2006 5th International Power Electronics and Motion Control Conference

Shanghai, China
13-16 August 2006

Volume 4 of 4

| IEEE Catalog Number: | 06EX1405 |
| ISBN: | 1-4244-0448-7 |

**Copyright © 2006 by The Institute of Electrical and Electronics Engineers, Inc.
All Rights Reserved**

Copyright and Reprint Permissions: Abstracting is permitted with credit to the source. Libraries are permitted to photocopy beyond the limit of U.S. copyright law for private use of patrons those articles in this volume that carry a code at the bottom of the first page, provided the per-copy fee indicated in the code is paid through Copyright Clearance Center, 222 Rosewood Drive, Danvers, MA 01923.

For other copying, reprint or republications permission, write to IEEE Copyrights Manager, IEEE Operations Center, 445 Hoes Lane, Piscataway, New Jersey USA 08854. All rights reserved.

IEEE Catalog Number: 06EX1405

ISBN: 1-4244-0448-7

Library of Congress: 2006925601

Additional Copies of This Publication Are Available from:

IEEE Service Center
445 Hoes Lane
Piscataway, NJ 08854
IEEE Service Center
445 Hoes Lane
Piscataway, NJ 08854
Phone: (800) 678-IEEE
            (732) 981-1393
Fax: (732) 981-9667
E-mail: customer-service@ieee.org

2006 5th International Power Electronics and Motion Control Conference

Shanghai, China
13-16 August 2006

IEEE Catalog Number: CFP06792-POD
ISBN: 978-1-42440-448-3

Table of Contents

Design Challenges For Distributed Power Systems 1
Fred C. Lee, Ming Xu, Shuo Wang, Bing Lu

A Smarter Grid for Improving System Reliability and Asset Utilization 16
D. Divan, H. Johal

Medium-Voltage Power Conversion Systems in the Next Generation 23
Hirofumi Akagi, Shigenori Inoue

Modern Electrical Drives: Design and Future Trends 31
R. W. De Doncker

Power Semiconductors development trends 39
L. Lorenz

Power Electronics in Wind Turbine Systems 46
F. Blaabjerg, Z. Chen, R. Teodorescu, F. Iov

Sustainable Energy and Mobility, and Challenges to Power Electronics 57
C.C.Chan

Wind farms with increased transient stability margin provided by a STATCOM 63
Marta Molinas, Jon Are Suul, Tore Undeland

A New Super Junction LDMOS with N+-Floating Layer 70
Baoxing Duan, Bo Zhang, Zhaoji Li

Unified Power Flow Controller: Comparison of Two Advanced Control Schemes and Performance Analysis for Power Flow Control 74
Liu Liming, Zhu Pengcheng, Kang Yong, Chen Jian

A New Analytical Model for the Surface Electrical Field Distribution of Double RESURF LDMOS 79
Qi Li, Zhaoji Li

A Novel Centralized HID Ballast System with Power-Bus 83
Xiaodong Lu, Bo Yang, Jiande Wu, Xiangning He

Research on a Novel Structure of SiGeC/Si Heterojunction Power Diodes 88
Liu Jing, Gao Yong, Ma Li

Gate driving of high power IGBT by wireless transmission 92
Stéphane Bréhaut, François Costa

The Characteristics of Thyristor Controlled Reactance Series Compensation by Adjustable Coupling 97
Guo-rong Zhu, Min-zu Li, Yong Kang

Dual-Side Cooled Novel IPM and Improved Capability of Inverter for Elevated-Temperature Operations ... 102
Jie (Jay) Chang, Changming Liao

An Improved Current-Doubler with Coupled Inductors 108
T.-F. Wu, C.-T. Tsai, W.-C. Lin, Y.-M. Chen

Monolithic Integration of Trench Power JFET with Schottky Diode 113
Yang Gao, Jie Chen, Alex. Q. Huang

Sequential Color LED Back-Light Driving System for LCD Panels 117
C.-C. Chen, C.-Y. Wu, P.-C. Lu, Y.-M. Chen, T.-F. Wu

Development of Large Capacity Programmable Harmonic Current Generator Based on Three-phase-four-wire Configuration 122
LIU Tao, ZHUO Fang, CHEN Bo, ZHAI Xi, WANG Zhao-an

A Universal Digital Platform and Software Library for Power Electronic Systems Integration 127
Haibing HU, Tianjun Jin, Wenxi YAO, Zhengyu LU, Zhaoming Qian

Table of Contents

Unipolar SiC Devices - Latest Achievements on the Way to a New Generation of High Voltage Power Semiconductors 132
Peter Friedrichs

Implementation of GA-trained GRNN for Intelligent Fast Charger for Ni-Cd Batteries 137
Panom Petchjatuporn, Noppadol Khaehintung, Khamron Sunat, Phaophak Sirisuk, Wiwat Kiranon

Modelling and Analysis of a Novel Transformer with Ability to Suppress Conducted interference 142
Zongxiang Chen, Pengsheng Ye, Junmin Pan

An Observer-Based Three-Phase Current Reconstruction using DC Link Measurement in PMAC Motors 147
Li Ying, Nesimi Ertugrul

Experiment Research of Chaotic PWM Suppressing EMI in Converter 152
R. Yang, B. Zhang , F.Li, J.J. Jiang

Emitter Size Effect in 4H-SiC BJT 157
Yan Gao, Alex Q. Huang, Sumi Krishnaswami, Anant K. Agarwal, Charles Scozzie

PSIM and SIMULINK Co-simulation for Threelevel Adjustable Speed Drive Systems 161
Zhang Yongchang, Zhao Zhengming, Baihua, Yuan Liqiang, Zhang Haitao

Three-Phase Z-Source AC-AC Converter for Motor Drives 166
Xu-Peng Fang

Construction and Application of Macro Model for ZVS Resonant Mode Controller MC34067 171
Wei Chen, Yilei Gu, Zhengyu Lu, Zhaoming Qian

Optimum Design of Hollow Conductor in Stator Winding for Large Evaporative Hydro-generator 175
Z. Wen , L. Ruan, G. Gu

Rotor Suspension Principle and Decoupling Control for Self-bearing Induction Motors 179
Tengchao Zhang, Huangqiu Zhu, Yuxin Sun

Field Oriented Control of Linear Induction Motor Considering Attraction Force & End-Effects 184
Jianqiang Liu, Fei Lin, Zhongping Yang, Trillion Q. Zheng

Series Resonant High Frequency Link Sine-wave Inverter System Modeling using Sampled Data 189
Jin Xiaoyi, Dong Wei, Sun Xiaofeng, Wu Weiyang

Maximal Power Point Tracking under Speed-Mode Control for Wind Energy Generation System with Doubly Fed Introduction Generator 194
Y. Zhao, X. D. Zou, Y. N. Xu, Y. Kang, J. Chen

Effective Mobility in Nano-Scaled n-MOSFETs 199
Yue-Hua Dai, Jun-Ning Chen, Dao-Ming Ke, Jia-E Sun

Investigation on the Factors Affecting Inrush Current of Transformers Based on Finite Element Modeling 204
M. Reza Feyzi, M. B. B. Sharifian

An Improved Support Vector Machine Method for Harmonic and Inter-harmonic Detecting 209
Ma Li, Liu Kaipei, Lei Xiao

A Common Mode and Differential Mode Integrated EMI Filter 214
Liu Nan, Yang Yugang

Electromagnetism Model and Characteristic Simulation of Novel Claw Pole Generator with Permanent Magnet Outer Rotor 219
Fengge Zhang, Haijun Bai, Shifu Zhang, Hans Pert Gruenberger, Eugen Nolle

An Improved Adaptive Filter for Voltage and Current Reference Extraction 224
A. Abedini, A. Nasiri

Simulation Analysis on Current SVM Algorithm of Matrix Rectifier 229
Xi-jun Yang, Peng-sheng Ye, Xiang Liu, Xing-hua Yang, Jian-quan Wang, Luan-guo Zhang

iv

Table of Contents

Study of Measurement Approach of Loop Gain of Converter 236
Weiping Zhang, Yunpeng Chen, Yuanchao Liu, Dongyan Zhang, Zheng Meng

A Stand-Alone Hybrid Generation System Combining Solar Photovoltaic and Wind Turbine with Simple Maximum Power Point Tracking Control 242
Nabil A. Ahmed, Masafumi Miyatake

Design Optimization of Industrial Motor Drive Power Stage Using Genetic Algorithms 249
F. Wang, W. Shen, D. Boroyevich, S. Ragon, V. Stefanovic, M. Arpilliere

FEM Based Simulation of a Permanent Magnet Synchronous Motor Performance Characteristics 254
L. Petkovska, G. Cvetkovski

Analytical Modeling of Semiconductor Losses in Matrix Converters 259
Bingsen Wang, Giri Venkataramanan

Nonlinear Robust Sliding Mode Control for PM Linear Synchronous Motors 267
Xi Zhang, Junmin Pan

Dynamic Analysis of PWM Switching DC-DC Converters 272
Liu Jian, Wang Yuanbin

A Novel LLC Resonant Converter Topology: Voltage Stresses of All Components in Secondary Side Being Half of Output Voltage 276
Yilei Gu, Zhengyu Lu, Zhaoming Qian

On the hybrid automaton models and control synthesis of a single inductor, double output boost converter 281
Sreekumar C, Vivek Agarwal

Complex Intermittency in Voltage-Mode Controlled Buck Converter 286
Zheng-Ping Li, Yu-Fei Zhou, Jun-Ning Chen

Dual Mode Control Multiphase DC/DC Converter for CPU Power 291
Li-Wei Lin, Chung-Hsing Chang, Huang-Jen Chiu, Shann-Chyi Mou

An Analog Implementation of Pulse-Width-Modulation Based Sliding Mode Controller for DC-DC Boost Converters 296
Siew-Chong Tan, Y. M. Lai, Chi K. Tse

Low Cost Electronic Ballast with Buck Converter as PFC Stage 301
Li Xiangrong, Xu Dianguo, Zhang Xiangjun

A New Converter Architecture for Future Generations of Microprocessors 306
Dodi Garinto

A Combined ZVS Converter with Naturally Sharing Input-Current and High Voltage Gain 311
Linbing Wang, Bo Yang

Matrix Coefficient Polynomial Description Model of DC-DC Converters Based on Switched Linear Systems 316
Yongping Zhang, Bo Zhang, Zongbo Hu, Dongyuan Qiu, Guiping Du

Development of DC-DC Multiple Converter based on Push-pull Forward Topology accomplished 321
Weihao Hu, Yunqing Pei, Zhaoan Wang

Voltage Fed and Current Fed Full Bridge Converter for the Use in Three Phase Grid Connected Fuel Cell Systems 325
M. Mohr, F.-W. Fuchs

Small-Signal Modeling of Asymmetrical Half Bridge Flyback Converter 332
Tso-Min Chen, Chern-Lin Chen

A DSP Based Controller for High Power Dual-Phase DC-DC Converters 337
Xin Guo, Xuhui Wen, Ermin Qiao

Table of Contents

Effective Load Resistance; A New Method to Evaluate DC/DC converters Efficiency 342
Alan Elbanhawy

Calculation of Power Loss in Output Diode of a Flyback Switching DC-DC Converter 346
Jiaxin Chen, Jianguo Zhu, Youguang Guo

A Multiple Output Forward Converter Adopting Weighted Time-Sharing Control and Switch-Linear Hybrid Scheme 351
Xiaodong Liu, Songqin Hu, Sizhou Sun

A Novel Soft-Switching PWM Full-Bridge DC/DC Converter with DC Busline Series Switch-Parallel Capacitor Edge Resonant Snubber Assisted by High-Frequency Transformer Leakage Inductor 356
Khairy Fathy, Toshimitsu Doi, Keiki Morimoto, Hyun Woo Lee, Mutsuo Nakaoka

High-Efficiency Cascode Forward Converter of Low Power PEMFC System 361
Jiann-Fuh Chen, Wei-Shih Liu, Ray-Lee Lin, Tsorng-Juu Liang, Ching-Hsiung Liu

Control of Bifurcation by Fuzzy Logic Controller for Current-mode Boost Converters 368
Noppadol Khaehintung, Phaophak Sirisuk, Anantawat Kunakorn

An Improved Three-Level Soft-Switching DC/DC Converter 373
Z. L. Lou, Z. S. Wang

A Novel Soft Switching Bidirectional DC/DC Converter and Design Consideration 378
Ma Gang, Qu Wenlong, Liu Yuanyuan

State-Variable Description and Analysis of a DC-Rail ZVT Inverter Feeding a Permanent Magnet Synchronous Motor 382
Ming Zhengfeng, Zhong Yanru

Analysis, Simulations and Experiments Of A Novel ZVS -ZCS Inverter With Pulse Current Feedback Transformer Auxiliary Commutation 386
Yaogang, Mahamnad Mansoor Khan, Chenchen

A Novel Eddy-Current Based Far-Infrared Rays Radiant Planner Heater using High-Frequency ZVT-PWM Inverter 392
Hisayuki Sugimura, Bishwajit Saha, Hideki Omori, Hyun Woo Lee, Mutsuo Nakaoka

3 Phases-3 Devices AC Voltage Regulator With Quasi-Zero Switching 397
Qianzhi Zhou, Wenhua Hu, Bin Wu

Study on Power Decoupling Control of Three Phase Voltage Source PWM Rectifiers 401
Wang Jiuhe, Yin Hongren, Zhang Jinlong, Li Huade

A Fully Digital Controlled 3KW, Single-Stage Power Factor Correction Converter Based on Full-Bridge Topology 406
HANG Li-jun, YANG Yue-feng, SU Bin, LU Zheng-yu, QIAN Zhao-ming

A New ZVT Power Factor Corrected Three-Phase AC-AC Converter with Single-Phase HF Link 411
T. H. Abdelhamid, A. Sabzali

Simple Bridge-Type AC/DC Converters with Natural Input-Current-Shaper 417
Hsing-Fu Liu, Chih-Yu Wu, Chin Sun, Lon-Kou Chang

Rough Controlling TSC for Reactive Current Compensation in Traction Substations 423
Hongsheng Su, Qunzhan Li

A Digitally Controlled 4-kW Single-Phase Bridgeless PFC Circuit for Air Conditioner Motor Drive Applications 428
Yong Li, Toshio Takahashi

Optimized Electrical Design for Single Phase PFC Active IPEM 433
Qiaoliang Chen, Xu Yang, Zhao-an Wang

Table of Contents

A Novel Topology of APFC with On-Line Half-Bridge UPS Controlled by DSP 438
Xuejun Ma, Xuezhi Hu, Hongxia Wu, XuWu Chen

Nonlinear Current Control of Single-Phase PFC Suitable for Mixed-Signal IC Implementation 442
Min Chen, Anu Mathew, Jian Sun

A Novel Detection Method for Three-Phase Reactive Current 449
Zong Ming, Wang Fengxiang, Hua Funian, Sun Yidan

Selective Harmonic Controlling for Three-Level High Power Active Front End Converter with Low Switching Frequency 453
Hui Zhang, Kaipei Liu,

A Unity Power Factor Three-Phase Buck Type SVPWM Rectifier Based on Direct Phase Control Scheme 458
LI Yabin, Li Heming, Peng Yonglong

3-Phase Current-Source SMES-UPS Based on TFSC and its Control Strategies Control Strategies 463
WANG Fu-sheng, LI Hong-mei

A novel control scheme of 230kA DC power source using thyristor, Phase-shifting rectifier transformer and On-load tap changer 468
Qiao Shutong, Jiang Jianguo, Zuo Dongsheng, Wu Xiaojie

Research on Control Method of Double-Mode Inverter with Grid-Connection and Stand-Alone 473
Herong Gu, Zilong Yang, Deyu Wang, Weiyang Wu

Power and Energy Management of a Dual- Energy Source Electric Vehicle - Policy Implementation Issues 478
P.C.K. Luk, L.C. Rosario

Study on Non Contact Automatic on-Load Voltage Regulating Distributing Transformer Based on Solid State Relay 484
Zhao-Yulin, Dong-Shoutian, Li-Jiahui, Yao-Xin, Zheng-Na, Liu-Xueli

The Principle of a Novel Arc-suppression Coil and its Implementation 489
Cheng Lu, Chen Qiaofu, Zhang Yu, Zhang Changzheng

Grid Connection to Stand Alone Transitions of Slip Ring Induction Generator During Grid Faults 494
G. Iwanski, W. Koczara

System Control of Power Electronics Interfaced Distribution Generation Units 499
D. Feng, Z. Chen

Test Loadability of Power Systems using A Networked Power Electronic Devices Control and Measurement System 505
Sheng Yang, Venkataramana Ajjarapu, Bo Zhang

Test-Bed of Doubly Fed Induction Generator for Variable-Speed Constant-Frequency Wind Power Generation 510
S. Y. Yang, X. Zhang, C. W. Zhang, R. X. Cao

Control strategy of Hybrid sources for Transport applications using supercapacitors and batteries 515
M.B. Camara, H. Gualous, F. Gustin, A. Berthon

Wind Generator Stabilization With Doubly-Fed Asynchronous Machine 520
Li Wu, Zhixin Wang

Design Consideration of a Novel Digital Bidirectional Constant Current Source Used in Hybrid Electric Vehicle 526
Qingbo Hu, Zhengyu Lü

A Single-Phase Grid-Connected Inverter System With Zero Steady-State Error 532
Guo Xiaoqiang, Zhao Qinglin, Wu Weiyang

DC Transformer with Line Frequency Ripple Cancellation 537
Sen Dou, Wilson Wu, Annabelle Pratt, Pavan Kumar

Table of Contents

A Novel PWM Method for Stacked Flying Capacitor Inverter 542
Gangui Yan, Gang Mu, Yafeng Huang, Wenhua Liu

Study on a New Method of Voltage-Source Induction Heating Load-Matched 549
Li Jin-gang, Zhong Yan-ru, Zhao Miao

An Alternating-master-salve Parallel Control Research for Single Phase Paralleled Inverters Based on CAN Bus 554
Zhang Chunjiang, Chen Guitao, Guo Zhongnan, Wu Weiyang

Analysis and Design of a Novel Dual Secondary Winding and Dual Power Bridge High Frequency Link Inverter 559
Zhang Zhe, Zhang Chunjiang, Wu Weiyang, Gu Herong, Shen Hong

Reduction of Common Mode EMI in a Full-Bridge Converter through Automatic Tuning of Gating Signals 564
Kai Zhang, Yunbin Zhou, Yonggao Zhang, Yong Kang

Phase Multilevel Inverter Fault Diagnosis and Tolerant Control Technique 569
Wang Baocheng, Wang Jie, Sun Xiaofeng Wu Junjuan, Wu Weiyang

Microcontroller-Based Single Phase Inverter Using a New Switching Strategy 574
K. Meghriche, O. Mansouri, A. Cherifi

Study of Stability Regions in Parallel Connected Boost Converters 580
Yuehui Huang, Chi. K. Tse

A Novel Analysis and Design Method for Integrated Magnetics 585
Zheng Feng, Weihao Hu, Pei Yun-qing

Investigation on the Space Vector PWM for Large Power Three-Level DC-Link Voltage Source Inverter Equipped with IGCTs 589
Wang Chengsheng, Li Chongjian, Li Yaohua, Zhao Xiaotan

Status and Opportunities of Photovoltaic Inverters in Grid-Tied and Micro-Grid Systems 593
Xiaoming Yuan, Yingqi Zhang

Adaptive Neuro-Fuzzy Control with Fuzzy Supervisory Learning Algorithm for Speed Regulation of 4-Switch Inverter Brushless DC Machines 597
A. Halvaei Niasar, H. Moghbelli, A. Vahedi

Combined Modulation and Harmonic Suppression 602
Cheng Weibin, Zhong Yanru, Jin Shun

Application Research of Maximum Wind-energy Tracing Controller Based Adaptive Control Strategy in WECS 607
Changhong Shao, Xiangjun Chen, Zhonghua Liang

Research on Synchrodrive Control Technology for Wind Turbine Adjustable-Pitch System Based on Adaptive decoupling Control 612
Hongche Guo, Qingding Guo

Limit-Trajectory Single- and Two-Mode Overmodulation Technology 617
Shun Jin, Yan-ru Zhong

Multiphase Permanent Magnet Motor Drive System Based on A Novel Multiphase SVPWM 622
Shan Xue, Xuhui Wen, Zhao Feng

Novel Random-Harmonic Elimination PWM Technique for Single-Switch Three-Phase AC-DC Buck Converter 627
Guang-Hui Tan, Wenchuan Ma, Yanchao Ji, Hongxiang Yu, Wancai Xu

Table of Contents

FPGA Based Multichannel PWM Pulse Generator for Multi-modular Converters or Multilevel Converters 632
Liqiao Wang, Weiyang Wu

Cascaded Multilevel Converters with Non-Integer or Dynamically Changing DC Voltage Ratios 637
Shuai Lu, Keith A. Corzine

Practical Thermal Design Considerations for IPEM-based Converter 642
Qiaoliang Chen, Xu Yang, Zhao-an Wang

Realization of an FPGA-Based Space-Vector PWM Controller 647
Zhou Yuan, Xu Fei-peng, Zhou Zhao-yong

Chaotifying Control of Permanent Magnet Synchronous Motor 652
Hai Peng Ren, Chong Zhao Han

Analysis of PMLSM Direct Thrust Control System Based on Sliding Mode Variable Structure 657
Junyou Yang, Guofeng He, Jiefan Cui

Carrier-based Pulse Width Modulation for Three-Level Inverters: Neutral Point Potential and Output Voltage Distortion 662
Jang-Hwan Kim, Seung-Ki Sul

AC Current Sensorless Control of Three-Phase Three-Wire PWM rectifiers under the Unbalanced Source Voltage 669
Jia-peng Xu, Yu-peng Tang

Waveform Library Control of Converter 674
Xiaofeng Sun, Bin Wang, Meng Lingjie, Weiyang Wu

d-model Adaptive Algorithm Based on Plant-Parameterization 679
Zhao Feng, Liu Weiguo

Dynamics and Control of Electronic Cascaded Systems 684
Wen Wei, Xu Haiping, Wen Xuhui, Shi Wenqing

The Controlling Strategy for Electronic Ballast of HID Lamps 688
Weiping Zhang, Xiaohan Guan, Xusen Zhao, Hongtao Li, Zhengang Liu

Voltage Spectra of Three-Level Inverters with Three-Phase Modulation 693
S. Halász, I. Varjasi

Design of Motion Control System Used for Filter Rod Production Machine 699
Yang Qingyu, Ge Sibo, Ye Kesong, Shi Ren

Magnetic Pole Identification for PMSM at Zero Speed Based on Space Vector PWM 703
Jiangang Hu, Longya Xu, Jingbo Liu

Study on Stagewise Control of Connecting DFIG to the Grid 708
Xueguang Zhang, Dianguo Xu, Yongqiang Lang, Hongfei Ma

Generalized Control Approach for Active Power Filters 713
Xiaoyu Wang, Jinjun Liu, Chang Yuan, Zhaoan Wang

Novel Circuit Configuration for Hybrid Reactive Power Compensator 718
H.L Jou, J.C Wu, J.J. Yang, W.P. Hsu

Shunt Active Power Filter with Sample Time Staggered Space Vector Modulation Based Cascade Multilevel Converters 724
Liqiao Wang, Weiyang Wu

Shunt Active Power Filter Synthesizing Resistive Loads by Means of Adaptive Inverse Control 729
Wu Yanfeng, Wu Zhengguo, Li Hua, Li Hui

Table of Contents

Single Neutral Element Self-Adaptive PID Controller Used In SVC...**734**
Zeng Guang, Ke Min-qian, Su Yan-min, Fu Qi-gang

A Novel Shunt Single-Phase Active Power Filter for High Voltage Application **739**
Zhang Changzheng, Chen Qiaofu, Zhao Youbin, Chen Yuda, Cheng Lu

Three-phase Active Power Filter Based on Space Vector and One-cycle Control...**744**
Wang Yong, Shen Songhua, Guan Miao

Implementation of a Shunt-Series Compensator for Nonlinear and Voltage Sensitive Load.........................**748**
Bor-Ren Lin, Chien-Lan Huang

Three-Phase Active Filter using a Single-Phase STATCOM Structure with Asymmetrical Dead-band Control..**753**
Seyyed Hossein Hosseini, Mehran Sabahi

Mitigation of Voltage Sag Using Adaptive Neural Network with Dynamic Voltage Restorer.......................**759**
M. R. Banaei, S. H. Hosseini, M. Darkalee Khajee

Mitigation of Current Harmonic Using Adaptive Neural Network with Active Power Line Conditioner **764**
M. R. Banaei, S. H. Hosseini

A direct control strategy for UPQC in three-phase four-wire system...**769**
Tan Zhili, Li Xun, Chen Jian, Kang Yong, Duan Shanxu

Three-Phase Harmonic Selective Active Filter Using Multiple Adaptive Feed Forward Cancellation Method...**774**
Lewei Qian, David Cartes, Qiang Zhang

Reactive Power Compensation in Distribution Networks with STATCOM by Fuzzy Logic Theory Application **779**
Seyyed Hossein Hosseini, Reza Rahnavard, Yousef Ebrahimi

A Distributed Fuel Cell Based Generation and Compensation System to Improve Power Quality...............**784**
Haimin Tao, Jorge L. Duarte, Marcel A. M. Hendrix

Parallel Control of Three-Phase Three-Wire Shunt Active Power Filters **789**
Xueliang Wei, Ke Dai, Xin Fang, Pan Geng, Fang Luo, Yong Kang

Study and Design of Noninductive Bus bar for high power switching converter **794**
Zhiling Qiu, Hongyan Zhang, Guozhu Chen

A New Minimum Torque-ripple and Sensorless Control Scheme of BLDC Motors Based on RBF Networks..**798**
Juan Wang, Hongwei Liu, Yuran Zhu, Bo Cui, Huijuan Duan

Improved Modelling and Calculation on Electromagnetic Transient of Power Transformer........................**802**
Chen Zhe, Wen Yuanfang, Lu Guojun

The Simulation and the Experimental Research of the Stator Bars' Evaporative Cooling System in the Three Gorges' Hydrogenerator **808**
Ruan Lin, Gu Guobiao, Tian Xindong, Yuan JiaYi

An Investigation of Multi-phase Transverse Flux Permanent Magnet Machine **813**
G.Q. Bao, J.K.Wang, D.Zhang, J.Z. Jiang

Suspension Principle and Digital Control for Bearingless Permanent Magnet Slice Motors **817**
Huangqiu Zhu, Liang Fang

The effect of parameter variations on the performance of indirect vector controlled induction motor drive **821**
A. Shiri, A. Vahedi, A. Shoulaie

Magnetic Field Analysis and Performance Calculation for New Type of Claw Pole Motor with Permanent Magnet Outer Rotor **826**
Fengge Zhang, Shifu Zhang, Haijun Bai, Eugen Nolle, Hans Pert Gruenberger

Table of Contents

Performance Analysis of a PM Claw Pole SMC Motor with Brushless DC Control Scheme 831
Youguang Guo, Jianguo Zhu, Jiaxin Chen, Jianxun Jin

Solving Induction Motor Equivalent Circuit using Numerical Methods for an In-Service and Nonintrusive Motor Efficiency Estimation Method 836
Bin Lu, Wei Qiao, Thomas G. Habetler, Ronald G. Harley

Fault Investigation of X-by-wire Permanent Magnet Synchronous Machine 842
L. Feng, A. Binder, A. Rentschler, A. Paweletz, D. Guenther

PLC-Based Speed Control of DC Motor 847
Ashraf Salah El Din Zein El Din

H8 Control of Adjustable-Pitch Wind Turbine Adjustable-Pitch System 853
Hongche Guo, Qingding Guo

The Motion Control Algorithm based on Quaternion Rotation for a Permanent Magnet Spherical Stepper Motor 857
Qun-jing Wang, Kun Xia

Research on Restraining Thrust Force Ripple for Permanent Magnet Linear Synchronous Motor 862
Cui Jiefan, Wu Hui, Sun Qing, Zhang Yi, Zhao Lijun

Using Recurrent Fuzzy Wavelet Neural Network to Control AC Servo System 866
Yan Tang, Wei Sun, Yaonan Wang, Xiaohua Zhai

new topology of multi - level - converter for harmonic reduction 870
Frank Grundmann, Jian Xie

PWM Based Sensing and Control of Magnetic Bearings 875
Zhuliang Yeic, Flalph Vansencc

Position Sensorless Direct Torque Control of Synchronous Reluctance with Permanent Magnet Motor 880
Jiang Dong, Zhao Zhengming, Duan Yao, Guo Wei

Counter-Rotating Permanent Magnet Brushless DC Motor for Underwater Propulsion 885
Jianqi Qiu, Cenwei Shi, Mengjia Jin, Ruiguang Lin

A Special Flux-weakening Control Scheme of PMSM - Incorporating and Adaptive to Wide-Range Speed Regulation 890
Song Chi, Longya Xu

Model-based Disturbance Attenuation for Linear Motor Servo System 896
Guiqiu Liu, Qingding Guo

A Fuzzy-Wavelet-Network-Based Position Control for PMSM 899
Wang Jun, Peng Hong, Xia Ling

Stability Analysis of Magnetic Bearing with Resonance Circuit 903
Zong Ming, Wang Fengxiang, Sun Yidan, Wang Jiqiang

Flux-Weakening Characteristics of Trapezoidal Back-EMF Machines in Brushless DC and AC Modes 908
Z.Q. Zhu, J.X. Shen, D. Howe

A Cost Effective Sensorless Control Method for Permanent Magnet Synchronous Motors Based on Average Terminal Voltage 913
Cheng-Hu Chen, Wei-Chih Tai, Ming-Yang Cheng

DSP-based Discrete-Time Reaching Law Control of Switched Reluctance Motor 918
Ge Baoming, Zhao Nan

Digital Control System on Bearingless Permanent Magnet-type Synchronous Motors 923
Jianming Deng, Huangqiu Zhu, Yang Zhou

Table of Contents

Practical Issues in Sensorless Control of PM Brushless Machines Using Third-Harmonic Back-EMF 928
J.X. Shen, Z.Q. Zhu, D. Howe

Switched Reluctance Motors Drive for the Electrical Traction in Shearer 933
H. Chen

Research on Three-level Inverter of Six-phase Synchronous Motor 937
Yao Wenxi, Hu Haibing, Lu Zhengyu, Xu Haijie

Doubly-Salient Permanent-Magnet Machine with Skewed Rotor and Six-State Commutating Mode 942
Yongbin Li, Chris Mi

Sensorless Control and PMSM Drive System for Compressor Applications 947
Dongsheng Li, Takahiro Suzuki, Kiyoshi Sakamoto, Yasuo Notohara, Tsunehiro Endo, Chikara Tanaka, Tatsuo Ando

Analysis and Experimental Study of Slot Effect in Synchronous Reluctance Permanent Magnet Motors 952
Wei Guo, Zhengming Zhao,Yingchao Zhang

A New BLDC Motor Drives Method Based on BUCK Converter for Torque Ripple Reduction 958
Zhang Xiaofeng, Lu Zhengyu

Performance Investigation of a Fault-Tolerant Brushless Permanent Magnet AC Motor Drive 962
Jingwei Zhu, Nesimi Ertugrul, Wen Liang Soong

Current sensorless integral variable structure controller of synchronous reluctance motor 967
Huann-Keng Chiang, Chien-An Chen, Bor-Ren Lin, Kai-Sheng Hsu

An Improved Sliding Mode Observer for Speed Sensorless Vector Control Drive of PMSM 972
K. Paponpen, M. Konghirun

Analysis of an AC fed direct converter for a switched reluctance machine in aerospace applications 977
S. J. Forrest, J. Wang, G. W. Jewell, C. M. Johnson, S.D. Calverley

Direct Torque Control of an Interior Permanent Magnet Synchronous Machine fed by a Direct AC-AC Converter 983
D. Xiao, M. F. Rahman

A Novel Modular Permanent Magnet Drive System Design 989
Wen Ouyang, Nicholas Lemberg, Ruoping Yao, T.A.Lipo

Research on Digital Control Systems for Large Power AC-DC-AC Converters with Synchronous Motor Load 995
Xiaotan Zhao, Chongjian Li, Weihui Sheng, Yaohua Li

About the Prediction of Undesired Higher Current and Torque Harmonics of Inverter Driven Motors with Numerical Methods 999
C. Grabner

A Method of Stator Voltage Error Compensation in MRAS Sensorless Vector Control of Induction Motor 1006
Wen Xuhui, Chen Guilan, Han Li

Systematic Design of Fuzzy Logic Based Hybrid On-Line Minimum Input Power Search Control Strategy for Efficiency Optimization of IM 1012
Zhang Liwei, Liu Jun, Wen Xuhui, Trillion Q. Zheng

Research on an AC Variable-frequency Power Dynamometer Based on PWM Rectifier and Fuzzy Direct Torque Control 1017
Jia-qiang Yang, Jin Huang

Characteristic Research of Bearing Currents in Inverter-Motor Drive Systems 1023
Xing Shancheng, Wu Zhengguo

Research on a New Motor Drive Control System for Electric Transit Bus 1027
SHAO Gui-xin, ZHANG Cheng-ning

xii

Table of Contents

New Micro-Drive Series For Induction Motors & Survey of Market Trends 1032
Henrik Rosendal Andersen, Ruimin Tan, Zhang Hui

Robust Backstepping Control of Induction Motor Drives Using Artificial Neural Networks 1038
J. Soltani, R. Yazdanpanah

Robust Nonlinear Control of Linear Induction Motor taking into account the Primary End Effects 1043
J. Soltani, M.A. Abbasian

A Novel Adaptive Scheme for Stator Resistance Estimation in Sensorless Induction Motor Drives 1049
Han Li, Wen Xuhui, Chen Guilan

Ripple-Free Sampling of Current Signals in Drives with Carrier-based PWM Patterns 1054
Haihui Lu, Qiang Yin, Russel J. Kerkman, Thomas A. Nondahl

Study of Speed Sensorless Control Methodology for Single Inverter Parallel Connected Dual Induction Motors Based on the Dynamic Model 1061
Shi Wei, Wang Ruxi, Wang Yue, He Yanhui, Wang Zhaoan, Liu Jinjun

ADC architecture with direct binary output for digital controllers of high-frequency SMPS 1066
Tao Zhou, Jianping Xu

Analysis and Evaluation of a High-Voltage AC Amplifier for Electrostatic Suspension 1071
F. T. Han, Q. P. Wu, K. Liu, Z. Y. Gao

Design and Development of a 50kW Z-Source Inverter for Fuel Cell Vehicles 1076
Miaosen Shen, Alan Joseph, Yi Huang, Fang Z. Peng, Zhaoming Qian

Identification and improvement of stray coupling effect in an L-C-L common mode EMI filter 1081
Junping He, Wei Chen, Jianguo Jiang

High Step-up Converter Associated with Soft-Switching Circuit with Partial Energy Processing for Livestock Stunning Applications 1086
S. -Y. Tseng, S.-H. Tseng, J. -Z. Shiang

A Computationally Intelligent Methodologies and Sliding Mode Control Based Traction control System for in-wheel driven EV 1091
Ming Zhengfeng, NI Guangzheng

A Low-Cost Gate Driver Design Using Bootstrap Capacitors for Multilevel MOSFET Inverters 1096
J. J. Graczkowski, K. L. Neff, X. Kou

An Effective Method to Suppress Resonance in Input LC Filter of a PWM Current-Source Rectifier 1101
Y.W. Li, B. Wu, N. Zargari, J. Wiseman, D. Xu

Topological and Modulation Design of Three-Level Z-Source Inverters 1107
P. C. Loh, F. Gao F. Blaabjerg

Investigation of Power Supplies for a Piezoelectric Brake Actuator in Aircrafts 1112
Rongyuan Li, Norbert Fröhleke, Hermann Wetzel, Joachim Böcker

A Line Power-Supply for LED Lighting using Piezoelectric Transformers in Class-E Topology 1117
F.E. Bisogno, S. Nittayarumphong, M. Radecker, A. V. Carazo, R. N. do Prado

Integrating Large Wind Farms into Weak Power Grids with Long Transmission Lines 1122
Richard Piwko, Nicholas Miller, Juan Sanchez-Gasca, Xiaoming Yuan, Renchang Dai, James Lyons

Turn-on Condition and Characteristics of Highpower Semiconductor Switch RSD 1129
Y. M. Zhou, Y. H. Yu, H. G. Chen, L. Liang

The analysis and simulation of power circuits for high voltage converter 1133
S. I. Volskiy, Y. Y. Skorokhod, V. V. Shergin

A novel IGCT-based Half-controlled Bridge Type Fault Current Limiter 1138
Wanmin Fei, Yanli Zhang

xiii

Table of Contents

Influence of Proton Irradiation dose on the Performance of Local Lifetime Controlled Power Diode with Proximity Gettering of Platinum 1143
B.D. Han, D.Q. Hu, S.S. Xie, Y.P. Jia, B.W. Kang

IMPLEMENTATION OF A HIGHER QUALITY DC POWER CONVERTER 1148
Barsoum, N.N., YII, M.L.

Design of a Digital Programmable Control IC for Single-Phase Controlled Rectifiers 1154
Ming-Fa Tsai, Fu-Jing Ke, Ying-De Lin, Jui-Kum Wang

Feasibility Study of AlGaN/GaN HEMT for Multimegahertz DC/DC Converter Applications 1159
Yang Gao, Alex Q. Huang

The Mechanism Analysis of IGBT Module Invalidation 1162
Xu Aide, Fan Yinhai, Wang Xinxin, Liu Yuanyuan

A New Injection Efficiency Controlled GTO 1167
Wang Cailin, Gao Yong, Zhang Ruliang

Implementation and Analysis of 3-phase Voltage Sourced Regenerative Rectifier 1171
Rui Chen, Qiongxuan Ge, Shijie Li

Design and Implementation of Electronic Ballast for Fluorescent Lamps with Low Lighting Flicker 1178
Yang-Sheng Lin, Chun-An Cheng, Jiann-Fuh Chen, Tsorng-Juu Liang, Wei-Shih Liu

A Floating-point Coprocessor Configured by a FPGA in a Digital Platform Based on Fixed-point DSP for Power Electronics 1183
Haibing HU, Tianjun Jin, Xianmiao Zhang, Zhengyu LU, Zhaoming Qian

An Analytical Model for 4H-SiC Super-Junction Devices 1188
L.C. Yu, K. Sheng

Architecture Implementation of Class-D Amplifiers Using Digital-Controlled Multiphase-Interleaved PWM Technique 1192
Yu-Tzung Lin, Chi-Yang Lee, Ying-Yu Tzou,

Integrated IC-like Thyristor–based Switching Structure for Pulse Current Generation to Electronic Ignition 1198
C. L. Zhang, K. S. Jeon, C. H. Ahn, J. D. Park, E. D. Kim, Na Zhi, Yong Gao

A Wide Bandwidth Current Probe Based on Rogowski Coil and Hall Sensor 1202
Dong Li, Guiyou Chen

Voltage Dip Detection Based on an Efficient Least Squares Algorithm for D-STATCOM Application 1207
Thip Manmek, Chathura P. Mudannayake, Colin Grantham

Optimal Design and Analysis on Bearingless Permanent Magnet-type Synchronous Motors Using Finite Element Method 1213
Chang Jiang , Huangqiu Zhu, Zhenyue Huang

The Restrain of Harmonic Circulating Currents between Parallel Inverters 1218
Yu Zhang, Shanxu Duan, Yong Kang, Jian Chen

Simulation of Permanent Magnet Synchronous Motor with Dual Closed Loop by Time-Stepping Finite Element Model 1223
Xinhua Liu, Jianzhong Jiang, Yu Gong, Ye Ding

Online Dynamic Parameter Estimation of Transformer Equivalent Circuit 1228
M. Reza Feyzi, Mehran Sabahi

Worst-Case Tolerance Analysis for a Power Electronic System by Modified Genetic Algorithms 1233
Toshiji Kato, Kaoru Inoue, Kazuya Nishimae

The Reduction of Force Ripples of PMLSM Using Field Oriented Control Method 1238
Yu-wu Zhu, Kun-seok Jung, Yun-hyun Cho

Table of Contents

Analysis and Design of Signal Stage AC/DC Converter with Resonant Model PFC 1243
Weiping Zhang, Liangrui Lin, Dongyan Zhang, Xusen Zhao

Low Frequency Model for the Metal Halide Lamp 1248
Weiping Zhang, Yuanchao Liu, Xiaoqiang Zhang, Hongtao Li, Wenji Liu

H8 Robust Controller Based on Local Feedback Recurrent Neural Network for Permanent Magnet Linear Synchronous Motor 1253
Junyou Yang, Naiguang Fa, Ruijuan Chen

Parameter Estimate Modeling of Electronic Transformer 1258
Jiaju Wu, Hidehiko Sugimoto, Changkun Wang

Analysis and Design of Boost DC-DC Converters for Intrinsic Safety 1267
Shu-Lin Liu, Jian Liu, Hong Mao

Modeling and Fuzzy Logic with Integrator Control for the ZVZCS PWM DC/DC Converter 1273
Shen Hong, Wan Jianru, Yang Xiaobo, Wu Weiyang, Wang Xiaohuan

ZVS DC-DC Converter with Parallel-Connected Current Doubler Rectifier 1278
Bor-Ren Lin, Shuh-Chuan Tsay, Chun-Sheng Yang, Chien-Lan Huang

Study on the Dynamical Model and Analytical Method for DC-DC Switching Converter 1283
Li-Li Wang, Yu-Fei Zhou, Jun-Ning Chen

A Novel Topology Family of Single-stage Parallel Mode Uninterruptible AC/DC Converter with PFC ... 1288
Xuejun Ma, Hongxia Wu, Congsheng Huang, Xuwen Huang

Analysis and Design of an Automatic-Current-Sharing Control Based on Average-Current Mode for Parallel Boost Converters 1293
Wenxun Xiao, Bo Zhang, Dongyuan Qiu

A Novel Digital Charge Control for DC-DC Converters 1298
Shi Wenqing, Xu Haiping, Wen Xuhui, Wen Wei

An Asymmetrical Switched Capacitor and Lossless Inductor Quasi-Resonant Snubber-Assisted ZCS-PWM DC-DC Converter with High frequency Link 1302
Khairy Fathy, Keiki Morimoto, Toshimitsu Doi, Hyun Woo Lee, Mutsuo Nakaoka

A Divided Voltage Half-Bridge High Frequency Soft-Switching PWM DC-DC Converter with High and Low Side DC Rail Active Edge Resonant Snubbers 1307
Khairy Fathy, Keiki Morimoto, Toshimitsu Doi, Hiroyuki Ogiwara, Hyun Woo Lee, Mutsuo Nakaoka

Dynamic Analysis of a Current Source Inductively Coupled Power Transfer System 1312
Wenqi Zhou, Hao Ma

A New Topology of Capacitor-Clamp Cascade Multilevel Converters 1318
Anees Abu Sneineh, Ming-Yan Wang, Kai Tian

Evaluation of Semiconductor Losses in Cryogenic DC-DC Converters 1323
C. Jia, A. J. Forsyth

Design and Performance Evaluation of a 10-kW Interleaved Boost Converter for a Fuel Cell Electric Vehicle 1328
G. Calderon-Lopez, A. J. Forsyth, D. R. Nuttall

Analysis of Abnormal Phenomenon in Common-Source-type Forward Converter with Self-driven Synchronous Rectifier 1333
Kentaro Fukushima, Takayoshi Hashimoto, Tamotsu Ninomiya, Takeshi Segawa

Power Quality Conditioning in Distributed Generation Systems 1338
R.K. Járdán, I. Nagy

Table of Contents

Active Clamp Forward Converter Combined with Dither Voltage Generator for Poultry Stunning Applications...1343
S. -Y. Tseng, H.-T. Wen, H.-H. Chang, J. -S. Kuo

A Novel Zero-Voltage Switching Resonant Pole Inverter 1348
Sanbo Pan, Junmin Pan

Analysis of Three-Level ZVS PWM Inverter for Induction Heating Applications 1353
A. Jangwanitlert, J. Songboonkaew, W. Thammasiriroj, J.C. Balda

Dual Duty Cycle Controlled Voltage Source Soft-Switching High Frequency Inverter with AC Load Side Reverse Blocking Switched Resonant Capacitor 1358
Khairy Fathy, Ju-Sung Kang, Hiroyuki Ogiwara, Bin Eiuo, Hideki Omori, Hyun Woo Lee, Mutsuo Nakaoka

A Switched-Capacitor Lossless Inductor ZCS Snubber-Assisted Series Load Resonant High Frequency Inverter with Dual Mode Pulse Modulation Scheme...1363
Khairy Fathy, Takaaki Okude, Hideki Omori, Hyun Woo Lee, Mutsuo Nakaoka

Topologies of Switch-Linear Hybrid Power Conversion & Special Operation States...................................1368
Lu-sheng Ge, Qian-zhi Zhou, Wu bin

Single Reverse Blocking Switch Type Pulse Density Modulation Controlled ZVS Inverter with Boost Transformer for Dielectric Barrier Discharge Lamp Dimmer...1372
Hisayuki Sugimura, Bishwajit Saha, Hideki Omori, Hyun-Woo Lee, Mutsuo Nakaoka

PDM Controlled Series Load Resonant Soft Switching High Frequency Inverter for Induction Heated Toner Fixing Outer Roller with Inner Cylindrical Working Coil Stator 1377
Hisayuki Sugimura, Hideki Omori, Hyun Woo Lee, Mutsuo Nakaoka

Zero-Voltage and Zero-Current Switching Two-Transformer Full-Bridge Converter Using the Output-Voltage-Doubler 1382
H.K. Yoon, E.S. Choi, S.K. Han, G.W. Moon, M.J. Youn

A Single-stage Boost-Flyback PFC Converter 1387
Zhao Qinglin, Wen Yi, Wu Weiyang, Chen Zhe

Control Bifurcation in PFC Boost Converter under Peak Current-Mode Control 1392
Yi-Jing Ke, Yu-Fei Zhou, Jun-Ning Chen

Analysis and Design of One-Cycle-Controlled Dual-Boost Power Factor Corrector 1397
Yue-feng Yao, Yuan-rui Chen

A Novel Single-phase Buck PFC Converter Based on One-cycle Control...1401
Chen Bing, Xie Yun-Xiang, Huang Feng, Chen Jiang-Hui

Modeling and Simulation of Three Phase High Power Factor PWM Rectifier factor correction....................1406
Yu Fang, Yong Xie, Yan Xing

Effect of the Ripple Current on Power Factor of CRM Boost APFC 1412
A. Abramovitz

Simulated Study of Three-Phase Single-Switch PFC Converter with Harmonic Injected PWM by MATLAB..1416
Zhanlong Li, Yupeng Tang

A Simple Digital Controller for Constant Instantaneous Input Power type Three-Phase Boost Rectifier under Unbalanced System..1421
Jin Ai-Juan, Li Hang-Tian, Li Shao-Long

An Improved and Digital Current Control Strategy for One Cycle Control Based Three-Phase Boost Rectifier under Unbalanced System..1426
Li Shao-Long, Jin Ai-Juan, Li Hang-Tian

Table of Contents

Control Method for Power Quality Compensation Based on Levenberg-Marquardt Optimized BP Neural Networks..1431
Zhou Ming, Wan Jian-Ru, Wei Zhi-Qiang, Cui Jian

A Nonlinear Method for Hybrid Electromagnetic Suspension..1436
Junwei Cui, Jianhui Wang

New topology of multi - level - converter for harmonic reduction 1442
Frank Grundmann, Jian Xie

Model Reference Adaptive Control based on Neural Network for Electrode System in Electric Arc Furnace...1447
Zhang Shi-feng, Zhang Shao-De, Li Kun, Zheng Xiao

STATCOM ETO Failure Analysis...1450
Zhong Du, Bin Chen, Chong Han, Zhaoning Yang, Wenchao Song, Subhashish Bhattacharya, Alex Q. Huang

Modeling and Control of Three-phase Voltage Source PWM Rectifier 1454
Yao Chen, Xin Min Jin

Mitigation of Electric Arc Furnace Voltage Flicker Using Static Synchronous Compensator...................1458
Y.F. Wang, J.G. Jiang, L.S. Ge, X.J. Yang

Design of Distributed FACTS Controller and Considerations for Transient Characteristics....................1463
Gaidi Ning, Shijie He, Yue Wang, Lei Yao, Zhaoan Wang

A Wind-Power Generation System Having a Function of Suppressing Line Voltage Deviation..................1468
Y. Nakayama, S. Fukuda, M. Futami, M. Ichinose, S. Ohara, H. Kita

A Novel Active Islanding Detection Method of Grid-connected Photovoltaic Inverters Based on Current-Disturbing..1473
Zhang Chunjiang, Liu Wei, San Guocheng, Wu Weiyang

Grid Connection of Doubly-Fed Induction Generators in Wind Energy Conversion System....................1477
Ahmed G. Abo-Khalil, Dong-Choon Lee, Se-Hyun Lee

Active and Reactive Power Control of DFIG for Wind Energy Conversion under Unbalanced Grid Voltage 1482
Jeong-Ik Jang, Young-Sin Kim, Dong-Choon Lee

A BASIC STUDY OF FUZZY-LOGIC-BASED POWER SYSTEM STABILIZATION WITH DOUBLY-FED ASYNCHRONOUS MACHINE...1487
Li Wu, Zhixin Wang

Quantitative Analysis on Different Modes of Energy Optimal Control for Series Power Quality Controllers...1492
Huang Xinming, Liu Jinjun, Zhang Hui

Resonance inverter power system for improving plasma sterilization effect 1497
Y.M Kim, J.Y Kim, M. C Jo, S.H Lee, S.P Mun, H.W Lee, S.K Kwon, K.Y Suh

Generic optimization for SMPS design with Smart Scan and Genetic Algorithm................................1502
Heidi H.T. Yeung, N. K. Poon, Stephen L. Lai

Novel Single-Stage Isolated Buck-Boost Inverter Based on Improved SPWM Control Method 1507
Guang-Hui Tan, Fanpeng Zeng, Yanchao Ji, Xi Chen, Hua Wang

On the Effects of Voltage Loop in Paralleled Converters Under Master-Slave Current Sharing.............1512
Yuehui Huang, Chi K. Tse

Improved Control for Parallel Inverter with Current-Sharing Control Scheme 1517
Zhao Qinglin, Chen Zhongying, Wu Weiyang

A Novel Digital Controlled battery charger for High power UPS application......................................1522
Fang Luo, Yong Kang, Shan Xu Duan, Xueliang Wei

xvii

Table of Contents

A Novel High Input Power Factor Single-Stage Single-Phase AC/AC Converter 1527
Chien-Ming Wang, Chien-Yeh Ho, Maoh-Chin Jiag

Research on the Power Sharing of the Parallel Inverters without Control Interconnection Basing on Droop Characteristic 1532
Kan Jiarong, Xie Shaojun

Analysis and Design of Repetitive controlled Inverter System with High Dynamic Performance 1537
Mingzhu Li, Zhongyi He, Yan Xing

Study on a large-volume high-performance programmable voltage disturbance source 1542
Zhan Qizhi, Zhuo Fang, Dong Wenjuan, Wang Zhao'an

1 KW Dual Interleaved Boost Converter for Low Voltage Applications 1546
Heinz van der Broeck, Ibrahim Tezcan

Control of Multilevel Flying Capacitor Inverters for High Performance 1551
L. Zhang, S. J. Watkins, Duan Qi Chang

Analysis of Harmonics in Input Line Current for Matrix Converter based on Double Input Line-toline Voltages 1557
Guo Yougui, Deng Wenlang, Zhu Jianlin

Research on Neutral-point Balancing Control for Three-level NPC Inverter Based on Correlation between Carrier-based PWM and SVPWM 1560
Wenxiang Song, Guocheng Chen, Xiaoyu Ding, Mantang Shu

Instantaneous Voltage Regulated Seamless Transfer Control Strategy for Utility-interconnected Fuel cell Inverters with an LCL-filter 1566
Guoqiao Shen, Dehong Xu, Xiaoming Yuan

An Anti-windup Design Method for Internal Model Control Based on H8 Optimization 1571
Hou Yansong, Li Hua

Study on Pwm Control Strategy of Photovoltaic Grid-connected Generation System 1576
Shi-cheng Zheng, Pei-zhen Wang, Lu-sheng Ge

Robust Sliding Model Control for Regenerative Braking of Electric Vehicle 1581
Min Ye, Zhifeng Bai, Binggang. Cao

A Self-adaptive Fuzzy Control Scheme of High Frequency Link SPWM Inverters 1585
Herong Gu, Deyu Wan, Weiyang Wu

Using Automatic Frequency Shifting Techniques for LLC-SRC Output Voltage Regulation 1590
Kuo-Kai Shyu, Ching-Ming Lai, Ko-Wen Jwo, Ming-Ho Pan, Chung-Ping Ku

Design and Test of Novel Programmable Digital Three Phases SPWM Chip 1595
Yang Yuan, Gao Yong, Chen Lijie

An Improved Performance of Five-Leg Inverter in Two Induction Motor Drives 1598
Ryuji Omata, Kazuo Oka, Atsushi Furuya, Shuji Matsumoto, Yusuke Nozawa, Kouki Matsuse

Adaptive Three Dimensional Space Vector Modulation in abc Coordinates for Three Phase Four Wire Split Capacitor Converter 1603
Xiao-bo Yang, Wei-yang Wu, Hong Shen

Inverters Parallel Operation Based on CAN 1608
Yong Wu, Xianglong Jiang, Jinbang Xu, Qingyi Wang, Shuyun Wan

EMI Reduction Method for a Single-Phase PWM Inverter by Suppressing Common-Mode Currents with Complementary Switching 1613
Toshiji Kato, Kaoru Inoue, Koji Akimasa

xviii

Table of Contents

Analysis and Design of a Novel Dual Secondary Winding and Dual Power Bridge High Frequency Link Inverter 1618
Zhang Zhe, Zhang Chunjiang, Wu Weiyang, Gu Herong, Shen Hong

Research of Complex Fuzzy Control on-off Magnetism Team Motor Speed-Adjusting System 1623
Zhao Ming-fu, Chen Yan, Zhang Zhi-yuan, Dong Chun, DongYu

A New BLDC Motor Drives Method Based on BUCK Converter for Torque Ripple Reduction 1626
Zhang Xiaofeng, Lu Zhengyu

Design of Wind Turbine Generator Control System 1630
Chen Guiyou, Zhou Li, Sun Tongjing, Wang Zhongmin

Non-touching Intelligent Control System of Water Intenerating Equipment Based on Sodion Exchange 1634
Chen Guiyou, Zhang Qingfan, Zhou Li, Luo Donghua

Investigation of Hybrid Modeling and Control for DC-DC Converters 1637
Hao Ma, Feng Qi, Wenqi Zhou

Effect of Peak Current Mode Control on Transient Response for VRM Application 1641
Seiya Abe, Tamotsu Ninomiya

Modulations for Voltage Source Rectification and Voltage Source Inversion Using Direct Space Vector Approach 1646
Keping You, M. F. Rahman

Synchronization of Voltage Waveforms in Basic Topologies of Dual Inverter-Fed Motor Drives 1651
V. Oleschuk, F. Profumo, A. Tenconi, R. Bojoi, A.M. Stankovic

Research on Fast Magnetic Valve Controllable Reactor 1657
Zhang Jian-wen, Cai Xu

Study and comparison of fault tolerant shunt threephase active filter topologies 1663
H. El Brouji, P. Poure, S. Saadate

Application of GA-BP in Fault Diagnosis of Power Circuit of SVC 1669
Zeng Guang, Xi Yu-fan, Su Yan-min, Zhang Jing-Gang

The Optimization-Sliding Mode Control For Three-Phase Three-Wire DSP-based Active Power Filter 1674
Zhou Wei-ping, Liu Da-ming, Wu Zheng-guo, Xia Li, and Yang Xuan-fang

Three-Phase DVR using a Single-Phase Structure with Combined Hysteresis/ Dead-band Control 1679
Seyyed Hossein Hosseini, Mehran Sabahi

Harmonic Detection Based on the TLS Estimation Algorithm 1684
Liu Kaipei, Zhang Junmin

Control Strategy Study of Hybrid Active Power Filter 1689
Jia Zhang, Guohong Zeng

Novel Harmonic Free Single Phase Variable Inductor Based on Active Power Filter Strategy 1693
Mu Xianmin, Wang Jianze, Ji Yanchao, Wei Xiaoxia, Fu Xiangyun

A Multi-Output Series Resonant Inverter with Asymmetrical Voltage-Cancellation Control for Induction-Heating Cooking Appliances 1697
S.H. Hosseini, A. Yazdanpanah Goharrizi, E. Karimi

Capacitor Voltage Control in a Cascaded Multilevel Inverter as a Static Var Generator 1703
M. Li, J. N. Chiasson, L. M. Tolbert

DC-link Pumping-up Voltage Suppression of a Series Active Voltage Regulator With Phase Shift Control 1708
G. C. Xiao, Z. L. Hu, C. H. Nan, Z. A. Wang

The Fuzzy Soft-startup Controller of Active Power Filter 1713
He Na, Wu Jian, Xu Dianguo

Table of Contents

A Novel Control Method for DSTATCOM Using Artificial Neural Network..1718
Yang Xiao-ping, Zhong Yan-ru, Wang Yan

A Detailed Analysis of Unexpected DC-side Voltage Boost in Series Power Quality Controllers 1722
Yuan Chang, Liu Jinjun, Wang Xiaoyu, Wang Zhaoan

Comparative Analysis of Popular Control Schemes for Parallel Active Power Filter and Experimental Verification..1726
Xiaoyu Wang, Jinjun Liu, Chang Yuan, Zhaoan Wang

Accurate Modeling of the Three Phase Induction Motor Including Saturation Effects................................1731
E. V. N. Souza, S. R. Naidu

A study on the reliability evaluation of driving parts for note handling units 1736
Joo Han Kim, Jung Kee Chung, Ha Kyeong Sung, Se Hyun Rhyu

Analysis on Toothless Permanent Magnet Machine with Halbach Array...1741
Xu Yanliang, Feng Kaijie

Improvement in Reliability of Doubly Salient Permanent Magnet Motor Drive.......................................1746
Wenxiang Zhao, Ming Cheng, Xiaoyong Zhu, Wei Hua, Jianzhong Zhang

A New Approach of Modeling the Saturated Induction and Synchronous Salient Pole Machines 1751
A. Câmpeanu, M. Badica

Inductance characteristics of 3-phase fluxswitching permanent magnet machine with doubly-salient structure 1758
Wei Hua, Cheng Ming

Performance Index Evaluations of a Micro Axialflux Switched-reluctance Motor.....................................1763
Cheng-Tsung Liu, Yen-Ming Chen, Da-Chen Pang

Study of Variable Frequency Operation of Induction Generator for Wind Power.......................................1768
Noriyuki Kimura, Mitsuhiro Hirao, Toshimitsu Morizane, Katsunori Taniguchi

Optimal Power Control Strategy of Maximizing Wind Energy Tracking and Conversion for VSCF Doubly Fed Induction Generator System 1773
H. Li, Z. Chen, John K. Pedersen

Design and Evaluation of a Dual Mechanical Port Machine and System 1779
Longya Xu, Yuan Zhang

Characteristic Analysis on Overhang Effect in Axial Flux PM Synchronous Motors with Slotted Winding 1784
WonYoung Jo, YunHyun Cho, YonDo Chun, DaeHyun Koo

Design and Analysis of a Double-Stator Cup-Rotor Directly Driven Permanent Magnet Wind Power Generator 1788
Dong Zhang, Shuangxia Niu, K. T. Chau, J. Z. Jiang, Yu Gong

Feasibility Analysis of Accelerometer Configuration of Non-gyro Micro Inertial Measurement Unit....................1793
Ding Mingli, Zhou Qingdong, Wang Qi, Wang Changhong

Design of Fractional-Order a PI Controller with two modes...1797
Wen Li, Yoichi Hori

Sliding Mode Robust Tracking Control Based on Learning Feedforward Compensation for High Precision Linear Servo System 1802
Zhu Guoxin, Guo Qingding, Zhao Ximei

Application of Fuzzy Self-learning Sliding Mode Variable Structure Control in Linear AC Servo System..............1806
Qing Hu, Shuo Jie, Dongmei Yu

Dynamics Research of Robot Manipulator 1811
Zhibing Shu, Caizhong Yan, Hairong Zhang

Table of Contents

Advanced Angle Control Schemes for Stator Hybrid Excited Doubly Salient Motor Drive 1815
Xiaoyong Zhu, Ming Cheng, Wenxiang Zhao, Wenguang Li

A Design Method of Reconfigurable Controller for AC Position Servo Systems 1820
Wu Qinmu, Qin Yi, Li Yesong

Position Sensorless Control of PMSM Based on a Novel Sliding Mode Observer over Wide Speed Range 1825
Song Chi, Student Member, Longya Xu,

Design of Motion Control System Used for Filter Rod Production Machine 1832
Yang Qingyu, Ge Sibo, Ye Kesong, Shi Ren

Analysis and Implementation of Sensorless Position Detection in a Permanent Magnet Generator 1836
Sebastian Rosado, Xiangfei Ma, Fred Wang, Jerry Francis, Dushan Boroyevich

Torque-Speed Characteristics of Interior-Magnet Machines in Brushless AC and DC Modes, with Particular Reference to Their Flux-Weakening Performance 1841
Y. F. Shi, Z. Q. Zhu, D. Howe

H8 Robust Control for Dual Linear Motors Servo System 1846
Zhao Ximei, Guo Qingding

Research on Linear Motor Driving System Based on Wavelet Transform 1849
Cui Jiefan, Zhao Lijun, Wang Hemin, Wan Junzhu, Jiang Lili

Study on Rotor Position Detection Error in Sensorless BLDC Motor Drives 1853
Li Qiang, Wang Ruixia

A New Scheme to Direct Torque Control of Interior Permanent Magnet Synchronous Machine Drives for Constant Inverter Switching Frequency and Low Torque Ripple 1858
Jun Zhang, M. Faz Rahman, Colin Grantham

A Modified Direct Toque Control for Interior Permanent Magnet Synchronous Motor Drive Without a Speed Sensor 1863
Yanping Xu, Yanru Zhong, Hui Yang

Direct Torque Control for Interior Permanent Magnet Synchronous Motors Using Matrix Converters 1867
D. Xiao, M. F. Rahman

A Neural Network Based Initial Position Detection Method To Permanent Magnet Synchronous Machines 1872
Mengjia Jin, P.C.K Luk, Jianqi Qiu, Cenwei Shi, Ruiguang Lin

A New Recurrent Fuzzy Neural Network Sliding Mode Position Controller Based on Vector Control of PMLSM Using SVM 1877
Junyou Yang, Ruijuan Chen, Naiguang Fa

DSP Implementation of Rotor Position Detection Method for Hybrid Stepper Motors 1882
M. Bendjedia, Y. Ait-Amirat, B. Walther, A. Berthon

An In-Wheel Switched Reluctance Motor for Electric Vehicles 1887
P.C.K. Luk, P. Jinupun

Speed Sensorless Vector Control of Induction Motor Based on Full-Order Flux Observer 1892
Shanshan Wu, Yongdong Li, Zedong Zheng

A Parameter Identification Method for General Inverter-fed Induction Motor Drive 1896
Xiaochun Jiang, Geng Yang, Yunfei Wang

Indirect Rotor Field Orientation Vector Control for Induction Motor Drives in the Absence of Current Sensors 1901
Z. S. WANG, S. L. HO

A Robust Adaptive Sliding-Mode Controller for Slip Power Recovery Induction Machine Drives 1906
J.Soltani, A. Farrokh Payam

xxi

Table of Contents

Identification of the Rotor Time Constant in Induction Machines without Speed Sensor............................1912
M. Li, J.N. Chiasson, M. Bodson, L.M. Tolbert

**Adaptive Control of Doubly Fed Field-Oriented Induction Machine Based On Recursive Least Squares
Method Taking the Iron Loss Into account**...1917
N. R. Abjadi, J. Askari, J. Soltani

Analysis and Design of PDM Converter with High Frequency Link for HEV Drive System........................1922
Ma Xianmin

**A Multi-Directional Power Converter for a Hybrid Renewable Energy Distributed Generation System
with Battery Storage**..1926
Mei Qiang, Wu Wei-Yang, Xu Zhen-lin

Four-bridge Multilevel Converters Based on Hybrid-clamped Techniques..1931
Xiaofeng Wang, Yan Deng, Xiangning He

**Standardization of Input/Output Impedance Specifications of Buck Converters Based on the System
Integration Concept**...1936
Tao Wu, Xinbo Ruan

Research on The Magnetic Integration in Three-Level ZCS Quasi-Resonant Buck Converter...................1942
Jiang Ying, Xiang Hui-jie, Yang Yu-gang, Liu Nan

Decoupling Control of Magnetically Levitated Induction Motor with Inverse System Theory...................1947
Yang Zhou, Huangqiu Zhu, Tianbo Li

Fault Detection and Accommodation for Nonlinear Systems Using Fuzzy Neural Networks.....................1952
H. Xue, J.G. Jiang

**A Novel Constant Power Control of High Frequency Electronic Ballast Applying the PLL Technique for a
Metal Halide Lamp**...1957
Chang-Hua Lin, Chung-Lun Ou, Tien-Shuo Liu, Ken-Chuan Hsu

The Voltage Stability Research of Ship Electric Power System...1962
Fanyinhai Zhaomin

Parasitic Gate Resistance and Switching Performance..1967
Alan Elbanhawy

**PWM Rectifier with DC Reverse-Blocking Diode for High-Reliability Generating Apparatus and Its
Application to Gas Heat Pump System**..1971
Akio Toba, Toshihiro Maeda, Kouetsu Fujita, Tomohiko Kato

A Novel Stator Section Crossing Method of Long Stator Linear Synchronous Motor for Maglev Vehicles..............1976
Qian Zhang, Fei Lin, Xiaojie You, Trillion Q. Zheng

Common Mode Current Suppression in Full-Bridge Converter Based on Simulated Annealing Algorithm.............1981
Yonggao Zhang, Kai Zhang, Yunbin Zhou, Yong Kang

Summary of Distance Measurement Based on Vision in Localization Technology..1986
Handong Zhang, Gang Wang, Yuwan Cen

**The studies of Single-phase Inverter Fault Diagnosis Based on D-S Evidential Theory and Fuzzy Logical
Theory**...1991
Wang Baocheng, Li Danhe, Sun Xiaofeng, Wu Weiyang

A Novel Single-Stage High-Power-Factor Electronic Ballast with Symmetrical Half-Bridge Topology.................1995
Chien-Ming Wang, Chien-Yeh Ho

**Smoothed-Power Output Supply System for Battery of Stand-alone Renewable Power System Using
EDLC**..2000
Y. Jia, R. Shibata, N. Yamamura, M. Ashida

Table of Contents

Supercapacitors characterization for hybrid vehicle applications 2005
F. Rafik, H. Gualous, R. Gallay, A. Crausaz, A. Berthon

Power Transfer Maximization and Di/Dt Based Extremum Tracking for a Swing Engine Based Portable Power System 2010
Satish Rajagopalan, Deepak M. Divan, Ronald G. Harley, J. Rhett Mayor

3D FEA of the Stator of the Linear Magnetic Flux Compression Generator 2015
Yanjie Cao, Chengxue Wang

The Effect of Current Control Strategies on Power Consumption of a Magnetically Levitated Turbomolecular Pump 2018
A.E. Hartavi, R.N. Tuncay, M.N. Sahinkaya

Direct Torque Control of an Interior Permanent Magnet Synchronous Machine fed by a Direct AC-AC Converter 2023
D. Xiao, M. F. Rahman

Control of Distributed Power Systems 2029
Z. Chen, Y. Hu, F. Blaaberg

xxiii

2006 5th International Power Electronics and Motion Control Conference

Volume 4 of 4

2006 5th International Power Electronics and Motion Control Conference

Research on the Power Sharing of the Parallel Inverters without Control Interconnection Basing on Droop Characteristic

Kan Jiarong   Xie Shaojun

Nanjing University of Aeronautics and Astronautics/Automation department, Nanjing, China

E-mail address: kanjr@163.com

Abstract—In the distributed AC supply system without control interconnection, the parameters unbalance because of nonuniformity between the unit inverters is inevitable. The supplier unit is difficult to share power properly. Based on the digital control inverter with instantaneous capacitor-current feedback, the affection to power sharing of the filter inductance and equivalent line impedance are analyzed. Theory and simulation prove that the active power can be shared averagely, the phase-shift between the units changes along with the load variation, the reactive power cannot be shared properly when the parameters are unbalanced, the sharing effect is affected by the load variation. Basing on the analysis, the strategy to reduce the unbalance of the power sharing is presented.

Keywords--Inverter; Parallel operation; Power sharing ; Parameters unbalance

I. INTRODUCTION

In recent years, there are a considerable increase in the using of uninterruptible power supplies (UPS) from the pattern of centralized system to distributed system. One of the crucial technique to the distributed power system (DPS) is paralleled technology. Compared to a single UPS with a big power, DPS has many desirable features, such as expandability, modularity, maintainability, increased reliability and redundancy.

Among many control methods, the frequency and voltage droop method [1-3] is effective. Contrast to the other control methods [4-6], using droop method really achieves the characteristic of redundancy. The digital double instantaneous value close loop controlled voltage source inverter is widely used because of its excellent performance. Therefore, this kind of inverter can be paralleled using the frequency and voltage droop method. However, the parallel inverters can't export equivalent voltage amplitude when difference lies in the elements, such as filter capacitor and filter inductance, which can cause the unequal power sharing. Moreover, the difference lying in line impedance difference between the inverter modules and the load bus also cannot leads to the proper power sharing. Literature [1] has referred to this problem, but the reason that causes this phenomenon

hasn't been analyzed. In the later literature [2-3], the reason hasn't expatiated either.

In this paper, the causation why the parallel inverters can't share the power properly when the component numerical value aren't equivalent is explained and the control strategy which can improve the power sharing between the parallel inverters is presented.

II. OUTPUT VOLTAGE CHARACTERISTIC OF A SINGLE INVERTER

Fig.1 shows the control diagram of a single digital double close-loop controlled inverter.

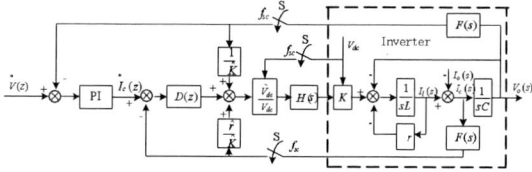

Fig.1 The control diagram of a single inverter

Where f_{sc} is digital sampling frequency.

From Fig.1, the closed-loop voltage gain G is derived.

$$|G(j\omega)| \approx \frac{\dfrac{k_p}{T_i}\dfrac{\hat{L}}{L}}{\left|\dfrac{k_p}{T_i}\dfrac{\hat{L}}{L} - \dfrac{m}{L} + \dfrac{\hat{L}m}{L^2}\right|} = \frac{1}{\left|1 + \left(\dfrac{\hat{L}}{L} - 1\right) * \dfrac{mT_i}{\hat{L}k_p}\right|} \cdots (1)$$

Where K_p is the proportion coefficient of the voltage loop; L is the actual value of the filter inductance; $\hat{L}$ is the estimating value of the filter inductance L, mT_i is the delay time for sampling and calculation of DSP. Equation (1) shows that output voltage characteristic is rigidity when $\hat{L}/L=1$. The characteristic curve will go up when $\hat{L}/L<1$ and the characteristic curves will go down when $\hat{L}/L>1$. In the Fig 2, the marking '*' curves express output voltage characteristics of a 1KVA/220V inverter according to different L value when $K_p=0.3$, $\hat{L}=1mH$, $\hat{C}=30\,uf$. From Fig 2, the error between L and $\hat{L}$ has a prodigious effect to the output voltage characteristic. If the coefficient K_p is increased, from equation (1), the condition that the curves deviate from each other will be alleviate. Homoplastically, the

1-4244-0448-7/06/$25.00 ©2006 IEEE   1532

marking ' ◇ ' curves show the output voltage characteristics when $K_p=1$, $\hat{L}=1mH$, $\hat{C}=30uf$. However, increasing K_p only can minish but not eliminate the deviation between characteristic curves because of the error between L and $\hat{L}$. Moreover, the biggish K_p will reduce the stability of the system.

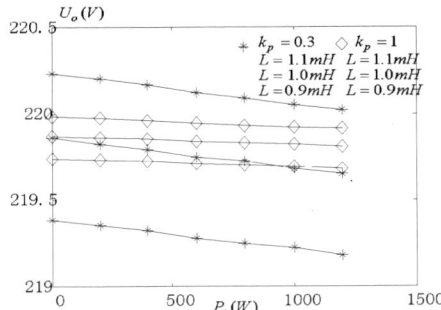

Fig.2 The output voltage characteristic diagram of

a single inverter

The other parameters of inverter affect a little to the output characteristic and the effect can be neglected.

III. THE ANALYSIS OF POWER SHARING

Fig 3 depicts the equivalent circuit of two inverters connected in parallel to a common load, by considering that line impedance is inductive.

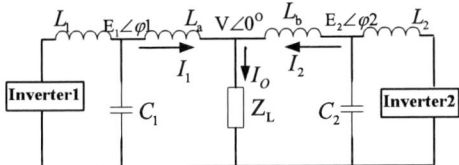

Fig3   The equivalent circuit model of the system

where L_i and C_i are the filter inductance and capacitor of inverters (i=1,2); L_a and L_b are the line impedance between inverter modules and common load bus respectively. The capacitor voltage is the controlled object of the inverter. Therefore, the load voltage can be set $V\angle 0°$ and the filter capacitors' voltage can be set $E_1\angle\varphi 1$ and $E_2\angle\varphi 2$ respectively. The giving voltage amplitude can be set as U_1 and U_2 and the output voltage amplitude of the inverters can be set as E_1 and E_2 because the controlled object of the voltage droop method is giving voltage amplitude but power sharing directly relates with the output voltage amplitude.

From literature [3], the inverters using the frequency and voltage droop method have following features:

1). The active power P_i of a single inverter mainly depends on the phase angle between the output voltage of inverter and the common connection point;

2). The reactive power Q_i of a single inverter hardly influenced by output voltage amplitude. (i=1,2)

A. THE UNBALANCE BETWEEN FILTER INDUCTANCE

In Fig 3, on the assumption that $L_a=L_b$ and $L_1\neq L_2$ (for instance $L_1<L_2$), but the estimating value of the filter inductance $\hat{L}_1=\hat{L}_2$, which will lead to the output voltage characteristic curves of the two inverters have a deviation according to Fig 2.

Fig 4(a) shows the giving voltage amplitude droop control method. $L_1<L_2$ can be regarded as the precondition for analyzing power sharing. $Q_1=Q_2$ is desired, which will result in $U_1=U_2$ according to the amplitude droop control method. Therefore, this will lead to $E_1<E_2$ on the principle of output voltage characteristic and $Q_1<Q_2$. At last, the two inverters will stably work at the condition of $U_1>U_2$, $E_1<E_2$, $Q_1<Q_2$. So the average power sharing will be unequal to each unit because a single inverter can't realize the zero deviation between giving voltage amplitude and output voltage amplitude when each inverter utilize the same voltage-amplitude/reactive-power droop curve.

Fig 4(b) shows the frequency/active-power droop curve. Because the phase angle between the output voltage of inverter and the common connection point is adjusted through changing the frequency, the two inverters will work at the same frequency at last. Or else, the parallel system would be at an unstable state. Thus, the active power P can be shared properly.

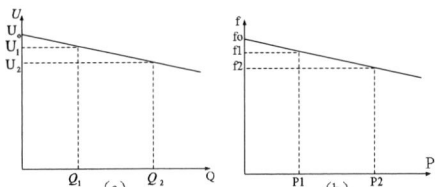

Fig4 Frequency-droop and Amplitude-droop control scheme of

inverters

B THE UNBALANCE BETWEEN LINE IMPEDANCE

In addition to the unbalance of filter inductance of inverters, the difference between line impedance can also cause unequal power sharing.

Equation (2) can be acquired from Fig (3) when $L_1=L_2$ and $L_a\neq L_b$(for instance $L_a<L_b$).

$$\begin{cases} I_1=\dfrac{E_1(\cos\varphi_1+j\sin\varphi_1)-V}{j\omega L_a} \\ I_2=\dfrac{E_2(\cos\varphi_2+j\sin\varphi_2)-V}{j\omega L_b}\cdots\cdots(2) \\ I_O=I_1+I_2=\dfrac{V}{Z_L} \end{cases}$$

Where ω is the foundational frequency of output voltage.

In the general condition, φ_i is very small, So $\sin\varphi_i=\varphi_i$ and $\cos\varphi_i\approx 1$ come into existence. The circular current $(I_1-I_2)/2$ should equal to zero. Equation (3) can be attained from equation (2).

$$\begin{cases} \dfrac{E_1\varphi_1}{\omega L_a} = \dfrac{E_2\varphi_2}{\omega L_b} \\ \dfrac{E_1 - V}{\omega L_a} = \dfrac{E_2 - V}{\omega L_b} \end{cases} \cdots\cdots(3)$$

It is apparent that $E_1 < E_2$ and $\varphi_1 \neq \varphi_2$ from equation (3) and $L_a < L_b$.

In addition, equation (4) is tenable in general condition.

$$\dfrac{|E_1 - E_2|}{E_1} << \dfrac{|L_a - L_b|}{L_a} \cdots\cdots(4)$$

According to equation (2), Fig 5 shows the vector diagram of the parallel system when the load is resistor and inductance.

Where $X_a = \omega L_a$, $X_b = \omega L_b$; I_O is the load current that can be disassembled I_L and I_R. I_L is composed of I_{L1} and I_{L2} and I_R is composed of I_{R1} and I_{R2}. It is obviously that the resistance component of the load not only can bring voltage amplitude difference ($E_{R1} < E_{R2}$) but also result in phase difference ($\varphi_1 < \varphi_2$) of the two inverters; the inductance component of the load only can bring voltage amplitude difference ($I_{L2}X_b > I_{L1}X_a$) but not engender the angle difference because the phase angle in the point of common connection and the voltage of line impedance are consistent. As a result, the angle difference and the voltage amplitude difference of the two inverters aren't equal to zero ($\varphi_1 < \varphi_2$ and $|E_1| < |E_2|$).

From the above analysis, the reactive power Q can't share equally because of $|E_1| < |E_2|$; The active power P is shared properly and its causation similar to the analysis in III A.

The condition that the load changes must be discussed because it is universal in our day using, for instance $Z_{L1} = nZ_{L2}(n>1)$. the angle difference and the voltage amplitude difference of the two inverters is $\Delta\varphi_1$ andΔE_1 when the load is Z_{L1} and $\Delta\varphi_2$ andΔE_2 when the load is Z_{L2}. From Fig 5, it can be attained $\Delta\varphi_1 = n\Delta\varphi_2$ and $\Delta E_1 = n\Delta E_2$. Therefore, there is a regulation process of amplitude and angle difference when the load changes.

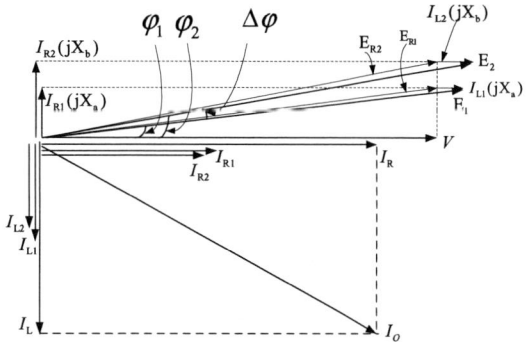

Fig5 Vector diagram of the parallel system

IV. CONTROL STRATEGY FOR IMPROVING THE POWER SHARING

For reducing the $\Delta Q(|Q_1 \text{-} Q_2|)$ caused by unbalanced element numerical value, a regulative tache must be added to the amplitude droop method besides increasing coefficient K_p. The giving voltage amplitude (U_i) changes along with regulating the voltage amplitude droop curve according the degree of the load, which will result in that the output voltage amplitude (E_i) of two inverter aren't equal and the reactive power Q is shared properly.

A. Using Different Voltage Amplitude At no Load s

Equation (5) shows that a regulating voltage tache is added to the voltage amplitude droop control scheme.

$$\begin{cases} U_1 = (1-k)U_0 - mQ_1 \\ U_2 = U_0 - mQ_2 \end{cases} \cdots(5)$$

Where U_O is giving voltage amplitude of inverter 2 at no load; k is voltage amplitude regulating coefficient of inverter 1 at no load. Fig 6 shows that k should be positive number according to equation (3) and equation (5) in the condition of $L_a < L_b$, which can reduce the unbalance of reactive power sharing. From the above analysis in III.B, k should vary along with the load changing. Only in this way, the reactive power can be shared properly, which is a shortcoming of this method.

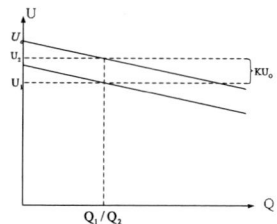

Fig6 The theory of sharing reactive power using different voltage amplitude of no load

B. Using Different Droop Coefficient

Equation (6) shows that different voltage amplitude droop coefficient is used in two inverters.

$$\begin{cases} U_1 = U_0 - n_1Q_1 \\ U_2 = U_0 - n_2Q_2 \end{cases} \cdots(6)$$

Where n_1 and n_2 is the voltage amplitude droop coefficient of inverter 1 and inverter 2 respectively.

In Fig 3, if $L_a < L_b$, $n_1 > n_2$ can be set, which can keep the diversity of giving voltage amplitude. From Fig 7, the reactive power can be shared properly because the giving voltage amplitude diversity changes along with the changing of load, which is needed to the unbalanced numerical value parallel system.

1534

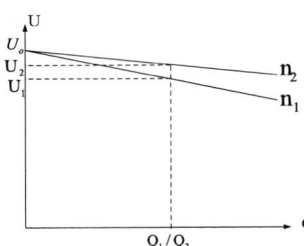

Fig7 Power sharing when the difference between the coefficient of voltage amplitude

V. SIMULATION REASEARCH

The unbalanced situation was simulated for a two-inverter paralleled system in order to verify the theoretical assumptions exposed above. The active power and reactive power calculation was presented in literature [2]. The parameters, correspondingly in Fig 3, listed in Table I.

Tab I

The simulation parameter table when the difference between filter inductance

Inverter1		Inverter2	
L_1	0.9mH	L_2	1.1mH
C_1	30uf	C_2	30uf
L_a	2mH	L_b	2mH
Starting f	50Hz	Starting f	50.1Hz
Starting U	314V	Starting U	311V

Equation (7) shows voltage amplitude and frequency droop control scheme. The voltage frequency can be limited at the range from 49.95Hz to 50.05Hz and the voltage amplitude can be limited at the range from 309V to 315V.

$$\begin{cases} U_i = 312 - (6/3000)Q_i \\ f_i = 50.025 - (0.1/2000)P_i \end{cases} \cdots(7)$$

Fig 8(a) shows the situation of reactive power sharing and Fig 8(b) depicts the situation of active power sharing when the load is equal to $(20+j20)\Omega$. As it can be seen, active power can be shared very well, but the reactive power that inverter 1 endure isn't equal to inverter 2 sharing, which is identical with the theoretical analysis.

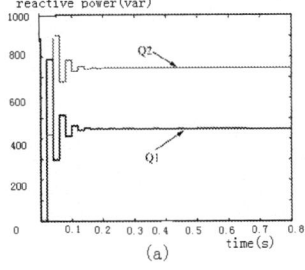

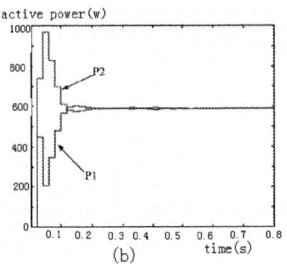

Fig8 The situation of the P/Q sharing when difference lying between filter inductance

Fig (9) shows simulation result when the difference lying in line inductance. The parameters, correspondingly in Fig 3, listed in Table II.

Tab II

The simulation parameter table when the difference between line inductance

Inverter1		Inverter2	
L_1	1mH	L_2	1mH
C_1	30uf	C_2	30uf
L_a	1.9mH	L_b	2.1mH
Starting f	50Hz	Starting f	50.1Hz
Starting U	314V	Starting U	311V

At the time 0.4s, The load of the parallel system changes from $(50+j50)\Omega$ to $(22+j22)\Omega$.

Fig 9(a) shows reactive power sharing condition changes along with changing of the load. Furthermore, the difference reactive power will increase when the load aggravates. Fig 9(b) depicts the phase angle between the two inverters which caused by the unbalanced numerical value of elements. Its magnitude is proportion to the degree of the load.

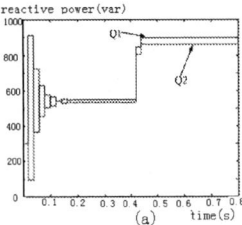

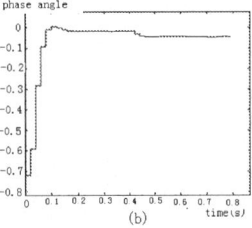

Fig9 The situation of the reactive power sharing and phase shift when difference lying in line inductance

In order to properly share reactive power, the voltage amplitude droop coefficients can be set unequal. Equation (8) is presented.

$$\begin{cases} U_1 = 312 - (6/2460)Q_1 \\ U_2 = 312 - (6/3000)Q_2 \end{cases} \cdots (8)$$

The parameters, correspondingly in Fig 3, listed in Table II. At the time 0.4s, The load of the parallel system changes from $(20+j40)\Omega$ to $(10+j20)\Omega$.

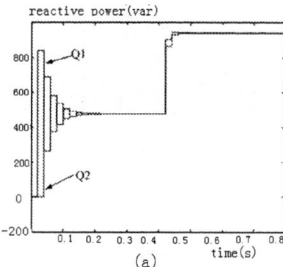

(a)

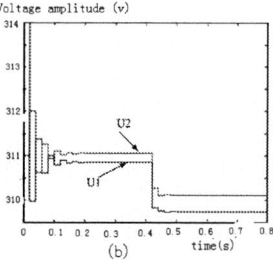

(b)

Fig10 The situation of the reactive power sharing and voltage amplitude when difference lying in voltage coefficient

Fig 10(a) shows the reactive power can be properly shared even if the corresponding parameters is unequal after adding the control tache to the droop method. The difference giving voltage amplitude of the two inverters changes big along with increasing of the load.

All of these results are identical with the theoretical analysis.

VI. CONCLUSIONS

In this paper, the power sharing of parallel inverters without control interconnection system has been analyzed when the difference lies between the actual value and the estimative value.

1). The active power can always be shared properly;

2). The reactive power sharing is unequal when the parallel inverters utilize the same voltage-amplitude/Q droop curve;

3). Adding the regulating tache to the voltage amplitude droop method can realize the reactive power sharing;

4). There are two kinds of methods to average reactive power sharing: one is regulating the voltage amplitude at no load; the other is regulating the coefficient of the inverter. The later can keep a difference between the giving voltage amplitudes, which can achieve the reactive power sharing properly.

REFERENCE

[1] Tuladhar. A，Jin.H，Unger. T，Mauch. K.Parallel operation of single phase inverter modules with no control interconnections. APEC '97 Volume 1，23-27 Feb. 1997 Page(s):94 - 100 vol.1

[2] Y.B.Byun，T.Y.Joe，E.S.Kim，J.I.Seo，D.H.Kim.Parallel operation of Three-Phase UPS Inverter by Wireless Load Sharing Control ,IEEE Trans.on Indus.Appl.,2000,29(1):526-532

[3] J.M.Guerrero，L.Garcia de Vicuna，J.Miret，M.Castilla.A Wireless Load Sharing Controller to Improve Dynamic Performance of Parallel-Connected UPS Inverters, Power Electronics Specialist Conference, 2003. PESC '03. IEEE 34th Annual

[4] Chen J F, Chu C L.Combination voltage controlled and Current controlled PWM inverter for UPS parallel operation IEEE Trans.on PE , 1995 , 10 (5)

[5] TAKAO KAWABATA , SHIGENORI HIGASHINO Parallel operation of voltage sourcr inverters.IEEE Trans on IA , 1988 , 24 (2)

[6] Wu T F , Huang Y H , Chen Y K, et al.A 3C strategy for multi-module inverters in parallel operation to achieve an equal current distribution ,IEEE Transaction IE , 2000 , 47(2)

2006 5th International Power Electronics and Motion Control Conference

Analysis and Design of Repetitive controlled Inverter System with High Dynamic Performance

Mingzhu Li[*], Zhongyi He[**], Yan Xing[**]

[*] Nanjing Design Center, Lite-On Technology Corp., Nanjing, China
[**] College of Automation Engineering, Nanjing University of Aeronautics & Astronautics, Nanjing, China
Email: frequence.li@liteon.com   hezhongyi@nuaa.edu.cn

Abstract—Although repetitive controlled PWM inverters present an excellent steady-state performance, they can not present a good dynamic performance under non-periodic disturbances. To resolve this problem, this paper analyzes the error characteristic of pulse width modulation (PWM) inverter system controlled by repetitive controller and voltage-current dual-loop controller, pointes out the necessary condition for repetitive controlled system to acquire good dynamic performance, and proposes a simple and effective control strategy to amend the demerit of conventional voltage-current dual-loop controlled inverter for excellent performance. At last, Simulation and experiment results in a TMS320LF2407 controlling 3KVA UPS are provided to testify the above conclusion.

Keywords- repetitive control; dynamic; UPS; inverter; PWM; instantaneous feedback

I. INTRODUCTION

Application of PWM inverter in serious ac power-conditioning systems, such as uninterruptible power supply (UPS) and programmable ac source (PAS), requests the better steady-state performance and faster transient response of inverter under different input reference signals and/or disturbances. Consequently, several instantaneous feedback control techniques [1]-[3], such as PID control and deadbeat control, have been introduced into the control of inverter system, and satisfactory results have been obtained for transient load disturbances. However, periodic distortions in the output waveform remain when the load disturbance is cyclic by nature. Because of these, repetitive controller, which has excellent restraining ability to periodic disturbance, attracts more research in [4]-[6] for single-phase inverter and [7] for three-phase rectifier.

Repetitive control theory [8]-[10] originates from the internal model principle. A repetitive controller can be viewed as a periodic waveform generator existing within the control loop of a control system, which is close-loop regulated by a feedback controller, so that the periodic errors can be eliminated. However, as repetitive controller is a learning controller that uses the information of the output error in the previous cycles to compute the repetitive action, the conventional repetitive

controller does not present good dynamic performance under non-periodic disturbances, such as sudden liner load changes. Considering the merit of transient control and repetitive control, and benefiting from the development of DSP technique, the multiple control strategy consists of transient control and repetitive control is becoming more and more familiar in inverter control system.

To find out the factor which may influence the dynamic performance of the multiple strategy controlling inverter system, this paper analyzes the error characteristic of the control system without the repetitive controller and the reason which may affect the regulating characteristic of the control system with repetitive controller, based on the inverter system controlled by repetitive controller and voltage-current dual-loop controller, and proposes a control strategy to amend the demerit of it to improve the whole performance.

II. INVERTER SYSTEM UNDER MULTIPLE CONTROL STRATEGY

A. Tracking Error of Repetitive Controlled System

Fig.1 presents a control diagram of repetitive controlled inverter system and the repetitive controller is just in the dashed-line frame, where $G_o(s)$ is the transfer function of the primary system without repetitive controller. $D(s)$ is the assumed disturbance.

From Fig.1, some basic equations [11] are shown in the following:

$$E(s) = R(s) - Y(s) \qquad (1)$$

$$Y(s) = G_0(s)V(s) + D(s) \qquad (2)$$

$$V(s) = E(s) + M(s)W(s) \qquad (3)$$

$$W(s) = Q(s)e^{-Ls}(W(s) + E(s)) \qquad (4)$$

Combining (1)(2)(3)(4), we can obtain (5):

$$E(s) = \frac{R(s) - D(s)}{1 + G_0(s)}(1 - Q(s)e^{-Ls})\frac{1}{1 - Q(s)e^{-Ls}(1 - G_m(s))} \qquad (5)$$

Where $G_m(s) = \dfrac{M(s)G_0(s)}{1 + G_0(s)}$, and $\dfrac{R(s) - D(s)}{1 + G_0(s)}$ is the tracking error expression of the system without repetitive

1-4244-0448-7/06/$25.00 ©2006 IEEE

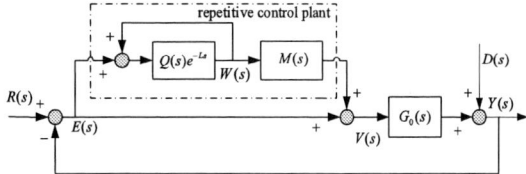

Figure 1. Inverter control system with repetitive controller

Figure 2. Open-loop block diagram of inverter with conventional dual-loop controller

controller. From (5), we can find that the tracking error between the reference signal and the output signal in the system without repetitive controller is an important factor which affects the tracking error of the inverter control system expressed in Fig.1.

Benefiting from the restraining ability of repetitive controller to periodic disturbance, the control system can achieve zero steady-state error in theory. But it can not track a step reference or sudden load changes in time, such as the change from empty load to full load or the reverse. Usually, the tracking error of repetitive controlled system is related to the system without repetitive controller when the above load changes happen. If the system without repetitive controller has big difference error-tracking characteristic in empty load and in full load, and the tracking error is obviously different, the output of the controlled inverter will come out distortion, such as up rush or fall for at least one output cycle, and the control system will have poor dynamic performance because of repetitive controller using the information of previous cycles to compute the repetitive action, when repetitive controller is introduced into the system shown in Fig.1.

B. Conventional Dual-loop Controller and Its Problem

Thinking about the good dynamic performance of voltage-current dual-loop controller and the excellent steady-state tracking characteristic of repetitive controller, we introduce the multiple control strategy consisting of the above two controller in our inverter control system, as can be seen from Fig.1. $G_o(s)$, the open-loop transfer function of the inverter controlled by voltage-current dual-loop controller, is expressed in Fig.2 in details, where C is the output filter capacitor, L is the output filter inductor, R_L is the resistance of the output filter inductor; $V_f(s)$ (equals to $Y(s)$) represents the output voltage of the inverter, I_o and I_L are the load current and inductor current, respectively.

The error of inductor current and load current is filter capacitor current, so the voltage-current dual-loop control scheme is similar to an inner filter capacitor current loop control scheme presented in [3]. But the voltage-current dual-loop control method has more stability and agility than the inner filter capacitor current loop control method. In Fig.2, K_v and K_I are the proportion-controller of voltage-loop and current-loop controller, respectively. K_{PWM} presents a ratio from regulating signal to the output of inverter. Though the value of the DC bus voltage directly affects K_{PWM}, proper methods can compensate for the changes in the DC bus voltage and make K_{PWM} be constant. Here, we set it to be 1. The introduced positive feedback loop of $V_f(s)$ is used to decouple the output voltage.

Setting $K_{PWM}=1$ and According to Fig.2, the open-loop transfer function of the inverter controlled by voltage-current dual-loop controller can be obtained:

$$G_o(s) = \frac{K_v K_I}{sC(sL+K_I)+\dfrac{sL+R_L}{R}} \qquad (6)$$

Equation (6) shows that R_L has important effect to the characteristic of the control system shown in Fig.2. Ideally, R_L equals to zero. The system is I model system no matter in empty load or in liner load, and it will have good tracking performance in the above two situations. But the output filter inductor of inverter always has resistance in natural and R_L does not be zero. That, the system will present as 0 model system in liner load, though it is I model system in empty load. The tracking performance of the system in liner load will be poor and the steady-state error in liner load will be bigger than the error in empty load. From the analysis in section A, we can see that the big difference of tracking error will result in the up rush or fall of output voltage when repetitive controller is introduced.

To eliminate the influence of R_L and improve the performance of multiple control strategy controlled inverter system, an algorithm is proposed in the next section.

C. Proposed Algorithm

It can be seen from (6) that finding method to eliminate the influence of R_L and make the inverter system shown in Fig.2 be I model system is the useful approach to improve the system performance. According to this, Fig.3 presents a resolve algorithm.

Setting $K_{PWM}=1$ and according to Fig.3, the same as Fig.2, the advanced open-loop transfer function of voltage-current dual-loop controlled inverter system can be shown in (7):

1538

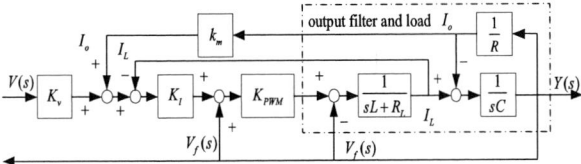

Figure 3. proposed block diagram

$$G_o(s) = \frac{K_v K_I}{sC(sL + K_I) + \dfrac{sL + R_L + (1 - k_m)K_I}{R}} \qquad (7)$$

According to (7), it is obvious that the following (8) is the sufficient condition to reduce the influence of R_L and ensure the dual-loop controlled inverter system to be I model system in liner load condition.

$$R_L + (1 - k_m)K_I = 0 \qquad (8)$$

That is to say, the introduced compensation modulus of load current k_m shown in (8) can eliminate the influence of R_L, make the inverter system have the similar tracking characteristic in empty load and in liner load, avoid the up rush and fall of output voltage in repetitive controlled inverter system shown in Fig.1, and improve the steady-state and dynamic performance.

III. SIMULATION AND EXPERIMENT RESULTS

The simulation and experimental verification is carried out on a 3KVA full-bridge single-phase UPS with output isolation transformer, controlled by DSP TMS320LF2407 from Texas Instruments, where the transfer ratio of the output transformer is 1:2.2, the equivalent filter inductor is 2mH, filter capacitor is 30uF, the DC bus voltage ranges from 180V to 350V, and the output ac voltage is 220Vrms/50Hz.

Fig.4 shows the Matlab simulation results, the response of the control system with conventional dual-loop controller (as shown in Fig.2) under two load changes, from full resistance load to empty load and reverse, where the resistance of filter inductor, R_L, is assumed to be $4\,\Omega$. Fig.4 (a) gives the output voltage of the inverter without repetitive controller. Fig.4 (b) gives the output voltage of the inverter with repetitive controller, as Fig.1 shows.

Fig.5 is still the Matlab simulation results, the response of the control system with advanced dual-loop controller (as shown in Fig.3) under two load changes, from full resistance load to empty load and reverse, where R_L is $4\,\Omega$, too. Fig.5 (a) gives the output voltage of the inverter without repetitive controller. Fig.5 (b) gives the output voltage of the inverter with repetitive controller.

Fig.4 (a) shows the steady-state error of the inverter system is big and the output voltage is lower than

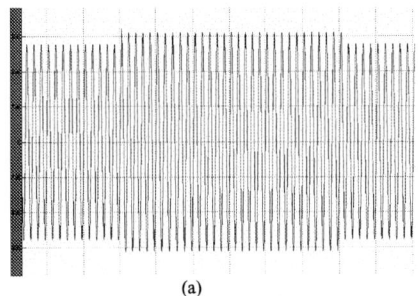

(a)

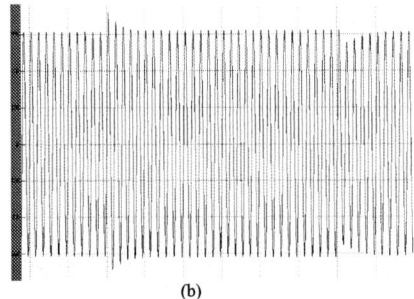

(b)

Figure 4. Simulation results. Transient response of the control system with conventional dual-loop controller under two load changes. (a) without repetitive controller. (b) with repetitive controller

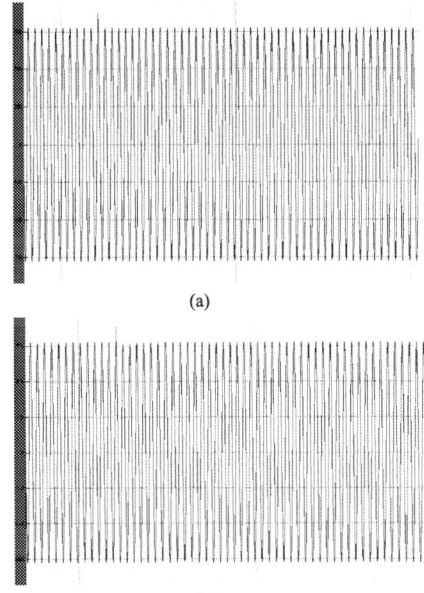

(a)

(b)

Figure 5. Simulation results. Transient response of the control system with advanced dual-loop controller under two load changes. (a) without repetitive controller. (b) with repetitive controller

reference when the system is 0 model system in full load; Fig.4 (b) shows repetitive controller can improve the steady-state performance and make the output in full load similar to the output in empty load, no matter the system is 0 model system or not. Moreover, the difference of Fig.4 and Fig.5 illustrates the transient performance of the control system with advanced dual-loop controller.

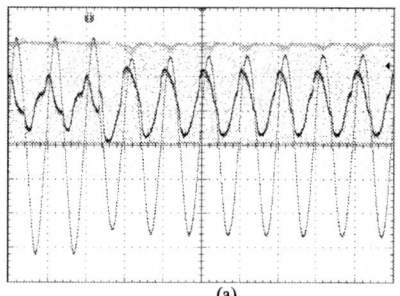 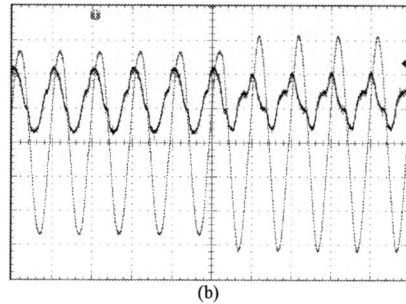

(a)   (b)

Figure 6. Experiment results. Transient response of the control system with conventional dual-loop controller and without repetitive controller under load changes. (a) From empty load to 2 electric bulb load. (b) From 2 electric bulb load to empty load. (red line is output voltage,100V/div ; blue line is the information of tracking error got from D/A converter)

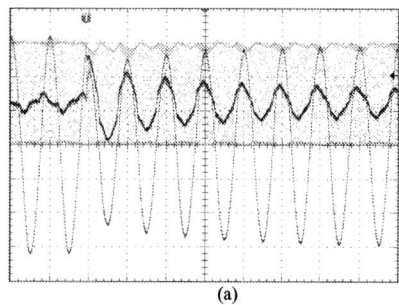 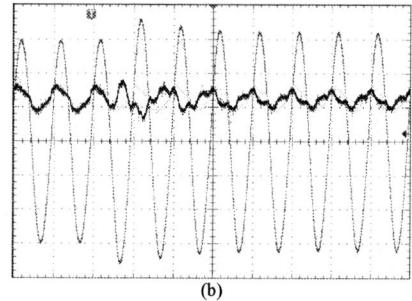

(a)   (b)

Figure 7. Experiment results. Transient response of the control system with conventional dual-loop controller and repetitive controller under load changes. (a) From empty load to 4 electric bulb load. (b) From 4 electric bulb load to empty load. (red line is output voltage,100V/div; blue line is the information of tracking error got from D/A converter)

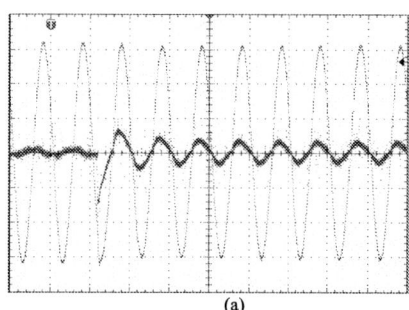 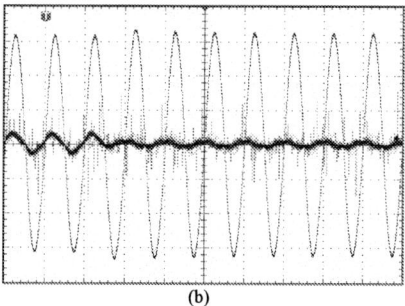

(a)   (b)

Figure 8. Experiment results. Transient response of repetitive controlled inverter with advanced dual-loop controller under load changes. (a) From empty load to 4 electric bulb load. (b) From 4 electric bulb load to empty load. (red line is output voltage,100V/div; blue line is the output filter inductor current (the current in the primary side of output transformer)), 45A/div.

Fig.6 and Fig.7 are the experiment results of the control system with the conventional dual-loop controller. Fig.6 is the result of the system without repetitive controller and Fig.7 is the result of the system with repetitive controller. Fig.6 (a) and Fig.6 (b) shows the transient response of the control system under a load change from empty load to 2 electric bulb load (one electric bulb is 200W and its cold resistance is very small) and reverse, respectively. Fig.7 (a) and Fig.7 (b) shows the transient response of the control system under a load change from empty load to 4 electric bulb load and reverse, respectively. The results in these figures are output voltage and tracking error got from D/A converter.

The difference of Fig.6 and Fig.7 presents that the error between output and reference in repetitive controlled system is smaller than that without repetitive controller. Moreover, Fig.7 shows the obvious up rush and fall of the output voltage under the load changes, which is similar to the simulation results illustrated in Fig.4.

Fig. 8 is the transient response of the repetitive controlled inverter with proposed dual-loop controller under load changes. Fig.8 (a) presents the output voltage and inductor current under a load change from empty load to 4 electric bulb load (the cold resistance of electric bulb is very small, which can be seen from this figure), and Fig.8 (b) shows the output voltage and inductor current under the load change from 4 electric bulb load to empty load. After comparing Fig.7 and Fig.8, we can easily find that the proposed control method is useful to improve the dynamic performance of the whole inverter system.

IV. CONCLUSION

Although the inverter system with repetitive controller presents a good steady-state performance when reference signals and disturbances are periodic, they usually have a poor transient response under non-periodic disturbances. Therefore, this paper analyzed the factor which may affect the dynamic performance of repetitive controlled inverter system, and pointed out the sufficient reason for repetitive controlled system to have good transient response. If the inverter system without repetitive controller has the similar tracking ability in empty load and in liner load, it will have transient response under the load change from empty load to liner load or reverse. Center of this paper is based on an inverter system controlled by multiple control strategy, which consists of repetitive controller and voltage-current dual-loop controller. An algorithm, introducing load current to eliminate the influence of the resistance in output filter inductor, is proposed to improve the dynamic performance of this inverter system. Simulation and experimental results verified the conclusion.

REFERENCES

[1] A. Kawamura and T. Yokoyama, "Comparison of five different approaches for real time digital feedback control of PWM inverter", in *IEEE Ind. Appl. Society Conf.*, vol. 2, 1990, pp.1000-1005.

[2] A. Kawamura, T. Haneyoshi, and R. G. Hoft, "Deadbeat control PWM inverter with parameter estimation using only voltage sensor", in *IEEE PESC Conf. Rec.*, 1986, pp.576-583.

[3] Ryan. M. J and Lorenz. R. D, "A high performance sine wave inverter controller with capacitor current feedback and "back-EMF" decoupling", in *IEEE PESC Conf. Rec.*, vol. 1, 1995, pp.507-513.

[4] Rech. C and Pinheiro. J. R, "New repetitive control system of PWM inverters with improved dynamic performance under nonperiodic disturbances", in *IEEE. PESC. 35th Annual*, vol.1, June. 2004, pp.54-60.

[5] Y. Y. Tzou, R. S. Ou, S. L. Jung, and M. Y. Chang, "High-performance programmable ac power source with low harmonic distortion using DSP-based repetitive control technique", *IEEE Trans. Power Electron.*, vol.12, pp.715-725, July 1997.

[6] Y. Y. Tzou, S. L. Jung, and H. C. Yeh, "Adaptive repetitive control of PWM inverters for very low THD AC-voltage regulation with unknown loads", *IEEE Trans. Power Electron.*, vol.14, pp.973-981, Sept. 1999.

[7] Keliang Zhou and Wang. D, "Digital repetitive controlled three-phase PWM rectifier", in *IEEE. PESC*. Vol.18, Jan. 2003, pp.309-316.

[8] S. Hara, Y. Yamamoto, T. Omata, and M. Nakano, "Repetitive control system: A new type servo system for periodical exogenous signals", *IEEE Trans. Auto. Contr.*, vol.33, pp.659-667, July 1988.

[9] H. L. Broberg and R. G. Molyet, "Reduction of repetitive errors in tracking of periodic signals: Theory and application of repetitive control", in *Proc. 1st IEEE Conf. Contr. Applicat.*, Dayton, OH, Sept, 1992, pp.1116-1121.

[10] M. Tomizuka, T. Tsao, and K. Chew, "Analysis and synthesis of discrete-time repetitive controllers", *Trans. ASME: J. Dyn. Syst., Meas., Contr.*, vol.110, pp.271-280, 1988.

[11] Broberg. H. L, and Molvet. R. G, "A new approach to phase cancellation in repetitive control", in *IEEE. IASAM. Conf. Rec.*, vol.3, Oct. 1994. pp.1766-1770.

2006 5th International Power Electronics and Motion Control Conference

Study on a large-volume high-performance programmable voltage disturbance source

Zhan Qizhi*, Zhuo Fang, Dong Wenjuan, Wang Zhao'an

School of Electrical Engineering, Xi'an Jiaotong University

28 West Xianning Road, Xi'an 710049, P. R. China

*E-mail: zmzqz@163.com

Abstract—Based on the separation of the fundamental parts and harmonics parts of a voltage waveform, this paper presents a novel scheme for a large-volume high-performance programmable voltage disturbance source (PVDS). The purposed equipment uses a serial main circuit to reduce the current flowing through the switches and increase the volume of the system. The system has a high stability and precision with a multiple feedback loop, which can simulate various voltage disturbance and supply a test voltage to the voltage quality equipments such as APF, SPQC, UPQC etc. The results of simulation and experiment indicate that the system has a good static and dynamic performance. The proposed method has been verified on a 300kVA experimental system. Since the system can simulate the three-phase unbalanced voltage, voltage sags, fluctuation and so on. So the proposed system can supply a full voltage test to the related power quality equipments to be tested.

Key words: multiple feedback loop, serial main circuit, and voltage harmonics

I. Introduction

POWER QUALITY problems encompass a wide range of disturbances such as voltage sags/swells, flicker, harmonic distortion, impulse transients , and interruptions [1].In recent years ,various power equipments such as APF(active power filter)、SPQC(series power quality controller) and so on have been widely used in the grid in order to improve the power quality and decrease the harmonic pollution. The volume of these equipments is between several kVA to MVA. To develop these equipments, people usually use programmable voltage source as test power. Recently the programmable voltage source is mainly small and medium-sized inverse source, which cannot satisfy the request and is also expensive. So it is very necessary to develop a high-volume voltage disturbance source to large-volume power quality equipments as test power.

This paper presents a novel scheme suiting for a superpower harmonic voltage source. Based on the principle of separating the fundamental components and harmonic components of the voltage, a series main circuit is adopted in this approach. The grid supplies the fundamental parts and the inverter supplies the harmonic parts. The harmonic parts are coupled to the main circuit through a step-down transformer; by this way the harmonic parts and the fundamental parts are composed of the needed voltage waveform for the equipments to be tested. It is clearly that the step-down transformer can effectively decrease the current flowing through the switch device and increase the volume of the system. A combined construct is adapted to the main circuit to supply an unbalanced three-phase voltage. A control strategy of output voltage feedback and filter's capacitor current inner loops is also adapted to the system to improve the accuracy. A simulation model is created to verify the correctness of the system, and experiments have been done at a 300kVA platform.

II. The control strategy of the PVDS

A. The principle and structure

The PVDS connected between the grid and the power quality equipments to be tested, shown as figure.1. In this figure the inverter can be equivalent to a controlled voltage source to supply harmonic voltage, which can be controlled in frequency、amplitude and phase. The harmonic voltage is coupled to the main circuit through a series injection transformer and added with fundamental component to get the required harmonic voltage source.

The inverter is the center part in harmonic voltage source, its performance directly affect the stability and accuracy of the whole system. As a general-purpose test platform, the harmonic voltage source requires to simulate all kinds of conditions. The requirements to the system mainly include: good performance in dynamic and static state, good stability, high accuracy to assure a excellent test work, high resistance ability of overload and load attack to improve the reliability of the system, these requirements will be discussed in this paper.

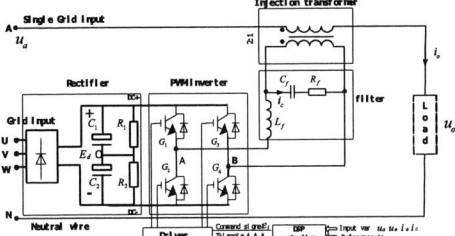

Figure.1 the main circuit of single-phase PVDS

B. The multiple feedback of load voltage and filter capacitor's current

In order to get a high accuracy and stability, the feedback control strategy is used in the purposed system. Since the output voltage waveform can be deviation of fundamental parts and harmonics parts in a main frequency cycle, so the accuracy of the inverter's output waveform is most important to the whole system. The load voltage feedback loop can get rid of the influence of various non-sine element and disturbance to improve the performance of the system[2][3].

The filtering capacitor current is the differential of the inverter's output voltage that a very small change of the voltage can create a big fluctuation in capacitor current, So the filtering capacitor current is a better variable than other signal such as the filtering inductor current, and the system may have better dynamic performance also[4][5]. That is to say, controlling the accuracy of filtering capacitor current can get high accuracy of output voltage waveform. Meanwhile, the inner loop of current can regulate the contained disturbances such as: fluctuation of output voltage、effect of dead-zone time, change in induction parameter, fluctuation in load current and so on, it can improve the static and dynamic performance、adaptability to the nonlinear etc.

The control block diagram of load voltage feedback with capacity current inner loop, shown as figure 2:

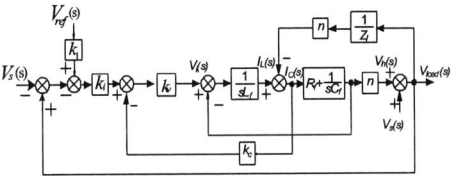

Figure.2 the control block diagram of load voltage and filtering capacity current feedback

In figure 2, V_{ref} is reference harmonic voltage, V_s is supply voltage, V_{load} is the load voltage, V_h is output voltage of the filter, V_i is output voltage of inverter, k_v is equivalent to a magnification of the inverter, k_i is adjustable parameter of voltage loop, k_c is adjustable parameter of current loop, L_f、C_f and R_f are respectively filtering induction、capacity and resistance of high frequency filter, Z_L is load resistance, n is the ratio of transformer and it is set as 1/2 in this paper. The transfer function expressions of two-loop feedback are given as follow.

The transfer function of the output voltage to the reference voltage as below:

$$G_{lr}(s) = \frac{n k_1 k_v k_i Z_l (R_f C_f s + 1)}{a_{11} s^2 + a_{21} s + a_{31}}$$

The transfer function of the output voltage to the source voltage as below:

$$G_{ls}(s) = Z_L \frac{L_f C_f s^2 + C_f (n k_v k_i R_f + k_c k_v + R_f) s + n k_v k_i + 1}{a_{11} s^2 + a_{21} s + a_{31}}$$

In these expressions, a_{11}, a_{21} and a_{31} are respectively given as below:

$$a_{11} = L_f C_f (n^2 R_f + Z_l) , \quad a_{31} = (n k_v k_i + 1) Z_L$$

$$a_{21} = n^2 L_f + C_f Z_L (R_f + k_c k_v + n k_v k_i R_f) ,$$

The expression of output voltage to reference voltage can be given as follow when the influence of load is ignored:

$$G_o(s) = \frac{n k_1 k_i k_v (s C_f R_f + 1)}{s^2 L_f C_f + s C_f (R_f n k_i k_v + k_v k_c + R_f) + n k_i k_v + 1}$$

It's easy to find that by adjusting current-loop parameters properly, the stability of the system can be enhanced, at the same time properly increase voltage-loop controller parameters to develop the response speed and steady-state accuracy.

Figure 3 shows zeros and poles distribution map with load voltage feedback loop and the capacitor current feedback loop when k_i is a fixed value and k_c is gradual increased. In the figure k_i is set at 4.2, increased from 0.001 to 0.397 with step length at 0.099. Spot A、B、C、D and E separately express the zero and pole distribution when k_c increased gradually. For the designed system, the further the distant of poles to virtual axis the better, the convergence speed will be accelerated and the stability will be enhanced. We can see from figure 3 when k_i is fixed, as k_c increased, the distant of closed-loop poles to virtual axis are further; hence

1543

the stability of the system is increased gradually.

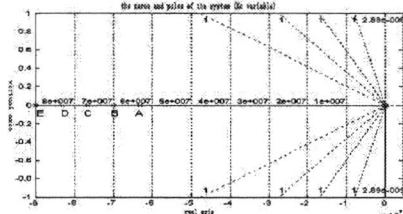

Figure.3 the zeros and poles distribution with multiple loops when k_i is 4.2,and

k_c is increased gradually

III. Simulation results

This paper created a simulation model for the purposed system; the simulation results verified the correctness of the system principle and the control strategy.

First the voltage harmonics is simulated, a fifth-order harmonic is created on a three-phase model, figure 4 shows phase-a load voltage wave, figure 5 shows the FFT analysis spectrum graph, fifth-order harmonic THD is set to be 1/5, by FFT analysis the result HRU_3=0.1976 is achieved, whose relative error is 0.1% compared to expect value.

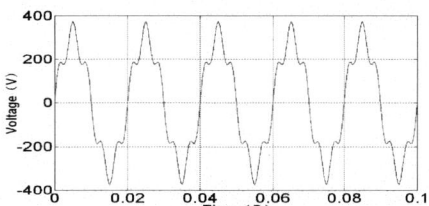

Figure.4 the load voltage wave of phase-A

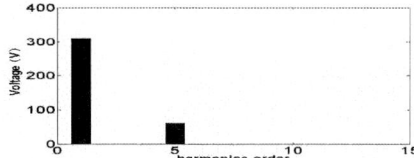

Figure.5 the FFT analysis result of load voltage

In order to test the harmonic voltage source output ability. We created a combine voltage source with triple-order, fifth order, seventh order, eleventh order and twenty-fifth order harmonics; Figure 6 shows the output wave of the system. Figure 8 shows the FFT analysis result.

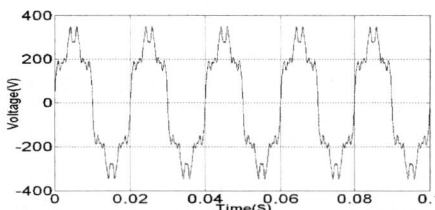

Figure .6 the load voltage wave of phase-A

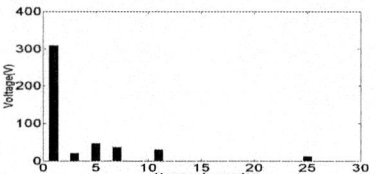

Figure.7 the FFT analysis result of load voltage

Table.1 lists the Fourier analysis results, in which the fundamental wave is enactment as 100%;

Table.1 each order harmonics setting value and detect value

Harmonics (order)	3	5	7	11	25
HRU_n(reference)	0.0667	0.15	0.12	0.1	0.04
HRU_n(practical)	0.0663	0.1487	0.1192	0.0990	0.0403
Relative error(%)	0.59	0.86	0.67	1.0	0.75

The result of the FFT analysis shows that the relative error of each harmonics is lower than 1.0 percent, which verifies the accuracy of the system.

Voltage sag is also simulated on the system; figure.8 shows the output voltage when voltage sag is 30%.

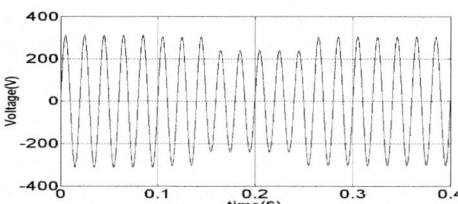

Figure.8 the load voltage when the command is voltage sag

When the command voltage is a pulse, the system can simulate the voltage pulse, the result is show as figure.9.

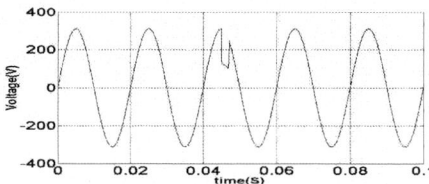

Figure.8 the load voltage when the command is voltage pulse

Unbalanced three-phase voltage can also be simulated on the system; the result is show as figure.9.

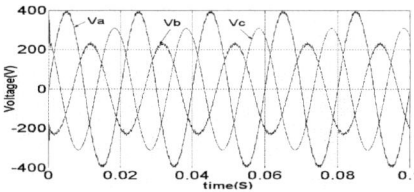

Figure.8 the three-phase load voltage when the command is unbalanced

voltage

IV. Experimental result

Based on the main circuit and control strategy discussed above, a 300kVA experimental platform has developed. Which can supply a test voltage to the voltage quality equipment to be tested. All functions of the system have been verified. An example is show as below when the platform supply a harmonics voltage source to a SPQC. Figure.9 shows the output voltage and current of the system, Figure.10 show the content of each harmonics when the fundamental voltage supposed to be 100%.

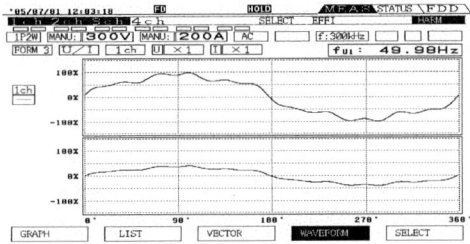

Figure.9 the wave of load voltage and current

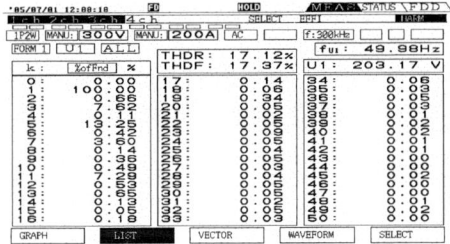

Figure.10 the content of each order harmonics

We can see from above results, each frequency harmonic output is basically the same, switching frequency harmonic has been filtered effectively, the difference between output value and setting value is under $\pm 1\%$, the output wave can completely satisfy the require of the equipment to be tested.

By control the output fundamental voltage wave of the inverter, the voltage sags can be simulated, figure 11 shows the output voltage wave with 30% sags when the supply voltage is 220V, the tested voltage sag amplitude is 29.3%, wave and accuracy all satisfy the requirement of the equipment to be tested.

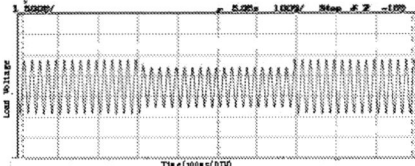

Figure.11 output voltage wave when voltage sag is 30%

V. Conclusions

This paper mainly studied the implementation method and control strategy of a high power programmable voltage disturbance source; the main circuit and control strategy are analyzed. Which have been verified by simulation and experiment. From the results we can see, series main circuit can improve the volume of the harmonic voltage source, the system has a good accuracy with the load voltage and filtering capacitor current feedback method, a 300kVA experimental platform have been designed successfully, which can efficiently supply the required voltage for related power quality equipments and simulate various voltage quality questions.

References

[1] H. Akagi, "New trends in active filters for power conditioning," *IEEE Trans. Ind. Applicat.*, vol. 32, pp. 1312–1322, Nov./Dec. 1996.

[2] Annette von Jouanne, Prasad N. Enjeti and Donald J. Lucas. DSP control of high-power UPS systems feeding nonlinear loads *[J] .IEEE Trans.Ind Electron.1996 ,43(1) :1212125.*

[3] A.Kawamura, R.G.Hoft. Instantaneous Feedback Controlled PWM Inverter with Adaptive Hysterics. *IEEE trans. on Ind. Vol.1, Apr. 1984, pp.769-755*

[4]M.Abdel-Rahim,E.Quaicoe, "Analysis and design of a multiple feedback loop control strategy for the single-phase voltage-source UPS inverters" *,IEEE Trans. On Power Electron., Vol.11,No.4,July 1996,pp.532-541*

[5]M.Queidat,D.Sadaruac, "Multiple feedback loop control strategy for UPS system having active filter ability",*INTELEC'96,1996, pp. 450-453f*

2006 5th International Power Electronics and Motion Control Conference

1 KW Dual Interleaved Boost Converter for Low Voltage Applications

Heinz van der Broeck, Ibrahim Tezcan

University of Applied Sciences Cologne, Faculty IME, Betzdorfer Str. 2 50679 Köln, Germany
Email: heinz.vdbroeck@fh-koeln.de

Abstract—This paper deals with a boost converter for low input voltages in battery or photovoltaic applications. The performance of a dual interleaved boost converter is investigated theoretically and experimentally. In order to increase the power density and to avoid EMI problems caused by the reverse recovery current of the diode, the discontinuous mode is considered only . At first, the steady state performance of the boost converter is investigation by a MathCAD simulation. It is shown that the input current ripple of a boost converter can substantially be reduced in the interleaved mode. A 1KW dual boost converter has been designed, realized and successfully tested in the laboratory. The integrated circuit UCC28220 from Texas Instruments has been used to control the converter. It employs fast current mode controllers and gate drivers for both sub converters. An error amplifier is inserted to stabilize and to limit the output voltage. The measured input current wave of the dual boost converter is in good agreement to the result of the simulation. The converter shows a high efficiency in the whole operation range.

Keywords: Dual boost converter, step up chopper, discontinuous mode, interleaved operation, UCC28220

1. Introduction

Step up choppers for DC to DC power conversion find widespread applications in different areas. They are used to boost a varying DC input voltage to a higher stabilised DC output voltage or to draw a certain current from a DC power source. These features can be used to supply a 42V power net from a 14V battery in automotive applications or to track the maximum point of power in photovoltaic systems.
The step up chopper or boost converter is a basic topology of power electronics and thus a well known circuit [1]. However, its operation mode and control scheme strongly depend on the specific application and the available components. In the last years, high power density, EMI compatibility and low cost are more and more important for many power electronics applications and thus also for the design of a boost converter. At low input voltages, typically given in automotive or photovoltaic applications, is also a challenge to achieve a high efficiency. This paper deals with the analysis and hardware realisation of dual interleaved boost converter aiming at an optimised design.

2. Operation modes of the boost converter

The boost converter shown in figure 1a can either be operated in the continuous or in the discontinuous mode. This is illustrated in figure 1b and 1c where both, the voltage at the transistor and the current in the choke are presented for the same operation point considering a large and a small choke.

a)

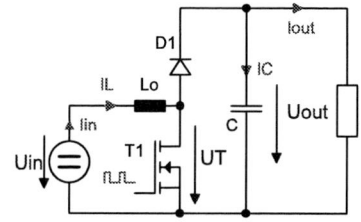

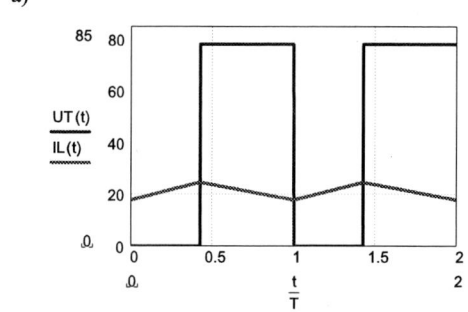

b) $\bar{I}_L = 21.1A$ $\hat{I}_L = 24.5A$ $\tilde{I}_L = 21.2A$ $\Delta I_L = 7A$

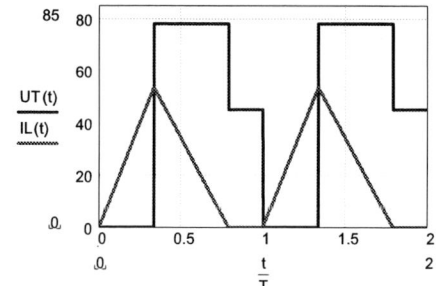

c) $\bar{I}_L = 21.1A$ $\hat{I}_L = 53.6A$ $\tilde{I}_L = 27.5A$ $\Delta I_L = 53.6A$

Figure 1 a) topology of a single boost converter
b) MCAD simulation: Uin = 45V Uout = 78V P = 950W f = 112 kHz Lo = 25μH
c) same as (b) however lower inductor Lo = 2.5μH

1-4244-0448-7/06/$25.00 ©2006 IEEE   1546

The voltage and current curves have been simulated by a MATHCAD program, which also calculates the peak, ripple, average and RMS values of all currents.

From the comparison of the two operation modes it becomes obvious that the continuous mode leads to a small current ripple and to the lowest RMS current (see fig. 1b), As a disadvantage of this mode, the reverse recovery current of the diode has to be mentioned. This is not considered in the simulation program but it has to be taken into account in the hardware. The reverse recovery current, occurring at the turn-off of the diode, is a large source for EMI problems. This effect can be avoided if the circuit operates in the discontinuous mode as shown in figure 1c. The higher conducting losses of this operation mode caused by the increased RMS current can be compensated by using MOSFETS with a low Rdson. Such power semiconductors are recently developed for low voltage applications. Several semiconductor manufacturers offer 100V MOSFETs in a TO220 package with a Rdson of only 10mΩ. Concerning the choke it is also evident that the smaller choke shows a lower DC resistance.

Hence, by using MOSFETs with a low Rdson the discontinuous mode has to be preferred for a boost converter because of the higher power density and the improved EMI compatibility.

As will be shown later, the discontinuous mode also allows galvanic isolated current sensing at low cost with almost no voltage drop. In order to prevent HF losses in the choke, its winding has to be realised by litze wire.

3. Interleaved operation

The remaining disadvantage of the discontinuous mode can be seen in the large current ripple which requires extra HF filtering at the input (not shown in fig. 1a). The effort for this low pass filter can substantially be reduced if the boost converter is split into two parallel connected units, which are controlled by a phase shifted switching function (interleaved operation) [3,4]. There are no additional semiconductors needed for the power part since most high current boost converters are already composed by a parallel connection of MOSFETs and diodes. The effect of the interleaved operation is illustrated in figure 2, which shows the topology, the currents in the subsystems (dotted line) and the resulting input current of the interleaved boost converter. Since the simulated current curves in figure 1c and 2 are based on the same parameters the improvement of the interleaved operation becomes obvious.

The choke currents I1(t) I2(t) of the sub boost converters still flow discontinuously while the superposed input current Iin(t) flows continuously with a lower current ripple $\Delta Iin = Iin_{MAX} - Iin_{MIN}$.

The waveforms of the currents have been calculated by the MathCAD program mentioned before.

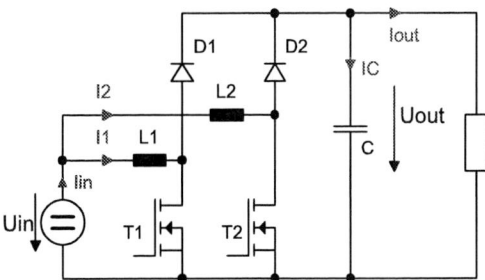

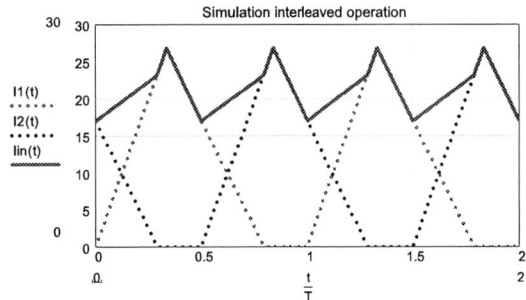

Figure 2 Interleaved boost converter. Simulated input current for Uin = 45V Uout = 78V P = 950W f = 112kHz
L1 = L2 = 5µH $\overline{Iin} = 21.1A$ $\hat{Iin} = 26.8A$ $\Delta Iin = 9.8A$

4. Analysis of the bost converter topologies

This chapter defines the input current range for a single or a dual boost converter operating in the discontinuous mode and it analyses the corresponding current ripples. For the analysis the equivalent topologies shown in figure 3 are used.

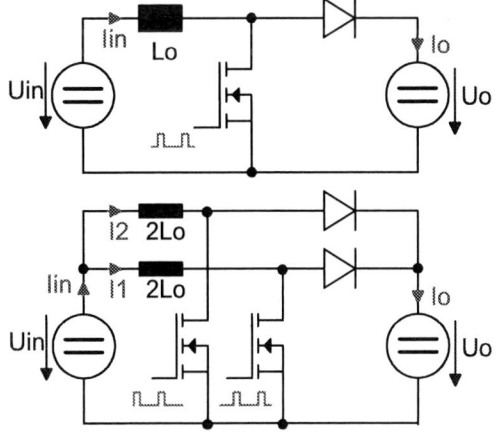

Figure 3 Single and dual boost converter

In both cases a constant output voltage of Uo is assumed and the converters are operated with a fixed switching frequency fs. In general the input voltage Uin may vary between: $0 \leq Uin \leq Uo$.

1547

The inductor Lo of the single boost converter is split into two inductors with an inductance of 2Lo and half the current capability for the dual boost converter.

If the average input current should always flow discontinuously it has to be limited to

$$\overline{Iin} \le \frac{Uo}{2 \cdot fs \cdot Lo} \cdot \frac{Uin}{Uo} \cdot \left(1 - \frac{Uin}{Uo}\right)$$

This is illustrated in figure 4 in a normalised way.

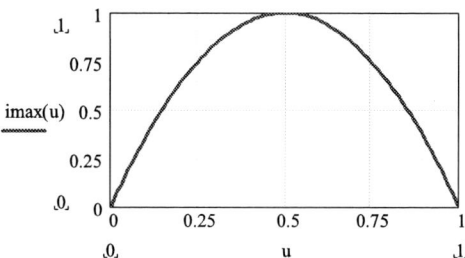

Figure 4   Average input current limit for discontinuous operation. $i_{MAX} = \overline{Iin} \cdot 8 \cdot fs \cdot Lo / Uo$ $u = Uin / Uo$

As can be seen in the fig. 4 the maximum discontinuous flowing current $\overline{Iin}\max = \dfrac{Uo}{8 \cdot fs \cdot Lo}$ .can be drawn from a voltage source Uin for $Uin = Uo/2$. For all other voltage conversion ratios the average current is lower. Considering an input voltage range of $Uo/4 \le Uin \le 3 \cdot Uo/4$ at least ¾ of the maximum current is allowed within the discontinuous mode.

While the average input current is the same for the single and the dual boost converter their current ripples ΔIin are different. This is illustrated in the following figures.

Figure 5 shows the currents in the filter chokes and the superposed current Iin(t) of both converters for $\overline{Iin} = \overline{Iin}\max$ and $Uin = Uo/2$. In this special case the current ripple of the single boost converter is twice the average input current $\Delta Iin,s = 2 \cdot \overline{Iin}\max$ and no current ripple occurs for the dual boost converter $\Delta Iin,d = 0$

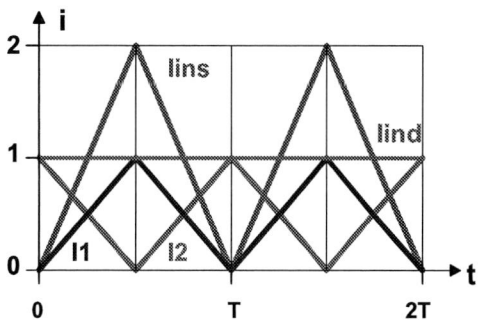

Figure 5 Performance of the single and dual boost converter for the critical conduction mode Uin=Uo/2

Fig. 6 presents the currents of the dual interleaved boost converter for different voltage conversion ratios and the average input current: $\overline{Iin} = \overline{Iin}\max \cdot \dfrac{1}{2} = \dfrac{Uo}{16 \cdot fs \cdot Lo}$.

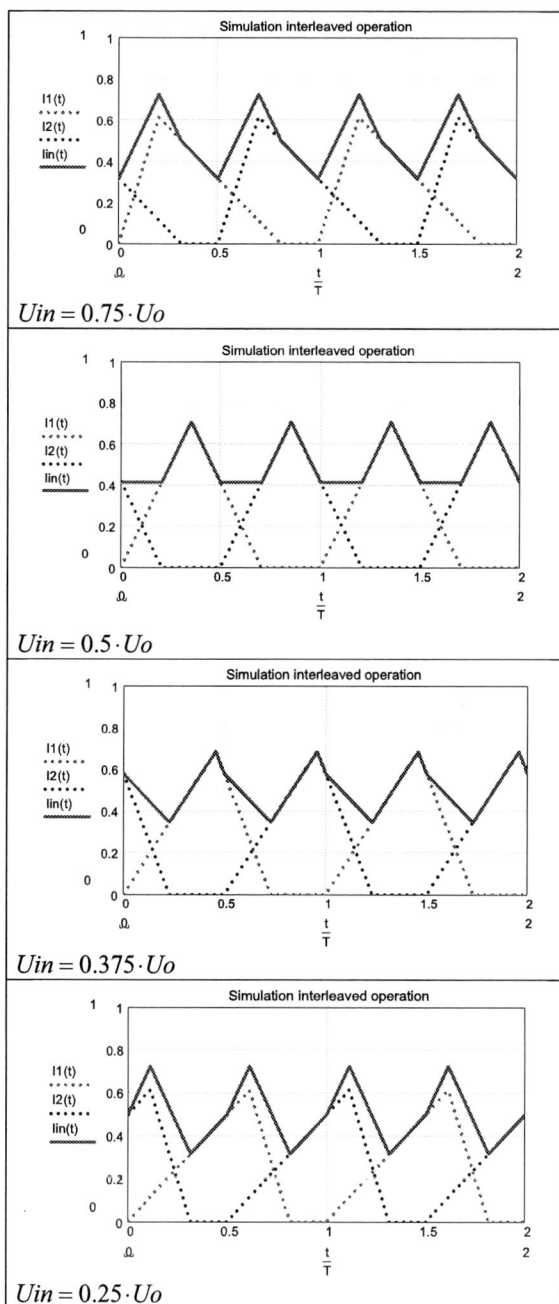

Figure 6 Currents in the interleaved boost converter
Normalised current $I_v = I_v(t) \cdot 8 \cdot fs \cdot Lo / Uo$

The dotted lines represent the current in the chokes and the solid line shows the input current which flows continuously for the chosen voltage conversion ratios.

1548

For $Uin = Uo/2$ the current ripple of the both boost converters can easily be determined as a function of the average input current $\overline{Iin}$. The calculation of the current ripple in a single boost converter yields:

$$\Delta Iin,s = 2 \cdot \sqrt{\frac{\overline{Iin} \cdot Uo}{8 \cdot fs \cdot Lo}}$$

In case of the dual interleaved converter one has to distinguish between two different operation ranges. For $\overline{Iin} < \overline{Iin}\,max/4$ the input current Iin(t) flows discontinuously leading to a current ripple of

$$\Delta Iin,d = \sqrt{\frac{\overline{Iin} \cdot Uo}{8 \cdot fs \cdot Lo}}$$

This is similar to the single boost converter but it benefits from the higher switching frequency which is doubled effectively. For higher input currents:

$\overline{Iin}\,max/4 < \overline{Iin} < \overline{Iin}\,max$ the input current Iin(t) flows continuously due to the overlapping of the choke currents. The corresponding current ripple is given by

$$\Delta Iin,d = \frac{Uo}{8 \cdot fs \cdot Lo} - \sqrt{\frac{\overline{Iin} \cdot Uo}{8 \cdot fs \cdot Lo}}$$

The calculated current ripples are evaluated in figure 7 in a normalised way.

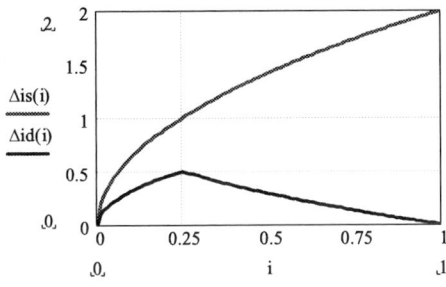

Figure 7 Normalised current ripple $\Delta i = \Delta I / \overline{Iin}\,max$ of the single and dual boost converter for Uin=Uo/2

From the comparison of the current ripples it becomes obvious that the interleave operation is a preferable solution since it combines all advantages of the continuous and the discontinuous mode.

5. Hardware realisation of the dual interleaved boost converter

In the past, interleaved boost converters could hardly be realised for low cost applications because of missing controller ICs. The problem was to optimal synchronize frequency and phase of two single controller ICs. It always required a very complex control circuitry.

The problem has been overcome by the controller IC UCC28220 from Texas Instruments [2], which contains all required functions needed for the optimum performance of a dual interleaved converter.

Based on the UCC28220 a 1 KW boost converter has been designed, realised and successfully tested for the

input voltage 30V...60V and a stabilised maximum output voltage of Uout = 80V. The boost converter and the control part of one sub system are presented in fig. 8.

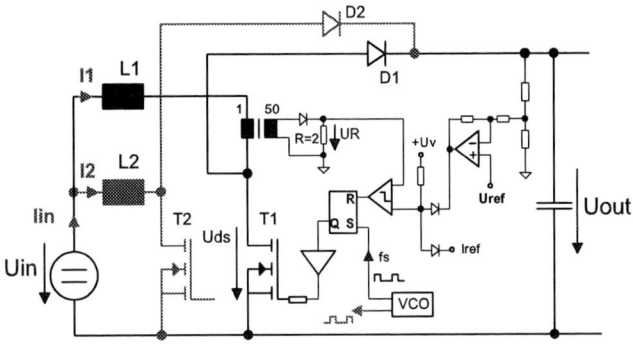

Figure 8 Interleaved dual boost converter & control part

As already mentioned before separate current mode controllers are applied for both subunits. The currents I1 and I2 in the two chokes (L1, L2) are sensed by means of current transformers. They provide galvanic isolation in the current feed back loop and avoid the use of shunts. Thus, less HF noise disturbs the controller IC and additional conducting losses are saved. An accurate current measurement via the transformers is only possible in the discontinuous mode. In general, the boost converter controls the peak values of the current in the chokes to be equal to a reference current Iref.

An additional control loop with an error amplifier prevents the output voltage to exceed an upper value. It overrules the current controller if the output voltage exceeds a certain threshold set by Uref

Figure 5 presents the realized hardware. Each sub converter consists of two parallel connected transistors IXUC120N10 in a TO220 package and one diode 60EPU02 . All power semiconductors are mounted below the PCB on the heatsink.

Figure 9 Realised 1kW dual interleaved boost converter

With the chosen switching frequency fs =112kHz and the small chokes 2Lo =L_1 = L_2 = 5 µH both sub converters never leave the discontinuous conduction mode.

1549

The performance of the developed interleaved boost converter has been successfully proved by several measurements. Figure 10 shows the performance of the sensor, where $U_R = I_{L1} \cdot R/50$ is proportional to the current $I_{L1} = Iz$ measured by a Tektronix current probe.

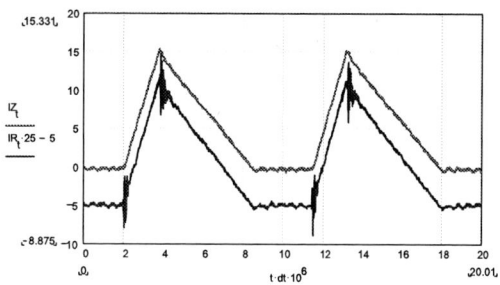

Figure 10 Choke current measured by a current transformer and by a Tek current probe

The characteristic voltage and current wave of one basic boost converter module can be seen in figure 11. The measurement has been taken at an input voltage of Uin = 40V and an output power of P = 2 x 500W =1000W

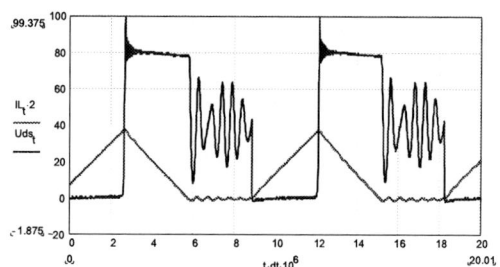

Figure 11 Current and voltage of one sub unit (Uin= 40V Uo=80V P = 2 x 500W)

Figure 12 presents the common input current of the interleaved dual boost converter at almost nominal power. This measurement confirms the results of the simulation shown in figure 2 , which are based on the same parameters.

At last, the efficiency of the dual boost converter has been measured as a function of the output power for three input voltages. The results are presented in figure 13. As could be expected the highest efficiency is achieved for an input voltage of Uin =60V.

6. References

[1] Ned Mohan; Tore M. Underland; William P. Robbins Power electronics: Converters, applications, and design New York : John Wiley & Sons , 1989

[2] Texas Instruments UCC28220 UCC28221 Datasheet SLUS544A, August 2004

[3] Perreault, D.J.; Kassakian, J.G.; Distributed interleaving of paralleled power converters IEEE Transactions on Circuits and Systems Vol. 44, Aug. 1997, pp 728 – 734

[4] Veerachary, M.; Senjyu, T., Uezato, K. Modeling and analysis of interleaved dual boost converter Proc. IEEE International Symposium on Industrial Electronics 2001 Volume 2, June 2001 pp 718 - 722

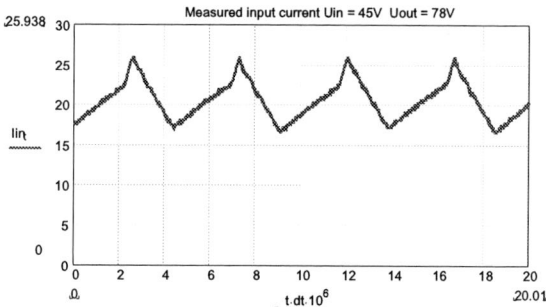

Figure 12 Measured input current   (Parameter Uin = 45V Uout = 78V   Pin = 950W   f = 112 kHz   L = 5μH)

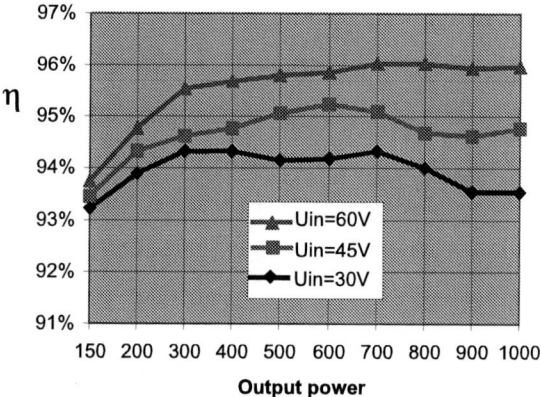

Figure 13 Efficiency of the dual interleaved boost converter (Uo = 78V = const.)

7. Conclusion

A 1kW dual interleaved boost converter has been investigated theoretically and experimentally. For high power density and EMV reasons the dual boost converter preferably operates in the discontinuous mode if the input voltage range is not larger than 1:3. Although the sub boost converters are always operating in the discontinuous mode the common input current of the converter flows continuously over a large current range. Thanks to the new IC UCC28220 all required control functions can be applied. This includes current mode control and equal current sharing. Another important advantage is the use of current transformers instead of shunts.

Control of Multilevel Flying Capacitor Inverters for High Performance

L. Zhang[*], S. J. Watkins[*] and Duan Qi Chang[**]

[*] School of Electronic and Electrical Engineering, University of Leeds, UK
[**] School of Automation, Chong Qing University, China

Abstract − The paper presents a switching pattern selection scheme for the control of a multilevel flying capacitor inverter. The scheme reduces capacitor voltage fluctuation without using voltage feedback. The method is developed for sinusoidal voltage generation using the selective harmonic elimination (SHE) technique and was compared favorably with the result when voltage feedback was used.. The result of the comparison is presented. Simulation and experimental results from a four-cell five level flying capacitor inverter validates the method.

Keywords − *Flying Capacitor Inverter, Selective Harmonic elimination, THD*

I. INTRODUCTION

The general function of multilevel converters is to synthesize a desired voltage from several levels of DC voltages. In contrast to their conventional two-level counterpart, their main advantage lies in reaching high voltage with low harmonics, using solid-state devices of lower voltage rating and much lower switching frequencies [1].

This paper explores the modulation strategies for flying-capacitor multilevel inverter. The four-cell five-level implementation of this circuit is shown in Fig. 1 [2, 3]. An important advantage of this circuit lies in having multiple switching combinations for the same voltage level. This gives flexibility in switching control strategies to achieve optimized output performance. Maintaining the correct voltage across the floating or cell-capacitor is the main challenge in ensuring proper operation of the inverter. Though it is desirable to use smaller capacitors in order to reduce cost and size of the inverter, the subsequent drawback is in the increased voltage swings on the intermediate voltage levels. This may cause excessive voltage stress on switching devices and the voltage ripple is counter-productive to producing a low harmonic content voltage waveform. A number of publications has addressed the issue of capacitor voltage balance [4,5,6].

The present paper investigates an optimal switching pattern selection mechanism for capacitor voltage balance without voltage feedback. The method is developed for sinusoidal voltage generation using the selective harmonic elimination (SHE) technique [7] and devices are switched only at or near fundamental frequency. The technique is intended for the control of high voltage static compensators. The paper discusses the principles applied for selecting the

switching sequences. Simulation studies of the selected optimal pattern applied to the four-cell inverter circuit are presented. Comparison of the method with a capacitor voltage feedback scheme is also given. Simulation results demonstrate that the balanced capacitor voltages are obtained, and hence output line voltage/current THDs are significant reduced.

II. REVIEW OF SELECTIVE HARMONIC ELIMINATION SCHEME

The selective Harmonic Elimination (SHE) scheme is a well known PWM technique [7] in which the switching angles of the inverter are determined to set the fundamental voltage at some specific magnitude and simultaneously suppress certain harmonics. A multi-level inverter has the increased flexibility to synthesis its voltage waveform offered by the multiple voltage levels. In its simplest control form, the resultant waveform for the phase voltage of the inverter in Fig. 1 has the shape of a staircase shown in Fig. 2.

Notice that the number of controlled components is equal to half the number of cells in the inverter due to the requirement for symmetry around the mean level, with each inverter limb operated in bipolar mode. With basic SHE control, the four-cell, five-level inverter under investigation can only be controlled with the fundamental regulated and one harmonic eliminated yielding equations as must be solved using the Newton-Raphson method to obtain the required control angles

$$\cos(\alpha_1) + \cos(\alpha_2) = \frac{\pi m_a}{2} \qquad (1)$$

$$\cos(5\alpha_1) + \cos(5\alpha_2) = 0 \qquad (2)$$

where α_1 and α_2 are the angles at which a level transition occurs. The Newton-Raphson method can be to solve the nonlinear equations (1) and (2) for the α_1 and α_2. Fig. 2 shows the resultant staircase shaped phase and line voltage waveforms when fundamental magnitude is set to 85%.

This method can be extended to higher orders of harmonic elimination, but obtaining a solution can be tedious and other approaches have been suggested [8].

The performance of the control schemes has been assessed using the total harmonic distortion (THD) and distortion factor 1 (DF1) [9] figures of merits. However as is known the balancing of a multilevel inverter can generate

1-4244-0448-7/06/$25.00 ©2006 IEEE

significant sub- and inter-harmonic frequency components at multiples of the fundamental divided by the number inverter cells per phase [9]. To ensure that these components are not ignored, the THD and DF1 formulae have been modified as

$$THD = \frac{100}{V_1} \sqrt{\sum_{i=1}^{n} V_{\frac{i}{m}}^2 - V_1^2} \ \% \qquad (3)$$

$$DF1 = \frac{100}{V_1} \sqrt{\sum_{i=1}^{n} \frac{m^2 V_{\frac{i}{m}}^2}{i^2} - V_1^2} \ \% \qquad (4)$$

to take into account the number of levels in a multilevel inverter.

In addition a parameter, the energy factor ξ, is defined in this study for assessing the performance of a control scheme with respect to the characteristics of a flying capacitor inverter

ξ relates the unit cell-capacitor stored energy to the total load power of the three-phase inverter and is given as

$$\xi = \frac{3}{4RC} \ \text{s}^{-1} \qquad (5)$$

Derivation of ξ is given in [10]. .In a practical system, the designer would aim to minimise the capacitor size leading inevitably to increased ripple on the capacitors voltages. So the parameter, ξ, needs to be as large as realistically possible, while maintaining safe operation of the inverter by minimising the peak switch blocking voltages and peak capacitor voltages. The ripple voltage will also be dependent on the amount of current flowing when the capacitors are in the load current path. Therefore, the current lag angle, ϕ, is also an instructive parameter when quantifying the inverter performance with the system design parameters.

III. SWITCHING PATTERN SELECTION

The rules for selecting the optimal switching patterns for voltage balance are as follows:

1. eliminate those switching patterns causing a capacitor to be charged and subsequently discharged, or vice versa, at peak current,

2. match the switching states in a cycle to ensure nil net charge accumulation in a capacitor.

The subsequent selection scheme is developed and tested using the four-cell five-level inverter as an example and can be extended to multilevel inverters of any number of levels.

A. Switching States and Sequences in a Four-Cell Inverter

In a four-cell chopper circuit shown in Fig. 3(a) which forms one phase limb of the three-phase flying capacitor inverter shown in Fig. 1, there are 16 distinct switching states. Table 1 lists all these switching states represented as 16 binary numbers and their corresponding net charging effect on each of the cell capacitors for a positive load current. The most significant bit of these binary numbers controls the outer complementary switch pair nearest the dc link (S_4). '1' indicates that the upper switch is in conduction and contributes to a net voltage level at the inverter output. As can be seen in the table, the switching

states for the levels -0.5 and 0.5 all lead to different charging effects in the three cell-capacitors, hence causing the voltage swings.

In the case of the level 0 switching states, these can be grouped in three complementary pairs, considering the charging characteristic for all capacitors in this inverter. The complementary pairs are states 3 & C, states 5 & A and states 6 & 9.

Synthesizing a sinusoidal cycle using the SHE staircase requires the inverter stepping through a set of three intermediate states, i.e. a switching sequence, from voltage levels -1 to +1. Even though there are a large number of possible switching sequences, the number is limited to those with only one switch state transition per level change. This is to minimize the switching losses in the power semiconductors of a real inverter. The number of allowable transitions between switching states in the four-cell inverter is illustrated in Fig. 3(b). The figure shows that there are four possible paths between level -1 and a level -0.5, three possible paths between any level -0.5 and a level 0, two possible paths between any levels 0 and a level +0.5 and one possible path between any level +0.5 and level +1.

Therefore, 24 different sequences of switching states can be used when stepping-up in voltage level between -1 and +1. For instance, the switching states to be used in one cycle may be [0000], [0001], [1001], [1101] and [1111], so 0, 1, 9, D and F in hexadecimal notation. This switching sequence is labeled as $\overset{D}{\underset{1}{9}}$.

Using the above labeling, the 24 possible switching sequences are as follows:

$$\overset{7}{\underset{1}{3}}, \overset{B}{\underset{1}{3}}, \overset{7}{\underset{1}{5}}, \overset{D}{\underset{1}{5}}, \overset{B}{\underset{1}{9}}, \overset{D}{\underset{1}{9}}, \overset{7}{\underset{2}{3}}, \overset{B}{\underset{2}{3}}, \overset{7}{\underset{2}{6}}, \overset{E}{\underset{2}{6}}, \overset{B}{\underset{2}{A}}, \overset{E}{\underset{2}{A}}, \overset{7}{\underset{4}{5}}, \overset{D}{\underset{4}{5}}, \overset{7}{\underset{4}{6}}, \overset{E}{\underset{4}{6}}, \overset{D}{\underset{4}{C}}, \overset{E}{\underset{4}{C}}, \overset{B}{\underset{8}{9}},$$
$$\overset{D}{\underset{8}{9}}, \overset{B}{\underset{8}{A}}, \overset{E}{\underset{8}{A}}, \overset{D}{\underset{8}{C}}, \overset{E}{\underset{8}{C}}$$

B. Switching Patterns and Their Selection

A pattern refers to a set of four switching state sequences for voltage balancing over four cycles. To analyze the pattern selection problem, it is assumed load current is sinusoidal and the phase current lags the phase voltage fundamental component. Fig. 4 shows the relationship between the phase voltage and current under staircase SHE control. The figure also shows the component of load current which can flow through the cell-capacitors for 0 and 0.5 voltage levels. As can be seen from level 0 operation, the current is of equal amplitude but of opposite polarity in the two halves of the staircase voltage cycle, hence the mean current is zero. This means that under ideal conditions, the same switching state could be used throughout for level 0 synthesis However this strategy can not be employed for the reasons explained below. As shown in Fig. 4 (c), the current waveforms for level 0.5 are totally different for the two halves. There is no symmetry in the load current to exploit for balancing purposes when operating at level +0.5, so all four switching states for this level as shown in Fig. 4 have to be cycled through over four cycles.

1552

One approach used in identifying valid balancing switching patterns is to keep the individual level switching states the same over one complete cycle, thus forming a sequence. According to three complimentary pairs of level 0 switching states, it is clear that there are groups of 6 pattern permutations made up of just 4 individual sequences, where the sequence order is varied. These groups of sequences are listed in Table 2, with the level 0 contributing states shown as reference. The basic requirement is that all four -0.5 and +0.5 level states are used, but the level 0 states can be taken from either one complementary pair set or two sets out of the six. Thus there are in fact 24 different groups of switching sequences, and by permutation each has 6 different patterns. So this gives a grand total of 144 valid balancing patterns that meet the minimum switching transition criteria. Taking the upper left group in Table 2 as an example, the 6 pattern permutations in this group are as follows:

$$\frac{7\,E\,D\,B}{36C9}, \frac{7\,E\,B\,D}{369C}, \frac{7\,D\,E\,B}{3C69}, \frac{7\,D\,B\,E}{3C96}, \frac{7\,B\,E\,D}{396C}, \frac{7\,B\,D\,E}{39C6}$$
$$\scriptstyle 1\,2\,4\,8 \quad 1\,2\,8\,4 \quad 1\,4\,2\,8 \quad 1\,4\,8\,2 \quad 1\,8\,2\,4 \quad 1\,8\,4\,2$$

Two rules have been devised to select a pattern out of 144 which will give the best performance. The first is to avoid those switching patterns which cause the same capacitor to be charged and subsequently discharged or vise versa at peak current. Fig. 5 shows that the phase current magnitude is maximum at levels -0.5 and +0.5. Hence these regions will have the largest levels of capacitor voltage variation. So if state 7[0111] is used at +0.5 level in the first sequence, voltage drop on C_3 would be the maximum due to the peak discharging current. On the other hand if the state 8[1000] is applied at -0.5 level in the subsequent sequence, C_3 would be charged with the most negative current causing further voltage drop. Thus if the switching state at level +0.5 gives the same polarity of voltage change as the next sequence's switching state at level -0.5, then this is to be avoided. Consequently preferred patterns should not include the next level -0.5 state being the 1's complement of the previous level +0.5 state. For instance, state 8 [1000] should not follow state 7 [0111]. According to this rule the following consecutive sequences pairs are not preferred: $\frac{7}{X}\frac{X}{8}, \frac{E}{X}\frac{X}{1}, \frac{D}{X}\frac{X}{2}, \frac{B}{X}\frac{X}{4}$

The second rule considers the charging/discharging of capacitors within one sinusoidal cycle due to the switching sequence used. The aim is to ensure that a particular capacitor is in the same current path over this cycle, so that there is no or as little as possible net charge accumulation. As shown in Table 1, when adjacent switches are in the opposite state, the capacitor between them lies in the load current path, and so its voltage would swing. Again, with regard to the peak and trough of the current occurring around level +0.5 and level -0.5 respectively, it is preferable if as many cells as possible are in the same state at both levels. Priority for this rule can be assigned to the higher voltage capacitor, i.e. C_3, which will see the largest voltage change. For example if at level -0.5 the switching state is 4 [0100], then state 7 [0111] is preferred for level +0.5. In both sequences C_3 lies in the load current path and is in discharging mode. Fig. 6 shows the resultant current

the cell-capacitor current and so the mean current is zero leading to no net change in cell-capacitor voltage. Applying the same rule, states 8 [1000] and B [1011] are also preferred within a sequence as C_3 is in charging mode for both voltage stepping-up and down cases. Another condition applicable to state transitions within a sequence is that of only switching one cell transistor pair at a time. This means that only four sequences would be preferred for minimising the voltage variation on C_3 over one sequence cycle, and in each case C_3 is in the same current path throughout the cycle. These are:

$$\frac{7}{5}, \frac{7}{6}, \frac{B}{9} \text{ and } \frac{B}{A}$$
$$\scriptstyle 4 \quad 4 \quad 8 \qquad 8$$

According to this rule there are 4 sequences, shown below, preferred within a pattern:

$$\frac{7}{5}, \frac{7}{6}, \frac{B}{9} \text{ and } \frac{B}{A}$$
$$\scriptstyle 4 \quad 4 \quad 8 \qquad 8$$

When applying this second rule to C_2, and to C_1, states in the adjacent cells around C_3 must not be complements, otherwise the above rule will be broken. Therefore, this means that the following 4 sequences are preferred for use in a pattern:

$$\frac{D}{5}, \frac{D}{C}, \frac{E}{6} \text{ and } \frac{E}{A}$$
$$\scriptstyle 4 \quad 4 \quad 2 \qquad 2$$

Applying the above two rules, one particular pattern for the four-cell five-level inverter is likely to be among those offering the best performance, viz:

$$\frac{7\,E\,D\,B}{36C9} \qquad \textbf{PATTERN 1}$$
$$\scriptstyle 1\,2\,4\,8$$

This pattern does not have any consecutive states which break the first rule governing the level +0.5 followed by level -0.5. It also contains 3 sequences from the second rule which have preferred state combinations for levels -0.5 and +0.5.

The various balancing patterns can also be screened for ones which are likely to cause poor overall performance. One such pattern which breaks the first rule only has one good sequence from the second rule for C_1 is as follows:

$$\frac{7\,B\,E\,D}{3AC5} \qquad \textbf{PATTERN 2}$$
$$\scriptstyle 2\,8\,4\,1$$

The same rule-based analysis can be done for other load characteristics. In the case of a leading phase current angle, the second rule governing a sequence still applies, the first rule's reasoning is the same but applied this time to a sequence level -0.5 state followed by the next sequence level +0.5 state.

The overall pattern, thus derived, is repeated once every four cycles (number of cells), with different switch states used at the same voltage level of each cycle.

IV. COMPARISON OF BALANCED SWITCHING PATTERNS

Simulation results for switching patterns 1 and 2 identified in the previous section are presented in Table 3. These are also compared with the ideal case where constant capacitor voltages are assumed. The dc link voltage was set to 400 V and control settings were $m_a = 1$ and a fundamental frequency of 50 Hz, using 5th harmonic elimination angles for staircase SHE control. The phase voltage fundamental is 244.9 V(rms). The load model used has component values R = 2.5 Ω and L = 7.958 mH, and the individual cell-capacitance is 10 mF. For these component values the energy factor, ξ, is 30 s^{-1} and the nominal output power for an equivalent ideal sinusoidal system is 12 kW with power factor 0.707.

As can be seen in Table 3, PATTERN 1, $\left(\begin{smallmatrix} 7 & E & D & B \\ 3 & 6 & C & 9 \\ 1 & 2 & 4 & 8 \end{smallmatrix}\right)$, offers lower output waveform harmonic distortion compared to PATTERN 2, $\left(\begin{smallmatrix} 7 & B & E & D \\ 3 & A & C & 5 \\ 2 & 8 & 4 & 1 \end{smallmatrix}\right)$. The table also shows that by optimum pattern selection the phase and line voltage THDs can be lowered compared to the ideal case. However, because of the additional sub- and inter harmonics at the low frequency, phase current THD is higher. The DF1 term indicates good correlation in predicting the THD of the phase current for this particular simple inductive load.

Fig. 7(a) shows the line voltage spectrum for the two cases, with the frequency components normalised to the fundamental. PATTERN 1 gives lower THD level than its counterpart. Notice that a 12.5 Hz component is present in the voltage spectrum for both patterns and its magnitude is higher for the inferior PATTERN 2. This component is due to the balancing pattern control strategy which repeats every four cycles, i.e. 12.5 Hz.

Fig. 7(b) shows the individual cell-capacitor voltages, plotted on the same zero axis for reference, of the inverter controlled by the two balancing pattern schemes. With PATTERN 1, voltage ripple is reasonably low with blocking voltage variations peaking at ±50 % of the nominal values. However for PATTERN 2 capacitor voltage ripple is significantly higher with a peak blocking voltage value around 100% of the nominal level. This is the primary reason for PATTERN 2's poorer waveform performance.

V. CLOSED-LOOP CAPACITOR VOLTAGE BALANCING

Capacitor voltage balance may be obtained by voltage feedback control method using voltage sensors. This is particularly valid in cases when the load parameters vary considerably within a cycle [11] There have been a number of proposed schemes for maintaining balanced regulation of the cell-capacitor voltages specifically under PWM control [12]. No work has been reported for balancing when using staircase SHE control and so an algorithm which will accomplish balanced operation is proposed in this section.

The proposed method employs comparators for an upper and lower voltage level band for each capacitor. Their output states are combined so that the signals for each capacitor are shown in Table III. This uses the relationship between the switch state of those mutually connected to a capacitor and charging/discharging behaviour dependency on current direction. For instance, an adjacent two cells' state of '10' with a positive phase current flowing will boost the voltage across the capacitor connected between the cells.

Using these comparator signals, a simple algorithm for a four cell inverter can be implemented to decide the

optimum switching state to apply to the inverter at a given level. A flow diagram of the algorithm is shown in Fig. 8. The algorithm input signals are the three comparator outputs C1 - 3 and the required output voltage level L. The state of the highest voltage capacitor is checked first. If it is within bands [0000], then the state of the middle capacitor is checked, otherwise the output firing switching state S is assigned to C3. If C2 is outside the regulation bands then S is assigned to C2. The comparator signal C1 for the lowest voltage capacitor is added to S only if the state of C2 is [0000]. This gives a potential switching state for the inverter control. If all three capacitor voltages are within the regulation bands, S = [0000], then the optimum pattern for the given voltage level demand is found using a look-up table. Otherwise, S is checked to ensure that it operate the inverter at the required voltage level. If this is not the case, S is either incremented of decremented until the switching state gives the appropriate voltage level.

Simulations were conducted under the same conditions as before and the resultant capacitor voltages are shown in Fig. 8, and compared with the optimum PATTERN 1 result. This shows that the voltage ripple has been reduced by applying feedback, but this appears to lead to a pseudo-random behaviour rather than a repeating waveform as before. The tolerance band was set at ±5%, and it was found that tighter tolerances caused instability in the simulation. The mean capacitor voltages were all kept within 3% of the target values, and the peak voltages were reduced compared with PATTERN 1. For instance, peak voltage on C$_3$ as a percentage of the unit cell-voltage was 318.2% compared to 332.3%. These show that the algorithm is achieving balanced voltage control with reduced ripple even compared with the optimum selected pattern. The maximum blocking voltage as a percentage of unit cell-voltage across any switch was also reduced to 156.4%, compared to 167.8% for PATTERN 1. This shows that some improvement can be gained in protecting against over-voltage conditions across the power switch, or it may be possible to increase output power for a given cell-capacitance.

The effect of the change in capacitor voltage on output power quality is shown in the plots of line voltage and phase current spectra. For comparison, the PATTERN 1 spectra are also shown in Figs. 9 and 10. The most interesting feature in the harmonics is a general increase in the sub- and inter-harmonics, but the 100 Hz component is noticeably reduced. This is due to the more random nature of the capacitor ripple waveform effectively spreading the spectral components. This is reflected in increased line voltage and phase current THD, but the values are still lower than a poorly selected open-loop balancing pattern. These results indicate that under steady-state conditions, the pre-selected balancing pattern is preferred due its lower harmonic distortion characteristic, but in a real system the closed-loop cell-capacitor voltage control would ensure better transient operation.

VI. CONCLUSIONS

The paper presented an optimal capacitor voltage balancing strategy using a selective harmonic elimination scheme. Switching states for a three-phase four-level flying capacitor inverter were investigated. Criteria have been identified to choose the switching sequence/patterns which would cause least disturbances to the capacitor voltages.

REFERENCES

[1] R. Teodorescu, F. Blaabjerg, J. K. Pedersen, E. Cengelci, S. U. Sulistijo, B. W. Woo and P. Enjeti, "Multilevel converters - a survey", Proceedings of EPE '99, 8th European Conference on Power Electronics and Applications, 7 - 9 September 1999, CD-ROM paper

[2] T. A. Meynard, and H. Foch, "Multi-level choppers for high voltage applications", EPE Journal, Vol. 2, No. 1, March 1992, pp. 45 – 50.

[3] G. Gateau, P. Maussion, and T. A. Meynard, "Fuzzy phase control of series multicell converters", Proceedings of the 6th IEEE International Conference on Fuzzy Systems, 1 - 5 July, 1997, Vol. 3, pp. 1627 – 1633.

[4] T. A. Meynard, M. Fadel and N. Aouda, "Modeling of multilevel converters", IEEE Transactions on Industrial Electronics, Vol. 44, No. 3, June, 1997, pp. 356 – 364.

[5] G. Gateau, M. Fadel, P. Maussion, R. Bensaid, and T. A. Meynard, "Multicell converters: active control and observation of flying-capacitor voltages", IEEE Transactions on Industrial Electronics, Vol. 49, No. 5, October 2002, pp. 998 – 1008.

[6] J. L. Thomas, S. Poullain, A. Donzel, and G. Bornard, "Advanced torque control of induction motors fed by a floating capacitor multilevel VSI actuator", IEE Seminar, 'Advances in Induction Motor Control', 23 May, 2000, pp. 5/1 - 5/5.

[7] F. G. Turnbull, "Selected harmonic reduction in static d-c – a-c inverters", IEEE Transactions on Communications and Electronics, Vol. 83, July 1964, pp. 374 – 378.

[8] Enjeti, P. and Lindsay, J.F., "Solving nonlinear equations of harmonic elimination PWM in power control", IEE Electronic Letters, Vol. 23, No. 12, June 1987, pp. 656 – 657.

[9] Fukuda, S. and Suzuki, K, "Using harmonic determination factor for harmonic evaluation of carrier-based PWM methods", Conference Record of IAS '97, IEEE Industry Applications Society 32nd Annual Meeting, 5 - 9 October, 1997, pp. 1541 – 1541.

[10] Steve Watkins, Optimal Control of Flying Capacitor Converters, PhD thesis, September 2005, Leeds University, UK

[11] Wilkinson, R.H., Meynard, T.A., Richardeau, F. and Enslin, J.H.R., "Dynamic control and voltage balance of multilevel converters: large signal one-cycle response", Proceedings of EPE '99, 8th European Conference on Power Electronics and Applications, Lausanne, France, 7 - 9 September, 1999.

[12] Escalante, M.F. and Vannier, J.-C., "Direct approach for balancing the capacitor voltages of a 5-level flying capacitor converter", Proceedings of EPE '99, 8th European Conference on Power Electronics and Applications, Lausanne, France, 7 - 9 September, 1999.

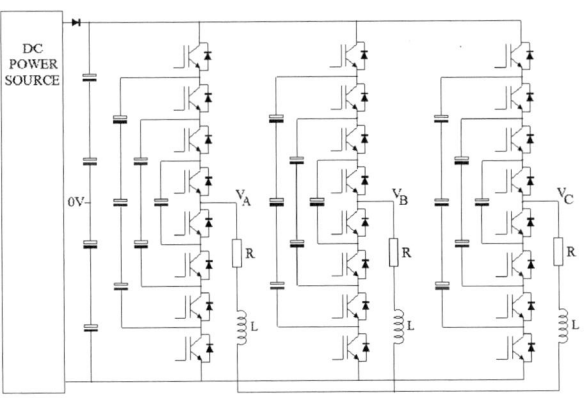

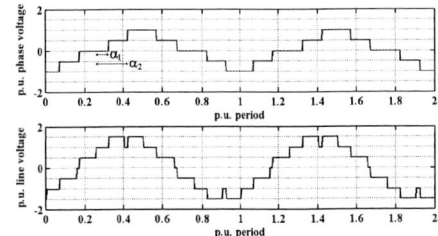

Figure 2. Idealized phase V/I waveforms with 45° lagging current

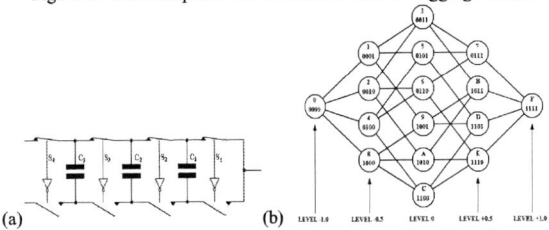

(a)   (b)

Figure 3. Simple four-cell chopper circuit and its allowable transitions between level

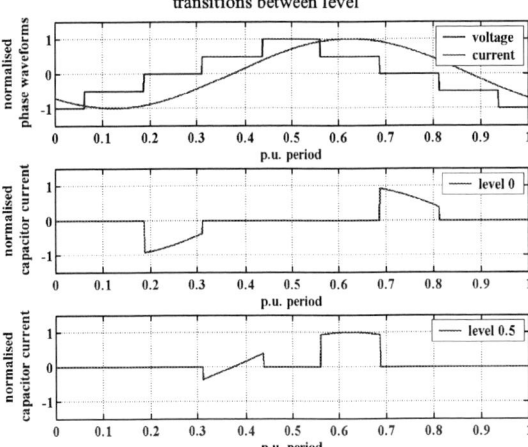

Figure 4. Idealised phase current and voltage waveform with 45° lagging current

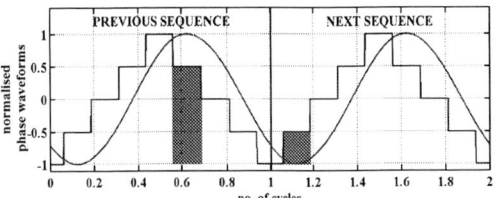

Figure 5. Switching patterns causing large voltage swings should be avoided

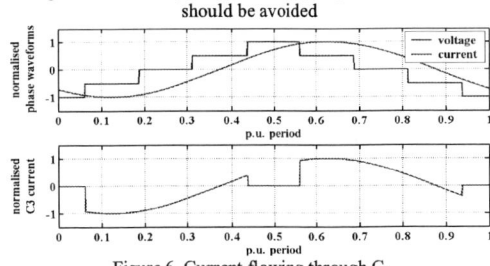

Figure 6. Current flowing through C_3 when using switching sequence 4,7,5

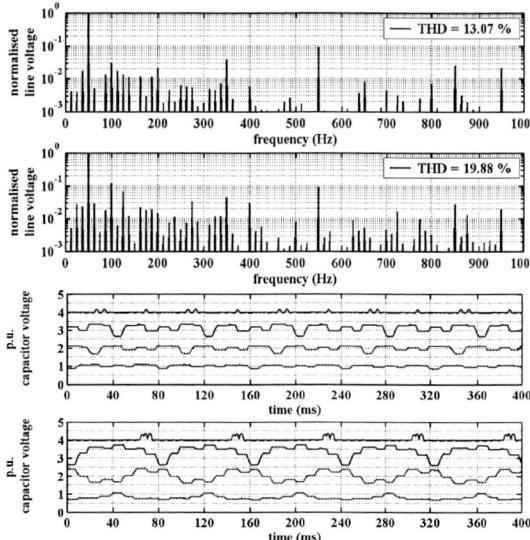

Figure 7. Comparison of Line voltage spectrum and capacitor voltages
PATTERN 1 (top) and PATTERN 2 (bottom)

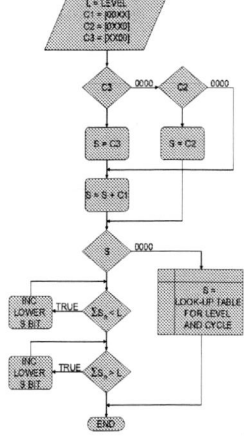

Figure 8. Flow chart for closed-loop voltage regulation

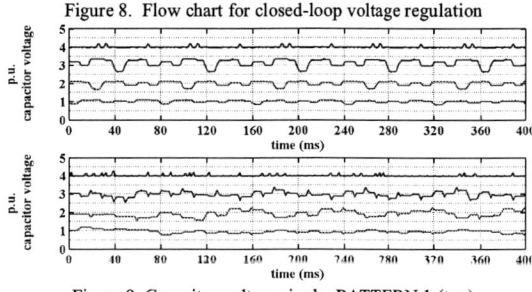

Figure 9. Capacitor voltage ripple, PATTERN 1 (top)
and closed-loop regulated capacitor voltage (bottom)

Figure10. Line voltage spectrum comparison, PATTERN 1 (top)
and closed-loop regulated capacitor voltage (bottom)

TABLE I
CAPACITOR VOLTAGE NET CHANGE FOR EACH INVERTER LIMB SWITCHING STATE

Switching State $S_4S_3S_2S_1$	Output Voltage Level	Change(+VE)/ discharge(-VE) C_3	Change(+VE)/ discharge(-VE) C_2	Change(+VE)/ discharge(-VE) C_1
0000 (0)	-1.0			
0001 (1)	-0.5			-VE
0010 (2)	-0.5		-VE	+VE
0100 (4)	-0.5	-VE	+VE	
1000 (8)	-0.5	+VE		
0011 (3)	0		-VE	
0101 (5)	0	-VE	+VE	-VE
0110 (6)	0	-VE		+VE
1001 (9)	0	+VE		-VE
1010 (A)	0	+VE	-VE	+VE
1100 (C)	0	+VE		
0111 (7)	+0.5	-VE		
1011 (B)	+0.5	+VE	-VE	
1101 (D)	+0.5		-VE	+VE
1110 (E)	+0.5			-VE
1111 (F)	+1.0			

TABLE II
24 GROUPS OF SWITCHING SEQUENCES

Level 0	Switching Sequence Groups			
3+C and 6+9	$\frac{7}{1},\frac{E}{2},\frac{D}{4},\frac{B}{8}$ 3, 6, C, 9	$\frac{B}{1},\frac{7}{2},\frac{E}{4},\frac{D}{8}$ 3, 6, C, 9	$\frac{D}{1},\frac{B}{2},\frac{7}{4},\frac{E}{8}$ 9, 3, 6, C	$\frac{E}{1},\frac{D}{2},\frac{B}{4},\frac{7}{8}$ 9, 3, 6, C
3+C and 5+A	$\frac{7}{1},\frac{B}{2},\frac{D}{4},\frac{E}{8}$ 3, A, 5, C	$\frac{B}{1},\frac{E}{2},\frac{7}{4},\frac{D}{8}$ 3, A, 5, C	$\frac{7}{1},\frac{B}{2},\frac{D}{4},\frac{E}{8}$ 5, 3, C, A	$\frac{D}{1},\frac{7}{2},\frac{E}{4},\frac{B}{8}$ 5, 3, C, A
6+9 and 5+A	$\frac{7}{1},\frac{B}{2},\frac{E}{4},\frac{D}{8}$ 5, A, 6, 9	$\frac{D}{1},\frac{E}{2},\frac{7}{4},\frac{B}{8}$ 5, A, 6, 9	$\frac{B}{1},\frac{7}{2},\frac{D}{4},\frac{E}{8}$ 9, 6, 5, A	$\frac{D}{1},\frac{E}{2},\frac{7}{4},\frac{B}{8}$ 9, 6, 5, A
3+C	$\frac{7}{1},\frac{B}{2},\frac{D}{4},\frac{E}{8}$ 3, 3, C, C	$\frac{7}{1},\frac{B}{2},\frac{E}{4},\frac{D}{8}$ 3, 3, C, C	$\frac{B}{1},\frac{7}{2},\frac{D}{4},\frac{E}{8}$ 3, 3, C, C	$\frac{B}{1},\frac{7}{2},\frac{E}{4},\frac{D}{8}$ 3, 3, C, C
5+A	$\frac{7}{1},\frac{B}{2},\frac{D}{4},\frac{E}{8}$ 5, A, 5, A	$\frac{7}{1},\frac{E}{2},\frac{D}{4},\frac{B}{8}$ 5, A, 5, A	$\frac{D}{1},\frac{B}{2},\frac{7}{4},\frac{E}{8}$ 5, A, 5, A	$\frac{D}{1},\frac{E}{2},\frac{7}{4},\frac{B}{8}$ 5, A, 5, A
6+9	$\frac{B}{1},\frac{7}{2},\frac{E}{4},\frac{D}{8}$ 9, 6, 6, 9	$\frac{7}{1},\frac{B}{2},\frac{E}{4},\frac{D}{8}$ 9, 6, 6, 9	$\frac{D}{1},\frac{7}{2},\frac{E}{4},\frac{B}{8}$ 9, 6, 6, 9	$\frac{D}{1},\frac{E}{2},\frac{7}{4},\frac{B}{8}$ 9, 6, 6, 9

TABLE III
TRUTH TABLE FOR CAPACITOR VOLTAGE COMPARATORS

Capacitor	Over-voltage Condition		Under-voltage Condition	
	Positive Current	Negative Current	Positive Current	Negative Current
C_3	0100	1000	1000	0100
C_2	0010	0100	0100	0010
C_1	0001	0010	0010	0001

2006 5th International Power Electronics and Motion Control Conference

Analysis of Harmonics in Input Line Current for Matrix Converter based on Double Input Line-to-line Voltages

GUO Yougui DENG Wenlang ZHU Jianlin
* Institute of Information and Engineering of Xiangtan University, Xiangtan 411105, China
e-mail: guoyougui@sina.com

Abstract—**Control strategy on double input line-to-line voltages for matrix converter was studied fewer compared with others. The relative theorem and inferences are put forward. That is the general composition rules and expressions of input line current is gotten. The simulated model was setup based on this modulation strategy. The simulated experiments have verified correctness of the theorem and inferences, and have gotten the harmonics distributed rule of input line current with harmonics in input voltages.**

Keywords-matrix converter; double line-to-line voltages; theorem; inference; general demonstration; simulation

I. INTRODUCTION

Matrix converter becomes a study hot spot of AC-AC variable frequency[1-4]. Japanese expert-A. Ishiguro and T. Furuhashi put forward double input line-to-line voltages composition as a control strategy for matrix converter[5]. Its main properties are that it can self-adjust the output line-to-line voltages to reach the demanded output line-to-line voltages while input three phase voltages are non-balanced; its voltage transfer ratio is higher, which respectively reaches 0.75 and 0.866 with two line-to-line voltages or three line-to-line voltages for composing; input power factor is 1 when three phase input voltages are symmetrical with a higher use efficiency; but its control is complex and realization more difficult.

Some experts get a few achievements in our country with double input line-to-line voltages composition as a modulation strategy. For example, professor MU Xinhua , etc. put forward a concept 'source key' and simplified the control strategy[6]-[7] in Nanjing University of Aeronautics & Astronautics. Professor CHEN Xiyou, etc. put forward a control strategy with adjustable input power factor and gave a analysis method of input current while asymmetric input voltages in Harbin university of technology on the basis of researches of Nanjing University of Aeronautics & Astronautics[8]-[9]. Professor WANG Yi, etc. put forward a closed-loop control strategy of double input line-to-line voltages' composition in Harbin university of technology[10], and gave a deep study to common mode voltage[11]. All their studies have very important sense to richen and develop achievements of matrix converter. But scholars in our country or aboard gave a careful analysis only for a kind of sector combination of between input

and output voltage; only studied asymmetrical input things in harmonics analysis. Actually, the combined rule and formula of output line-to-line voltages and input line currents are different in each sector combination; actual grid makes input supply asymmetrical but also have many harmonics because of one phase or two phase loads, asymmetrical, nonlinear loads or quality of converters, and so on. So research on general combined rule and formula of input line currents and effects of abnormal input supply on combination of input line currents has some theoretical values and realistic senses.

II. DOUBLE INPUT LINE-TO-LINE VOLTAGES CONTROL PRINCIPLE FOR MATRIX CONVERTER

The topology of matrix converter is shown by figure one.

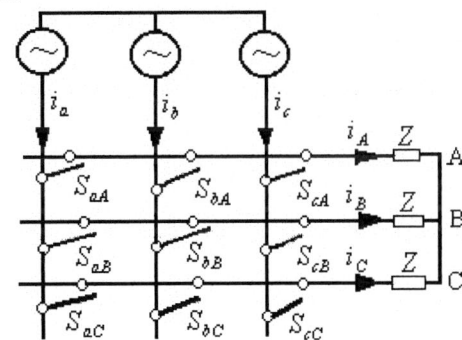

Figure.1 The topology of matrix converter

Input voltage and output voltage are respectively divided into 6 sectors in one cycle[5]. So they have 36 switching combinations in a switching period. Reference [6] put forward a concept of 'source key', and further concluded 9 things, each of them corresponds to a different 'source key', which simplified its control strategy. Double input line-to-line voltages composition principle is to get demanded output line-to-line voltages through vector composition of corresponding cycle ratios and two suitable input line-to-line voltages for each case in each switching period in these 9 cases.

III. THEOREM AND INFERENCES

The theorem and inference are gotten by deep analysis of input line-to-line local average composition in each sector combination of all these 9 cases.

1-4244-0448-7/06/$25.00 ©2006 IEEE

Theorem: the input line current formula for three-phase to three-phase AC-AC matrix converter with control strategy using double input line-to-line voltages to compose output line-to-line voltages:

$$i_n = P_L \cdot \psi \cdot (u_{ng} + u_{nh})$$

(1)

Where $\psi = 1/(u_{ab}^2 + u_{bc}^2 + u_{ca}^2)$, P_L is average power of load, $n \in \{a, b, c\}$, $g \in \{b, c, a\}$, $h \in \{c, a, b\}$, and n, g, h have a relation of ordered one to one.

Inference: the control strategy with double input line-to-line voltages composing into output line-to-line voltages cannot self-adjust input line currents.

Take one-phase to explain, suppose a-phase. When input three-phase voltages are symmetrical, then $u_{ab} + u_{ac} = 3u_a$, input line current $i_a = 3P_L\psi \cdot u_a$ according to the theorem. Under this case, P_L and ψ are constants. So input current depends on input voltage. Its variable rule is the same as that of input voltage, and the input power factor is 1. Obviously, it cannot be self-adjusted.

When abnormal input line-to-neutral voltage makes corresponding three input line-to-line voltages abnormal, ψ in the theorem is not a constant any more. But $u_{ab} + u_{ac}$ is the same frequency waveform. Obviously, this item is abnormal. So i_a is not only affected by two line-to-line voltages, but also be affected by ψ. One is one order, another is two order, which cannot be counteracted through the relationship between numerator and denominator. It concluded that any abnormal of input line-to-neutral voltage will result in abnormal of input line current. That is to say, this control strategy cannot self-adjust input line current.

IV. SIMULATION EXPERIMENT

Simulation[12]-[15] conditions: suppose the demanded output voltages are symmetrical three-phase sine waveforms, input frequency is 50Hz, the voltage transfer ratio is 0.866, output frequency is 100Hz, switching frequency is 2.56kHz, load is resistance and inductive, that is R=233.4 Ω, X_L=134.75 Ω.

(1) symmetrical three-phase input voltages

Simulation waveform is shown as figure 2. It shows that the input line current is sine waveform, which verifies the correctness of the theorem.

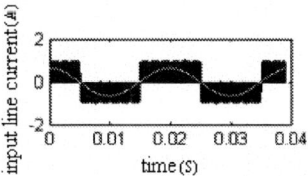

Figure.2 instantaneous input current and its average

(2) asymmetrical three-phase input voltages with harmonics

Suppose three-phase input voltages as follows:

$$\begin{bmatrix} u_a \\ u_b \\ u_c \end{bmatrix} = \sqrt{2} \cdot \begin{bmatrix} 220\cos(\omega_i t) + 0.2 * 220\cos(5\omega_i t) \\ 235\cos(\omega_i t - 120°) + 0.2 * 235\cos(5\omega_i t) \\ 210\cos(\omega_i t + 120°) + 0.2 * 210\cos(5\omega_i t) \end{bmatrix}$$

Simulation waveform is shown by figure 3. The modulation strategy of double input line-to-line voltage composition cannot self-adjust input line current. Similarly, the same conclusion is gotten through the simulation analysis for things with even harmonics and other odd harmonics and different amplitude. So the correctness of the theorem was verified.

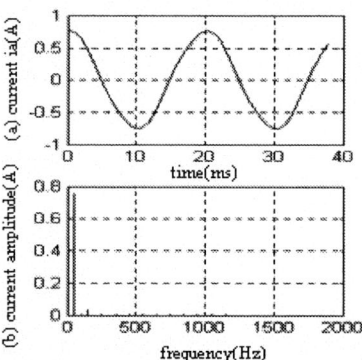

Figure.3 input line current waveform and its spectrum analysis

(3) Three-phase input voltages respectively have n-times harmonics

Suppose three-phase input voltages as follows:

$$\begin{bmatrix} u_a \\ u_b \\ u_c \end{bmatrix} = 220\sqrt{2} \cdot \begin{bmatrix} \cos(\omega_i t) + 0.2\cos(5\omega_i t) \\ \cos(\omega_i t - 120°) + 0.2\cos(5\omega_i t) \\ \cos(\omega_i t + 120°) + 0.2\cos(5\omega_i t) \end{bmatrix}$$

Simulation result shows that input line current is sinusoidal waveform. Similarly, the same conclusion is gotten through the simulation analysis for things with even harmonics and other odd harmonics and different amplitude. So three-phase input line-to-neutral voltages respectively have the same harmonics has no effect on input line current.

(4) Two-phase or one-phase input line-to-neutral voltage has n-times harmonics

① Two-phase input line-to-neutral voltage has n-times harmonics

Suppose three-phase input line-to-neutral voltages as follows:

$$\begin{bmatrix} u_a \\ u_b \\ u_c \end{bmatrix} = 220\sqrt{2} \cdot \begin{bmatrix} \cos(\omega_i t) + 0.2\cos(5\omega_i t) \\ \cos(\omega_i t - 120°) + 0.2\cos(5\omega_i t) \\ \cos(\omega_i t + 120°) \end{bmatrix}$$

Simulation waveform is shown as figure 4. input line current has 5 ± 2 times harmonics. Further simulation has gotten that the input line current mainly has $n\pm2$ times harmonics when two-phase input line-to-neutral n-times harmonics(no matter what n is odd or even).

② one-phase input line-to-neutral voltage has n-times harmonics

Similarly, the result has been gotten that input line current mainly has n±2 times harmonics when n is odd; that it mainly has 7 or 11 times harmonics when n is even.

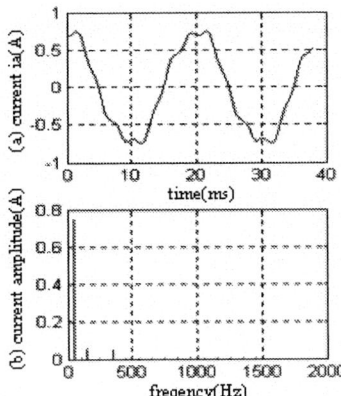

Figure.4   The waveform and spectrum of input line-to-line current

V.   CONCLUSION

The corresponding conclusions can be gotten through analysis and simulation experiments of control strategy of double input line-to-line voltages composition:

(1) the input line current of matrix converter can be shown in a general mathematical model, which is suitable for any sector combination;

(2) control strategy of double input line-to-line voltages composition has no adjustability to input line current;

(3) the input line current is not impacted when three-phase input line-to-neutral voltages respectively has the same times harmonics;

(4) the input line current has some rules when two or one-phase input line-to-neutral has harmonics.

REFERENCES

[1]   L. Huber, D. borojevic. Space vector modulated three-phase to three-phase matrix converter with input power factor correction[J]. IEEE Transactions on industry application, 1995, 31(6):1234~1246.

[2]   Ding Wei, Zhu Jianlin, Li Zhiyong, etc. Simulation model of matrix converter with space vector modulated control strategy[J]. Natural Science Journal of Xianntan University, 2002, 24(3): 100-103.

[3]   Ding Wei, Zhu Jianlin, Fan Lv, etc. Matrix converter and its research situation[J]. Natural Science Journal of Xianntan University, 2002, 24(2): 185-187.

[4]   Lan Zhiyong, Zhu Jianlin. Simulation research of three phase to three phase sparse matrix converter[J]. Natural Science Journal of Xianntan University, 2005, 27(3): 110-115.

[5]   Akio Ishiguro, Takeshi Furuhashi, Shigeru Okuma. A novel control method for forced commutated cycloconverters using instantaneous values of input line-to-line voltages[J]. IEEE Transactions on industrial electronics, 1991, 38(3): 166-172.

[6]   Mu Xinhua, Zhuang Xinfu. Two-voltage control principle and waveform synthesis of AC-AC matrix converter [J]. Journal of Nanjing University of Aeronautics & Astronautics, 1997, 29 (2): 151~157.

[7]   Mu Xinhua, Zhuang Xinfu, Chen Huaiya. The switch state analysis and simulation of matrix converter using double line-to-line voltages control technique[J]. Transactions of China Electrotechnical Society, 1998，13(1): 46~50.

[8]   Chen Xiyou, Chen Xueyun. The control of reactive power and the harmonic elimination of input current for matrix converter based on   double line-to-line voltage synthesis[J]. Electric Drive, 2001, (1): 11-15

[9]   Chen Xiyou, Cong Shujiu, Chen Xueyun. The analysis of the relationship between the voltage sectors and the performances of matrix converter based on two-voltage synthesis{J}. Proceedings of the CSEE, 2001, 21(9): 63-67.

[10]   Wang Yi, Chen Xiyou, Xu Dianguo. Research on closed-loop control for matrix converter based on double line-to-line voltages synthesis{J}. Proceedings of the CSEE, 2002, 22(1): 74-79.

[11]   Liu Hongchen, Chen Xiyou, Feng Yong, etc. A research on common-mode voltage for matrix Converter based on two line voltage synthesis[J]..Proceedings of the CSEE, 2004, 24(12)：182-186

[12]   Zhou Xu, Zhu Jianlin, Xiang Ga. Simulation research on simulink in integration language matlab[J]. Natural Science Journal of Xianntan University, 2000, 22(1): 99-102.

[13]   Luo Zhenzhong, Zhu Jianlin, Tan Ping'an, Lan Zhiyong. Computer simulation of squirrel-cage induction motor based on the basic simulink module[J]. Natural Science Journal of Xianntan University, 2004, 26(2): 85-87.

[14]   Tang Guangdi, Mei Baishan, Zhu Jianlin, Yi Lingzhi. Rearch on simulation a new method of direct torque control[J]. Natural Science Journal of Xianntan University, 2003, 25(2): 95-98.

[15]   Yi Lingzhi, Zhu Jianlin, Zhang Linting, Deng Wenliang, Li Weiping, et al. The simulation design of three-phase digital cycloconverter[J]. Natural Science Journal of Xianntan University, 2003, 25(2): 90-94.

2006 5th International Power Electronics and Motion Control Conference

Research on Neutral-point Balancing Control for Three-level NPC Inverter Based on Correlation between Carrier-based PWM and SVPWM

Wenxiang Song, Guocheng Chen, Xiaoyu Ding, Mantang Shu
Shanghai Key Laboratory of Power Station Automation Technology
Shanghai University
Shanghai, P. R. China, 200072
gchchen@mail.shu.edu.cn

Abstract — Based on inherent correlation between the carrier-based PWM and SVPWM for three-level inverter, this paper presents a new modulation approach for the control of the neutral-point voltage variation in the three-level neutral-point-clamped voltage source inverter. The new modulation approach realizes the control of inverter neutral-point current by modifying redundant small vectors pairs' distribution factor, only requiring the information of dc-link capacitor voltages and three-phase load currents, which is convenient to apply and is compatible of digital computer realization. The effectiveness of proposed control approach is verified by simulation and experiment results.

Key Words : neutral-point-clamped inverter, neutral-point voltage control, carrier-based PWM, SVPWM

I. INTRODUCTION

As is known to all, there exists the fluctuation of neutral-point voltage in a three-level neutral-point-clamped (NPC) inverter. It has attracted popular attentions and many schemes have been presented[1-5]. Fig.1 shows the topology of a NPC inverter that each phase terminal can be connected either positive (P), negative (N) or neutral (O) terminal of the DC link. All the switch states can be illustrated by the space vectors diagram as shown in Fig.2. These space vectors can be classified into zero vectors, small vectors (the vertexes of inner hexagon), medium vectors (the mid-points of sides of outer hexagon) and large vectors

(the vertexes of outer hexagon). Both zero vectors and small vectors have redundant switching states. For the small and medium vectors, one or two phases of the load are connected to the NP, which results in a neutral current that disturbs NP potential balance. For each pair of redundant small vectors, the line-line voltages are identical, but NP currents have opposite polarity. So the key to neutral-point voltage control lies in the use of appropriate small vector states to generate an average neutral current of the desired polarity to balance the two capacitor voltages.

By now, all the solutions for NP potential ripples are based on triangle carried-wave PWM[1-3],[6] or SVPWM[4-5],[7]. Although the space vectors based neutral-point potential behavior analysis in [4] is comprehensive, it does not lead to a simple/efficient control scheme. The SVPWM scheme proposed in [5] is the most effective method for voltage balancing control; however, it is considerably complex. In [6], a zero-sequence voltage is added to the commanded voltages to drive the voltage unbalance to zero and the NP-fully-controllable region region is given. But it is not in favor of digital computer realization because of its complexity.

In this paper, first the influence of three-phase load current to flow in/out of neutral point is analytically investigated for the neutral-point potential, and presents an average NP currents model. Then, based on essential correlations between the triangle carrier based PWM and

---

Spported by Shanghai Leading Academic Discipline Project, Project Number:T0103

1-4244-0448-7/06/$25.00 ©2006 IEEE

the SVPWM, this paper presents a novel NP balancing scheme. The present control algorithm only requires the information of the capacitor voltages and three phase load currents and it's easy to digitally achieve with computer. At last, simulation and experiment results verify the effectiveness of the method.

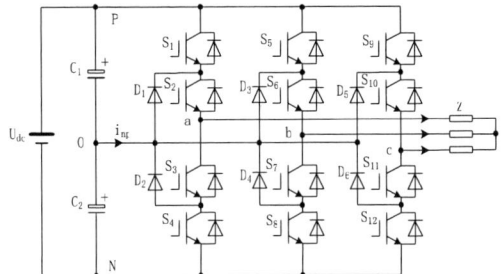

Fig.1 Scheme diagram of three-level NPC inverter

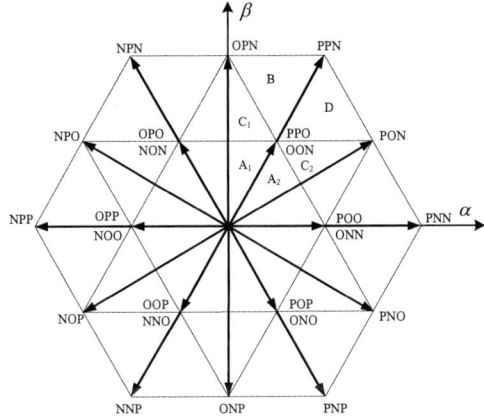

Fig.2 Space voltage vector with their switching states

for three-level inverter

II. ANALYSIS OF THE NP CURRENTS

For small and medium vectors, one or two phases can be connected to the NP, it must lead to NP currents (i_{np}) that disturbs the NP potential. Because of one-phase current's connection to the NP, medium vectors make NP potential dependent in part on the load conditions. They are the most important source of the NP potential unbalance[4].

Every arm of the three-phase inverter has three output states: positive (P), negative (N) and neutral (O). Assume the inverter switching states as:

$$V_s = [S_a, S_b, S_c]^T \qquad (1)$$

In equation (1), S_x=-1,0,1 , x=a,b,c

To any arm, when it connects to the NP, which means

$S_x(t)$= 0, the current passing by it flows into the NP through the clamping diodes. The instantaneous current could be expressed as:

$$i_{np}(t) = [1-abs(S_a)] \cdot i_a + [1-abs(S_b)] \cdot i_b + [1-abs(S_c)] \cdot i_c$$

$$= -abs(S_a) \cdot i_a - abs(S_b) \cdot i_b - abs(S_c) \cdot i_c \qquad (2)$$

In equation (2), $abs(\bullet)$ is the absolute function.

The aforementioned NP current can generate the DC capacitor voltage ripple by passing through them. Therefore the method has to make the change of DC capacitors to be zero during every control period T_s by controlling the average NP currents to be zero within T_s.

III. NP POTENTIAL CONTROL ALGORITHM

A. Inherent Correlation between Carrier-Based PMM and SVPWM

For conventional two-level three-phase-three-wire inverter using SPWM, when appropriate three-order harmonics, which is a kind of zero-sequence component, is injected into the sine modulation wave, it produces the same output line-to-line voltage as SVPWM does. On the other hand, SVPWM corresponds to the result of the regular sampling of the sine modulation wave which is added the zero-sequence component on it. But the zero-sequence component doesn't exactly equate the three-order harmonics. Especially during the dynamic system regulation, the phase angle and amplitude change. Sometimes the modulation wave even distorts into non-sine wave. The similar conclusion is achieved in the three-level VSI and further analysis is shown as follows.

Assume the operation states in one triangle carrier-based PWM period are shown in Fig.3. V_a^*, V_b^* and V_c^* are three-phase reference modulation waves. V_{c1} and V_{c2} are triangle carried-waves. The switching states of the three phase arms, which correspond to the action sequence of the three phase arms, result from the comparison between the triangle carried-wave and the reference wave. Given careful analysis on the three-phase-arm action sequence from the point of SVPWM, it's easy to find that this pattern agrees with the SV synthesis action sequence in triangle B of section 2 in Fig.2(assume the α axis in Fig.2 as the start and define counterclockwise and respectively every spam of 60° as section 1, section 2, ..., section 6). It reflects certain relation qualitatively between carrier-based PWM and SVPWM in the three-level VSI [7].

Note that all the parameters in Fig.3 are normalized.

Assume the operation times of three space vectors that synthesize the desired reference vector, which are PPO (OON), PPN and OPN, are separately $2T_0$, $2T_1$ and $2T_2$. The coefficient k is the distribution factor of the time T_0. When the aforementioned parameters are known, the connotative three-phase modulation functions for SVPWM are

$$V_a^* = (k \cdot T_0 + T_1)/T \qquad (3)$$

$$V_b^* = (k \cdot T_0 + T_1 + T_2)/T \qquad (4)$$

$$V_c^* = k \cdot T_0/T - 1 \qquad (5)$$

Suppose V_a, V_b and V_c are the given sine reference voltage sign. According to the regular sampling principle and the volt-second balance principle, the times are

$$T_1 = (V_a - V_c - 1) \cdot T \qquad (6)$$

$$T_2 = (V_b - V_a) \cdot T \qquad (7)$$

$$T_0 = T - T_1 - T_2 \qquad (8)$$

Substitute (6), (7) and (8) into (3), (4) and (5) and a new series formulas are

$$V_a^* = V_a + V_z \qquad (9)$$

$$V_b^* = V_b + V_z \qquad (10)$$

$$V_c^* = V_c + V_z \qquad (11)$$

$$V_z = (k-1) \cdot V_c - k \cdot V_b + 2k - 1 \qquad (12)$$

In (12), the distribution factor k can change from 0 to 1, which leads to the variation of V_z in the triangle carried-wave modulation. k can be a constant or a time-based variable. The difference of k leads to different PWM patterns. Equation (12) is almost exactly equal to the one for two-level VSI that illustrates the relation between triangle carried-wave PWM and SVPWM[8]. The only difference is the meaning of zero sequence vector. According to aforesaid conclusion, three-level SVPWM could be implemented by injecting appropriate

zero-sequence component into triangle carried-wave modulation. On the other hand, triangle carried-wave modulation could be realized by choosing reasonable

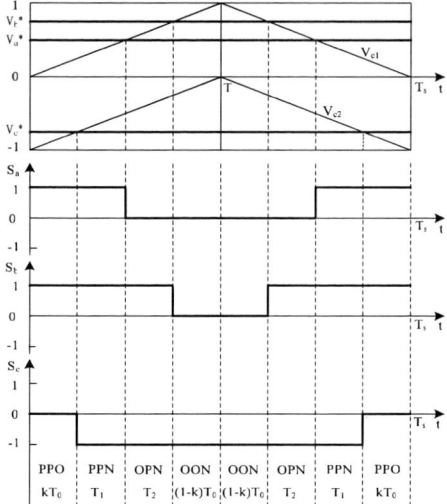

Fig 3 Timing of switching of SVPWM and triangle comparison PWM for three-level inverter

distribution factor of the redundancy vectors in SVPWM. There is an inherent and inevitable correlation between them.

B. Control Algorithm for the Average Zero NP Current

The most popular and employed scheme is adjusting the relative operation time of small vectors in pairs by hysteresis band control, which is a rude and qualitative modulation method. Its control effect depends much on the load power factor (PF). The lower PF is, the worse NP control effect is.

The redundancy small vectors exist always in pairs. Each pair of vectors makes the same line-to-line voltages, but results in different NP currents directions, hence different effects to the NP potential. To regulate the operation times of the small vectors means regulating the distribution factor of durations of the redundant small vector pairs. According to the foregoing section, this regulation of distribution factor for SVPWM could be achieved by injecting proper zero-sequence component into the triangle carried-wave. Thus the control task of NP potential balancing can be translated into finding appropriate zero-sequence voltage component to realize the NP potential stabilization by regulating accurately the operation time of the redundancy small vectors.

According to the analysis of Section 2, during one

control period T_s , the average NP current flowing into and out of the NP must be zero in order to make the variation of capacitor voltage be zero. Based on equation (2), considering the three-phase-arm switching states and the assumption that all three-phase currents are constant as I_a, I_b and I_c respectively during one control period T_s, it can be concluded that when the desired reference vector lies in Triangle B of Section 2, the average NP current in a T_s is

$$I_{np} = -(V_{ac} \cdot I_a + V_{bc} \cdot I_b - V_{cc} \cdot I_c) \qquad (13)$$

And

$$V_{ac} = V_a^* + V_0 \qquad (14)$$

$$V_{bc} = V_b^* + V_0 \qquad (15)$$

$$V_{cc} = V_c^* + V_0 \qquad (16)$$

Substitute equation (14), (15) and (16) into (13) and define $I_{np}=0$, a new equation is

$$V_0 = \frac{V_a^* \cdot I_a + V_b^* \cdot I_b - V_c^* \cdot I_c}{2I_c} \qquad (17)$$

Equation (17) is an average NP potential balancing equation. If the zero-sequence components out of equation (17) are injected into reference wave V_a^*, V_b^* and V_c^*, the precise operation time modulation of the redundancy small vectors will be achieved and the NP potential will be balanced.

Equation (17) is gained on the condition that the reference vectors lies in triangle B of section 2. It should be noted that when the reference vector falls into triangle A or C, one phase modulation wave (here is V_b) of the three phases may flip its sign. In this case, the zero-sequence component should be calculated according to the final status of flipping sign. When the reference vector locates in other sections, the corresponding NP potential balancing equations can be achieved.

C. *NP Balancing Algorithm during Dynamic Process*

In reality, the NP potential may departures the balance point or the initial capacitor voltages differ with each other. In this case, proper NP zero-sequence voltages should be injected to control the average NP current so that the NP potential comes toward the balance point as fast as possible within one T_s.

Suppose that in each T_s the error between the two DC capacitor voltages is:

$$\Delta U_{c12} = U_{c1} - U_{c2} \qquad (18)$$

Then the correlation between the average injected NP current and the capacitor voltage variation is:

$$\Delta U_{c1} = -\Delta U_{c2} = \frac{1}{C_1} \frac{I_{np0}}{2} T_s \qquad (19)$$

In (19), C_1 is the volume of the DC capacitor as shown in Fig.1.

Here the NP current under control is not exactly equal to zero, but to

$$I_{np0} = -\frac{C_1 \cdot \Delta U_{c12}}{T_s} \qquad (20)$$

Define $I_{np}=I_{npo}$ and according to (13), an new equation is

$$V_0 = \frac{(V_a^* \cdot I_a + V_b^* \cdot I_b - V_c^* \cdot I_c) + I_{np0}}{2I_c} \qquad (21)$$

Equation (21) is the average NP potential balancing equation in practice. When the reference vector locates in other sections, there will be corresponding equations.

D. *Restriction Condition for NP Potential Balancing*

When the zero-sequence component V_0 for NP potential balancing, which results from the assumption $I_{np}=0$, is immitted into the reference modulation waves, it's restricted by the modulation wave amplitude and has to satisfy the limit condition as follows[6]:

$$\left| V_x^{**} \right| = \left| V_x^* + V_0 \right| \leq 1 \qquad (22)$$

In (22), x=a,b,c.

When the reference vector locates in section 2, the reference modulation voltage is from equation (3),(4),(5) while the zero-sequence component is from equation (17). Based on these equations, an analysis about the limit condition can be given analytically.

When the zero-sequence component satisfying the limit condition is injected into the reference modulation wave, the average NP current could be made to zero so that the NP potential fluctuation is restrained fully. The zero-sequence component unsatisfying the limit condition should be as large as possible within the limit condition, so that the average NP current is as small as possible and in this case, the NP potential fluctuation is diminished, although not completely eliminated.

1563

IV. SIMULATION AND EXPERIMENT RESULTS

The proposed method is simulated for a three-phase RL load to verify the effectiveness. The parameters of the three-phase RL load include R=12.6 Ω , L=20mH, $C_1=C_2$=220uF. The DC link voltage equals to 2000V. The control period T_s is 400us. The basic frequency is 50Hz.

As shown in Fig.4, the results of simulations are executed in the case of following conditions: the output frequency f=40Hz, the modulation index m=0.8. To compare the NP potential variation, the NP balancing control is unimplemented before t=0.04S while applied after that. It's easy to learn that the NP potential variation is large without balancing control and is effectively restrained with it.

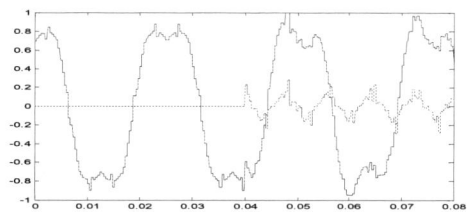

(a) Reference voltage and zero-sequence voltage

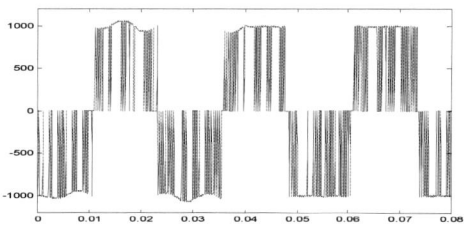

(b) Output PWM phase voltage

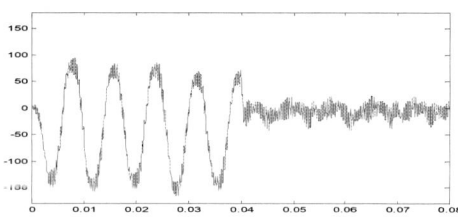

(c) NP potential variation

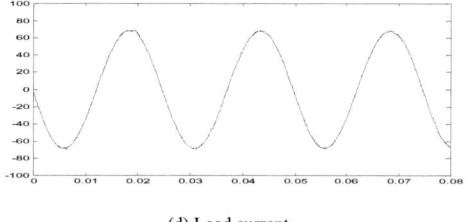

(d) Load current

Fig.4 Simulation results for NP balancing control when M=0.8, f=40Hz

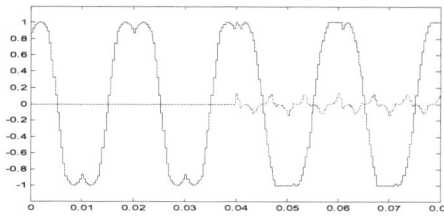

(a) Reference voltage and zero-sequence voltage

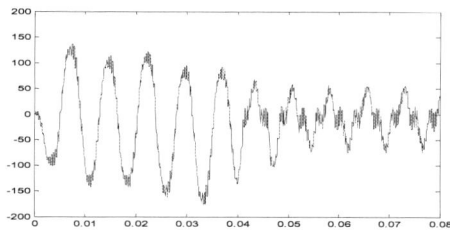

(b) NP potential variation

Fig 5 Simulation results for NP balancing control when M=1.0,f=50Hz

In the case of f=50Hz and m=1.0, the simulation wave with the NP balancing control is shown in Fig.5. The reference wave is flat-roofed. It means the compensation current is restrained within the limit condition. So the NP balance algorithm can diminish the NP fluctuation, although can not eliminate it.

We employed a digital signal processor (DSP) TMS320LF2407A as the core control chip for the experimental work. The dc bus consists of two capacitors in series with 200uF each, and the rating parameters of the induction motor used in the experiment consist of 380V, 1430min[-1], "Y" connected, 3kW. The experimental wave of the motor line-to-line voltage and current is shown as Fig.6, and NP voltage variation before and after employing the algorithm is shown as Fig.7.

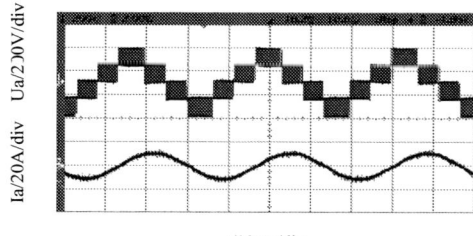

t/10ms/div

Fig.6 Line-to-line Voltage and current for NP balancing control when

M=0.8,f=40Hz

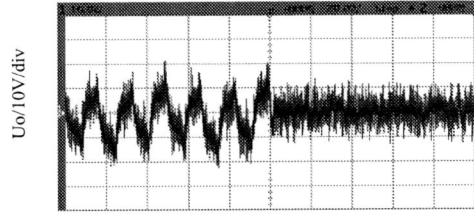

t/20ms/div

Fig.7 Experimental results of NP variation comparison without and

with NP balancing control when M=0.8,f=40Hz

All the foregoing results verify the effectiveness of the NP balance algorithm proposed in this paper.

V. CONCLUSION

This paper investigates the inherent correlation between triangle carrier-based PWM and SVPWM for the three-level inverter and comes to a conclusion that three-level SVPWM could be achieved by injected proper zero-sequence component into triangle carrier-based PWM. Based on the correlation, this paper proposes a novel NP potential balancing algorithm and deduces mathematic equations of the required zero-sequence component to deal with the NP potential fluctuation of three-level diode-clamped VSI. The scheme realizes the control of the NP current, only requiring the information of dc-link capacitor voltages and three-phase load currents. The effectiveness of proposed balancing algorithm is verified by the simulation and experiment results.

REFERENCE

[1]   J.K. Steinke, "Switching frequency optimal PWM control of a

three-level inverter," IEEE Transactions on Power Electronics, vol. 7, no. 3, pp. 487-496, 1992.

[2]   Dong-Hyun Kim et. al., "The analysis and comparison of carrier-based PWM methods for 3-level inverter" Industrial Electronics Society, 2000. IECON 2000. 26th Annual Confjerence of the IEEE Volume 2, 22-28 Oct. 2000 pp:1316 - 1321

[3]   S. Ogasawara and H. Akagi, "Analysis of variation of neutral point potential in neutral-point-clamped voltage source PWM inverters," *Conference Record of the IEEE Industry Applications Society*, vol. 2, pp. 965-970, 1993.

[4]   N. Celanovic and D. Boroyevich, "A comprehensive study of neutral-point voltage balancing problem in three-level neutral-point-clamped voltage source PWM inverters," *IEEE Transactions on Power Electronics*, vol. 15, no. 2, pp. 242-249, 2000.

[5]   K. Yamanaka *et. al.*, "A novel neutral point potential stabilization technique using the information of output current polarities and voltage vector," *IEEE Transactions on Industry Applications*, vol. 38, no. 6, pp.1572-1580, 2002.

[6]   Q. Song, W. Liu, Q. Yu, X. Xie and Z. Wang .A neutral-point potential balancing algorithm for three-level NPC inverters using analytically injected zero-sequence voltage [C]. *Proceedings of the Applied Power Electronics Conference*, vol. 1, pp.228-233, 2003.

[7]   H. Y. Wu, X. L. He. Relationship between multilevel carrier-based PWM and SVPWM and its applications[J] *Proceedings of the CSEE*, Vol.22,No.5, pp.10-15, 2002

[8]   Vladimir Blasko. A hybrid PWM strategy combining modified space vector and triangle comparison methods PESC'96, Baveho, (VB), Italy, 1996, 1872~1878

2006 5th International Power Electronics and Motion Control Conference

Instantaneous Voltage Regulated Seamless Transfer Control Strategy for Utility-interconnected Fuel cell Inverters with an LCL-filter

Guoqiao Shen, Dehong Xu

College of Electrical Engineering, Zhejiang University,

Yugu Lu, Hangzhou, P. R. China, 310027

xdh@cee.zju.edu.cn

Xiaoming Yuan

General Electric Corporate R&D-Shanghai

200233, Shanghai, China

xiaoming.yuan@geahk.ge.com

Abstract—This paper proposes a new control strategy for the transition between grid-tied mode and off-grid mode of the utility interconnected fuel cell inverters with an LCL filter. During the transition from grid-tied mode to stand-alone mode, the inverter control is switched to voltage-controlled mode from the beginning of the transition, using inverter instantaneous output voltage regulation to reduce the output grid current of the inverter quickly, so that the static transfer switch is turned off immediately and the inverter maintain a continuous voltage output. It will result in a short transition time and minimum voltage fluctuation on the load. The algorithm is described along with their transition performances. Simulations and experiments on a 5kW fuel cell inverter prototype are provided to validate the algorithm.

I. INTRODUCTION

Fuel cell generation now is regarded as one of the most competitive energy source for the distributed energy generations due to its advantages of higher efficiency, lower emissions, and higher power density. Most of the utility interactive fuel cell inverters have the capability to operate in two modes, a) utility interactive mode for peak shaving to reduce the overall cost of power by generating during peak hours, b) stand-alone mode to provide power during utility outages until service can be restored.[1]

In the case of sensitive and mission critical loads at the user end, a continuous uninterrupted AC power is of utmost importance. So it is very valuable for fuel cell inverters to be able to operate in both grid-tied and off-grid modes with seamless transitions between the two modes. When the inverter is grid-connected, it is operated in current-controlled mode. In the stand-alone mode, the inverter is operated in the voltage-controlled mode. Heretofore, many researches have made on these two modes of inverters [2], but no enough attention has been paid to the issue of a seamless transition between the two modes. The control algorithm reported in [3] switches the inverter from current-controlled mode to voltage-controlled mode after the grid-disconnecting switch (SCR) current goes to zero free-wheeling, so the output voltage on the load is uncontrolled from the instant of utility fault to the time of the first zero crossing of the

grid current. To overcome the turn-off delay and reduce the voltage uncontrolled time during the transfer, the voltage amplitude regulation algorithm (VAR) and the voltage phase regulation algorithm (VPR) are introduced to reduce the grid current promptly [4]. However, they both depend on the in-phase running for the grid voltage and current, which in some case, will be of a problem.

Focusing on the seamless transition between grid-tied and off-grid modes of operation, this paper proposes a new control algorithm based on voltage-controlled method for utility-interactive fuel cell inverters with an LCL filter, the instantaneous voltage regulation algorithm (IVR), which is introduced to implement forced current commutation between the grid and the inverter. The algorithm and its transition performances are described and discussed in this paper. Simulations results on a 5kVA single-phase inverter are presented using PSPICE. The results show the feasibility of the proposed approaches and the effectiveness of the algorithms in minimizing voltage transients across the inverter and loads, even at the moment of utility fault. A 5kVA DSP controlled inverter prototype has been made, and the present control strategy is validated.

II. TOPOLOGY AND RUNNING MODES OF THE INVERTER

Fig.1 shows the system topology for the utility interactive fuel cell inverter. The topology comprise a fuel cell stack, a DC to DC converter, a PWM inverter, a low-pass LCL filter and a grid-disconnecting transfer switch (STS). Due to the advantages of short turn-off time, large current capacity and low cost, SCR is selected as the static transfer switch to enable the inverter disconnect from the grid rapidly. The LCL form of low-pass filter offers the potential for improved harmonic performance at lower switching frequencies [5], which is a significant advantage in higher-power applications, (e.g. several hundreds kilowatt inverters). For sensitive and mission critical loads at the user end, a continuous uninterrupted AC power is required when the utility is abnormal.

1-4244-0448-7/06/$25.00 ©2006 IEEE

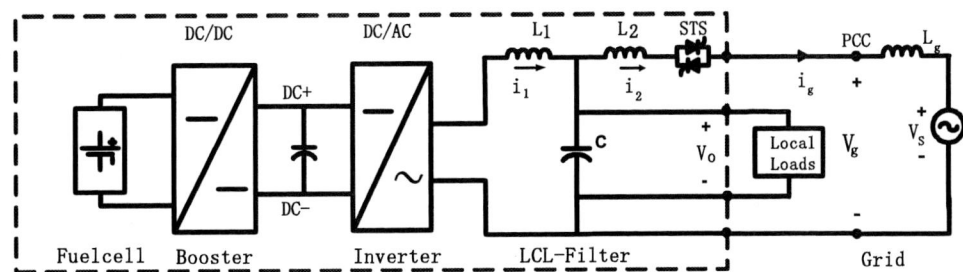

Fig.1   System topology for the utility interactive fuel cell inverter

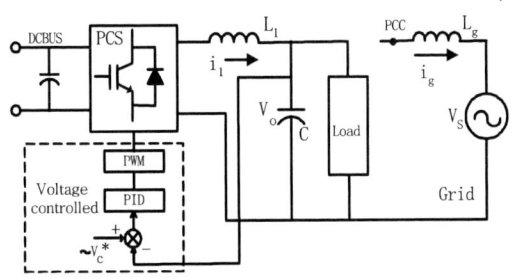

（a）Voltage-controlled stand-alone mode

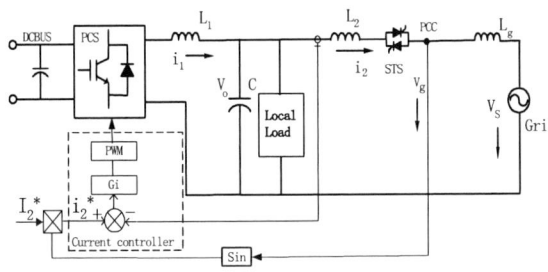

（b）Current-controlled utility-interactive mode

Fig.2 Schematic diagram of the converter in different modes

Fig.2a shows the schematic diagram of the converter when it is disconnected from the grid. The inverter is operated in voltage-controlled mode. Voltage feedback is used to regulate the voltage across the load.

Fig.2b shows the schematic diagram of the converter when it is grid-connected. The Inverter is operated in current-controlled mode at this time. It is controlled through current feedback to regulate the current injected into the grid. The utility is assumed to be relatively strong and maintains the voltage across the load.

For the transition from stand-alone to grid-connected, it is relatively easy and well known as in the uninterruptible power supply (UPS), so no more descriptions is given here. For the transition from grid-connected mode to stand-alone mode, there are two essential work should be done for the control algorithm: a)

Disconnect from the grid by the static switch (SCR).Because SCR has no self turn-off capability, the control algorithm should ensure appropriate conditions for the SCR current to go down to zero. b) Change inverter control method from current-controlled mode to voltage-controlled mode. In order to maintain a continuous uninterrupted AC power across the load, quickly turning-off of the SCR and earlier voltage controlling of the inverter output are required.

III. PRINCIPLE AND PERFORMANCE OF TRANSFER STRATEGIES

A. Control algorithm with grid current free-wheeling (CFW)[3]

For the transfer from utility-interactive mode to stand-alone mode, the inverter is initially current-controlled and the grid maintains the voltage at the PCC. When the inverter starts to switch to the stand-alone mode, the drive of the SCR is removed firstly. The grid current is freewheeling through the SCR, as it has no self turn-off capability. The grid is disconnected until the line current goes to zero, and the SCR has been turned off. At this instant, the PWM inverter is shifted from current controlled mode to the voltage controlled mode. Thus there is no current spike during the transition. But in the case of a utility fault, the voltage across the load may be dropped within half a line cycle (in addition to the time required to detect the fault) due to the delay of switching inverter to voltage-controlled mode.

B. Control Algorithm with VAR and VPR [4]

With VAR transfer control, the grid current is in phase with the grid voltage before switching, because a unity power factor is required. The drive of the SCR is removed at the beginning of the operation mode transition. At the same time, the inverter is shifted to the voltage-controlled mode with the desired output voltage Vo being lower or higher than the grid voltage Vs (Fig.3a), introduces a voltage drop V_L across the line inductance and filter inductance L_2, just being opposite to the grid current Ig . So the grid current is forced to decrease quickly. Once the grid current goes to zero, the SCR is turned off, and the output voltage is changed to the rated value soon. Consequently, it takes less time to complete the transfer, and the voltage transition across the load is minimized (Fig.3).

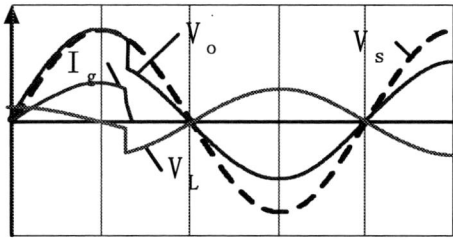

(a) Current and voltage waveforms during transition

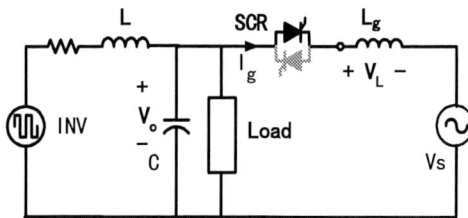

(b) Topology of the utility-interactive inverter

Fig.3 Control algorithm with voltage amplitude regulation (VAR)

For the VPR control algorithm, instead of the voltage amplitude, the voltage phase of the inverter output voltage is regulated according to the current waveform, so a opposite voltage is introduced upon the grid side inductance to transfer the grid current to the inverter.

It can be seen that both VPR and VAR method are depend upon the waveform of the grid voltage and grid current, in-phase running are required, so they are limited to be used. For the moment of utility fault, the waveforms of the grid voltage and current are always distorted.

C. Proposed Control Algorithm with the Instantaneous Voltage Regulation (IVR)

The proposed control algorithm is an improvement of VAR. Here, the instantaneous voltage of the inverter output is controlled according to the direction of the grid current with a fixed voltage difference to the grid voltage, so that a voltage opposite to the grid current is introduced

on the grid side inductance in despite of the waveforms of the grid voltage and current. Fig. 4a shows the waveforms of the grid voltage Vg, the corresponding desired output voltage Vo of the inverter for the proposed IVR transition control, and consequently the voltage drop V_L introduced on the grid side inductance L2 by the IVR control. Whenever the transition is carried out, and no matter the grid voltage is distorted, the voltage across the inductance L2 is same because the inverter output voltage is controlled to follow the grid voltage with a constant voltage difference. Fig.4b shows the waveforms of the grid voltage and current and the inverter output voltage before and after the transition. Before the transition, the inverter was running in current-controlled mode as shown in Fig.4b, so the inverter output voltage was nearly the same as the grid voltage. When the transition began at t0, the inverter turned to voltage-controlled mode, and the inverter output voltage was controlled to follow the grid voltage with a fixed difference, as shown in Fig.4a, so that the grid current fell down to zero at t1, after a delay, at the time of t2, the inverter output voltage turned to trace the standard waveform.

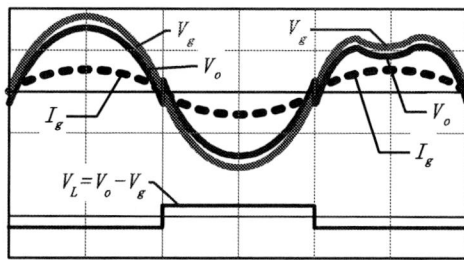

(a) Grid voltage and the desired inverter output voltage proposed for IVR control transition

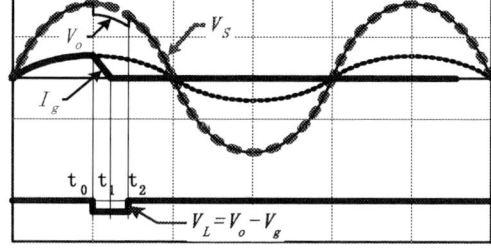

(b)Waveforms during the IVR control transition, the inverter output voltage from t0 to t2 is controlled to decrease , as shown in (a)

Fig.4 Control algorithm with instantaneous voltage regulation (IVR)

In this control algorithm, the grid current falling

1568

time can be expressed as Δt,

$$\Delta t = t_1 - t_0 = L \frac{i_g}{V_L}$$

For a fixed V_L value, the grid current falling time is proportional to the instant current value. When the grid current is given, then the grid current falling time under different control voltage difference V_L can be calculated, as shown in fig.5a. For normal utility condition transition, the maximum grid current falling time is occur at the current peak, so it can be derived as:

$$\Delta t = \frac{1}{\omega\beta(1-\alpha)} \qquad (1)$$

where β is the radio of grid voltage amplitude to the voltage amplitude across the grid side inductance L2 caused by the grid current in normal operating condition,

$$\beta = \frac{V_S}{V_L} = \frac{V_{SM}}{\omega L_2 I_{SM}}$$

$\alpha = Vo/Vs$ is the amplitude radio of the desired inverter output voltage to the grid voltage during the transition.

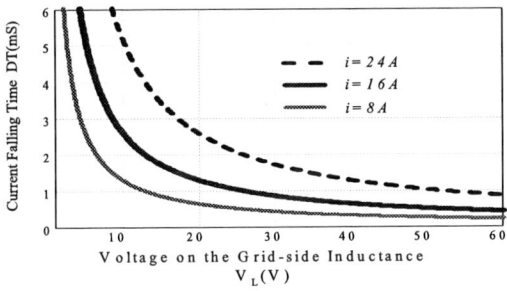

a) Grid current falling time under different voltage difference (V_L)

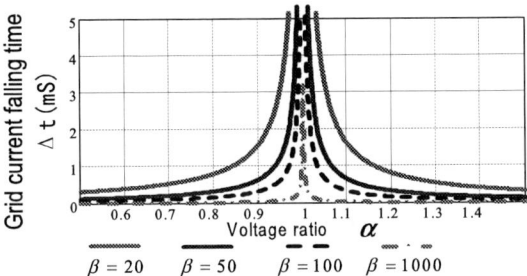

b) Maximum grid current falling time for normal grid condition

Fig.5 Grid current falling time vs voltage difference

Fig.5b shows the maximum grid current falling time vs. the voltage phase difference between the inverter output voltage and the grid voltage for different β values. The grid side inductance is 1.6mH. The grid current falling time is decreasing as enlarges. For $\beta=20$, 15% voltage difference between the grid voltage and the

inverter output voltage will result in 1ms less falling time of the grid current and a short transition period.

IV. SIMULATION AND EXPERIMENT RESULTS

The simulation is realized by PSPICE on a single-phase half-bridge inverter and a six-switch 3-phase inverter, as shown in Fig.6a and Fig.6b. The rated power of each phase is around 5kVA. The voltage of DC bus is 800V, the current injected into grid is 0~30A, the switching frequency is 20 kHz. The parameters of the filter are: L1=1.6mH, C=12μF, the grid side inductance is assumed to be 10μH~2.0mH, including the additional inductance.

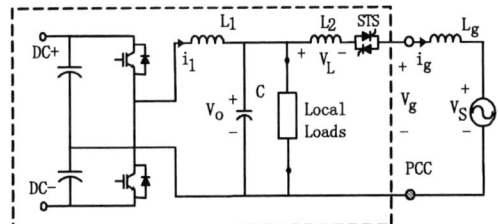

a) Schematic diagram of a single-phase half-bridge inverter

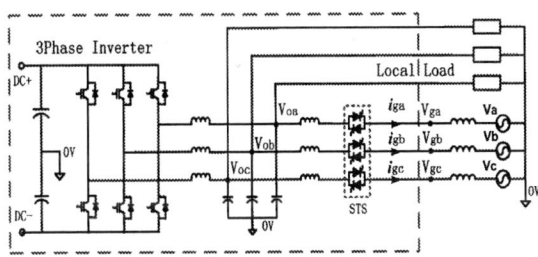

b) Schematic diagram of a six-switch 3-phase inverter

Fig.6 Schematic diagram of grid-tied inverters for simulation

The voltage and current waves of the inverter during the transition from grid-connected to stand-alone in the case of a utility fault are shown in Fig.7. Where, Fig.7a is under the traditional grid current free-wheeling control. Fig.7b is under the voltage amplitude regulation control. The utility fault takes place at the time of t0, and grid voltage has fallen to 50% of the normal value. The detecting time of fault is 1ms. The result shows that the duration of voltage drop under the traditional grid current free-wheeling control is comparative longer (7ms), whereas the duration of voltage drop under the IVR control is only 1.2ms.

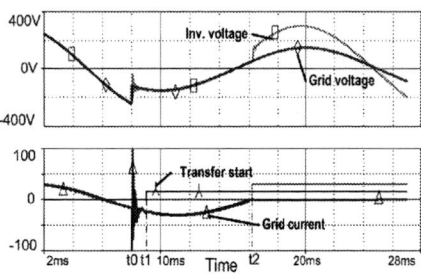

a) Under traditional grid current free-wheeling control

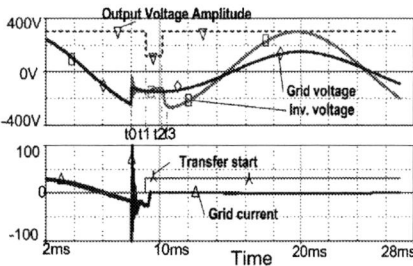

b) Under IVR control

Fig.7 Transition in the case of a fault on the grid

A 5kVA DSP controlled inverter prototype has been made. The experimental results are shown as Fig.8 to Fig.9. Fig.8 shows the grid current (curve1#) and the load voltage (curve2#) under IVR when the transitions start near current zero-cross points. Fig.9 shows the waveforms when the transitions start at current peak points. Whenever the transition is carried out, the transition time is very short, and the voltage on load is continuous.

V. CONCLUSIONS

A novel control algorithm for the transition between grid-tied and off-grid modes of the utility-interactive inverters is proposed. The principle and performances have been discussed. The simulation and experimental results show that the output instantaneous voltage regulation algorithm can provide seamless transfers between the two modes for the inverter, avoiding the temporarily uncontrolled output voltage of the inverter，which occurs during the time of grid current free-wheeling under the traditional control algorithm. The algorithm presented in this paper is valid even in the event of a fault on the grid.

ACKNOWLEDGEMENGT

The authors would like to acknowledge the support of

Delta Power Electronic Technology and Education Fund, and the support of GE (China) Research & Development Center Co., Ltd.

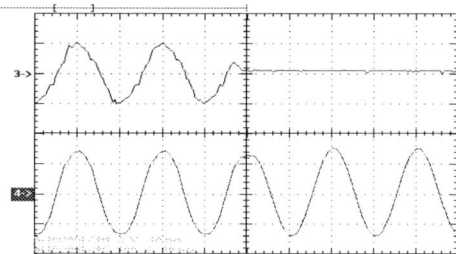

(3_ Grid current, 20A/div.   4_Voltage on local load)

Fig.8 Experimental results for transition from grid-tied mode to off-grid mode under IVR near current zero-cross

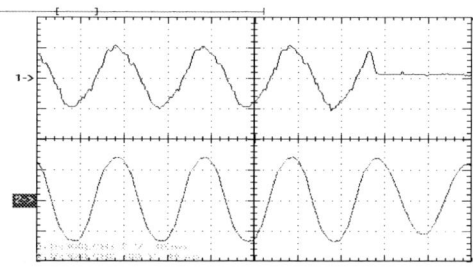

(1_ Grid current, 20A/div.   2_Voltage on local load)

Fig.9 Experimental results for transition from grid-tied mode to off-grid mode under IVR at current peak point.

REFERENCES

[1] Coles, L.R.; Chapel, S.W.; Iamucci, J.J.; "Valuation of Modular Generation, Storage, and Targeted Demand-side Management", Energy Conversion, IEEE Transactions on , Volume: 10 Issue: 1 , Mar 1995 ,Page(s): 182 –187

[2] O'Sullivan, G.A.;"Fuel Cell Inverters for Utility Applications", Power Electronics Specialists Conference, 2000 , Volume: 3 , Page(s): 1191 -1194 vol.3

[3] Tirumala, R.; Mohan, N.; Henze, C., "Seamless Transfer of Grid-connected PWM Inverters between Utility-interactive and Stand-alone modes", IEEE APEC2002. Volume: 2 , pp.1081 -1086.

[4] Guoqiao Shen, Dehong Xu, Danji Xi, "Novel seamless transfer strategies for fuel cell inverters from grid-tied mode to off-grid mode", APEC2005, March 2005, Pages:109-113

[5] Guoqiao Shen, Dehong Xu, Danji Xi, "An Improved Control Strategy for Grid-Connected Voltage Source Inverters with a LCL Filter", APEC2006, Session 29.1, March 2006

2006 5th International Power Electronics and Motion Control Conference

An Anti-windup Design Method for Internal Model Control Based on H_∞ Optimization

Hou Yansong, Li Hua

School of Information and Electrical Engineering, Lanzhou Jiao Tong University, Lanzhou, 730070, China
houyans@163.com

Abstract—Performance and stability of internal model control can deeply deteriorate when plant control input is subject to saturation. Study is made on the reason why two-degree-of-freedom internal model control has anti-windup characteristics. From this consideration, a new modified three-degree-of-freedom internal model control scheme is proposed. This method can realize the decoupling of anti-windup, reference tracing and disturbance rejection. Anti-windup controller synthesis based on H_∞ optimization theory. Small gain theorem guarantees the global stability. Simulation results show the method has better ability of anti-windup and disturbance rejection.

Keywords—saturation; anti-windup; internal model control (IMC); degree-of-freedom; H_∞ optimization

I. Introduction

A good deal of the control system design concerns itself with the problem designing linear controllers based on linear model of plant. In practice, there often exist nonlinear element between controller and plant. For example, maximum actuator movement limits control input for the plant; also, actuators have a restriction to permit voltage input to save its own circuit. Because of limitations, the actual plant input will be different from the output of the controller. When this happens, the controller output does not drive the plant and as a result, the states of the controller are wrongly updated. This effect is called windup. Since the linear controller is designed ignoring actuator nonlinearities, the adverse effect caused by windup is not only badly performance deterioration but also the system stability does not be guaranteed. Windup is usually interpreted as an inconsistency between the controller output and the states of the controller.

Many papers have reported various methods to the anti-windup in the unity feedback framework [1] ~ [5]. Typically, these schemes use additional local feedback to correct the difference between the actual plant input and controller output. In fact, these methods are condition techniques based on linear controller. However, stability of closed loop is difficult to analyze for its nonlinear essence.

This work was supported by the Natural Science Foundation of Gansu province under Grant 3ZS042-B25-039.

Internal model control (IMC) is well known as a new control scheme that has advantages of easy design procedure, easy stability analysis. Internal model control also guarantees the stability if include the nonlinear into internal model. A numerical example shows the above internal model control has anti-windup characteristics [6]. However, the above scheme lost robust servo characteristics subjected to control input saturation if plant contains a pole on the origin [7]. Although stability is guaranteed, provided that there is no mismatch between plant and model, controller state is still inconsistency with the plant input. Considering that internal model control controller itself also contains an internal model, which is still driven by the unsaturated control, a new scheme was proposed in [8]. A similar method was given in [9]. Disturbance rejection is seldom considered except Yamada [10]. In [11], this takes into account nonlinear as a part of plant, given a method based on approximately inverse of nonlinear. However, input saturation mainly occurs as a result of large fast changes in external signal. This anti-windup mechanism does not vanish if unsaturated and long response time will occur consequently.

This paper first analyzes the reason why two-degree-of-freedom internal model control (2-DOFIMC) has anti-windup characteristics. From this consideration, a new modified three-degree-of-freedom internal model control scheme is proposed. This method can realize the decoupling of anti-windup, reference tracing and disturbance rejection. Anti-windup controller synthesis based on H_∞ optimization theory. Simulation results show the method has better ability of anti-windup and disturbance rejection.

II. Two-degree-of-freedom IMC with Input Saturation

In this section, the reason why two-degree-of-freedom internal model control has anti-windup characteristics is analyzed. For simply, the plant is focused on single input and single output system.

Fig.1 shows the two-degree-of-freedom internal model control with control input saturation. Here $G(s)$ is the plant and $G_0(s)$ is the nominal model of the plant, N is saturation nonlinear, $Q_1(s)$ and $Q_2(s)$ are controllers.

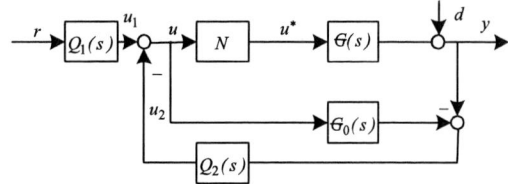

Figure 1. 2-DOFIMC with input saturation

r is the reference input, u is controller output, u^* is plant input, y is plant output and d is disturbance. If $Q_1(s)$ is equal to $Q_2(s)$, the scheme is equivalent to one-degree-of-freedom internal model control. Assume $G_0(s)$ is equal to $G(s)$ and both stable, $d=0$. When saturation does not occur, two-degree-of-freedom internal model control equal to open loop control and $Q_1(s)$ plays the role of a feed forward controller designed for reference tracing. While, $Q_2(s)$ is introduced for disturbance rejection and robustness consideration.

The nonlinear element depicted in Fig.1 can be regarded as a system with perturbation depicted in Fig.2. The saturation function is given by:

$$u^* = \begin{cases} +u_{sat} & u > +u_{sat} \\ u & -u_{sat} \le u \le +u_{sat} \\ -u_{sat} & u < -u_{sat} \end{cases} \quad (1)$$

A single static saturation nonlinearity function defined by (1) can be characterized by:

$$0 < \frac{u^*}{u} \le 1 \quad (2\text{-}a)$$

$$\frac{u^*}{u} - 1 = \frac{u^* - u}{u} = W(s)\Delta(s) \quad (2\text{-}b)$$

$$\|\Delta(s)W(s)\|_\infty \lessdot 1 \quad (2\text{-}c)$$

The saturation nonlinearity represented by (2) is said to be bounded by the conic sector $(0,1]$. $W(s)$ is frequency dependent weighting function. Also, the following equations can be derived from Fig.2:

$$u = u_1 - u_2 \quad (3)$$

and

$$u_2 = Q_2(s)[G_p(s) - G_0(s)]u = Q_2(s)G_0(s)W(s)\Delta(s)u \quad (4)$$

Substituting (3) into (4) and rearranging gives:

$$u_2 = \frac{Q_2(s)G_0(s)W(s)\Delta(s)}{1 + Q_2(s)G_0(s)W(s)\Delta(s)}u_1 \quad (5)$$

and

$$y = u_1 G_0(s) + u_1 G_0(s)W(s)\Delta(s) - u_2 G_p(s) \quad (6)$$

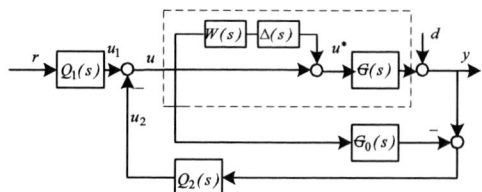

Figure 2. 2-DOFIMC with perturbation

From (5) and (6), consider the 'loop shifting' diagram shown in Fig.3.

Because $G_0(s)$ is equal to $G(s)$ and both stable, the necessary and sufficient condition that the equivalent scheme shown in Fig.3 is internal stable is determined by local feed back. From the small gain theorem [12], the system is internal stable if and only if

$$\|Q_2(s)G_0(s)W(s)\Delta(s)\|_\infty < 1 \quad (7)$$

If one factors $G_0(s)$ as:

$$G_0(s) = \frac{B_0(s)B_{0u}(s)}{A_0(s)} \quad (8)$$

where $B_0(s)$ and $B_{0u}(s)$ are the stable and unstable factors in the numerator, respectively, with $B_{ou}(0) = 1$. Then a suitable choice for $Q_2(s)$ would be

$$Q_2(s) = \frac{A_0(s)}{B_0(s)}F_2(s) = \frac{A_0(s)}{B_0(s)}\frac{1}{(1 + a_2 s)^n} \quad (9)$$

and (7) is equal to

$$\left\|\frac{B_{0u}(s)}{(1 + a_2 s)^n}W(s)\Delta(s)\right\|_\infty < 1 \quad (10)$$

From Fig.3, design freedom $Q_2(s)$ can modify the control valuable u when saturation occur. For this consideration, if we take into account saturation as perturbation, large value of a_2 will enhance system robust stability while good robust performance can be obtained for small value of a_2. Therefore, it can design a suitable small a_2 to obtain good performance at the same time guarantee robust stability. This characteristics just like anti-windup compensation.

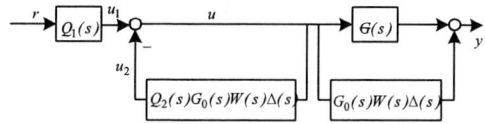

Figure 3. Equivalent 2-DOFIMC with perturbation

1572

When the error between $G_0(s)$ and $G(s)$ exists, in this case, a_2 does not be designed too small for robust stability consideration. In addition, anti-windup characteristics of two-degree-of-freedom internal model control will deeply be deteriorated.

III. THREE-DEGREE-OF-FREEDOM IMCScheme For Anti-windup

In this section, we propose a three-degree-of-freedom design method of the anti-windup compensation for internal model control. The idea of this method is obtained using previous discussion.

We propose a three-degree-of-freedom internal model control for anti-windup design in Fig.4 and additional feed back achieve anti-windup compensation. Here $G(s)$ is the plant and $G_0(s)$ is the nominal model of the plant, N is saturation nonlinear, $Q_1(s)$, $Q_2(s)$ and $Q_3(s)$ are controllers. r is the reference input, u is controller output, u^* is plant input, y is plant output and d is disturbance. Assume $G_0(s)$ is equal to $G(s)$ and both stable, $d=0$. No matter saturation occurs or not, system is often working in open style. $Q_1(s)$ is designed for reference tracing and $Q_2(s)$ for disturbance rejection, while $Q_3(s)$ is introduced for anti-windup. On the other hand, when saturation does not occur, that is $Q_3(s)$ does not work, then Fig.4 is standard two-degree-of-freedom internal model control.

Because $G_0(s)$ is equal to $G(s)$ and both stable, $Q_1(s)$ is also stable; the necessary and sufficient condition that the scheme shown in Fig.4 is internal stable is determined by local feedback. Consider the 'loop shifting' diagram shown in Fig.5(a). From the small gain theorem [12], the necessary and sufficient condition of system is internal stable is

$$\|Q_3(s)\|_\infty < 1. \tag{11}$$

If the error exists between $G(s)$ and $G_0(s)$ $d=0$. The nonlinear can be take account as in Fig.2 with $W(s)=1$, consider the 'loop shifting' diagram shown in Fig.5(b),

$$M(s) = \frac{Q_3(s) + Q_2(s)(G(s) - G_0(s))}{1 + Q_2(s)(G(s) - G_0(s))} . \tag{12}$$

From (2-c) and small gain theorem, the condition that the control system in Fig.4 is internal stability is

$$\|M(s)\|_\infty < 1 . \tag{13}$$

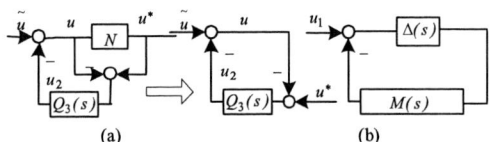

Figure 5. (a) Loop shifting $(G(s)=G_0(s))$ (b) Loop shifting $(G(s)\neq G_0(s))$

IV. CONTROLLER SYNTHESIS

In this section, a synthesis procedure of three-degree-of-freedom internal model control in Fig.4 is given.

Assume $G_0(s)$ is equal to $G(s)$ and both stable, the closed loop formed by the $Q_1(s)$ and $Q_3(s)$ can be drawn as shown in Fig.6 (a). The design problem then can be considered to compute a controller $Q_3(s)$, which satisfy the (11) and in the case when $u^* \neq u$ correct the u so that the controller output u is forced to become equal to u^*. From Fig.6 (a) it follows that

$$u - u^* = S(s)Q_1(s)r - S(s)u^* \tag{14}$$

where $S(s) = (1 - Q_3(s))^{-1}$. In order to formulate a sensible H_∞ optimization problem, it is suitable to minimize both the signal $u - u^*$ and the signal $u_2 = Q_3(s)(u - u^*)$, expressed formally as

$$\min_{stable Q_3(s)} \left\| \begin{matrix} W_1(s)S(s)Q_1(s) & -W_1(s)S(s) \\ W_2(s)Q_3(s)S(s)Q_1(s) & -W_2(s)Q_3(s)S(s) \end{matrix} \right\|_\infty \tag{15}$$

where $W_1(s)$ and $W_2(s)$ are frequency dependent weighting functions.

Formulate the standard H_∞ optimization problem as in Fig.6 (b), the exogenous inputs are $w = [r, u^*]^T$; the signals to be made small are $z = [W_1(s)(u - u^*), W_2(s)Q_3(s) (u - u^*)]^T$; the plant is $Q_1(s)$; and the controller $Q_3(s)$ is driven by $(u - u^*)$. In practice, correct signal u_2 must be vanish when $u = u^*$, that is $Q_3(s)$ is stable. However, the H_∞ optimization does not guarantee it. This condition can be satisfy to select appreciate weighting functions. The rest of work can be done MATLAB tools. We mainly focus on the synthesis of $Q_3(s)$, $Q_1(s)$ and $Q_2(s)$ can be designed by standard internal model control.

$$P = \begin{bmatrix} -W_1(s) & W_1(s)Q_1(s) & W_1(s) \\ 0 & 0 & W_2(s) \\ -1 & Q_1(s) & 1 \end{bmatrix} \tag{16}$$

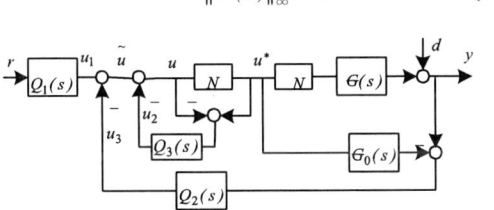

Figure 4. 3-DOFIMC scheme for anti-windup

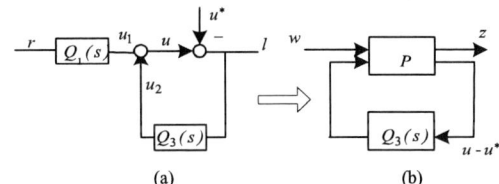

Figure 6. (a) Feedback for anti-windup (b) Standard H∞ optimization control

1573

V. NUMERICAL EXAMPLE

Example1. This example taken from [11], the plant model contains two second-order Butterworth filters in series, with a pair of lightly damped poles. The transfer function of the plant is given by

$$G_0(s) = 0.2 \frac{s^2 + 2\varsigma_1\omega_1 s + \omega_1^2}{s^2 + 2\varsigma_1\omega_2 s + \omega_2^2} \frac{s^2 + 2\varsigma_2\omega_1 s + \omega_1^2}{s^2 + 2\varsigma_2\omega_2 s + \omega_2^2} \quad (17)$$

$\varsigma_1 = 0.3872, \varsigma_2 = 0.9239, \omega_1 = 0.2115, \omega_2 = 0.0473$.

Let saturation limits set at ± 1, reference input $r = 5$ at time 0sec, input disturbance $d_i = 0.1$ at time 200sec, output disturbance $d_o = 0.1$ at time 300sec. Fig.7 shows output response. Here, the solid line shows the output of IMC in Fig.1 with $Q_1(s)=Q_2(s)=(s+1)/(16s+1)$ and $N=0$. The dot line shows the output of IMC in Fig.1 with $Q_1(s)=Q_2(s)=(s+1)/(16s+1)$ and with saturation. The dot-dash line shows the output of IMC in Fig.1 with $Q_1(s)=(s+1)/(16s+1)$, $Q_2(s)=(s+1)/(2s+1)$. Obviously, it has characteristic to anti-windup. The thin line shows the output of the method in [11]. The dash line shows the output of modify internal model control in Fig.4 with $Q_1(s)=(s+1)/(16s+1)$, $Q_2(s)=(s+1)/(2s+1)$ and $Q_3(s)=1/(0.01s+1)$ with $\| Q_3(s) \|_\infty=0.998$. Comparing the method in [11], it has better anti-windup characteristic and disturbance rejection.

Example2. This example taken from [9], the plant model is given by

$$G_0(s) = \frac{2}{100s + 1} \quad (18)$$

Let saturation limits set at ± 1, reference input r = 1 at time 0sec, input disturbance $d_i = 1$ at time 500sec, output disturbance $d_o = 0.2$ at time 800sec. Fig.8 shows output response. Here, the solid line shows the output of internal model control in Fig.4 with $Q_1(s)=(100s+1)/(66s+2)$, $Q_2(s)=(100s+1)/(15s+2)$ and $Q_3(s) = 1/(0.1s+1)$. The dot line shows the output of the method in (9).

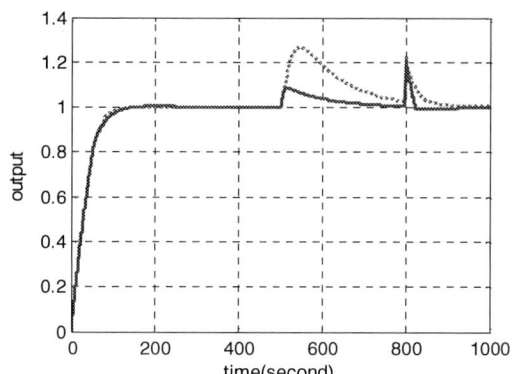

Figure 8.   Output response

From this numerical example, it is shown that the disturbance rejection characteristics of present method are better than the method in (9). While, the anti-windup characteristics are quite comparable.

Fig.9 shows the control input, the dot line shows the control input without saturation, the dash line shows the control input with saturation, the solid line shows the control input with anti-windup compensation.

VI. CONCLUSIONS

There are mainly two group techniques to deal with control input saturation, one is well known model predictive control and the other is anti-windup method. The later is based on the linear controller and have many advantages to the designer, such as easy implement, short computational times. It is often used in the control of mechanical systems and aircraft. A key problem yet to be solved is control input saturation when internal model control is used in the control of mechanical systems. In practice, input saturation mainly occurs as a result of large fast changes in external signal. So consider saturation effect as a special disturbance; take account anti-windup problem as a robust disturbance rejection problem. An optimal strategy has been proposed to solve anti-windup controller based on H_∞ optimization framework.

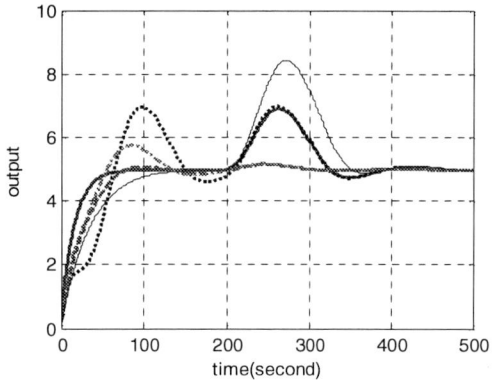

Figure 7.   Output response

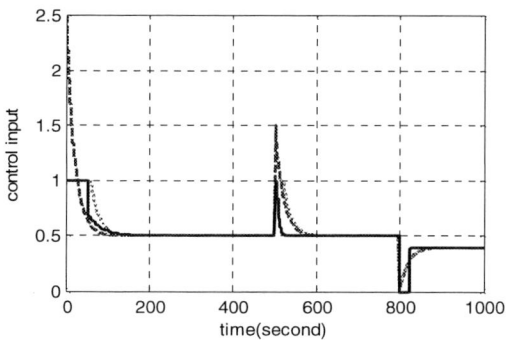

Figure 9.   Control input

From this consideration, a new modified three-degree-of-freedom internal model control scheme is given in this paper. It has been shown through simulation results that the additional feedback controller $Q_3(s)$ can give a fast compensation when saturation occurs. Also, disturbance rejection can be achieved by second design freedom $Q_2(s)$. In addition, the stability of system is guaranteed via small gain theorem.

VII. REFERENCES

[1]   M. V. Kothare, P. J. Campo, M. Morari and C. N. Nett, "A unified framework for the study of anti-windup design", *J. Automatica,* vol. 30, pp. 1869-1883, 1994.

[2]   R. Hanus, M. Kinnaert and J. L. Henrotte, "Conditioning technique, a general anti-windup and bumpless transfer method", *J. Automatica,* vol. 23, pp. 729-739, 1987.

[3]   C. Edwards, I. Postlethwaite, "Anti-windup and bumpless-transfer schemes", *J. Automatica,* vol. 34, pp. 199-210, 1998.

[4]   C. Edwards, I. Postlethwaite, "An anti-windup scheme with closed-loop stability considerations", *J. Automatica,* vol. 35, pp. 761-765, 1999.

[5]   P. F. Weston, I. Postlethwaite, "Linear conditioning for systems containing saturating actuators", *J. Automatica,* vol. 36, pp. 1347-1354, 2000

[6]   Y. Funami, K. Yamada, "An anti-windup control design method using modified internal model control structure", *Proc. IEEE Int. Conf. on Systems, Man, and Cybernetics, Tokyo*: IEEE Press, 74-79, 1999.

[7]   K. Yamada, "Robust internal model servo control with control input saturation", *Proc. American Control Conference,* 3685-3686,1998.

[8]   G. C. Goodwin, S. E. Graebe and W. S. Levine, "Internal model control of linear systems with saturating actuators" *Proc. European control Conference,* 1072-1077, 1993.

[9]   A. Zheng, M. V. Kothare and M. Morari, "Anti-windup design for internal model control" *J. Int. J. control,* vol. 60, pp. 1015-1024, 1994.

[10]   K. Yamada, Y. Funami, "A design method of anti-windup control system with unknown external signal" *Proc. Int. Workshop on Advanced Motion Control,* 205-210, 2000.

[11]   C. S. Ling, M. D. Brown, P. F. Weston and C. Roberts, "Gain tuned internal model control for handling saturation in actuators", *Proc. American Control Conference,* 4692-4697, 2004.

[12]   S. W. MEI, T. L. SHEN and K. Z. LIU, *Theory and Application of Modern Robust Control.* Tsinghua University Press, 2003 (in Chinese).

Study on Pwm Control Strategy of Photovoltaic Grid-connected Generation System

Shi-cheng Zheng, Pei-zhen Wang and Lu-sheng Ge

Anhui University of Technology school of electrical engineering and information
Ma'anshan China, 243002
e-mail:zsc108@ahut.edu.cn

Abstract—The principle of photovoltaic grid-connected generation system is analyzed, and the research on PWM control strategy is done based on the power output side of photovoltaic grid-connected generation system. The direct current control strategy of photovoltaic grid-connected inverter is presented. The control fashion of applying fixed switch frequency and the feedforward compensation control of utility grid voltage is introduced. The current response speed is proved to be faster, and the grid-connected current waveform is sine and the power factor is unity.

Keywords-photovoltaic grid-connected generation system; direct current control strategy; feedforward compensation control

I. INTRODUCTION

When each country of world is developing the green recyclable energy, the solar energy is valued uniformly by its advantages, and the photovoltaic generator of using solar energy is transiting from the supplemental energy to substituting energy gradually. Photovoltaic grid-connected generation system will be the trend of solar energy application. At present, the USA and Germany have all produced the large-scale roof photovoltaic grid-connected generation plan, and the law of recyclable energy is also brought into effect in China on January 1, 2006. The recyclable energy generation will hold enormous development space. This paper characterizes the photovoltaic grid-connected generation system, and researches the direct current control strategy based on the main circuit of three phases photovoltaic grid-connected generation system. The grid-connected current waveform of system is improved effectively, and the faster dynamic response speed is obtained.

II. MAIN CIRCUIT STRUCTURE AND PRINCIPLE

Fig.1 is main circuit structure of the system.

In Fig.1, the solar cell array (is also called as photovoltaic array) converts the solar energy from the sun into DC power immediately, and the DC power is connected to the input of the grid-connected inverter. Then, after energy conversion, the inverter outputs the high frequency SPWM waveform, and its fundamental waveform is sine. After going through the filter inductance, the inverter feeds sine current into the utility

grid which its frequency and phase is the same as utility grid. Thus, the unity power factor and sine grid-connected current waveform are realized. The three phases inverter is also named as current control voltage inverter, briefed as CCVI. The solar cell array is the DC power input of inverter, it is neither constant voltage source nor constant current source, and its V-I character curve and P-V character curve have intense nonlinearity [1], varying as the solar radiance and surroundings temperature, thus in order to make the solar cell array work steadily and output its maximum power, its working voltage must be stabilized firstly. Accordingly, its working voltage of maximum power point is attained by the method of software searching.

In main circuit, the aim of adopting the series-parallel connection of the capacitor of C1, C2, R1 and R2 is to stabilize the DC high voltage and average voltage. Resistance (R), Capacitor (C) and Diode (D) compose the absorbing and snubber circuit of three phases inverter and this circuit functions to reduce the di/dt and dv/dt arising from the IGBTs when they are switched frequently. Thus the IGBTs will be protected effectively. The three phases transformer functioned as promoting voltage and isolating the main circuit from utility grid, so the reliability of system is improved. The filter inductance is important component, and its value relates to DC voltage, the voltage of utility grid, the switch frequency of system and the amplitude of grid-connected current. If its value is too small, the effect of filter is worse and the waveform of grid-connected current will contains more harmonic components. Contrarily, if its value is too large, the voltage drop of inductance will increase. Moreover, the damping and time-delay will increase also, and the power of feed-in utility grid will be influenced. According to the calculations based on stabile model and experiments, the value of filter inductance is taken as about 2.6mH.

III. CONTROL STRUCTURE OF SYSTEM

Fig.2 is the control structure of system.

The double closed loop control structure is adopted in the system, the loop of voltage lies in outer loop and the loop of current is inner loop. The outer loop functions to stabilize the DC voltage of photovoltaic array. The aim of inner loop is to track the given current signal. The output of voltage loop is treated as the given reference amplitude signal I_f of inner loop, and the value of multiplied by that discrete sine tables will be the given real-time signal of inner loop. After the regulation of current loop, the track control of grid-connected current is realized. By regulating the parameters of current loop, tracking speed can be advanced and tracking error can be reduced. The stability of working voltage of photovoltaic array must be attained. Known from the characteristic of photovoltaic array, the choice of working voltage may affect the output power of photovoltaic array. The Maximum Power Point Tracking (MPPT)[2] control is used to get the best working voltage

of photovoltaic array Vd*. In addition, intelligent control of system is realized by judging the output power, DC voltage of photovoltaic array and so on. In Fig.2, AVR is voltage regulator and its output is the input given of the inner loop. ACR is current regulator, and its output added by the feed-forward compensation output will be the real-time modulating signal.

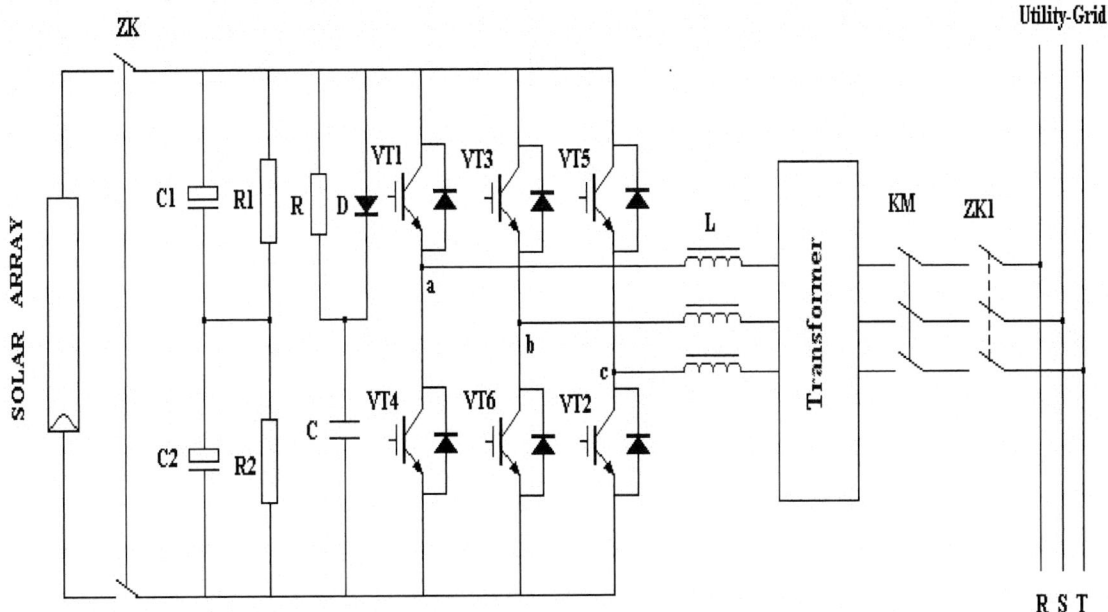

Fig.1 Main circuit structure of the system

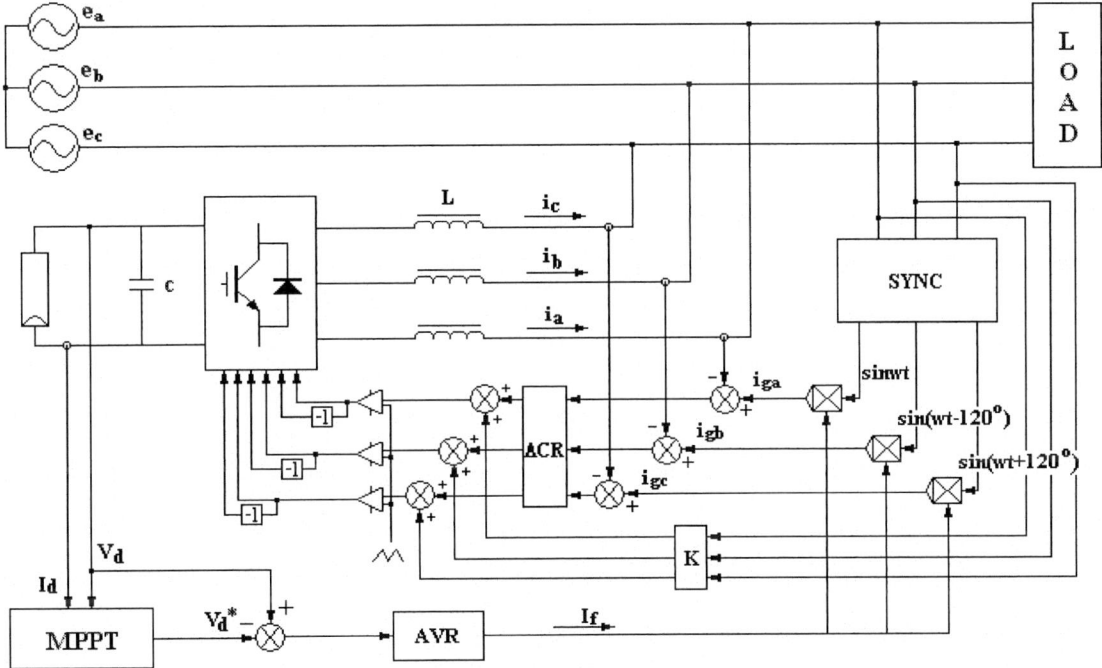

Fig.2 Control structure of the system

IV. CONTROL STRATEGY OF DIRECT CURRENT

In the grid-connected generation system, the load is utility grid whose capacity is infinite, so the grid-connected fashion of current type is adopted to make the grid-connected current waveform be ideal sine. Then the output side of inverter presents the characteristic of controlled current source. In previous current control strategy, allowing for the speed influence of microcomputer, the indirect current control strategy, which is so called control of amplitude and phase, is taken. That is to say, the control of grid-connected current is realized by controlling the amplitude and phase of output voltage of voltage source type inverter. As the dynamic response speed of indirect current control is slow and this method is sensitive to the parameters variation, it is replaced by the direct current control strategy gradually when many kinds of microprocessor of high speed and high capability are produced. The advantages of direct current control strategy are that the dynamic response speed is fast and this kind of control strategy is insensitive to the parameters variation. So, the output waveform of grid-connected current is better and is easier to meet national standards. The frame of direct current control strategy is shown as Fig.3.

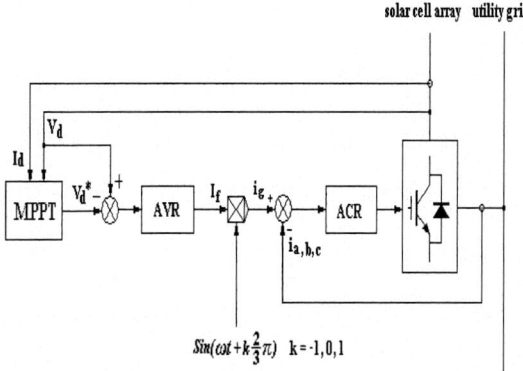

Fig.3 Frame of direct current control strategy

A. Feedforward Compensation Control

In Fig.2, taking the A phase as example, i_a is assumed as state parameter which flowing through filter inductance L[3], the voltage equation of inverter output side from the Fig.2 is obtained as equation (1)

$$V_a = \frac{e_a}{n} + L\frac{di_a}{dt} + i_a r \tag{1}$$

After the transform of Laplas, $I_{(S)}$ is get as

$$I_a(s) = \frac{1}{sL+r}\left[V_a(s) - \frac{e_a(s)}{n}\right] \tag{2}$$

Therein, V_a is the output SPWM waveform of A phase of inverter and it isn't filtered. n is the transformer ratio of

three phases transformer. e_a is A phase voltage amplitude of utility grid. R is equal inner resistance of filter inductance L, transformer and circuit etc.

When the switch frequency is higher, if the influence of time-delay and power devices and the nonlinear influence of deadtime are neglected, the inverter controlled by SPWM fashion may be approximated to be a magnification part with coefficient K_{pwm}. That is expressed as

$$G(s) = K_{pwm} \tag{3}$$

Based on the above details, the closed loop control frame of grid-connected current may be shown as Fig.4.

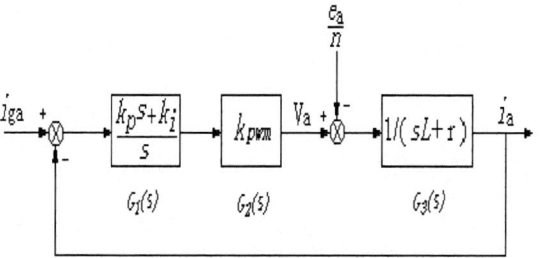

Fig.4 Closed loop control frame of grid-connected current

Differing from the passive inverter, the grid-connected inverter is active inverter, and its load is utility grid. As the voltage amplitude of common connection point can change suddenly when other alternative loads vary, accordingly, the current waveform of grid-connected may distort. Analysis is as follow

From Fig.4, when the voltage of utility grid varies suddenly (assuming from e_a to $e_a + \Delta e_a$), the variation of output grid-connected current of system (Δi_a) will be expressed as

$$\Delta i_a = -\frac{G_3(s)}{1+G_1(s)G_2(s)G_3(s)}\frac{\Delta e_a}{n} \tag{4}$$

That is to say, when the voltage of utility grid increases suddenly, the grid-connected current amplitude will reduce, then the deviation occurs. Two methods may be adopted to restrain the disturbance influence to the grid-connected current arising from the instantaneous variation of utility grid.

(1) From equation (4), if the following demands can be met

$$\left|G_1(s)G_2(s)G_3(s)\right| \gg 1 \text{ and } \left|G_1(s)G_2(s)\right| \geq \left|G_3(s)\right| \tag{5}$$

The influence of utility grid will vanish completely, in other words, $\Delta i_a = 0$.

(2) Feedforward compensation control of utility grid voltage is used.

Although the method (1) can eliminate the nonlinear disturbance influence to the grid-connected current arising from the instantaneous variation of utility grid voltage, the open loop gain of circuit is higher, so the more EMI and noise will be brought to the system, and the system can be unstable. Based on the above analyses, in order to get the better effect of restraining the disturbance influence from utility grid, the feedforward compensation control of utility grid voltage is used to counteract the disturbance, and make the system be a passive tracking system. So the control frame of system is simplified and the control effect is improved. From the control theory [4], the feedforward compensation control is applying the open loop control fashion to compensate the measurable disturbance signal in practice. So the feedforward control fashion won't change the characteristic of system. From the effect of restraining the disturbance, the feedforward control may lighten the burden of feedback control. Thus, the gain of feedback control may be reduced, and the stability of system will be better. The Fig.5 is the control frame of feedforward compensation.

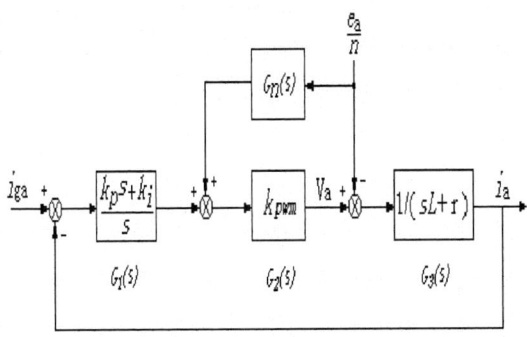

Fig.5 Control frame of feedforw and compensation

From Fig.5, the analogy analysis is expressed as follows

When the voltage of utility grid varies suddenly (assuming from e_a to $e_a + \Delta e_a$), the variation of output grid-connected current of system (Δi_a) will be shown as equation (6)

$$
\begin{aligned}
\Delta i_a &= \Delta i_{a1} + \Delta i_{a2} \\
&= -\frac{G_3(s)}{1+G_1(s)G_2(s)G_3(s)}\frac{\Delta e_a}{n} + \frac{G_2(s)G_n(s)G_3(s)}{1+G_1(s)G_2(s)G_3(s)}\frac{\Delta e_a}{n} \\
&= \frac{G_3(s)[G_2(s)G_n(s)-1]}{1+G_1(s)G_2(s)G_3(s)}\frac{\Delta e_a}{n}
\end{aligned}
$$

(6)

From equation (6), when the transfer function of feedforwrd compensation control part $G_n(s)$ is shown as equation (7)

$$
G_n(s) = \frac{1}{G_2(s)} = \frac{1}{Kpwm}
$$

(7)

The disturbance influence from the utility grid may be eliminated utterly. That is to say, when the given grid-connected current signal is zero, corresponding duty ratio to counteract the utility grid voltage is calculated out by the feedforward compensation control, and doing like so can make the system be a passive tracking system completely.

B. SPWM control of fixed switch frequency

In the direct current control strategy, there are mainly two kinds of control fashions being used. One is the hysteresis loop current control, which can track the given current signal at the fast speed. The other is SPWM control applying fixed switch frequency and the feedforward compensation control of utility grid voltage[5].The hysteresis loop current control means to treat the deviation of given current value subtracted by feedback current value as the input of hysteresis loop comparator. The output signal of hysteresis loop comparator is isolated and amplified to drive the power devices of main circuit, and the tracking control of grid-connected current is realized. If the width of hysteresis loop is wider, the harmonics components of grid-connected current will be much. On the contrary, if the width of hysteresis loop is narrower, the harmonic components of grid-connected current are less, however the switch frequency must be very high. In addition, the switch frequency isn't fixed and varies intensely during the current tracking, so the spectrum of output grid-connected current is very broad, thus, it is very difficult to design the filter. According to the above analyses, the SPWM control fashion of applying the fixed switch frequency and the feedforward compensation control of utility grid voltage in this system is adopted, and the real-time tracking of grid-connected current is achieved.

V. CONCLUSION

According to the above control strategy, the experiments of photovoltaic grid-connected generation system are done. Its main circuit is shown as Fig.1. The power devices are IGBTs produced by Mitsubishi Company and its drive circuit is made up of special chip of M57962L produced by Mitsubishi Company too. The switch frequency is 10KHz. The high speed and high capability chip of TMS320LF2407A is used as the control core of system [6]. This chip has 16 high speed A/Ds, and its fastest conversion time is only 0.5 us. The processing speed of chip meets the function demands of system completely. The experiments results are shown as Fig.6 and Fig.7. The waveforms of Fig.6 are grid-connected currents waveforms of A phase and B phase. The waveforms of Fig.7 are the grid-connected current waveform and utility grid voltage waveform of A phase. From Fig.6, the phase difference between A phase current and B phase current is 120°. From Fig.7, the phase of A phase grid-connected current is synchronized with A phase utility grid voltage's and the power factor of grid-connected inverter equals unity.

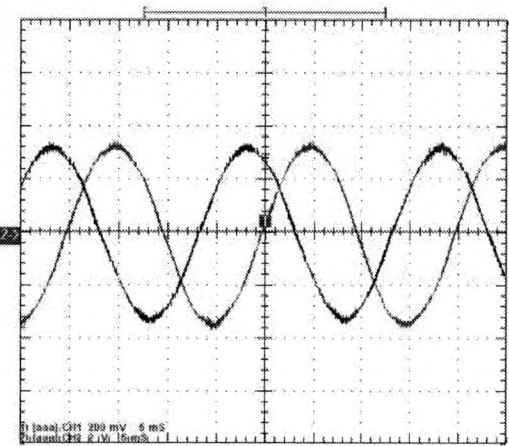

Fig.6 Grid-connected currents waveforms of A phase and B phase

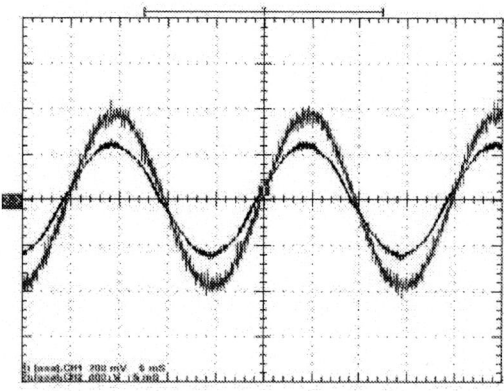

Fig.7 Waveforms of grid-connected current and utility grid voltage

ACKNOWLEDGMENT

This work is supported by the natural science foundation of Anhui Province Education Bureau under Project 2001kj040zd.

REFERENCES

[1] Wei Zhao. Study on the photovoltaic grid-connected generation system of solar energy. Doctoral dissertation of Hefei University of technology, April,2003.

[2] Bodur M, Ermis M. Maximum power point tracking for low power photovoltaic solar panels. IEEE on Electronics Technical Conference Proceedings,Feb,1994, p758~761.

[3] Shicheng Zheng. Study on photovoltaic generation system and its control. Doctoral dissertation of Hefei University of technology, Mar,2005.

[4] Dejin Xia. Automatic control theory. mechanical industry press, July,2002.

[5] Chongwei Zhang, xing Zhang. PWM rectifier and its control. Mechanical Industry Press, Oct, 2003, p06~08.

[6] Heping Liu.TMS320LF240X DSP structure, theory and application. Beijing University of aeronautics and astronautics press, April, 2002.

Robust Sliding Model Control for Regenerative Braking of Electric Vehicle

Min. Ye, Zhifeng. Bai and Binggang. Cao

Research and development center for electric vehicles,
Xi'an Jiaotong University, Xi'an 710064, China.

Abstract— A novel sliding mode control based on H$_\infty$ theory has been developed to meet the nonlinear control of regenerative braking of Electric Vehicle(EV). By combining the sliding phase and the hitting phase, the parameter of the sliding mode controller is designed by H$_\infty$ optimization procedure. In the scheme the sliding mode is ensured from the beginning of the response without the reaching problem and the traditional clustering is reduced. In addition, the gain matrix of the system guarantees the robust stability and disturbance rejection specification. Comparison is made between conventional PI control and the proposed scheme. Simulated results showed that the robustness and superior dynamic performance of the proposed control system.

Keywords-Sliding mode control; H$_\infty$ norm; regenerative braking; EV.

I INTRODUCTION

In a world where environment protection and energy conservation are growing concerns, the development of the electric vehicles has taken on an accelerate pace with the integration of electric motor drive, electronic, and controls [1]. Now the limitation of driving range is the key restriction for the development of EV. Regenerative braking is an affective approach to extend the driving range of EV, at the same time it plays an important role in energy economizing. So regenerative braking is being a hot topic recently around the world [2]. Regenerative braking means that part of the kinetic energy of EV can be transformed to the electric energy which is stored in the energy cell by the boost circuit. The recycle energy can be utilized to accelerate the vehicle. At one hand the boost circuit works at PWM mode, which is a seriously nonlinear system. On the other hand the voltage of the battery, the state of the road, and the driving range of vehicles vary largely under the presence of regenerative braking. All these lead to the invalidation of the traditional control method for regenerative braking. Various attempts to satisfy the control performance for regenerative braking have been presented in literatures [3-6], such as fuzzy PID control in [3], variable structure control in [4], and H$_\infty$ control in [5], [6]. It should be noted, however, that a fairly concise model is usually required for the above controller design. Since the linear design model does not express the exact behavior and the unmodelled dynamics, in this paper a robust controller is designed to guarantee the robustness for both stability and performance based on the concept of variable structure control. By combining the sliding phase and the hitting phase, the hitting controller is firstly proposed to guarantee the existence of the sliding mode for regenerative braking. The parameter of the controller is designed by H$_\infty$ norm theory to ensure that the uncertainty satisfies the matching condition. The advances in both H$_\infty$ control and variable structure control guarantee the more tolerance to achieve the desired goals of the system.

The paper begins with modeling the mathematic model of the regenerative braking system, then focusing on the design of the robustness controller; subsequent to the simulation of the regenerative braking by Matlab/Simulink, it discusses in detail the advantage of the sliding mode controller based on H∞ norm; finally the conclusion is summarized.

II SYSTEM MODEL OF REGENERATIVE BRAKING

A. Dynamic model of EV

On level ground, the resistance of EV consist of the air resistance F_w, the rolling resistance F_f, the electric braking force F_e and the mechanical braking force F_m during the deceleration or braking. They can be presented as

$$\begin{cases} F_w = \frac{1}{2}C_d A \rho V_r^2 \\ F_f = fmg\cos\alpha \\ F_e = \dfrac{K_g K_t I_m}{R_w} \\ F_m = K_m F_0 \end{cases} \tag{1}$$

where C_d=the air resistance coefficient; A=the front area of EV; ρ=the air density; V_r=the relative velocity of EV to the air; f=the rolling resistance coefficient; m=the vehicle mass; K=the transfer ratio from the motor to the wheel; K_t=the electromagnetic thrust coefficient; I_m=the current of the motor armature; R_w=the radius of the wheel. Assume that the mechanical braking force is proportional to the displacement of the braking pedal, the displacement Km is $Km \in [0 \quad 1]$. F_0 denotes the maximum braking force with the maximum displacement of K_m. F_0 can be estimated by the initial velocity, the braking time and the braking distance of EV, which is rational to the study of this paper.

The dynamic model of EV can be presented as

1-4244-0448-7/06/$25.00 ©2006 IEEE   1581

$$m\frac{dv}{dt} = -(F_w + F_f + F_e + F_m) \tag{2}$$

where v =the velocity of EV(km/h). The back-electromagnetic of the motor F_e is

$$v_M = \frac{vKe}{7.2\pi Rw} \tag{3}$$

where K_e =the coefficient of the back-electromagnetic of the motor.

B. Circuit model of regenerative braking

The buck-boost converter incorporates the buck and boost converters into one converter being able to operate both when the output voltage is higher or lower than the source voltage. The control of the converter is accomplished by altering the pulse width of the modulated valve form. The buck-boost converter of regenerative braking is depicted in figure 1, which shows the scheme of the circuit. While regenerative braking, the converter works at boost mode: T1 switches off, T2 switches on PWM mode. The boost converter consisting of T2, L and the diode increases the back-electromagnetic voltage of the motor to charge the battery.

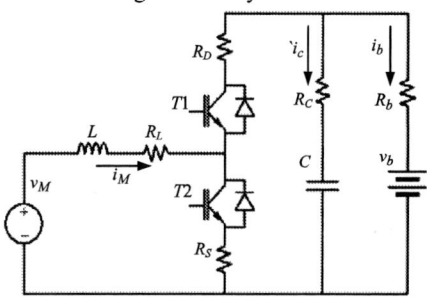

Figure.1 Circuit scheme of regenerative braking

In Fig.1, L is the inductance of the motor armature, C is the filter capacitor of the main circuit, R_L, R_D, R_S and R_C denote the resistance of every circuit lead, R_b is the inter resistance of the battery. I_b is the average current which charges the battery during one PWM period. Take the current direction of Fig.1 as the normal direction. Then the boost circuit of regenerative braking can be characterized by the following tow sets of equations:

$$\begin{cases} \dot{x} = A_1 x + B_1 u & switch\,on \\ \dot{x} = A_0 x + B_0 u & switch\,off \end{cases} \tag{4}$$

where $A_1 = \begin{pmatrix} -\dfrac{R_m + R_s}{L} & 0 \\ 0 & -\dfrac{1}{R_m + R_s} \end{pmatrix}, B_1 = \begin{pmatrix} -\dfrac{1}{L} & 0 \\ 0 & \dfrac{1}{R_L + R_s} \end{pmatrix},$

$A_0 = \begin{pmatrix} -\dfrac{1}{L}(R_b + R_m + \dfrac{R_b R_c}{R_b + R_c}) & 0 \\ \dfrac{R_b}{C(R_b + R_c)} & \dfrac{-1}{C(R_b + R_c)} \end{pmatrix},$

$B_0 = \begin{pmatrix} \dfrac{1}{L} & \dfrac{-R_c}{C(R_b + Rc)} \\ 0 & \dfrac{1}{C(R_b + Rc)} \end{pmatrix}, x = [i_m \quad V_c]^T, u = [V_m \quad i_b]^T,$

According to the state space averaging method[7], the average system behavior is described by the following small-signal equation

$$\begin{cases} \dot{\tilde{x}} = (A + \Delta A)\tilde{x} + B\tilde{u} + D\omega \\ \tilde{y} = C\tilde{x} \end{cases} \tag{5}$$

where

$A = \dfrac{A_1 + A_0}{2}, \Delta A = A_1(\delta - \dfrac{1}{2}) - A_0(\dfrac{1}{2} - \delta), B = \dfrac{B_1 + B_0}{2},$

$C = [\dfrac{1}{2} \quad \dfrac{-L}{R_c}], D = (A_1 - A_0)X + (B_1 - B_0)U$

(X,U) is the steady state working point. δ is the perturbation of the duty cycle of T2 at steady point with the external disturbance. The sign $\sim$ denotes the perturbation of variable near the steady point.

III SLIDING MODE CONTROL BASED ON H∞ NORM

The sliding surface for the system (5) can be chosen as

$$S(t) = Fx(t) - \int_0^t (FA + FBG)x(\tau)d\tau \tag{6}$$

where $x(t)$ is the solution of equation(5), $s(t) = [s_1(t) \quad s_2(t)]^T \in R^2$. F and G are time-invariant matrices to be designed. The selection of F is constrained to satisfy that FB is nonsingular and $FD = 0$; The selection of G should satisfy that the time response of the system is confined within the predefined specification when the system is on the sliding surface. With $FD = 0$ using (5), the derivative of (6) is

$$\dot{S}(t) = F\Delta Ax(t) + FBu(t) - FBGx(t). \tag{7}$$

On the sliding surface, with $S(t) = 0$ and $\dot{S}(t) = 0$ the equivalent control $u_{eq}(t)$ can be presented as

$$u_{eq}(t) = Gx(t) - (FB)^{-1}F\Delta Ax(t) + Du(t). \tag{8}$$

Substituting (9) in (5), then the sliding dynamic equation is

$$X(t) = (A + BG)x(t) + (I - B(FB)^{-1}F)\Delta Ax(t) + Du(t) \tag{9}$$

where I denotes the unit matrix. For the controllable system, the unmodelled difference satisfies that $Rank[B : \Delta A] = Rank[B]$. Moreover assume $FM = 0$ and combine (5), the closed-loop system can be rewritten as

$$\begin{cases} x(t) = A_c x(t) + \Delta Ax(t) + Du(t) & A_c = A + BG \\ y(t) = Cx(t) \end{cases} \tag{10}$$

It is obvious that the solution F of equations $FM = 0$ and $FD = 0$ is existent and not unique. Furthermore it is true that FB is nonsingular. The closed-loop transfer function $H(s)$ from the system disturbance input $\omega(t)$ to system $y(t)$ is

$$H(s) = C(SI - A_c - A)^{-1}D. \tag{11}$$

Now the project is just to design $u(t)$ to ensure that the existence of the sliding mode. Moreover on the sliding surface, the closed system (5) satisfies the performance specification as follows: 1.the system is stable and the stability regain is limited within α ; 2.$\|H(s)\|_\infty \leq \gamma$, where $\|H(s)\|_\infty = \underset{u \in R}{SUP}\,\sigma[H(ju)]$ denotes the maximum singular value of the equivalent matrix of H(s). γ is a given positive constant.

Theorem 1:

The Lyapunov function for system (5) can be defined as

$$V(S(t)) = S^T(t)S(t) = S_1^2(t) + S_2^2(t)\,. \tag{12}$$

If the switching function is equation (6), with $FM = 0$ and $FD = 0$, the following equation hold that

$$\dot{V}(S(t)) = S^T(t)\dot{S}(t) \tag{13}$$

The validation can be seen in detail in literature [8]. The theorem shows that selection of matrix F does not affect the existence of the sliding surface. So it is random to select F under the situation of that $FD = 0$ and $FM = 0$.

Theorem 2:

Assume that, the condition holds that $FM = 0$ and $FD = 0$ for system(5) and FB is nonsingular, then the control can be selected as

$$u(t) = Gx(t) - (FB)^{-1}[K\|KC\|_\infty \|x(t)\|_\infty + \alpha]Sgn(S(t)) \tag{14}$$

where $k \geq \|\Delta A\|$, $Sgn(S(t)) = [Sgn(S_1(t)) \quad Sgn(S_2(t))]$, $\|x(t)\|$ denotes the standard Eucliden norm on $x(t)$, $\|\Delta A\|$ and $\|C\|$ denote the induced norm of H_2 norm. At the same time the system state can tend to the sliding surface at the possibility of 100%. During the tending process, parameter α can adjust the velocity of the sliding mode tending to the sliding surface.

Theorem 1 and 2 show that if the selected matrix F satisfies $FM = 0$, $FD = 0$ and FB is nonsingular, it's always true that the system state will tend to the sliding surface under the control of the designed controller. Moreover the sliding mode equation (5) can be gotten.

Theorem 3:

Now we introduce theorem 3 to give out the expression of feedback matrix G and its necessary and sufficient condition for the existence.

Assume matrix Q is symmetric and proper, if and only if

$$(I - BB^+)(AQ + QA)^T + 2\alpha I + \gamma^2 QC^TCQ + MM^T + QN^TNQ + DWD^T)(I - BB^+) = 0 \tag{15}$$

, the matrix equation has solution G . The expression of G is

$$G = -\frac{1}{2}B^+(AQ + QA)^T + 2\alpha I + \gamma^2 QC^TCQ + MM^T$$
$$+ QN^TNQ + DWD^T - \hat{Z}) + (I - BB^+)Z \tag{16}$$

where $\hat{Z}$ is slope symmetric matrix, Z is compatible dimension matrix, and B^+ denotes the Moore-penrose psuedoinverse of matrix B.

IV SIMULATION AND ANALYSIS

The simulation model of the regenerative is presented in Fig.2 based on the MATLAB/SIMULINK software. In Fig.2 the vehicle unit denotes the dynamic model of EV during regenerative braking, whose inputs are the displacement of braking pedal and the current of the armature and output is the back-electromagnetic voltage of the motor. The controlled feed-back current of the battery is proportional to the displacement of the braking pedal.

The required parameter of the simulation refers to the EV XJTUEV-II of Xi'an Jiaotong University, which are as follows:$Cd=0.5$, $A=2.43m^2$, $f=0.0112$, $\rho= 1.225Ns2/m4$, $m=1500kg$, $Kg=4.7$, $Kt=0.4Nm/A$, $Ke= 0.0421Vs/r$, $Rw=0.287m$.

First, under the initial condition of the velocity $V=60$ km/h and the displacement of the braking pedal $Km=0.85$, the parameter of PI controller is tuned to $Kp=0.004$, and $Ki=1.8$. The time response of PI control is shown in Fig. 3. Then the comparative simulation are conducted on PI and the proposed control, the results are shown in Fig.4 and 5.

The input of these simulation is $\begin{cases} K_m = 0.85 & 0 \leq t \leq 0.5 \\ K_m = 0.25 & 0.5 < t \end{cases}$

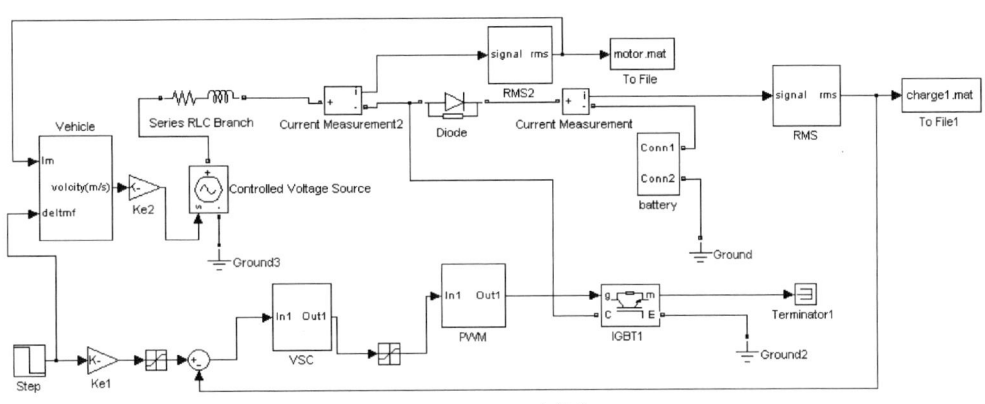

Figure.2 Simulink model of regenerative braking

In Fig.3, 4 and 5, the continuous line means the current of the armature of the motor and the dashed line means the controlled feed-back current of the battery. From Fig.3 we can see that the PI control is satisfactory. However once the braking pedal steps up suddenly, which is equivalent to the parameter perturbation and the disturbance of the system, the PI control is bad in Fig.4. Fig.4 shows the greater overshoot, the long setting time and bad tracking ability. But we can conclude from Fig.5 that the dynamic performance of the proposed scheme is highly satisfactory. The proposed control system has fast response, no steady state difference, strong robustness to the disturbance and it's stable.

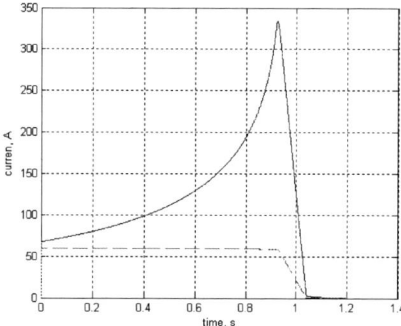

Figure.3 Transient response of PI without disturbance

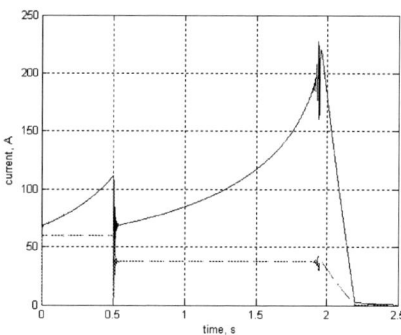

Figure 4. Step response of PI with disturbance

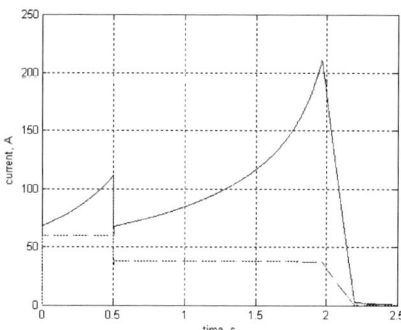

Figure 5. Step response of the proposed scheme with disturbance

V   CONCLUSION

Based on the mathematical model of the regenerative braking of EV, a new sliding mode controller is developed with the integration of the structure control and H∞ norm theory. The dynamic performance and robustness performance of the system was significantly enhanced comparing with only one method is mentioned. The special features of the new control scheme lies in the solution of the traditional clustering without the decreasing of the dynamic performance. The comparing simulations show the proposed scheme can ensure the robustness performance and the superior dynamic performance under the presentation of the unmodelled difference and additional disturbance. The proposed scheme is widely suitable for the nonlinear control of EV.

REFERENCE

[1]. C.C Chan, "the state of the art of the electric and hybrid vehicles," in Proc. IEEE, Vol.90, No.2, Feb.2002, pp.247-275.

[2]. B.G. Cao, Z.F. Bai, W. Zhang, "research on control for regenerative braking of electric vehicle," in Proc. IEEE ICVES'2005, pp.92-97.

[3]. J. Paterson, and M. Ramsay, "Electric vehicle braking by fuzzy logic control," in Proc. IEEE IASAM'93, pp.2200-2204.

[4]. S.K. Chung, J.H. Lee, and T.S. Ko, "Robust speed control of brushless direct-drive motor using integral variable structure control," IEE Proc. Electr. Power Appl., Vol.142, No.6, Nov.1995, pp.361-370.

[5]. Z.F. Bai, "H_∞ robust control for driving and regenerative braking of electric vehicle," Journal of Xi'an Jiaotong University, Vol.39, No.3, Mar.2005, pp.256-260.

[6]. C.W. Zhang, "Study on regenerative braking of Electric vehicle," in Proc. IEEE IPEMC'2004, pp.836-39.

[7]. P. Mattavell, L. Rossetto, and G. spiazzi, "Small-signal analysis of DC-DC converters with sliding mode control," IEEE Trans. on Power Electron., Vol.12, No.1, Jan.1997, pp.96-102.

[8]. D.P. Ye and L.L. Cui, "H_∞ norm and variance-constrained variable structure control for linear stochastic uncertain system," College Mathmatics, Vol.20, No.6, Dec.2004, pp.93-97.

2006 5th International Power Electronics and Motion Control Conference

A Self-adaptive Fuzzy Control Scheme of High Frequency Link SPWM Inverters

Herong Gu, Deyu Wan and Weiyang Wu
YanShan University, Qin Huangdao, P.R. China
Email: ydghr@ysu.edu.cn

Abstract—As the High frequency link (HFL) SPWM inverter possesses highly nonlinear property and variable parameters due to its special structure, the perfect control can not be obtained with conventional control methods especially under nonlinear load. A Self-adaptive Fuzzy Control scheme in this paper is proposed for a single-phase high frequency isolated DC\AC\AC inverter, in which a novel assistant Fuzzy Controller is designed to improve the self-adaptive and robust property by adjusting the scaling factor on-line. It takes a role to provide stronger control action for a large voltage error and a smoother control action for a small voltage error considering synchronously the error change. This in turn ensures that high quality output voltage is always maintained on HFL while the transient response improved regardless of the loading property. Simulation and experiment results of 1.5kW prototype are shown in the paper. Comparison between Self-adaptive Fuzzy controller based on scaling factor self-adjusting and conventional FC verifies the proposed control method.

Keywords-fuzzy control; self-adaptive control; high frequency link inverter; rectifier load; DSP

I. INTRODUCTION

High frequency link inverter is well known to be a kind of popular circuit topologies these days due to low power loss, high system efficiency and compact bulkness. And it is at the same time a complicated system with multi-stage power converting, serious nonlinear property and variable parameters. So an accurate mathematic model is hard to obtain. The general requirement of inverter systems is to maintain high quality output voltage with total harmonic distortion (THD) typically specified to be less than 5% regardless of the loading conditions. However, with most of today's load being nonlinear and subject to variations, conventional control methods can not fulfill requirement. Fuzzy control (FC) attracts enough attention among various digital control methods because it can effectively estimate and deal with complicated process [1]. As one of the rules based intelligent control method, fuzzy control is composed of fuzzy mathematics, fuzzy logic rational rules and special knowledge expression mode of fuzzy language. Fuzzy logic control can furnish faster and more robust performance under parameter variations and load

disturbances. It is a closed loop digital control system with feedback passage supported by computer control technology. Furthermore, because of little occupying of processor for looking up the FC table [2] [3], high sample ratio can be adopted to compensate the lack of Fuzzy rules and actual experience.

However, there is some limitation in conventional FC. Control precision of Steady-state using FC system is worse than that of conventional control strategy. In addition, establishment of fuzzy control rules depends on the expert experience and it can't deal with durative changes [4]. In this paper, a novel scaling factor self-adjusting based self-adaptive Fuzzy control (SAFC) scheme is designed to control the HFL inverters. The assistant Fuzzy controller is introduced to distinguish from system state and then generate modifying information to correct parameters of the primary Fuzzy controller in order to eliminate the influence of the perturbations. The system can achieve outstanding steady-state and dynamic performance. In this work, SAFC is developed for a single-phase HFL inverter. A 1.5KW DC\AC\AC HFL SPWM inverter is implemented, and the effectiveness is demonstrated by detailed simulation and experiment results.

II. SELF-ADAPRATIVE FUZZY CONTROL FOR HFL INVERTERS

The fuzzy control system consists of five parts: fuzzy controller, interface of input and output, control object, execute institution and sensors [5], which actually translates the knowledge and experience of expert into the control law. Because SAFC is not designed based on the mathematical analysis of a process model, MATLAB/ SIMULINK and Fuzzy Logic Toolbox are used for simulation analysis. The actual instantaneous output voltage is sensed, sampled and compared with sinusoidal reference value to create the error voltage. Discrete error voltage $E(k)$ and its change of error $Ec(k)$ are processed by the FC through fuzzification, fuzzy inference and defuzzification operations. The change of control signal $\Delta u(k)$ as the output variable of primary FC is added to the control signal, $U(k)$ to give a updated value of switching angles to compensate properly with any loading variation. In the FC system, weight factor Ke, Kc and scaling factor Ku are very important to the static and dynamic performance. The effect of the gain settings for a conventional PI controller in a closed loop system is related to the scaling factors adjustment in the SAFC, which can be approximated as a actual PI controller. Referring to the conventional integral digital PI controller.

$$\Delta u(k) = K_P \Delta e(k) + K_I e(k) \qquad (1)$$

Referring to FC [6].

$$K_U \Delta u(k) = K_C Ec(k) + K_E E(k) \qquad (2)$$

Relating (1) and (2),

$$K_P \equiv \frac{K_C}{K_U} \quad \text{and} \quad K_I \equiv \frac{K_E}{K_U} \qquad (3)$$

Referring to PI control experience, the following conclusions can be gained:

1) A higher value of Ke will cause a long transition progress or even an overshoot. On the contrary, a smaller Ke will result in a poor dynamic response and a large system error.

2) Increscent Kc can make the controller sensitive, which will avoid overshoot of the system output and bring out a slow dynamics. Otherwise, a large overshoot or even a surge will occur.

3) The value of Ku will influence directly the output scale of controller. By reducing Ku, a steady output can be achieved. On the other hand, a large Ku makes for a proper dynamics. A compromise should be carried when selecting an appropriate Ku.

In Fig1, a fuzzy controller based on scaling factor self-adjusting online is designed. The assistant fuzzy controller is introduced to modify the scaling factor of primary fuzzy controller for better robust depending on various load features. The basic rules of assistant fuzzy controller are shown as follows. Scaling factor Ku should be reduced for a small output overshoot and short rise-time when error E is large and has the inverse sign with the change of error, EC. On the contrary, Ku should be increased when E and

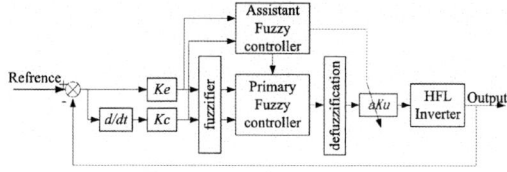

Figure1. Block diagram of SAFC

TABLE I.
PRIMARY FC RULES

EC/E	NB	NM	NS	ZE	PS	PM	PB
NB	NB	NB	NB	NB	NM	NS	ZE
NM	NB	NB	NB	NM	NS	ZE	PS
NS	NB	NB	NM	NS	ZE	PS	PM
ZE	NB	NM	NS	ZE	PS	PM	PB
PS	NM	NS	ZE	PS	PM	PB	PB
PM	NS	ZE	PS	PM	PB	PB	PB
PB	ZE	PS	PM	PB	PB	PB	PB

TABLE II
ASSISTANT FC RULES

EC/E	NB	NM	NS	ZE	PS	PM	PB
NB	VB	VB	VB	B	M	SM	VS
NM	VB	VB	B	B	BM	SM	S
NS	VB	BM	NM	B	VB	S	S
ZE	SM	M	BM	VS	M	BM	SM
PS	S	SM	S	VB	B	BM	VB
PM	S	SM	BM	B	B	VB	VB
PB	VS	SM	M	B	VB	VB	VB

EC has the same sign, because system output is drawing away with reference at this point. On the other hand, when E has a large value, scaling factor Ku should have a large change range. It should be reduced if system output draws away from reference after drawing on it. This method is in fact a gain-variable nonlinear controller which is unattached with model [7]. Rules of primary FC and assistant FC are shown in Table I and Table II respectively.

III. DESIGN OF SYSTEM

In this paper a laboratorial prototype of 1.5KW DC\AC\AC HFL inverter based on DSP (TMS320LF2407A) is built up, which is shown in Fig2. The circuit consists of four parts: full-bridge converter, HF transformer, cyclo-converter Bridge and LC low-pass filter. Power bidirectional flow and former-stage ZVS function are accomplished by means of bipolar phase-shifted modulation.

A. Circuit system parameters

System parameters are specified as Table III.

TABLE III
SYSTEM PARAMETERS

Input battery voltage	Vi =40-48V
Output AC voltage	Vo= 110V (50Hz)
Output power	Po=1.5KW
Switching frequency	F=33 kHz

B. Controller structure and parameters

The controller is designed as a double closed-loop of output voltage instantaneous SAFC and inductor current P control, which is shown in Fig.2. As the input variations

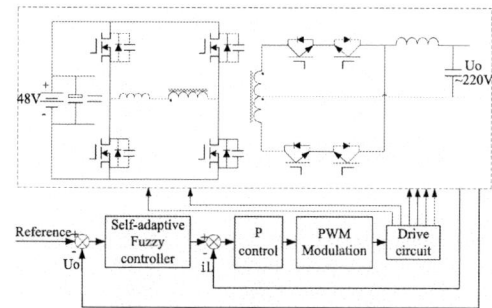

Figure2. Diagram of HFL inverter system

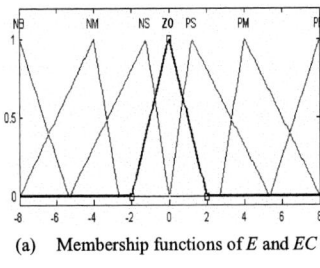

(a) Membership functions of *E* and *EC*

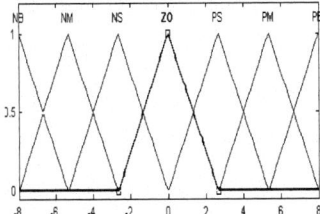

(b) Membership function of Δu

Figure3. Shape of membership functions for SAFC

of PI-type primary fuzzy controller and PD-type assistant fuzzy controller, *E* and *EC* have the definition of error and the change of error respectively [8]. They are all divided into seven fuzzy subsets from [-8, +8]: {NB, NM, NS, Z, PS, PM, PB}. As shown in Fig.3, their membership functions have the same geometry of asymmetric Triangle [9]. Δu as the output variation of primary fuzzy controller is also divided into seven subsets, which is not like α, the output of assistant fuzzy controller. α is divided into seven fuzzy subsets from [0,1], {VS, S, SM, M, BM, B, VB}. But they have the same membership functions of symmetrical triangle.

The inference logic of MAX-MIN method is employed in two fuzzy controllers, similarly the defuzzification method of centroid formula is shown in (4) [9].

$$u = \frac{\sum_{i=1}^{n} \mu(u_i) \times u_i}{\sum_{i=1}^{n} \mu(u_i)} \quad (4)$$

Fig.4 shows a 3-dimension surface plot of the control output with the membership functions above mentioned.

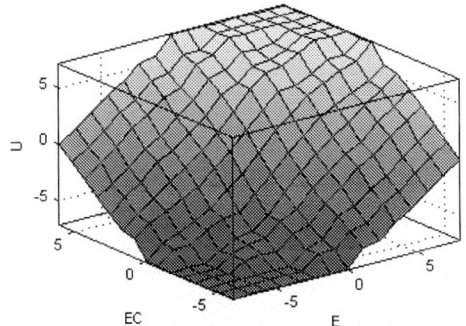

Figure4. 3-D surface plot of SAFC output

The characteristic of a FC controller is exactly a linear plane while that of the SAFC shows some plateaus which governed that variations and uncertainty can be accommodated with the proposed SAFC. The size of the plateaus depends on the appropriate selection of membership functions and it can be readily determined according to the desired system output performance. This is a great modified advantage compared with FC.

IV. SIMULATION AND EXPERIMENT RESULTS

The SAFC performance was verified theoretically for the HFL inverter via simulation utilizing MATLAB FZ Tools. The simulation results are shown in Fig.4 and Fig.5.

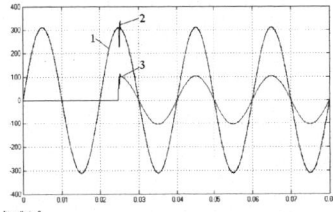

Figure4. Simulation waveforms of output voltage and current under a step-up load. 1-Vref, 2-Vo, 3-Io
(Y-axis:50V/div; 20A/div X-axis: 10ms/div)

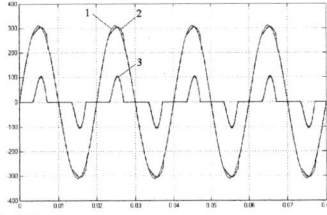

Figure5. Simulation waveforms of output voltage and current under a Rectifier load. 1-Vref, 2-Vo, 3-Io
(Y-axis:50V/div; 20A/div X-axis: 10ms/div)

Rapid transient response of controller can be found in Fig.4 and Fig.5 by the transitory modulation. The experiment waveforms of load voltage and current under inductive load are shown in Fig.6 and its corresponding harmonic analysis in Fig.7, in which THD of the voltage wave is as high as 0.65%.

SAFC dynamic response under a step change load is given in Fig.8 and Fig.9, which accords with the simulation results. Short transition and small overshoot illustrate fast dynamic response.

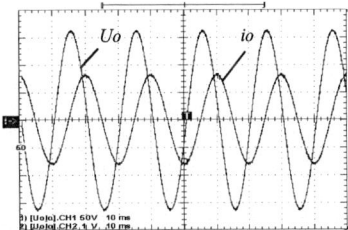

Figure6. SAFC inductor load
(Y-axis:50V/div; 1A/div X-axis: 10ms/div)

1587

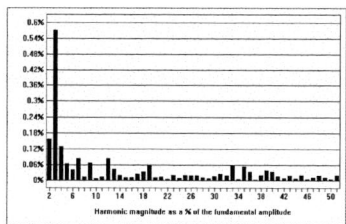

Figure7. Harmonic analyse histogram of voltage

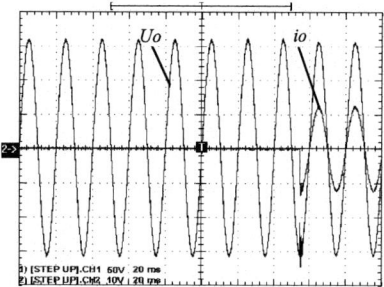

Figure8. SAFC dynamic response under a step load
(Y-axis:50V/div; 10A/div X-axis: 20ms/div)

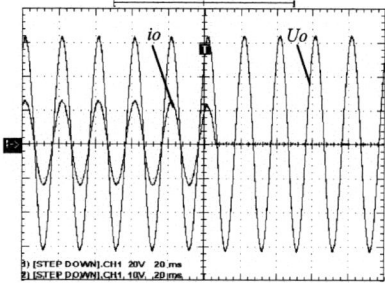

Figure9. SAFC dynamic response under a drop load
(Y-axis:50V/div; 10A/div X-axis: 20ms/div)

For comparison purpose both the FC and SAFC operations of HFL inverter are implemented in the experiment. Output voltage and current waveforms with rectifier load under SAFC are shown in Fig.10. THD of the corresponding voltage wave is as high as 3.31% in Fig.11 of harmonic analyzing histogram, crest factor: 3.51 and Po: 730W.

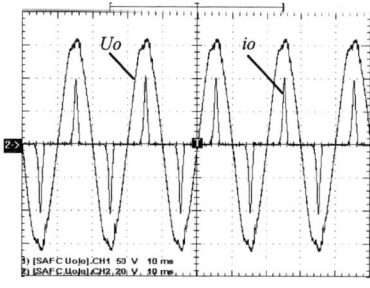

Figure10. SAFC output waves under rectifier load
(Y-axis:50V/div; 10A/div X-axis: 10ms/div)

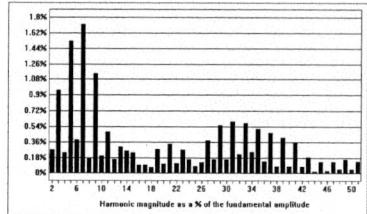

Figure11. Harmonic analyse histogram of voltage

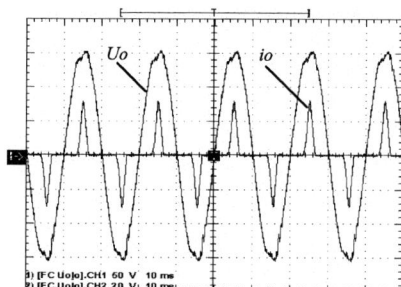

Figure12. FC output waves under rectifier load
(Y-axis:50V/div; 10A/div X-axis: 10ms/div)

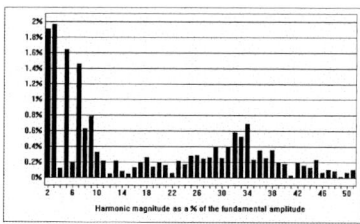

Figure13. FC output waves under rectifier load

Comparatively, output voltage and current waveforms with the same rectifier load under FC are given in Fig.12. Fig.13 is the corresponding harmonic analyses histogram, in which THD is as high as 4.0%, crest factor: 3.22 and Po: 676W. It can be seen above that SAFC can maintain better robust with various loads.

TABLE IV
EXPERIMENT RESULTS OF SAFC AND FC

Load voltage	SAFC	FC
RMS/V	109.42	108.55
THD	3.31%	4%
crest factor	3.51	3.22
Po/W	730	676

V. CONCLUSIONS

SAFC presented in this paper has a simple design progress no need of exact mathematics model of HFL inverter and can operate automatically without man-made effect. SAFC provides insensitivity to plant-parameter variations and external distribution such as making for overcoming perturbations caused by rectifier load. System control performance can be optimized depending on the modification of weight factor and scaling factor during the process of simulation and experiment. Different combinations of input and output scaling factors, shape or location of input and output membership functions are possible to obtain the optimal corresponding output performance. Simulation and experiment results show that: control parameter modified continually by assistant FC makes for better stability and dynamics than conventional FLC. The prominent improvement of system robust verified the proposed control method. Therefore, SAFC is suitable for DC/AC/AC inverter to generate a high quality AC voltage power source.

ACKNOWLEDGEMENT

This work was supported by the National Natural Science Foundation of China, NO. 50237020.

REFERENCES

[1] XIE Li-hua, SU Yan-min. POWER ELECTRONICS, 2001,35 (6): 52-55.

[2] Tzou Ying-Yu, et al. Fuzzy Control of a Close-Loop Regulated PWM Inverter under Large Load Variation[C]. IEEE-IECON'93,1993,(l):267-272

[3] Guang Da. Chen, Wei You. Cai, Tian Fu. Cai, Hai Feng. Liu, A hybrid fuzzy current regulator for three-phase voltage source PWM-inverter Machine Learning and Cybernetics, 2002. Proceedings. 2002 International Conference on Volume 2,4-5 Nov. 2002 (2):919-923

[4] Lin C T and Lu Y C. A Neural Fuzzy System with Fuzzy Supervised Learning. IEEE Trans. Syst, Man, and Cybern,2001,26(5):754-763

[5] Luo Ling, Zhou Yong-peng; Xu Jin-bang, Wan Shu-yun, Parameters self-adjusting fuzzy PI control with repetitive control algorithms for 50 Hz on-line UPS controlled by DSP Industrial Electronics Society, 2004. IECON 2004. 30th Annual Conference of IEEE Volume 2,2-6 Nov. 2004 (2):1487-1491

[6] Azli, N.A. Ning, W.S. Application of fuzzy logic in an optimal PWM based control scheme for a multilevel inverter Power Electronics and Drive Systems, 2003. PEDS 2003. The Fifth International Conference on Volume 2, 17-20 Nov. 2003 Page(s): 1280-1285,Vol.2

[7] SUN Dan, HE Yi-kang, ZHI Da-wei. Self-adapting fuzzy control of SRM based on its nonlinear model. ADVANCED TECHNOLOGY OF ELECTRICAL ENGINEE RING AND ENERGY,2001(4):9-13

[8] Bulale, Y.I. Salami, M.J.E. Sulaiman, M. Design and development of DSP-based hybrid controller for servo driver applications Control Applications, 2003. CCA 2003. Proceedings of 2003 IEEE Conference on Volume 1, 23-25 June 2003 Page(s):773 - 778 vol.1

[9] En-Chih Chang, Tsorng-Juu Liang; Jiann-Fuh Chen, Ray-Lee Lin, A sliding-mode controller based on fuzzy logic for PWM inverters Circuits and Systems, 2004. Proceedings. The 2004 IEEE Asia-Pacific Conference on Volume 2, 6-9 Dec. 2004 Page(s): 965 - 968 vol.2

2006 5th International Power Electronics and Motion Control Conference

Using Automatic Frequency Shifting Techniques for LLC-SRC Output Voltage Regulation

Kuo-Kai Shyu, Ching-Ming Lai, Ko-Wen Jwo, Ming-Ho Pan* and Chung-Ping Ku**

National Central University/Department of Electrical Engineering, Taiwan, China
*Ching Yun University /Department of Electrical Engineering, Taiwan, China
**Industrial Technology Research Institute/Energy and Environment Research Laboratories, Taiwan, China

Abstract— **In this paper, an automatic frequency shifting technique is developed for achieving output voltage regulation of LLC series resonant converter (LLC-SRC). Furthermore, the operating modes and dc characteristics are discussed for designing the steady state operating point. Finally, an LLC-SRC, rated about 90W with operating frequency range from 92KHz to 102KHz is provided to verify the control strategy. As theoretical analysis, the proposed system gives a satisfactory result within the overall load range. This technique is very promising in resonant converter applications.**

Keywords- Frequency shifting technique, LLC -SRC

I. INTRODUCTION

Pulse width modulation (PWM) switching technique is widely used for power electronics control in recent years. For achieving higher efficiency and, to reduce the overall size of the power converters, the switching frequency is stepping up. However, for traditional hard-switched power converters, an increase in operating frequency results in large EMI, RFI and switching losses. To solve the above-mentioned problem, the resonant converter was proposed and widely used nowadays [1]. Due to the advantages of zero voltage switching (ZVS) or zero current switching (ZCS), the resonant converter can be designed to operate at a high frequency and maintaining high efficiency. Within the various resonant topologies, series resonant converter (SRC), parallel resonant converter (PRC) and series-parallel resonant converter (SPRC/LCC) were discussed a lot. The SRC offers excellent part load efficiency but the light load output voltage regulation is always a problem for the control design [2]. For the PRC, it provides satisfactory voltage controllability but circulating energy will damage the light load efficiency. Combining the good characteristics of SRC and PRC is the SPRC or LCC resonant converter [3]. It is with smaller circulating energy and not so sensitivity for load change. Unfortunately, the SPRC requires large range of frequency variation if controlled by changing the operating frequency of the resonant converter. Therefore, the switching devices can not be best utilized [4].

Recently, owing to the benefits of simple structure, ZVS capability within extensive load range, low turn off loss, high efficiency and high power density, the LLC-SRC has been successfully applied in the distributed power system (DPS) and voltage regulator modules

(VRM) [5]. Although its performance is very attractive, the LLC-SRC belongs to three-element resonant structures. For this reason, the analysis and design of the controller are too complex and lack of discussion [6].

To overcome the above-mentioned problems, an automatic frequency shifting technique is developed for achieving output voltage regulation of the LLC-SRC in this paper. Furthermore, the operating modes and dc characteristics are discussed for designing the steady state operating point. Experimental results show the accuracy performance of the proposed control strategy, the ZVS turn on capability and low turn off loss for switches are performed. A rated practically 90W LLC-SRC prototype is implemented with high efficiency about 97% under rating condition.

II. ANALYSIS OF LLC-SRC

Fig. 1 shows the system diagram of the proposed LLC-SRC. The integrated magnetics (leakage inductance L_r, magnetizing inductance L_m) and the dc blocking capacitor C_r compose a resonant tank. The splitting capacitors C_1, C_2 and two switches Q_1, Q_2 form a symmetrical half bridge inverter. The half bridge inverter gives the voltage V_{AB} applied to the resonant tank to be half of V_{in} in the form of square waveform with symmetrical duty cycle.

As a result of the low voltage stress on secondary rectifier, the secondary side is a center-tapped rectifier followed by a capacitive filter. Furthermore, the automatic frequency-shifting technique is adopted for developing a variable frequency controller. The variable frequency controller improves the overall system stability and the output voltage regulation.

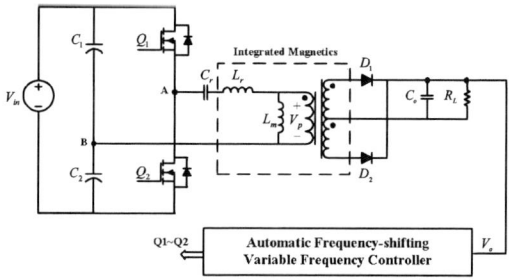

Figure 1.   System diagram of the proposed LLC-SRC

1-4244-0448-7/06/$25.00 ©2006 IEEE   1590

A. Operating principle

The detailed operating modes of the proposed LLC-SRC is shown in Fig. 2 will be described as follows.

(a) Mode I ($t_0 < t < t_1$):

This mode begins when Q_1 and D_1 are turned on at $t = t_0$. The voltage of magnetizing inductance L_m is clamped to nV_o. At this moment, magnetizing current i_m is negative and increases linearly. Within this operating mode, leakage inductance L_r and capacitance C_r resonate. The resonant current i_r is a sine-wave current. This mode ends when Q_1 is turned off.

(b) Mode II ($t_1 < t < t_2$):

At $t = t_1$, Q_1 is turned off. i_r goes through body diode of Q_2, which creates a ZVS condition for Q_2. When Q_1 is turned off, since the resonant frequency is larger than the switching frequency ($f_o > f$), the resonant current i_r is larger than the magnetizing current i_m ($i_r > i_m$) and D_1 maintains turned on yet. Furthermore, the resonant current i_r decreases rapidly simultaneously.

(c) Mode III ($t_2 < t < t_3$):

At $t = t_2$, Q_2 is turn on with ZVS and the resonant current i_r decreases rapidly. At $t = t_3$, the resonant current i_r is equal to the magnetizing current i_m ($i_r = i_m$) hence D_1 is turned off.

(d) Mode IV($t_3 < t < t_4$):

At $t = t_3$, the resonant current i_r is still equal to the magnetizing current i_m and D_1 is turned off. When the i_r is smaller than i_m ($i_r < i_m$), D_2 is turned on hence the voltage of the magnetizing inductance L_m is clamped to $-nV_o$. The magnetizing current i_m will decrease linearly and the resonant capacitor discharges. Only the leakage inductance L_r and capacitance C_r resonate during the mode. At $t = t_4$, this mode ends when Q_2 is turned off.

(e) Mode V ($t_4 < t < t_5$):

At $t = t_4$, Q_2 is turned off and the resonant current i_r goes through body diode of Q_1, which creates ZVS condition for Q_1. The leakage inductance L_r and capacitance C_r resonate, hence D_2 is still turned on. Besides, the resonant current i_r decreases rapidly.

(f) Mode VI ($t_5 < t < t_6$):

At $t = t_5$, Q_1 is turned on with ZVS and D_2 is still maintaining turned on. At $t = t_6$, the i_r is equal to i_m ($i_r = i_m$) hence D_2 is turned off. This mode ends followed by the next operating cycles starts.

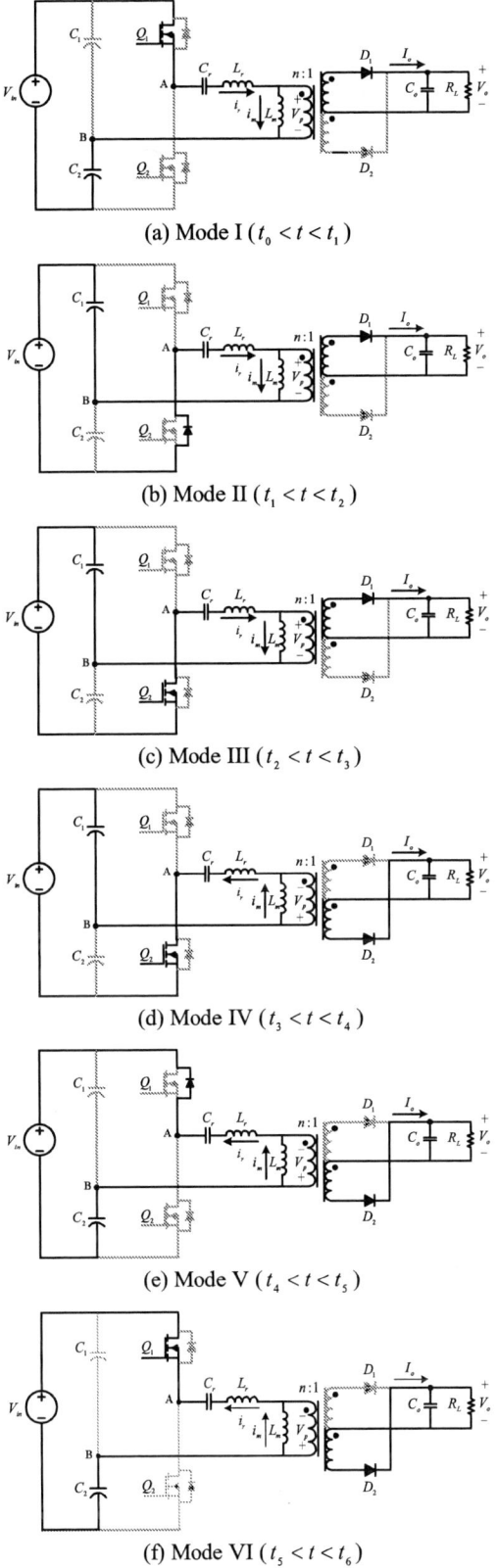

(a) Mode I ($t_0 < t < t_1$)

(b) Mode II ($t_1 < t < t_2$)

(c) Mode III ($t_2 < t < t_3$)

(d) Mode IV ($t_3 < t < t_4$)

(e) Mode V ($t_4 < t < t_5$)

(f) Mode VI ($t_5 < t < t_6$)

Figure 2.   The operating modes of the proposed LLC-SRC

B. DC characteristics

Fig. 3 shows the ac equivalent circuit with fundamental element simplification (FES) method, which is used for the analysis of LLC-SRC [8].

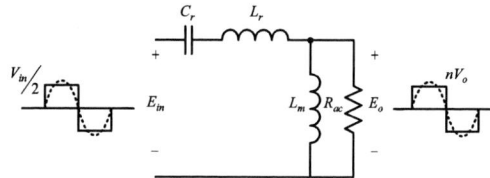

Figure 3.   The ac equivalent circuit of LLC-SRC

The ac voltage gain can be analyzed from Fig. 3 as follows

$$G_{ac} = \frac{R_{ac} \ // \ j\omega L_m}{\frac{1}{j\omega C_r} + j\omega L_r + j\omega R_{ac} \ // \ j\omega L_m} \tag{1}$$

$$|G_{ac}| = \frac{E_o}{E_{in}} = \frac{1}{\sqrt{\left[1 + \frac{1}{k}\left(1 - \frac{f_o^2}{f^2}\right)\right]^2 + \left[\left(\frac{f}{f_o} - \frac{f_o}{f}\right) \cdot \frac{\pi^2 Q}{8n^2}\right]^2}} \tag{2}$$

where $k = \frac{L_m}{L_r}$, $f_o = \frac{1}{2\pi\sqrt{L_r C_r}}$, $Q = \sqrt{\frac{L_r}{C_r}}\frac{1}{R_L}$, $R_{ac} = \frac{8n^2}{\pi^2} \cdot R_L$.

Spreading the voltage V_{AB} by Fourier-series, ones adopts only fundamental component. Thus, the maximum value of the fundamental component of the voltage V_{AB} is

$$E_{in(max)} = \frac{4}{\pi} \cdot \frac{1}{2} V_{in} = \frac{2}{\pi} V_{in} \tag{3}$$

and its RMS value of the fundamental component is

$$E_{in} = \frac{2}{\pi} V_{in} \cdot \frac{1}{\sqrt{2}} = \frac{\sqrt{2}}{\pi} V_{in} \tag{4}$$

One has the fundamental component RMS value of transformer primary side

$$E_o = \frac{2\sqrt{2}}{\pi} \cdot n V_o \tag{5}$$

From equations (3) to (5), the DC voltage gain of LLC-SRC is as following:

$$G_{dc} = \frac{V_o}{V_{in}} = \frac{1}{2n} \cdot \frac{E_o}{E_{in}} = \frac{1}{2n}|G_{ac}| \tag{6}$$

$$G_{dc} = \frac{1}{2n \cdot \sqrt{[1 + \frac{1}{k} \cdot (1 - \frac{f_o^2}{f^2})]^2 + [(\frac{f}{f_o} - \frac{f_o}{f}) \cdot \frac{\pi^2 \cdot Q}{8n^2}]^2}} \tag{7}$$

The DC characteristics of LLC-SRC is shown in Fig. 4. From Fig. 4, since the assumed input voltage is 48V and output voltage is 28V, the DC voltage gain is 0.583. The design parameters are $L_r = 21\mu H$, $L_m = 125\mu H$ $C_r = 0.167\mu H$, $n = 1$ and $Q = 1.31$ at 3A load current.

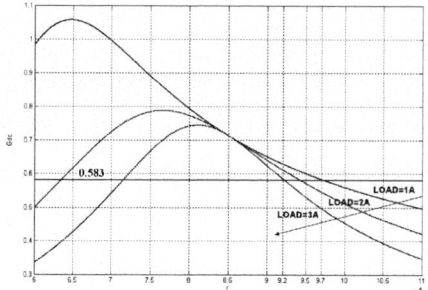

Figure 4.   DC characteristics of the proposed LLC-SRC

III.   AUTOMATIC FREQUENCY SHIFTING TECHNIQUES

From Fig. 4, one can clearly see that the converter gain decreases when the switching frequency deviates from the resonant point. Using the above-mentioned concept, an automatic frequency shifting technique is adopted for achieving the output voltage regulation of LLC-SRC. The sketching chart of control locus is shown in Fig. 5. The switching frequency of power MOSFETs is designed to be higher than the steady state resonant frequency of the resonant tank resulting in ZVS on power switches within full load range. When the system operates at heavy load, the controller has to decrease the switching frequency automatically for supplying the larger output power for load. In contrast with the system operating at light load, the controller increases the switching frequency efficaciously for supplying smaller output power to load. In conclusion, the proposed automatic frequency shifting technique can adjust the output power for conforming voltage regulation to the specifications.

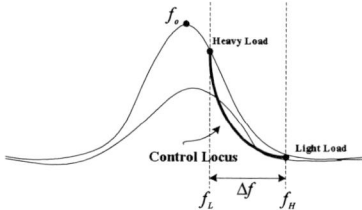

Figure 5.   The sketching chart of the control locus

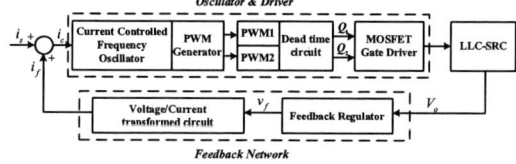

Figure 6.   Automatic frequency shifting techniques for the LLC-SRC

The automatic frequency shifting technique for the system is shown in Fig. 6 and will be described as follows.

A.   Oscillator and PWM circuit

The oscillator consists of a current controlled frequency oscillator (CCFO) and a PWM generator. The CCFO oscillates a variable frequency saw-toothed signal for PWM generator.

The PWM generator compares the saw-toothed signal with a setting reference voltage to produce two PWM signals. An accomplished example of CCFO is shown in Fig. 7. The proposed oscillator offers benefits of simple structure, low cost and widely range variable frequency.

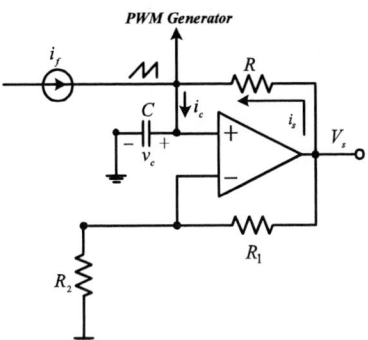

Figure 7. An accomplished example of CCFO

Fig. 8 shows the sketching chart of the charge current i_c, the voltage of capacitor v_c and switching frequency f_s of MOSFET Q_1 with load change, respectively. From Fig. 8, the output voltage V_o of LLC-SRC via feedback regulator to generate the feedback voltage v_f, and then v_f is converted into the corresponding current i_f by voltage/current transformed circuit. The lower v_f brings the smaller charge current and the switching frequency f_L of Q_1 is equal to 90KHz. By contrast, the higher v_f brings the larger charge current i_c and by the switching frequency f_H is adjusted to reaches 120KHz gradually.

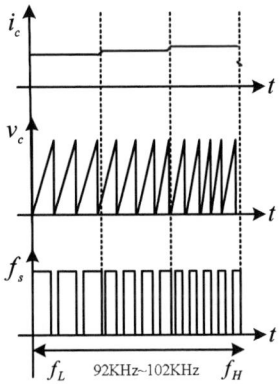

Figure 8. The sketching chart of i_c, v_c and f_s with load change

where the charge current i_c is described

$$i_c = i_s + i_f \tag{8}$$

the derived voltage of capacitor v_c is

$$v_c \cong \frac{i_c}{C} t = \frac{i_s + i_f}{C} t \tag{9}$$

The charge stage of capacitor C can be analyzed as follows.

State 1: Positive slope charge ($f_L \sim f_H$)

The feedback voltage v_f raises and the corresponding current i_f charges the capacitor C resulting in i_c increases rapidly. For this reason, the switching frequency of power switches deviates from the original operating frequency f_L of the system. Consequently, the switching frequency varies from f_L to f_H gradually. In this state, the system can raise the output power for load rapidly and the output voltage increases.

State 2: Negative slope charge ($f_H \sim f_L$)

In this state, the feedback voltage v_f falls and the current i_c decreases rapidly. The switching frequency deviates from the original operating frequency f_H of the system, hence the switching frequency varies from f_H to f_L gradually. The system can reduce the output power for load rapidly and the output voltage decreases.

Therefore the continuous variable frequency behavior of the proposed system can be provided by modulating the charge current i_c of capacitor. When the steady state output feedback, the CCFO can output a switching frequency f

$$f = \frac{1}{T} = \frac{1}{2RC \ln(1 + \frac{2R_1}{R_2})} \tag{10}$$

The deviation frequency f' resulted by feedback regulation can be described

$$f' \propto (i_s + i_f)' \tag{11}$$

From equation (10), the key parameter is the time constant value $T_c = RC$. This value influences the shifting-frequency characteristic. When T_c is too large, the variable frequency range may be narrowed.

B. Feedback network

For output voltage regulation of the proposed LLC-SRC, a simple voltage control mode is adopted. The feedback regulator circuit is shown in Fig. 9.

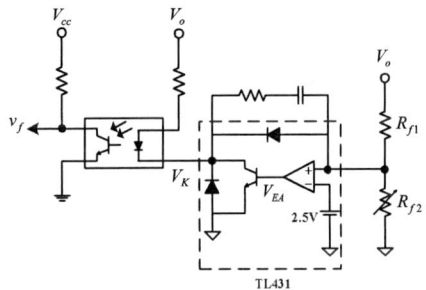

Figure 9. Feedback regulator circuit

1593

IV. Experimental Results

Fig. 10 shows the experimental V_{GS}, V_{DS}, v_{C_r} and V_o with half load and full load, respectively. From Fig. 10, one can clearly see that the proposed system has benefits of ZVS characteristic and low turn off loss. When the load changes from half load to full load, the system can shift the switching frequency automatically for output voltage regulation.

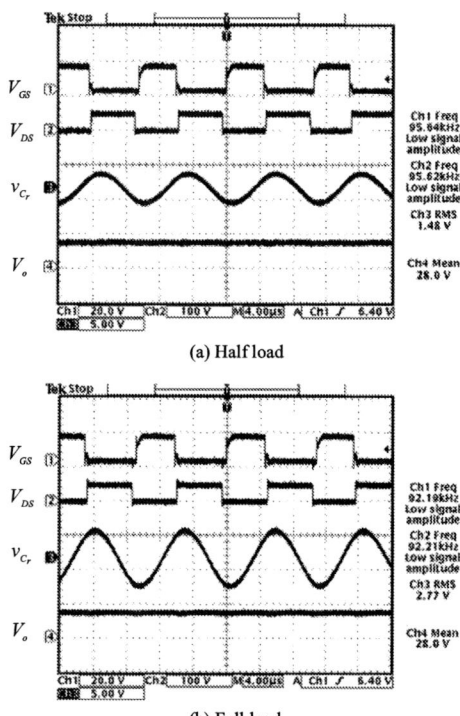

(a) Half load

(b) Full load

Figure 10. Experimental waveforms of the proposed LLC-SRC

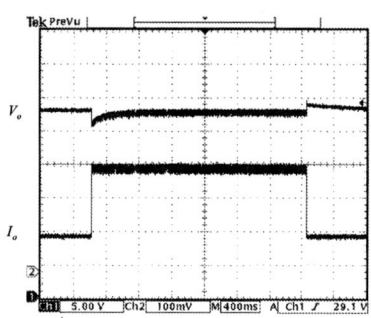

Figure 11. Transient response waveform of output voltage

Furthermore, the transient response waveform of output voltage with load change from 0.5A to 3A is experimented is shown in Fig. 11. The transient response time is about 0.2s, the experimental result is provided to verify the control strategy. The implemented prototype and measured efficiency of the LLC-SRC is shown in Fig. 12 and Fig. 13.

The prototype of LLC-SRC is implemented with high efficiency about 97% under rating condition.

Figure 12. The implemented prototype of LLC-SRC

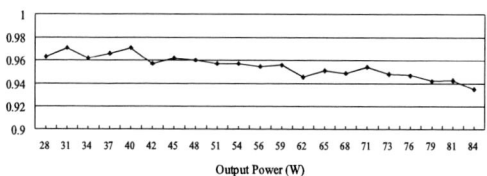

Figure 13. Measured efficiency

V. Conclusion

In this paper, an automatic frequency shifting techniques is adopted for achieving output voltage regulation of the LLC-SRC. The operating modes and dc characteristics are discussed for designing the steady state operating point. Experimental results show the accuracy performance of the proposed control strategy, ZVS capability and low turn off loss for switches are performed. As theoretical analysis, the proposed system gives a satisfactory result within the overall load range.

Acknowledgment

This research is sponsored by National Science Council under contract NSC-94-2622-E008-001-CC3.

References

[1] R. L. Steigerwald, "A comparison of half-bridge resonant converter topologies," *IEEE Trans. on Power Electronics*, Vol. 3, No. 2, pp. 174-182, 1988

[2] C. Fernandez, O. Garcia and J. A. Cobos, "Design guidelines of the series resonant converter for very low current or very high frequency applications," *Proc. IEEE Power Electronics Specialists Conference*, pp. 2454-2460, 2004

[3] J. Forsyth, G. A. Ward and S.V. Mollov, "Extended fundamental frequency analysis of the LCC resonant converter," *IEEE Trans. on Power Electronics*, Vol. 18, No. 6, pp. 1286-1292, 2003

[4] C. Chakraborty, M. Ishida and Y. Hori, " Novel half-bridge resonant converter topology realized by adjusting transformer parameters," *IEEE Trans. on Industrial Electronics*, Vol. 49, No. 1, pp. 197-205, 2002

[5] B. Yang and F. C. Lee, "LLC resonant converter for front end DC/DC conversion," *Proc. IEEE Applied Power Electronics Conference*, pp. 1108-1112, 2002

[6] M. Z. Youssef and P. K. Jain, "A review and performance evaluation of control techniques in resonant converters," *Proc. Annual Conference of IEEE Industrial Electronics Society*, pp. 215-221, 2004

1594

2006 5th International Power Electronics and Motion Control Conference

Design and Test of Novel Programmable Digital Three Phases SPWM Chip

Yang Yuan, Gao Yong, Chen Lijie

Department of Electronic Engineering, Xi'an University of Technology, Xi'an, 710048, China

Abstract—**In order to improve the sinusoidal pulse width modulation (SPWM) control system's real time capability and reduce the system's complexity, a high precision programmable digital three phases SPWM chip based on variable sampling frequency method is designed. The chip can be "programmed" by MCU to set the needed parameters with great flexibility and can be interfaced with multiplexed and non-multiplexed data bus mode. To implement the high precision for any frequency sinusoidal power wave, an improved method is presented by changing the sampling clock frequency for different power frequency while keeping the sampling points constant. In addition, the pipeline structure also has improved the speed of the system. The chip is fabricated with 0.35μm CMOS technology and the core area is 1755×1746μm². The test results show that the chip has reached its specifications. The control resolution for any power frequency is up to 16 bits, the system clock can reach 50MHz, and the ROM operation period is up to 1.68μs.**

Keywords-SPWM; chip; design; test

I. INTRODUCTION

In recent years, SPWM technology has been widely used in variable speed AC induction motors, static and uninterruptible power supplies and other forms of electronic equipments[1,3]. Using DSP to generate the SPWM control signal is hard to satisfy the system's real time capability[2,3], and using other logic circuit to realize that increases the system's complexity. In this paper we designed a programmable digital three phases SPWM chip. There are many functions for our chip such as

narrow pulse deletion, "programmed" by MCU, and so on. To keep the high precision for any power frequency, we use an improved method by changing the sampling clock frequency for different power frequency while keeping the sampling points constant. In addition, the pipeline structure and three phases sharing ROM improved system speed and reduced the chip area. We fabricated the chip using chartered 0.35μm CMOS technology and the core area is 1755×1746μm². The functions and specifications are tested. The test results show that the chip has reached its specifications. The design, implementation and test results of the chip are described in the following parts.

II. SYSTEM ARCHITECTURE BASED ON VARIABLE SAMPLING FREQUENCY

The system is composed of three main modules: the sine wave generator, triangle wave generator and six outputs generator as shown in Fig.1. In order to reduce the chip area and power, most of the modules for the three phases are sharing, and the phase control logic is used to control each module operate at the time division mode for the three phase The system clock SYSCLK is divided into two signals by 2^{n+1} frequency divider, one for generating triangular carrier waveform, the other for generating three-phase sinusoidal AC signal. The n was decided by carrier waveform frequency control word *CFS*. The triangular carrier waveform is generated by a 7-bit plus-subtract counter. The sine wave generator is the key of the system. We use ROM to store fixed sine wave sampling data. The most important for this module is to produce sine waveform at any frequency with high precision and to

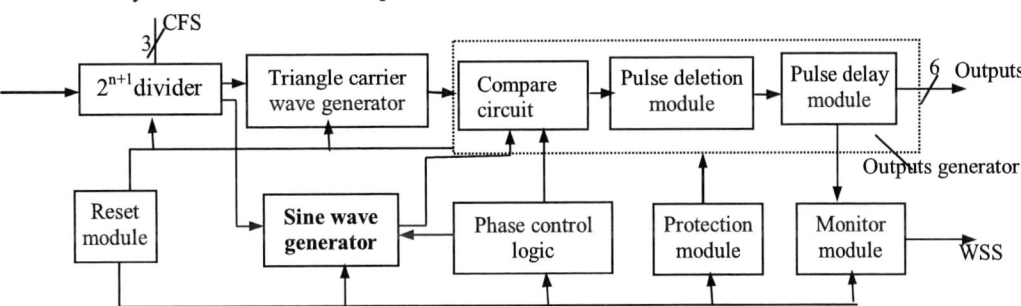

Figure1 System architecture

1-4244-0448-7/06/$25.00 ©2006 IEEE

reduce the ROM size. In traditional way, the sampling data number will decrease with the power frequency *FCW* increases, so the precision of the sine wave decreases and the ROM store a complete period data[4,5]. In our chip, an improved method with variable frequency sampling is presented as shown in Fig.2.

In this structure, the clock to control the ROM phase address is not a constant frequency but a variable frequency relative to the *FCW*, which keeps the sine wave constant number of sample data in a cycle so keep the same precision for any frequency. And through adding a sign bit and an adder/substrate counter, the ROM size of the sine wave reduces to one fourths of the traditional one. Only the sine sampling points(384 points) between $0 \sim \pi /2$ are memorized in LUT. Using the well-known quarter-wave symmetry technique, the sine wave samples for the full range of 2π are generated from 0 to $\pi/2$ rads of sine information stored in the ROM and the sign bit.

Under the control of clock f_0, the 23-bit phase accumulator execute the sum calculation of the FCW and the 16-bit register which is the result of the adder from the last calculation in every period, the high 7 bits are used as the counter of the low 16-bit register overflows. Through the selector one bit of the high 7 bits are used for the clock of a 9-bit address up/down counter adding or subtracting 1. The up/down counter is taken as an adder when address counter is 0 and as a subtracter when address counter is 383. When the address counter is 0, the 1-bit sign-flag-bit is in reverse. The addresses derived from the address counter are considered as the ROM look-up table's address. After reading the sample data memorized in the ROM, a 1/2 quantized sine wave is achieved when addressing ROM one time from 0 to 383 to 0. In the end, when the radium of the sine wave is between $0 \sim \pi$, the sign flag-bit is 1, when between $\pi \sim 2\pi$, the sign flag-bit is 0. So a complete sine wave is obtained by synthesizing the sign flag-bit register.

The sine wave frequency is derived as follows:

$$ f_{\sin e} = \frac{f_0 \times FCW}{2^{16}} \times \frac{1}{2^{7-m} \times 384 \times 4} \quad (1) $$

From formula (1), we can conclude that the 16-bit frequency control word can control the output frequency with 16-bit resolution for any power frequency.

The six outputs generator is obtained by comparing the sine wave output and the triangle carrier wave. In this module, the narrow pulse deletion circuit and pulse delay circuit are also included to reduce the switch loss and avoid transient short circuit between two devices at the same arm, the minimum pulse width and the pulse delay time can be set by parameters.

In addition to the three main modules, some other function modules are also designed in this chip, such as watchdog circuit, reset circuit, monitor circuit and protection circuit. The watchdog module can monitor the communication between the SPWM chip and MCU. If during the set time the MCU don't reset watchdog, the watchdog will reset the whole chip. There are two kinds of reset way in the chip: one is software way, the other is hardware way. The monitor circuit provides the user the state of the chip, whether the chip is working normally or not. The protection module can shut off the outputs pulse and give an alarm when emergency happen. It makes the chip shut off outputs override MCU, so decrease the reflection time. All the control parameters can be written into registers of the chip by MCU. The chip has been designed to operate with multiplexed and non-multiplexed data bus mode.

In the chip, in order to satisfy the system speed performance, pipeline structure is designed.

III. TEST RESULT OF THE CHIP

The chip was fabricated by CHARTERED 0.35μm CMOS technology and the core area is $1755 \times 1746 \mu m^2$. Fig.3 is the layout of the chip.

In order to test the fabricated chip, we developed a PCB with the control unit of AT89S8252. Each function and specification has been passed the test. MCU send control information to the chip, by changing the software of MCU, we change the specification and function to be tested. During the test, we use the Logic Analyzer TEKTRONIX TLA601 to capture the test result. The test results show that the operation frequency of the chip can be up to 50MHz. The power supply can be 2.7V~5.5V. When the power supply is 3.3V, the power of the chip is 42.32mW. Following is part of the test results.

Fig.4 is the six outputs, where power frequency was set to 3.9kHz and the carrier frequency is 24kHz, the amplitudes of three phases are different, for the red,

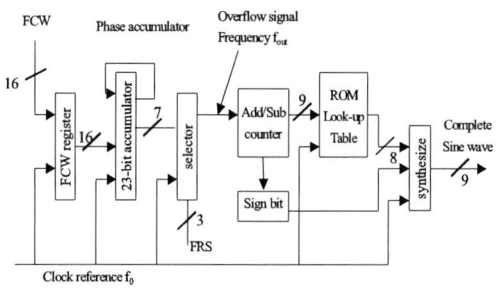

Figure 2 The improved architecture to generate sine wave

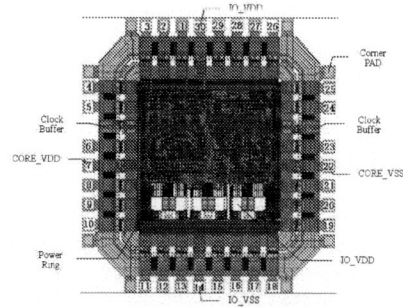

Figure 3 Layout of the chip

1596

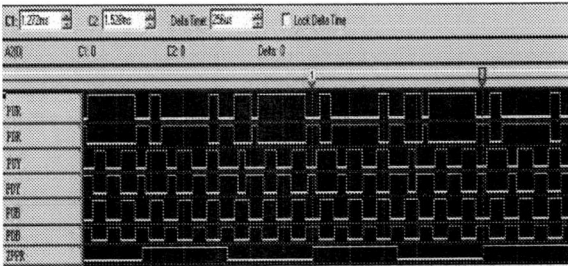

Figure 4 Outputs waveform and power frequency

yellow, blue phase are 100%, 50%, 0% separately. Outputs are divided into three-phases six-output pulses. Every phase is divided into two output pulses. In the figure, from up to down, the outputs are red phase up arm *PUR*, red phase down arm *PDR*, yellow phase up arm *PUY*, yellow phase down arm *PDY*, blue phase up arm *PUB*, blue phase down arm *PDB* and the power frequency monitor signal *ZPPR* which has the same frequency as that of the modulation waveform, separately. From the figure, we can see that the six SPWM outputs are the SPWM waves, the pulse widths for three phases are different also and the period of ZPPR is 256μs so the frequency is 3.90625kHz, the control error is 0.16%.

In order to verify the narrow pulse deletion function of the chip, from the software we set minimum pulse width to 10μs. We captured the minimum output pulse as shown in Fig.5, from which we can see that it is conformed to the value we set.

Fig.6 shows the delay function of the chip, where we set the pulse delay time to 1μs. From the figure we can see that it is 1μs after the red phase down arm "off", the up arm change to "on".

Fig.7 shows the watchdog function. We set the watchdog time to 20ms. From the figure we can see that after 20.04ms the system doesn't reset the watchdog the

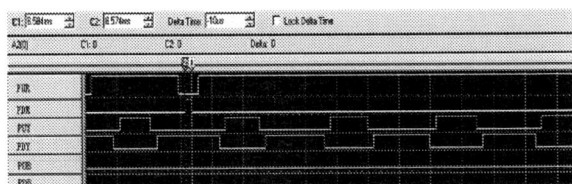

Figure 5 Minimum pulse width waveform

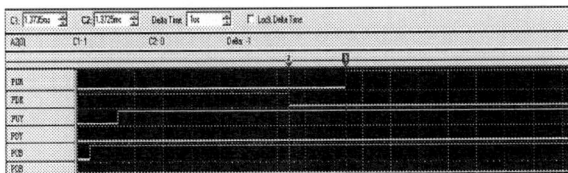

Figure 6 Pulse delay function waveform

outputs are turned off. The control error is 0.2%.

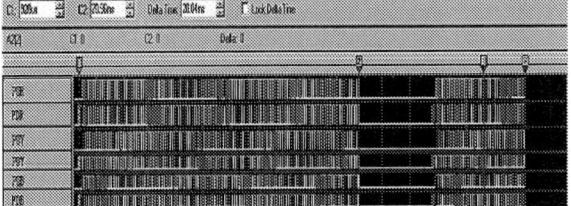

Figure 7 Watchdog function

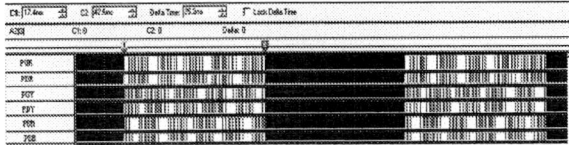

Figure 8 Reset function

Fig.8 shows the reset function. We use the software reset function to reset the system for 25ms after it works for 25ms. From the figure we can see that the work time is 25.2ms. The control error is 0.8%.

IV. CONCLUSION

A programmable three phase digital SPWM generator chip is designed and fabricated. By using an improved structure to generate sine waveform and the time division structure for three phases, high control precision has been obtained and the chip area has been reduced much. The pipeline structure satisfied the system speed performance. The test results show that the chip is fabricated successfully, the functions and specifications has been reached, system can work up to 50MHz.

REFERENCES

[1] K Zhang, Y Kang, J Xiong, J Chen., *Direct repetitive control of SPWM inverter for UPS purpose*, IEEE Trans. Power Electronics ,Vol.18,No.3, 2003

[2] LUO Xiaowei; LI Shuguang, *The Implementation of Modulation Signals in a Spwm Mode Based on TMS320F240*, Microcomputer applications,VOL.25, NO.5, 2004

[3] WANG Hong-wei; LIANG Hui, *Variable Speed System of Low Power Induction Motor based on DSP Control*, Power Electronics, Vol.39, No.3, 2005

[4] J.M.P Langlois,D.Al-Khalili, *A Low Power Direct Digital Frequency Synthesizer with 60 dBc Spectral Purity*, GLSVLSI'02, April, 2002, New York, USA

[5] J.Vankka, *Methods of mapping from phase to sine amplitude in direct digital synthesis*, in Proc.1996 IEEE Int. Frequency Control Symp.pp.942-950.

An Improved Performance of Five-Leg Inverter in Two Induction Motor Drives

Ryuji Omata[*], Kazuo Oka[*], Atsushi Furuya[*], Shuji Matsumoto[*], Yusuke Nozawa[*] and Kouki Matsuse[*]

[*] Meiji University / Dept. of Electrical and Electronics Engineering, Kawasaki, JAPAN

E-mail: ce66012@isc.meiji.ac.jp

Phone/Fax: +81-44-934-7298

A Number of Scope is 9.

Abstract— This paper presents an improved method of voltage utility factor from 50% to 86.6% for a five-leg inverter when two induction motors are driven in vector control and in a condition that a frequency of two motors isn't extremely different, and the improved method of voltage utility factor is theoretically shown. The five-leg inverter is a single inverter that can drive two motors independently and consists of five legs. *U* and *v* phases of both motors are connected in each leg respectively whereas *w* phase of both motors is connected in a common leg. In the five-leg inverter, because *w* phase of the motors are connected in the common leg, it causes difference from a switching pattern of w phase in two motors. For this reason, the modulation methods for a three leg inverter can't use for the five-leg inverter. Many modulation methods for the five-leg inverter have been proposed, but voltage utility factor is 50% in these methods. From the reason a maximum voltage utility factor of the three leg inverter is 100%, the voltage utility factor of the five-leg inverter is lower than one of the three leg inverter

Keywords; Five-Leg Inverter, Voltage Utility Factor,

Two Induction Motor Drives

I. INTRODUCTION

This paper presents an improved method of voltage utility factor (VUF) from 50% to 86.6% for a five-leg inverter when two induction motors are driven in vector control and in a condition that a frequency of two motors isn't extremely different. VUF is defined at the rate of the maximum output voltage in an inverter and a direct current (DC) -link voltage. VUF had better have as high as possible. Because the DC-link voltage required to output a same output is smaller on condition that a stator current root mean square (RMS) for the motor equals. In other words, an inverter capacity to output the same output is smaller. Using the five-leg inverter, it is possible to control the position, speed and torque of motors independently [1],[2],[3],[4]. However, since w phase of both motors are connected in a common leg, it causes difference from a switching pattern of w phase in both motors. Therefore, the modulation methods for the three leg inverter can't use for the five-leg inverter. Though

many modulation methods for the five-leg inverter have been proposed, VUF in these methods is 50%. From the reason a maximum VUF of the three leg inverter is 100%, VUF of the five-leg inverter is lower than one of the three leg inverter.

II. FIVE-LEG INVERTER

The five-leg inverter is a single inverter that can drive two motors independently. Figure.1 shows a structure of the five-leg inverter [1]-[4], [6]. The five-leg inverter consists of five legs. *U* and *v* phases of both motors are connected in each leg respectively whereas *w* phase of two motors is connected in a common leg (Leg5).

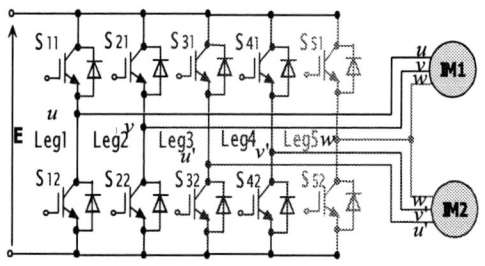

Figure.1 Structure of five-leg inverter

III. EXPANDED TWO-ARM MODULATION

An improved method of VUF in this work is based on the ETAM. Therefore, we introduce the ETAM in this section.

As mentioned above, the modulation methods of the three-leg inverter can't use for the five-leg inverter.

As one of the modulations method for the five-leg inverter, an expanded two-arm modulation (ETAM) has been proposed [3],[4]. Since a carrier-wave comparison method is applicable for the ETAM, it is an easier modulation method than other methods for the five-leg inverter.

In the ETAM, signal waves in each leg are given by the equations (1).

$$\begin{cases} v_{2arm\_uk}^* = v_{umk}^* - v_{wmk}^* \\ v_{2arm\_vk}^* = v_{vmk}^* - v_{wmk}^* \\ v_{2arm\_uk}^* = v_{wmk}^* - v_{wmk}^* = 0 \end{cases} \quad (k=1,2) \qquad (1)$$

Where, v_{imk}^* ($i=u,v,w$) is a command of i phase voltage and $v_{2arm\_ik}^*$ is signal waves of i phase in motor k ($k=1,2$).

The ETAM is the modulation method giving signal waves based on w phase voltage of each motor. Giving signal waves as equations (1), signal waves of w phase in both motors are zero. This means that we can give a zero command as the signal wave in the common leg. Adapting the carrier-wave comparison method, the independent control of two motors is possible.

IV. IMPROVED METHOD OF VOLTAGE UTILITY FACTOR

A. Voltage Utility Factor

We assume that a virtual neutral point is a central point of the DC link voltage E and a height of a carrier wave is 1/2 times of the DC-link voltage. Based on this assumption, the height of the carrier wave equals to amplitude of the phase voltage command in both motors, we have a merit that an analysis can be done easily.

There, using modulation index a_k of motor k, An amplitude of the signal wave of each phase are expressed by $a_k E/2$.

VUF is defined at the rate of the maximum output voltage for the inverter and the DC-link voltage. Therefore, adapting the assumption, it is calculated by the equation (2).

$$VUF(\%) = \frac{\sqrt{3} \cdot 1/2 \cdot a_{max} E}{E} \cdot 100 = \frac{\sqrt{3}}{2} a_{max} \cdot 100 \qquad (2)$$

Where, a_{max} is a maximum modulation index and a maximum value of the signal wave is $E/2$.

VUF had better have as high as possible. Because the DC-link voltage required to output a same output is smaller on condition that a stator current RMS for the motor equals.

In the general carrier-wave comparison method used for the three-leg inverter, since the phase voltage command of both motors is given as the signal waves, inequality (3) is obtained.

$$0 \le \frac{1}{2} a_k E \le \frac{1}{2} E \qquad (3)$$

From inequality (3), a_{max} is 1 and VUF is 86.6% from equation (2). Otherwise, in the ETAM, because the amplitude of signal waves equals to it of a line voltage for the motor, inequality (4) is obtained from equations (1).

$$0 \le \sqrt{3} \cdot \frac{1}{2} a_k E \le \frac{1}{2} E \qquad (4)$$

From inequality (4), a_{max} is $1/\sqrt{3}$ and VUF is 50% from equation (2).

As mentioned above, if a_{max} equals 1, we can improve VUF to 86.6%.

B. Improved Method of Voltage Utility Factor

For the improved method of VUF in this work, we assume that two induction motors are driven in vector control and in the condition that the frequency of two motors isn't extremely different. Moreover, initial phases of voltages and the frequency for both motors must correspond. As the initial phase of u phase voltage for both motors is zero at starting, only the frequency for them must correspond. In fact, we can equal the frequency of both motors with compensating a rotor flux command on condition that the frequency of two induction motors isn't extremely different and they are driven by vector control [5].

In the proposed method, we assume that signal waves of each phase in two motors are given in equations (5)

$$\begin{cases} v_{uk}^* = v_{umk}^* - \left(v_{wmk}^* - v_w^*\right) = v_{2arm\_uk}^* + v_w^* \\ v_{vk}^* = v_{vmk}^* - \left(v_{wmk}^* - v_w^*\right) = v_{2arm\_vk}^* + v_w^* \\ v_{wk}^* = v_{wmk}^* - \left(v_{wmk}^* - v_w^*\right) = v_w^* \end{cases} \quad (5)$$

v_w^* is a signal wave synchronizing to w phase, having the amplitude that the modulation index always equals 1.

Figure.2 shows a vector diagram of signal waves in two motors. From equation (2) and Figure.2, signal waves for the proposed method are given by the following inequality constraints (6) [6].

$$\begin{cases} v_{uk}^{*2} = v_{2arm\_uk}^{*2} + v_w^{*2} - 2 \cdot v_{2arm\_uk}^* \cdot v_w^* \cos(\pi/6) \le \left(\frac{1}{2}E\right)^2 \\ v_{vk}^{*2} = v_{2arm\_vk}^{*2} + v_w^{*2} - 2 \cdot v_{2arm\_vk}^* \cdot v_w^* \cos(\pi/6) \le \left(\frac{1}{2}E\right)^2 \end{cases} \quad (6)$$

Using the assumption for v_w^*, signal waves of u and v phase are the same formation from the inequality constraints (6). This is remarkable point.

If a_k has the range from 0 to 1, the maximum voltages of v_{uk}^* and v_{vk}^* are $E/2$. For this reason, the inequality (6) is satisfied on condition that a_k has the range from 0 to 1. Therefore, a_{max} is 1, it is possible to improve VUF to 86.6% by the proposed method.

1599

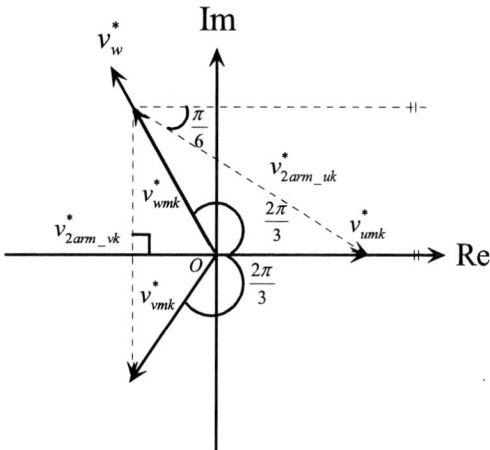

Figure.2 Vector diagram of signal waves in two induction motors

From equations (1) and (5), if v^*_w is zero, signal waves giving to each leg in the proposed method are the same as ETAM. For this reason, we can understand that the ETAM is an especial case of the proposed method.

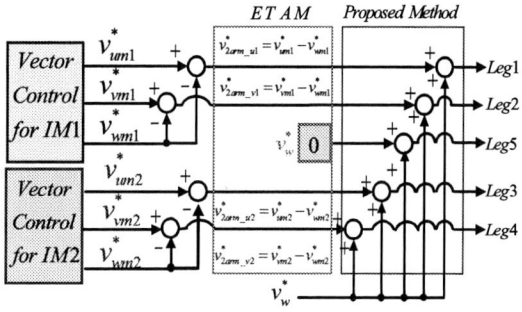

Figure.3 Signal waves giving to each leg

Figure.3 shows a relationship between signal waves giving to each leg and phase voltage command of both motors. As you can see from Figure.3, adding v^*_w to signals waves in the ETAM, them in the proposed method are given. Therefore, since a carrier-wave comparison method is applicable for the proposed method, a switching algorithm for the proposed method is easy.

C. Between Voltage Utility Factor and Common Mode Voltage

In the ETAM, common mode voltage (CMV),which is a voltage between a neutral point of the motor and an earthing point, is given by a following equation.

$$v_{ck\_etam} = \frac{1}{3}\left(v^*_{2arm\_uk} + v^*_{2arm\_vk} + v^*_{2arm\_wk}\right) \quad (7)$$

Where, v_{ck_etam} is CMV of the motor k in adapting ETAM. Since a sum of the phase voltage for the motor is zero, an equation (8) is obtained.

$$v^*_{umk} + v^*_{vmk} + v^*_{wmk} = 0 \qquad (8)$$

From the equation (1), (7), (8), CMV of the motor k equals the w phase voltage command of the motor k.

$$v_{ck\_etam} = -v^*_{wmk} \qquad (9)$$

Therefore, amplitude of the v_{ck_etam}, V_{ck_etam}, is $a_k E/2$.

$$V_{ck\_etam} = \frac{1}{2} a_k E \quad (10)$$

Otherwise, CMV of the motor k in adapting the proposed method is given by the following equation.

$$v_{ck\_pro} = \frac{1}{3}\left(v^*_{uk} + v^*_{vk} + v^*_{wk}\right) = v^*_w - v^*_{wmk} \quad (11)$$

Where, v_{ck_pro} is CMV of the motor k in adapting the proposed method. From the equation (11), amplitude of the v_{ck_pro}, V_{ck_pro}, is given by the equation (12).

$$V_{ck\_pro} = \frac{1}{2} E - \frac{1}{2} a_k E = \frac{1}{2}\left(1 - a_k\right) E \qquad (12)$$

And, we express that a voltage between each phase of the motor k and the earthing point is v_{iok} and a maximum value of it is v_{iokmax}. In this case, since v_{iokmax} equals to $E/2$, an inequality constraint is obtained.

$$\left[v_{imk} + v_{ck}\right]_{Amp\_\max} \le v_{iok\max} = \frac{E}{2}$$

$$\left[x\right]_{Amp\_\max} \to maximum\ amplitude\ of\ x \qquad (13)$$

Where, v_{ck} is CMV. Using the ETAM, v_{ck} equals v_{ck_etam} and using the proposed method, v_{ck} equals v_{ck_pro}. Moreover, adapting inequality (13) about v_{imk} and v_{ck} on condition that modulation index equals a_{max}, we get.

$$\left[v_{imk(a\max)} + v_{ck(a\max)}\right]_{Amp\_\max} \le v_{iok\max} = \frac{E}{2} \qquad (15)$$

Where $v_{imk(amax)}$ is v_{imk} and $v_{ck(amax)}$ is v_{ck} on condition that modulation index equals a_{max}. From the equation (2), VUF is decided in $v_{imk(amax)}$. Therefore, you can understand that the $v_{ck(amax)}$ is smaller, the VUF is higher. We assume that $v_{ck(amax)}$ in the ETAM is $v_{ck\_etam(amax)}$ and $v_{ck(amax)}$ in the proposed method is $v_{ck\_pro(amax)}$. In this case, $v_{ck\_etam(amax)}$ and $v_{ck\_pro(amax)}$ are as the follows.

$$\begin{cases} v_{ck\_etam(amax)} = \frac{1}{2}a_{max}E = \frac{1}{2\sqrt{3}}E \\ v_{ck\_pro(amax)} = \frac{1}{2}(1-a_{max})E = 0 \end{cases} \quad (16)$$

CMV of the proposed method varies according to modulation index. Especially, when modulation index equals a_{max}, CMV of the proposed method is zero. This is a remarkable point.

D. Compensation of motor frequenvy

For the proposed method, it is required that the frequency of both motors has the same value. In this work as shown Figure.4, the frequency commands of both motors are given as the input signal of Proportional-Integral (PI) controller, and the value adding the output signal of PI controller to the rotor flux command for motor 1 is the new rotor flux command for motor 1

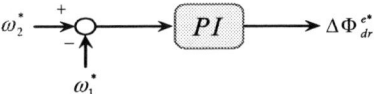

Figure.4 Compensatory method of motor frequency

In this Figure.4, ω^*_i (i=1,2) is a frequency command of motor i and $\Delta\Phi^{e^*}_{dr1}$ is a output signal of PI controller, compensatory value of rotor flux in the motor 1.

V. SIMULATION RESULTS

We simulated to confirm a validity of the proposed method. Table1. is the specification of induction motor using for simulations. The specification of both motors is the same. Table2. shows the speed and load torque commands. The values in parentheses are at the rate for a rated torque. DC-link voltage is the constant value of 150(V). The Simulation results of the ETAM cases are also shown to compare the proposed method of the ETAM. Figure.5 shows the results for the proposed method.

Inverter output voltage can be outputted up to 75(V) in the ETAM and up to 150(V) in the proposed method. As you can see from Figure.5 (a) and (b) and Figure.6 (a) and (b), using the proposed method, output voltage is about 130(V). The validity of the proposed method is shown from these simulation results.

TABLE1. SPECIFICATION OF INDUCTION MOTOR

Output	2.2 [kW]	$Rs = 0.7185\,[\Omega]$
Voltage	180 [V]	$Rr = 0.5965\,[\Omega]$
Current	10 [A]	$Ls = 63.38\,[mH]$
Speed	1750 [rpm]	$Lr = 63.38\,[mH]$
Poles	4	$M = 61.28\,[mH]$

TABLE2. SPEED AND LOAD TORQUE COMMANDS

	Time [s]	0	0.5	0.8	1	1.5	1.6	2
IM1	Speed [rpm]		235.0			485.0		585.0
	Torque [N·m]		10.0(83%)			14.0(117%)		
IM2	Speed [rpm]		250.0			500.0		600.0
	Torque [N·m]		12.0(100%)			16.0(133%)		

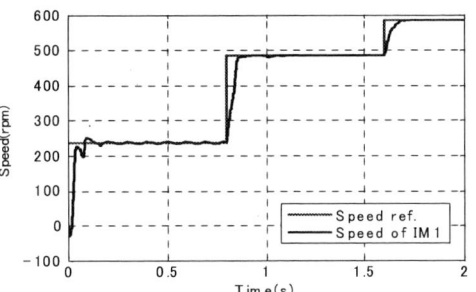

Figure.5 (a) Speed of induction motor1

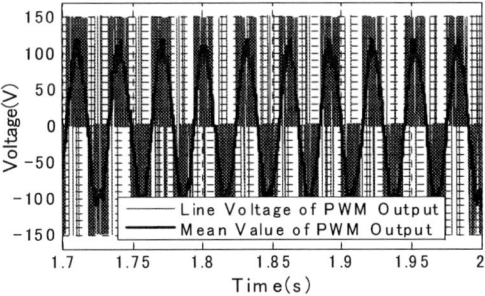

Figure.5 (b) Speed of induction motor2

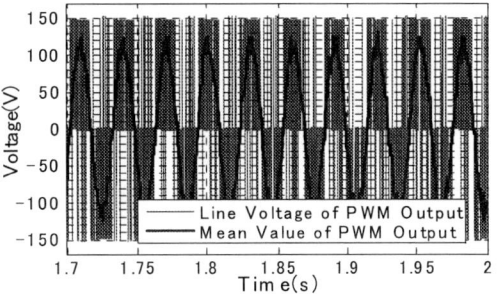

Figure.6 (a) Line Voltage between u and v phase of motor1

Figure.6 (b) Line Voltage between u and v phase of motor2

VI. CONCLUSION

This paper presents that voltage utility factor can be improved from 50% to 86.6% for the five-leg inverter if the frequency of two induction motors isn't extremely different and they are driven in vector control.

Moreover, the validity of the proposed method is shown theoretically and by simulation results.

REFERENCES

[1] Francois B., Bouscayrol A.: "Control of Two Induction Motors Fed by a Five-Phase Voltage-Source Inverter", ELECTRIMACS'99, pp.313-318 (1999)

[2] M.Hizume,S.Yokomizo.K.Matsuse: "Independent Vector Control of Parallel-Connected Two Induction Motors by a Five-Leg Inverter", EPE2003, CD-ROM (2003)

[3] Y.Kimura, M.Hizume, K.Oka.and K.Matsuse: "Independent Vector Control of Two Induction Motors with Five-Leg Invereter by the Expanded Two Arm PWM Method", The 2005 International Power Electronics Conference, p.613-616 (2005)

[4] Y.Nozawa, M.Hizume, Y.Kimura, K.Oka, K.Matsuse: "Independent Position Control of Two Permanent Magnet Synchronous Motors with Five-Leg Invereter By the Expanded Two Arm Modulation Method", JIASC2005, I-79-80 (2005)

[5] Enrique Ledezma, Brenden McGrath, Alfred Munoz, and Thomas A.Lipo : "Dual AC-Drive System With a Reduced Switch Count", IEEE TRANSACTIONS ON INDUSTRY APPLICATIONS, Vol.37, No.5, pp.1325-1333 (2001)

[6] K.Oka, M. Hizume, Y.Kimura, Y.Nozawa, K.Matsuse: "Improved Method of Voltage Transfer Ratio for Five-Leg Inverter", JIASC2005, I-75-78 (2005)

2006 5th International Power Electronics and Motion Control Conference

Adaptive Three Dimensional Space Vector Modulation in abc Coordinates for Three Phase Four Wire Split Capacitor Converter

Xiao-bo Yang[*], Wei-yang Wu[*] and Hong Shen[* **]
[*]Yanshan University, Qinhuangdao, China
[**]Tianjin University, Tianjin, China
yxb@ysu.edu.cn

Abstract—a new three dimensional space vector modulation (3D-SVM) strategy in abc coordinates for the three phase four wire converter with split-capacitor is presented. Among various three phase four wire converters, the center split converter is suit to low cost, low power applications for its fewer switches. Compare to the 3D-SVM in $\alpha\beta\gamma$ coordinates, the strategy in abc coordinates is more clearly and simpler, and therefore the complexity of modulation algorithm and the computational burden are dramatically reduced. An adaptive 3D SVM is employed to compensate the voltage variation of split capacitors. A 3kVA prototype is established and the experimental results are presented to validate the method.

Keywords-three phase four wire coverter; center split; 3D SVM; abc coordinates

I. INTRODUCTION

In recent years, the three phase four wire PWM converter has been used widely, such as distributed power generators, active power filters, phase PWM rectifiers and DVR[1]. Three phase voltage-source inverters normally have three ways of providing a neutral connection for three-phase four-wire systems:

1) using split dc link capacitors and tying the neutral point to the mid-point of the dc link capacitors;

2) using a D/y0 connected 50Hz transformer to estabilish the neutral point as return ground[2];

using a four-leg converter topology and tying the neutral point to the mid-point of the fourth leg[1, 3].

The three phase four-leg converter topology has been reported as the best alternative to implement a three phase four wire devices[4], especially the three phase four-leg inverter for unbalance loads and three phase four-leg PWM rectifier improving fault tolerant capability. The demerit for the split capacitor converter is that the large and expensive dc link capacitors are needed to maintain an acceptable voltage ripple level across the dc link capacitors in case of a large zero sequence current due to unbalanced or nonlinear load. In addition, it suffers from an insufficient utilization of the dc link voltage, the maximum modulation index is only 0.866. Despite the demerits mentioned above, the center split three phase four wire converter is still attractive: with the split-

capacitor approach, the three-phase four wire converter essentially becomes three single-phase half-bridge converters and permits each of the three legs to be controlled independently, making its current tracking control simpler than phase four-leg converter; the number of switches are reduce to six and easy to control; without the 50 Hz output transformer, the cost is much lower. Thus, there are still some interesting solutions that use a conventional three-leg converter with split capacitor to implement a three phase four wire shunt active power filter, especially for some low cost, low power APF applications. A typical two level three phase four wire split capacitor converter as APF is shown in Fig.1.

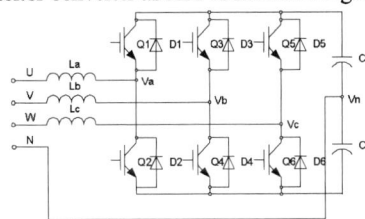

Figure 1. Split capacitor three phase four-leg converter as active filter

Because of the independency of the three legs, the control of split capacitor converter is usually under three dimension space. There are many papers have discussed the control schemes for the split capacitor three phase four converter, such as Hysteresis Control strategies[5] and the 3D-SVM strategies[6]. Most papers dealing with the 3D-SVM use a representation of vectors in $\alpha\beta\gamma$ coordinates instead of abc coordinates. This representation offers interesting information about the zero sequence component of both current and voltages, however, it has some drawbacks: the change of reference frame implies complex calculations; the switch vectors in $\alpha\beta\gamma$ coordinates is difficult to understand; the 3D-SVM method in $\alpha\beta\gamma$ need to determine the "sextant" in which the desired voltage vector is included, which leads to many complicated operations[3].

The 3D-SVM algorithm is first proposed in [7] for a three phase four-leg inverter. A 3D-SVM for multilevel is discussed in[8] . In this paper, a very simple and fast 3D-SVM algorithm in abc coordinates for the three phase four

1-4244-0448-7/06/$25.00 ©2006 IEEE   1603

wire split capacitor converter is proposed. The adaptive SVM strategy is first proposed in [9] for a component-minimized voltage source inverter to compensate DC-link voltage ripple. However, the method is also effective for the 3D SVM strategy discussed in this paper to compensate the voltage variation of split capacitors.

After Introduction, the paper shows the principle of the 3D-SVM algorithm in abc coordinates in Section II, the switch vectors in abc coordinates are illustrated also. Next, the method for choosing the tetrahedron containing a given voltage is presented in Section III. In Section IV, the 3D-SVM under the capacitor voltage variation is discussed; an adaptive 3D SVM strategy is employed to compensate the voltage variation. Finally, in Section V, the simulation and experimental results are presented to validity the proposed method.

II. THE SWITCH VECTORS IN THREE DEMENSIONAL SPACE

By extending two dimensional space to three dimensional space, the three legs of the split capacitor converter is decoupled. The three phase-neutral voltages, defined as $[V_{an} \ V_{bn} \ V_{cn}]^T$, can be express in the three dimensional space. The switching vectors present a very simple and straightforward expression:

$$\begin{bmatrix} V_{an} \\ V_{bn} \\ V_{cn} \end{bmatrix} = \begin{bmatrix} V_a - V_n \\ V_b - V_n \\ V_c - V_n \end{bmatrix} \quad (1)$$

As a traditional three-leg converter, the split capacitor three phase converter has eight possible switching combinations. However, with the split capacitors, the decoupled three phase-neutral voltages extend to three dimensional space rather than a 2D $\alpha\beta$ plane. The switching combinations can be represented by the ordered sets [Sa, Sb, Sc], where Sj=1 (j=a, b, c) denotes upper switch of the j leg is closed and Sj=0 denotes lower switch of the j leg is closed. Note that the dead band times are not considered in this method.

TABLE I shows the phase to neutral voltages for all eight switching combinations in abc coordinates. To simplify the notation, the vectors are normalized by V_{dc}, the voltage of DC link. Considering the voltage variation of the split capacitors, define:

$$V_n = \varepsilon V_{dc}, \ (0 < \varepsilon < 1) \quad (2)$$

Thus, the normalized value of Vn is ε.

The eight switching vectors can be plotted in the three dimensional space. If the voltage variation of the split capacitors is neglected, we have $\varepsilon = 0.5$ and all the switching vectors are equal length, as shown in Fig. 2. It is easy to notice that the vectors are all in the vertices of a cube. Note that unlike the traditional 2D SVM, the zero vectors, V1 and V7, also contribute to synthesize the reference vector. In fact, the zero vector represents the zero sequence component.

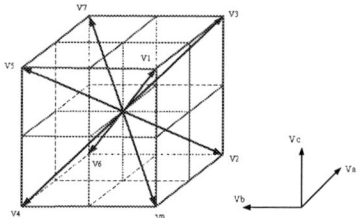

Figure 2. Switching vectors in the abc coordinate

TABLE I.
SWITCHING STATES, AC TERMINAL VOLTAGES AND SWITVHING VECTORS IN ABC COORDINATES

state	Sa	Sb	Sc	V_{an}	V_{bn}	V_{cn}	Vector
1	0	0	0	$-\varepsilon$	$-\varepsilon$	$-\varepsilon$	V0
2	0	0	1	$-\varepsilon$	$-\varepsilon$	$1-\varepsilon$	V1
3	0	1	0	$-\varepsilon$	$1-\varepsilon$	$-\varepsilon$	V2
4	0	1	1	$-\varepsilon$	$1-\varepsilon$	$1-\varepsilon$	V3
5	1	0	0	$1-\varepsilon$	$-\varepsilon$	$-\varepsilon$	V4
6	1	0	1	$1-\varepsilon$	$-\varepsilon$	$1-\varepsilon$	V5
7	1	1	0	$1-\varepsilon$	$1-\varepsilon$	$-\varepsilon$	V6
8	1	1	1	$1-\varepsilon$	$1-\varepsilon$	$1-\varepsilon$	V7

Joining the corresponding three vertices of the cube, the region including in this regular hexahedron is complete equivalent to the one that appears in the $\alpha\beta\gamma$ representation[7], but the distribution of the switching voltages appear more clearly, allowing the development of a simpler 3D SVM algorithm[3].

Different from the 3D-SVM in $\alpha\beta\gamma$ coordinates, the planes that define the control region, or in another words, the over modulation boundaries, in which the switching vectors will be included, present very simple expressions: The all six boundary are parallel to the coordinate planes, expressed by (3).

$$\begin{cases} 2|V_a| = 1 \\ 2|V_b| = 1 \\ 2|V_c| = 1 \end{cases} \quad (3)$$

III. CALCULATION OF THREE DIMENSIONAL SPACE VECTOR MODULATION IN ABC COORDINATES

Similar to the 2D-SVM, the reference vector V_{ref} is synthesized by the switching vectors shown in Fig.2. The synthesis of the reference vector in abc coordinates follows the same procedure as a classical 3D-SVM strategy[1]:

1) *Selection of switching vectors;*
2) *Projection of the reference vector onto selected switching vectors.*

Instead of selecting a region that is usually a sextant in $\alpha\beta\gamma$ coordinates, the tetrahedron composed of three vectors point to its corners has to be determined. Different from classical 3D SVM strategy, the selecting of zero vectors should be considered deliberately. Because the zero vectors contribute to the reference vector, each of them will have a certain effect. Thus, there are four switching vectors should be selected once the tetrahedron is determined: the three vectors point to the tetrahedron, including two non-zero switching vectors and one zero switching vectors; and another zero switching vector is in addition. An example is shown in Fig.3. The reference vector is in the tetrahedron composed of three switching vectors: V0, V4 and V6. V0 is a "negative" zero switching vector, V4 and V6 are the others two non-zero switching vector. With the addition of V7, the "positive" zero switching vector, there are total four switching vectors, V0,V7,V4 and V6 contribute to the synthesis the reference vector. The restriction will be discussed later.

A. Sellection of switching vectors

As illustrated in Fig.3, the given tetrahedron is surrounded by four planes. By the same dividing method, the control cube can be split into six tetrahedrons. There are nine dividing planes to define the six tetrahedrons:

- Six of them are the control boundaries paralleling to the coordinate planes, see (3);

- The other three planes are vertical to the coordinate planes and through the origin point of the coordinates. Their equations are shown in (4).

$$\begin{cases} V_a - V_b = 0 \\ V_b - V_c = 0 \\ V_c - V_a = 0 \end{cases} \quad (4)$$

Note that all the tetrahedrons are equal in size, providing a symmetrical division of the control region. It is easy to determine the tetrahedron in which the reference vector is dwelling. An efficient way has discussed in[10] for a three phase four-leg inverter. For the proposed 3D SVM in this paper, the method is still effective and easier.

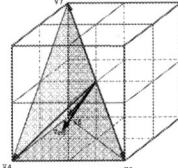

Figure 3. Selection of switching vectors

Define three indexes, C_i (i=1,2,3), each of them representing the position relative to the plane equations in (5):

$$\begin{cases} C_1 = \text{Sign}(V_{aref} - V_{bref}) \\ C_2 = \text{Sign}(V_{bref} - V_{cref}) \\ C_3 = \text{Sign}(V_{cref} - V_{aref}) \end{cases} \quad (5)$$

Where $[V_{aref} \quad V_{bref} \quad V_{cref}]^T$ are the reference voltage vector; Sign(x) extracts the sign of x, being 1 if x is positive, being 0 if x is negative.

Define a region pointer RP to represent the tetrahedron in which the reference vector is included. The RP can be calculated as

$$RP = 4C_1 + 2C_2 + C_3 \quad (6)$$

Table II shows the six tetrahedrons with their RP values. The all six tetrahedrons and the corresponding switching vectors are shown in Figure 4.

TABLE II.
REGION POINTERS AND THE SWITCHING VECTORS TO COMPOSE TETRAHEDRONS

RP	Vd1	Vd2	Vd2	Vd4
1	V0	V1	V5	V7
2	V0	V4	V6	V7
3	V0	V4	V5	V7
4	V0	V2	V3	V7
5	V0	V1	V3	V7
6	V0	V2	V6	V7

Each tetra has two non-zero switching vectors and two zero switching vectors. Synthesizing the reference vector by using adjacent switching vectors of a tetrahedron leads to minimal circulating energy and current ripple. Different from traditional 2D SVM or 3D SVM for three phase four-leg inverters, which have two class sequence schemes according to the select of zero switching vectors[11], the 3D SVM in this paper has only class I sequence scheme: both of the zero sequence vectors should be used unless over modulation occurs.

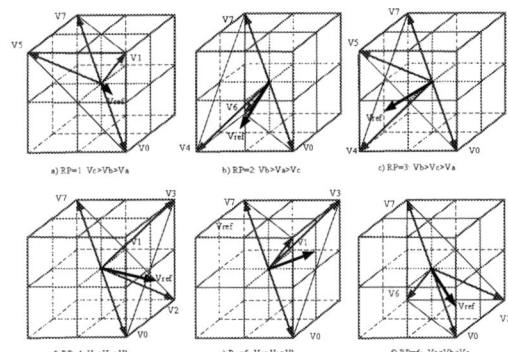

Figure 4. Tetrahedron identification

B. Projection of the reference vector onto selected switching vectors

The time duration of the selected switching vectors can be easily computed by projecting the reference vector onto the switching vectors. For example, if the reference voltage vector $V_{ref} = [V_{aref} \quad V_{bref} \quad V_{cref}]^T$ is in the region of RP=2, and the available switching vectors are Vd1=V0, Vd2=V4, Vd3=V6 and Vd4=V7. The duty ratios d= $[d_1 \ d_2 \ d_3 \ d_4]^T$ of the corresponding switching vectors are giving by

$$\mathbf{V_{ref}} = d_1 \mathbf{Vd1} + d_2 \mathbf{Vd2} + d_3 \mathbf{Vd3} + d_4 \mathbf{Vd4} \quad (7)$$

and

$$d_1 + d_2 + d_3 + d_4 = 1. \quad (8)$$

From (5) and (6) we have:

$$\begin{bmatrix} \mathbf{V_{ref}} \\ 1 \end{bmatrix} = \begin{bmatrix} \mathbf{Vd1} & \mathbf{Vd2} & \mathbf{Vd3} & \mathbf{Vd4} \\ 1 & 1 & 1 & 1 \end{bmatrix} \cdot \mathbf{d} \quad (9)$$

The duty ratios can be obtained:

$$\mathbf{d} = \mathbf{M_d}^{-1} \begin{bmatrix} \mathbf{V_{ref}} \\ 1 \end{bmatrix} \quad (10)$$

where

$$\mathbf{M_d} = \begin{bmatrix} \mathbf{Vd1} & \mathbf{Vd2} & \mathbf{Vd3} & \mathbf{Vd4} \\ 1 & 1 & 1 & 1 \end{bmatrix} \quad (11)$$

IV. US ADAPTIVE 3D SVM TO COMPONSATE THE VOLTAGE VARIATION OF THE SPLIT CAPACITORS

In the split capacitor three phase converter topology, the zero sequence current flows through C1 and C2 so the mid-point voltage of the two DC link capacitors changes inevitably. When the DC voltage change occurs, the Vdc1 $\neq$ Vdc2 or $\varepsilon \neq 0.5$ in (2). The lengths of the switching vectors are not equal any more. However, it is easy to verify that all the switching vectors are still in the vertices of a cube. The new cube in the abc coordinates is shown in Figure 5.

Reference [6] has propose a method to deal with the DC voltage variation in $\alpha\beta\gamma$ coordinates. In this paper, a new method, based on the adaptive 3D SVM is proposed.

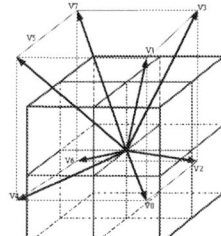

Figure 5. Switching vectors under DC voltage variation of split capacitors

By sampling the mid-point voltage of the split capacitors, the ac terminal voltages in Table I. can be acquired in real-time. For example, when the reference vector is in the tetrahedron composed of three switching vectors V0, V4 and V6, as shown in Fig.4(b), the corresponding transform matrix $\mathbf{M_d}$ is given by

$$\mathbf{M_d} = \begin{bmatrix} \mathbf{V0} & \mathbf{V4} & \mathbf{V6} & \mathbf{V7} \\ 1 & 1 & 1 & 1 \end{bmatrix} = \begin{bmatrix} -\varepsilon & 1-\varepsilon & 1-\varepsilon & 1-\varepsilon \\ -\varepsilon & -\varepsilon & -\varepsilon & 1-\varepsilon \\ -\varepsilon & -\varepsilon & 1-\varepsilon & 1-\varepsilon \\ 1 & 1 & 1 & 1 \end{bmatrix}. \quad (12)$$

The duty ratios $\mathbf{d}$ under voltage variation is

$$\mathbf{d} = \mathbf{M_d}^{-1} \begin{bmatrix} \mathbf{V_{ref}} \\ 1 \end{bmatrix} = \begin{bmatrix} -1 & 0 & 0 & 1-\varepsilon \\ 1 & 0 & -1 & 0 \\ 0 & -1 & 1 & 0 \\ 0 & 1 & 0 & \varepsilon \end{bmatrix} \begin{bmatrix} V_{aref} \\ V_{bref} \\ V_{cref} \\ 1 \end{bmatrix} \quad (13)$$

In fact, the calculation procedure of duty ratios using $\mathbf{M_d}^{-1}$ with variable ε under voltage variation is just the same as using a constant matrix, except for the real-time calculation of ε is needed. The additional time spending on a DSP is negligible.

The transform matrix $\mathbf{M_d}^{-1}$ for the six tetrahedrons can be calculated beforehand, as shown in Table III. The sequence scheme is shown in Fig.6, where the RP is equals to 5.

The switching times of the gating signals for upper switches t1, t2 and t3 can be obtained by the selected sequence scheme. As shown in (14).

$$\begin{cases} t1 = V_{aref} + \varepsilon \\ t2 = V_{bref} + \varepsilon \\ t3 = V_{cref} + \varepsilon \end{cases} \quad (14)$$

Note that it is just the same as the carrier based SPWM with mid-point voltage compensation.

TABLE III.
SUMMARY OF TRANSFORM MATRIX $\mathbf{M_b}^{-1}$ FOR DUTY RATIOS CALCULATION

RP	$\mathbf{M_d}^{-1}$	RP	$\mathbf{M_d}^{-1}$
1	$\begin{bmatrix} 0 & 0 & -1 & 1-\varepsilon \\ -1 & 0 & 1 & 0 \\ 1 & -1 & 0 & 0 \\ 0 & 1 & 0 & \varepsilon \end{bmatrix}$	4	$\begin{bmatrix} 0 & -1 & 0 & 1-\varepsilon \\ 0 & 1 & -1 & 0 \\ -1 & 0 & 1 & 0 \\ 1 & 0 & 0 & \varepsilon \end{bmatrix}$
2	$\begin{bmatrix} -1 & 0 & 0 & 1-\varepsilon \\ 1 & -1 & 0 & 0 \\ 0 & 1 & -1 & 0 \\ 0 & 0 & 1 & \varepsilon \end{bmatrix}$	5	$\begin{bmatrix} 0 & 0 & -1 & 1-\varepsilon \\ 0 & -1 & 1 & 0 \\ -1 & 1 & 0 & 0 \\ 1 & 0 & 0 & \varepsilon \end{bmatrix}$
3	$\begin{bmatrix} -1 & 0 & 0 & 1-\varepsilon \\ 1 & 0 & -1 & 0 \\ 0 & -1 & 1 & 0 \\ 0 & 1 & 0 & \varepsilon \end{bmatrix}$	6	$\begin{bmatrix} 0 & -1 & 0 & 1-\varepsilon \\ -1 & 1 & 0 & 0 \\ 1 & 0 & -1 & 0 \\ 0 & 0 & 1 & \varepsilon \end{bmatrix}$

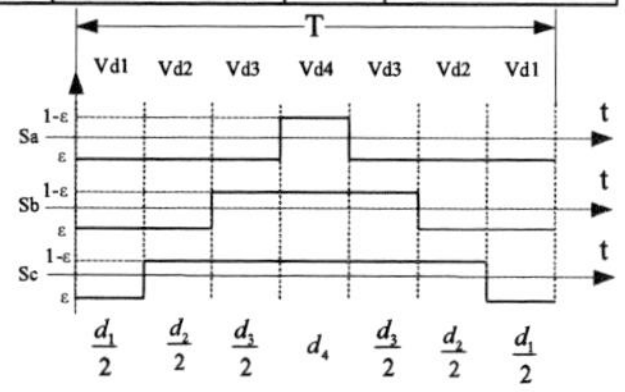

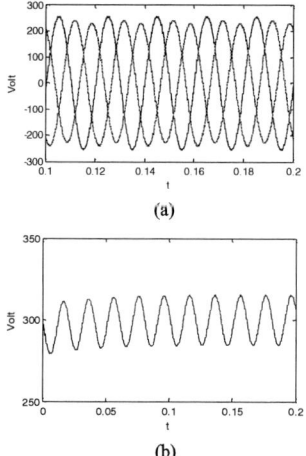

Figure 6. Sequence scheme for the proposed method (RP=5)

V. EXPERIMENTAL AND SIMULATION RESULTS

A 3kVA prototype for the proposed converter has been built. A DSP (TMS320F2812@150M) is used to realize the proposed algorithm.

Simulation results on Matlab/Simulink for a three phase four wire split capacitor converter with unbalanced loads are shown in Fig.7 and Fig.8. The dc link voltage is Vdc=600V and modulation index is 0.8. Unbalanced resistance loads are used in the simulation: Ra=220Ω, Rb=220Ω and Rc=22Ω.

Fig.7 shows the phase to neutral output voltages under the mid-point variation without compensation. Fig.8 shows the same output voltages with voltage compensation by employing adaptive 3D SVM. The symmetry of output voltages are improved greatly by the proposed method.

Figure 7. (a) Three phase output voltage of the three phase four wire split capacitor converter and (b) mid-point voltage: without voltage variation compensation.

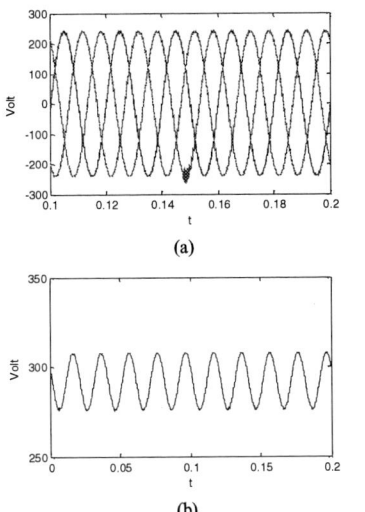

(a)

(b)

Figure 8. (a) Three phase output voltage of the three phase four wire split capacitor converter and (b) mid-point voltage: with voltage variation compensation.

Fig. 9 is the driver signals of the three legs, only the gate signals of the upper switches are displayed.

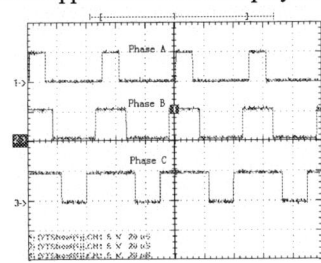

Figure 9. Driver signals of the upper switches

CONCLUTION

This paper presents a new 3D SVM in abc coordinates for the three phase four wire converter with split-capacitor. Compare to the 3D SVM in $\alpha\beta\gamma$ coordinates, the proposed method is more clear and easy to employ. The 3D SVM strategy under split capacitor voltage variation is also discussed and an adaptive SVM strategy is proposed. The calculation procedure of adaptive SVM under voltage variation is just the same as under the constant mid-point voltage except for the real-time acquisition of the mid-point voltage is needed. The additional time spending on a DSP is negligible. The equivalent carrier of the 3D SVM is deduced, which implies that the 3D SVM strategy for a three phase four wire spilt capacitor converter is equivalent to the carrier based SPWM with mid-point voltage compensation, either under abc coordinates or $\alpha\beta\gamma$ coordinates.

ACKNOWLEDGMENT

This work was supported by the Key Programs of the National Natural Science Foundation of China. (No. 50237020)

REFERENCE

[1] Zhang, R., et al., Three-dimensional space vector modulation for four-leg voltage-source converters. Power Electronics, IEEE Transactions on, 2002. 17(3): p. 314-326.

[2] Bai Dan, et al., Study on the unbalanced load of a three phase four leg inverter (In chinese). Automation of electronic power systems 2004. Vol.28 (No.9).

[3] Perales, M.A., et al. New controllability criteria for 3-phase 4-wire inverters applied to shunt active power filters. IECON 2002, pt. 1, p 638-43 vol.1.

[4] Aredes, M., J. Hafner, and K. Heumann, Three-phase four-wire shunt active filter control strategies. Power Electronics, IEEE Transactions on, 1997. 12(2): p. 311-318.

[5] Man-Chung, W., et al. Theoretical study of 3 dimensional hysteresis PWM techniques. Conference Proceedings IPEMC 2004, pt. 3, p 1635-40 Vol.3.

[6] Ning-Yi, D., W. Man-Chung, and H. Ying-Duo. Three leg center-split inverter controlled by 3D SVM under DC variation. Conference Proceedings IPEMC 2004, pt. 3, p 1362-7 Vol. 3

[7] Zhang, R., et al. A three-phase inverter with a neutral leg with space vector modulation. APEC, 1997, p 857-863.

[8] Prats, M.M., et al., A 3-D space vector modulation generalized algorithm for multilevel converters. Power Electronics Letters, IEEE, 2003. 1(4): p. 110-114.

[9] Blaabjerg, F., D. Neacsu, and J.K. Pedersen. Adaptive SVM to compensate DC-link voltage ripple for component minimized voltage source inverters. PESC 1997, p 580-9 vol. 1.

[10] Perales, M.A., et al., Three-dimensional space vector modulation in abc coordinates for four-leg voltage source converters. Power Electronics Letters, IEEE, 2003. 1(4): p. 104-109.

[11] Prasad, V.H., D. Borojevic, and R. Zhang. Analysis and comparison of space vector modulation schemes for a four-leg voltage source inverter. APEC 1997, p 864-71 vol. 2.

2006 5th International Power Electronics and Motion Control Conference

Inverters Parallel Operation Based on CAN

Yong Wu, Xianglong Jiang, Jinbang Xu, Qingyi Wang and Shuyun Wan

Huazhong University of Science and Technology / Department of Control Science and Engineering, Wuhan, China

yngww@163.com

Abstract - A novel digital current sharing scheme for inverters parallel operation is proposed and discussed. The new scheme is based on Controller Area Network (CAN). This scheme eliminates the circumfluence which is produced by the difference of inverters' voltage amplitudes. A key feature of the scheme is that it uses CAN to transmit parameters among inverters. It is important in high reliability inverters parallel operation system where system operation can be maintained in noisy environments. The mathematic model of the scheme is deduced and some representative experimental results are also presented. The results show that it can improve the capacity, reliability and redundancy of inverters parallel operation system.

Keywords–inverter; parallel operation; CAN; circumfluence

I. INTRODUCTION

Inverters parallel operation is believed to be a key technology to achieve a high, reliable and flexible power ac source. It can promote the development of inverter technology. However, the parallel operation of inverters is much more complicated than dc parallel operation because the output voltage of all the parallel inverters must be strictly synchronized in frequency, phase and amplitude to guarantee the output power sharing. Types of inverters parallel operation system configuration, control methods of parallel operation, matters that demand special attentions and means of protection against failure are discussed in [1].

For the inverters parallel operation system, a crucial problem is the circumfluence among inverters that is produced by the differences of phases and amplitudes of inverters. Many models have been proposed to eliminate the circumfluence. In master-slave model [2] and [3], an inverter is specified as master module and other inverters are slave modules. The master decides current sharing of each slave module. This model is easily implemented, but when the master is fault the inverters parallel operation system is breakdown. So it has poor redundancy. Reference [4] and [5] achieves real and reactive power sharing among inverters by controlling two independent quantities – the power angle and fundamental inverter voltage amplitude. This model does not need communication of control signals among inverters, but it needs complicated calculation and leads to great delay. Most inverters parallel operation models have poor stability in noisy environments.

In this paper, a novel instantaneous digital current

sharing scheme based on CAN is proposed. It eliminates circumfluence which is produced by the difference of the amplitudes through CAN when the phases of inverters in parallel operation have synchronized. The scheme shares current among inverters through digital communication and improves the reliability and effect of inverters parallel operation system. Because of high reliability of CAN in noisy environments, inverters parallel operation based on CAN has high stability. This scheme has been successfully used in 3KVA inverters parallel operation system and has been verified good effect.

In the following, firstly the principle of inverters parallel operation is given a brief overview. Secondly, the current sharing scheme based on CAN is described and its mathematic model is deduced. Then, the circuit configuration of inverters parallel operation system is described. Finally, some experimental results demonstrate the accuracy of the proposed scheme and the conclusion is given.

II. PRINCIPLE OF INVERTERS PARALLEL OPERATION

Circumfluence among inverters should be given extremely careful consideration. The ac output of inverter can be represented by frequency, amplitude and phase. Only when these three parameters of all inverters are equal, the circumfluence among inverters can be eliminated and inverters can parallel operate. In these three parameters of inverter, through carefully designing and controlling inverter, the stability and the precision of the frequency can get very high degree. Therefore, the difference of frequencies can be taken no account of when we design inverters parallel operation system. So we consider that the frequencies of all inverters are the same and mainly consider the influence of the phase and amplitude. Fig. 1 shows the principle of two inverters parallel operation.

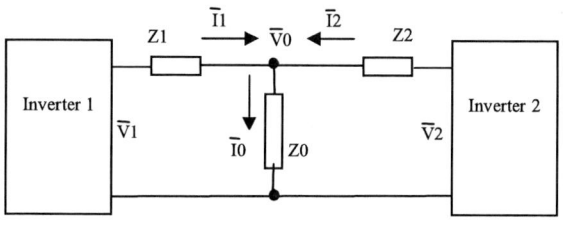

Fig.1 Parallel operation of two inverters

1-4244-0448-7/06/$25.00 ©2006 IEEE

Where Z1 and Z2 are the reactance of connection, Z0 is reactance of load.

The circumfluence between two inverters is defined as follows.

$$\bar{I}_H = (\bar{I}1 - \bar{I}2)/2 \qquad (1)$$

The voltages of load in Fig. 1 is illustrated as

$$\bar{V}0 = (1 - Z1/(Z1 + Z0 // Z2)) * \bar{V}1 + (1 - Z2/(Z2 + Z0 // Z1)) * \bar{V}2 \quad (2)$$

The output currents of inverter 1 (I1) and inverter 2 (I2) are

$$\bar{I}1 = (\bar{V}1 - \bar{V}0)/Z1 \qquad (3)$$

and

$$\bar{I}2 = (\bar{V}2 - \bar{V}0)/Z2 \qquad (4)$$

respectively.

The paralleled inverters must be same designed and the parameters of the inverters are equal.

It is assumed that the reactance of the connection is equal and in practical it is far less than the reactance of load. That is

$$Z1 = Z2 = Z, Z << Z0. \qquad (5)$$

Replacing (5) in (2), (3) and (4), we get

$$\bar{V}0 = (\bar{V}1 + \bar{V}2)/2 \qquad (6)$$

and

$$\bar{I}_H = (\bar{V}1 - \bar{V}2)/(2 * Z) = \Delta \bar{V}/(2 * Z). \qquad (7)$$

In Fig.1, take the voltage of the load as the reference voltage, that is $\bar{V}0 = V0 \angle 0^0$, where V0 is the amplitude of the load voltage. When all inverters in the parallel operation system are synchronized, the phases are the same. From (6), the phases of two inverters' output voltages in parallel operation are equal to the voltage of load. So $\bar{V}1 = V1 \angle 0^0$, $\bar{V}2 = V2 \angle 0^0$. Replacing in (7), we get the circumfluence between two inverters as follows.

$$\bar{I}_H = (V_1 - V_2) \angle 0^0/(2 * Z) = \Delta V \angle 0^0/(2 * Z) \quad (8)$$

Where V1 and V2 are voltage amplitudes of two inverters respectively, Δ V is the amplitude difference of output voltages. Analyzing (8), in inverters parallel operation system, we can conclude:

1) When inverters are synchronized, the circumfluence is produced by the difference of amplitudes of the inverters' output voltages (ΔV).

2) The inverter whose amplitude is higher than the average voltage will output circumfluence, the inverter whose amplitude is less than the average voltage will input circumfluence.

3) Circumfluence doesn't pass through load.

From the conclusion above, when inverters are synchronized, in order to eliminate the circumfluence in inverters parallel operation system, we must eliminate the difference of the amplitudes of inverters.

III. CURRENT SHARING BASED ON CAN

A. General principle

CAN is a serial communication protocol that it is used efficiently for applications of distributed systems control in real time with high level of safety. The CAN specification defines the physical and data link layers of ISO/OSI reference model. It has many characteristics such as low cost, high reliability in noisy environments, priority-based bus arbitration, fail nodes automatic disconnection, high transmission rate (1Mbit/s) and standard ISO [6]-[8]. Applying CAN in inverters parallel operation system can easily increase the number of the inverters and make system reliable, efficiency and low cost.

In inverters parallel operation system, when all inverters are synchronized, in order to eliminate the circumfluence that is produced by the difference of amplitudes of inverters, it's very importance to know the average voltage of the inverters parallel operation system. For an inverter of the parallel operation system, if its output voltage is higher than the average voltage, the inverter will output circumfluence; if its output voltage is less than the average voltage, the inverter will input circumfluence.

In the digital current sharing scheme based on CAN, the average voltage is calculated through CAN. CAN has the function of broadcast that we can transmit an inverter's parameters such as output voltage, output current and output power to other inverters. In every 50HZ ac periods, every inverter transmits its parameters (output voltage, current and power) to other inverters through CAN, and synchronously it receives parameters (output voltage, current and power) that other inverters transmit. After the inverter has received the parameters of all inverters, it calculates the average voltage of the parallel operation system and then adjusts its output voltage to the average voltage. When all inverters' output voltages equal to the average voltage, the difference of the output voltage of inverters is eliminated. That is Δ V=0. From (8), the circumfluence among inverters equals to zero. $\bar{I}_H = 0$.

B. Analyzing of mathematic model

Applying CAN in inverters parallel operation system to eliminate the circumfluence among inverters equals to introduce negative voltage feedback in inverters to eliminate the voltage amplitude difference of inverters. Fig.2 is the model of a single inverter and Fig.3 is the

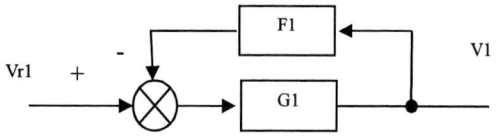

Fig.2 Model for a single inverter

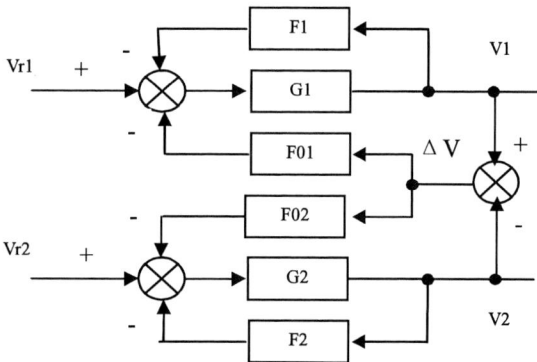

Fig.3 Model for parallel operation of two inverters

model of the parallel operation of two inverters.

Where Vr1 and Vr2 are the reference input voltages of inverters respectively, V1 and V2 are the output voltages of inverters respectively, ΔV is the difference of output voltage, G1 and G2 are the open loop transfer functions of inverters respectively, F1 and F2 are feedback voltages in inverters respectively, F01 and F02 are feedback voltages between inverters respectively.

Fig.2, the transfer function of a single inverter 1 is obtained as

$$V1/Vr1 = G1/(1+F1*G1). \qquad (9)$$

Fig.3, the transfer function of an inverter 1 in parallel operation is obtained as

$$V1/Vr1 = (1-\Delta V/Vr1*F01)*G1/(1+F1*G1). \quad (10)$$

In (10), because $\Delta V << Vr1$, if propriety select F01, we get

$$1-\Delta V/Vr1*F01 \approx 1. \qquad (11)$$

So the transfer function is approximately equal to

$$V1/Vr1 \approx G1/(1+F1*G1). \qquad (12)$$

Comparing (9) with (12), it is seen that the transfer function of parallel operation is approximate equal to that of single operation. This indicates the output voltage of the inverter can still follow the reference voltage after we introduce the average voltage negative feedback to an inverter. And because we introduce the average voltage

negative feedback in an inverter, the output voltage of all inverters can follow the average voltage. So it eliminates the difference of voltage amplitude of inverters and eliminates the circumfluence that is produced by the difference of the amplitudes of inverters.

IV. CIRCUIT CONFIGURATION OF INVERTERS PARALLEL OPERATION SYSTEM

Fig.4 shows circuit configuration of the 3KVA inverters parallel operation system.

In Fig.4, the inverters parallel operation system eliminates difference of phases and amplitudes through synchronization line and CAN respectively. After the system has synchronized, every inverter transmits its parameters (output voltage, current and power) to other inverters and synchronously receives parameters of other inverters in every 50HZ ac periods through CAN. Then it calculates the average voltage of the system and adjusts its output voltage to average voltage. When output voltages of all inverters equal to average voltage of the system, the difference of voltage amplitudes is eliminated and circumfluence is eliminated.

Because of introducing CAN to share current among inverters in inverters parallel operation system, all inverters adjust their output voltages to average voltage and the system can efficiently share current. When an inverter joins the inverters parallel operation system that is running, it can receive and transmit parameters with other inverters through CAN after its phase has synchronized. Because of high data transmission rate of CAN, it can transmit all parameters in a 50HZ ac periods. So it can eliminate circumfluence and all inverters share current in inverters parallel operation system in a short time. When an inverter in inverters parallel operation system is power-off or fault, it automatic disconnects from the CAN bus and does not influence the running of the parallel operation system. So this inverters parallel operation system has perfect current sharing effect and reliability.

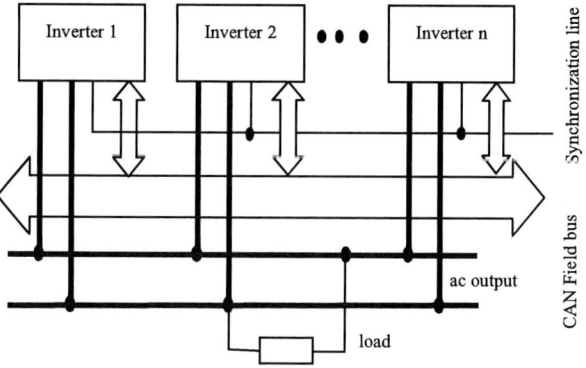

Fig.4 Configuration of inverters parallel operation

V. EXPERIMENTAL RESULTS

The proposed current sharing control scheme of inverters parallel operation based on CAN is verified by 3KVA inverters parallel operation system. In the 3KVA inverters parallel operation system, the output voltage of inverter is 220V/50HZ and output power is 3KVA.

In the first experiment, the load inductor is 2mH and resistance is 16Ω. Fig.5 shows the experimental voltage waveform of two inverters parallel operation. The output currents of two inverters are 13.6A and 13.6A respectively.

The result shows that digital current sharing based on CAN does not influence the waveform of output voltage.

In the second experiment, when two inverters parallel operation system mentioned above is running, the third inverter is power on. Fig.6 shows the voltage waveform from two inverters parallel operation to three inverters parallel operation. The output currents of three inverters are 8.9A, 9.1A and 9.2A respectively.

The result shows that suddenly changing the number of inverters does not influence the synchronization and current sharing of inverters parallel operation system.

In the third experiment, the load capacitor is 470μF and resistance is 16Ω. Fig.7 shows the waveform of load

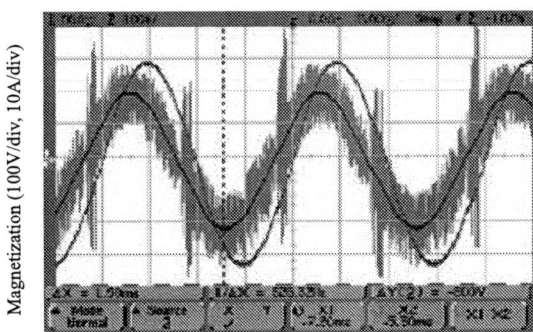

Time (20ms/div)

Fig.7 Experimental waveform of output voltage and current of three

inverters parallel operation

voltage and current of three inverters parallel operation system. The output currents of three inverters are 5.8A, 5.8A and 5.9A respectively.

The experimental results above show that the proposed scheme based on CAN has perfect current sharing effect (13.6A and 13.6A in two inverters parallel operation. 8.9A, 9.1A and 9.2A in three inverters parallel operation). The current sharing time is less than one second and the difference of output current amplitudes among inverters is less than 0.5A.

VI. CONCLUSION

This paper has described a method to effectively control inverters parallel operation with CAN. This method has a highly modular structure. It can easily increase or reduce the number of inverters in inverters parallel operation system. When an inverter is power-off or fault, it automatic disconnects from inverters parallel operation system and doesn't influence the running of the parallel operation system. This feature enables easy modification of the structures to meet the requirements of different ac power system and the system has high stability.

Analyzing of mathematic model of inverters parallel operation based on CAN proves that the digital current sharing method proposed in this paper equals to introduce negative voltage feedback among inverters and efficiently shares current among inverters. The experimental results presented indicate that the scheme effectively achieves the goals of current sharing in inverters parallel operation system. It not only increases the capacity of inverter power system, but also improves the reliability and fault-tolerance of inverter power system. It's very important to improve the modularization and enhance the redundancy of inverter power system.

REFERENCES

[1] Takao kawabata and Shigenori higashino, "Parallel operation of voltage source inverters," IEEE Trans. Industry Applications, vol. 24, no.2, pp.281-287, Mar./Apr. 1988.

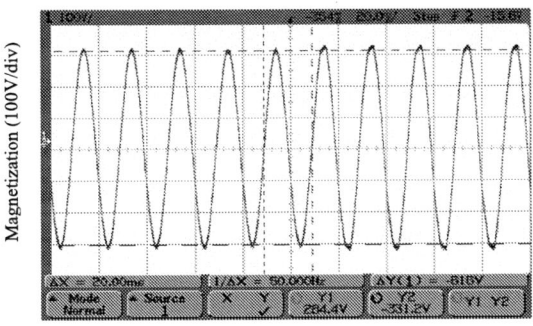

Time (20ms/div)

Fig.5 Experimental voltage waveform of two inverters parallel

operation

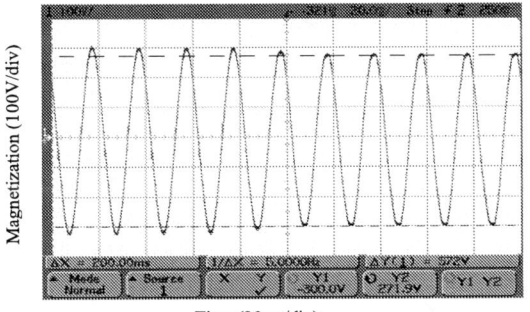

Time (20ms/div)

Fig.6 Experimental voltage waveform from two inverters parallel

operation to three inverters parallel operation

[2] Jiangfu Chen and Chinglung Chu, "Combination of voltage-control and current-controlled PWM inverters for UPS parallel operation," IEEE Trans. on PE, vol.10, no.5, pp547-558, 1995.

[3] Y. Xing, L.P. Huang and Y.G. Yan, "A decoupling control method for inverters in parallel operation," Power System Technology. Proceedings. PowerCon 2002. International *Conf.* vol. 2, pp.13-17, Oct., 2002.

[4] Mukul C. Chandorkar, Deepakraj M. Divan and Rambabu Adapa, "Control of parallel connected inverters in standalone ac supply systems," IEEE Trans. Industry applications, vol. 29, no.1, pp.136-143,Jan./Feb. 1993.

[5] Youichi Ito and Osamu Iyama, "Parallel Redundant Operation of UPS with Robust Current Minor Loop," Power Conversion Conference, vol. 1 3-6 Aug. 1997 pp.489-494.

[6] Lamia Chaari, Nouri Masmoudi and Lotfi Kamoun, "Electronic control in electric vehicle based on CAN network," SMC 2002 IEEE International *Conf.* vol.7, pp.6-.9.

[7] Liming He, Zuohua Tian, "Time division multiple access based on CAN," Proceedings of the 4th World Congress on Intelligent Control and Automation (Shanghai, P.R.China), pp.10-14, Jun., 2002.

[8] Scott, A. V., Buchanan, W. J., "Truly distributed control systems using fieldbus technology," ECBS 2000 Proceedings, Seventh IEEE International Conference and Workshop on the 3-7 April 2000 pp.165-173.

2006 5th International Power Electronics and Motion Control Conference

EMI Reduction Method for a Single-Phase PWM Inverter by Suppressing Common-Mode Currents with Complementary Switching

Toshiji Kato, Kaoru Inoue, Koji Akimasa
Department of Electrical Engineering, Doshisha University
Kyotanabe, Kyoto, 610-0321, JAPAN
E-mail: tkato@mail.doshisha.ac.jp

Abstract—With advancement of recent high-speed switching devices, PWM inverters are operated in high frequencies and often generate EMI noises due to common-mode leakage currents. There are mainly two types of leakage currents. One is inverter leakage currents due to parasitic capacitances between switching devices and heat sinks. The other is load leakage currents due to parasitic capacitances between the load and the earth. This paper proposes an elimination method which is effective both to the leakage currents in a single-phase full-bridge PWM inverter. The principle is so simple that the inverter is operated only complementarily between two arms. This eliminates neutral voltage drifts of the load by keeping the common-mode voltage zero. Furthermore parasitic capacitor currents circulate only inside the inverter body because the switching times are synchronous between the two arms. This is valid even when there are dead-times in arm switching. These effects are validated through simulation by Saber and experiment by a DSP-based system with switching frequency of 10kHz.

Keywords—EMI, PWM Inverter, High-Frequency Leakage Current, Common-Mode Voltage

I. INTRODUCTION

With recent development of high-speed switching devices, switching frequencies of converters have been becoming higher. This tendency generates more high-frequency leakage currents of converters which cause electro-magnetic interference (EMI) noises. Various techniques by passive filters have been proposed to reduce the EMI noises[1] − [3]. These passive techniques are effective. However increases of system volumes and costs are inevitable. They cause other problems on magnetic saturations of filter inductor cores which need careful design with full inverter operation modes. Other techniques by active filters with switching devices have been alternatively reported[4] − [6]. They have advantages in volume and cost aspects. However they have limits on response speeds which are not effective in a frequency region over tens of MHz.

This paper proposes a complementary switching technique which can reduce leakage currents both from loads and switches without any special filters. This principle is so simple that it eliminates a common-mode voltage component by complementary switching between two arms of a single-phase inverter to reduce its load leakage current. It has another additional advantage that it can also reduce leakage currents which circulate only inside the inverter and do not go out because the switches are operated on synchronous timings by the complementary switching. It has an advantage in volume and cost because it does not need

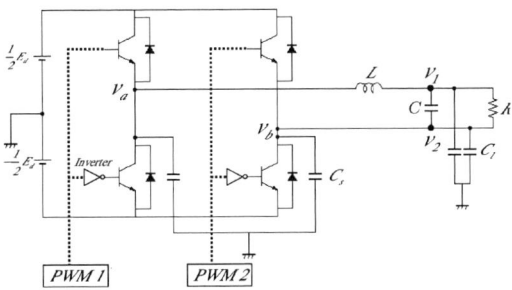

Figure 1. Single-phase PWM inverter circuit.

Figure 2. Conventional generation of PWM control signals.

TABLE I
INVERTER CIRCUIT PARAMETERS.

parameters	symbol	value
input voltage	E_d	40V
switching frequency	f_s	10kHz
filter inductance	L	3mH
filter capacitance	C	25μF
load resistance	R	30Ω
parasitic capacitance	C_l	2000pF
parasitic capacitance	C_s	1000pF

any special filter nor devices. The principle is investigated for its leakage current reduction effects first by simulation by Saber simulator and then by experiment.

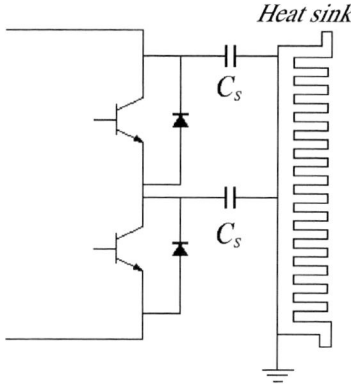

Figure 3. Parasitic capacitances between switches and heat sinks.

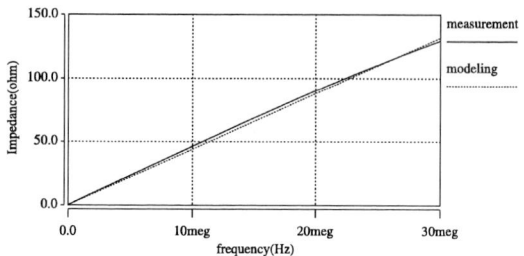

Figure 4. Frequency characteristic of the line impedance.

II. HIGH-FREQUENCY LEAKAGE CURRENTS IN SINGLE-PHASE PWM INVERTER

A. System Configuration

The conventional control method of a single-phase PWM inverter is shown in Fig.1 where L is a filter inductor, C a filter capacitor, R a load resistor, C_l, C_s parasitic capacitors of the load and switches. Typical control gate signals are PWM signals which are shifted mutually by 180 degrees in Fig.2 and they are inputted into the arms respectively.

High-frequency leakage currents which are discussed in this paper have mainly two routes. One is from the load, through the parasitic capacitor C_l, and to the earth. The other is the switches, through the parasitic capacitors C_s, and to the earth. They are generally called as common-mode currents. They cause EMI problems like interferences of inverter control and malfunctions of other devices.

B. Leakage Currents from the Load

An inverter load has a parasitic capacitance to the earth as in Fig.1. A common-mode voltage v_{common} of the inverter output varies in a step-wise shape whose break points are synchronous with arm switch timings due to PWM operations. It is equivalent to the neutral voltage of the load and is expressed with the following equation.

$$v_{common} = \frac{1}{2}(v_1 + v_2) \qquad (1)$$

This common-mode voltage is applied across the above parasitic capacitor and a high-frequency leakage current flows along this common-mode route, the load, through the parasitic capacitor, to the earth.

C. Leakage Currents from Switches

Generally switches are equipped with a heat sink as in Fig.3 and there are parasitic capacitors between them. The heat sink is connected to the inverter earth. The capacitances are expressed with the following equation where S is an area size of the sink and d is a thickness of the insulator between the sink and the switches.

$$C_s = \frac{\mu_0 \mu_r S}{d} \qquad (2)$$

A phase voltage is always applied to the parasitic capacitors. When it is changed in a step-wise shape due to a switching, a high-frequency leakage current is generated along a common-mode route, from the switch, through the capacitor, to the earth. This may cause an EMI problem.

D. Line Impedance along Leakage Route

The line impedance increases as frequency increases. It should be considered in leakage current analysis. The upper frequency limit of conductive-mode EMI is 30MHz whose wave length λ is 10m. It is necessary to consider the line as a distributed circuit when its length is longer than 2.5m which equals to $\lambda/4$. The main purpose of this paper is to show effectiveness of the proposed method and the length of all lines between the inverter and the load is set to 1m length. The line is modeled with lumped elements synthesized from a measured frequency characteristic in Fig.4 and it is used for simulation.

First the impedance of a line of 1m length is measured in the conduction-mode frequency band from 100kHz to 30MHz with an impedance meter(Agilent4294A). The characteristic is shown in Fig.4 and is almost linearly increased. It is approximated with an usual $L - R$ series model($R = 58m\Omega, L = 800nH$) and is synthesized with the impedance meter. The model characteristic is also shown in Fig.4 and it almost coincides with the measured one. This line model is used for all simulation cases in this paper.

E. Leakage Current Analysis by Simulation with Saber

Leakage currents from the load and the switches are simulated with the model for three 1m length lines between the inverter and the load by Saber. Simulation parameters are listed in Table 1.

A computed common-mode voltage and a leakage current from the load are shown in Fig.5(a). The current waveform is synchronous with the voltage change timings. They are enlarged in Fig.5(b) and their oscillation amplitude and frequency are 1.2A and 1.6MHz respectively.

Two phase voltages and their leakage current from switches are shown in Fig.5(c). The current waveform is also synchronous with the voltage change timings. It is enlarged in Fig.5(d) and its oscillation amplitude and frequency are 0.55A and 1.7MHz respectively.

1614

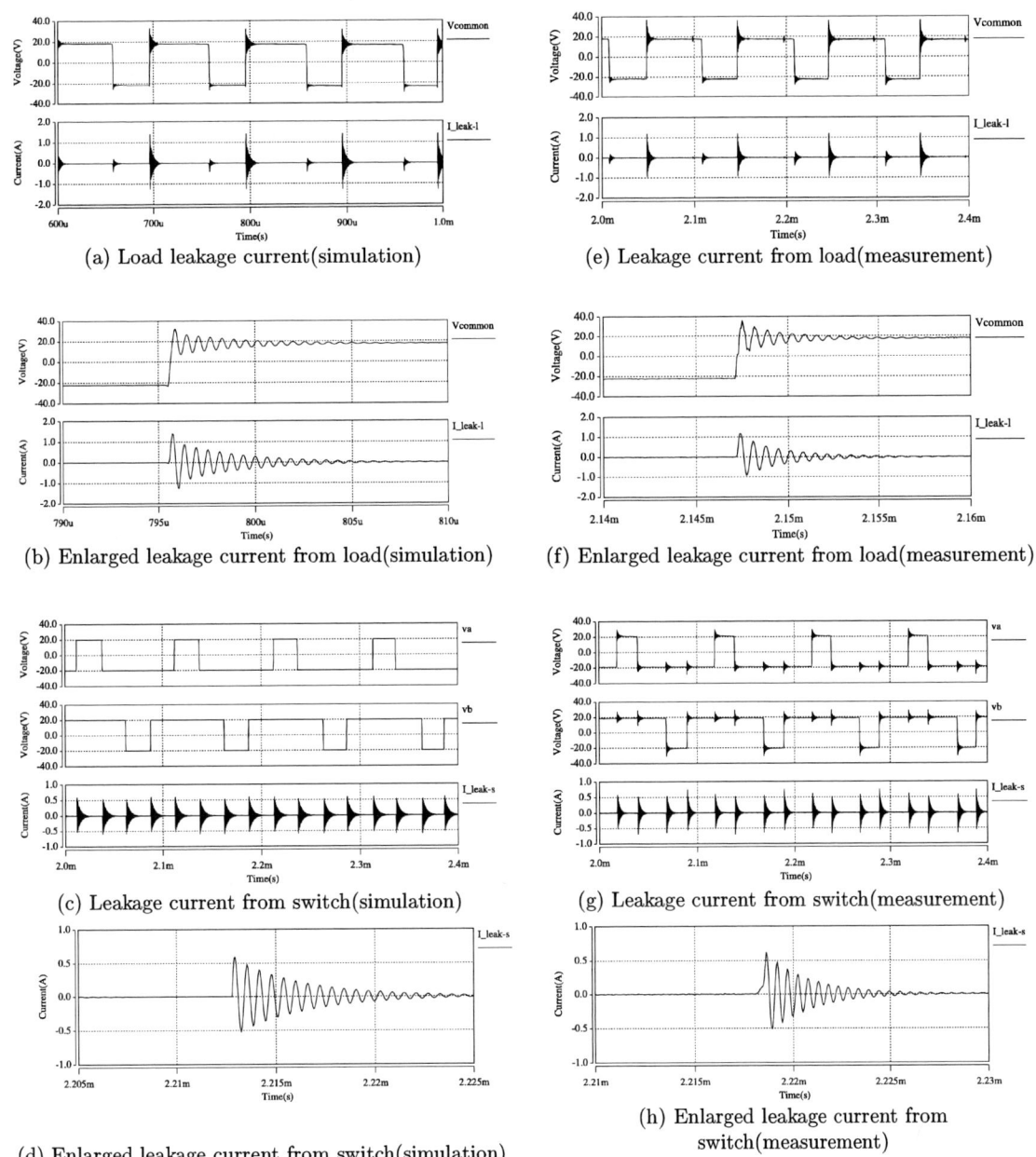

(a) Load leakage current(simulation)

(b) Enlarged leakage current from load(simulation)

(c) Leakage current from switch(simulation)

(d) Enlarged leakage current from switch(simulation)

(e) Leakage current from load(measurement)

(f) Enlarged leakage current from load(measurement)

(g) Leakage current from switch(measurement)

(h) Enlarged leakage current from switch(measurement)

Figure 5. Comparison between simulated and measured leakage currents.

III. LEAKAGE CURRENT ANALYSIS BY EXPERIMENT

A. Experimental System Setup

The conventional inverter system in Fig.1 is experimentally analyzed where L is an inductor, C a filter capacitor, R a load resistor, C_l a parasitic load capacitor, and C_s a parasitic switch capacitor. The two parasitic capacitors are selected to be large enough as ($C_l = 2000pF, C_s = 1000pF$) and are externally connected to emphasize their effects. Their currents are considered to be the leakage ones.

A measured leakage current from the load is shown in Fig.5(e),(f). Its oscillation maximum amplitude and frequency are 1.2A and 1.55MHz which are close to those by simulation. This shows validity of the simulation models including the line model.

A measured leakage current from the load is shown in Fig.5(g),(h). Its oscillation maximum amplitude and frequency are 0.55A and 1.75MHz which are close to those by simulation.

1615

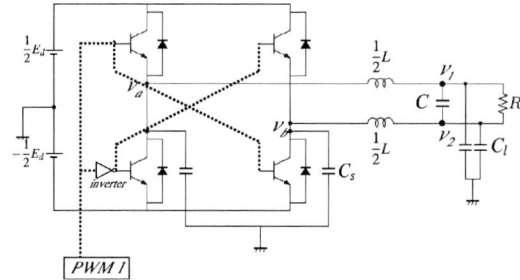

Figure 6. The proposed complementary switching technique.

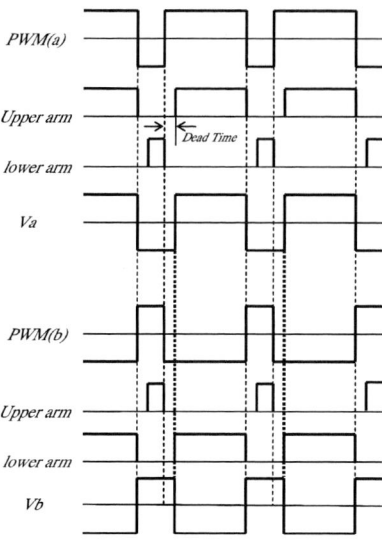

Figure 8. Relation between dead-times and phase voltages.

Figure 7. Leakage current cancellation from switches.

IV. LEAKAGE CURRENT REDUCTION BY COMPLEMENTARY ARM SWITCHING

A. Complementary Arm Switching

A switch control technique for a single-phase PWM inverter which can reduce the two types of leakage currents from a load and switches is proposed in this paper. The principle is so simple that two arms of an objective inverter are switched complementarily. The gate signals are shown in Fig.6 which fires switches in the arms synchronously. The filter inductor L is divided into two parts to keep its voltage potential balance.

B. Reduction of Leakage Current from Load

According to the conventional switching control, the common-mode voltage in (1) is not zero. This voltage is applied to the parasitic load capacitor and a high-frequency leakage current is generated. The proposed control tries to eliminate the voltage and the leakage current by keeping sum of the two arm voltages zero with the complementary switching.

C. Reduction of Leakage Current from Switches

The proposed complementary switching keeps the common-mode voltage zero originally to eliminate a load leakage current. Furthermore it makes it possible to eliminate leakage currents from switches at the same time. The

principle is explained in Fig.7. When one arm voltage is set to be the upper voltage level, the other arm voltage is set to the lower voltage level at the same time. This makes it possible to cancel one current with the other of the opposite sign and to eliminate leakage current. This technique has advantages in cost and size because it does not need any additional element, for example, a link transformer in [3].

D. Effects of Dead-Times over Output Phase Voltages

The proposed principle is absolutely based on simultaneous switching of two arms. However actual switches need dead-times which prevent arm short currents. It is necessary to investigate what kind of effects these dead-times have over output phase voltages.

Typical PWM gate signals and phase voltages with the dead-times are shown in Fig.8 where one gate signal is denoted as PWM(a) and the other as PWM(b). When one phase with PWM(a) has a positive current flow to the load, it becomes to the upper voltage level after the dead-time. On the contrary, the other phase with PWM(b) has a negative current flow to the load, it becomes to the lower voltage level after the dead-time. When the phase(a) has a negative current flow, the situation is quite opposite. In both cases the two phases operates complementarily and leakage currents from switches are suppressed.

V. EXPERIMENTAL REULTS

A. Output Voltage and Current Waveforms

Experimentally measured waveforms of phase voltages v_1, v_2, an output voltage $v = v_1 - v_2$, a filter inductor current i_L are compared between the conventional result in Fig.9 and the proposed one in Fig.10. The proposed result has a very little zero voltage component and noise components are considerably reduced. However only rip-

1616

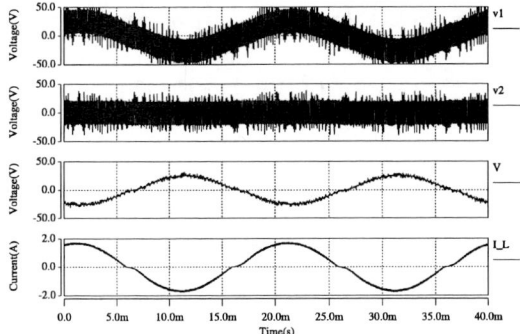

Figure 9. Measured load voltage and current (conventional).

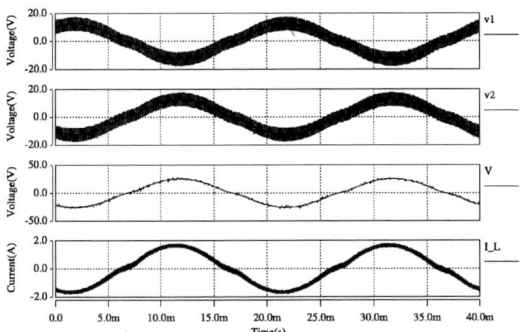

Figure 10. Measured load voltage and current (proposed).

ple components of the current are twice because the output voltage does not have any zero voltage level due to the complementary switching. These should be reduced by other measures; increase of switching frequency or filter component values.

B. Reduction of Leakage Current from Load

Measured common-mode voltage and leakage current waveforms by the proposed method are shown in Fig.11. Comparing with the conventional result of the pulse shape in Fig.5(e), the common-mode voltage v_{common} is considerably suppressed. The leakage current is almost zero, but not perfectly zero. This is caused by a small unbalance of the divided filter inductors. However the suppression effect is enough for the leakage current from the load.

C. Reduction of Leakage Currents from Switches

Measured phase voltages and leakage current waveforms from switches are shown in Fig.12. Comparing the conventional result in Fig.5(g), the leakage current is considerably reduced. Also in this case, the leakage current is almost zero, but not perfectly zero. This is caused by small differences of turn-on and turn-off timings of the IGBTs. However the suppression effect is enough for the leakage current from the switches.

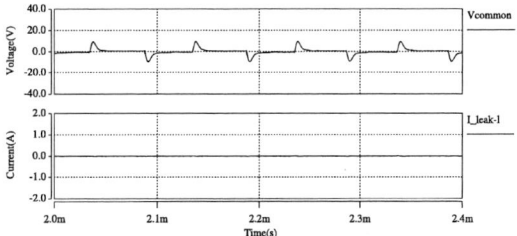

Figure 11. Common-mode voltage and leakage current (proposed).

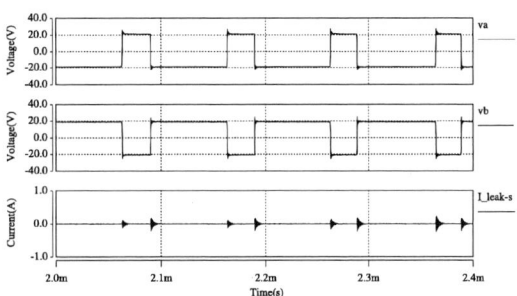

Figure 12. Phase voltage and switch leakage current (proposed).

VI. CONCLUSIONS

This paper proposed a complementary arm switching technique which can suppress leakage currents from a load and switches in a single-phase PWM inverter. The principle is so simple that the inverter is operated only complementarily between two arms. This eliminates neutral voltage drifts of the load by keeping the common-mode voltage zero. Furthermore parasitic capacitor currents circulate only inside the inverter body because the switching times are synchronous between the two arms. This is valid even when there are dead-times in arm switching. These effects were validated through simulation by Saber and experiment by a DSP-based system with switching frequency of 10kHz.

REFERENCES

[1] H. Hasegawa, T. Doumoto, H. Akagi : "A three-phase voltage-source PWM inverter system characterized by sinusoidal output voltage with neither common-mode voltage nor normal-mode voltage -design and performance of a passive EMI filter- ," *T. IEE Japan*, Vol.122-D, No.8, pp.845-852 (2002)

[2] W. Khan-ngern, Y. Prempaneerach : "Reduction emission for USPs using passive EMI filters," *IEEE Trans. Ind. Applications*, pp.189-192 (1998)

[3] D. Cochrane, D.Y. Chen:"Passive cancellation of common-mode noise power electronic circuits," *IEEE Trans. on Power Electronics*, Vol.18, No.3, pp.756-763(2003)

[4] S. Ogasawara, H. Ayano, H. Akagi : "An active circuit for cancellation of common-mode voltage generated by a PWM inverter," *IEEE Trans. Power Electronics*, Vol.13, No.5, pp.835-841(1998)

[5] A. Rao, T.A. Lipo :"A modified single phase inverter topology with active common mode voltage cancellation, " *IEEE Power Electronics Specialist Conference Record*, Vol.30, No.2, pp.850-854(1999)

[6] C. Yo, K. Seung : "A new active common-mode EMI filter for PWM inverter", *IEEE Trans. Power Electronics*, Vol.18, No.6, pp.1309-1314(2003)

2006 5th International Power Electronics and Motion Control Conference

Analysis and Design of a Novel Dual Secondary Winding and Dual Power Bridge High Frequency Link Inverter

Zhang Zhe, Zhang Chunjiang, Wu Weiyang, Gu Herong and Shen Hong
Yanshan University, Qinhuangdao, China
zhangzhe@ysu.edu.cn

Abstract—a novel dual secondary winding and dual power bridge high frequency link (HLF) inverter is discussed in this paper Single-stage power conversion, standard half-bridge connection of devices, soft-switching for all the power devices (ZVS or ZCS), low conduction loss, simple bipolar combined phase-shifted control, and high efficiency are among the salient features of the HLF inverter. The principles of circuit operation, PWM control and synthesis, topological extension are discussed. The theoretical analysis is presented and the realization of power stage and close loop control is described. The experimental results are given to validate the effectiveness of analytical result and operation principle of the circuit.

Keywords-inverter; high frequency link; soft-switching

I. INTRODUCTION

The bi-directional high frequency link inverters are used widely in the UPS system, battery-backup stand-alone inverter systems, and alternative energy systems such as photovoltaic applications, specially the power supply systems of recreational vehicle and marine boats that need the inverters having the features of higher reliability, smaller volume, higher efficiency and lower cost [1].

In the past decade, the bi-directional HFL inverters have undergone a great development worwide. The existing HFL inverters can be classified into three major categories from the view of power processing. One is single stage bi-directional circuit, which is mainly based on cycloconverter topology [2-3]. It has the feature of just single power conversion stage with the bidirectional switches, and without the large energy storage elements. But it has a inherent problem of high voltage surge on the bidirectional switches [4]. The second one is the two stage power converter configuration, which usually has two or three power converters in cascade and needs the DC-link [5]. The more power stages relatively leads to more power loss and lower reliability. The third one is quasi-single-stage bi-directional inverter/charger proposed by Virginia Power Electronics Center in 1998, and with the voltage clamp schemes all switch devices can achieve soft-

---
[a] This paper is supported by the Key Program National Nature Science Foundation of China (No. 50237020).

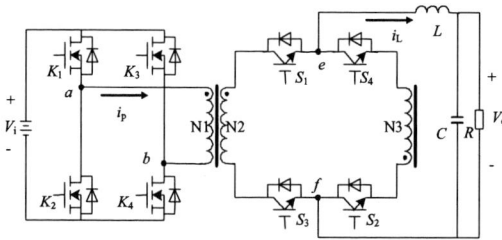

Figure 1. Dual Secondary Winding and Dual Power Bridge HFL Inverter

switching [6].

In this paper a novel soft-switched single-stage bi-directional HFL inverter topology is discussed with the bipolar combined phase-shifted SPWM control schemes. Analysis of the operation principles in four quadrants is performed. Based on the analytical result, the design consideration is given and also illustrated in a design. The close loop control for the whole system is is also discussed and described. Finally, the experimental results from prototype (24 V dc input and 110 V ac output, 500 W output power) are presented.

II. ANALYSIS OF THE TOPOLOGY

A. Dual secondary winding and dual power bridge HFL inverter topology

The HFL inverter topology is shown in Fig.1. It consists of a full bridge high frequency inverter, two standard half-bridges to achieve the high frequency voltage polarity inversion to get the sinusoidal output voltage on the ac load and a high frequency transformer with two independent secondary windings. Because of the two secondary winding, the secondary side of the transformer increases a new degree of freedom, and the circuit structure of the secondary side is more flexible. So the proposed topology can be extended to a topology family according to the different requirements.

The proposed dual secondary winding and dual power bridge high frequency link inverter has the following salient features compared to other existing circuit topologies furnishing the same functionality:

• Simplified SPWM pattern and control;
• Simplified the circuit structure and the number of

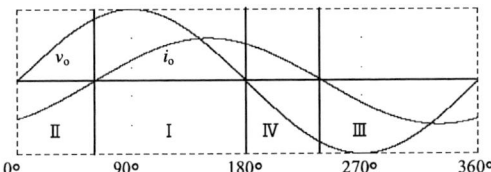

Figure 2. Four-quadrant operation

switches is small;
•Standard half-bridge circuit connection and easily achieved modularization;
•Bi-directional switches are not necessary due to two independent secondary windings;
•Easier circuit layout.

B. Operational Principles

The high-frequency operation principles of the dual secondary winding and dual power bridge HFL inverter in each operation mode will be discussed in this section. Because the load can assume any power factor, either leading or lagging, the circuit needs to operate in all the four quadrants in the V_o-I_o plane during an output line cycle as indicated in Fig. 2. In this paper the bipolar combined phase-shifted control method is used to make the inverter realize four-quadrant operation. It is clear that the output current and voltage are both positive in quadrant I, and that v_o is positive and i_o is negative. Circuit operations in quadrants III and IV are exact replicas of those in quadrants I and II. So only the operations in quadrants I and II need to be analyzed.

In the following analyses, it is assumed that all the power devices are ideal, and that the inductor of output filter, L_f, is much higher than leakage inductor of the transformer, L_k, so the inductor current, I_L, can be considered constant during circuit commutations.

1) Operating Principles in Quadrant I: The PWM pattern and the main operating waveforms are shown in Fig. 3 for operation in quadrant I. The equivalent circuits in each interval during a half high frequency cycle are drawn in Fig. 4.

[t_0~t_1]: Before t_0, switches K_1 and K_4 are on, the primary current, i_p, flows through the K_1, the primary winding of the transformer N_1 and K_4. I_L flows through the secondary winding N_2, S_1 and the antiparallel diode of S_3, D_3. The voltage on the output filter, u_{ef}, is positive, the dc source delivers the energy to the load.

At t_0, S_2 is closed under ZCS and K_1 and K_4 are turned-off. under ZVS. The u_{ab} is:

$$u_{ab} = V_i - \frac{nI_L}{C}t. \qquad (1)$$

Where $C=C_1+C_2$, and n is turn ratio of the transformer.

[t_1~t_2]: At t_1, u_{ab} reduces to zero and increases continually in negative direction. Because the I_L can reflect to the primary side, $i_p=nI_L$, is approximately constant.

[t_2~t_3]: C_1 and C_4 are charged to V_i, so C_2 and C_3 are discharged to zero at t_2. i_p starts to freewheel through D_{K2}, D_{K3} and N1. I_L flows through N_2, S_1 and D_3 so that

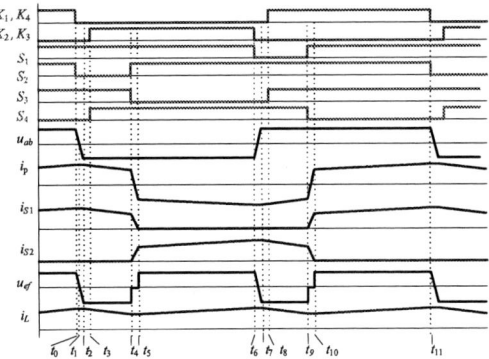

Figure 3. PWM pattern and key waveforms in quadrant I.

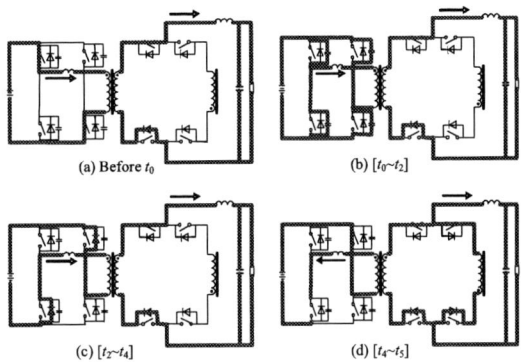

Figure 4. Equivalent circuits in a half-frequency cycle in quadrant I

$u_{ef} = nu_{ab} = -nV_i$ the load feedbacks the energy to dc source.

[t_3~t_4]: Both K_2 and K_3 are turned on under ZVS at t_3 without interruption to the freewheeling path.

[t_4~t_5]: At $t4$, S_3 is turned off under ZVZCS. At the same time S_2 is turned on under ZCS, and according to the voltage polarity of N_3 the current in the secondary side begins to commutate from S_1 to S_2. So u_{ef}=0, u_{ab}=0. The dc voltage is exerted on the transformer leakage inductance L_k oppositely and i_p starts to ramp down, described as:

$$\frac{di_p}{dt} = -V_i / L_k. \qquad (2)$$

when i_p reverses its direction, it starts to flow through K_2 and K_3.

After the current in S_1 is reduced to zero and the current in S_2 is reached to I_L the next half high frequency cycle begins.

2) Operating Principles in Quadrant II: The PWM pattern and high frequency waveforms are shown in Fig. 5, while the equivalent circuits in each interval are shown in Fig. 6.

[t_0~t_1]: Before t_0, are on, u_{ab}=-V_i, and i_p is negative and flows through K_2, N_1 and K_3. While I_L flows through N_2, D_1 and S_3. At t_0, K_2 and K_3 are turned off under ZVS, and i_p starts to charge C_1, C_4, and discharge C_2, C_3.

1619

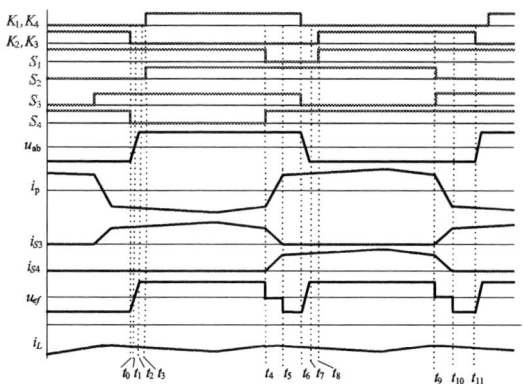

Figure 5.   PWM pattern and key waveforms in quadrant II.

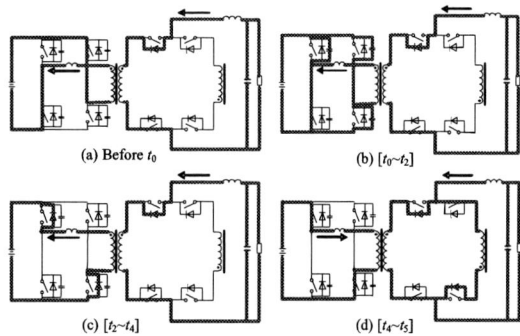

(a) Before t_0 (b) [t_0~t_2]

(c) [t_2~t_4]   (d) [t_4~t_5]

Figure 6.   Equivalent circuits in a half-frequency cycle in quadrant II

[t_1~t_2]: At t_1, u_{ab} increased to zero and resumed to increase in positive direction. The voltage on S_2 begins to increase from zero and the current paths are not changed.

[t_2~t_3]: C_1 and C_4 are charged to V_i, so C_2 and C_3 are discharged to zero at t_2. i_p starts to freewheel through D_{K2}, D_{K3} and N1. I_L flows through N_2, S_1 and D_3 so that $u_{ef} = nu_{ab} = -nV_i$ the load feedbacks the energy to dc source.

[t_3~t_4]: Both K_2 and K_3 are turned on under ZVS at t_3 without interruption to the freewheeling path.

[t_4~t_5]: Turns off S_1 under ZVZCS. S4 is turned on under ZCS, and the current in S_4 begins to increase and the current in S_3 starts to reduce. So i_p begins to reset by the dc voltage. Finally, i_p reverses its direction, and i_p starts to flows through K_1, K_4 and N1. During this interval the secondary side of the HF transformer is equivalent to be short-circuit condition, so $u_{ef} = 0$.

After the current in S_4 is reached to I_L the next half high frequency cycle begins.

C.   SPWM Control Method

The bipolar combined phase-shifted SPWM with the signal of output current polarity ensures transition between different quadrants as shown in Fig. 7. The modulation wave is the output of the current regulator rectified and biased, and the carrier signal is output of the

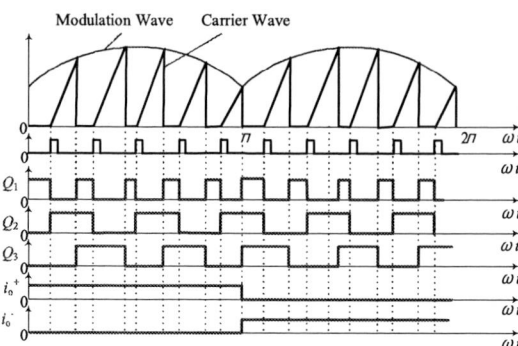

Figure 7.   Bipolar combined phase-shifted control scheme

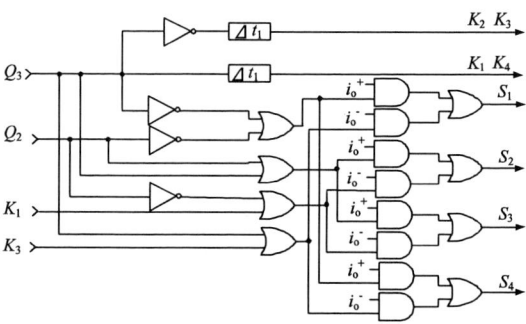

Figure 8.   Synthesis of logic signals

constant-frequency integration. Q_2 and Q_3 are modulated by the SPWM phase-shifted scheme. The driving signals for all the main switches can be synthesized with simple circuits as shown in Fig. 8. The signal of Q_3 is obtained by dividing trailing edge of the carrier signal and is 50% fixed duty cycle. Moreover, 50% fixed duty-cycle switching signals are applied to all the switches on the high frequency inverter bridge with proper delay, Δt_1, inserted between the complementary switch-pairs. With some simple logic operations of Q_2, Q_3, K_1 and K_3 we can generate the driving signals for switches on secondary side.

III.   SYSTEM DESIGNS

The block diagram of the whole HFL inverter system is shown in Fig. 9.

A.   Designs of the Power Stage

1)   Transformer turn ratio n: For the high frequency transformer turn ratio should guarantee the voltage range. To ensure appropriate inverter output voltage scaling, the turns ratio n of the transformer needs to be selected such that $n(V_{imin} - V_{FET})D_M \geq V_o^p = \sqrt{2}V_o$, where D_M is the maximum duty ratio, V_{imin}, V_{FET}, and V_o^p are the minimum input battery voltage, on-drop voltage on the primary MOSFET switches, and peak output ac voltage, respectively. If D_M=0.9, V_{FET}=0.5 V, and V_o=110 V, then $n \geq 8$. In reality, $n = 10$ is selected.

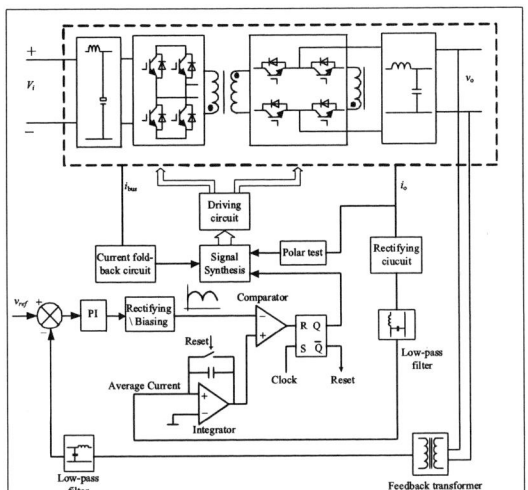

Figure 9. Block diagram for the whole system

2) Election of the switch devices: The high voltage justify the use of IGBT devices for the secondary side switches. Because of the structure of two independent secondary windings of the transformer, standard low cost half-bridge IGBT modules can be used. On the primary side, the switches are switched under ZVS and low voltage MOSFETs can be adequately used.

3) Output filter: The output filter of the power inverter is used to smooth out the waveforms generated from the DC/AC stage. The signal contains many unwanted harmonic frequencies including multiples modulation switching frequency. The cut-off frequency f_c is described as:

$$f_c = \frac{1}{2\pi\sqrt{LC}} = \frac{R}{2\pi L} \tag{3}$$

Where $R = \sqrt{L/C}$.

So from (3) we can get:

$$L = \frac{R}{2\pi f_c}. \tag{4}$$

$$C = \frac{L}{R^2} = \frac{1}{2\pi f_c R}. \tag{5}$$

When the fundamental frequency of output voltage is 50 Hz, f_c is chosen to be 400~1000 Hz generally. And R is $(0.5\sim0.8)R_L$, where R_L is the load resistance. If $R = 0.5R_L = 0.5 \times 16.1 \approx 8\,\Omega$ and f_c=1 kHz, we can get: L_f=1.4 mH and C=10 μ F:

B. Close loop control scheme

This paper proposed a novel control strategy with output voltage and inductance current double loop, shown in Fig. 9. Because the one cycle control can restrain the dynamic and steady-state errors in one on-off cycle, it be used to the inner loop of the inductance current and to realize the PWM scheme. A instantaneous outer voltage feedback loop with PI controller is used to stabilize the sinusoidal voltage.

IV. EXPERIMENTAL RESULTS

A laboratorial prototype based on dual secondary winding and dual power bridge HFL Inverter is designed according to the following specifications:

Input battery voltage V_i=20~28 V (nominal value V_i=24 V);

Output ac voltage V_o = 110 V, 50 Hz;

Output capacity P_o = 500 W;

Output voltage THD<5%.

The main power stage parameters are as follows:

$K_1\sim K_4$: FQA65N20, 200V/65A, TO-3P, MOSFET;

$S_1\sim S_4$: Toshiba MG50J2YS50 600V/50A IGBT Module;

The transformer T: PC40 materials, ETD-59 core, N_1 : N_2 : N_3 = 1 : 10 : 10;

Switching frequency: f_s =30 KHz;

Primary leakage inductance L_k=1 uH.

Fig. 10 shows output of constant-frequency integrator is low level while the Q_1 is high. Fig. 11 shows the logic relationship between output of outer loop regulator and output of constant frequency integrator in a large time range. From Fig. 12 it is clear that voltage surge on collector-emitter of switch in secondary side is restrained effectively. The bipolar SPWM voltage on the output LC filter, u_{ef}, is shown in Fig. 13. The output voltage is shown in Fig. 14

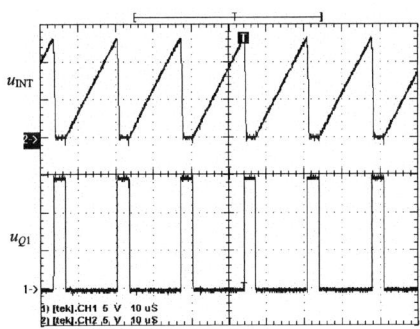

Figure 10. Waveforms of u_{INT} and u_{Q1} (5 V/div)

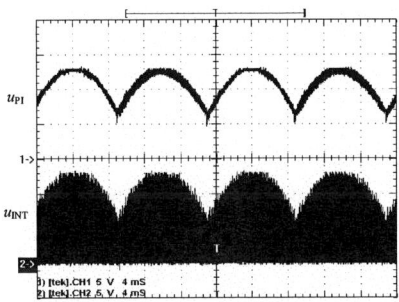

Figure 11. Waveforms of u_{PI} and u_{INT} (5 V/div)

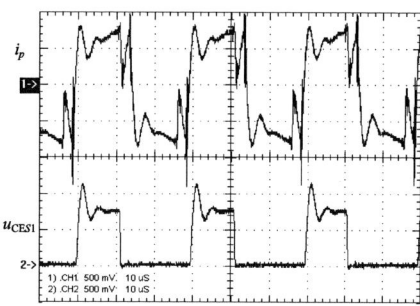

Figure 12.   Waveforms of i_p and u_{CES1} (CH1: 20 A/div;
CH2: 250 V/div)

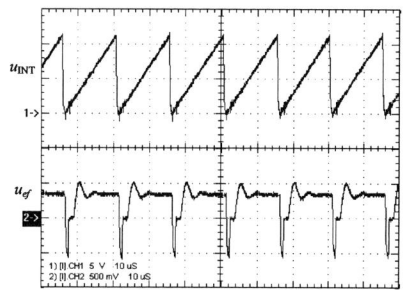

Figure 13. Waveforms of u_{INT} and u_{ef} (CH1: 5V/div; CH2:
250V/div)

V.   CONCLUSION

A novel single-phase soft-switched single-stage bi-directional high frequency link inverter topology is discussed in this paper. With the bipolar combined phase-shifted control method it can realize four-quadrant operation. The main circuit structure, detailed operation modes, PWM control scheme and synthesis, the closed-loop control are investigated. The feasibility for the novel HFL inverter with the proposed control scheme has been verified

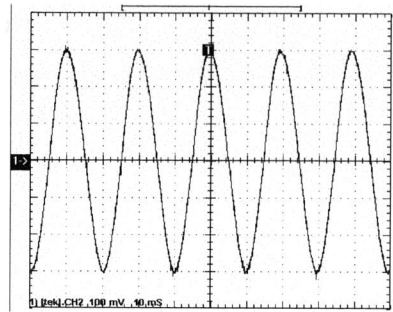

Figure 14.   Waveforms of output voltage in resistant load
(50 V/div)

by the experimental results on a on a laboratorial prototype.

REFERENCES

[1]   W. Wu, C. Zhang, X. Sun and X. Yang, "Research on the topologies of medium and small power high frequency link inverter," Proceedings of Ninth Chinese Power Electronics Seminar, Sept., 2004, pp. 24-27

[2]   I.Yamato, N.Tokunage, Y.Matsuda, H.Amano and Y.Suzuki, "High frequency link DC-AC converter for a new voltage clamper," *IEEE PESC*, 1990, pp. 749-756.

[3]   H. Pinherio, P. Jain and G. Joos, "Zero voltage switching series resonant based UPS", *IEEE PESC*, 1998, pp. 1879-1885.

[4]   Tazume K., Aoki T., Yamashita T., "Novel method for controlling a high-frequency link inverter using cycloconverter techniques," *IEEE PESC*, 1998, pp. 497-502.

[5]   Salam Z., Ramli, Z., "A DC-DC type bidirectional high frequency link inverter using center-tapped active rectifier," *IEEE IECON, 30th Annual Conference of IEEE*, 2004, pp. 47-50.

[6]   Kunrong Wang, Fred C. Lee, and Wei Dong, "A new soft-switched quasi-single(QSS) bi-directional Inverter/Charger", *Proceedings of the Sixteenth Annual VPEC Seminar*, Sept., 1998, pp. 55-62.

2006 5th International Power Electronics and Motion Control Conference

Research of Complex Fuzzy Control on-off Magnetism Team Motor Speed-Adjusting System

Zhao Ming-fu, Chen Yan, Zhang Zhi-yuan, Dong Chun, DongYu

Department of Electronic Engineering Chongqing Institute of Technology

Chongqing 400050 P.R. China

zmf@cqit.edu.cn

Abstract—Usually on-off magnetism team motor speed-adjusting adopts traditional linear control algorithm (such as PID). In modern control system, it is nonlinear, multivariable, multi-coupling, high-dimension and the object is time-varying, uncertain and difficult to describe by math model because of the complex control object. The traditional algorithm is difficult to meet the need of on-off magnetism team motor speed-adjusting. The paper takes the fuzzy control technique into on-off magnetism team motor speed-adjusting system, designs a complex fuzzy control motor speed-adjusting system. The system composed to be a real-time data collection and control system by computer, A/D, D/A and sensor. It adopts on-off magnetism team motor (rating power 11kw, rating rotate speed 1500rpm), analog load (0～10kw) to take a platform-frame experiment and make the system precision and dynamic performance achieve the need. Actually it is turn out to be feasible to use motor speed-adjusting, at the same time, have a good effect. After we take it, the control precision has been improved and had a good effect on improving the system quality.

Keywords-On-off magnetism team motor; Complex fuzzy control; speed-adjusting system

I. INTRODUCTION

Usually On-off magnetism team motor speed-adjusting adopts traditional linear control algorithm (such as PID). In modern control system, it is nonlinear, multivariable, multi- coupling, high-dimension and the object is time-varying, uncertain and difficult to describe by math model because of the complex control object. The traditional algorithm is difficult to meet the need of on-off magnetism team motor speed-adjusting.

The paper takes the fuzzy control technique into on-off magnetism team motor speed-adjusting system, designs a complex fuzzy control motor speed-adjusting system.

The system composed to be a real-time data collection and control system by computer, A/D, D/A and sensor. Actually it is turn out to be good effect.

II. SYSTEM WHOLE STRUCTURE

Intelligence speed fuzzy controller; analog current adjuster; signal drive; power amplifier; electromotor; speed current sensor compose double closed-loop speed-adjusting system which have rotate speed and current. It is different from plain double closed-loop speed-adjusting system. This system's speed-adjuster adopts an intelligence digital adjuster.

Fig1 is the system's control structure, namely system adopts double closed-loop speed-adjusting system structure which have rotate speed and current. Using current adjusting loop as system's inner loop; adopting fuzzy control and warp complex of fuzzy digital adjuster

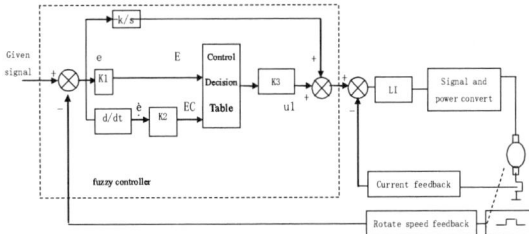

Figure 1. Control system block diagram

to realize control; with control –electromotor compose speed loop.

Considering fuzzy control is a droop control system essentially, can't eliminate error, so the control precision will be effect. This system adopts complex control structure and adds proportion integral part. The controller contains basic fuzzy control and a warp integral control, the output of controller equal to the sum of two components, namely:

$$u = u_1 + u_2$$

Bring complex structure in order to reduce swing around zero, bring integral part in order to improve precision of system, so eliminate error comes from interrupt.

III. STRUCTURE AND ARITHMETIC OF BASIC FUZZY CONTROL

Fig2 is the system structure of basic fuzzy control. The system is the typical double-input fuzzy control system. Its input contains a warp which between given speed signal and feedback signal and the rate of warp change, according to the fuzzy knowledge base, we carry out fuzzy reasoning, then fuzzy judge, at last get the precision control variable.

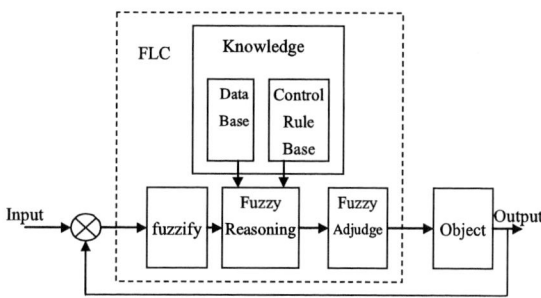

Figure 2. Structure of basic fuzzy control system

A. Fuzzifying

Considering the large warp bound of electromotor rotate speed and high precision, we order the discoursing universe of warp variable and warp change rate to be 15 ranks（±7，±6，±5，±4，±3-，±2，±1，0）。

Suppose the change range of input x= [a，b]，

Integer discoursing universe

N={-n，-n+1，…,-1,0,1，…,n}，

So the calculate formula quantization gene k1、k2 is:

$$k = 2n/(b-a)$$

The fuzzy controller is two dimensions input controller, namely the warp e which between given signal r and speed feedback signal and the change rate of warp ec, after fuzzifying the corresponding fuzzy vector E、EC is the discoursing universe of e and ec mapped to the fuzzy subclass individually. Namely:

$$E = k_1 \times e$$
$$EC = k_2 \times ec$$

Ensure membership function and linguistic variable

We make the fuzzy linguistic variable of wrap variable, wrap change rate and control quantity divided into 7 ranks（NB、NM、NS、ZE、PS、PM、PB）；the shape of the fuzzy subclass' membership function adopts triangle wave and evaluate E、EC、U. The definition of in-out fuzzy subclass' membership function is indicated in fig3.

B. Design of fuzzy control rule

Based on the expert or runner's experience and skill, we conclude the control rule of fuzzy controller, there are many ways can produce the control rule, here we use artificial control experience, namely conclude by system's step-response signal. Look at the table 1; it is the fuzzy control rule.

TABLE I.
FUZZY CONTROL RULE

U EC / E	NB	NM	NS	ZE	PS	PM	PB
NB	PB	PB	PB	PB	PM	ZE	ZE
NM	PB	PB	PB	PB	PM	ZE	ZE
NS	PM	PM	PM	PM	ZE	NS	NS
ZE	PM	PM	PS	ZE	NS	NM	NM
PS	PS	PM	ZE	NM	NM	NM	NM
PM	ZE	ZE	NM	NB	NB	NB	NB
PB	ZE	ZE	NM	NB	NB	NB	NB

Table I indicates a set of control rule; show the fuzzy relationship of fuzzy system. We can get the fuzzy relationship matrix R according to the control rule. It can be like this:

$$IF \ A_i \ AND \ B_j \ THEN \ C_{ij}$$

A_i、B_j、C_{ij} is the corresponding linguistic variable of e、ec and u. The control rule base all has 49.

C. Fuzzy reasoning

There are many ways of fuzzy reasoning arithmetic; usually we adopt CRI's Mandani, according to the control rule, we can get fuzzy relation matrix, then use different synthesizing measures to calculate the result. Table1 indicates a set of control rule, it shows the fuzzy relationship, and we can get the system's fuzzy output according to the matrix operation and synthesizing operation.

D. Fuzzy judgment

Suppose the output's change range u= [p，q]，Integer discoursing universe

N= {-n，-n+1，…,-1, 0, 1, …, n}，then proportion gene: $k_3 = (q-p)/2n$

This system adopts method of barcenter namely weighted average method to realize fuzzy judgment Suppose the discoursing universe gene of the linguistic variable's output control quantity U: $u_1, u_2, \cdots, u_n$，$\mu(i)$ is the subjection degree of output element, then the control quantity is（U）

$$u = \frac{\sum_i^n \mu(i) \times u_i}{\sum_i^n u_i}$$

Exact control quantity: $u_\Delta = k_3 \times U$

IV. ARITHMETIC OF COMPLEX FUZZY CONTROL SYSTEM'S REAL TIME CONTROL

According to the fuzzy relationship between control rule and arithmetic language we can get fuzzy relation matrix R, fuzzify aimed at warp e and differential coefficient of warp ec, then according to the fuzzy reasoning synthesizing principia we can get the fuzzy ensemble U of electro motor's control quantity. Then

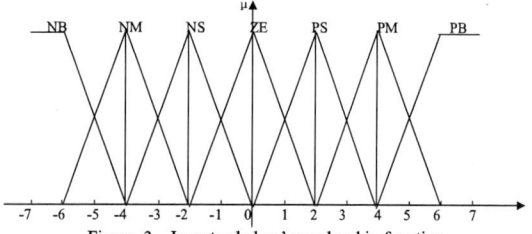

Figure 3. In-out subclass' membership function

adopt method of barcenter to calculate, at last get the fuzzy control decision-making table. After testing debug the table in practice and more improve, we put the table into EPROM. In real time control, the tested rotate speed wrap and rotate speed wrap variable multiply their quantization gene K1、K2 respectively to get the e and ec's fuzzy quantity E and EC, according to the E and EC, we can get the fuzzy quantity U by consulting control decision-making table. The basic fuzzy control quantity u be got by U multiply K3.

We get the control quantity u2 by proportion and integral of speed warp, the basic fuzzy control quantity u adds u2 is complex control quantity u. It is difficult to calculate control table by off-line way. We can use software to program and set a fuzzy reasoning to calculate. The fuzzy exploitation setting which is made by Institute of Mechanics is used in this system, set the structure of fuzzy controller, calculate control table by off-line.

V. STRUCTURE DESIGN OF SYSTEM

A. The structure of control system

The system adopts on-off magnetism team motor as its execute structure. The control circuit contains starting circuit, given circuit, given integral, speed-adjusting circuit, current--adjusting circuit, pulse-modulating, logical controlling, drive-amplifier, power amplifier, speed sensor tester, current sensor tester. As Fig4 shows:

B. The structure and function of complex fuzzy controller

The complex fuzzy controller as the system's core adopts digital computer system with one SCM. According to the arithmetic and D/A transition, it gets the analogical control signal. It is easy to realize the complex control.

The program of controller mainly contains initialization, data-collecting, data-processing, data-outputting, etc.

Using T0 as timer/counter, setting count value by interrupt mode which makes overflow one time every

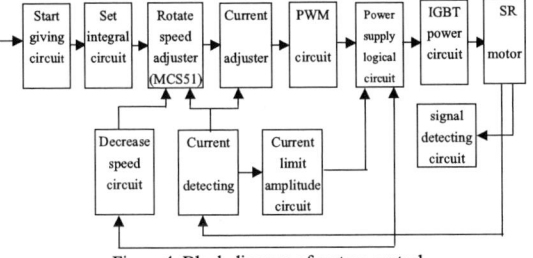

Figure 4. Block diagram of system control

10ms and produced one interrupt, starting A/D ,then finish data-collecting, data-processing, data-outputting, etc mainly by main program and interrupt subprogram. The main program firstly sets work unit for given current and rotate speed signal in memory, modifies INT0 interrupt serve subprogram, allows interrupt and initializes. User

interrupt subprogram finishes data-collecting, data-processing, and data-outputting.

Software adopts data wave-filtrating, dead-setting to eliminate signal disturb and remove vibration of system's zero-returning. It also adds watch dog circuit and protection program in case of special circumstances

VI. CONCLUSION

The system adopts on-off magnetism team motor (rating power 11KW, rating rotate speed 1500rpm), analogical carrying load （0～10kw） to start frame-platform experiment. The system precision and dynamic capability will meet to the need by adjusting k1、k2、k3、k. Fig5 is the given step signal. Apparently, the system's response is better than third-order best system which is adjusted by PID.

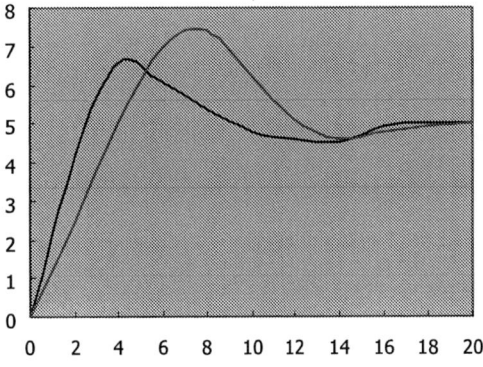

Figure 5. System response under step signal

It is turn out to be doable to use fuzzy control on speed-adjusting of motor. But the basic fuzzy control system is a droop system and can't eliminate steady-state error, so the high-powered controller must improve their system structure. This complex fuzzy control system structure improved the system quality.

REFERENCES

[1] Wan Peilin, Fuzzy technology basic and application, Chongqing, Chongqing University publishing company, 1995, 10

[2] Zhang Zhiyuan, using fuzzy NN to finish design and simulation, Journal of Chongqing Institute of Technology,2000, 1

[3] Yu Yongquan, Zeng Bi. Fuzzy logical control of SCM, Bejing University of Aeronautics and Astronautics publishing company, 1995

[4] Zhang Yigang , Apply design of SCM MCS-51, Harbin Institute of Technology publishing company, 1997

2006 5th International Power Electronics and Motion Control Conference

A New BLDC Motor Drives Method Based on BUCK Converter for Torque Ripple Reduction

Zhang Xiaofeng ,Lu Zhengyu

Department of Electrical Engineering

Zhejiang University, Hangzhou,310027, P.R.China

Zxfeng001@163.com

Abstract—**This paper presents a comprehensive analysis on torque ripples of brushless dc motor drives in conduction region and commutation region. A novel method for reducing the torque ripple in brushless dc motors with a single current sensor has been proposed by adding BUCK converter in the front of 3-phase inverter.In such drives, torque ripple suppression technique is theoretically effective in commutation region as well as conduction region. Effectiveness and feasibility of the proposed control method is verified through experiments.**

Keywords—*Brushless dc motor; torque ripple; conduction region; commutation region*

I .INTRODUCTION

The brushless direct current (BLDC) motor has high torque,compact size, and high efficiency. Therefore, the BLDC motor is widely used in computers, household, industrial products and automobiles.However, the BLDC motor has a disadvantage of high cost compared with the direct current (DC) motor because it is necessary to use an inverter and controller to remove a brush of DC motor.

As known to all,brushless dc motor with trapezoidal back-EMF have been widely used due to their high power density and easy control method. Further, basic trapezoidal brushless dc motors make it possible to use a single dc-link current sensor to regulate the phase current flowing through two motor phases.But torque ripple generated in conduction region and commutation region is the main drawback of BLDCM, however, which deteriorates the precision of BLDCM[1].In this paper, the proposed control scheme eliminates torque ripples in conduction region and attenuates torque ripples in commutation region .Meanwhile, this scheme improves the veracity of a single dc-link current sensor.

II .PRINCIPLE AND ANALYSIS

Generally a BLDCM has two operation region: conduction region and commutation region.In the conduction region,with position of rotor selected 2-phases are conducted.On the other hand commutation region is to be transient region which converts from the current conduction into next one,is relatively shorter than

conduction region,and 3-phases(rising phase,decaying phase, non-commutation phase)are all conducted. Conduction and commutation appears six times per one electrical rotating of the rotor[2].

The novel proposed circuit consists of the step-down BUCK converter that regulates the amplitude of the current and the inverter which is controlled in such a way as to supply 3-phase rectangular current with a pulse width of $120°$ electrical degree to the motor. Fig.1shows the conventional and new proposed circuit configuration respectively[3][4][5]. The motor is assumed symmetrical and salient effect is neglected. The phase inductance denoted by L_a , L_b , L_c are constant. E_a , E_b , E_c represent three phase back EMF respectively. i_a,i_b,i_c are stator current in phase A,B,C respectively. R_a ,R_b , R_c are phase resistance respectively. N denotes the neutral node of the motor windings with reference to groud. R_{sample} is the current sampling resistance[6].

Fig.2 shows the normal and new type of PWM pulse patterns in 2-phase feeding scheme waveforms.The following analysis is based on the region of 6-1 shown in Fig.2 .

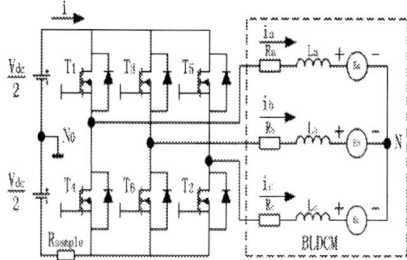

Fig1(a). The conventional circuit configuration

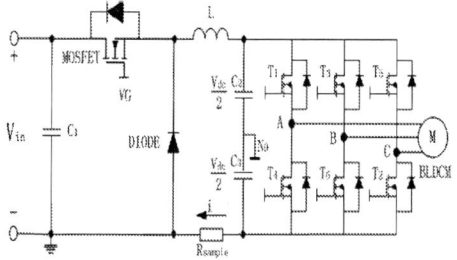

Fig1(b). The new proposed circuit configuration

1-4244-0448-7/06/$25.00 ©2006 IEEE   1626

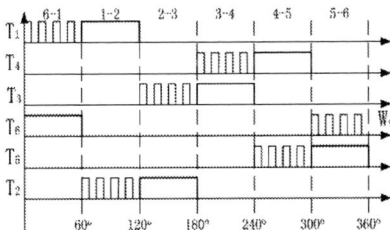

Fig2(a). The normal modulation of PWM-ON pattern

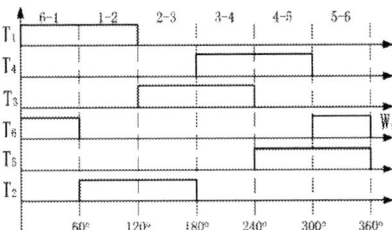

Fig2(b). The new modulation pattern

The torque is given by Eq.(1).

$$T_e = \frac{1}{\omega_m}\left(e_a i_a + e_b i_b + e_c i_c\right) \qquad (1)$$

where T_e is the belectromagnetic torque of motor, ω_m is mechanical angular velocity of rotor. e_a, e_b, e_c are the BEMF of 3-phase, i_a, i_b, i_c are the current of 3-phase. In order to only show off the torque ripple due to supply, ideal trapezoidal EMFs with a 120° constant plateau are considered.

According to the switching conditions of the inverter switches, the voltage equations related to the switch-on and switch-off intervals in the normal mode can be described as Eq.(2).

$$\begin{bmatrix} V_{ka} \\ \dfrac{V_{dc}}{2} \\ V_{kc} \end{bmatrix} = \begin{bmatrix} R & 0 & 0 \\ 0 & R & 0 \\ 0 & 0 & R \end{bmatrix} \bullet \begin{bmatrix} i_a \\ i_b \\ i_c \end{bmatrix} + \frac{d}{dt}\begin{bmatrix} L & 0 & 0 \\ 0 & L & 0 \\ 0 & 0 & L \end{bmatrix}\begin{bmatrix} i_a \\ i_b \\ i_c \end{bmatrix} + \begin{bmatrix} e_a \\ e_b \\ e_c \end{bmatrix} + \begin{bmatrix} V_{NN_0} \\ V_{NN_0} \\ V_{NN_0} \end{bmatrix}$$

$$(2)$$

Where V_{dc} is the dc-link voltage, $L = L_s - M$ (L_s is self inductance and M is mutual inductance),V_{ka}, V_{kc} are the voltages of phase A and C,S is the switching function(1 denotes switch-on,0 denotes switch-off), N_0 and N are shown in Fig.1.V_{NN_0} is derived from (2) as follows.

$$V_{NN_0} = \frac{1}{3}\left(V_{ka} - V_{kc} + \frac{1}{2}V_{dc}\right) - \frac{1}{3}\left(e_a + e_b + e_c\right)$$

$$(3)$$

Combining (2) and (3),the currents of 3-phase can be described as Eq.(4).

$$\begin{cases} i_a = \left(\dfrac{2}{3}V_{ka} + \dfrac{1}{3}V_{kc} - \dfrac{1}{6}V_{dc} - \dfrac{4}{3}E_m\right)\dfrac{t}{L} + i_{a0} \\[2mm] i_b = \left(-\dfrac{1}{3}V_{ka} + \dfrac{1}{3}V_{kc} + \dfrac{1}{3}V_{dc} + \dfrac{2}{3}E_m\right)\dfrac{t}{L} + i_{b0} \\[2mm] i_c = \left(-\dfrac{1}{3}V_{ka} - \dfrac{2}{3}V_{kc} - \dfrac{1}{6}V_{dc} + \dfrac{2}{3}E_m\right)\dfrac{t}{L} + i_{c0} \end{cases}$$

$$(4)$$

Where E_m is BEMF constant and i_{a0}, i_{b0}, i_{c0} are steady-state value of phase-current in the conduction region. It is assumed that the motor winding resistance is neglected,and e_a maintains the value of E_m and e_b, e_c hold $-E_m$.

In the conduction region(taking"6-1" for example),the torque equation is derived from(1).In principle, the torque in BLDCM is proportional to the current amplitude in the non-commutation phase winding.

$$T_e = \frac{1}{\omega_m}\left[E_m \bullet I_a + (-E_m)\bullet(-I_a) + e_c \bullet 0\right] = \frac{2E_m}{\omega_m}I_a$$

$$(5)$$

In the commutation region(taking from "6-1" to "1-2" for example), the torque equation is expressed as Eq.(6) derived from (1) and (4). Correspondingly the torque ripples which are caused by the inductance L is Eq.(7).

$$T_e = \frac{2E_m I_a}{\omega_m} + \frac{2E_m t}{3\omega_m L}\left(2V_{ka} + V_{kc} - \frac{1}{2}V_{dc} - 4E_m\right)$$

$$(6)$$

$$\Delta T_e = \frac{2E_m t}{3\omega_m L}\left(2V_{ka} + V_{kc} - \frac{1}{2}V_{dc} - 4E_m\right) \qquad (7)$$

Where t is time of commutation region.

In conduction region, the dc-link current sensor can not reflect the real phase-current when the switching device turns off in the conventional PWM-ON modulation method, then the torque ripples are generated because of these.However in the new proposed method, BUCK converter which adopts PWM modulation transforms V_{in} into V_{dc},accordingly changing constant voltage-source into quasi-current-source which can provide the satisfactory waveform matched to the induced EMF waveform in the stator windings.So, the dc-link current sensor can reflect the phase-current exactly,conclusively eliminating the torque ripples in conduction region.

In commutation region, the conventional control method adopts PWM-ON modulation.The torque ripples are expressed as Eq.(8) during commutation interval (S_a =1, S_c =0 or 1).

$$
\begin{cases}
\Delta T_e = \dfrac{2E_m t}{3\omega_m L}\left(V_{dc} - 4E_m\right)\ldots\ldots S_C = 1 \\[4mm]
\Delta T_e = \dfrac{2E_m t}{3\omega_m L}\left(-4E_m\right)\ldots\ldots S_c = 0
\end{cases}
\tag{8}
$$

The expression shows the commutation torque ripples in trapezoidal BLDCM, including torque spikes in the low speed range($V_{dc} > 4E_m$) and torque dips in the high speed range($V_{dc} < 4E_m$).

But using the new proposed method that adds Buck converter between the battery and the inverter,the converter can regulate the amplitude of the out-voltage and the current.The duty ratio is denoted by D(t),the following equation is gained:

$$
V_{dc} = D(t) \bullet V_{in}
\tag{9}
$$

So torque spikes decreases with decreasing the duty ratio D in the low speed range($V_{dc} > 4E_m$) and torque dips decreases with increasing the duty ratio D in the high speed range($V_{dc} < 4E_m$).Because of $S_a = S_c = 1$ in the proposed scheme,the torque ripples are expressed as Eq.(10) during commutation interval.In this way,the case of $V_{dc} < 4E_m$ exists over the entire speed range , consequently attenuating the commutation torque ripples.

$$
\Delta T_e = \frac{2E_m t}{3\omega_m L}\left[D(t) \bullet V_{in} - 4E_m\right]
\tag{10}
$$

In the low speed range, the torque ripple ΔT_e can be eliminated by making the $D(t) \bullet V_{in} = 4E_m$.

In the high speed, the torque ripple ΔT_e can be attenuated by making the $D(t) = 1$.

In summary,the new proposed method makes drive to produce smooth torque and linear torque with current.

III.EXPERIMENTAL RESULTS

Experiments are carried out to verify the feasibility and effectiveness of the proposed method. The parameters of BLDCM prototype are shown in table.1.

TABLE1:PARAMETERS OF BLDCM PROTOTYPE

U_{rated} (V)	24
P_{rated} (W)	146
Poles	4
$R_{phase}(\Omega)$	0.402
$L_{phase}(mH)$	0.185
$f_{buck}(KHz)$	30

To implement the new control method, the TMS320LF2407A DSP is employed in the prototype. Digtal PID speed control, constant current control and constant voltage control operate every 33.3 μs sampling time.Fig.3 and Fig.4 show the experiment

waveforms of two-phase currents and commutation currents respectively in the case of conventional control method and in the case of the new proposed scheme.Fig.3 shows that torque ripples during conduction region are eliminated effectively by the new proposed control technique.Fig.4 notes that the commutation current slopes of the incoming and outgoing phases balanced, so the resulting commutation torque ripples are effectively suppressed with the help of the new proposed scheme.

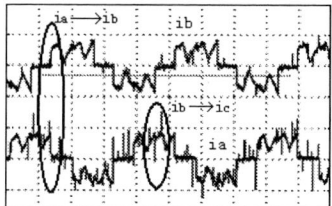

i: 10A/div   time: 3ms/div

Fig.3.(a) The 2-phase current-waveforms of conventional modulation mode

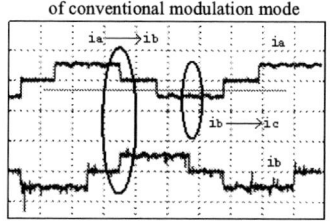

i: 10A/div   time: 2ms/div

Fig.3.(b) The 2-phase current-waveforms of new proposed modulation mode

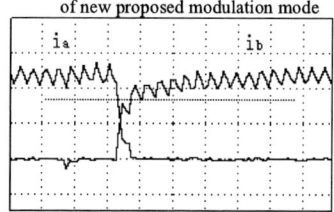

i: 2A/div   time: 100us/div

Fig.4.(a)The commutation current-waveforms of conventional modulation mode

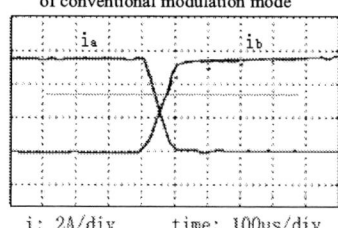

i: 2A/div   time: 100us/div

Fig.4.(b)The commutation current-waveforms of new proposed modulation mode

IV.CONCLUSIONS

In this paper,a new torque ripple reduction method based on buck converter has been proposed for brushless dc motor drives using a single dc current sensor.In such control method, the dc-link current sensor can give correct information corresponding to the motor phase currents to eliminate torque ripples in conduction region. Meanwhile, torque ripples have been attenuated effectively during commutation region.Subsequently effectiveness and feasibility of the proposed control method are verified through experiments.

REFERENCE

[1] Joong-Ho Song and Ick Choy, "Commutation torque ripple reduction in brushless DC motor drives using a single DC current sensor,"*IEEE Trans. on Power Electronics*,vol. 19, No.2 ,pp.312-319,March 2004.

[2] Byoung-Hee Kang,Choel-Ju Kim,Hyung-Su Mok and Gyu-Ha Choe, "Analysis of torque ripple in BLDC motor with commutation time,"*Proceedings of IEEE*,vol.2,pp.1044-1048, June 2001.

[3] Carlson R,Lajoie-Mazenc M and Fagundes J.C.d.S, "Analysis of torque ripple due to phase communtation in brushless DC machines,"*IEEE Trans. on Industry Applications*,vol.28,no.3, pp.632-638,May-June 1992.

[4] Luk P.C.K and Lee C.K, "Efficient modeling for a brushless DC motor drive,"*International Conference on Industrial Electronics,Control and Instrumentation*,vol.1,pp.188-191, September 1994.

[5] Lei Hao,Toliyat,H.A, "BLDC motor full speed range operation including the flux-weakening region,"*IEEE-IAS Annual Meeting*,vol.1,pp.618-624, Octorber 2003.

[6] Wei Kun,Lou Zhenli,Zhang Zhongchao, "Estimate of rotor position of BDCM based on the third harmonic component"*Power Electronics and Motion Control Conference,2004, IPEMC* ,vol.3,pp.1306-1310, August 2004.

Design of Wind Turbine Generator Control System

Chen Guiyou, Zhou Li, Sun Tongjing, Wang Zhongmin
School of Control Science and Engineering
Shandong University
Ji'nan, Shandong, P.R. China
chenguiyou@sdu.edu.cn

Abstract—**The principle of wind turbine generator (WTG) and its control system based on programmable logic controller (PLC) are presented. The wind energy is converted into electric energy by WTG. Because of the uncertainties of the speed, the direction of the wind, and the large inertia of the wind turbine of WTG, reliable control strategies are adopted to assure the WTG to run normally under serious conditions. According to the special requirement of the control system of 55kW WTG, a gate method is used to get the precise measurement of the generator's speed. The generator can be merged into or disengaged from the power system safely by the control system. The crosswind protection of the wind turbine and casting loose automatically are also realized. The practical running results show that the control system is reliable.**

Keywords- wind turbine generator; programmable logic controller; gate method; control system

I. INTRODUCTION

With the development of the economy and the technology, countries all over the world become more concerned about the exploitation of the green energy. The wind energy is a kind of green energy. And it is abundant. Therefore more and more attentions are paid to its exploitation. Wind energy is becoming one of the most appealing energy today as described in Ref. [1]. An efficient method of using the wind energy is to convert it into the electric energy. This kind of device or system is called the wind turbine generator (WTG). After the first WTG was developed in Denmark in 1980's, every country strives for its development as described in Ref. [2]. Only after several years, the commercialization and industrialization of the WTG was realized. The huge WTG industry was formed in the middle of 1990s. The largest capability of WTG in the world was 20000MW in Oct. 2001, and it will break through 100GW in 2010 according to Ref.[3].

Because of the uncertainty of the speed, the direction of the wind, and the large inertia of the wind turbine of WTG, it is difficult to control WTG. In general, the running environment of the WTG is very bad and complex. So the automatic running and no-person on duty is needed. Therefore the mergence into or disengagement from the power net of WTG is required strictly. And it has become the key technique of the control system of WTG whether the generator can merge into or disengage from the power net reasonably. It is

obvious that high reliability is very important in the control system of WTG. In order to improve the measurement accuracy of the generator speed, a gate method is introduced. The programmable logic controller (PLC) is used as the main control unit of the control system referring to [4]. Reliable and feasible control strategies are used in the main control unit. The control system introduced in the paper is used in the middle-sized WTG.

II. PRINCIPLE AND STRUCTURE OF WTG

A. The structure of WTG

WTG is a generator having a horizontal axis. The wind energy is converted into the electric energy by WTG. The electric energy produced by WTG is merged into power system by the control system which is described in Ref.[5]. The WTG is made up of three glass-steel laminas, which is placed windward and adjusted by a fixed oar when it is out of speed in Ref.[6]. The impeller of the generator is driven by the speedup gear box. Two servo motors are installed symmetrically on the chassis of the engine room. They are used to adjust impeller to the windward direction or the crosswind direction and to cast loose automatically. The impeller is the part that absorbs the wind power and changes the wind power into mechanical energy. The laminas are driven by the revolving torque which is formed under the function of the certain velocity of wind and the angle between the wind and laminas. The generator is driven by speedup case after the wind power is changed into mechanical energy.

As the braking process is completed by the leaf pinch speed limiting board and hydraulic pressure arrester, the stopping operation can be realized quickly and reliably. For the convenience of maintenance in the engine room, a control box is installed. The impeller, the speedup box, the generator, the direction adjusting machine, the wind's speed and direction sensors, the engine room control box, and so on, are installed on the top of the tower frame of the engine room. The whole engine room can be turned along the tower tray horizontally. The structure of the tower is a taper pipe. The main control unit is installed in the control box at the bottom. The hanging control box is installed on the tower wall side of control ark. The equipments in the engine room are connected to the control system at the bottom by cable. The structure of WTG is shown in Fig. 1.

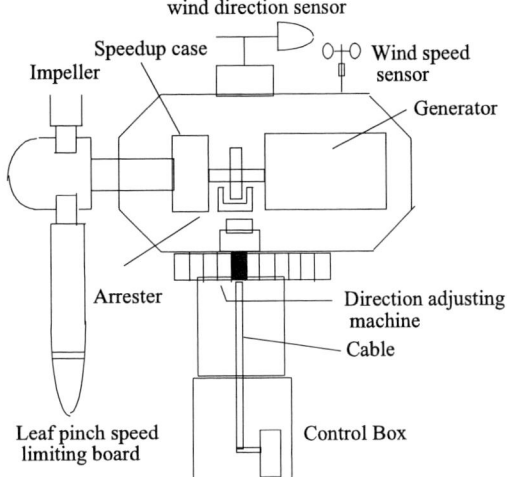

Figure 1. The structure of WTG

B. The requirement of control system of WTG

The power that the impeller absorbed from the wind can be denoted as following which is described in Ref.[7].

$$p = kCPV\cos\alpha \qquad (1)$$

Where, k is the coefficient which is determined by the density of wind and the area that the impeller has scanned; CP is the using coefficient of wind energy, which is determined by the shape and the number of the lamina and the distance of two lamina; α is the angle between the impeller normal direction and the wind direction; V is the wind speed of the for-forth frontage of the impeller.

As to the 55kw WTG, whose impeller is composed of three leaves, the coefficient k and CP can be considered as constants. So the power absorbed by the impeller depends on the speed and the direction of the wind.

There are two WTGs in the whole system, 55kw and 11kw. The impeller is connected to the 55kw generator though the speedup box. The speedup ratio is 1 to 20 according to Ref.[7]. The 11kw is connected to the 55kw generator though the strap. The connection speed ratio is 1.25 to 1. The direction adjustment system is made up of direction sensors and drivers, which is used to stop the impeller running. The synchronization speed of the two generators is 1000r/min.

The impeller starts at low wind speed and turn to the windward direction automatically when the wind speed is in the range of 3.5m/s to 7m/s. When the speed of impeller is up to 40r/min, the 11kw generator will overload. The generator will disengage from the power system after it overloads for specified period. When the speed of the impeller is up to 50r/min, the 55kw generator will be at the status of synchronization speed and merge to the power system. The control system must make the precise measurement of the speed and power of

the generator, switch the two WTGs merge to and disengage from the power system automatically according to the variety of the wind speed.

The value of $\cos\alpha$ varies little near zero when the direction of the wind changes. So the impeller does not adjust the direction to follow the wind direction, and the absorptivity of the impeller was commonly requested in the 95%of the most absorb power and was not adjusted the direction to follow the wind direction, while the value range of α is from -18 to 18 according to Ref.[7]. When the value of α exceeded the range above, the impeller should be adjusted the direction to follow the wind direction according to the direction of the wind direction sensor. The adjust value is at least 20, which should estimate the real-time torsion of the cable in order to loose the cast in proper time and prevent the cable from being turned over the limitation that can destroy the cable.

III. ACCURATE MEASUREMENT AND RELIABLE CONTROL USED IN THE CONTROL SYSTEM

A. Using Programable Controller as the Main Control Unit Units

A programmable controller is used as the main control unit of WTG control system. An 85VAC-264VAC wide range input power is equipped in the programmable controller. There are 32 input ports whose power supply is DC 24V and are insulated by photo-electricity coupling in the programmable controller. It also has a counting input port with 2KHz high speed and 28 output ports with relay insulation. Enough data section, soft node and timer/counter are included in the controller. It has many application instructions. Its control program executes in a closed loop, and the inputs and outputs refresh periodically. It can run without failure for a long time. The reliability of the control system depends on the programmable controller.

OMRON CPM2A PLC is used as the controller, which has the features of three usable output and two usable power supply, 120 I/O points at most, quick interruption input function, quick response input function and stable input filter etc.

B. Gating Method for the Precise Measurement of the Generator's Speed

It is difficult to merge the WTG into the power net at its synchronization speed. If the speed exceeds the synchronization speed by 8 turns, great strike current to the net will occur when the WTG merge into the power net. Not only may it have bad influence on the power net quality, but also it can damage the main loop control equipments and the engine of the WTG. So, how to control the generator to merge into or disengage from the power net is the key technology for the safe running of the WTG. The precise measurement of the generator's speed is the necessary condition of the control to the WTG. So, the chief task is to achieve the precise measurement of the generator's speed. The gating method is utilized. The method is to install an electromagnetic tray having 60 alveolars on the high speed axis, on which a high speed adjacent switch is installed to generate induction counting square waves. The frequency of the square wave pulses equals to the speed of generator. The speed of generator will be measured by introducing the induction pulses into the high speed counter input of the programmable controller. In practical application, there is about 2% error between the measurement of the square-wave pulses and the real speed of generator. By analyzing, it is caused mainly by the affection of the programmable controller's scan period when reading the high speed counter in the application program. The solution is to control the induction pulses using 1:0.125 gating signal before it is introduced into the high speed counter input. In one second the gate is opened, the square wave is allowed to enter the high speed counter input to be counted; in 0.125 second the gate is closed, the square wave is not allowed to enter the programmable controller, and the reading of the prior 1 second's counting value and the reset operation of the counter are finished. The gating signal is produced by a 4.194304MHz crystal with frequency division by CD4521. It is precise and stable, and can meet the need of the precise measurement of the generator's speed. The measurement circuit of the generator's speed shown in Fig. 2.

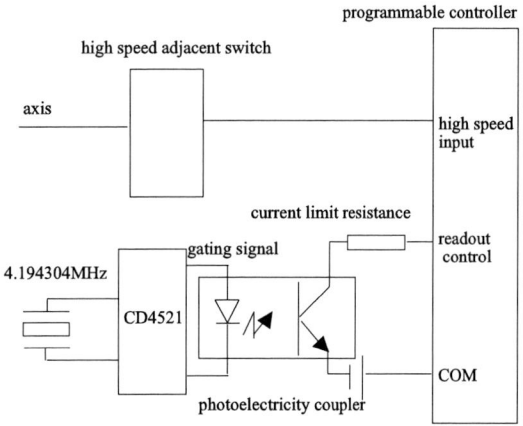

Figure 2. The measurement circuit of the generator's speed

C. Control of the Two Generators' Mergence into or Disengagement from the Power Net

Within the wind speed range from 3.5m/s to 23m/s, the control system can merge or disengage the generator into/from the power net automatically. For the convenience of description, the action of the mergence into the power net of the generator is called on-line, the action of the disengagement from the power net of the generator is called off-line. The consequence of these two actions or the status of the generator is called ON-LINE and OFF-LINE separately.

In order to increase the efficiency, two different power capacity generators are used. When the wind speed is low, smaller generator should be used. However, when the smaller generator's power exceeds by 20% of its rating, it should be disengaged from the power net, and the larger generator should be merged into the power net. When the wind speed becomes low again, it is required to switch back to the small generator. Thus, if the wind speed changes between 6m/s and 7m/s repeatedly, the action of the mergence into or disengagement from the power net

of the generators will occur frequently. The back difference control method is used to avoid such case occurring. When the back difference value is between 3kw and 8kw, the control engine will finish self-tuning dynamically according to the switch situation. The simplified control rules set described in Ref. [6] are as followings:

- **IF** (STARTED=1) AND $\{[(V_i \geqslant 800)$ AND $(G_l$ is OFF-LINE)] OR $(800 \leqslant V_i \leqslant 840)\}$ **THEN** G_s on-line
- **IF** G_s is ON-LINE AND $\{(V_i \geqslant 820, T_1 \geqslant 10s)$ OR $(P1_c \geqslant 120\% P1_r, T_2 \geqslant 4s)\}$ **THEN** G_s off-line
- **IF** G_s is OFF-LINE AND $\{[(V_i \in V_n)$ AND $(T \to 0)$] OR $(V_i \geqslant 1000)\}$ **THEN** G_l on-line
- **IF** G_l is ON-LINE AND $(P2_c \leqslant P_{ts}, T_3 \geqslant 6s)$ **THEN** G_l off-line
- **IF** $V_i < 800$ **THEN** G_s off-line

Where, STARTED is the flag of the system starting working. Vi is the current speed of the generator. G_l stands for the larger generator. G_s stands for the smaller generator. 'OFF-LINE', 'ON-LINE', 'off-line' and 'on-line' are described as above. T_1 is the duration time of the mergence into the power net, T_2 is the time of the 20% over load lasting. $P1_c$ is the current power of the smaller generator. Pl_r is the rating power. V_n is the speed closed to the synchronization rotating speed corresponding to the wind speed in the knowledge base. T is the delayed time of the mergence into the power net corresponding to the closed synchronization rotating speed in the knowledge base. $T \to 0$ denotes the end of delay. $P2_c$ is the current power of the larger generator. T_3 is the delayed time of the larger generator disengaging from the power net. P_{ts} is the threshold of the back difference control of the disengagement from the power net of the big generator, which can be determined by the Equation (2),

$$P_{ts} = 6 - (N_s - 4)/2 \qquad (2)$$

Where, N_s is the switch times of WTG per hour.

D. Protection against Over-speed

The wind turbine of the WTG will absorb the wind energy to rotate when it is windward. If it reaches the synchronization rotating speed of the generator, the generator merges into the power net to generate electricity. The wind turbine's speed does not exceed 53r/min in general. However, in case of something wrong or failure during the mergence into the power net, the wind turbine will speed up quickly under the empty load condition. If it is not dealt with on time, the wind turbine will be out of control and the bad accident will occur. The timing method to measure the rotating speed of the wind turbine in real time is used to avoid this situation. The wind turbine generates an induction square wave using an adjacent switch, which is connected to an input of the programmable controller. A timer of the programmable controller begins to work at the declining edge of the square wave, whose timing unit is 0.1 second, and the set value is 21. If the contactor of the timer turns on in the period of the wind turbine rotating two circles, the time between the two circles is greater than or equal to 2.1 seconds, and the speed of the wind turbine is within the

1632

limit of safe rotating speed of 57r/min. If the contactor of the timer keeps OFF in the period of the wind turbine rotating two circles, the wind turbine must be over-speed. Then the hydraulic brake system must be driven to startup by the controller immediately to stop the wind turbine by holding the brake of the low speed axis braking tray. At the same time the direction adjusting machine is started up to adjust the short cabin to the right 90 degrees, so the crosswind protection is realized.

E. Casting Loose

The generator, the direction adjusting motor, the wind direction sensor and the wind speed sensor, the speed measuring sensor, and so on, are all installed on the short cabin at the top tower, which is connected to the control ark at the bottom of the tower using a cable. The cable will retort following the rotation of the short cabin when it is adjusting the direction to follow the wind direction. If the torsion exceeds by two circles, the short cabin should turn in the reversed direction to cast loose. There is an adjacent switch installed on the decelerating machine of the direction adjusting servo motor. The sensor will generate a square wave pulse when the short cabin rotates 1/8 circle in one direction, and the square wave pulse is counted by a reversed counter of the programmable controller. The set value of the counter is 16. If 16 pulses in one direction are detected, the cable has retorted two circles in this direction. In appropriate time, the wind turbine of WTG stops rotating through holding the brake, and the direction adjusting system turns 2.5 circles in the reversed direction automatically to cast loose. The control system also sets the safeguard of the failure stopping engine at 4 circles of the cable in order to cast loose on time when the cable retorts more than two circles.

IV. CONCLUSION

Reliable controls of WTG are presented in this paper. The programmable controller is used as the main control unit in the control system of WTG. The practical running denotes that all kinds of measurements and controls are reasonable and practical, which can guarantee the WTG to run without faults for long time. According to the special accuracy requirement of the control system of 55kW WTG, a gate method is used to get the precise measurement of the generator's speed. The practical running results show that the control system is reliable.

REFERENCES

[1] Basic Aspects for Application of Wind Energy, *Commission of the European Communities Directorate -General for Energy* ,pp:5-9, 1994.

[2] Jens Peter Molley, General Aspects on Wind Energy Development. *DEWI*,Wilhelmshaven,1995.

[3] Ye Hangye. The Control Technology of Wind Power Generator Set, *China Mechine Press*, Beijing, 2002.

[4] Xu De, Sun Tongjing, Chen Guiyou. The Application Technology of Programmable Controller (PC), *2nd Edition. ShanDong Science and Technology Publishing House*, ShanDong, 2002.

[5] Henry Seifert. Principles of Wind Energy Extraction. *Eldorado*, 1995.

[6] Sun Tongjing, Xu De, The Main Points of Control and Technique Measures of 55kW WTG, *Electrical Dirve Automation*, 1997.5.

[7] Sun Tongjing, Xu De, The PLC control system of 55/11kw wind turbine generator, *Inductrial control computer*, No.6, pp:24-26,1997

Non-touching Intelligent Control System of Water Intenerating Equipment Based on Sodion Exchange

Chen Guiyou, Zhang Qingfan, Zhou Li, Luo Donghua
School of Control Science and Engineering
Shandong University
Ji'nan, Shandong, P.R. China
chenguiyou@sdu.edu.cn

Abstract—**A non-touching intelligent control system of water intenerating equipment is designed. The combination of power electronic technology and microcontroller technology is realized. Moc3061 is used as photoelectric coupler and KS3-5 is used as TRIAC. The reliability of the whole system has been greatly improved by using the non-touching intelligent control technology. The principle, structure and software design of the system are introduced. Practical running results show that the system is not difficult to build up, featuring the low price, but the high performance and high reliability. It can be used as a universal non-touching control platform. So it has great spread value.**

Keywords- water intenerating equipment; power electronic technology; non-touching control; TRIAC

I. INTRODUCTION

Long time running practice of boilers shows that the quality of the feedwater to boilers is one of the most important factors that influences the safety, stabilization and economical operation of the boilers and hot power system. There are many impurities in the non-purified water, especially the calcium and magnesium hydroniums. These hydroniums do a lot of harm to the boiler system. Some examples are listed in the following.

- The scale, a hard mineral coating that forms on the inside surface of boilers in which water is repeatedly heated, can lead to the asymmetric heat of the boiler and can destroy the metal.
- Thermal efficiency will decline and the energy consumption will increase.
- The medicament would be added to clean scales and the operation cost would be increased.
- These hydroniums can result in the rot of metal.
- The quality of steam is easy to be deteriorated.

Therefore, it is necessary to intenerate the water before it enters the boiler. The water is intenerated using the exchanger.

The rigidity of water is the content of salty substances that dissolve in the water, that is to say the content of calcium (Ca^{2+}) and magnesium hydronium (Mg^{2+}). The principle and feature of the water intenerating equipment

are described in Ref.[1]. The more hydroniums are included in water the larger is the rigidity of water. On contrary, the less hydroniums are included in water the smaller is the rigidity of water. When the original water that contained rigid hydronium gets through the colophony exchanger, the Ca^{2+} and Mg^{2+} are exchanged with the natrium hydronium (Na^+), and the Na^+ enters the water. Thus, the water that comes out from the exchanger is the intenerated water. This course is called intenerating process. With the uninterrupted operation of exchanging process, the colophony loses the exchangeable capacity after the natrium hydroniums are all exchanged, so the Ca^{2+} and Mg^{2+} that are adsorbed in colophony must to be replaced by using Nacl solution, and the colophony resumes the exchangeable capacity after absorbing the Na^+ again. This course is called regeneration. After many times of intenerating process and regeneration, the colophony should be cleaned periodically in order to improve the efficiency of hydronium exchange.

II. SYSTEM STRUCTURE AND WORK PROCESS

The system structure is showed in Figure1.

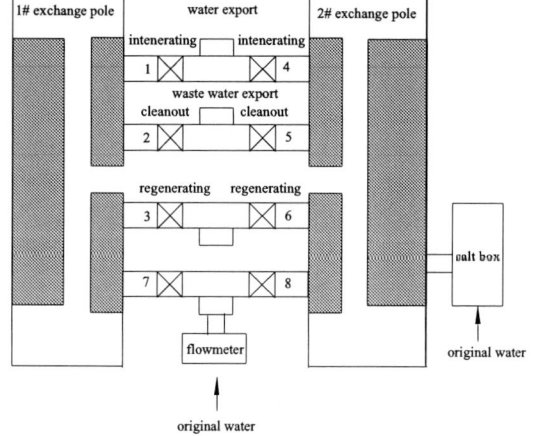

Figure1. The structure of the water intenerating equipment

In order to improve the work efficiency of the system, two water exchange poles are installed in the water intenerating equipment. The two poles work alternately, and the work time of each working procedure can be adjusted automatically according to requirements. In Figure1, the 1,2,3,7 electromagnetic valves are connected to the 1# exchange pole, and the 4,5,6,8 electromagnetic valves are connected to the 2# exchange pole. The process of intenerating water includes intenerating process, cleaning process and regeneration. 1, 2, 3, 4, 5, 6 electromagnetic valves are normally closed valves, achieve the intenerating process, cleaning process and regeneration. The 7 and 8 electromagnetic valves are normally open valves, which control the feedwater.

The time of intenerating process, cleaning process and regeneration is 30 minutes, 15 minutes, 10 minutes respectively as in [2]. The two exchange poles work alternately in the water intenerating equipment, and the work scheduling is showed in Figure2.

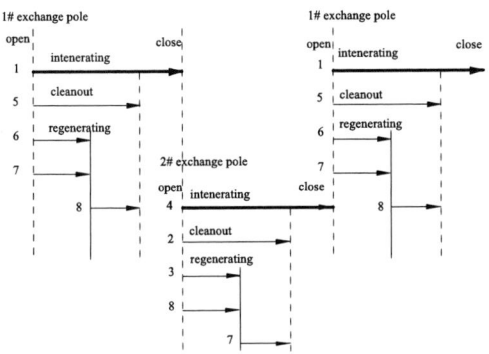

Figure2. The work scheduling of the water intenerating equipment

In the intenerating process of 1# exchange pole, the flow direction of each solution is shown in Figure3.

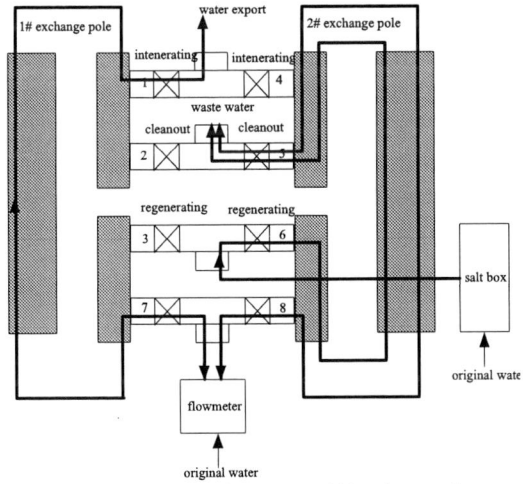

Figure3. The intenerating process of 1# exchange pole

In the intenerating process of 2# exchange pole, the flow direction of each solution is shown in Figure4.

In order to realize the automatic operation of the water intenerating equipment, the automatic control system is designed to control each electromagnetic valve. According to the specified scheduling which is set in advance, the automatic control system makes the on-off control of the electromagnetic valves. The output of traditional control system is commonly realized by using relay. The life of this relay control system is limited and the reliability is low. In order to solve the problem, a non-touching intelligent control system based on power electronic and microcontroller technology is designed, which combined the power electronic with microcontroller technology, the reliability and the life of the system are improved.

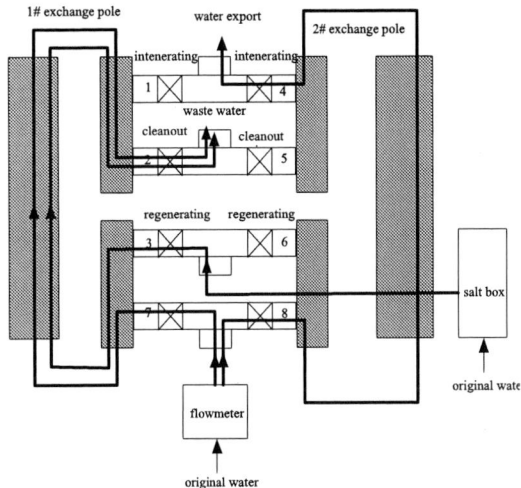

Figure4. The intenerating process of 2# exchange pole

III. HARDWARE DESIGN

The microcontroller µPSD3251 is used as the CPU of the control system. It combines a flash PSD architecture with an 8032 microcontroller core. The description about uPSD3251 is shown in [3]. The µPSD3251 features dual banks of Flash memory, SRAM, general purpose I/O and programmable logic, supervisory functions and access via ADC, DAC and PWM channels, and an on-board 8032 microcontroller core, with two UARTs, three 16-bit Timer/Counters and two External Interrupts. As with other Flash PSD families, the µPSD3251 is also in-system programmable (ISP) via a JTAG ISP interface. The hardware structure of the non-touching system is showed in the Figure5.

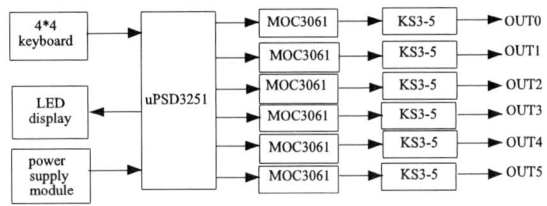

Figure5. The hardware structure of the control system

In Figure5, The isolation driver between the microcontroller and TRIAC is realized by using the photoelectric coupler MOC3061, which also realizes the non-touching control combined with the TRIAC. The 4×4 keyboard is designed to setup time of the process of intenerating, cleaning and regeneration. The LED display is designed to show the setting up time in the time setting up period, and to display the operation time in the working phase of the equipment.

The MOC3061 consists of a GaAs infrared emitting diode optically coupled to a monolithic silicon detector performing the function of a zero voltage crossing bilateral TRIAC driver as in [4]. It is used as the interface between TRIAC and the microcontroller to equipment powered from 115/240 VAC lines, such as solid-state relays, industrial controls, motors, solenoids and consumer appliances, etc. Its features are showed as following.

- Simplifies logic control of 115/240 VAC power
- Zero voltage crossing
- dv/dt of 1000 V/μs guaranteed
- 600 μs guaranteed
- VDE recognized
- ordering option V
- Underwriters Laboratories (UL) recognized

One of the six outputs is showed in detail in Figure 6.

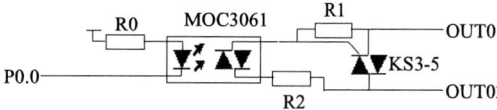

Figure6. One of the six outputs

When the output of P0.0 is "0", the emitting diode will emit lights and the TRIAC in the MOC3061 will works, then the "G" pole and "A" pole of the KS3-5 are turned on, which is called the phase strong triggering as in [5]. Where, R1 is the diffluent resistance of Gate pole and R2 is the limitative resistance that limits the current of the KS3-5.

IV. THE SOFTWARE DESIGN

The flow of the whole system is showed as Figure7.

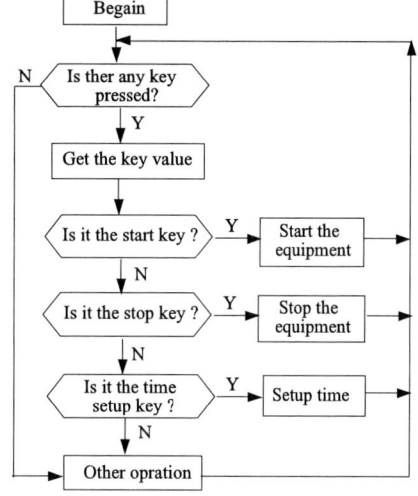

Figure7. The flow of the system

The work flow of 1# exchange and 2# exchange pole is shown in Figure 8. These codes are designed in the timer interrupt routine.

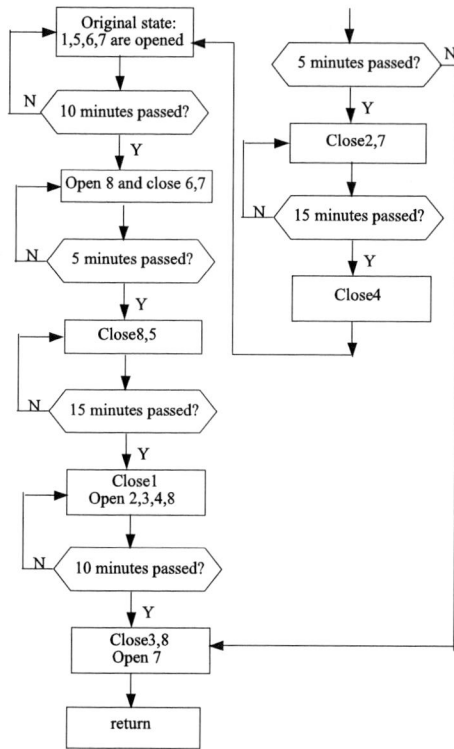

Figure8. The work flow of the1 and 2 exchange pole

The control program is not so complicated, so all the programs running in the uPSD3251 are written in ASM.

V. CONCLUSIONS

A non-touching intelligent control system of water intenerating equipment is designed. Power electronic technology and microcontroller technology are used in the system. An interface between TRIAC and microcontroller is introduced. Practical running results show that the system has creativity and practicability, which fills the blanks of non-touching control in the water process field and is worth being spread.

REFERENCES

[1] The work principle and feature of the water intenerating equipment. http://www.gxinvest.gov.cn/gqnr.php?id=3690

[2] Lu Jingjun, "The automatic control of sodion watersoftening system", *Journal of Nantong Vocational College,* Vol.13 No.2, pp. 49-52, Jun.1999.

[3] μPSD325X, Flash Programmable System Devices with 8032 Microcontroller Core, *STMicroelectronics*, March, 2003.

[4] MOC3061 datasheet, Fairchild Semiconductor Corporation, 2003.

[5] Wang Taoan and Huang Jun, power electronic technology, *China Machine Press,* Beijing, 2000.

2006 5th International Power Electronics and Motion Control Conference

Investigation of Hybrid Modeling and Control for DC-DC Converters

Hao Ma, Feng Qi, Wenqi Zhou
College of Electrical Engineering, Zhejiang University, Hangzhou, China
mahao@cee.zju.edu.cn

Abstract—This paper proposed uniform hybrid system models for second-order DC-DC converters working under continuous current mode. Contrast to the averaged and linearized models, no approximation is introduced to the modeling. The Direct Method of Lyapunov is used to analyze stability of the system and a new sliding mode control strategy is derived from the analysis. Simulation and experiment results are presented to validate the strategy.

Keywords-hybrid system; modeling; DC-DC converters

I. INTRODUCTION

Power semiconductor components are widely used in power converter since they were introduced in the 1950's. But owing to their nonlinear characteristic [1], modeling and analysis of the converters are tough tasks contrast to the linear circuits. For a long time, state-space averaging approach [2], which is implemented to derive the small-signal averaged equations of the switching converters, is considered the key step in modeling. Nevertheless, when facing the large disturbance of load or input voltage etc, the small-signal based method seems not so effective due to the approximation of the model.

Hybrid dynamical systems are characterized by interacting continuous and discrete dynamics. Power electronic converters can be viewed as a kind of hybrid systems due to their natural hybrid behavior. The basic structure of hybrid system for the converters is showed in Fig. 1, where $x(t)$ is the vector of state variable and $s(t)$ stands for the switch states. Here, the "controller" can be designed and control algorithm such as PID, current programmed mode (CPM) and sliding mode are widely used. Since there is no approximation when modeling

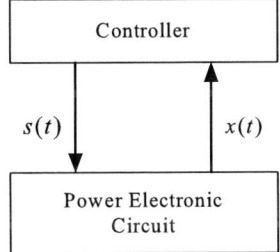

Figure 1. Internal structure of hybrid control system for power electronic circuit

hybrid systems, modes and analysis of power electronic converters based on the hybrid system theory can be more precise than that on the averaged ways. In [3], the circuit is modeled as composition of hybrid automata, guards are designed and a numerical algorithm is given to compute the maximum radius of safe ball connected with the system stability. The deadbeat operation and controllability for DC-DC converters by use of the concept of hybrid system is studied and extending deadbeat controllability to higher order systems is proposed in [4]. This paper proposed uniform hybrid system models for second-order DC-DC converters working under continuous current mode (CCM), applies the Direct Method of Lyapunov (DML) to propose result for stability, from which a new control algorithm is derived.

In section II modeling in the style of hybrid system and the Lyapunov sufficient stability condition are introduced. In section III a new sliding-mode-like control algorithm in order to satisfy the condition is proposed and the simulation of the Boost circuit results are presented. Experimental results and discussions are shown in section IV. Brief conclusion and further research directions are discussed in section V.

II. MODEL OF THE SYSTEMS IN HYBRID SYSTEM STYLE AND STABILITY SYNTHESIS

A. Model of the DC-DC systems

Second-order DC-DC circuits worked under CCM mode, such as Buck, Boost, Buck-Boost, Flyback and Forward, can be modeled in hybrid systems form [5] as

$$\dot{x} = F[(J(s(t)) - R(s(t)))x(t) + B(s(t))u(t)] \quad (1)$$

$$y = Cx + Du(t) \quad (2)$$

$$s(t^+) = \phi(x(t), m(t)) \quad (3)$$

Where, $x(t) \in R^n$ is the continuous-time state and is usually the vector of inductance current and capacitance voltage. F, J and R are all $n \times n$ matrices, the first is related with the parameter of the circuit, the second is skew symmetric matrix, meaning $J = -J^T$, the third is nonnegative matrix associated with the load. $u(t)$ is a

1-4244-0448-7/06/$25.00 ©2006 IEEE   1637

continuous control input. $s(t) \in \{0,1\}$ is a Boolean value representing the switching function that determine $\dot{x}$. $m(t)$ is a "discrete event" input, and $\phi(\cdot,\cdot)$ is a discontinuous function behind which lies the control algorithm.

B. Equilibrium point and Lyapunov asymptotical stability condition

Using the conception of average method where the control s is considered continuous, $x(t) = x_{ref}$ of the DC-DC circuits is called the equilibrium point when there exits $s(t) = s_{eq}$ and $0 \le s_{eq} \le 1$. Such that constraint (4) is satisfied.

$$F[(J(s_{eq}) - R(s_{eq}))x_{ref} + B(s_{eq})u] = 0 \qquad (4)$$

In order to apply the DML, a common Lyapunov function V should be proposed to help analyzing the stability of system (1)-(3). If the function V around x_{ref} satisfies that

1) $V(x, x_{ref}) > 0$ anywhere excepted in x_{ref} where $V(x_{ref}, x_{ref}) = 0$,
2) $V(x, x_{ref})$ is continuous without jump and radially unbounded,
3) for any x, a control signal s can be selected so that $\dot{V}(x, x_{ref}) < 0$.

If such a control algorithm is applied, then the system will stabilize on x_{ref} asymptotically.

In practice, the function V can be defined as Lyapunov function of the system.

$$V(x, x_{ref}) = \frac{1}{2}(x - x_{ref})^T F^{-1}(x - x_{ref}) \qquad (5)$$

It is positive and represents the total energy of DC-DC system. It is the combination energy of the inductance and the capacitance. The time derivative of V can be denoted by $\dot{V}_s$ [5].

$$\begin{aligned} \dot{V}_s &= (x - x_{ref})^T F^{-1}\dot{x} \\ &= (x - x_{ref})^T [(J(s) - R(s))x + B(s)u] \end{aligned} \qquad (6)$$

Using the skew symmetry property of $J(s)$ and the property of the equilibrium point (4), (6) can be rewritten as

$$\begin{aligned} \dot{V}_s = &-(x - x_{ref})^T R(s)(x - x_{ref}) + (x - x_{ref})^T \{[(J(s) - R(s)) \\ &-(J(s_{eq}) - R(s_{eq}))]x_{ref} + (B(s) - B(s_{eq}))u\} \end{aligned} \qquad (7)$$

Obviously, the first term of the expression is nonnegative because $R(s)$ is a nonnegative matrix. By applying the control law, we can make $\dot{V}_s < 0$ so as to achieve stability of the system.

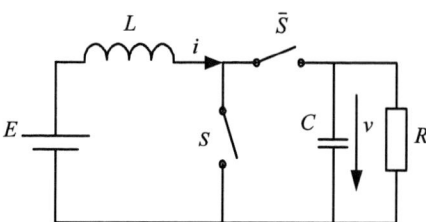

Figure 2. A simplified Boost converter

III. EXAMPLE AND SIMULATION RESULTS

Fig. 2 shows a simplified circuit of Boost converter working under CCM. We choose inductor current i and the capacitor voltage v as the state vector, namely $x = \begin{bmatrix} i & v \end{bmatrix}^T$. The equivalent series resistor (ESR) of the capacitor is neglected so the capacitor voltage is the same as the output voltage. The state matrices corresponding to (1)-(3) are

$$F = \begin{bmatrix} \dfrac{1}{L} & 0 \\ 0 & \dfrac{1}{C} \end{bmatrix}, \quad J = \begin{bmatrix} 0 & -(1-s) \\ 1-s & 0 \end{bmatrix}, \quad R = \begin{bmatrix} 0 & 0 \\ 0 & \dfrac{1}{R} \end{bmatrix},$$

$$B = \begin{bmatrix} 1 \\ 0 \end{bmatrix}, \quad u = E, \quad C = \begin{bmatrix} 1 & 0 \end{bmatrix}, \quad D = 0$$

The uniform state equation is

$$\begin{bmatrix} \dot{i} \\ \dot{v} \end{bmatrix} = \begin{bmatrix} 0 & -\dfrac{1-s}{L} \\ \dfrac{1-s}{C} & -\dfrac{1}{RC} \end{bmatrix} \begin{bmatrix} i \\ v \end{bmatrix} + \begin{bmatrix} \dfrac{1}{L} \\ 0 \end{bmatrix} E \qquad (8)$$

When $s = 1$, the switch S is ON while $\bar{S}$ is OFF and vice versa.

The equilibrium point according to (4) is

$$v_{ref} = \frac{1}{1 - s_{eq}}E, \quad i_{ref} = \frac{v_{ref}}{(1 - s_{eq})R}$$

According to (5), the proposed Lyapunov function is

$$V(x, x_{ref}) = \frac{1}{2}L(i - i_{ref})^2 + \frac{1}{2}C(v - v_{ref})^2 \qquad (9)$$

The time derivative referred to (7) is

$$\dot{V}_s = -\frac{(v - v_{ref})^2}{R} + (vi_{ref} - iv_{ref})(s_{eq} - s) \qquad (10)$$

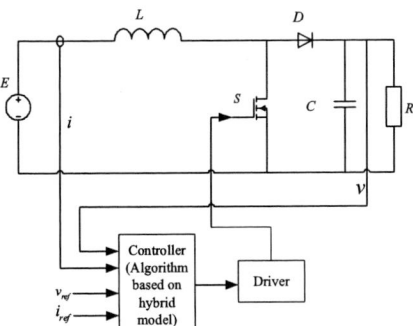

Figure 3. Topology and control block scheme of Boost

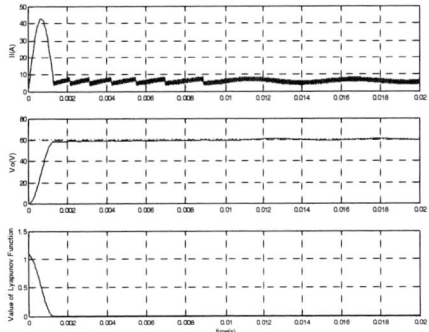

Figure 4. Inductance current i, output voltage v and the value of proposed Lyapunov function V

The first term of the expression (10) is nonnegative and likely negative for most of time since the trajectories of periodically switched DC-DC converters in steady-state are limit cycles [4]. If we can adopt the switching law that makes the second term of (10) equal to zero, the condition $\dot{V}_s < 0$ will be satisfied and the system will stabilize asymptotically.

To implement the control algorithm, we denote the function below as the sliding mode surface function

$$P(x,t) = vi_{ref} - iv_{ref} \qquad (11)$$

And the switching law can be described as:

$$s = \begin{cases} 1, \text{ when } \quad P(x,t) > 0 \\ 0, \text{ when } \quad P(x,t) < 0 \end{cases} \qquad (12)$$

when $P(x,t) > 0$, if the second term of (10) is expected to stay negative, the multiplier $(s_{eq} - s)$ must be negative, so $s = 1$ is the choice. Then, the inductor current i will increase and the output voltage v will decrease at the same time, which will make $P(x,t)$ decrease and eventually negative. At that time, the second term of (10) will become positive and $\dot{V}_s$ would be positive accordingly, so the output of the controller will let $s = 0$ so as to prevent $\dot{V}_s$ from being positive. The whole system is showed in Fig. 3.

In order to avoid the Zeno problem in simulation of the control algorithm, the controller will sample the value of state variable at a fixed frequency. The parameters are as following

Sampling frequency $f_{sample} = 40\text{kHz}$

Input voltage $E = 30\text{V}$

Output voltage $V_{out} = 60\text{V}$

Inductance $L = 300\mu\text{H}$

Output capacitance $C = 600\mu\text{F}$

Load resistance $R = 20\Omega$

The simulation is completed by MATLAB, and the results are showed in Fig. 4 and Fig. 5. Fig. 6 shows the limit cycle between the inductance current and the output voltage in the stable state.

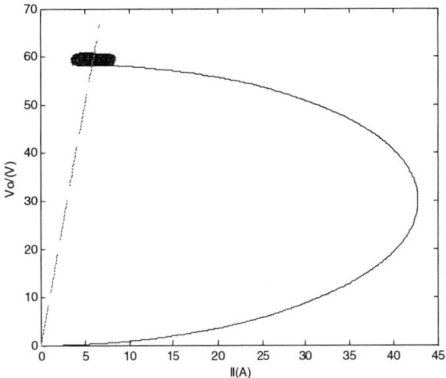

Figure 5. Trajectory of inductance current i and output voltage v (solid line) and the sliding mode curve (dash line)

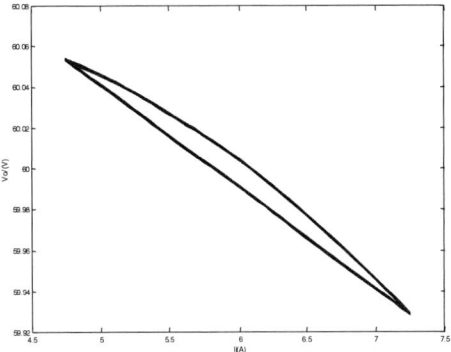

Figure 6. Stable limit cycle between inductance current i and output voltage v

IV. EXPERIMENTAL RESULTS AND FURTHER IMPROVEMENT

We implement the control algorithm mentioned above in a Boost converter and the controller is DSP TMS240LF2407. Inductor current, output voltage and the

1639

drive signal are shown in Fig. 7. Experimental stable limit cycle between inductor current and output voltage are shown in Fig. 8.

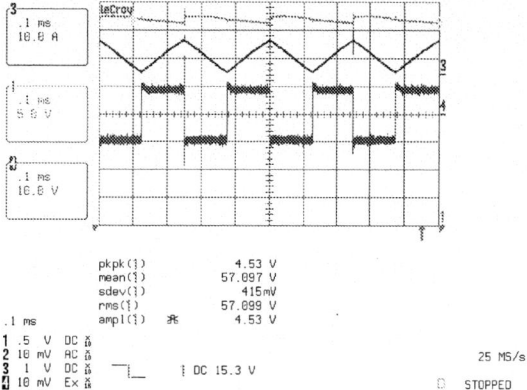

Figure 7.   Inductor current, output voltage and the drive signal

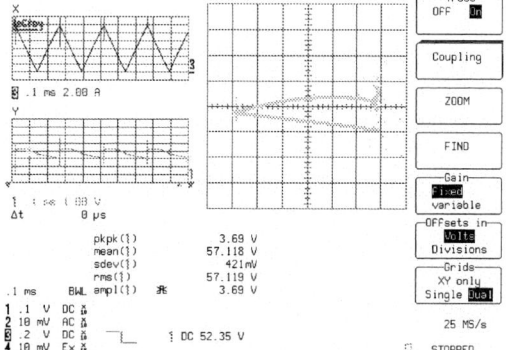

Figure 8.   Experimental stable limit cycle between induce current and output voltage (offset is made to the right diagram by the oscilloscope)

Experimental results above are identical with simulation. However, this algorithm needs to know the reference inductor current before it works, which limits its feasibility in practical use. To solve the problem, we can change equation (11) as

$$P(x,t) = (v(t) - v_{ref})i_{ref} - (i(t) - i_{ref})v_{ref} \qquad (13)$$

let $\varepsilon_i = i(t) - i_{ref}$, it can be viewed as the high harmonic component of the inductor current and obtained by high pass filter (HPF) as shown in Fig. 9. The whole system can be designed as that shown in Fig. 10.

V.   CONCLUSIONS

In this paper, hybrid modeling for DC-DC converters under CCM is proposed. DML is used to analyze stability of the model and a new sliding mode control strategy is derived. Simulations by MATLAB and experiments by DSP are done to validate this method. Hybrid systems techniques are more conformable than the averaged approach because there is no approximation in the analysis. Also, hybrid systems theory gives totally new perspective to power electronics.

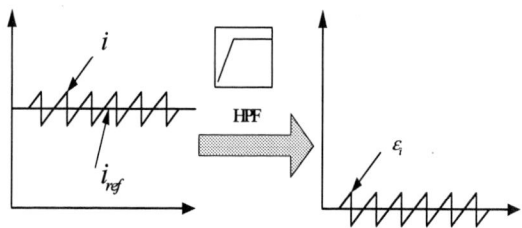

Figure 9.   Principle of HPF in the system

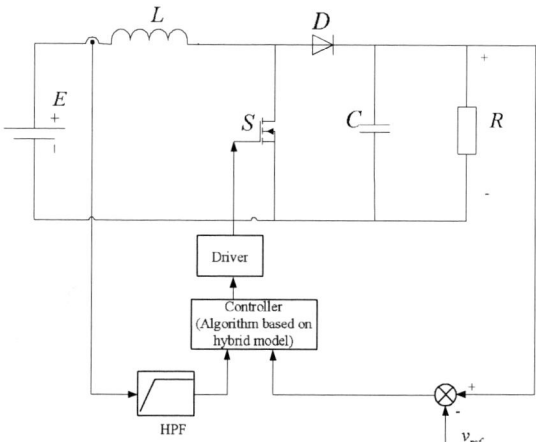

Figure 10. Block diagram of the whole system

REFERENCES

[1]   Xenofon D. Koutsoukos, Panos J. Antsaklis, James A. Stiver and Michael D. Lemmon, "Supervisory Control of Hybrid Systems," *Proceeding of the IEEE*, vol. 88, No. 7, pp. 1026–1049, July 2000.

[2]   R. D. Middlebrook and S Cuk, "A general unified approach to modelling switching power converter stages," *IEEE Power Electronics Specialists Conference*, June 8-10, Record. pp. 18–34. 1976.

[3]   Senesky, M. Eirea and G. Koo, T, "Hybrid modeling and control of power electronics," *Proc. 6th International Workshop on Hybrid Systems: Computation and Control*, vol. 2993 of Lecture Notes in Computer Science. Springer pp. 450–465, 2003.

[4]   J. T. Mossoba and P. T. Krein, "Exploration of Deadbeat Control for Dc-dc Converters as Hybrid Systems," *IEEE Power Electronics Specialists*, 36th Conference on June 12, pp. 1004–1010, 2005.

[5]   Jean Buisson. Pierre-Yves Richard and Herve Cormerais, "On the Stabilisation of Switching Electrical Power Converters," *Proc. 8th International Workshop on Hybrid Systems: Computation and Control*, vol. 3414 of, Lecture Notes in Computer Science. Springer pp. 184–197, 2005.

2006 5th International Power Electronics and Motion Control Conference

Effect of Peak Current Mode Control on Transient Response for VRM Application

Seiya Abe and Tamotsu Ninomiya
Kyushu University, Fukuoka, Japan
abe@ees.kyusyu-u.ac.jp

Abstract— **The dynamics is improved to approximately first order by using peak current mode control, because of reducing a system order due to the inductor current is restrained as a control signal. However, the voltage mode control is mainly used, and there are few applications for which peak current mode control is used in VRM application. This paper presents, the transient response of VRM with peak current mode control is examined analytically and experimentally.**

Keywords-component; peak current mode control; voltage mode control; transient response

I. INTRODUCTION

With recent progress of semiconductor technology, the circuit pattern used in LSIs has become much denser. Therefore, the driving voltage of LSI has been decreased to suppress the power loss due to charging and discharging of parasitic capacitances and the breakdown by electric field generating between patterns. Moreover, increase of load current is also remarkable by advanced function of LSIs. In order to meet these demands, a specified power supply with low voltage and high current output is needed. Moreover, high-efficiency and low-noise features are required as usual. Furthermore, in the case of operating mode changes, the load current variation of its power supply occurs suddenly. In this instant, the output voltage of the power supplies changes. However, it is required to suppress this output voltage variation within the limits, and to prevent malfunction of LSIs. To realize the above-mentioned power requirements, the fast transient response power supply is needed, it is usually called Voltage Regulator Module (VRM) [1-4], and many researchers have recently investigated VRM. The chief advantage of peak current mode control is the approximately first order dynamics. Therefore, the bandwidth of loop gain can be extended easily without any phase-lead compensation. Extending bandwidth decreases the voltage drop at rapid load change. However, the voltage mode control is mainly used, and there are few application for which peak current mode control is used in VRM application[5-7]. This paper presents, the transient response of voltage mode control and peak current mode control are examined analytically and experimentally.

II. VOLTAGE DROP ON TRANSIENT TIME

Figure 1 shows the synchronous buck converter used for standard VRM application. The rise time of inductor current falls behind load current when the load current rapid change, and the charge of an output capacitor is overdue and the discharge becomes large, as shown in Fig. 2. This causes the output voltage variation occurrence in transient time. Moreover, the voltage variation reaches a peak value in the time when inductor current reaches the desired value for the first time as shown in Fig. 3. To suppress this unbalance charge of the output capacitor, the fast response of inductor current is needed, and then the output voltage drop is reduced.

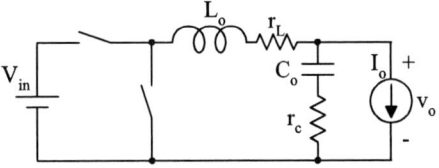

Figure 1.   Synchronous buck converter.

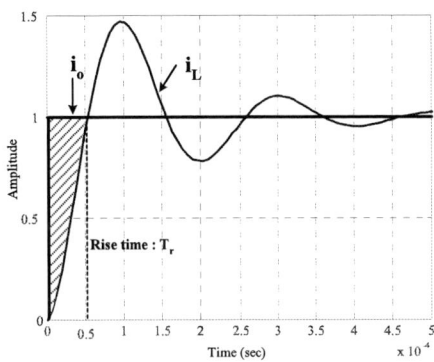

Figure 2.   Step response of load current and inductor current.

1-4244-0448-7/06/$25.00 ©2006 IEEE   1641

III. VOLTAGE MODE CONTROL

Applying the stage space averaging method can derive the dynamic response[8-12]. Figure 4 shows the block diagram of buck converter with feedback loop, and the transfer function of bullocks are defined follows.;

Gdv(s): Transfer function of control to output voltage

Gdi(s): Transfer function of control to inductor current

Zo(s): Output impedance

Gii(s): Transfer function of output current to inductor current

Fm: Frequency modulation

H(s): Voltage sensor gain

Gc(s): Transfer function of compensator

Kiv: Inductor current sensing gain

Hereafter, the transfer functions (loop gain, current transfer function and output impedance) are analyzed which are related to the transient response.

Loop gain;

$$T_v(s) = G_{dv}(s)H(s)G_c(s)F_m \quad (1)$$

Closed-loop output impedance

$$Z_{o\_close}(s) = \frac{Z_o(s)}{1 + T_v(s)} \quad (2)$$

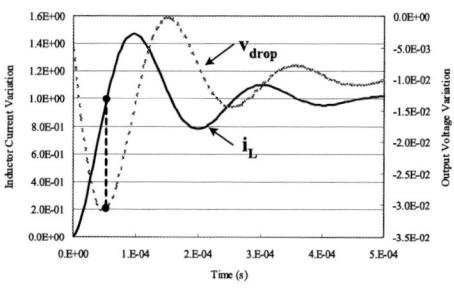

Figure 3. Relationship between inductor current and output voltage drop on transient time.

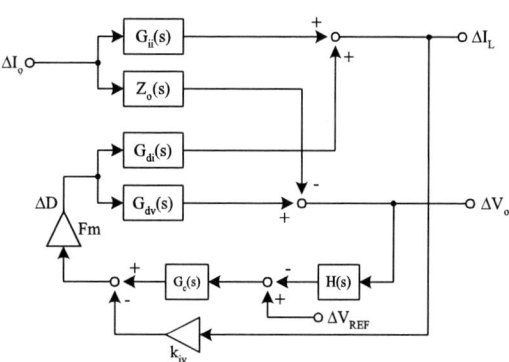

Figure 4. Block diagram.

Closed-loop current transfer function;

$$G_{iic}(s) = G_{ii}(s) + H(s)G_c(s)F_m G_{di}(s)Z_{o\_close}(s) \quad (3)$$

where;

$$G_{dv}(s) = \frac{(sCrc + 1)RV_{in}}{P(s)} \quad (4)$$

$$G_{di}(s) = \frac{sCV_{in}}{P(s)} \quad (5)$$

$$G_{ii}(s) = \frac{sCrc + 1}{P(s)} \quad (6)$$

$$Z_o(s) = \frac{s^2 LC + s(L + Cr_L r_c) + r_L}{P(s)} \quad (7)$$

$$P(s) = s^2 LC + sC(r_L + r_c) + 1 \quad (8)$$

From Eq. (3), the rise time Tr of inductor current can be derived as follows;

$$T_r = \frac{\pi}{2\omega_c} \quad (9)$$

(ωc : bandwidth of loop gain)

The rise time is decreased by extending bandwidth as shown in Eq. (9). Figure 5 shows the experimental results of relation of bandwidth and output voltage drop. The voltage drop is smaller as extension of bandwidth. This is because, the output impedance peak is reduced by extending bandwidth as shown in Fig. 6.

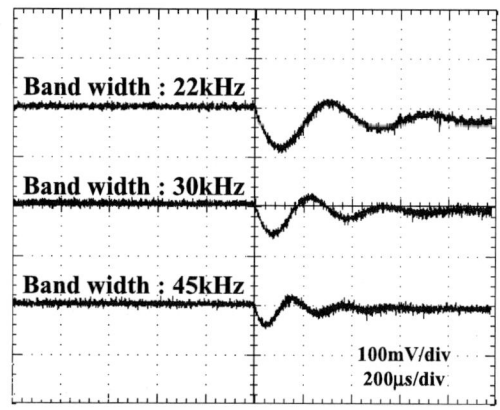

Figure 5. Relation between bandwidth and output voltage variation.

IV. PEAK CURRENT MODE CONTROL

The transfer function can be derived for peak current mode control similarly by applying the stage space averaging method as following equations;

Loop gain;

$$T_c(s) = \frac{F_m G_{dv}(s)}{1 + k_{iv} F_m G_{di}(s)} H(s) G_c(s) \quad (10)$$

Closed-loop output impedance

$$Z_{o\_close}(s) = \frac{Z_o(s) + \dfrac{k_{iv} F_m G_{dv}(s) G_{ii}(s)}{1 + k_{iv} F_m G_{di}(s)}}{1 + T_c(s)} \quad (11)$$

Closed-loop current transfer function;

$$G_{iic}(s) = \frac{G_{ii}(s) + H(s) G_c(s) F_m G_{di}(s) Z_{o\_close}(s)}{1 + k_{iv} F_m G_{di}(s)} \quad (12)$$

In peak current mode case, the current transfer function Gii(s) is changed by influence of current loop as follows;

$$G_{ii}(s) = \frac{sCrc + 1}{P(s)\left(1 + k_{iv} F_m G_{di}(s)\right)}$$
$$= \frac{sCrc + 1}{s^2 LC + sC\left(r_L + r_c + k_{iv} F_m V_{in}\right) + 1} \quad (13)$$

The dumping factor may be over 1 depending on circuit parameter by influence of current loop. In this case, the conjugate complex root changes to two real roots. Figure 7 shows the analytical results of loop gain. As shown in Fig. 7, the resonance peak is disappeared due to the two real poles, and this system can be treated as first order system approximately. From open loop current transfer function, the rise time Tr of inductor current can be derived as follows;

$$T_r = \frac{4}{\omega_{f1}} \quad (14)$$

($\omega f1$: inflection point of lower frequency)

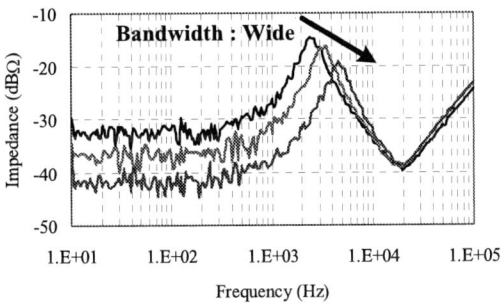

Figure 6. Output impedance of voltage mode control.

Next, a closed loop is considered. The transfer function Giic(s) means the closed loop transfer function of Gii(s). The frequency response of Giic(s) is shown in Fig. 8. As shown in Fig. 8, the frequency of a inflection point of Giic(s) is well agreed with crossover frequency of loop gain. In this case, the rise time is well agreement with Eq.(14), and the relationship between rise time of inductor current and crossover frequency is given by the following equation;

$$T_r = \frac{4}{\omega_c} \quad (15)$$

(ωc : bandwidth of loop gain)

The rise time is decrease by extending bandwidth, as shown in Eq. (15).

Here, the technique of extending bandwidth is considered. There are two methods as a technique for extending bandwidth. There are two methods to extend the bandwidth, one is enlargement DC gain, and the other is used phase compensation. The rise time can be reduced by both methods.

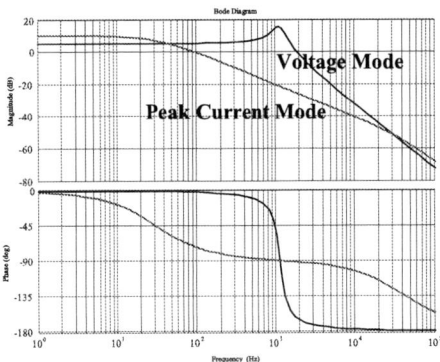

Figure 7. Frequency response of Loop gain.

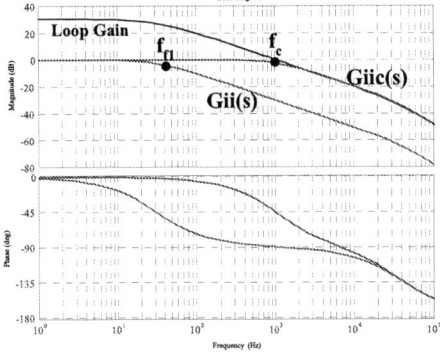

Figure 8. Frequency response of the closed-loop current transfer function.

However, a difference comes out in the transient response. When DC gain is enlarged, the transient response is improved as shown in Fig. 9. On the other hand, in phase compensation case, although bandwidth is wider, the transient response is not changed. This is because the output impedance changes. Since the steady state value Zo(0) of the output impedance becomes small when DC gain is enlarged, the steady-state deviation of output voltage becomes small. On the other hand, when the phase compensation is used, the steady state value Zo(0) of the output impedance dose not change as shown in Fig. 10. Therefore, the steady-state deviation of output voltage does not change and the transient response does not improve. This is because the transient response of POL with peak current mode control becomes the first-order response, and there is advantage that the over shot does not appear in transient response. When the steady state value Zo (0) of the output impedance becomes small and steady-state deviation becomes small and, the transient response is improved. Therefore, it turns out that it does not depend for the transient response of POL with peak current mode control on bandwidth, but is dependent on the steady state value Zo (0) of the output impedance. However, the output impedance of peak current mode control becomes very large compared with voltage mode control as shown in Fig. 11, and the reduction of the output impedance is needed. It is possible to use an integrator as a method of redacting output impedance. In this case, the peak is appeared in output impedance, and the transient response has the second order characteristic and become the same as the case of voltage mode control as shown in Fig. 12. Therefore, it is desirable to enlarge DC gain to reduce the output impedance for transient response improvement. Figure 13 shows the experimental results of relation between DC gain and steady-state value Zo(0) of output impedance. As shown in Fig. 13, enlargement of DC gain reduce the output impedance. Figure 14 shows the experimental results of relation between steady-state value Zo(0) of output impedance and output voltage drop. As shown in Fig. 14, decreasing output impedance reduces the output voltage drop. Moreover, there is no over shoot in transient time.

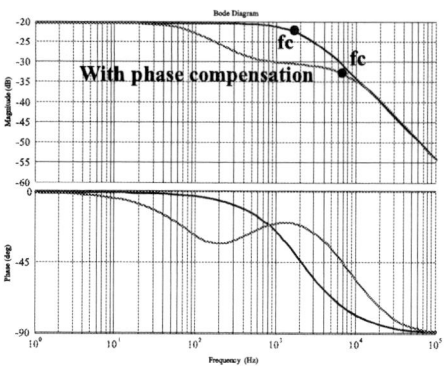

Figure 10. Output impedance with phase compensation.

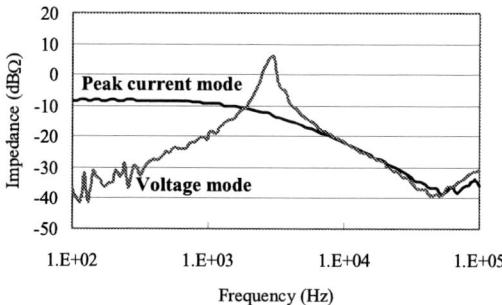

Figure 11. Output impedance.

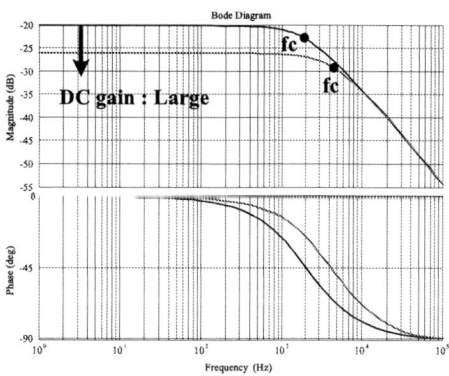

Figure 9. Output impedance with large DC gain.

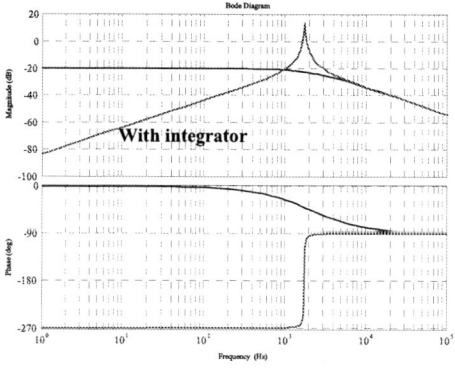

Figure 12. Output impedance with integrator.

V. CONCLUSION

This paper presents the effect of peak current mode control on transient response. The dynamic characteristics are analyzed and it is conformed by experimentally. As a result, in peak current mode control, it is desirable to enlarge DC gain to reduce the output impedance, and decreasing output impedance reduces the voltage drop. Moreover, it turns out that it does not depend for the transient response of POL with peak current mode control on bandwidth, but is dependent on the steady state value Zo (0) of the output impedance.

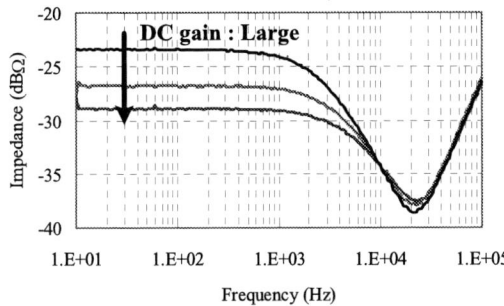

Figure 13. Output impedance of peak current mode control.

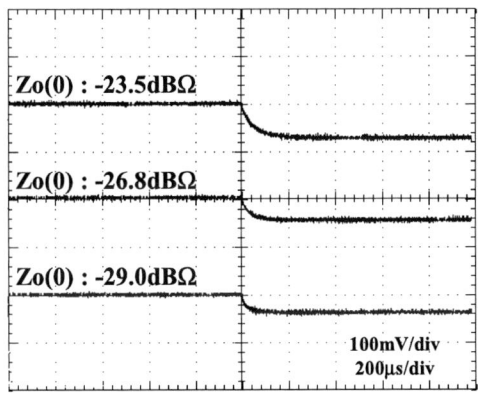

Figure 14. Relation between Zo(0) and output voltage variation.

REFERENCES

[1] M. T. Zhang, M. M. Javanovic, F. C. Lee, "Design Considerations for Low-Voltage On-Board DC/DC Modules for Next Generations of Data Processing Circuits, IEEE Transactions on Power Electronics, Vol. 11, No. 2, pp328-337 , 1996

[2] K. Yao, Y. Meng, F. C. Lee,"Control Bandwidth and Transient Response of Buck Converters," IEEE PESC'02 Record

[3] P. Leong, F.C. Lee, P. Xu, K. Yao,"Critical Inductance in Voltage Regulatro Modules," IEEE Transactions on Power Electronics Vol. 17. No. 4. pp. 485-492, 2002

[4] J. Sun, "Control Design Considerations for Voltage Regulator Modules," IEICE/IEEE INTELEC'03 Proceeding, pp.84-91

[5] R.D. Middlebrook, "Topics in Multiple-Loop Regulators and Current-Mode Programming," IEEE PESC'05 Record, pp. 716-732

[6] R. B. Ridley, "A New, Continuous-Time Model for Current-Mode Control," IEEE Transactions Power Electronics, Vol. 6, No.2, pp271-280

[7] A. Fontan, S. Ollero, E. D. L. Cruz, J. Sebastian, "Peak Current Moede Control Applied to the Forwerd Converter with Active clamp," IEEE PESC'98 Record, pp45-51

[8] R.D. Middlebrook, S. Cuk, "A General Unified Approach to Modeling Switching-Converter Power Stages," IEEE PESC'76 Record, pp. 18-34

[9] B. H. Cho, J. R. Lee, F. C. Lee, "Large-Signal Stability Analysis of Spacecraft Power Processing Systems, IEEE Transactions on Power Electronics, Vol. 5, No. 1, pp110-115, 1990

[10] A .F. Witulski, R. W. Erickson, "Extension of State-Space Averaging to Resonant Switches and Beyond, IEEE Transactions on Power Electronics, Vol. 5, No. 1, pp98-109, 1990

[11] T. Ninomiya, M. Nakahara, T. Higashi, K. Harada, "A Unified Analysis of Resonant Converters," IEEE Transactions Power Electronics, Vol. 6, No. 2, pp260-270 , 1991

[12] S. R. Sanders, J. M. Noworolski, X. Z. Lui, G. C. Verghese, "Generalized Averaging Method for Power Conversion Circuits, IEEE Transactions on Power Electronics, Vol. 6, No. 2, pp251-259, 1991

2006 5th International Power Electronics and Motion Control Conference

Modulations for Voltage Source Rectification and Voltage Source Inversion Using Direct Space Vector Approach

Keping You[*] and M. F. Rahman[**]

School of Electrical Engineering and Telecommunications

University of New South Wales

Sydney 2052, Australia

*: youkeping@ieee.org **:f.rahman@unsw.edu.au

Abstract- **Two arbitrary zero-placement (discontinuous) PWM strategies are proposed in this paper for DC-AC three-phase voltage source-inversion and three-phase AC-DC voltage-source rectification respectively. They are based on the general space vector modulation approach derived from matrix converter theory. The two PWM strategies are sharing the same algorithm in calculating space vector duty-cycles.**

I. INTRODUCTION

This paper will develop two PWM algorithms by exploring the general direct space vector modulation algorithm (G-SVM) approach derived from matrix converter theory in order to operate two-to-three and three-to-two reduced matrix converter for DC-AC Voltage Source Inversion (VSI) and AC-DC Voltage Source Rectification (VSR) respectively.

Originally, for three-to-three AC-AC matrix converters, Alesina-Venturini optimum PWM algorithm (A-V PWM)[1][2], fictitious DC-Bus indirect space vector PWM algorithm (FB-SVM)[4], and the general direct space vector modulation algorithm (G-SVM)[5][6] were developed.

The A-V PWM is a continuous PWM method, while the G-SVM is a discontinuous or arbitrary zero-placement approach.

The reduction of three-to-three A-V PWM into three-to-two and into two-to-three ones were reported by previous work [3][7]. But no scientific reduction work for the discontinuous G-SVM approach has been reported.

The work of this paper will present the reduction for G-SVM, introducing a new option of modulation approach for the practical engineering design and hence offering the intrinsic advantages of G-SVM over A-V PWM in, for example, the controllable three-phase input current displacement, extra freedom of zero-placement.

II. REVIEW OF THE GENERAL SPACE VECTOR MODULATION (G-SVM) FOR THREE-TO-THREE MATRIX CONVERTER

The basic method of the general space vector modulation will be represented here with minor modification.

The basic configuration of three-to-three matrix converter, in Fig. 1, has 21 effective switching configurations (18 active and 3 zero ones) listed in Table I.

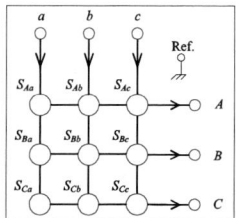

Fig. 1 Basic structure of three-to-three matrix converter

Table I
Switching Configurations of Three-to-three Matrix Converter

No.	Name	A	B	C	v_o	α_o	i_i	β_i
1	+1	a	b	b	$(2/3)v_{ab}$	0	$(2/\sqrt{3})i_A$	$-\pi/6$
2	-1	b	a	a	$-(2/3)v_{ab}$	0	$-(2/\sqrt{3})i_A$	$-\pi/6$
3	+2	b	c	c	$(2/3)v_{bc}$	0	$(2/\sqrt{3})i_A$	$\pi/2$
4	-2	c	b	b	$-(2/3)v_{bc}$	0	$-(2/\sqrt{3})i_A$	$\pi/2$
5	+3	c	a	a	$(2/3)v_{ca}$	0	$(2/\sqrt{3})i_A$	$7\pi/6$
6	-3	a	c	c	$-(2/3)v_{ca}$	0	$-(2/\sqrt{3})i_A$	$7\pi/6$
7	+4	b	a	b	$(2/3)v_{ab}$	$2\pi/3$	$(2/\sqrt{3})i_B$	$-\pi/6$
8	-4	a	b	a	$-(2/3)v_{ab}$	$2\pi/3$	$-(2/\sqrt{3})i_B$	$-\pi/6$
9	+5	c	b	c	$(2/3)v_{bc}$	$2\pi/3$	$(2/\sqrt{3})i_B$	$\pi/2$
10	-5	b	c	b	$-(2/3)v_{bc}$	$2\pi/3$	$-(2/\sqrt{3})i_B$	$\pi/2$
11	+6	a	c	a	$(2/3)v_{ca}$	$2\pi/3$	$(2/\sqrt{3})i_B$	$7\pi/6$
12	-6	c	a	c	$-(2/3)v_{ca}$	$2\pi/3$	$-(2/\sqrt{3})i_B$	$7\pi/6$
13	+7	b	b	a	$(2/3)v_{ab}$	$4\pi/3$	$(2/\sqrt{3})i_C$	$-\pi/6$
14	-7	a	a	b	$-(2/3)v_{ab}$	$4\pi/3$	$-(2/\sqrt{3})i_C$	$-\pi/6$
15	+8	c	c	b	$(2/3)v_{bc}$	$4\pi/3$	$(2/\sqrt{3})i_C$	$\pi/2$
16	-8	b	b	c	$-(2/3)v_{bc}$	$4\pi/3$	$-(2/\sqrt{3})i_C$	$\pi/2$
17	+9	a	a	c	$(2/3)v_{ca}$	$4\pi/3$	$(2/\sqrt{3})i_C$	$7\pi/6$
18	-9	c	c	a	$-(2/3)v_{ca}$	$4\pi/3$	$-(2/\sqrt{3})i_C$	$7\pi/6$
19	zero	a	a	a	0	0	0	0
20	zero	b	b	b	0	0	0	0
21	zero	c	c	c	0	0	0	0

And their corresponding switching state space vectors in term of the output voltages and input currents are presented by space vector diagrams in Fig. 2, where input phase *a* is the zero space angle reference axis.

The basic idea of the general space vector modulation for three-to-three matrix converter is, by combination of four switching configurations within one switching period, to synthesize volt-second average voltages as close as possible to the expected output fundamental

1-4244-0448-7/06/$25.00 ©2006 IEEE

phase voltages in form of space vector (both the magnitude v_o and space angle α_o) and simultaneously to keep the input current displacement angle β_i as close as possible to the expected one.

Table II
Duty-cycle Distribution for three-to-three matrix converter

		output voltage sectors												
		1 or 4					2 or 5					3 or 6			
input current sectors	1 or 4	(1 1) or (4 4)	+9	-7	-3	+1	(4 2) or (1 5)	+6	-4	-9	+7	(1 3) or (4 6)	+3	-1	-6	+4
		(4 1) or (1 4)	-9	+7	+3	-1	(1 2) or (4 5)	-6	+4	+9	-7	(4 3) or (1 6)	-3	+1	+6	-4
	2 or 5	(5 1) or (2 4)	+8	-9	-2	+3	(2 2) or (5 5)	+5	-6	-8	+9	(5 3) or (2 6)	+2	-3	-5	+6
		(2 1) or (5 4)	-8	+9	+2	-3	(5 2) or (2 5)	-5	+6	+8	-9	(2 3) or (5 6)	-2	+3	+5	-6
	3 or 6	(3 1) or (6 4)	+7	-8	-1	+2	(6 2) or (3 5)	+4	-5	-7	+8	(3 3) or (6 6)	+1	-2	-4	+5
		(6 1) or (3 4)	-7	+8	+1	-2	(3 2) or (6 5)	-4	+5	+7	-8	(6 3) or (3 6)	-1	+2	+4	-5
		$(K_i\,K_v)$	1	2	3	4	$(K_i\,K_v)$	1	2	3	4	$(K_i\,K_v)$	1	2	3	4

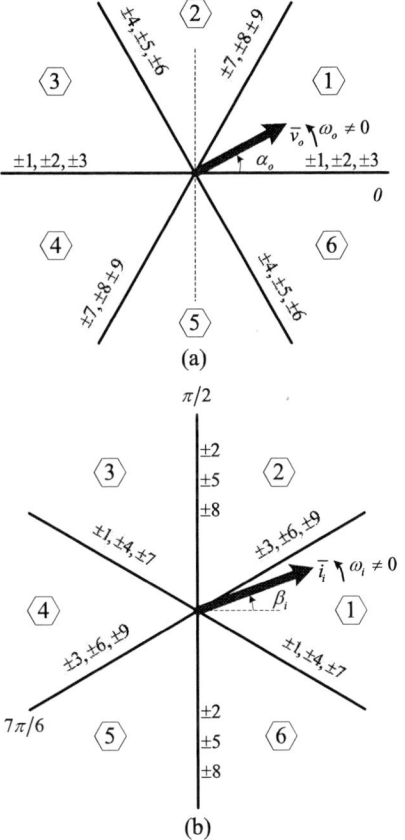

(a)

(b)

Fig. 2 Output voltage and input current hexagons by the general space vector modulation approach for three-to-three matrix converter.

The vector duty-cycles for the abovementioned four switching configurations are calculated in equation group(1), then are converted into switch duty-cycles by aid of Table II.

The Table II, unlike its original form in [6], specifies the combination of sector-pair for every four active switching configurations, leaving the calculation of duty-cycles in (1) without considering the signs and resulting to straight use of the table in programming.

$$d_1 = \frac{2}{\sqrt{3}} q \, \frac{\cos(\tilde{\alpha}_o - \pi/3)\cos(\tilde{\beta}_i - \pi/3)}{\cos\varphi_i}$$
$$d_2 = \frac{2}{\sqrt{3}} q \, \frac{\cos(\tilde{\alpha}_o - \pi/3)\cos(\tilde{\beta}_i + \pi/3)}{\cos\varphi_i} \tag{1}$$
$$d_3 = \frac{2}{\sqrt{3}} q \, \frac{\cos(\tilde{\alpha}_o + \pi/3)\cos(\tilde{\beta}_i - \pi/3)}{\cos\varphi_i}$$
$$d_4 = \frac{2}{\sqrt{3}} q \, \frac{\cos(\tilde{\alpha}_o + \pi/3)\cos(\tilde{\beta}_i + \pi/3)}{\cos\varphi_i}$$
$$d_0 = 1 - (d_1 + d_2 + d_3 + d_4)$$

$$q = \frac{v_o}{v_i} \tag{2}$$

where, $-\frac{\pi}{6} < \tilde{\alpha}_o < \frac{\pi}{6}$, $-\frac{\pi}{6} < \tilde{\beta}_i < \frac{\pi}{6}$, and $-\frac{\pi}{2} < \varphi_i < \frac{\pi}{2}$, $\tilde{\alpha}_o$ and $\tilde{\beta}_i$ are the angles with respect to the bisecting lines of target sectors respectively; φ_i is the phase displacement of input current; v_o, v_i are the amplitude of phase output and phase input voltage space vector respectively.

III. SIMPLIFIED GENERAL SPACE VECTOR APPROACH FOR DC-AC OPERATION

The two-to-three or three-to-two form of the conventional three-to-three matrix converter can be obtained under the constraint of matrix converter theory by means of adjustment of modulation and topological configuration. This section will focus on the issue of two-to-three configuration.

If the voltage source terminals of a three-to-three matrix converter is supplied by the three-phase balanced sinusoidal signals and all switches are PWM controlled by G-SVM scheme, then the general space vector modulation is still naturally adaptive if the input voltage space vector by any way is stopped at an arbitrary angle. And any stopped symmetric balanced voltage space vector is always physically mapped to three unique voltage potentials, as in(3), at the three terminals where the voltage space vector was physically rotating. And we assume the existence of the input current space vector as in(4).

$$v_{a,b,c} = [v_i \cos\alpha_i \;\; v_i\cos(\alpha_i - 2\pi/3) \;\; v_i\cos(\alpha_i + 2\pi/3)]^T \tag{3}$$

$$i_{a,b,c} = [i_i\cos(\alpha_i + \varphi_i) \;\; i_i\cos(\alpha_i + \varphi_i - 2\pi/3) \;\; i_i \cos(\alpha_i + \varphi_i + 2\pi/3)]^T \tag{4}$$

where α_i is the space angle at which the input voltage space vector is stopped, and φ_i is input current displacement angle, v_i, i_i are peak phase voltage and current respectively.

The three phase terminals can be electrically "merged" into two terminals if and only if two of the three terminals have exactly the same voltage potential.

By trigonometric manipulation one can easily prove that the situation of equal voltage potential among any two terminals can be obtained if and only if the expression of (5) is true.

$$\alpha_i = zero \;\; or \;\; 2\pi/3 \;\; or \;\; 4\pi/3. \tag{5}$$

One is free to select any value in (5), and hereafter we let $\alpha_i = 0$ for further discussion, the input voltages then become

$$v_{a,b,c}=[v_i \ (-1/2)v_i \ (-1/2)v_i]^T \qquad (6)$$

Thus two terminals b and c have equal voltage potential leading to that the input line-to-line voltages equivalently become an identical DC voltage, and consequently the input current space vector is expected to be stopped in the same manner as in(7). This is shown in Fig. 4.

$$i_{a,b,c}=[i_i cos(\varphi_i) \ i_i \ cos(\varphi_i - 2\pi/3) \ i_i \ cos(\varphi_i + 2\pi/3)]^T \quad (7)$$

Replacing the voltage-source terminals (a, b, c) by (p, n, n) in Fig. 1, one may obtain Fig 3, and have 12 active switching configurations remained, which can be divided into six groups, each having one identical output voltage switching state space vector and two different input current switching-state space vectors, as shown in Table III.

For the case in (3) and (4), only sector 1 and 4 of input current hexagon are involved in the synthesis of input DC current. The duty-cycle distribution coding table then may be simplified as that in Table IV. The whole procedure can be illustrated by space vector diagrams in Fig. 4. And the calculation of space vector duty-cycles is in (8) and (9).

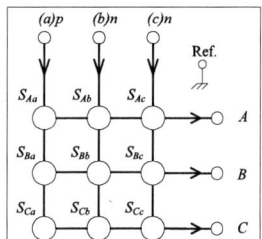

Fig 3 One instance of forming the two-to-three configuration

Table III
Switching Configurations of Two-to-three Matrix Converter

No.	Group	Name	A	B	C	v_o	α_o	i_i	β_i
1	1	+1	p	n	n	$(2/3)Vdc$	0	$(2/\sqrt{3})i_A$	$-\pi/6$
2		-3	p	n	n	$(2/3)Vdc$	0	$-(2/\sqrt{3})i_A$	$7\pi/6$
3	2	+3	n	p	p	$-(2/3)Vdc$	0	$(2/\sqrt{3})i_A$	$7\pi/6$
4		-1	n	p	p	$-(2/3)Vdc$	0	$-(2/\sqrt{3})i_A$	$-\pi/6$
5	3	+4	n	p	n	$(2/3)Vdc$	$2\pi/3$	$(2/\sqrt{3})i_B$	$-\pi/6$
6		-6	n	p	n	$(2/3)Vdc$	$2\pi/3$	$-(2/\sqrt{3})i_B$	$7\pi/6$
7	4	+6	p	n	p	$-(2/3)Vdc$	$2\pi/3$	$(2/\sqrt{3})i_B$	$7\pi/6$
8		-4	p	n	p	$-(2/3)Vdc$	$2\pi/3$	$-(2/\sqrt{3})i_B$	$-\pi/6$
9	5	+7	n	n	p	$(2/3)Vdc$	$4\pi/3$	$(2/\sqrt{3})i_C$	$-\pi/6$
10		-9	n	n	p	$(2/3)Vdc$	$4\pi/3$	$-(2/\sqrt{3})i_C$	$7\pi/6$
11	6	+9	p	p	n	$-(2/3)Vdc$	$4\pi/3$	$(2/\sqrt{3})i_C$	$7\pi/6$
12		-7	p	p	n	$-(2/3)Vdc$	$4\pi/3$	$-(2/\sqrt{3})i_C$	$-\pi/6$
13	7	zero	p	p	p	0	0	0	0
14	8	zero	n	n	n	0	0	0	0

$$d_1 = d_2 = (1/\sqrt{3})q\cos(\tilde{\alpha}_o - \pi/3)$$
$$d_3 = d_4 = (1/\sqrt{3})q\cos(\tilde{\alpha}_o + \pi/3) \qquad (8)$$
$$d_z = 1-(d_1 + d_2 + d_3 + d_4)$$

where, $-\dfrac{\pi}{6} < \tilde{\alpha}_o < \dfrac{\pi}{6}$; the subscript z represents the *zero* switching configuration, and

$$q = \frac{v_o}{(2/3)V_{dc}} \qquad (9)$$

It is easy to prove that the proposed modulation algorithm for two-to-three matrix converter is the general form of the well-known space vector modulation for traditional VSI.

Table IV
Duty-cycle distribution for Two-to-three Matrix Converter

		output voltage sectors													
		1 or 4				2 or 5				3 or 6					
Stopped $ie^{j\beta}$	1 or 4	(1 1) or (4 4)	+9	-7	-3	+1	(4 2) or (1 5)	+6	-4	-9	+7	(1 3) or (4 6)	+3	-1	-6	+4
		(4 1) or (1 4)	-9	+7	+3	-1	(1 2) or (4 5)	-6	+4	+9	-7	(4 3) or (1 6)	-3	+1	+6	-4
		$(K_i \ K_v)$	1	2	3	4	$(K_i \ K_v)$	1	2	3	4	$(K_i \ K_v)$	1	2	3	4

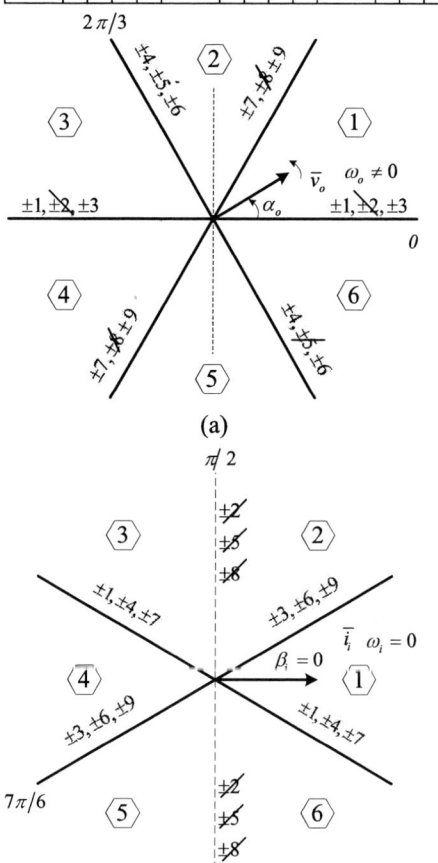

Fig. 4 Output voltage and input current hexagons by the general space vector modulation approach after the DC-AC two-to-three simplification.

IV. SIMPLIFIED GENERAL SPACE VECTOR APPROACH FOR AC-DC OPERATION

For the reduction from three-to-three matrix converter to the so-called three-to-two AC-DC bidirectional matrix converter, the required modulation method must satisfy the following three constraints simultaneously:

(1) Existence theme of matrix converter theory in [1],

(2) Fundamental output voltage space vector must be stopped at certain space position, and

(3) Full use of the input AC line-line voltage in synthesis of the voltage between the two terminals to be used as DC terminals.

With the three-phase balanced signals at output terminals (A, B, C), one is free to select which two terminals is going to be used as DC terminals, and one may also easily find that, if, for example and hereafter, terminal A and C are selected, the required angle of stopped output voltage space vector satisfying the above constraint (3) must be $\alpha_o = 30^o \pm k90^o$, $k=0,1,2,...$ referring to the zero space angle axis located at input phase a.

Let $k=0$, then the angle of the stopped output voltage space vector is 30°, and the current-source terminal voltages become

$$[v_A\ v_B\ v_C]^T = [(\sqrt{3}/2)v_o\ \ 0\ \ -(\sqrt{3}/2)v_o]^T \quad (10)$$

where the v_o is the amplitude of the sinusoidal phase voltage space vector.

The two terminals having reversed signs but equal magnitude of voltage can then be treated as DC terminals with voltage drop V_{dc} from A to C, knowing (10), one has

$$V_{dc} = v_A - v_C = \sqrt{3}\,v_o \quad \text{or} \quad v_o = (1/\sqrt{3})V_{dc} \quad (11)$$

, and the rest zero-voltage terminal becomes the potential mid-point between the foresaid two ones, can then be ignored in physical implementation.

Replacing terminal names A and C by p and n respectively in Fig. 1, one may obtain the Fig 5, and find immediately only 12 of the total 18 active switching configurations can contribute to the synthesis of the output DC voltage.

This can be illustrated by both the two shadowed parts of Table V and the space vector diagrams for the output voltages and input currents in Fig. 6. The reduced form of duty-cycle distribution scheme can be obtained in Table VI. The calculation of the corresponding duty-cycles can then be simplified in(12).

$$d_1 = d_3 = (1/\sqrt{3})q\cos(\tilde{\beta}_i - \pi/3)/\cos\varphi_i$$
$$d_2 = d_4 = (1/\sqrt{3})q\cos(\tilde{\beta}_i + \pi/3)/\cos\varphi_i \quad (12)$$
$$d_0 = 1-(d_1+d_2+d_3+d_4)=1-(2d_1+2d_2)$$
$$-\frac{\pi}{6}<\tilde{\beta}_i<\frac{\pi}{6}, \text{ and } -\frac{\pi}{2}<\varphi_i<\frac{\pi}{2}$$

$$q = (1/\sqrt{3})V_{dc}/v_i = V_{dc}/v_{il-l} \quad (13)$$

where, $\tilde{\beta}_i$ and φ_i are of the same definition in (1); v_{il-l} is input line-line voltage amplitude, which is normally used in most practical engineering design.

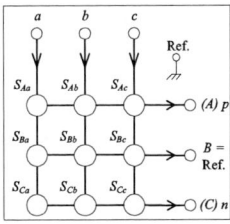

Fig 5 One instance of forming the three-to-two configuration

Table V
Switching Configurations of Three-to-two Matrix Converter

No.	Name	A	B	C	v_o	α_o	i_i	β_i
1	+1	a	b	b	$(2/3)v_{ab}$	0	$(2/\sqrt{3})i_A$	$-\pi/6$
2	-1	b	a	a	$-(2/3)v_{ab}$	0	$-(2/\sqrt{3})i_A$	$-\pi/6$
3	+2	b	c	c	$(2/3)v_{bc}$	0	$(2/\sqrt{3})i_A$	$\pi/2$
4	-2	c	b	b	$-(2/3)v_{bc}$	0	$-(2/\sqrt{3})i_A$	$\pi/2$
5	+3	c	a	a	$(2/3)v_{ca}$	0	$(2/\sqrt{3})i_A$	$7\pi/6$
6	-3	a	c	c	$-(2/3)v_{ca}$	0	$-(2/\sqrt{3})i_A$	$7\pi/6$
7	+4	b	a	b	$(2/3)v_{ab}$	$2\pi/3$	$(2/\sqrt{3})i_B$	$-\pi/6$
8	-4	a	b	a	$-(2/3)v_{ab}$	$2\pi/3$	$-(2/\sqrt{3})i_B$	$-\pi/6$
9	+5	c	b	c	$(2/3)v_{bc}$	$2\pi/3$	$(2/\sqrt{3})i_B$	$\pi/2$
10	-5	b	c	b	$-(2/3)v_{bc}$	$2\pi/3$	$-(2/\sqrt{3})i_B$	$\pi/2$
11	+6	a	c	a	$(2/3)v_{ca}$	$2\pi/3$	$(2/\sqrt{3})i_B$	$7\pi/6$
12	-6	c	a	c	$-(2/3)v_{ca}$	$2\pi/3$	$-(2/\sqrt{3})i_B$	$7\pi/6$
13	+7	b	b	a	$(2/3)v_{ab}$	$4\pi/3$	$(2/\sqrt{3})i_C$	$-\pi/6$
14	-7	a	a	b	$-(2/3)v_{ab}$	$4\pi/3$	$-(2/\sqrt{3})i_C$	$-\pi/6$
15	+8	c	c	b	$(2/3)v_{bc}$	$4\pi/3$	$(2/\sqrt{3})i_C$	$\pi/2$
16	-8	b	b	c	$-(2/3)v_{bc}$	$4\pi/3$	$-(2/\sqrt{3})i_C$	$\pi/2$
17	+9	a	a	c	$(2/3)v_{ca}$	$4\pi/3$	$(2/\sqrt{3})i_C$	$7\pi/6$
18	-9	c	c	a	$-(2/3)v_{ca}$	$4\pi/3$	$-(2/\sqrt{3})i_C$	$7\pi/6$
19	zero	a	a	a	0	0	0	0
20	zero	b	b	b	0	0	0	0
21	zero	c	c	c	0	0	0	0

Table VI
Duty-cycle Distribution for Three-to-two matrix converter

		Fixed output voltage Space Vector			
		1 or 4			
Rotating input current Space Vector	1 or 4	(1 1) or (4 4)	+9	-7	-3	+1
		(4 1) or (1 4)	-9	+7	+3	-1
	2 or 5	(5 1) or (2 4)	+8	-9	-2	+3
		(2 1) or (5 4)	-8	+9	+2	-3
	3 or 6	(3 1) or (6 4)	+7	-8	-1	+2
		(6 1) or (3 4)	-7	+8	+1	-2
		$(K_i\ K_v)$	**1**	**2**	**3**	**4**

V. CONSIDERATIONS ON IMPLEMENTATION

This section focuses on the brief considerations in implementing the proposed simplified three-to-two G-SVM for VSR. The implementation for the case of two-to-three variety is ignored because it is identical to that for conventional SVM for VSI.

The microelectronic circuitry, usually a combination of a microprocessor or digital signal processor (DSP) and a sequential logic circuitry, is needed to implement the proposed G-SVM for VSR. Since the G-SVM algorithm does not require any deliberate hardware circuitry for the synchronization to the three-phase inputs, the input voltage sector is captured by calculating the discrete sampled input voltages. And here we assume the speeds of all programmable devices are sufficient. Generally speaking three main tasks are involved in sequence:

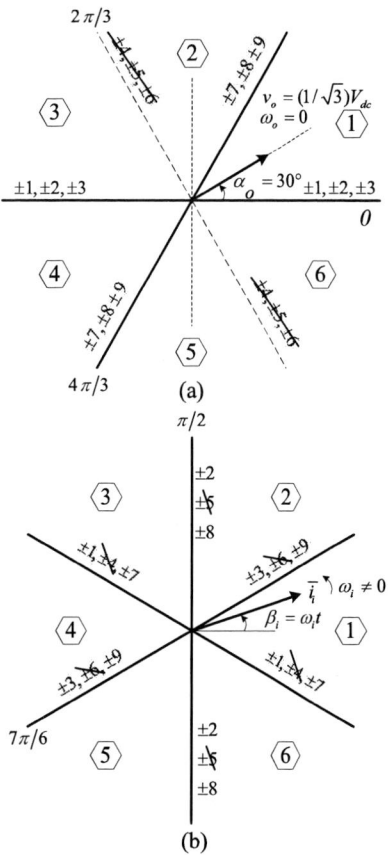

(a)

(b)

Fig. 6 Hexagons by the general space vector modulation approach after AC-DC three-to-two simplification

(a) calculation of expected vector duty-cycles

(b) synthesis and allocation of switching duty-cycles for topological switches

(c) design of sequencers for the right current commutation among physical switches

Task (a) is usually a software design on the platform of a given microprocessors or DSP, since it almost always deeply involve in the system control scheme. The task (c) is usually a pure hardware design of a finite state machine (FSM).

Task (b) is a grey area between the software design of DSP and the hardware design of logic circuitry.

Table V and Table VI can be used directly in software programming for the task (a).

If, for example, the task (b) is supposed to be implemented by DSP as well, the program merely needs an extra lookup table and an extra matrix multiplication.

Allocating task (b) in the microprocessor stage allows the flexible selection between symmetric and unsymmetric pattern of PWM by swapping the relevant lookup tables in program coding.

If the burden of task (b) is shifted into the hardware design of the logic circuitry, the inputs of the logic circuit are four vector duty-cycles and one address for locating the switching duty-cycles.

The microprocessor must have four independent PWM timers and a usable interrupt source for synchronization of the four independent PWM timers.

VI. CONCLUSION

General space vector modulation algorithms for three-to-two AC-DC VSR and two-to-three DC-AC VSI are proposed and theoretically justified. And it is feasible to use the simplified general space vector modulation in engineering design for AC-DC VSR and DC-AC VSI.

REFERENCE:

[1] Marco Venturini, Aleberto Alesina, "The Generalised Transformer: A New Bidirectional Sinusoidal Waveform Frequency Converter With Continuously Adjustable Input Power Factor", IEEE Power Electrn. Specialist Conf. (PESC) Rec., pp. 242-252, 1980.

[2] Alesina, A.; Venturini, M.;"Intrinsic amplitude limits and optimum design of 9-switches direct PWM AC-AC converters";Power Electronics Specialists Conference, 1988. PESC '88 Record., 19th Annual IEEE,11-14 April 1988,Pages:1284 - 1291 vol.2

[3] Holmes, D.G.; "A new modulation algorithm for voltage and current source inverters, based on AC-AC matrix converter theory"; Industry Applications Society Annual Meeting, 1990., Conference Record of the 1990 IEEE , 7-12 Oct. 1990; Page(s): 1190 -1195 vol.2

[4] Huber, L.,Borojevic, D.; "Space vector modulated three-phase to three-phase matrix converter with input power factor correction"; Industry Applications, IEEE Transactions on , Volume: 31 , Issue: 6 , Nov.- ec. 1995; Pages:1234 - 1246

[5] Casadei, D.; Grandi, G.; Serra, G.; Tani, A.; "Space vector control of matrix converters with unity input power factor and sinusoidal input/output waveforms," Power Electronics and Applications, 1993., Fifth European Conference on , 13-16 Sep 1993,Pages:170 - 175 vol.7

[6] Casadei, D.; Serra, G.; Tani, A.; Zarri, L.; "Matrix converter modulation strategies: a new general approach based on space-vector representation of the switch state," Industrial Electronics, IEEE Transactions on , Volume: 49 , Issue: 2 , April 2002 Pages:370 - 381

[7] Holmes, D.G.; Lipo, T.A.; "Implementation of a controlled rectifier using AC-AC matrix converter theory;" Power Electronics, IEEE Transactions on , Volume: 7 , Issue: 1 , Jan. 1992;Pages:240 – 250

2006 5th International Power Electronics and Motion Control Conference

Synchronization of Voltage Waveforms in Basic Topologies of Dual Inverter-Fed Motor Drives

V. Oleschuk*,**, F. Profumo**, A. Tenconi**, R. Bojoi** and A.M. Stankovic***

*Academy of Sciences of Moldova, Institute of Power Engineering, Kishinau, Republic of Moldova
**Politecnico di Torino, Department of Electrical Engineering, Turin, Italy
***Northeastern University, Department of Electrical and Computer Engineering, Boston, USA

Abstract — This paper presents results of application of novel methods of synchronized pulsewidth modulation (PWM) to the open-end winding induction motor drives on the base of dual two-level voltage source inverters, with both single and separate dc voltage sources. The proposed control algorithms provide synchronous symmetrical control of the motor phase voltage, together with elimination of the common-mode voltages in the systems, during the whole control range including the zone of overmodulation. Spectra of the phase voltage of motor drives with synchronized PWM do not contain even harmonics and sub-harmonics (combined harmonics), which is especially important for the systems with increased power rating. Simulations give the behavior of basic versions of synchronized PWM in these systems.

Keywords – induction motor drive; dual two-level inverters; synchronized pulsewidth modulation

I. INTRODUCTION

Multilevel converters foe power adjustable speed drives are a subject of increasing interest in the last years due to some advantages compared with standard inverters. Ones of the interesting and perspective topologies of power converters are now cascaded (dual) two-level [1]-[4] and three-level [1],[5] converters which utilize two standard three-phase voltage source inverters.

The structure of adjustable speed drive system based on the cascaded converters is constructed by splitting the neutral connection of the induction motor and connecting both ends of each phase coil to an inverter. In this case the cascaded converters are capable of producing voltages which are identical to those of multilevel converters [2].

Between the most important problems for dual inverter-fed drive systems are problem of minimization or elimination of the common-mode voltages in the systems, and also problem of synchronization of operation of two inverters in dual inverter-fed drive system. In order to provide synchronization of the output voltage waveforms for drive converters, novel methods (methodology) of synchronized PWM have been recently proposed [6]-[8].

This paper presents results of dissemination of new methodology of PWM for control of dual inverter-fed drive systems on the base of two-level voltage source inverters, allowing providing both synchronization of the phase voltage waveforms and elimination of the common-mode voltages in these systems.

II. DUAL INVERTER-FED DRIVES ON THE BASE OF TWO-LEVEL INVERTERS WITH SINGLE DC VOLTAGE SOURCE

Fig. 1 presents the basic structure of a dual inverter-fed open-end winding induction motor drive with two-level inverters, where INV1 and INV2 are standard two-level voltage source inverters. The single power supply is used for both inverters in this case, and elimination of the common-mode voltages is provided by the specialized schemes of pulsewidth modulation.

A. Control Scheme with Mutual Compensation of the Common-Mode Voltages

Fig. 2 shows the switching state vectors of two inverters, which provide elimination of zero sequence currents in drive system [3]. The conventional definition for the switching state sequences (voltage vectors) for the switches of the phases of *ABC* of each individual inverter is used here. In particular, for INV1: **1** – 100; **2** – 110; **3** – 010; **4** –011; **5** – 001; **6** – 101, **0** – 000, **7** - 111 (1 - switch-on state, 0 – switch-off state); and the same definition are used for INV2: **1'** – 1'0'0'; **2'** – 1'1'0'; **3'** – 0'1'0'; **4'** –0'1'1'; **5'** – 0'0'1'; **6'** – 1'0'1'; **0'** – 0'0'0'; **7'** – 1'1'1', where 1' - switch-on state of switches of INV2, and 0' – switch-off state in INV2.

Zero switching state sequences 0 and 7 for INV1, and **0'** and 7' for INV2 provide notches in the output voltage waveform of each inverter. Each inverter generates in this case alternating common-mode voltage, but these common-mode voltages do not cause zero sequence currents in the machine phase winding, because these voltages compensate each other [3].

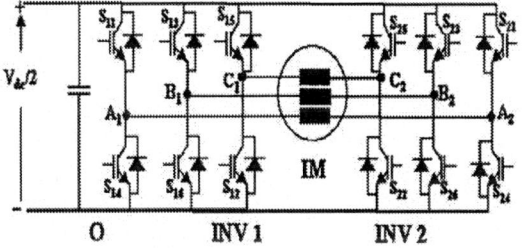

Figure 1. Topology of dual inverter-fed drive on the base of two-level inverters with single dc-link voltage source [3]

1-4244-0448-7/06/$25.00 ©2006 IEEE

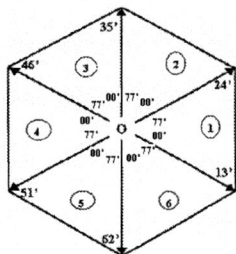

Figure 2. Voltage space-vector combinations, providing mutual compensation of the common-mode voltages [3]

In order to avoid asynchronism of conventional versions of voltage space-vector modulation, novel method (methodology) of direct synchronized PWM can be used for independent control of each inverter [6]. Figs. 3-4 show continuous (CPWM, Fig. 3) and discontinuous (DPWM, Fig. 4) schemes of synchronized PWM for a quarter-period of the output voltage of inverter. The upper curves are here the switching state sequence, then - control signals for the cathode switches of the phases a, b and c. The lower curves in Figs. 3-4 show the corresponding quarter-wave of the line output voltage of inverter. Signals βj represent total switch-on durations during the switching period τ, signals γ_k are generated on the boarders (Fig. 3) or in the centers (Fig. 4) of the corresponding β. Widths of notches λ_k represent duration of zero sequences.

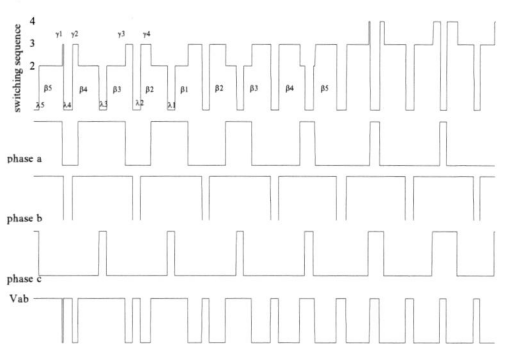

Figure 3. Control and output signals for quarter-period of three-phase inverter with continuous PWM (CPWM)

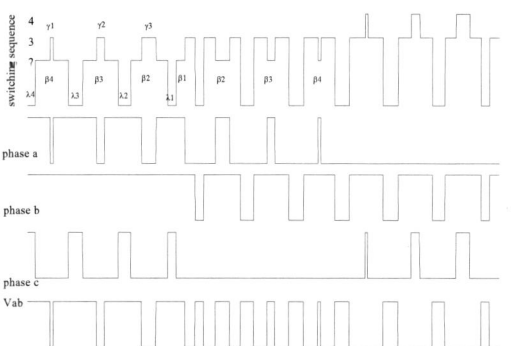

Figure 4. Control and output signals for quarter-period of three-phase inverter with discontinuous PWM (DPWM)

Special signals λ' (λ_5 in Fig. 3, λ_4 in Fig. 4) with the neighboring β'' (β_5 in Fig. 3, β_4 in Fig. 4) are formed in the clock-points (0^0, 60^0, 120^0..) of the output curve of inverters with synchronized PWM. They are reduced simultaneously till close to zero value at the boundary frequencies F_i between control sub-zones, providing a continuous adjustment of voltage waveforms with smooth pulses-ratio changing. Control correlations for the method of synchronized PWM are presented in [6].

Fig. 5 presents the pole voltages of the phase A V_{A10} and V_{A20} of the inverters INV1 and INV2, zero sequence voltages of each inverter V_{01} and V_{02} and their difference V_{01} - V_{02}, and phase voltage of the system $V_{A1A2} = V_{A10}$ - V_{A20}, for continuous (CPWM) scheme of synchronous modulation, applied for control of a dual inverter-fed open-end winding motor drive with scalar V/F control at the fundamental frequency $F = 40\ Hz$ (modulation index $m = 0.8$ in this case). Switching frequency $F_s = 1\ kHz$.

All control and output signals of the inverters INV1 and INV2 have mutual phase shift in 120^0 in accordance with the scheme of space-vector PWM (see Fig. 2), which provides elimination of the common-mode voltages in the phase voltage of dual inverter-fed drive system.

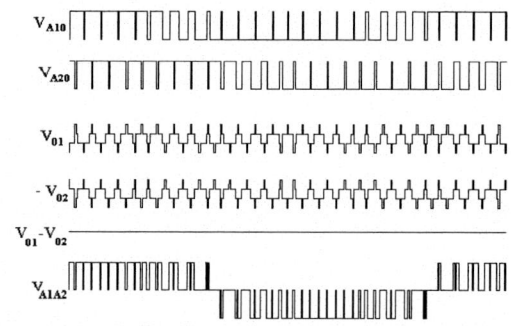

Figure 5. The pole and common-mode voltages and their difference, and phase voltage of drive system with synchronized CPWM ($F = 40\ Hz$)

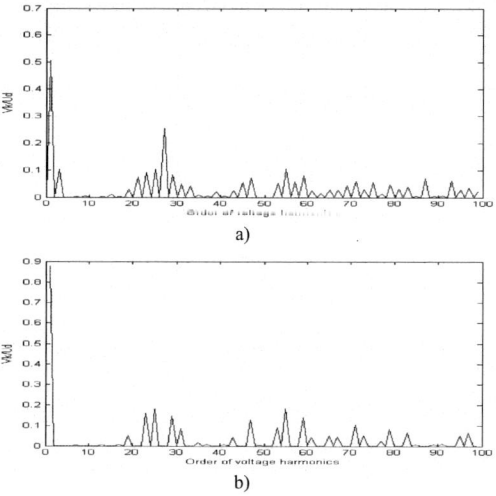

Figure 6. Spectrum of the pole (a) and phase (b) voltages of the system with synchronized CPWM ($F = 40\ Hz$)

So, difference $V_{01} - V_{02}$ of zero sequence voltages of each inverter V_{01} and V_{02} is equal to zero during the whole control range (these voltages compensate each other, see the corresponding curves in Fig. 5). Fig. 6 shows the corresponding spectra of the pole and phase voltages. Spectrum of the pole voltage (Fig. 6,a) of each inverter includes odd harmonics, including triplen order harmonics. But spectrum of the phase voltage of the drive system (Fig. 6,b) includes only odd (non-triplen) voltage harmonics, and all triplen order harmonics and all even harmonics are lacking here due to the described algorithm.

Fig. 7 presents the corresponding voltage waveforms of the drive system with discontinuous synchronized PWM (DPWM) at the fundamental frequency $F = 40Hz$, modulation index $m = 0.8$ in this case. Fig. 8 shows the spectra of the pole and phase voltages of the system. The spectra of the phase voltage of the system, presented in Fig. 8,b, contain only odd harmonics without the triplen components.

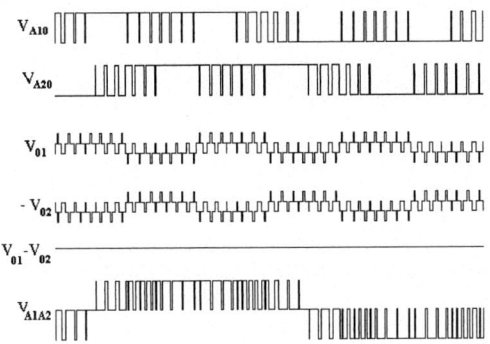

Figure 7. The pole and common-mode voltages and their difference, and the phase voltage of the drive system with synchronized DPWM
($F = 40 Hz$)

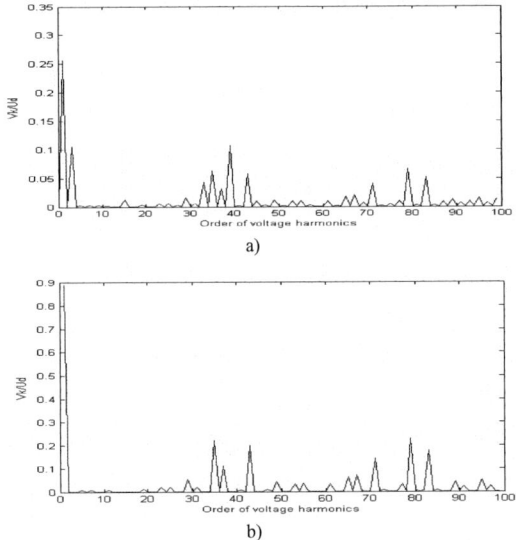

a)

b)

Figure 8. Spectrum of the pole (a) and phase (b) voltages of the system with synchronized DPWM ($F = 40 Hz$)

B. Control Scheme with Full Elimination of the Common-Mode Voltages

Full cancellation of the common-mode voltages in dual inverter-fed drives can be achieved by the use of the specialized control schemes. Fig. 9 shows the switching state vectors of two inverters, which provide full cancellation of the common-mode voltages generated by the individual inverters [4]. In order to eliminate generation of the common-mode voltages by individual inverters in dual inverter-fed drive, only odd switching sequences (combinations of voltage space-vectors **1**, **3**, **5** and **1'**, **3'**, **5'**, see Fig. 9) are used in this scheme of PWM. So, the possibility of undesirable bearing currents and leakage currents are fully avoided by this scheme.

Fig. 10 – Fig. 13 present some results of simulation of a dual inverter-fed open-end winding motor drive with full elimination of the common-mode voltages. Curves in Figs. 10-11 correspond to continuous synchronized PWM, Figs. 12-13 correspond to the discontinuous scheme of synchronized PWM. Operation of the drive system is here under standard scalar *V/F* control. Average switching frequency F_s of each inverter is equal to *1 kHz*. Figs. 10 and 12 present switching state sequences for INV1, pole voltages of two inverters V_{A10} and V_{A20}, phase voltages V_{A1A2} and V_{B1B2} and their difference $V_{A1A2} - V_{B1B2}$ (line-to-line voltage). Figs. 11 and 13 show the corresponding spectra of the pole and phase voltages.

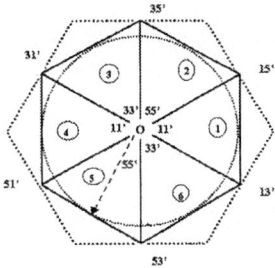

Figure 9. Voltage space-vector combinations, providing full elimination of the common-mode voltages in dual inverter-fed drive [4]

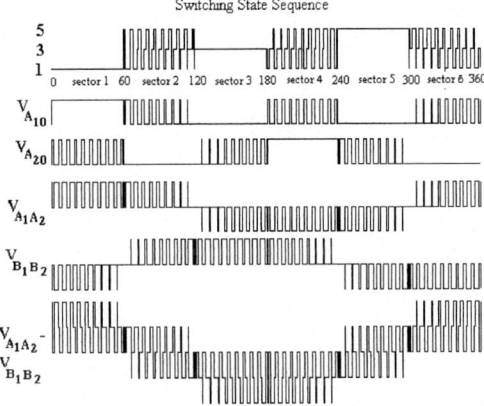

Figure 10. Switching state sequence, pole voltages, phase voltages, and their difference for the system with CPWM at $F = 30 Hz$ ($m=0.6$)

1653

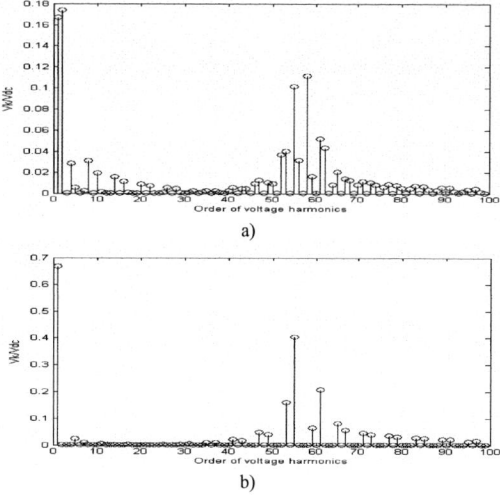

a)

b)

Figure 11. Spectrum of the pole (a) and phase (b) voltages of the system with synchronized CPWM (*F = 30 Hz, m=0.6*)

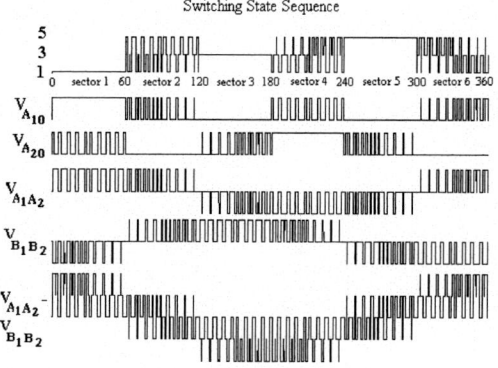

Figure 12. Switching state sequence, pole voltages, phase voltages, and their difference for the system with DPWM at *F =30 Hz (m=0.6)*

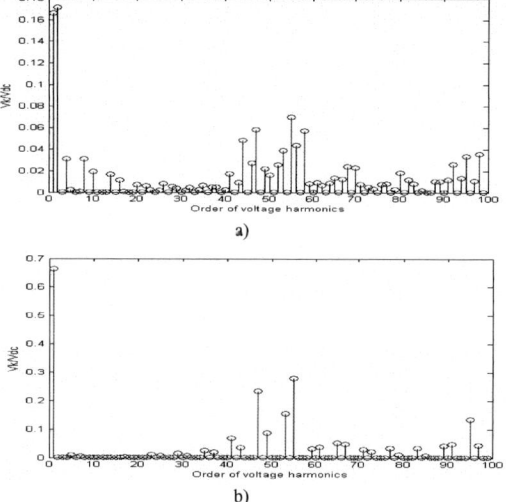

a)

b)

Figure 13. Spectrum of the pole (a) and phase (b) voltages of the system with synchronized DPWM (*F = 30 Hz, m=0.6*)

All waveforms of the phase and line-to-line voltages, presented in Figs. 10 and 12, have full quarter-wave symmetry during the whole control range including the zone of overmodulation, and the spectra of the phase voltage (see Figs. 11,b and 13,b) do not contain even harmonics and combined harmonics (sub-harmonics).

In order to compare behavior of the drive systems with the presented schemes of modulation, a comparative analysis of the spectra of the motor phase voltage has been executed based on computer simulation. Weighted Total Harmonic Distortion factor ($WTHD = (1/V_1)\sqrt{\sum_{i=2}^{n}(V_i/i)^2}$, [6]) was used for determination of its quality.

Fig. 14 presents averaged results of calculation of *WTHD* factor versus modulation index for the motor phase voltage V_{A1A2} of the drive systems with average switching frequency of each inverter $F_s = 1kHz$, for both continuous (CPWM) and discontinuous (DPWM) schemes of synchronized modulation, for the two described in parts *II.A* and *II.B* control algorithms. Control mode corresponds here to standard scalar *V/F* control, and spectral characteristics of motor phase voltage have been analyzed until the zone of overmodulation (modulation index *m = 0.3–0.9*).

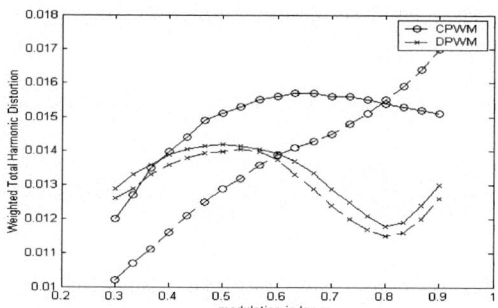

Figure 14. *WTHD* factor versus modulation index for the systems with full elimination of the common-mode voltages (—), and with mutual compensation of the common-mode voltages (- - -)

Presented in Fig. 14 results of analysis of spectral composition of motor phase voltage show, that at low modulation indices *m*, *WTHD* factor is better for the drive system with continuous scheme of synchronized PWM with mutual compensation of the common mode voltages, and when *m>0.6*, both versions of discontinuous schemes of synchronized PWM provide better values of *WTHD*.

III. DUAL TWO-LEVEL INVERTER-FED MOTOR DRIVES WITH SEPARATE DC-LINK SOURCES

Fig. 15 presents basic structure of a dual inverter-fed open-end winding induction motor drive with two standard voltage source inverters, which are supplied by two separate dc-link sources with voltages V_{dc1} and V_{dc2} [2]. Separate dc supply is used for each inverter to block the flow of third harmonic currents, and it is achieved using isolated transformers in providing the dc-link supply for both inverters [1],[2].

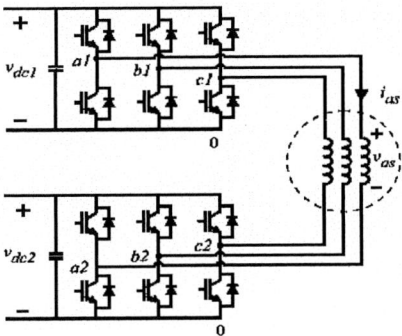

Figure 15. Basic topology of dual inverter-fed open-end winding induction motor drive with two separate dc-link sources [2]

Different dc-sources voltage ratios can be in these dual inverter-fed drives. In particular, if $V_{dc2} = 0$, the voltage vector plot is the same as that of a two-level standard inverter (lower inverter is like a short circuit in this case). If $V_{dc2} = 0.5V_{dc1}$, voltage space vector patterns are in this case the same as that of a four-level converter. If $V_{dc2} = V_{dc1}$, voltage patterns are the same as that of a three-level converter [2].

Synchronous symmetrical control of the output voltage waveforms of each inverter in accordance with basic PWM algorithm provides synchronous symmetrical regulation of voltage in the induction machine phase windings. Rational phase shifting between output voltage waveforms of the inverters is equal in this case to one half of the switching sub-cycle τ (is equal to $0.5\,\tau$) [1].

In the case, when both dc-link sources have the same voltage ($V_{dc2} = V_{dc1}$), resulting voltage space-vectors are equal to space-vector patterns of conventional three-level inverter. Fig. 16 present pole voltages of two inverters V_{a10} and V_{a20} , zero sequence (triplen harmonic components) voltage V_0, motor phase voltage V_{a1a2}, and motor phase voltage without triplen harmonic components (actual motor phase voltage) V_{a1a2} - V_0. Both inverters are here under control in accordance with continuous synchronized PWM with standard V/F control mode. Average switching frequency is equal to $1\ kHz$. Curves in Fig. 16 correspond to the fundamental frequency F equal to $40\ Hz$ (modulation index $m=0.8$). Fig. 17 shows the corresponding spectra of the pole and phase voltages of the system for this control regime.

Figs. 18-19 present analogous characteristics for the system with discontinuous synchronized PWM with V/F control mode. Average switching frequency is equal to $1\ kHz$, fundamental frequency $F = 40\ Hz$, $m=0.8$. Actual motor phase voltages without triplen harmonics (lower curves in Figs. 16, 18) are similar to the voltage wave-forms of the conventional three-level converter. So, when the open-end winding induction motor is fed by two two-level inverters with half the dc-link voltage ($V_{dc}/2$), compared to the dc-link voltage of standard three-level converter, a three-level inverter structure is realized [1],[3].

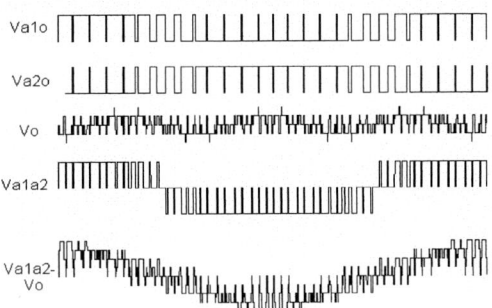

Figure 16. Voltage waveforms of the drive system with continuous synchronized PWM ($F = 40\ Hz$, $m=0.8$)

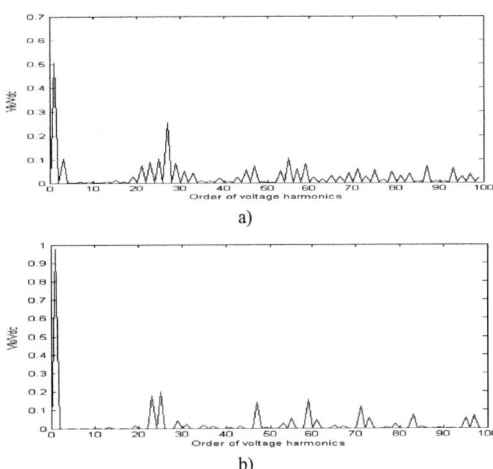

Figure 17. Spectrum of the pole (a) and phase (b) voltages (CPWM, $m=0.8$)

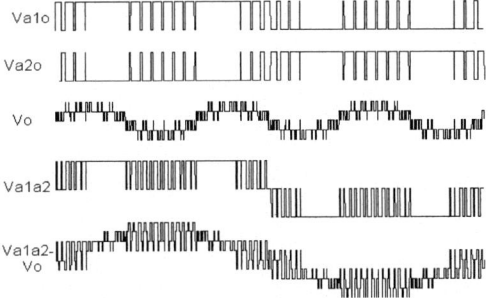

Figure 18. Voltage waveforms of the drive system with discontinuous synchronized PWM ($F = 40\ Hz$, $m=0.8$)

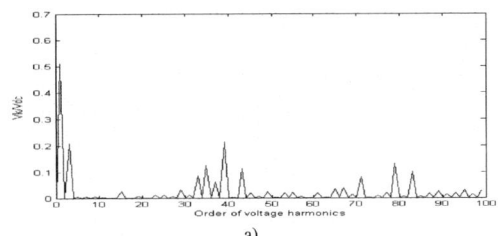

a)

1655

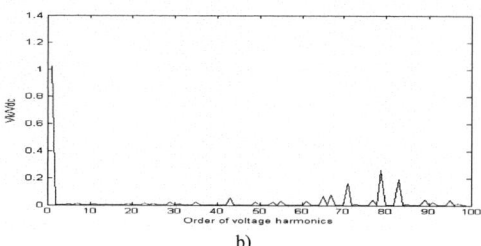

b)

Figure 19. Spectrum of the pole (a) and phase (b) voltages (DPWM, $m=0.8$)

Fig. 20 presents averaged results of calculation of Weighted Total Harmonics Distortion factor (*WTHD*, see part *II*) for actual motor phase voltage ($V_{a1a2}-V_o$) for the analyzed continuous (CPWM) and discontinuous (DPWM) versions of synchronized modulation during standard scalar *V/F* control until overmodulation (modulation index *m = 0.3–0.9*). Average switching frequency of each inverter is equal to *1 kHz* for both versions of PWM. But average motor phase switching frequency is equal to the double switching frequency of each inverter, and is equal to *2 kHz* in this case. Dotted lines in Fig. 20 show results of calculation of *WTHD* factor for the line-to-line output voltage of each inverter ($V_{a1b1} = V_{a2b2}$) for the mentioned above schemes of synchronized PWM.

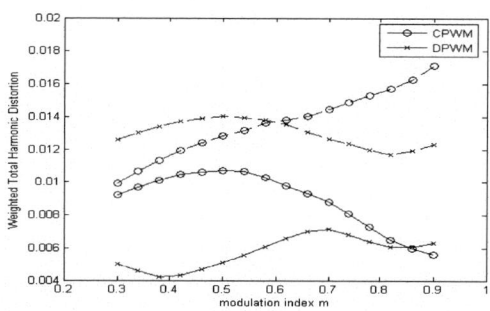

Figure 20. *WTHD* of actual motor phase voltage (—), and of the line-to-line output voltage of each inverter (- - -)

Presented in Fig. 20 results of analysis of spectral composition of actual motor phase voltage for dual inverter-fed open-end winding motor drive show that Weighted Total Harmonic Distortion factor of the actual motor phase voltage is much better for the drive systems with discontinuous (DPWM) version of synchronized PWM during almost all undermodulation control region.

Described in parts II - III methods and techniques of synchronized modulation are well suited for high quality linear control of the fundamental voltage of both each inverter and dual inverter-fed drive systems during overmodulation. The detailed description of the control process in this zone for standard inverters with algorithms of synchronized PWM is in [9].

IV. CONCLUSION

Novel methods of synchronized PWM have been disseminated for control of dual inverter-fed open-end winding motor drives on the base of the cascaded two-level inverters with both single and separate dc voltage sources. General control strategy for these systems is based on the corresponding phase shifting of control and output signals of two standard inverters, providing elimination of undesirable common-mode voltages in the load. Spectra of the motor phase voltage of the drive systems with algorithms of synchronized PWM do not contain even harmonics and sub-harmonics, which is especially important for the systems with increased power rating. Comparison of spectral characteristics of the motor phase voltages in systems shows some advantage of discontinuous schemes of synchronized PWM at higher values of modulation index.

ACKNOWLEDGMENT

This research has been supported partly by Marie Curie Int'l Fellowships Award of FP6 Program of EC, and by Award MOE2-2612 of the US Civilian R&D Foundation (CRDF).

REFERENCES

[1] H. Stemmler and P. Guggenbach, "Configurations of high power voltage source inverter drives", *Proc. of EPE'93 Conf.*, pp.7-12.

[2] K.A. Corzine, S.D. Sudhoff and C.A. Whitcomb, "Performance characteristics of a cascaded two-level converter", *IEEE Trans. on Energy Conversion*, vol.14, no.3, pp.433-439, 1999.

[3] V.T. Somasekhar, K. Gopakumar, E.G. Shivakumar and S.K. Sinha, "A space vector modulation scheme for a dual two level inverter fed open-end winding induction motor drive for the elimination of zero sequence currents", *EPE Journal*, vol.12, no.2, pp.12-36, 2002.

[4] M.R. Baiju, K.K. Mohapatra, R.S. Kanchan and K. Gopakumar, "A dual two-level inverter scheme with common mode voltage elimination for an induction motor drive", *IEEE Trans. on Power Electr.*, vol.19, no.3, pp.794-805, 2004.

[5] K.A. Corzine, M.W. Wielebski, F.Z. Peng and J. Wang, "Control of cascaded multilevel inverters", *IEEE Trans. on Power Electr.*, vol.19, no.3, pp.732-738, 2004.

[6] V. Oleschuk and F. Blaabjerg, "Direct synchronized PWM techniques with linear control functions for adjustable speed drives", *Proc. of the IEEE APEC'2002*, pp.76-82.

[7] V. Oleschuk and F. Blaabjerg, "Three-level inverters with common-mode voltage cancellation based on synchronous pulsewidth modulation", *Proc. of the IEEE PESC'02*, pp.863-868.

[8] V. Oleschuk, F. Profumo, A. Tenconi. and R. Bojoi, "Synchronised PWM for control of inverter-fed open-end winding motor drives", *CD-ROM Proc. of the PELINCEC'2005 Conf.*, 8p.

[9] V. Oleschuk, V. Ermuratski and E.M. Chekhet, "Drive converters with synchronized PWM during overmodulation", *Proc. of the IEEE ISIE'2004*, pp.1339-1344.

2006 5th International Power Electronics and Motion Control Conference

Research on Fast Magnetic Valve Controllable Reactor

Zhang Jian-wen and Cai Xu

Shanghai Jiaotong University Electrical engineering department   Shanghai, China

icebergzjw@sjtu.edu.cn

Abstract: The paper analyzes the principle and structure of the magnetic valve controllable reactor, and presents a practical mathematic model. The simulation and the experimental results in this paper verify that the dynamic response of the general magnetic valve controllable reactor is slow and unpractical. A novel accelerative magnetic valve controllable reactor is proposed, and both the theory analysis and experiments prove that it can speedup the reactor's dynamic response and make the response time less than 2 line cycles.

Key words: Magnetic valve type, controllable reactor, fast

I. Introduction

The controllable reactor is applied in many practical fields. There are mainly four kinds of controllable reactor: regulative reactors, tapped coil, adjustable air gap, controllable saturated type and thyristor controlled reactor(TCR). For tapped coil, it can't be adjusted the inductance continuously. For adjustable air gap reactor, its precision is low because of some mechanical reasons. The magnetic saturation type reactors and TCR can be electrically controlled continuously. Without mechanical components in structure, they possess much higher reliabilities and wider adjustable ranges.

Due to the limitation of thyristor manufacture, high power thyristors are very expensive; therefore TCR is seldom used in super-high power network. Moreover complicated protection circuits are usually needed in TCR for over-voltage and over-current, and additional filters for abundant harmonics.

The controllable saturated reactor overcomes the problems of the TCR. The magnetic valve controllable reactor is one kind of the controllable saturated reactors. The reactor makes use of power source itself to convert voltage through self-coupling windings and to gain exciting source by thyristor rectification without additional windings. It combines working windings with controlling windings organically to reduce energy loss and gain structure simplification, and which has an approximately linear characteristic of voltage /fundamental current.

In practical application, owe to its flexible control and low cost, the magnetic valve controllable reactor is used to adjust the reactive power of power system continuously, limit the operation over-voltage effectively and enhance the stability of system. Take the var-compensation as an example; it is widely applied in the electrical railway power system and the power system of high power promotion. The var-compensation requires fast response, which a fast response reactor fits. From the theory analysis and the experimental result, this paper verifies that the dynamic response of the ordinary magnetic valve controllable reactor is too slow to satisfy the requirement for fast response reactor. And according to its characteristic, an acceleration approach, which can enhance the speed of the dynamic response process, is proposed. The prototypical test result shows that this method can cause the reactor's dynamic response process to be less than two periods of working frequency.

II. The responding characteristic of general magnetic valve controllable reactor

A. The control principle and structure of the magnetic valve controllable reactor.

The circuit of the magnetic valve controllable reactor is shown in detail as fig.1. And its structure

1-4244-0448-7/06/$25.00 ©2006 IEEE   1657

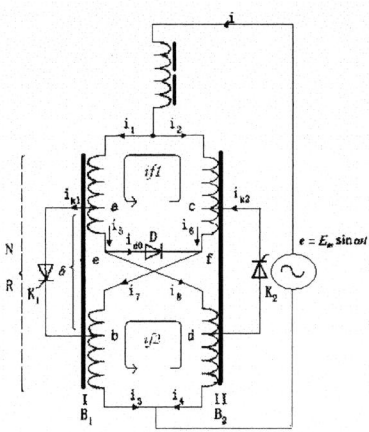

is shown in fig.2. The main iron core of the reactor contains two parallel limbs, which have two stages of varied transverse sections for the limitation of harmonics, and a middle column. A winding is placed on the middle column and two windings are mounted individually on the upper and lower part of left and right limb.

There is a tap with tapping ratio δ in each winding. The upper and lower windings on the different limbs are crossly connected and then two branches of the series windings are connected in parallel .Two thyristors and one diode are connected to the taps of the windings. These switching devices constitute the control circuit. During the positive period and negative periods of working frequency, the thyristor K_1 and K_2 are triggered cyclically and conducted in turn to provide the reactor with DC biased current if_1, if_2, the biased current is drawn by the self-coupling of the reactor. The diode is used to dredge the current caused by inductive load during the blocking of the thyristors. The current passing through the windings of the reactor is divided into two parts: one is the working current, and the other is for excitation. By controlling the conducting angle of the thyristor, it is possible to control the exciting current and hence to control the saturation degree of the reactor. Thus the working current of the reactor can be adjusted continuously.

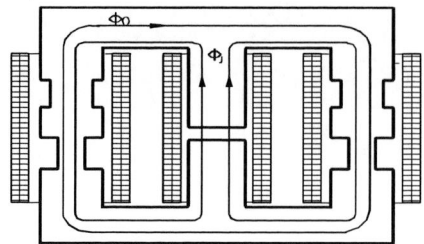

Fig.2 the structure of self-arc-suppression

B. The mathematic model of the magnetic valve controllable reactor

It is supposed that there are two windings containing $N/2$ circles of coils on left and right limbs of the iron core and there is a tap with tapping ratio δ in each winding. R denotes the total resistance of the N circle coil; other denotations are shown in fig.1.

According to the conducting states of thyristors K_1, K_2 and diode D, five various statuses are obtained. Consequently the following five electromagnetism equations are listed.

（1） K_1 On, K_2 and D off

$$\begin{cases} F_1 = (1-\delta)Ni_1 + \delta Ni_2 \\ F_2 = Ni_2 \\ i_{k1} = i_1 - i_2 \end{cases} \quad (1)$$

$$\begin{cases} \dfrac{d\Psi_1}{dt} = \dfrac{1}{1-\delta}e - Ri_1 \\ \dfrac{d\Psi_2}{dt} = \dfrac{1-2\delta}{1-\delta}e + \delta Ri_1 - (1+\delta)Ri_2 \end{cases} \quad (2)$$

（2） K_1 and D on, K_2 off

$$\begin{cases} F_1 = Ni_1 - \delta Ni_{k1} \\ F_2 = Ni_2 \\ i_{k1} = \dfrac{1}{R}\dfrac{d\Psi_1}{dt} + i_1 \\ i_{do} = i_1 - i_2 - i_{k1} \end{cases} \quad (3)$$

$$\begin{cases} \dfrac{d\Psi_1}{dt} = \dfrac{1}{1-\delta}e - Ri_1 \\ \dfrac{d\Psi_2}{dt} = e - Ri_2 \end{cases} \tag{4}$$

（3）D on, K_1 and K_2 off

$$\begin{cases} F_1 = Ni_1 \\ F_2 = Ni_2 \\ i_{do} = i_1 - i_2 \end{cases} \tag{5}$$

$$\begin{cases} \dfrac{d\Psi_1}{dt} = e - Ri_1 \\ \dfrac{d\Psi_2}{dt} = e - Ri_2 \end{cases} \tag{6}$$

（4）K_2 on, K_1 and D off

$$\begin{cases} F_2 = (1-\delta)Ni_2 + \delta Ni_1 \\ F_1 = Ni_1 \\ i_{k2} = i_1 - i_2 \end{cases} \tag{7}$$

$$\begin{cases} \dfrac{d\Psi_1}{dt} = \dfrac{1-2\delta}{1-\delta}e - (1+\delta)Ri_1 + \delta Ri_2 \\ \dfrac{d\Psi_2}{dt} = \dfrac{1}{1-\delta}e - Ri_2 \end{cases} \tag{8}$$

（5）K_2 and D on, K_1 off

$$\begin{cases} F_2 = Ni_2 - \delta Ni_{k2} \\ F_1 = Ni_1 \\ i_{k2} = \dfrac{1}{R}\dfrac{d\Psi_2}{dt} + i_1 \\ i_{do} = i_1 - i_2 + i_{k2} \end{cases} \tag{9}$$

$$\begin{cases} \dfrac{d\Psi_1}{dt} = \dfrac{1-2\delta}{1-\delta}e - (1+\delta)Ri_1 + \delta Ri_2 \\ \dfrac{d\Psi_2}{dt} = \dfrac{1}{1-\delta}e - Ri_2 \end{cases} \tag{10}$$

C. the characteristic of magnetic valve controllable reactor

According to the mathematic model described in the last section, the dynamic response of reactor can be analyzed. The simulated waveform of dynamic

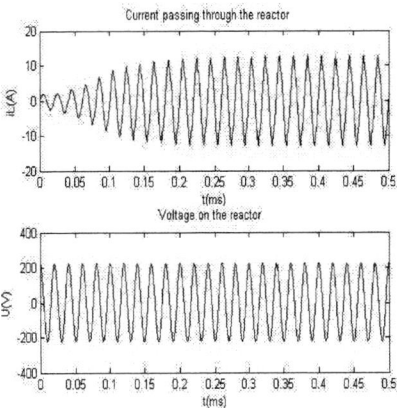

Fig.3 the simulated waveform of the dynamic response reactor

response is presented in fig.3. And the experimental waveform obtained in the same condition as the simulation is shown in fig.4. From the results showed above, it can be found that the current will be stable after 9-10 periods of working frequency after the power is applied.

The reactor's dynamic response process can be analyzed theoretically from the simplified mathematical model. When the reactor's magnetic core is not saturated, the coil's resistance is much less than the inductance, and can be neglected during analysis. Moreover it is mainly concerned to the response time from the idle to the rated state. It is supposed that the triggering angle is $\alpha = 0^0$, only the K1 or K2 which are in "on" status is necessary for analysis, so these two equations of simplified state are obtained:

$$\begin{cases} \dfrac{d\Psi_1}{dt} = e + K(t)\dfrac{\delta}{1-\delta}e \\ \dfrac{d\Psi_2}{dt} = e - K(t)\dfrac{\delta}{1-\delta}e \end{cases}$$

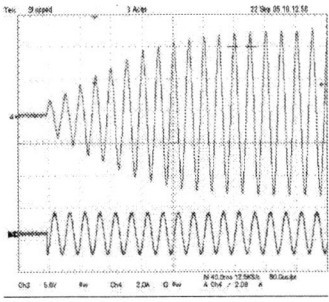

Fig.4 the dynamic response of the magnetic valve controllable

$$K(t) = \begin{cases} -1 & \text{during positive period } K_1 \text{ On} \\ 1 & \text{during negtive period } K_2 \text{ On} \end{cases} \quad (11)$$

Its magnetic flux includes DC and AC components, the DC component is $\Psi_d(t) = (\Psi_1 - \Psi_2)/2$, and AC component is $\Psi_s(t) = (\Psi_1 + \Psi_2)/2$, then:

$$\Psi_d(t) = \frac{2\delta E_m t}{2\pi(1-\delta)} \quad (14)$$

$$\Psi_s(t) = -\frac{E_m \cos(\omega t)}{2\pi} T \quad (15)$$

And one parameter that reflects the degree of reactor's saturation is necessary: the saturated degree, which can be expressed as

$$K_s = (B_0 - B_s)/B_m \quad (16)$$

Here B_s denotes the silicon steel plate saturated magnetic-flux density, B_0, B_m denotes biased magnetic-flux density (magnetic-flux density DC component) and working magnetic-flux density (magnetic-flux density AC component), actually only $-1 < K_s < 1$ is valid for the magnetic saturation reactors. Practically, in order to use the iron core fully, it requires $K_s = \pm 1$ when achieving rated maximum capacity. Since $B_m = B_s$ and $B_0 = 2B_s$,so the maximum of AC and DC components of magnetic flux are:

$$\Psi_{sm} = \frac{E_m}{2\pi} T = B_s N S_A , \quad (17)$$

$$\Psi_{dm} = \Psi_d(T_s) = \frac{2\delta E_m}{2\pi(1-\delta)} T_s = B_s N S_A . (18)$$

(T_s is the time that the DC component achieves the stable status)

According to equation (16) (17), we can have:

$$T_s = \frac{1-\delta}{2\delta} T , \quad (19)$$

Thus, it can be shown that the tap δ determines the dynamic response time. In the actual model $\delta = 0.05$, and according to (19) we can get $T_s = 0.95T$, which is accorded with the result of simulation.

But in practice, the response time of 9-10 periods of working frequency is still too long. It is necessary to speed up the reaction of the magnetic valve controllable reactor.

III. Fast response magnetic valve controllable reactor

A. Structure analysis as well as control strategy

According to the analysis to the dynamic response of the magnetic valve controllable reactor, it can be known that increasing the value of the tap δ can reduce T_s , but the increment of the tap δ makes the adjustment characteristic worse. The adjustment ratio $\Delta i_L / \Delta \alpha$ is so large that it is difficult to control reactor, so variable tap is needed. Suppose that the tap δ is large during the transient period and small in the stable state. The method is realized as fig.5, with auxiliary switches K_3, K_4 and the diodes D_1, D_2 .When the voltage on the reactor became extremely large abruptly (suppose during the positive period), the main and auxiliary SCR K_1, K_3 will be triggered to conduct. Because of self-coupling of the reactor, the electric potential

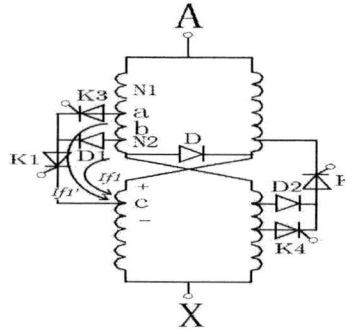

Fig.5 the circuit of the speeded reactor

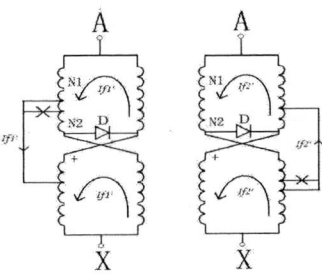

Fig.6 the excitation current of reactor in speeded mode

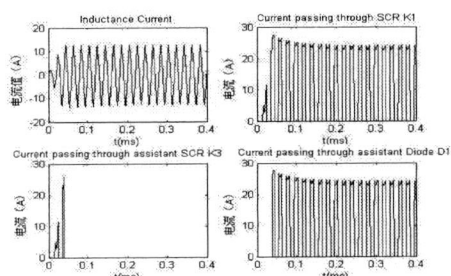

Fig.8 the simulated waveform of the dynamic response of speeded arc-suppression

of the point a is higher than the electric potential of the point c, so the K_1, K_3 is conducted and make the potential of an equal to point c, but higher than the point b. So the diode D_1 is not conducted due to the reversal voltage. Then the direction of excitation current flows is shown as the fig.6.

During the negative period, the K_2, K_4 are triggered to conduct and the D_2 is blocked and the direction of excitation current flows is shown as fig.7. Combining these two processes, it is equal to increasing the tap δ during the transient period.

According to the analysis above, it can seen that increasing δ may speed up the dynamic response of the reactor. When the acceleration process is completed, the excitation current satisfies the demand, so the K_3, K_4 will no longer triggered. And when the main SCR K_1 is triggered at certain conducted angle, the potential of point b is higher

than point c, the diode D_1 is conducted, so as K_1 and D_2, and the tap δ is changed to be small value.

B. *The simulation analysis and the experiment verifications*

This acceleration method is verified by the numeric simulation, and the simulation waveform is shown as the fig.8. The first sub-fig shows the situation of the inductance current, the current achieves the rated value within two periods of working frequency, the second one shows the current passing through K_1, it increases dramatically. The third one shows the current through K_3, it is active during the first two periods. The last one shows the current in D_1. Then with this topology, the reactor is tested in the actual model, the experimental waveform is shown as the fig.9. The curve on the top of the fig is the experimental waveform of the inductance current's dynamic response. The next curve is waveform of the K_1, K_3 's trigger pulse, and the third curve is the waveform of the K_2, K_4 's trigger pulse, the last one is the waveform of the voltage on the K_1. So the result from the experiment is in accordance with the one from simulation. It proves that the dynamic response time of the reactor in this structure is less than 2 period of working frequency.

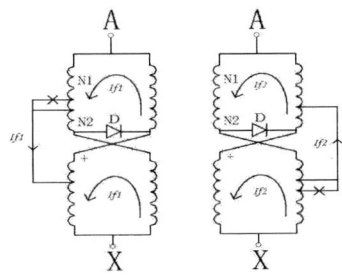

Fig.7 the excitation current of reactor in general mode

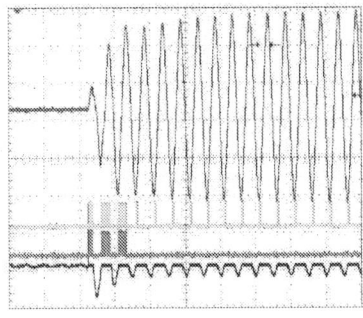

Fig9 the experimental waveform of the dynamic response of speeded arc-suppression

1661

IV. Conclusions

This article proposes one acceleration method for the magnetic valve controllable reactor. According to the result of the numeric simulation analysis and the practical model experiment, the acceleration method can control the reactor's dynamic response time in less than 2T (T is working frequency), and make the reactor satisfy the requirements to fast reactor used in the voltage quality control system.

REFERENCES

[1] Yin Zhongdong, Chen Weixian, Chen Jianye, Wang Zhonghong; Wong Manchung, "Research on the magnetic valve thyristor controlled reactor", Power Electronics Specialists Conference, 1998. PESC 98 Record. 29th *Annual IEEE* Volume 2, 17-22 May 1998 Page(s):2131 - 2136 vol.2[Digital Object Identifier 10.1109/PESC.1998.703478]

[2] Liu Hong; Yin Zhongdong, Chen Weixian; "Research on voltage regulation of magnetic valve controlled reactor based on thyristor", Power System Technology, 1998. Proceedings. POWERCON '98. 1998 International Conference on Volume 1, 18-21 Aug. 1998 Page(s):664 - 667 vol.1 [Digital Object Identifier 10.1109/ICPST.1998.729048]

[3] Baichao Chen; Kokernak, J.M., "Thyristor controlled two-stage magnetic-valve reactor for dynamic VAR-compensation in electric railway power supply systems", Applied Power Electronics Conference and Exposition, 2000. APEC 2000. Fifteenth *Annual IEEE* Volume 2, 6-10 Feb. 2000 Page(s):1066 - 1072 vol.2 [Digital Object Identifier 10.1109/APEC.2000.822820]

[4] ДОРОЖКОЛИ, "Compare analysis with different structure of controllable reactors". *Электрическч естанцич* (Power Plant),1988,volume 4: 19~24

[5] Cai xu, "A new arc-suppression coil with magnetic bias and its control". *Automation of Electric Power System*,2002, volume 10(8),32-35

[6] Liu hong, Yi zhong-dong, Chen bai-chao, "The study on a new appliance of automation Arc-suppression", *Automation of Electric Power System,*2002, volume 10(8),32-35

[7] Chen bai-chao, "The theory and application of a new controllable reactor (the first version)", Wuhan, *Wuhan university of Hydr. & Electric Engineering*

[8] Cai xu, Xie gui-lin, "Study on arc-suppression coil with magnetic bias", *Journal of North China Electric Power University.* 2001, volume 28(1),10-14

[9] Chen bai-chao, Chen wei-xian, "A mathematic model of the magnetic valve type controllable reactor and its characteristic", *Wuhan university of Hydr. & Electric Engineering* 1995, volume 28(3),293-298

2006 5th International Power Electronics and Motion Control Conference

Study and comparison of fault tolerant shunt three-phase active filter topologies

H. El Brouji*, P. Poure** and S. Saadate*

*Groupe de Recherches en Electrotechnique et Electronique de Nancy, GREEN-UHP, CNRS UMR 7037
** Laboratoire d'Intrumentation Electronique de Nancy, LIEN, EA 3440
Universite Henri Poincaré de Nancy I, BP 239, 54506 Vandoeuvre les Nancy cedex France
Email adress : philippe.poure@lien.uhp-nancy.fr

Abstract-This paper studies 3 fault tolerant shunt three phase active filter topologies. It mainly details converter reconfiguration and uses an optimised fast and reliable fault diagnosis method, avoiding spurious semiconductor fault detection due to semiconductor switching.
The fault tolerant converter topologies are based on classical three-leg active power filter topology but includes bi-directional switches for converter reconfiguration after fault detection. The power filter structure, after reconfiguration, could either be a three-leg or a two-leg topology.
An unique control is applied before and after fault detection, which avoids any controller reconfiguration. Simulation results obtained with Saber CAD tools validate the theoretical study.

Keywords- active power filter, fault tolerant converter topology, fault detection, converter reconfiguration, fault tolerant control

I. INTRODUCTION

The reliability of power electronic equipments becomes extremely important in industry applications. The fault mode behaviour of static converters, protection and fault tolerant control of voltage source inverter systems has been covered in a large number of papers. Most of them are focused on induction motor drive applications.

D. Kastha and B. K. Bose considered various fault modes of a voltage source PWM inverter system for induction motor drive [1]. However, they do not propose to reconfigure the inverter topology.

E. R. C. Da Silva and all investigated fault detection of open-switch damage in voltage source PWM motor drive systems [2]. They mainly focused on detection and identification of the power switch in which the fault has occurred. In an other paper, they investigated the utilization of a two-leg based topology when one of the inverter legs is lost. Then, the machine operates with only two stator windings [3]. They proposed to modify PWM control to allow continuous free operation of the drive.

More recently, E. R. C. Da Silva and al studied fault tolerant active power filter system [4], [5]. They proposed to reconfigure power converter and PWM control and examined a fault identification algorithm.

The present paper study 3 fault tolerant shunt three phase active filter topologies. It mainly details converter reconfiguration and uses an optimised fast and reliable fault diagnosis method, avoiding spurious semiconductor fault detection due to semiconductor switching.

II. FAULT TOLERANT TOPOLOGIES

Fig. 1 presents a classical three-leg shunt power active filter. It is composed by a grid (e_{si} for i = {1, 2, 3}), a non-linear load and a voltage source converter. The load is a three-phase diode rectifier feeding an (R, L) series load. The grid is supposed to be balanced with equal series resistances r_s and inductances l_s for each phase. The static converter is a voltage source inverter with equal series resistances r_f and inductances l_f for each phase.

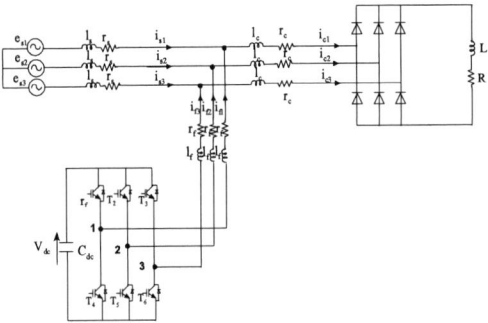

Figure1.   Classical three-leg shunt active filter topology

The output currents of the shunt active filter are controlled to provide reactive power and harmonic currents generated by the non-linear load to ensure filtering. The capacitor C_{dc} of the DC filter side is an energy storage capacity.

1-4244-0448-7/06/$25.00 ©2006 IEEE   1663

In this paper, we studied more particularly 3 fault tolerant shunt active filter schemes. These structures use connecting and isolating devices; they allow reconfiguring the power converter topology after fault detection of one of the semi-conductor devices. These three fault tolerant filter structures are based on the classical and "healthy" structure presented in fig. 1.

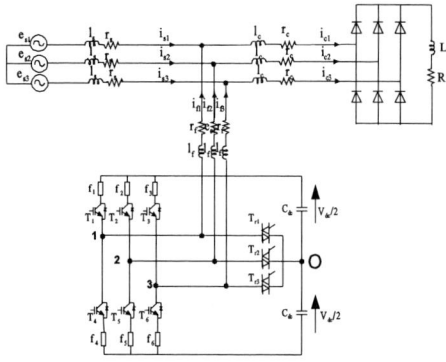

Figure 2.   Fault tolerant three-leg shunt active filter topology

The first one presented figure 2 is a three legs topology. The three legs are used in "health" condition. This structure includes bi-directional devices T_{ri} (i = {1, 2, 3}) (triacs in this case to simplify control) allowing the connection between the grid (for the phase number k having a faulty semiconductor device) and the middle point O of the capacitor C_{dc}. By this way, for the leg number k, if one of the semi-conductor T_i or its associated switch driver is faulty, the leg including T_i could be isolated and the point k will be directly connected to the middle point 0 by switching on the triac T_{rk}. Consequently the post fault topology is a two legs structure with faulty phase directly connected to the point O.

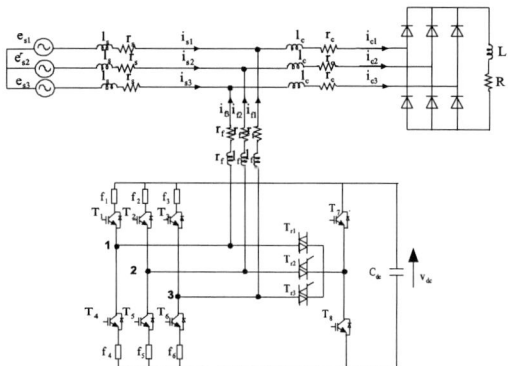

Figure 3.   Fault tolerant four-legs shunt active filter topology

The second structure is a "four-leg" filter topology, presented in fig. 3. The fourth leg composed of devices T_7 and T_8 is a redundant leg. This leg will replace if necessary the faulty one of the three other legs. Bi-directional devices allow connecting the faulty phase number k to the leg constituted by T_7 and T_8. After fault detection, the filter structure remains a classical three-legs topology as presented in figure 1.

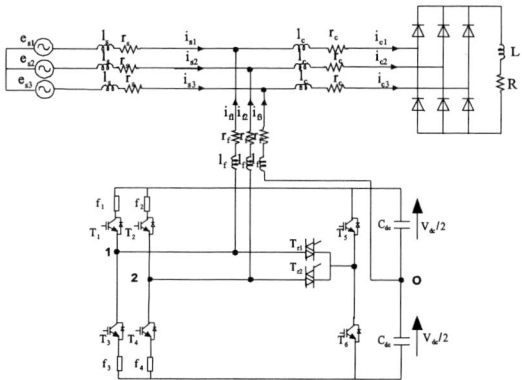

Figure 4.   Fault tolerant three-leg shunt active filter topology with one redundant leg

The third and last structure fig.4 is a three-leg topology, including a redundant leg composed of devices T_5 and T_6. The "healthy" structure is a two-leg topology with phase number three connected to the middle point O of the capacitor. The third leg is dedicated to replace the leg connected to phase one or two in fault case. To do this, only 2 bi-directional devices T_{r1} and T_{r2} are needed. The post fault topology remains a two-leg topology with phase number 3 connected to the point O.

In general, several faulty cases can occur : power switch or power switch driver can be faulty. In each case, it results in the following models :

- a switch is open instead of being normally closed. It results in an open phase. Only triacs are useful and allow to select the leg to be isolated;

- a switch is closed instead of being normally open. It results in a short-circuit of the capacitor, increasing i_{sk} current and decreasing to zero regulated capacitor voltage v_{dc}. To isolate the faulty switch as fast as possible, one can also place, despite of cost considerations, isolated devices such as fast active fuses in series with semiconductor devices used in healthy mode.

In summary, in a fault case on the leg number k, the compensation is achieved by the following steps :

- detection of the faulty switch of the leg number k (detailed in next section);

- removing the commands of the 2 switching drivers of the leg number k;

- replacing the faulty leg by the redundant one (cases of structure fig.3 and fig. 4) or connecting the faulty DC

1664

side phase k to the point O of the capacitor energy storage case of fig. 2);

- applying the suited control strategy.

III   FAULT DETECTION AND CONTROL STRATEGY

Fault detection is based on the comparison between measured and estimated V_{kO} voltages for $k = \{1, 2, 3\}$, respectively noted V_{kOm} and V_{kOth}. Voltage V_{kOth} can be expressed as :

$$V_{kOth} = (2.C_k-1).V_{dc}/2$$

with $C_k = \{O, 1\}$ depending on the conducting state of the top semi-conductor of the leg number k.

However, semi-conductors switching disturb measured V_{kOm} voltage. Consequently the error voltage signal defined by $\varepsilon_k = V_{kOm} - V_{kOth}$ for $k = \{1, 2, 3\}$ is constituted of picks induced by each switching on k arm and during about 0.1 µs if IGBT are used.

To avoid spurious fault detection due to power semi-conductors switching, we think of transforming the "voltage" signal $\varepsilon_k (t)$ in a "time" signal $int_k(t)$. From signal $\varepsilon_k (t)$, we defined the signal $int_k(t)$, constituted of picks having as maximal value the time during which $\varepsilon_k(t)$ is different from zero.

The calculation of $int_k(t)$ is achieved for each phase by first taking the absolute value of $\varepsilon_k(t)$, applying the results to an hysteresis comparator and integrating the comparator output as presented in fig.5. The output of the hysteresis comparator is equal to 0 if $\left|\varepsilon k(t)\right| = 0$ and equal to 1 if $\left|\varepsilon k(t)\right| \neq 0$. The maximal value after integration of this comparator output square signal results in the time during which V_{kOm} and V_{kOth} are different, if integration is initialised to 0 after each square waveform integration. Consequently, we detect the default using a "time based" default criterion instead of "voltage based" default criterion. To do this, we applied the integration result signal to a second hysteresis comparator having a band width several times larger than switching time of the semi-conductor. The final result is noted x_k. By this way, we avoid spurious default detections due to semiconductor switching.

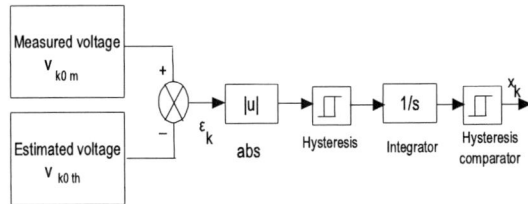

Figure 5.   Block diagram of the detection algorithm for the phase k

When a default is detected in a leg number k, the previous switching control orders C_k and $not(C_k)$ for the 2 semiconductors of this leg must be set to '0'. For the leg number k and in case depicted in fig.3, we respectively note "Com_high$_k$" and "Com_low$_k$" the new logic switching orders of the top and the bottom switches T_7 and T_8 including fault tolerant control. They are defined by the logic equations :

Com_high$_k$ = C_k *and* (*not*(s_k))
Com_low$_k$ = (not(C_k)) *and* (*not*(s_k))

IV   ACTIVE FILTER CONTROL

Fig. 6 presents a block diagram of the proposed control system. This shunt active filter control is available for three-legs and two-legs structures. The major advantage of this control principle is to be suited if there is fault or not. Consequently, no control reconfiguration is necessary.

The task of this control is to determine the current harmonic reference to be generated by the active filter. They are defined using classical active and reactive power method proposed by Akagi [7], associated with 2 "multi variable" filters.

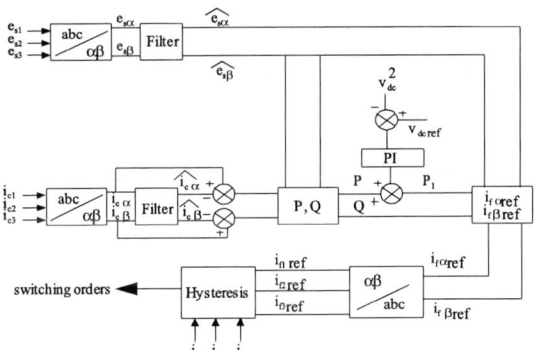

Figure 6.   Block diagram of the control system

(α, β) components of voltages ($e_{s\alpha}$ and $e_{s\beta}$) and currents ($i_{c\alpha}$ and $i_{c\beta}$) are defined by the classical Concordia transformation :

$$\begin{bmatrix} x\alpha \\ x\beta \end{bmatrix} = \sqrt{\frac{2}{3}} \begin{bmatrix} 1 & -\frac{1}{2} & -\frac{1}{2} \\ 0 & \frac{\sqrt{3}}{2} & -\frac{\sqrt{3}}{2} \end{bmatrix} \begin{bmatrix} x1 \\ x2 \\ x3 \end{bmatrix}$$

The originality of this control is that it extracts the harmonic components directly from the $\alpha\beta$ axis by using multi-variable filter, as given by the expressions:

$$\hat{x}\alpha = \frac{k}{s}[x\alpha(s) - \hat{x}\alpha(s)] - \frac{\omega}{s}\hat{x}\beta(s)$$

$$\hat{x}\beta = \frac{k}{s}[x\beta(s) - \hat{x}\beta(s)] + \frac{\omega}{s}\hat{x}\alpha(s)$$

This filter and its principle are detailed in [6]

The power P and Q are calculated by :

$$P = v_\alpha.i_\alpha + v_\beta.i_\beta$$
$$Q = v_\alpha.i_\beta - v_\beta.i_\alpha$$

(α, β) current references are calculated (see fig. 6) by :

$$i_{f\alpha}^{ref} = \frac{v_\alpha}{v_\alpha^2 + v_\beta^2} P1 - \frac{v_\beta}{v_\alpha^2 + v_\beta^2} Q$$

$$i_{f\beta}^{ref} = \frac{v_\beta}{v_\alpha^2 + v_\beta^2} P1 + \frac{v_\alpha}{v_\alpha^2 + v_\beta^2} Q$$

Finally filter current references are defined by :

$$\begin{bmatrix} i_{f1}^{ref} \\ i_{f2}^{ref} \\ i_{f3}^{ref} \end{bmatrix} = \sqrt{\frac{2}{3}} \begin{bmatrix} 1 & 0 \\ -\frac{1}{2} & \frac{\sqrt{3}}{2} \\ -\frac{1}{2} & \frac{\sqrt{3}}{2} \end{bmatrix} \begin{bmatrix} i_{f\alpha}^{ref} \\ i_{f\beta}^{ref} \end{bmatrix}$$

This control with the same reference voltage V_{dcref} can be used for before and after fault topologies in the cases of fig. 3 and fig.4 because topologies before and after converter reconfiguration are the same.

In the case fig. 2, the topology before reconfiguration is a three legs topology and becomes a two legs topology after reconfiguration. One can also chose between applying the same V_{dcref} value before and after reconfiguration or changing V_{dcref} value simultaneously with converter reconfiguration. In the second case, the new V_{dcref} reference should be about twice greater.

All these cases will be studied by simulation in next section.

V COMPARISON STUDY AND SIMULATION RESULTS

This section presents simulation results obtained with Saber CAD simulator for the 3 fault tolerant topologies. General simulation parameters are given in appendix 1.

A. Case of the three-leg topology fig.2

Fig. 7 presents results in an open circuit case (fault of the bottom switch of the leg number 3), introduced at t = 200 ms. The value of the capacitor voltage reference V_{dc}^{ref} is set to 1600V for both before fault three-leg topology and post fault two-leg topology. The value of l_f is 300μF and the hysteresis band is equal to 30 A. Results presented in fig. 7 show that proposed fault tolerant system preserved the main performance features after fault compensation. More, the mean switching frequency

is equal to 8 kHz before fault and equal to 8.4 kHz after fault.

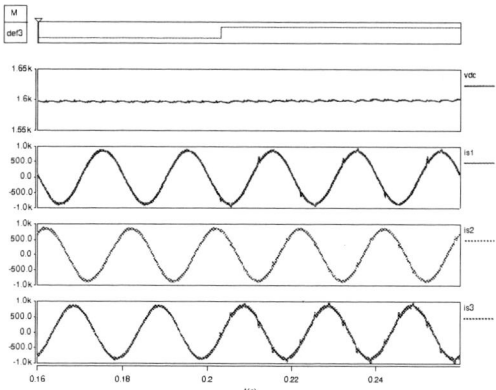

Figure 7.   Case of an open circuit with an unique reference voltage V_{dc}^{ref} =1600 V

Fig. 8 presents results in an open circuit case (fault of the bottom switch of the leg number 3), introduced at t=200 ms. The value of the capacitor voltage reference V_{dc}^{ref} is set to 800 V for before fault three-leg topology and set to 1300 V for post fault two-leg topology. The value of l_f is 150μF and the hysteresis band equal to 20 A. The proposed fault tolerant system preserved the main performance features after fault compensation. More, the mean switching frequency is equal to 7 kHz before fault and equal to 9'5 kHz after fault.

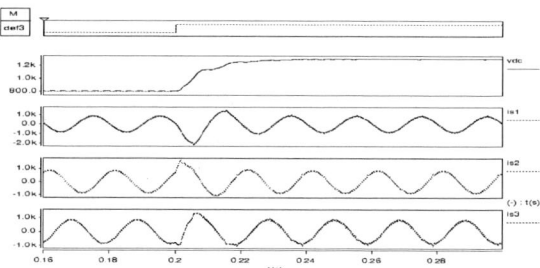

Figure 8.   Case of an open circuit with 2 reference voltages

B. Case of the "four-leg" topology fig.3

Fig. 9 presents results in an open circuit case (fault of the bottom switch of the leg number 3), introduced at t = 200 ms. The value of the capacitor voltage reference V_{dc}^{ref} is set to 700V. The value of l_f is 300μF and the hysteresis band is equal to 30 A. Performances for waveforms figures 9 are the same before and after reconfiguration. The mean switching frequency remains equal to 7 kHz.

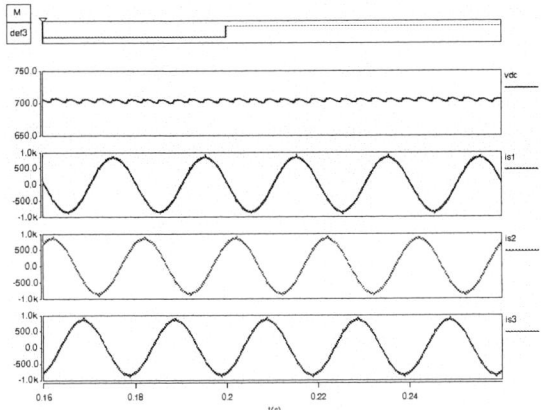

Figure 9.   Case of an open circuit with V_{dc}^{ref} =700 V

Fig. 10 presents results in an short circuit case (fault of the bottom switch of the leg number 3), introduced at t = 200 ms. The value of the capacitor voltage reference V_{dc}^{ref} is set to 700V. The value of l$_f$ is 150µF and the hysteresis band is equal to 20 A. Performances are the same as the previous one.

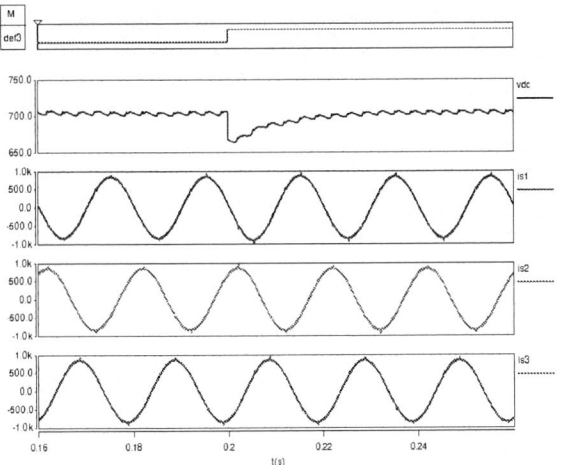

Figure 10.   Case of a short circuit with V_{dc}^{ref} =700 V

C.   Case of the three legs topology fig.4

Fig. 11 presents results in an open circuit case (fault of the bottom switch of the leg number 3), introduced at t = 200 ms. The value of the capacitor voltage reference V_{dc}^{ref} is set to 1600V. The value of l$_f$ is 300µF and the hysteresis band is equal to 30 A. The main performance features are preserved after topology reconfiguration.

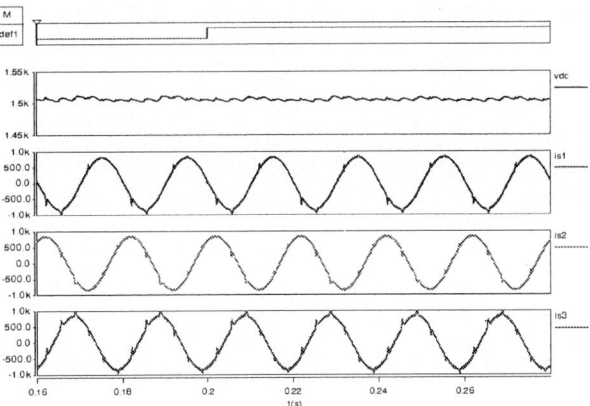

Figure 11.   Case of an open circuit V_{dc}^{ref} =1600 V

Fig. 12 presents results in an short circuit case (default of the bottom switch of the leg number 3), introduced at t = 200 ms. The value of the capacitor voltage reference V_{dc}^{ref} is set to 700 V. The value of l$_f$ is 150µF and the hysteresis band is equal to 20 A. Performances for waveforms figures 10 are the same as the one for fig.11.

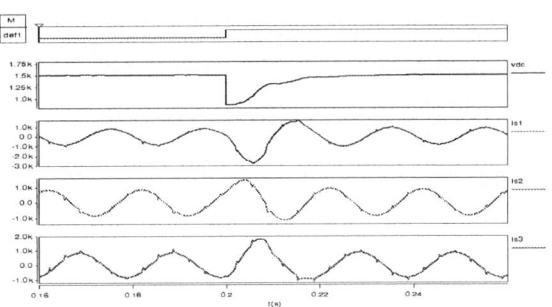

Figure 12.   Case of a short circuit with V_{dc}^{ref} =700 V

VI   CONCLUSION

In this paper, we studied 3 fault tolerant active power filter topologies and associated fault diagnosis method, control and converter reconfiguration.

This paper has demonstrated the possibility of designing fault tolerant active filter topologies. These structures used fault detection and fault compensation. They can achieve continuous free operation even if a complete loss of one of the converter legs has happened.

The fault is compensated by reconfiguring the converter topology using connecting bi-directional devices. In the cases of structures depicted in fig. 3 and fig. 4, the faulty leg is replaced by a redundant one. In the other case (fig. 2), faulty leg is isolated. Then, the corresponding phase is connected to middle point of the capacitor C$_{dc}$.

1667

When comparing the 3 proposed topologies, we demonstrated that, in fault case, performances can be guaranteed for each of them. Structure depicted in fig. 2 do not need a redundant leg but however requires a higher dc-bus reference voltage to preserve the main performance features after fault compensation.

The configuration fig.4 is the most cost effective; however, filter parameters must be optimised to achieve high performances, more particularly for the phase number 3. Finally, configuration presented in fig. 2 preserves the main performance features after fault compensation, even if two values of V_{dc}^{ref} are used.

REFERENCES

[1] Investigation of fault modes of voltage-fed inverter system for induction motor drive, Kastha, D.; Bose, B.K., *IEEE Transactions on Industry Applications*, Vol. 30, Issue 4, July-Aug. 1994 pp. 1028 – 1038.

[2] Fault detection of open-switch damage in voltage-fed PWM motor drive systems, De Araujo Ribeiro, R.L.; Jacobina, C.B.; da Silva, E.R.C.; Lima, A.M.N., *IEEE Transactions on Power Electronics*, Vol. 18 , Issue: 2 , March 2003, pp. 587 – 593.

[3] An induction motor drive system with improved fault tolerance, Beltrao de Rossiter Correa, M.; Brandao Jacobina, C.; Cabral da Silva, E.R.; Nogueira Lima, A.M., *IEEE Transactions on Industry Applications*, Vol. 37, Issue 3, May-June 2001, pp. 873 – 879.

[4] Improved fault tolerance of active power filter system, Jacobina, C.B.; Correa, M.B.R.; Pinheiro, R.F.; Lima, A.M.N.; da Silva, E.R.C., *Power Electronics Specialists Conference,* 2001. PESC. 2001 IEEE 32nd Annual , Vol. 3, 17-21 June 2001.

[5] Control of a three-phase four-wire active filter operating with an open phase, Jacobina, C.B.; Pinheiro, R.F.; de R. Correa, M.B.; Lima, A.M.N.; da Silva, E.R.C., *Industry Applications Conference, 2001. Thirty-Sixth IAS Annual Meeting. Conference Record of the 2001 IEEE* , Vol.: 1 , 30 Sept.-4 Oct. 2001.

[6] Generalized theory of the instantaneous reactive power filter, H. Akagi, Y. Kanazawa, A. Nabae , Proceeding *International power electronics conference.* Tokyo, Japan, PP. 1375-1386, 1983.

[7] A new robust experimentally validated phase locked loop for power electronic control , M.C.Benhabib and S.Saadate,. *EPE journal*, vol. 15, no. 3, 2005.

APPENDIX – SIMULATION PARAMETERS

Grid : 230V, 50 Hz
Non-linear load : R = 0,6 Ω, L = 2,5 mH,
l_c = 15µH, r_c = 0,4 mΩ
Filtering capacitor : C_f = 17,6 mF

2006 5th International Power Electronics and Motion Control Conference

Application of GA-BP in Fault Diagnosis of Power Circuit of SVC

Zeng Guang*, **, Xi Yu-fan**, Su Yan-min* and Zhang Jing-Gang**
*Xi'an Jiaotong University, Xi'an, 710049, China.
**Xi'an University of Technology, Xi'an, 710048, China.
E-mail: g-zeng@mail.xaut.edu.cn

Abstract-The multi-layer feed-forward neural network is in essence a dynamic system including a large number of interconnected processing elements (neurons) working in unison to solve special problems, and this characteristic makes it suitable for the fault diagnosis. If we regard fault signs as inputs of the network and fault causes as outputs, we can construct a network to map the complicated relationship between inputs and outputs. However, the unavoidable shortcomings of the Back-Propagation (BP) algorithm typically adopted by feed-forward neural network limits its use. The main problem is that gradient methods employed by classical BP algorithm find only a local optimum, the local optimum found depends on the starting point and the goal function must be smooth. From the viewpoint of mathematics, Genetic Algorithms (GA) is a kind of technique for searching optimal solutions. What a perfect method of combination of GA and BP! In this paper, we optimize the BP network weights by genetic algorithms, and then the GA-BP network is applied to the fault diagnosis of power circuit of the Static Var Compensator (SVC) based on Digital Signal Processor (DSP) TMS320F240. The experiment shows the performance of the system is excellent.

Keywords- on-line fault diagnosis; SVC; neural network; genetic algorithms; DSP controller

I. INTRODUCTION

With the rapid development of the power electronic technology, the power electronic equipment is increasingly used in such areas as electroplating, induction heating, DC power transmission, AC adjustable speed motor, electric power, locomotive engine , SVC and Uninterrupted Power Supply (UPS) etc., in the light of its advantages of high efficiency, flexible control and easy realization, and therefore the fault diagnosis of the power electronic equipment arises naturally.

Many efforts have been done in analyzing and performing the fault diagnosis of the power electronic equipment since artificial neural network and genetic algorithm appeared. A approach for diagnosing single open-circuit faults in the inverter of the advanced static var generator (ASVC) using the feed forward network is presented in [1]. Fault diagnosis of three-phase rectifier

based on neural network theory and frequency analysis is presented in [2]. In [3], a hybrid algorithm that combines improved genetic manipulation with artificial network is applied to fault diagnosis of power electronic three-phase rectifier circuits. In [4], the author discuss several advanced genetic algorithms that have proved to be efficient in solving difficult design problems such as complex nonlinear function, then give an overview of applications of genetic algorithms to different domains of engineering design. A dynamic genetic algorithm based on continuous neural networks for a kind of non-convex optimization problems is proposed in [5].

Power electronic device plays a very important role in the power electronic circuits. According to the realistic running of the power electronic circuits, most of the faults resulted from the power electronic device failure. Especially in Thyristor Control Reactor (TCR) of high power SVC circuit, many thyristors in parallel or in series make the power circuits more complex and the control circuit as well. So the fault diagnosis of TCR is of importance. In this paper, fault diagnosis of power circuit of SVC is fulfilled based on DSP TMS320F240 of TI corporation using GA-BP network. The realistic running indicates that the system has great capabilities in both on-line fault detection and accurate fault analysis and diagnosis.

II. GA-BP NETWORK

A. Introduction of Artificial Neural Network

The multi-layer feed-forward artificial neural network is in essence a dynamic system including a large number of interconnected processing elements (neurons) working in unison to solve special problems, and its behavior is strongly influenced by two aspects, one is its connecting topology including the number of neurons and the way they connect to each other, the other is the connection weights. The Kolmogarav theorem has proved that a three-layer Back-Propagation (BP) network can approximates to the nonlinear bounded function in any given operational precision with appropriate structure and proper weights. The Artificial Neural Network have been widely used in fault diagnosis during the recent years. In this paper, a three-layer BP network illustrated in Fig. 1 is used.

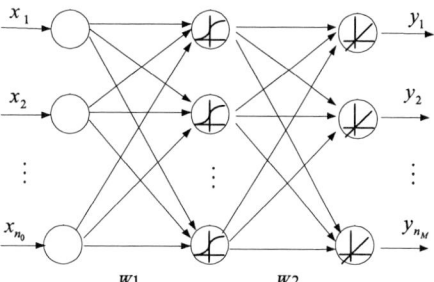

Figure 1. Topology of three-layer BP neural network

The operating function of middle layer called hidden layer here is Sigmoid given below, and the output layer's is linear function:

$$y = \frac{1}{1 + \exp(-x)} \qquad (1)$$

Assuming that there are N groups of inputs and corresponding outputs all together called training patterns for M-layer networks (excluding input layer), and there are n_L neurons in the L-th layer, the objective function, the mean square error between the actual output of a multi-layer feed-forward network and the desired output, should be as minimal as possible given below:

$$E = \frac{1}{2} \sum_{p=1}^{N} \sum_{j=1}^{n_M} (t_{pj} - y_{pj})^2 \qquad (2)$$

where, the t_{pj} presents the desired output of the j-th output neuron at the p-th pattern, and W1、W2、θ_1、θ_2 presents the weights matrix and the biases matrix from input layer to hidden layer and from hidden layer to output layer respectively.

B. Steps of GA-BP Algorithms

The layer number and the neuron number of multi-layer feed-forward network is very important to determine its structure. The neuron number of input and output layer is limited by the realistic fault event. Hunting for the proper number of nodes for the hidden layers is difficult and can be determined only by experimentation. Generally, the network performance may be improved by increasing the number of hidden nodes. However, the addition of more hidden nodes may actually degrade the performance due to the difficulty in training and the increased potential of over-fitting the trained data. It is generally believed that the best network contains the least number of necessary neurons to accomplish the input/output relationships. The most common algorithm for feed-forward network is BP (back propagation) which adjust the weights and the biases of the network using the gradient steepest descent method, nevertheless, it is known to us all that the unavoidable disadvantages of BP restricts its use

sometimes. The basic problem is that gradient methods find only a local optimum, and no information is available on how good it is compared to the global one. Moreover, the local optimum found depends on the starting point; to improve results the computation is usually repeated for a number of starting points. The goal function must be smooth, and a procedure is needed to compute gradients (analytically, or at least numerically). In real design problems—with complicated or possibly discontinuous goal functions, and discrete variables—these conditions are in general not straightforward to fulfill.

From the viewpoint of mathematics, Genetic Algorithms (GA) is a kind of technique for searching optimal solutions. GA strongly differ in conception from other search methods. The basic difference is that while other methods always process single points in the search space, genetic algorithms maintain a population of potential solutions. Considering the shortcomings of the BP algorithm, this paper adopted a method called GA-BP which is done as follows to get the optimal weights and biases for the feed-forward network. GA-BP combine the advantages of both GA and BP:

Step 1. Determine the coding method and create the initial population of individuals. Here we adopt real-coded method, it is said that each element of the individual must be a real number, and we choose the beginning weights and biases in $(-3/\sqrt{F},\ 3/\sqrt{F})$ stochastically, there F express the number of input neurons connected to the weight matrix.

Step 2. Valuate the fitness function of every individual according to some certain rule, now we use the function 1/E.

Step 3. Reproduction. We present here the most commonly used selection schemes for reproduction, fitness-proportional selection, it select the new individuals in direct proportion to the fitness value of every individual. It means that the higher is the value of the individual, the more directed the selection is towards the individual, and in addition, we preserve the best one of previous group of individuals.

Step 4. Crossover. Operate with mutation operator in a certain probability, we use linear crossover method, for example:

Assuming that the individual v1 and v2 are selected to crossover, v1=[p11,p12,…,p1i,…p1m]
v2=[p21,p22,…,p2i,…p2m]
after crossover they become as
v1'=[p11,p12,…,p1i',…p1m']
v2'=[p21,p22,…,p2i',…p2m']
there   p1j'= p1j -a0*(p1j − p2j)
p2j'= p2j +a0*(p1j - p2j)
and a0∈ (0,1), is a random decimal fraction, j∈ [i,m] , i∈ [1,m], i is also selected stochastically.

Step 5. Mutation. Modify one or more selected element of an individual to produce a new individual, we use dynamic mutation operator presented as follows:

Supposing that we select an individual to mutate after crossover, v={p1,p2,...,pi,...pm}, the new individual after mutation is v={p1,p2,...,pi',...pm} if the i-th element is mutated,

$$\begin{cases} pi' = pi + (bi - pi)(1 - r^{\lambda T}) & r < 0.5 \\ pi' = pi - (pi - ai)(1 - r^{\lambda T}) & r \geq 0.5 \end{cases}$$

there $T = 1 - f_i / f_{max}$, f_i is the fitness value of v, f_{max} is the maximum fitness value of all individuals. [aj,bj] is the area to which pi should be subject , λ is in general an integer from 2 to 5

Step 6. Repeat Step 2 and Step 5 until some preconditions are satisfied., here if the fitness of the best individual satisfies the inequation $E \leq \varepsilon$, then stop ,and there ε express the desired value of the function E.

III. SIMULATION AND EXPERIMENT

A. Introduction of power circuit and control circuit of SVC (FC+TCR)

The power circuit of SVC (FC+TCR) is as shown in Fig. 2, it works mainly by adjusting the time when the thyristor be triggered and the current flow through the power inductance in order to adjust the reactor power of the circuit and correct the power factor of the circuit. And the TCR in general consists of many thyristors in series as shown in fig. 3, it is supposed that there are 10 pair of thyristors in every phase, we encode the positive phase components as No. 1 to No. 10 and the negative phase as No. 11 to No. 20. It is clear that thyristor No. 1 and No. 11 are reversely in parallel, and the cathodes of the thyristor No. 1 and No. 12 are joined together.

Trigger circuit board are equipped near the power circuit, each board offers the trigger pulse to the two thyristors which are reversely in parallel and receives signals that indicate the thyristor is triggered by normal trigger or by emergency trigger of the two components of which the cathodes are joined together. In the trigger circuit board,

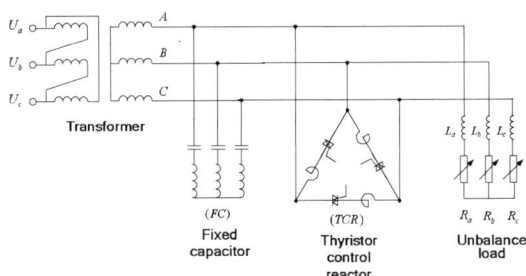

Figure 2. Power circuit of SVC (FC+TCR)

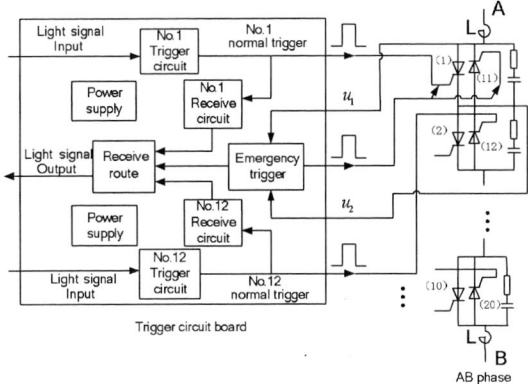

Figure 3. AB phase of TCR and trigger circuit board

power supply come from the voltage u_1 to u_2 , emergency trigger sent a trigger pulse to the thyristor when the voltage u_1 to u_2 is higher than a given value which approximates the highest voltage from anode to cathode the thyristor could bear. The input light signal denote the trigger pulse of the thyristor calculated and sent through the fibre-optics by power circuit controller, and the output light signal denote the received signals sent by the receive circuit or by emergency trigger after they trigger the thyristor, this is the gist by which we can judge if the circuit and the thyristor work properly or not. The trigger circuit play a part in transferring the light signal to electrical signal and amplifying the signal.

B. Fault dignosis and simulation of neural network

We can easily get the table of the relationship between fault signs and fault causes according to the realistic running of the system as shown in Table I. The trigger pulse is sent through fibre-optic and is transferred to the gate of the thyristor in normal state, at the same time, the signal which indicates the component is triggered by the normal trigger pulse is sent by receive circuit to receive route. If the input light signal doesn't trigger the thyristor normally, the voltage u_1 to u_2 may rise rapidly, which make the emergency trigger sent out and meanwhile the signal which indicates the component is triggered by emergency trigger pulse is sent to receive route.

We judge if the thyristor No.1 work normally using the receive signal synthetically from thyristor No.1, No.11 and No.12, so we can regard these receive signals as inputs of a neural network and the fault causes as outputs, map the complex relationship between the inputs and outputs. The number of input layer is six and the output layer is five in the light of the number of the received signals and the kind of fault causes, and we use six neurons in the hidden layer, we have 64 (2^6) groups of training patterns all together. At last we determine the parameters of the network as given in equation (3) to equation (6).

TABLE Ⅰ. RELATIONSHIP BETWEEN FAULT SIGN AND FAULT CAUSE

Received signal						Fault cause of No.1				
From No.1		From No.11		From No.12		Normal	Trigger circuit	Receive circuit	Receive Route	Damage
Normal	Emergency	Normal	Emergency	Normal	Emergency					
0	0	0	0	×	×					√
0	0	0	1	0	0				√	
0	0	1	0	0	0				√	
0	0	1	1	0	0				√	
0	1	×	×	×	×		√	√		
1	0	×	×	×	×	√				
1	1	×	×	×	×		√			
others								√		

$$W1 = \begin{bmatrix} 14.615 & -18.053 & -0.05013 & -0.05013 & 0.0112 & 0.0112 \\ 5.3755 & -5.872 & 0 & 0 & 0 & 0 \\ -37.898 & 10.058 & 7.3406 & 7.3406 & 0.073 & 0.073 \\ 30.441 & 30.605 & 7.1136 & 7.1136 & 20.381 & 20.381 \\ -26.181 & 4.1441 & 1.8931 & 1.8931 & -0.0228 & -0.0228 \\ 27.502 & 27.913 & -7.587 & -7.587 & -0.675 & -0.675 \end{bmatrix}$$

$$(3)$$

$$\theta_1 = \begin{bmatrix} -5.5338 \\ 1.2252 \\ 1.5639 \\ -20.332 \\ 7.15 \\ 0.5236 \end{bmatrix} \quad (4)$$

$$W2 = \begin{bmatrix} 0.4253 & -1.3118 & 0 & 0 & 0.1274 & 0 \\ 0 & 0 & 0.0057 & 0 & -1.0054 & 0 \\ -1.1214 & 6.5442 & 14514 & 1.0023 & -10162 & 3.9924 \\ 1.5016 & -4.6316 & -7.0343 & -1.0023 & 3.9543 & -3.5308 \\ 0.1964 & -0.6058 & -7.4836 & 0 & 7.0808 & -0.46185 \end{bmatrix}$$

$$(5)$$

$$\theta_2 = \begin{bmatrix} 0.8848 \\ 0.9992 \\ -9.4087 \\ 7.657 \\ 0.8705 \end{bmatrix} \quad (6)$$

C. Frame of diagnosis system and experiment

With its remarkable advantages such as flexible instruction set, high-speed performance, innovative parallel architecture, cost effectiveness etc., DSP (Digital signal processor) have been extensively utilized resolve diverse areas of science and engineering problems. Except these advantages the following features of DSP are useful in adopting neural network:

a) High-speed central processing unit (CPU) allows the digital designer to process algorithms in real time rather than approximate results with look-up tables.

b) Single cycle instruction MAC following repeat (RPT) execution.

c) Large program-memory space is well-suited for saving neural network weights and data tables to calculate the non-linear function

d) To support multi-processor (multi-bus and bus-arbitrate, high-speed series I/O port) is suitable for dealing with large complex neural network.

The weights of the neural network are in general some exact decimal fraction, thus it is easier to realize it using the floating-point processor than using fixed-point processor at the cost of operating speed. This paper adopted 16-bit fixed-point processor, TMS320F240 of TI corporation, in order to raise the operation speed. The frame of the diagnosis system is illustrated in Fig. 4. It can be devided into three parts, the first part is high voltage segment including the power circuit and trigger circuit board, the second part is the power circuit controller, and the part which core is DSP is fault diagnosis portion. Here, the light signal is sent by fibre-optic, the computer monitoring the fault records real time are named as monitor, HD7279 serves keyboard and LED (Light Emitting Diode), DS12887 supply the exact time when any fault just happens. The calculation of non-linear function is fulfilled through the following function:

$$y = y(index) + (x - x(index)) \times \frac{dy}{dx}(index) \quad (7)$$

here, $\frac{dy}{dx}(index)$ means the slope of No. index becline

which are made to approximate the nonlinear function between two close dots .

It takes the controller about 37.5us to calculate neural network, in addition to other time, the system may diagnose as many as 170 thyristors in a AC electricity cycle (20ms, if it has a frequency of 50Hz), in other words, there could be no less than 28 pair of thyristors in each phase. With simulating all kinds of fault state and many times of experiments, it is strongly proved that the system can give the right cause of the fault, correct rate is up to100%.

IV. CONCLUSIONS

The high-performance of feed-forward neural network in approximating the complex non-linear function with multi inputs and outputs is well-suited for the fault diagnosis system. Its main idea is to map the relationship between fault sign and fault cause, train the network in terms of typical fault patterns. The GA search for the optimum solution without the gradient information and analytic differential of the function so long as the optimum solution exists in given constrained condition, and it consequentially get global optimum solution at last. If we use GA to optimize the neural network weights, there is an additional advantage of better fault tolerance than using classical BP algorithm.

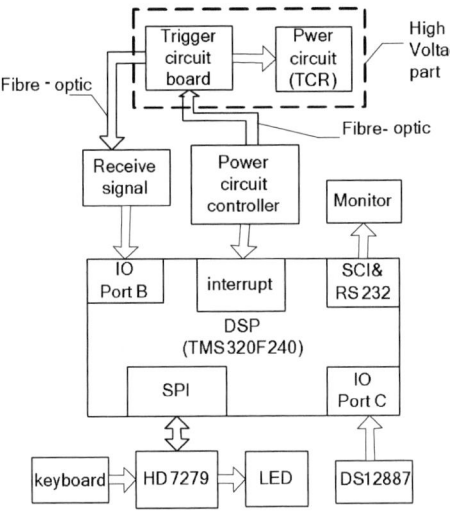

Figure 4. Frame of diagnosis system

The high-speed operation of fixed-point DSP is well-suited for real-time on-line fault detection and diagnosis. In this paper we diagnose the thyristors of the SVC power circuit with GA-BP neural network. The experiment results show the performance of the system is excellent. We hope this method and system are helpful to other users in similar areas.

REFERENCES

[1] SU P S, WANG Z Q, and JANG J G, " Fault Diagnosis of Power Electronic Circuits using artificial neural networks," *J Tsinghua Uni (Sic&Tech). Beijing,* vol. 39, No 3,1999, pp. 19-22.

[2] MA HAO,XU De-hong, and BIAN Jing-ming, "Fault diagnosis of power electronic circuits based on neural network and frequency

analysis," *Journal of Zhejiang university (engineering Science). China,* Vol. 33, No 6, Nov. 1999 ,. pp. 19-22.

[3] CAI J D, TU J, and WANG S F，"Fault Diagnosis of Power Electronic Circuit Based on Genetic Manipulation and Artificial Neural Network，" *High Voltage Engineering, China,*Vol. 30, No 9, Sep. 2004, pp. 3-5.

[4] Ga´bor Renner, Aniko´ Eka´rt, "Genetic algorithms in computer aided design," *Computer-Aided Design. Hungary,* Vol. 35, 2004, pp. 709-726.

[5] Qing Tao, Xin Liu, and Meisheng Xue, "A dynamic genetic algorithm based on continuous neural networks for a kind of non-convex optimization problems," *Applied Mathematics and Computation.* Vol. 150, 2004, pp. 811-820.

[6] Adriana R. Garcez Castro, Vladimiro Miranda, "An interpretation of neural networks as inference engines with application to transformer failure diagnosis," *Electrical Power and Energy Systems.* Vol. 27, 2005, pp. 620-626.

[7] R. Rengaswamya, Dinkar Mylaraswamy, "A comparison of model-based and neural network-based diagnostic methods," . *Engineering Applications of Artificial Intelligence.* Vol. 14, 2001, pp. 805-818.

[8] Jiantao Liu, Hongbing Chang, T.Y. Hsu (Xu Zuyao), Xueyu Ruan, "Prediction of the flow stress of high-speed steel during hot deformation using a BP artificial neural network," *Journal of Materials Processing Technology.* Vol. 103, 2000, pp. 200-205.

[9] Patricia R.S. Jota , Syed M Islam , Tony Wu , "Gerard Ledwich. A class of hybrid intelligent system for fault diagnosis in electric power systems," *Neurocomputing.* Vol. 23, 1998, pp. 207-224.

[10] L.A. Snider,Y.S. Yuen, "The artificial neural-networks-based relay algorithm for the detection of stochastic high impedance faults," *Neurocomputing.* Vol. 23, 1998, pp. 243-254.

[11] C.S.Chang, Z.Xu, A.Khambadkone, "Enhancement and laboratory implementation of neural network detection of short circuit faults in DC transit system," *IEE Proc -Electr Power Appl,* Vol 150, No 3, May 2003, pp. 344–350.

[12] Tsong-Lin Lee, "Back-propagation neural network for long-term tidal predictions," *Ocean Engineering.* Vol. 31, 2004, pp. 225–238.

[13] Sang-Joon Lee, Hyosung Kim, Seung-Ki Sul, "A Novel Control Algorithm for Static Series Compensators by Use of PQR Instantaneous Power Theory," *IEEE Transaxtions on Power Electronics,* vol. 19, No 3, May 2004, pp. 814–827.

[14] Fang Zheng Peng, George W. Ott, Jr, and Donald J. Adams, " Harmonic and Reactive Power Compensation Based on the Generalized Instantaneous Reactive Power Theory for Three-Phase Four-Wire Systems," *IEEE Transaxtions on Power Electronics,*vol. 13, No 6, Nov.1998, pp. 1174–1179.

The Optimization-Sliding Mode Control For Three-Phase Three-Wire DSP-based Active Power Filter

Zhou Wei-ping, Liu Da-ming, Wu Zheng-guo, Xia Li, and Yang Xuan-fang
College of Electrical and intelligent, Naval University of Engineering
Wuhan, Hubei, China
E-mail: cnwhwp@126.com

Abstract—**Based on the time-domain analysis, a time-domain mathematical model of three-phase three-wire active power filter (APF) current control loop is established; Predictive control equation and its optimal feasible solution with minimum objective function value are gotten, and the sliding-mode control strategy of current control loop is deduced. A novel optimization–sliding mode control strategy is proposed, with this control strategy the current tracking error can be reduced quickly and the chattering can be reduced effectively. The control performance under the proposed control strategy is analyzed. The proposed control strategy can improve current tracking speed and accuracy, provide both robustness response and a good dynamic response of the current tracking control, and has the characteristics of simple computation and easy application. The simulation and experiment results show its validity and feasibility.**

Keywords-active power filter; harmonic and reactive power compensation; mathematical model; optimization control; Sliding mode control; Variable structure control;

I. INTRODUCTION

Power electronic equipment usually introduces harmonic currents and reactive power, which result in problems such as low power factor, low efficiency, power system voltage fluctuations, EMI and communications interference. Passive filters have been used to improve power factor and to eliminate line current harmonics. However passive filters have some disadvantages such as resonance, fixed compensation character, harmonic amplification, possible overload and large size. So active power filters (APF) have been presented as a current-harmonic compensator for reducing the total harmonic distortion of the current and correcting the power factor[1~13].

Fig. 1 shows the configuration of a three-phase three-wire parallel active power filter. Its main power circuit is composed of a pulse-width-modulation (PWM) converter and a DSP-based control unit that generate the PWM control signals, the operation principle of the active power filter is to generate compensating currents into power system for canceling the current harmonics in the nonlinear load currents and compensating the reactive power. The performance of parallel active power filter with voltage source converter highly depends on the current tracking control capacity, which includes tracking speed and tracking precision.

In this paper, a time-domain mathematical model of three-phase three-wire active power filter current control loop is established, the sliding mode control law is deduced, and a novel optimization–sliding mode control approach is proposed. The proposed control approach can improve current tracking speed and accuracy, and the chattering can be reduced effectively with this method. Also it has the characteristics of simple computation and can provide both robustness response and a good dynamic response of the current tracking control.

II. TIME-DOMAIN MODEL ESTABLISHMENT

Consider Fig. 1, where U_{AO}, U_{BO} and U_{CO} represent the phase voltages of a power system at the point of common coupling (PCC), i_s and i_c represent the line current of power system and the input current of the active power filter.

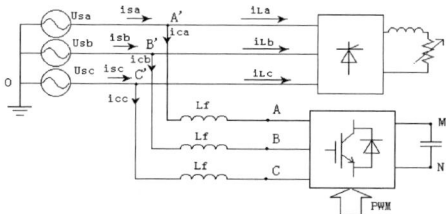

Figure 1. Main power circuit of three-phase three-wire APF

Six power transistors of a typical 3-phase power converter are controlled by PWM signals. When an upper transistor is switched on, the lower transistor is switched off. Thus, the on and off states of the upper transistors (j_a, j_b, j_c) is sufficient to evaluate the outputs of the APF. The voltage at nodes A, B and C referred to the negative rail of bridge "*N*" can be written as

$$\begin{cases} U_{AN} = j_a \cdot U_{dc} \\ U_{BN} = j_b \cdot U_{dc} \\ U_{CN} = j_c \cdot U_{dc} \end{cases} \quad . \quad (1)$$

where $j_a, j_b, j_c \in \{0, 1\}$, and U_{dc} is the dc-rail voltage. The electric circuit equations of the three-phase three-wire APF in time domain are as follows:

$$\begin{cases} U_{A'O} - U_{AO} = L_f \dfrac{d\,i_{ca}}{d\,t} \\[2mm] U_{B'O} - U_{BO} = L_f \dfrac{d\,i_{cb}}{d\,t} \\[2mm] U_{c'o} - U_{co} = L_f \dfrac{d\,i_{cc}}{d\,t} \end{cases} \quad (2)$$

Where L_f is the inductance of APF input inductor, i_{ca}, i_{cb} and i_{cc} are the input currents of the APF. For a three-phase three-wire system, it holds that

$$i_{ca} + i_{cb} + i_{cc} = 0. \qquad (3)$$

This leads to

$$U_{AO} + U_{BO} + U_{CO} = 0. \qquad (4)$$

Combination of （1）and（4）yields

$$U_{NO} = -\frac{U_{dc}}{3}(j_a + j_b + j_c). \qquad (5)$$

Substituting（5）into（1）results in

$$\begin{cases} U_{AO} = \dfrac{U_{dc}}{3}(2j_a - j_b - j_c) \\[2mm] U_{BO} = \dfrac{U_{dc}}{3}(-j_a + 2j_b - j_c) \\[2mm] U_{co} = \dfrac{U_{dc}}{3}(-j_a - j_b + 2j_c) \end{cases} \quad (6)$$

And the reference-voltages of U_{AO}, U_{BO}, U_{CO} are

$$\begin{cases} U_{AO}^{*} = U_{A'O} - L_f \dfrac{d\,i_{ca}^{*}}{d\,t} \\[2mm] U_{BO}^{*} = U_{B'O} - L_f \dfrac{d\,i_{cb}^{*}}{d\,t} \\[2mm] U_{co}^{*} = U_{c'o} - L_f \dfrac{d\,i_{cc}^{*}}{d\,t} \end{cases} \quad (7)$$

Where i_{ca}^{*}, i_{cb}^{*} and i_{cc}^{*} are reference currents.

Based on equidistant sampling with the sampling period T_s,（7）can be written approximately as

$$\begin{cases} U_{AO}^{*}(k) = U_{A'O}(k) - L_f \dfrac{i_{ca}^{*}(k) - i_{ca}^{*}(k-1)}{Ts} \\[2mm] U_{BO}^{*}(k) = U_{B'O}(k) - L_f \dfrac{i_{cb}^{*}(k) - i_{cb}^{*}(k-1)}{Ts} \\[2mm] U_{CO}^{*}(k) = U_{c'o}(k) - L_f \dfrac{i_{cc}^{*}(k) - i_{cc}^{*}(k-1)}{Ts} \end{cases} \quad (8)$$

Where $U_{AO}^{*}(k), U_{BO}^{*}(k)$ and $U_{CO}^{*}(k)$ are reference voltages at the kth sampling instant. The control goal can be expressed as: by changing the switching patterns several times during the control period of $(t_k, \; t_{k+1})$, to make the average value of $U_{mO}(t)$ be equal to the average value of $U_{mO}^{*}(t)$ at any period of time as short as possible, it also holds during the control period of (t_k, t_{k+1}). Assuming the sampling frequency is high enough, and the change of $U_{A'O}$ is slow enough to be considered as a constant during the period of $(t_k, \; t_{k+1})$, when $t \in [t_k, t_{k+1}]$, that is:

$$\begin{cases} \overline{U_{AO}(t)} = U_{AO}^{*}(k+1) \\[2mm] \overline{U_{BO}(t)} = U_{BO}^{*}(k+1) \\[2mm] \overline{U_{CO}(t)} = U_{CO}^{*}(k+1) \end{cases} \quad (9)$$

There are eight possible combinations of on and off patterns of (j_a, j_b, j_c) for the three upper power transistors that feed the three phase power converter. According to the magnitude of the binary number of $j_a j_b j_c$, from small to big, the corresponding pattern is denoted as k0 to k7, and the corresponding controlled voltage vector, $\mathbf{U}(i) = [U_{AO}(i), U_{BO}(i), U_{CO}(i)]^{T}$, can be obtained from (6). Within the sampling period of $(t_k, \; t_{k+1})$, the respective durations for each switching patterns are denoted as $\Delta t(i)$. The time-domain mathematical model of three-phase three-wire active power filter current control loop is obtained from (9).

$$\frac{1}{Ts}[\mathbf{U} \cdot \mathbf{\Delta t}] = \begin{bmatrix} U_{AO}^{*}(k+1) \\ U_{BO}^{*}(k+1) \\ U_{CO}^{*}(k+1) \end{bmatrix}. \quad (10)$$

where: $\mathbf{U} = [\mathbf{U}(0), \mathbf{U}(1), \cdots, \mathbf{U}(6), \mathbf{U}(7)]$

$\mathbf{\Delta t} = [\Delta t(0), \Delta t(1), \cdots, \Delta t(6), \Delta t(7)]^{T}$

III.   OPTIMIZATION-SLIDING MODE CONTROL

A. Current Tracking optimization Control

In proposed optimization control approach, the objective function of current tracking performance optimization control is defined as

$$J = \sum_{m=A,B,C} \int_{t(k)}^{t(k+1)} [i_{cm}^{*}(t) - i_{cm}(t)]^2 \, dt. \quad (11)$$

Where $i_{cm}^{*}(t)$ and $i_{cm}(t)$ can be written as

$$\begin{aligned} i_{cm}^{*}(t) &= i_{cm}^{*}(k) + [\widetilde{i_{cm}^{*}}(k+1) - i_{cm}^{*}(k)](t - t_k)/T_s \\ &= i_{cm}^{*}(k) + \Delta i_{cm}^{*}(k+1)(t - t_k)/T_s \end{aligned} \quad (12)$$

$$i_{cm}(t) = i_{cm}(k) + \frac{1}{L_f} \int_{t_k}^t [U_{A'O'}(t) - U_{AO'}(t)]dt \qquad (13)$$

where $m \in \{A,B,C\}$, $t_k \leq t < t_{k+1}$. So (11) can be written as:

$$J = \sum_{m=A,B,C} \int_{t(k)}^{t(k+1)} [i_{cm}^*(t) - i_{cm}(t)]^2 dt \qquad (14)$$

$$= \sum_{m=A,B,C} \int_{t(k)}^{t(k+1)} [\Delta i_{cm}(k) + \Delta i_{cm}^*(k+1)\frac{(t-t_k)}{T_s}$$

$$- \frac{1}{L_f} \int_{t_k}^t (U_{m'O'}(t) - U_{mO'}(t))dt]^2 dt$$

In order to get the optimal feasible solution of the time-domain mathematical model, the optimal feasible solution is discussed first under the assumption of $\hat{U}_{AO}^*(k+1) > -\hat{U}_{BO}^*(k+1) \geq -\hat{U}_{CO}^*(k+1) > 0$. And the optimal feasible solution of (10) in this case can be solved as follows[6]:

$$\begin{cases} \Delta t(4) = \frac{T_s}{U_{dc}}(U_{AO}^*(k+1) - U_{CO}^*(k+1)) \\ \Delta t(5) = \frac{T_s}{U_{dc}}(U_{CO}^*(k+1) - U_{BO}^*(k+1)) \\ \Delta t(0) = T_s - \Delta t(4) - \Delta t(5) \end{cases} \quad .(15)$$

According to the approach, the optimal feasible solutions of (10) are obtained as in Table I

TABLE I.   OPTIMAL FEASIBLE SOLUTION OF PREDICTIVE CONTROL EQUATION

case	U_{AB}^*	U_{BC}^*	U_{CA}^*	Optimal feasible solution
I	+	+	-	k4×t(A,B)+k6×t(B,C)
II	-	+	-	k2×t(B,A)+k6×t(A,C)
III	-	+	+	k2×t(B,C)+k3×t(C,A)
IV	-	-	+	k1×t(C,B)+k3×t(B,A)
V	+	-	+	k1×t(C,A)+k5×t(A,B)
VI	+	-	-	k4×t(A,C)+k5×t(C,B)

where $t(m,n) = \frac{Ts}{U_{dc}}[U_{mO}^*(k+1) - U_{nO}^*(k+1)]$, m,n

$\in \{A,B,C\}$; and switching pattern K0 for $\Delta t(0)$ duration of time is inserted for the rest of the PWM period.

B. Sliding Mode Control of APF

The switching surface functions are defined as

$$\begin{aligned} S_a &= i_{ca}^*(t) - i_{ca}(t) \\ S_b &= i_{cb}^*(t) - i_{cb}(t) \\ S_c &= i_{cc}^*(t) - i_{cc}(t) \end{aligned} \qquad (16)$$

Where i_{ca}^*, i_{cb}^*, i_{cc}^* are reference currents. For a three-phase three-wire power system, there are only two independent equations of (16), and the control signals should be regulated to satisfy the reaching mode operation[7-13]:

$$S_a \frac{dS_a}{dt} = S_a[\frac{di_{ca}^*}{dt} - \frac{1}{L_f}U_{sa} + \frac{1}{L_f}U_{dc}U_{a-con}] < 0 \qquad (17)$$

$$S_b \frac{dS_b}{dt} = S_b[\frac{di_{cb}^*}{dt} - \frac{1}{L_f}U_{sb} + \frac{1}{L_f}U_{dc}U_{b-con}] < 0$$

Where U_{a-con}, U_{b-con} are the control signals of phase A and phase B. the control signals of the following form can satisfy the requirements of the reaching mode:

$$U_{a-con} = -k_a sign(S_a) \qquad (18)$$

$$U_{b-con} = -k_b sign(S_b)$$

Where $k_a > |L_f d i_{ca}^* / dt - U_{sa}| / U_{dc}$

$$k_b > |L_f d i_{cb}^* / dt - U_{sb}| / U_{dc} \qquad (19)$$

C. Optimization-Sliding Mode Control

There is a high current tracking accuracy in proposed optimization control approach, but it usually can't track the rapidly changed reference currents very well, and this leads the current spines of the compensated line currents.

The sliding mode control has a good dynamic and can easily eliminate the current spines of the compensated currents, but it has the drawback of chattering phenomena.

Here, a novel optimization-sliding mode control approach combined with the optimization control and sliding mode control is proposed.

$$U_{m\_con} = \begin{cases} U_{m\_con\_SMC} & \max|S_{a,b,c}| > \Delta \\ U_{m\_con\_OPT} & \max|S_{a,b,c}| \leq \Delta \end{cases} \quad (20)$$

Where $U_{m\_con\_SMC}$ is the control signal of sliding mode control which satisfy Equ.(17), $U_{m\_con\_OPT}$ is the control signal of the optimization control, and Δ is the appropriate value of determining the control action of them on or off.

IV.   SIMULATION RESULTS

To demonstrate the performance of the proposed approach, some computer simulations were carried out. Fig.2 shows the load current and its spectrum of phase A. Fig.3 shows the line current and its spectrum of phase A after compensation using the optimization control approach, it is shown that the currents tracking accuracy is high, but current spines can be found.

Fig.4 shows the line current and its spectrum of phase A after compensation using the sliding mode control approach. it can be found that the current spines of line current are eliminated, but it has the chattering phenomena. Fig.5 shows the line current and its spectrum of phase A after compensation using the proposed optimization-sliding mode control approach. It is shown that the load current is well compensated, the

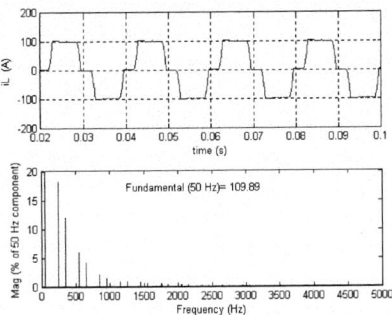

Figure 2. load current (up) and its spectrum (down) of phase A

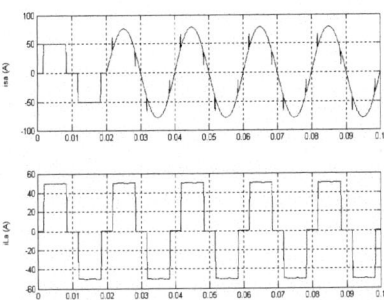

Figure 3. compensated current (up) and load current (down) of phase A using optimization control

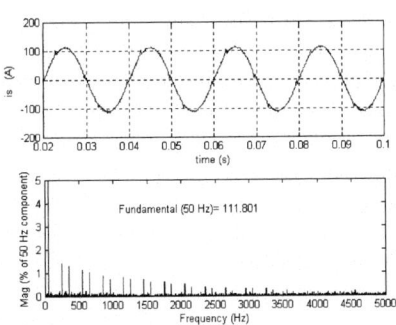

Figure 4. compensated current (up) and its spectrum (down) of phase A using sliding mode control

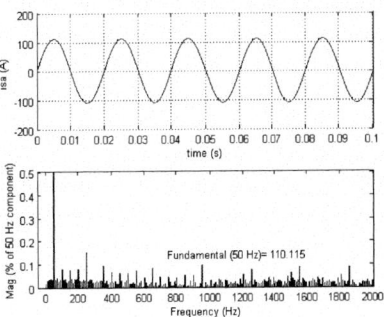

Figure 5. compensated current (up) and its spectrum (down) of phase A using optimization-sliding mode control

chattering phenomenon is suppressed effectively, and a high current tracking precision was gotten with the proposed method.

V. EXPERIMENTAL RESULTS

To demonstrate the performance of the proposed approaches, an experiment was conducted. The prototype three-phase three-wire APF was built using an IPM (PM75CSA120) as Fig.1. The control strategy is carried out based on TI-DSP of TMS320LF2407A.

Fig.6 shows the experimental result of the optimization control approach, it is shown that the load current is well compensated and source current is in-phase with supply voltage. The currents tracking accuracy is high, but current spines can be found.

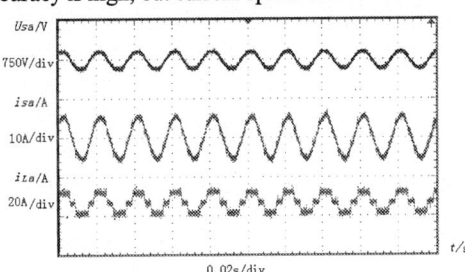

Figure 6. Experiment waveforms of phase A using optimization control method
a: voltage b: current after compensation c: load current

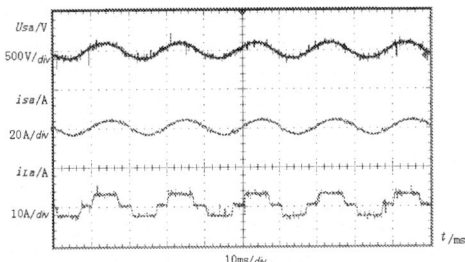

Figure 7. Experiment waveforms of voltage and currents of phase A using sliding mode control method
a: voltage b: current after compensation c: load current

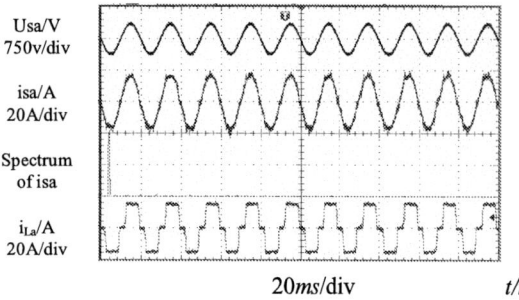

Usa/V
750v/div

isa/A
20A/div

Spectrum
of isa

i$_{La}$/A
20A/div

20*ms*/div   *t/ms*

Figure 8.   Experiment waveforms of voltage and currents of phase A
using proposed method
a: voltage b: current after compensation c: load current

Fig.7 shows the experimental result of the sliding mode control approach, it can be found that the current spines of line current are eliminated, but it has the chattering phenomena.

Fig.8 shows the experimental result of the proposed optimization-sliding mode control approach. It is shown that the load current is well compensated, the chattering phenomenon is suppressed effectively. It shows that the proposed approach can get a high current tracking precision.

VI. CONCLUSION

A novel optimization-sliding mode control approach of parallel active power filter is proposed combined with the optimization control and sliding mode control. The control performance is analyzed. The proposed optimization control approach can improve current tracking accuracy, and the optimal solutions can be gotten by simple judgments and computations, but it usually can't track very well under large change rate of reference current. The sliding mode control has a good dynamic, but it has the drawback of chattering phenomena. The load current can be well-compensated using proposed novel optimization-sliding mode control approach, the chattering phenomena is suppressed effectively, and a high current tracking precision is gotten with this method. The source current is in-phase with supply voltage and there are low current harmonics.

The proposed approach avoids complex co-ordinate transformations, which can significantly decrease computational complexity, and has the characteristics of simple computation and is much suitable for APF control application based on DSP.

REFERENCES

[1] Chongming Qiao, Keyue Ma Smedley, "Three-phase bipolar mode active power filters", *IEEE trans. Ind. Applica.*, vol. 38, no.1, pp. 149-158, Jan/Feb 2002.

[2] Wang Qun, Yao Wei-zheng, Liu Jin-jun, et al. "Harmonic source and compensation characteristics of active power filters"[J]. Proceeding of the CSEE, 2001,21(2):16-20. in Chinese

[3] Luowei Zhou and Zichang Li , "A novel active power filter based on the least compensation current control method", IEEE trans. Power Electron., Vol.15, NO.4, pp655-659, July 2000.

[4] Weiping Zhou, Li Xia, Zhengguo Wu, Wenhua Li. High-Accuracy Digitalization Method of phase and frequency Detection for power fundamental waves and its application. The 5th ISTM/2003, Shenzhen, China, pp1592-1595.

[5] Chen Guo-zhu. Liu Zheng-yu, Qian Zhao-ming. The general principle of active filter and its application[J], Proceeding of the CSEE, 2000,20(9)：17-21.

[6] Zhou Weiping, Wu Zhengguo, Xia Li, et. al. A current tracking performance optimization control approach in three-phase three-wire active power filter. IPEMC'04, conference proceedings vol.2, pp430-434. Xi'an,China,2004.

[7] B.R.Lin,S.C.tsay and M.S. Liao, "Integrated power factor compensator based sliding mode controller", IEE P.A.,Vol.148,NO.3,May 2000.pp237-244.

[8] V.M.Cardenas. et al, "Analysis and evaluation of control techniques for Active Power Filters: sliding mode control and proportional-integral control ", IEEE PESC'1999. pp649-654.

[9] Suttichai saetieo, rajesh devaraj, david a. torrey. "The design and implementation of a three-phase active power filter based on sliding mode control", IEEE . trans. on IA, 1995(31)5: 993-999.

[10] Severine guffon. et al, "Indirect sliding mode control of a three-phase power filter", IEEE PESC'1998. pp1408-1414.

[11] V.Carenas. et al, "Sliding mode control applied to 3 Φ shunt Active Power Filter using compensation with instantaneous reactive power theory", IEEE PESC'1998,pp236-241.

[12] Claudia Hernandez. et al, "Sliding mode control for a singer phase Active Filter", IEEE CIEP' 1998. pp171-176.

[13] NASSAR Mendalek, Kamal Al-Haddad, Farhat Fnaiech, et al. "Sliding Mode Control of 3-Phase Shunt Active Power Filter in d-q Frame", PESC'02 , pp369-375, Queenland Australia, June 23-27, 2002.

2006 5th International Power Electronics and Motion Control Conference

Three-Phase DVR using a Single-Phase Structure with Combined Hysteresis/Dead-band Control

Seyyed Hossein Hosseini and Mehran Sabahi
Department of Electrical Engineering, Islamic Azad University of Tabriz, Tabriz, Iran
hosseini@tabrizu.ac.ir

Abstract. **This paper proposes a dynamic voltage restorer for low or medium power applications with a single phase inverter (or STATCOM to acts as a series active filter) structure to compensate the line voltage unbalancing and distortions on the three phase load. The number of required switches of the conventional series three phase circuits is reduced from six to four. A new and simple control routine is applied to eliminate requirements for fast data samplers. In addition a new combined hysteresis/dead-band control with fixed time steps is used to proper charging of batteries (or capacitors) and minimizes energy transferring between line power and batteries by optimum reference voltage estimation, between the standard limits. A circuit model has been simulated with PSCAD/EMTDC software and the results are presented which verify the analysis.**

Keywords- Dynamic voltage restorer - Active filter – Single phase STATCOM

I. INTRODUCTION

Nowadays medium and low power dynamic voltage restores (DVR) are widely used for voltage restoration of sensitive loads and electronic devices. The requirements for power quality become more and more important to keep safety of the electric devices and costumer satisfaction. The growth of the nonlinear loads like the devices with switching power supplies have increased the current harmonics, EMI problems, unnecessary reactive power and power losses which causes distortion, harmonics, flicker phenomena, sag and swell conditions on the line voltages and other problems. The static VAR compensators are presently used in the power systems to improve power quality but they haven't fast enough time responding to the instantaneous or non-sinusoidal line voltage variations. Fast switching devices like CMOS or IGBT transistors provide implementation of the full bridge inverters to serve as a real time series compensators by bidirectional energy flow to control, compensate non-standard voltage wave forms and harmonics. Several circuits and control routines have been proposed to act as a dynamic voltage restore [1-9].

In this paper the main goal is proposing a cheap single phase circuit topology to act as a dynamic voltage restorer for compensating non-sinusoidal three phase line voltage on the loads by a new combined hysteresis/dead-band control technique. Using the proposed method, it is not only possible to correctly charge the batteries (or capacitors) but also to stop charging at the full charge conditions of the batteries. Also there is minimum energy transferring between batteries and utility. In fact energy transfer between batteries and line is done only in the case of sag or swell conditions of the line voltages. Required data to control the DVR is achieved from capacitors DC voltage, zero crossing detector of the input power for reference synchronization and voltage comparators which maybe contain some fast Op-Amps. There are not any mathematical computation blocks or DSP processors. Main tuning parameters are obtained by means of the conventional PI controllers. A circuit model has been simulated with PSCAD – EMTDC software [10-11]. The results show the good performance of proposed circuit and verify the analysis.

II. CIRCUIT CONFIGURATION

Fig.1 illustrates the power circuit diagram of the proposed DVR based on the single phase inverter structure. The load has not null connection and if that is required a delta to star transformer, just before the load, maybe added. Main structure of the circuit has a full bridge inverter with a voltage divider which consist capacitors C_1 and C_2. Each legs of this inverter can make independent voltages with respect to the midpoint of capacitors by proper gate pulsations of Q_1 to Q_4.

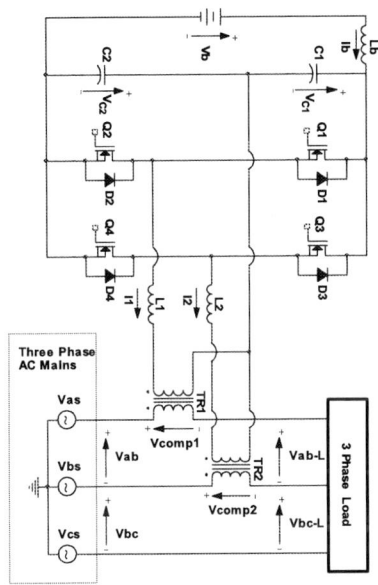

Figure 1. Proposed DVR circuit diagram.

1-4244-0448-7/06/$25.00 ©2006 IEEE   1679

Inductance L_b is used to eliminate instantaneous currents of batteries. Inductances L_1 and L_2 are used to act as low pass filters to eliminate high frequency components of PWM voltage outputs of the inverter legs and provide smooth compensation voltage at the primary sides of the isolation transformers. In order to voltage injection in series by the line voltages two isolation transformers TR_1 and TR_2 are required which is clear in Fig. 1 respectively, so the average values of the injection voltages v_{comp1} and v_{comp2} are zero. The midpoint of the capacitors is connected to common heads of the isolation transformers. This midpoint of the capacitors acts as a reference point to the inverter circuit, so two independent compensation voltages can be provided. Compensation of the voltages $v_{ab}(t)$ and $v_{bc}(t)$ lead to compensate $v_{ac}(t)$. Main condition to proper operation of the circuit is the amount of charge on the DC link capacitors, which is provided from batteries or from the utilities. At the normal conditions the amount of compensation voltages is zero so average energy transferring between batteries and line will be zero except for charging the batteries and power losses of the inverter. If the sag and swell conditions of the line voltages is not to be considered, it is possible to eliminate batteries and using a STATCOM structure to modify line voltages as a series active filter.

III. CONTROL STRATEGY

In symmetrical conditions we can suppose:

$$C_1 = C_2 \; , [V_{c1}]_{Avr.} = [V_{c2}]_{Avr.} = \frac{V_b}{2} \qquad (1)$$

If capacitor voltages are assumed to remain constant (equal to $V_b/2$), during a switching period and total losses of the inverter are considered to be zero, then the average amount of compensation voltage during one switching interval of the inverter is given by (2):

$$[V_{comp1,2}]_{Avr.} = \frac{n_2}{n_1} \frac{V_b}{2} (2D_{1,2} - 1)$$
$$where: \; D_{1,2} = \frac{t_{1,2}}{T_s} \quad , \quad 0 < D_{1,2} < 1 \qquad (2)$$

Where V_b is the voltage of batteries, n_2/n_1 is transformers turns ratio, $t_{1,2}$ shows conducting time interval of $Q_{1,2}$ and D is duty cycle of the switching period. The value of capacitors is depending to load currents which determine the amount of charging or discharging it during one cycle of utility voltage. If we assume the average value of one phase current during its half period time interval to $I_{Lavr.}$, and V_{Ripple} as the allowed voltage variation of the capacitor during same interval, then the following approximation can be derived:

$$C_i \frac{\Delta V_{ci}}{\Delta t} \approx I_{Lavr} \Rightarrow C_i \approx \frac{2I_{Lavr}}{V_{Ripple} f_{Line}} \quad , \quad i = 1,2 \quad (3)$$

Where, f_{Line} is the main utility frequency. In the next step proper charge voltage of capacitors must be found. If we assume maximum absolute amount of the compensation voltage to $\Delta V_{compMax}$, then (4) is obtained:

$$V_b \geq 2 \frac{n_1}{n_2} \Delta V_{compMax} \qquad (4)$$

If maximum effective value of the line current is represented by $I_{LrmsMax}$ then the rating of isolation transformer can be achieved approximately by (5):

$$S[TR_i] \approx I_{LrmsMax} \frac{\Delta V_{compMax}}{\sqrt{2}} \quad [VA], i = 1,2 \qquad (5)$$

The operation modes are as follows:

a) Charging mode:

In this case switches and inductances act like a boost converter to transfer either the stored energy in L_1, L_2 or line current to C_1 (or C_2). To create this mode a dead-band between gate pulses of transistors pair Q_1 - Q_2 and Q_3-Q_4 are required. The width of this dead-band determines dead time interval of the transistors and hence determine the amount of the energy pumping to the capacitors. It is possible to consider two separate dead bands to independent control of the energy flow to C_1 or C_2. First case in the charging mode control will be shown by (6):

$$i_1 < 0 \quad ; \quad D_1 \; is \; conducting \quad ;$$
$$v_{c1}(t^+) = v_{c1}(t^-) - \frac{1}{C_1} \int_{t^-}^{t^+} i_1(t) dt \qquad (6)$$

By assuming constant amount of inductor current during the switching interval equals to I_1, (7) will be achieved:

$$v_{c1}(t^+) \approx v_{c1}(t^-) + \frac{1}{C_1} I_1(t^+ - t^-)$$
$$\Rightarrow v_{c1}(t^+) > v_{c1}(t^-) \qquad (7)$$

Second case in the charging mode control is similar to (6) and (7) which is shown by (8):

$$i_2 < 0 \quad ; \quad D_3 \; is \; conducting \quad ;$$
$$v_{c1}(t^+) \approx v_{c1}(t^-) + \frac{1}{C_1} I_2(t^+ - t^-) \qquad (8)$$
$$\Rightarrow v_{c1}(t^+) > v_{c1}(t^-)$$

Third and 4th cases of the charging mode control are applied to charge capacitor C_2 with similar equations which are given briefly by (9) and (10):

$$i_1 > 0 \quad ; \quad D_2 \; is \; conducting \quad ;$$
$$v_{c2}(t^+) \approx v_{c2}(t^-) + \frac{1}{C_2} I_1(t^+ - t^-) \qquad (9)$$
$$\Rightarrow v_{c2}(t^+) > v_{c2}(t^-)$$

Also

$$i_2 > 0 \quad ; \quad D_4 \; is \; conducting \quad ;$$
$$v_{c2}(t^+) \approx v_{c2}(t^-) + \frac{1}{C_2} I_2(t^+ - t^-) \qquad (10)$$
$$\Rightarrow v_{c2}(t^+) > v_{c2}(t^-)$$

b) Compensation mode:

In this mode of operation, both diodes and transistors may be conducting relative to amount and direction of the compensation current. First case in compensation mode is given by (11):

$$to \ have \ \Delta V_{comp1} > 0 \Rightarrow Q_1 \ turns \ on$$

$$V_{comp1}(t) = \frac{n_2}{n_1}\left(L_1 \frac{di_1(t)}{dt} + V_{c1}(t)\right) \tag{11}$$

Due to small switching interval, by assuming constant capacitor voltage during that, the amount of compensation voltage will be approximated by (12):

$$V_{comp1}(t_0 + T_s) \approx \frac{n_2}{n_1}\left(\frac{V_b}{2} + L_1 \frac{\Delta I_1}{T_s}\right) \tag{12}$$

$$\frac{V_b}{2} \gg L_1 \frac{\Delta I_1}{T_s} \Rightarrow \Delta V_{comp1} > 0$$

Second, third and 4th cases case of the compensation mode is given similarly by (13):

$$to \ have$$
$$\Delta V_{comp1} < 0 \Rightarrow Q_2 \quad turns \ on$$
$$\Delta V_{comp2} > 0 \Rightarrow Q_3 \quad turns \ on \tag{13}$$
$$\Delta V_{comp2} < 0 \Rightarrow Q_4 \quad turns \ on$$

It is clear that to correct control of the compensation current, proper values should be chosen for capacitor voltages, inductors and time interval.

IV. ESTIMATION OF REFERENCE VOLTAGE VALUE

The energy equilibrium of the whole DVR system may be presented by (14):

$$W_S = W_{Load} + W_{Loss} + W_{C1} + W_{C2} + W_{Batt.} \tag{14}$$

Where W_s is the input energy from the utility, W_{Loss} is wasted energy of the inverter elements due to the losses, W_C and W_{Batt} are the stored energy in the capacitors and batteries respectively. Optimum control of DVR circuit is achieved when average energy circulation between batteries and utility decreases to zero. First in a given time interval and in absence of the batteries, it is possible to suppose average amounts of W_{Load} and W_{Loss} as constant values, an approximate relation as (15) will be obtained:

$$\Delta W_S[avr] = \Delta W_{C1}[avr] + \Delta W_{C2}[avr] \tag{15}$$

$$\Delta P_S[avr] \approx \frac{1}{2\Delta T} C_1 \left[V_{c1}(T^+)^2 - V_{c1}(T^-)^2 \right]$$
$$+ \frac{1}{2\Delta T} C_1 \left[V_{c2}(T^+)^2 - V_{c2}(T^-)^2 \right] \tag{16}$$

In (16), ΔT presents long time interval duration. If V_{LL} and I_s are assumed as the effective value of the main harmonic amplitude of the input line to line voltage and line current respectively, (17) will be obtained as follows:

$$\Delta P_S[avr] = \sqrt{3} I_s V_{LL} \cos\varphi \tag{17}$$
$$\Rightarrow \Delta P_S[avr] = k\Delta I_s$$

Comparing (16) and (17) yields to drive (18):

$$\Delta V_{LL} \propto \Delta V_{cap1} + \Delta V_{cap2} \quad where$$
$$\Delta V_{cap1} = V_{c1}(T^+)^2 - V_{c1}(T^-)^2 \quad ; \tag{18}$$
$$\Delta V_{cap2} = V_{c2}(T^+)^2 - V_{c2}(T^-)^2$$

If main harmonics of the voltages remain between standard values, it is possible to arrange a conventional PI controller by considering (18) to track sum of the capacitors voltage variations regarding to a specific DC value (which is taken equal to normal full charge of batteries i.e. E_{ch}), it is used as the input signal to the PI controller. Also effective value of the estimated reference voltage is obtained as the output signal of the PI controller. In the stable conditions with proper amount of the reference voltage value, the variations of the capacitor voltages tend to zero. Energy transferring occurs only in the case of charging batteries, non-standard sag or swell of the line voltage conditions. The PI controller output is limited; the allowed variations of the line voltages are assumed about 3 % of the nominal voltage.

V. COMBINED HYSTERESIS/ DEAD-BAND CONTROLLER

In this method first a conventional hysteresis zone for switches is considered. Utility voltages v_{ab} and v_{bc} are compared with a reference which has a proper value of the hysteresis band, then required gate pulses are provided from a logic circuit. In addition another zone is assumed between mentioned hysteresis bands which we call it as a dead-band. If the line to line voltage v_{ab} takes place between dead zone, then the gate pulses of Q_1 and Q_2 are stopped. The width of the dead zone and its symmetry relative to reference voltages determine charging mode time intervals to the C_1 and C_2. The following equations offer detailed explanation for symmetrical conditions. By considering a positive sequence phase voltage respect to null point, equals to $V_{an} = V_s cos(\omega t)$, for proper compensation of the voltage harmonics and distortions, the reference voltages is defined by (19):

$$V_{refAB} = \hat{V}_{mref} \cos\left(\omega t + \frac{\pi}{6}\right)$$
$$V_{refBC} = \hat{V}_{mref} \cos\left(\omega t - \frac{\pi}{2}\right) \tag{19}$$

Where, V_{mref} is the estimated reference voltage which obtains from PI controller output which is explained in the pervious section. Upper and lower hysteresis and dead-band zone limits are given by (20) and (21) respectively:

$$V_{upAB\_h} = V_{refAB} + \alpha = \hat{V}_{mref} \cos\left(\omega t + \frac{\pi}{6}\right) + \alpha \quad ; \quad \alpha > 0$$

$$V_{downAB\_h} = V_{refAB} - \alpha = \hat{V}_{mref} \cos\left(\omega t + \frac{\pi}{6}\right) - \alpha$$

$$V_{upBC\_h} = V_{refBC} + \alpha = \hat{V}_{mref} \cos\left(\omega t - \frac{\pi}{2}\right) + \alpha \tag{20}$$

$$V_{downBC\_h} = V_{refBC} - \alpha = \hat{V}_{mref} \cos\left(\omega t - \frac{\pi}{2}\right) - \alpha$$

$$V_{upAB\_d} = V_{refAB} + \beta = \hat{V}_{mref} \cos\left(\omega t + \frac{\pi}{6}\right) + \beta \quad ; \quad \beta > 0$$

$$V_{downAB\_d} = V_{refAB} - \beta = \hat{V}_{mref} \cos\left(\omega t + \frac{\pi}{6}\right) - \beta$$

$$V_{upBC\_d} = V_{refBC} + \beta = \hat{V}_{mref} \cos\left(\omega t - \frac{\pi}{2}\right) + \beta \tag{21}$$

$$V_{downBC\_d} = V_{refBC} - \beta = \hat{V}_{mref} \cos\left(\omega t - \frac{\pi}{2}\right) - \beta$$

Logic of the gate pulses is obtained from (22):

In each time step:

if $v_{ab}(t) > V_{upAB\_h}$; *Positive voltage injection*

$\Rightarrow Q_1 = ON$; $Q_2 = OFF$

if $v_{ab}(t) < V_{upAB\_h}$; *Negative voltage injection*

$\Rightarrow Q_1 = OFF$; $Q_2 = ON$

if $V_{downAB\_d} < v_{ab}(t) < V_{upAB\_d}$; *charging mode*

$\Rightarrow Q_1 = OFF$; $Q_2 = OFF$

if $V_{upAB\_d} < v_{ab}(t) < V_{upAB\_h}$; *keep previous state*

if $V_{downAB\_h} < v_{ab}(t) < V_{downAB\_d}$; *keep previous state*

$$(22)$$

if $v_{bc}(t) > V_{upBC\_h}$; *Positive voltage injection*

$\Rightarrow Q_3 = ON$; $Q_4 = OFF$

if $v_{bc}(t) < V_{upBC\_h}$; *Negative voltage injection*

$\Rightarrow Q_3 = OFF$; $Q_4 = ON$

if $V_{downBC\_d} < v_{bc}(t) < V_{upBC\_d}$; *charging mode*

$\Rightarrow Q_3 = OFF$; $Q_4 = OFF$

if $V_{upBC\_d} < v_{bc}(t) < V_{upBC\_h}$; *keep previous state*

if $V_{downBC\_h} < v_{bc}(t) < V_{downBC\_d}$; *keep previous state*

Fig.2 shows combined hysteresis/dead-band control diagram for compensation of v_{ab}. To avoid of high frequency chattering on the border lines, a fixed frequency switching is applied. The value of β determines the charging interval of the C_1, C_2 and batteries. Charging intervals increase by increasing β and cause more energy storing on capacitors or batteries. To determine proper values for β, the difference between capacitors voltages (or the current of batteries, i_b) is used as input signal for a conventional PI controller. Final value of β is depended on the inverter losses and charging mode required time interval. Ideal conditions are yield when width of dead-band limits to zero.

Fig. 3 illustrates control diagram of the proposed circuit. It is clear that the used parts in this diagram can be provided easily, with cheap structure. There are not any fast analog to digital converters or DSP processors. All of the required functions may be obtained by conventional analog circuits, CMOS or TTL logic gates. So it is possible to obtain fast online responding of the proposed AF system. E_{ch} in Fig. 3 is the proper full charge voltage of the batteries.

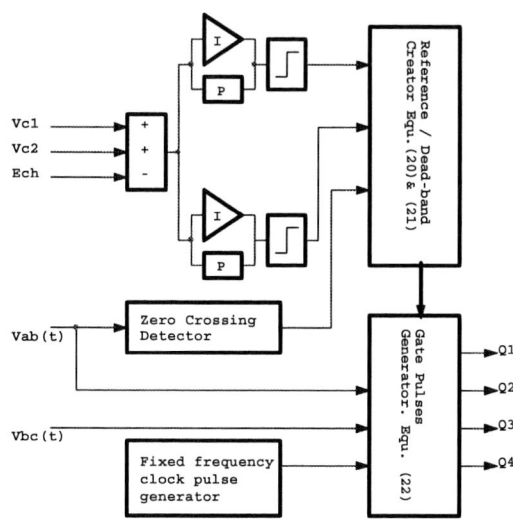

Figure 3. Control diagram

VI. THE RESULTS OF SIMULATION

The following values are selected in order to simulation of the proposed circuit. Switching frequency is considered about 10 [KHz], V_{LL} =380V RMS, C_1=C_2=10000 [μF], four 24 volts series batteries so E_{ch} is assumed equal with 96 volts, L_1=L_2=2 [mH], an inductive symmetric three phase load with power consumption about 1500 [W] and about 450 [VAR] reactive power. Fig. 4 shows wave forms of the load voltages without compensation for each phase (with respect to the star point of the load). Fig. 5 shows harmonic spectrum of the load voltages respectively. By activating of proposed DVR the load voltages are corrected as is shown in Fig. 6. The harmonic spectrum of resulting wave forms is given by Fig.7 for comparison with previous harmonic spectrum of uncompensated load voltages. DVR circuit modifies load voltages to the standard desired wave forms. In the next step the effect of DVR circuit on the sag or swell conditions of line voltage is simulated. Fig. 8 shows one of the input phase voltages and resulting phase voltage respectively. Table I compares the total harmonic distortion values, THD, for load voltages and illustrates the good performance of the proposed DVR.

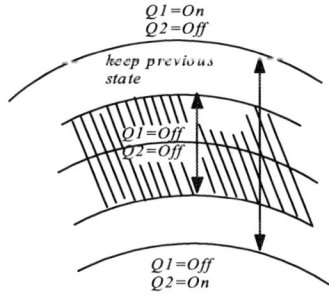

Figure 2. Combined hysteresis / dead-band control (for Q_1 and Q_2)

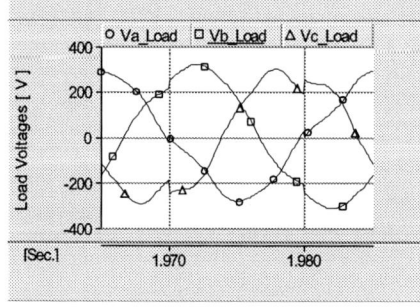

Figure 4. Load voltages before compensation

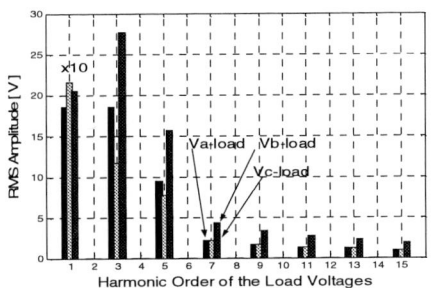

Figure 5. Harmonic spectrum of load voltages before compensation

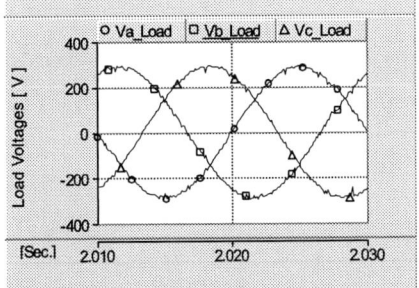

Figure 6. Load voltages after compensation

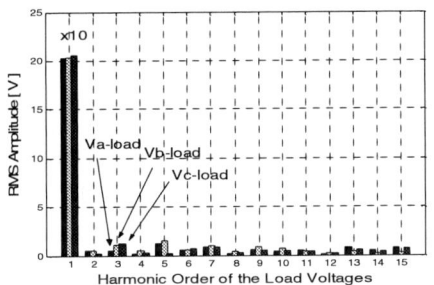

Figure 7. Harmonic spectrum of load voltages after compensation

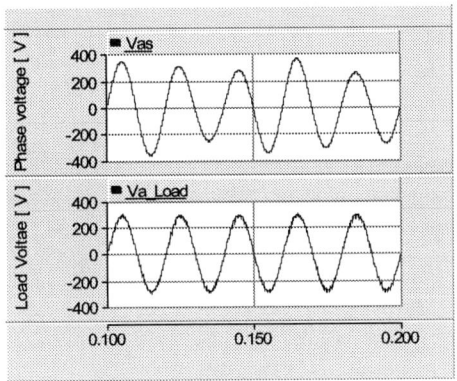

Figure 8. Sag and swell compensation

TABLE I.
Comparing THD results for each phase

Corporation of total harmonic distortion (DVR mode)					
Without Compensation			With Compensation		
THD_A	THD_B	THD_C	THD_A	THD_B	THD_C
0.1141	0.0669	0.1591	0.012	0.013	0.011

VII. CONCLUSION

A single phase inverter (or STATCOM) topology, as a three phases DVR (or series active filter) to compensation of the non-liner, unbalanced, sag or swell three phase input power voltages, has been proposed. The number of required switches of conventional parallel active filters is reduced from six to four. Also the number of required isolating transformers is reduced. A combined hysteresis/dead-band controller for load voltage proper compensation and minimum energy transferring is applied. Optimum reference voltage within the standard limits is estimated to minimized losses and energy transferring between batteries and power line. Furthermore this method can be charge the batteries automatically and stop charging at the full charge voltage. By disconnecting of batteries proposed DVR it is changed to a series active filter to correct input power line wave forms to sinusoidal shape on the loads. Also provides sufficient charging of the DC link capacitors and required optimum reference voltage value estimation which causes minimum energy flow between power system and active filter circuit. Furthermore proposed control system maybe structured with conventional cheap devices with fast online time response. A circuit model has been simulated with PSCAD/EMTDC software and the results verify the analysis.

REFRENCES

[1] S.S Choi,B.H. Li,D. M. Vilathgamuwa, "Dynamic Voltage Restoration with Minimum Energy Injection" IEEE Trans. On Power systems, Vol. 15, No. 1, Feb. 2000

[2] J.G.Nielsen, M.Newman, H.Nielsen, F.Blaabjerg, "Control and testing of a dynamic voltage restorer (DVR) at medium voltage level" , PESC, IEEE 3-4th annual Conf., pp 1248-1253, June 2003

[3] Y.H.Cho, S.Sul, "Controller design for dynamic voltage restorer with harmonic compensation function", Indust. App. Conf., 39th IAS Annual Meeting, pp 1452-1457, Oct. 2004

[4] N.H.Woodley, P.E., "Field experience with dynamic voltage restorer systems", Pow. Eng. Society Winter Meeting, pp 2864-2871, 2000

[5] Y.H.Chang, G.H.Kwon, T.B. Park, K.Y.Lim, "Dynamic voltage regulator with solid state switched tap changer", Quality and Security of Elec. Pow. Del. Systems, IEEE PES International Symp., pp 105-108,Oct. 2003

[6] M.El-Habrouk, M.K.Darvish, P.Mehta, "Analysis and design of a novel active power filter configuration", IEE Proc.-Electr. Power Appl., Vol. 147, No. 4, pp 320-328, July 2000

[7] J.H.Sung, S.Park, K.Nam, "New hybrid parallel active filter configuration minimising active filter size"; IEE Proc.-Electr. Power Appl., Vol. 147, No.2, pp 93-97, March 2000

[8] J.M. Carrasco, E. Galvan, M.Perales, G.Escobar, A.M. Stankovic, P.Mattavelli, " Direct current control: A novel control strategy for harmonic and reactive compensation with active filters under unbalanced operation", IECON'01, 27th Annual Conference of the IEEE Ind. Electr. Society, pp 1138-1143, 2001

[9] B.N.Singh, A.Chandra, K.Al-Haddad, " DSP- based indirect-current-controlled STATCOM , Part 1: Evaluation of current control techniques", IEE Proc.-Electr. Power Appl., Vol. 147, No.2, pp 107-112, March 2000

[10] Manitoba HVDC research center Inc., "PSCAD/EMTDC Version 4", 2003

[11] O.Anaya, E.Acha, "Modeling and analyses of custem power systems by PSCAD/EMTDC", IEEE Trans. On Pow. Delivery, Vol. 17, No. 1, pp 266-272, Jan. 2002

Harmonic Detection Based on the TLS Estimation Algorithm

Liu Kaipei ,ZHANG Junmin

School of Electrical Engineering, Wuhan University Wuhan 430072

e-mail: forzhanghua@sohu.com

Abstract—Based on the Newton learning algorithm, a new algorithm for TLS estimation and its application in harmonic detection is presented in this paper. It is shown that the LMS algorithm and Constrained Anti-Hebbian algorithm are two examples of the proposed algorithm in this paper. To solve the contravention between the detecting precision and speediness, analog detecting architectures are studied. The simulation results show that its noise rejection capability is superior to those of the LMS and Constrained Anti-Hebbian algorithms and the new algorithm gives faster convergence and better precision than another two methods.

Keywords- Newton algorithm; TLS estimation; adaptive; harmonic detection

I. INTRODUCTION

Harmonic and reactive current detecting play an important role in active power filters (APF) for harmonic and reactive current compensation of power system [1]. The detecting approaches of harmonic and reactive current mainly include the Fryze's time domain analysis method [2~3], and the filtration method by fixed frequency filters, in whose circuit there are fixed frequency filters such as low-pass or high-pass filter, and the composition method of the p-q and p-q-r based on the instantaneous reactive power theory [4~6], and the adaptive filter methods based on the LMS [7~9] and the neuron method [10]. Although each method has its own advantages, there are some problems that are difficult to overcome. In the adaptive and neuron harmonic detecting methods the algorithm are simple and easy to implement. The filtering characteristic is not sensitive to the element parameters. The shortcomings of these methods mentioned above are that their detecting precision cannot be consistent with characteristic of transient process when the amplitudes or frequency of the load currents are changed because of disturbance in loads and power system.

Usually we employed the least mean square method estimate parameters of a system. However, in recent years, there has widespread interest in the total least square method and the Hebbian learning algorithm [11~12] and its application for curve and surface fitting parameter estimation, and adaptive filtering. In LMS method we consider only the noise of the input in adaptive filtering or output of the system in modeling of a system. In many tasks there are not only input noise but

also output noise. The TLS method attempts to solve these problems. In this paper, we propose a new neural learning algorithm that is based on the TLS learning rule. We will find the LMS and the anti-Hebbian method are two special cases of the new algorithm. And analog detecting architectures are studied.

II. TLS ALGORITHM AND THE RELATIONSHIP OF THREE ALGORITHMS

A. TLS algorithm based on Newton algorithm

According to Newton algorithm, the learning rule of weight vector is [13]:

$$\psi(k+1) = \psi(k) - \mu' R^{-1} \nabla(k)$$
$$= \psi(k) - \mu'' R^{-1} \eta(k)\xi(k) \quad (1)$$

Where: R is the input self-correlation matrix.

Usually R is unknown and unfixed, so the computation of its inverse matrix can spend a lot time. Therefore many improved algorithms have been studied. Assume R is a diagonal matrix, so is its inverse matrix. We can obtain:

$$\psi(k+1) = \psi(k) - \mu\eta(k)\xi(k) \quad (2)$$

Where $\mu = \mu'' R^{-1} = \begin{bmatrix} \mu_1 & 0 & \cdots & 0 \\ 0 & \mu_2 & \cdots & 0 \\ \vdots & \vdots & \vdots & \vdots \\ 0 & 0 & 0 & \mu_{n+1} \end{bmatrix}$ is a diagonal matrix.

In order to get the least whole square error, the ideal value can be regardede as one of the input signal. In order to avoid all weight vectors are zero, assume $\psi_{n+1}(k) = -1$. Based on (2) we have:

$$\psi_{n+1}(k+1) = \psi_{n+1}(k) - \mu_{n+1}\eta(k)\xi_{n+1}(k)$$
$$= -1 - \mu_{n+1}\eta(k)\xi_{n+1}(k) \quad (3)$$

If $\psi_{n+1}(k+1)$ keeps in value -1, we obtain:

$$1 = 1 + \mu_{n+1}\eta(k)\xi_{n+1}(k) \quad (4)$$

Then (2) is divided by (4):

$$\psi_i(k+1) = \frac{\psi_i(k) - \mu_i\eta(k)\xi_i(k)}{1 + \mu_{n+1}\eta(k)\xi_{n+1}(k)} \quad (5)$$

If the coefficient μ_i is small enough such that $\left|\mu\eta(k)\xi_{n+1(k)}\right| < 1$, then (5) can be expanded as a Taylor series in μ:

$$\psi_i(k+1)=\psi_i(k)-\mu_i\eta(k)\xi_i(k)+\mu_{n+1}\eta(k)\psi_i(k)\xi_{n+1}(k)$$
$$=\psi_i(k)-\mu_i\eta(k)\left[\xi_i(k)+\frac{\mu_{n+1}}{\mu_i}\psi_i(k)\xi_{n+1}(k)\right] \quad (6)$$
$$==\psi_i(k)-\mu_i\eta(k)\left[\xi_i(k)+K_i\psi_i(k)\xi_{n+1}(k)\right]$$

Where $K_i=\dfrac{\mu_{n+1}}{\mu_i}$.

Matrix form is:
$$\psi(k+1)=\psi(k)-\mu\eta(k)\left[\xi(k)+K\psi(k)\xi_{n+1}(k)\right] \quad (7)$$

Where $K=\begin{bmatrix} k_1 & 0 & \cdots & 0 \\ 0 & k_2 & \cdots & 0 \\ \vdots & \vdots & \cdots & \vdots \\ 0 & 0 & \cdots & k_{n+1} \end{bmatrix}$ is a diagonal matrix.

Theorem: In algorithm
$$\psi(k+1)=\psi(k)-\mu\eta(k)\left[\xi(k)+K\psi(k)\xi_{n+1}(k)\right] \quad (8)$$
if $\psi_{n+1}(0)=-1$, **then** $\psi_{n+1}(k)=-1$ for all t.

Proof: From (8) we know

$$\psi_{n+1}(k+1)=\psi_{n+1}(k)-\mu_{n+1}\eta(k)\left[\xi_{n+1}(k)+K_{n+1}\psi_{n+1}(k)\xi_{n+1}(k)\right]$$
$$=\psi_{n+1}(k)-\mu_{n+1}\eta(k)\left[\xi_{n+1}(k)+\frac{\mu_{n+1}}{\mu_{n+1}}\psi_{n+1}(k)\xi_{n+1}(k)\right]$$
$$=\psi_{n+1}(k)-\mu_{n+1}\eta(k)\left[\xi_{n+1}(k)+\psi_{n+1}(k)\xi_{n+1}(k)\right]$$

Suppose $k=N$, we have $\psi_{n+1}(N)=-1$, then
$$\psi_{n+1}(N+1)=-1-\mu_{n+1}\eta(k)\left[\xi_{n+1}(k)+(-1)\xi_{n+1}(N)\right]$$
$$=-1$$

By using the mathematical induction, if $\psi_{n+1}(0)=-1$, then $\psi_{n+1}(k)=-1$ for all t.

B. the relationship of three algorithms

From reference [14] the weight vector incremental term of the LMS algorithm is:
$$\psi(k+1)=\psi(k)-2\mu\eta(k)\xi(k) \quad (9)$$

Where $\xi(k)$ ----input vector

$\quad\quad\quad\quad \eta(k)$ ----output vector

$\quad\quad\quad\quad \psi(k)$ ---- weight vector

$\quad\quad\quad\quad \mu$ ----incremental coefficient

The anti-Hebbian learning rule [15] is:
$$\psi(k+1)=\psi(k)-\mu\eta(k)\left[\xi(k)+\xi_{n+1}(k)\psi(k)\right] \quad (10)$$

Obviously, (10) has one more item $-\mu\eta(k)\xi_{n+1}(k)\psi(k)$ than (9). From discuses below we can know the anti-Hebbian is optimal solution [3]. From (8) we know:

(1) when $K=0$, i.e. $\mu_{n+1}=0$, the algorithm is:
$$\psi(k+1)=\psi(k)-\mu\eta(k)\xi(k) \quad (11)$$

(2) When $K=1$, i.e. μ_i is same, the algorithm is:
$$\psi(k+1)=\psi(k)-\mu\eta(k)\left[\xi(k)+\eta(k)\psi(k)\right] \quad (12)$$

From analysis above, the LMS algorithm and anti-Hebbian algorithm are two examples of the TLS algorithm. In order to simple the TLS algorithm, suppose $\mu_1=\mu_2=\cdots=\mu_n$, then in (8), there are two coefficients: μ and K.

III. THE APPLICATION OF ANALOG ADAPTIVE FIR FILTER IN HARMONIC DETECTION

Adaptive FIR filter has been widely used in the signal process and auto-control fields. The structure of the typical adaptive FIR filter is shown in figure 1.

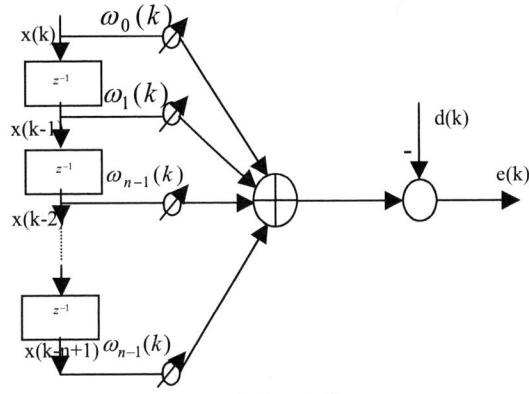

Figure 1. Adaptive FIR filter

Where $\overline{X}(k)=\left[x(k),x(k-1),\cdots,x(k-n+1)\right]^T$ is input vector, $\overline{\omega}(k)=\left[\omega_0(k),\omega_1(k),\cdots,\omega_{n-1}(k)\right]^T$ is weight vector, $d(k)$ is the ideal input. The output of the filter is:
$$y(k)=\overline{\omega}^T(k)\overline{X}(k) \quad (13)$$

The error is:
$$e(k)=y(k)-d(k)$$

For LMS algorithm, we have:
$$\overline{\omega}(k+1)=\overline{\omega}(k)-2\mu e(k)\overline{X}(k) \quad (14)$$

For anti-Hebbian algorithm, we have:
$$\overline{\omega}(k+1)=\overline{\omega}(k)-\mu e(k)\left[\overline{X}(k)+d(k)\overline{\omega}(k)\right] \quad (15)$$

For TLS algorithm, we have:
$$\overline{\omega}(k+1)=\overline{\omega}(k)-\mu e(k)\left[\overline{X}(k)+Kd(k)\overline{\omega}(k)\right] \quad (16)$$

Equations (14), (15) and (16) are similar structurally. When K is 0 and 1 in (16), the equation is (14) and (15) respectively.

Usually the adaptive method is only used in digital system. But many researches show that real biologic neuron system works by analog mode. So analog circuit is more feasible than digital circuit. And the structure of analog circuit is simple; the response speed of it is rapid.

Formula (14), (15) and (16) are divided by sample period T respectively, then we can obtain:
$$\frac{\overline{\omega}(k+1)-\overline{\omega}(k)}{T}=\frac{\mu e(k)\overline{X}(k)}{T} \quad (17)$$
$$\frac{\overline{\omega}(k+1)-\overline{\omega}(k)}{T}=\frac{\mu e(k)\left[\overline{X}(k)+d(k)\overline{\omega}(k)\right]}{T} \quad (18)$$
$$\frac{\overline{\omega}(k+1)-\overline{\omega}(k)}{T}=\frac{\mu e(k)\left[\overline{X}(k)+Kd(k)\overline{\omega}(k)\right]}{T} \quad (19)$$

If the sample period T is small enough such that a discrete variable can be regarded as continuous one. The discrete variable k is replaced the continuous time variable t; and let $\eta = \dfrac{\mu}{T}$; then the analog forms of three formulae above are:

$$\omega(t) = \int \eta e(t) x(t) dt \qquad (20)$$

$$\omega(t) = \int \eta e(t)(x(t) + d(t)\omega(t)) dt \qquad (21)$$

$$\omega(t) = \int \eta e(t)(x(t) + Kd(t)\omega(t)) dt \qquad (22)$$

When the only input of the system is voltage reference input $u_s(t)$ that has same frequency and phase with power voltage. And the input has no delay unit. The load current $i_L(t)$ is ideal input, i.e. $d(t)$ in the equations above. A kind of analog circuit based on adaptive harmonic current detection method is shown in Figure2.

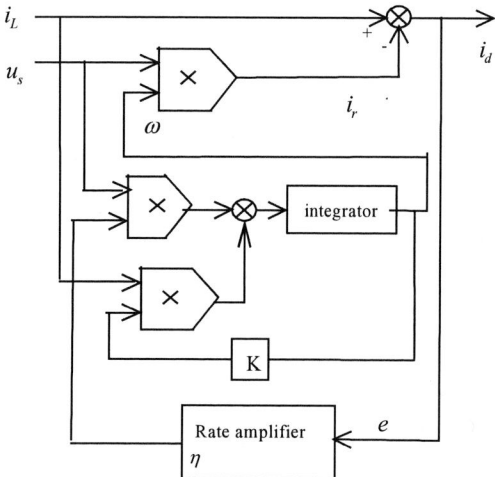

Figure 2.   Analog circuit diagram of adaptive detecting

In Figure2 when coefficient K=0, the diagram shows detection method based on LMS algorithm; when K=1, it shows the method of anti-Hebbian algorithm; when K is random value, it shows the method of the TLS algorithm.

When K=0[16], **_Theorem 1_**: If the signals in Figure 2 satisfy relations as follows

$$u_s = \sin(100\pi t) \qquad (23)$$

$$i_L = \sum_{n=1}^{N}(a_n \sin(100\pi nt) + b_n \cos(100\pi nt)) \qquad (24)$$

$$\omega = \int_0^t u_s i_d \eta \, dt \qquad (25)$$

$$i_d = i_L - u_s \omega \qquad (26)$$

Where a_n, b_n, η are constants, then ω is not constant.

$$\omega = \int_0^t u_s(i_L - u_s\omega)\eta \, dt$$

$$= \eta \int (a_1 - \omega)\sin^2(100\pi t)dt + \frac{\eta b_1}{2}[\frac{1 - \cos(200\pi t) + \sin(200\pi t)}{200\pi}]$$

$$+ \eta \sum_{n=2}^{N}[\frac{a_n}{2}(\frac{\sin(100(n-1)\pi t)}{100(n-1)} - \frac{\sin(100(n+1)\pi t)}{100(n+1)})$$

$$+ \frac{b_n}{2}(\frac{1 - \cos(100(n-1)\pi t)}{100(n-1)} - \frac{1 - \cos(100(n+1)\pi t)}{100(n+1)})$$

From theorem 1 we can come to the conclusion that parameter ω is not constant. It shows that the iterative element η must be very small, or else ω may bring errors, which may be not ignored. The smaller is the parameter η, the smaller is the errors of ω. It will be convergent to the ideal value, but iterative process may be long. The fluctuation of ω means that the parameter i_r is not sine wave, nor is i_r the active current or i_d the compensation current accurately.

When K is not zero, the only difference is:

$$\omega = \int_0 \eta i_d(u_s + K i_L(t)\omega(t))dt \qquad (27)$$

So we can obtain the similar conclusion. The proving is the same as the theorem 1.The improved adaptive detecting of the harmonic for power system can be described in Figure 3.

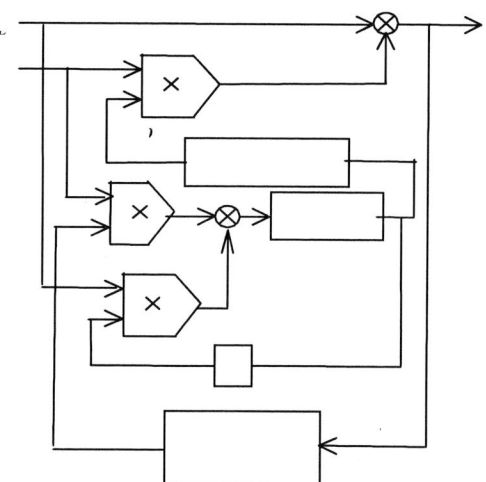

Figure 3. Improved Analog circuit diagram of adaptive detecting

According to the analysis above, we choose one kind of inductance load as compensation object. Suppose the voltage in power system is sinusoidal signal, the waveform of the current is shown in Figure3. The current can be decomposed of high harmonic except $50Hz$ fundamental frequency current. If this kind of current is injected into power system, harmonic voltage will be brought and the power quality may be worse. Therefore we must compensate harmonic current to maintain good power quality.

1686

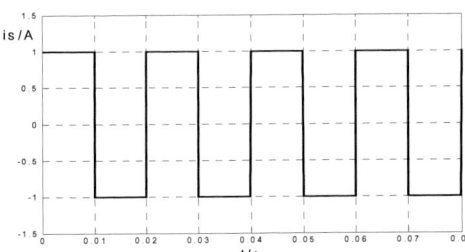

Figure 3. Harmonics current

Next three harmonic current methods based on three algorithms respectively are used to detect harmonic current shown in Figure3. Their following performances are shown in Figure4, Figure5 and Figure6 respectively.

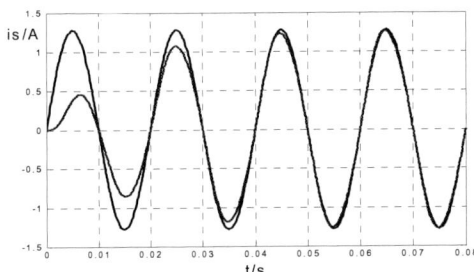

Figure 4. Simulation for the LMS algorithm

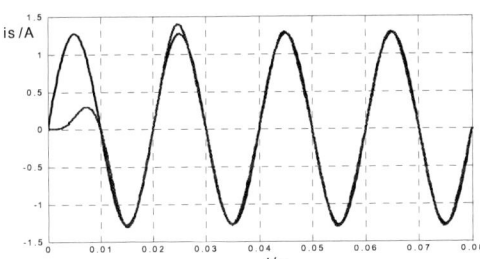

Figure 5. Simulation for the anti-Hebbian algorithm

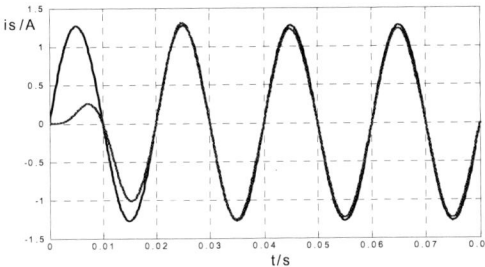

Figure 6. Simulation for the TLS algorithm

According to the figures above, the dynamic time of the LMS algorithm is more two periods; the dynamic time of the anti-Hebbian is probably one and half period; the dynamic time of the TLS is maybe one period.

IV. CONCLUSION

In this paper, we propose a new adaptive learning algorithm that is based on the TLS learning rule. We will find the LMS and the anti-Hebbian method are two special cases of the new algorithm.

To solve the contravention between the detecting precision and speediness, analog detecting architectures are studied. The improved adaptive leaning methods with lowpass filter are proposed to decrease the errors.

The simulation results show that its noise rejection capability is superior to those of the LMS and Constrained Anti-Hebbian algorithms and the new algorithm gives faster dynamic time and better precision than another two methods.

REFERENCES

[1] Akagi H, Kanazawa Y , Nabae A, "Instantaneous reactive power compensator comprising switching devices without energy storage components". IEEE Trans on IA 1984 20(3) 625-630.

[2] S.Fryze, Wirk, Blind, "Und Scheinleistung in Elektrischen Stromkreisen mit nichtsinusoidalem Verlauf von Strom und Spannung". Elektrotech Zeitschrift. Vol.53, No.53, 1932, pp.596-599.

[3] S.Fryze, "Wybrane Zagadnienia Elektrotechniki". Warsaw, Poland: PWN, 1996.

[4] H. Akagi, Y. Kanazawa, and A. Nabae, "Instantaneous reactive power compensators comprising switching devices without energy storage components," *IEEE Trans. Ind. Applicat.*, vol. IA-20, no. 3, pp. 625–633, 1984.

[5] A. Nabae and T. Tanaka, "A new definition of approach of instantaneous active-reactive current and power based on instantaneous space vectors on polar coordinates in three-phase circuits," *IEEE Trans. Power De-livery*, vol. 11, pp. 625–633, July 1996.

[6] H. Akagi, "New trends in active filters for power conditioning," *IEEE Trans. Ind. Applicat.*, vol. 32, pp. 1312–1322, Nov./Dec. 1996.

[7] Liu Kaipei, Zhang Junmin, Xuan Yang, "Hamonics detection for three-phase circuits based on resampling theory and mean filtering", Proceedings of the CSEE, 23(9):78-82, 2003.

[8] H. Kim, F. Blaabjerg, "Instantaneous power compensation in three-phase systems by using p-q-r theory", IEEE TRANSACTIONS ON POWER ELECTRONICS, VOL.17, NO.5, pp701-710, SEPTEMBER 2002.

[9] H. Kim, Hi Akagi, "The instantaneous power theory on the rotating p-q-r reference frames", IEEE 1999 International Conference on Power Electronics and Drive System, PEDS'99, July 1999, Hong Kong:422-427.

[10] Chen Yanhui, Liu Kaipei, Guo Shufang, "Power system harmonic detection by p-q-r method and its limitation, Power System Technology", 2004, 28(13): 88-91.

[11] M.A. Rahnon and K. B.Yu, "TLS approach for frequency estimation using linear prediction". IEEE Transactions on ASSP, vol.35, No.5, 1987.

[12] Keqin Gao, M. Omair Ahmad, M. N. S. Swanmy, "A constrained anti-Hebbian learning algorithm for total least-squares estimation with applications to adaptive FIR and IIR filtering". IEEE Transactions on CS. vol.41, No.11, Nov.1994.

[13] Shen Fuming, *Adaptive signal processing*. Publishing House of Xidian University. Mar.2001.

[14] Simon Haykin, *Adaptive filter theory*. Publishing House of Electronics Industry, Aug.1998.

[15] Scott C. Douglas, "Analysis of an Anti-Hebbian Adaptive FIR filtering algorithm". IEEE Transactions on CS. vol.43, No.11, Nov.1966.

[16] Liu kaipei, Tan Qian, Sun Junfu, Zhanghui, "Time domain analysis do adaptive detecting approaches of harmonic and reactive current for APF and its improvement". 2005 IEEE/PES Transmission and Distribution Conference & Exhibition: Asia and Pacific Dalian, China.

2006 5th International Power Electronics and Motion Control Conference

Control Strategy Study of Hybrid Active Power Filter

Jia Zhang, Guohong Zeng

College of Electrical Engineering Beijing Jiaotong University
Mailing address: Beijing Jiaotong University 403# (100044)
Email address: dqzhangj@master04.bjtu.edu.cn

Abstract—A novel hybrid active power filter (HAPF) is proposed in this paper, which has the advantages of both active power filter (APF) and passive filter (PF), and overcomes their defects. A review of the present development of harmonic suppressing and reactive power compensation is presented in the first part. Principle and structure of the proposed HAPF is discussed in the second part, and an optimized switch controlling strategy is introduced, which can control the 3-phase 3-wire APF more effactually and accurately. A reference current generating method named as v_p–v_q is deduced in the third part, which can solve the problem of the existing methods under asymmetrical and distorted source-voltage. Software simulation result is presented in the last part, which can demonstrate the validity and effect of the proposed HAPF.

Keywords—active power filter (APF); passive filter (PF); reactive power compensation; harmonic suppressing

I. INTRODUCTION

At present, Harmonic and Reactive Power have an increasing influence on power quality and electronic devices. Therefore, harmonic suppressing and reactive power compensation has become important researching direction in the domain of improving the quality of electric power. Although the widely adopted SVC which controlled by SCR has achieved certain effects in compensating harmonic and reactive power, it still has such defects as fixed working place, large size and easiness of bringing about syntony with power line, making it fail to meet the need of dynamic compensation.

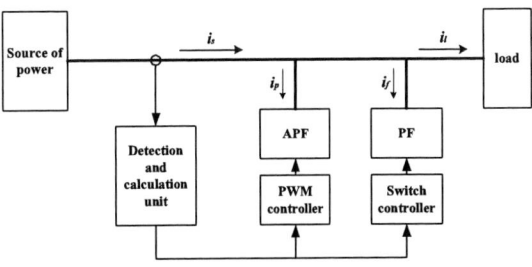

Fig. 1 Principle Diagram of HAPF

Hybrid active power filter, composed of passive filter and active power filter, merges the passive filter's advantage of great capacity with that of active power filter—possibility to realize dynamic compensation, so as to meet the requirements of wide range of dynamic compensation. Consequently, it has gained great attention from experts in the domain of harmonic suppressing and reactive power compensation and has a promising marketing prospect.

HAPF is connected with load in parallel (Fig.1), which mainly includes three units: reactive and harmonic current detection and calculation unit, APF unit and PF unit. In this strategy, PF is used to compensate a majority of low-order harmonic and high-order harmonic and most reactive power, while APF is for the residual low-order harmonic and a little reactive power which flow into HAPF.

II. MAIN CIRCUIT STRUCTURE OF HAPF

Fig.2 shows the main circuit structure of HAPF. Tran stands for system isolation transformer; Load indicates

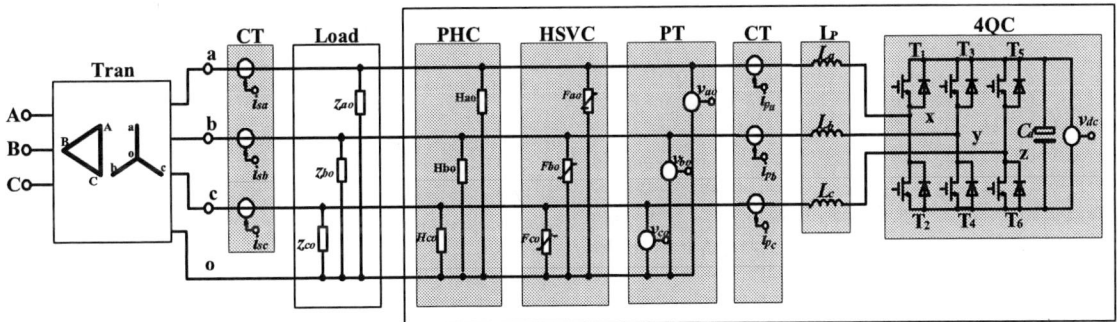

Fig. 2 Main Circuit Structure of HAPF

1-4244-0448-7/06/$25.00 ©2006 IEEE

asymmetrical and distorted load which generates reactive and harmonic current. The stuff in the block is HAPF. PHC represents Passive Harmonic Compensator and HSVC is Hybrid Static Var Compensator.

Fig.3 shows the single phase reactive power compensation circuit which represents one element of HSVC and PHC in Fig.2 respectively. In order to adjust reactive power smoothly, thyristor switching capacitor (TSC) in HSVC is divided into three separately controlled parts, whose capability are 200KVA, 200KVA and 100KVA respectively. The capability of each part is considered according to the whole capability and the APF capability; Thyristor controlled reactor (TCR) is also utilized to compensate three phase unbalance load. PHC is made up of triple, quintuple, septuple and high-order harmonics resonance circuit.

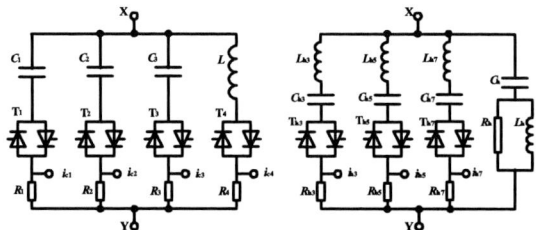

Fig.3 Main Circuit Structure of Reactive Power Compensation Network

Reactive power compensation: Detect current and voltage signals, and calculate three phase load in real time. Choose the seemly compensating capability of HSVC network based on the whole reactive power capability. APF compensates the remainder reactive power after HSVC network compensation.

Harmonic compensation: PHC compensates most triple, quintuple and septuple harmonics and high-order harmonic, while APF compensates the remainder harmonics. When harmonic current goes beyond the capacity limit, HAPF closes PHC and APF to protect them.

III. ARITHMETIC OF REFERENCE CURRENT

Based on the principle of instantaneous reactive power, the *p-q* method, first proposed by H. Akagi in 1983, has been utilized widely in the APF field. But it requires the source voltage to be sinusoidal and symmetric. When the source voltage is distorted by harmonic, big error will occur in the result. If the phase-lock-loop (PLL) is used to generate the synchronous signals: $\sin\omega t$ and $\cos\omega t$, the reference current is same as what is generated by *p-q* method. That is the basic idea of $i_p{-}i_q$ method. Its principle is as follows: The load current signals are transformed into $\alpha{-}\beta$ coordinate from *a-b-c* coordinate at first, and then are disintegrated into i_p and i_q in *p-q* coordinate which is cycling synchronously with source voltage. Because different components in i_p and i_q can be made by the load's active, reactive, negative and harmonic current, it is possible to eliminate some parts of these components from i_p and i_q, to use the reverse

transformation, and get the demanded APF's reference current. This method has the advantage of quick response and can be applied to all the compensating purpose. In the case of unbalanced source voltage, $i_p{-}i_q$ method can generate correct reference current in the harmonic and negative fundamental current compensation system; in the positive fundamental current compensation system, $i_p{-}i_q$ method will have some error.

From above statements, when the source voltage is unbalanced and distorted, reference current of APF can be generated accurately if we can get the positive fundamental component of the source voltage. So the key to answer is how to get the positive fundamental component from the distorted and unbalanced source voltage accurately.

The analysis of $i_p{-}i_q$ method has disclosed the fact that even if the load current contains the load's active and reactive current, there would be no error in the result of load positive fundamental component. This property means that if the source voltage takes the place of the load current as the input, the output should be the source positive sequence voltage without error. Therefore, a positive sequence voltage filter method $v_p{-}v_q$ can be designed similarly to the $i_p{-}i_q$ method.

The structure of $v_p{-}v_q$ method is shown in Fig. 4:

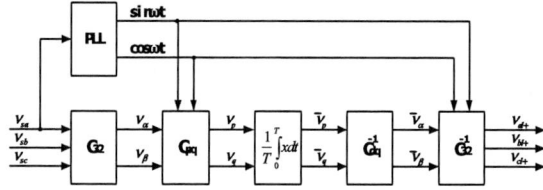

Fig. 4 Principle Diagram of $V_p{-}V_q$ Filter

In *p-q* coordinate, v_p and v_q can be deduced as follows:

$$\begin{bmatrix} v_d \\ v_q \end{bmatrix} = \frac{1}{\sqrt{2}} \begin{bmatrix} \sqrt{3}v_{sa}\sin\omega t - (v_{sb} - v_{sc})\cos\omega t \\ \sqrt{3}v_{sa}\cos\omega t + (v_{sb} - v_{sc})\sin\omega t \end{bmatrix} \quad (1)$$

One of the three-phase source voltages is used to generate a set of sine and cosine orthogonal signal with PLL. Assume $\bar{v}_p$ and $\bar{v}_q$ are the constant component of v_p and v_q respectively, the source positive fundamental voltage is:

$$\begin{bmatrix} v_{a1+} \\ v_{b1+} \\ v_{c1+} \end{bmatrix} = \frac{1}{\sqrt{6}} \begin{bmatrix} 2 & 0 \\ -1 & \sqrt{3} \\ -1 & -\sqrt{3} \end{bmatrix} \begin{bmatrix} \bar{v}_p\sin\omega t + \bar{v}_q\cos\omega t \\ \bar{v}_q\sin\omega t - \bar{v}_p\cos\omega t \end{bmatrix} \quad (2)$$

With the result of $v_p{-}v_q$ method as the source voltage, the $i_p{-}i_q$ method can be used to create the reference current without considering the result error. This paper only discusses the situation of HAPF, so that, for easy implementation, the synchronous detection (SD) algorithm can be adopted for the reference current calculation. The signal flow chart is shown in Fig.5.

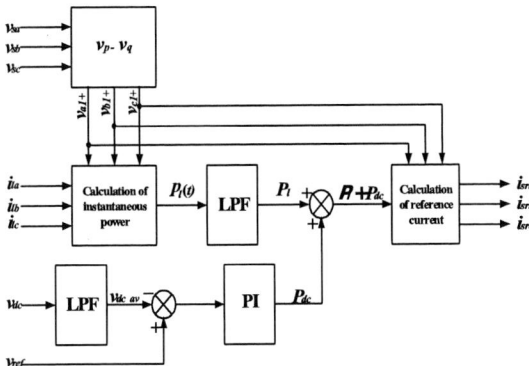

Fig. 5 Generation Principle Diagram of Reference Current

The amplitude of supply voltage U_m and load's average active power P_l is calculated by using the source's positive sequence voltage output from v_p-v_q method, a pure sinusoidal fundamental waveform. The equations for calculation are:

$$P_l = \frac{1}{T} \int_0^T \left[v_{a1+}(t) \cdot i_{la}(t) + v_{b1+}(t) \cdot i_{lb}(t) + v_{c1+}(t) \cdot i_{lc}(t) \right] dt \quad (3)$$

$$U_m^2 = \frac{2}{3} \left[v_{a1+}^2(t) + v_{b1+}^2(t) + v_{c1+}^2(t) \right] \quad (4)$$

The source will only supply load's positive sequence active current if APF works properly. But in fact, APF will consume some active power, which causes the decline of capacitance voltage on the side of DC. So the source must supply some other active power to APF in order to keep the voltage of this capacitance. As is shown, P_{dc} is just the active power which is consumed by APF, which can be calculated by the equation as follows:

$$P_{dc} = \frac{1}{2T} C \left(v_{ref}^2 - v_{dc\_av}^2 \right) \quad (5)$$

Here, C indicates the capacity of capacitor in DC side. And v_{ref} is the reference DC voltage, v_{dc_av} is average DC voltage.

The source current should be symmetrical and in-phase with the positive sequence voltage. Therefore, the source's reference current should be:

$$\left[i_{sra} \quad i_{srb} \quad i_{src} \right]^T = \frac{2(P_l + P_{dc})}{U_m^2} \cdot \left[v_{a1+} \quad v_{b1+} \quad v_{c1+} \right]^T \quad (6)$$

IV. SIMULATION RESULT

Simulation based on SABER software has been done to verify the proposed reference current generating method. The waveform of source's voltage and current (shown in Fig.6 and Fig.7) shows the waveform of load's

current and the source reference current. The waveform of the compensated source current shows that three-phase current is of symmetric fundamental sine wave. This result approved the proposed reference generating method.

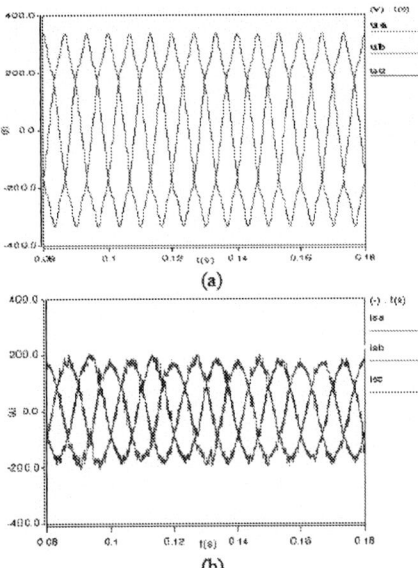

Fig.6 (a) Source voltage wave after compensated
(b) Source current wave after compensated

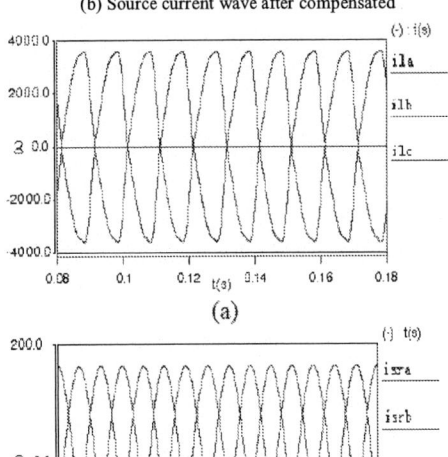

Fig.7 (a) Load current.
(b) Source reference current.

V. CONCLUSION

The theoretic analysis and the result of simulation have verified validly and feasibility of the proposed reference current generating method. Based on the method, APF can reduce reactive and harmonic current efficiently.

Combining the advantages of APF and PF, HAPF can satisfy the need of large capability and high efficiency very well in reality.

REFERENCES

[1] Guohong Zeng, "Research of a novel railway traction supply system based on 3/1 phase balanced converter." A dissertation submitted to academic degrees evaluation committee of Northern Jiaotong University for the degree of doctor of philosophy in engineering, Nov. 2002.

[2] Senini.S, Wolfs P J. "Hybrid active filter for harmonically unbalanced three-phase three-wire railway traction loads[J]." IEEE Trans. Power Electronics, July 2000, pp.702-710.

[3] Bhavaraju V B, Enjeti P N. "Analysis and design of active power filter for balancing unbalanced loads[J]." IEEE Trans. Power Electronics, Oct. 1993, pp.640-647.

[4] Akagi H. "Active and hybrid filters for power conditioning[C]." Proceedings of 2000 IEEE International Symposium on Industrial Electronics, 2000, Vol.1:TU26-TU36.

[5] Singh B, Al-Haddad K, Chandra A. "A review of active power for power quality improvement[J]." IEEE Trans. Industrial Electronics, 1999, pp.960-971.

[6] Adil M A, David A T. "A Passive Series, Active Shunt Filter for High Power Applications[J]." IEEE Trans. Power Electronics, Jan. 2001 pp.101-109.

2006 5th International Power Electronics and Motion Control Conference

Novel Harmonic Free Single Phase Variable Inductor Based on Active Power Filter Strategy

Mu Xianmin, Wang Jianze, Ji Yanchao, Wei Xiaoxia and Fu Xiangyun

Electrical Engineering Department of Harbin Institute of Technology, Harbin 150001, Heilongjiang, P.R.China

Abstract—This paper presents a variable inductor for single phase system which employs a new control method based on active power filter schemes. The novel variable inductor combines an active power filter circuit and a control circuit. The active power filter circuit is based on a single phase inverter with four controllable switches, a standard H-bridge inverter. The AC side of the inverter is connected to an aid harmonic elimination winding of the variable inductor. Its DC side is connected to a filter capacitor and the secondary winding of variable inductor through the control circuit based on BUCK circuit. The inverter switches are controlled to shape the current through the filter inductor such that the primary current is in the same shape as the input voltage, and in the same time supply the active power consumer of control circuit. The controller is based on a conceptually simple control scheme, the energy balance concept. The advantages of new variable inductor are simplicity of control circuits, low cost and good response. In theory, the harmonic current component of the variable inductor is zero. The feasibility of this theory is verified by using Matlab/Simulink simulation results.

Keywords-variable inductor; harmonic elimination; PWM converter; energy balance

I. INTRODUCTION

Recent year, increase of nonlinear load cause a great demand for reactive power generation and control in electric power systems. In electric power systems, the shunt capacitor is commonly used as the var compensator to control reactive power flow. It performs well under the circumstance of low voltage and insufficient reactive power. However, modem transmission systems are regularly encountered with large voltage variation and great capacity requirement. Synchronous condenser, which consumes the active power for generating the reactive power, has been used for var compensation. Drawbacks of this device are decrease of power efficiency and increase of cost because it also consumes active power. A Static VAR Compensator (SVC) constructed with a thyristor, linear inductor and capacitor is one of the devices for regulating and stabilizing transmission lines. SVC can be tuned, by varying the inductance in thyristor controlled reactor (TCR), to control reactor power smoothly. Several magnetic devices for variable inductor have also been presented [1]–[4]. Direct current

controllable reactor (DCCR) is one kind of magnetically controlled reactor. DCCR regulates the equivalent inductor by changing current of control winding.

However, in all these systems SVC has a problem of harmonic currents arising. The harmonics of the output current can not be neglected for large power applications. For harmonic reduction, there are two approaches. First one is the application of a high frequency PWM switching technique to the SVC system. Another effective solution is utilization of a linear variable inductor with sinusoidal output.

In this paper, the authors propose a new variable inductor based on conventional shunt DCCR which two primary windings are shunt connected. The secondary DC control winding is connected to the primary AC winding through a voltage source inverter (VSI) and a switch. The primary current is easily controlled with the secondary current. The primary current is sinusoidal in a wide control region. The operating principle is described in this paper.

II. CONVENTIONAL DCCR

A. Circuit Configuratuion

Fig. 1 shows the schematic diagram of a conventional shunt DCCR using tow core. The turns of tow primary windings are n_1 and n_2 respectively, and secondary one is n_3. The currents of the three windings are i_1, i_2 and i_3 respectively. The dashed curves Φ_1 and Φ_2 illustrate the fluxes in the tow core. The control current i_3 can be varied by changing the variable resistor R . Because of nonlinear B-H characteristics of the core, the effective inductance of the primary winding is continually decreased by increasing the secondary excitation.

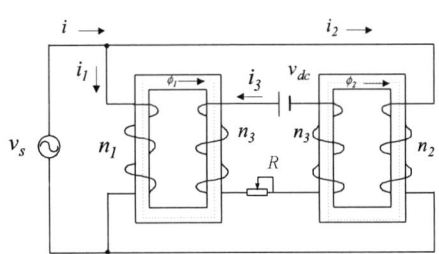

Figure 1. Conventional shunt DCCR configuration scheme

B. Hamonic Analysis

Fig.2 shows the magnetic equivalent circuit of the conventional shunt DCCR.

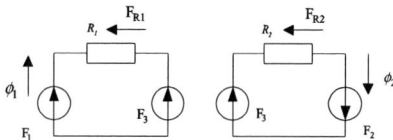

Figure 2. Magnetic equivalent circuit of shunt DCCR

F_1 and F_2 are MMF's of primary winding, F_3 is MMF of secondary winding, and F_{R1} and F_{R2} are the MMF's of the core magnetic reluctances, and Φ_1 and Φ_2 the fluxes. The relations held in the magnetic circuit are:

$$F_1 = F_{R1} + F_3 \ , \qquad F_2 = F_{R2} - F_3 \qquad (1)$$

For simplification, assume the nonlinear characteristics of the magnetic reluctances as: $F_{R2}=k\Phi_1{}^3$ and $F_{R2}=k\Phi_2{}^3$. Then the total MMF's primary winding is given by:

$$F_1 + F_2 = k\phi_1{}^3 + k\phi_2{}^3 \qquad (2)$$

Equation (2) shows that the total current of primary winding is increased by the secondary flux. When the primary voltage is $u_1= u_2=V_s\sin\omega t$, the flux can be expressed to $\Phi=\Phi_m\cos\omega t+\Phi_d$. Here $\Phi_m=V_s/\omega n$, n is turns of the primary winding, assumed turns of tow primary winding are equal $n_1= n_2= n$. Φ_d is the flux excited by F_3. Substituting the fluxes into (2), the total MMF's can be obtained as follow:

$$F_1 + F_2 = \left(k/2\right)\left[3\phi_m\left(\phi_m{}^2 + 4\phi_d{}^2\right)\sin\omega t - \phi_m{}^3\sin3\omega t\right] \ (3)$$

Due to $F_1+F_2=n(i_1+i_2)$, the primary current of primary windings $i=i_1+i_2$ can be obtain from (3). The 3rd harmonic current to fundamental one is calculated as follows:

$$\Gamma_3 = 1/\left[3+12\left(\phi_m/\phi_d\right)^2\right] \qquad (4)$$

Fig.3 shows the simulation results using Matlab/Simulink when V_s =100V, V_{dc} =10V, R=0.5Ω, $n_1= n_2= n_3=100$. The 3rd harmonic component of primary current is about 40% of the fundamental one. The total harmonic distortion (THD) is 41.9%.

III. THE PROPOSED SHUNT DCCR

A. Configuration of New Shunt DCCR

Fig.4 shows the configuration of proposed shunt DCCR based on conventional one. At the every primary winding side add one harmonic compensate winding respectively. They are connected to a voltage source inverter. The capacitor of DC side of the inverter is connected to secondary control winding through a BUCK circuit. One

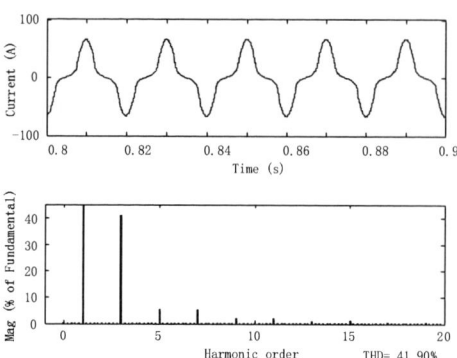

Figure 3. Current and harmonic component of conventional shunt DCCR

controller is configured to control the current of secondary control winding. And at the same time the controller controls the current of inverter to compensate the harmonic component of current of primary winding. Thus harmonic component of the DCCR is eliminated and the primary current is sinusoidal.

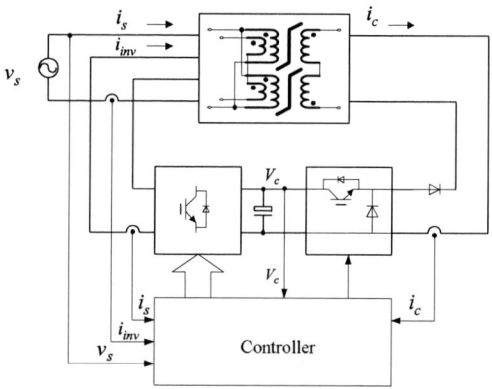

Figure 4. Configuration of new shunt DCCR

B. System Description

The proposed shunt DCCR is divided into main body and controller part which includes inductor control and harmonic control function. Controller controls the current of secondary winding to adjust the equivalent inductance of reactor using a simple PI link. Fig. 5 shows the control schemes. Where, G(s) is the transfer function of the shunt DCCR. Rms link calculates the fundamental component of the primary current. The proposed shunt DCCR eliminated harmonic employed active power filter technique. Controller detects the primary current and calculates the fundamental component and the harmonic one. Then calculate the command current of inverter. The voltage source inverter output the compensate current according to the calculated current.

IV. CONTROL PRINCIPLE OF PROPOSED SHUNT DCCR

Due to control scheme of inductance of reactor is simpler compare with harmonic elimination, the control stratagem of harmonic elimination will be discussed mostly. Be differed from common extract method of command current, the reference current of proposed shunt DCCR is not include the fundamental component of reactor power. The inverter compensates the harmonic component. Consider the previous obtain method of command current of single phase active power filter [5-8], method of total compensation is invalid and one based on instantaneous reactor power theory by construct a three phase system is complexity.

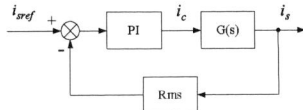

Figure 5. Control scheme of adjust inductance

For simplify the control scheme, we proposed a simple method based on power balance concept [5]. From Fig. 4 it can be found that when assume turns of primary winding and the second one is equaled the primary current can be represent as

$$i_s = i_{inv} + i_R \qquad (5)$$

Where i_s is primary current, i_{inv} is the inverter output current, and i_R is a virtual reactor current which cannot be measured, and $i_R = i_s$ when $i_{inv} = 0$.

$$\begin{cases} i_s = i_{sa} + i_{sr} + i_{sh} \\ i_{inv} = i_{inva} + i_{invr} + i_{invh} \\ i_R = i_{Ra} + i_{Rr} + i_{Rh} \end{cases} \qquad (6)$$

Where i_{sa}, i_{sr} and i_{sh} is the real part, reactive part and harmonic part of primary current i_s respectively, and i_{inva}, i_{invr} and i_{invh} is the real part, reactive part and harmonic part of inverter output current i_{inv} respectively, and i_{Ra}, i_{Rr} and i_{Rh} is the real part, reactive part and harmonic part of reactor current i_R respectively.

A. Reactive power balance

Assuming the mains voltage is a pure sinewave, it is represented as

$$v_s(t) = V_s \sin \omega t \qquad (7)$$

Primary current i_s when it is not compensated can be represented as

$$i_s = \sum_{n=1}^{\infty} I_n \sin(n\omega t + \theta_n) \qquad (8)$$

It can be subdivided into the fundamental and harmonic components as

$$i_s = I_1 \sin(\omega t + \theta_1) + \sum_{n=2}^{\infty} I_n \sin(n\omega t + \theta_n) \qquad (9)$$

When primary current i_s is compensated by inverter current i_{inv}, primary current can be represented as

$$i_s = I_1 \sin(\omega t + \theta_1) \qquad (10)$$

According to the Fourier algorithm its real part is $I_1 \cos\theta_1$, the system real power current is represented as

$$i_{sa} = I_1 \cos\theta_1 \sin \omega t \qquad (11)$$

And the reactive power current is

$$i_{sr} = I_1 \cos\theta_1 \cos \omega t \qquad (12)$$

Similarly the reactive part of inverter output current i_{inv}, can be represented as

$$i_{invr} = I_{inv1} \cos \theta_{inv1} \cos \omega t \qquad (13)$$

Where, I_{inv1} and θ_{nv1} is fundamental amplitude and initial phase angle of i_{inv} respectively.

B. Real power balance

During the transient, for instance, when adjusted the inductance of DCCR, the real power balance would be lost. From Fig. 4, if the real current of primary current is i_s with amplitude of I_s, then the real power of system can be represented as

$$\frac{1}{2} V_s I_s = P_L + P_{inv} \qquad (14)$$

Where V_s is the amplitude of power source voltage, P_L is the real power consumed by the DCCR primary winding, and P_{inv} is the real power supplied from the inverter. Consider the variation of real power, Equation (14) can be rewritten as

$$\Delta P_{inv} = \Delta P_L - \frac{1}{2} V_s I_s \qquad (15)$$

And the energy variation of capacitor can be represented as

$$\frac{1}{2} C (V_c + \Delta V_c)^2 - \frac{1}{2} C V_c^2 = \int_0^t \Delta P_{inv}(t) dt \qquad (16)$$

Simplified (16),

$$\frac{1}{2} C (2 V_c \Delta V_c + \Delta V_c^2) = \int_0^t \Delta P_{inv}(t) dt \qquad (17)$$

Neglected ΔV_c^2, the high order of voltage variation of capacitor, the voltage of inverter DC side capacitor can reflect the real power variation.

As the mention above, the primary current harmonic component i_{sh} can be extracted from the equation

$$i_{sh} = i_s - i_{sa} - i_{sr} \qquad (18)$$

Hence the calculated compensation current can be obtained. The block diagram of the compensation current circuit is showed in Fig. 6. In the block diagram, there are tow PI controller. One is adjusting the real power balance and another is adjusting the reactive power balance.

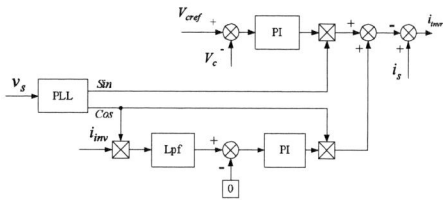

Figure 6. Block diagram of the compensation current calculate circuit for the proposed control strategy

V. SIMULATION RESULTS

To verify the performance of the proposed shunt DCCR, a simulation model was be setup employed Matlab/Simulink software. The main parameters of the model are as follows: DC busbar voltage V_c =250V, DC capacitor C= 5000 μF, power supply voltage V_s =100V, turns of primary windings and secondary one $n_1= n_2= n_3$=100. The simulation results were shown in Fig. 7. From the primary current waveform, it can be seen that the primary current is nearly a sinewave, and the total harmonic distortion (THD) is smaller than 4%.

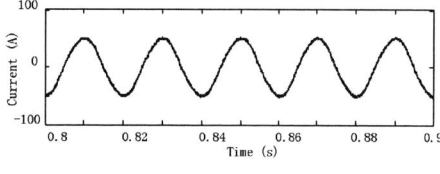

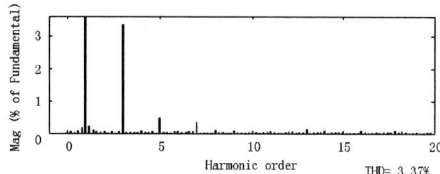

Figure 7. Current and harmonic component of novel proposed shunt DCCR

VI. CONCLUSIONS

In this paper conventional shunt DCCR has been analyzed employing magnetic equivalent circuit method. The conventional shunt DCCR included a mount of harmonic component in the output current. This paper proposes a new shunt DCCR, employing an active power filter technique to eliminate the harmonic. For simplify the control strategy, proposed a simple scheme based on energy balance concept. The novel shunt DCCR has tow important features as follows:

(a) It can force the primary current to be a shape with sinewave.

(b) Its new topology simplified the control scheme, increased credibility of the system.

Above all, the proposed single phase shunt DCCR is superior to conventional shunt DCCR.

REFERENCES

[1] Nakamura, K. and Ichinokura, O., et al., "Analysis of orthogonal-core type linear variable inductor and application to VAr compensator,", *IEEE Trans. on Magnetics*, vol.36, pp.3565 – 3567, Sept 2000.

[2] M. V. Biki, Yu. L. Chizhevskij, et al., "Controlled Shunting reactors for high-voltage electric network," Electrotekhika, No.9, Sept. 1994.

[3] J. Tellinen and J. Jarvik, "Minimization of higher harmonics in high power saturated reactors for AC power transmission lines," *Power Quality '98, IEEE*, pp.133-138, 1998.

[4] Baichao Chen and J. M. Kokernak, "Thyristor controlled two-stage magnetic valve reactor for dynamic var-compensation in electric railway power supply systems," *IEEE 2000 International Conference on Power Electronics and Drive Systems*, vol. 2, pp.1066-1072. Sept, 2000.

[5] Jou, H.-L., Wu, J. C. and Chu, H. Y., "New single-phase active power filter", *Electric Power Applications, IEE Proceedings-*, vol. 141, pp.129 – 134, May 1994.

[6] Chin Yuan Hsu, and Horng Yuan Wu, "A new single-phase active power filter with reduced energy storage capacitor," *PESC '95, 26th Annual IEEE*, vol.1, pp.202 – 208, June 1995.

[7] Wu, J. C. and Jou, H. L., "Simplified control method for the single-phase active power filter," *Electric Power Applications IEE Proceedings-*, vol.143, pp.219 – 224, May 1996.

[8] Torrey, D.A. and Al Zamel, A.M.A.M. "A single-phase active power filters for multiple nonlinear loads," *IEEE Trans. on Power Electronics*, vol.10, pp.263 – 272. May 1995.

2006 5th International Power Electronics and Motion Control Conference

A Multi-Output Series Resonant Inverter with Asymmetrical Voltage-Cancellation Control for Induction-Heating Cooking Appliances

S.H. Hosseini, A. Yazdanpanah Goharrizi and E. Karimi
Faculty of Electrical and Computer Engineering, University of Tabriz, Tabriz, Iran
e-mail: hosseini@tabrizu.ac.ir

Abstract—This paper presents and analyzes the asymmetrical voltage-cancellation (AVC) control, for a multi-output series resonant inverter. It is applied to the multiple-burner induction-heating cooking appliances. Some common approaches use several single-output inverters or a single-output inverter multiplexing the loads along the time periodically. By specifying a multi-output series-resonant high frequency inverter, a new inverter is obtained fulfilling the requirements. The synthesized converter can be considered as a multi-output extension of a full-bridge technology. It allows the control of those outputs, simultaneously and independently, up to their rated power, saving component count compared with the same-output, convert solution and providing a higher utilization of electronics. A three output inverter is presented and analyzed. All simulation results are provided with *PSCAD/EMTDC* software.

Keywords-asymmetrical voltage-cancellation; multiple-burner; induction-heating cooking

I. INTRODUCTION

Induction cookers are low-power induction-heating systems with a maximum output power usually less than 3-kw per load. An induction-heating cooking appliance is basically made up of a flat-type inductor coil on which the pan to be heated is placed, between the pan and the coil an insulator (usually a ceramic glass) is placed. The heat is generated at the bottom of the pan due to eddy currents and hysteresis losses. These induced currents are caused by an alternating magnetic field generated by a medium frequency (20-100 kHz) current through the coil. Pan-inductor coupling is usually modeled as the series connection of an inductor and a resistor, based on the transformer analogy. The values of the equivalent inductance and resistance depend on the operating frequency and the required maximum power. A typical arrangement of an induction cooker is shown in Fig.1

Induction Cookers take the energy from the main voltage, which is rectified by a bridge of diodes .A bus filter is designed to allow a big voltage ripple getting a resultant input power factor close to one. Then inverter supplies high-frequency current to the induction coil. Due to this ripple all components have to be designed for the voltage and current peak values (Fig.1). The main inverter topologies used in induction cookers are resonant inverters, including full-bridge [2], half–bridge [3], [7], and single–switch inverters [8], [9].

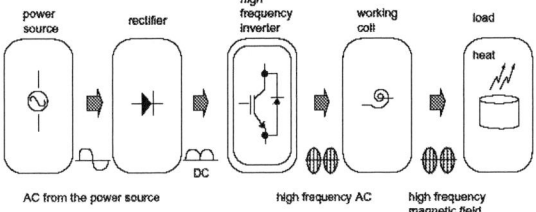

Figure 1. Typical arrangement of induction-heating cooking appliance.

Multi-burner induction–heating cooking appliances including two or four inductors are the most common solution. In a multiple–burner induction cooker it is possible to use one inverter per burner as traditional approaches or one inverter for two or more burners. In the last case, a common technique uses a single-output inverter multiplexing the loads along the time periodically by means of electromechanical switches [4], causing a very low frequency switching with power distribution and acoustic noise not completely satisfactory. Another approach [5], is based on the use of several resonant capacitors connected with loads by electromechanical switches activated to change the power division. The operating frequency is not fixed; actually it depends on power distribution and load conditions. The use of multiple-output inverters has clear benefits for multiple-burner cookers; higher utilization ratio of electronics (the converter is configured to supply either both inductors are operated), higher maximum power (when only one of inductors is active, all of the rated power of the converter can be applied to the load, obtaining a quick heating function) and finally, it is possible to share some components of the power converters. The purpose of this work is to analyze three inverter for using it in power electronic converters mainly for induction cooking, even though it could be extended to other applications. We centralized our study to three output inverter, then by using the results of two-output and three output inverters,

---

1-4244-0448-7/06/$25.00 ©2006 IEEE   1697

can predict some properties for n-output inverter. For this purpose, the basic required specifications are: a three-output series–resonant inverter, with variable output power and a fixed frequency control. The synthesized converter should allow the control of the three output, simultaneously and independently, up to their rated powers saving component count compared with the three-output converter solution. The required fixed–frequency control has some additional advantages, as reducing the electromagnetic noise spectrum and avoiding the acoustic noise due to different operating frequencies which cause low–frequency interferences amplified by the pans. A synthesis method proposed for switched converters [8] is applied here as a tool for generating the new inverter. In addition, the synthesis process also considers some studies about fixed-frequency control techniques, in particular the asymmetrical voltage–cancellation control [10].

This paper is organized as follows. In section II the three-output series–resonant inverter is synthesized from the two-output series–resonant inverter which discussed in [1]. In section III a three-output series–resonant inverter is simulated and results are provided with PSCAD/EMTDC software. Finally some conclusions of the work are provided in section IV.

A.   Fundamental Structure.

The synthesis method applied here to the previously specified converter considers a generic switched converter composed of independent sources, resistive or reactive loads, and a conversion block consisting of non dissipative components, such as switches, inductors, and capacitors. We suppose the following basic specifications for the converter synthesis: a three-output series-resonant inverter, with variable output power and a fixed-frequency control. Three inductive loads should be supplied, simultaneously and independently, up to their rated output powers, without losing the mentioned benefits of fixed-frequency control strategies. Additionally, a soft-switching operation for the active devices is recommended taking into account efficiency performances. The basic structure for the desired converter to be synthesized is shown in Fig. 2. The formulation of the energy state equation of a generic converter expressed in [8] can be particularized for the desired converter as

$$\begin{bmatrix} L & 0 \\ 0 & C \end{bmatrix} \begin{bmatrix} \dot{i}_L \\ \dot{v}_C \end{bmatrix} = \begin{bmatrix} -R & I \\ -I & 0 \end{bmatrix} \begin{bmatrix} i_L \\ v_C \end{bmatrix} + \begin{bmatrix} f_j \\ 0 \end{bmatrix} U \qquad (1)$$

Where i_L and v_C vectors o the three inductor currents and the three capacitor voltages, respectively.

$$i_L = \begin{bmatrix} i_{L_1} \\ i_{L_2} \\ i_{L_3} \end{bmatrix}, \; v_C = \begin{bmatrix} v_{C_1} \\ v_{C_2} \\ v_{C_3} \end{bmatrix} \qquad (2)$$

L, C and R diagonal matrices whose diagonal Elements are the inductances, capacitances and resistances, respectively

$$L = \begin{bmatrix} L_1 & 0 & 0 \\ 0 & L_2 & 0 \\ 0 & 0 & L_3 \end{bmatrix}, C = \begin{bmatrix} C_1 & 0 & 0 \\ 0 & C_2 & 0 \\ 0 & 0 & C_3 \end{bmatrix} R = \begin{bmatrix} R_1 & 0 & 0 \\ 0 & R_2 & 0 \\ 0 & 0 & R_3 \end{bmatrix} \qquad (3)$$

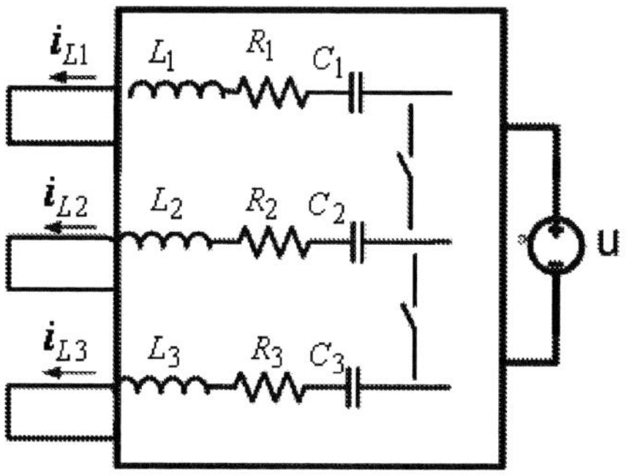

Figure 2.   Basic structure of the two-output series resonant inverter to be synthesized

f_j : Vector which describes the connections between the network components (all serial RLC Blocks) and the input voltage U for each circuit configuration j of the converter

$$f_j = \begin{bmatrix} f_{1j} \\ f_{2j} \\ f_{3j} \end{bmatrix} \qquad (4)$$

The possible values of the three elements of f_j are either 1 (forward connection), 0 (no connection) or -1 (reversed connection).

From (1)

$$L \frac{di_L}{dt} = -Ri_L + v_C + f_j U \qquad (5)$$

$$C \frac{dv_C}{dt} = -i_L \qquad (6)$$

According to (5) and (6), the unique function controlled by the switches in the converter is f_j . So it has to be the control function, representing the connection of the input voltage source to the three loads. From (5)

$$f_j = \left(L \frac{di_L}{dt} + Ri_L - v_C \right) \frac{1}{U} \qquad (7)$$

As a consequence, assuming that i_L and v_C are periodic function, the control function f_j has to be also periodic. The possible values for the three elements of f_{ij} are 1, 0,-1. Only this range of values can be used to obtain periodic sequences to control the final converter. Taking into account that a fixed–frequency control is required, and considering the generalized control strategy presented in [10], all possible values for each element f_{ij} of the vector f_j to satisfy (6) will produce a quasi–square waveform shown In Fig.3 (a).This generalized control strategy (with control of the parameters: α_+, α_- and β) is called asymmetrical voltage–cancellation (AVC) Control and it is typically achieved with full-bridge topologies. As a particularly case analyzed in [10] considering efficiency performances, Fig.3 (b) shows the proposed technique for the optimum AVC control with ZVS operation above resonance and output power variations. In this particular case, β =180 is constant and only the control parameters α_+ or α_- are individually varied to reduce the output voltage and power. The control angle α_+ is considered in Fig.3 (b). If a further reduction of the output power is needed; the control angle should be varied.

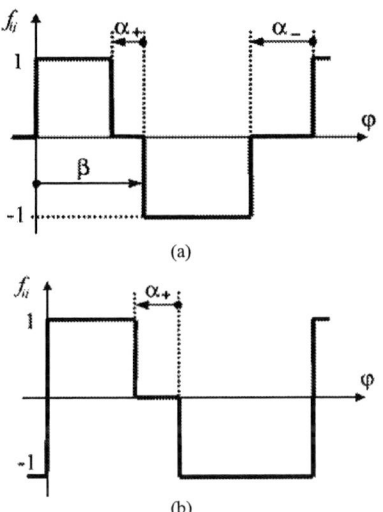

(a)

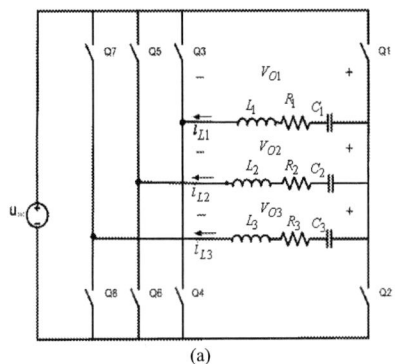

(b)

Figure 3. Possible periodic sequence of values for each element of :(a) generalized AVC control and (b) proposed optimum AVC control for ZVS operation.

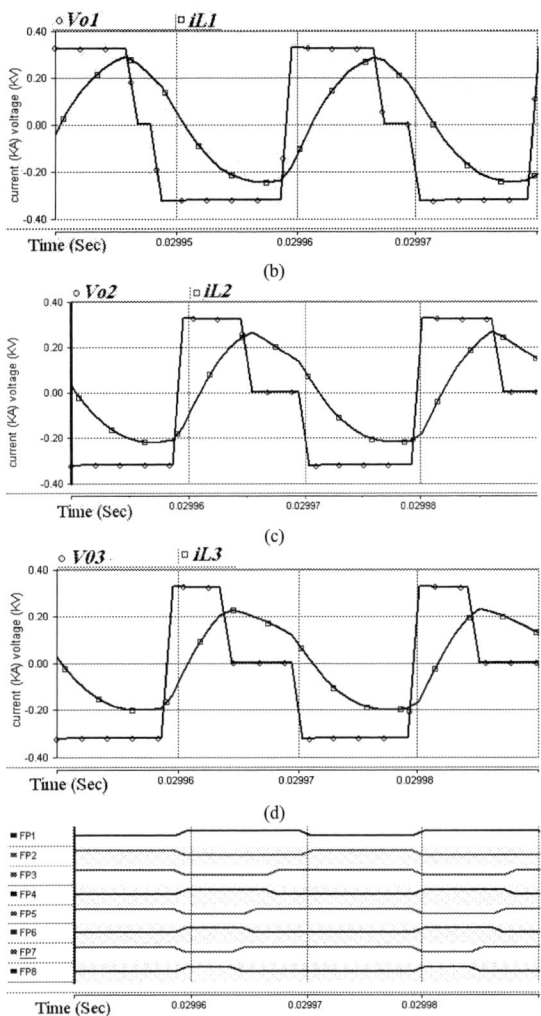

Figure 4. Synthesized three-output series-resonant inverter with some typical waveforms: (a) Vo1, iL1, (b) Vo2, iL2, (c) Vo3, iL3, for the selected AVC all currents amplitude multiplied by 10 and (e) gate control signals.

B. Proposed Three-Output Inverter

According to the previous results for the basic structure and control of the converter, a possible synthesis approach could be a one kind of full-bridge topology, with a fixed–frequency control of each output. Fig.4 (a) shows the used full-bridge structure with some typical waveforms for the selected AVC control strategy including the sequences of conduction of switches, the quasi-square output voltages (v_{O1}, v_{O2}, v_{O3}), and the output currents (i_{L1}, i_{L2}, i_{L3}).The vector of output voltages is:

$$v_O = \begin{bmatrix} v_{O1} \\ v_{O2} \\ v_{O3} \end{bmatrix} = f_j U \qquad (8)$$

Considering the sequence of conduction of switches for the selected fixed-frequency control strategy, we can observe that the corresponding switches of right legs of the bridges are activated at the same time with a fixed phase lag of 180. So the right legs are merged into one common leg and, as a result, the obtained three-output inverter is shown in Fig.4, where some typical waveforms are included, This converter has one common leg of switches and three independent legs. It displays a saving of three switches compared with the same full-bridge solution. All switches have bidirectional currents and unipolar voltages (transistors with anti parallel diodes).The states of switches for each configuration j appear in Table I. vector f_j shows the value of each output voltage in p.u. Fig.5 shows the complete periodic sequence of configurations to obtain the required output power in each output. Notice that this complete periodic sequence depends on the values of the control angles $\alpha_{+1}, \alpha_{+2}\ \alpha_{+3}$ and β defined as in Fig.3 (a). For usual continuous output power ranges of the considered application, just the control angles α_{+1}, α_{+2} and α_{+3} are normally used represented in the typical waveforms in Fig.4. As a consequence, the circuit configurations of the converter are usually restricted to $j \langle 7$. Anyway, the other control angles and circuit configurations would be used for further reductions of the output power in continuous mode.

To improve the final performance of the converter, the final design (Fig.6 (a)) includes three additional low cost relays (electromechanical switches) SW1, SW2 and SW3 for paralleling both independent legs when only one or three output is required. These relays are only activated to change the power level, without variation of the switching frequency of the main devices.

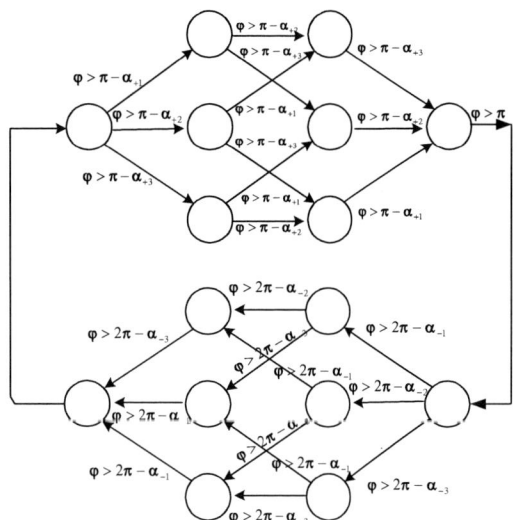

Figure 5.   Periodic sequences of configurations for the three-output inverter

So the output power can be increased for the same efficiency. Available maximum power in outputs varies depending on the activation states of topology relays. From the topology in Fig.6 (a), the activation states shown in Table II can be deduced. This strategy allows

the quick heating function as well as an optimum utilization of rated power. There are three activation states of relays (normally open (no), normally closed (nc)) to supply all burners up to rated power, so there is not a unique relationship between individual legs and inductors. The three-output series-resonant inverter has been implemented using IGBT transistors with ant parallel diodes as switches (Fig. 6 (a)).

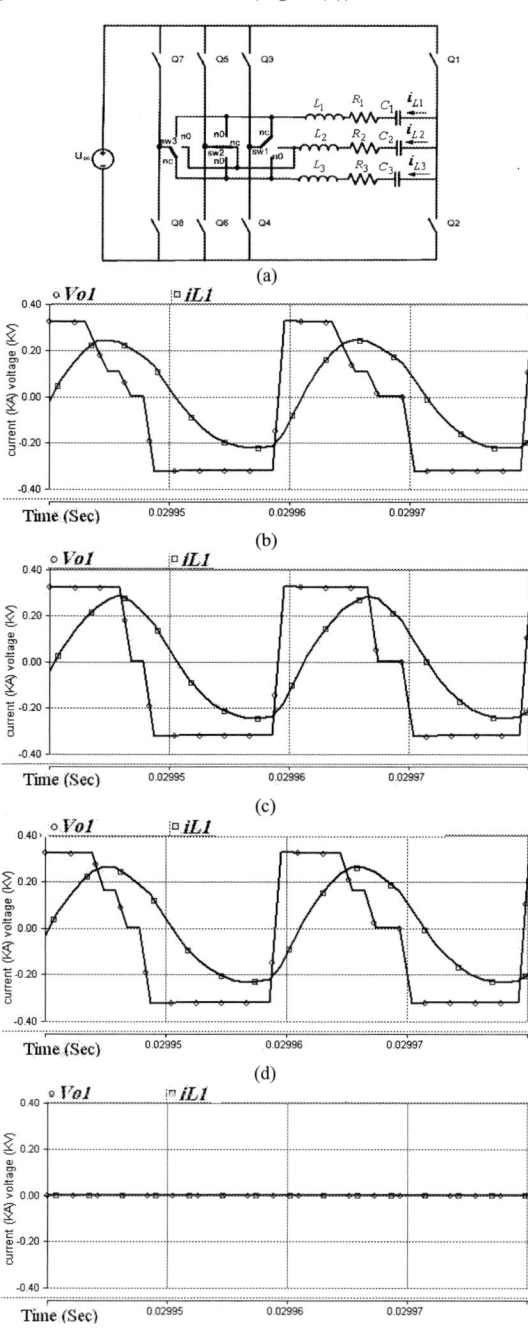

Figure 6.   (a) Implemented three-output series-resonant inverter, (b) nc, no, no, (c) no, no, nc, (d) no, nc, no, (e) no, nc, nc.

II. DESINGN AND SIMULATION

Although not shown in Fig. 6 (a) for simplicity, a snubber capacitor of 10 nF is connected to each transistor to reduce turn-off losses. The two loads consist of flat inductor coils coupled to standard cooking pans. Model parameters of loads are approximately R=9.5 Ω and L=60H. The series resonant capacitors are C=450 nF. The inverter is operated with a peak input voltage of 325 V and a switching frequency of 48 kHz, with a maximum overall output power of 3.2 kW. An asymmetrical voltage-cancellation control technique is used because, as analyzed in [10], this control technique achieves better efficiency performances than conventional fixed-frequency control techniques, while considering ZVS operation, output power and load variations. So the asymmetrical voltage-cancellation control, initially proposed for one-output full-bridge topologies, has been extended to our three-output inverter. The three-output inverter configuration is shown and simulated and voltages and currents of the first output for different relays situation are shown in fig. 6 (b, c, d and e). This figure shows the current and voltage waveform of the first load, for simplicity and condition which has same output power was neglected.

By observing the circuit configurations for one-output inverter, two-output inverter and three-output inverter ,we can conclude that for one-output, according to full-bridge topology, it needs 4 power switches, for two-output as previously presented in [1], It needs 6 power switches, and finally for 3-output :8 power switches. Besides per n-output according to full-bridge topology, it needs 4n switches, so by merging the right legs into one common leg, explained in section II two switches are eliminated. In general 2*1+2 switches are need for one-output inverter and for two-output inverter 2*1+2+2*1=2*2+2=6 and for three-output inverter: 2*3+2=8 switches are need respectively. As a result 2*n+2 switches for the inverter of n-output is desired. Consequently, 2n-2 switches or 2n-2 control circuits are eliminated.

III. CONCLUSION

In this paper, a three-output series-resonant inverter has been obtained from some specifications, using a synthesis method and the asymmetrical voltage-cancellation control strategy. The synthesized converter allows supplying three inductive loads up to their rated values simultaneously and independently with only one converter, saving two transistors compared with the full-bridge alternative solution. The final addition of three low-cost relays allows a quick heating function as well as an optimum utilization of rated power. The three-output inverter has been simulated with PSCAD/EMTDC software. Finally the three output inverter generalized in to n-output Series-Resonant Inverter with saving switching devices i.e. 2n+2 switching device instant of 4n switch device is required.

TABLE I. ELEMENTS OF f AND STATES OF SWITCHES FOR EACHCONFIGURATION j OF THE THREE-OUTPUT INVERTER

j	f_{1j}, f_{2j}, f_{3j}	Q_1	Q_2	Q_3	Q_4	Q_5	Q_6	Q_7	Q_8
$j=0$	(1,1,1)	ON	OFF	OFF	ON	OFF	ON	OFF	ON
$j=1$	(0,1,1)	ON	OFF	ON	OFF	OFF	ON	OFF	ON
$j=2$	(1,0,1)	ON	OFF	OFF	ON	ON	OFF	OFF	ON
$j=3$	(1,1,0)	ON	OFF	OFF	ON	OFF	ON	ON	OFF
$j=4$	(0,0,1)	ON	OFF	ON	OFF	ON	OFF	OFF	ON
$j=5$	(0,1,0)	ON	OFF	ON	OFF	OFF	ON	ON	OFF
$j=6$	(1,0,0)	ON	OFF	OFF	ON	ON	OFF	ON	OFF
$j=7$	(0,0,0)	ON	OFF	ON	OFF	ON	OFF	ON	OFF
$j=8$	(-1,-1,-1)	OFF	ON	ON	OFF	ON	OFF	ON	OFF
$j=9$	(0,-1,-1)	ON	OFF	ON	OFF	ON	OFF	ON	OFF
$j=10$	(-1,0,-1)	OFF	ON	ON	OFF	OFF	ON	ON	OFF
$j=11$	(-1,-1,0)	OFF	ON	ON	OFF	ON	OFF	OFF	ON
$j=12$	(0,0,-1)	ON	OFF	ON	OFF	OFF	ON	ON	OFF
$j=13$	(0,-1,0)	ON	OFF	ON	OFF	ON	OFF	OFF	ON
$j=14$	(-1,0,0)	OFF	ON	ON	OFF	OFF	ON	OFF	ON
$j=15$	(0,0,0)	ON	OFF	ON	OFF	OFF	ON	OFF	ON

TABEL II. AVAILABLE MAXIMUM POWER IN THE OUTPUTS AS FUNCTION OF THE STATES OF RELAYS SW1, SW2, SW3.

Sw1,Sw2,Sw3	no,no,no	no,no,nc	no,nc,no	no,nc,nc	nc,no,no	nc,no,nc	nc,nc,no	nc,nc,nc
P1	100% or 3.1 Kw	0.0 or 3.1 Kw	0.0 or 3.1 Kw	0%	>100% or 4.9 Kw	3.1 or 4.9 Kw	0%	100% or 3.1 Kw
P2	100% or 3.1 Kw	0.0 or 3.1 Kw	>100% or 4.9 Kw	3.1 or 4.9 Kw	0.0 or 3.1 Kw	0%	3.1 or 4.9 Kw	100% or 3.1 Kw
P3	100% or 3.1 Kw	>100% or 4.9 Kw	0.0 or 3.1 Kw	3.1 or 4.9 Kw	0.0 or 3.1 Kw	3.1 or 4.9 Kw	3.1 or 4.9 Kw	100% or 3.1 Kw

REFERENCES

[1] José M. Burdío, Fernando Monterde, José R. García, Luis A. Barragán, and Abelardo Martínez, "A two-output series-resonant inverter for induction-heating cooking appliances," IEEE Trans. Power Electron., vol. 20, no. 4, pp. 815-822, Jul. 2005.

[2] F. P. Dawson and P. Jain, "A comparison of load commutated inverter system for induction heating and melting applications," *IEEE Trans. Power Electron.*, vol. 6, no. 4, pp. 430–441, Jul. 1991.

[3] L. Hobson, D. W. Tebb, and D. Turnbull, "Dual element induction cooking unit using power MOSFETs," *Int. J. Electron.*, vol. 59, pp.747–757, Jun. 1985.

[4] H. W. Koertzen, J. D. van Wyk, and J. A. Ferreira, "Design of the half bridge series resonant converter for induction cooking," in *Proc. IEEE Power Electronics Specialists Conf. (PESC)*, 1995, pp. 729–735.

[5] M. Kamli, S. Yamamoto, and M. Abe, "A 50–150 kHz half-bridge inverter for induction heating applications," *IEEE Trans. Ind. Electron.*, vol. 43, no. 1, pp. 163–172, Feb. 1996.

[6] S. Wang *et al.*, "Induction-heated cooking appliance using new quasi-resonant ZVS-PWM inverter with power factor correction," *IEEE Trans. Ind. Apl.*, vol. 34, no. 4, pp. 705–712, Jul./Aug. 1998.

[7] Y. S. Kwon, S. Yoo, and D. Hyun, "Half-bridge series resonant inverter for induction heating applications with load-adaptative PFM control strategy," in *Proc. IEEE Applied Power Electronics Conf. (APEC'99)*, 1999, pp. 575–581.

[8] H. Omori and M. Nakaoka, "New single-ended resonant inverter circuit and system of induction-heating apparatus," *Int. J. Electron.*, vol. 67, pp. 277–296, Feb. 1989.

[9] H.W. Koertzen, J. A. Ferreira, and J. D. vanWyk, "A comparative study of single switch induction heating converters using novel component effectively concepts," in *Proc. IEEE Power Electronics Specialists Conf. (PESC'92)*, 1992, pp. 298–305.

[10] J. M. Burdío, L. A. Barragán, F. Monterde, D. Navarro, and J. Acero, "Asymmetrical voltage-cancellation control for full-bridge series resonant inverters," *IEEE Trans. Power Electron.*, vol. 19, no. 2, pp. 461–469, Mar. 2004.

2006 5th International Power Electronics and Motion Control Conference

Capacitor Voltage Control in a Cascaded Multilevel Inverter as a Static Var Generator

M. Li*, J. N. Chiasson*, L. M. Tolbert*
*The University of Tennessee, ECE Department, Knoxville, USA

Abstract—The widespread use of non-linear loads and power electronics converters has increased the generation of non-sinusoidal and non-periodic currents and voltages in power systems. Reactive power compensation or control is an important part of a power system to minimize power transmission losses. Given a modulation index, the switch times can be chosen to achieve the fundamental while eliminating specific harmonics. However, the resulting total harmonic distortion (THD) depends on the modulation index (see [1][2]). This work considers the control of the DC capacitor voltage in such a way that one can operate at the modulation index which results in the minimum THD. This paper presents the development of specific control algorithms for a cascaded multilevel inverter to be used for static var compensation.

Index Terms—Multilevel Inverter, Static Var Generator (SVG), Cascade inverter.

I. INTRODUCTION

Multilevel inverters have gained much attention in recent years as an effective solution for various high power and high voltage applications. A multilevel inverter is a power electronic device built to produce ac waveforms from small voltage steps by utilizing isolated dc sources or a bank of series capacitors. The multilevel inverter is ideal for connecting distributed dc energy sources (solar cells, fuel cells, the rectified output of wind turbines) to an existing three phase power grid [3].

Multilevel inverter structures have been developed to overcome shortcomings in solid-state device ratings so that they can be applied to high-voltage, high power electrical systems. As pointed out in [3][4][5], the advantage of the cascaded multilevel inverter includes: (1) its active devices switch at (or nearly) the fundamental frequency drastically reducing the switching losses, (2) it eliminates the need for a transformer to provide the requisite voltage levels, (3) packaging is much easier because of the simplicity of structure and lower component count, and (4) as there are no transformers, it can respond much faster.

It is widely acknowledged that a major concern in any power system is power quality, and especially to have low harmonic content. This is because of the effects harmonics have on the energy efficiency of the power system as well as the detrimental effect they have on the reliability of the equipment connected to it. Because the multilevel inverter is switching at the fundamental frequency, its generated harmonics are much lower in frequency than high-carrier frequency based PWM systems. As a result, a major concern in designing a static var compensator based on the multilevel inverter is to ensure that its total harmonic distortion is within allowable standards.

Previous work in [1][2] has shown the switching angles in the multilevel inverter are found so as to produce the required fundamental voltage while at the same time not generate higher order harmonics. However, for 3-level multilevel inverter, if modulation index is out of the range 1.18 through 2.5, there exists no set of switching angles such that the fundamental can be controlled while at the same time completely eliminating the 5th and 7th order harmonics. In this work, a control strategy is presented to vary the level of the DC capacitor voltage so that use of the staircase switching scheme (with its inherent low switching losses).

II. CASCADED H-BRIDGES

A cascaded multilevel inverter is made up from a series of H-bridge (single-phase full bridge) inverters, each with their own isolated dc bus. This multilevel inverter can generate almost sinusoidal waveform voltage from several separate dc sources (SDCSs), which may be obtained from solar cells, fuel cells, batteries, ultracapacitors, etc. Figure 1 shows a single-phase structure of an M-level H-bridges multilevel cascaded inverter. Each level can generate three different voltage outputs $+V_{dc}, 0$ and $-V_{dc}$ by connecting the dc sources to the ac output side by different combinations of the four switches.

The output voltage of an M-level inverter is the sum of all of the individual inverter outputs. It is clear from Figure 1 that to have an M-level cascaded multilevel inverter we need $\left(\frac{M-1}{2}\right)$ H-bridge units in each phase. An example phase voltage waveform for a 7-level cascaded multilevel inverter with three dc sources and three full bridges is shown in Figure 2. The output phase voltage is given by $v_{an} = v_{a1} + v_{a2} + v_{a3}$.

As Figure 2 illustrates, each of the H-bridge's active devices switches only at the fundamental frequency, and each H-bridge unit generates a quasi-square waveform by phase-shifting its positive and negative phase legs' switching timings. Further, each switching device always conducts for 180° (or 1/2 cycle) regardless of the pulse width of the quasi-square wave so that this switching method results in equalizing the current stress in each active device.

1-4244-0448-7/06/$25.00 ©2006 IEEE

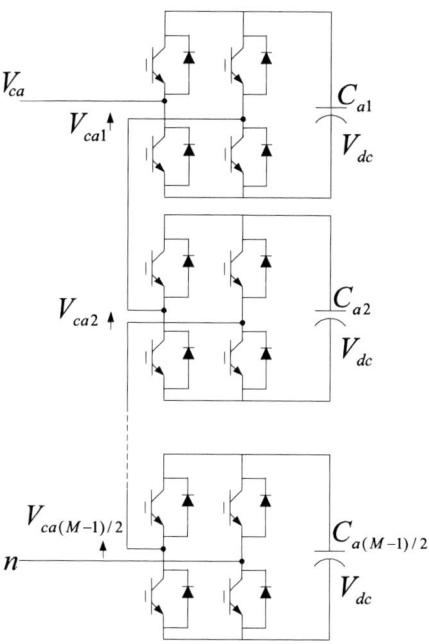

Fig. 1. Single-phase structure of a m-level H-bridges multilevel cascaded inverter.

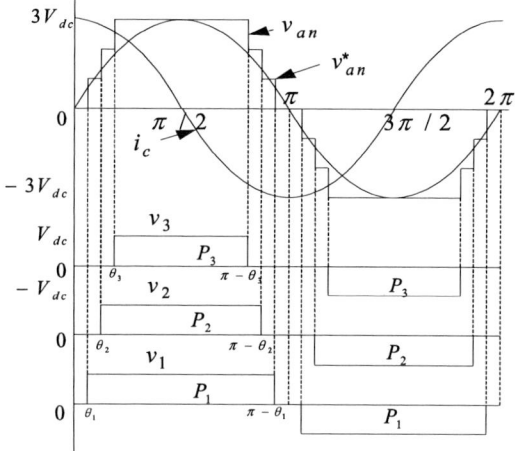

Fig. 2. Output waveform of a 7-level cascade multilevel inverter.

III. SVG SYSTEM CONFIGURATION AND OPERATION

Figure 3 shows the system configuration and control block diagram of a Static Var Generator (SVG) using a cascaded multilevel inverter, where L_c is the inverter interface inductance, v_s represents the source voltage, I_c^* (or q_c^*) is the reactive current (or reactive power) reference, and V_{dc}^* is the dc link voltage reference (see [6]). The switching pattern table shown in Figure 3 generates the switching gate signals by given modulation index and phase angles through a look-up table. The look-up table is made from the switching angles

which are computed off-line to minimize harmonics for each modulation index m. θ is the phase angle of the source voltage. α_c is phase-shift angle of the output voltage.

Here the modulation index m is defined by

$$m = s \frac{V_c^*}{V_{c\max}}, \tag{1}$$

where V_c^* is the magnitude reference of the inverter output voltage. Using the techniques in [3][6],

$$V_c^* = \sqrt{v_{ca}^{*2} + v_{cb}^{*2} + v_{cc}^{*2}}. \tag{2}$$

$V_{c\max}$ is the maximum obtainable magnitude of voltage when all the switching phase angles are zero:

$$V_{c\max} = \sqrt{\frac{3}{2}} \frac{4}{\pi} s V_{dc}, \tag{3}$$

where s is the number of sources.

Figure 4 shows the equivalent circuit of the SVG system (see [6]). A leading reactive current (capacitive current) is drawn from the line when the amplitude of the output voltage V_C is larger than the source voltage's amplitude which means vars are generated. A lagging reactive current (inductive current) is drawn from the line when the amplitude of the output voltage V_C is smaller than the source voltage's amplitude which means vars are absorbed. Since phase current i_{ca} is leading or lagging the phase voltage v_{can} by 90° as shown in Figure 2, the average charge on each dc capacitor will be zero which means there is no net real power exchange between the multilevel inverter and the utility line. To compensate the switching device loss and capacitor loss, the multilevel inverter should be controlled so that some real power is delivered to the dc capacitor. In principle, each dc capacitor voltage can be controlled to be exactly the dc desired voltage, V_{dc}^*.

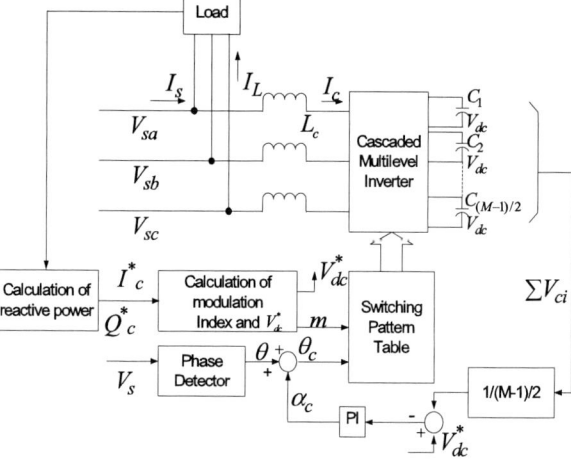

Fig. 3. SVG system configuration using the cascaded multilevel inverter.

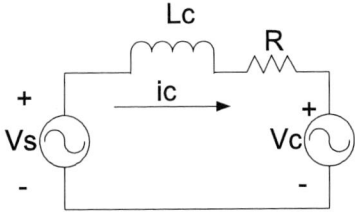

Fig. 4. Equivalent circuit of the SVG system.

IV. DYNAMIC MODELS OF SVG SYSTEM

Following [6] the source voltage v_s, output voltage of the multilevel inverter v_c, and SVG system current i_c can be represented in the $\alpha\beta$-frame using the $abc - \alpha\beta$ transformation

$$C = \sqrt{\frac{2}{3}} \begin{bmatrix} 1 & -1/2 & -1/2 \\ 0 & \sqrt{3}/2 & -\sqrt{3}/2 \end{bmatrix} \quad (4)$$

matrix, then by using the synchronous reference frame transformation

$$T = \begin{bmatrix} \cos\theta & \sin\theta \\ -\sin\theta & \cos\theta \end{bmatrix} \quad (5)$$

v_s can be represented by dq-coordinate expressions. Thus the equivalent circuit of the SVG system can be represented by

$$L_c \frac{d}{dt} \begin{bmatrix} I_{cd} \\ I_{cq} \end{bmatrix} + \omega L_c \begin{bmatrix} -I_{cq} \\ I_{cd} \end{bmatrix} + R \begin{bmatrix} I_{cd} \\ I_{cq} \end{bmatrix} = \begin{bmatrix} V_{sd} - V_{cd} \\ V_{sq} - V_{cq} \end{bmatrix} \quad (6)$$

and

$$v_s = \begin{bmatrix} V_{sd} \\ V_{sq} \end{bmatrix} = \begin{bmatrix} V_s \\ 0 \end{bmatrix}, \quad (7)$$

where V_s is the rms value of the line-to-line voltage, and θ is the phase angle.

The instantaneous active power P_c flowing into the SVG, and instantaneous reactive power Q_c drawn by the SVG can be represented by

$$P_c = V_s I_{cd} \quad \text{and} \quad Q_c = V_s I_{cq}, \quad (8)$$

where I_{cd} and I_{cq} are the active current and reactive current of SVG respectively.

Based on equation (6), in order for the SVG system to generate the desired the active current and reactive current, the modulation index should be given by the following

$$\begin{bmatrix} V_{cd}^* \\ V_{cq}^* \end{bmatrix} = \begin{bmatrix} V_{sd} + \omega L_c I_{cq}^* - \left(L_c \frac{d}{dt} I_{cd}^* + R I_{cd}^* \right) \\ V_{sq} - \omega L_c I_{cd}^* - \left(L_c \frac{d}{dt} I_{cq}^* + R I_{cq}^* \right) \end{bmatrix} \quad (9)$$

$$V_c^* = \sqrt{V_{cd}^{*2} + V_{cq}^{*2}} \quad \text{and} \quad m = \frac{V_c^*}{\sqrt{\frac{3}{2}} \frac{4}{\pi} V_{dc}}. \quad (10)$$

V. CONTROL SCHEME OF SVGS

A cascaded multilevel inverter is used as a static var generator to minimize the non-active power/current, which is shown in Figure 3. In this work, an RL load is used. The desired reactive current to be injected by SVG is obtained by

$$Q_c^* = 3V_{sph} I_{sph} \left| \sin\left(\theta_V - \theta_I \right) \right| \quad (11)$$

$$I_{cd}^* = 0 \quad \text{and} \quad I_{cq}^* = \frac{Q_c^*}{\sqrt{3} V_{sph}} \quad (12)$$

where V_{sph} and I_{sph} are the rms value of the phase-to-phase voltage and current of voltage source. θ_V and θ_I are the phase angles of V_{sph} and I_{sph} separately.

The modulation index m is obtained by equation (9) and (10). For each m, switching angles are computed off-line to eliminate the 5th and 7th harmonics (see [1][2]) and are plotted in Figure 5. Figure 6 shows the THD out to the 49th harmonic. However, one may note that outside the range $m = 1.18$ through $m = 2.5$ and some intervals between $m = 2.4$ and $m = 2.5$, there exists no set of switching angles such that the fundamental can be controlled while at the same time completely eliminating the 5th and 7th order harmonics. So for modulation indices outside this interval, other switching schemes can be used, however, they will typically result in a larger THD.

A control method is proposed here so that m is operated close to the value that gives the minimum THD. By equation (10), it can be seen that in order to generate the desired output voltage (or desired reactive power) with smallest THD, changing the dc link voltage of each level can also force the modulation index to be in the range 1.18 through 2.4 where a solution exists that eliminates the lower order harmonics. In other words, one would not regulate the capacitor voltage to a constant value, but rather they would be changed according to the steady-state operating conditions.

Given the Q_c^* (or I_c^*), modulation index m is computed by equations (9) and (10). If m is in the range 1.18 through 2.4, then

$$V_{dc}^* = V_{dc}. \quad (13)$$

If m is out of the range $1.18 - 2.4$, fix $m = 2.0$, then

$$V_{dc}^* = \frac{V_c^*}{\sqrt{\frac{3}{2}} \frac{4}{\pi} m} = \frac{V_c^*}{\sqrt{\frac{3}{2}} \frac{4}{\pi} 2.0}. \quad (14)$$

A PI controller is used to control each capacitor voltage equal to V_{dc}^*. The control principle can be explained with the aid of Figure 7. In Figure 7, v_s is the source voltage, i_c is the current flowing into the inverter, and v_c is the multilevel inverter output voltage. v_c is controlled so that it lags or leads v_s by α_c, then the total real power P_i flowing between the multilevel inverter and the utility line is

$$P_i = \frac{V_s V_c \sin\alpha_c}{X_{Lc}} \quad (15)$$

where X_{Lc} is the impedance of interface inductor. If v_c lags v_s by α_c, and P_i flows into the multilevel inverter, and the capacitor is charged. If v_c lags v_s by α_c, and P_i flows from the multilevel inverter to the utility line, the capacitor is discharged. By controlling the charging and discharging of the capacitor voltage, and the capacitor voltage is kept equal to v_{dc}^*.

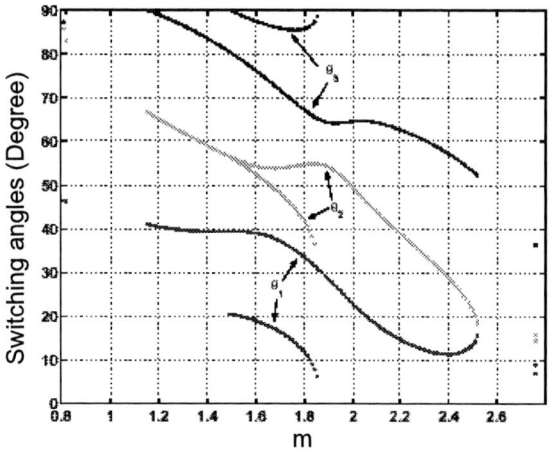

Fig. 5. Switching angles vs modulation index m for 3 dc sources multilevel inverter.

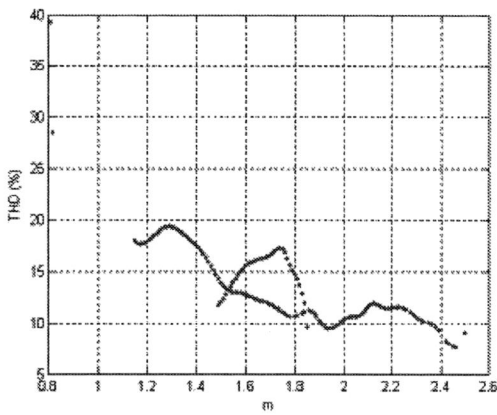

Fig. 6. THD vs modulation index m for 3 dc sources multilevel inverter.

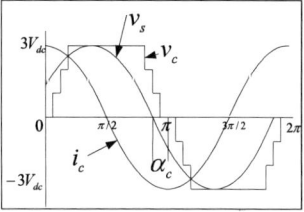

Fig. 7. Control principle for the capacitor voltage of multilevel inverter.

The switching angles are computed in the work [1][2] assuming the dc capacitor voltage of each source of multilevel

inverter is the same. To keep the dc voltage balanced between the capacitors of each inverter, the rotated switching scheme using fundamental frequency switching is used, where the switching patterns are rotated every cycle. Figure 8 shows the control logic scheme of rotating the switching patterns (see [7]). By rotation of the switching patterns, all dc capacitors are equally charged and discharged, as well as each of the switching devices having the same switching and current stresses.

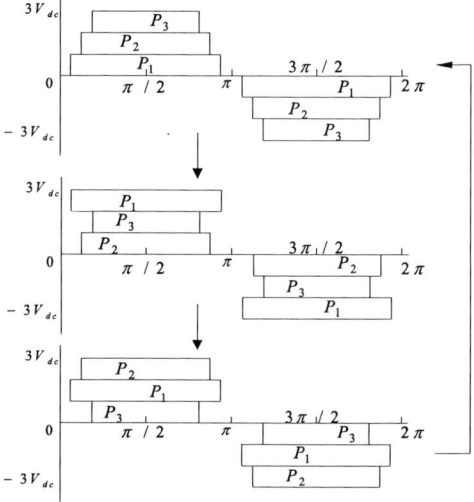

Fig. 8. Rotated switching pattern.

VI. SIMULATION RESULTS

A mathematical model of a 7-level cascaded multilevel inverter is built using Matlab/Simulink. A SVG system and the control system is modeled. In this work an RL load is used, source voltage (rms value of the line-to-line voltage) $V_s = 240$ V, DC link voltage (initial capacitor voltage) $V_{dc} = 70$ V, interface inductance $L_C = 32$ mH, total ac resistance $R = 1.0$ Ω, and fundamental frequency $f = 60$ Hz.

Figure 9 shows the simulation results. By equation (11) and (12), the reactive power Q_c^* or equivalently the reactive current I_c^* needed to be injected into the utility system is computed. This gives $Q_c^* = 520.8$ var or desired reactive current $I_{cd}^* = 0$, $I_{cq}^* = 2.170$ A. In Figure 9, the multilevel inverter is connected to the utility line at $t = 100T = 1.667$ S. It can be seen that the voltage and current sources are out of phase before the multilevel inverter is connected. The modulation index m is computed according to (10), results in $m = 2.32$. Since m is in the range 1.18 through 2.4, $V_{dc}^* = V_{dc} = 70$ V will suffice. A PI controller is used to keep each capacitor voltage at 70 V. The PI gain is chosen as $K_p = K_I = 0.001$. From Figure 9, it can be seen after 1 or 2 cycles, the source voltage v_s and the i_s are in phase.

Figures 10 and 11 show the simulation results when the load is changed. The total reactive power $Q_c^* = 262.8$ var or desired reactive current $I_{cd}^* = 0$, $I_{cq}^* = 1.095$ A is needed for injection into the utility line. In Figure 10, the multilevel inverter is connected to the utility line at $t = 100T = 1.667$ S. It can be seen that the voltage and current sources are out of phase before the multilevel inverter is connected. The modulation index m is again obtained using (10), giving $m = 2.439$. Since m is not in the range 1.18 through 2.4, then fix $m = 2.0$ and $V_{dc}^* = 85.35$ V (by equation (14)). A PI controller is again used to change each capacitor voltage equal to V_{dc}^*, where the PI gain is chosen as $K_p = K_I = 0.001$. From Figures 10 and 11, after 3 seconds, the source voltage v_s and the source current i_s are in phase.

VII. CONCLUSIONS

A cascaded multilevel inverter has been presented for static var compensation/generation application. This paper has introduced a control strategy to vary the level of the DC capacitor voltage so that use of the staircase switching scheme (with its inherent low THD) can be applicable for a wider range of modulation indices. The simulation results corresponded well with the predicted results.

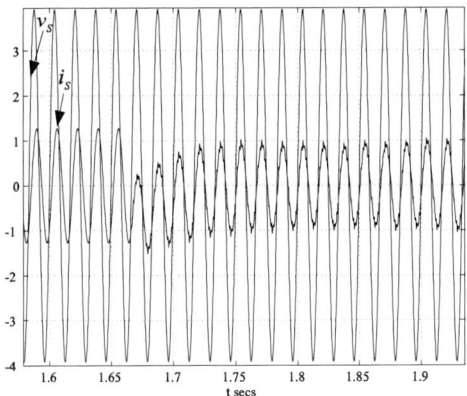

Fig. 9.   Source voltage (scaled 0.02) v_s and source current i_s.

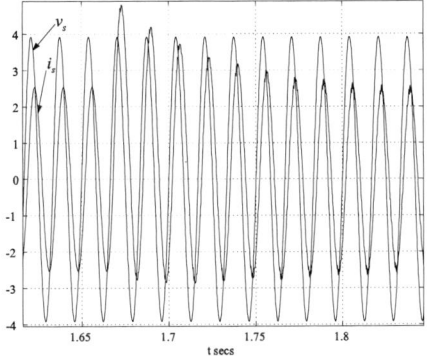

Fig. 10.   Source voltage (scaled 0.02) v_s and source current i_s.

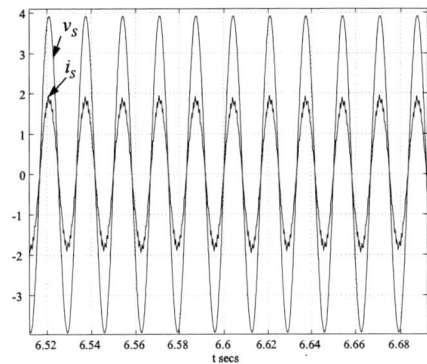

Fig. 11.   Source voltage (scaled 0.02) v_s and source current i_s.

REFERENCES

[1] J. Chiasson, L. M. Tolbert, K. McKenzie, and Z. Du, "A unified approach to solving the harmonic elimination equations in multilevel converters," *IEEE Transactions on Power Electronics*, vol. 19, pp. 478–490, March 2004.

[2] J. Chiasson, L. M. Tolbert, K. McKenzie, and Z. Du, "Control of a multilevel converter using resultant theory," *IEEE Transactions on Control System Technology*, vol. 11, pp. 345–354, May 2003.

[3] F. Z. Peng, J. S. Lai, J. W. McKeever, and J. VanCoevering, "A multilevel voltage-source inverter with separate dc sources for static var generation," *IEEE Transactions on Industry Applications*, vol. 32, pp. 1130–1138, September/October 1996.

[4] D. E. Soto-Sanchez and T. C. Green, "Voltage balance and control in a multi-level unified power flow controller," *IEEE Transactions on Power Delivery*, vol. 16, pp. 732–738, Oct. 2001.

[5] L. M. Tolbert and F. Z. Peng, "Multilevel converters as a utility interface for renewable energy systems," in *IEEE Power Engineering Society Summer Meeting*, pp. 1271–1274, July 2000. Seattle, WA.

[6] F. Z. Peng and J. S. Lai, "Dynamic performance and control of a static var generator using cascade multilevel inverters," *IEEE Transactions on Industry Applications*, vol. 33, pp. 748–755, May 1997.

[7] L. M. Tolbert, F. Z. Peng, T. Cunnyngham, and J. Chiasson, "Charge balance control schemes for cascade multilevel converter in hybrid electric vehicles," *IEEE Transactions on Industrial Electronics*, vol. 49, pp. 1058–1064, October 2002.

2006 5th International Power Electronics and Motion Control Conference

DC-link Pumping-up Voltage Suppression of a Series Active Voltage Regulator With Phase Shift Control

G. C. Xiao, Z. L. Hu, C. H. Nan, and Z. A. Wang

School of Electrical Engineering, Xi'an Jiaotong University , Xi'an, P.R. China

e-mail: xgc@mail.xjtu.edu.cn

Abstract—**The dc-link voltage pumping-up phenomenon is analyzed for a series active voltage regulator in the paper. In order to avoid dc-link voltage pumping-up, a phase shift compensation strategy is proposed, and the relationship between the source overvoltage ratio, load power factor, phase shift angle and compensation voltage provided by the inverter is presented in detail. Taking a series active voltage regulator used in low voltage distribution for example, the parameters aforementioned are calculated. A group practical design curves for suppressing dc-link voltage pumping-up with phase shift control algorithm are obtained. At last, the validity and the effectiveness of the proposed novel method is verified by some computer simulations and the laboratory experimental results.**

Keywords-overvoltage; dc voltage pumping-up; phase shift

I. INTRODUCTION

The series active voltage regulator proposed in this paper is a new power electronic equipment which is designed to mitigate the voltage disturbance in forms of voltage swell, voltage sag, severely distorted voltage waveforms and frequency deviation on sensitive loads. And it is gradually used in industrial processes to deal with field voltage quality problem.

When the source voltage is long duration overvoltage, the phase of the compensation voltage will be opposite to the source voltage. In this way, the compensator will absorb energy through the coupling transformer. But the diode rectifier can not feed the energy back to the source; it will lead to dc-link voltage pumping-up so as to endanger the equipment's safety. Therefore necessary measures have to be taken to regulate the dc-link voltage.

For series active voltage regulator, many literatures have studied its operation principle, reference value detecting and its control strategy[1-5], but few of them focus on the dc side voltage pumping-up phenomenon, especially adopting the shift phase control algorithm to suppress the dc-link voltage pumping-up. This paper present the relationship between the source overvoltage ratios, load power factor, shift phase angle and compensation voltage provided by inverter in detail under phase shift control method. Taking a series active voltage regulator used in low voltage distribution for example, the parameters aforementioned are calculated. A group of practical design curves for suppressing dc-link voltage

pumping-up with phase shift control algorithm are obtained. The effectiveness of this novel control strategy is verified by some simulations and experimental results .

II. SERIES ACTIVE VOLTAGE REGULATOR'S PRINCIPLE AND CONFIGURATION

The series active voltage regulator controls the voltage across the sensitive load by injecting an appropriate voltage through the series-connected coupling transformer. In this case the series active voltage regulator can be considered as a controlled voltage source. As is shown in Fig. 1, the main circuit of series active voltage regulator consists of inverter VT, coupling transformer TB, energy storage capacitor C and rectifier VD. The regulator detects the voltage deviation of the power system and generates reference signals in the control circuit. Then generates PWM signals to control the inverter. The inverter output is filtered before being fed to the coupling transformer in order to mitigate switching frequency harmonics generated in the inverter. The rectifier is used to charge the dc-link capacitor.

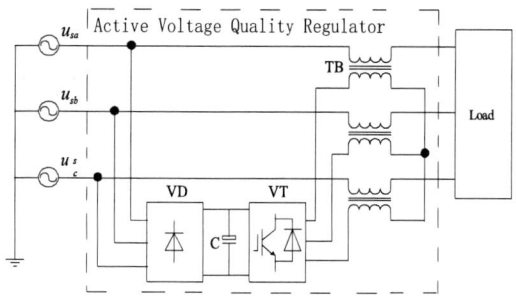

Figure 1. Main circuit of the series active voltage regulator

III. THE DC-LINK VOLTAGE PUMPING-UP PHENOMENON

When inverter injects compensation voltage into power system, there is certain energy exchanged between the power system and the series active voltage regulator. When source voltage is lower than the rated load voltage(sag), the series active voltage regulator is required to inject active power into distribution line; hence the dc-link voltage will not any rise. When source voltage is overvoltage(swell), the phase of the compensation voltage will be opposite to the source. The energy would inject into the inverter through coupling transformer from distribution line. But the diode rectifier can not feed the energy back to the distribution line. Hence the dc-link voltage will pumping-up. And this will endanger the equipment's safety, so it is important to regulate dc voltage in an appropriate limit value.

Ordinarily, there are three methods to deal with the pumping-up voltage in dc side.

● Energy-consumed. This method is to make the redundant energy consumed by a resistor.

● Energy-feedback. That is mean to make the redundant energy feedback to the power distribution through a special inverter.

● Phase shift control. In this way, the compensation voltage is not in-phase with the source. The distinct of this method is that the energy supplied by the source will decrease, so it can suppress dc-link voltage pumping-up phenomenon. But this method will make the compensation voltage and the equipment capacity increase as well as make the phase of the output voltage jump at the beginning of compensation.

All of the three methods have both merits and demerits, so we should analyze them in order to determine which one is the most suitable for a particular situation.

IV. PHASE SHIFT CONTROL FOR INDUCTIVE LOAD

Generally, we can use in-phase compensation technique and phase shift compensation technique to control the series active voltage regulator. In-phase compensation means that the phase of the compensation voltage is the same (undervoltage) as or opposite (overvoltage) to the source voltage. The compensated load voltage is in phase with the source, too. Phase shift means that there is a phase displacement between the compensation voltage and the source voltage. The load voltage will jump at the beginning of compensation.

The phasor diagram during overvoltage is depicted in Fig. 2 for most inductive load. As a symmetrical and balanced power system, $\dot{U}_S$, $\dot{U}_N$ and φ are the supply voltage, load-side voltage and the load power factor angle, respectively.

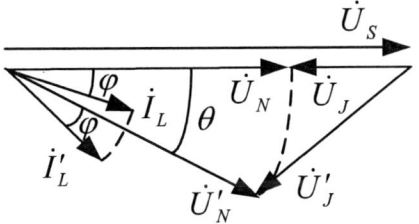

Figure 2. Phasor diagram of power distribution system during overvoltage

Using in-phase technique, let $\dot{U}_J$, $\dot{I}_L$ represent the compensation voltage, load current respectively. P_S and P_L are the input power from the source and the load power respectively, then

$$P_L = 3U_N I_L \cos\varphi \qquad (1)$$

$$P_S = 3U_S I_L \cos\varphi \qquad (2)$$

Let ΔP be the real power inject into the inverter from the source, then from (1) and (2):

$$\Delta P = P_S - P_L = 3(U_S - U_N)I_L \cos\varphi \qquad (3)$$

Unlike the in-phase voltage injection technique considered above, the proposed phase shift compensation technique is realized by the adjustment in θ, as show in Fig. 2. Let $\dot{U}'_N$, $\dot{I}'_L$ and $\dot{U}'_J$ represent the load voltage, load current and compensation voltage respectively. The magnitude of $\dot{U}'_N$ and $\dot{I}'_L$ are equal to $\dot{U}_N$ and $\dot{I}_L$. And the magnitude of the compensation voltage is larger than that based on in-phase injection method. In this case, the load power is still P_L, but the output power of the supply can be written as:

$$P_S = 3U_S I_L \cos(\varphi + \theta) \qquad (4)$$

The real power injected into the inverter can be given by:

$$\Delta P = P_S - P_L = 3I_L[U_S \cos(\varphi + \theta) - U_N \cos\varphi] \qquad (5)$$

The compensation voltage can be written as:

$$U'_J = \sqrt{U_S^2 + U_N^2 - 2U_S U_N \cos\theta} \qquad (6)$$

One distinct advantage of the proposed scheme is that less real power would inject into the inverter. The dc-link voltage pumping-up phenomenon would be suppressed consequently. The possibility of operating at $\Delta P = 0$ during voltage swell is an interesting proposition. From (5)

$$U_S \cos(\varphi + \theta) = U_N \cos\varphi \qquad (7)$$

In this case, as it is shown in Fig.3, the phase angle between the load current $\dot{I}'_L$ and compensation voltage $\dot{U}'_J$ is 90°. No real power would inject into the inverter, and the dc-link voltage will not rise any more. The magnitude of the compensation voltage can be calculated as:

$$U'_J = U_S \sin(\varphi + \theta) - U_N \sin\varphi \qquad (8)$$

Using equation (7) and do some mathematical manipulations, the relationship between compensation voltage and source voltage, load rated voltage, load power factor angle is given by

$$U'_J = \sqrt{U_S^2 - U_N^2 \cos^2\varphi} - U_N \sin\varphi \qquad (9)$$

And the corresponding phase shift angle is

$$\theta = \cos^{-1}\left(\frac{U_N}{U_S}\cos\varphi\right) - \varphi \qquad (10)$$

1709

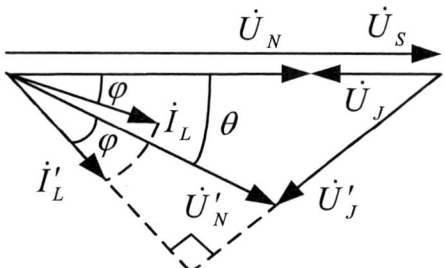

Figure 3. No pumping-up for inductive load under phase shift compensate

Obviously, when $\varphi = 0$, we can obtain the phase shift angle and the compensation voltage under pure resistive load from equ. (9) and (10). They are given by:

$$U'_J = \sqrt{U_S^2 - U_N^2} \qquad (11)$$

$$\theta = \cos^{-1}\frac{U_N}{U_S} \qquad (12)$$

V. COMPENSATION VOLTAGE AND PHASE SHIFT ANGLE CALCULATION

Under the proposed phase shift control technique, phase shift angle θ and compensation voltage U'_J are two important parameters. Phase shift angle θ and compensation voltage are related to the amplitude of source overvoltage and load power factor. The source overvoltage ratio δ is given by

$$\delta = \frac{U_S - U_N}{U_N} \times 100\% \qquad (13)$$

For low voltage distribution, assuming the rated load voltage is 220V. When the source overvoltage ratio δ is 5%,10%,15%,20%,25%,30% and the power factor is $0.5\sim1.0$, using (9)and (10), we can obtain the curves of the compensation voltage U'_J versus the power factor, shown in Fig. 4, and the curves of the phase shift angle θ versus the power factor, shown in Fig. 5.

From Fig. 4 and Fig. 5, it can be seen that the compensation voltage and the phase shift angle achieve the maximum value as $\cos\varphi = 1$ (pure resistive load). Take $\delta = 20\%$(the source voltage is 264V) for example. Under in-phase control, $U_J = 44$ V and Under phase shift control, $U'_J = 146$, $\theta = 33.6$. With the decrease of the load power factor, the compensation voltage and the phase shift angle will decreases. And with the increase of the source overvoltage ratio δ , the compensation voltage and the phase shift angle will increase too.

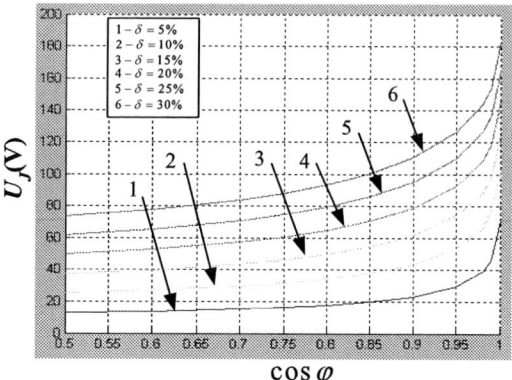

Figure 4. Compensation voltage versus the load power factor and the overvoltage ratio

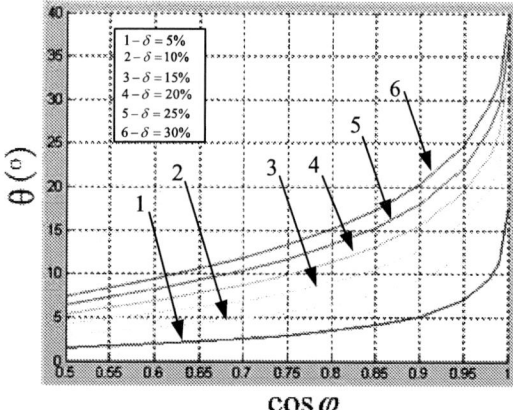

Figure 5. Phase shift angle versus the load power factor and the overvoltage ratio

It should be point out that the aforementioned formulas do not take the loss of the inverter in consideration. If considering the loss (for example the switch loss), the practical compensation voltage and the phase shift angle will be smaller than the theoretic calculation under phase shift control.

VI. SIMULATION AND EXPERIMENTAL RESULTS

A detailed simulation of series active voltage regulator with phase shift control using MATLAB/SIMULINK program as well as an experimental investigation is carried out in order to verify the effectiveness of the proposed control method.

In the simulation, the load is three-phase balance resistive load and we also take the loss of the inverter operation into consideration. The main parameters are as follows:
- Capacity of the regulator   15kVA
- Over voltage ratio δ 5 %
- Frequency   50Hz

If we don't consider the loss, the phase shift angle and the compensation voltage are 17.7° and 70V. But they are only 14.4° and 57V when consider the loss.

We adopt in-phase control before 0.08s, and then adopt phase shift control after 0.08s. Fig.6 shows the varieties of the dc-link voltage. From Fig.6, we can see that the dc-link voltage is pumping-up under in-phase control. But it gradually decreases to a steady value under phase shift control. Fig.7 shows the waveforms of the source voltage, the load voltage and the compensation voltage.

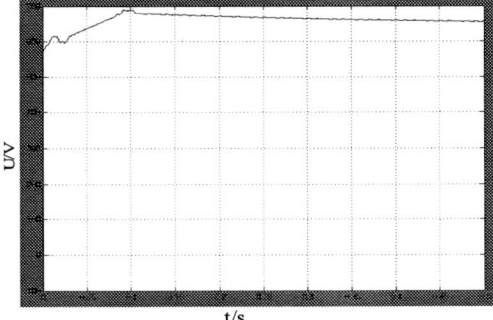

Figure 6. Dc-link voltage

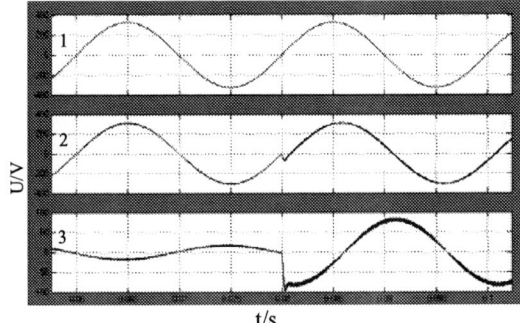

Figure 7. The waveforms of the source voltage, load voltage and compensation voltage(1: the source voltage, 2: the load voltage ,3: the compensation voltage)

The experimental results are also got on a 15KVA/220V, 50Hz prototype.

We use in-phase control firstly, then the phase shift control when DC-link voltage increases above the permitted level. Fig.8 shows the varieties of the dc-link voltage. Fig.9 shows the waveforms of the compensation voltage. From Fig.8 and Fig.9, we can see that the varieties of the DC-link voltage and the compensation voltage are the same as simulations, as well as the theoretic analysis too.

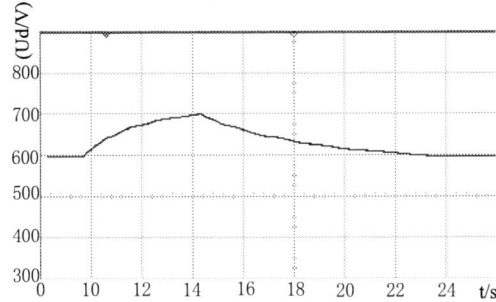

Figure 8. The experimental result of Dc-link voltage

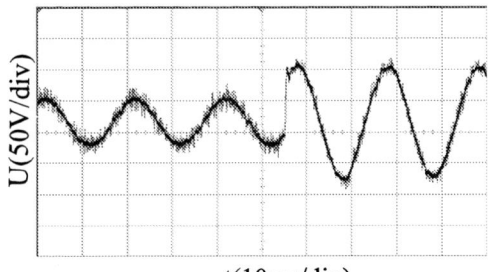

t(10ms/div)

Figure 9. The experimental result of compensation voltage

We also obtain the experimental data about the ratio, phase shift angle and compensation voltage against the source overvoltage under the condition that the load is 3000W/220V. They are shown in TABLE Ⅰ.

TABLE Ⅰ

THE EXPERIMENTAL DATA ABOUT THE RATIO, PHASE SHIFT ANGLE AND COMPENSATION VOLTAGE AGAINST THE SOURCE

Over-Vol. ratio (％)	Compensation Voltage(V)			Phase shift angle(deg)		DC-link Voltage(V)	
	In-phase	Phase shift		Theo-retic	Exper-iment	In-phase	Phase shift
		Theo-retic	Exper-iment				
10	22	100.8	75.8	24.62	17.2	653	549
15	33	124.9	112.2	29.59	25.2	726	624
20	44	145.9	135.4	33.56	31.2	801	653
25	55	165	155.3	36.86	34.8	-------	763

Note: '-------' denote the DC-link voltage will exceed the capacitor permitted voltage.

VII. CONCLUSION

This paper analyzes the pumping-up phenomenon in dc-link voltage for a series active voltage regulator. In order to avoid dc-link voltage pumping-up, a phase shift compensation approach is proposed, and the relationship between the source overvoltage ratio, load power factor, phase shift angle and compensation voltage provided by inverter is presented in detail.

Taking a series active voltage regulator used in low voltage distribution for example, the parameters aforementioned are calculated. A group of practical design curves for suppressing dc-link voltage pumping-up with phase shift control algorithm are obtained. The effectiveness of this novel control strategy is also verified by some simulation results and experimental results. And this paper provides the theoretic basis restraining dc-link voltage pumping-up, designing the main circuit and control system for the series active voltage regulator.

REFERENCES

[1] Sang-Joon Lee, Hyosung Kim, Seung-Ki Su. A novel control method for the compensation voltages in dynamic voltage restorers. Applied Power Electronics Conference and Exposition, 2004.

APEC '04. Nineteenth Annual IEEE Volume 1, 2004，Page(s):614-620

[2] Young-Hoon Cho，Seung-Ki Sul. Controller design for dynamic voltage restorer with harmonics compensation function. Industry Applications Conference, 2004. 39th IAS Annual Meeting. Conference Record of the 2004 IEEE, Volume 3,3-7 Oct. 2004，Page(s):1452 - 1457

[3] Hyosung Kim, Seung-Ki Sul. Compensation voltage control in dynamic voltage restorers by use of feed forward and state feedback scheme. IEEE Transactions on Power Electronics.Volume 20, Issue 5,Sept. 2005 ,Page(s):1169 - 1177

[4] J auch Thomas, Kara Alexander, Rahmani Mohamed, et al. Power Quality Ensured by Dynamic Voltage Restorer. ABB Review, 1998，4: 25-36

[5] Mahinda Vilathgamuwa, A A D Ranjith Perera, S S Choi. Performance Improvement of the Dynamic Voltage Restorer With Closed-Loop Load Voltage and Current-Mode Control. IEEE Transactions on Power Electronics，2002，17(5):824-83

The Fuzzy Soft-startup Controller of Active Power Filter

He Na, Wu Jian, Xu Dianguo

Department of Electrical Engineering, Harbin Institute of Technology, Harbin, 150001, CHINA

E-mail: hena1012@hit.edu.cn

Abstract—The study of a fuzzy logic controlled, three-phase shunt active power filter to improve power quality by compensating harmonics and reactive power is presented. The advantage of fuzzy controller with self-adjustable factor proposed in this paper is that it is based on a linguistic description and does not require a mathematical model of the system. Furthermore the fuzzy control rules can be modified according to the real system states automatically and instantaneously, which can improve the dynamic performance and robustness of the system. Finally the fuzzy controller and conventional PI controller are combined to build a combined controller which absorbs the merits of PI controller and fuzzy controller simultaneously. The performance of the combined controller is compared with a conventional PI controller. The combined controller can restrain voltage overshoot, reduce inrush current when APF is plunged into, which realizes soft-startup. Simulation results verify the good dynamic performance, transient stability and strong robustness of the combined controller. Furthermore compensation of reactive power and harmonics is found to be satisfactory.

Keywords- fuzzy control. self-adjustable factor. APF(Active Power Filter) . robustness.

I. INTRODUCTION

In recent years, the usage of modern electronic equipment is increasing rapidly. As a byproduct, these appliances impose nonlinear loads to the power system. And these nonlinear loads inject harmonic and reactive current into the power system, thus contributing to the degradation in the power quality. The shunt active power filter (APF) shows a strong vitality in eliminating harmonics and reactive power, which has received extensive concern in power electronic filed [1,2].

Most of the using APFs are based on voltage source inverter. According to PWM control laws, the DC-link voltage of inverter must be kept constant in order that APF can compensate harmonics and reactive power effectively. Since APF itself can not produce power to maintain the DC-link voltage, some actions must be taken to do this[3]. Generally conventional PI controller is applied to control DC-link voltage by adding an active component to the source current reference [4,5]. However, the PI controller requires precise linear mathematical models, which are difficult to obtain , and fails to perform satisfactorily under parameter variations, nonlinearity, load disturbance, etc. It will cause DC voltage overshoot and inrush source current which will

lead to protection or even equipment damage when APF is plunged into the system. The voltage overshoot and inrush current have been the bottleneck which restricts the development of APF[6].

Recently, fuzzy logic controllers have received a great deal of interests in APFs. The advantages of fuzzy controllers over conventional controllers are that they do not need an accurate mathematical model, can work with imprecise inputs, can handle non-linearity, and are more robust than conventional controllers[7]. Several fuzzy controllers of DC-link voltage are designed separately in paper [8-10]. But the structures of those controllers are very complex and the control rules are not adjustable, which results in poor control performance in real complex system.

A fuzzy controller with self-adjustable factor is proposed in this paper to realize the soft-startup of DC-link voltage. The distinctive characteristic of this controller is that its fuzzy control rules can be modified according to the real system states automatically and instantaneously which can conquer the nonlinearity and uncertainty of power system. Furthermore this feature is achieved by using simplified control technique thereby enhancing the system reliability.

Finally conventional PI controller and fuzzy controller are combined to build a combined controller to control DC-link voltage. This combined controller absorbs the merits of PI controller which has high accuracy in steady state and fuzzy controller which has strong robustness and excellent self-adaptability in transient state. That is the combined controller can ascertain good operating performance both in transient state and steady state. It can not only restrain voltage overshoot and inrush current to realize soft startup but also eliminate the steady error. Section II analyses the principles of APF and section III proposes its whole control scheme. Fuzzy logic is analyzed and then a novel fuzzy controller is designed in section IV. Section V designs the combined controller. The simulations and conclusions are given in Section VI and VII, respectively.

II. THE WORKING PRINCIPLES OF APF

Fig.1 shows the main circuit of three-phase APF system. The nonlinear load is implemented by a three-phase uncontrolled diode bridge rectifier feeding a R-L load and APF is implemented by a voltage source inverter.

From fig.1, it can be obtained that:

$$i_s(t) = i_L(t) + i_c(t) \tag{1}$$

The source voltage is given by

$$u_s(t) = U_m \sin \omega t \tag{2}$$

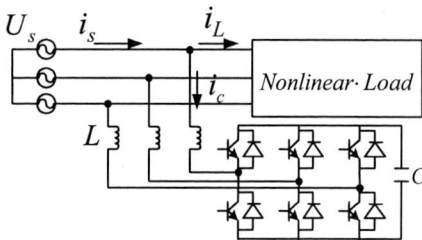

Fig.1 The main circuit of three-phase APF

The nonlinear load current includes fundamental current i_{Lf} and harmonic current i_{Lh}. And the fundamental current i_{Lf} includes active and reactive component i_{Lfp} and i_{Lfq}, as follows,

$$i_L(t) = \sum_{n=1}^{\infty} I_n \sin(nwt + \phi_n)$$

$$= I_1 \sin(wt + \phi_1) + \sum_{n=2}^{\infty} I_n \sin(nwt + \phi_n)$$

$$= I_1 \sin wt \cdot \cos \phi_1 + I_1 \cos wt \cdot \sin \phi_1 + \sum_{n=2}^{\infty} I_n \sin(nwt + \phi_n)$$

$$= i_{Lfp} + i_{Lfq} + I_{Lh}$$

The compensated current is the sum of i_{Lh} and i_{Lfq}.

When APF exports a current i_c with same amplitude and opposite phase of the compensated current, the source current i_s will become sinusoidal which only includes the fundamental active component of load current. This is the working principle of APF expressed in following formulas.

If $\qquad i_c = -(i_{Lh} + i_{Lfq})$ $\qquad$ (3)

Then $\qquad i_s = i_{Lfp}$ $\qquad$ (4)

III. THE WHOLE CONTROL SCHEME OF APF

Fig.2 is the control scheme of the whole system. The instantaneous reactive power theory is adopted to detect the source current references. The abc-dq transformation circuit converts three-phase supply currents i_{sa}, i_{sb}, i_{sc} into the instantaneous active current

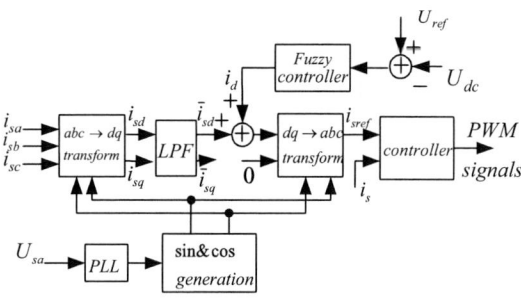

Fig.2 The control scheme of APF

i_{sd} and instantaneous reactive current i_{sq}. The fundamental components in i_{sa}, i_{sb}, i_{sc} correspond to dc components in i_{sd} and i_{sq}, and harmonic components to ac components. Two first-order low-pass filters extract dc components $\bar{i}_{sd}$, $\bar{i}_{sq}$ from i_{sd} and i_{sq}. The dc-link voltage control current i_d should be added to $\bar{i}_{sd}$ to maintain dc-link voltage. Then the dq-abc transformation produces the source current references. In this detection loop, the LPFs (Low-pass filters) are adopted which can decrease the delay time and improve the system stability over the method which adopts high-pass filters[11]. Because only fundamental-frequency components flow through dq-abc transformation circuits whereas high-frequency components are bypassed form them. Since the fundamental-frequency components are in a low-frequency range, a delay-time does not cause any significant phase delay. This small delay time also makes no contribution to causing instability in a high-frequency range.

From Fig.2 we can see that the whole system contains two controllers. The outer-loop fuzzy voltage controller controls DC-link voltage to follow the voltage reference; the inner-loop PI current controller, using generalized integrators, decides the switching signals for the devices of the PWM converter, thus forcing the source current to follow the current reference. The adopted current controller is shown in fig.3.

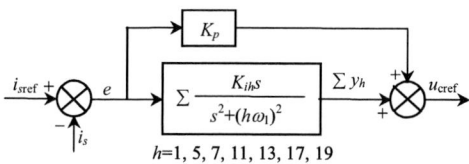

h=1, 5, 7, 11, 13, 17, 19

Fig.3 PI current controller using generalized integrators

u_{cref} is the voltage reference of the inverter, K_p the proportional coefficient, K_{ih} the coefficient of generalized integrators, h the harmonic order, y_h the output of generalized integrators.

In order to avoid saturation of the generalized integrators, a self-tuned strategy is adopted to modify the integrator coefficient as shown in (5),

$$k_{ih} = \begin{cases} \alpha k_{ih}, & |y_h| > y_{h\max} \\ k_{ih}, & |y_h| \le y_{h\max} \end{cases} \qquad \alpha \in (0,1) \qquad (5)$$

$y_{h\max}$ is the output limit of generalized integrators. After disposed in the current controller, the signals will be sent to the space vector pulse width modulation section to get the PWM signals.

IV. THE DESIGN OF FUZZY CONTROLLER WITH SELF-ADJUSTABLE FACTOR

A. The Fuzzy Logic Analysis

In a fuzzy logic controller, the control action is based on experience and control engineering knowledge so that mathematical models are not necessary to develop the control action. Thus the principles of the control system are characterized by an extremely simply and intuitive approach, explained as follows.

The time step response of a stable closed loop system should have a shape as shown in fig.4. The input variables of the fuzzy controller are the error E and its varying ratio $\dot{E}$. The output is control signal. The time response has been divided into four regions Z_1, Z_2, Z_3, Z_4. The index used for identifying the response area is defined as follows:

Z_1: if $E > 0$ & $\dot{E} < 0$; Z_2: if $E < 0$ & $\dot{E} < 0$;

Z_3: if $E < 0$ & $\dot{E} > 0$; Z_4: if $E > 0$ & $\dot{E} > 0$;

From the step response we can obtain that the corresponding rule for the region Z_1 should have the effect of shortening the rise time. Rule for region Z_2 should decrease the overshoot of the system response. Similarly, rules for other regions can be formed. Based on this the rules can be obtained from an understanding of the filter behaviour and modified by simulation performance.

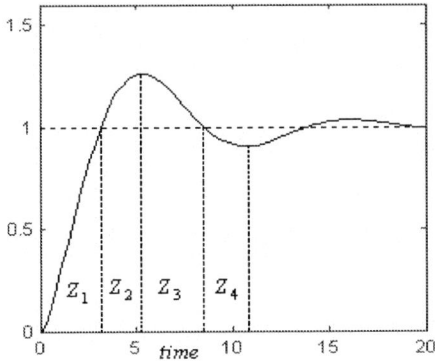

Fig.4   Step response

B. Proposed F uzzy Controller

Based on the aforementioned discussion, a fuzzy controller with self-adjustable factor is designed in this paper. The self-adjustable factor can regulate in the whole region and adjust the control strategies according to the real state of system instantaneously, thus optimization can be realized at the same time.

In the proposed fuzzy controller, the fuzzy values E^* and $\dot{E}^*$ of error signal E and its varying ratio $\dot{E}$ are regarded as inputs, and the fuzzy control variable U is the output. The method of non-uniform quantization is used during the process of fuzzification and defuzzification[12], as shown in table 1. That is when the exact values of E and $\dot{E}$ are small, we divide them finely; conversely when they are large, we divide them coarsely. And probability statistic method has been adopted in ascertaining the dividing sections. We can omit the process of ascertaining membership function because the dividing method itself is a process of ascertaining membership function. Then the fuzzy controller can be designed with the fuzzy input values of

E^* and $\dot{E}^*$.

Table I
THE FUZZIFICATON OF E (the positive part only)

E	≥ 30	10-30	5-10	2-5	1-2	0.5-1	0.2-0.5	-0.2-0.2
E^*	7	6	5	4	3	2	1	0

The fuzzy control rules adopted are shown in (5):

$$\begin{cases} U = -(\alpha E^* + (1-\alpha)\dot{E}^*) \\ \alpha = \dfrac{1}{N}(\alpha_s - \alpha_0)|E^*| + \alpha_0 \end{cases} \quad (5)$$

In (5), α is a self-adjustable factor, N is the classification degree, U is the control variable, and $0 \leq \alpha_0 \leq \alpha_s \leq 1$ are constants.

The characteristic of the control rules shown in (5) is that the factor α can regulate in the whole region. When the error E is very big, it indicates that the system is far from balance state. So the controller should enlarge its control variables in order to make the system attain balance state as soon as possible. In this case, the error E plays a more important role in control variables. However, when the error E is very small, it indicates that the system is close to the balance state. In order to make the system to steady state and restrain the oscillations as soon as possible, the most important thing is to decrease the overshoot. Then the role of E in the control variables is small while the role of $\dot{E}$ becomes large. The self-adjustable factor can regulate in the whole region similar to the thinking characteristics of human being in the process of controlling. And as a result it can realize optimization.

Based on the aforementioned theory, the DC-link voltage fuzzy controller is designed as shown in fig.5. In order to decrease the voltage overshoot and inrush current better, the voltage reference U_{ref} is set as a slope-function which increases gradually till to the rated value.

The DC-link voltage is detected and compared with a reference U_{ref} to achieve the error ΔU, and the varying ratio $\Delta \dot{U}$ can also be calculated out. Then pass them to the fuzzy controller to produce the control variable ΔI_d.

According to (5), the fuzzy control rules of APF are shown in (6):

$$\begin{cases} \Delta I_d = -(\alpha \Delta U^* + (1-\alpha)\Delta \dot{U}^*) \\ \alpha = \dfrac{1}{N}(\alpha_s - \alpha_0)|\Delta U^*| + \alpha_0 \end{cases} \quad (6)$$

ΔU^* and $\Delta \dot{U}^*$ are the fuzzy values of ΔU and $\Delta \dot{U}$ separately.

And then $\qquad I_d(k) = I_d(k-1) + \Delta I_d(k) \qquad (7)$

That is the output of the fuzzy controller is the change in the amplitude of active current I_d which is used to maintain the DC-link voltage. This is equivalent to an integral function which can contribute to eliminating steady error.

1715

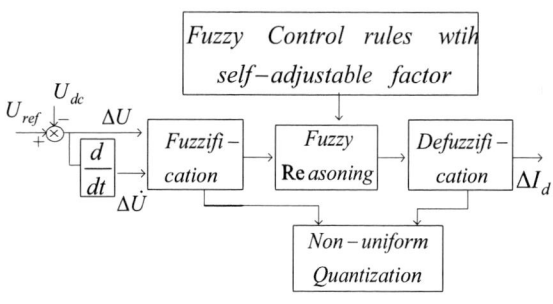

Fig.5 Structure of DC-link fuzzy controller

V. THE DESIGN OF COMBINED CONTROLLRE

In order to absorb the merits of fuzzy controller and conventional PI controller, a combined controller is then proposed to control the DC-link voltage in steady and transient states, as shown in fig.6. At the first transient stage the fuzzy controller takes actions to control the DC-link voltage in order to improve the dynamic response; when the voltage error is very small and the system is almost in steady state, the PI controller is plunged into and takes the place of fuzzy controller to eliminate the steady error and improve the steady performance of system. The switch of the two controllers is decided by the voltage error.

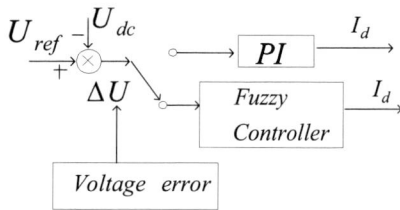

Fig.6 Structure of combined controller

VI. SIMULATION RESULTS

Based on the above discussion, the whole system is simulated in MATLAB. The source voltage is 100V, DC-link voltage reference is set 220V, load current is 8.7A, $L = 4mH$, $C = 3400uF$.

At the first stage, all IGBT devices are turned off so the DC-link is charged by the parallel diode, which is equivalent to three-phase bridge rectifier. And then the second transient stage, the fuzzy controller is plunged into to control the DC-link voltage to a rated value gradually without overshoot. When the DC voltage error is less than a valve value, it goes to the third steady stage and PI controller is switched into to eliminate the steady error.

Form Fig.7 and Fig.8 we can see that compared to conventional PI controller, the combined controller proposed in this paper can improve the system dynamic response, restrict the voltage overshoot and reduce the inrush current, which realizes the soft-startup of APF.

Fig.9 shows the source current curves before and after the APF is plunged into. It can be seen that APF can

compensate the harmonics and reactive power effectively, which makes the source current sinusoidal.

Fig.10 shows the curves of transient experiments in which the load decreases to a half at some time. From the curves we can see that the combined controller has a better dynamic performance in tracing load current compared to PI controller.

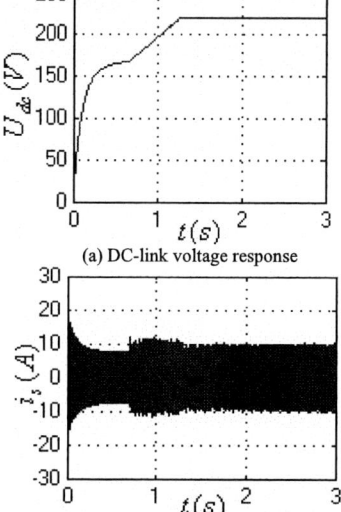

(a) DC-link voltage response

(b) Source current response
Fig.7 Response of combined controller

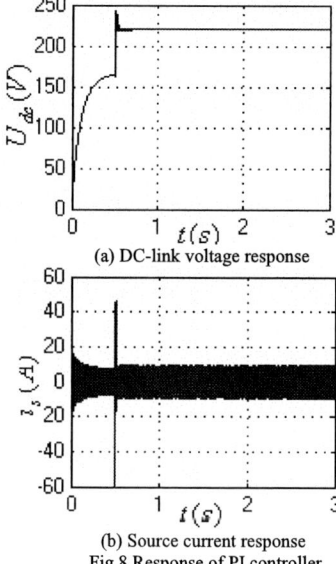

(a) DC-link voltage response

(b) Source current response
Fig.8 Response of PI controller

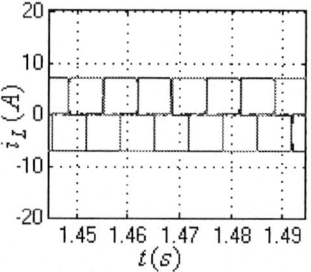

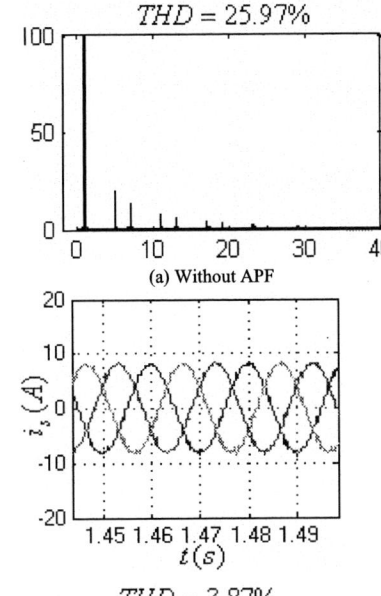

(a) Without APF

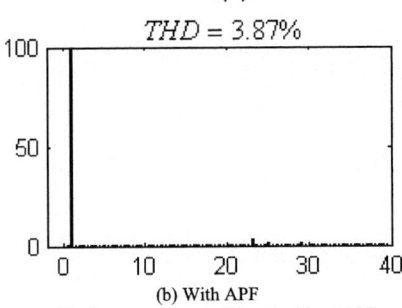

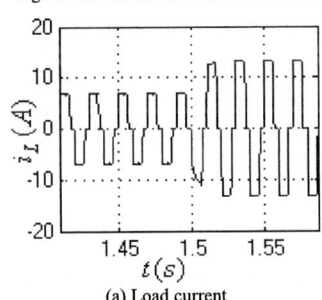

(b) With APF

Fig.9 source current with and without APF

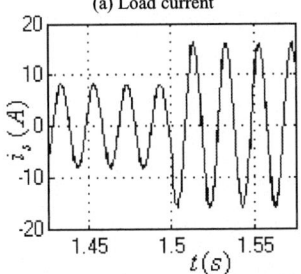

(a) Load current

(b) Source current of combined controller

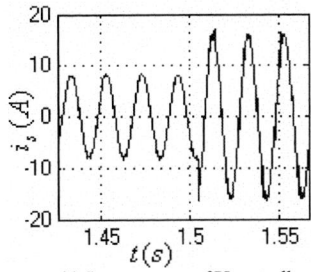

(c) Source current of PI controller
Fig.10 Transient response of current

VII. CONCLUSIONS

In order to control the DC-link voltage of APF better and realize soft-startup, a combined controller of fuzzy controller and PI controller is proposed in this paper. The structure of the fuzzy controller with self-adjustable factor is simple, the calculating methods are convenient and the effect is significant. The combined controller can improve system performance in both steady state and transient state. Simulation results show that the combined controller has strong robustness and excellent dynamic response compared to PI controller. It can greatly restrict the voltage overshoot and inrush current and compensate the harmonic and reactive current well.

REFERENCES

[1] Wang Zhaoan, Yang Jun, Liu Jinjun. Harmonics Restraint and Reactive Power Compensation[M]. Bei Jing, China Machine Press, 1998.

[2] Gu Jianjun, Xu Dianguo, Liu Hankui, etal, "Active Power Filter Technology and its Development". Electric Machines and Conrol, vol.7, 2003, pp. 126-132.

[3] Ding Kai, Chen Yunping, Wang Xiaofeng, etal, "Investigation on Relevant Problems for DC side Voltage of Parallel Active Power Filter". Electric Engineering, vol.10, 2002, pp. 27-29..

[4] Zhang Jianhui, Jiang Qirong, Zhao DI, etal, "Design of Active Power Filter Controller". Power System Technology, vol.26, 2002, pp. 48-52.

[5] You Xiaojie, Li Yongdong, Victor Valouch, etal. "SAPF Control Strategy under the Condition of Non-ideal Source Voltages". Proceedings of the CSEE, vol.24, 2004, pp. 55-60.

[6] Ruan Lifei, Zhang Wanfeng, Ye Pengsheng, "Study on Control Strategy for Three-phase Shunt Active Power Filter". Electric Drive, vol.3, 2002, pp. 36-38.

[7] Li Shiyong. Fuzzy Control .Neuro Control and Intelligent Control. Harbin: Harbin Institute of Technology Press, 1996.

[8] S.K.Jain, P.Agrawal, H.O.Gupta, "Fuzzy logic controlled shunt active power filter for power quality improvement". IEE Proc.-Eletrc. Power Apply, vol.149, 2002 , pp.317-328.

[9] Juan Dixon, Jose Contardo, Luis Moran, "DC link Fuzzy control for an active power filter, sensing the line current only". IEEE 1997, pp. 1109-1114.

[10] Xu Wanfang, Luo An, Wang Lina, etal, "Development of Hybrid Active Power Filter Using Intelligent Controller". Automation of Electric Power Systems, vol.27, 2003, pp. 49-52.

[11] Sunt Srianthumrong, Hideaki Fujita, Hirofumi Akagi. Stability Analysis of a Series Active Filter Integrated with a Double-Series Diode Rectifier. IEEE TRANSACTIONS ON INDUSTRIAL APPLICACTIONS, VOL.17, NO.1, JANUARY 2002.

[12] Liu Rui-ye, "The Research on Fuzzy Variable Structure Control of Tubro-Generator". A Dissertation for the PH.D. Degree of Harbin Institute of Technology, 1999

2006 5th International Power Electronics and Motion Control Conference

A Novel Control Method for DSTATCOM Using Artificial Neural Network

YANG Xiao-ping[*], ZHONG Yan-ru[**], WANG Yan[*]
[*]Xi'an University of Technology/Institute of Water Resources and Hydro-electrical Engineering,
[**]Xi'an University of Technology/Department of Electrical Engineering
Xi'an, China

Abstract—In this paper, a new control method for Distribution Static Compensator (DSTATCOM) using Artificial Neural Networks (ANN) instead of the conventional direct controller is presented. A feedforward three-layer neural network is adopted to control the output compensating currents. This is accomplished by producing the appropriate switching patterns of the IGBTs. A backpropagation algorithm trains this network. A practical case with PSCAD/EMTDC is presented to check the proposed control performance. The simulation results show a better source current waveforms and a near unity power factor operation is achieved. This technique offers an alternative to the multi-pulse techniques that require complex magnetic circuit arrangements as well as to the conventional PI controller that the parameters are hard to determine.

Keywords-DSTATCOM, Artificial Neural Network(ANN), Back propagation, PSCAD/EMTDC, Simulation.

I. INTRODUCTION

As a typical flexible AC transmission system (FACTS) device, DSTATCOM is one of the most effective reactive power compensation devices to improve power transfer capability [1]. To obtain the best compensation effects, DSTATCOM must be controlled properly. In recent years most of the papers have suggested many methods for designing DSTATCOM controllers using linear control techniques, in which the system equations are linearized at a specific operating point and based on the linearized model, the PI controllers are tuned in order to have the best possible performance. The drawback of such PI controllers is that their performance degrades as the system operating conditions change. Nonlinear adaptive controllers on the other hand can give good control capabilities over a wide range of operating conditions, but they have a more sophisticated structure and are more difficult to implement compared to linear controllers. In addition, they need a mathematical model of the system to be controlled [2].

Artificial Neural networks (ANN) offer a solution to this problem, they are able to identify and model such nonlinear system. Nowadays, this technique is considered as a new tool to design DSTATCOM control system. The

ANN presents two principal characteristics. It's not necessary to establish specific input-output relationships but they are formulated through a learning process or through an adaptive algorithm. Moreover, it can be trained online without requiring large amounts of offline data [3].

In this paper, a novel design of a DSTATCOM control method based on ANN is presented. The ANN block will be used instead of the conventional direct control. The performance of the method is demonstrated via transient simulation using the electromagnetic transients simulation program PSCAD/EMTDC.

II. CONTROL SYSTEM OF DSTATCOM

Fig.1 shows the complete control system with DSTATCOM. The ac source is three-phase sine wave. The load current I_l is the sum of the source current I_S and the compensation current I_n. The objectives are to get a source current without harmonic and reactive components and to get a constant value of the DC voltage. The ac side of the converter is connected to the power supply through a synchronous link reactor L, Fig.2, which also performs as a low pass filter. The main losses are reduced to R. The simplified main circuit of DSTATCOM is shown in Fig.3.

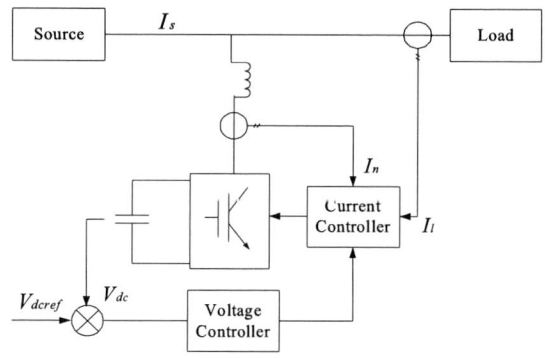

Fig.1 Control system with DSTATCOM

1-4244-0448-7/06/$25.00 ©2006 IEEE

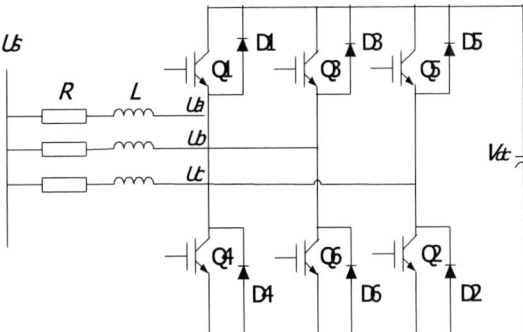

Fig.2 The main circuit

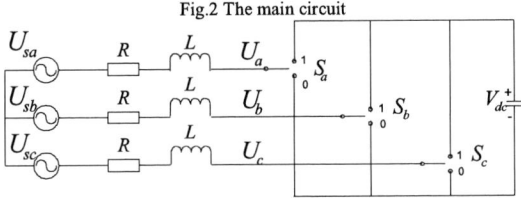

Fig.3 Simplified main of DSTACOM

III. CONVENTINAL CONTROL METHOD OF DSTATCOM

Fig.4 shows the conventional direct control of DSTATCOM. V_{dc} is the DC side voltage, I_{na}, I_{nb}, I_{nc} are the output currents of the inverter, and I_{la}, I_{lb}, I_{lc} are the load currents [4].

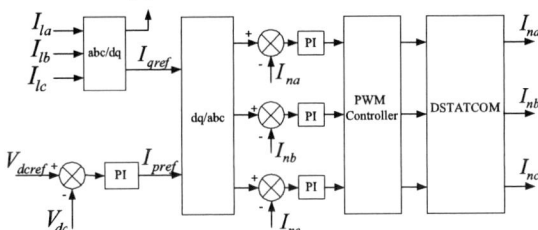

Fig.4 The conventional direct control of DSTATCOM

A. Generation of The Reactive Reference Current

The reactive reference current I_{qref} is generated by the load current, using dq transformations. The relationship between the peak ac input current and the mean -axis values is as follow:

$$\begin{bmatrix} I_d \\ I_q \end{bmatrix} = \sqrt{\frac{2}{3}} \begin{bmatrix} \sin\omega t & \sin(\omega t - \frac{2\pi}{3}) & \sin(\omega t + \frac{2\pi}{3}) \\ -\cos\omega t & -\cos(\omega t - \frac{2\pi}{3}) & -\cos(\omega t + \frac{2\pi}{3}) \end{bmatrix} \begin{bmatrix} I_a \\ I_b \\ I_c \end{bmatrix} \quad (1)$$

$$\begin{bmatrix} I_a \\ I_b \\ I_c \end{bmatrix} = \sqrt{\frac{2}{3}} \begin{bmatrix} \sin\omega t & -\cos\omega t \\ \sin(\omega t - \frac{2\pi}{3}) & -\cos(\omega t - \frac{2\pi}{3}) \\ \sin(\omega t + \frac{2\pi}{3}) & -\cos(\omega t + \frac{2\pi}{3}) \end{bmatrix} \cdot \begin{bmatrix} I_d \\ I_q \end{bmatrix} \quad (2)$$

B. Generation of Tthe Active Reference Current

Although the voltage of the DC side should be maintained at a pre-specified design value, voltage fluctuations cannot be avoided. Voltage control of the DC side is achieved by adjusting a small amount of real power flowing into the DC capacitor. Therefore a DC voltage control loop is included in the DSTATCOM. The error between the actual DC voltage and its reference value is treated in a proportional integral (PI) controller, and the output is the active reference current I_{pref}. The parameters of the PI controller must be chosen carefully to attain the desired stability and dynamics.

C. Generation of Trigger Signals

The DC/AC inverter is a full bridge, three-phase inverter made of six Insulated Gate Bipolar Transistors (IGBTs) controlled by Pulse Width Modulation (PWM) technique. The IGBT is selected as switching component because it is voltage controlled and easy to drive, with relatively low on-state voltage drop. In simulations, the PWM inverter is modeled with related snubber circuits attached to the IGBTs. A current error is observed and fed to the PI control block. The output of the PI controller is the voltage-modulating signal in which its magnitude and phase are controlled. In the PWM technique, the triangular carrier signal is compared with the voltage-modulating signal so as to obtain the firing signals of the IGBTs.

IV. THE PROPOSED ANN CONTROLLER

The disadvantage of direct control method is that harmonics are big. Moreover, the parameters of PI controllers are hard to determine, which influence the performance of the controller.

This paper introduces another option to reduce the ac current harmonics that uses an ANN. The ANN quickly computes the necessary firing instants for providing the necessary reactive power as well as for eliminating only the necessary harmonics, Fig.5.

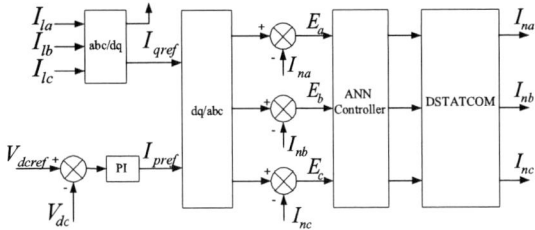

Fig.5 The proposed controller

A. The Architecture of ANN

The ANN controller used is a feed-forward one, comprising three neuron layers, the input layer, the hidden layer and the output layer. The input layer offers connection point to transmit the input signal to the hidden layer. The latter begins the learning process and the output layer continues the learning process and provides

1719

outputs. The network topology is shown in Fig.6.The hidden-layer neurons have a tan-sigmoid transfer function, and the output layer neurons have a linear transfer function. The ANN has three inputs that are the three-phase current errors of the DSTATCOM. It also has three outputs that are the three switching functions of the inverter legs.

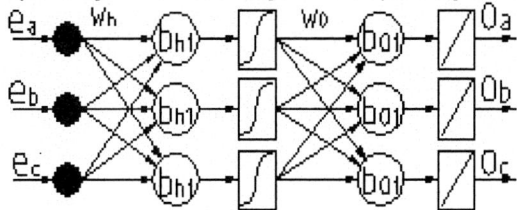

Fig.6 The architecture of ANN

The control objective of the ANN is to provide the wanted proper gating patterns of the PWM inverter, leading to adequate tracking of the DSTATCOM reference phase currents and constant DC voltage. The number of neurons in the hidden layer is specified as the minimum number that produces the permitted training criterion.

On introducing the input vector e_j and $e_j = [e_a, e_b, e_c]^T$, the equations associated with the signals flowing from each layer to the next are

$$I_j = f(W_{hj}^T e_j^T + B_{hj}^T) \qquad (3)$$

$$O_j = g(W_{oj}^T I_j^T + B_{oj}^T) \qquad (4)$$

where I_j is the hidden layer output vector, O_j is the output vector of the ANN, f is the transfer function of the hidden layer, g is the transfer function of the output layer; W_h is the weight matrix of the hidden layer; W_o is the weight matrix of the output layer; B_h is the bias vector of the hidden-layer neurons; and B_o is the bias vector or the output layer. In this paper, the functions of f and g are as follows:

$$f(x) = \frac{1}{1 + e^{-\lambda x}} \qquad (5)$$

$$g(x) = \begin{cases} 0 & x < 0 \\ 1 & x \geq 0 \end{cases} \qquad (6)$$

Where λ is the up ratio of the function, which is between 0 and 1[5].

B. Training Algorithm

The ANN is trained by changing the weights W_{ij} and the biases B_j. The training criterion is taken as the mean square error of the ANN output with a value of 0.0001and the error function is defined as:

$$E = \frac{1}{2} \sum_{i=1}^{N} e(i) \qquad (7)$$

Where N is the number of output neurons and $e(i)$ is the instantaneous error between the actual and estimated values of the output. The training is finished when the value of E is less than 0.0001[6].

Sufficient input-output training samples with different compensation modes are obtained by using the triangular carrier modulation technique. The backpropagation (BP) algorithm is used for training. The BP algorithm was developed by Paul Werbos in 1974. It applies a weight correction to the neural network connection weights, which is proportional to the partial derivative of the error function. BP learning algorithm is the most commonly used technique for updating neural network weight parameters [7]. The adjustment to the weights is in the negative direction of the gradient of the error. The neuroidentifier weights are adjusted according to:

$$\Delta W_{ij} = -\eta \frac{\partial E}{\partial W_{ij}} \qquad (8)$$

Where η is the learning rate parameter. A large learning rate might lead to oscillations in the convergence trajectory, while a small learning rate provides a smooth trajectory at the cost of slow convergence speed.

V. SIMULATION RESULTS

The detailed DSTATCOM model with a three-phase bridge converter connected into a distribution system is modeled by using PSCAD/EMTDC. The compensator was connected to the network through an inductance. A list of the system parameters considered in the simulation is given in table I. The ANN technique is used for the switching pattern generation. Learning process of ANN is developed in MATLAB 6.5, helped by the toolbox Neural Network.

TABLE I.
SYSTEM PARAMETERS USED IN SIMULATION

Power source	Line voltage = 380V Frequency = 50.0 Hz
DSTATCOM	R=0.074 Ω
	L=0.7mH
	C=1000 μF
	V_{dc} =760V
	Capacity = $\pm$ 100kvar

Simulation results are presented in Fig. 7-8, where V_{as} is the system voltage (phase A), I_{as} is the system current (phase A), and V_{dc} is the DC side capacitor voltage. The DSTATCOM device is put into service at 0.2s. From the waveforms we can see that I_{as} comes in phase with V_{as} quickly after compensation and V_{dc} becomes constant with a short-time transition. As a consequence of the action of the switches, high frequency harmonics arise in the capacitor voltage and system phase currents. A second harmonic is excited in the capacitor voltage and a large third harmonic component is present in the system phase currents. Changing the inductance value or install a filter in the power circuit can reduce the harmonics.

1720

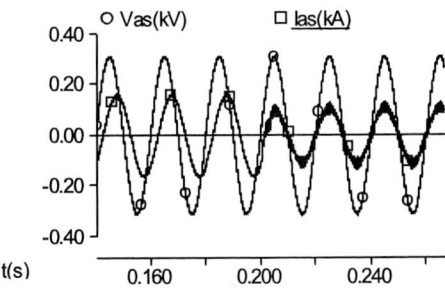

Fig.7 The system voltage and current waveforms of phase A before and after compensation

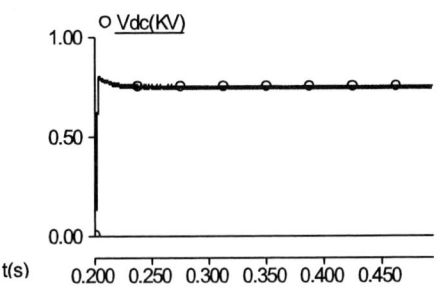

Fig.8 The DC side voltage

VI. CONCLUSIONS

A DSTATCOM based on an artificial neural network has been proposed and analyzed in this paper. Load currents and the DC voltage are sensed, the ANN block calculates the power circuit control signals from the reference compensation currents, and the power circuit injects the compensation current into the power system. The viability of the system has been proved through simulation results. The device has been able to shape the system current into nearly a sinusoidal waveform in phase with the system voltage. The simulation results show that the proposed control strategy is effective.

REFERENCES

[1] Chen Xianming, Xu Heping et al, "Development of a ± 500kvar STATCOM", *Automation of Electric Power System*, pp: 53-57, December 2001.

[2] LI Chun, J IANG Qirong, and XIU Lincheng, "Reactive Power Control of STATCOM Using PID Auto-Tuning Technique and Fuzzy Set", *Control Theory and Applications*, China, 6th, vol.17, pp.856-860, December 2000.

[3] J. R. Vazquez, P.R.Salmeron, "Three-phase Active Power Filter Control Using Neural Networks", 10th *Mediterranean Electrotechnical Conference*, MEleCon, Vol.III, 2000, pp. 924-927.

[4] Su Chen, Géza Joós, "Direct Power Control of DSTATCOMs for Voltage Flicker Mitigation", *IEEE2001*, pp. 2683-2690.

[5] LIU Guiying, SU Shiping, "A Parallel-Connected Three-phase-four-Active Power Filter and its Control Based on Neural Network", *Hunan Electric Power*, vol.23, 3rd, 2003.

[6] Salman Mohaghegi, et al, "A Comparison of PSO and Backpropagation for Training RBF Neural Networks for Identification of A Power System With STATCOM", *IEEE2005*, pp. 1-4.

[7] S.S. Haykin, "Neural Networks- A Comprehensive Foundation", *Prentice-Hall*, 2nd, 1998, ISBN 0-1327-3350-1.

2006 5th International Power Electronics and Motion Control Conference

A Detailed Analysis of Unexpected DC-side Voltage Boost in Series Power Quality Controllers

Yuan Chang, Liu Jinjun, Wang Xiaoyu, Wang Zhaoan
School of Electrical Engineering, Xi'an Jiaotong University
Xi'an, Shaanxi Province, P. R. China
Chang.Yuan@ieee.org

Abstract—Series power quality controller (SPQC) can effectively suppress power quality interferences. Yet the normal operation of whole system based on the maintenance of DC-side voltage. This paper presents an unexpected voltage boost of DC side in SPQC and analyzes it in detail. The experimental results verified the analysis.

Keywords- DC-side; power quality; dead-time

I. INTRODUCTION

Series power quality controller (SPQC) based on self-commutated power converters has been drawing more and more attentions since late 1980s [1-7]. In the beginning it was mainly proposed to suppress the input current harmonics of voltage-source type harmonics-generating loads, such as capacitor-filtered diode rectifier circuits, so that the current flowing into the electric power network could be almost sinusoidal [1-4]. More recently, most publications in this field have been on using it as a series voltage-source type compensator for power quality interferences such as voltage harmonics and sags coming from the electric power network side, so that the quality of the load voltage can be guaranteed [5-7].

As shown in Figure 1, a series power quality controller, connected in series with the electric power supply source and the load, usually consists of a voltage-source inverter using IGBTs or other fully-controlled power semiconductor devices and a coupling transformer. The single phase rectifier is used to offer the DC-side voltage. For the normal operation of whole system, the maintenance of DC-side voltage of the series power

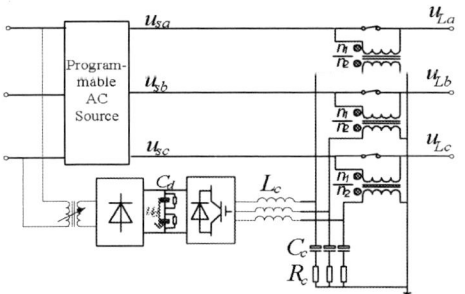

Figure 1. System configuration of SPQC

quality controller has been a significant issue all the while. The DC-side voltage not only affects the compensating performance, but also relates to the safe operation of whole system.

This paper presents an unexpected voltage boost in DC side of SPQC. Both qualitative analysis and quantitative analysis are proposed, and the experimental results verified the analysis.

II. THE UNEXPECTED DC-SIDE VOLTAGE BOOST

A prototype 3kVA SPQC using IGBT devices is developed to compensate voltage harmonics and sags coming from the electric power network side. The configuration of main stage is shown as Fig.1.

Fig.2 shows the unwonted raise of the DC-link voltage. In normal state, the DC-link voltage should be determined by the AC side voltage of the rectifier. The Fig.2 gives out the waveform of the DC-side voltage during the start-up of the SPQC. The reference of the compensating voltage is set to be zero to predigest the analysis. We can see that the DC-link voltage raise

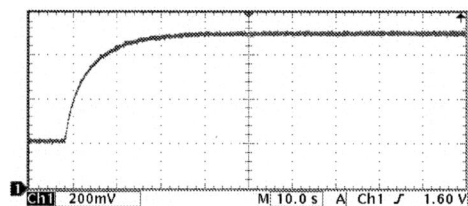

Figure 2. The waveform of the DC-side voltage during the start-up

rapidly after the SPQC is started, and the final value (340V) is much larger than the expected value (100V). Even the source voltage is less than 50 V, the DC-link voltage can reach several hundreds volts.

III. THEORETICAL ANALYSIS

In the control of the inverter in SPQC, it is necessary to include a "dead time" in switching signals in order to avoid the simultaneous conduction of the devices on the same electrical trajectory or leg. It results in a momentary loss of control, and the output voltage deviates from the reference voltage. This effect is well known as the dead-time effect [8,9]. After a detailed analysis, we found the dead time injection is the real reason of DC-side voltage boost.

1-4244-0448-7/06/$25.00 ©2006 IEEE

A. From the View Point of Circuit Operation

The voltage reference is set to zero to emphasize the effect of dead-time injection, so the driving signals are square waveforms with duty cycle 50%. If the converter is working without dead-time injection, the output line voltage of the converter will be zero at anytime (as shown in Fig.3), and the current flow through the DC-side capacitor will be zero. Then the capacitor voltage will keep constant if the power losses are negligible.

Considering the dead-time injection, the only difference comparing with the previous situation is the dead time transient. During the dead time, both the upper leg and lower leg are turned off, so the current path is decided by the current directions before the dead time. Assuming the three-phase utility voltages are balanced and synchronous, the current of the DC-side capacitor will always flow from high potential terminal to low potential terminal (as shown in Fig.4), so the capacitor voltage will boost. In a practical system, power losses are unavoidable, so the DC-side voltage couldn't keep rising. The final value depends on the power losses and the dead time.

B. From the View Point of the equivalent AC voltage error

Fig.4 illustrates the analysis from the view point of the equivalent AC voltage error. Taking phase a as an example, Fig.5.a shows the single phase circuit. Fig.5.b illustrates error voltage generated by the dead-time injection. T1 and T2 are the driving signals for the upper and lower switch respectively. The dead-time injection generates one error voltage pulse in each switching period. The polarity of this error voltage pulse is decided by the input current. The error voltage pulse series can be replaced by an equivalent square voltage (Fig.5.c) when the switching noise can be ignored, and the equivalent square voltage is in phase with the input current. Then we found the injection of dead-time will always leads to the absorbing of electric power, and there are no other energy restorer in the inverter but DC-side capacitor, so the capacitor voltage will boost. As mentioned before, the final value depends on the power losses and the dead time.

ΔU_d is used to represent the magnitude of this square voltage, V_{cc} denotes the DC-link voltage, T_s and T_d represent the switching period and the dead-time respectively. It's easy to obtain: $\Delta U_d = V_{CC} \cdot \dfrac{T_D}{T_S}$

Assume the peak value of the fundamental component of the square waveform is U_{cm}, we can get that:

$$U_{CM} = \frac{4}{\pi} \cdot \Delta U_d = \frac{4}{\pi} \cdot V_{CC} \cdot \frac{T_D}{T_S}$$

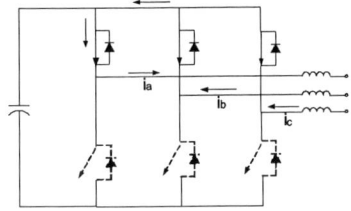

Figure 3. Working state of the converter without dead-time injection

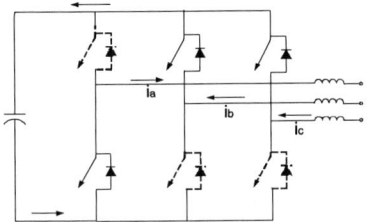

a) Current flow into the converter through two phases and flow out through one phase

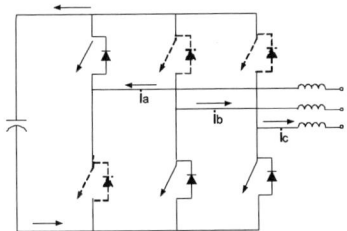

b) Current flow into the converter through one phase and flow out through two phases

Figure 4. Working state of the converter with dead-time injection

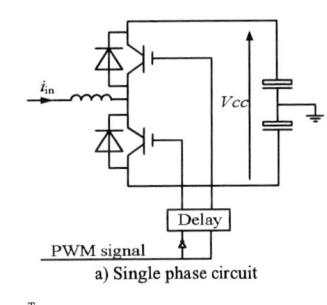

a) Single phase circuit

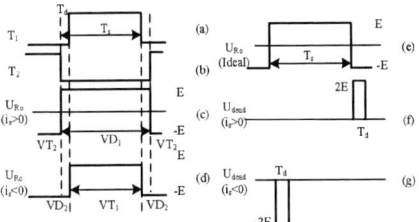

b) The error voltage generated by dead-time injection

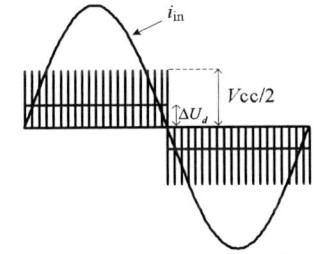

c) The relationship between the input current and the voltage error

Figure 5. Illustration of dead-time error voltage

1723

IV. QUANTITATIVE ANALYSIS

Based on the previous analysis, we can calculate the final value of DC-side voltage, and the agreement between the quantitative results and experimental results will strength the theoretical analysis greatly.

Both the absorbed power and the power loss of the inverter are affected by the DC-link voltage, the final value of DC-link voltage can be calculated by the power balance equation: $P_{in} = P_{loss}$

A. The input power

Assume the peak value of the source voltage is U_{SM}, the load resistor is R, the peak value of input current is I_{CM}, and the peak value of the fundamental component of the equivalent error voltage is U_{CM}. We can obtain:

$$I_{CM} = \frac{U_{SM} - U_{CM}}{R}$$

$$U_{CM} = U_{SM} - I_{CM} \cdot R$$

So, the absorbed power of one phase is:

$$P'_{in} = \frac{1}{2} \cdot I_{CM} \cdot U_{CM} = \frac{1}{2} \cdot I_{CM} \cdot (U_{SM} - I_{CM} \cdot R)$$

B. The power losses

The power losses of an inverter were discussed a lot, so we referred some conclusion directly. [10] The total power losses are made of three parts: forward losses, switching losses and recovery losses. The calculation of each part is shown below.

1) Forward losses

The relationship between the collector-to-emitter voltage (v_{CE}) and the collector current (i_C) can be approximated by the following leaner equation:

$$v_{CE} = \frac{V_{CEN} - V_{CEO}}{I_{CN}} \cdot i_C + V_{CEO}$$

where I_{CN} and V_{CEN} are the rated current and the collector-to-emitter voltage at the rated current.

Assume the current to be sinusoidal:

$$i_C = I_{CM} \cdot \sin \alpha$$

We can calculate the forward losses both in IGBT and diode:

$$
\begin{aligned}
P_{fi} &= \frac{1}{T} \cdot \sum \Delta E_f \\
&= \frac{1}{T} \cdot \int_0^{T/2} v_{CE} \cdot i_C \cdot \frac{T_S - 2T_D}{2T_S} \, dt \\
&= \frac{1}{2\pi} \cdot \int_0^{\pi} v_{CE} \cdot i_C \cdot \frac{T_S - 2T_D}{2T_S} \, d\alpha \\
&= \frac{T_S - 2T_D}{2T_S} (\frac{1}{4} \cdot \frac{V_{CEN} - V_{CEO}}{I_{CN}} \cdot I_{CM}^2 + \frac{1}{\pi} \cdot V_{CEO} \cdot I_{CM})
\end{aligned}
$$

$$P_{fd} = \frac{T_D}{T_S} (\frac{1}{4} \cdot \frac{V_{FN} - V_{FO}}{I_{CN}} \cdot I_{CM}^2 + \frac{1}{\pi} \cdot V_{FO} \cdot I_{CM})$$

2) Switching losses

Switching losses can be divided into two parts: turn-on losses and turn-off losses.

Turn-on losses can be obtained by the following equation:

$$
\begin{aligned}
P_{on} &= \frac{1}{2\pi} \int_0^{\pi} \Delta E_{on} \cdot F_S \, d\alpha \\
&= \frac{1}{8} V_{CC} \cdot t_{rN} \cdot \frac{I_{CM}^2}{I_{CN}} \cdot F_S
\end{aligned}
$$

where t_{rN} is the rated raise time.

Assume the fall time can be approximated by the following equation:

$$t_f \approx (\frac{2}{3} + \frac{1}{3} \cdot \frac{i_c}{I_{cN}}) t_{fN}$$

where t_{fN} is the rated fall time.

Turn-off losses can be calculated by the following equation:

$$
\begin{aligned}
P_{off} &= \frac{1}{2\pi} \int_0^{\pi} \Delta E_{off} \cdot F_S \, d\alpha \\
&= V_{CC} \cdot I_{CM} \cdot t_{fN} \cdot F_S (\frac{1}{3\pi} + \frac{1}{24} \cdot \frac{I_{CM}}{I_{CN}})
\end{aligned}
$$

3) Recovery losses

Recovery losses can be obtained by the following equation:

$$
\begin{aligned}
P_{rr} &= \frac{1}{2\pi} \int_0^{\pi} \Delta E_{rr} \cdot F_S \, d\alpha \\
&= V_{CC} \cdot F_S \{ [0.28 + \frac{0.38}{\pi I_{CN}} I_{CM} + 0.015 (\frac{I_{CM}}{I_{CN}})^2] \cdot \frac{I_{rrN} t_{rrN}}{2} \\
&\quad + I_{CM} \cdot t_{rrN} (\frac{0.8}{\pi} + 0.05 \cdot \frac{I_{CM}}{I_{CN}}) \}
\end{aligned}
$$

where t_{rrN} and I_{rrN} are the rated recovery time and rated recovery current respectively.

From the previous analysis, we can get the power balance equation:

$$3P'_{in} = 6(P_f + P_{on} + P_{off} + P_{rr})$$

Substituting the system parameters to the equations shown above, we can get the final value of the DC-side voltage easily

V. EXPERIMANTAL RESULTS

From the equations above, we can obtain the final value of DC-side voltage. When the correlative parameters are determined (Table. I), the curve in Fig.5.a shows the relationship between the final value of the DC-

TABLE I.
SYSTEM PARAMETERS

F_S	T_s	T_d	I_{cN}	I_{rrN}	t_{rN}	t_{fN}	t_{rrN}
12.8 kHz	78 us	3 us	25 A	5 A	1 us	2 us	0.15 us

side voltage and the peak value of source voltage when the load impedance is 100 Ω. The curve in Fig.7.b shows the relationship between the final value of the DC-side voltage and the load impedance when the peak value of source voltage is 20V. The triangular markers in Fig.8 show the experimental results. When the peak value of source voltage is 20V, 40V and 60V, the final value of DC-side voltage is 100V, 300V and 480V respectively (Fig.6.a). When the load impedance is 20 Ω, 50 Ω and 100 Ω, the final value of DC-side voltage is 195V, 160V and 100V respectively (Fig.6.b). The experimental results agree with the previous two curves very well. It strongly supports the previous analysis.

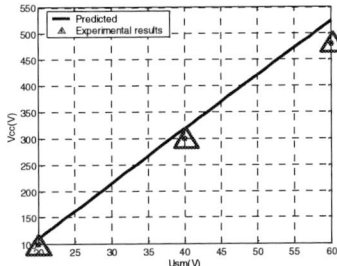

(a) Relationship between U_{sm} and V_{cc}

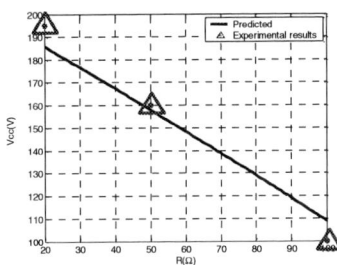

(b) Relationship between R and V_{cc}
Figure 6.   Experimental verification

VI.   CONCLUSIONS

The unexpected boost of the DC side voltage of SPQC systems is analyzed in detail from two view points. The agreement between the quantitative analysis and the experimental results verified the theoretical analysis. Since the existence of dead-time causes the boost of DC-side voltage, there are several effective solutions to this dead-time effect problem:

1)   *Feed forward compensation*
  Calculate the equivalent dead-time voltage error, and subtract it from the output voltage reference.
2)   *DC-side voltage control*
  Control the inverter to export real power to maintain the DC-side voltage.
3)   *Feedback control of output voltage*
  The feedback loop will attenuate the voltage error effectively if the compensator is well designed.

The last two methods are widely used in practical, but in a complex system specially employing full digital controller, the first method has its own advantages-you don't have to design the feedback loop compensator, which is always time-consuming and arduous.

REFERENCES

[1]   F. Z. Peng, H. Akagi, and A. Nabae,   "Compensation characteristics of the combined system of shunt passive and series active filters," IEEE Trans. Industry Applications, Vol. 29, No. 1, pp. 144-152, 1993

[2]   F. Z. Peng, and D. J. Adams, "Harmonic sources and filtering approaches," Proceedings of IEEE IAS Annual Meeting, 1999

[3]   Z. Wang, Q. Wang, W. Yao, and J. Liu, "A series active power filter adopting hybrid control approach," IEEE Trans. Power Electronics, Vol. 16, No. 3, pp. 301-310, 2001

[4]   L. A. Moran, I. Pastorini, J. Dixon, and R. Wallace, "A fault protection scheme for series active power filters," IEEE Trans. Power Electronics, Vol. 14, No. 5, pp. 928-938, 1999

[5]   Woo-Cheol Lee, Taeck-Kie Lee, Chang-Su Ma, and Dong-Seok Hyun, "fault protection scheme for series active compensators," Proceedings of IEEE PESC 2002, pp. 1217-1222, 2002

[6]   Su Chen, G. Joos, L. Lopes, Wennan Guo, "A nonlinear control method of dynamic voltage restorers," Proceedings of IEEE PESC 2002, pp. 23-27, 2002

[7]   R. Li, A. T. Johns, M. M. Elkateb, "Control concept of unified power line conditioner," in Proceedings of IEEE PES Winter Meeting 2000, pp.2594-2599.

[8]   Seung-Gi Jeong; Min-Ho Park, "The analysis and compensation of dead-time effects in PWM inverters", Industrial Electronics, IEEE Transactions on Volume 38, Issue 2, April 1991 Page(s):108 - 114

[9]   Ben-Brahim, L., "The analysis and compensation of dead-time effects in three phase PWM inverters", Industrial Electronics Society, 1998. IECON '98. Proceedings of the 24th Annual Conference of the IEEE,Volume 2, 31 Aug.-4 Sept. 1998 Page(s):792 - 797 vol.2

[10]   Casanellas, F, "Losses in PWM inverters using IGBTs", Electric Power Applications, IEE Proceedings-Volume 141, Issue 5, Sept. 1994 Page(s):235 – 239

Comparative Analysis of Popular Control Schemes for Parallel Active Power Filter and Experimental Verification

Xiaoyu Wang, Jinjun Liu, Chang Yuan, and Zhaoan Wang
School of Electrical Engineering
Xi'an Jiaotong University
28 West Xianning Road, Xi'an, Shaanxi 710049 China
e-mail: xywang@ieee.org

Abstract—**Active Power Filter (APF) is an effective solution to the compensation of power quality problems, especially harmonics. To achieve better performance, the control method is crucial for APF. Currently, there are many control schemes available for APF, which may be classified as two types: one needs harmonics detection from load current, while the other does not need harmonics detection but does need detecting and controlling of the DC-link voltage. This paper presents a study on the popular control schemes comparatively, and leads to the conclusion that the latter type, the control scheme without harmonics detection, is just a reduced version of the broadly used regular load current harmonics detection control, and by nature has the similar essential control principles in the system point of view. The latter type is a simple feedback control, while the regular control scheme with load harmonics detection is a composite control, with both feedback of DC-link voltage and feedforward of the fundamental load current which results in better dynamic performance. This clear disclosure of the internal relationship between the two types of the control schemes would be very helpful to choose the most appropriate control scheme effectively. The above conclusion is illustrated with signal-flow graphs and corresponding analysis, and verified by both time-domain computer simulations and experimental tests on hardware prototypes.**

Keywords- Parallel Active Power Filter

I. INTRODUCTION

Power quality (PQ) problems have been drawing more and more attentions in these years, especially with the development of modern electronics industry and the continuous proliferation of nonlinear type of electric load. To solve these problems, passive power filters were used at the beginning, and the active power filter (APF) is now widely researched and used [1, 2]. In the following, only parallel active power filters (PAPF) is discussed because it is the first and the most popular topology of APF. Besides, much of the analysis and principle of PAPF can be applied to the other topologies. Fig. 1 shows the basic

construction of PAPF, in which i_S is the source current, i_L is the load current, and i_C is the compensation current.

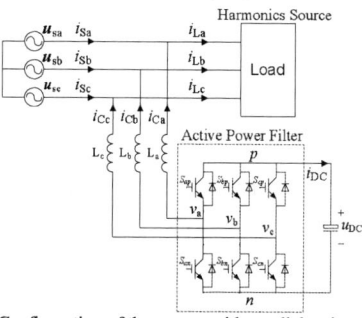

Figure 1. Configuration of the system with parallel active power filter.

APF is a quite effective solution to enhance the power quality, yet the performance and compensation results depend much on the control schemes, which is very important and a hot research spot since the beginning. Currently, there are many control schemes available for APF, which may be classified as two types: one needs harmonics detection from load current, while the other does not need harmonics detection but does need detecting and controlling of the DC-link voltage. Fig. 2 shows the categories of popular control schemes of PAPF.

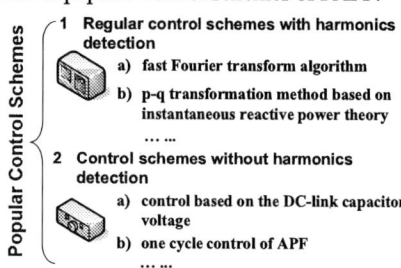

Figure 2. Categories of popular control schemes of PAPF.

Regular control schemes need harmonics detection, in which the PAPF acts as a controlled harmonics current source, injecting current which is inverse equivalent to the load harmonics in order to prevent the load from drawing harmonic current from the source. How to detect the harmonics from the load current quickly and exactly is a key issue. Lots of harmonics detection algorithms were proposed such as fast Fourier transform algorithm (FFT), *p-q* transformation method based on instantaneous reactive power theory. They are affective, but need much calculation, which usually implemented with digital signal processor (DSP), hence the cost of the control system is much high. In late 1990s, some simpler control schemes which don't need harmonics detection are proposed and draw much attention [3-6]. Many advantages were reported such as implementation simplicity at low cost and comparable compensation characteristics with the regular control scheme. Among these control methods, the control based on the DC-link capacitor voltage and the one-cycle control are two typical representatives [7-9].

In our practice of power quality control with active power filters, the authors feel puzzled to choose the best one from so many control schemes, which motivates us to research on the internal relationship and essential compensating mechanism of them. What's the difference between the two main kinds, and what are their internal relationships? Focusing on the control variable and signal flow, this paper presents a study on the popular control schemes comparatively, and leads to the conclusion that the latter type, the control scheme without harmonics detection, is just a reduced version of the broadly used regular load current harmonics detection control, and by nature has the similar essential control principles in the system point of view. The latter type is a simple feedback control, while the regular control scheme with load harmonics detection is a composite control, with both feedback of DC-link voltage and feedforward of the fundamental load current which results in better dynamic performance. The above conclusion is illustrated with signal-flow graphs and corresponding analysis, and verified by both time-domain computer simulations and experimental tests on hardware prototypes.

II. REVIEW OF POPULAR CONTROL METHODS OF APFs

A. Regular Control Scheme with Harmonics Detection

The most popular control scheme of PAPF is based on harmonics detection-and-compensation strategy. Fig. 3a) shows the equivalent circuit of the system and Fig. 3b) shows its corresponding control scheme. The controller has to detect the load current i_L and extract its harmonics out with a high performance high pass filter (HPF) as the current reference to PAPF. Following this reference, the PAPF is controlled as a harmonic current source. A feedback loop of DC-link capacitor voltage is added to the reference as a fundamental part based on the principle of power balance to maintain the DC-link voltage of the power converter constant.

In regular control, the performance of the low pass filter (LPF) is very important, and there are currently lots of algorithm proposed, such as fast Fourier transform algorithm (FFT), *p-q* method based on instantaneous reactive power theory, etc. Most of them need large

computation workload, and commonly implemented with high-cost DSP. Fig. 3c) shows one commonly used algorithm of harmonics extraction based on instantaneous reactive power theory or *p-q* transformation, which is very popular in industry application due to its quick dynamic characteristics and simple implement.

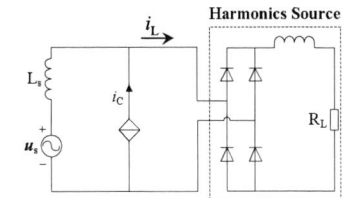

a) Equivalent circuit of system with PAPF.

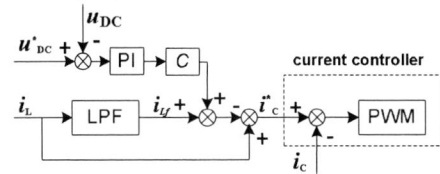

b) Regular control scheme.

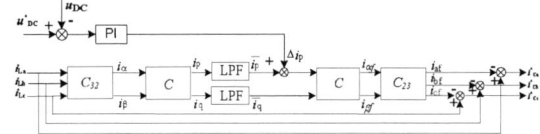

c) Harmonics extraction algorithm in regular control scheme based on instantaneous reactive power theory or *p-q* transformation

Figure 3. Regular control scheme of PAPF.

B. Control Scheme Based on DC Capacitor Voltage

In late 1990s, a kind of simpler control scheme was proposed with many of attractive advantages proclaimed such as:

1) This control method requires minimum measurement with only source currents and DC-link capacitor voltage.
2) The compensation is achieved without harmonics extraction, hence simplifying the control and reducing the cost.
3) The control logic and the associated hardware are simple, thereby enhancing the system reliability.

Fig. 4 shows the basic principle of this kind of control scheme.

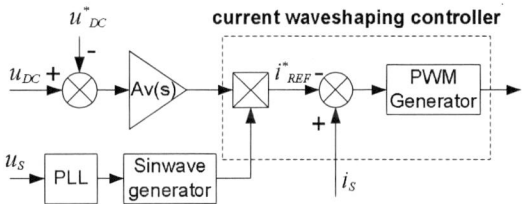

Figure 4. Control schemes of PAPF based on DC capacitor voltage.

Unlike regular controls, the reference here comes directly from the DC-link capacitor voltage, which contains the information of the real power flow, thus the amplitude control of the source current can be obtained by using a DC side voltage regulation circuit [3-6]. To maintain a constant DC-link capacitor voltage, the PAPF must not consume the average power from the source; hence in steady state the current of PAPF is purely harmonics which could exactly compensate the harmonics part of load current.

The well-marked advantages for this kind of control schemes based on the DC-link capacitor voltage over the regular control are:

1) It is not necessary to calculate the harmonics of the load, which is much simpler and hence more reliable than the regular control method.

2) It detects source current instead of both load current and injected current i_C, which is much less than the regular control method.

3) The calculation result in this method is the reference of source current, while it is the reference of converter output current or compensating current in regular methods

C. Control of PAPF with One Cycle Control

To control the PAPF based on the idea of one cycle control is one of the popular control schemes without harmonics detection, which has drawn much attention now. It is a nonlinear method for switch circuit control, which was first put forward in 1990s, and applied into the control of APF from the year of 2000 [7-9].

Fig. 5 illustrates the basic control scheme of one-cycle control of PAPF. It contains a PI controller, an integrator with a reset switch, a comparator, and a flip-flop. The voltage of DC-link capacitor is sensed and its error is fed to a controller of Av(s) which usually contains a PI controller. This reference leads the source current i_S, which is sensed with a resistor R_S at the source side, to sinusoidal waveform to achieve the goals of active power filter.

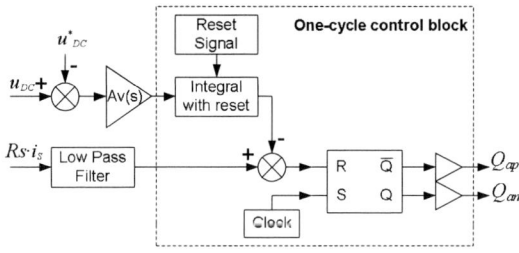

Figure 5. Control scheme of PAPF with one cycle control.

In this control scheme, a nonlinear technique of one-cycle control is used to generate PWM signals instead of regular PWM techniques, which makes it quite different at first appearance. Actually it has the same control loop and hence essential principle with previously discussed control scheme without harmonics extraction based on DC-link capacitor voltage.

Compared to the regular control method, one cycle control has the similar characteristics as the previous control without harmonics detection, such as no need to sense the load current, no complicated computation for generating harmonics and reactive current reference, and hence higher reliability and simpler implementation [7-9].

III. COMPARATIVE ANALYSIS OF CONTROL ALGORITHMS

Although the control methods mentioned above are proposed for a long time, each control scheme is reported separately and focuses only on the advantages of its own features. Most acclaimed advantages are not proved or verified strictly. The authors felt puzzled to select the best from so many control schemes, which motivates us to research on the internal relationship and essential compensating mechanism of them. After lots of detailed analysis and derivation, we found that all these omnifarious control schemes have similar control loops, and thus similar control principle in the view of system level.

To illustrate the author's conclusion and simplify the complicated analysis, the signal-flow graphs are found very useful because it gives a simplified schematic overview of the system, which is helpful to prevent us from falling in the detail discussion of each control method.

A. Analysis of Regular Control Scheme with Signal-Flow Graph

Signal-flow graph is an alternative schematic method to represent the flow and processing of signals in system much clearer than block diagram. Fig. 6 a) shows the signal-flow graph of regular control method, compare to Fig. 3 b) in block diagram.

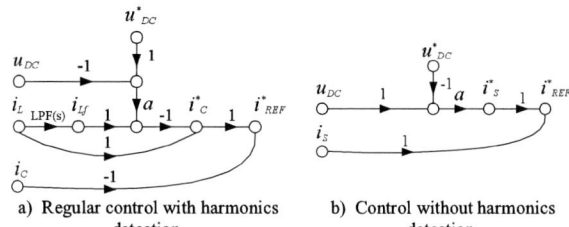

a) Regular control with harmonics detection.

b) Control without harmonics detection.

Figure 6. Signal flow graphs of the main popular control methods.

The signal-flow graph is composed of the input nodes i_L (load current), i_C (compensate current), u_{DC} (DC-link capacitor voltage), and the output node i^*_{REF} (reference of compensation current), properly linked with the branch gain LFP (low pass filter), and a (the composite of the PI compensator and the transformation from DC to AC synchronizing with source voltage).

In a similar way, Fig. 6 b) shows the signal-flow graph of the control method without harmonics detection. It can be easily seen that this control method need less computation, and is much simple to implement.

B. Evolutional Reduction of Regular Control Method

To find out the internal relationship between the regular control and those without harmonics detection, the signal-flow graph can be reduced by signal-flow graph reduction techniques to simplified schematics.

By splitting the input node of i_L, Fig. 6 a) can be reduced to Fig. 7 a). Considering the fact that

$$i_S = i_L - i_C \tag{1}$$

a simplified signal-flow graph of the regular can be get as Fig. 7 b) with totally information of Fig. 6 a).

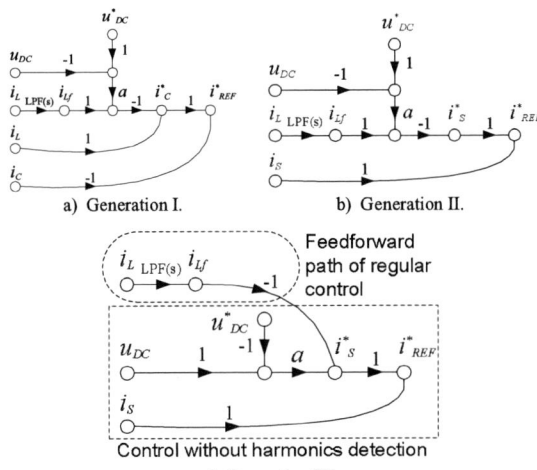

a) Generation I.   b) Generation II.

c) Generation III.

Figure 7. Signal flow graphs revolution of the regular control method.

C. Internal Relationship between the Popular Control Schemes

Fig. 7 b) can be represented as Fig. 7 c), which is much similar to Fig. 6 b). By comparing the signal-flow graphs of Fig. 7 c) and Fig. 6 b), the essential relationships between the control methods are revealed as the following:

1) The control scheme without harmonics detection is just a reduced version of the broadly used regular load current harmonics detection control, and by nature has the similar essential control principles in the system point of view. It is a simple feedback control, while the regular control scheme with load harmonics detection is a composite control, with both feedback of DC-link voltage and feedforward of the fundamental load current which results in better dynamic performance.

2) The control without harmonics detection is simple and easy to implement, hence is suitable with low cost application. While the regular control is more complicated, and should be chosen when dynamic performance is concerned more.

3) One cycle control is proposed in a fresh view and seems quite different from normal control schemes, yet it is actually one kind of implementation of the control scheme without harmonics detection. Based on similar main control loop, one cycle control generates PWM signal with nonlinear modulation method to achieve a better dynamic response.

4) Since this relationship is not unique to PAPF, it is possible for the control scheme without harmonics detection to spread to the other topologies and applications.

IV. SIMULATION VERIFICATION

Plenty of time-domain computer simulations were held comparatively to verify the pervious analysis and conclusion. Fig. 8 illustrates some simulation results of three-phase PAPF controlled with different kinds of schemes. When there is any disturbance at the load, regular control scheme with harmonics detection has a better dynamic response than those without harmonics detection.

V. EXPERIMENTAL VERIFICATION

Experiments were carried out on hardware prototypes to verify above analysis and conclusion. Fig. 9 shows the results of PAPF controlled with regular method, while Fig. 10 shows those without harmonics detection.

Obviously, in steady state, both methods are effective, but the regular control has a faster start up. When there is any disturbance at the load, the regular method can sustain the DC-link voltage perfectly, while the control without harmonics detection is not so good.

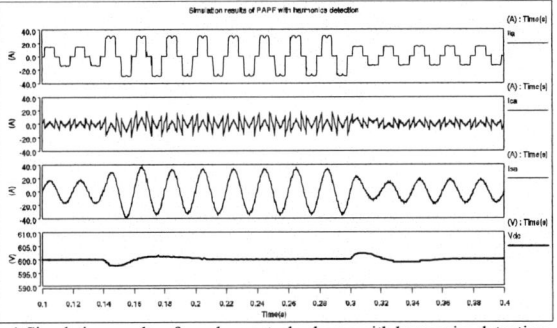

a) Simulation results of regular control scheme with harmonics detection

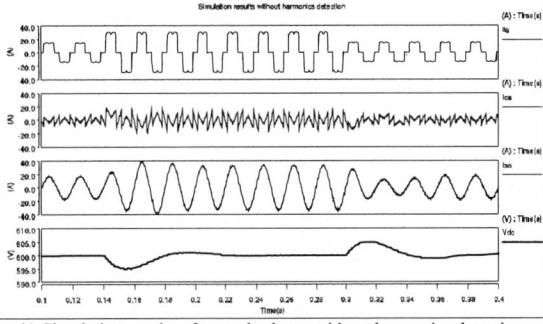

b) Simulation results of control scheme without harmonics detection

(1) i_L: Load current; (2) i_C: Compensated current; (3) i_S: Source current; (4) u_{DC}: DC voltage.

Figure 8. Simulation results of three-phase PAPF controlled with different kind of schemes.

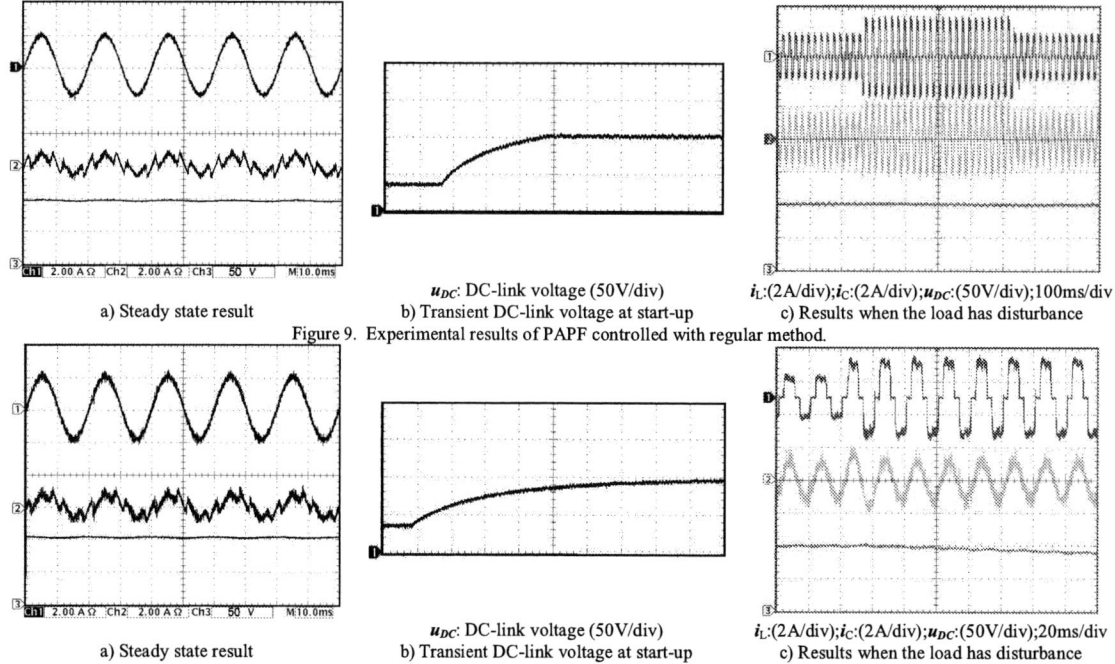

| a) Steady state result | u_{DC}: DC-link voltage (50V/div)
b) Transient DC-link voltage at start-up | i_L:(2A/div);i_C:(2A/div);u_{DC}:(50V/div);100ms/div
c) Results when the load has disturbance |

Figure 9. Experimental results of PAPF controlled with regular method.

| a) Steady state result | u_{DC}: DC-link voltage (50V/div)
b) Transient DC-link voltage at start-up | i_L:(2A/div);i_C:(2A/div);u_{DC}:(50V/div);20ms/div
c) Results when the load has disturbance |

Figure 10. Experimental results of PAPF controlled without harmonics detection.

VI. CONCLUSION

Active Power Filter (APF) is an effective solution to the compensation of power quality problems, especially harmonics. To achieve better performance, the control method is crucial for APF. Currently, there are many control schemes available for APF, which may be classified as two types: one needs harmonics detection from load current, while the other does not need harmonics detection but does need detecting and controlling of the DC-link voltage. This paper presents a study on the popular control schemes comparatively, and leads to the conclusion that the latter type, the control scheme without harmonics detection, is just a reduced version of the broadly used regular load current harmonics detection control, and by nature has the similar essential control principles in the system point of view. The latter type is a simple feedback control, while the regular control scheme with load harmonics detection is a composite control, with both feedback of DC-link voltage and feedforward of the fundamental load current which results in better dynamic performance. Although one cycle control is proposed in a fresh view and seems quite different, it is actually one kind of implementation of the control schemes without harmonics detection. Based on similar main control loop, one cycle control generates PWM signal with nonlinear modulation method to achieve a better dynamic response.

This clear disclosure of the internal relationship between the two types of the control schemes would be very helpful to choose the most appropriate control scheme effectively. What's more, different recombination of each part in the control schemes may improve the compensation characteristics, since these controls may differ in some implementation. The above conclusion is illustrated with signal-flow graphs and corresponding analysis, and verified by both time-domain computer simulations and experimental tests on hardware prototypes.

REFERENCES

[1] Akagi, H., "New trends in active filters for power conditioning," *Industry Applications, IEEE Transactions on* , vol.32, no.6pp.1312-1322, Nov/Dec 1996.

[2] Akagi, H., "Active harmonic filters," *Proceedings of the IEEE* , vol.93, no.12pp. 2128- 2141, Dec. 2005.

[3] Wu, J.-C.; Jou, H.-L., "Simplified control method for the single-phase active power filter ," *Electric Power Applications, IEE Proceedings-* , vol.143, no.3pp.219-224, May 1996.

[4] Casadei, D.; Grandi, G.; Reggiani, U.; Rossi, C., "Control methods for active power filters with minimum measurement requirements," *Applied Power Electronics Conference and Exposition, 1999. APEC '99. Fourteenth Annual* , vol.2, no.pp.1153-1158 vol.2, 14-18 Mar 1999.

[5] K. Chatterjee, B. G. Fernandes, and G. K. Dubey. "An instantaneous reactive volt-ampere compensator and harmonic suppressor system". *Power Electronics, IEEE Transactions on*, 14 (2):381-392, 1999.

[6] S. K. Jain, P. Agarwal, and H. O. Gupta. "A control algorithm for compensation of customer-generated harmonics and reactive power". *Power Delivery, IEEE Transactions on* 19:357-366, 2004.

[7] Qiao, C.; Smedley, K.M., "Three-phase active power filters with unified constant-frequency integration control," *Power Electronics and Motion Control Conference, 2000. Proceedings. PIEMC 2000. The Third International* , vol.2, no.pp.698-705 vol.2, 2000.

[8] Zhou, L.; Smedley, K.M., "Unified constant-frequency integration control of active power filters," *Applied Power Electronics Conference and Exposition, 2000. APEC 2000. Fifteenth Annual IEEE* , vol.1, no.pp.406-412 vol.1, 2000

[9] K. M. Smedley, L. Zhou, and C. Qiao, "Unified constant-frequency integration control of active power filters-steady-state and dynamics" *Power Electronics, IEEE Transactions on*, vol. 16, no. 3, pp. 428-436, 2001.

2006 5th International Power Electronics and Motion Control Conference

Accurate Modeling of the Three Phase Induction Motor Including Saturation Effects

E. V. N. Souza[*] and S. R. Naidu[**]

[*] Centro Federal de Educação Tecnológica de Alagoas/Industry Department, Palmeira dos Índios, AL, Brazil
[**] Universidade Federal de Campina Grande/ Elecrical Engineering Department, Campina Grande, PB, Brazil
vidal@dee.ufcg.edu.br, naidu@dee.ufcg.edu.br

Abstract—**A model has been proposed for including the effects of saturation in the simulation of the three phase induction motor with delta-connected stator windings. The model combines the well known model for the machine in a rotating reference frame attached to the air-gap flux with magnetic harmonic functions, obtained experimentally, that describe the nonlinear behavior. The model also takes into account the spatial third harmonic flux distribution which arises from saturation in the main flux path. The proposed model has been validated by comparison with measurements.**

Keywords-**air-gap flux; harmonic function; third harmonic; saturation.**

NOMENCLATURE

v_{sd}, v_{sq}	instantaneous source voltages applied to the dq-axes equivalent circuits.
i_{sd}, i_{rd}	d-axis stator and rotor currents.
i_{s3}, i_{r3}	third harmonic current induced in the stator and rotor windings, respectively.
i_{md}	d-axis magnetizing current.
$\lambda_{ms3}, \lambda_{mr3}$	spatial third harmonic air-gap flux linking a stator and a rotor winding.
λ_{md}	d-axis magnetizing linkage flux.
ω_r	rotor angular velocity, in electrical rad/s.
r_s, r_r	resistances of stator and rotor phase windings.
R_m	iron losses equivalent resistance.
P, F_c, J	number of pole pairs, friction coefficient and moment of inertia, respectively.

Superscript '*f*' refers to variables in a rotating reference frame attached to the air-gap flux.
Superscript '*c*' refers to variables in a rotating reference frame attached to the stator current.

I. INTRODUCTION

Magnetic saturation in the main flux path affects the air-gap flux in two ways. In the first instance, the peak value of the spatial flux distribution is no more linearly related to the magnetizing current, even though this flux distribution could remain sinusoidal around the rotor. This effect is referred to as the saturation of the spatial fundamental component. A second effect of magnetic saturation is the distortion in the spatial flux distribution, which becomes flattened and deviates from the sinusoidal waveform. The distortion is predominantly due to the appearance of the third harmonic component in the spatial flux distribution. Incorporation of the spatial harmonic content of the air-gap flux into the machine's equations is not simple.

Several models have been proposed that have presented magnetic saturation by nonlinear magnetizing inductances in the dq-axes equivalent circuits [1,2]. Alignment of the d-axis with the main flux vector has been proposed in [2]. In this model, the alignment has been obtained by closed loop control. In another model, saturation in the leakage flux has also been included besides the saturation in the spatial fundamental component of the air-gap flux [3]. A small signal analysis that uses two magnetizing inductances in a two-axes model has been presented to study the dynamic stability of an induction motor, as in [4]. The models mentioned above are all based on the two-axes formulation and do not take into account the distortion in the spatial distribution of the air-gap flux.

More recently, attention has been given to the spatial harmonic distortion of the air-gap flux. Ref. [5] presents a dq-axes induction motor model that takes the distortion into account empirically by means of a variable air-gap length, and requires detailed knowledge of the machine design parameters. Modeling of the induction machine directly in the phase domain has been proposed in [6]. Saturation is taken into account by means of "magnetic harmonic functions" that are obtained experimentally by measuring the voltages and currents at the machine terminals during the no-load test. The phase domain model is well established and experimentally validated.

Modeling of the induction machine in the dq-coordinates, that takes into account the spatial harmonics in the air-gap flux distribution due to saturation effects, is lacking in the literature. This paper proposes a dq-axes model for the induction motor that includes saturation in the main flux path. The spatial fundamental and third harmonic components of the air-gap flux distribution have been incorporated in the model. The model utilizes

1-4244-0448-7/06/$25.00 ©2006 IEEE

magnetic harmonic functions" similar to those proposed in [6], and also aligns the d-axis with the air-gap flux vector. Saturation of the leakage flux path has not been considered. Experimental and simulated results for a delta-connected, three-phase, induction motor are presented to validate the proposed model.

II. PROPOSED MODEL

A. Saturation in the Spatial Fundamental Component of the Air-gap Flux

We will first consider only one of the effects of saturation, i.e., saturation of the spatial fundamental component of the air-gap flux distribution. Inclusion of the spatial third harmonic air-gap flux distribution will be considered later. The proposed model uses a rotating reference frame attached to the air-gap flux, by aligning the d-axis of the rotating frame with this flux. The angular velocity ω_f of the reference frame is an unknown variable to be determined at each time step. Since the d-axis is aligned with the air-gap flux, the q-axis air-gap flux component, λ_{mq}^f, equals to zero. It results that the q-axis voltage component is given by equation (1).

$$v_{mq}^f = d\lambda_{mq}^f \big/ dt = 0 , \qquad (1)$$

It follows that $i_{rq}^f = -i_{sq}^f$. The dq-axes circuits are shown in Fig. 1. The equations for the electrical variables are:

$$v_{sd}^f - r_s i_{sd}^f - L_{ls} di_{sd}^f \big/ dt + \omega_f L_{ls} i_{sq}^f - v_{md}^f = -\delta v_{sd}^f = 0 \quad (3)$$

$$v_{md}^f + r_r i_{rd}^f + L_{lr} di_{rd}^f \big/ dt + (\omega_f - \omega_r)L_{lr} i_{sq}^f = \delta v_{rd}^f = 0 \quad (4)$$

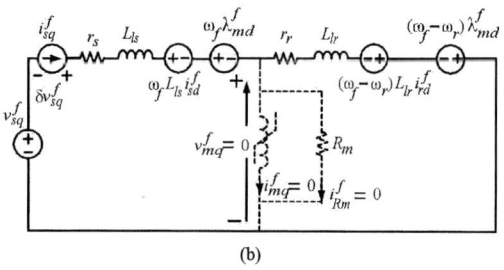

(a)

(b)

Figure 1. Equivalent circuits attached to the air-gap flux: (a) d-axis circuit: q-axis circuit.

$$i_{sd}^f + i_{rd}^f - v_{md}^f \big/ R_m - i_{md}^f = \delta i_{md}^f = 0 \qquad (5)$$

$$i_{md}^f = F_1(\lambda_{md}^f) \qquad (6)$$

$$v_{md}^f = d\lambda_{md}^f \big/ dt \qquad (7)$$

$$v_{sq}^f - r_s i_{sq}^f - L_{ls}\frac{di_{sq}^f}{dt} - \omega_f L_{ls} i_{sd}^f - \omega_f \lambda_{md}^f = v_{mq}^f = 0 \quad (8)$$

$$v_{sq}^f - (r_s + r_r)i_{sq}^f - (L_{ls} + L_{lr})di_{sq}^f \big/ dt$$
$$+ (\omega_f - \omega_r)L_{lr} i_{rd}^f - \omega_f L_{ls} i_{sd}^f - \omega_r \lambda_{md}^f = -\delta v_{sq}^f = 0$$

(9)

Equations (3) to (9) are solved iteratively in each time step for the electrical variables $i_{sd}^f, i_{rd}^f, i_{sq}^f$ and λ_{md}^f. A technique for solving the nonlinear equations is described in [7]. The phase domain components of any variable are obtained by applying the transformation from the dq-axis to the phase-domain. The function $F_1(\lambda_{md}^f)$ in equation (6) is the spatial fundamental magnetic function, and it is obtained experimentally for the machine under consideration. The angular velocity of the rotating frame is computed using equation (8) at each time step, and this computed value is used in the next time step. Finally, the fundamental component of the electromagnetic torque is written as

$$T_{el} = P\lambda_{md}^f i_{sq}^f \qquad (10)$$

Up to this point, the model is equivalent to that described in [2].

B. Inclusion of the Spatial Third Harmonic Flux Distribution Effects

When the machine enters deeper into magnetic saturation, the spatial distribution of the air-gap flux is no longer sinusoidal. The flattening of the flux distribution may be described by the appearance of the spatial third harmonic component in this flux distribution. The spatial third harmonic flux linking each of the stator windings is identical and it is given by

$$\lambda_{ms3}(t) = -F_3(\lambda_{md}^f)\cos(3\omega_f t) \qquad (11)$$

The function $F_3(\lambda_{md}^f)$ is the spatial third harmonic magnetic function, and it is yet another function which is determined experimentally for the given machine. The spatial third harmonic flux linking each rotor winding is given by:

$$\lambda_{mr3}(t) = -F_3(\lambda_{md}^f)\cos[3(\omega_f - \omega_r)t] \qquad (12)$$

Equations (11) and (12) show that the spatial third harmonic flux is a zero-sequence flux for the machine windings. At this point, it becomes necessary to take into

1732

account the type of stator winding connection in order to compute the effects of the third harmonic flux linkages.

Therefore, for a delta-connected stator winding, the spatial third harmonic air-gap flux will induce circulating currents in the rotor as well as in the stator. The third harmonic current induced in each stator winding is computed by numerical integration of the following equation:

$$r_s i_{s3}(t) + L_{ls}\, di_{s3}(t)/dt = d\lambda_{ms3}(t)/dt \qquad (13)$$

This current is added to each current of the stator phases obtained in section *II.A*.

The current induced in each rotor phase by the spatial third harmonic air-gap flux satisfies equation (14), i.e,

$$r_r i_{r3}(t) + L_{lr}\, di_{r3}(t)/dt = d\lambda_{mr3}(t)/dt \qquad (14)$$

Equation (14) is solved numerically for the current $i_{r3}(t)$. This current is added to each of the rotor phase currents that have been obtained using the procedure described in section *II.A*. The third harmonic component of the electromagnetic torque may be written as [8]:

$$T_{e3}(t) = 3P[\lambda_{mr3}(t)3i_{r3}(t)] \qquad (15)$$

Having obtained all the electrical variables, and assuming that the load torque T_m is known, the mechanical equations (16), (17) and (18) are used for numerically determining the angular velocity ω_r. The velocities ω_{r1} and ω_{r3} are the fundamental and third harmonic components of the angular velocity ω_r.

$$(J/P)\, d\omega_{r1}(t)/dt + F_c/P\,\omega_{r1}(t) = T_{e1} - T_m \qquad (16)$$

$$(J/P)\, d\omega_{r3}(t)/dt + F_c/P\,\omega_{r3}(t) = T_{e3}(t) \qquad (17)$$

$$\omega_r(t) = \omega_{r1}(t) + \omega_{r3}(t) \qquad (18)$$

Finally, the angular velocity ω_f of the rotating reference frame is updated using equation (8) and, together with the rotor velocity ω_r given by equation (18), it is used in the computations of the next time step.

C. Computational algorithm

The computational procedure that emerges for simulating the saturated induction machine is outlined below:

1. Using ω_f and ω_r of the previous time step, solve equations (3), (4), (5), (6), (7) and (9) iteratively for the variables i_{sd}^f, i_{rd}^f, i_{sq}^f and λ_{md}^f. The fundamental magnetic function $F_1(\lambda_{md}^f)$ is required in this step.

2. Obtain the peak value of the third harmonic air-gap flux distribution using the third harmonic magnetic function $F_3(\lambda_{md}^f)$.

3. Compute the electromagnetic torques and solve the mechanical equations for ω_{r1} and ω_{r3}.

4. Update ω_f and ω_r. Proceed to the next time step.

III. HARMONIC MAGNETIC FUNCTIONS MEASUREMENTS

The harmonic magnetic functions are obtained from the no-load test performed on the machine and also from its model. Fig. 2 shows the test circuit for the delta-connected stator. The induction machine is driven at synchronous speed by a DC motor coupled to its shaft. The stator windings are then fed with balanced, sinusoidal voltages from a three-phase voltage source with controlled output voltages.

The instantaneous values of the voltages and currents shown in Fig. 2 are measured and stored in the memory of a data-acquisition system (DAQ). Since the induction machine is being driven at synchronous speed, currents are not induced in the rotor. Moreover, all sinusoidal variables are reduced to constant quantities in a synchronously rotating reference frame. Fig. 3 shows the equivalent circuits when the rotating frame is attached to the stator current.

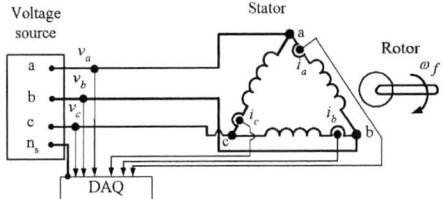

Figure 2. No-load test for determining the magnetic harmonic function of a delta-connected motor.

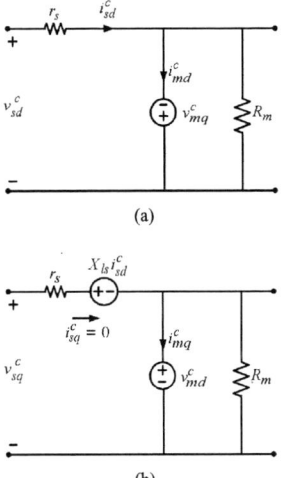

Figure 3. Equivalent circuits in a rotating reference frame attached to the stator current: (a) d-axis circuit; (b) q-axis circuit.

The governing equations for the flux linkage are:

$$v_{md}^c = \omega_f \lambda_{sd}^c = v_{sq}^c - X_{ls} i_{sd}^c \qquad (19)$$

$$v_{mq}^c = \omega_f \lambda_{sq}^c = -(v_{sd}^c - r_s i_{sd}^c) \qquad (20)$$

$$i_{md}^c = i_{sd}^c - v_{mq}^c / R_m \qquad (21)$$

From the above equations, the amplitude of the rotating flux linkage vector is

$$\lambda_{md}^f = \sqrt{(\lambda_{sd}^c)^2 + (\lambda_{sq}^c)^2} , \qquad (22)$$

and the current i_{md}^f aligned with the flux linkage vector is given by

$$i_{md}^f = \frac{\lambda_{sd}^c}{\lambda_{md}^f} i_{md}^c + \frac{\lambda_{sq}^c}{\lambda_{md}^f} i_{mq}^c \qquad (23)$$

$$i_{md}^f = F_1(\lambda_{md}^f) \qquad (24)$$

Let λ_{m3p} be the peak value of the spatial third harmonic flux linking each of the stator windings. Then, the function $\lambda_{m3p} = F_3(\lambda_{md})$ is the third harmonic magnetic function for the machine under consideration. This function is shown in Fig. 5. The fundamental magnetic function obtained from the no-load test on the delta-connected machine is shown in Fig. 4.

The third harmonic magnetic function for the delta-connected induction motor is obtained by monitoring the third harmonic component of the zero-sequence current circulating in the closed delta-connected stator windings. This current is limited only by the leakage reactances and resistances of the stator windings.

At high levels of saturation, the circulating current leads to saturation of the leakage reactances.

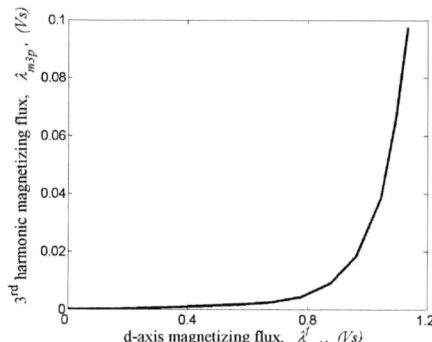

Figure 5. Third harmonic magnetic harmonic function.

Therefore, the third harmonic magnetic function obtained from tests on the delta-connected machine is different from the curve which can be obtained for the star-connection. The study of leakage reactance saturation falls outside the scope of this paper. In addition, the procedure for obtaining the "magnetic harmonic functions" for the star-connected induction motor is the subject of another paper under construction.

IV. EXPERIMENTAL VALIDATION

In the laboratory, a 1.5 hp, three-phase induction motor was tested with the stator windings in delta connection. The motor parameters will be given in an Appendix. The voltage applied to the machine terminals was 120% of rated voltage. With the unloaded motor running in the steady state, a load torque was applied to the motor. The motor stator phase currents and voltages were measured and stored by the data-acquisition system. Figs. 6 and 7 show the measured stator phase current for the unloaded and loaded steady state, respectively. The simulated results using the proposed model are also shown in these figures, and a satisfactory agreement with the measured results is observed. The simulated rotor phase currents shown in Fig. 8 are flattened due to the spatial third harmonic component of the air-gap flux originated by the magnetic saturation.

The larger deviations between the measured and simulated curves are attributed to saturation of the leakage reactances.

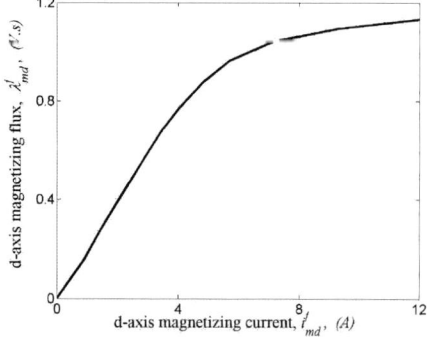

Figure 4. Fundamental magnetic harmonic function.

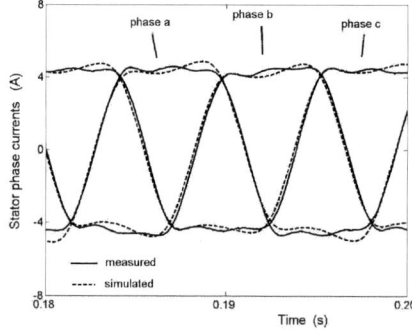

Figure 6. Model validation for the unloaded, delta-connected motor.

1734

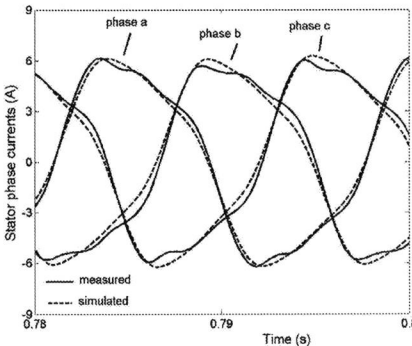

Figure 7. Model validation for loaded, delta-connected motor.

The proposed model includes the effects of the third harmonic in the spatial air-gap flux distribution. Other triple harmonics, such as 9th, 15th and 21st have not been included. The model also assumes a symmetrical machine with smooth rotor and stator surfaces.

V. CONCLUSIONS

A model has been proposed for simulating the three phase delta-connected induction motor, including the effects of saturation. The model takes into account the spatial third harmonic flux distribution that arises from saturation in the main flux path. The proposed model combines the model for the machine in a rotating reference frame attached to the air-gap flux and magnetic harmonic functions that describe the nonlinear behaviour.

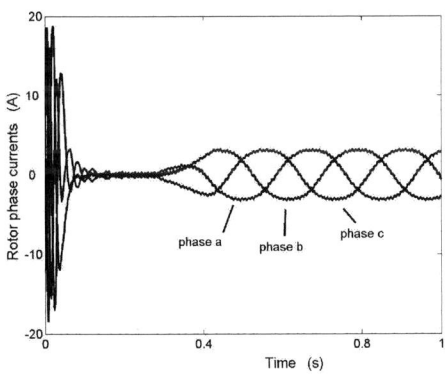

Figure 8. Simulated rotor phase currents for the delta-connected motor.

Experimental determination of the magnetic harmonic functions has been described.

The proposed model has been validated by measurements on a 1.5 hp motor. Satisfactory agreement between simulation and measurements has been observed. The model does not include saturation of the leakage reactances.

ACKNOWLEDGMENT

The authors thank Brazil's CNPq (Conselho Nacional de Pesquisa e Desenvolvimento) for its continued support.

APPENDIX

The studied three-phase, delta-connected induction motor data are given as they are shown below.

- *Nameplate data*:

Three-phase 1.5 HP squirrel cage induction motor, 220/380V (Δ/Y), 6.5/3.8A (Δ/Y), 60 Hz, 860 rpm;

- *Mechanical parameters*:

$P = 4$, $J = 0.0116$ Kg.m^2, $F_C = 0$;

- *Electrical parameters*:

r_s=6 Ω, r_r=4.2 Ω, L_{ls}= L_{lr} = 0.023 H, R_m = 0.023 H

REFERENCES

[1] J. E., Brown, K. P. Kovacs and P. Vas, "A Method of Including the effects of Main Flux Path Saturation in the Generalized Equations of A.C. Machines," *IEEE Transactions on Power Apparatus and Systems*, vol. PAS-102, no. 1, pp. 96–103, January 1983.

[2] Y. He and T. A. Lipo, "Computer Simulation of an Induction Machine with Spatially Dependent Saturation," *IEEE Transac. on Power Apparatus and Systems*, vol. PAS-103, no. 4, pp. 707–714, April 1984.

[3] T. A. Lipo and A. Consoli, "Modelling and Simulation of Induction Motors With Saturable Leakage Reactances," *IEEE Transactions on Industry Applications*, vol. IA-20, no. 1, pp. 180-189, January/February 1984.

[4] J. A. A. Melkebeek, "Magnetizing-Field Saturation and Dynamic Behavior of Induction Machines. Part 1: Improved Calculation Method for Induction Machine Dynamics," *IEE Proceedings, vol. 130 - part B*, no. 1, pp. 1-9, January 1983.

[5] J. C. Moreira and T. A. Lipo, "Modeling of Saturated A.C. Machines Including Air Gap Flux Harmonic Components", *IEEE Transactions on Industry Applications*, vol. 28, no. 2, pp. 343-349, March/April 1992.

[6] D. Bispo, L. M. Neto, J. T. Resende and D. A. Andrade, "A New Strategy for Induction Machine Modeling Taking Into Account the Magnetic Saturation", *IEEE Transactions on Industry Applications*, vol. 37, no. 6, pp. 1710-1719, november/december 2001.

[7] S. R. Naidu and A. M. N. Lima, "A New Approach for the Simulation of Power Electronic Circuits", *IEEE Transactions on Circuits and Systems-I: Fundamental Theory and Applications*, vol. 49, no. 9, pp. 1317-1324, September, 2002.

[8] P. Vas, *Electrical Machines and Drives - A Space Vector Theory Approach*, New York, USA, Oxford University Press Inc, 1992.

A study on the reliability evaluation of driving parts for note handling units

Joo Han Kim*, Jung Kee Chung, Ha Kyeong Sung, Se Hyun Rhyu

Korea Electronics Technology Institute (KETI)

203-101 B/D 192 Yakdae-Dong, Wonmi-Gu Puchon-Si, Kyunggi-Do, 420-140, Korea

kimjh@keti.re.kr

Abstract— ATM is element that reliability, stability is important in relation with customer by financial agency used device. To embody high reliability and stability, reliability estimation technology development is important. When new product is developed, performance and reliability evaluation of product are essential element. In this paper, is treating contents on reliability estimation of stepping motor, BLDC motor and solenoid that is main driving source of note handling units.

Keywords-component; *Automatic Teller Machine, BLDC motor, Stepping Motor, Solenoid, Note Handling Units*

I. Introduction

Importance of automation device (ATM) in banking service is increased gradually by automation request and integration banking service of banking service by financial industry competitive power strengthening. Therefore, point technology of each part materials that make core of banking automation device is raised on main constituent of financial industry competitive power. Note handing units are ATM(automatic teller machine) point parts that achieve Bank's function of depositing, paying, transferring, returning current, saving etc.

ATM is element that reliability, stability is important in relation with customer by financial agency used device. To embody high reliability and stability, reliability estimation technology development is important. When new product is developed, performance and reliability evaluation of product are essential element. In this paper, is treating contents on reliability estimation of stepping motor, BLDC motor and solenoid that is main driving

source of note handling units. Stepping motor was measured accuracy of motor. Velocity fluctuating rate was measured in BLDC motor, Achieved study of static characteristics and analysis in solenoid.

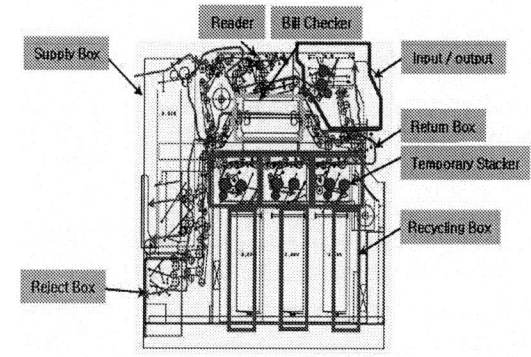

Fig. 1 Note Handling Units structural drawing

II. Accuracy estimation of stepping motor

Repeatability of stepping motor measured in using laser interferometer. Specifications of stepping motor are table 1

Table 1

Specs of stepping motor

Item	Specs	Unit
Model	34HY0802-01	
Step angle	1.8	Degree
Number of lead wire	8	Number
Current phase	3.6	A
Resistance/phase	0.45	Ohm
Inductance/phase	1.6	mH

Accuracy estimation of stepping motor is achieved repeatability test. Repeatability examination means degree of how each position points are situated in the fixed point in case of located several times vehicle repeatedly about some control target point. Examination of repeatability is testing agreement expedition of position of when decide repeat position according to same method under uniformity condition about each of each control axis departments. Measurement system composes with Fig 2 and tested.

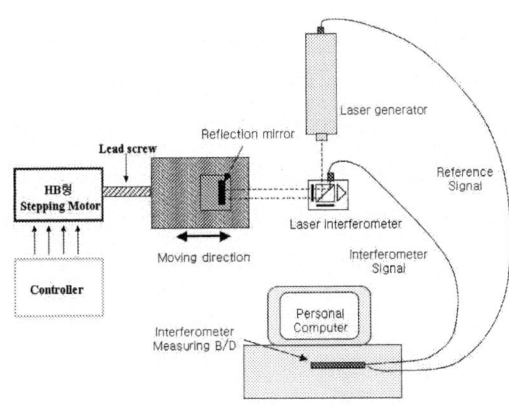

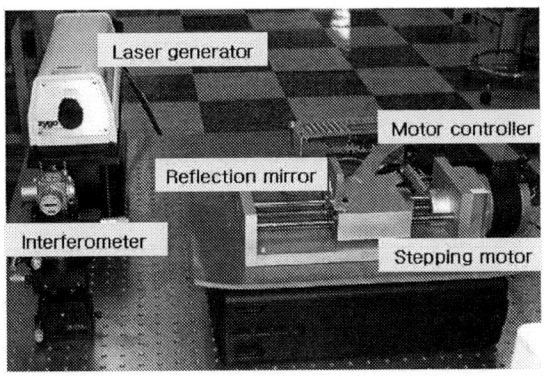

Fig. 2 Stepping motor accuracy measurement

Examination item and measuring method have 1/2 of indication value maximum variation in measuring holding point repeating positioning 7 times to direction that is same in any one point. This measurement does at central and two end each position of movement distance in principle. Measurement value does by absolute maximum. We got experiment result such as Fig 3, 4

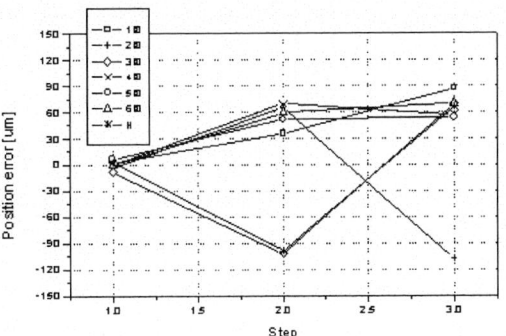

Fig. 3 Position error of stepping motor

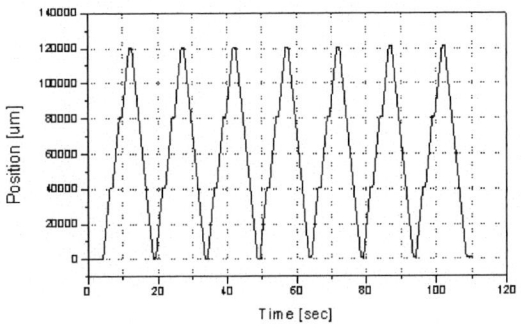

Fig. 4 Position value of stepping motor

- Repeatability accuracy of stepping motor

Step	1	2	3
Repeatability accuracy (µm)	+-7.4	+-85.6	+-97.4

III. Velocity fluctuating rate estimation of BLDC motor

We measured Velocity fluctuating rate quality in given fixed load state of BLDC motor used as main drive motor of ATM device. Specification of BLDC motor that test is same table 2

Table 2 Specs of BLDC motor

Item	Specs
Rate voltage	24V
Output	150W
Velocity control	3 Phase PWM

We designed and manufactured estimation device especially for Velocity fluctuating rate estimation. Manufactured estimation device is consisted to give fixed load by inertia whirl, hysteresis powder break that have about 10 times of rotation inertia moment of motor. And we connected motor to use torque meter so that can measure this load size and speed. Output of torque that came out in torque meter did to connect by indicator and know size by digital numerical value. We did it so that analysis may be available after storing data by computer connecting analog output that came out in indicator at the same time to data acquisition system separately. Also, Velocity signal displayed in torque meter as square wave was 60 pulse/rev. We acquired data to use signal conditioner that change this frequency to analog voltage.

It is that signal conditioner that possess present is inputted to 500Hz. We make reduce by 1/8 using logic integrated circuit and used this, because exceed extent speed of measured motor is 2000 rpm (2000/60rps x 60pulse/rev =2000Hz). We processed data to use DASYLab. Composition of estimation device is fig 4

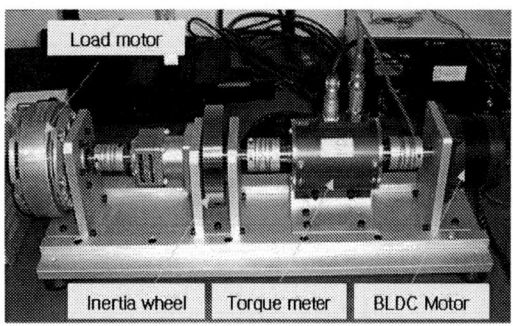

Fig. 5 Velocity fluctuating rate measurement of BLDC motor

Fig 6 is results of measurement for BLDC motor. Above graph displays speed of revolution (unit : rpm) and load torque (Kgf ㎝) that lower part graph takes to motor.

Controller of BLDC motor can know that keep the speed that is set even if load changes because there is given speed control function. First, when motor operate, motor

load rise to 7.5kgfcm by inertia. Motor rotates about 0.75kgfcm loads if come in stable state. This time, the motor speed is 1996 rpm.

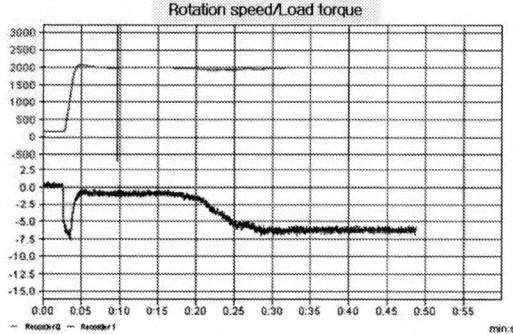

Fig. 6 Velocity fluctuating rate of BLDC motor

We increased load to 6.22kgfcm for an after this experiment. Motor speed of this state occurred change of the speed when load changes, but can know that return by stable state if load is fixed. Speed of motor was exposed by 2005 rpm. Therefore, Velocity fluctuating rate is as following.

$$SRF = \frac{\Delta \omega_{rpm}}{\omega_{mean}} \times 100\% = \frac{2005 - 1996}{(2005 + 1996)/2} \times 100\% = 0.45\%$$

Ⅳ. Solenoid static characteristics estimation and analysis

To evaluate characteristics of solenoid that is main driving source of ATM device, we measured sucking force of solenoid that is static test characteristic, and compared with that test sucking force at analytical solution using finite element method. Table 3 is importance specification of solenoid to test.

Table 3 Specs of Solenoid

Item	Specs	Units
Model	G-1253(JU WON)	
Rate voltage	DC30±10%	V
Rate current	833±10%	mA

Diameter of coil	0.31	mm
Number of winding	1880	turns
Resistance of winding	14.6	Ω

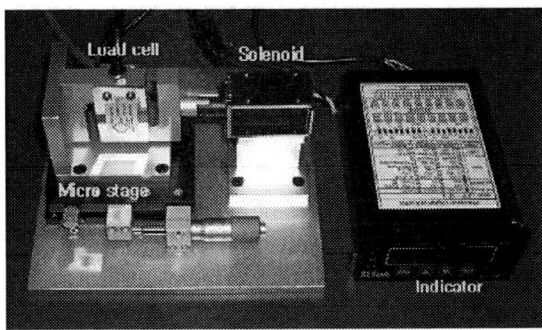

Fig. 7 Static quality measurement of solenoid

In this experiment, composed experiment device that use micro stage and load cell to measure sucking force by Solenoid Stroke. Fig 8 is result that measure solenoid sucking force.

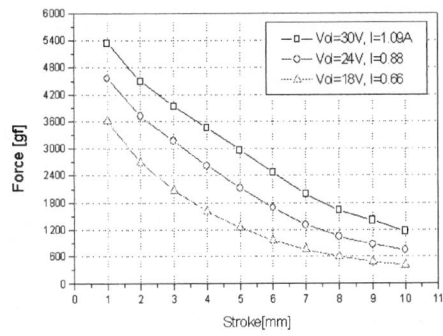

Fig. 8 sucking force of solenoid

Fig 10,11 are result that analyzes solenoid sucking force using finite element method. We behaved analysis that use two-dimension finite element method for quality analysis of solenoid

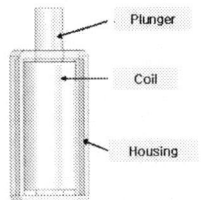

Fig. 9 Analysis model of solenoid

We supposed stator frame in circle shape for swiftness of analysis. And we did two-dimension axial symmetry analysis.

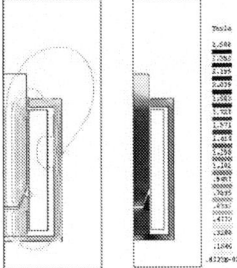

Fig.10 Magnetic Flux & Flux density of solenoid
(Stroke 1mm)

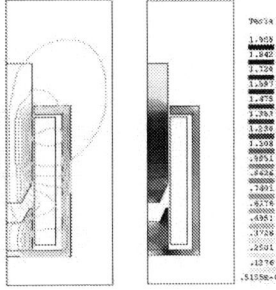

Fig. 11 Magnetic Flux & Flux density of solenoid
(Stroke 7mm)

Fig 12 is picturing that compare experimental value with analysis result of solenoid. Although experimental value is similar to analysis results, aspect of sucking force by position displayed a little different value at initial position place than analysis wave. This is assumed for cause by magnetic leakage and quality of the material and approximation of shape that is in joint of frame core part.

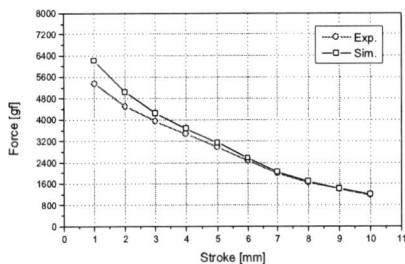

Fig. 12 Measurement value & Analysis value of solenoid
sucking force

V. Conclusion

In this paper, is treating contents on reliability estimation of main driving source of note handling units as ATM devices point parts. We got following result through this research and development.

(1)We estimated accuracy of stepping motor that is main driving source.

(2)We estimated velocity fluctuating rate of BLDC motor that is main driving source.

(3)We achieved static characteristics of solenoid and analysis that is main driving source.

REFERENCES

[1] TJE Miller, Design of Brushless Permanent Magnet Motors, OXFORD, 1994

[2] B.C Kuo, Theory and application of step motor, WEST PUBLISHING CO.1974

[3] BRUND LEQUESNE, " Dynamic Model of solenoid under impact excitation", IEEE Vol. 1996

[4] Huh, U., Lee, J., "A Torque Control Strategy of the Brushless DC Motor With Low Resolution Encoder" Proceedings of 1995 International Conference on Power Electronics and Drive Systems. PEDS'95. IEEE N.Y.1995

[5] Franklin,G.F., J.D.Powell, and M.L.Workman, "Digital control of Dynamic Systems", 2nd ed. Addison-Wesley, 1990.

2006 5th International Power Electronics and Motion Control Conference

Analysis on Toothless Permanent Magnet Machine with Halbach Array

Xu Yanliang[*]   Feng Kaijie[**]

[*]School of Electric Engineering, Shandong University, Jinan, P.R.China
[**] Wendeng Allwin Motor Co., Ltd, Wengdeng, P.R.China

Abstract- **Toothless permanent magnet (PM) machines with Halbach magnet array and normal one are studied comparatively in this paper. First of all, the analytic model is established for these two kinds of PM machines, then, their magnetic field distributions, magnitudes of magnetic flux at different locations are analyzed, laying emphasis on their different magnetic field variations along with the magnet thickness. These two kinds of PM machines without iron rotor are especially studied. One toothless stator, three kinds of rotor named as Halbach array iron one, Halbach array ironless one and normal array ironless one respectively are manufactured to structure three different prototyped machines, which have been used to verify the theoretical analysis successfully.**

Keywords- Halbach magnet array, toothless machines, ironless rotor, permanent magnet machines

I. INTRODUCTION

Toothless permanent magnet (PM) machine has found a wide application in high-speed drive such as energy storage flywheel and spindle system especially when suspended by magnetic bearing because of its elimination of tooth iron loss, very low torque ripple and, most important, very little effect on magnetic bearing system[1]. However it has a large effective air-gap because of its toothless characteristic, which results in a lower air-gap magnetic flux density and then a lower torque to volume ratio. On the other hand, permanent magnet synchronous motor, a kind of PM machine, when employing the normal permanent magnet array, must adopt a winding configuration with short pitch and distribution to satisfy its ideal demand of sinusoidal electromotive force, which undoubtedly lowers its torque ratio. Fortunately, a new kind of permanent magnet array named as Halbach one has such advantages as follows compared with the normal one when used in electric machine[2]: (1) Ideally sinusoidal air-gap flux density distribution, (2) Larger air-gap magnetic flux, (3) Good magnetic shielding.

In this paper, taking the example of toothless permanent magnet brushless direct current machine (BDCM) which was intended in application of a satellite attitude control/energy storage flywheel with angular moment of 15Nms at the speed of 30,000 rpm, the Halbach array and the normal one have been studied comparatively, especially in the different air-gap flux density variation with the thickness of the permanent magnet and in the

different influence when the machine employs a rotor with or without a iron core. The analystic model is established first of all, and finally, the prototyped machines were manufactured to verify the theoretical analysis.

II. ANALYTIC MODEL

A 6-pole toothless PM electric machine is used to study comparatively as shown in Fig.1, where, the permanent magnet can be the normal radially magnetized array, the normal parallel magnetized one and Halbach one shown respectively in Fig.2 (a), (b) and (c), and the rotor yoke can be made of iron or non-iron. The Halbach array may have three segments per pole as shown in Fig.2 (c) or any other amount of segments and has fixed direction of magnetization in each individual magnet segment by:

$$\theta_m = (1-p)\theta_i \qquad (1)$$

where θi is the angle between $\theta=0$ and the center of ith magnet segment (Fig.3).

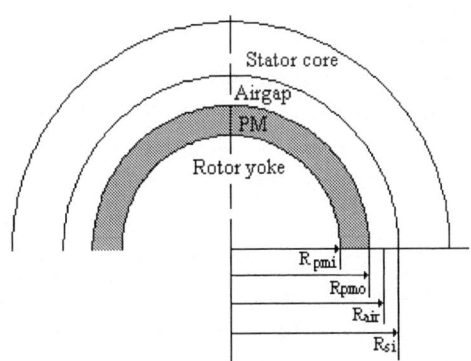

Fig.1 Structure diagram of toothless PM machine

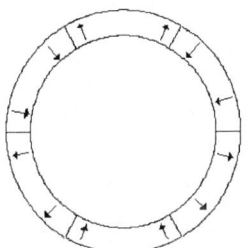

(a) Normal radially magnetized array

1-4244-0448-7/06/$25.00 ©2006 IEEE

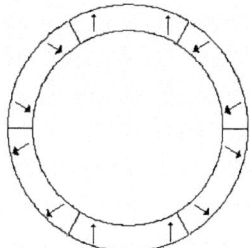

(b) Normal parallel magnetized array

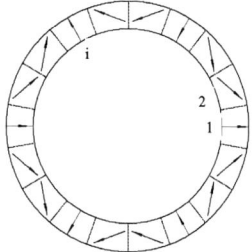

(c) Halbach magnet array (three segments per pole)

Fig.2 Different magnet array

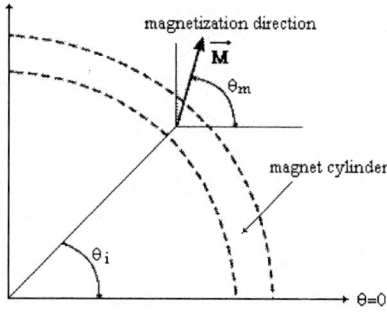

Fig.3　Angular relationship between θm and θi

III. COMPARISON OF MAGNETIC FIELD FOR PM MACHINES WITH DIFFERENT MAGNET ARRAY

A. PM machines with iron rotor

Under the same dimensions, four kinds of PM machines with normal parallel magnetized array, normal radially magnetized array, Halbach magnet array with 3 segments per pole and Halbach array with 5 segments per pole respectively are studied comparatively. Fig.4, Fig.5 and Fig.6 give the magnetic field distributions, air-gap flux density distributions and the harmonic magnitudes of air-gap flux density for the four machines respectively, and Table I gives the magnetic flux at different locations for the different machines. It can be shown that from Fig.4, Fig.5, Fig.6 and Table I.

1) PM machine with Halbach array features sinusoidal distribution of air-gap flux density much better than that with normal magnetized array and the more for its segments per pole, only a little better its sinusoidal distribution. Considering the more complication when the

Halbach array with more segments per pole, the following analysis takes the example of Halbach array with three segments per pole only.

2) PM machine with Halbach array enjoys a very low flux density in rotor yoke, as a result, the rotor yoke can have a very small area and even be replaced by non-iron materials, as is very beneficial to reduce the machine's weight and volume, reduce the inertia, and enhance its torque to volume ratio.

3) PM machine with normal parallel magnetized array features a larger air-gap flux, larger air-gap fundamental flux density, and a lower flux in the location of inner permanent magnet than that with normal radially magnetized array. It means that, for 6-pole PM machine, parallel magnetization is prior to radial one, and then the normal parallel magnetized array is designated specially to be used to compare with Halbach array. In fact, the superiority of normal magnetized PM machine with parallel or radial magnetization is determined mainly by its pole number [3].

From Table I, the PM machine with Halbach array produces a lower air-gap magnetic flux than that with normal magnetized array, just because the machine adopts a small magnet thickness. In the next section, the influence of magnet thickness on the air-gap flux will be involved.

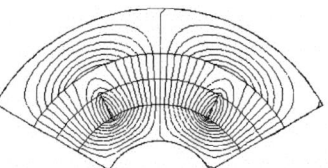

(a) Normal radially magnetized array

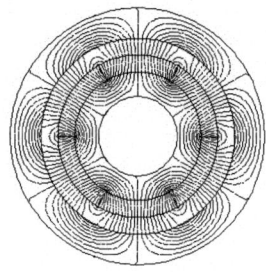

(b) Normal parallel magnetized array

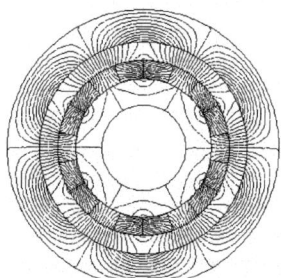

(c) Halbach array (three segments per pole)

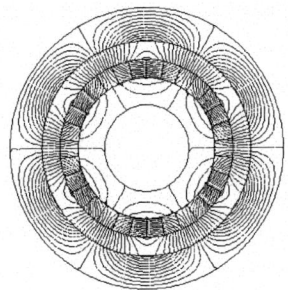

(d) Halbach array (five segments per pole)

Fig.4　Magnetic field distribution for different PM machines

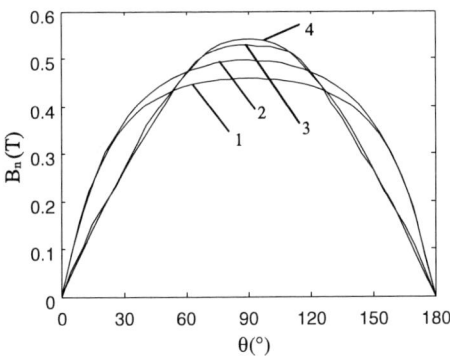

Fig.5　Air-gap flux density distribution for different PM machines

1—Normal radially magnetized array, 2—Normal paraller magnetized array, 3—Halbach array(three segments per pole), 4—Halbach array(five segments per pole)

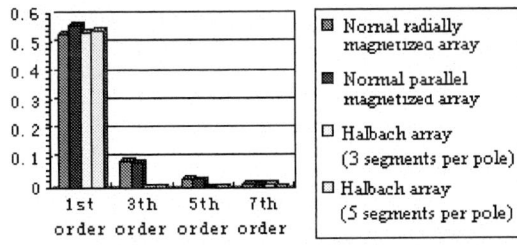

Fig.6　Harmonic magnitude of air-gap flux density waveform for different PM machines (Unit:T)

Table I　Magnetic flux at different locations for different PM machies

Different PM machines	Inner radial of PM(R_{pmi})	Outer radial of PM(R_{pmo})	Air gap(R_{air})	Inner radial of stator core (R_{si})
1	1.459	1.158	0.8966	0.847
2	1.082	1.195	0.9457	0.8954
3	0.2988	0.993	0.8528	0.8204
4	0.2386	0.983	0.8615	0.83

1—Normal radially magnetized array, 2—Normal parallel magnetized array, 3—Halbach array(3 segments per pole), 4—Halbach array(5 segments per pole)

B. *PM machines without iron rotor*

In order to reduce the volume, weight and inertia of the PM machine, and at the same time to keep a larger air-gap magnetic field, the advantage of magnetic shielding for Halbach array should be much taken, which can make the PM machine rotor yoke non-magnetic as mentioned above. Fig.7 and Table II give the relative results for PM machines when with different magnet array and when with different rotor core. It is well known that the normal magnetized-array PM machine with a ironless rotor core results in a very low air-gap magnetic field compared with that with iron rotor core. However for PM machine with Halbach array, it is very different, it can be shown from Fig.7 and Table II that, the PM machine with Halbach array just leads to a very little reduction of the air-gap magnetic field when the rotor core is replaced by non-iron material.

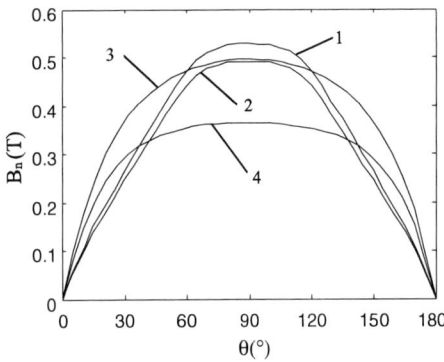

Fig.7　Air-gap flux density distribution when the machine with or without rion rotor core

1—Halbach array with iron rotor core, 2—Halbach array without iron rotor core, 3—normal array with iron rotor core, 4—normal array without iron rotor core

Table II　Magnetic flux at different locations when the machine with or without iron rotor (Unit:Wb/mm×10^{-5})

Different PM machines		Inner radial of PM(R_{pmi})	Outer radial of PM(R_{pmo})	Air gap(R_{air})
Halbach array	With iron rotor	0. 2988	0. 993	0. 8528
	Without iron rotor	0. 141	0. 928	0. 7955
Normal array	With iron rotor	1.082	1.195	0.9457
	Without iron rotor	0.49	0.927	0.711

IV. INFLUENCE OF MAGNET THICKNESS ON AIR-GAP MAGNETIC FIELD FOR DIFFERENT PM MACHINES

The influences of magnet thickness on air-gap magnetic field for PM machines with normal array and Halbach one are shown in Fig.8 and Fig.9, where h_{pm} is the thickness of magnet. It is shown that, when there is a smaller h_{pm}, the

1743

PM machine with Halbach array produces a lower air-gap flux and density than that with normal array, but along with the enlargement of h_{pm}, the former provides a larger air-gap magnetic field increment than the latter, and when h_{pm} is enlargement to some magnitude, the air-gap magnetic field in the former machine is stronger than that in the latter. Therefore, in order to enhance the air-gap magnetic field in the toothless PM machine and then to enhance the torque to volume ratio, it is a good choice to adopt Halbach array with a large magnet thickness especially when the machine employs a ironless rotor.

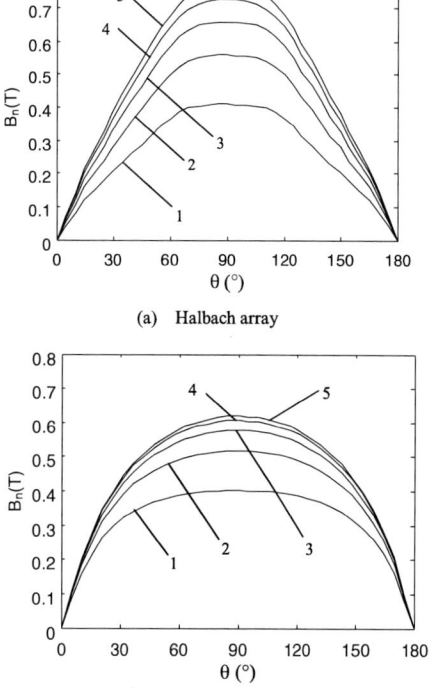

(a)   Halbach array

(b)   Normal array

Fig.8   Variation of flux density distribution with magnet thickness
for machines with different magnet array

1—$h$pm=3mm，2—$h$pm=5mm，3—$h$pm=7mm，4—$h$pm=9mm，5—$h$pm=11mm

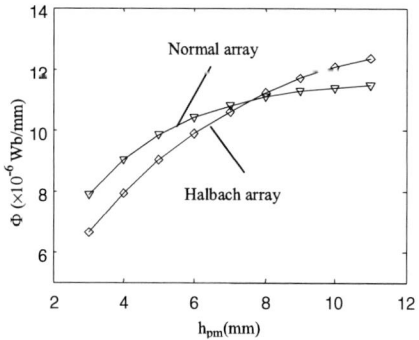

Fig.9   Variation of air-gap flux with magnet thichness

V. PROTOTYPED MACHINES AND EXPERIMENTS

To verify the above analyses, three prototyped PM machines are manufactured with a same toothless stator, three different rotor named as iron rotor with Halbach array, ironless rotor with Halbach array and iron rotor with normal array respectively, some components of which are given in Fig.10. Fig.11 gives the EMF waveforms for different PM prototyped machines. It is shown from Fig.11 that the PM machines with Halbach array make only a little difference in the EMF waveform whether or not employing a iron rotor, and both feature a sinusoidal EMF waveform compared with the PM machine with normal array, as are well conformed to the above analyses.

The prototyped toothless stator and the prototyped rotor with Halbach array and iron core has been structured as a permanent magnet BDCM in application of a satellite attitude control and energy storage flywheel with angular moment of 15Nms at speed of 30,000 rpm successfully.

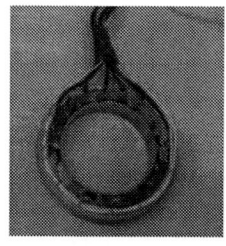 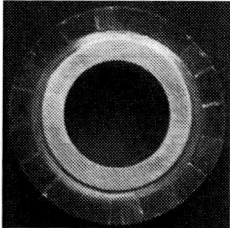

(a) Toothless stator   (b) ironless Rotor with Halbach array

Fig.10 Some Components of prototyped machines

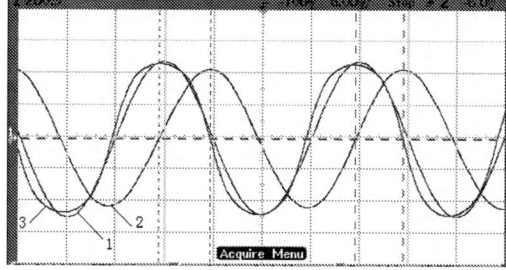

Fig.11 EMF waveform for different PM prototyped machines

1—iron rotor with Halbach array, 2——ironless rotor with Halbach array, 3——iron rotor with normal array

VI. CONCLUSION

After theoretical analysis and experimental study between toothless PM machines with different magnet arrays and with different rotor yoke materials, such conclusions as follows are reached:

1) PM machine with Halbach array features sinusoidal distribution of air-gap flux density much better than that with normal magnetized array and the more for its segments per pole, only a little better its sinusoidal distribution, so three segments per pole for Halbach array is a moderate choice between its complexity and advantages.

2) PM machine with Halbach array presents a larger air-gap magnetic field increment than that with normal array along with the enhancement of the magnet thickness. It means that, the former machine can produce a larger air-gap magnetic field than the latter one only when the magnet thickness exceeds a definite value.

3) PM machine with Halbach array makes only a little difference in the air-gap magnetic field when it is replaced from a iron rotor to a ironless one.

As a whole, for a toothless PM machines, especially when the rotor is equipped with ironless core, the Halbach magnet array with a definitely large magnet thickness is much recommended.

REFERENCES

[1] H. Hofmann and S. R. Sanders, "Synchronous reluctance motor/alternator for flywheel energy storges systems," *IEEE Workshop on Power Electronics in Transportation*, Dearborn, MI, October 24-25, 1996, pp.199-206

[2] Z.Q.Zhu and Howe, "Halbach permanent magnet machines and applications: a review", *IEE Proc.-Electr. Power Appl.*, vol.148, No.4, pp.299~308, July 2001.

[3] Wang Cheng, Zhou Meiwen, Guo Qingjing, "Vector control and alternative current servo motor", Beijing: Mechenical Inductrial Press, 1994.

Improvement in Reliability of Doubly Salient Permanent Magnet Motor Drive

Wenxiang Zhao*,**, Ming Cheng*, Xiaoyong Zhu*,**, Wei Hua*, Jianzhong Zhang*

*Department of Electrical Engineering, Southeast University, Nanjing 210096, P.R. China
**School of Electrical and Information Engineering, Jiangsu University, Zhenjiang 212013, P.R. China
Phone: 86-25-83794152 Fax: 86-25-83791696 E-mail: mcheng@seu.edu.cn

Abstract—The doubly salient permanent magnet (DSPM) motor is a novel type of AC brushless machine and it offers advantage of high power density, wide speed range and simple structure. To improve the reliability and fault-tolerant performance, a new dual-channel 12/8-pole DSPM motor speed-adjustable system is proposed in this paper. With the comprehensive introduction for the proposed redundant DSPM motor, a close loop control scheme is designed. The motor drive is developed with speed, position and current feedback control using a digital signal processor (DSP) TMS320F2812. The speed and position feedback are obtained from an absolute optical encoder, whereas Hall transducer is implemented to sense the current. The operation performance simulation is conducted based on a prototype DSPM motor and the results show that the dual-channel DSPM motor drive offers higher reliability and better operation characteristics than the traditional single channel motor drive at the same fault condition.

Keywords—Doubly salient motor; Dual-channel; Reliability; Fault

I. INTRODUNTION

It has been a long-live objective to develop an advanced adjustable speed drive that combines the advantages of the DC motor of good speed controllability as well as the AC motor of high reliability, robustness, and maintenance free. Recently, a new type of brushless machine, termed the doubly salient permanent magnet (DSPM) motor, has been developed, and the recent literature has shown that the DSPM motor is of high efficiency, high power density and simple structure [1]. The DSPM motor incorporates the merits of both permanent magnet (PM) brushless motor and the switched reluctance (SR) motor. First, PMs locating in stator, DSPM motor can eliminate the problem of irreversible demagnetization and mechanical instability, and retain the merits of high efficiency and high power density. Secondly, the rotor of DSPM motor is the same as that of a SR motor, adopting the advantages of simple configuration and mechanical robustness.

In the development of the DSPM motor drive, there were a number of publications [2]-[8]. However, most of

Projects supported by NSFC (50337030 and 50377004)

them were biased on the analysis and the control, rather than fault-tolerance characteristic and the way to improve reliability, of the DSPM drive. For drive applications, such as aerospace and electric vehicle application, the high degree of reliability and fault-tolerance is required. Because of the magnetic independence of the motor phases and the circuit independence of the inverter phases, the DSPM motor drive has inherent fault-tolerant capability for some type of faults, but not for others. To achieve higher reliability, a dual-channel DSPM motor drive is developed in this paper. With redundant structure, a digital signal processor TMS320F2812 is implemented to a digital dual-channel DSPM motor speed-adjustable system. The proposed effective control system is conceived to be high reliability, simple system structure, and minimum requirement of additional hardware.

II. MOTOR STRUCTURE AND PRINCIPLE ANALYSIS

Similar to the SR motor, Ps/Pr=6/4, 8/6 and 12/8 are possible configurations of the DSPM motor. Compared with the 3-phase 6/4-pole one, the 3-phase 12/8-pole DSPM motor possesses shorter flux paths and traditional shape instead of football shape in cross-section. Because of the fact that the circumference of stator split into four segments by 4 PMs, less magnetic potential drop and iron losses are achieved in 12/8-pole motor. Moreover, because the flux per magnetic pole is halved in 12/8-pole motor, the width of both stator yoke and teeth is almost one-half of those of a 6/4-pole machine, which allows greater inner stator diameter and greater rotor diameter. Therefore, higher torque density can be achieved. Furthermore, less width of stator teeth results in shorter end part of phase windings, leading to less copper consumption and resistance of windings. Based on the above fact, higher efficiency can be expected in the 12/8 pole DSPM motor.

Under the normal operation, the reliability rate of the DSPM motor drive can be evaluated based on reliability rate of the individual components in the system. As a series system, the reliability function of the system can be calculated by the product of the prime component reliability rate, as follows:

$$R_s(t) = \prod_{i=1}^{n} R_i(t) \qquad (1)$$

where Ri(t) is the reliability rate of the prime component in the DSPM motor drive, Rs(t) the reliability rate of the motor drive.

In the dual-channel DSPM motor drive, the system is consist of two subsystem in parallel. Therefore, reliability rate of the redundant system is calculated as follows:

$$R(t) = 1 - \prod_{j=1}^{2}[1 - R_{sj}(t)] = 1 - \prod_{j=1}^{2}[1 - \prod_{i=1}^{n} R_i(t)] \quad (2)$$

If reliability rate of the subsystem is 0.9, reliability rate of the dual-channel DSPM motor will be improved to 0.99 with the redundant structure, and the system reliability has improved greatly.

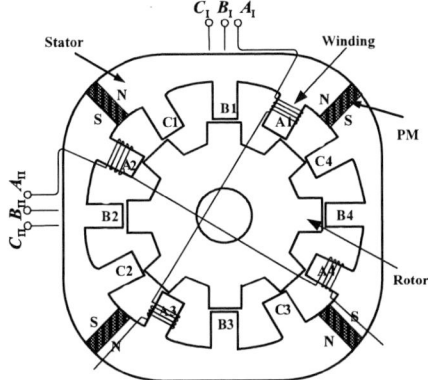

Fig. 1 Cross section of dual-channel DSPM motor.

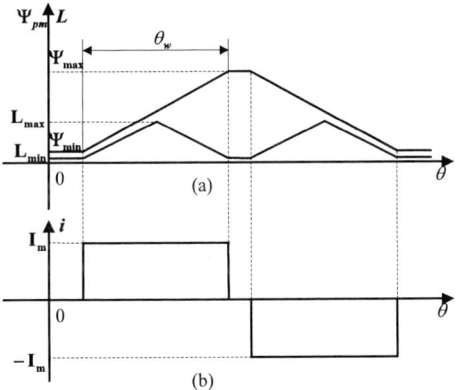

Fig. 2 Theoretical flux and current waveform.

Based on the above analysis, the redundant structure of DSPM motor is proposed. As shown in Fig. 1, for a 3-phase 12/8-pole motor, there are four parts of each phase, so there are two combinations to achieve dual-channel coil arrangement for increased redundancy. Here A-phase is explained as the example, and other phases are like that. If phase-leg A1 and phase-leg A2 are connected in series as one phase in a channel, the phase is likely to produce unbalanced magnetic radial forces and different airgap lengths on opposite sides of the machine under the fault condition. The forces result in a displacement of the rotor from its center location, and the diametrically opposite poles cause stresses on the rotor bearings. Therefore, the opposite pole-coils, phase-leg A1 and phase-leg A3, are connected in series, and the dual-channel DSPM motor winding configuration is developed by two parallel

connections. Fig. 1 shows the cross section of the proposed dual-channel 12/8-pole DSPM motor, and the corresponding theoretical waveform of PM flux Ψ_{pm} and phase current i are shown in Fig. 2. The parallel connection of pole coils is operated in two channels under the normal condition. Under the fault condition, there is only one channel in work while the fault channel is switched off, whatever short-circuit fault or open-circuit fault.

Under the normal condition, the matrix equation describing the 12/8-pole DSPM motor is as follows:

$$u_p = R_p i_p + e_p = R_p i_p + \frac{d\psi_p}{dt} \quad (3)$$

And

$$\psi_p = [L_p] \cdot [i] + \psi_{pmp} \quad (4)$$

where $p=A$, B, C. By neglecting mutual induction, the model is simplified as:

$$\begin{aligned} e_p &= \frac{d\psi_p}{dt} = L_p \frac{di_p}{dt} + i_p \frac{dL_p}{dt} + \frac{d\psi_{pmp}}{dt} \\ &= L_p \frac{di_p}{dt} + i_p \frac{dL_p}{d\theta}\omega + \frac{d\psi_{pmp}}{d\theta}\omega \\ &= L_p \frac{di_p}{dt} + e_r + e_{pm} \end{aligned} \quad (5)$$

According to the principle of virtual work, the electromagnetic torque of the DSPM motor is obtained as:

$$T_{ep} = \frac{1}{2} i_p^2 \frac{\partial L_p}{\partial \theta} + i_p \frac{\partial \psi_{pmp}}{\partial \theta} = T_{rp} + T_{mp} \quad (6)$$

With variation of stator winding inductance, the reluctance torque T_{rp} has a zero average value, and the reaction torque T_{mp}, the dominant torque, can be produced by applying either a positive current to a phase windings when its flux linkage is increasing (e_{pm}>0) or a negative current when the flux linkage is decreasing (e_{pm}<0) as shown in Fig. 2.

III. SYSTEM DESCRIPTION

According to the characteristic of the proposed redundant DSPM motor, a full digital control system for dual-channel DSPM motor has been developed. The circuit frame of this system is given in Fig. 3. The whole control system is composed of a multiprocessor system built up by compatible industrial PC (IPC) as the host and a DSP TMS320F2812 as the slave CPU, a position and speed sensor based on optical encoder, a current sensor based on Hall-effect, and a PWM-controlled power inverter based on intelligent power module (IPM). The task control assignment has been distributed between IPC and DSP, so that the first executes man-machine interface and program compiler, while the second executes detecting signal and real-time calculation. Furthermore, the DSP processor executes PWM output.

A. Redundant Motion Controller

The DSP TMS320F2812, a member of the TMS320C28x™ DSP generation, is a highly integrated, high-performance solution for demanding control applications. Compared with former generation DSP TMS320F240, TMS320F2812 is integrated with three 32-bit CPU-Timers, enhanced ADC module, more communication interfaces, etc. Especially, there are two Event Managers (EV), EVA and EVB, which provide a broad range of functions and features that are particularly useful in redundant motor control applications. Each EV is capable of controlling three Half-H bridges, when each bridge requires a complementary PWM pair for control. Each EV also has two additional PWMs with no complementary outputs. As the motion controller in the dual-channel DSPM motor drive, EVA and EVB individually offer the PWM output for the each channel of the redundant DSPM motor. The redundant motion controller provides the possibility to achieve higher reliability.

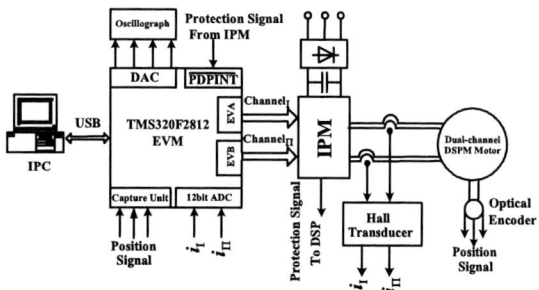

Fig. 3 Hardware circuit frame of control system.

B. Detection of Rotor Position and Speed

In accordance with the operation principle of the dual-channel DSPM motor, each of the three phase windings should be turn on or off at the specific rotor position. The shaft position information is provided using an 8-slot slotted disk connected to the rotor shaft and three opto-couplers mounted to the stator housing as shown in Fig. 4.

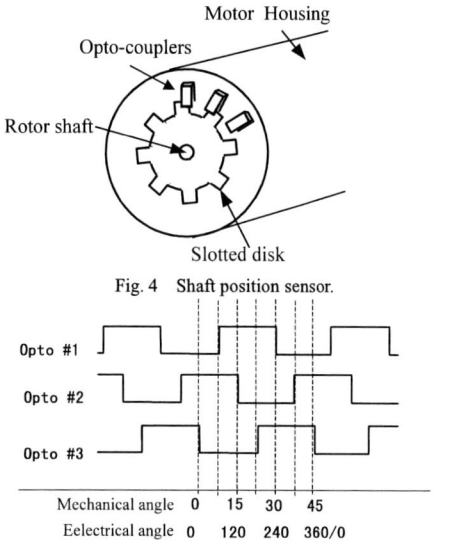

Fig. 4   Shaft position sensor.

| Mechanical angle | 0 | 15 | 30 | 45 |
| Eelectrical angle | 0 | 120 | 240 | 360/0 |

Fig. 5 Opto-coupler output signal vs. rotor angle.

The opto-couplers are nominally located 30° apart from each other along the circumference of the disk. The output signal waveforms of the sensor are shown in Fig. 5.The sensor generates an opto-coupler edge for every 7.5° of mechanical rotation. For every 45° of mechanical rotation the signal pattern repeats, corresponding to one electrical cycle of the dual-channel DSPM motor, of which there are 8 cycles per shaft revolution.

In this application, both mechanical angle and electrical angle are referenced. Mechanical angle is useful when considering velocity control of the dual-channel DSPM motor, and electrical angle is convenient when considering commutation. At each edge, the speed is calculated by:

$$n = \frac{\Delta\theta}{\Delta t} = \frac{60\Delta\theta \cdot f_{clk}}{N} \qquad (7)$$

where n is the speed in rpm, $\Delta\theta$ the distance between opto-coupler edges in revolution, Δt the time between edges in minutes, N the number of clock counts between edges, and f_{clk} the clock frequency.

Position measurements are identified when opto-coupler transition occurs, using the DSP's capture units. From Fig. 5, it can be seen that there are six possible combinations of the opto-coupler output states per electrical cycle of the dual-channel DSPM motor. Corresponding to each state, the six power switches have the fixed conducting state.

C. Current Sensing

The control strategy of the dual-channel DSPM motor drive consists of two basic schemes, namely current chopping control (CCC) and angle position control (APC), respectively for constant torque operation at speeds below the base speed and constant power operation at speeds above the base speed. At low speed, the current rises very fast to a very high level when the phase winding is turned on due to low back EMF and reactance. To perform the current chopping as well as over-current protection, the phase current of the dual-channel DSPM motor must be detected instantaneously. Hence, a current sensor should be provided for each phase in each channel.

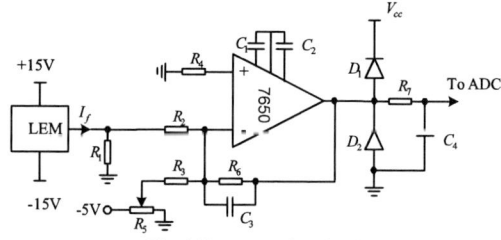

Fig. 6 Current sensing circuit.

In the application, current-voltage transducer based on Hall effect is implemented to sense the currents, using ADC embedded in TMS320F2812. The current sensor output signal needs to be rearranged and scaled so that it can be sent to DSP. The output signal of the current sensor may be either positive or negative. Hence the signal must be translated by the analog interface into a range of (0, 5V)

1748

in order to allow the single voltage ADC module to read both positive and negative values. To achieve the above objective, the current sensing circuit is designed as Fig. 6.

D. Power Converter

Unlike traditional electric machine, such as induction and synchronous machine, the DSPM motor can not work from industrial power supply directly. Hence, a power converter is indispensable in the DSPM motor control system.

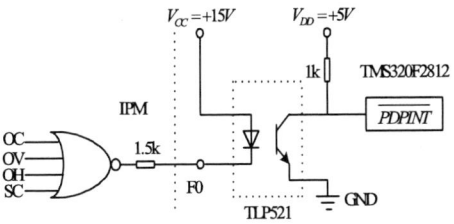

Fig. 7 Interface circuit of $\overline{PDPINT}$ protection.

As a slave CPU, TMS320F2812 produces current references to two IPMs that each is packed by six IGBT power transistors in the converter. For the safe operation of the dual-channel motor drive, an external interrupt in each EV, which is maskable interrupt, is generated when the device pin $\overline{PDPINT}$ (power-drive protection interrupt) is pulled low. This maskable interrupt can put the timers and the PWM output pins in high-impedance states, and then inform the CPU in case of motor drive abnormalities such as over-voltage, over-current, and excessive temperature rise. In the proposed redundant motor drive, each channel has its own independent converter. There are two separate control systems because the two converters can remain isolated from each other, while the pin $\overline{PDPINT}$ acts as the switch of the dual-channel. When any fault is detected, the fault channel is switched off, and the motor drive runs at single channel.

As shown in Fig. 7, the protection signals, namely over over-current signal "OC", over-voltage signal "OV" over-heat signal "OH", and IGBT short-circuit signal "SC", are fed to the interrupt pin $\overline{PDPINT}$. Thus, any action in the four protections results in a high-to-low transition on pin $\overline{PDPINT}$, generating an external interrupt to turn off all of the power switches in the fault channel.

IV. SYSTEM SIMULATION

To assess the fault-tolerant performance of the motor drive under the fault condition, operation performance simulation is conducted on a available DSPM prototype machine as shown in Fig. 8. The parameters used in simulation have been deduced from nonlinear finite element analysis, which have been verified by the measurements on the machine. Fig. 9 gives the back EMF and self-inductance characteristics of the motor.

The principal electromagnetic faults considered here are those taking place inside the motor or in the converter.

Without magnet, brush or windings in the rotor, there is no electrical fault in the rotor, but there may be electrical faults in the stator. On the other hand, faults may occur in power converter since the main switches and the freewheeling diodes in the converter may be damaged by over-voltage, over-current, or excessive temperature rise. Furthermore, the DSPM motor may not be commutated normally if the rotor position detector is in fault, and the switches in the phase will not be triggered. In summary, there are four types of prime fault in motor drive: armature phase open-circuit, armature phase short-circuit, converter switch open-circuit and converter switch short-circuit. In this paper, the open-circuit fault condition of one phase is simulated to illustrate higher reliability of the dual-channel DSPM motor drive. When the fault is detected, the pin $\overline{PDPINT}$ generates the interrupt signal, and the corresponding EV will not produce PWM output. With the fault channel switched off, the control system runs at single channel.

Fig. 8 12/8-pole DSPM prototype machine

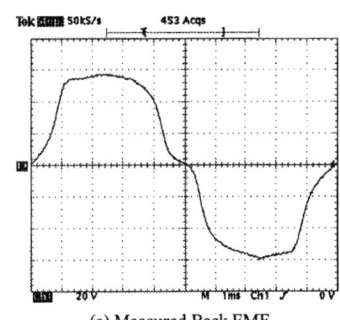

(a) Measured Back EMF

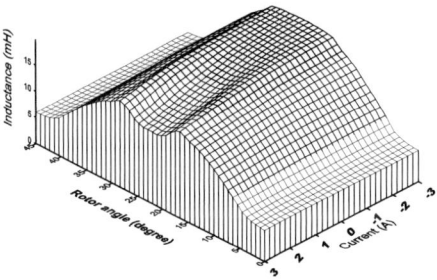

(b) Predicted self-inductance

Fig. 9 Static characteristics of DSPM motor

Under the fault and the normal conditions, the waveforms of total output torque in both motor drives, traditional DSPM motor and dual-channel DSPM motor, are shown in Fig. 10. Fig. 10 (a) shows torque waveform of the

1749

traditional DSPM motor drive under the one phase open-circuit fault in which the corresponding fault phase can not produce the torque, so the DSPM motor drive operation is provided by the healthy phases. It is seen that continued operation under the fault is possible, but the torque ripple is much bigger than the normal condition. Fig. 10 (b) illustrates the output torque of the dual-channel DSPM motor drive under the same fault when the control system switches off the fault channel, and the dual-channel DSPM motor drive operation is provided by the healthy channel.

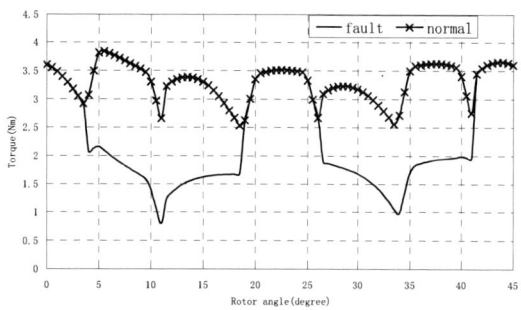

(a) Traditional DSPM motor

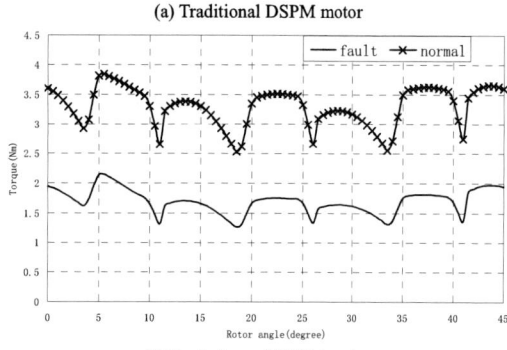

(b) Dual-channel DSPM motor

Fig. 10 The torque waveforms under the normal and the fault condition.

To scale the grade of the torque ripple, a torque ripple factor is defined as follows:

$$k = \frac{T_{\max} - T_{\min}}{T_{avg}} \qquad (8)$$

where T_{max} is the maximum of the output torque, T_{min} the minimum of the output torque, and T_{avg} the average of output torque. Under the normal condition, T_{avg} is 3.306Nm, and k is 39.5% in both the motor drivers. Under the fault condition, T_{avg} is 1.679Nm, and k is 170% in the traditional DSPM motor drive, whereas T_{avg} is 1.702Nm, and k is 51.8% in dual-channel DSPM motor drive. It can be observed that the dual-channel DSPM motor drive can offer not only higher average torque but also much less torque ripple than single channel one under the same fault condition. Therefore, fault-tolerance operation with good performance is possible, though the output torque is nearly halved.

V. CONCLUSION

In this paper, a digital dual-channel DSPM motor drive

has been developed for the improvement of reliability. Based on the analysis of the construction and operation principle of the dual-channel DSPM motor, a redundant motor is developed, and a digital control system for the motor has been designed by using a 32-bit high speed DSP TMS320F2812. With the redundant motion controller, the two channels are offered to the motor drive. The simulation of operation performance of the motor drives has been conducted, showing that less torque ripple and higher average torque can be achieved in the dual-channel DSPM motor drive than in the traditional DSPM motor drive under the fault condition. By switching off the fault channel, the fault-tolerant characteristic and the reliability in the DSPM motor drive have improved greatly. The developed digital dual-channel DSPM motor drive lays a foundation for the further research and development of evolving suitable fault handling control strategies for not only DSPM motor but also other doubly salient motors such as hybrid excited doubly salient motor.

REFERENCES

[1] Y. Liao, F. Liang, T. A. Lipo, "A novel permanent magnet motor with doubly salient structure," *IEEE Transactions on Industry Applications*, vol. 31, pp. 1069–1078, Sept.-Oct. 1995.

[2] M. Cheng, K.T. Chau and C.C. Chan, "A new doubly salient permanent magnet motor," *Proceedings of IEEE 2nd International Conference on Power Electronics Drives and Energy System for Industrial Growth*, Vol. 1, Perth, Australia, Dec. 1998, pp. 2–7.

[3] M. Cheng, K.T. Chau and C.C. Chan, "Static characteristics of a new doubly salient permanent magnet motor," *IEEE Transactions on Energy Conversion*, vol. 16, pp. 20–25, March 2001.

[4] M. Cheng, K.T. Chau, C.C. Chan and Q. Sun, "Control and operation of a new 8/6-pole doubly salient permanent-magnet motor drive," *IEEE Transactions on Industry Applications*, vol. 39, pp. 1363–1372, Sept.-Oct. 2003.

[5] B.C. Mecrow, "New winding configurations for doubly salient reluctance machines," *IEEE Transactions on Industry Applications*, vol. 32, pp. 1348–1356, Nov.-Dec. 1996.

[6] A.R.C.S. Babu, K.R. Rajagopal, "FE analysis of multiphase doubly salient permanent magnet motors," *IEEE Transactions on Magnetics*, vol. 41, pp. 3955–3957, Oct. 2005.

[7] M. Cheng, Q. Sun, "Two-phase operation of 8/6-pole doubly salient permanent magnet motor with minimum torque ripple", *Proc. ICEMS*, Jeju, Korea, Nov. 1-3, 2004, CD-ROM.

[8] M. Cheng and E. Zhou, "Analysis and control of novel split-winding doubly salient permanent magnet motor for adjustable speed drive," *Science in China (Series E)*, vol. 44, pp. 353–364, June 2001.

[9] C.M. Stephens, "Fault detection and management system for fault tolerant SR motor drives", *IEEE Transactions on Industry Applications*, vol. 27, pp. 1098–1102, Nov.-Dec. 1991.

[10] A.G. Jack, B.C. Mecrow, and J. Haylock, "A comparative study of PM and SR motors for high performance fault tolerant applications", *IEEE Transactions on Industry Applications*, vol. 32, pp. 889–895, July-Aug. 1996

2006 5th International Power Electronics and Motion Control Conference

A New Approach of Modeling the Saturated Induction and Synchronous Salient Pole Machines

A. Câmpeanu [*], M. Bădică [**]

[*] University of Craiova, Craiova, Romania
[**] Automobile Dacia S.A., Mioveni, Romania

Abstract—**In this paper there are obtained mathematical models of the saturated induction and synchronous machines. It is shown that various sets of state variables have a common theoretical background what simplifies the deduction of the saturated models. In case of salient-pole synchronous machines a basically new computational a basically new computational assumption aiming to a higher fidelity representation of the main field saturation process is stated. A corrected model of the saturation is considered via a specific assumption preservation of the main flux linkage-combined with the introduction of the computational smooth axis-gap machine.**

Under these assumptions specific computations are performed and there are written the state equations of the electromagnetic processes for several combinations of the state variables.

Keywords-component; formatting; style; styling; insert

I. INTRODUCTION

The high precision simulation of the behaviour of alternative current machines during dynamic processes (start, self-excitation) having large variations of magnetic stress is closely connected to a correct mathematical model. Within this framework the literature displays an increased interest in considering saturation of the main flux linkage.

In [1], [3], [4], [16] a.o. there are detailed and analyzed nonlinear dynamic models for the induction machine. As a rule the analyses of the saturated machines is performed with the stator and rotor currents [1],[11] or flux linkages [7],[12] as state variables. In [13],[14] the method of space phasors is applied with winding currents as state variables, while in [9] the equations of some start processes [14] are solved using Simulink techniques. Mixed combinations of currents and flux linkages as state variables were also taken into account [4]. In the following a unique computational algorithm has been obtained.

The paper [5] is dedicated to the modelling of the magnetic saturation for the smooth air-gap synchronous machine. The same problem for the salient-pole synchronous machine appears as much more complicated since usually the magnetizing curve in the q-axis is not known. When this curve is known it is possible to describe saturation in the two (d, q) axes by introducing two saturation coefficients (see, e. g. [10]).

Other more or less recent research (e. g. [8], [18]) shows that one could accept the idea of the same saturation degree in both d, q axes. In the line of this assumption the paper [6], [8] considers an isotropic machine for which there are deduced the equivalent expressions of the main flux linkage ψ_μ and of the magnetizing current i_μ; nevertheless these are only computational quantities and do not define saturation for given ampere-turns of the windings.

In the following a unique saturation degree is assumed and the considered synchronous machine is replaced by a smooth air-gap one with the conservation of the saturation degree.

The computation assumptions are specified and it is shown that they ensure a higher degree of precision of the main field saturation. With respect to this it has to be mentioned that the considered type of rotor (with constant air-gap and non-magnetic gap along d-axis) ensure preservation of the fundamental harmonic of the magnetic field.

This harmonic is of basic interest since in practice only its effect, are mainly captured at machine's terminals. Within this framework there are deduced the non-linear dynamic models for various combinations of the state variables. Under these assumptions there are deduced the nonlinear dynamic models for various combinations of the state variables.

II. MATHEMATICAL MODELS

A. The saturated induction machine

The mathematical model of the induction machine saturated on the main flux way is considered to have the form

$$U_{dq} = A\frac{dX_{dq}}{dt} + BX_{dq}$$

$$U_{dq} = \begin{vmatrix} u_{sd} & u_{sq} & 0 & 0 \end{vmatrix}^{\mathrm{T}},$$

$$X_{dq} = \begin{vmatrix} X_{sd} & X_{sq} & X_{rd} & X_{rq} \end{vmatrix} \tag{1}$$

where $X_{sd}, X_{sq}, X_{rd}, X_{rq}$ are the d,q projections of the currents and flux linkages taken as state variables.

To the above equations the motion equation is added

$$M - M_r = \frac{J}{p}\frac{d\omega}{dt} \quad M = \frac{3}{2}p\,\mathrm{Re}\big[j\underline{\psi}_s \underline{i}_s^*\big], \tag{2}$$

The notations are the usual ones. The quantities are related to the stator. It is assumed that the rotor is short-circuited.

Saturation occurs in the expression of the main flux linkage

$$\underline{\psi}_m = L_m\left(\underline{i}_s + \underline{i}_r\right) = L_m\left(i_m\right)\underline{i}_m \qquad \underline{i}_m = i_m e^{j\varphi}$$

L_m is the magnetization inductance and $\underline{i}_m$ is the space phasor of the magnetizing current. The iron losses are neglected and consequently $\underline{\psi}_m$ and $\underline{i}_m$ have the same direction. The possible combinations of the state variables are grouped in two classes for which are established direct calculus algorithms.

1) The winding currents – flux linkages as state variables.

In [4] there are deduced the mathematical models for mixed combinations currents - flux linkages associated with the introduction of a generalized space phasor $\underline{i}_m$, which may or may not have a real existence in the machine and has the same orientation as the phasors of the magnetizing current i_m. For the given state variables, ψ is a linear combination of them.

In what follows it will be seen that regardless the state variables choice (combinations winding currents - flux linkages), the deduction of the mathematical models is always associated with a term of the form $\underline{\psi}\dfrac{d}{dt}\left(\dfrac{1}{L'_m}\right)$

which may be expressed in a general form; $L'_m = \psi/i_m$ is an inductance associated to the saturation state of the magnetic circuit. All is needed is that ψ and L'_m be defined for each choice of the state variables. Further it is showed how to define and express in detail this term thus obtaining a result that is valid for all choices of the state variables.

It is computed first $\dfrac{d}{dt}\left(\dfrac{1}{L'_m}\right) = -\dfrac{1}{L'^2_m}\dfrac{dL'_m}{dt}$.

We have [16]

$$\frac{dL'_m}{dt} = \frac{dL_m}{dt} = \frac{1}{i_m}\left(L_{mt} - L_m\right)\frac{di_m}{dt} \qquad (3)$$

where $L_{mt} = \dfrac{d\psi_m}{di_m}$ is the differential magnetizing inductance associated to L_m.

We have further

$$\frac{d\underline{\psi}}{dt} = \frac{d}{dt}\left(L'_m \underline{i}_m\right) = L'_{mt} e^{j\varphi}\frac{di_m}{dt} + jL'_m\frac{d\psi}{dt}\underline{i}_m \quad (4)$$

where we denoted

$$L'_{mt} = L_{mt} + \left(L'_m - L_m\right) \qquad (5)$$

L'_{mt} being the differential inductance associated to L'_m

For the computation of $\dfrac{d\varphi}{dt}$ it is taken into account that ψ and ψ_m are in phase and it follows

$$\frac{d\varphi}{dt} = \frac{1}{\psi^2}\left(\psi_d\frac{d\psi_q}{dt} - \psi_q\frac{d\psi_d}{dt}\right) = \frac{1}{\psi}\left(\frac{d\psi_q}{dt}\cos\varphi - \frac{d\psi_d}{dt}\sin\varphi\right),$$
$$(6)$$

Finally taking into account (4), (5), (6) and introducing

$$\underline{i}_m = \frac{1}{L'_m}\underline{\psi} = \frac{1}{L'_m}\psi e^{j\varphi} \qquad (7)$$

the term containing saturation is

$$\frac{1}{L'^2_m}\underline{\psi}\frac{dL'_m}{dt} = \left(\frac{1}{L'_m} - \frac{1}{L_{dd}}\right)\frac{d\psi_d}{dt} - \frac{1}{L_{dq}}\frac{d\psi_q}{dt} + $$
$$+ j\left[\left(\frac{1}{L'_m} - \frac{1}{L_{qq}}\right)\frac{d\psi_q}{dt} - \frac{1}{L_{dq}}\frac{d\psi_d}{dt}\right]. \qquad (8)$$

The computation inductances are defined:

$$\frac{1}{L_{dd}} = \frac{\sin^2\varphi}{L'_m} + \frac{\cos^2\varphi}{L'_{mt}}, \quad \frac{1}{L_{qq}} = \frac{\cos^2\varphi}{L'_m} + \frac{\sin^2\varphi}{L'_{mt}},$$
$$\frac{1}{L_{dq}} = \left(\frac{1}{L'_m} - \frac{1}{L'_{mt}}\right)\sin\varphi\cos\varphi. \qquad (9)$$

From (6) also results the transient angular speed $\omega_\psi(t)$ of the main flux linkage

$$\omega_\psi(t) = \frac{d\varphi}{dt} + \omega_B \qquad (10)$$

ω_B - the angular speed of the general reference frame d,q.

Expressions (8), (9), (10) are valid for any combinations of the state variables discussed above. The general flux linkage ψ and the inductances L'_m, L'_{mt} for various pairs of currents and flux linkages are relatively easily computed.

Let $\underline{i}_s, \underline{\psi}_r$ be state variables.

After performing of the calculus [16] it is obtained $L'_m = L_{r\sigma} + L_m = L_r$ and the general flux linkage $\underline{\psi} = \psi_d + j\psi_q = L'_m\underline{i}_m = \underline{\psi}_r + L_{r\sigma}\underline{i}_s$.

We remark that ψ, i_m, ψ_m are co-linear and ψ depends linearly of the state variables i_s, ψ_r.

The components ψ_d, ψ_q of (8) are defined. Separating the real and imaginary components in the mathematical model (1) where $X_{dq} = \left|i_{sd} \quad i_{sq} \quad \psi_{rd} \quad \psi_{rq}\right|^T$

the matrices A, B have the form

$$A = \begin{vmatrix} L_\sigma - \dfrac{L^2_{r\sigma}}{L_{dd}} & -\dfrac{L^2_{r\sigma}}{L_{dq}} & 1 - \dfrac{L_{r\sigma}}{L_{dd}} & -\dfrac{L_{r\sigma}}{L_{dq}} \\ \dfrac{L^2_{r\sigma}}{L_{dq}} & L_\sigma - \dfrac{L^2_{r\sigma}}{L_{qq}} & -\dfrac{L_{r\sigma}}{L_{dq}} & 1 - \dfrac{L_{r\sigma}}{L_{qq}} \\ 0 & 0 & 1 & 0 \\ 0 & 0 & 0 & 1 \end{vmatrix}$$

$$B = \begin{vmatrix} R_s & -\omega_B\left(L_\sigma - \dfrac{L_{r\sigma}^2}{L_r}\right) & 0 & -\omega_B\dfrac{L_m}{L_r} \\[2mm] \omega_B\left(L_\sigma - \dfrac{L_{r\sigma}^2}{L_r}\right) & R_s & \omega_B\dfrac{L_m}{L_r} & 0 \\[2mm] -R_r\dfrac{L_m}{L_r} & 0 & \dfrac{R_r}{L_r} & -(\omega_B-\omega) \\[2mm] 0 & -R_r\dfrac{L_m}{L_r} & \omega_B-\omega & \dfrac{R_r}{L_r} \end{vmatrix}$$

where $L_\sigma = L_{s\sigma} + L_{r\sigma}$.

The electromagnetic torque M in the coordinates of the state variables i_s, ψ_r is obtained from (2) by eliminating $\underline{\psi}_s$. We have

$$M = \frac{3}{2} p \frac{L_m}{L_r}\left(\psi_{rd} i_{sq} - \psi_{rq} i_{sd}\right)$$

2) The current i_m - currents (flux linkages) of the windings as state variables.

Everywhere in equations where i_m occurs as state variable in combination with the currents or the flux linkages of the windings, the term $\dfrac{d}{dt}\left(L_m \underline{i}_m\right)$ has to be computed. This remark is valid also for the state variables (i_s, i_r) satisfying $i_s + i_r = i_m$. The following expression is used [14]

$$\frac{d}{dt}\left(L_m \underline{i}_m\right) = L_{dd}\frac{di_{md}}{dt} + L_{dq}\frac{di_{mq}}{dt} + j\left(L_{qq}\frac{di_{mq}}{dt} + L_{dq}\frac{di_{md}}{dt}\right) \quad (11)$$

where the computation inductances are:

$$L_{dd} = L_{mt}\cos^2\varphi + L_m\sin^2\varphi, \quad L_{qq} = L_{mt}\sin^2\varphi + L_m\cos^2\varphi$$
$$L_{dq} = (L_{mt} - L_m)\sin\varphi\cos\varphi$$

$$(12)$$

The angular velocity of the phasor ψ_m in the reference frame B has the form [16]

$$\frac{d\varphi}{dt} = \frac{1}{i_m}\left(\frac{di_{mq}}{dt}\cos\varphi - \frac{di_{md}}{dt}\sin\varphi\right)$$

$$(13)$$

Equations (11)÷(13) are generally applicable regardless the state variables associated to i_m.

Let i_m, ψ_s be state variables.

After straightforward computation [16] and taking also into account (11) in equation (1) where

$$X_{dq} = \begin{vmatrix} i_{md} & i_{mq} & \psi_{sd} & \psi_{sq} \end{vmatrix}^{\mathrm{T}}$$

it follows

$$A = \begin{vmatrix} 0 & 0 & 1 & 0 \\ 0 & 0 & 0 & 1 \\ L_{r\sigma}+K_\sigma L_{dd} & K_\sigma L_{dq} & 1-K_\sigma & 0 \\ K_\sigma L_{dq} & L_{r\sigma}+K_\sigma L_{qq} & 0 & 1-K_\sigma \end{vmatrix}$$

$$B = \begin{vmatrix} -R_s\dfrac{L_m}{L_{s\sigma}} & 0 & \dfrac{R_s}{L_{s\sigma}} & -\omega_B \\[2mm] 0 & -R_s\dfrac{L_m}{L_{s\sigma}} & \omega_B & \dfrac{R_s}{L_{s\sigma}} \\[2mm] R_r\dfrac{L_s}{L_{s\sigma}} & (\omega_B-\omega)(L_{r\sigma}+K_\sigma L_m) & \dfrac{R_r}{L_{s\sigma}} & -(\omega_B-\omega)(1-K_\sigma) \\[2mm] (\omega_B-\omega)(L_{r\sigma}+K_\sigma L_m) & R_r\dfrac{L_s}{L_{s\sigma}} & (\omega_B-\omega)(1-K_\sigma) & \dfrac{R_r}{L_{s\sigma}} \end{vmatrix}$$

where $k_\sigma = 1 + L_{r\sigma}/L_{s\sigma}$.

As previously, the electromagnetic torque is obtained from the expression

$$M = \frac{3}{2} p \frac{L_m}{L_{s\sigma}}\left(\psi_{sq} i_{md} - \psi_{sd} i_{mq}\right)$$

B. The saturated anisotropic synchronous machine.

The mathematical model (1) specified for the synchronous salient poles machine is presented in [17], where

$$\psi_d = L_{s\sigma} i_d + \psi_{md}, \quad \psi_q = L_{s\sigma} i_q + \psi_{mq}$$

$$\psi_E = L_{E\sigma} i_E + L_{DE\sigma} i_D + \psi_{md}$$

$$\psi_D = L_{D\sigma} i_D + L_{ED\sigma} i_E + \psi_{md}, \quad \psi_Q = L_{Q\sigma} i_Q + \psi_{mq}$$

$$\psi_{md} = L_{md} i_{md}, \quad \psi_{mq} = L_{mq} i_{mq}$$

$$i_{md} = i_d + i_E + i_D, \quad i_{mq} = i_q + i_Q$$

The usual notations are used. The rotor values are also related to the stator. The damping windings D,Q are short-circuited.

For a salient poles machine the expression for the main inductance in d-axe have the form [15]

$$L_{md} = \frac{2\mu_0}{\pi^2} m_1 (N_1 K_{B1})^2 \frac{\tau_i}{p\delta_0''} K_{ad} \quad (14)$$

where $\quad \delta_0'' = \delta_0 K_C K_S \quad (15)$

is the computational d-axis air-gap. K_C is the Carter coefficient and K_S – the saturation coefficient, assumed to be the same for all main flux lines. In this conditions

$$K_{ad} = \frac{2}{\tau}\int_0^\tau \frac{\delta_0''}{\delta''(x)}\sin^2\frac{\pi}{\tau}x = \frac{2}{\tau}\int_0^\tau \frac{\delta_0}{\delta(x)}\sin^2\frac{\pi}{\tau}x = const. \quad (16)$$

A analogous expressions are obtained for L_{mq}, K_{aq}. According to (16) the coefficients $K_{ad,q}$ depend on machine geometry only; they are given in [15] a.o. For a smooth air-gap machine with the air-gap $\delta(x) = \delta_0$ it follows that $L_{md} = L_{mq} = L_m$ where L_m has the form (4) with $K_{ad} = 1$.

The inductances L_{md}, L_{mq} correspond to a machine with constant air-gap and non-magnetic gap along the d axis (cross – hatched in fig. 1).

This machine displays for all lines of the direct field ψ_{md} the computational air–gap $\delta_0^{//} / K_{ad}$ and for all lines of the quadrature field ψ_{mq} the computational air–gap $\delta_0^{//} / K_{aq} = \delta_0^{//} / K_{ad}) + + (\delta_0^{//} / K_{aq} - \delta_0^{//} / K_{ad}$; obviously the thickness of the non–magnetic gap is $\delta_0^{//} (1 / K_{aq} - 1 / K_{ad})$. The machine with this computational rotor having harmonically distributed windings considers saturation and space fundamental harmonics of the field. The inductances L_{md}, L_{mq} depend on saturation via $\delta_0^{//}$ given by (15).

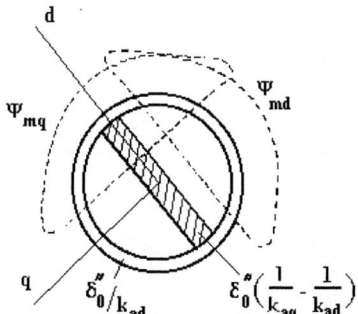

Figure 1. The computational rotor

1) The introduction of the computational smooth air-gap machine The main flux linkages are given by

$$\psi_{md} = L_{md} i_{md}, \ \psi_{mq} = L_{mq} i_{mq} \qquad (17)$$

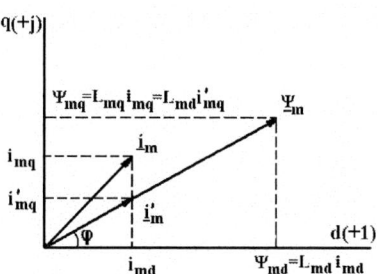

Figure 2. The current and the flux linkages

One may also write

$$\psi_{mq} = L_{md} \frac{L_{mq}}{L_{md}} i_{mq} = L_{md} i_{mq}^{'} \qquad (18)$$

Since it is assumed the same saturation degree for both d-and q-axes, we have

$$\frac{L_{md}}{L_{mq}} = \frac{K_{ad}}{K_{aq}} = m^2 = const. \qquad (19)$$

and therefore $\quad i_{mq}^{'} = m^2 i_{mq} \qquad (20)$

The currents i_{md}, i'_{mq} correspond to a smooth air-gap machine with the main inductance L_{md}. In the following we denote $L_{md} = L_m$, the unique magnetizing inductance.

In fig. 2 there are represented the currents i_{md}, i_{mq}, the resulting magnetizing current i_m of the salient pole

machine and the corresponding flux-linkages $\underline{\psi}_{md}$, $\underline{\psi}_{mq}$, $\underline{\psi}_m$.

Since $L_{md} \neq L_{mq}$, $\underline{i}_m$ and $\underline{\psi}_m$ are spatially displaced.

For the adjusted current $i_{mq}^{'}$ it is obtained the resulting magnetizing current $i_m^{'}$ of the smooth air gap machine that sets the saturation level superposed to $\underline{\psi}_m$.

The following relations are immediate

$$\underline{\psi}_m = L_m (i_m^{'}) \underline{i}_m^{'}, \underline{i}_m^{'} = i_m^{'} e^{j\varphi}$$

$$i_m^{'} = \sqrt{i_{md}^2 + m^4 i_{mq}^2} = \sqrt{i_{md}^2 + i_{mq}^{'2}} \qquad (21)$$

$$\varphi = arctg \frac{i_{mq}^{'}}{i_{md}}, \ \psi_m = \sqrt{\psi_{md}^2 + \psi_{mq}^2}$$

Equations (21) are structurally different from those given in [6], [8]; therefore the results are different. We remark that the magnetically isotropic machine requires to re-evaluate the magnetizing current in the q-axis in order to preserve the main flux linkage – a basic condition. By using the smooth air-gap computational rotor the magnetic asymmetry problem is eliminated. It is now clear that saturation is taken into account through the unique magnetizing inductance L_m in the first equation of (12); this inductance is supported to be current-dependent i.e. $L_m = L_m (i_m^{'})$. This dependence is accomplished through air-gap thickness $\delta_0^{//}$, more precisely through k_s – the unique saturation coefficient. The main magnetizing characteristic $\psi_m (i_m^{'})$ may be computed or obtained experimentally. By introducing i'$_m$, according to (21), for the deduction of the mathematical models corresponding to various combinations of state variables, it is possible to make use of the general relations valid for the induction machine. For a direct deduction of the non-linear dynamical models for the saturated synchronous machine in various state representations, a computational algorithm is used, as in the case of the induction machine; in the following some general recurrence relations are defined.

It is computed the derivative

$$\frac{d}{dt} \left(\frac{\underline{\psi}_m}{L_m} \right) = \frac{1}{L_m} \frac{d\underline{\psi}_m}{dt} + \underline{\psi}_m \frac{d}{dt} \left(\frac{1}{L_m} \right) \qquad (22)$$

From (8), with the notation $\psi = \psi_m$, $L'_m = L_m$, $i_m = i'_m$ we obtain

$$\underline{\psi}_m \frac{d}{dt} \left(\frac{1}{L_m} \right) = \left(\frac{1}{L_{dd}} - \frac{1}{L_m} \right) \frac{d\psi_{md}}{dt} + \frac{1}{L_{dq}} \frac{d\psi_{mq}}{dt} +$$

$$+ j \left[\left(\frac{1}{L_{qq}} - \frac{1}{L_m} \right) \frac{d\psi_{mq}}{dt} + \frac{1}{L_m} \frac{d\psi_{md}}{dt} \right] \qquad (23)$$

where the computational inductance are

$$\frac{1}{L_{dd}} = \frac{\sin^2 \varphi}{L_m} + \frac{\cos^2 \varphi}{L_{mt}} \quad \frac{1}{L_{qq}} = \frac{\cos^2 \varphi}{L_m} + \frac{\sin^2 \varphi}{L_{mt}}$$

$$\frac{1}{L_{dq}} = \left(\frac{1}{L_{mt}} - \frac{1}{L_m} \right) \sin \varphi \cos \varphi \qquad (24)$$

1754

$\cos\varphi = \dfrac{\psi_{md}}{\psi_m}$, $\sin\varphi = \dfrac{\psi_{mq}}{\psi_m}$, $L_{mt} = \dfrac{d\psi_m}{di'_m}$ - the differential

cyclic inductance associated to L_m.

Separating the real and imaginary parts in (22), (23) it follows

$$\frac{d}{dt}\left(\frac{\psi_{md}}{L_m}\right) = \frac{1}{L_{dd}}\frac{d\psi_{md}}{dt} + \frac{1}{L_{dq}}\frac{d\psi_{mq}}{dt}$$
$$\frac{d}{dt}\left(\frac{\psi_{mq}}{L_m}\right) = \frac{1}{L_{qq}}\frac{d\psi_{mq}}{dt} + \frac{1}{L_{dq}}\frac{d\psi_{md}}{dt}$$
(25)

It is computed [9] with the same substitution the derivative

$$\frac{d\underline{\psi}_m}{dt} = \frac{d}{dt}\left(L_m i'_m\right) = L_{dd}\frac{di_{md}}{dt} + L_{dq}\frac{di'_{mq}}{dt} +$$
$$+ j\left(L_{qq}\frac{di'_{mq}}{dt} + L_{dq}\frac{di_{md}}{dt}\right)$$
(26)

where the computational inductance are

$$L_{dd} = L_{mt}\cos^2\varphi + L_m\sin^2\varphi$$
$$L_{qq} = L_{mt}\sin^2\varphi + L_m\cos^2\varphi$$
$$\cos\varphi = \frac{i_{md}}{i'_m}\ \sin\varphi = \frac{i_{mq}}{i'_m}$$
$$L_{dq} = (L_{mt} - L_m)\sin\varphi\cos\varphi$$
(27)

We have obviously

$$\frac{d\psi_{md}}{dt} = L_{dd}\frac{di_{md}}{dt} + L_{dq}\frac{di'_{mq}}{dt}$$
$$\frac{d\psi_{mq}}{dt} = L_{qq}\frac{di'_{mq}}{dt} + L_{dq}\frac{di_{md}}{dt}$$
(28)

The speed of the main magnetic field ψ_{sh} during the dynamic processes will be

$$\omega_\psi = \frac{d\varphi}{dt} + \omega$$
(29)

Taking into account (21) it follows

$$\frac{d\varphi}{dt} = \frac{1}{\psi_m}\left(\frac{d\psi_{mq}}{dt}\cos\varphi - \frac{d\psi_{md}}{dt}\sin\varphi\right) =$$
$$= \frac{1}{i'_m}\left(\frac{di'_{mq}}{dt}\cos\varphi - \frac{di_{md}}{dt}\sin\varphi\right)$$
(30)

2) The mathematical models for various combinations of the state variables

The state variables $\underline{i}_s$, i_E, $\underline{\psi}_m$ The non-linear model (1) is considered, and

$U_{dq} = |u_{sd}\ u_{sq}\ u_E\ 0\ 0|^T$, $X_{dq} = |i_{sd}\ i_{sq}\ i_E\ \psi_{md}\ \psi_{mq}|^T$

By eliminating i_D, i_Q, i_E, ψ_D, ψ_Q from the equations of the synchronous machine in Park co-ordinates [6],[15] there are obtained the terms $\dfrac{d}{dt}\left(\dfrac{\psi_{md}}{L_m}\right)$, $\dfrac{d}{dt}\left(\dfrac{\psi_{mq}}{L_m}\right)$ which

are replaced according to (25). After some transformations for given state vector X_{dq} the matrices A, B have the form

$$A = \begin{vmatrix} L_{s\sigma} & 0 & 0 & 0 & 0 \\ 0 & L_{s\sigma} & 0 & 0 & 1 \\ -L_{DE\sigma} & 0 & L_{E\sigma}-L_{DE\sigma} & 1+\frac{L_{DE\sigma}}{L_{dd}} & \frac{L_{DE\sigma}}{L_{dq}} \\ -L_{D\sigma} & m^2 L_{dq} & L_{DE\sigma}-L_{D\sigma} & 1+\frac{L_{D\sigma}}{L_{dd}} & \frac{L_{D\sigma}}{L_{dq}} \\ 0 & -L_{Q\sigma} & 0 & \frac{L_{Q\sigma}}{m^2 L_{dq}} & 1+\frac{L_{Q\sigma}}{m^2 L_{qq}} \end{vmatrix}$$

$$B = \begin{vmatrix} R_s & -\omega L_{s\sigma} & 0 & 0 & -\omega \\ \omega L_{s\sigma} & R_s & 0 & \omega & 0 \\ 0 & 0 & R_E & 0 & 0 \\ -R_D & 0 & -R_D & \frac{R_D}{L_m} & 0 \\ 0 & 0 & 0 & 0 & \frac{R_Q}{m^2 L_m} \end{vmatrix}$$

The electromagnetic torque is

$$M = \frac{3}{2}p\left[\psi_{md}i_{sq} - \psi_{mq}i_{sd}\right]$$

The state variables $\underline{i}_s$, i_E, i_D, i_Q In the voltage equations appear the terms $\dfrac{d\psi_{md}}{dt}$, $\dfrac{d\psi_{mq}}{dt}$ which are given by (28) where $i_{md}=i_{sd}+i_E+i_D$, $i'_{mq}=m^2(i_{sq}+i_Q)$. To the state vector $X_{dq}=|i_{sd}\ i_{sq}\ i_E\ i_D\ i_Q|^T$ there correspond the matrices

$$A = \begin{vmatrix} L_{s\sigma}+L_{dd} & m^2 L_{dq} & L_{dd} & L_{dd} & m^2 L_{dq} \\ L_{dq} & L_{s\sigma}+m^2 L_{qq} & L_{dq} & L_{dq} & m^2 L_{qq} \\ L_{dd} & m^2 L_{dq} & L_{E\sigma}+L_{dd} & L_{DE\sigma}+L_{dd} & m^2 L_{dq} \\ L_{dd} & m^2 L_{dq} & L_{DE\sigma}-L_{D\sigma} & L_{D\sigma}+L_{dd} & m^2 L_{dq} \\ L_{dq} & m^2 L_{qq} & L_{dq} & L_{dq} & L_{Q\sigma}+m^2 L_{qq} \end{vmatrix}$$

$$B = \begin{vmatrix} R_s & -\omega\left(L_{s\sigma}+m^2 L_m\right) & 0 & 0 & -\omega m^2 L_m \\ \omega\left(L_{s\sigma}+L_m\right) & R_s & \omega L_m & \omega L_m & 0 \\ 0 & 0 & R_E & 0 & 0 \\ 0 & 0 & 0 & R_D & 0 \\ 0 & 0 & 0 & 0 & R_Q \end{vmatrix}$$

The electromagnetic torque is

$$M = \frac{3}{2}pL_m\left[(i_E+i_D)i_{sq} - m^2 i_Q i_{sd} + (1-m^2)i_{sd}i_{sq}\right]$$

and for non-salient poles (m=1)

$$M = \frac{3}{2}pL_m\left[(i_E+i_D)i_{sq} - i_Q i_{sd}\right]$$

III. SIMULATION RESULTS

In order to emphasize the saturation in accordance with the carried out theory, the dynamic reversal process of an asynchronous motor with star connection and rated voltage $U_n = 220$ V is considered. The motor's parameters are:

$R_s = 1{,}16 \; \Omega \; R_r = 4.1 \; \Omega$,

$L_{s\sigma} = 0{,}024 \, \mathrm{H}, L_{r\sigma} = 0{,}0034 \, \mathrm{H}, \mathrm{p} = 2, \mathrm{J} = 0{,}024 \, \mathrm{kgm^2}$.

The magnetic characteristic $\psi_m(i_m)$ is approximated by:

$$\psi_m = \left| \begin{array}{ll} 0{,}4 i_m & i_m < 0{,}75A \\ 0{,}407 \cdot 0{,}921^{i_m} \cdot i_m^{0,81} & i_m > 0{,}75A \end{array} \right.$$

The curves M(t), ω_ψ(t), ω(t) have been plotted at U=300V, in order to ensure a significant saturation level. The exact program ($L_{mt}(i_m) \neq L_m(i_m); L_{dq} \neq 0$) and the classical one ($L_{mt} = L_m(i_m); L_{dq} = 0$) have been used.

The representations from the fig. 3, 5 respective 4, 6 are not essentially different except the final part of the analyzed dynamic process, when the saturation of the main field also occurs and where the exact method emphasizes damped electromechanical oscillations. They are justified essentially by the dynamic variation of the saturation.

It must be noticed (fig. 4 and 6) the evolution of the main field transient speed ω_ψ(t) where occur oscillations overlap, including sense reversals, in tight connection with the rotor transient speed ω(t). By considering $L_{mt}(i_m) \neq L_m(i_m)$ (in general $L_{dq} \neq 0$), the exact method takes into account the dynamic saturation and it appears to be the only one capable to present in a correct manner the entire evolution of a complicated dynamic process. For low saturation, both programs provide results which practically overlap. Other quantitative results are presented in [13], [15] a.o.

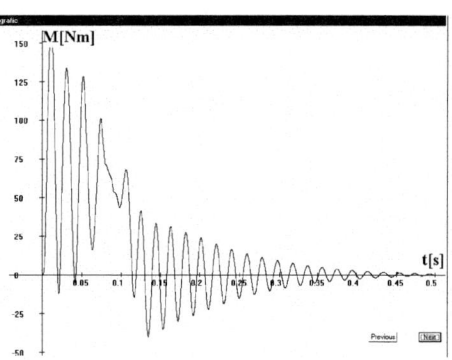

Figure 3. M(t) for $L_{dq} \neq 0$

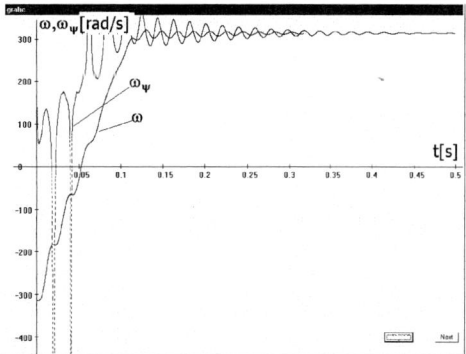

Figure 4. ω(t), ω_ψ(t) for $L_{dq} \neq 0$

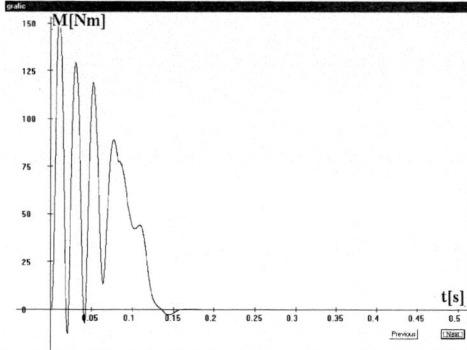

Figure 5. M(t) for $L_{dq} = 0$

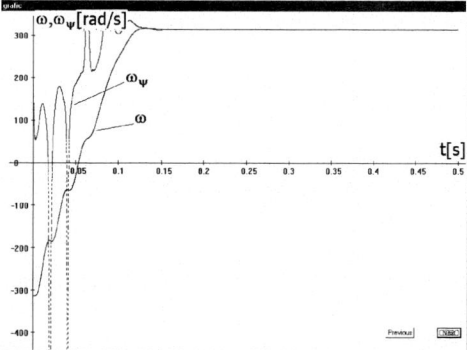

Figure 6. ω(t), ω_ψ(t) for $L_{dq} = 0$

IV. CONCLUSIONS

In the paper there are deduced the mathematical models of the saturated induction machine for various combinations currents - flux linkages. There are analyzed almost all combinations of state variables with the computational inductances of the form (9) or (12), according to the presence of (8) or (11) respectively. The only omitted combination is (ψ_s, ψ_r) when A becomes the identity matrix.

Equation (4) shows that saturation may be included in the theory of the space phasors. It appears thus necessary to consider jointly with the static inductance L'$_m$ the differential inductance L'$_{mt}$, and also the rotation speed of

$\psi(\psi_m)$ in the considered d, q system - via $\dfrac{\mathrm{d}\varphi}{\mathrm{d}t}$.

The paper presents and explains the passage from the anisotropic machine to the non-salient pole machine with the preservation of the fundamental harmonic of the main flux linkage when a unique saturation coefficient is considered.

The equations (1)-(2) remain valid and relations (21) are used. For the sake of generality it is considered also the cross-leakage between E and D windings.

It is shown that, as it should be expected, the expressions of the computational inductance L_{dd}, L_{dq}, L_{qq} of the magnetically isotropic machine (where the equivalent magnetizing current is introduced) are the same with those of the induction machine. The analysis corrects the theory of the non-linear models as it is given in [6], [8] a.o.

The equations (25), (28) are quite general and simplify deduction of the mathematical models for various combinations of state variables.

The equations (6), (10), (13), (29), (30) contain valuable information about the velocity ω_ψ of the main rotating magnetic field during the transient processes.

The numerical simulation points-out that in the presence of saturation the simplified models do not reveal in a correct way the complete evolution of the dynamic electromechanical processes. The dynamic variation of the saturation ($L_{dq} \neq 0$) essentially justifies the forced electromechanical oscillations.

We insist finally on the computational character of the inductances L_{dd}, L_{qq}, L_{dq} which depend on saturation, reference frame and state variables.

V. REFERENCES

[1] K. P. Brown, P. Kovacs, P. Vas, "A method of including the effects of main flux path saturation in the generalized equations of a. c. machines", *IEEE Trans. on* PAS-102, No. 1, 1983.

[2] A. N. El-Serafi, J. Wu, "Saturation representation in synchronous machine models", *Elec. Machines and Power Systems*, Vol. 20, pp. 355-369, 1992.

[3] R. I. Kerkman, "Steady-state and transient analysis of an induction machine with saturation of the magnetizing branch", *IEEE Trans. on Ind. Appl.*, IA-21,No. 1, 1985.

[4] E. Levi, Z. Krzeminski, "Main flux saturation modelling in axis models of induction machines using mixed current - flux state-space models*", ETEP*, No. 6, 1996.

[5] E. Levi, "Modeling of magnetic saturation in smooth air-gap synchronous machines", *IEEE Trans. On Energy Conversion*, Vol. 12, pp151-156, 1997.

[6] E. Levi, "Saturation modeling in D-Q axis models of salient pole synchronous machines", *IEEE Trans. On Energy Conversion*, Vol. 14, No.1, pp 44-50, March 1999.

[7] T. A. Lipo, A. Consoli, "Modelling and simulation of induction motors with saturated leakage inductances", *IEEE Trans. on Ind. Appl.*, IA - 20, No.1, 1984.

[8] L.Pierrat, E. Dejaeger, M.S. Garrido, "Models unification for the saturated synchronous machines", *Proc. Int. Conf. on Evolution and Modern Aspects of Synchronous Machines*, Zürich, Switzerland, pp.44-48, 1991.

[9] I. M. Rousseau, G. Hennuy, "Numerical treatment of the transient behavior of an induction machine taking into account the saturation effect", *OPTIM '96*, Brasov.

[10] S. A. Tahan, I. Kamwa, "A two-factor saturation model for synchronous machines with multiple rotor circuits", *IEEE Trans. On Energy Conversion*, Vol. 10, No. 4 pp. 609-616, 1995.

[11] P. Vas, "Generalized analysis of saturated a.c. machines" *A. f. E.*,Vol. 64, No. 1-2, 1981.

[12] P. Vas, *Electric Machines and Drives*, Clarendon Press, Oxford, 1992.

[13] A. Câmpeanu, "About the saturation effects in the alternative current machine equations", Revue Roum. Sci. Techn-Electrotech. et Energ.,Vol. 38, No.4, 1993.

[14] A. Câmpeanu, "Transient performance of the saturated induction machine", *Electrical Engineering (A. f. E)*, Vol. 78, No. 4 , 1995.

[15] A. Câmpeanu, *Introduction in AC Electrical Machines Dynamics* (in Romanian), Editura Academiei Romane, Bucuresti, 1998.

[16] A. Câmpeanu, "Nonlinear dynamical models for the saturated induction machine", *Revue Roum. Sci. Techn-Electrotech. et Energ.*, Vol 46, No. 1, pp 89-99, January-March 2001.

[17] A. Câmpeanu, "Saturation and saliency in synchronous machines: A corrected model", *OPTIM 2004*, Brasov, Romania.

[18] I. Iglesias, L. Garcia Tabares, I. Tamaret, "A D-Q Model for the Self Commutated Synchronous Machine Considering the Effect of Magnetic Saturation", *IEEE Trans. Energy Conv.*, **7**, *4*, 1992, pp. 768-776.

2006 5th International Power Electronics and Motion Control Conference

Inductance characteristics of 3-phase flux-switching permanent magnet machine with doubly-salient structure

Wei Hua, Cheng Ming

Department of Electrical Engineering, Southeast University, Nanjing, China
huawei1978@seu.edu.cn, mcheng@seu.edu.cn

Abstract—This paper investigates the inductance characteristics of a novel 3-phase 12/10 (12-stator-tooth/10-rotor-pole) flux-switching permanent magnet (FSPM) machine with doubly-salient structure based on finite element (FE) analysis. Firstly, the topology and operation principle of the machine are presented. Secondly, the apparent (static) and incremental (dynamic) inductances are defined. Then, the conventional calculation method of transforming the 3-phase inductances in stator frame into d/q-axes rotor frame is performed, which consumes lots of time. To avoid it, a newly simplified and fast method is proposed to obtain the d-axis and q-axis inductances directly and accurately at only two special rotor positions. The proposed method is validated by the comparison of the predicted results in two ways.

Keywords-flux-switching; doubly-salient; permanent magnet (PM) machine; inductance; finite element (FE) analysis

I. INTRODUCTION

Conventional permanent magnet (PM) brushless machines usually have magnets on the rotor. However, brushless machines having magnets on the stator [1], namely doubly-salient permanent magnet (DSPM) machines [2]-[4], flux-reversal permanent magnet (FRPM) machines [5], and flux-switching permanent magnet (FSPM) machines [6][7], have recently been the subject of considerable research. Since inductances are essential parameters for the dynamic performance predictions of PM brushless machines, which significantly affect the output torque/power and the field-weakening capability, it is necessary to predict inductance accurately for the control system design by taking saturation-effect into account due to the inductances gradually reduce with the increase of armature current. For those novel machines, both the self-inductance inductance and mutual-inductance are mostly calculated by the conventional method step by step for each phase based on finite element (FE) analysis [3][5][6], which consumes much time and a large quantity of data need to be dealt. In this paper, in addition to the conventional way of transforming the inductance in stator frame into rotor frame, a newly simplified and fast method is proposed which can directly

calculate d-axis and q-axis inductances at two special rotor positions avoiding the transforming procedure, significantly saving time and workload. The proposed method is validated by the comparison of the predicted results from two methods with acceptable accuracy.

II. TOPOLOGY AND OPERATION PRINCIPLE

Fig. 1 shows the cross-sections of a 3-phase, 12/10-pole FSPM machine. It can be found that the rotor of the machine is similar to that of a switched reluctance (SR) motor. In addition, the concentrated windings, also same to SR motors, are employed, which leads to low copper consumption and low copper loss due to short end-windings. In the FSPM machine the concentrated coil is wound around the two adjacent teeth with a piece of magnet in the middle. Compared to SR motors, the main difference lies in the configuration of magnets in the stator, containing 12 segments of "U"-shape magnetic cores, between which 12 pieces of magnets are inset pre-magnetized circumferentially in alternative opposite directions. Unlike the conventional PM machines having magnets in the rotor, the placement of both magnets and windings in the stator is favorable for cooling and is desirable for the aerospace and EV applications where the ambient temperature of the machine may be high. The operation principle of FSPM machine can be described in Fig. 2. Obviously, both the value and polarity of the phase PM flux-linkage vary versus rotor position [7]. Due to the PM flux-linkage and consequently the back-emf are essentially sinusoidal [6][7], it makes the FSPM machine an excellent candidate for brushless AC drive operation. The detailed analysis of the structure of the FSPM machine is presented in [7].

III. DEFINITION OF INDUCTANCE

In generally, inductance can be characterized as the property of a circuit element by which energy is capable of being stored in a magnetic field [8]. The terminal voltage u of a general inductor can be expressed as,

$$u = Ri + \frac{d\psi}{dt} = Ri + \frac{d\psi}{di}\frac{di}{dt} = Ri + L_i\frac{di}{dt} \qquad (1)$$

where,

$$L_i = d\psi/di, \qquad (2)$$

1-4244-0448-7/06/$25.00 ©2006 IEEE

and R is resistance of the inductor, ψ is flux-linkage and L_i is inductance. In the absence of ferro-magnetic material, ψ is directly proportional to current i for all values and the inductance can be simply expressed as,

$$L_a = \psi/i \qquad (3)$$

(3) means that inductances are independent of the current and depended only on the topology of the inductor. In order to distinguish the different definitions of inductance, (3) is referred to as the apparent or static inductance, whilst (2) is referred to as the incremental or dynamic inductance [8]. Fig. 3 shows the apparent and incremental inductances for a typical ψ-i curve of an electromagnetic system. Obviously, the apparent inductance will equal or very close to the incremental inductance when the electromagnetic system does not include a magnetic material or the magnetic material is not saturated. However, both the apparent and incremental inductances will gradually decrease as the material becomes saturated, consequently, incremental inductance will be less significantly than the corresponding apparent inductance.

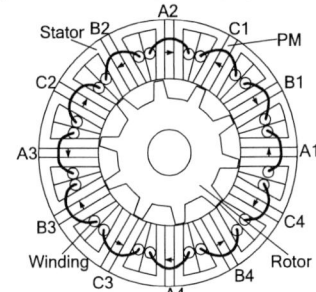

(a) 3-phase 12/10-pole FSPM machine

(b) Configuration of PMs and windings

Figure 1.   Topologies of FSPM machine

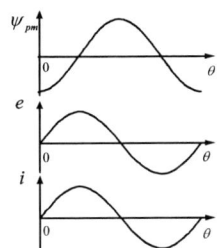

Figure 2.   Operation principle of FSPM machine

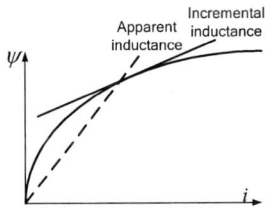

Figure 3. ψ-i curve

IV.   APPARENT INDUCTANCE

A.   Inductance calculation in stator frame

1)   Unsaturated condition

According to (3), the self- and mutual-inductance of the FSPM machine is performed based on two steps to consider the saturation-effect. Firstly, the unsaturated inductance is investigated assuming the PMs being absent, dealt as free space. Whilst, a DC armature current density of J_s=5A/mm^2 is injected into each phase winding respectively. In Fig. 1(a) there are totally four coils contributing to one phase, e.g. coil A_1-A_4 for phase A. Due to the topology symmetry, coil A_1 and A_3 are identical from the viewpoint of magnetic circuit, and so do coil A_2 and A_4. Supposing only one turn wound on the corresponding tooth belonging to each coil, i.e. the winding turns of one coil n_{coil}=1. Thus, for coil A_1,

$$L_{a1} = \frac{\psi_{a1}}{i_a} = \frac{\psi_{a1}}{J_s A_{slot} k_s / n_{coil}} = n_{coil}^2 \frac{\phi_{a1}}{J_s A_{slot} k_s} = \Lambda_{a1} \quad (4)$$

where, L_{a1}, ψ_{a1}, ϕ_{a1}, and Λ_{a1} is the self-inductance, flux-linkage, flux, and self-permeance of the coil A_1, respectively; i_a and J_s is the injected DC armature current and current density; A_{slot} is the half area of one slot; k_s is the slot packing or filling factor. Fig. 4(a) shows the self-inductances per turn of coil A_1 and coil A_2 as well as their sum under unsaturated field, i.e. neglecting the existence of magnets. Due to the asymmetry of magnetic circuit for coil A_1 and A_2 as the rotor rotates in a period of $36^\circ_{\text{mech.}}$[7], there is periodical difference between L_{a1} and L_{a2} as shown. Hence, if the four coils are connected serially to compose one phase, e.g. phase A, the relationship between phase inductance L_a and coil inductances (L_{a1}~L_{a4}) are expressed as follows,

$$L_a = L_{a1} + L_{a2} + L_{a3} + L_{a4} = 2(L_{a1} + L_{a2}) \qquad (5)$$

Obviously, the combined phase inductance is more sinusoidal than the individual coil one, which indicates further that the FSPM machine is more suitable for AC operation. For mutual-inductances, similarly, Fig. 4(b) shows the results of all mutual-inductances of three phase windings. Apparently, M_{ab} is identical to M_{ba}, for example. It is noted that the mean value of mutual-inductance is almost half of that of self-inductance due to the special machine topology. The relationship between the self-inductance and the mutual-inductance is shown in Fig. 5. Fig. 4(c) shows the unsaturated phase inductance including the self- and mutual-inductance, respectively. Fig. 4(d) is the unsaturated 3-phase self-inductances, indicating the perfect symmetry between them with 60°_{elec} phase shift. Hence, the apparent inductances are obtained without taking the saturation-effect into account.

2)   Saturated condition

To investigate the saturation-effect on the inductance characteristics, the existence of magnets is considered, i.e. both the PM excitation and the armature excitation take effect simultaneously. In this case, besides the magnetic field induced by armature currents, the PMs field is included and (4) can be re-written as follows,

1759

$$L_{a1} = \frac{\psi_{a1} - \psi_m}{i_a} = \frac{\psi_{a1} - \psi_m}{J_s A_{slot} k_s / n_{coil}} = n_{coil}^2 \frac{\phi_{a1} - \phi_m}{J_s A_{slot} k_s} \quad (6)$$

where, ψ_m and ϕ_m are the flux-linkage and flux per coil produced by magnets, respectively. For the armature reaction may strengthen or weaken the PM magnetic field, so two cases are considered here by injecting the current density of 5A/mm^2 and -5A/mm^2 respectively into one phase windings, respectively. Fig. 4(e) and (f) shows the comparison results of self- and mutual-inductance under three loadings, i.e., only armature field, pro-magnetized field and de-magnetized field, respectively. Obviously, due to the saturation effect, both the self- and mutual-inductance are significantly reduced compared to the unsaturated results. However, in both the saturated fields, the average value of inductances are almost the same, i.e. the inductances under PM strengthening action is similar to that under PM weakening action due to bipolar PM flux-linkage, which is different from the unipolar PM flux-linkage in DSPM machine [3]. Moreover, the waveforms of saturated inductances including self- and mutual-inductances are quite different from those on the unsaturated condition with the reversal rotor position of peak and bottom values. It should be noted that this difference will play an important role in the relative values of d-axis inductance L_d and q-axis inductance L_q as presented in the followings.

B. Inductance calculation in rotor frame

The apparent inductance in stator frame reference has been obtained. However, since the 3-phase FSPM machine is proposed as a synchronous motor with the sinusoidal inductances, the machine can be analyzed and controlled in the rotor frame, i.e. the classic dq-axes model shown in Fig. 6. Considering this, due to $L = \psi/i$, through the famous Park-transformation, the L_d and L_q can be obtained as follows,

$$L(d,q,0) = P^* \psi(a,b,c)^* i(a,b,c)^{-1} * P^{-1} = P^* L(a,b,c)^* P^{-1} \quad (7)$$

where,

$$P = 2/3 * \begin{bmatrix} \cos\theta & \cos(\theta - 120^\circ) & \cos(\theta + 120^\circ) \\ -\sin\theta & -\sin(\theta - 120^\circ) & -\sin(\theta + 120^\circ) \\ 1/2 & 1/2 & 1/2 \end{bmatrix} \quad (8)$$

$$\psi(a,b,c) = \begin{bmatrix} \psi_a \\ \psi_b \\ \psi_c \end{bmatrix} \quad (9) \text{ and } i(a,b,c) = \begin{bmatrix} i_a \\ i_b \\ i_c \end{bmatrix} \quad (10)$$

According to (7), the self- and mutual-inductance can be transformed into L_d and L_q. Fig. 4(g) shows the transformed unsaturated L_d and L_q based on the unsaturated $L(a,b,c)$. It can be seen that the variation of L_d and L_q in terms of the rotor position is so small that they can be regarded as constants, and the other components including L_{dq}, L_{qd}, L_{d0}, L_{q0} are almost zero, satisfying exactly the request of the d-q axes model. Similarly, the L_d and L_q in rotor frame is also affected by the armature field as that does on self- and mutual-inductance in stator frame. Fig. 4(h) shows the saturated L_d and L_q based on the Park-transformation. It is interesting that the relative values between L_d and L_q are reversely changed, i.e.,

under unsaturated condition: $L_d > L_q$; under saturated conditions: $L_q > L_d$. This unique characteristic should be paid more attention in the design and control of the machine since the relative values between L_d and L_q affect the control algorithm and the flux-weakening capability of the machine significantly [8]. The small value of L_d will reduce the flux-weakening capability of the FSPM machine with the fixed PM flux-linkage ψ_m, and rated current I_a, as shown in

$$k_{fw} = L_d I_d / \psi_m \quad (11)$$

where k_{fw} is the factor to reflect the flux-weakening capability of the machine. For the FSPM machines adopting ferrite material, the flux-weakening capability can reach infinite speed in theory [6] due to lower open-circuit flux-linkage ψ_m. However, in the case of NdFeB material, the merit will discount due to higher saturated ψ_m and smaller L_d with the same rated current I_a.

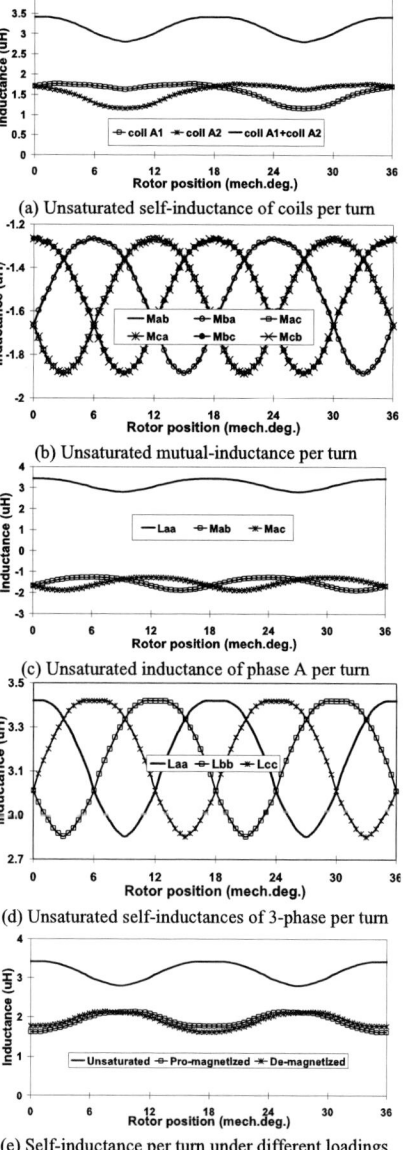

(a) Unsaturated self-inductance of coils per turn

(b) Unsaturated mutual-inductance per turn

(c) Unsaturated inductance of phase A per turn

(d) Unsaturated self-inductances of 3-phase per turn

(e) Self-inductance per turn under different loadings

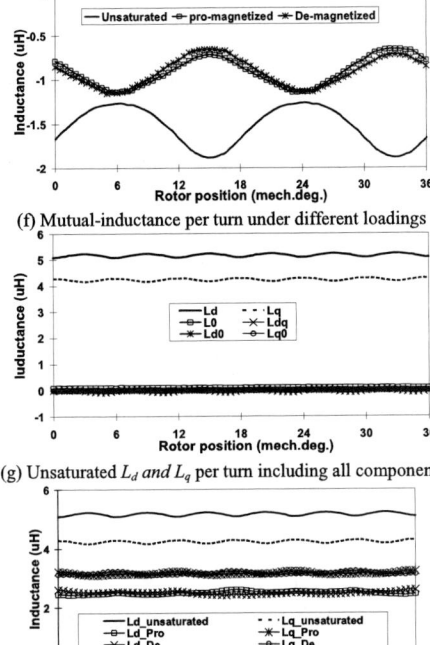

(f) Mutual-inductance per turn under different loadings

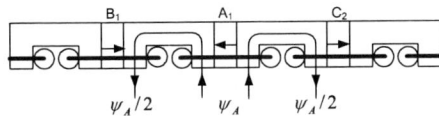

(g) Unsaturated L_d and L_q per turn including all components

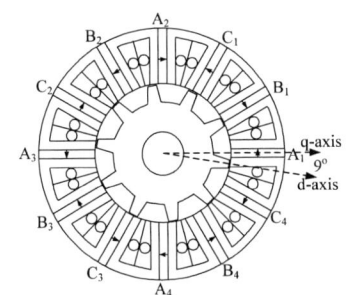

(h) L_d and L_q per turn under different loadings

Figure 4.   Apparent inductances of FSPM machine

Figure 5.   Relationship of self- and mutual-inductances

Figure 6.   Definition of *dq*-axes of FSPM machine

C.   The simplified calculation: two-position method

The d- and q-axis synchronous inductance L_d and L_q can be deduced from the calculated self- and mutual-inductance $L(a,b,c)$ as discussed above. However, both $L(a,b,c)$ and L_d and L_q are dependent on the working point of both hard and soft magnetic materials when saturation is considered. As a fact, it will cost too much effort and computing time to calculate self- and mutual-inductances in the stator frame because the finite element (FE) analysis has to be repeated over a whole rotor pole pitch step by step to get the complete waveforms of $L(a,b,c)$ before transforming them into L_d and L_q. Thus, a

simplified and fast approach to avoiding the time-consuming procedure is proposed based on the *dq*-axes theory and the equations in the following sections, in which the d- and q-axis inductances can be directly obtained according to only two special rotor positions, so it is called two-position method.

When the rotor position $\theta_r=0$, i.e. the d-axis is displaced from the center of magnet of coil A_1, shown in Fig. 6. According to (8), if the machine is supplied with the DC currents as $I_a=I$, $I_b=-I/2$, and $I_c=-I/2$, then only the d-axis current exists, i.e. $I_d=I$ and $I_q=0$. Thus, the corresponding d-axis flux-linkage can be calculated:

$$\psi_d = \frac{2}{3}(\psi_a - \frac{1}{2}\psi_b - \frac{1}{2}\psi_c) \qquad (12)$$

Considering the defined (3) of apparent inductance,

$$L_d = \frac{\psi_d - \psi_m}{I_d} \qquad (13)$$

Where, ψ_m is d-axis PM flux-linkage transformed into rotor frame, equal to the peak value of the PM flux-linkage per phase in stator frame. Similarly, keeping supplied armature currents unchanged, i.e. $I_a=I$, $I_b=I_c=-I/2$, rotating the rotor $9°_{mech.}$, i.e. $\theta_r=-90°_{elec}$ (the rotor pole number is 10), then only the q-axis current exists, i.e. $I_d=0$, and $I_q=I$. Hence, the q-axis inductance can be calculated in the same way, which is only caused by I_q

$$L_q=\psi_q/I_q \qquad (14)$$

By (13) and (14), the d- and q-axis inductance can be obtained only by these two special positions avoiding the repeated calculations of inductance in the stator frame step by step. Table I compares the corresponding predicted results under unsaturated condition (only armature field), as well as saturated conditions (pro-magnetized field and de-magnetized field), respectively by the conventional method and the proposed two-position method. It is obviously that the predicted L_d and L_q are in good agreements by two methods under three conditions, effectively validating the two-position method.

V.   INCREMENTAL INDUCTANCE

In order to further accurately predict the dynamic behaviour and performance of electromagnetic systems with saturation, a more accurate knowledge of the value of the incremental inductance ($L_i=d\psi/di$), rather than the apparent inductance ($L_a=\psi/i$) is required [8]. This is because the inductance voltage terms $d\psi/dt$ encountered in such models can be readily expressed as $\dfrac{d\psi}{di}\dfrac{di}{dt}$ when saturation is an important factor. Since iron saturation has to be considered, the d- and q-axis synchronous inductances can be defined in terms of the apparent and incremental inductances respectively as follows,

$$L_{di} = \frac{d\psi_d}{dI_d} \qquad (15a) \text{ and } L_{qi} = \frac{d\psi_q}{dI_q} \qquad (15b)$$

$$L_{da} = \frac{\psi_d - \psi_m}{I_d} \qquad (16a) \text{ and } L_{qa} = \frac{\psi_q}{I_q} \qquad (16b)$$

where, L_{di} and L_{qi} is the incremental *dq*-axes inductances, respectively; L_{da} and L_{qa} is the apparent *dq*-axes inductances relative to the open-circuit working point, respectively. Both apparent and incremental inductances need to be used in order to account for saturation and to predict the dynamic behavior and performance accurately in the *dq*-axes system. To calculate the incremental inductance, on the base of the results above, which are obtained from a DC armature currents density $Js=5\text{A/mm}^2$ field, simultaneously to keep the PMs work points almost the same, a new incremental DC armature field with a current density of 6A/mm^2 is applied. Thus, Equation (15) can be modified as,

$$L_{di} = \frac{d\psi_d}{dI_d} \approx \frac{\Delta\psi_d}{\Delta I_d} = \frac{\psi_{d2}-\psi_{d1}}{I_{d2}-I_{d1}} \qquad (17a)$$

$$L_{qi} = \frac{d\psi_q}{dI_q} \approx \frac{\Delta\psi_q}{\Delta I_q} = \frac{\psi_{q2}-\psi_{q1}}{I_{q2}-I_{q1}} \qquad (17b)$$

The incremental inductance in *d-q* frame can be obtained by (17) adopting two-position method. Besides, to compare with the results based on two-position, the incremental inductance in stator frame is also calculated as,

$$L_{kk} = \frac{\psi_{k2}-\psi_{k1}}{I_{k2}-I_{k1}} \quad (18a) \text{ and } M_{kj} = \frac{\psi_{j2}-\psi_{j1}}{I_{k2}-I_{k1}} \quad (18b)$$

Here, k, j denotes either of (a,b,c). After all incremental inductances in stator frame are calculated, they are transformed into the incremental inductances in rotor frame as done in the section of apparent inductance. Table II compares the results from the two methods, showing a good agreement between them. It can also be found that the incremental inductances are close or slightly lower than the corresponding apparent values, in consistent with the theory analysis in Section III.

VI. CONCLUSION

This paper investigates the inductance characteristics of a 3-phase (12-stator-tooth/10-rotor-pole) flux-switching permanent magnet (FSPM) machine based on finite element (FE) analysis. The apparent and incremental inductances are defined firstly. The conventional calculation method of transforming the inductance in

stator frame into rotor frame is performed. Since the conventional method cost too much computation time, a newly simplified and fast method is proposed to directly calculate the d- and q-axis inductances based on two special rotor positions. The comparison between the results obtained by conventional and proposed methods verifies the effectiveness and accuracy of the proposed two-position method. The proposed method lays a foundation for the further performance analysis, design and control of the FSPM motor.

REFERENCES

[1] S. E. Rauch and L. J. Johnson. Design principles of flux-switching alternators [J]. *AIEE Trans.* 1955, 74III: 1261-1268.

[2] Y. Liao, F. Liang and T. A. Lipo. A novel permanent magnet machine with doubly saliency structure [J]. *IEEE Trans. Industry Applications*, 1995, 3(5): 1069-1078.

[3] M. Cheng, K.T. Chau, and C. C. Chan. Static characteristics of a new doubly salient permanent magnet machine [J]. *IEEE Trans. Energy Conversion*, 2001, 16(1): 20-25.

[4] Ming Cheng, K.T. Chau, C. C. Chan and Qiang Sun. Control and operation of a new 8/6-pole doubly salient permanent magnet motor drive [J], *IEEE Transactions on Industry Applications*, September/October 2003, 39(5): 1363-1371.

[5] R. P. Deodhar, S. Andersson, I. Boldea, et al. The flux-reversal machine: a new blushless doubly-salient permanent-magnet machine [J]. *IEEE Trans. Industry Applications*, 1997, 33(4): 925-934.

[6] E. Hoang, A. H. Ben-Ahmed, J. Lucidarme. Switching flux permanent magnet polyphased machines [J]. *7th European Conf. Power Electronic and Applications*, 1997, 3: 903-908.

[7] W. Hua, Z. Q. Zhu, M. Cheng, et al. Comparison of Flux-Switching and Doubly-Salient Permanent Magnet Brushless Machines [J]. *8th International Conf. on Electrical Machines and System*, 2005. 1: 165-170.

[8] Chen, Y. S., Motor topologies and control strategies for permanent magnet brushless AC drives [D]. PhD Thesis, UK: the University of Sheffield, 1999.

TABLE I.   APPARENT INDUCTANCES PREDICTION

Inductance (μH)	L_{da} (abc-dq)	L_{da} ($\theta_r=0°$)	L_{qa} (abc-dq)	L_{qa} ($\theta_r=-9°_{mech}$)
Unsaturated	5.081	5.071	4.165	4.167
Pro-magnetized	2.488	2.345	3.133	3.128
De-magnetized	2.559	2.671	3.138	3.128

(note: L_{da} and L_{qa} means the apparent d- and q-axis inductance, respectively.)

TABLE II.   INCREMENTAL INDUCTANCES PREDICTION

Inductance (μH)	L_{di} (abc-dq)	L_{di} ($\theta_r=0°$)	L_{qi} (abc-dq)	L_{qi} ($\theta=-9°_{mech}$)
Unsaturated	5.027	5.019	4.123	4.115
Pro-magnetized	2.410	2.109	3.138	3.106
De-magnetized	2.529	2.548	3.150	3.062

(note: L_{di} and L_{qi} means the incremental d- and q-axis inductance, respectively.)

2006 5th International Power Electronics and Motion Control Conference

Performance Index Evaluations of a Micro Axial-flux Switched-reluctance Motor

Cheng-Tsung Liu[*], Yen-Ming Chen[*], and Da-Chen Pang[**]

[*] Department of Electrical Engineering, National Sun Yat-Sen University, Kaohsiung, Taiwan, China
[**] Department of Mech. Eng., National Kaohsiung University of Applied Sciences, Kaohsiung, Taiwan, China

Abstract—For supplying adequate driven force with high operational precision in semiconductor manufacturing industry, construction of a low cost device that can be integrated with the micro-electromechanical system (MEMS) fabrication scheme has been investigated. With both the degraded material property after electroplating process and the geometric constraints of MEMS constructions being considered, magnetic characteristics of a micro axial-flux switched-reluctance motor (μAFSRM) will be carefully analyzed by both analytical modeling and three-dimensional finite element analysis. Supported by a laboratory prototype, the generated torques and attract forces at different pole gaps and overlapped areas of the motor can be evaluated to provide thorough performance indices for relative micro-machine designs and operations.

Keywords-axial flux; finite element analysis; micro-electromechanical system; performance index

I. INTRODUCTION

With its simpler mechanical structure and larger torque supply, even in the miniaturized scale application like micro-electromechanical system (MEMS), the competence of switched-reluctance motor (SRM) is still preserved. For those mass application requirements on micro positioning stage, micro fluidic devices, minimally invasive surgery, relays, and switches, more and more attentions have been addressed on the related device constructions by employing such machine concepts. Based on the similar fabrication schemes as being used on semiconductor associated processes, a micro switched-reluctance motor (μSRM) is thus expected to fulfill these operational objectives.

By using the x-ray lithography and electroplating processing (LIGA processing) scheme, a three-phase switched-reluctance motor was first fabricated and tested by Guckel *et al.* [1], [2] in the early '90s. The rotor of this motor can be levitated in the air gap while providing a maximum speed of 34,000 rpm and an output torque of 10^{-8} N·m. Based on this successfully fabrication process, a three-phase micro linear switched-reluctance motor

(μLSRM), which can provide a propulsive force of 4 mN with a operational power loss less than 200 μW, was then constructed by Ohnstein *et al.* [3] for high-pass optical filter application with a traveling range of 0.78 mm.

As the material properties for micro machine system constructions will be significantly changed after the electroplating process, the commonly used analyzing schemes for electromagnetic motion devices can not be directly applied. Through detailed evaluations on various processing techniques that are used for micro magnetic device fabrications [4], a general observation showed that the magnetic materials made by electroplating will have much lower permeability. The other design and fabrication constraints are the overhead cost and geometric limitations for the associated motor constructions. From these investigations, it is understood that the μLSRM should equipped with phases as few as possible at a designated operational step length, hence a larger slot for winding allocation and thermal dissipation can be achieved. By considering such material and physical constraints in fabrication process, Liu and Chiang evaluated the performance and saturation effects of a bi-directionally operated μLSRM that is operated with only two phase of windings [5]. Detailed comparisons of motor generated propulsive/normal force ratios at various pole shapes and winding current levels were investigated and an optimal pole shape combination under the design constraints was also identified.

A micro axial flux switched-reluctance motor (μAFSRM) that is operated with four-phase stator windings has been proposed in [6], along with the detailed comparisons of motor generated torque and normal force ratios at various pole arrangements through a commercial three-dimensional (3-D) finite element analysis (FEA) package [7]. However, such 3-D FEA will only provide the information for certain specific structures and operational conditions, an analytical model that can provide systematic performance index evaluations of the proposed μAFSRM for associated applications will thus be devised in this paper to support the validity of the proposed machine.

1-4244-0448-7/06/$25.00 ©2006 IEEE

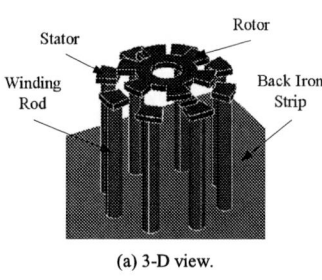

(a) 3-D view.

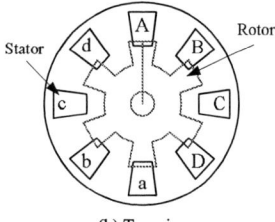

(b) Top view.

Figure 1. Conceptual structure of the proposed micro axial-flux switched-reluctance motor (μAFSRM).

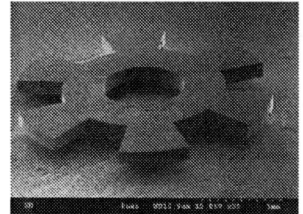

(a) Part of the rotor base.

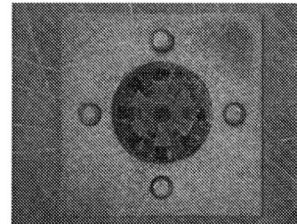

(b) Top view of the assembly.

Figure 2. Laboratory prototype of the constructed μAFSRM for preliminary investigations.

II. THE MICRO AXIAL-FLUX SWITCHED-RELUCTANCE MOTOR

The conceptual structure of the proposed μAFSRM is illustrated in Fig. 1 and its associate physical dimensions are provided in Table I. With 8 stator poles and 6 rotor poles, it is expected that this motor can be conveniently operated by applying appropriate 4-phase currents to its stator windings. Since the rotor of this motor has the tendency of moving to the minimum magnetic resistance positions, with a stator tooth pitch of 45° and a rotor tooth pitch of 60°, it is clear that the step length (SL) of this motor is 15°.

Based on the fundamental arrangement, a laboratory prototype of the μAFSRM for preliminary investigations is depicted in Fig. 2. Due to physical space and material

TABLE I
PHYSICAL DIMENSIONS OF THE μAFSRM

Part	Item	Dimension
Stator	Pole Numbers	8
	External diameter	5.0 mm
	External diameter of teeth	4.6 mm
	Internal diameter of teeth ($2r_2$)	2.8 mm
	Depth (Thickness)	0.2 mm
Rotor	Pole Numbers	6
	External diameter of teeth ($2r_1$)	3.2 mm
	Internal diameter of teeth	2.0 mm
	Internal diameter of bearing	1.0 mm
	Depth (Thickness)	0.2 mm
Winding Rod	Radius	0.5 mm
	Length	4.0 mm
Back Iron Strip	Length	10.0 mm

constraints, in addition to the basic stator and rotor, the stator windings are wound on the winding rod along with a back iron strip to provide the return path for the magnetic flux.

The materials selected for electroplating/fabricating the motor are pure nickel (Ni) and iron (H60 series). For Ni, it has a fairly small relative permeability (μ_r) of about 250 and a saturation flux density level (B_{sat}) of 0.6 T if the anneal temperatures is below 500°C. While with a higher anneal temperature, such magnetic characteristics will be improved to $\mu_r \approx 400$ and $B_{sat} \approx 1.2$ T [1]. For H60 series materials, which can generally provide higher relative permeability ($\mu_r \approx 6000$) and saturation flux density level ($B_{sat} \approx 1.75$ T), certain parts of the μAFSRM that will not affect the main fabrication process can be employed to enhance the designed flux paths. Based on such arrangement evaluation, a general guidance of fabricating the entire μAFSRM system is to select Ni for assembling the stator and rotor, while H60 series materials for the winding rod and back iron strip.

III. ANALYTICAL MODEL DEVELOPMENTS

Since the source is applied sequentially to every one phase of the stator winding, based on the relative arrangements of stator and rotor, it is obvious that the individual rotor pole will only be attracted by the electromagnetic force generated by the corresponding set of stator poles. By first assuming that the air-gap reluctances are much larger than those of the machine poles and winding rods, an equivalent magnetic circuit of such conducting state can be formulated as shown in Fig. 3(a), among which the subscript A denotes the positive winding while a denotes the negative ones. Due to magnetic symmetry, the magnetic circuit can be further

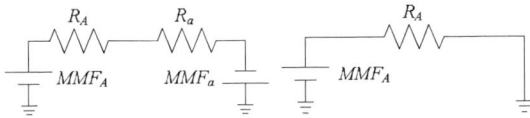

(a) Flux path at one conducting state. (b) Simplified flux path.

Figure 3. Equivalent magnetic circuit at one conducting state of the µAFSRM system.

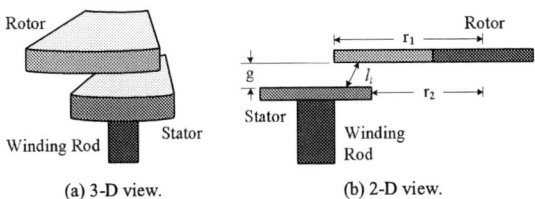

(a) 3-D view. (b) 2-D view.

Figure 4. Geometrical relations between stator and rotor poles.

and using the correction factors α and β to represent the 3-D unalignment of poles, the equivalent air-gap flux path lengths at two of the extreme conducting states can be respectively represented as in

$$l_0 = \left(g^2 + (\alpha \times g)^2 + [\beta \times (\frac{r_1 + r_2}{2} \times SL)]\}\right)^{1/2} . \quad (1)$$

$$l_{SL} = \left(g^2 + (\alpha \times g)^2 \right)^{1/2} . \quad (2)$$

Where l_0 is the equivalent length with the corresponding stator pole first energized, and l_{15} is the equivalent length with the corresponding stator pole is aligned with the rotor pole. The related area at these two flux paths can be derived as in

$$A_0 = (\pi/32) \times (r_1^2 - r_2^2) . \quad (3)$$

$$A_{SL} = (\pi/2)(1/12 + 1/16) \times (r_1^2 - r_2^2) . \quad (4)$$

Hence the resultant equivalent inductances L_0 and L_{SL} at these two distinct positions can be easily devised, and the inductances of the system can then be approximated as in

$$L_\theta = L_0 + (L_{SL} - L_0) \cdot \cos[6(SL - \theta)] . \quad (5)$$

Based on the above formulations, the estimated torques and normal forces can be derived and the operational performance of this µAFSRM at various depth and gap combinations can be illustrated in Fig. 5.

IV. VERIFICATIONS BY THREE-DIMENSIONAL FINITE ELEMENT ANALYSES

To   confirm   the   preceding   performance   index

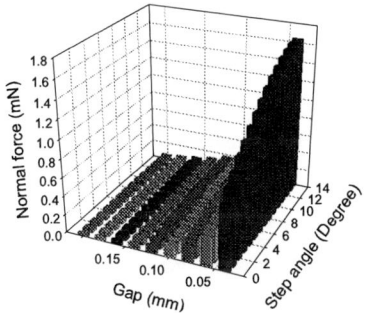

(a) Normal forces at various depth conditions.

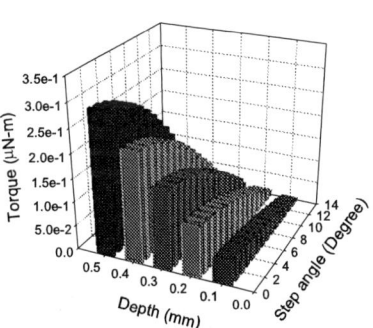

(b) Torques at various depth conditions.

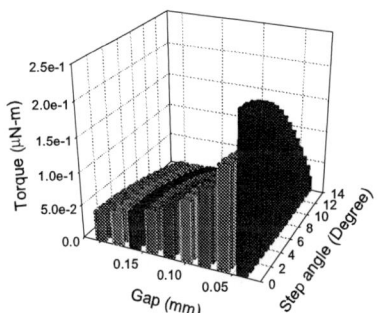

(c) Normal forces at various gap conditions.

(d) Torques at various gap conditions.

evaluations as obtained from the analytical models, 3-D FEA will be thoroughly performed. Since the magnetic property of Ni is relative low compared with those materials commonly used for electromagnetic motion devices, no symmetric simplification in the Cartesian coordinate will be applied for the desired μAFSRM 3-D analyses. As the performance of μAFSRM will be highly dependent on the permeability of selected materials, not only those applicable material properties must be carefully evaluated, but also the related meshes for 3-D FEA at those pole tips of the motor must be precisely addressed for related analyses.

By taking those specific constraints into accounts, parts of the 3-D meshes for the basic type μAFSRM are shown in Fig. 6 and their associated sizes are provided in Table II. With the flux flowing among corresponding stator and rotor poles mainly in the motor axial direction, the generated torques and normal forces within one step length (*SL*) are respectively depicted in Fig. 7 to support a further verification. Clearly it can be confirmed that the devised analytical model is quite applicable for the desired μAFSRM performance index evaluations.

V. CONCLUSION

For light-load semiconductor industry applications in the miniature aspects, the μAFSRM is desired to provide adequate and smooth torques in its rotating directions that can overcome the generated frictions contributed from the axial-direction normal forces of the motor. Especially at the positions where the stator and rotor poles are near-aligned and the motor is quasi-stalled. As can be observed from Fig. 7, the resulting torques at such conditions are much smaller than the values when this stator pole pairs are first energized. Since the gaps and overlapped areas of stator and rotor poles will greatly affect the generated axial normal forces, a concise and accurate enough analytical model that can provide the required operational characteristics of the proposed μAFSRM at various stator and rotor pole geometrical arrangements has been developed. Confirmed by 3-D FEA results, it is convinced that the analytical model will provide the desired performance indices for relative design and operational objectives.

ACKNOWLEDGMENT

The authors wish to express their gratitude to the valuable technical suggestions and assistances from Mr. T.-S. Chiang for system setups and formulations.

REFERENCES

[1] H. Guckel, T. R. Christenson, K. J. Skrobis, T. S. Jung, J. Klein, K. V. Hartojo, and I. Widjaja, "A first functional current excited planar rotational magnetic motor," *Proc. 1993 IEEE Micro Electro Mechanical Systems*, FL, U.S.A., Feb. 1993, pp. 7-11.

[2] H. Guckel, T. R. Christenson, K. J. Skrobis, J. Klein, and M. Karnowsky, "Planar rotational magnetic micromotors with integrated shaft encoder and magnetic rotor levitation," *Proc. 2nd*

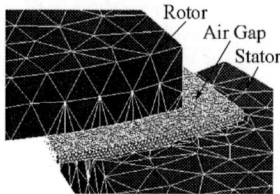

Figure 6. Three-dimensional meshes of the μAFSRM.

TABLE II

PARTITIONED MESH SIZES OF THE μAFSRM

Item	Size
External air area	2.50 mm
Rotor	0.10 mm
Stator	0.10 mm
Air gap and pole tips	0.01 mm
Winding rod and back iron strip	1.00 mm

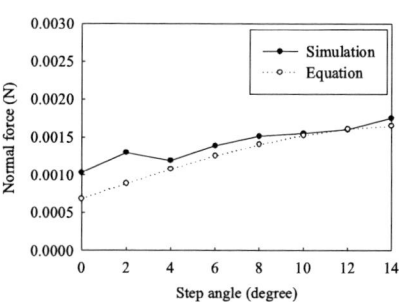

(a) Operational normal forces.

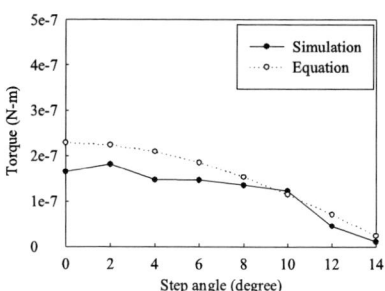

(b) Operational torques.

Figure 7. Operational performance verifications of the μAFSRM by 3-D FEA at fixed depth and gap conditions (depth = 0.4 mm, gap = 0.04 mm).

[3] T. R. Ohnstein, J. D. Zook, and J. B. Starr, *Micromechanical Stepper Motor*, US Patent No. 5,929,542, 1999.

[4] J. Y. Park and M. G. Allen, "Development of magnetic materials and processing techniques applicable to integrated micromagnetic devices," *J. Micromechanics and Microengineering*, vol. 8, 1998, pp. 307-316.

vol. 40, no. 4, July 2004, pp. 2861-2863.

[6] C.-T. Liu, Y.-M. Chen, and D.-C. Pang, "Optimal Design of a Micro Axial Flux Switched-reluctance Motor," *Proc. 2005 Int. Electric Machines and Drives Conference*, San Antonio, TX, U.S.A. May 2005.

[7] Magsoft, *FLUX 3D User's Guide, Version 3.20*, Troy, NY, U.S.A.: Magsoft Corp., Dec. 2000.

Study of Variable Frequency Operation of Induction Generator for Wind Power

Noriyuki Kimura*, Mitsuhiro Hirao*, Toshimitsu Morizane*, Katsunori Taniguchi*

* Osaka Institute of Technology/Faculty of Eng., Electrical and Electronic Systems Eng. Dept., Osaka, Japan
E-Mail: kimura@ee.oit.ac.jp

Abstract— **This paper investigated the reactive power increase when the cage rotor induction machine is connected to the utility system through the full rating VSC link system. The calculation using equivalent circuit indicates that the increase in the reactive power consumption may be suppressed by the decrease in the reactive power consumption of the secondary circuit. The combination of the cage rotor induction generator and the full rating converter link system is not more disadvantageous compared with the synchronous machine and the converter system. More investigation in detail is attractive.**

Keywords-wind power; voltage source converter; equivalent circuit ; cage rotor induction machine;

I. INTRODUCTION

It is often said that wind energy will be the most cost effective source of electrical power in the near future[1,2]. In fact some good case can be said that it already has achieved this status. The actual life cycle cost of fossil fuels is not exactly known, however it is certainly far more than the current wholesale rates.

To make the wind power generation more popular, infinite refinements and improvements are required. Especially the reduction of the cost and the increments of the output is the fundamental issue. We have studied to have more power from the induction generator by using a voltage source converter(VSC) for exchanging the power with utility system.

The induction generator is the most popular and the lowest cost machine for the wind power generation. However it cannot generate power at low speed. The synchronous generator using the permanent magnets can generate power at any speed, but costs high. Hence we have proposed to use the induction generator with VSC system [3].

In this paper, we have studied the characteristics of the induction generator with VSC in more detail.

II. INVESTIGATED WIND POWER GENERATION SYSTEM

Several types of generators can be coupled to the wind turbine. Most popular one is the cage rotor induction machine shown in Fig. 1(a). It needs no power converter to connect to the utility system. However, it has one large

demerit that it cannot generate power at the lower speed than the synchronous speed. Another problem is the large reactive power consumption, which sometimes cause large voltage change in the connected utility system. Meanwhile, the permanent magnet type synchronous machine is often used to compensate these demerits. It can generate power at any speed of the rotor and needs only small amount of reactive power. However, it needs power converter at full power rating, since the generated power has various frequency and cannot be connected directly to the utility system, as shown in Fig.1 (b).

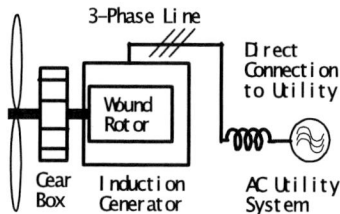

(a) Cage rotor induction machine

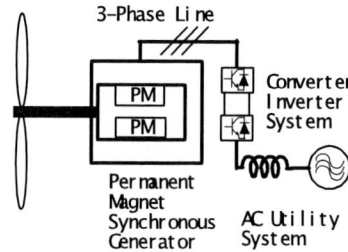

(b) Permanent magnet type synchronous machine

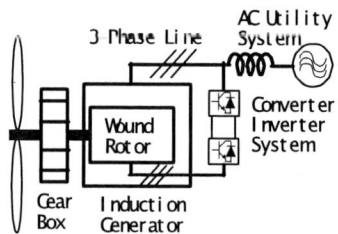

(c) Doubly fed wounded rotor induction machine

Figure 1.   Generator types for wind power generation

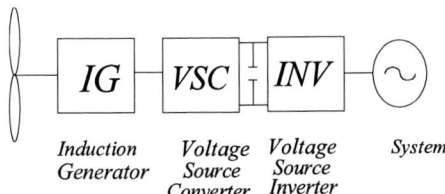

Figure 2. Suggested wind power generation system using cage rotor induction machine (IG: inductions generator)

The cost of machine and additive power converter system is pretty large. To deteriorate the increase of the cost, the doubly fed wounded rotor induction machine is one of the alternative solution. Although it also needs the power converter system in the rotor exciting circuit as shown in Fig.1 (c), the rating of the converter system can be reduced to 1/3 of the full rating necessary for the synchronous machine. However, the cage rotor induction machine is the lowest cost both in manufacturing and maintenance. We suggest to use the cage rotor machine with the converter system of the full rating as shown in Fig. 2. In this configuration, the voltage source converter connected to the induction generator can supply variable frequency ac voltage depending on the rotor speed which is changed by the wind power change. Therefore the induction generator can supply power at any rotor speed. Usually the large amount of reactive power consumption in the induction machine is considered to be big disadvantage compared to the synchronous machine. However the power factor of the induction generator in steady state operation is larger than 0.9. This fact means the increments in the converter rating is 10% compared to the converter for the synchronous machine. Another anxiety is the increase of the reactive power consumption when the frequency is decreased. We have mainly focused on this matter in this paper.

III. Equivalent Circuit of Induction Machine

Output power from the induction generator can be analyzed by using the equivalent circuit. We show some important characteristics of the induction generator by using the equivalent circuit.

A. Equivalent Circuit and its simplification

Single phase diagram of the equivalent circuit of the induction machine is depicted as Fig. 3. A standard equivalent circuit is shown in Fig. 3(a). It is rather complicated to express the output active and reactive power, P and Q, we use simpler L-type circuit shown in Fig. 3(b-d).

The output active and reactive power, Pb and Qb of Fig. 3(b) are calculate by the following equations.

$$P_b = g_0 V_s^2 + \left(r_1 + \frac{r_2}{s} \right) \left(\frac{V_s}{z} \right)^2 \quad ...(1),$$

$$Q_b = b_0 V_s^2 + \left(x_1 + x_2 \right) \left(\frac{V_s}{z} \right)^2 \quad ...(2),$$

$$\text{where, } z = \left(r_1 + \frac{r_2}{s} \right)^2 + \left(x_1 + x_2 \right)^2 \quad ...(3).$$

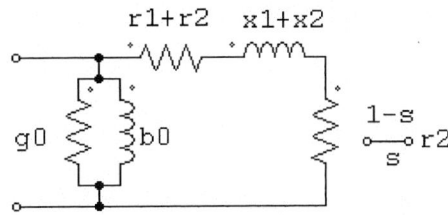

(a) Equivalent circuit (A) : standard

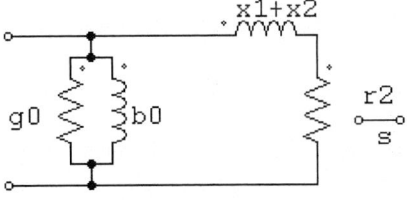

(b) Equivalent circuit (B) : L-type

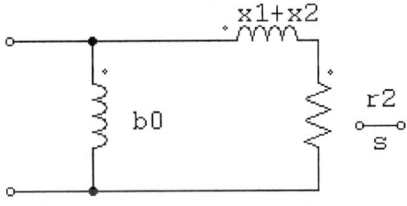

(c) Equivalent circuit (C) : L-type

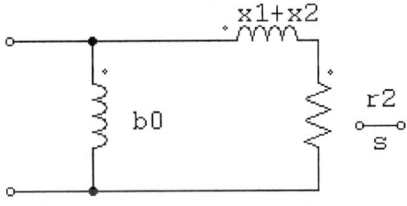

(d) Equivalent circuit (D) : L-type

Figure 3. Various equivalent circuit of induction machine

1769

Resistance in primary winding in Fig. 3(b) can be neglected and then the equivalent circuit is simplified to Fig. 3(c). Moreover the iron loss in the exciting circuit represented by g0 can also be neglected and the equivalent circuit is simplified to Fig. 3(d).

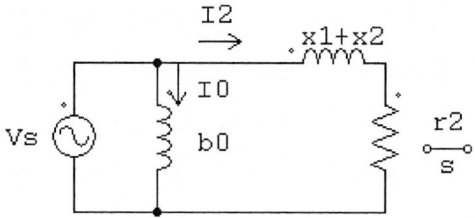

Figure 4.　Simple equivalent circuit for analysis

The currents in the equivalent circuit are calculated by using the circuit in Fig. 4. The output active and reactive power, Pb and Qb of Fig. 3(d) are calculate by the following equations.

$$P_b = \frac{r_2}{s}\left(\frac{V_s}{z}\right)^2 \quad ...(11),$$

$$Q_b = b_0 V_s^2 + (x_1 + x_2)\left(\frac{V_s}{z}\right)^2$$
$$= \frac{1}{\omega L_0} V_s^2 + \omega(L_1 + L_2)\left(\frac{V_s}{z}\right)^2 \quad ...(12),$$

$$\text{where, } z = \left(\frac{r_2}{s}\right)^2 + \omega^2 L^2$$
$$b_0 = 1/\omega L_0, \quad x_1 + x_2 = \omega(L_1 + L_2) \quad ...(13).$$

If slip s is small, active power Pb has little dependency on the frequency, since (r_2/s) can be much larger than ωL. On the other hand, reactive power has dependency on the frequency as follows.

$$Q_b = \left(\frac{A}{\omega} + \omega B\right) V_s^2$$

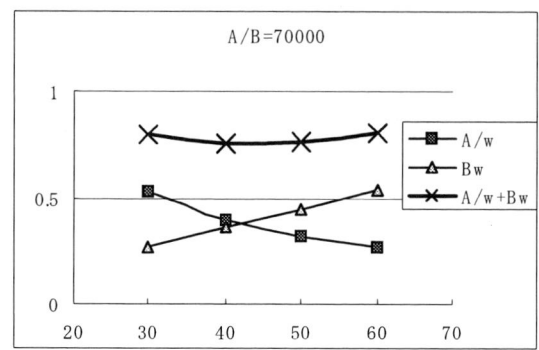

Figure 5.　Desirable parameter condition of induction generator

If the ratio of parameter A and B, that is A/B, equals 70000, the reactive power consumption has little dependency on the frequency between 30[Hz] and 60[Hz] as shown in Fig.5.

IV. ANALYSIS OF INDUCTION GENERATOR

We have investigated the output power and required reactive power for the wind power generator listed in the Table 1. In this case, A/B=38000 approximately.

Table 1. Parameter of induction generator

Rated voltage	690[V]
Rated Power	1000[kW]
Apparent Power	1111[kVA]
Resistance (Primary+Secondary)	0.0114+0.015[ohm]
Reactance (Primary+Secondary)	0.126+0.07[ohm]
Exciting susceptance	0.233[Siemens]

We have calculated the output active power and reactive power when the voltage is kept constant at the rated value, 690[V]. The results are shown in Fig.6. As one can expect, the reactive power is increased when the frequency is decreased.

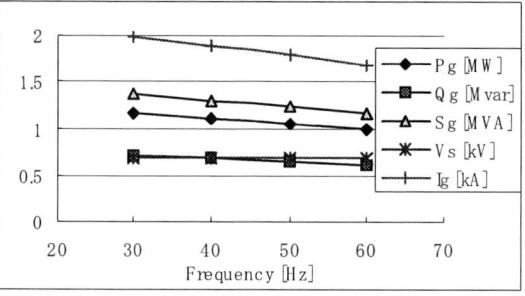

Figure 6.　PQ output diagram of 1MW induction generator

To avoid overcurrent in the VSC current, Ig, the output voltage of the VSC must be decreased. In motor drive, the V/f=constant operation is often used to get the constant torque condition. Fig.7 shows the output active power and reactive power when the V/f=constant operation is used. It is clear that the output active power is decreased substantially, and

1770

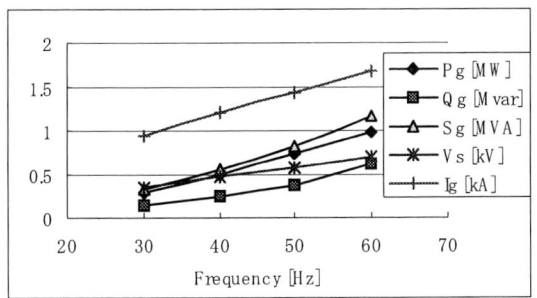

Figure 7.   PQ output diagram of 1MW induction generator

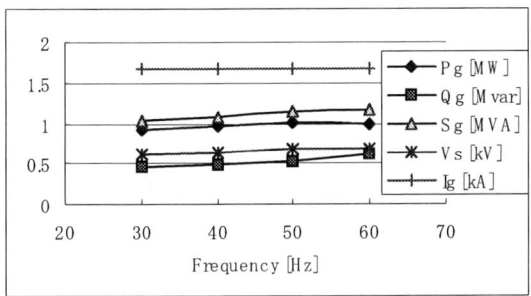

Figure 8.   PQ output diagram of 1MW induction generator

The limitation of the VSC rating is determined by the current in the switching device of the VSC, Ig. By increasing the output voltage of the VSC, Ig reaches its limit and the output power is increased certain amount as shown in Fig.8. This case means the decrease of the output power depending on the output frequency of the VSC is not so large as anxiously expected.

V.   P-Q DIAGRAM OF INDUCTION GENERATOR

In this section, we compare two link converter systems. To investigate the better cost performance, decrease of the rating of the VSC1 is desirable and the system shown in Fig. 9 is investigated. The VSC1 in Fig. 9(a) is replaced by the combination of the VSC1 and the diode rectifier in Fig. 9(b). Usually the diode rectifier costs much lower than the VSC. But the chopper must be installed to match the output voltage of the rectifier to the common dc capacitor voltage. "DI-REC + chop" in Fig. 9(b) means series connected diode rectifier and chopper. In this system, the VSC1 is used to supply only the exciting current of the induction generator. All the generated real power from the induction generator is absorbed by the diode rectifier. Hence, the rating of VSC1 is decreased and the cost reduction is also expected.

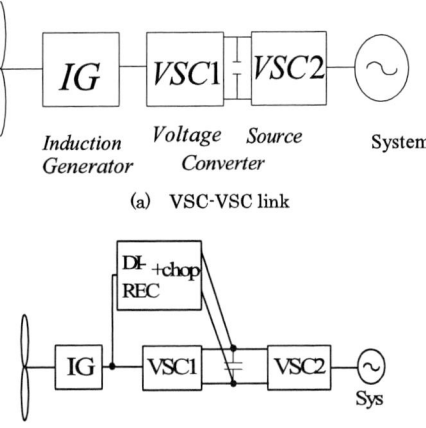

(a)   VSC-VSC link

(b)   VSC&Diode REC-VSC link

Figure 9.   Studied wind power generation link system

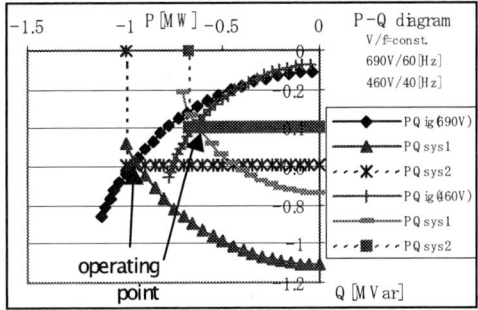

(a)   Output diagram when V/f=constant

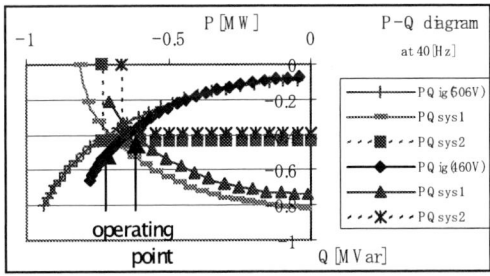

(b) Output change depending on VSC voltage when f=40[Hz]

Figure 10.   PQ output diagram of 1MW induction generator

P-Q diagram are changed as shown in Fig.10(a) when V/f=constant. The output active power and reactive power are decreasing almost proportionally to the frequency. However, as seen in the previous section, increase a certain amount in the output voltage of the VSC1 does not violate the overcurrent of the VSC1. Fig. 10(b) shows that the output active power can be increased near to its rated value. This is realized because the increase in the reactive power is not so large as expected for increasing the voltage.

1771

VI. CONCLUSIONS

We have investigated the reactive power increase when the cage rotor induction machine is connected to the full rating VSC link system. The calculation using equivalent circuit indicated that the increase in the reactive power consumption may be suppressed by the decrease in the reactive power consumption of the secondary circuit. The calculation indicated that compensation between 30[Hz] and 60[Hz] are complete when the parameter ration A/B=70000. The analysis about the real induction machine with the parameter ration A/B=38000 is performed and results show that the system can output almost 90[%] of its ratings.

We have compared the VSC-VSC link system and VSC+Diode REC – VSC link system. The maximum output from both system is almost same. The rating of VSC compared to the diode rectifier is 4/6 or 40[%] of using VSC-VSC link system.

The combination of the cage rotor induction generator and the full rating converter link system is not more disadvantageous compared with the synchronous machine and the converter system. More investigation in detail is attractive.

REFERENCES

[1] S. Meier, et.al, "New Voltage Source Converter Topology for HVDC Grid Connection of Offshore Wind Farms", EPE-PEMC-2004 Riga, Latvia. (European Association of Power Electronics and Applications - Power Electronics and Motion Control conference), Paper No.A131501, 2004.9.2-4.

[2] L. Leclercq, et.al, "Grid Connected or Islanded Operation of Variable Speed Wind Generators Associated with Flywheel Energy Storage Systems", EPE-PEMC-2004 Riga, Latvia. (European Association of Power Electronics and Applications - Power Electronics and Motion Control conference).

[3] Noriyuki Kimura, Toshimitsu Morizane, Katsunori Taniguchi, Takanori Oono, Mitsuhiro Hirao, Takuma Hayashi, "Basic Study of Power Conditioner for Wind Farm Power Generation", Proceedings of The International Conference on Electrical Engineering 2005, Paper No. ICEE-F0516, Kunming, China July 10-14, 2005.

2006 5th International Power Electronics and Motion Control Conference

Optimal Power Control Strategy of Maximizing Wind Energy Tracking and Conversion for VSCF Doubly Fed Induction Generator System

H. Li.*[†], Z. Chen[†], *Senior Member, IEEE*, John K. Pedersen[†], *Senior Member, IEEE*
* The Key Lab. of High Voltage Eng. & Elec. New Technology, Ministry of Education, Elec. Eng. College of
Chongqing University, Chongqing, China
[†] Aalborg University, Aalborg, Denmark
E-mail Lih@ iet.aau.dk; zch@ iet.aau.dk; jkp@ iet.aau.dk

Abstract—This paper focuses on the development of maximum wind power extraction strategies for variable speed constant frequency (VSCF) grid-connected wind power generation systems with a doubly fed induction generator (DFIG). A new optimal control method is proposed by controlling the generator stator active and reactive power, which is based on the condition of the system operation for not only the extracted maximum power of the wind turbine below the rated wind speed but also the higher generator efficiency. Based on the DFIG mathematical models, the optimal stator reactive power value is derived for minimal machine copper losses. According to wind turbine power characteristics and generator power flow equations, the optimal stator active power reference value is also obtained for capturing maximal output power from wind turbines. A dual-passage excitation control strategy is applied to control the active and reactive power independently. Detailed simulation results have confirmed the feasibility and performance of the optimal control strategy.

Keywords- wind power generation; doubly fed induction generator; maximum wind energy extraction; copper loss-minimization; efficiency; excitation control

I. INTRODUCTION

Recently, wind generation systems are attracting great attention and interest as clean, safe and economical renewable sources [1-4]. Due to the highly variable wind velocity, variable speed constant frequency (VSCF) wind generation system has distinct advantages than the traditional constant speed constant frequency (CSCF) wind generation system, such as more effective power capture, lower mechanical stress and less power fluctuation [2]. Three main types of VSCF wind power conversion system are usually adopted in the world [2], such as directly driven synchronous generators, squirrel cage induction generators and doubly fed induction generators (DFIG). The stator of DFIG connects the grid directly and provides for variable speed operation by using a partially rated converter on the rotor side, so it is a common choice for its smaller converter capacity than

other VSCF generators by a number of manufactures of large offshore wind turbines [2-5]. In order to fully realize the benefits of VSCF in the doubly fed wind power generator system, it is critical to develop optimal control strategies to most effectively utilize the wind power energy as wind speed varies.

In order to achieve the maximum power point tracking (MPPT) control, some control schemes have been researched. Some of these strategies depend on measuring or estimating the wind speed [5,6], which is very difficult to do with highly variable wind conditions and adds to system cost in practical implementation. Some methods require more knowledge of wind turbines by using mechanical output power feed back control [7]. To overcome the aforementioned drawbacks, some control strategies have been proposed to continuously search for the peak output power of the wind turbine with the optimum algorithm [8-10], such as hill-climb searching (HCS) control strategy. It usually requires to real-time measurement of the wind turbines speed which may be varying or unstable with wind condition, moreover the interval is usually only determined by some specificity experiments, which influences the peak point calculation of wind turbine output power. Also, above these methods almost neglect to incorporate the DFIG generator efficiency [11-13], that is to say, though it captures the maximal wind energy, the wind power generator system may not inject a maximum value of output active power into the grid, because the DFIG efficiency may not be optimized [14]. Though the control method of optimized power flow with high efficiency of DFIG is proposed in [14], but its optimal solution is based on the simplified steady state equations neglecting the rotor and stator resistances.

In this paper, the main goals of the proposed optimal strategy of DFIG are to choose the operation points depending on the conditions of both the extracted maximum power and copper loss-minimized operation of DFIG. The maximum mechanical power is obtained by the stator active power control without the difficult and complex wind speed measurement. The copper losses are minimized by optimal stator reactive power. Furthermore, a dual-passage excitation control strategy based on accurate DFIG model is applied to control the stator active and reactive power. The optimum performances of DFIG wind generation system have been studied by simulation with Matlab/Simulink.

II. WIND TURBINE CHARACTERISTICS

The mechanical power extracted from a wind turbine, P_T, is dependent on the power coefficient, C_p, for the given turbine operation conditions and is given by

$$P_T = \frac{1}{2} C_p A \rho v^3 \tag{1}$$

where C_p, A, ρ and v are the power coefficient (maximum value Betz's limit 0.593), the area swept by the blades, the specific density of air and the wind speed, respectively. Therefore, if the air density, swept area and wind speed are constant, the output power of the wind turbine will be a function of C_p. However, C_p is a function of the pitch angel of turbine rotor blades, β, and the tip-speed ratio, λ, which is defined as

$$\lambda = \frac{\omega_T R_T}{v} \tag{2}$$

where ω_T and R_T are the rotor angular speed and the radius of the turbine blade. A typical C_p curve is shown in Fig. 1.

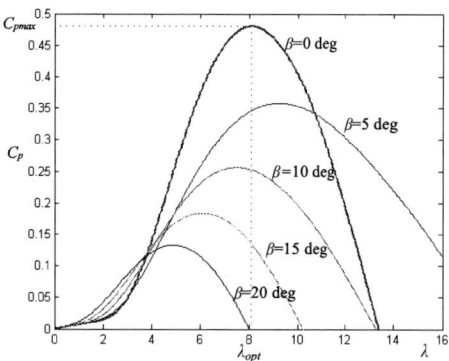

Fig.1. Power coefficient versus tip-speed ratio

The parameters of wind turbine for this calculation are shown in Appendix A Table I. The same parameters are also adopted in simulation of section V. As it can be seen, there is a value of λ for which C_p value is maximized (C_{pmax}) by a given pitch angle β value, and the C_{pmax} value is reduced with the β value increased. To extract the maximum wind energy, pitch angle is usually around zero when the wind speed below the rated value [1].

The per unit (p.u.) system for wind turbine and DFIG is adopted in this paper. The mechanical power equation (1) in p.u. can be written as equation (3) and the power characteristics of wind turbines for different wind speed are shown in Fig. 2.

$$P_{T\_pu} = k_p C_{p\_pu} v_{pu}^3 \tag{3}$$

where C_{p_pu} is power coefficient in p.u. of the maximum value of C_p (where β =0), v_{pu} is wind speed in p.u. of the base wind speed, which is the mean value of the expected wind speed in m/s. This base wind speed produces a mechanical power, which is usually lower than the turbines nominal power [1]. So here k_p is power gain for C_{p_pu}=1 p.u. and v_{pu}=1 p.u., k_p is usually less than or equal to 1. (where k_p is 0.73, and the generator rational speed is 1.2 p.u. at base wind speed 12m/s).

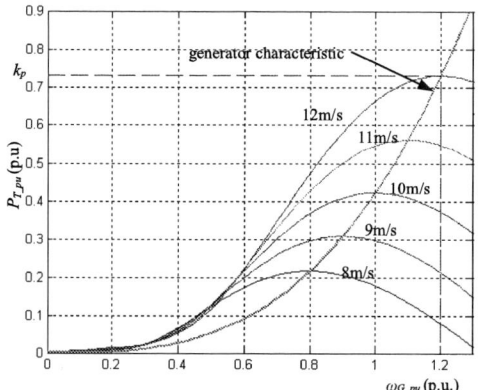

Fig.2. Power characteristics of a wind turbine for various wind speed

As shown in Fig. 2, it can be seen that every wind speed has a generator speed that gives a maximum output power. To capture the maximal power from wind turbines, it is desirable to have a power characteristic that will follow the peak point power curve by regulating generator speed. This maximum power value at various wind speed may be derived by using equations (2) and (3).

$$P_{T\max\_pu} = k_p v_{pu}^3 = k_p \omega_{T\_pu}^3 = k_{opt} \omega_{G\_pu}^3 \tag{4}$$

where $k_{opt} = k_p / 1.2^3$

III. OPTIMAL CONTROL SCHEME OF DFIG

A. Mathematic Model of DFIG

It is well known that the d-axis of the synchronous rotating d-q reference frame is taken as the direction of the stator voltage vector, and the stator and rotor side are taken as the generator convention and motor convention respectively, the steady state voltage equations of DFIG can be expressed as [3] [15] [16] (all of variables are in per unit system in the following expression)

$$\begin{bmatrix} U_s \\ 0 \\ U_{rd} \\ U_{rq} \end{bmatrix} = \begin{bmatrix} -R_s & -X_s & 0 & X_m \\ X_s & -R_s & -X_m & 0 \\ 0 & -sX_m & R_r & sX_r \\ sX_m & 0 & -sX_r & R_r \end{bmatrix} \begin{bmatrix} I_{sd} \\ I_{sq} \\ I_{rd} \\ I_{rq} \end{bmatrix} \tag{5}$$

Where subscripts s and r denote stator and rotor respectively, subscripts d and q denote d-axis and q-axis components. U is voltage, I is current, R is resistance, X_s, X_r and X_m are respectively the self reactance and mutual reactance of the stator and rotor windings respectively, s is slip, U_s is the phase voltage of the stator winding.

B. Minimal Copper Losses Control Strategy

The copper losses in DFIG depend on the machines coil resistances and currents, as described in equation (6)

$$
\begin{aligned}
p_{cu} &= p_{cus} + p_{cur} \\
&= R_s(I_{sd}^2 + I_{sq}^2) + R_r(I_{rd}^2 + I_{rq}^2)
\end{aligned}
\tag{6}
$$

where p_{cus} and p_{cur} are stator and rotor copper loss, respectively, p_{cu} is total copper loss.

Based on equation (5), stator currents can be derived as a function of rotor currents and machine parameters.

$$
\left.\begin{aligned}
I_{sd} &= B[A(X_m I_{rq} - U_s) + X_m I_{rd}] \\
I_{sq} &= B[(X_m I_{rq} - U_s) - A X_m I_{rd}]
\end{aligned}\right\}
\tag{7}
$$

where $A = R_s / X_s$, $B = X_s /(X_s^2 + R_s^2)$

In order to minimization copper loss, the above expressions are substituted into equation (6), the expressions of the copper loss are differentiated with respect to the q-axis current components of the rotor, I_{rq}, for $\partial p_{cu} / \partial I_{rq} = 0$, the optimal I_{rqopt} can be expressed as follows (if $\partial p_{cu} / \partial I_{rd} = 0$, the optimal I_{rdopt} value is zero, which is obviously unreasonable.)

$$
I_{rqopt} = \frac{(1+A^2)B^2 R_s X_m}{R_r + (1+A^2)B^2 X_m^2 R_s} U_s
\tag{8}
$$

To achieve this current value in steady state regulation, the optimal q-axis current component of the stator, I_{sqopt}, can be derived as follows

$$
I_{sqopt} = -\frac{U_s + R_s i_{sd}}{X_s} + \frac{X_m}{X_s} I_{rqopt}
\tag{9}
$$

Equations describing stator active and reactive power from such an induction generator are given as follows

$$
\left.\begin{aligned}
P_s &= U_s I_{sd} \\
Q_s &= U_s I_{sq}
\end{aligned}\right\}
\tag{10}
$$

From equations (9) and (10), the stator reactive power can be obtained for minimizing copper loss.

$$
Q_{sopt} = \left(\frac{-U_s}{X_s} + \frac{X_m}{X_s} I_{rqopt}\right) U_s
\tag{11}
$$

From the above equations, It can be seen that I_{rqopt} value is only related to the machine parameters for a fixed terminal voltage, the optimal stator reactive power value, Q_{sopt}, is constant with active power variation for the copper loss-minimization.

C. Maximal Wind Energy Extraction Strategy

If the stator iron loss is ignored, the power flow equations of DFIG can be expressed as follows [3] [15] [16]

$$
\left.\begin{aligned}
P_s &= P_e - p_{cus} \\
P_e &= \frac{P_m}{1-s} \\
P_{outgrid} &= P_s - P_r
\end{aligned}\right\}
\tag{12}
$$

where P_e is electromagnetic power; P_m is net input mechanical power from the rotor shaft; P_r is active power of the rotor side, $P_{outgrid}$ is the output power into the grid from DFIG system. The block diagram of various active and reactive power flows is marked as Fig. 3.

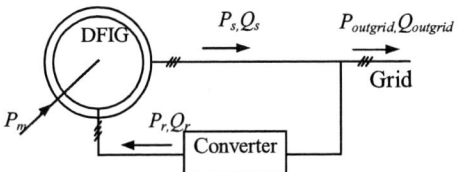

Fig. 3. Configuration of power delivers for DFIG

In order to track the maximal mechanical power of wind turbines, P_m in equation (12) can be substituted by $P_{Tmax\text{-}pu}$ in equation (4) (where the mechanical loss is neglected), the optimal stator reactive power reference value can be derived as

$$
P_{sopt} = \frac{P_{T\max\_pu}}{1-s} - p_{cus} = \frac{k_{opt}\omega_{G\_pu}^3}{1-s} - p_{cus}
\tag{13}
$$

Therefore the proposed optimal control schemes of maximal wind energy tracking and conversion for DFIG wind generation system are described in equations (11) and (13), As a result, the wind power energy can be most effectively utilized, and the maximum generated power is obtained.

IV. DUAL-PASSAGE EXCTION CONTROL STRATEG

A. Control Principle

From the steady state voltage equation (5), the current expressions can be obtained as follows

$$
\begin{bmatrix} I_{sd} \\ I_{sq} \\ I_{rd} \\ I_{rq} \end{bmatrix} = \begin{bmatrix} Y_{11} & Y_{12} & Y_{13} & Y_{14} \\ Y_{21} & Y_{22} & Y_{23} & Y_{24} \\ Y_{31} & Y_{32} & Y_{33} & Y_{34} \\ Y_{41} & Y_{42} & Y_{43} & Y_{44} \end{bmatrix} \begin{bmatrix} U_s \\ 0 \\ U_{rd} \\ U_{rq} \end{bmatrix}
\tag{14}
$$

where Y_{11}, $Y_{12}, \ldots Y_{44}$ are respectively elements of admittance matrix of the voltage equations. the d-axis and q-axis current components can be described as

$$
\left.\begin{aligned}
I_{sd} &= Y_{11}U_s + Y_{13}U_{rd} + Y_{14}U_{rq} \\
I_{sq} &= Y_{21}U_s + Y_{23}U_{rd} + Y_{24}U_{rq}
\end{aligned}\right\}
\tag{15}
$$

So, the equation (10) can be written as

$$
\left.\begin{aligned}
P_s &= Y_{11}U_s^2 + Y_{13}U_s U_{rd} + Y_{14}U_s U_{rq} \\
Q_s &= Y_{21}U_s^2 + Y_{23}U_s U_{rd} + Y_{24}U_s U_{rq}
\end{aligned}\right\}
\tag{16}
$$

It can be seen that stator active and reactive power is related to the d-q-axis rotor voltage, which will be changed with d-axis or q-axis rotor voltage variation. So

1775

it is difficult to decouple control of stator active and reactive power as equation (16).

Applying the admittance matrix of voltage equation (14), the above expressions can be written as [15]

$$P_s = Y_{11}U_s^2 + YU_sU_r \cos(\alpha + \delta) \atop Q_s = Y_{21}U_s^2 + YU_sU_r \sin(\alpha + \delta) \Big\} \quad (17)$$

where

$$Y = \sqrt{Y_{13}^2 + Y_{14}^2} = \sqrt{Y_{24}^2 + (-Y_{23})^2} \,, U_r = \sqrt{U_{rd}^2 + U_{rq}^2}$$

$$\cos\delta = \frac{Y_{13}}{Y}, \sin\delta = \frac{-Y_{14}}{Y}, \cos\alpha = \frac{-U_{rd}}{U_r}, \sin\alpha = \frac{U_{rq}}{U_r},$$

If $U_r \cos(\alpha + \delta)$, $U_r \sin(\alpha + \delta)$ can be considered as the rotor voltage components upon the new synchronous rotating d'-q' reference frame [15], the stator active and reactive powers can be written as

$$P_s = Y_{11}U_s^2 + YU_sU'_{rd} \atop Q_s = Y_{21}U_s^2 + YU_sU'_{rq} \Big\} \quad (18)$$

This result shows that rotor voltage d'-axis and q'-axis components are decoupled for the stator active and reactive power, respectively. In addition, it is derived by the accurate mathematical model, however, the tradition vector control strategy is usually based on neglecting the stator resistance. Thus, the desired stator active and reactive power can be controlled independently based the voltage components upon the d'-q' reference frame. The relationship between the new synchronous rotating d'-q' reference frame and traditional synchronous rotating d-q reference frame is shown in Fig. 4.

$$U_{rd} = -U'_{rd}\sin\delta + U'_{rq}\cos\delta \atop U_{rq} = U'_{rd}\cos\delta + U'_{rq}\sin\delta \Big\} \quad (19)$$

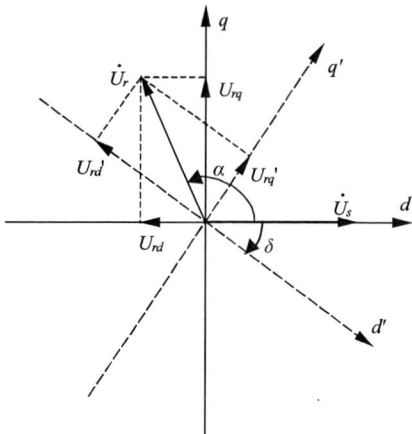

Fig. 4. Diagram of relationship between d'-q' and d-q reference frame

B. Excitation Control Model

The excitation control model based on the d'-q' reference frame can be expressed as [15,16]

$$U'_{rd} = U'_{rd0} + K_p\Delta P_s + K_s\Delta s + K_{is}\int \Delta s dt \atop U'_{rq} = U'_{rq0} + K_q\Delta Q_s + K_{iq}\int \Delta Q_s dt \Bigg\} \quad (20)$$

Where U'_{rd0}, U'_{rq0} is the steady state value of the rotor voltage d'-q' components, respectively, which can be calculated in equation (18); K_p, K_s, K_{is}, K_q and K_{iq} are excitation controller parameters, respectively. ΔP_s, ΔQ_s and Δs are described from the optimal control strategy.

$$\Delta P_s(k) = P_{sopt}(k) - P_s(k-1) \atop \Delta Q_s(k) = Q_{sopt}(k) - Q_s(k-1) \atop \Delta s(k) = s(k) - s(k-1) \Bigg\} \quad (21)$$

The optimal control strategy block of DFIG wind power generation system is shown in Fig. 5. It can be seen that the maximal wind energy is captured by the active power loop control; the copper loss-minimization is controlled by the reactive power loop.

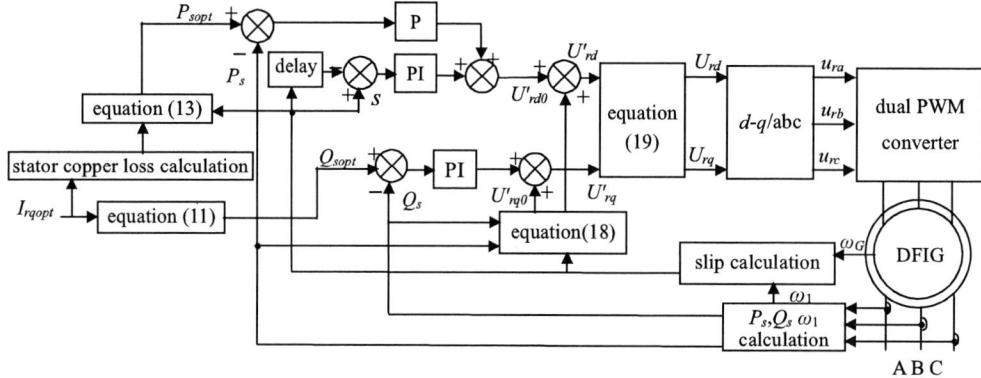

Fig. 5. Optimal control strategy block of maximal wind energy tracking and conversion for DFIG

V. SIMULATION RESULTS

To testify the proposed optimal control strategy, some simulations of dynamic performances for DFIG wind power generating system are implemented.

The wind speed is set as 9m/s at the beginning of the simulation. As a step input, the wind speed jumps to 12m/s at 6 seconds, and then drops to 10m/s at 10 seconds, finally drops to 9m/s at the time of 12 seconds. Due to the wind speed variation, the generator speed is also changed for optimal operation, which is shown in Fig. 6. It can be seen that the generator speed is regulated fast and stably, the stable generator speed in p.u. is 0.8992, 0.99912 and 1.1989, respectively, which is nearly close to theoretic optimal value (0.9, 1.0 and 1.2, respectively). So it may be concluded that the optimal control strategy of tracking maximal wind energy is feasible and correct. The maximum generated active power $P_{outgrid}$, mechanical power $P_{T\text{-}pu}$, stator active and reactive power P_s, Q_s, and rotor active power P_r are respectively shown in Fig. 7. As it can be seen, the maximal mechanical power can be extracted for a given wind speed, the stator active and reactive power can be control independently, the control strategy has good robustness.

In addition, the cooper losses, active powers, reactive powers and efficiency are calculated respectively as different stator reactive power values. When the wind speed is 12m/s (DFIG speed is above synchronous operation) and 9m/s (DFIG speed is below synchronous operation), the results are shown as Table I and Table II, respectively (where * denotes the optimal reactive power value). As it can be seen, the generator efficiency can be significantly improved by controlling the copper loss-minimization based on maximal wind energy tracking, and its advantage is more obvious below the base wind speed.

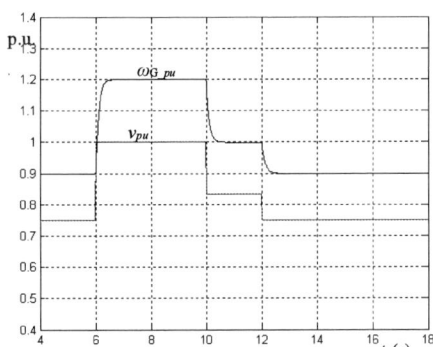

Fig. 6 ω_{G_pu} response with v_{pu} variation

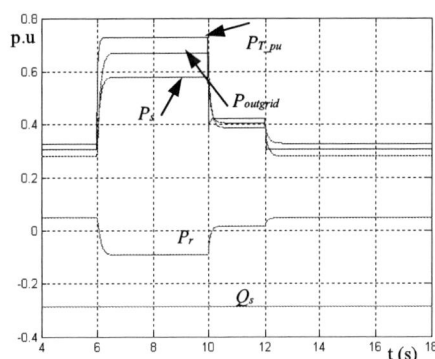

Fig. 7. P_{T_pu}, $P_{outgrid}$, P_s, Q_s and P_r response

TABLE I
THE COPPER LOSSES, POWERS AND EFFICIENCY FOR DIFFERENT STATOR REACTIVE POWER Q_s (V=12m/s)

Q_s	p_{cus}	p_{cur}	p_{cu}	P_s	P_r	Q_r	$P_{outgrid}$	$Q_{outgrid}$	η (%)
-0.2851*	0.0283	0.0292	0.0575	0.5791	-0.0919	-0.0879	0.6710	-0.1972	91.92
0	0.0232	0.0460	0.0692	0.5841	-0.0751	-0.1552	0.6592	0.1552	90.30
-0.7	0.0541	0.0223	0.0764	0.5532	-0.0988	-0.0006	0.6520	-0.6994	89.32
-0.85	0.0688	0.0249	0.0937	0.5385	-0.0963	0.0279	0.6348	-0.8779	86.96
0.7	0.0541	0.1283	0.1824	0.5532	0.0071	-0.3456	0.5461	1.0456	74.81
0.85	0.0688	0.1535	0.2223	0.5385	0.0324	-0.3911	0.5061	1.2411	69.33

TABLE II
THE COPPER LOSSES, POWERS AND EFFICIENCY FOR DIFFERENT STATOR REACTIVE POWER Q_s (V=9m/s)

Q_s	p_{cus}	p_{cur}	p_{cu}	P_s	P_r	Q_r	$P_{outgrid}$	$Q_{outgrid}$	η (%)
-0.2851*	0.0128	0.0139	0.0267	0.3287	0.0484	0.0376	0.2803	0.3227	91.01
0	0.0076	0.0304	0.0380	0.334	0.0650	0.0376	0.2690	-0.0376	87.34
-0.7	0.0395	0.0076	0.0471	0.3021	0.0422	-0.0064	0.2599	0.6936	84.40
-0.85	0.0547	0.0107	0.0654	0.287	0.0452	-0.0206	0.2418	-0.8294	78.52
0.7	0.03951	0.1136	0.15311	0.3021	0.1481	0.1683	0.154	0.5317	50.00
0.85	0.0547	0.1394	0.1941	0.287	0.1739	0.1914	0.1131	0.6586	35.72

VI. CONCLUSION

Based on analyzing the characteristics of wind turbines and the mathematic models of DFIG, a new optimal control strategy of maximal wind energy tracking and conversion is proposed and testified by simulation, the predominance of the control algorithm is to achieve the optimal operation for not only the maximal wind energy extraction but also the generator copper loss-minimization. The dual-passage excitation control strategy is applied to decouple control of the stator active power and reactive power. Simulation results have shown that the stator active and reactive power is controlled independently with varying wind speed conditions; VSCF doubly fed wind generator control system has good robustness. In addition, the wind power energy can be most effectively utilized and the higher generator efficiency is obtained by the proposed optimal power control strategy.

APPENDIX A
PARAMETERS OF DFIG WIND POWER SYSTEM

TABLE I
WIND TURBINE PARAMETERS

MAIN PARAMETERS	VALUE
maximal power coefficient	0.48
optimal tip-speed ratio	8.1
base wind speed (m/s)	12
maximal output power (p.u) at the base wind speed	0.73
generator speed (p.u.) at maximal output power for the base wind speed	1.2
wind turbine speed (p.u.) at maximal output power for the base wind speed	1

TABLE II
DFIG PARAMETERS

MAIN PARAMETERS	VALUE
rated power (kW)	7.5
rated voltage (V)	380
pairs of poles	3
rated frequency (Hz)	50
stator resistance (p.u.)	0.06797
rotor resistance (p.u.)	0.05765
stator leakage inductance (p.u.)	0.09225
rotor leakage inductance (p.u.)	0.09225
mutual inductance (p.u.)	1.61455
The system total moment of inertia (s)	3.694

APPENDIX B
POWER COEFFICIENT CHARACTERISTIC

The power coefficient is approximated as [1]

$$C_p(\lambda,\beta)=0.5176(\frac{116}{\lambda_i}-0.4\beta-5)^{\frac{-21}{\lambda_i}}+0.0068\lambda$$

where

$$\frac{1}{\lambda_i}=\frac{1}{\lambda+0.08\beta}-\frac{0.035}{\beta^3+1}$$

REFERENCES

[1] Siegfried Heier, "Grid integration of wind energy conversion systems," John Wiley & Sons Ltd, 1998, ISBN 0-471-97143-X.

[2] Holdsworth L., Wu X. G., Ekanayake J. B. and Jenkins N., " Comparison of fixed speed and doubly-fed induction wind turbines during power system disturbances," IEE Proc. Generation, Transmission and Distribution, Vol.150, No.3, 2003, pp. 343–352.

[3] Muller S., Deicke M. and De Doncker R. W., "Doubly fed induction generator systems for wind turbine," IEEE Industry Applications Magazine, Vol.8, No. 3, 2002, pp.:26-33.

[4] R. Pena, J. C. Clare and G. M. Asher, " Doubly fed induction generator using back-to-back PWM converts and its application to variable speed wind-energy generation," IEE Proceedings Electrical Power Application, Vol. 143, No.3, pp. 231-241, May, 1996.

[5] T. Thiringer and J. Linders, "Control by variable rotor speed of a fixed-pitch wind turbine operating in a wide speed range," IEEE Trans. Energy Conv., Vol.EC-8, 1993, pp. 520-526

[6] Teodorescu R. and Blaabjerg, F, "Flexible control of small wind turbines with grid failure detection operating in stand-alone and grid-connected mode," IEEE Transactions on Power Electronics, Vol. 19, Sept. 2004, pp.1323-1332.

[7] R. Chedid, F. Mrad, and M. Basma, " Intelligent control of a class of wind energy conversion systems," IEEE Trans. Energy Conv., Vol.EC-14, 1999, pp. 1597-1604.

[8] Rajib Datta and V. T. Ranganathan, " A method of tracking the peak power points for a variable speed wind energy conversion system," IEEE Trans. Energy Conv.,Vol. 18, No. 1, 2004, pp. 163-168.

[9] Ahmed G. Abo-Khail, Dong-choon Lee and Jul-Ki Seok, " Variable speed wind power generation system based on fuzzy logic control for maximum output power tracking," 35th Annual IEEE Power Electronics Specialists Conference, Germany, 2004.

[10] Esmaili, R., Xu, L. and Nichols, D.K., "A new control method of permanent magnet generator for maximum power tracking in wind turbine application," IEEE Power Engineering Society General Meeting 2005, June 12-16, 2005, pp. 1162 – 1167.

[11] Mohamed Orabi, Tarek Ahmed, Mutsuo Nakaoka, M. Z. Youssef, "Efficient performances of induction generator for wind energy utilization," The 30th Annual Conference of the IEEE Industrial Elec-tronics Society, Korea, 2004, pp. 838-843

[12] Tapia, A. Tapia, G. Ostolaza, J.X. Saenz, J.R., "Modeling and control of a wind turbine driven doubly fed induction generator," IEEE Transactions on Energy Conversion, Vol. 18, June 2003, pp.194 – 204

[13] Datta, R. and Ranganathan, V.T., "Variable-speed wind power generation using doubly fed wound rotor induction machine-a comparison with alternative schemes," IEEE Transactions on Energy Con-version, Vol. 17, Sept. 2002, pp.414 – 421

[14] B. Rabelo and W. Hofmann, "Control of an optimized power flow in wind power plants with doubly-fed induction generators", IEEE 34th Annual Power Electronics Specialist Conference, 2003. PESC '03, Vol. 4, 15-19 June 2003, pp:1563-1568 vol.4

[15] LIAO Yong and YANG Shunchang, "The selection principles of the feed back coefficient for the dual-passage excitation generators", Proceedings of the CSEE, Vol. 19, No. 1, 1999, pp. 52-55.

[16] LI Hui and YANG Shunchang, "Study of control strategies for hydroturbine generating system with doubly fed generators based on CMAC neural network", Proceedings of the CSEE, Vol. 24, No. 12, 2004, pp. 187-192.

2006 5th International Power Electronics and Motion Control Conference

Design and Evaluation of a Dual Mechanical Port Machine and System

Longya Xu and Yuan Zhang

Dept. of Electrical and Computer Engineering, The Ohio State University, Columbus, OH 43210, USA
xu.12@osu.edu, zhang.564@osu.edu

Abstract—A Dual Mechanical Port (DMP) machine is a brand new type of electric machine, promising in many applications such as multi-source hybrid traction, integrated starter and generator, variable gearboxes and so forth. In this paper, the dynamic model is established and analyzed. Further, a prototype machine is designed and evaluated by finite element analysis. Several important operation modes are highlighted and the whole control system block diagram is configured for dynamics investigation. The control algorithms with the overall system operation performance are evaluated by computer simulation.

Keywords—electric machine; integrated starter and generator; dual mechanical port; FEA; system simulation

I. INTRODUCTION

A conventional electric machine may have one or two electric ports but only one mechanical port for the purpose of electromechanical energy conversion. A Dual Mechanical Port (DMP) machine is a brand new electric machine, which has more than one mechanically movable part. The prototype to be discussed in this paper has two mechanically movable parts: a PM rotor and an inner wound rotor as shown in Fig. 1. The outer most part of DMP is designed to be stationary, acting as a purely electrical port. The PM rotor is a purely mechanical port since it has no windings. The inner wound rotor functions as both a mechanical and electrical ports. The DMP machine has many different operation modes and these modes of our most interests are promising in applications like multi-source hybrid traction, integrated starter and generator, variable gearboxes, and other high power variable speed applications.

In this paper, the dynamic model is established and analyzed briefly. Further, a prototype machine is designed and evaluated by finite element analysis. Several interesting operation modes are discussed and the whole control system block diagram is configured for dynamic investigation. The control algorithms with the overall system operation performance are evaluated by computer simulations.

II. TRANSIENT MODEL OF DMP MACHINE

The detailed derivation of the transient model of the

DMP machine is given in [1]. Here, only some major equations are presented and discussed. Eqs. (1) through (4) are the voltage equations for the stator and the inner rotor windings.

$$V_{qs}^e = i_{qs}^e r_s + \frac{d\lambda_{qs}^e}{dt} + \omega\lambda_{ds}^e \tag{1}$$

$$V_{ds}^e = i_{ds}^e r_s + \frac{d\lambda_{ds}^e}{dt} - \omega\lambda_{qs}^e \tag{2}$$

$$V_{qr}^e = i_{qr}^e r_r + \frac{d\lambda_{qr}^e}{dt} + (\omega - \omega_r)\lambda_{dr}^e \tag{3}$$

$$V_{dr}^e = i_{dr}^e r_r + \frac{d\lambda_{dr}^e}{dt} - (\omega - \omega_r)\lambda_{qr}^e \tag{4}$$

where r_s is the resistance of the stator windings and r_r the resistance of the inner rotor windings. ω and ω_r represent the rotating speeds of the PM-rotor and the inner rotor. Eqs. (5) through (8) are the flux linkage equations for stator and inner rotor windings.

$$\lambda_{qs}^e = L_s i_{qs}^e + L_m i_{qr}^e \tag{5}$$

$$\lambda_{ds}^e = \lambda_m^e + L_s i_{ds}^e + L_m i_{dr}^e \tag{6}$$

$$\lambda_{qr}^e = L_r i_{qr}^e + L_m i_{qs}^e \tag{7}$$

$$\lambda_{dr}^e = \lambda_m^e + L_r i_{dr}^e + L_m i_{ds}^e \tag{8}$$

where L_s is the self-inductance of the stator windings. L_m is the mutual inductance between the stator and the inner rotor windings. λ_m^e is the flux linkage produced by the PM-rotor.

Eqs. (9) through (11) are the three torque equations for the DMP machine.

$$T_{e,stator} = \frac{3}{2}\frac{P}{2}(\lambda_{dr}i_{qs} - \lambda_{qr}i_{ds}) \tag{9}$$

$$T_{e,pm-rotor} = \frac{3}{2}\frac{P}{2}\lambda_m(i_{qs} + i_{qr}) \tag{10}$$

$$T_{e,wd-rotor} = \frac{3}{2}\frac{P}{2}(\lambda_{ds}i_{qr} - \lambda_{qs}i_{dr}) \tag{11}$$

Eq. (10) indicates that the torque on the PM-rotor is composed of two parts. The first part is due to the interaction between the permanent magnet and the stator current. Likewise, the second part is due to the interaction between the permanent magnet and the inner rotor current.

III. PROTOTYPE DESIGN AND FEM EVALUATION

A 125kw, six-pole prototype DMP electric machine is

designed. Some main dimensions of the prototype are given in Table I.

TABLE I.   MAJOR DIMENSIONS OF DESIGNED DMP

	Stator	Outer rotor	Inner rotor
OD	350	214.2	185.2
ID	215	186	60
Stack L	350	350	350
#of turns	12	PM	12

The FEA geometry model is shown in Fig. 1.

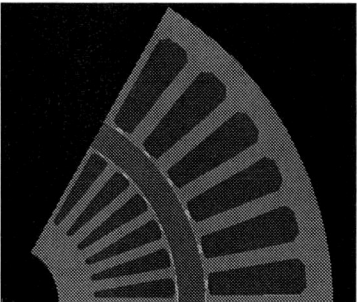

Figure 1. Geometry model for FEA

For purpose of comparison, three different PM-rotor structures are examined as shown in Fig. 2. For the first full PM structure in Fig. 2(a), the north poles and the south poles are alternatively distributed on the circle and each permanent magnet spans a full pole pitch, 60° mechanical. For the second PM rotor structure in Fig. 2(b), the north poles and the south poles are also alternatively distributed on the circle, but each magnet spans only 2/3 of a full pole pitch, 40° mechanical. The space between a pair of magnets is air. The third PM rotor structure is similar to the second one except that the space between two magnets is iron instead of air, shown as in Fig. 2(c).

For the full PM structure case in Fig. 2(a), if both rotors are assumed not rotating, and windings are fed with a set of three-phase currents, the torque on the PM rotor is approximately a sinusoidal wave as shown in Fig. 3. In the winding excitation, the peak values of the stator and the inner rotor windings currents are 250A and 100A respectively. The maximum torque generated on the PM rotor is about 460Nm. For the three different PM rotor structures designed above, the maximum torque on the PM rotor with different amount of current excitation are plotted in Fig. 4. The peak values of currents in stator and inner wound rotor windings vary from 250A/100A to 1250A/500A.

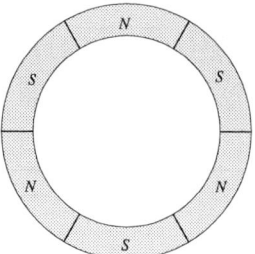

(a) Full PM structure

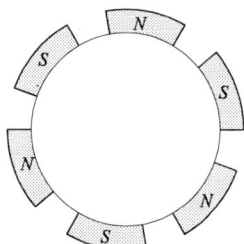

(b) PM-air structure

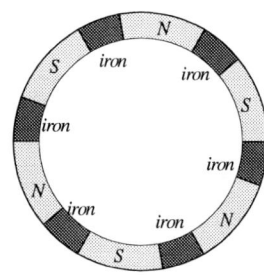

(c) PM-iron structure

Figure 2. Three types of PM rotor structures

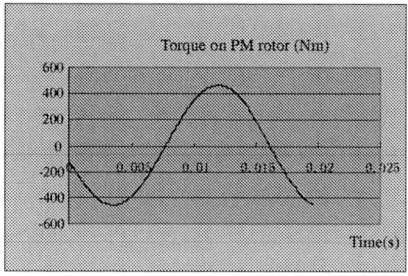

Figure 3. Torque on stationary PM rotor

In Fig. 4, the blue, purple and yellow lines represent the maximum torque produced by full PM structure (Fig. 2(a)), the PM-air structure (Fig. 2(b)) and the PM-iron structure (Fig. 2(c)) respectively. The maximum torque generated by the full PM structure is always larger than that produced in the PM-air structure. When the peak current in the stator winding is small, maximum torque produced by the PM-iron structure is lower than those produced by the other two structures. However, the maximum torque in PM-iron structure increases faster than that in the other two structures with increasing winding currents.

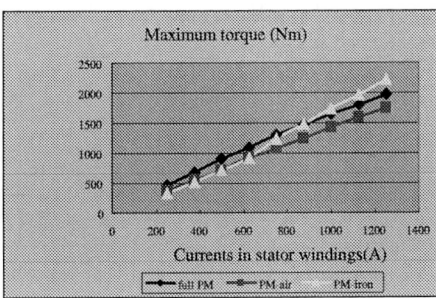

Figure 4. Torque in three different PM rotor structures

IV. SYSTEM CONTROL CONFIGURATION AND EVALUATION

Because of the two mechanical and two electrical ports, the DMP machine has many possible operation modes, which should all satisfy Eq. (12),

$$\pm P_{e,stator} \pm P_{m,pm-rotor} \pm P_{e,wd-rotor} \pm P_{m,wd-rotor} = 0 \qquad (12)$$

where "±" represent the power flow in and out. Subscripts "e" and "m" represent "electrical" and "mechanical" respectively. From mathematical point of view, there are 16 different solutions to the above equation. But here we choose to discuss and explain two solutions to Eq. (12) that we are most interested in.

a. $-P_{m,pm-rotor} + P_{m,wd-rotor} = \mu P_{e,stator} \pm P_{e,wd-rotor} = 0$ (13)

In this case, additional constrain is imposed, that is, not only is (12) satisfied but also are the electrical and mechanical subset of the equations balanced and equal to zero. The reason that this mode is called variable ratio gearbox mode is as follows. Since the two rotor speeds are different with the power equal to each other, the torque of the two rotors will be inversely proportional to the speeds. Therefore, we need to involve in the stator and inner wound rotor electrical power. In the process, the power taken by one electrical port from one rotor will be sent to another rotor in such a way that the electrical power is always balanced. We may call the DMP machine functioning as a variable E-gearbox because we use electromechanical device to achieve variable gearbox results.

b. $-P_{m,pm-rotor} + P_{m,wd-rotor} = \mu P_{e,stator} \pm P_{e,wd-rotor} = C$ (14)

Case "b" is almost the same as Case "a" except that "0" is replaced by a constant "C". In this case, the mechanical power of the two rotors is not balanced any more and the remaining can be absorbed by the electrical terminals and sent to the common DC bus, like a battery bank, for energy storage or draining. This operation mode is much like a hybrid traction system for electric vehicles.

As shown in the control block diagram in Fig. 5, the DMP system control consists of 2 independent speed loops, one for outer PM rotor and another inner wound rotor. In each speed loop, there are two current loops, one for current regulation in the q-axis and another d-axis. Also to be noticed is that the two mechanical ports - the PM and inner wound rotors, are attached to external

mechanical loads and an engine respectively. To highlight the DMP uniqueness in operation, the entire simulation process is divided into 6 major intervals in which each corresponds to an application feature.

a. Interval I - PM Rotor Acceleration From 0 to 2500 rpm

In the first interval, the stator windings are powered from the DC power supply through a DC/AC inverter and the PM rotor and the associated mechanical loads are accelerated from 0 to 2500rpm with vector control. The inner wound rotor is not excited nor driven by any external mechanical torque. Thus, there is no electromagnetic torque between the inner and PM rotors - the inner rotor remains standstill. Through controller, the stator electrical power flows to the PM rotor and is converted into mechanical power for its acceleration. We have assumed some mechanical frictions on the PM rotor. So, the electromagnetic torque generated by the stator currents is balanced by those for acceleration and overcoming frictions. The simulation results are shown in Fig. 6. From the top to bottom, plotted are PM rotor speed, inner rotor speed, stator torque, PM rotor torque and inner rotor torque. As indicated by the simulation results, it takes about 6 seconds to accelerate the PM rotor from 0 to 2500rpm.

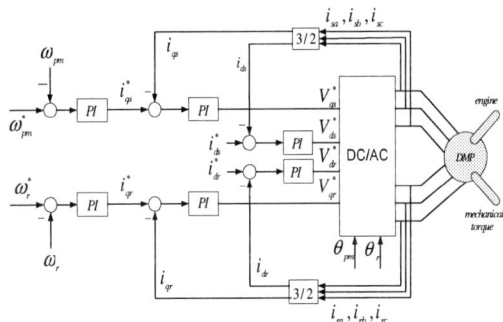

Figure 5. DMP control system block

b. Interval II - Inner Rotor Acceleration from 0 to 3500rpm

The second interval starts in 10 sec and lasts for about 0.5 sec. At the beginning of this interval, the PM rotor is already at constant 2500 rpm. Then, a step speed command is applied to the inner wound rotor to accelerate it from standstill to 3500rpm. Notice that the inner rotor acceleration torque is given by the outer PM rotor, which, in turn, is from the stator. It can be observed that the PM rotor is slowed down a little bit by the inner rotor dragging force. To stabilize the PM rotor speed, the PM rotor speed loop responds with more stator currents for more torque so that the PM rotor remains at constant 2500rpm. The simulated stator current result clearly shows the boosted stator current and torque for the inner rotor acceleration.

During inner rotor acceleration, it is most interesting to

1781

examine the winding current of the inner rotor before and after the speed of 2500 rpm. Before reaching 2500 rpm, the inner rotor actually receives more than the power needed for acceleration from the outer PM rotor. The surplus amount of power then is recovered by the windings located on the inner rotor to the power supply. At 2500rpm, synchronized with the PM rotor, and thereafter, the inner rotor will not have enough power to go to higher speed. Therefore, the slip power control now has to reverse the direction, sending power to the inner rotor so that the rotor speed can reach 3500rpm eventually. The slip power reversal and current sequence change in Interval II are clearly displayed in Fig. 7.

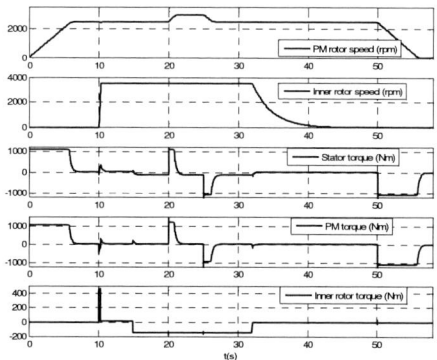

Figure 6. Simulation results of DMP speed and torque

The DMP system power balance is shown in Fig. 8. The simulated current and power confirm the analysis stated above.

c. Interval III – The Engine Attached to the Inner Rotor Starts and Delivers 150 Nm Torque

In Intervals III, we start the engine attached to the inner rotor and command it to deliver a mechanical torque of 150 Nm. This interval covers the period from 15s to 32s. In this period, the engine dictates the inner rotor speed at 3500rpm. Consequently, the inner rotor has to output an electromagnetic torque of 150 Nm to balance the mechanical torque. The dragging electro-magnetic torque imposed by the outer rotor is actually given by the stator currents acting on the PM rotor. Therefore, the stator is operated as a generator. On the other hand, the stator field and PM rotor have a synchronous speed of 2500 rpm, lower than the inner rotor mechanical speed of 3500 rpm, which forces the surplus power provided by the engine absorbed by the slip power windings on the inner rotor. In this interval, both the stator and inner rotor are sending power to the electrical source, originated from the engine attached.

d. Interval IV – PM Rotor Goes to 3000 rpm from 2500 rpm and Returns

With the engine outputting 55kw (3500 rpm and 150 Nm) while both the stator and inner rotor sending power

back to the electrical power supply, speed control of the PM rotor is investigated in Interval IV. This interval starts at 20s and ends at approximately 25s, nesting inside Interval III. As shown, the speed control is composed of three stages: acceleration, stable speed, and deceleration. At acceleration stage, a step speed command from 2500rpm to 3000rpm is applied to the PM rotor and the stator sends the needed power and torque to the PM rotor, which offsets the power generation due to the engine attached to the inner rotor. At the stage of the PM rotor speed stable at 3000 rpm, more power will be generated to the stator from the inner rotor and simultaneously, the slip power recovery by the inner rotor windings is reduced proportionally for reduced slip. At the final stage, with the PM rotor speed return from 3000 to 2500 rpm, the stator has to absorb the both the PM rotor kinematical energy and the one originated from the engine. The simulation results in Interval IV clearly confirm all of these.

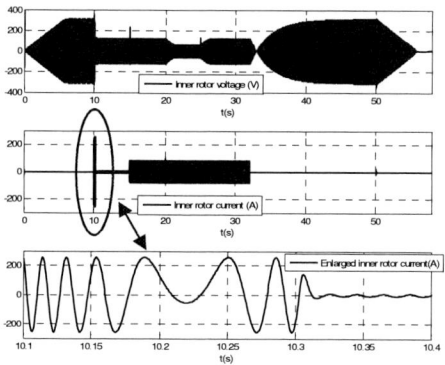

Figure 7. Simulation results of inner rotor i_a and v_a

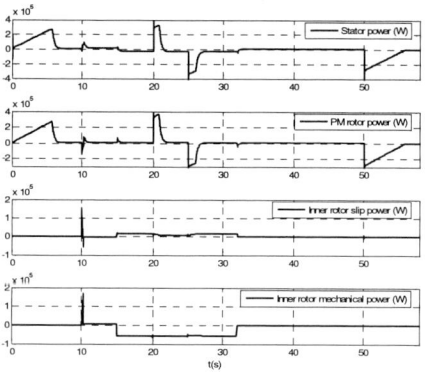

Figure 8. Simulation results of DMP power balance

e. Interval V – Engine Shut Down and Inner Rotor Windings De-excited

At 32s, in Interval V, we shut down the engine and remove electrical excitation to the inner rotor windings. Therefore, there is no electromagnetic and mechanical torque on the inner rotor. From simulation result it shows that the inner rotor speed drops to zero gradually because of the friction torque. For the PM rotor, the speed control loop is still active to command the stator current and electromagnetic torque, maintaining the PM rotating at 2500rpm. In this case, the electromagnetic torque generated by the stator current is balanced by the friction torque on the PM rotor.

f. Interval VI - PM is Decelerated to 0 rpm

In the last interval, after the engine and the inner rotor stop, we give a negative step speed command to the PM rotor. The electromagnetic process in this interval is of the reverse in Interval I. In the last interval, we recover the mechanical energy given to the PM rotor and its associated mechanical load during acceleration. In this interval the DMP functions as a generator. Similar to what happened in Interval I, it takes about 6 seconds to slow down the PM rotor to zero speed and the peak power shoots to as high as 280kw. The stator power plot also shows that the energy recovered (area under power curve) in this interval equals to approximately that in Interval I.

V. Summary

DMP machine is a brand new type of electric machine, promising to be applied in many fields as multi-source hybrid traction, integrated starter and generator, variable gearboxes and so forth. However, little research in DMP machine design and system control has been done in this area. In this paper, the dynamic model is established and explained briefly. Further, a prototype machine is designed and evaluated by finite element analysis. Several interesting operation modes are discussed and the whole control system block diagram is configured for dynamic simulation. The control algorithms with the overall system operation performance are evaluated by computer simulation.

References

[1] L. Xu, "A New Breed of Electrical Machines – Basic Analysis and Applications of Dual Mechanical Port Electric Machines", in *Proc. 8th Int. Conf. on Electrical Machines and Systems*, Aug. 2005, vol. 1, pp. 24-31.

[2] P. Hammond, "*Energy Methods in Electromagnetism,*" (Book), Clarendon Press, Oxford, 1981.

[3] S. A. Nasar, L. E. Unnerhr, "*Electromechanics and Electric Machines,*" (Book), John Wiley and Sons, 1983.

[4] L. Xu and W. Cheng, "Torque and Reactive Power Control of a Doubly-Fed Induction Machine by Position Sensorless Scheme", IEEE Transactions on Industry Applications, Vol. 31, No. 3, May/June 1995, pp.636-642.

[5] Martin J. Hoeijmakers, Jan A. Ferreira, "The Electrical Variable Transmission", IEEE-IAS Annual Meeting Proceedings, Oct. 2004, Seattle.

[6] Yuusuke Minagawa, Masaki Nakano, Minoru Arimitsu and Kan Akatsu, "New Concept Motor with Dual Rotors Driven by Harmonic", *SAE TECHNICAL, PAPER SERIES* 2002-01-2857, NISSAN Motor Co., LTD

2006 5th International Power Electronics and Motion Control Conference

Characteristic Analysis on Overhang Effect in Axial Flux PM Synchronous Motors with Slotted Winding

WonYoung Jo[*], YunHyun Cho[*] YonDo Chun[**] and DaeHyun Koo[**]

[*] Department of Electrical Engineering/Dong-A University, Busan, Korea
[**] Korea Electrotechnology Research Institute/Mechatronics Group, Changwon-si, Korea
jo6554@donga.ac.kr, yhcho@dau.ac.kr

Abstract— **This paper deals with an overhang effect of permanent magnet (PM) in the new type axial flux permanent magnet synchronous motor (AFPMSM) with a double-sided air-gap. 3-D electromagnetic finite element method (FEM) is used to calculate the air-gap magnetic flux density, back electromagnetic force (EMF) and the cogging torque according to the variation of the overhang angle. From the results, we can confirm the overhang effect quantitatively which improves the performance of the AFPM motor.**

Keywords- AFPMSM; overhang effect; linkage flux; leakage flux; cogging torque

I. INTRODUCTION

Recently, axial flux permanent magnet synchronous motors (AFPMSMs) have been used increasingly for various application such as electric ship, electric vehicle, and air-plain propulsion due to its compact construction and high power density [1-3].

In general, permanent magnet (PM) motors have the overhang structure of which PM length is longer than the stator's stack length. The overhang effect has been generally used to enhance the linkage flux in the motor. Some researches have been reported about radial type brushless motor and linear PM motor [4-6]. However, there is few and far between about the overhang model of AFPM motor with a double-sided air-gap.

This article quantitatively investigates the influences of the overhang on the new type AFPM motor performances such as back electromotive force (EMF), air-gap magnetic flux density, cogging torque and torque. 3-D finite element method (FEM) is used to simulate the characteristics of AFPM motor. From the results, we can confirm the overhang effect and select the proper overhang angle of PM which improves the performance of the AFPM motor.

II. THE AFPMSM CONSTRUCTION

The AFPMSM was designed with a single rotor and double-sided airgaps and stators. The anisotropic rotor with permanent magnets is located in the middle of both the double-sided stators with exciting coils as shown in

Fig. 1. The stator back yokes are attached themselves to the lateral case covers of the motor. The stator teeth are fixed as each piece to the stator disk. The fan-shaped magnets of rotor are inserted in holes of the rotor disk without the back yokes. In order to increase the power density of this motor, Neodymium-Iron-Boron is selected due to its high energy product. Fig. 2 shows the structure of the AFPMSM.

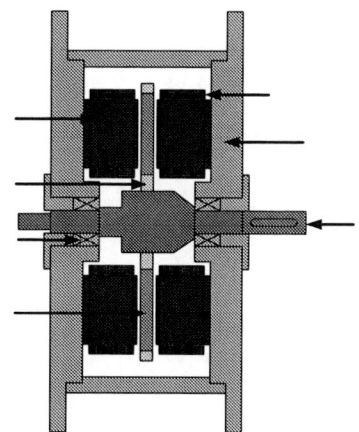

Figure 1. Cross section view of the AFPMSM

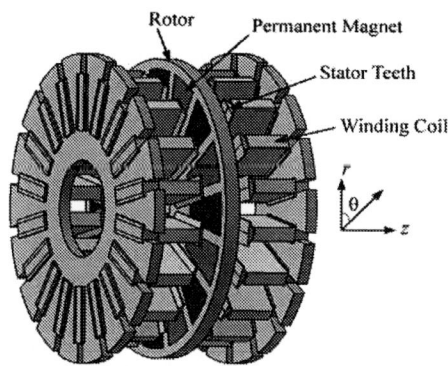

Figure 2. The structure of the AFPMSM.

Table I shows the specifications of the AFPM motor. We proposed the innovative teeth structure. The teeth are segmented structures and assembled together with wound coils. It enables us to manufacture the motor simply and easily.

TABLE I.
MAIN SPECIFICATION OF THE AFPMSM

Rated	Output power, kW	15
	Voltage, V	380
	Current, A	4 × 5.87
	Speed, rpm	1800
	Torque, Nm	76.9
Stator	Slot number	18
	Phase	3
	Phase resistance, Ω	0.735
	Turns per phase	95
	Inner radius, mm	151
	Outer radius, mm	262
	Air-gap length, mm	2
Rotor	Pole number	16
	Inner radius, mm	124
	Outer radius, mm	268
	Axial length of PM, mm	12.5
	Remanent of PM, T	1.2
	PM material	Nd-Fe-B
	Coercivity, kA/m	970
Winding connection		4-Y

III. OVERHANG EFFECT CONCEPT

Fig. 3 shows the overhang concept of the permanent magnet in a slot with exciting winding coil. In order to investigate the overhang effect of permanent magnet, the tangential degree, of PM is increased from 0 to 6 degree and we carried out 3-D finite element analysis of magnetic fields with the aid of ANSYS package. Each overhang angle of PM is 0(Case 1: no overhang), 1.5(Case 2), 3(Case 3), and 6(Case 4), degree respectively.

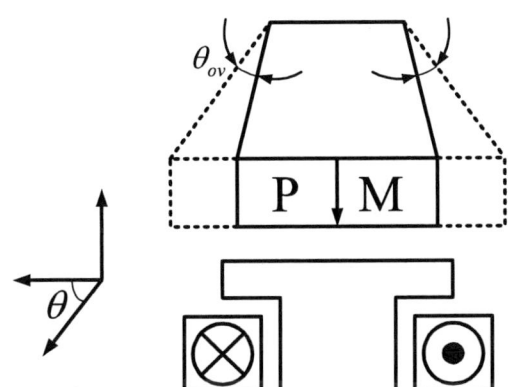

Figure 3. Overhang concept of the PM in a slot

IV. 3-D FEM ANALYSIS

A. Fundamental theory

It is assumed that eddy currents are confined to the rotor cage. Therefore, in the finite element models, only

the part of the meshing that represents the rotor cage is modeled as conducting. Fields in this conducting region can be modeled using the magnetic vector potential , and the electrical scalar potential . The equations to be solved are given by:

$$\nabla \times (\frac{1}{\mu}\nabla \times A) + \sigma(\frac{\partial A}{\partial t} + \nabla V) \qquad (1)$$

$$\nabla \bullet \sigma(\frac{\partial A}{\partial t} + \nabla V) = 0 \qquad (2)$$

The electric scalar can be implicitly included, so that a simpler formulation that includes the rotor velocity u can be used.

$$\nabla \times \frac{1}{\mu}\nabla \times A = \sigma[\frac{\partial A}{\partial t} + (u \bullet \nabla)A + A \times \omega) \quad (3)$$

where, ω is the angular velocity.

But in the non-conducting region, the total and reduced magnetic scalar potential, φ and ψ can be used. Both will lead to the following Laplacian type of equation which has to be φ and ψ can be used. Both will lead to the following Laplacian type of equation which has to be solved:

$$\nabla \bullet \mu\nabla \psi = 0 \qquad (4)$$

B. Analysis Model

The 2-dimension analysis is generally used because of the benefit of calculating time reduction and simplifying of modeling procedure for cylindrical motors. In the other hand, the 2-dimension analysis has to be performed under the condition that the geometrical and physical quantity in the vertical direction of the slice must be constant. So 2-dimension analysis is not applied to the AFPMSM with axial arrangement structures, and there is necessity for using the 3-dimension analysis.

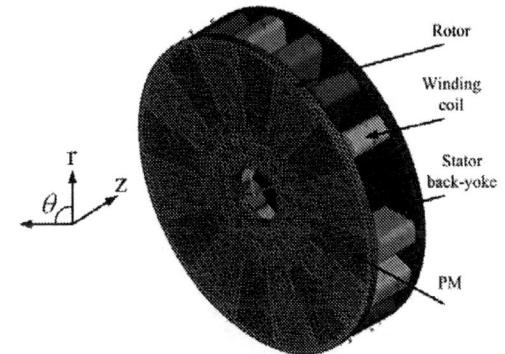

Figure 4. 3D mesh for a half analysis model

By taking advantage of machine symmetry, a half analysis model with 536125 elements and 101108 nodes is considered as Fig. 1. The elements in air-gap is divided as 4 stairs for precision of analysis, the torque and electromagnetic force characteristics are calculated by using Maxwell Stress Tensor equation. If the AFPMSM is operating using the inverter with 3-phases PWM, in order to understand and analyze the driving characteristics exactly, the switching pattern and voltage equation of driving circuit and free-wheeling diode must be considered. But a time of applying to 3-dimension analysis in this paper, analysis model is considered as static magnetic field: input current of each phase is 11.39A, using 3-phases operating method.

C. FEM Analysis

In order to investigate quantitatively the overhang effect, the vector distribution of the linkage flux and leakage flux of PM without exciting current according to variable overhang angles is shown as Fig. 4. In case of no overhang(Case 1), the fringing and leakage flux are produced at the edges, and the air-gap flux density at the edges of the stator teeth decreases due to the leakage flux at the tangential axis edges of the stator, as shown in Fig. 4 (a). According to the increment of overhang angle, the air-gap flux density at the edges also increases owing to the rise of linkage flux at the edges as shown in Fig. 4 (b). From these results we know that in case of the gradual increment of overhang angle the fluxes of the edges also increase gradually owing to the linkage of a portion of flux generated at a overhang section of PMs. Therefore the overhang structure has the effect of increasing the linkage flux when the leakage components at the edges of θ direction are rising.

Fig. 7 shows the flux density distribution in air-gap according to variable overhang angles. As increasing of θ_{ov}, the flux density at the edges in air-gap is high due to overhang effect. But in case of 9[deg] overhang angle, on the contrary the flux density is lower than others. This result is solved as the surface flux density of PM is higher at the middle than at the edges. Therefore from these results we can estimate that the PM with 6[deg] overhang angle is suitable.

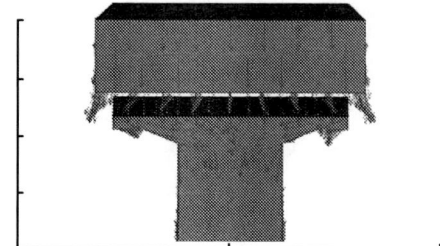

(b) θ_{ov} =1.5[deg]

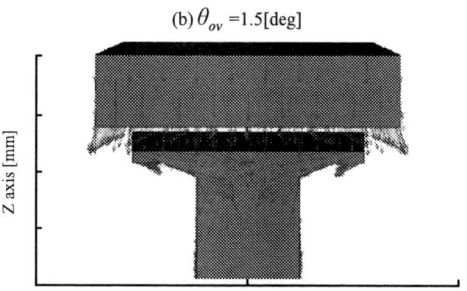

(c) θ_{ov} =3[deg]

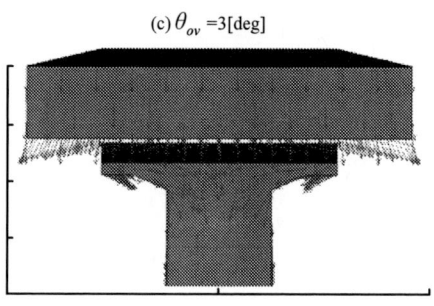

(d) θ_{ov} =6[deg]

Figure 6. Vector distribution of flux density according to overhang angle variations.

Fig. 8 shows the back electromotive force (EMF) of each phase according to increment of overhang angle when the speed is under no-load and 1800 rpm at the static state operating. As the linkage flux is increasing according to overhang angle, we can know that the back EMF is also increasing in proportion to the linkage flux. Fig. 9 shows the characteristics of cogging torque as variable overhang angles. So we can know that the overhang effect don't influence the characteristics of cogging torque. From this result we can expect that overhang effect don't bring on a change of the characteristics of torque. Fig. 10 shows the flux density distribution for 3-dimension analysis. The peak value of this result is 1.05[T], so this result satisfies the required motor design that has the high flux density in air-gap.

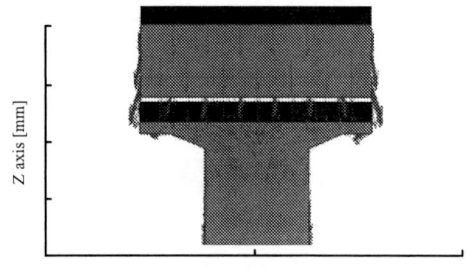

(a) θ_{ov} =0[deg]

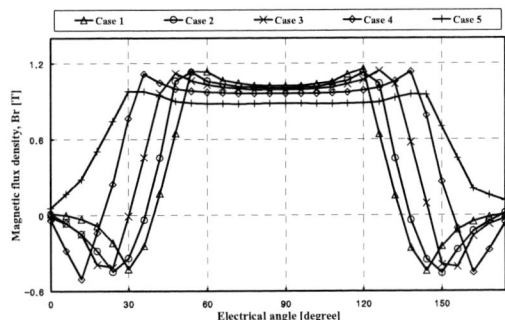

Figure 7. The flux density distribution at air-gap according to variations of overhang angle

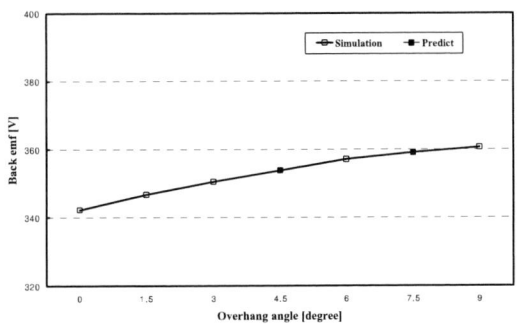

Figure 8. Characteristics of back EMF at no-load according to variations of overhang angle

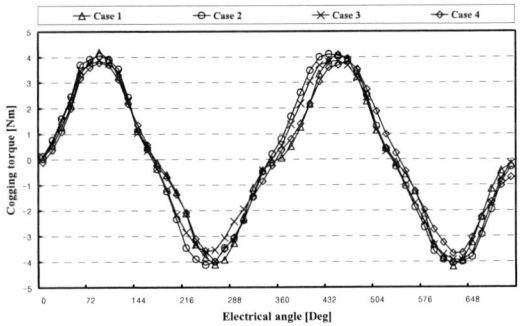

Figure 9. Characteristics of cogging torque according to variations of overhang angle

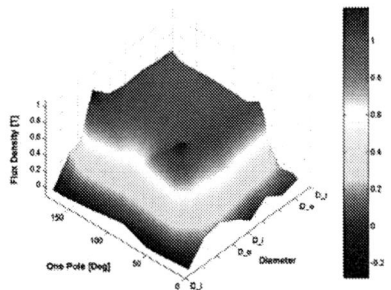

Figure 10. Flux density distribution for one-pole in air-gap

V. CONCLUSION

This article describes the overhang effect of an new type AFPM motor with a double sided air-gap. The motor characteristics such as back EMF, linkage flux, cogging torque are investigated according to the variation of the overhang angle. 3D FEM method is used for the efficient analysis of the magnetic field. From the result, we can confirm that the increase of overhang angle enhances the linkage fluxes of the stator in AFPM. Also, the cogging torque is not much influenced by an overhang effect of PM.

ACKNOWLEDGMENT

This work was supported by the KEMCO.

REFERENCES

[1] R. J. Hill-Cottingham, P. C. Coles, J. F. Eastham, F. Profumo, A. Tenconi, G. Gianolio, "Multi-disc axial flux stratospheric propeller drive", Proc. of IEEE IAS Annual Meeting Confrence 2001, vol. 3 pp. 1634-1639, 2001.

[2] F. Caricchi, F. Crescimbini, O. Honorati, Modular, "Axial-flux permanent magnet motor for ship propulsion drives", IEEE Trans. on Energy Conversion, vol. 14 pp. 673-679, 1999.

[3] F. Profumo, Z. Zhang, A. Tenconi, "Axial flux machines drvies a new viable solution for electric cars", IEEE Trans. on Industrial Electronics, vol. 44 no. 1 pp. 39-45, 1997.

[4] Surong Huang, Jian Luo, Franco Leonardi, Thomas A. Lipo, "A Comparison of Power Density for Axial Flux Machines Based on General Purpose Sizing Equations", IEEE trans on Energy Conversion, Vol. 14, No. 2, June 1999.

[5] M. Aydin, S. Huang, T.A. Lipo, "Optimum design and 3D finite element analysis of nonslotted and slotted internal rotor type axial flux PM disc machines", Power Engineering Society Summer Meeting, 2001. IEEE, Vol. 3 pp.1409 - 1416, July 2001.

2006 5th International Power Electronics and Motion Control Conference

Design and Analysis of a Double-Stator Cup-Rotor Directly Driven Permanent Magnet Wind Power Generator

Dong Zhang[*], Shuangxia Niu[**], K. T. Chau[**], J. Z. Jiang[*], Yu Gong[*]

[*] College of Mechatronics Engineering and Automation, Shanghai University, Shanghai, China

[**] Department of Electrical and Electronic Engineering, Hong Kong University, Hong Kong

Abstract—This paper presents a new double-stator cup-rotor directly driven permanent magnet machine for wind power generation. The machine configuration is so unique that it can significantly improve the torque density with small cogging torque. Because of the nature of double-stator windings, the machine can flexibly change their connections, hence providing a constant output voltage over a wide speed range for battery changing. Computer simulation is conducted by using the time-stepping finite element method. Both simulation and experimental results are given to verify the validity of the proposed machine.

Keywords-double-stator cup-rotor permanent magnet machine; time-stepping finite element method; connection

I. INTRODUCTION

In recent years, with the increasing concern on the wind power generation, a lot of researches about the wind power generator are implemented; many kinds of machines began to work for it. Just because of the permanent magnet generator has a lot of advantages, such as high efficiency, high power factor and high power density, it has been the focus of research [1] [2] [3]. But the conventional permanent magnet (PM) generator, which works indirectly coupled to the wind turbine, is not applicable for the variable wind turbine speed, and has low efficiency with the gearbox. It is important to find a low speed and high efficient generator, which can change terminal voltage flexibly to different speed and be directly driven by the wind turbine.

The purpose of this paper is to propose a new 3-phase 22-pole double-stator cup-rotor PM generator (DSCR-PMG) for wind power generation, so it can work at low speed, its axis can directly couple to the wind turbine. The novel design of the pole-to-slot ratio can lead to a very low torque ripple, and more sinusoidal EMF waveform. The windings in the two stators can flexibly be connected, so the machine can offer the good controllability for the constant terminal voltage in a wide speed range for battery charging. The time-stepping finite element method (TS-FEM) model of the generator is developed, and the performance of the machine is simulated by TSFEM. The prototype is made and

experiments were carried out to validate the proposed design and analysis.

II. MACHINE DESIGN AND ANALYSIS

A. Machine Configuratioin

The Structure of the proposed DSCR-PMG is shown in Fig. 1. This machine has two concentric stators, and one cup-type rotor. Firstly, it has the high power density due to two stators. That means we can put more windings into the machine, and we have two airgaps to contribute for the output torque. So the DSCR-PMG can produce double

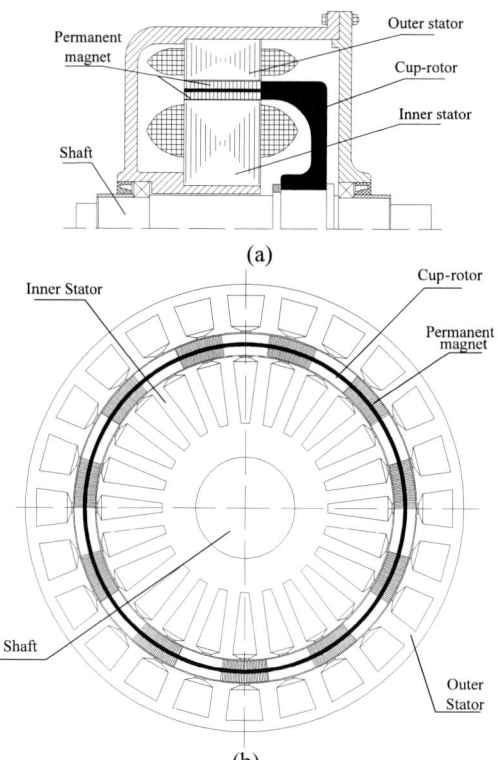

Figure 1. Proposed Machine. (a) Structure. (b) Configuration

power density than the single stator machine with the same armature diameter.

Secondly, the fractional number of stators per pole per phase enhances the sinusoidal EMF waveform and reduces the torque ripple. Especially the number of slots closes to number of poles, which virtually fulfills the zero cogging torque without skew, and this will be discussed later in this paper.

Thirdly, the high number of poles requires less yoke material in the two stators, then it will enhance the power density.

Fourthly, coils of each phase wound on adjacent teeth, which enable short end-windings material, so it can improve the utilization of copper, which will increase the power density.

Finally, the rotor core is designed like a cup with PMs mounted on its inner and outer surface. This cup-type

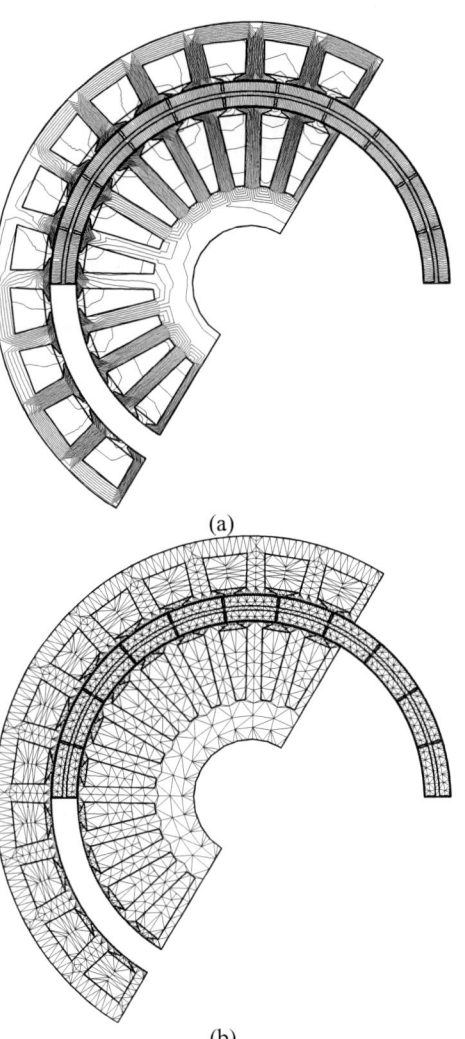

(a)

(b)

Figure 2. Magnetic field analysis using TS-FEM. (a) Mesh diagram. (b) Magnetic field distribution at full load.

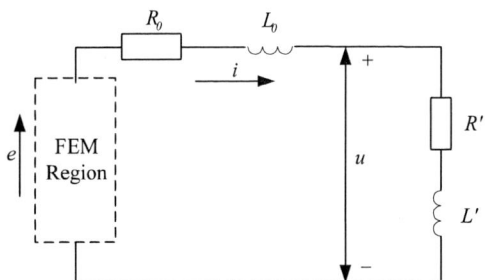

Figure 3. The circuit of the generator system diagram.

rotor can effectively shorten the magnetic circuit length and hence improving the controllability.

B. Analysis with Time-stepping finite-element method

The two dimensional TS-FEM is applied to analyze the transient and steady-state performance of the proposed machine [4] [5].

Supposing the machine works as a generator in a constant speed, the mathematical model of the system is composed of two equations. One is finite element equation of the electromagnetic field of the machine and the other is circuit equation of the double-stator windings and load. Both of them are the function of time and space, too.

Thus, the governing equation for electromagnetic field is represented by Maxwell's equation as

$$\frac{\partial}{\partial x}\left(\upsilon\frac{\partial A}{\partial x}\right) + \frac{\partial}{\partial y}\left(\upsilon\frac{\partial A}{\partial y}\right) = -J + \sigma\frac{\partial A}{\partial t} - \upsilon\frac{\partial B_{ry}}{\partial x} + \upsilon\frac{\partial B_{rx}}{\partial y} \quad (1)$$

Where A is the magnetic vector potential of the z axis component, υ is the reluctivity, J is the current density, σ is electrical conductivity, it exists in the solid rotor iron core and PM where eddy currents cannot be ignored, and B_{rx} and B_{ry} are the remnant flux density vector of the PM in x axis component and y axis component, respectively.

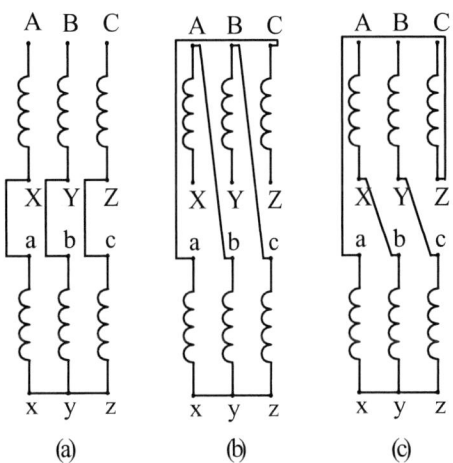

Figure 4. Different connection modes. (a) Same phases in series. (b) Adjacent anti-phases in series. (c) Adjacent phases in series.

1789

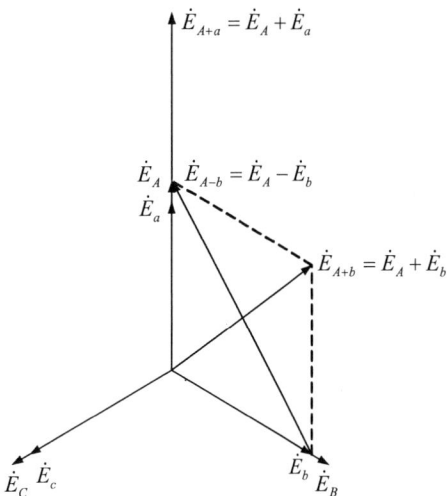

Figure 5. Vector diagrams of different connection modes.

Figure 7. Machine prototype.

Then the circuit diagram of the machine for generating mode is given in Fig. 3, u is the terminal voltage of the generator system. So the circuit equation of the stator windings and outer load for the generator is derived as

$$(R_0 + R')i + (L_0 + L')\frac{di}{dt} - \frac{l}{S}\iint_{\Omega}\frac{\partial A}{\partial t}d\Omega = 0 \quad (2)$$

Where R_0 is the inner resistance of one phase winding of the generator, R' is the outer resistance load, L_0 is the inductance of the end windings, L' is the outer inductance load.

By using the circuit-field-torque coupled time-stepping finite-element method (TS-FEM), both the transient and steady-state performances of the proposed machine can be analyzed. Fig. 2 shows the finite-element mesh and the corresponding magnetic field distribution at full load. In order to meet the requirement of constant output voltage over a wide range of speeds for battery charging, the double-stator windings have versatile connection modes as shown in Fig. 4. Below the base speed ω_b, the same phases are connected in series (Fig. 4(a)). Between ω_b and $(2/\sqrt{3})\omega_b$, the adjacent anti-phases are in series

(Fig.4(b)). Between $(2/\sqrt{3})\omega_b$ and $2\omega_b$, the adjacent phases are in series (Fig. 4(c)). The corresponding vector diagrams are illustrated in Fig. 5. For fine tuning or beyond $2\omega_b$, the conventional flux weakening control can be further employed to adjust the output voltage.

III. SIMULATION AND EXPERIMENTAL RESULTS

Due to the fractional-slot winding design, the cogging torque is simulated as shown in Fig. 6, which verifies that the cogging effect of the proposed machine is insignificant.

For experimental verification, the proposed machine is prototyped as shown in Fig. 7. The corresponding key data is listed in Table I. The no-load EMF waveforms at 400rpm under different connection modes are first simulated by using the TS-FEM, and then measured from experimentation. As shown in Fig. 8, the measured waveforms closely match with the simulated waveforms, hence verifying the validity of the proposed machine.

Then we make the load experiment of three connection modes and give the efficiency curves of DSCR-PMG, which is shown in Fig. 9. So we can find that the machine has high efficiency, and the simulated efficiency curves are closely matched. They confirm that the generator design is successful.

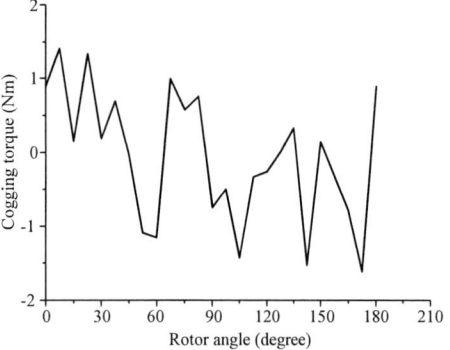

Figure 6. Simulated cogging torque.

TABLE I.
SPECIFICATIONS OF PROPOSED MACHINE

Rated speed	400 rpm
Rated power	2 kW
Rated phase voltage	48 V
Inner stator inner diameter	92 mm
Inner stator outer diameter	165 mm
Outer stator inner diameter	191 mm
Outer stator outer diameter	245 mm
Stack length	50 mm
Airgap length	0.6 mm
Inner stator slot number	24
Outer stator slot number	24
Pole number	22

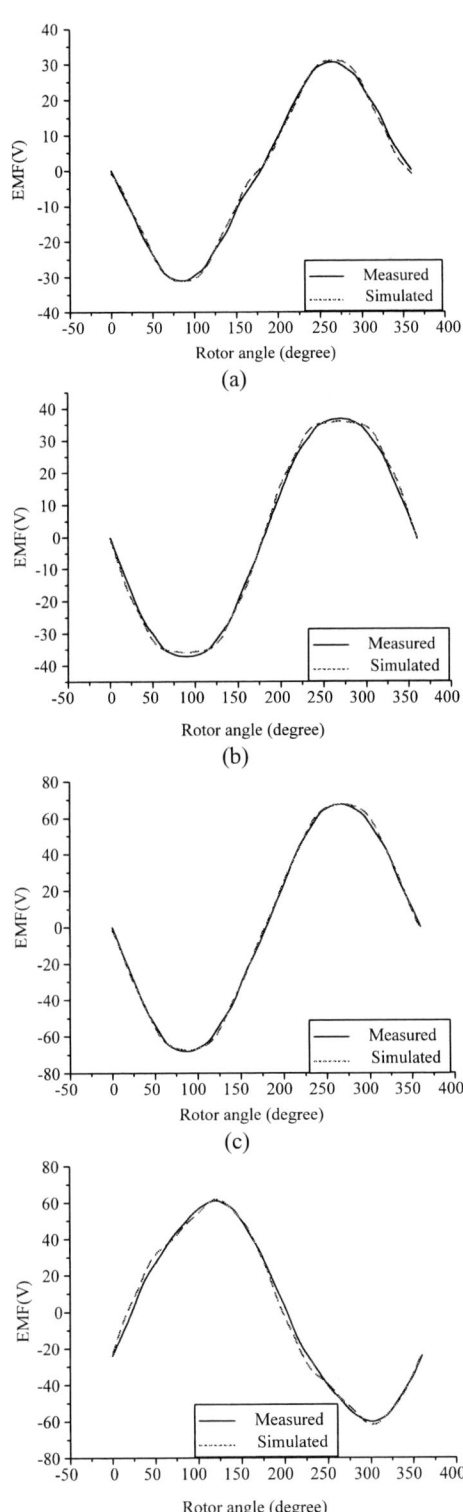

(a)

(b)

(c)

(d)

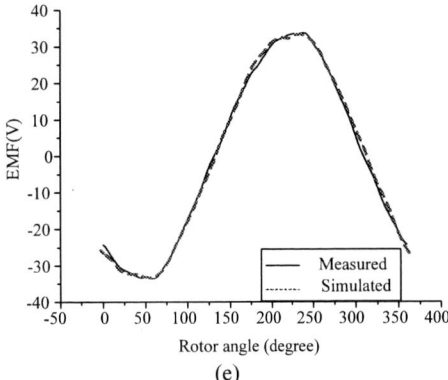

(e)

Figure 8. Different no-load EMF waveforms at 400rpm. (a) Inner stator only. (b) Outer stator only. (c) Same phases in series. (d) Adjacent anti-phases in series. (e) Adjacent phases in series.

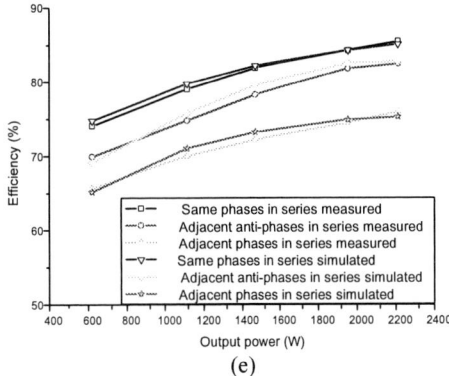

(e)

Figure 9. The efficiency curves of different connection modes.

IV. CONCLUSION

In this paper a new double-stator cup-rotor PM wind power generator is proposed and analyzed. Because of its special configuration design, the DSCR-PMG can achieve a high torque density. Due to the structure of double-stator winding, a flexible winding connection method is proposed to obtain a constant output voltage over a wide speed range for battery charging. Both the computer simulation and the experimental results confirm the validity of the proposed double-stator cup-rotor directly driven permanent magnet wind power generator application.

ACKNOWLEDGMENT

The first author wishes to thank Dr. K. T. Chau and Department of Electrical and Electronic Engineering, Hong Kong University, for providing the facilities , financial resources and the test results.

REFERENCES

[1] K.T. Chau, Ying Fan, Ming Cheng; "A novel three-phase doubly salient permanent magnet machine for wind power generation," *Industry Applications Conference, 2004. 39th IAS Annual Meeting.*

[2] Jianyi Chen, Chemmangot V. Nayar and Longya Xu, "Design and Finite-Element Analysis of an Outer-Rotor Permanent-Magnet Generator for Directly Coupled Wind Turbines," *IEEE Trans. On Magn.*, vol. 36, pp. 3802–3809, September 2000

[3] E. Spooner, A. C. Williamson, "Direct coupled, permanent magnet generators for wind turbine applications," *IEE proc.-electr. Power Appl.*, vol. 143, pp. 1–8, January 1996

[4] S. J. Salon, *Finite Element Analysis of Electrical Machines.* Norwell, MA: Kluwer, 1995

[5] Jinyun Gan, K. T. Chau, Yong Wang, C. C. Chan and J. Z. Jiang, "Design and analysis of a new permanent magnet brushless DC machine," *IEEE Trans. On Magn.* Vol. 36, pp. 3353–3356, September 2000

2006 5th International Power Electronics and Motion Control Conference

Feasibility Analysis of Accelerometer Configuration of Non-gyro Micro Inertial Measurement Unit

DING Mingli[*], ZHOU Qingdong[*], WANG Qi[*] and WANG Changhong[**]
[*] Department. of Automatic Test and Control, Harbin Institute of Technology, Harbin, China
[**] Department. of Control Science and Engineering, Harbin Institute of Technology, Harbin, China
E-mail: dingml@hit.edu.cn

Abstract—A sufficient condition is given to determine if a configuration of accelerometers of non-gyro micro inertial measurement unit (NGMIMU) is feasible. NGMIMU uses only accelerometers to compute the motion of a moving body. If the condition is satisfied, the angular acceleration and the linear acceleration in three directions can be expressed by the linear combination of the angular velocity and the accelerometer output. And the posture of a moving body can be estimated using a numerical iteration algorithm. Two configurations of accelerometers are analyzed in detail and the results verify the correctness of the condition. This research serves as an important criterion for the research of other problems of NGMIMU.

Keywords-Measurement Unti; Accelerometer; Feasibility; Configuration

I. INTRODUCTION

Most of current inertial navigation systems use liner accelerometer and gyroscopes to sense linear accelerations and angular velocity, respectively. The gyroscope has the disadvantage of complicate manufacture technique, high cost, high power consumption, and large volume for missile. In addition, gyroscope is hard to bear the impact of high acceleration and angular acceleration. We hope to use the small accelerometers with low cost to replace the gyroscopes to realize navigation for the missile. So the non-gyro micro inertial measurement unit (NGMIMU) using accelerometers, is studied extensively to determine the posture of a rigid body.

The idea of replacing gyroscopes with accelerometer to realize navigation has been presented by foreign scholars for some years. In 1965, DiNapoli proposed the idea of non-gyroscope scheme to measure the angular velocity of a rigid body for the first time in his master thesis as in [1]. In 1967, Alfred R. Schuler presented many kinds of accelerometer-based configurations as in [2]. In 1994, Chen presented a novel design of six-accelerometer configuration to compute the rotational and translational acceleration of a rigid body as in [3]. In 2001, Tan gave a sufficient condition for the feasibility of configuration of accelerometers of NGMIMU. But in fact, the condition

given by Tan is only adapted to the design of six accelerometers. If the number of accelerometers is larger than six, the algorithm contributed to the condition cannot compute the navigation parameters.

In light of the facts mentioned above, this work proposes a sufficient condition to determine if a configuration of accelerometers of NGMIMU is feasible based on the conventional configurations. Two examples of configuration of accelerometers are also discussed to verify the correctness of the theory.

II. FEASIBILITY OF ACCELEROMETER CONFIGURATION OF NGMIMU

An inertial frame and a rotating moving body frame are exhibited in Fig. 1, where b represents the moving body frame and I the inertial frame.

The acceleration of point Q is given by
$$a = \ddot{R}_I + \ddot{r}_b + \dot{\omega} \times r + 2\omega \times \dot{r}_b + \omega \times (\omega \times r) \qquad (1)$$
where $\ddot{r}_b$ is the acceleration of point Q relative to body frame. $\ddot{R}_I$ is the inertial acceleration of O^b relative to O^I. $2\omega \times \dot{r}_b$ is known as the Coriolis acceleration, $\omega \times (\omega \times r)$ represents a centripetal acceleration, and $\dot{\omega} \times r$ is the tangential acceleration owing to angular acceleration of the rotating frame.

If Q is fixed in the b frame, the terms $\dot{r}_b$ and $\ddot{r}_b$ vanish. And (1) can be rewritten as
$$a = \ddot{R}_I + \dot{\omega} \times r + \omega \times (\omega \times r) \qquad (2)$$

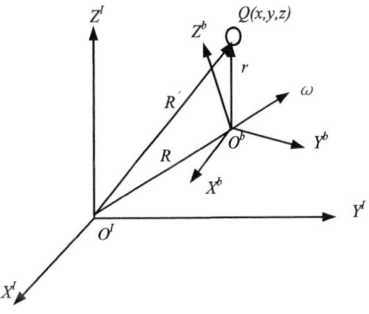

Figure 1.   Geometry of body frame (*b*) and inertial frame (*I*)

Thus the accelerometers rigidly mounted at location r_i on the body with sensing direction θ_i produce A_i as outputs.

$$A_i = [\ddot{R}_I + \dot{\Omega} r_i + \Omega\Omega r_i] \cdot \theta_i \qquad (i=1,2,...,N) \quad (3)$$

where

$$\Omega = \begin{bmatrix} 0 & -\omega_z & \omega_y \\ \omega_z & 0 & -\omega_x \\ -\omega_y & \omega_x & 0 \end{bmatrix}, \ddot{R}_I = \begin{bmatrix} \ddot{R}_{Ix} \\ \ddot{R}_{Iy} \\ \ddot{R}_{Iz} \end{bmatrix} \quad (4)$$

In (4), ω_x, ω_y and ω_z represent the angular velocity along x, y and z axis, respectively.

Consider N accelerometer distributed at locations $r_1,...,r_N$ with sensing directions $\theta_1,...,\theta_N$, respectively. The pair (r_i, θ_i) is expressed in the body frame.

Let $\omega = \begin{bmatrix} \omega_x & \omega_y & \omega_z \end{bmatrix}^T$. Considering the skew-symmetric vector Ω, for any $N = \begin{bmatrix} n_x & n_y & n_z \end{bmatrix}^T$, we have

$$\Omega \cdot N = \begin{bmatrix} 0 & -\omega_z & \omega_y \\ \omega_z & 0 & -\omega_x \\ -\omega_y & \omega_x & 0 \end{bmatrix} \begin{bmatrix} n_x \\ n_y \\ n_z \end{bmatrix} = \begin{bmatrix} \omega_y n_z - \omega_z n_y \\ \omega_z n_x - \omega_x n_z \\ \omega_x n_y - \omega_y n_x \end{bmatrix} = \omega \times N \quad (5)$$

Given $\omega \in R^3$, the cross-product $\omega \times N$ is a linear operator, and ω has a matrix representation given by the skew-symmetric vector Ω.

Using (3) and $\omega \leftrightarrow \Omega$, we have

$$A_i = (\ddot{R}_I + \dot{\Omega} r_i + \Omega\Omega r_i) \cdot \theta_i$$
$$= \theta_i^T \ddot{R}_I + (r_i \times \theta_i)^T \dot{\omega} + \theta_i^T \Omega^2 r_i \quad (6)$$
$$= \begin{bmatrix} (r_i \times \theta_i)^T & \theta_i^T \end{bmatrix} \begin{bmatrix} \dot{\omega} \\ \ddot{R}_I \end{bmatrix} + \theta_i^T \Omega^2 r_i$$

From (4), we get

$$\Omega^2 = \begin{bmatrix} -\omega_y^2 - \omega_z^2 & \omega_x\omega_y & \omega_x\omega_z \\ \omega_x\omega_y & -\omega_x^2 - \omega_z^2 & \omega_y\omega_z \\ \omega_x\omega_z & \omega_y\omega_z & -\omega_x^2 - \omega_y^2 \end{bmatrix} \quad (7)$$

From (7), the accelerometer outputs are

$$A_i = M_1 \begin{bmatrix} \dot{\omega}_x \\ \dot{\omega}_y \\ \dot{\omega}_z \\ \ddot{R}_{Ix} \\ \ddot{R}_{Iy} \\ \ddot{R}_{Iz} \end{bmatrix} + M_2 \begin{bmatrix} \omega_x^2 \\ \omega_y^2 \\ \omega_z^2 \\ \omega_y\omega_z \\ \omega_x\omega_z \\ \omega_x\omega_y \end{bmatrix} \quad (8)$$

In (8), $M_1 = \begin{bmatrix} (r_i \times \theta_i)^T & \theta_i^T \end{bmatrix}$ is a $N \times 6$ matrix,

and $\theta_i^T = \begin{bmatrix} \theta_{ix} & \theta_{iy} & \theta_{iz} \end{bmatrix}$, $r_i = \begin{bmatrix} r_{ix} \\ r_{iy} \\ r_{iz} \end{bmatrix}$. Then

$$M_2 = \begin{bmatrix} -\theta_{iy}r_{iy} - \theta_{iz}r_{iz} \\ -\theta_{ix}r_{ix} - \theta_{iz}r_{iz} \\ -\theta_{ix}r_{ix} - \theta_{iy}r_{iy} \\ \theta_{iz}r_{iy} + \theta_{iy}r_{iz} \\ \theta_{iz}r_{ix} + \theta_{ix}r_{iz} \\ \theta_{iy}r_{ix} + \theta_{ix}r_{iy} \end{bmatrix}^T \quad (9)$$

From (8), the posture of a moving body in three-dimensional space can be represented by linear motions($\ddot{R}_{Ix}, \ddot{R}_{Iy}, \ddot{R}_{Iz}$) and rotational motions($\dot{\omega}_x, \dot{\omega}_y, \dot{\omega}_z$). If $\dot{\omega}_x, \dot{\omega}_y, \dot{\omega}_z, \ddot{R}_{Ix}, \ddot{R}_{Iy}, \ddot{R}_{Iz}$ can be represented by the linear combinations of A_i and $\omega_x^2, \omega_y^2, \omega_z^2, \omega_y\omega_z, \omega_x\omega_z, \omega_x\omega_y$, the configuration of accelerometers of NGMIMU is considered feasible.

Sufficient Condition: In (8), if the order of the $N \times 6$ matrix M_1 is equal to or larger than 6, i.e.,

$$R(M_1) \geq 6 \quad (10)$$

the configuration of accelerometers is considered feasible. The condition given by Ref. [4] is that the $N \times 6$ matrix M_1 has a left inverse. When $N = 6$, or M_1 is a 6×6 matrix, the condition can meet the computation requirement; on the other hand, when $N > 6$, the algorithm of computing the inverse of M_1 is not applicable. So the conclusion given by Ref. [4] can be only used when $N = 6$.

This paper generalizes the condition for the feasibility of configuration of accelerometers of NGMIMU by removing the restrain of the number of the accelerometer. When $R(M_1) \geq 6$, $\dot{\omega}_x, \dot{\omega}_y, \dot{\omega}_z, \ddot{R}_{Ix}, \ddot{R}_{Iy}, \ddot{R}_{Iz}$ can be expressed only by the linear combination of $\omega_x^2, \omega_y^2, \omega_z^2, \omega_y\omega_z, \omega_x\omega_z, \omega_x\omega_y$ and A_i. And the posture of a moving body can be computed using the accelerometer outputs A_i and the initial angular velocity value ω_0.

III. EXAMPLES FOR VERIFYING THE FEASIBILITY CONDITION

A. A Nine-accelerometer Configuration

Nine-accelerometer configuration is a typical configuration in the research of NGIMU. The Configuration C as in [2] is as follows. The locations and the sensing directions of the nine accelerometers in the body frame are shown in Fig. 2. Each arrow in Fig. 2 points to the sensing direction of each accelerometer.

The locations and sensing directions of the nine accelerometers are

$$[r_1,\cdots,r_9] = l \begin{bmatrix} 0 & 0 & 1 & -1 & 0 & 0 & 0 & 0 & 1 \\ 1 & -1 & 0 & 0 & 1 & -1 & 0 & 0 & 0 \\ 0 & 0 & 0 & 0 & 0 & 0 & 1 & 1 & 0 \end{bmatrix} \quad (11)$$

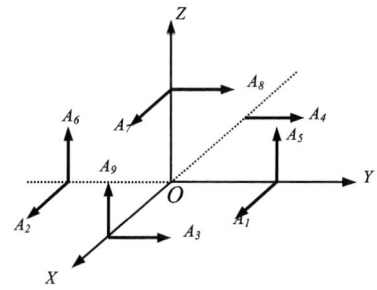

Figure 2. Nine-accelerometer NGMIMU Configuration

where l is the distance between the accelerometer and the origin of the body frame.

$$[\theta_1,\cdots,\theta_9]=\begin{bmatrix} 1 & 1 & 0 & 0 & 0 & 0 & 0 & 1 & 0 & 0 \\ 0 & 0 & 1 & 1 & 0 & 0 & 0 & 1 & 0 \\ 0 & 0 & 0 & 0 & 1 & 1 & 0 & 0 & 1 \end{bmatrix} \quad (12)$$

By the definition of M_1,

$$M_1=\begin{bmatrix} 0 & 0 & -l & 1 & 0 & 0 \\ 0 & 0 & l & 1 & 0 & 0 \\ 0 & 0 & l & 0 & 1 & 0 \\ 0 & 0 & -l & 0 & 1 & 0 \\ l & 0 & 0 & 0 & 0 & 1 \\ -l & 0 & 0 & 0 & 0 & 1 \\ 0 & l & 0 & 1 & 0 & 0 \\ -l & 0 & 0 & 0 & 1 & 0 \\ 0 & -l & 0 & 0 & 0 & 1 \end{bmatrix}, \quad (13)$$

and the order of (13) is

$$R(M_1)=6 \quad (14)$$

So the nine-accelerometer configuration is feasible according to the sufficient condition presented above. Then we compute the linear expressions of the navigation parameters. Substitute (11) and (12) into (9) and we obtain

$$M_2=\begin{bmatrix} 0 & 0 & 0 & 0 & 0 & l \\ 0 & 0 & 0 & 0 & 0 & -l \\ 0 & 0 & 0 & 0 & 0 & l \\ 0 & 0 & 0 & 0 & 0 & -l \\ 0 & 0 & 0 & l & 0 & 0 \\ 0 & 0 & 0 & -l & 0 & 0 \\ 0 & 0 & 0 & 0 & l & 0 \\ 0 & 0 & 0 & 0 & l & 0 \\ 0 & 0 & 0 & 0 & l & 0 \end{bmatrix} \quad (15)$$

Using (8), the accelerometer outputs are

$$A_i=\begin{bmatrix} 0 & 0 & -l & 1 & 0 & 0 \\ 0 & 0 & l & 1 & 0 & 0 \\ 0 & 0 & l & 0 & 1 & 0 \\ 0 & 0 & -l & 0 & 1 & 0 \\ l & 0 & 0 & 0 & 0 & 1 \\ -l & 0 & 0 & 0 & 0 & 1 \\ 0 & l & 0 & 1 & 0 & 0 \\ -l & 0 & 0 & 0 & 1 & 0 \\ 0 & -l & 0 & 0 & 0 & 1 \end{bmatrix}\begin{bmatrix} \dot{\omega}_x \\ \dot{\omega}_y \\ \dot{\omega}_z \\ \ddot{R}_{lx} \\ \ddot{R}_{ly} \\ \ddot{R}_{lz} \end{bmatrix}+\begin{bmatrix} 0 & 0 & 0 & 0 & 0 & l \\ 0 & 0 & 0 & 0 & 0 & -l \\ 0 & 0 & 0 & 0 & 0 & l \\ 0 & 0 & 0 & 0 & 0 & -l \\ 0 & 0 & 0 & l & 0 & 0 \\ 0 & 0 & 0 & -l & 0 & 0 \\ 0 & 0 & 0 & 0 & l & 0 \\ 0 & 0 & 0 & 0 & l & 0 \\ 0 & 0 & 0 & 0 & l & 0 \end{bmatrix}\begin{bmatrix} \omega_x^2 \\ \omega_y^2 \\ \omega_z^2 \\ \omega_y\omega_z \\ \omega_x\omega_z \\ \omega_x\omega_y \end{bmatrix}$$

$$(16)$$

With (16), the linear expressions are

$$\dot{\omega}_x=\frac{1}{4l}(A_3+A_4+A_5-A_6-2A_8), \quad (17a)$$

$$\dot{\omega}_y=\frac{1}{4l}(-A_1-A_2+A_5+A_6+2A_7-2A_9), \quad (17b)$$

$$\dot{\omega}_z=\frac{1}{2l}(-A_1+A_2+A_3-A_4) \quad (17c)$$

$$\ddot{R}_{lx}=\frac{1}{2}(A_1+A_2), \ \ddot{R}_{ly}=\frac{1}{2}(A_3+A_4), \ \ddot{R}_{lz}=\frac{1}{2}(A_5+A_6) \quad (17d)$$

In (17), the navigation parameters are all expressed by the accelerometer outputs. And we can compute the

posture of a moving body using a numerical iteration algorithm. The analysis above demonstrates the feasibility of the nine-accelerometer configuration.

B. A Six-accelerometer Configuration

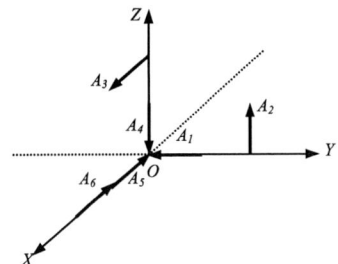

Figure 3. Six-accelerometer NGIMU configuration

The locations and the sensing directions of nine accelerometers in the body frame are shown in Fig. 3. Using the similar method mentioned above, we are easy to get

$$[r_1\times\theta_1,\cdots,r_6\times\theta_6]=l\begin{bmatrix} 0 & 2 & 0 & 0 & 0 & 0 \\ 0 & 0 & 2 & 0 & 0 & 0 \\ 0 & 0 & 0 & 0 & 0 & 0 \end{bmatrix} \quad (18)$$

so

$$M_1=\begin{bmatrix} 0 & 0 & 0 & 0 & -1 & 0 \\ 2l & 0 & 0 & 0 & 0 & 1 \\ 0 & 2l & 0 & 1 & 0 & 0 \\ 0 & 0 & 0 & 0 & 0 & -1 \\ 0 & 0 & 0 & -1 & 0 & 0 \\ 0 & 0 & 0 & -1 & 0 & 0 \end{bmatrix} \quad (19)$$

and the order of (19) is

$$R(M_1)=5 \quad (20)$$

The configuration is not feasible because the order of M_1 is 5 and that implies there exists only five accelerometer output equations, which is not enough for the six navigation parameters (three angular accelerations and three linear accelerations) in the navigation computation. The solution for the equations is not unique and the navigation computation cannot be implemented for this six-accelerometer configuration. So the six-accelerometer configuration of NGMIMU is not feasible.

In addition, some conventional configurations of accelerometers of NGMIMU as in [6, 7] are analyzed using the condition and all these configurations meet the condition.

IV. CONCLUSIONS

This paper proposes a sufficient condition for the configuration of accelerometers of NGMIMU. The condition $R(M_1)\geq6$ indicates that the navigation parameters can be linearly expressed by the accelerometer outputs and the initial conditions. And we can compute the posture of the moving body using a numerical iteration algorithm. Analysis is performed over some conventional configurations of accelerometers of NGMIMU using the condition in this paper and all these configurations meet the condition.

The sufficient condition can serve as an important criterion for the design of NGMIMU. In this paper, the navigation parameter errors attributed to the computational accumulation errors are not fully considered. The future work will consider the optimality of calculation.

REFERENCES

[1] L.D. DiNapoli, "The measurement of angular velocities without the use of gyros," *Master Thesis*, Philadelphia: The Moore School of Electrical Engineering, University of Pennsylvania, 1965, pp. 34-41.

[2] Alfred R. Schuler, "Measuring rotational motion with linear accelerometers", *IEEE Trans. on AES*. 1967, Vol. 3, No. 3, pp. 465-472.

[3] Jeng-Heng Chen, S. C. Lee, Daniel B. DeBra, "Gyroscope free strapdown inertial measurement unit by six inear accelerometers",

Journal of Guidance, Control and Dynamics. 1994, Vol. 17, No. 2, pp. 286-290.

[4] Chin-Woo Tan and Sungsu Park, "Design of gyroscope-free navigation systems", *In: 2001 IEEE Intelligent Transportation Systems Conference Proceedings*, USA: 2001 IEEE, 2001, pp. 286-291.

[5] Sou-Chen Lee and Yu-Chao Huang, "Innovative estimation method with measurement likelihood for all-accelerometer type inertial navigation system", *IEEE Trans. on AES*. 2002, Vol. 38, No. 1, pp. 339-346.

[6] Ma Shutian, Chen Shiyou and Li Yanmei, "Nongyroscopic strapdown inertial navigation system", *Chinese Journal of Aeronautics*, 1997, Vol. 18, No. 4, pp. 484-488.

[7] Shi Zhen, "Accelerometer allocation scheme in gyroscope free strapdown inertial navigation system", *Journal of Chinese inertial technology*, 2002, Vol. 10, No. 1, pp. 15-19.

2006 5th International Power Electronics and Motion Control Conference

Design of Fractional-Order PI^α Controller with two modes[*]

Wen Li[**], Yoichi Hori[***]

[**]Dept. of Electrical Engineering, Dalian JiaoTong University, Dalian, China
E-mail: lw6017@vip.sina.com, dllw2004@mail.goo.ne.jp
[***]Rm. Ce-501, Institute of Industrial Science, the University of Tokyo, Tokyo, Japan
E-mail: hori@iis.u-tokyo.ac.jp, y.hori@ieee.org

Abstract — A kind of fractional-order PI^α controller is considered. Controller structure with two modes is presented, active status of two modes and mode switching are described. The Tustin operator is used for digital realization of the fractional-order PI^α controller. Different approximation order is discussed to get a tradeoff between approximation precision and complexity of discretization expression. Comparison of fractional-order PI^α controllers with two modes and without two modes is demonstrated by experimental examples of torsion vibration suppression in the two cases. Some conclusions are given.

Keywords-Fractional-order controller with two modes; discretization; performance comparison

I. INTRODUCTION

Early developments of Fractional Order control (FOC) in 1950s and 60s were presented [1]-[3]. However, because the lack of sufficient mathematical knowledge and the limited computational power available at that time, applying fractional-order calculus to dynamic systems control is just a recent focus of interest [4]-[6]. The main topics in control field are to expand the existing control theory, based on integer-order differential equations into Fractional Order System (FOS). In theory, the control systems can include both the fractional order dynamic system or plant to be controlled and the fractional order controller. There are two research hot spots for FOC. One is model identification of parameters of fractional-order models, and the other is fractional-order controller design and implementation. General real systems are better described by fractional-order models than described by classical integer order models. However, usually the plant model may have already been obtained as an integer-order model in classical senses, in control

practice it is more common to consider fractional -order controllers. Fractional-order $PI^\alpha D^\beta$ ontroller is an example of fractional-order controllers and it is an extension of integer order PID controllers. The transfer function of $PI^\alpha D^\beta$ controller can be given by $K_p + T_i s^{-\alpha} + T_d s^\beta$, where α and β are positive real numbers, K_p, T_i and T_d are proportional gain, integration constant and the differentiation constant respectively. The $PI^\alpha D^\beta$ controller may improve system control performance due to more tuning parameters introduced. When $\alpha = 1$ and $\beta = 1$, $PI^\alpha D^\beta$ controller becomes traditional PID controller.

More tuning parameters make controller be able to better fit controlled system. But more tuning parameters make it difficult how to match those parameters of $PI^\alpha D^\beta$ controller. The purpose of this paper is to propose a fractional-order PI^α controller with two modes (FOPI for short). As regards two modes, one is control mode; the other is regulating mode.

We focus on the structure design of the FOPI, active status of two modes and mode switching method firstly. The Tustin operator is used for digital realization of the FOPI. Different approximation order is discussed to get a tradeoff between approximation precision and complexity of discretization expression. Then, comparison experiments are demonstrated. These experiments were conducted under two cases for testing control properties of the FOPI based on an experiment system of torsion vibration suppression. Finally, some conclusions are given.

II. DESIGN OF CONTROLLER STRUCTURE

In this section, a structure of the FOPI is proposed to get an easy-to-used method of parameter configuration. This FOPI consists of two components, one is general fractional-order PI^α controller, and the other is parameter regulating unit. The FOPI has two control modes, control mode and regulating mode. The control mode is used during steady state. It is well

---

[*] This paper is supported by China Scholarship Council and Science and Technology Fund of Ministry of Education P.R.C

1-4244-0448-7/06/$25.00 ©2006 IEEE

known that parameter match of $PI^\alpha D^\beta$ controller will play an important role for control performance. In real world, lots of systems usually are fractional order systems. Although a good way to more efficient control of fractional order systems is to use fractional-order controllers, parameter selection still is an important problem needed to be considered. For a little complex system, it is difficult to determine a group of better matching parameters for fractional-order $PI^\alpha D^\beta$ controller by experience. Thus we add a parameter regulating unit in the FOPI to complete parameter matching task instead of trial method. The unit consists of a neural network to realize parameter matching automatically. Structure of the FOPI is shown in Fig. 1.

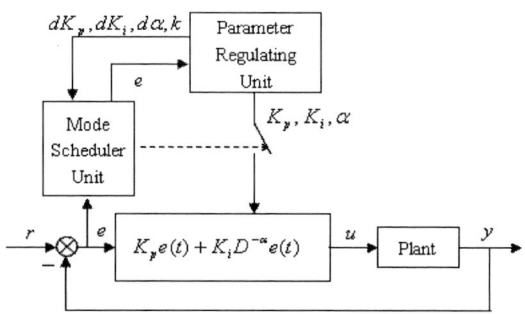

Fig. 1   Block diagram of controller

A   Mode Switching Strategy

In Fig.1, the mode scheduler unit is used to manage the switching of modes. Mode switching strategies are designed according to specific demand. Here, the strategies of scheduler are set as following.

The strategy of switching from control mode to regulating mode only has one rule, i.e. if $|e| > e_{set}$, then switching on the parameter regulating unit.

There two strategies of switching from regulating mode to control mode. They are:

if $|e| \le e_{set}$, then switching off the parameter regulating unit; or if $f(dK_p, dK_i, d\alpha, q)$=True, then switching off the parameter regulating unit.

In above rules, e_{set} is a threshold for mode switching; the $f(dK_p, dK_i, d\alpha, q)$ is a logical expression, it denotes following logical relation:

$|dK_p| < \varepsilon_p$ and $|dK_i| < \varepsilon_i$ and $|d\alpha| < \varepsilon_\alpha$ and $q > Q$, where dK_p, dK_i and $d\alpha$ are changes of parameters K_p, K_i and α , q is the number of iterative times, and $\varepsilon_p > 0$, $\varepsilon_i > 0$, $\varepsilon_\alpha > 0$ and $Q > 0$ are setting values.

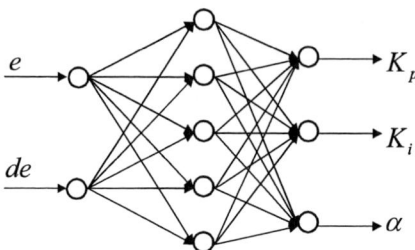

Fig.2   Parameter regulating unit

B   Parameter Regulating Unit

In order to realize self-tuning parameters of the FOPI, learning function should be introduced. Here, a neural network is adopted to get learning performance for the parameter regulating unit. Fig.2 shows structure of the parameter regulating unit. It is implemented using a BP neural network with 2-5-3 network architecture. From Fig.2, it is easy to be understood that the number of nodes in output layer is 3, because there are three parameters, K_p, K_i and α . In order to simplify the neural network, two input nodes are set. About the number of hidden layer nodes, Lippmann provided comprehensive geometrical arguments and reasoning to justify why the maximum number of nodes in a single hidden layer should be equal to M(N+1), where M is the number of output nodes and N is the number of input nodes[7]. In this paper, 5 nodes are used in the single hidden layer. Due to restriction of paper length there is not scope to tell more about the learning algorithm of parameter configuration for the neural network unit.

III.   DISCRETIZATION OF THE FOPI

A   Discretization Methods

Reference [8] points out that the key step in digital implementation of a FOC is the numerical evaluation or discretization of the fractional-order differentiator. In general, there are two discretization methods: direct discretization and indirect discretization. In indirect discretization methods [9], two steps are required, i.e., frequency domain fitting in continuous time domain first and then discretizing the fit s-transfer function. Other frequency-domain fitting methods can also be used but without guaranteeing the stable minimum-phase discretization. There are several direct disretization methods, such as the finite memory length expansion, the direct power series expansion (PSE) of using Euler operator. In addition, literature [10] introduces two methods expanding the Tustin

1798

operator into a rational function. The first one is based on Muir-recursion applied to numerator and denominator of the Tustin operator and the second one is using continued fraction expansion (CFE) technique. For Al-Alaoui operator [8], above two expanding method also can be applied.

In this paper, the Tustin rule is considered as generating function for the digital realization of the FOPI. The following is the transfer function of the FOPI,

$$G(s) = \frac{U(s)}{E(s)} = K_p + \frac{K_i}{s^\alpha} = G_1(s) + G_2(s). \quad (1)$$

The Tustin operator or so-called generating function is

$$\left(\omega(z^{-1})\right)^{\pm\alpha} = \left(\frac{2}{T}\frac{1-z^{-1}}{1+z^{-1}}\right)^{\pm\alpha}. \quad (2)$$

According to Muir-recursion formula, there is following equation for positive $\alpha \le 1$,

$$s^\alpha = \left(\frac{2}{T}\frac{1-z^{-1}}{1+z^{-1}}\right)^\alpha \approx \left(\frac{2}{T}\right)^\alpha \frac{A_n(z^{-1},\alpha)}{A_n(z^{-1},-\alpha)}. \quad (3)$$

Therefore,

$$G_2(z) = \frac{K_i}{s^\alpha} = \lim_{n\to\infty} \frac{K_i}{\left(\frac{2}{T}\right)^\alpha \frac{A_n(z^{-1},\alpha)}{A_n(z^{-1},-\alpha)}}, \quad (4)$$

Where A_n denotes polynomial with truncation order n, in the variable z^{-1}. $A_n(z^{-1},\alpha)$ is gotten by recursion formula,

$$A_0(z^{-1},\alpha) = 1,$$
$$A_n(z^{-1},\alpha) = A_{n-1}(z^{-1},\alpha) - c_n z^{-n} A_{n-1}(z,\alpha), \quad (5)$$

and

$$c_n = \begin{cases} \alpha/n, & n \text{ is odd}, \\ 0, & n \text{ is even}. \end{cases} \quad (6)$$

Thus, for a specific truncation order n, equation (4) can be expressed by

$$\frac{K_i}{s^\alpha} \approx \frac{K_i}{\left(\frac{2}{T}\right)^\alpha \frac{A_n(z^{-1},\alpha)}{A_n(z^{-1},-\alpha)}} = \frac{K_i A_n(z^{-1},-\alpha)}{\left(\frac{2}{T}\right)^\alpha A_n(z^{-1},\alpha)}. \quad (7)$$

B Approximation Order

From equation (4), it is can be seen that fractional-order integrator $\frac{K_i}{s^\alpha}$ is an infinite dimensional linear filter. Thus the realization of this fractional-order integrator is an approximation problem of equation (7) in a considered frequency bandwidth domain actually. This subsection will discuss the relation between approximation order and approximation precision. So-called approximation order is the truncation order. According to recursive formula (5) and (6), the expressions of $A_n(z^{-1},\alpha)$ can be obtained for truncation order n = 1, 2, ..., 9. Therefore corresponding approximation expressions of $G_2(s)$ are gotten also, here they are listed as follows:

$$G_{21}(s) = \frac{K_i}{\left(\frac{2}{T}\right)^\alpha} \frac{z+\alpha}{z-\alpha} = \frac{K_i}{K} \frac{z+\alpha}{z-\alpha},$$

$$G_{23}(s) = \frac{K_i}{K} \frac{z^3 + \alpha z^2 + \frac{\alpha^2}{3}z + \frac{\alpha}{3}}{z^3 - \alpha z^2 + \frac{\alpha^2}{3}z - \frac{\alpha}{3}},$$

$$G_{25}(s) = \frac{K_i}{K} = \frac{z^5 + \alpha z^4 + \frac{2\alpha^2}{5}z^3 + \left(\frac{\alpha}{3} + \frac{\alpha^3}{15}\right)z^2 + \frac{\alpha^2}{5}z + \frac{\alpha}{5}}{z^5 - \alpha z^4 + \frac{2\alpha^2}{5}z^3 - \left(\frac{\alpha}{3} + \frac{\alpha^3}{15}\right)z^2 + \frac{\alpha^2}{5}z - \frac{\alpha}{5}},$$

$$G_{27}(s) = \frac{K_i}{K} \times \frac{z^7 + \alpha z^6 + \frac{3}{7}\alpha^2 z^5 + \left(\frac{1}{3}\alpha + \frac{2}{21}\alpha^3\right)z^4 + \left(\frac{26}{105}\alpha^2 + \frac{1}{105}\alpha^4\right)z^3 + \left(\frac{1}{5}\alpha + \frac{2}{35}\alpha^3\right)z^2 + \alpha^2 z + \frac{\alpha}{7}}{z^7 - \alpha z^6 + \frac{3}{7}\alpha^2 z^5 - \left(\frac{1}{3}\alpha + \frac{2}{21}\alpha^3\right)z^4 + \left(\frac{26}{105}\alpha^2 + \frac{1}{105}\alpha^4\right)z^3 - \left(\frac{1}{5}\alpha + \frac{2}{35}\alpha^3\right)z^2 + \alpha^2 z - \frac{\alpha}{7}},$$

$$G_{29}(s) = \frac{K_i}{K} \times$$

$$z^9 + \alpha z^8 + \frac{4}{9}\alpha^2 z^7$$
$$+\left(\frac{1}{3}\alpha + \frac{1}{9}\alpha^3\right)z^6 + \left(\frac{17}{63}\alpha^2 + \frac{1}{63}\alpha^4\right)z^5$$
$$+\left(\frac{1}{5}\alpha + \frac{16}{189}\alpha^3 + \frac{1}{945}\alpha^5\right)z^4$$
$$+\left(\frac{34}{189}\alpha^2 + \frac{2}{189}\alpha^4\right)z^3$$
$$\underline{+\left(\frac{1}{7}\alpha + \frac{1}{21}\alpha^3\right)z^2 + \frac{1}{9}\alpha^2 z + \frac{\alpha}{9}}$$
$$z^9 - \alpha z^8 + \frac{4}{9}\alpha^2 z^7$$
$$-\left(\frac{1}{3}\alpha + \frac{1}{9}\alpha^3\right)z^6 + \left(\frac{17}{63}\alpha^2 + \frac{1}{63}\alpha^4\right)z^5$$
$$-\left(\frac{1}{5}\alpha + \frac{16}{189}\alpha^3 + \frac{1}{945}\alpha^5\right)z^4$$
$$+\left(\frac{34}{189}\alpha^2 + \frac{2}{189}\alpha^4\right)z^3$$
$$-\left(\frac{1}{7}\alpha + \frac{1}{21}\alpha^3\right)z^2 + \frac{1}{9}\alpha^2 z - \frac{\alpha}{9}$$
.

Obviously truncation order n affects discretization effect, i.e. approximation precision of equation (1) depends on that of $G_2(s)$. Fig. 3 give magnitude and phase curves with different truncation order n, α =0.5 and sampling period T=0.001 second.

Fig.3 shows that the best approximating frequency bandwidth is [200, 2000] when truncation order n=1, 3,

5, 7, 9. It can be found that no matter how many n is, errors of magnitudes and phases are bigger at high frequency for this approximating method. More detailed discuss of this problem will be given in other paper.

IV. EXPERIMENTAL EXAMPLES

Considering the approximating precision and the complexity of discretization expression, truncation order n=5 was determined. The FOPI is realized by C program. The testing of control performance was conducted based on an experiment system of torsional vibration suppression. This experiment system includes a 2-inertial component. The two inertias are linked by a long thin shaft and equivalent model can be simplified as shown in Fig.4.

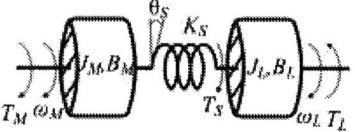

Fig.4   Simplified 2-inertia model

Two cases were designed for testing performance of the FOPI.

Case 1. the controller adopted is a general fractional-order PI^α controller without parameter regulating unit. Three parameters were set as follows:

$$K_p = 1.0, \quad K_i = 0.28, \quad \alpha = 0.1.$$

Firstly, an additional friction was applied to the experiment system, corresponding time response curve is the upper one in Fig.5. Then, maintaining controller parameters and removing the additional friction from experiment system, the time response obtained is the lower one in Fig.5. Clearly, the control performance became worse because parameters could not be regulated automatically.

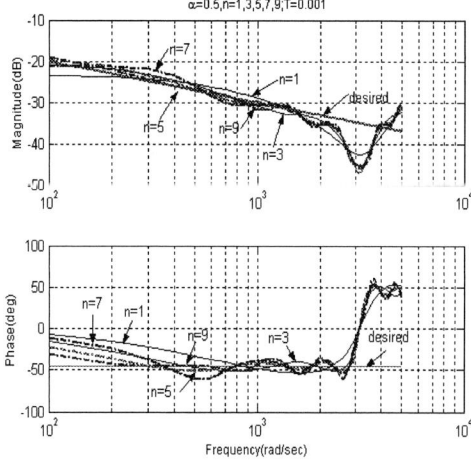

Fig.3   Approximating curves with different n

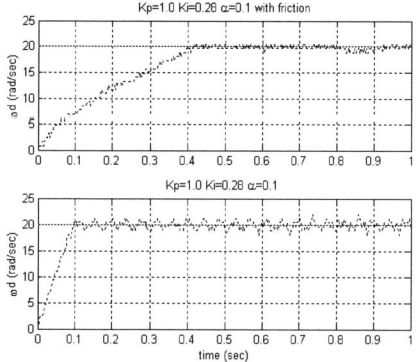

Fig.5   Response curves for case 1

Case 2. the FOPI is adopted. Similar process was repeated. When an additional friction was put on the experiment system, firstly, the regulating mode was started during 0-0.33 second. Then system run into steady state, the FOPI worked at control mode. Time response curve and parameters changing curves are given in Fig.6. When the additional friction was canceled, the controller parameters were re-tuned during regulating mode. These updated parameters guaranteed satisfied control performance. Corresponding curves are shown in Fig.7.

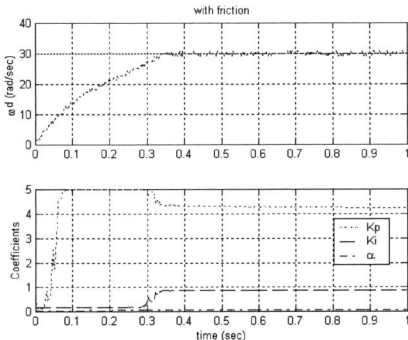

Fig.6   Response curves with friction

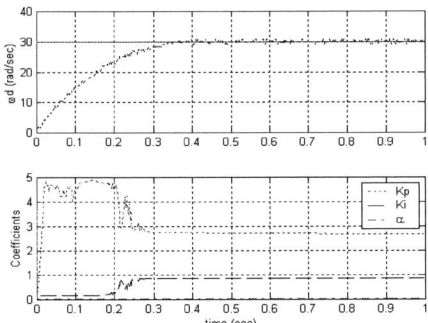

Fig.7   Response curves without friction

V.   CONCLUSION

Summarizing the above discussion, several conclusions can be obtained. Fractional-order controller can fit fractional order systems in real world better. Combining some intelligent approaches, fractional-order controllers can have better control characteristics. Experiments show that the controller proposed in this paper has better adaptive property for system condition changing due to introducing two modes.

REFERENCE

[1]   K.Hashimoto, S.Takahashi, and H.Amano, "Realization of non-integer integral computing element," JIEE, vol. 89, no. 974, pp. 2213-2220, 1969.

[2]   S.Manabe, "The non-integer integral and its application to control systems," JIEE, vol. 80, no. 860, pp. 589-597, 1960.

[3]   S.Manabe, "The system design by the use of non-integer integral and transport delay," JIEE, vol. 81, no. 878, pp. 1803-1812, 1961.

[4]   I.Podlubny, "Fractional-order systems and PI^λ D^μ -controllers," IEEE Trans. Automat. Contr., vol. 44, no. 1, pp. 208-214, Jan 1999.

[5]   A.Oustaloup, J.Sabatier, and P.Lanusse, "From fractal robustness to CRONE control," Fractional Calculus and Applied Analysis, vol. 2, no. 1, pp. 1-30, 1999.

[6]   H.-F.Raynaud and A.Zergalnoh, "State-space representation for fractional order controllers," Automatica, vol. 36, no. 7, pp. 1017-1021, 2000.

[7]   R.P.Lippmann, "An introduction to computing with neural nets," IEEE Acoustics, Speech and Signal Processing Magazine, 4-22, Apr. 1987.

[8]   Y.Q.Chen, K.L.Moore, "Discretization Schemes for Fractional-Order Differentiators and Integrators," IEEE Trans. Circuits and Systems-I: Fundamental Theory and Applications, vol. 49, no.3, pp. 363-367, Mar. 2002.

[9]   A.Oustaloup, et al, "Frequency-band complex noninteger differentiator: characterization and synthesis," IEEE Trans. On Circuit and Systems-I: Fundamental Theory and Applications, vol.47, no.1, pp. 25-39, 2000.

[10]   B.M.Vinagre, I.Petras, et al, "Two Digital Realization of Fractional controllers: Application to Temperature control of a Solid," in:Proceedings of the European Control Conference(ECC2001), pp. 1764-1767, Porto, Portugal, 2001.

2006 5th International Power Electronics and Motion Control Conference

Sliding Mode Robust Tracking Control Based on Learning Feedforward Compensation for High Precision Linear Servo System

Zhu Guoxin, Guo Qingding and Zhao Ximei
School of Electric Engineering, Shenyang University of Technology
Shenyang,110023, Liaoning Province
ghost0303@sohu.com

Abstract—The permanent magnet linear servo system has some advantages of high speed, high response and direct drive, but the servo performance is influenced by the load disturbances, end-effects, nonlinear friction and the parameters variations. In order to eliminate the influences of the above-mentioned uncertain factors on the basis of ensuring better tracking performance, a robust tracking control strategy combining variable structure control (VSC) with B spline nervous network (BSNN) is presented in this paper. The variable structure control has advantages of the fast response and the invariability of the uncertain factors, but its chattering phenomenon will influence the stationarity of the linear servo system and tracking precision. To reduce the "chattering" input, a BSNN is adopted to eliminate the above-mentioned uncertain factors, thus the steady state accuracy of the system is enhanced further. The simulation results show that the solution not only has stronger robustness to the uncertainty of the linear servo system, but also has better tracking performance.

Keywords - Permanent magnet linear synchronous motor; slide mode variable structure control; B spline nervous network; end-effects; chattering; robust tracking

I. INTRODUCTION

The permanent magnet linear synchronous motor (PMLSM) has the advantages of higher energy density, reliability and efficiency; recently it is widely used in feed drive of high precision NC machine tools as in [1]. However, PMLSM adopts the direct drive method, so the load variation and the external disturbances will influence the performance of the linear servo system directly. Furthermore, the parameters variation, the nonlinear friction, end-effects, cogging effects and flux linkage harmonic and so on will affect the servo performance of the linear system and the machining accuracy of the machine tools as in [2]. To eliminate the influences of the above-mentioned uncertain factors on the base of ensuring tracking performance, a robust tracking control strategy combining variable structure control with B spline neural network (BSNN) is presented in this paper.

The variable structure control has advantages of the fast response and the invariability of the uncertain factors. But the switch control to deal with the uncertain factors will generate chattering phenomenon, and the amplitude of the chattering is proportional to the range of system parameters variation and the amplitude of external

disturbances as in [3]. The chattering will influence the steady performance and positioning accuracy, and increase energy loss. The common method to reduce the chattering is that the switch control is substituted by the continuous saturated control so that the discontinuous switch control becomes smooth. However, the stability of system is guaranteed only outside the critical layer, and the tracking error is related to the width of the critical layer. Although the method can eliminate chattering, robustness of VSC is destroyed, and the stability in the critical layer can not be guaranteed as in [4-5]. Ref.[4] has presented a compensation method based on the linear disturbance observer to reduce chattering. However, uncertain disturbance of the end-effects and friction force are usually time-varying nonlinear, so the effect of the compensation is not satisfactory.

To weaken the influences of chattering to the linear servo system, a BSNN is introduced to constitute the function approximator that is used to eliminate the uncertain factors caused by the end-effects, parameters changes, friction, and external load in the paper. BSNN is a local approximated network, and has advantages that rapidity convergence and avoids local extremum. By the compensation of the BSNN, both the stronger robustness and tracking precision are guaranteed.

II. THE MATHEMATICAL MODEL OF PMLSM

A. The Mode of End-effect

PMLSM is a thrust device that can provide linear motion directly. The end-effects caused by two cutting off terminals will generate thrust ripple which will influence the servo performance. The low frequency component is the end-effects force, while the high frequency component is cogging-effects force. The end-effects force is the main factor, and is periodic function with regard to position. The amplitude of the thrust ripples is also related to the magnitude of the current, but the end-effects without load are the main cause of the thrust ripples. The simplified mathematical model is

$$F_d = F_{dm} \cos(\frac{s}{\tau} 2\pi + \theta_0) \qquad (1)$$

Where, F_d is end-effects force; s is mover's position; F_{dm} is the amplitude of end-effects force; τ is polar pitch; θ_0 is electrical angle of initial phase as in [6].

B. The Mode of PMLSM

When only fundamental component is considered, the d-q axis model can be used. PMLSM voltage equations and flux linkage equations are as follows.

$$u_d = R_s i_d + p\lambda_d - v\lambda_q \tag{2}$$

$$u_q = R_s i_q + p\lambda_q + v\lambda_d \tag{3}$$

$$\lambda_d = L_d i_d + \lambda_{PM1} \tag{4}$$

$$\lambda_q = L_q i_q \tag{5}$$

When $i_d = 0$ control scheme is applied in the current inner loop, the expression of electromagnetic thrust is

$$F_e = \frac{3\pi}{2\tau}\lambda_{PM} i_q = K_f i_q = M\frac{dv}{dt} + Bv + F_d + F_l \tag{6}$$

Where, u_d, u_q; i_d, i_q; L_d, L_q; λ_d, λ_q are voltages, currents, inductances, the flux linkages of d-axis and q-axis respectively; R_s is the resistance of mover, λ_{PM} is excitation flux linkage caused by permanent magnet; v is speed of mover; B is viscous friction coefficient; M is mass of the moving part including the load; F_e is the electromagnet thrust; K_f is the electromagnetic thrust coefficient; F_l is the load resistance.

Defining state variables as $x(t) = [x_1(t) \ x_2(t) \ x_3(t)]^T$, then the obtained state equations of the PMLSM are

$$\dot{x}_1(t) = x_2(t)$$
$$\dot{x}_2(t) = x_3(t)$$
$$\dot{x}_3(t) = -a(t)x_2(t) + b(t)u(t) - d(t) \tag{7}$$

Where $u = i_q$ is control input; $a(t) = B/M$, $b(t) = K_f/M$ are time-varying parameters; $d(t) = (F_l + F_d)/M$ is time-varying disturbance.

III. DESIGN OF SLIDING MODE VARIABLE STRUCTURE CONTROLLER

The block diagram of the system with variable structure control based on BSNN is shown in Fig.1.

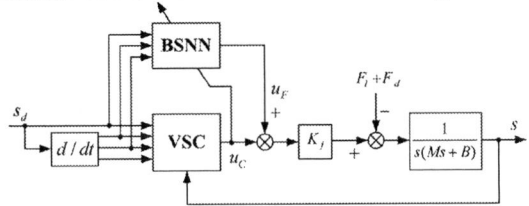

Fig.1 Block diagram of PMLSM system with VSC based on BSNN

In Fig.1, BSNN is trained by the output u_C of the feedback controller (VSC) in order to reduce the tracking error e. After fully trained, control signal u_F is generated by BSNN, which is used to track reference path. The design of sliding model variable structure control concludes two parts: one is the selection of sliding model switching hyper-plane that is stable and has ideal dynamic character; the other is definition of sliding model control law, such that the state can reach near the

sliding model switching hyper-plane in finite time to ensure the hyper-plane to slide to the stable point.

Assumption 1 :

1) $\{\dddot{s}_d(t), \ddot{s}_d(t), \dot{s}_d(t), s_d(t)\}$ are known bounded and continuous;
2) The state $x(t)$ is obtainable;
3) Time-varying parameters and disturbances are bounded, i.e.

$$a(t) = a_n + \Delta a(t), \quad |\Delta a(t)| \le \alpha(t) \quad \forall t$$
$$b(t) = b_n + \Delta b(t), \quad |\Delta b(t)| \le \beta(t) \quad \forall t$$
$$|d(t)| < \gamma(t) \quad \forall t$$

Where, a_n, b_n are the nominal parameters of the system; $\Delta a(t)$, $\Delta b(t)$ are the uncertain parts of the system parameters, $\gamma(t)$ is the upper bound of load disturbance.

Defining variable $e(t) = s_d(t) - s(t)$. Introducing An integral element into the switching function $\sigma(t)$ to eliminate the steady error of the system. The equation of the sliding mode switching hyper-plane is

$$\sigma(t) = \ddot{e}(t) + c_1\dot{e}(t) + c_2 e(t) + c_3\int_0^t e(\tau)d\tau \tag{8}$$

The dynamic character of $\sigma(t)=0$ will be stable and have ideal characteristic root by selecting $c_i (1 \le i \le 3)$ properly.

The dynamic Eq.(8) of the sliding model switching hyper-plane can be rewritten as follows.

$$\dot{E}(t) = CE(t) + D\sigma(t) \tag{9}$$

Where,

$$E(t) = \begin{bmatrix} \int_{t_0}^t e(\tau)d\tau \\ e(t) \\ \dot{e}(t) \end{bmatrix}, C = \begin{bmatrix} 0 & 1 & 0 \\ 0 & 0 & 1 \\ -c_3 & -c_2 & -c_1 \end{bmatrix}, D = \begin{bmatrix} 0 \\ 0 \\ 1 \end{bmatrix} \tag{10}$$

The solution of Eq.(9) is

$$E(t) = \exp[C(t-t_0)]E(t_0) + \int_{t_0}^t \exp[C(t-\tau)]D\sigma(\tau)d\tau \quad t \ge t_0 \tag{11}$$

Theorem 1: if two positive constants k_1, k_2 exist such that the inequality (12) holds, then Eq.(9) is stable.

$$\|\exp[C(t-t_0)]\| \le k_1 \exp[-k_2(t-t_0)]$$
$$\forall t \ge t_0 \ge 0, \ k_2 = -\max_{1 \le i \le 3} \Re e\{\lambda_i[C]\}. \tag{12}$$

Theorem 1: If sliding model switching hyperplane equation meets the inequality $|\sigma(t)| < \delta\|E(t)\|$ $(t \ge t_0, 0 < \delta < k_2/k_1)$ and the initial condition $E(t_0)$ is bounded, then, when $t \ge t_0$, $e(t)$ is approached to zero at the speed of $(k_2 - \delta k_1)$ as in [7].

According to Cayley-Hamilton theorem, the matrix $\exp[C(t-t_0)]$ can be denoted as 3×3 matrix. If the poles of C matrix lie in the left half plane, it is the more left, the value of σ is less. We can solve k_1, k_2 according to inequality (12) and $0 < \delta < k_2/k_1$.

The variable structure control law adopts the

equivalent control method, the control input is

$$u_C(t) = u_{eq}(t) + u_s(t) \tag{13}$$

Where, u_{eq} is equivalent control used to control the certain part of PMLSM; while the switching control u_s is used to control the uncertain part so as to ensure robustness. According to the conditions of the sliding model equivalent control $\sigma(t) = 0$ and $d\sigma(t)/dt = 0$, the equivalent control is derived by Eq.(7),(8),(13).

$$u_{eq}(t) = \frac{1}{b_n}\{\ddot{s}_d(t) + a_n x_2(t) + c_1[\dot{s}_d(t) - x_3(t)]$$
$$+ c_2[\dot{s}_d(t) - x_2(t)] + c_3[s_d(t) - x_1(t)]\} \tag{14}$$

The switching control is designed as follows.

$$u_s(t) = K(x,t)\mathrm{sgn}(\sigma) \tag{15}$$

Where

$$K(x,t) = [\alpha(t)|x_2(t)| + \beta(t)|u_{eq}(t)| + \gamma(t)]/[b_n - \beta(t)] \tag{16}$$

According to Assumption 1, the state $x_1(t)$ can approximately follow the given position $s_d(t)$, and the sliding model control input $u_c(t)$ is bounded.

IV. THE DESIGN OF BSNN

To weaken the influences of chattering to the linear servo system, a BSNN is introduced to constitute the function approximator that is used to simulate the uncertain factors caused by the end-effects, parameters variation, friction, external load. It is used to implement the function approximation of the feed-forward part. BSNN adopts the piecewise polynomial basis function, that is so called B spline. Two orders B spline is adopted in the paper. B spline is distributed into the input area of the BSNN and the sum of the membership degree is equal to one. The output of BSNN is the weighted sum of each B spline basis function, i.e.

$$u_F(x) = \sum_{i=1}^{N}\mu_i(x)w_i \tag{17}$$

where, μ_i and w_i are the primary function and weighting value of the ith B spline, N is the number of the B spline.

Block diagram of BSNN is shown in Fig.2.

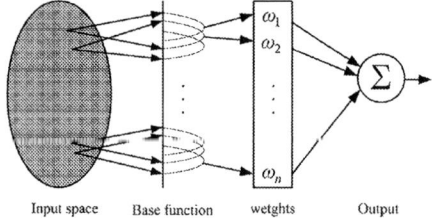

Fig.2 Block diagram of the BSNN

The dynamic equation of PMLSM is obtained by Eq.(6)

$$\ddot{s} = \frac{1}{M}\left(-B\dot{s} - F_d + K_f u\right) \tag{18}$$

Where, s is the position of the linear motor, $u = i_q$ is controlled variable. The state equation of PMLSM is

$$\begin{bmatrix} \dot{s} \\ \ddot{s} \end{bmatrix} = \begin{bmatrix} 0 & 1 \\ 0 & -\dfrac{B}{M} \end{bmatrix}\begin{bmatrix} s \\ \dot{s} \end{bmatrix} + \begin{bmatrix} 0 \\ \dfrac{K_f}{M} \end{bmatrix}u + \begin{bmatrix} 0 \\ -\dfrac{F_d}{M} \end{bmatrix} \tag{19}$$

The desired control signal of the PMLSM is as follows.

$$u_d = \frac{1}{K_f}\left(M\ddot{s}_d + B\dot{s}_d + F_d(s_d)\right) \tag{20}$$

Therefore selecting $\{s, \dot{s}, \ddot{s}\}$ serves as the input of the feed-forward controller (BSNN).

To obtain the minimal width of the B spline, defining

$$\varphi = \arccos\left[-0.0147\frac{\big||-T(j\omega)|\big|_{\infty}}{\min\limits_{\omega \in R|\cos(\varphi)\leq 0}\big|-T(j\omega)\big|}\right] \tag{21}$$

Where, φ is the phase angle of close-loop transfer function. $\varphi = \arccos(-0.0257) = -256[\deg]$ can be obtained according to frequency character of PMLSM, when phase frequency is -256 [deg], the frequency is $\omega = 220\big[rads^{-1}\big]$. According to $\omega = 2\pi/d$, the minimal width of the B spline is obtained

$$d_{\min} = \frac{2\pi}{\omega_{\max}} = \frac{2\pi}{220} = 2.855 \times 10^{-2}\,[s] \tag{22}$$

Selecting the objective function as

$$J_d = \frac{1}{2}\sum_{j=0}^{t/T}(u_C(jT))^2 = \frac{1}{2}\sum_{j=0}^{t/T}(u(jh) - u_F(jT))^2 \tag{23}$$

Where, J_d is the number of the junctions, t is the motion time, T is sample period, u_c is the output of the VSC, u_F is the output of the BSNN.

The gradient descent method is adopted to adjust the weights.

$$\Delta w_i = -\gamma\frac{\partial J_d}{\partial w_i} = -\gamma\frac{\partial J_d}{\partial u_F}\frac{\partial u_F}{\partial \omega_i}$$
$$= \gamma\sum_{j=0}^{t/T}\mu_i(jT)(u(jT) - u_F(jT)) \tag{24}$$
$$= \gamma\sum_{j=0}^{t/T}\mu_i(jT)u_C(jT)$$

where, Δw_i is the regulated quantity of the ith weights, γ is the rate of the learning, $0 < \gamma < 1$.

V. SIMULATION RESULTS AND ANALYSIS

For PMLSM, the normal parameters are $M_n - 11.0Kg$, $B_n = 2.0N\ s/m$, $K_{fn} - 25N/A$, $F_{en} = 200N$, $v_n = 2.0m/s$. For the sake of comparison, traditional VSC is also simulated. The desired position path of the system is $s_d(t) = 1\sin(2\pi t)$ m, the parameters change range is $|\Delta a(t)| \leq |a_n(t)| = \alpha(t)$ and $|\Delta b(t)| \leq 0.5|b_n(t)| = \beta(t)$, the upper bound of the external disturbance is $\gamma(t) = |d(t)|$, $F_d = 30\sin(0.035x/2\pi)$ is the end-effects force. The path input of the BSNN is used to compensate for the end-effects, the period of the PMLSM end-effects is about 3.5[cm], the width of the B spline is 0.0035[m]. Choosing the width of the B spline as

$0.2[ms^{-1}]$, γ is 0.1, the sample period T is 0.5[ms].

The position tracking character curves of the traditional VSC and BSNN-VSC system are shown in Fig.3 and Fig.4. Where □ and □ are ideal and practical position curves respectively. Compared Fig.3 with Fig.4, we can see that the tracking performance of the traditional VSC becomes worse and occur to chattering in the condition of having great uncertain factors. While the BSNN-VSC strategy proposed in the paper has not only stable tracking accuracy, but also has satisfactory transient state performance, which is because BSNN catches the major character of the uncertain factors.

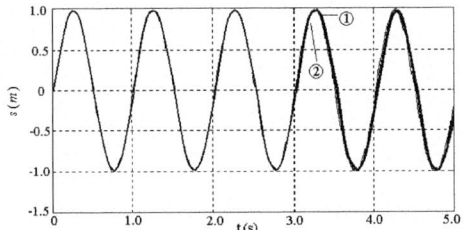

Fig.3 Curves of position tracking characteristic of the system with traditional VSC

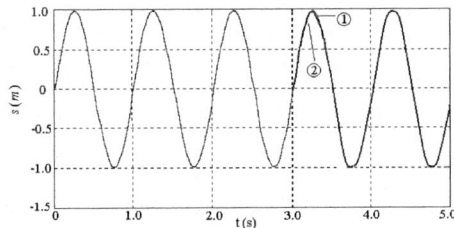

Fig.4 Curves of position tracking characteristic of the system with BSNN-VSC

VI. CONCLUSION

In the paper, a sliding mode variable structure control strategy based on BSNN is presented for direct drive high precision permanent magnet linear synchronous motor servo system. It can compensate for the uncertain factors of the end-effects, parameters variation, friction and external load, so the influence of chattering input to system is eliminated fundamentally. The simulation results show the control scheme has better fast tracking performance and robustness.

REFERENCES

[1] Boldea I. Nasar S. A. *Linear Electric Actuators and Generators.* Cambridge University Press, 1997.

[2] Q.D. Guo and C.Y. Wang, *Precision control technology for linear AC servo system.* Beijing: Machine Industry Press, 2000.

[3] F.Y. Wang. *Sliding mode variable structure control.* Beijing: Machine Industry Press, 1995.

[4] Y.B. Sun, Q.D. Guo and L.M. Shi. "Sliding mode variable structure control based on thrust observer for AC linear servo system". *Transactions of China Electrotechnical Society,* vol. 13 no. 2, pp. 1-5, 1998.

[5] C.L. Hwang, "Sliding mode control using time-varying switching gain and boundary layer for electro hydraulic position and differential pressure control". Proc. *IEE Control Theory Application,* vol. 143, no. 4, pp. 325~332, July 1996.

[6] Q.L. Li and X.K. Wang. "Design of permanent-magnet AC linear synchronic motor position servo control system". *China mechanical engineering,* vol. 12, no. 5, pp. 577~581, Oct. 2001.

[7] C.L. Hwang, "Neural-network-based variable structure control of electrohydraulic servosystems subject to huge uncertainties without persistent excitation", *IEEE Trans. Mechatronics,* vol. 4, no. 1, pp. 50-59, January 1999.

Application of Fuzzy Self-learning Sliding Mode Variable Structure Control in Linear AC Servo System

Qing Hu, Shuo Jie and Dongmei Yu

Shenyang University of Technology / School of Electrical Engineering, Shenyang, 110023, China
aqinghu@163.com

Abstract—A sliding mode variable structure control (SMVSC) method based on fuzzy self-learning for a direct-drive AC linear servo system is presented. SMVSC is a special nonlinear control strategy, which has strong robustness against parameter variations, load disturbances and uncertainty of system. This paper proposes that the chattering can be effectively minimized by introducing fuzzy self-learning in conventional sliding mode control, without sacrificing the strong robustness of SMVSC. Simulations show that this kind of method is effective. Simultaneously, the servo system promises good static and dynamic performance.

Keywords-sliding mode variable structure control; fuzzy control; linear servo system; fuzzy self- learning; chattering

I. INTRODUCTION

This paper presents the application of fuzzy self-learning sliding mode variable structure in linear AC servo system. It has the ability of online self- learning and adjusting the output accurately and timely. It has a vice-control effect, which ensures the kinematic state parameters to surpass the super surface of the sliding mode at a low speed, and thus provides the possibility to minimize the chattering, at least in theory.

Sliding mode variable structure control is a special discontinuous nonlinear control strategy, which has strong robustness against parameter variations, load disturbances and uncertainty of system, with the advantage of rapidness and easy realization. However, the control precision and stability of the system are affected by the chattering. The normal method to minimize the chattering is to replace the switch control with continued saturated nonlinear control to smooth the discontinuous variables. This method can eliminate the chattering, but at the same time minimizes the robustness of the sliding mode structure control system [4]. Another method is to control the tendency rate of the sliding mode, which minimizes the chattering through controlling the approaching speed of the state variables to the switch line of sliding modes without the loss of the robustness of the system. But it is difficult to decide the approaching speed in practice. Ref. [5] introduced the fuzzy theory to minimize the chattering,

but it is hard to establish the exact fuzzy rule. This paper presents a kind of fuzzy self-learning sliding mode control strategy, and it proceeds on-line learning through fuzzy deduction for uncertainty factor (parameter variable and external disturbance), also it proceeds real-time adjustment for direction and range of the switch control. Without the influence on system's robustness and rapid tracking performance, strike speed that the kinematic state of system traverse sliding mode line, is reduced, thereby the chattering of the system. is consumedly minimized.

Due to permanent magnet linear synchronous motor (PMLSM) needs no mechanical transmissions from rotational motion to line motion, PMLSM becomes the best choice of the high precision and micro-input servo system. However, owing to the direct driving of the load, the variations of the load and external disturbance will directly affect the performance of the servo system. The parameter variations of the motor under different working states and the nonlinear factors will have bad influence on the servo system. Classical PID control unable to adapt to the requirement of high-powered linear direct-drive AC servo system, this paper presents a new type of sliding mode variable structure control strategy in allusion to plant characteristics and limitation of classical control strategy.

II. MATHEMATICAL MODEL OF LINEAR AC SERVO SYSTEM

The basic structure of AC permanent magnet linear synchronous motor and the structure and principle figure of linear servo system are shown in Ref. [1].

The inner circle of the system adopts the field-oriented vector control, let current component of mover $i_d = 0$, we can obtain the simplified model of the plant

$$F_e = K_F i(t) , \qquad (1)$$

$$M\frac{dv}{dt} + Bv = F_e - F_L , \qquad (2)$$

$$s = \int v dt , \qquad (3)$$

where F_e, i, F_L, and v refer to electromagnetic thrust, dc component I_q in d-q coordinate transformation, equivalent resistance caused by the end effect and linear speed of mover, respectively; M, s, K_F, and B refer to mass of mover, mechanical displacement of mover, thrust coefficient and viscous frictional coefficient, respectively.

III. SLIDING MODE CONTROL

Consider the affine linear system below

$$
\begin{cases}
\dot{x}_1 = x_2 \\
\dot{x}_2 = x_3 \\
\cdots \\
\dot{x}_n = f(x) + b(x)u + d(t)
\end{cases}
$$

$$y = x_1, \tag{4}$$

where $x = [x_1, x_2, \cdots, x_n]^T = [x, \dot{x}, \cdots, x^{n-1}]^T$ is the state vector of system. $f(x)$, $b(x)$ are nonlinear functions which their precise values are unknown. u, y stand for the control input and output of system, respectively. The aim of the control is that when $f(x)$, $b(x)$ and $d(t)$(bounded disturbance) of (4) exist uncertainties, y can asymptotically tracks the desired output x_d. Take tracking error vector:

$$e = x_d - x = [e, \dot{e}, \cdots e^{(n-1)}]^T. \tag{5}$$

Substitute e into (4), we get error state equation of the system:

$$e^{(n)} = x_d^{(n)} - x^{(n)} = -(f(x) + b(x)u + d(x)) + x_d^{(n)}. \tag{6}$$

Choose sliding mode switch function as:

$$s = c^T e = c_1 e + c_2 \dot{e} + \cdots + c_n e^{(n-1)} \quad (c_i > 0). \tag{7}$$

From $\dot{s} = 0$ we can obtain sliding mode equivalent control:

$$u_{eq} = \frac{1}{b}\left(-f - d + x_d^{(n)} + \sum_{i=1}^{n-1} c_i e^{(i)}\right). \tag{8}$$

When b, f and d are precisely known, from (8) we directly obtain u_{eq}, according to basic theory of the sliding mode control [7], the structure of the control law is:

$$u = u_{eq} + u_s, \tag{9}$$

where u_s is nonlinear control, which function results in the sliding mode.

But in practice, b, d and f often exist some uncertainties, such as parameter uncertainties, uncertainties caused by nonlinear and so on, and makes it difficult to obtain the equivalent control. In this case,

based on Ref. [3], fuzzy system is adopted to approach equivalent control u_{eq}, then (9) can be rewritten as

$$u = u_f + u_s, \tag{10}$$

where u_f is the output of fuzzy control .

IV. FUZZY SLIDING MODE CONTROL DESIGN

A. Sliding Mode Control Strategy

In order to design sliding mode controller, redefine state variables: Suppose s^* is known, s for the actual displacement of the mover. Suppose $x_1 = s^*$-s, from (1), (2) and (3), error state equation of PMLSM can be obtained

$$
\begin{bmatrix} \dot{x}_1 \\ \dot{x}_2 \end{bmatrix} = \begin{bmatrix} 0 & 1 \\ 0 & a \end{bmatrix} \begin{bmatrix} x_1 \\ x_2 \end{bmatrix} + \begin{bmatrix} 0 \\ b \end{bmatrix} u + \begin{bmatrix} 0 \\ d \end{bmatrix} f, \tag{11}
$$

where $a = -B/M$, $b = -K_F/M$, $d = 1/M$. $u = I_q$ is control component; a, b, d are limited time-varying parameters, their rated values are a_r, b_r, d_r, respectively; $f = E_L + \Delta M dv/dt + \Delta B v$ is the defined generalized loading disturbance. Here ΔM, ΔB are offset of coefficient parameter M and B, respectively, take switch function of sliding mode variable structure as:

$$\sigma(x) = cx_1 + x_2, \tag{12}$$

where $x = [x_1, x_2]^T$ is state vector of error system. After the system enters into the state of sliding mode, the main performance of the system is decided by sliding mode line $\sigma(x) = 0$.

Adopt sliding mode equivalent control strategy, the output of controller is

$$u = u_{eq} + u_{fz}, \tag{13}$$

where u_{eq} is equivalent part of sliding mode control, i.e. the necessary known part of the control system when $d\sigma/dt = 0$ and $f = 0$. The sliding switch control u_{fz}, through high frequency switch control, makes the system state approach to the super surface of the sliding mode and ensures the system state point approaches the stable point along the super surface. Therefore, the kinematic state will not be affected by the uncertain parts of the system and the external disturbance, and thus the system shows strong robustness. In this paper the design of u_{fz} is accomplished through self-learning fuzzy controller.

B. Fuzzy Basis Functions

In order to solve the problem of the low convergent speed of learning algorithm, a special fuzzy control—fuzzy basic function is introduced in this paper. The system is treated as a nonlinear function. It is expanded based on a series of basic functions to accomplish self-learning through the correction of the weight of the basic function according to the error between the input and feedback. The convergence is superior to the neural network (NN)[2]. The fuzzy control (FC) consists of fuzzy, rule base, fuzzy deduction and anti-fuzzy as shown in Fig. 1.

The distance between the kinematic state point $P(x)$ and the super surface of sliding mode $\sigma(x)=0$ is $R=\sigma(x)/(1+c)^{1/2}$, its derivative $R=\dot{\sigma}(x)/(1+c)^{1/2}$ is just the speed at which the point $P(x)$ approaches the super surface. So we use switch function $\sigma(x)$ and its derivative as the input of the fuzzy controller and substituted the sliding switch u_{fz} with the output of fuzzy control. Assuming that the input of the fuzzy controller has seven level language variables, its membership function is shown in Fig. 2.

Two-input and one-output structure is adopted in fuzzy controller. The relation between $\sigma(x)$, $\dot{\sigma}(x)$ and u_{fz} can be seen as a nonlinear function with two input and one output and naturally can make fuzzy basic function expansion. Supposing that the input variables $\sigma(x) = s_1$, $\dot{\sigma}(x) = s_2$, the fuzzy rule in the rule base can be expressed as

$$\text{Rule } j: \text{If } s_1 \text{ is } A_1^j \text{ and } s_2 \text{ is } A_2^j \text{ then } Z \text{ is } B^j , \quad (14)$$

where $j = 1, ..., m$ ($m = 7 \times 7 = 49$) is the number of the fuzzy rule, s_i ($i = 1,2$) is the input of the fuzzy controller. $Z = u_{fz}$ is the output, A_i^j and B^j are the input and output language variables with the membership function $\mu_{A_i^j}(s_i)$ and $\mu_{B^j}(z)$, respectively. The corresponding expression is

$$\mu_{A_1^j \times A_2^j \to B^j}(s_1, s_2, z) = \mu_{A_1^j}(s_1) * \mu_{A_2^j}(s_2) * \mu_{B^j}(z), \quad (15)$$

where $*$ means multiplying , not getting minimum as in the traditional fuzzy controller. (15) can be transformed into

$$\mu_{A_i \circ R_j}(z) = \mu_{B^j}(z) * \prod_{i=1}^{2} \mu_{A_i^j}(s_i) . \quad (16)$$

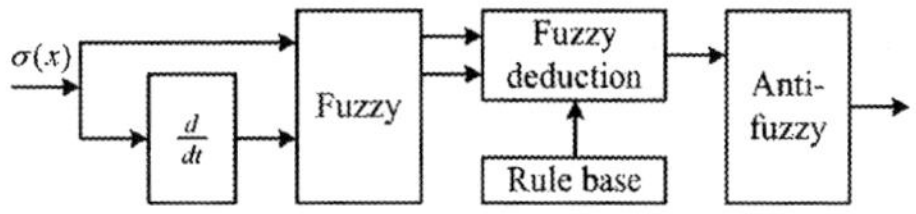

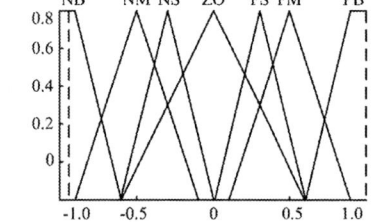

Figure 1. Structure of fuzzy controller

Figure 2. Membership function of input variables

The role of the anti-fuzzy is to map from the fuzzy aggregate in output space R to the exact point in the R. When the centroid defuzzification algorithms is used, we have

$$Z = \sum_{j=1}^{m} Z^j \mu_{A_s \circ R_j}(\underline{Z^j}) / \sum \mu_{A_s \circ R_j}(\underline{Z^j}), \quad (17)$$

where Z^j is Z value of maximum $\mu_B(Z)$. In general, let $\mu_B(Z^j)=1$. Substituting (16) into (17), the expansion form of fuzzy controller output is

$$u_{fz} = f(S) = \sum_{j=1}^{m} p_j(S)\omega_j , \quad (18)$$

where

$$p_j(S) = \prod_{i=1}^{n} \mu_{A_i^j}(s_i) / \sum_{j=1}^{n} \prod_{i=1}^{n} \mu_{A_i^j}(s_i) . \quad (19)$$

in which $n = 2$ is the number of input, $\omega_j = \underline{Z^j}$ is the weight decided by the following learning algorithm. (19) is called the fuzzy basic function (FBF). The expansion of (18) can approaches arbitrary nonlinear function with function with arbitrary accuracy [6].

C. Learning Algorithm

The task of the learning algorithm is to justify the weight ω_j to ensure the state point sliding on the super surface. According to the sliding mode condition $\sigma \cdot \dot{\sigma} < 0$ and the principle of minimizing the chattering, the justification of the weight should make $\sigma \cdot \dot{\sigma}$ decrease, that is, under the condition of sliding mode, the switch function converges to zero and the state point slides to the stability point along the sliding mode super surface. Gradient descend method is used to justify the weight ω_j

$$\Delta\omega_j = -\Gamma \nabla_{\omega j} \sigma(x)\dot{\sigma}(x) = -\Gamma \frac{\partial \sigma(x)\dot{\sigma}(x)}{\partial \omega_j(t)}, \quad (20)$$

where $\Gamma > 0$ is the rate of learning. (20) can be written as

$$\Delta\omega_j = -\Gamma \frac{\partial \sigma(x)\dot{\sigma}(x)}{\partial u(t)} \frac{\partial u(t)}{\partial \omega_j(t)}$$

$$= -\Gamma \frac{\partial \sigma(x)\dot{\sigma}(x)}{\partial u(t)} \frac{\partial (u_{eq} + \Delta u)}{\partial \omega_j}. \quad (21)$$

From (8) we can see that u_{eq} is decided by the state variable x, but irrelated to weight ω_j, so $\partial u_{eq}/\partial \omega_j = 0$. According to (11) and (12)

$$\frac{\partial \sigma(x)\dot{\sigma}(x)}{\partial u(t)} = \sigma(x)\frac{\partial \dot{\sigma}(x)}{\partial u(t)} = \sigma(x)\frac{\partial}{\partial u}(c\dot{x}_1 + \dot{x}_2)$$

$$= \sigma(x)\frac{\partial}{\partial u}(cx_2 + ax_2 + bu + df) = b\sigma(x) . \quad (22)$$

From (21) and (22), we get the learning rule of the weight ω_j

$$\Delta\omega_j = -\Gamma[b\sigma(x)]\frac{\partial u_{fz}(t)}{\partial\omega_j(t)} = -\Gamma[b\sigma(x)\frac{\partial\sum p_j\omega_j}{\partial\omega_j}]$$

$$= \Gamma b\sigma(x)p_j(\sigma,\dot\sigma) ,\qquad (23)$$

where $j = 1, ..., m$.

The block diagram of linear servo system with self-learning fuzzy sliding mode control (SFSMC) is given in Fig. 3. The dash part is self-learning fuzzy sliding mode control (SFSNC), including sliding mode control (SMC) and fuzzy self-learning control (FSLC).

V. SIMULATION RESULT

Based on the PMLSM servo system, we simulated the self-learning fuzzy sliding mode control by using MATLAB6.5 (fuzzy toolbox, control toolbox, simulink box). The parameters are:

M_n = 11.0 kg, B_n = 8.0 N·S/m K_{Fn}= 28.5 N/A, F_{en} = 100 N, v_{en} = 1.0 m/s. The simulations of the traditional PI and sliding mode control are also carried out. The parameters of PI controller are K_p = 46, K_I = 5.6 and the parameter of sliding mode control is c = 12. Let the dimensional factors of the input parameters of self-learning fuzzy mode control $\sigma(x)$, $\dot\sigma(x)$ be k_1 = 12, k_2 = 50, respectively, the proportional factor of the output parameter u_{fz} is K_3=10, learning rate is Γ = 1000, sample period is T=0.5ms.

A. Constant Disturbance

As constant disturbance F_L = 50N (t>0.6ms), $M = M_n$. The simulation curves of PI and self-learning fuzzy sliding mode control are shown in Fig.4, marked with 2 and 1, respectively. It can be seen clearly that the self-learning fuzzy sliding mode control has good tracking ability.

B. Variable Disturbance

Add F_L =35sin(25t), $M = M_n$. The simulation results are shown in Fig. 5(a) and 5(b). From the figures we can know that conventional sliding mode control has great adaptability to the disturbance and fuzzy self-learning control minimizes the chattering remarkably and has strong robustness to disturbance.

C. Parameter Perturbance

Supposing that the mass of mover, one of parameters of linear servo system, is $M = 2.5M_n$. The simulation curve of PI control and fuzzy self-learning sliding modecontrol are shown in Fig. 6, respectively. The curves marked with 1, 2 and 1', 2' are the speed responding curves when $M = M_n$ and $M = 2.5M_n$. From Fig. 6, we can see clearly that PI control can't fit this variation when the parameters are perturbing. However,

the fuzzy self-learning sliding control is almost not affected. That means it has strong ability against parameter perturbance.

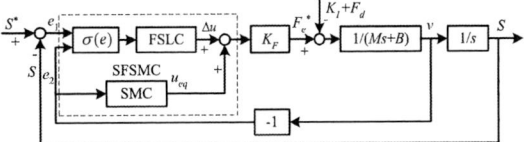

Figure 3. Block diagram of linear servo system with self-learning fuzzy sliding mode control

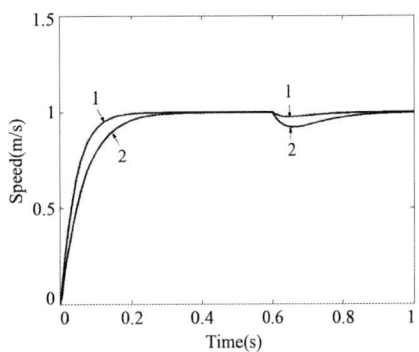

Figure 4. Speed responses of the servo system with constant disturbance

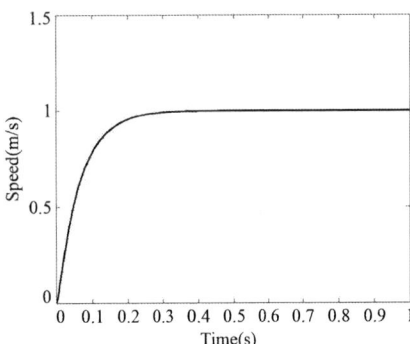

(a) Conventional SMC

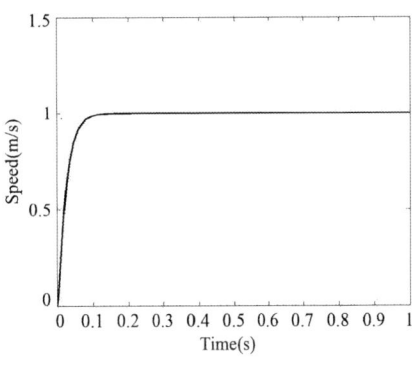

(b) SFSMC

Figure 5. Speed responses of the servo system with variational disturbance

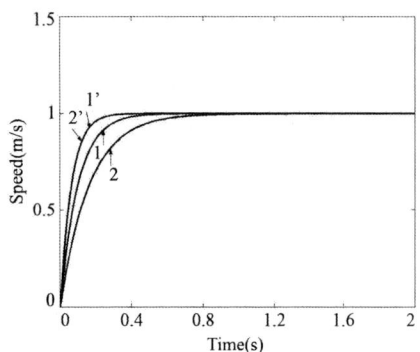

Figure 6. Speed responses of the servo system with parameter disturbance

1, 2 — PI; 1', 2' — SFSMC

VI. CONCLUSION

The simulation results indicate that the fuzzy self-learning sliding mode variable structure control strategy proposed in this paper is effective. It is superior to PI and traditional sliding mode control. It has strong robustness to the parameter perturbance and external disturbance, and minimizes the chattering remarkably. Simultaneously, it also improves the static and dynamic performance of the servo system.

REFERENCES

[1] S. Yibiao, G. Qingding, S. Limei, "Thrust observer-based sliding mode variable structure control for linear AC servo system," *Trans. of China Electrontechnical Society*, vol. 3, pp. 2-3, April 1998.

[2] Y. Bin, X. Dianguo and W. Zongpei, "Control of reluctance motor via fuzzy learning system," *Electric Machine and Control*, vol.2, pp. 100-103, 1998.

[3] W. LiXin, "Control design and stability analysis of adaptive fuzzy system, "*IEEE trans. on international Elect*, 1995.

[4] Z. Jin, W. Shunyan, S. Xiaopeng, "A survey to control strategies of AC servo systems based on sliding mode control," *Electric Drive*, pp. 2-6, 1996.

[5] K. Duk-Heon, K. Hack-Seong, R. Hong-Je, "Position control system for induction motor using variable structure system with fuzzy sliding surface," *Proceedings ICPE' 95 [C], Seoul*, pp. 200-210, 1995.

[6] W. LiXin and Mendel Jerry M., "Fuzzy basis functions universal approximation and orthogonal least-square learning," *IEEE Trans on Neural Network*, vol. 5, pp. 808-814, March 1992.

[7] R.A.DECARLO, "Variable structure of nonlinear multivariable systems, " *A tutorial Proc, IEEE*, Vol. 76, pp. 212-232, 1988.

2006 5th International Power Electronics and Motion Control Conference

Dynamics Research of Robot Manipulator

Zhibing Shu , Caizhong Yan,Hairong Zhang
Automation College
Nanjing University of Technology, NJUT
Nanjing, China
shuzhibing@163.com, czyan_22@sina.com

Abstract—**The main strategy in analyzing the motion of a rigid body is to split the motion into the linear motion of the CM and the angular motion of the body about its CM. This is because all particles composing the body show the same relative angular motion about the CM. In other words, one can describe the motion of the rigid body as a whole rather than those of the particles individually. The physical characteristics of a rigid body can be described by its inertial properties: mass and moment of inertia. For this reason, a major portion of discussion will be dedicated to the inertial properties of the rigid body: moment of inertia & inertia tensor.**

Keywords- **Planar Robot; Dynamics; self-tuning PI control**

I. INTRODUCTION

For analytical convenience, human body segments are considered as rigid bodies. A rigid body is similar to a system of particles in the sense that it is composed of particles. Therefore most of the equations for a system of particles are usable in the dynamics of a rigid body. The main difference is of course that the rigid body is rigid. There is no migration of mass within a rigid body. As a result, the relative positions among the particles composing a rigid body do not change. This will further simplify the equations obtained from a system of particles.

II. EXAMPLE OF DYNAMICS OF A PLANAR ROBOTS

Let us illustrate the applicability of Lagrange's equations of motion in robotics with the help of the simple planar robot given in fig 1

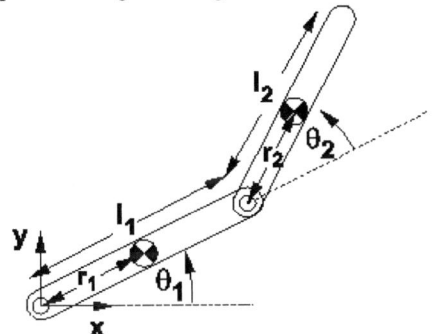

Figure1.   Two-link Planar Manipulator

For simplification each link i of this robot is modeled as a homogeneous rectangular beam of mass m1 and with a corresponding inertia tensor I:

$$T_i = \begin{bmatrix} I_{x,t} & 0 & 0 \\ 0 & I_{y,i} & 0 \\ 0 & 0 & I_{z,i} \end{bmatrix}$$

where: $i = 1,2$.

The inertia tensor is related to a body frame that is fixed in the center of masses of each link; again the coordinate axes are in parallel with the body axes.

If $v_i \in R^3$ is the translational velocity of the center of masses of the i-th link and $\omega_i \in R^3$ is the angular velocity of the same body, then we find the kinetic energy of the robot as:

$$T\left(\Theta, \overset{\bullet}{\Theta}\right) = \frac{1}{2} m_1 \|v_1\|^2 + \frac{1}{2} \omega_1^T T_1 \omega_1 + \frac{1}{2} m_2 \|v_2\|^2 + \frac{1}{2} \omega_2^T T_2 \omega_2$$

Because the motion of the robot is restricted to the x-y-plane, $\|v_i\|$ is equivalent to the magnitude of the xy-velocity of the center of masses, and the vector ω_i is always pointing in z-direction. Here $\|\omega_1\| = \dot{\theta}_1$ and $\|\omega_2\| = \dot{\theta}_1 + \dot{\theta}_2$ are valid.

Solving for the kinetic energy is carried out using generalized coordinates P. We assume that in this planar case the position of the i-th center of masses is defined as

$$p_i = [x_i, y_i, 0]^T .$$

Furthermore let r1, r2 be the distance from the previous joint to the following center of masses. This situation is presented here in figure 1

The position of the center of masses for each link is given by:

$$x_1 = r_1 \cos \theta_1$$
$$y_1 = r_1 \sin \theta_1$$
$$x_2 = l_1 \cos \theta_1 + r_2 \cos(\theta_1 + \theta_2)$$
$$y_2 = l_1 \sin \theta_1 + r_2 \sin(\theta_1 + \theta_2)$$

and their velocities are:

$$\dot{x}_1 = -r_1 \sin \theta_1 \dot{\theta}$$

$$\dot{y}_1 = r_1 \cos \theta_1 \dot{\theta}$$

$$\dot{x}_2 = -(l_1 \sin\theta_1 + r_2 \sin(\theta_1+\theta_2))\dot{\theta}_1 - r_2 \sin(\theta_1+\theta_2))\dot{\theta}_2$$

$$\dot{y}_2 = (l_1 \cos\theta_1 + r_2 \cos(\theta_1+\theta_2))\dot{\theta}_1 - r_2 \cos(\theta_1+\theta_2))\dot{\theta}_2$$

Thus the kinetic energy function can be reformulated into:

$$T\left(\theta,\dot{\theta}\right) = \frac{1}{2}m_1\left(\dot{x}_1^2 + \dot{y}_1^2\right) + \frac{1}{2}T_{21}\dot{\theta}_1^2$$
$$+ \frac{1}{2}m_2\left(\dot{x}_2^2 + \dot{y}_2^2\right) + \frac{1}{2}T_{22}\left(\dot{\theta}_1+\dot{\theta}_2\right)^2$$

This corresponds to:

$$T\left(\theta,\dot{\theta}\right) = \frac{1}{2}\begin{bmatrix}\dot{\theta}_1 \\ \dot{\theta}_2\end{bmatrix} M \begin{bmatrix}\dot{\theta}_1 \\ \dot{\theta}_2\end{bmatrix}$$

where:

$$M = \begin{bmatrix} M_{11} & M_{12} \\ M_{21} & M_{22} \end{bmatrix}$$

and

$$M_{11} = T_{21} + T_{22} + m_1 r_1^2 + m_2\left(l_1^2 + r_2^2\right) + 2m_2 l_1 l_2 \cos\theta_2$$

$$M_{12} = T_{22} + m_2 r_2^2 + m_2 l_1 l_2 \cos\theta_2$$

$$M_{21} = T_{22} + m_2 r_2^2 + m_2 l_1 l_2 \cos\theta_2$$

$$M_{22} = T_{22} + m_2 r_2^2$$

Substituting the Lagrangian $L = T_i$ and building the equations of motion leads to:

$$M_A\begin{bmatrix}\ddot{\theta}_1 \\ \ddot{\theta}_2\end{bmatrix} + M_B\begin{bmatrix}\ddot{\theta}_1 \\ \ddot{\theta}_2\end{bmatrix} = \begin{bmatrix}\tau_1 \\ \tau_2\end{bmatrix}.$$

Here we find:

$$M_A = \begin{bmatrix} M_{A,11} & M_{A,12} \\ M_{A,21} & M_{A,22} \end{bmatrix}$$

Where:

$$M_{A,11} = T_{21} + T_{22} + m_1 r_1^2 + m_2\left(l_1^2 + r_2^2\right) + m_2 l_1 l_2 \cos\theta_2$$

$$M_{A,12} = T_{22} + m_2 r_2^2 + \frac{1}{2}m_2 l_1 l_2 \cos\theta_2$$

$$M_{A,21} = T_{22} + m_2 r_2^2 + \frac{1}{2}m_2 l_1 l_2 \cos\theta_2$$

$$M_{A,22} = T_{22} + m_2 r_2^2$$

and:

$$M_B = \begin{bmatrix} -\frac{1}{2}m_2 l_1 l_2 \sin\Theta_2\,\dot{\theta}_2 & -\frac{1}{2}m_2 l_1 l_2 \sin\Theta_2\left(\dot{\theta}_1+\dot{\theta}_2\right) \\ \frac{1}{2}m_2 l_1 l_2 \sin\Theta_2\,\dot{\theta}_1 & 0 \end{bmatrix}$$

The first term of the equation of motion describes inertial forces for joint acceleration, the second part combines curious forces and centrifugal forces, and the right hand side gives the applied torques.

III. MULTI-BODY DYNAMICS

A. Overview

During the last four decades, the field of multibody dynamics has developed from classical analytical mechanics to an independent and important branch in mechanics to satisfy the growing needs arising in complex practical application problems. It comprises aspects like the mechanical modeling of systems with respect to their dynamics, formalization and structuring of the dynamical equations, and their numerical implementation and evaluation. Multibody dynamics thus extends from analytical to computational mechanics with strong emphasis on applied mechanics.

Classical mechanics relies on the differentiability of mappings and thus denies a proper treatment of discontinuity events as impacts or stick-slip transitions. Non-smooth dynamics, a new area of basic research that is developing and spreading at high speed, takes care of such events by fully incorporating them into general concepts on how to deal with inequalities.

B. Geometrical Interpretations of Impacts

The best known example of an impact in mechanics is the collision of two rigid bodies. There is a velocity jump at the impact due to the geometric impenetrability condition, but also an impulsive force caused by the assumption of rigidity. These two properties, velocity jumps and impulsive forces, characterize impacts. Impact theory, however, is not restricted to rigid bodies.

Impact laws can be regarded as the constitutive equations of the impact. In classical mechanics local formulations like Newton's impact law are used. Such approaches work well as long as highly dissipative collisions are investigated. In fact, impacts are to be understood as global processes, including all components of the participating system. For the general theoretical framework, however, geometrical concepts are needed to characterize and identify the impact parameters in an invariant and meaningful way.

The emphasis is a local formulation of impact laws and their dependence to global characteristics, and verification in experiments.

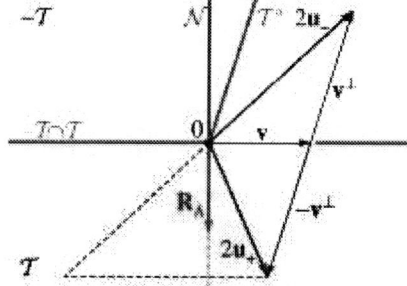

Figure2. geometric analysis and proposed solution for the impact: A reflecting hyper-plane is used to determine the post-impact velocities U$_+$ from U$_-$

C. Design of Under Actuated Robot

In order to verify the theoretical understanding of non-smooth dynamics experimental data has to be obtained and evaluated. The research within this project will be dealing with the design of a modular robot containing actuated (motors etc.) and under-actuated joints (frictional clutches etc.), which can be interchanged in any possible manner. A possible configuration could consist of two translational joints as shown in the picture below.

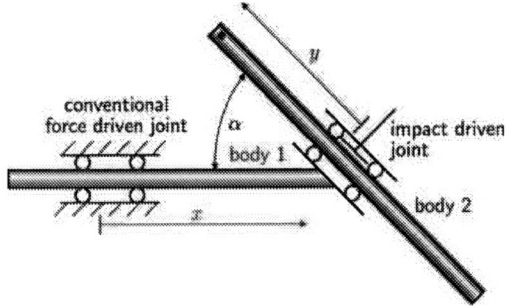

Figure3.   Design of a Simple Multi-body Robot Manipulator

Since constraints between the various bodies of a system can change from an activated state to a deactivated state due to single events like induced velocity jumps, such systems are often called "structure variant". Furthermore necessary hardware and software-interfaces have to be developed. They will ensure proper actuation and access to the robot in the context of later control efforts. Finally there will be first experiments in trajectory planning and the implementation of optimal control.

Future applications of structure variant robots may be found in the field of Biomechanics (replacements) and/or general robot-manipulators.

IV.   NEURAL-NETWORK-BASED SELF-TUNING PI CONTROL SYSTEM

To increase the robustness of fixed-gain PI controllers, we propose a new neural-network-based self-tuning PI control system in Fig.3. In this new approach, a well-trained neural network supplies the PI controller with suitable gain according to each operating condition pair (torque, angular velocity, and position error) detected. To demonstrate the advantages of our proposed neural-network-based self-tuning PI control technique, both computer simulations and experiments were executed in this research. During the computer simulation, the Direct Experiment Method was adopted to better model the problem of hysteresis in the ac servo motor. In real experiments, a PC-based controller DEC4DA was used to carry out the control tasks. Results of both computer simulations and experiments show that the newly developed dynamic PI approach outperforms the fixed PI scheme in rise time, precise positioning, and robustness.

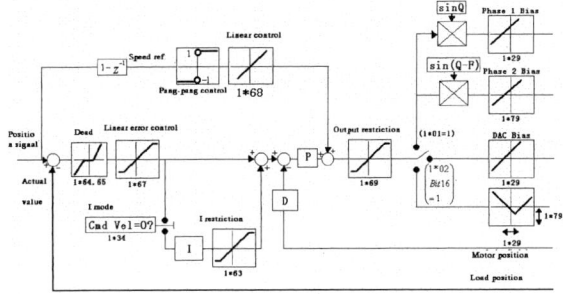

Figure4.   Structure of a new neural-network-based self-tuning PI control system

Dynamic response is one of the toughest dimensions of performance to define. But once clearly understood, it helps clarify many ambiguities of an application.

Dynamic response defines how quickly the speed of the motor and drive (amplifier) can stabilize after a load disturbance. Each manufacturer specifies dynamic response somewhat similarly, but common wording might read, "velocity regulation of 0.5% with a 90% load change and 2 Hz to stabilize the motor speed." Translated, this should mean that the motor speed will remain constant to 0.5% (of rated speed or set speed depending on the manufacturer) when there is a 90% load change, and the system should recover set speed within a half second. (Such a specification, by the way, indicates pretty good performance.)

Dynamic response might best be explained through a simple example. Suppose a 50-lb bag of dog food drops on an empty conveyor with a 50-lb. capacity. This changes the load instantly from 0 to 100%. The conveyor should slow momentarily and then return to the set speed. The time the conveyor takes to reestablish its set speed is a function of the drive dynamic response.

At a conveyor speed of 20 ft/min. (considered fairly slow by modem standards), the task of starting and stopping with an accuracy of 0.25 in. turns out to be more difficult than it might seem. At 20 ft/min., stopping in 0.25 in. incurs a 62-msec deceleration (corresponding to about 16 Hz). If the drive powering the application carries a 2 to 10-Hz rating, find a different solution.

If torque output is proportional to current input in the input-output model of the motor and drive, then dynamic response resembles the current rate-over-time (dEdt) used to specify the operating limits of power transistors. This region of operation, also called Safe Operating Area, is a limit which cannot be exceeded safely. Transistors that do exceed it tend to fail, sometimes catastrophically.

It is not uncommon for servo drives to exhibit current rate-of-rise that is 50 to 100 times that of ordinary motor drives. It is this current in-rush that primarily sets the speed with which the servodrive can accelerate or decelerate a load.

Dynamic response is also analogous to the behavior observed when a simple ac motor switches on. How many times can the motor start and stop under starting load conditions before the system might fail? In starting a standard ac motor, the torque variation is 100% over typically no more than 10 starts per minute. In fact, if the rotor is not stationary when the restart takes place, the electrical load could be a dead short.

Many variable-frequency drives have the same problem. Suppliers offer a special feature to synchronize the variable frequency drive to the rotor at the zero crossing point of the ac sine wave.

REFERENCES

[1] Th. Doersam and Th. Fischer, Aspects of Controlling a Multifingered gripper[J], International Conference on Conventional and Knowledge-based Intelligent Electronic Systems, 1997

[2] Th. Doersam and P. Dürrschmied, Compensation of friction in mechanical drives for a three fingered robot gripper[J], Proc. of the 1996 IEEE/RSJ Int. Conf. on Intelligent Robots and Systems, IROS, Osaka, Japan,Nov. 1996

[3] R. Menzel, Konstruktion und Regelung einer Hand, Fortschritt-Berichte VDI-Reihe 8 Nr.451,1995

[4] J.K. Salisbury, Articulated Hands: Force Control and Kinematics Issues[J], Ph.D thesis, Stanford University, 1982

[5] G. Hirzinger, Mechatronik-Konzepte nicht nur für die Raumfahrt, Deusche Forschungsanstalt für Luft- und Raumfahrt, Hannover Messe, 1996

[6] W.Paetsch, Exemplarische Untersuchungen zu mehrfingrigen Robotergreifern: Aufbau-Regelung- Systemintegration, Fortschritt-Berichte VDI-Reihe 8 Nr. 363, Düsseldorf, 1993.

[7] B. Magnussen, Infrastruktur für Steuerungsund Regelungssysteme von robotischen Miniatur- und Mikrogreifern, Fortschritt-Berichte VDI Reihe 8, Nr.567, Düsseldorf: VDI-Verlag, 1996

2006 5th International Power Electronics and Motion Control Conference

Advanced Angle Control Schemes for Stator Hybrid Excited Doubly Salient Motor Drive

Xiaoyong Zhu[*,**], Ming Cheng[*], Wenxiang Zhao[*], Wenguang Li[*]

[*]Department of Electrical Engineering, Southeast University, Nanjing 210096, China
[**]School of Electrical and Information Engineering, Jiangsu University, Zhenjiang, 212013, China
Phone: 86-25-83794152 Fax: 86-25-83791696 E-mail: mcheng@seu.edu.cn.

Abstract—A novel stator hybrid excited doubly salient (HEDS) brushless motor is presented in this paper. The topology and operation principle of the proposed motor are introduced. As supply to the HEDS motor drive system, a power converter with split capacitors is used and a basic current control method for this converter is explained. To achieve the maximum torque, an advanced angle control scheme is developed. Moreover, due to the asymmetry of the motor's electromagnetic parameters such as inductances, an unequal voltage distribution across the capacitors with the basic current control may occur. And the offset voltage of the centre-point may results in a poorer dynamic performance and low efficiency. To minimize the problems, a charge control strategy is developed. The two different strategies are implemented in an adjustable-speed drive and the steady-state and dynamic performances are analyzed. The results show that both control strategies improve the performance of the drive.

Keywords—hybrid excited motor; doubly salient motor; permanent magnet; angle control; charge control

I. INTRODUCTION

A viable brushless topology, named as the doubly salient permanent magnet (DSPM) motor, has been introduced in the past ten years [1]. This DSPM motor essentially adopts the same structure as a switched reluctance (SR) motor but with PMs placed in the stator. Recent literature has already illustrated that the DSPM machine is of high efficiency, high power density, and simple structure [2-5]. In [4-5], an 8/6 pole and a 6/4 DSPM motors have been well explored. However, similar to most permanent magnet brushless motors, this DSPM motor still suffers from a limited constant power operation range due to the fact that the field control capability of PM excitation is much more difficult to achieve than that of wound field excitation. Hence a novel topology to extend the constant-power operation range of the DSPM motor, namely a stator hybrid excited doubly salient (HEDS) motor has been proposed recently, which combines the advantage of PM machines with the possibility of controllable magnetic flux by auxiliary DC windings [6].

Projects supported by NSFC (50337030 and 50377004).

On the other hand, an energy optimizing control strategy for a current controlled DSPM was introduced in [7], where the turn-on angle is adjusted to minimize the power consumption. However, very little advanced control methods are discussed about control of this kind of HEDS motor. Due to the nonlinear characteristic of the HEDS motor, an advanced strategy for current control the motor is also needed to obtain the maximum torque and reduce the power losses by adjusting the turn on and turn-off angles on-line.

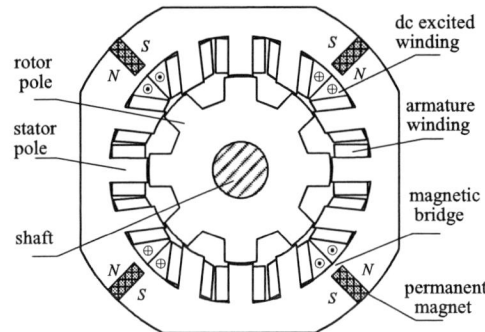

Fig. 1. Cross-section of the HEDS machine.

The main purpose of the paper is to present the details of the advance control methods to the proposed motor. The topology and operation principle of the proposed motor are introduced in section II. Next, a simulation model is developed in section III. Two control methods are presented in section IV, one is the torque-to-current ratio maximum scheme and the other is the charge control strategy of the split capacitors in the inverter. The two different strategies are implemented in an adjustable-speed drive in section V. The results have shown that the average torque can be significantly increased both in constant-torque operation and constant-power region. And the torque ripple is also dramatically reduced. Finally, it is concluded that both control strategies improve the performance of the drive.

II. MOTOR TOPOLOGY AND OPERATION PRINCIPLE

Fig. 1 shows the proposed topology, which is a three-phase 12/8-pole machine. It consists of two types of stator windings, a three-phase armature winding and a dc excitation winding. The function of the armature winding is the same as that for a DSPM machine, whereas the dc field winding not only works as an electromagnet but also as a tool for flux weakening operation [8].

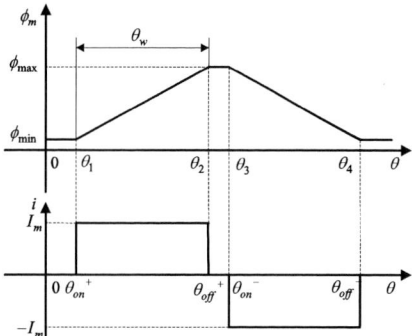

Fig. 2. Theoretical flux and current waveforms.

The theoretical waveforms of PM flux ϕ_m and phase current i with respect to the rotor position are shown in Fig. 2. When a rotor pole is entering the region occupied by a conductive phase, the flux is increasing. If a positive current is applied to the winding, a positive torque will be produced. When the rotor pole is leaving the stator pole from the aligned position, the flux is decreasing and also a positive torque will be produced if a negative current is applied to the winding. Thus, two possible torque producing zones are fully utilized.

III. MODELING OF THE MOTOR

The phase voltage equations of a three phase version of this motor can be expressed as follows:

$$\begin{bmatrix} u_a \\ u_b \\ u_c \end{bmatrix} = \begin{bmatrix} r_a & 0 & 0 \\ 0 & r_b & 0 \\ 0 & 0 & r_c \end{bmatrix} \begin{bmatrix} i_a \\ i_b \\ i_c \end{bmatrix} + \frac{d}{dt} \begin{bmatrix} \psi_a \\ \psi_b \\ \psi_c \end{bmatrix} \quad (1)$$

The flux linkage equation can be interpreted as:

$$\begin{bmatrix} \psi_a \\ \psi_b \\ \psi_c \end{bmatrix} = [L] \begin{bmatrix} i_a \\ i_b \\ i_c \end{bmatrix} + \begin{bmatrix} \psi_{ma} \\ \psi_{mb} \\ \psi_{mc} \end{bmatrix} = [L] \begin{bmatrix} i_a \\ i_b \\ i_c \end{bmatrix} + \begin{bmatrix} \psi_{pma} + L_{fa}i_f \\ \psi_{pmb} + L_{fb}i_f \\ \psi_{pmc} + L_{fc}i_f \end{bmatrix} \quad (2)$$

The total flux linkages and their time derivatives are:

$$\frac{d}{dt}\begin{bmatrix} \psi_a \\ \psi_b \\ \psi_c \end{bmatrix} - [L]\frac{d}{dt}\begin{bmatrix} i_a \\ i_b \\ i_c \end{bmatrix} \mid \frac{d[L]}{dt}\begin{bmatrix} i_a \\ i_b \\ i_c \end{bmatrix} \mid \begin{bmatrix} e_{ma} \\ e_{mb} \\ e_{mc} \end{bmatrix} \quad (3)$$

and the hybrid excited back EMF can be derived as:

$$\begin{bmatrix} e_{ma} \\ e_{mb} \\ e_{mc} \end{bmatrix} = \begin{bmatrix} \dfrac{d\psi_{ma}}{dt} \\ \dfrac{d\psi_{mb}}{dt} \\ \dfrac{d\psi_{mc}}{dt} \end{bmatrix} = \begin{bmatrix} \dfrac{d\psi_{pma}}{dt} \\ \dfrac{d\psi_{pmb}}{dt} \\ \dfrac{d\psi_{pmc}}{dt} \end{bmatrix} + \begin{bmatrix} \dfrac{dL_{fa}}{dt}i_f \\ \dfrac{dL_{fb}}{dt}i_f \\ \dfrac{dL_{fc}}{dt}i_f \end{bmatrix} + \begin{bmatrix} l_{fa}\dfrac{di_f}{dt} \\ l_{fb}\dfrac{di_f}{dt} \\ l_{fc}\dfrac{di_f}{dt} \end{bmatrix} \quad (4)$$

According the equations from (1) to (4), the dynamic equations of the motor can be expressed by:

$$\frac{d}{dt}\begin{bmatrix} i_a \\ i_b \\ i_c \end{bmatrix} = [L]^{-1}\{[U]-[E]\} - [L]^{-1}\left\{[R]+\frac{d[L]}{d\theta_r}\omega_r\right\}\begin{bmatrix} i_a \\ i_b \\ i_c \end{bmatrix} \quad (5)$$

The overall electrical magnetic torque is:

$$T_e = \begin{bmatrix} i_a & i_b & i_c \end{bmatrix}\begin{bmatrix} \dfrac{d\psi_{ma}}{d\theta} \\ \dfrac{d\psi_{mb}}{d\theta} \\ \dfrac{d\psi_{mc}}{d\theta} \end{bmatrix} + \frac{1}{2}\begin{bmatrix} i_a & i_b & i_c \end{bmatrix}\frac{d[L]}{d\theta_r}\begin{bmatrix} i_a \\ i_b \\ i_c \end{bmatrix} \quad (6)$$

where

$$[U]=\begin{bmatrix} u_a \\ u_b \\ u_c \end{bmatrix}, \; [E]=\begin{bmatrix} e_{ma} \\ e_{mb} \\ e_{mc} \end{bmatrix}, \; [R]=\begin{bmatrix} r_a & 0 & 0 \\ 0 & r_b & 0 \\ 0 & 0 & r_c \end{bmatrix}, \text{ and}$$

$$[L]=\begin{bmatrix} L_{aa} & L_{ab} & L_{ac} \\ L_{ba} & L_{bb} & L_{bc} \\ L_{ca} & L_{cb} & L_{cc} \end{bmatrix}.$$

The parameters in (5) and (6), namely flux linkage, self-inductance and mutual inductance, etc, can be obtained by finite element analysis. Fig. 3 and Fig. 4 give the flux linkage and self-inductance characteristics of the motor derived from 2D nonlinear finite element analysis in which the magnetic saturation, leakage flux are taken into account. By using (5) and (6), both the steady-state and dynamic behavior of the motor drive can be analyzed.

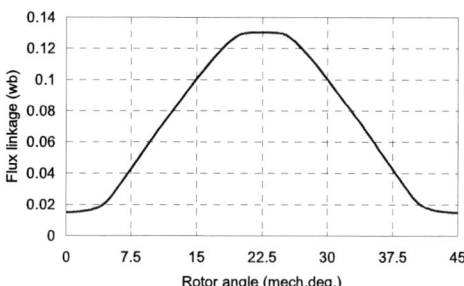

Fig. 3 Flux linkage versus rotor angle.

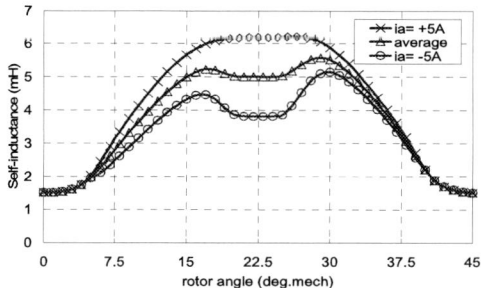

Fig. 4. Self-inductances at armature current of 5 A.

IV. STRUCTURE OF THE DRIVE SYSTEM

To supply the HEDS motor, a power converter topology with bi-directional current operation is preferable. Therefore, there are basically two converter topologies possible for bi-directional current operation of the motor, namely the full-bridge converter and the half-bridge converter with split capacitors, whose centre-point is connected to the neutral of the motor. The half-bridge converter is chosen in this paper because it uses the minimum semiconductors. As shown in Fig. 5, a half-bridge converter needs only two power switches per phase, one-half of those of the full-bridge converter. Two capacitors are connected in series across the DC input and their junction is ideally at mid-potential with a voltage of $1/2U_{dc}$ across each capacitor. From Fig. 5, there is additional field drive circuit in the drive system, which is specially designed to drive the HEDS motors' field current to achieve the flux-weakening operation for the whole drive system. For simplicity, the basic control method and advanced angle control scheme are mainly discussed in this paper without considering the flux-weakening operation or the field winding current i_f is set to a constant.

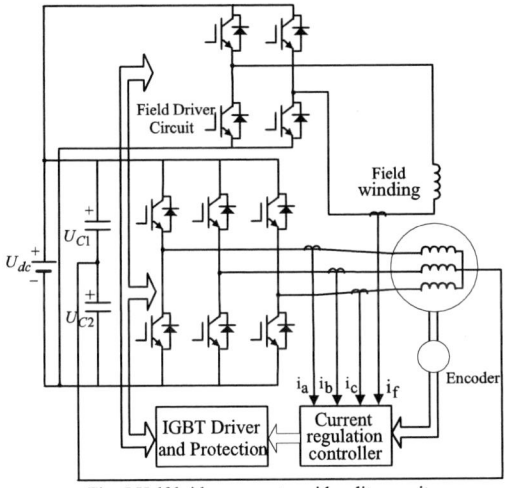

Fig. 5 Half-bridge converter with split capacitors.

The mean values of the developed torque from the motor can be controlled by two basic methods, namely current control or voltage control. The most obvious method is current control where the current impressed to the stator winding is adjusted to the desired torque. In this paper the focus is on current control realized as a hysteresis control.

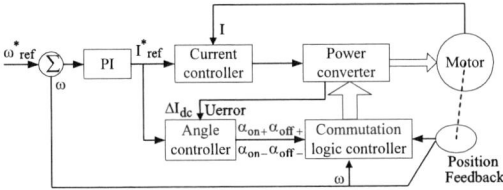

Fig. 6 Speed control of HEDS motor.

Fig. 6 shows the HEDS motor drive. The speed is controlled by a PI-controller and the torque command is executed by regulating the current in the inner loop. The reference current I^* for a given operation point is determined from the load characteristics. The controller needs current feedback information from each phase of the motor. Besides the basic method of current hysteresis control, an advanced angle controller is also adopted in the drive system in order to improve the dynamic performance and increase the efficiency.

V. ADVANCED ANGLE CONTROL

Assuming the speed can be kept constant, the different advanced control strategies can be implemented. Two methods are proposed: one is a torque-to-current ratio maximum scheme and the other is a charge control strategy of the split capacitors. Both the strategies are based on angle control of the motor. The angles used in the control are defined in Fig.7.

Compared with the SR motor, additional two angles, namely α_{on-} and α_{off-}, are controllable in the HEDS motor when the current is negative. Hence four angles are available which can give a very complicated control strategy. A simple but efficient angle control scheme, however, is proposed where α_{on+} controls the charge in the split capacitors, α_{off+} and α_{on-} control torque producing, and α_{off-} normally keeps fixed. Because of asymmetrical inductance characteristics which results in a different rise slope of the negative current from the positive current, the charge control is necessary. Otherwise, the capacitors will be charged and discharged differently in a period and consequently the voltage across the capacitors will be quite different, resulting in a poorer dynamical performance and more torque ripples. In the following detailed explanation of the two control methods, an angle control combination $\alpha_{on+}=5°$, $\alpha_{off+}=20°$ and $\alpha_{on-}=25°$, $\alpha_{off+}=40°$ is referred to as the standard angle control.

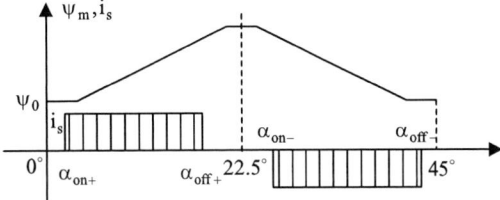

Fig. 7 Angle definitions used in the advanced control strategies.

A. Torque-to-current ratio maximum control

The standard angle control is not optimum, because the shape of the current does not give maximum torque from the PM flux linkage. The current in Fig. 8 is simulated at standard angle combination with a given reference current $I^*=5A$ and $n=1500$ rpm. The same conditions are used in an optimized angle control ($\alpha_{on+}=2°$, $\alpha_{off+}=20.5°$ and $\alpha_{on-}=20.5°$, $\alpha_{off+}=43°$) and the current is shown in Fig. 9. Fig. 8 and Fig. 9 also show the waveforms of the back EMF at no-load which indicates where most torque is produced. It

1817

is seen from Fig. 8 that the rise of the positive current is not the same as the rise of the negative current, which is caused by the difference in inductance and voltage. The positive current has reached its peak value in 60% of the time and the negative current in 20% of the time where the slope of the PM flux linkage is the maximum. The torque will be improved by ensuring that the current has reached its peak value before the maximum back EMF at no-load. The torque-to-current ratio is then maximized.

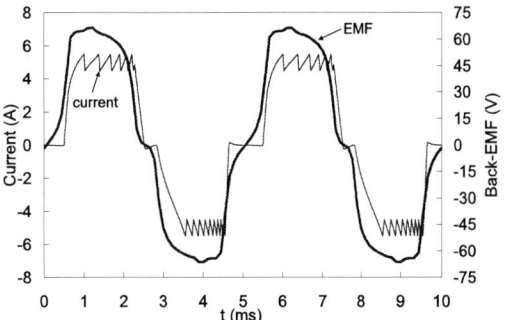

Fig. 8 Phase current of standard angle control.

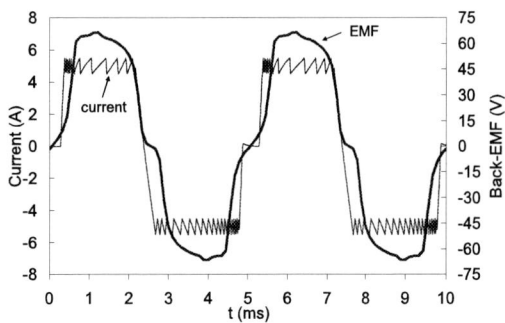

Fig. 9 Phase current of advanced angle control.

On the contrary, when the load characteristic is given in a drive system, the current can be lowered with the result that the copper losses decrease and the efficiency is higher. This is achieved by moving the four angles, namely α_{on+}, α_{off+} and α_{on-}, α_{off+}. Fig. 10 and Fig. 11 show the overall torque and output power capabilities with respect to different combination of α_{on+}, α_{off+} and α_{on-}, α_{off+} of the motor. It can be found that the maximum torque and power capabilities can be obtained when α_{on+} is selected to be 2°, α_{off+} and α_{on-} is set to be 20.5° and α_{off+} is kept 43°.

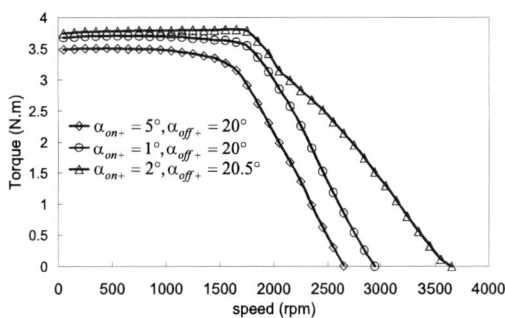

Fig. 10 The average torque of different angles control

B. Charge control

Due to the asymmetrical inductance for positive current and negative current, the voltage at the centre-point of the capacitors will be deflected from the mid-potential. By moving one of the angle α_{on+} on line, it is possible to stabilize the centre-point of the two capacitors by charge control. Fig. 12 shows the different α_{on+} may results in a distinct offset voltage of the centre-point. When the turn-on angle α_{on+} reaches to 2°, the voltage is nearly equal to zero. The difference between $1/2U_{dc}$ and the centre-point voltage U_{C2} is measured by a simple circuit as shown in Fig. 13, where $U_{error}=1/2U_{dc}-U_{C2}$.

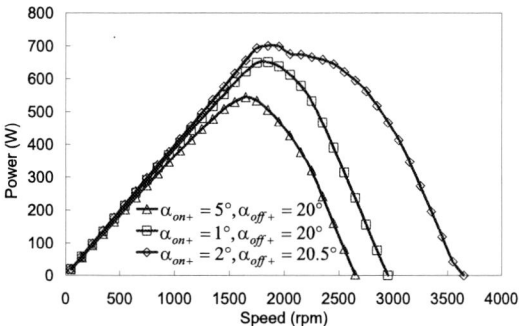

Fig.11 The output power of different angles control

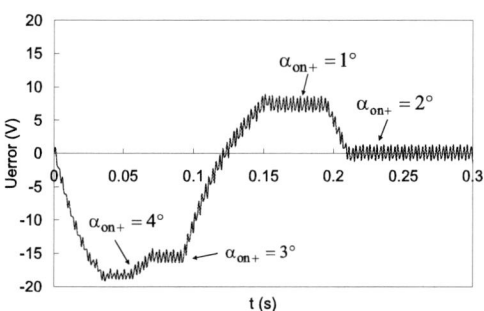

Fig.12 The offset voltage of the centre-point at different α_{on+}.

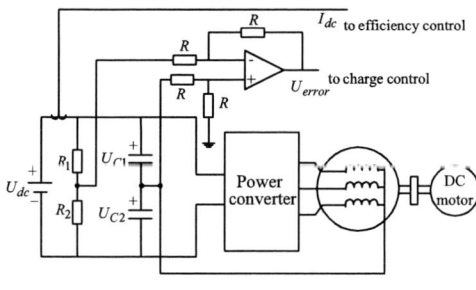

Fig. 13 Measurement circuit of the offset voltage of centre-point.

VI. Implementation and Performance

The advanced angle control schemes are developed as a combination of the torque-current-ration maximum control and charge control scheme. Fig. 14 shows the flow

1818

chart of the algorithm. For the torque-current-ration maximum control, if the power flow has decreased, $\Delta I_{dc}(k)=I_{dc}(k)-I_{dc}(k-1)<0$, α_{off+} and α_{on-} are moved further in the same direction (the sign of $\Delta\alpha_{off+}$ is unchanged), if $\Delta I_{dc}(k)>0$, the sign of Δ_{off-} is changed. For the charge control, if $U_{error}>0$, then α_{on+} is advanced ($\alpha_{on+}=+\Delta\alpha_{on+}$), else α_{on+} is backed off ($\alpha_{on+}=-\Delta\alpha_{on+}$).

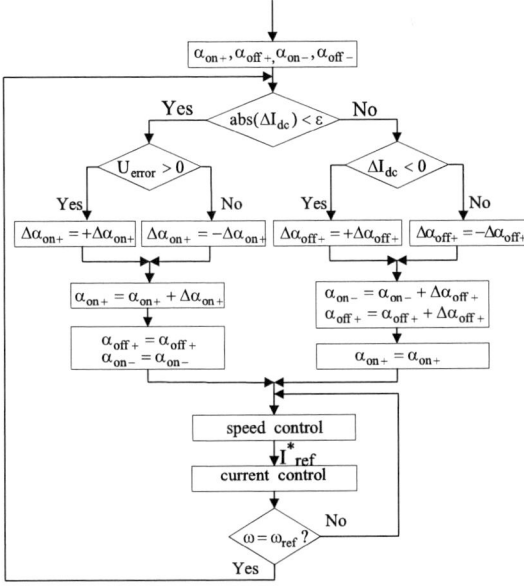

Fig. 14 Flow-chart of the advanced angle control

Fig. 15 shows the starting response from standstill to the speed of 1500 r/min. It can be seen the motor drive response quickly and there is almost no overshoot. Moreover, Fig. 15 also gives the dynamic responses of speed under a sudden change of load torque from 1.5 N.m to 4 N.m at 0.8s and a sudden decrease of load torque from 4 N.m to 0.5 N.m at 1.0s. It can be found that the transient change in speed is very small and the speed regulation of the motor drive is good.

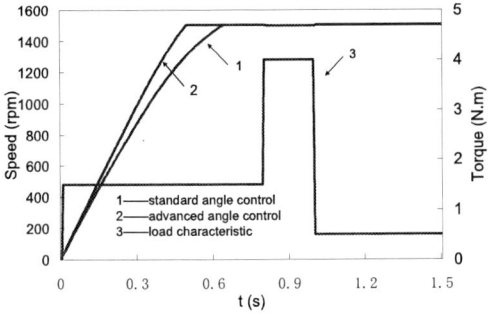

Fig.15 Step response on the speed with sudden load

Some performance parameters are listed in Table I. From the table, when the reference current I^*_{ref} is set to 5 A, the average torque reaches to 3.81 N.m in advanced angle control which is nearly 20% higher than that 3.27 N.m in standard angle control. And the torque ripple is dramatically reduced from 64.8% to 27.6%.

TBALE I
PERFORMANCE OF THE MOTOR AT 1500 R/MIN

I_{ref}^*=5A	T_{av} N.m	K_T	U_{c1}(V)	U_{c2}(V)	ΔU(V)
Standard angle control	3.27	64.8%	90.1	99.9	9.8
Advanced angle control	3.81	27.6%	94.9	95.1	0.2

VII. CONCLUSIONS

A new stator hybrid excitation doubly salient motor drive system is presented in this paper. Advanced angle control strategy with a torque-current-ratio maximum control and a charge control of the split capacitors are investigated and implemented in the adjustable-speed drive system. Compared with the traditional standard angle control, the results have shown that higher average torque and less torque ripple are achieved which are based on changing the turn-on and turn-off angles on-line. Furthermore, the dynamic response is also improved greatly. It is concluded that both control strategies improve the steady and dynamic performance of the HEDS motor drive system.

REFERENCES

[1] Y. Liao and T.A. Lipo, "A newly doubly salient permanent magnet motor for adjustable speed drives", *Electric Machines and Power Systems*, vol. 22, no. 2, 1995, pp. 259-270.

[2] M. Cheng and E Zhou, "Analysis and Control of Novel Split-winding Doubly Salient Permanent Magnet Motor for Adjustable Speed Drive," *Science in China (Series E)*, vol. 44, no. 4, 2001, pp. 353-364.

[3] M. Cheng, K.T. Chau, C.C. Chan and Q. Sun, "Control and operation of a new 8/6-pole doubly salient permanent magnet motor drive", *IEEE Trans. on Industry Applications*, vol. 39, no. 5, 2003, pp. 1363-1371.

[4] K.T. Chau, M. Cheng, Chan C.C., "Performance Analysis of 8/6-pole Doubly Salient Permanent Magnet Motor", *Electric Machines and Power Systems*, vol. 27, no. 10, 1999, pp. 1055-1067.

[5] Y. Li, and T.A. Lipo, "A doubly salient permanent magnet motor capable of field weakening," *Proceedings of IEEE Power Electronic Specialist Conference*, Atlanta, USA, June 18-22, 1995, pp. 565-571.

[6] X. Zhu and M. Cheng, "An overview of hybrid excitation permanent magnet machines," *Proc. ICEMS*, Jeju, Korea, Nov. 1-3, 2004, 513-M06-064.

[7] Frede Blaabjerg, Leif Christensen, Peter O. Rasmussen. "New advanced control methods for doubly salient permanent magnet motor". *Proc. Thirtieth IAS Annual Meeting*, IAS '95, USA, 8-12 Oct. 1995. Vol. I, pp. 222 – 230.

[8] X. Zhu, M. Cheng, and W Li, "Design and analysis of a novel stator hybrid excited doubly salient permanent magnet brushless motor", *Proc. ICEMS*, Nanjing, China, Sep. 27-29, 2005, Vol. I, pp. 401-406.

2006 5th International Power Electronics and Motion Control Conference

A Design Method of Reconfigurable Controller for AC Position Servo Systems

Wu Qinmu   Qin Yi   Li Yesong

Huazhong University of Science and Technology/Department of control science and engineering

Wuhan 430074, china. wuqingmu@21cn.com

Abstract - **Aiming at a wide application range of controller, a design method of reconfigurable controller was presented for AC position servo systems (ACPSS) based on permanent magnet synchronous motor (PMSM). In this controller, each of control loops adopts several control strategies. Their characteristics and parameter's calculation methods were analyzed in detail in this paper. The closed-loop bandwidth's range of each control loop, anti-disturbance ability and precision of position servo in ACPSS were also analyzed, and a classification table about them has been obtained. On the base of these analyses, a method that regroups control strategies to form new controller was proposed. Some simulation results show that the application range of this controller is wider than that of conventional controller.**

Keywords - *PMSM; Parameter calculation; Strategy selection; Reconfigurable controlle; AC position servo systems.*

I. INTRODUCTION

With the development of the technology of power electric and large scale integration circuit, high-performance ACPSS can be widely employed in industry application. In the usual way, the ACPSS adopts two or three loops' configuration. A good configuration is to enclose the velocity loop within the position loop, and the current loop is included in the velocity loop, as shown in Fig 1. The position controller feeds its output to velocity controller; the velocity controller output feeds into the current controller that creates motor's torque; and the system's position signal is fed back from a position senor, such as an encoder, most modern servo systems estimate velocity from the position signal too. The current loop and the velocity loop are

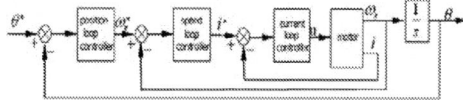

Figure 1. Structure of ACPSS

very important in this configuration, which decide the performance of position servo systems. In the conventional design method of controller, each control loop only adopts

single control strategy, which leads to a narrow application range. In the position servo area, different ACPSS have different characteristics and performance requirements. For example, in light industry position servo systems, the moment of inertia of plant (MIP) is small and the system's response is very fast; in robot position servo systems, the MIP is apt to change and the system should has better robustiveness; in mechanical given servo systems, the MIP is big and is prone to change. In order to obtain the expected performance in all the servo systems mentioned above, a conventional controller can't do it. Consequently, a design method of reconfigurable controller was proposed. Each of control loops adopts several control strategies in this controller, these control strategies form a strategy libaray, and it can be saved in host PC and local controller. So these control strategies can be flexibly regrouped to form new controller for anticipant performance requirements according to the method proposed. We arrange this paper as follows. The model of PMSM is analyzed in Section II; some control strategies whose realization is easy are analyzed and their characteristics and parameter's calculation methods are given in Section III; a classification table about characteristic and performance requiremeats of ACPSS and a method of reconfiguring controller are given in Section IV; finaly, correctness of this design method was verified by simulation.

II. PM SYNCHRONOUS MOTOR

The stator d, q equations of the PMSM in the rotor rotating reference frame are [1,5]:

$$v_q = Ri_q + p\psi_q + \omega_s\psi_d \tag{1}$$
$$v_d = Ri_d + p\psi_d - \omega_s\psi_q \tag{2}$$

and

$$\psi_q = L_s i_q \tag{3}$$
$$\psi_d = L_s i_d + \psi_f \tag{4}$$
$$\omega_s = n_p \omega_r \tag{5}$$

where v_d and v_q are the d, q -axis stator voltages, i_d, i_q are the d,

q-axis stator currents, L_s is d, q-axis stator inductance, ψ_d and ψ_q are the d, q-axis flux linkages, while R and ω_e are the stator resistance and inverter frequency, respectively. ψ_f is the flux linkage due to the rator magnets linking the stator, ω_r is the rator angle velocity, and n_p is the number of pole pairs.

The electric torque is

$$T_e = 3n_p \psi_f i_q / 2 \qquad (6)$$

And the equation for the motor dynamics is

$$\dot{\omega}_r = (T_e - T_L - B\omega_r)/J \qquad (7)$$

Where T_L is the load torque, B is the damping coefficient, and J is the MMI.

III. CHARACTERISTICS AND PARAMETER CALCULATION METHODS OF STRATEGIES

In AC position servo control area, a number of control schemes have been used, such as self-adapt control, nerval network control, fuzzy control and prediction control. However these schemes have some common defeats, namely complicated computation and algorithm. So hardware equipment must have high-speed computation ability and large memory space for realizing them, which is the reason why they can't be widely adopted. In this paper, some simple and easy realization control strategies are used.

A. Current Control Loop

Three control strategies were used in this loop, and they are the proportional-integral control combined with current feedback decouple (PICFD) [1], the proportional-integral control combined with voltage feed-forward decouple (PICFFD) [1] and the minimum step control combined with current feedback decouple (MSCFC). Transfer function of PICFD and PICFFD controller is the same, and is:

$$G_{Ic}(s) = K_p + K_i \frac{1}{s} \qquad (8)$$

where $K_p = w_{ib} L_s$, $K_i = w_{ib} R$, $w_{ib} = 5 w_{nb}$, w_{ib} and w_{nb} is the closed-loop bandwidth required of current loop and velocity loop, respectively. PICFD can apply to those systems with middle frequency closed-loop bandwidth requirement in current control loop, and it only need small computation steps. PICFFD can support low or middle frequency closed-loop bandwidth's servo systems, but need more computation steps. MSCFC is a control strategy accurately depending on parameters of motor, it can make servo systems acquire fast current response and high frequency closed-loop bandwidth, but which depents on sample time of system [3].

B. Velocity Control Loop

Four control strategies were adopted in this loop, they are the proportional-integral control (PIC), the PDFF control, the pole placement control (PPC) and the robust control (RBC).

1) The transfer function of the PIC controller is

$$G_{V_{PI}}(s) = K_{sk}(1 + \frac{1}{T_{sk}s}) \qquad (9)$$

where $K_{sk} \approx cos(\gamma_{sm})\omega_t/K_{so}$, $T_{sk} \approx ctg(\gamma_{sm})/\omega_t$, $K_{so} = 1.5 n_p \psi_f$, $\omega_t \leq \omega_{vp}/1.4$, $\omega_{vp} = 5\omega_{bb}$, ω_t, γ_{sm} is the opened-loop truncateion frequency and the phase margins of velocity loop, and ω_{pb} is the closed-loop bandwidth of the prosition loop [10]. The control strategy is simple and supports rapid response, but the velocity loop realized by it has a tendency to overshoot and a feeble anti-disturbance ability. It adapts to those ACPSS with low or middle frequency closed-loop bandwidth requirement.

2) The PDFF is a such controller that extends PI by modifying the control algorithm, its structure is shown in Fig. 2, it has

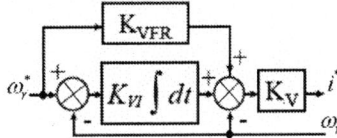

Figure 2. stucture of PDFF controller

greater anti-disturbance ability than PI, but it comes at the cost of reduced responsiveness. The PDFF controller [7] with different coefficient K_{VFR} has different performance. In design process, at first, K_{VI} and K_V shoule be found out according to the lightly higher performace of velocity loop than real requirement that, then adjusting K_{VFR} to satisfy anti-disturbance ability and responsiveness.

3) The PP controller [8] can has velocity loop obtain high frequency closed-loop bandwidth, and has better stability and robustness, non-overshoot. Its transfer function is

$$G_{Vpp}(s) = \frac{b}{s^2 + \alpha s + \beta} \qquad (10)$$

So the closed-loop transfer function of velocity loop is $G_o(s) = m\omega^2/((s+m)(s^2+2\zeta\omega s + \omega^2))$, where $b = m\omega^2/k_o$, $\alpha = m + 2\zeta\omega$, $\beta = 2m\zeta\omega + \omega^2$, $m \approx \omega_{ib}$, $\zeta = 0.5$, ω is lightly bigger than ω_{ib}.

4) The Robust controller [4,6] is designed based on slide control theory. its algorithm is

$$s(k) = ce(k) + \Delta e(k) \qquad (11)$$
$$\Delta e(k) = (e(k) - e(k-1))/T_s \qquad (12)$$
$$u(k) = u(k-1) + T_s \sum (Ks(k) + K_{eq}\Delta e(k)) \qquad (13)$$

where $K \leq T_s\alpha/(1+cT_s)\beta$, $K_{eq} = ((1+cT_s)\alpha-1)/((1+cT_s)\beta)$, $\beta = k_t(1-\alpha)/B$, $c = \omega_{ib}$, $\alpha = exp(-BT_s/J)$, $k_t = 3n_p\psi_f/2$, T_s is sample period of velocity loop, The velocity loop with the robust control strategy has better robustiveness, its closed-loop bandwidth is apt to be adapted from low to middle frequency.

C. Position Control Loop

Two control strategies is used in position loop, they are the proportion control (PC) and the hybrid control (HC) (proportion control plus feed-forword control). When PC is used in position loop of ACPSS, the servo systems has rapid position response, but dynamic servo error is big. It adapts to static position servo systems. Proportion coefficient of controller $K_p = c/(4*\zeta^2)$, where $c = 5*\omega_{pb}$, $\omega_{pb} \geq 1.4*(6\sim10)/t_s$ [2], $\zeta = 1.2$.

When HC [3] is adopted in position loop of ACPSS, the sevo systems has rapid position response, and dynamic servo error is smaller than that of system with PC controller. It adapts to both static and dynamic position servo systems. Proportion control part's parameter K_p is solved by the same method like above-mentioned proportion control. In this control strategy, transfer function of feed-forward section is

$$G_{PF}(s) = (k_{pf}s)/(1 + T_{pf}s) \qquad (14)$$

where $k_{pf} = 0.7 \sim 0.9$, T_{pf} is decided by acceleration of reference signal.

IV. PERFORMANCE CLASSIFICATION TABLE OF ACPSS AND METHOD OF REGROPINGG STRATEGIES

In industry application, the performance requirements of ACPSS are: 1) responsive speed, the closed-loop bandwidth of position loop can embody it; 2) antidisturbance ability; 3) Robustiveness; 4) Servo precision. In order to obtain a method of reconfiguring controller, a classification table of performance requirement of ACPSS is given and shown in table 1.

TABLE 1 CLASSIFICATION TABLE OF PERFORMANCE AND CHARACTERISTIC OF ACPSS

	classification	Value		
bandwidth	Low(Hz)	1-10	10-25	25-40
	Middle(Hz)	5-50	50-125	125-200
	High(Hz)	25-250	250-625	625-1000
disturbance	strong	1	(have to consider it)	
	feeble	0	(don't consider it)	
parameter variation of plant	large	1	(have to consider it)	
	small	0	(don't consider it)	
servo type	static position servo			
	dynamic position servo			

According to table 1 and characteristics of each control strategy, a flow procedure figure about reconfigureing controller is shown in Fig. 3.

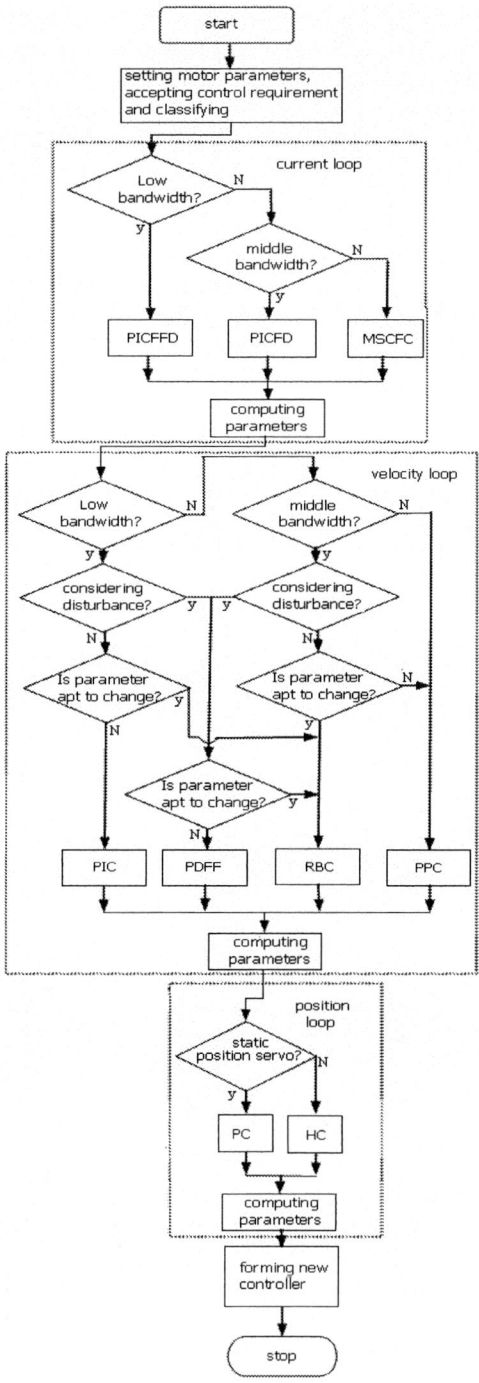

Figure 3. Flow chart of selection control strategy

V. SIMULATION ANALYSIS

In robot position servo systems, the MIP is apt to change and there is slow response, namely these systems should have low frequency closed-loop bandwidth. According to the method about reconfiguring controller, the position loop, the velocity loop and the current loop should adopt PC, RBC and PICFFD, respectively. Parameters of PMSM are: *L=0.0085mh, R=2.875Ω J=0.0008kg·m², n_p=4, $ψ_f$=0.175wb, B=0.000056 N·m·s/rad*. Position reference value is 1cm, the setting time required is 3s, the MMI varies in J ~50J, the system with reconfigurable controller is simulated, control strategies's parameter of each loop is shown in table 2, and position servo output of this ACPSS are shown in Fig. 4, when the MIP is J, 30J and 50J. If control strategies are fixed and the position loop,

TABLE 2 PARAMETERS OF CONTROL STRATEGIES

		position loop	velocity loop	current loop
recongiuration controller	control strategies	PC	RBC	PICFFD
	parameter value	k_p = 4.5	k = 0.0502 k_{eq} = 0.042 c = 25	k_p = 1.0625 k_i = 359.375
controller used by fixed control strategies	control strategies	PC	PIC	PICFFD
	parameter value	k_p = 4.5	k_p = 0.0128 k_i = 0.0831	k_p = 1.0625 k_i = 359.375

the velocity loop and the current loop use PC, PIC and PICFFD, respectively. The system controlled by them is simulated too, and position servo output of this ACPSS is shown in Fig 5 when the MIP is J, 30J and 50J. According to Fig 4 and 5, the system controlled by the reconfigurable controller has better robustiveness, regardless whether the MIP varies in J~ 50J, its position servo output is nearly fully same. But the system

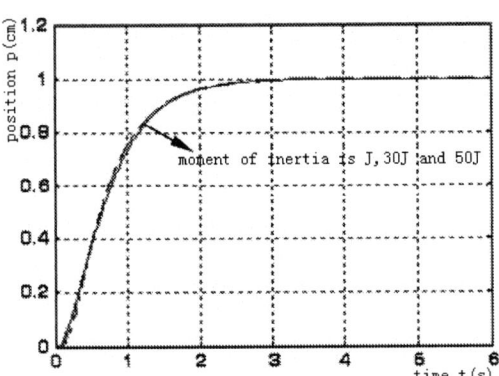

Figure 4. Position output of system with reconfigurable controller

when J'=J, 30J, 50J

controlled by fixed control strategies is not that case, only when the MIP is equal to J, the system has anticipant servo performance. But when the MIP is 30J or 50J, the system's position output yields serious oscillation.

In light industry position servo systems, the MIP is small and

has only very small variation, namely it can be considered as fixed value; rapid response is requied, that is to say, these systems should

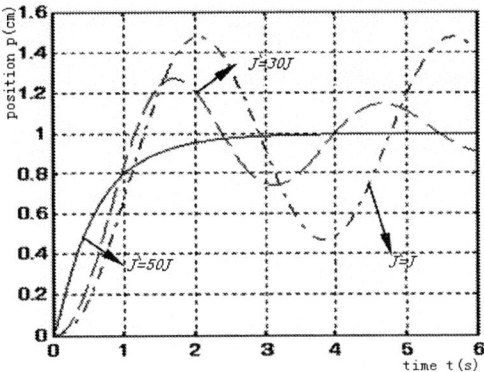

Figure 5. Position output of system with fixed control strategies

when J'=J, 30J, 50J

has high frequency closed-loop bandwidth. Supposing position reference value is 1cm, seting time required is 0.1s, and there exists a disturbance $ω$ in feedback path of velocity loop, $ω$=sin(400t). According to the method about reconfiguring controller, the position loop, the velocity loop and the current loop should use PC, PPC and MSCFC, respectively. Systems controlled by reconfigureation controller and fixed control strategies (position loop, velocity loop and current loop should be select PC, PIC and PICFFD, respectively) all are simulated, control strategies's parameter of each loop is shown in table 3, and servo system's position output are shown in Fig. 6. Due to disturbance, influence of position and velocity output are shown

TABLE 3 PARAMETERS OF CONTROL STRATEGIES

		position loop	velocity loop	current loop
reconfigurable controller	control strategies	PC	PPC	MSCFC
	parameter value or transfer function	k_p = 50	b = 39366 $α$= 798 $β$ =321625	$\dfrac{z^2-1.982z+0.98}{0.006z^2-0.012z+0.006}$
controller used by fixed control strategies	control strategies	PC	PIC	PICFFD
	parameter value	k_p = 50	k_p = 0.1534 k_i = 12	k_p =12.75 k_i = 4312.5

in Fig 7, 8. According to Fig 6, position output of two systems is same in large scale. From partial figure enlarged during stable output of Fig 7 and 8, the anti-disturbance ability of two systems are quite different, and the system controlled by reconfigurable controller outperform that controlled by fixed control strategies. Consequently, the reconfigurable controller can adapt to more ACPSS in industry application, namely its application range is the wider than conventional controller's.

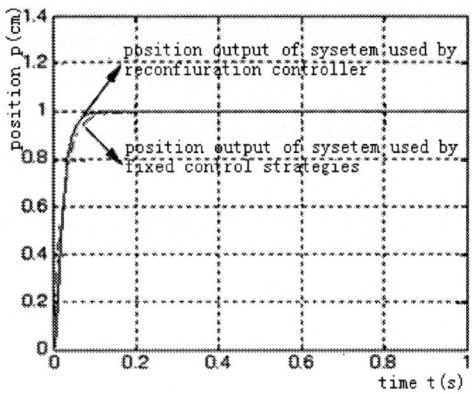

Figure 6. Position output of two systems

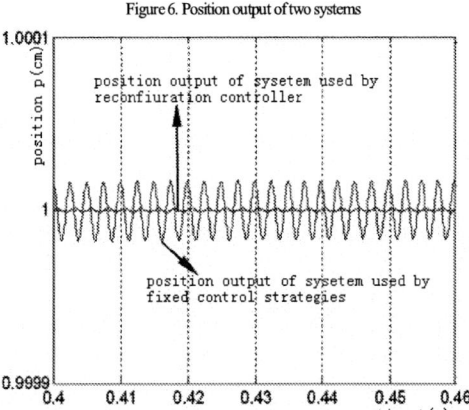

Figure 7. Partial figure enlarged during steady output of position

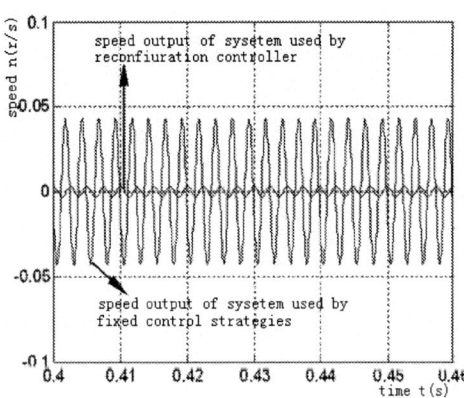

Figure 7. Partial figure enlarged during steady output of velocity

VI. CONCLUSION

On the base of these analyses about characteristics and parameter's calculation methods of control strategies and performance requirements of ACPSS in industry application, a design method of reconfigurable controller was proposed. In this method, the control strategies selected are simple and apt to realization, because of high-speed compution ability of microcontroller. The control strategy libaray formed by those control stratgegies selected can be saved in the host computer or the local controller, because of the mature network technology and the big memory space of micro-controller. New control strategies can be downloaded to controller throught network bus or flexibly regrouped in controller according anticipant performance requirements of ACPSS. This controller has wider application range than the conventional controller. The simulation results also verify it. Our method can enlarge application range of the controller with no additional cost, which is very advantageous to the end user.

REFERENCE

[1] Qin Yi, Li Ye Song etc., Modern AC servo systems, Huazhong University of Science and Technology, 1995.

[2] Jing Yu etc., Design guidance of servo systems, BeiJing University of Theory and Industry Publication, 2000.

[3] Gene F. Franklin, J. David Powell. Digital control of dynamic systems, Tsinghua University Publication : 2001.

[4] Wen-Chun Yu, Gou-Jen Wang, "Discrete sliding mode control with forgetting dynamic sliding surface", *Mechatronics* ,14, pp.737-755 , 2004.

[5] Paul Stewart, Visakan Kadirkamanathan, "Dynamic model reference PI control of permanent magnet AC motor drives". *Control Engineering Practice*, 9, pp. 1255-1263 , 2001

[6] Ge Baoming, Zheng Hongtao, "Robust control methods for PM synchronous motor", Proceedings of the 5th word Congress on Intelligent Control and Automation, pp.4437-4440 , 2004

[7] George Ellis, "PDFF: An Evaluation of a Velocity loop Control Method", Kollmorgen.

[8] SERVOSTAR S-and CD-Series Velocity Tuning Algorithms, Danaher Motion Kollmorgen, 2002.

[9] Li Ye Song, "An study on self-adjustmen and self-recognition control for full digital AC servo systems", PhD Dissertation, huazhong university of science and technology, 1996.

[10] Deng Zhong Hua, "An study on high-performance full digital AC servo systems", PhD dissertation, huazhong university of science and technology, 1995.

Position Sensorless Control of PMSM Based on a Novel Sliding Mode Observer over Wide Speed Range

Song Chi, Student Member, and Longya Xu, Fellow IEEE
Dept. of Electrical and Computer Engineering
The Ohio State University
2015 Neil Avenue
Columbus, OH 43210 USA
chi.36@osu.edu, xu.12@osu.edu

Abstract— A position sensorless control scheme of permanent magnet synchronous machines (PMSM) without saliency is proposed. The rotor position angle of PMSM is estimated by a novel sliding mode observer over a wide speed range including flux-weakening region. A feedback of the equivalent control is applied in the sliding mode observer. Compared to conventional sliding mode observers (SMO), the proposed approach features the flexibility to design the sliding mode observer with broad operating range. By properly selecting equivalent control feedback gains, the estimation error of rotor position angle can be reduced in the low-speed range and fast convergence guaranteed in the high-speed range. Computer simulation and experimental results are presented to verify the proposed sensorless control algorithm.

Keywords-sensorless; PMSM; sliding mode observer; wide speed range

I. INTRODUCTION

Permanent magnet synchronous machines (PMSM) have been increasingly applied for drive applications due to their high power density, torque to inertia ratio and high efficiency. To achieve the field-oriented control of a PMSM, knowledge of rotor position is required. Usually the rotor position angle is measured by a shaft encoder, resolver, or Hall sensors. However, the presence of such sensors increases the cost and encumbrance of the overall system as well as reduces its robustness and reliability. Furthermore, it is difficult to install and maintain a position sensor due to the limited space and/or rigid environment with severe shaft vibration. Therefore, it is desirable to eliminate position sensors in vector control PMSM drives.

For this purpose, researches have been conducted widely in the past two decades. Several main techniques of sensorless control have been extensively studied for PMSM drives, which can be broadly categorized into two groups: 1) magnetic-saliency-based and 2) observer-based estimation. In the first group, the rotor position can be estimated by using inductance variations due to the

magnetic saturation and/or saliency of PMSM. Techniques based on this idea, e.g. the "INFORM" method [1] and the high-frequency signal injection method [2], are playing an important role on the sensorless control of PMSM drives requiring standstill and low-speed operation. Unfortunately such approaches only work at low speeds for PMSMs with saliency and are too complicated to be easily implemented in real-time. The second group includes the Luenberger observer [3], full-order state observer with feedback linearization of non-linear state [4], D-state observer [5], and so on. However, the poles and zeros of the system transfer function may vary due to parameter variations, and model uncertainties always degrade the performance of these observers.

Previous studies show that sliding mode observers (SMO) have attractive advantages of robustness to disturbances and low sensitivity to parameter variations when the sliding mode truly happens. In principle, sliding mode approaches can only be achieved by discontinuous control and switching at infinite frequency. In reality, however, no such sliding mode will take place in implementation due to the limited switching frequency and sampling rate. As a result, the discretization chattering problem normally exists. The boundary solution has been used to solve the chattering problem by replacing the discontinuous control with a saturation function which approximates the sign function in a boundary layer of sliding mode manifold. In such a way, the invariance property of sliding mode is partially preserved in the sense that the state trajectories are confined to a small vicinity of the manifold. However, within the boundary layer, the state behavior and further convergence to zero cannot be guaranteed [6].

For SMO based sensorless control of PMSM, two challenges have to be dealt with properly: first, the very small magnitude of the back-EMF at low speeds and the second, the sufficient high switching gain satisfying necessary conditions for SMO convergence in the high-speed range. It is known that the minimum operating speed and the quality of the estimated rotor position at low speeds depend on the quantization error of discrete-time controller. On the other hand, the high switching gain causes large ripples on the estimated component, resulting in large estimation error.

Peixoto et al proposed a speed control system of a PMSM drive, in which the sliding mode observer was built to estimate the induced back-EMF, rotor position and speed. The back-EMF information was obtained from the filtered switching signals relative to the current estimation error. To design the switching gain over a wide speed range may be a challenging job using this observer model [7].

Han et al presented a method to estimate the speed of a PMSM based on sliding mode observer. Lyapunov functions were chosen for determining the adaptive law for the speed and stator resistance estimator. It can be found that the existence condition of sliding mode was not easily guaranteed for the convergence of the SMO. Also the integration of speed may bring more error on the estimated rotor position angle [8].

Elbuluk et al investigated a SMO for estimating the position and speed of a PMSM. Instead of directly using filtered switching signals by a low-pass filter, an observer was designed to undertake the filtering task for the estimated back-EMF. It is stated that the observer has the structure of an extended Kalman filter and is expected to have high filtering properties [9]. Unfortunately no experimental results were given to show the feasibility of the proposed method. Similar technique has also been discussed in [6].

Kang et al proposed an iterative sliding mode observer for the estimation of back-EMF of a PMSM without saliency and thus rotor position angle in high-speed range. By iterating the general SMO recursively several times within a sample period of PI current regulators, chatting components of the estimated currents and back-EMF can be reduced. However, this method doesn't help much for the low-speed operation [10].

To cope with the foregoing problems, this paper presents a sensorless vector control algorithm for PMSMs without saliency. The rotor position angle is estimated by a novel sliding mode observer over a wide speed range including flux-weakening region. A feedback method of equivalent control is proposed. By this method the estimation error of rotor position angle can be reduced in the low-speed range and fast convergence guaranteed in the high-speed range without chattering problem. In addition, the stability analysis is conducted and selection of the feed back gain discussed. The proposed sensorless control algorithm is implemented in a fully digital speed control system based on a digital signal processor (DSP), TMS320F2812. Computer simulation and experimental results are used to verify its validity.

II. MATHEMATICAL MODEL OF PMSM IN THE STATIONARY REFERENCE FRAME

The dynamic equations of a PMSM without saliency in the stationary reference frame can be expressed in the matrix form as

$$\dot{\vec{i}}_s = A \cdot \vec{i}_s + B \cdot (\vec{v}_s - \vec{e}_s) \qquad (1)$$

where
$$A = \begin{bmatrix} -R_s/L_s & 0 \\ 0 & -R_s/L_s \end{bmatrix} \qquad B = \begin{bmatrix} 1/L_s & 0 \\ 0 & 1/L_s \end{bmatrix}$$

$$\vec{i}_s = \begin{bmatrix} i_{\alpha s} \\ i_{\beta s} \end{bmatrix} \qquad \vec{v}_s = \begin{bmatrix} v_{\alpha s} \\ v_{\beta s} \end{bmatrix}$$

$$\vec{e}_s = \begin{bmatrix} e_{\alpha s} \\ e_{\beta s} \end{bmatrix} = K_e \cdot \omega_r \cdot \begin{bmatrix} -\sin(\theta_r) \\ \cos(\theta_r) \end{bmatrix}$$

L_s, R_s and K_e refer to the parameters of stator inductance, resistance and back-EMF constant.

III. SLIDING MODE OBSERVER

The sliding mode observer is designed as

$$\dot{\hat{\vec{i}}}_s = A \cdot \hat{\vec{i}}_s + B \cdot (\vec{v}^*_s + l \cdot \vec{Z}_{eq} + \vec{Z}) \qquad (2)$$

$$l > -1, \quad \vec{Z} = -K \cdot sign(\hat{\vec{i}}_s - \vec{i}_s)$$

where
$$A = \begin{bmatrix} -R_s/L_s & 0 \\ 0 & -R_s/L_s \end{bmatrix} \qquad B = \begin{bmatrix} 1/L_s & 0 \\ 0 & 1/L_s \end{bmatrix} \qquad K = \begin{bmatrix} k & 0 \\ 0 & k \end{bmatrix}$$

$$\vec{i}_s = \begin{bmatrix} i_{\alpha s} \\ i_{\beta s} \end{bmatrix} \qquad \vec{v}^*_s = \begin{bmatrix} v^*_{\alpha s} \\ v^*_{\beta s} \end{bmatrix} \qquad \hat{\vec{i}}_s = \begin{bmatrix} \hat{i}_{\alpha s} \\ \hat{i}_{\beta s} \end{bmatrix}$$

$$sign(\hat{\vec{i}}_s - \vec{i}_s) = \begin{bmatrix} sign(\hat{i}_{\alpha s} - i_{\alpha s}) \\ sign(\hat{i}_{\beta s} - i_{\beta s}) \end{bmatrix}$$

In (2), l is the feedback gain of the equivalent control Z_{eq}, and k, normally positive ($k > 0$), is the switching gain of the discontinuous control Z,. The superscript '*' indicates a command variable.

The equivalent control Z_{eq} can be obtained as in

$$\vec{Z}_{eq} = \begin{bmatrix} Z_{eq\alpha} \\ Z_{eq\beta} \end{bmatrix} = \begin{bmatrix} -k \cdot sign(\hat{i}_{\alpha s} - i_{\alpha s}) \cdot \dfrac{\omega_c}{s + \omega_c} \\ -k \cdot sign(\hat{i}_{\beta s} - i_{\beta s}) \cdot \dfrac{\omega_c}{s + \omega_c} \end{bmatrix} \qquad (3)$$

ω_c : freq_cutoff

A low-pass filter is used in (3). It is noticed that its cutoff frequency should be designed properly corresponding to the fundamental frequency of currents.

Fig. 1 shows a block diagram of the proposed sliding mode observer.

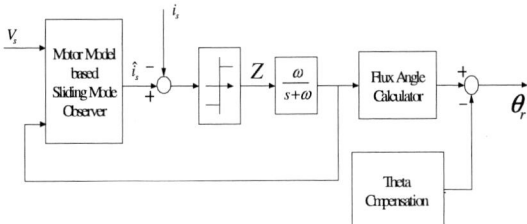

Figure 1.   Block diagram of the proposed sliding mode observer

To solve the chattering problem, the sign function is replaced by a saturation function as shown in Fig. 2. When the magnitude of current error is less than E_0, the control Z changes to a saturation function as

$$\vec{Z} = -k_s \cdot (\hat{\vec{i}}_s - \vec{i}_s)$$

where $k_s = k / E_0$.

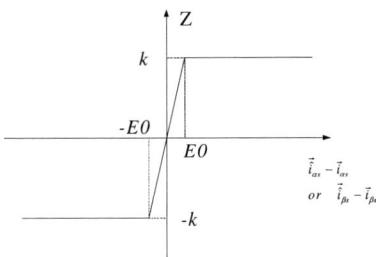

Figure 2.   Diagram of the saturation function

Considering (1) and (2), we can obtain the dynamic sliding mode motion equation as expressed in

$$\dot{S} = A \cdot S + B \cdot (\vec{e}_s + l \cdot \vec{Z}_{eq} + \vec{Z})$$
$$where \quad S = \hat{\vec{i}}_s - \vec{i}_s \tag{4}$$

If the switching gain of Z, k, is large enough to guarantee the following [2]

$$\dot{S}^T \cdot S < 0 \tag{5}$$

then sliding mode will exist and we get

$$\vec{e} = \begin{bmatrix} e_{\alpha s} \\ e_{\beta s} \end{bmatrix} = -(1 + l)\vec{Z}_{eq} \tag{6}$$

Furthermore, the rotor position angle $\hat{\theta}_r$ can be estimated by using the equivalent control Z_{eq} as follows.

$$\hat{\theta}_r = -\tan^{-1}(\frac{\hat{e}_{s\alpha}}{\hat{e}_{s\beta}}) = -\tan^{-1}(\frac{Z_{eq\alpha}}{Z_{eq\beta}}) \tag{7}$$

It is noticed that the estimated rotor position angle $\hat{\theta}_r$ should be added by π when the rotation direction is opposite.

IV.   STABILITY ANALYSIS

Let the positive definite function,

$$V = \frac{1}{2} \cdot S^T \cdot S > 0 \tag{8}$$

be a Lyapunov function candidate. Its time derivative along the system trajectories is of the form

$$\dot{V} = S^T \cdot \dot{S} = S^T AS + S^T B \cdot (\vec{e}_s + l \cdot \vec{Z}_{eq} + \vec{Z}) \tag{9}$$

Considering (4) and $\mu = 1/\omega_c$, we get

$$\dot{V} = S^T \cdot \dot{S}$$
$$\Rightarrow S^T AS + S^T B \cdot \vec{e}_s + k(\frac{l}{\mu s + 1} + 1)S^T B \cdot sign(S) \tag{10}$$
$$= F1 + F2$$

where

$$F1 = S^T AS$$

$$F2 = \frac{1}{L_s} \left\{ \bar{i}_{s\alpha} \left[e_{s\alpha} - k \cdot \frac{\mu s + 1 + l}{\mu s + 1} sign(\bar{i}_{s\alpha}) \right] \right.$$
$$\left. + \bar{i}_{s\beta} \left[e_{s\beta} - k \cdot \frac{\mu s + 1 + l}{\mu s + 1} sign(\bar{i}_{s\beta}) \right] \right\}$$
$$= \frac{1}{L_s} \left\{ \begin{array}{l} \bar{i}_{s\alpha} \cdot \left[e_{s\alpha} \mp k \cdot \frac{\mu s + 1 + l}{\mu s + 1} \right] (\substack{\bar{i}_{s\alpha} > 0 \\ \bar{i}_{s\alpha} < 0}) \\ + \bar{i}_{s\beta} \cdot \left[e_{s\beta} \mp k \cdot \frac{\mu s + 1 + l}{\mu s + 1} \right] (\substack{\bar{i}_{s\beta} > 0 \\ \bar{i}_{s\beta} < 0}) \end{array} \right\}$$

From (1), we know that A is negative and B positive definite. Consequently F1 is negative.

Assuming μ is very small with regard to the high cutoff frequency ω_c of the low-pass filter, F2 is negative if the following satisfies.

$$\begin{cases} k \cdot \frac{\mu s + 1 + l}{\mu s + 1} > |e_{s\alpha}| \\ k \cdot \frac{\mu s + 1 + l}{\mu s + 1} > |e_{s\beta}| \end{cases} \Rightarrow k \cdot (1 + l) > |\vec{e}_s|_{,max} \tag{11}$$

1827

It is implied that the feedback gain l must be larger than –1 with any positive switching gain k, i.e. $l > -1$.

Therefore, the time derivative of V is negative with enough large positive switching gain k, which testifies to convergence to $S(t) = 0$ with finite time and thereby the existence of sliding mode.

V. SELECTION OF FEEDBACK GAIN OF EQUIVALENT CONTROL

A. Below the base speed: $0 > l > -1$

From (6), we can see that the magnitude of the equivalent control Z_{eq} is always larger than that of the back-EMF with $0 > l > -1$. The rotor position of PMSM is calculated from Z_{eq}. Although the back-EMF is small at low speed, the estimated Z_{eq} is enlarged by selecting the feedback gain satisfying $0 > l > -1$. Therefore, with the same quantization limits in the real-time implementation, the proposed observer is able to work at lower speeds, or improve the quality of the estimated rotor position angle, equivalently extending the minimum operating speed.

B. Above the base speed, i.e. flux-weakening region: $l > 0$

From (11), it can be seen that the effective switching gain of the control Z is enlarged by selecting the feedback gain $l > 0$. This property results in faster convergence rate of the proposed asymptotic observer.

On the other hand, the switching gain k can be designed with smaller value than those in conventional SMOs under same conditions. By this approach, using same low-pass filters, the ripples on the equivalent control Z_{eq} can be reduced, and thus ripples on the estimated position angle are very small.

In the paper, $l = -0.5$ below the base speed and $l = 1$ above the base speed are selected for the computer simulation and experimental testing.

VI. OVERALL CONTROL SYSTEM

Fig. 3 shows a block diagram of the overall control system, which consists of a speed PI regulator, a flux-weakening controller, 2 current PI regulators, a speed calculator implemented by a Phase-Locked Loop (PLL) and a rotor position estimator by the sliding mode observer. In addition, the conventional modules for vector control such as Clark, Park and inverse Park transformation, space vector PWM generation module, a 3-phase power inverter and the PMSM are included. When the speed of the PMSM is above its base speed, the flux-weakening controller will be activated, which not only takes over the speed regulation but also automatically generates the required demagnetizing current for the flux-weakening operation.

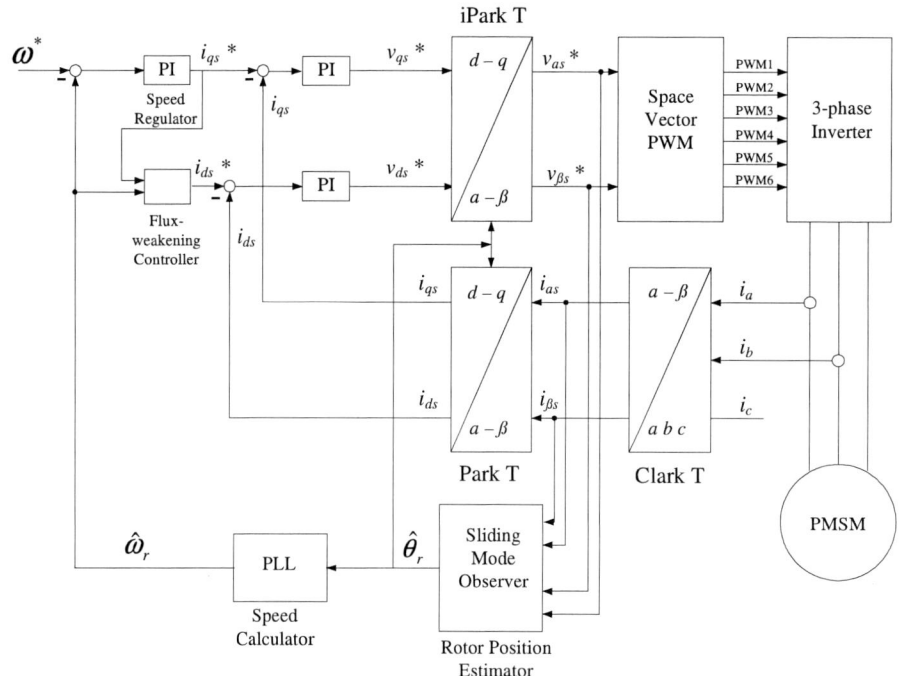

Figure 3.   Block diagram of overall control system

VII. SIMULATION AND EXPERIMENTAL RESULTS

A. Simulation results

The proposed position sensorless control algorithm based on the sliding mode observer was simulated by Simulink/MatLab. The parameters of PMSM were: R_s = 16 ohm, L_s = 60 mH and K_e = 0.22 Vs/rad. The base speed was 250 rpm. The dc bus voltage of the power inverter was 310V. The space vector PWM algorithm was applied and updated every 50 μs with respect to the switching frequency of 20kHz. Variables in the speed and current regulators were in per unit. The speed base was defined as 1250 rpm and current base was 7A. The switching gain k of the sliding mode observer was 800. The cutoff frequency of the low-pass filter for obtaining the equivalent control was 4000π rad/s while the maximum fundamental frequency of current was 1000π rad/s.

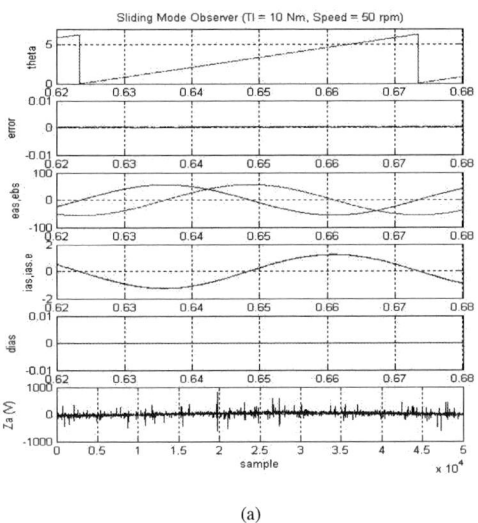

(a)

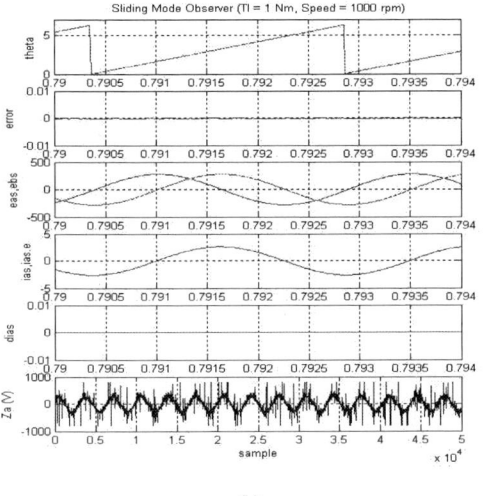

(b)

Figure 4. Actual and estimated angle (up) and estimation error (2nd), estimated back-EMF (3rd), measured and estimated current i_{qs} (4th) and error (5th), and sliding mode control Za (bottom) (a) l=-0.5, (b) l=1

Figs.4 shows the simulation results when the motor was running at 50rpm with l = -0.5 in (a) and at 1000rpm with l = 1 in (b). In the two cases, actual and estimated waveforms of both rotor position angle and stator current almost overlap together with very small errors (much smaller than 0.01). In the figure, 0.01 of angle error represents 0.57 electrical degree. In addition, the control Z_a was sampled in real time at minimum frequency of 1 GHz. Other signals were sampled at fixed 100 kHz. It is noticed that Z_a is not aligned with other signals in time and plotted in sample dots.

B. Experimental results

An experimental PMSM drive system was set up to verify the proposed sensorless control, which included: 1) a 48-pole non-saliency outer-rotor PMSM with its base speed of 250 rpm, 2) a DSP controller based on an eZdspF2812 DSP board, 3) a three-phase power inverter and 4) a dynamometer coupled with the shaft of the PMSM as load. The switching frequency of power inverter was 20kHz. Space vector PWM was used for the PWM generation. The dc bus voltage of the power inverter was 310V and the maximum current was 7A. The sampling frequency of the current and voltage measurement was 20 kHz. The parameters of PMSM were: R_s=16 ohm, L_s=60 mH and K_e=0.22 Vs/rad, same as in the simulation.

In Figs. 5 and 6, the real rotor position is measured by the encoder ($\theta_{encoder}$) and the estimated one ($\hat{\theta}_{SMO}$) by the proposed sliding mode observer. Hall sensor signal *Hall_A* and phase current i_a and i_b are also displayed.

Figs.5 shows the experimental results when the motor was running at 50 rpm (20% base speed) with a constant load of 10 Nm (30% maximum torque) and Fig.6 at 1000 rpm (400% base speed) with a constant load of 1 Nm (50% available maximum torque according to speed). We can see the well-behaved rotor position estimation from the SMO observer and the well-regulated sinusoidal current waveforms. The estimated angle $\hat{\theta}_{SMO}$ aligns with $\theta_{encoder}$ and *Hall_A* well. The output rotor position signals in Fig. 6 are rounded on the bottom corner due to the limited bandwidth of monitor circuits. The details of the measured and estimated rotor position angle were recored by real-time data logging and shown in Fig. 7.

Figs.8 through 10 show the transient response when the motor was running at 50, 500 and 1000 rpm with step-on and -off load of 10 Nm, 5 Nm and 2 Nm respectively. As observed, the speed is well regulated over a wide range regardless of load disturbance.

Figs.11 shows the experimental results when the motor was accelerated from 50 to 1025 rpm and stayed for 25 s and then decelerated to 50 rpm. The dc bus voltage varied between 320V and 280V due to operating conditions. The load was added with 0.5 Nm. The feedback gain k changed at 180 rpm with the onset of the flux-weakening operation. We can observe the automatically generated demagnetizing current i_{ds} and the good performance of speed control within the whole speed range including the flux-weakening region.

All the experimental results show that the proposed sensorless control algorithm is valid and the real-time implementation is successful.

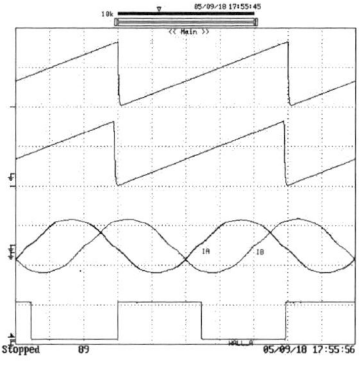

Figure 5. Rotor position angle $\theta_{encoder}$ (up), $\hat{\theta}_{SMO}$ (2nd), i_a and i_b (1.0 A/div, 3rd) and *Hall_A* (bottom)

Figure 6. Rotor position angle $\theta_{encoder}$ (up), $\hat{\theta}_{SMO}$ (2nd), i_a and i_b (2.0 A/div, 3rd) and *Hall_A* (bottom)

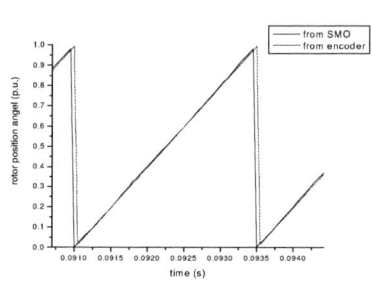

Figure 7. Rotor position angle $\theta_{encoder}$ (up), $\hat{\theta}_{SMO}$ (bottom) from real-time data logging

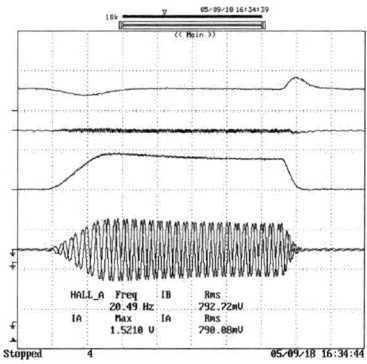

Figure 8. Rotor speed $n_{encoder}$ (*75 rpm/div*, up), i_{ds} (*1.7A/div*, 2nd), i_{qs} (*1.7A/div*, 3rd) and i_a and i_b (2.0 A/div, bottom)

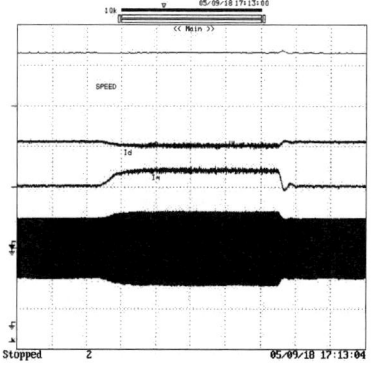

Figure 9. Rotor speed $n_{encoder}$ (380 *rpm/div*, up), i_{ds} (*1.7A/div*, 2nd), i_{qs} (*1.7A/div*, 3rd) and i_a and i_b (2.0 A/div, bottom)

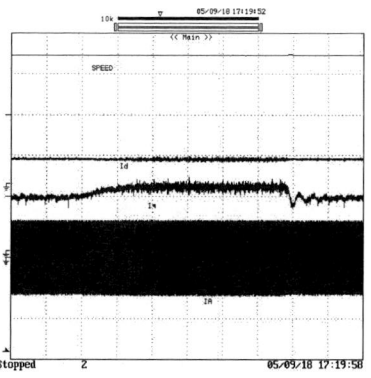

Figure 10. Rotor speed $n_{encoder}$ (*757 rpm/div*, up), i_{ds} (*1.7A/div*, 2nd), i_{qs} (*1.7A/div*, 3rd) and i_a and i_b (2.0 A/div, bottom)

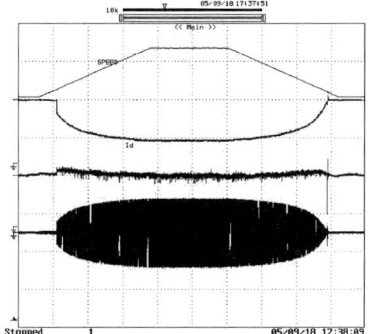

Figure 11. Speed command and feedback (757 rpm/div, up), i_{ds} (1.7 A/div, 2nd), i_{qs} (1.7 A/div, 3rd) and stator phase current i_{as} (1.0 A/div, bottom), 10 s/div

VIII. Conclusions

In this paper, a sensorless vector control scheme is presented for PMSMs without saliency using a novel sliding mode observer. The rotor position angle is estimated based on the equivalent control signals over wide speed range including flux-weakening region. By a feedback method of the equivalent control, the estimation error of rotor position angle is reduced in the low speeds and fast convergence guaranteed in the high speeds. Stability of the proposed sliding mode is proved by a Lyapunov function. The speed of PMSM is well regulated in the whole speed range regardless of load disturbance. The validity of the proposed control algorithm has been demonstrated by both computer simulation and experimental results.

Acknowledgment

This work was supported by the Research and Engineering Center of Whirlpool.

References

[1] M. Schroedl, "Sensorless control of AC machines at low speed and standatill based on the "INFORM" method," in *Conf. Rec. IEEE-IAS Annu. Meeting*, vol. 1, 1996, pp. 270-277.

[2] M. J. Corley and R. D. Lorenz, "Rotor position and velocity estimation for permanent magnet synchronous machine at standstill and high speed," in *Conf. Rec. IEEE-IAS Annu. Meeting*, vol. 1, 1996, pp. 36-41.

[3] K. W. Lim, K. S. Low and M. F. Rahman, "A position observer for permanent magnet synchronous motor drive," *IECON Annu. ConferenceRecord*, pp. 1004–1008, 1994.

[4] Y. Yamamoto, Y. Yoshida and T. Ashikaga, "Sensorless control of PM motor using full order flux observer," *IEEJ Trans. Ind. Applt.*, vol. 124, pp. 743-749, Aug. 2004.

[5] S. Shinnaka, "New "D-state-observer"-based vector control for sensorless drive of permanent-magnet synchronous motors," *IEEE Trans. Ind. Applicat.*, vol. 41, pp. 825-833, May./June. 2005.

[6] Vadim Utkin, J. Guldner and Jingxin Shi, "Sliding mode control in electromechanical systems," 1ˢᵗ Edition, Taylor&Francis, 1999.

[7] Z. M. Peixoto, et al., "Speed control of permanent magnet motors using sliding mode observers for induced emf, position and speed estimation," in *Conf. Rec. IEEE-IAS Annu. Meeting*, vol. 2, 1995, pp. 1023–1028.

[8] Y. S. Han, J. S. Choi and Y.S. Kim, "Sensorless PMSM drive with a sliding mode control based adaptive speed and stator resistance estimator," *IEEE Trans. Magnetics,* vol. 36, pp. 3588-3591, Sept. 2000.

[9] M. Elbuluk and C. Li, "Sliding mode observer for wide-speed sensorless control of PMSM drives," in *Conf. Rec. IEEE-IAS Annu. Meeting*, vol. 1, 2003, pp. 480–485.

[10] K. Kang, J. Kim and et al, "Sensorless control of PMSM in high speed range with iterative sliding mode observer," in *Conf. Rec.IEEE- APEC'04*, vol. 2, 2004, pp. 1111–1116.

2006 5th International Power Electronics and Motion Control Conference

Design of Motion Control System Used for Filter Rod Production Machine

Yang Qingyu, Ge Sibo, Ye Kesong and Shi Ren
Department of Automation, Xi'an Jiaotong University, Xi'an, P.R.China
yangqingyu@mail.xjtu.edu.cn

Abstract—The control system of filter rod production machine has characters of high control precision and high reliability, its performance affects quality of filter rod directly. How to improve its integrative control level is the key problem of control system design. Therefore, a suit of motion control system based on Profibus is presented in this paper. In this control system, some new technologies such as fieldbus and AC servo drive control are employed. System structure and software framework are issued. The key technologies are researched in depth. This system has already been implemented on the product line in a large-scale tobacco group successfully. The performance results show that the motion control system improves outputs and quality of products and integrative control level, saves energy, and obtains favorable economic and social benefits.

Keywords-filter rod production machine; Fieldbus; motion control; PLC

I. INTRODUCTION

Filter rod production machine used to make cigarette filter tip is the important device in tobacco industry. Carbonic acid fibre is incised into cigarette filter tip through a series of technics including slack control, insufflating glycerin, packaging papers, circle control, and so on. In this process, slack control performance of Carbonic acid fibre affects "draw resistance" directly, which is the key parameter used to judge the quality of filter rod and content of tar in cigarette. Today, the traditional method is mechanism drive when controlling filter rod production machine in tobacco industry in china. But this method has disadvantages of large noise, high energy wasting, and unstable quality of products. Recent years, along with the development of cigarette technologies and high demand of automation level and management information integration in cigarette industry, more higher and stricter control demand about control of filter rod production machine are brought forward[1].

Therefore using some new technologies, a suit of motion control system based on Profibus that is used for filter rod production machine is presented in this paper, and the key technologies are researched. This paper is structured as follows. Section II introduces shortly the technics and control demand of filter rod production machine. Section III describes the network architecture

and software framework of the motion control system. Section IV is the detailed design, such as AC servo drive control, data communication and advanced program design, etc. Section V is implementation results. Section VI summarizes this paper.

II. TECHNICS AND CONTROL DEMAND OF FILTER ROD PRODUCTION MACHINE

The technics process of filter rod production machine is shown as Fig.1. Normal length of the filter rod is 120mm. Firstly carbonic acid fibre is pulled into fibre rod with same thickness. In order to control slack performance, the fibre rod passes through slack roller 1-2 and output roller 3. Then glycerin is insufflated into the fibre rod, and the content of glycerin is 49mg/120mm. Moreover the fibre rod is packaged with twist papers after insufflating glycerin. Finally, the fibre rod passes through cooling, circle control and length incising to form the filter rod with the length of 120mm.

Control demands of filter rod production machine include three points. Firstly, the speed of main motor can be controlled according to multi-modes, which decides the production speed of the whole machine. The running mode of the main motor includes starting speed, normal production speed and meeting paper speed. Secondly, the speed of slack motors and glycerin motor must correspond with the ones of the main motor. In order to change glycerin content of filter rod, the speed of glycerin motor should be adjusted independently. There is a ruler that speed of the other motors cannot be changed as long as the speed of main motor is invariable, which is to keep the reliability of glycerin content. The most important control portion is slack roller 1-2 and output roller 3, which affects the slacking effects of fibre and the draw resistance of filter rod directly. Finally, the whole control system must have high reliability.

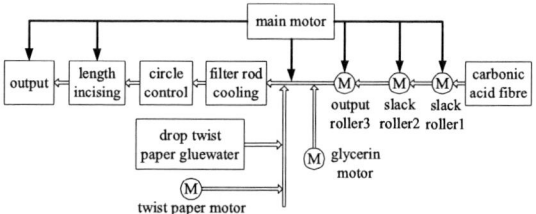

Fig. 1 Technics process of filter rod production.

1-4244-0448-7/06/$25.00 ©2006 IEEE

III. NETWORK ARCHITECTURE AND SOFTWARE FRAMEWORK

Network architecture of control system not only affects the exertion of system performance, but also decides the integration automatic level of the system in some sense. How to design reasonable system architecture is the chief problem during developing control system.

A. Network Architecture of Motion Control System

In order to satisfy the demand of integration automation, the motion control system designed for filter rod production machine adopts the structure of distributed control system, and is composed of operator stations (OS), local control stations (CS) and sensors/actuators. The three levels communicate with digital buses including fieldbus and RS485. The network architecture of the whole motion control system is depicted in Fig.2.

As shown in Fig.2, the OS is the center of the control system, its mission is to monitor the whole production line. Furthermore the OS provides an interface for operators to set parameters. According to the demand of high reliability, two suits of operator station are redundant and achieve redundant control of slack control segment. In order to save invest, multiple function panel MP370 is used as main station, and ordinary touch panel TP170A is used as redundant station. Communication between two redundant stations and PLC is realized through Profibus-DP. Process monitoring and parameters setting are implemented on MP370. TP170As is used as adjusting of slack motor, which can control two slack rollers, one output roller and glycerin motor independently.

CS responds for complicated control functions, and realizes the communication between OS and transducers at the same time. Siemens S7-300 PLC is chosen as the CS, which has the advantages of high reliability and high respond speed. The signals from sensors/actuators enter into I/O modules of PLC directly.

The control of main motor adopts technology of AC frequency conversion. CT/UNI transducer is adopted. The control mode of slack rollers change from mechanism drive to AC servo drive, and Germany LENZE drivers are selected. Panasonic NAIS driver is used to control glycerin motor. All the drivers and transducers communicate with PLC through RS485 protocol.

B. Software Framework

We use the thinking of modularization software design when realizing monitoring program. We design PLC program using Step7 language, and design OS HMI software using ProTool-SIMATIC HMI configuration language[2]. The functions on OS have device and process monitor, parameters setting, report and data statistic, fault alarming, providing correlative information for users, and so on. The software framework is represented in Fig.3. If there were alarm and fault in control system, the current menu changes to alarm and fault display menus

automatically. After alarm and fault are eliminated, the current menu will redisplay automatically.

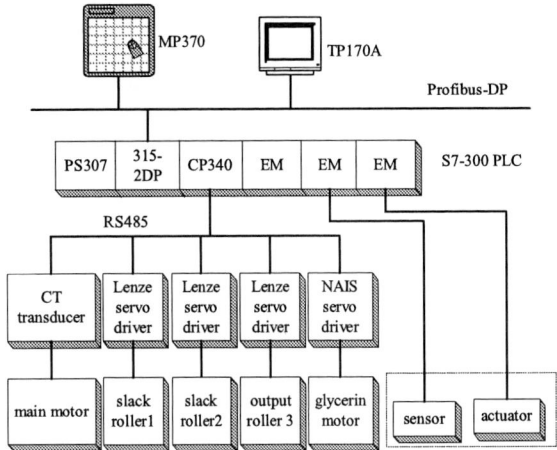

Fig. 2 Network architecture of motion control system.

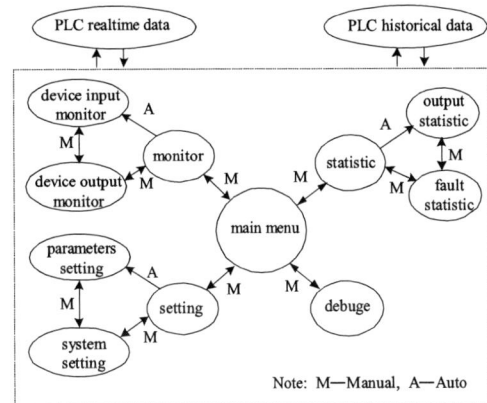

Fig. 3 Software framework of motion control system.

IV. KEY TECHNOLOGIES OF SYSTEM REALIZATION

Besides PLC control program and general HMI menus, the key technologies of the control system used for filter rod production machine are AC drive control, communication among all parts and advanced program design, which are also difficult and important. The realization of these key technologies is described as follows.

A. AC Servo Drive Control

As shown in Fig.1, the most important portion in the control of filter rod production machine is speed control of motors. The speed relation among main motor, slack motors and glycerin motor must be proportional strictly. Therefore we choose transducers and AC servo drive control subsystem to realize speed control.

Among these motors, slack and output roller motors have the highest control precision, which affect draw resistance directly. The speed of roller motors is very high,

1833

and the unit of speed changing is 1 rpm. The speed of the three motors has also proportional relation strictly, which is 892:1258:1146 (r/m) during normal production. Therefore three Lenze drivers are chosen to control slack roller 1-2 and output roller 3. Lenze drivers can realize complicated control functions and precise position control through embedded phase controller. The three drivers are connected into the form of digital frequency degree[3][4].

The control precision of the glycerin motor is lower relatively. The speed of this motor is not high and about 80 rpm. In stead of expensive Lenze driver, economic Panasonic driver is chosen to control glycerin motor. The configuration and setting of Panasonic driver are easier than Lenze driver.

The main motor should provide two suits of normal production speed and meeting paper speed at the same time according to the technics demand. So CT transducer is selected to control the main motor. CT transducer has abundant inside parameters and functions, and can realize complicated control strategy and high precise digital setting[5]. In addition, CT transducer has four work modes, which are open loop, closed vector loop, rebirth and user interface. According to control demand, we configure the transducer into open loop mode in this motion control system.

B. Data Communication

Data communication includes between not only PLC and drivers/transducers, but also PLC and OS. Reliability of communication affects the system performance directly. When designing the control system, we try our best to make data communication easy and reliability.

1) Data Communication Between Drivers/Transducers and PLC: RS485 protocol is used to realize communication between drivers/transducers and PLC. On PLC we use CP340 RS422/485 communication processor to extend RS485 interface. Because Lenze drivers and CT transducers haven't RS485 interface, UD71 and RS232/485 module 2102 are used to realize RS485 communication respectively. Reliable communication between PLC and drivers/transducers is achieved via steps above. Please note that the configuration data is downloaded to drivers using RS232 protocol, using RS485 protocol when system running.

2) Data Communication Between OS and PLC: The OS MP370 and TP170A communicate with PLC through Profibus-DP. For PLC and OS, communication configuration must be done in Step7 and Protool software respectively. The key configuration portion is the setting of baud rate and device bus address. All the devices on one Profibus must have the same baud rate and different address.

C. Advanced Program Design -Eembedded VB Script

When developing HMI using ProTool on MP370[6], some complicated functions such as the computation of production efficiency can not be developed using general function block of Protool. So we use embedded VB script to realize these functions.

We design the computation of production efficiency using embedded VB script according to real-time production data. The data includes outputs, runtime and stop-time because of fault. In addition, the VB script program must be trigged by some concrete event. Because real-time production is variable, the change of production is selected as trigger. The production efficiency will be computed one time as long as production varieties, then the efficiency display will be refreshed. Adopting this method, we design the computation subprogram of production efficiency and other complicated function modes. Please note the relationship between project variables and temporary ones in scripts when designing VB scripts to realize complicated control algorithm.

V. IMPLEMENTATION RESULTS

The motion control system that is presented in this paper has already been implemented on the product line in a large-scale tobacco group successfully since 2004. The control system has abundant functions. The speed of slack motors can be decreased or increased in 1 rpm unit. In addition, control mode of the system has changed from mechanism drive to servo drive, and decreased noise greatly. Table I and Table II is draw resistance of filter rod with the length of 120 mm before and after adopting the motion control system respectively. Here normal range of draw resistance with is 380 ± 30 mmH$_2$O.

Table I and II show that quality of draw resistance is improved greatly, at the same time stability and control precision increase. According to the results and documents from consumer, change range of draw resistance is very large, and even reaches 60 mmH$_2$O before the motion control system is implemented. Furthermore operators must verify and adjust system parameters every 10 minutes manually. After the motion control system is implemented, change range of draw resistance decreases to 20 mmH$_2$O, and the quality and output of production increase greatly. In addition, total automation is achieved without operator's participation, and good economic benefits are obtained.

TABLE I
DRAW RESISTANCE FROM OLD CONTROL SYSTEM (BEFORE)

No. Unit	1	2	3	4	5	6	7	8	9	10
mmH$_2$O	395	401	375	362	368	352	412	385	392	378

TABLE II
DRAW RESISTANCE FROM MOTION CONTROL SYSTEM (NOW)

No. Unit	1	2	3	4	5	6	7	8	9	10
mmH$_2$O	375	392	372	388	380	383	387	385	392	375

VI. CONCLUSIONS

Using some new technologies such as fieldbus and AC servo drive control, a suit of motion control system based on Profibus that is used for filter rod production machine is presented in this paper, and the key technologies are researched in depth. The performance results show that the motion control system improves output and quality of products and integrative control level, and saves energy and invest. Design thinking and application experience of the motion control system are useful for us to establish fieldbus control system in other industrial process, such as slack control system in spin industry.

REFERENCES

[1] X.C. Gu, "Application of AC servo drive system in tobacco production machine," *China west science and technology*, no.10, pp.12-14, 2004.

[2] ProTool User Configuration Manual Based On Windows. Germany Siemens Ltd., 1999.

[3] *Lenze Manual*. Germany Lenze GmbH & Co KG, 1998.

[4] Global Driver Control Getting Started. Germany Lenze GmbH & Co KG, 2000.

[5] *Unidrive Modle No.1-5 User Manual*. England Control Techniques Drives Ltd., 1999.

[6] Multi Panel MP370 Device Manual. Germany Siemens Ltd., 2001.

2006 5th International Power Electronics and Motion Control Conference

Analysis and Implementation of Sensorless Position Detection in a Permanent Magnet Generator

Sebastian Rosado, Xiangfei Ma, Fred Wang, Jerry Francis, and Dushan Boroyevich

Center for Power Electronic Systems
Virginia Tech, Blacksburg, VA, USA
e-mail: srosado@vt.edu

Abstract— **Driven by the requirement of knowing the rotor position in a permanent magnet generator (PMG), this work evaluates the performance of different position detection algorithms in a new domain of application were their use has not been reported before. Computer simulations and experimental tests are used to evaluate the characteristics and robustness of algorithms originally proposed for permanent magnet motors (PMM). Based on that evaluation, one algorithm is selected for implementation in a DSP, which requires some enhancements and additional features. The overall results obtained in this work show the appropriateness of the approach for the application under study.**

Keywords- permanent magnet machine; rotor position detection; sensorless; observer

I. INTRODUCTION

For using d-q model based control in synchronous generator excitation regulation it is required to know the position of the rotor shaft during operation. The generator under control is actually a set of three coupled rotating machines. Figure 1 shows the configuration of the system under study. The primary power for the generator excitation is taken from a permanent magnet generator (PMG) that feds the exciter, which is a synchronous machine, through a regulated DC/DC converter. This converter is the point where the excitation is controlled. Because all three rotating machines: PMG, exciter, and generator, are coupled on the same shaft it is possible to base the rotor position on any one of them. Due to its relative simplicity, the PMG is selected for the rotor position detection. Because of cost and reliability reasons it is desired that the position detection is done sensorless.

This work evaluates the performance of previously proposed rotor position detection algorithms in a new domain of application that creates new challenges given the signals that the sensorless algorithm must process at the operation conditions of the system.

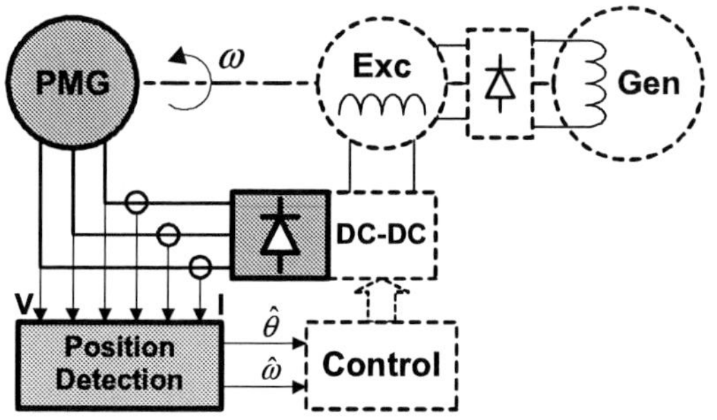

Figure 1 System general configuration. This paper concentrates on the position detection at the PMG

Several methods have been proposed in the literature for the rotor position detection of permanent magnet motors. Traditionally those methods were grouped in two main categories according to their working principles: reluctance techniques [1], and state observers [2]. The first group principle is based on determining the response of the machine to an injected electric signal. In addition, it is generally applied where the detection is required to operate at standstill or low speeds. On the other side, the observer principle relies on the state determination based on the measurement of the available magnitudes and the machine model. For this purpose, a proper system model is required [3]-[5]. Because the application under study that does not involve low speed operation, it is considered more suitable for the observer type approach. Many different observation techniques have been proposed and used for rotor position detection in permanent magnet motors [6]-[14]. After a wide review of those techniques, three of them were selected for a detailed evaluation in computer simulation. This evaluation is presented in section II of this paper. The simulation concluded in the selection of one of the algorithms for experimental testing, which is presented in section III. This last algorithm was then enhanced for its implementation in a DSP, aspect that is discussed in section IV. The results obtained probe the appropriate operation of the implemented detection algorithm.

This work made use of ERC Shared Facilities supported by the National Science Foundation under Award Number EEC-9731677.

1-4244-0448-7/06/$25.00 ©2006 IEEE   1836

II. EVALUATION OF POSITION DETECTION ALGORITHMS

A. PMG Modeling and Detection Principle

The advantage of the rotor position detection at the permanent magnet machine is the simplicity of its model [3]-[5]. The voltage at the machine three-phase terminals is given by:

$$V_{a,b,c} = R_s I_{a,b,c} + \frac{d\lambda_{a,b,c}}{dt} \qquad (2\text{-}1)$$

Where R_s is the resistance of the armature windings and $\lambda_{a,b,c}$ the total flux in the three phases. By decomposing the flux in its leakage and linkage flux components, the previous equation can be written as:

$$V_{a,b,c} = R_s I_{a,b,c} + L_s \frac{dI_{a,b,c}}{dt} + \frac{d\phi_{a,b,c}}{dt} \qquad (2\text{-}2)$$

The second term on the left represents the voltage drop due to the flux leakage and the third term is usually called the machine back-emf, *e*. For a given magnetic flux distribution in the machine which is created by the permanent magnet, the back-emf is produced by the time variation of the flux linkage seen by the machine armature windings. Therefore, the back-emf at each of the three-phases can be expressed as:

$$e = \frac{d\phi}{dt} = \frac{k_e}{n_p} f_e(\theta) \frac{d\theta_e}{dt} \qquad (2\text{-}3)$$

Where $f_e(\theta)$ gives the flux linkage distribution as function of the rotor angle; k_e is called the electromagnetic constant of the machine, and n_p the pole pair number of the machine.

Because of the direct relation between the back-emf and the rotor angle expressed in equation (2-3); it is possible to calculate the rotor position from the back-emf. This is the basis of the observer methods where the rotor angle is calculated from the back-emf obtained from the observer. Several different techniques of observer formulation have been proposed based on this principle; many of them were reviewed as part of this work. The review and analysis of some of techniques is discussed in the following sub-section.

B. Review and Analysis of Position Detection Techniques

The observer based techniques can be classified from both the point of view of the observer itself and the filtering method. Observers can be of deterministic or stochastic type as well as the filtering techniques. In any case the challenge is to overcome errors that are mainly created by parameter uncertainty or measurement errors and noise. The machine model used in the observer can be formulated in the stationary reference frame a-b-c, like equation (2-2), in the α-β-o coordinates, or in the rotating reference frame d-q-o. The models in the different references are obtained from (2-2) by application of standard matrix transformations.

Among the three methods preliminary selected, one of them uses a stochastic observer while the other two are based on back-emf estimation using different deterministic machine modeling. The stochastic observer uses a sliding mode observer and an extended Kalman filter. The idea of this method is to use a sliding mode observer (SMO) over an expanded state of the PM machine for a robust current detection [6], [7]. The Kalman filter is used for filtering measurement noise and SMO oscillation of the signals [8], [9], [10]. The second method [13], [14], calculate s the back-emf from the flux linkage change as expressed in equation (2-3). In order to improve the estimation and make the method less sensitive to the noise amplification due to the use of time step increments, the procedure averages the three-phase measurements. The rotor is then obtained from:

$$\Delta\theta = \frac{n_p}{k_e} \frac{\Delta\phi_a f_{eb}(\theta) + \Delta\phi_b f_{ec}(\theta) + \Delta\phi_c f_{ea}(\theta)}{f_{ea}(\theta) f_{eb}(\theta) + f_{eb}(\theta) f_{ec}(\theta) + f_{ec}(\theta) f_{ea}(\theta)} \qquad (2\text{-}4)$$

The last approach is based on an observer built on the d-q-o model of the machine [11]-[12]. The technique used is described in [11], where the difference between the observed and measured voltages is used to iteratively correct the position estimation.

C. Simulation Test Results

The preliminary evaluation was done in simulation using a detailed model of the PMG, which included non-linearities produced by non-uniform coil disposition and the rectifier load. The interest was to test the robustness of the algorithms for: machine modeling inaccuracies, parameter deviation due to miscalculation or changes in the operation conditions, non-linearity of the load circuit connected to the PMG, and noise in the voltage and current measurements. Figures 2, 3 and 4 show the results for the SMO + EKF, flux linkage, and d-q observer algorithms respectively.

The method based on the SMO and EKF presents the larger error in ideal conditions, but it is the most robust against disturbances. The flux linkage method presents a very small angle estimation overall, but it is sensitive to the electromagnetic constant value. The method based on the observation in the d-q coordinates is quite sensitive to disturbances. The results are summarized in Table I.

Table I Summary of Simulation Results

	SMO + EKF	Flux linkage	d-q observer
Typical angle error	Large, about 1.5 deg	Very small, < 0.1 deg	Very small, < 0.1deg.
Parameter deviation	Low	R_s, L_s low, k_e high	Low
Noise handling	Good	Good	Fair
Load non-linearities	Very small effect	θ small, ω large effect	Considerable effect

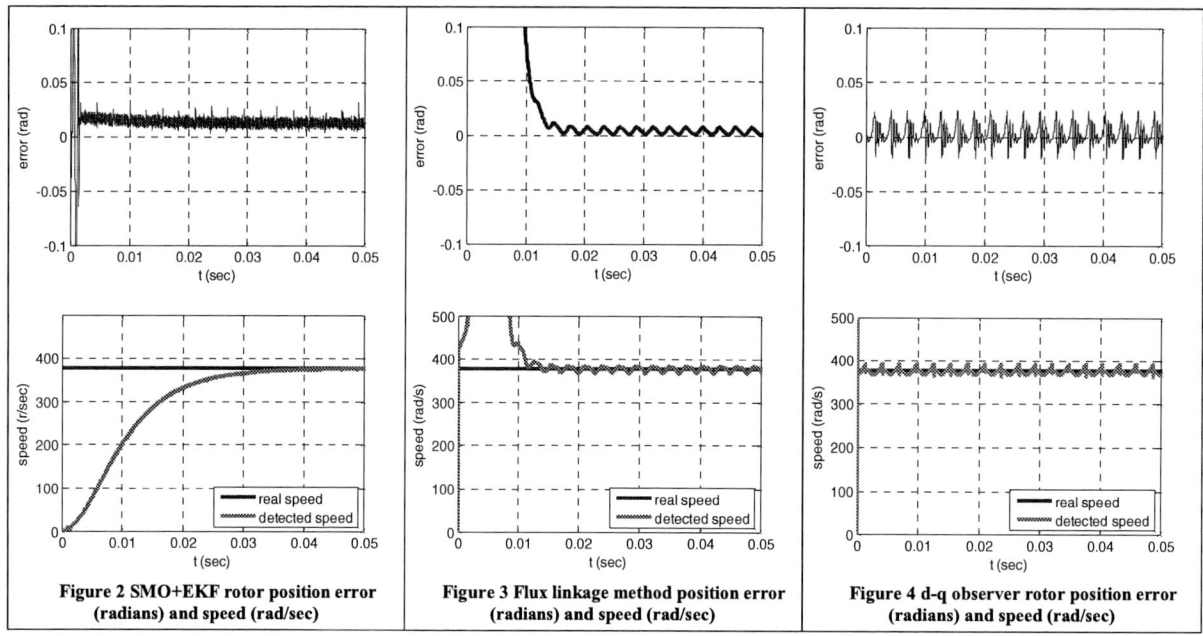

Figure 2 SMO+EKF rotor position error (radians) and speed (rad/sec)

Figure 3 Flux linkage method position error (radians) and speed (rad/sec)

Figure 4 d-q observer rotor position error (radians) and speed (rad/sec)

III. EXPERIMENTAL TEST OF THE FLUX LINKAGE METHOD

A. Machine Parameter Identification

Because of the good results produced by the flux linkage method, it was selected for experimental test. For that, the machine parameters R_s, L_s, and k_e must be determined. Moreover, as it is noted in Table I, the method is rather sensitive to the value of the electromagnetic constant k_e in order to produce good detection results. The value of k_e was calculated from the voltage-frequency characteristic of the machine, which was obtained experimentally, and it is shown in Figure 5. The linear characteristic of k_e shown in the Figure can be expressed by the relation:

$$k_e = 0.0042 \frac{V}{2\pi f} \left(\frac{V}{rad \cdot sec} \right)$$

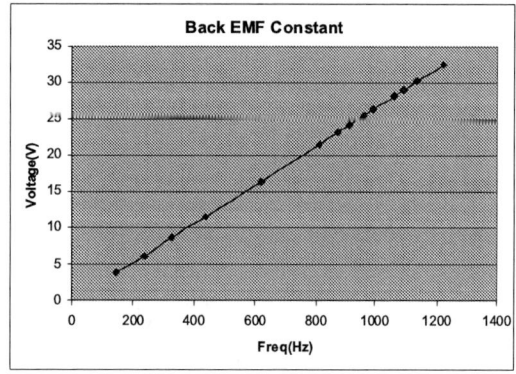

Figure 5 Voltage-frequency characteristic of the PMG

B. Response in case of Linear Load

In the studied machine, the back-emf has a THD of less than 4%, which is mostly fifth harmonic. Experimental waveforms for the phase voltage are shown in Figure 6. Therefore, in the detection algorithm, the function $f_e(\theta)$, the back-emf as function of the rotor angle was considered as a pure sine wave. Under this assumption, the denominator of equation (2-4) becomes a constant equal to 1.5 and the expression can be simplified to:

$$\Delta\theta = \frac{n_p}{k_e} \frac{\Delta\phi_a f_{eb}(\theta) + \Delta\phi_b f_{ec}(\theta) + \Delta\phi_c f_{ea}(\theta)}{1.5} \quad (3\text{-}1)$$

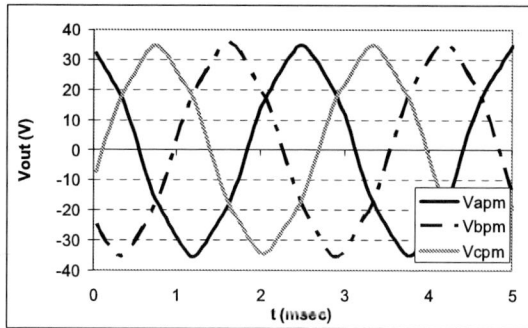

Figure 6 PMG voltage waveforms in case of a resistive load

Using equation (3-1) in the detection algorithm produced a total error in the range of 0.02 radians (or 1.2 degrees). The position detection error along one electrical revolution (2π electric radians) is shown in Figure 7. It is also worth mentioning that the tested algorithm has phase lock-loop characteristics being able to track the rotor angle in less than one electric turn. This is also observed in the simulations shown in Figure 3.

1838

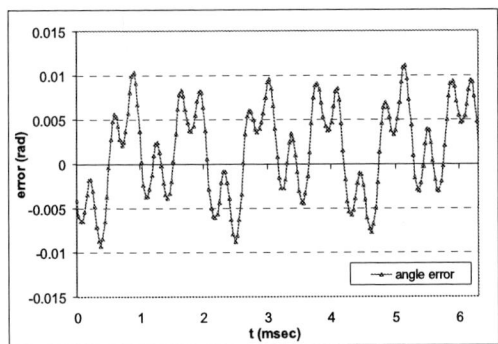

Figure 7 Rotor position error obtained for the flux linkage method during experimental tests

C. Response in Case of Non-linear Load

For the application under study it was required to know the response in case of non-linear load. Therefore, a diode rectifier with capacitive filter, which in the final application will supply the power for excitation, was connected to the PMG. Figure 8 shows the circuit used to test the algorithm response for this case.

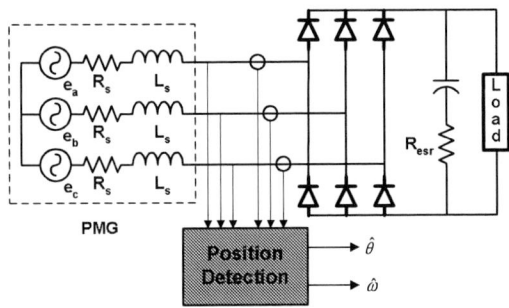

Figure 8 Schematic of the circuit used to test the position algorithm response in case of nonlinear load

The voltage and current waveforms at the PMG terminals, used for detection, are shown in Figure 9. Two line to line voltages were measured in the circuit of Figure 8 from where the three phase to neutral magnitudes were calculated. Additionally, the three line currents were also measured in the circuit.

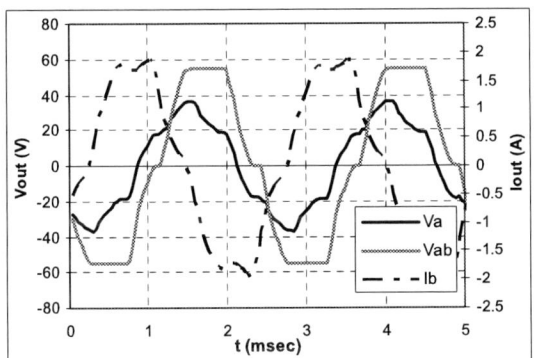

Figure 9 Waveforms at the PMG with the rectifier load; line and phase voltages , and line current

The harmonic contents of the voltage and current magnitudes are shown in Figure 10. In this case the total harmonic distortion (THD) in both voltage and current increased to about 12%.

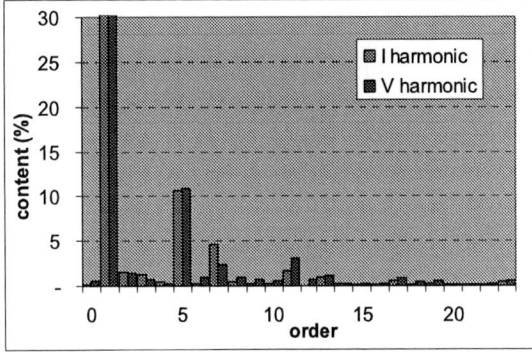

Figure 10 Harmonic spectra for the voltage and current used for the position detection

The response of the position detection scheme changes significantly under the condition being described. The angle error obtained is shown in Figure 11 that also includes the case of resistive load for comparison purposes. The angle error increased to 0.08 radians (or 4.6 degrees) in the rectified load case.

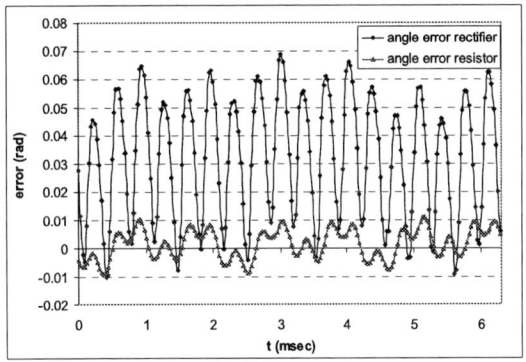

Figure 11 Rotor position error obtained for rectified and resistive loads

Although the considerable change in the response, the detected angle is still accurate enough for the application under analysis.

IV.   DSP IMPLEMENTATION OF POSITION DETECTION

A. PMG- Generator angle and synchronization

The real interest of this work is to detect the rotor position of the generator rather than the PMG. Although coupled on the same shaft, there exist a discrepancy between the PMG angle and the generator angle. The sources of this discrepancy are two; one due to the fact that the machine rotors may not be aligned, and the other given by the different number of pole pairs of the machines. To account for the mechanical misalignment, the generator position must be one-time detected. This can be done under no load steady-state condition where the generator output voltage is equal to its back-emf. The different pole pair ratio requires not only to scale the detected angle properly, but also to match the same generator and PMG poles every time the machine runs; otherwise, the generator angle will be displaced. In order to match the two machine poles, a synchronization routine runs every time the PMG starts operation. This routine generates a set of periodic waveforms, as many as the pole pair ratio, corresponding to the back-emf of the machine with larger number of poles, which in our case is the PMG. One of these waveforms, and only one, is in phase with the other machine no load voltage. After the phase agreement is detected the algorithm is ready to determine the main generator rotor angle.

B.   Offset Cancellation and Measurement Noise

The DSP implementation of the algorithm as it was presented in section II did not produce the desired results due to the integration of off-set errors originated at the measurements. If this off-set is not compensated, the flux integration will continually increase in absolute value diverging from its real value. Therefore, a term that cancels the flux offset was introduced in the flux linkage update equation. The flux linkage increment at each phase is then calculated as:

$$\Delta\phi_{a,b,c} = -\left(V_{a,b,c} - R_s I_{a,b,c}\right)\Delta t - L_s \Delta I_{a,b,c} - k_c \tanh\left(\phi_{a,b,c}\right)$$

The sampling rate at the application was 64 kHz, which probed to work well even at the maximum frequency of the application, which is 1600 Hz. A third order FIR was used for the filtering the voltage and current measurements. Figure 12 shows the results obtained applying the implemented algorithm for the calculation of the evolution of the I_d, I_q current components during a transient at the main generator.

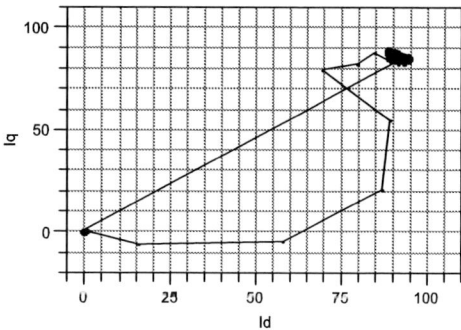

Figure 12 I_d, I_q evolution for load connection at main generator

V.   SUMMARY AND CONCLUSIONS

This paper presented the analysis and solution for the problem of determining the rotor position in a PMG connected to a non-linear load. The results from simulation and experiment show that accuracy of the algorithms is affected by the operation conditions of the PMG. However, the implementation of flux linkage method produces results that are appropriate for the application under study. It is also worth mentioning that a synchronous machine voltage controller implemented using the d-q component calculation based on the angle detected by the method described in this paper produced good results at all operation conditions.

ACKNOWLEDGMENT

The authors are grateful to Thales Avionics Electrical Systems (TAES) for their financial support to this work.

REFERENCES

[1]   S. Ostlund, M. Brokemper, "Sensorless Rotor-position Detection from Zero to Rated Speed for an Integrated PM Synchronous Motor Drive", IEEE Trans. Ind. Apps, Vol. 32, No. 5, Sep. 1996

[2]   L. Jones, J. Lang, "A State Observer for the Permanent-Magnet Synchronous Motor", IEEE Trans. Ind. Elec, Vol. 36, No. 3, Aug. 1989

[3]   P. Pillay, R. Krishnan, "Modeling of Permanent Magnet Motor Drives", IEEE Trans. Ind Electronics, Vol. 35, No. 4, Nov. 1988

[4]   V. Petrovic, M. Stankovic, "Modeling of PM Synchronous Motors for Control and Estimation Tasks", Proc. 40th IEEE Conf. on Decision and Control, Part vol. 3, pp.2229-34, Dec. 2001

[5]   P. Krause, O. Wasynczuk, S. Sudhoff, *Analysis of Electric Machinery and Drive Systems*, 2nd ed., IEEE Press, J. Wiley, 2002

[6]   T. Furuhashi, S. Sangwongwanich, S. Okuma, "A position-and-velocity sensorless control for brushless DC motors using an adaptive sliding mode observer", IEEE Trans. Ind. Elec., vol.39, no.2, pp.89-95, April 1992

[7]   S. Brock, J. Deskur, K. Zawirski "Modified sliding-mode speed controller for servo drives", Proc. of IEEE International Symposium on Industrial Electronics, Part vol.2, pp.635-40, 1999

[8]   A. Germano, F. Parasiliti, M. Tursini, "Sensorless speed control of a PM synchronous motor drive by Kalman filter", ICEM 94, Int. Conf. on Electrical Machines, Part vol.2, pp.540-4, 1994

[9]   F. Parasiliti, R. Petrella, M. Tursini "Rotor speed and position detection for PM synchronous motors based on sliding mode observer and Kalman filter", European Power Elec, EPE'99, pp.10

[10]   F. Parasiliti, R. Petrella, M. Tursini, "Sensorless speed control of a PM Synchronous motor based on sliding mode observer and extended Kalman Filter", IEEE 36th IAS Annual Meeting, Vol. 1, pp. 533-540, 2001

[11]   N. Matsui, "Sensorless PM Brushless DC Motor Drives", IEEE Trans. Industrial Electronics, Vol. 43, pp. 300-308, April 1996

[12]   M. Naidu, B.K. Bose, "Rotor Position Estimation Scheme of a Permanent Magnet Synchronous Machine for High Performance Variable Speed Drive", IEEE Ind. App. Soc. Annual Meeting 1992, Vol. 1, pp. 48-53

[13]   L. Ying, N. Ertrugul, "A New Algorithm for Indirect Position Estimation in Permanent Magnet AC Motors", IEEE Power Electronics Specialists Conf. PESC 2002, Vol. 1, pp. 289 -294

[14]   L. Ying, N. Ertrugul, "A Novel, Robust DSP-Based Indirect Rotor Position Estimation for Permanent Magnet AC Motors without Rotor Saliency", IEEE Trans. Power Elec, Vol.18, pp. 539-546

2006 5th International Power Electronics and Motion Control Conference

Torque-Speed Characteristics of Interior-Magnet Machines in Brushless AC and DC Modes, with Particular Reference to Their Flux-Weakening Performance

Y. F. Shi, Z. Q. Zhu, D. Howe

Department of Electronic and Electrical Engineering, University of Sheffield, Mappin Street, Sheffield S1 3JD, UK
Email: Z.Q.Zhu@sheffield.ac.uk

Abstract— The torque-speed characteristics of a 3-phase permanent magnet brushless machine, having an interior permanent magnet rotor and an essentially sinusoidal back-emf waveform, are determined experimentally when the motor is operated in brushless AC (BLAC) mode, with 3-phase, sinusoidal current waveforms, and in brushless DC (BLDC) mode with both 2-phase, 120° conduction and 3-phase, 180° conduction rectangular current waveforms. The performance is compared on the basis of (i) the same peak phase current, (ii) the same torque in the constant torque operating region, and (iii) the same RMS phase current. It is shown that while 2-phase, 120° BLDC operation results in the highest torque capability for the same peak current and BLAC operation results in the highest specific torque per rms current when operating below the base-speed, 3-phase, 180° BLDC operation generally results in the best performance in the flux-weakening region. Hence, although the IPM motor has an essentially sinusoidal back-emf waveform, it may be advantageous to operate it in a hybrid mode, i.e. BLAC in the constant torque region and 3-phase, 180° BLDC in the flux-weakening region.

Keywords- brushless ac, brushless dc, flux weakening, permanent magnet machine.

I. INTRODUCTION

Permanent magnet (PM) brushless motors are now used extensively in applications as varied as servo drives and automotive power-trains, and are generally classified as being either brushless DC, with essentially rectangular phase current waveforms, or brushless AC, with essentially sinusoidal phase current waveforms. Generally, BLDC operation is preferred for motors having trapezoidal back-emf waveforms, with BLAC operation being preferred for motors having sinusoidal back-emf waveforms, since this reduces the torque ripple. The steady-state performance of both BLAC and BLDC machines, in both the constant torque and constant power (flux-weakening) regions, has been studied extensively [1-6]. Vector control is readily applied to BLAC machines to achieve maximum torque per ampere below base-speed

and optimum flux-weakening performance above base-speed [2][5-8]. Similarly, for BLDC machines, maximum torque per ampere and extended speed operation is realized by advancing the commutation angle, in both 2-phase, 120° (BLDC-120) and 3-phase, 180° (BLDC-180) conduction modes [3][4]. However, the investigations in [3][4] are restricted to surface-mounted magnet, trapezoidal back-emf machines. A comparison of the relative merits of employing the forgoing modes of operation to a PM brushless machine having an interior permanent magnet rotor and a sinusoidal back-emf waveform, particularly in the flux-weakening region, have not been reported in the literature. Thus, in this paper, the torque-speed characteristics of a machine having an interior permanent magnet (IPM) rotor and an essentially sinusoidal back-emf waveform are determined experimentally when it is operated as a BLAC, BLDC-120, and BLDC-180 motor in the constant torque and flux-weakening modes, on the basis of (i) the same peak phase current, (ii) the same torque in the constant torque operating region, and (iii) the same rms phase current.

II. BRUSHLESS AC AND DC OPERATION

It is well known [5][6] that IPM brushless machines offer potential advantages over other permanent magnet brushless machine topologies, in terms of exhibiting a saliency torque, having a higher demagnetization withstand capability, and facilitating extended speed operation. The IPM motor on which the three different operational modes have been evaluated has a sinusoidal back-emf waveform, Fig. 1. The parameters which are given in Table I and its base-speed is around 1750rpm.

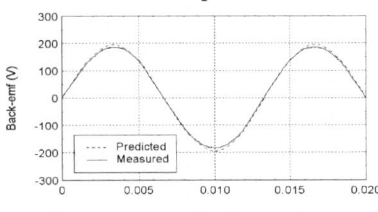

Figure 1. Finite element predicted and measured line back-emf waveform, 1500rpm.

1-4244-0448-7/06/$25.00 ©2006 IEEE

TABLE I. PARAMETERS OF IPM BRUSHLESS MACHINE

DC-link voltage (V_{dc})	285 V
Rated torque (T_e)	4.0 Nm
Current (peak) ($I_{a\max}$)	4.0 A
Phase resistance (R)	8.5 Ω
d-axis inductance (L_d)	$31.2 - 0.7I_d$ mH
q-axis inductance (L_q)	$55 - 3.7(I_q - 1.5)$ mH
Number of pole-pairs (p)	3
Flux-linkage ($\psi_m = E/\omega_e$)	0.227 W_b

(a) Brushless AC operation, BLAC

In the constant torque mode, optimal d-q axis current profiles for maximum utilization of the saliency torque are imposed for maximum torque per ampere, whilst in the constant power mode optimal current profiles are imposed according to the supply voltage and current constraints [6].

The torque-speed characteristics which result both with and without flux-weakening control are shown in Fig. 2, together with the optimal phase advance angle β_0. Figs. 3 and 4 show typical phase current waveforms when the machine is running below and above base-speed, both with and without flux-weakening control. As can be seen, below base-speed, a fixed phase advance angle (13°) is employed to utilize the saliency torque (this is also applied above base-speed even when 'without flux-weakening', Figs. 2 and 4(a)). As expected, above base-speed, the current and torque performance is significantly improved by employing flux-weakening control.

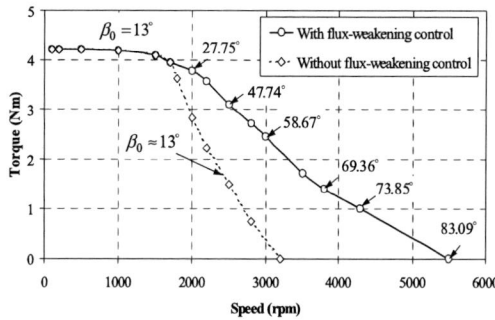

Figure 2. Measured torque-speed characteristics for BLAC operation.

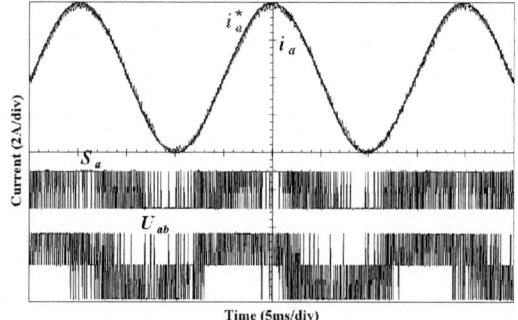

Figure 3. Phase current waveforms for BLAC operation below base-speed, 1000rpm. S_a :gate drive signal; U_{ab} : line voltage; i_a^* : demand Phase A reference current; i_a : actual Phase A current.

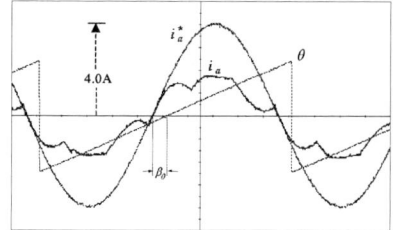

(a) Without flux-weakening

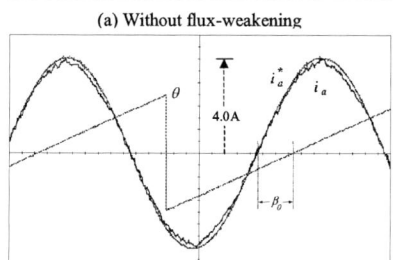

(b) With optimal flux-weakening

Figure 4. Phase current waveforms for BLAC operation above base-speed, 2500rpm. i_a^* : demand Phase A reference current; i_a : actual Phase A current.

(b) Brushless DC operation with 2-phase, 120° conduction, BLDC-120

Fig. 5 shows typical measured phase current waveforms for different commutation advance angles θ_c when running below and above base-speed, the maximum current being limited to 4A. As shown in Fig. 6, θ_c =0° corresponds to the instant at which the phase winding is switched on lagging 30° behind the zero-crossing of the back-emf waveform (θ=0). Thus, θ_c =30° corresponds to the commutation coinciding with the zero-crossing of the back-emf. In general, at low speed, Fig. 5(a), due to 2-phase conducting operation mode, the phase current is discontinuous. However, during the commutation period, all three phases conduct due to the circulating current via the freewheel diode. The phase current waveform may become continuous at high speed, particularly under flux-weakening control, Fig. 5(b). Hence, unlike in BLAC operation mode, the angle, β_0 , by which the zero-crossing of phase current waveform leads the zero-crossing of the back-emf waveform, may be different from the commutation advance angle, θ_c , as illustrated in Fig. 6. The optimal commutation advance angle for maximum torque is speed dependent as shown on Fig. 7, and can be obtained experimentally (or by simulation).

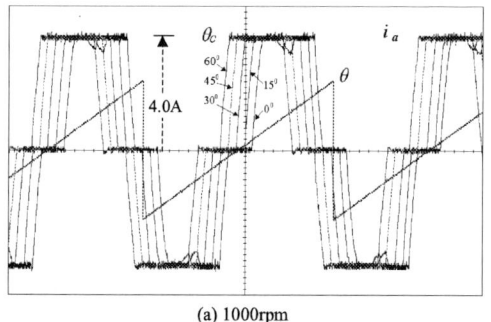

(a) 1000rpm

1842

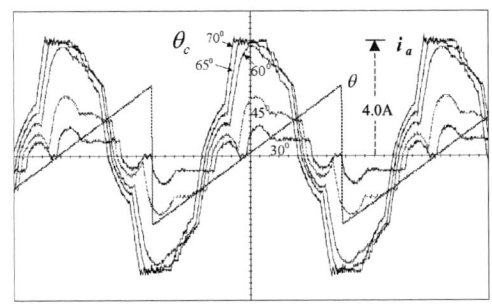

(b) 2500rpm

Figure 5. Measured phase current waveforms at different commutation advance angles for BLDC-120 operation.

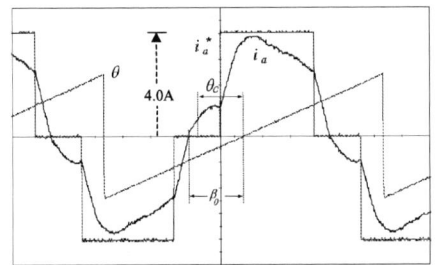

Figure 6. Demand and actual phase current waveforms and rotor position. BLDC-120, 2500rpm, θ_c =60°, $\beta_0 \approx 72°$.

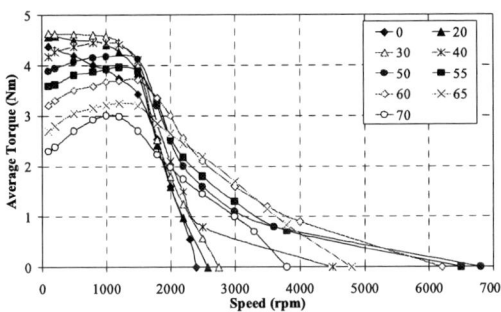

(a) Influence of commutation advance angle θ_c

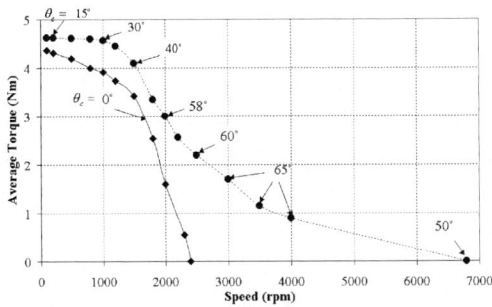

(b) Optimal commutation advance angle and maximum torque-speed

Figure 7. Measured torque-speed characteristics for BLDC-120 operation.

(c) Brushless DC operation with 3-phase, 180° conduction, BLDC-180

In this case, three phases are always conducting. θ_c =0° now corresponds to a phase winding being excited at the zero-crossing of its back-emf waveform (θ=0). Fig. 8 shows measured phase current waveforms for different commutation advance angles, again when the motor is

running below and above base-speed and with the maximum current limited to 4A. Similar to the case of BLDC-120 operation mode, the angle, β_0, by which the zero-crossing of phase current waveform leads the zero-crossing of the back-emf waveform, may be different from the commutation advance angle, θ_c, as illustrated in Fig. 9. The optimal commutation advance angle and the resulting maximum torque per ampere-speed characteristic is shown in Fig. 10.

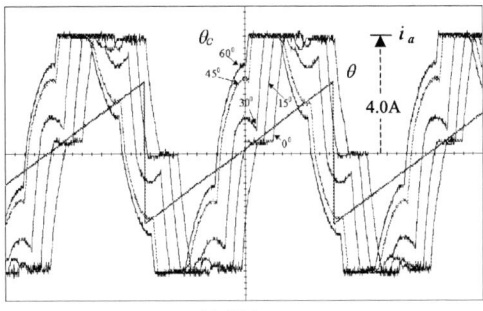

(a) 1000rpm

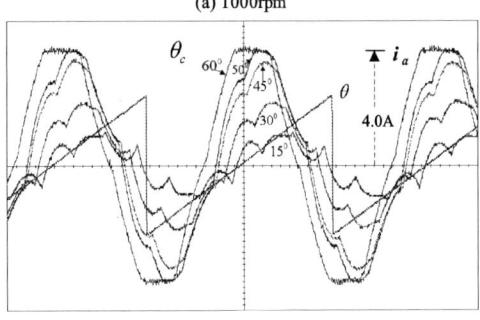

(b) 2500rpm

Figure 8. Measured phase current waveforms at different commutation advance angles for BLDC-180 operation.

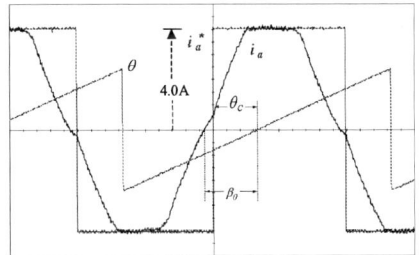

Figure 9. Demand and actual phase current waveforms and rotor position. BLDC-180, 2500rpm, θ_c =60°, $\beta_0 \approx 75°$.

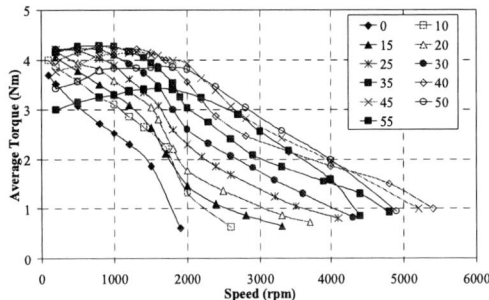

(a) Influence of commutation advance angle θ_c

1843

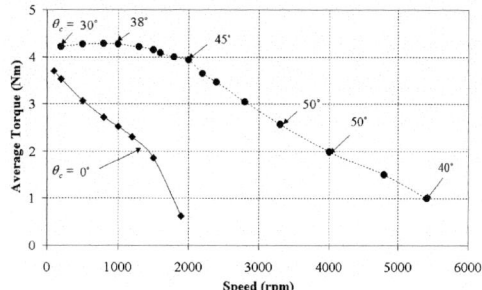

(b) Optimal commutation advance angle and maximum torque-speed

Figure 10. Measured torque-speed characteristics for BLDC-180 operation.

As will be evident, in both BLDC operating modes the phase current waveforms are far from being rectangular or sinusoidal, only the fundamental component contributing to the average torque and harmonics resulting in torque ripple, since the back-emf waveform is sinusoidal. However, since the zero-crossing of the current waveform varies with speed, the phase angle β_0, which is defined similarly as for a BLAC drive, cannot be controlled directly, since, for a given θ_c, β_0 varies with both the speed and the torque. Further, while the optimal value of θ_c increases with the speed up to ~4000rpm, above this speed the optimal value of θ_c reduces due to the varying phase angle β_0 between the current and the back-emf, as will be evident from Figs. 5 and 8, in which it can be seen that at high speed with optimal flux-weakening control the phase current waveforms in both BLDC operating modes become more sinusoidal since the high order current harmonics are suppressed by their higher reactance as the speed increases. In addition, at high rotational speeds, over-currents are observed in the phase current waveforms, as can be seen in Figs. 11 and 12. These are uncontrollable since they are due to circulating currents which flow via the freewheel diodes during commutation.

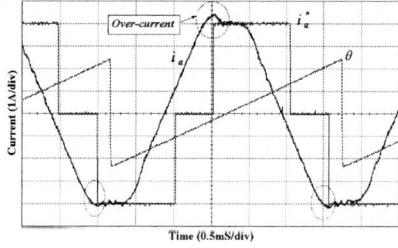

Figure 11. Phase currents for BLDC-120 operation, 6500rpm, θ_c =50°.

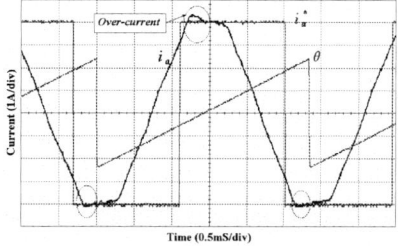

Figure 12. Phase currents for BLDC-180 operation, 7050rpm, θ_c =40°.

III. PERFORMANCE COMPARISON

The torque-speed characteristics which result when the motor is operated in BLAC, BLDC-120 and BLDC-180 modes are compared on the basis of (i) the same peak phase current (Case #1), (ii) the same torque in the constant torque region (Case #2), and (iii) the same rms phase current (Case #3).

Fig. 13 shows the torque-speed characteristics for BLAC and BLDC-120 operation. As will be seen from Fig. 13(a), BLAC operation results in superior flux-weakening performance, while the maximum torque is higher with BLDC-120 operation for the same peak phase current (Case #1), since the fundamental current is the highest. However, if the performance is judged on the basis of the specific torque per rms current, BLAC operation is superior throughout the operating speed range.

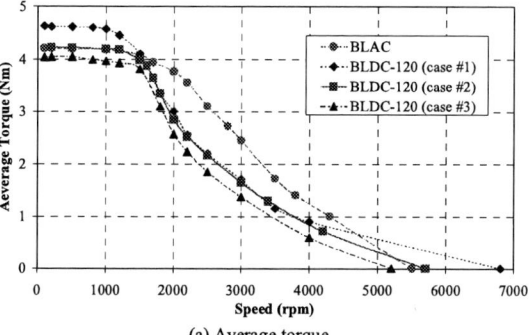

(a) Average torque

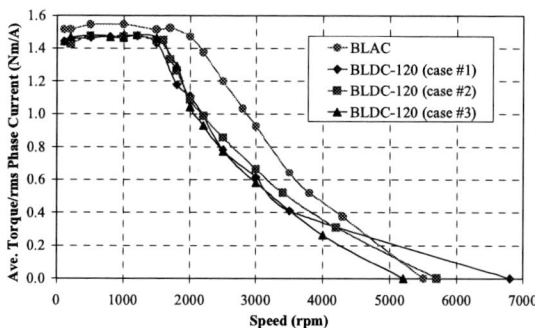

(b) Average torque / measured RMS current

Figure 13. Torque-speed characteristics for BLAC and BLDC-120 operation.

Fig. 14 compares the torque-speed characteristics which result with BLAC, BLDC-120 and BLDC-180 operation and the same peak phase current (Case #1). As will be seen from Fig. 14(a), due to its high fundamental current, BLDC-120 operation again results in the highest torque capability below base-speed, while BLDC-180 operation results in the highest torque in the flux-weakening region since its DC link voltage utilization is the highest. However, when the performance is judged on the basis of the specific torque per rms current, BLAC operation is superior below base-speed, while BLDC-180 operation remains superior in the flux-weakening region.

1844

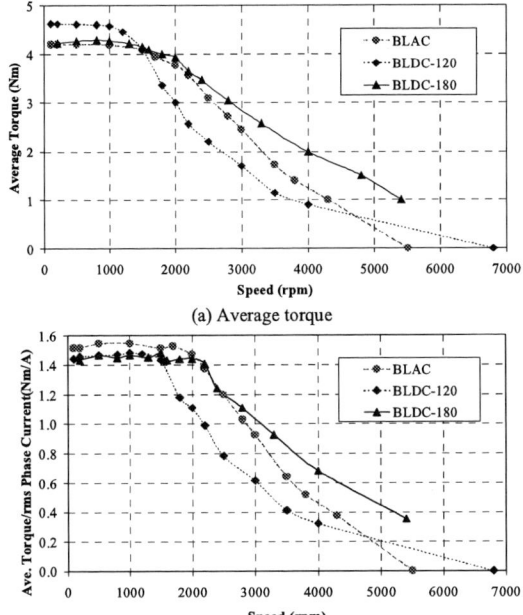

(a) Average torque

(b) Average torque / measured RMS current

Figure 14. Torque-speed characteristics for brushless BLAC and BLDC operation (Case #1).

Fig. 15 shows representative phase current waveforms when the motor is operated in all three modes, in both the constant torque and flux-weakening regions, with the phase currents constrained to have the same peak value (Case#1). As will be seen, in the BLDC-120 and BLDC-180 operating modes the current waveforms become more sinusoidal as the speed is increased since the high order current harmonics are suppressed by their higher reactance, as mentioned before, while in the flux-weakening region the fundamental supply voltage and current component is highest in the BLDC-180 operating mode due to the six-step voltage operation. Consequently, it results in the highest torque per rms current capability.

IV. CONCLUSIONS

The torque-speed characteristics of an IPM brushless machine when operated as a brushless AC motor and a brushless DC motor, with both 2-phase, 120° conduction and 3-phase, 180° conduction, has been investigated. It has been shown that (a) BLDC-120 operation yields the highest torque capability for the same peak phase current below base speed; (b) BLAC operation yields the highest specific torque per rms current below base-speed; (c) BLDC-120 operation exhibits the poorest flux-weakening performance; (d) BLDC-180 operation generally results in the best performance in the flux-weakening region. Hence, even for IPM motors which have essentially sinusoidal back-emf waveforms, it may be advantageous to employ BLDC-180 operation in the flux-weakening region, e.g. to employ a hybrid operation mode – BLAC operation in the constant torque region and BLDC-180

operation in the flux-weakening region, for which a transition method proposed in [7][8] may be employed. However, other aspects of performance, in terms of efficiency and torque ripple, should also be considered. The performance of a PM brushless motor having surface-mounted magnets and a trapezoidal back-emf waveform, when operated in both BLDC and BLAC modes, is described in a companion paper [9].

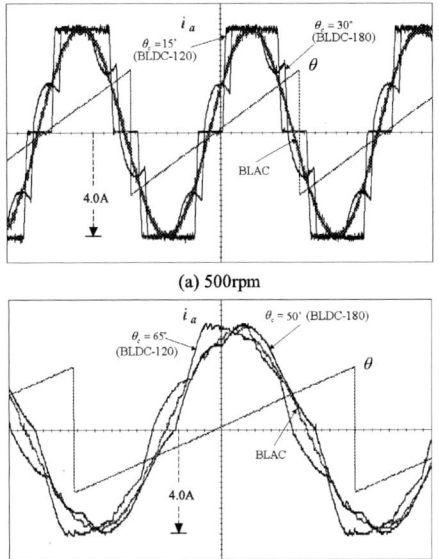

(a) 500rpm

(b) 3000rpm

Figure 15. Phase current waveforms for BLAC and BLDC operation (Case #1).

REFERENCES

[1] P. Pillay and R. Krishnan, "Application characteristics of permanent magnet synchronous and brushless dc motors for servo drives," *IEEE Trans. Ind. Appl.*, vol. 27, no. 5, pp. 986-996, 1991.

[2] T. M. Jahns, "Motion control with permanent-magnet ac machines," *IEEE Proc.*, vol. 82, no. 8, pp. 1241-1252, 1994.

[3] S. K. Safi, P. P. Acarnley and A. G. Jack, "Analysis and simulation of the high-speed torque performance of brushless dc motor drives," *IEE Proc.-Electr. Power Appl.*, vol. 142, no. 3, pp. 919-200, 1995.

[4] T. M. Jahns, "Torque production in permanent-magnet synchronous motor drives with rectangular current excitation," *IEEE Trans. Ind. Appl.*, vol. 20, no. 4, pp. 803-813, 1984.

[5] T. M. Jahns, "Flux-weakening regime operation of an interior permanent-magnet synchronous motor drive," *IEEE Trans. Ind. Appl.*, vol. 23, no. 4, pp. 681-689, 1987.

[6] S. Morimoto, M. Sanada and Y. Takeda, "Wide-speed operation of interior permanent magnet synchronous motors with high-performance current regulator," *IEEE Trans. Ind. Appl.*, vol. 30, no. 4, pp.920-926, 1994.

[7] B.K. Bose, "A microcomputer-based control and simulation of an advanced IPM synchronous machine drive system for electric vehicle propulsion," *IEEE Trans. Ind. Elect.*, vol.35, no.4, 1988, pp.547-559.

[8] B.K. Bose, "A high-performance inverter-fed drive system of an interior permanent magnet synchronous machine," *IEEE Trans. Ind. Appl.*, vol.24, no.6, 1988, pp.987-997.

[9] Z.Q. Zhu, J.X. Shen, D. Howe, "Flux-weakening characteristics of trapezoidal back-emf machines in brushless DC and AC modes", *Proc. Int. Power Electronics and Motion Control Conf., IPEMC 2006*, 13-16 August, 2006, Shanghai, China.

H∞ Robust Control for Dual Linear Motors Servo System

Zhao Ximei, Guo Qingding
Shenyang University of Technology, Shenyang, P.R.China
E-mail:zhaoximei79@yahoo.com.cn

Abstract—The gantry-moving milling machining tools that adapts dual linear motors drive is a high precision and fast motion system, the factors that influence the synchronous performances of the system are not only unknown bounded disturbances but also have parameters variety, unmodelled dynamic and the nonlinear uncertainty. These factors will make the performance of the servo system worse, so they must be considered in the servo drive system of high precision and tiny-feed numerical control machine, thus the satisfying control effect can be achieved. It is difficult to meet the requirement of the synchronous performance only by the position loop; the dynamic synchronous control in the velocity inner loop should be implemented. The problem of H∞ control for synchronous drive of dual linear servo motors used in gantry-moving milling machining center is researched in this paper. A synchronous drive controller is designed based on H∞ control theory. When loads or disturbances are changed suddenly, the H∞ synchronous controller can regulate rapidly so as to ensure the synchronous precision of the dual linear motors servo system. The H∞ feedback controller guarantees synchronous precision of the synchronous linear servo system and eliminate the inferences that are the unknown and bounded disturbance, parameter uncertainties and unmodelled dynamics. The system will automatically tune the parameters of the controller so that the influences can be suppressed effectively. The simulation results show that the proposed scheme is reasonable and effective.

Keywords-gantry-moving machining tool; linear motor; H∞ robust control

I. INTRODUCTION

Since middle of 1990s, linear drive technology has extensive application in precise positioning system. Especially, it request to implementation high velocity and super velocity, precision and super precision, it more requires support of linear servo drive technology with high speed response. Because it can eliminate a series of effects of traditional mechanically-driven chains, so it can increase high speed response competence and motion accuracy of the feed system, linear drive becomes the one most representative advanced technology of the new generation numerical control machine.

With the development of the industry, the higher requirements of the mechanical property and product quality are presented; it isn't feasible to control only one motor at some occasions. It requests to coordinately control more one motor. Synchronous drive control is a common one in the coordinate control, it requests several motors run at same setting value, ensure velocity, angle of rotation or position with consistency.

For the control problem to the high-accuracy single linear servo system, there are many research results on all kinds of control strategies as in [1~6]. But it is still a problem for the high precision synchronous drive feed technology in heavy-duty machine tools. At present, the most representative products use the mode with two same rotation motors plus ball and screws. In this paper, dual linear servo motors with so call "zero drive" mode are used as high speed, high-accuracy feed drive units in gantry-moving milling machine tools. And the position synchronous drive of the two linear servo motors is researched.

In synchronous drive, if the compensation control is only in the speed external loop, the synchronous performance of the servo system will become worse because of the inertia of the controlled plant when load changes constantly. In the paper, a synchronous feedback controller is designed based on H∞ control theory with regard the dual linear motors as one whole controlled plant. Because the compensating control is added to the current loop, when loads or disturbances are changed suddenly, the H∞ synchronous controller can regulate rapidly so as to ensure the synchronous precision of the dual linear motors servo system.

II. THE MODEL OF PMLSM AND THE CONTROLLER STRUCTURE

The synchronous drive system use two permanent magnet linear synchronous motors (PMLSM) as the actuators PMLSM is a thrust device that can exchange the electric energy into linear motion directly. It's equation of motion is:

$$F_d = K_f i_q = F_L + F_d + Dv + M\frac{dv}{dt} \qquad (1)$$

$$s = \int vdt \qquad (2)$$

Where, M is the mass of mover; D is viscous friction coefficient; i_q is q-axis current; F_L is the resistance of load; F_e is electromagnetism thrust; F_d is the resistance produced by end-effect of the PMLSM; K_f is thrust coefficient; s is displacement of mover; v is speed of mover.

From equation (1) and (2), we can obtain the transfer function model of the PMLSM as following

$$P(s) = \frac{1}{Ms + D} \qquad (3)$$

The block diagram of the position synchronous servo system is shown in Fig. 1. ω_1 and ω_2 are external disturbances of the two linear motors respectively.

The linear motor has one's own independent controller so as to ensure the position following performance of each axis. While the H∞ feedback controller is regarded as the synchronous controller used to suppress various kinds of perturbations so as to ensure the synchronous precision of the dual linear motors drive system. Therefore, the whole controllers include two major parts, IP position controller and H∞ feedback controller.

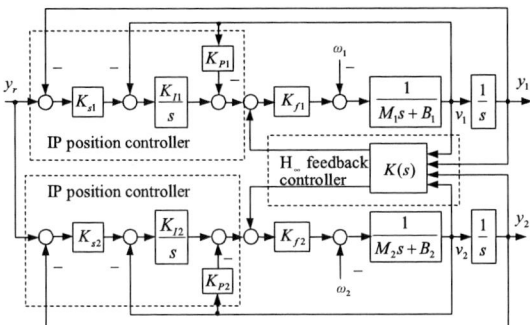

Fig.1. Block diagram of the position synchronous drive system

III. Design Of The Synchronous Controller

The high precision position servo system is required to follow the position command rapidly, no stable state error and have strong robustness to the disturbances and the parameters variety. Therefore, the conventional PI controller is hard to meet the request of following performance and robustness together. Here, two IP controllers are used as the position controllers of each sub-system to ensure the performance of the close loop systems, because it has quicker response ability to the reference signals and stronger resistance ability to the load disturbances while choosing higher integral gain. In Fig.1, the position controller adopts the proportional component, the speed controller adopts IP component, which is called IP position controller. Parameters of the two IP position controllers are designed respectively by the normal model of the controlled sub-system as in [4]. H∞ feedback controller is designed to compensate the influence of various kinds of disturbances so as to ensure the synchronous precision. To translate the H∞ feedback controller design issue to H∞ standard design problem, constituting a generalized plant by Fig.1. The block diagram of generalized system is described in Fig.2.

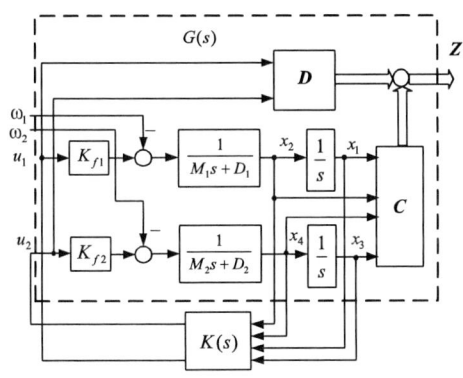

Fig.2. Block diagram of the general system.

Where, u_1 and u_2 are the output control signals of the H∞ feedback controller; ω_1 and ω_2 are the external disturbances; Z is valued signals, which has independent weights of the state variables and the control variables; $r_1 > 0$ and $r_2 > 0$ are used to limit oversize drive signal; the coefficient of the weights $q_1 > 0$ and $q_2 > 0$ are used to adjust permitted synchronous error; $G(s)$ is generalized controlled plant, including the actual controlled plant and the assumed weight function to describe the design index.

According to the state variables defined in Fig.2 and based on LPMSM motion equation, the state space expression of the generalized controlled plant is as follows.

$$\begin{cases} \dot{x} = Ax + B_1\omega + B_2u \\ Z = Cx + Du \end{cases} \tag{4}$$

Where,

$$x = \begin{bmatrix} x_1 & x_2 & x_3 & x_4 \end{bmatrix}^T ; \omega = \begin{bmatrix} \omega_1 & \omega_2 \end{bmatrix}^T ; u = \begin{bmatrix} u_1 & u_2 \end{bmatrix}^T ;$$

$$A = \begin{bmatrix} 0 & 1 & 0 & 0 \\ 0 & -\dfrac{D_1}{M_1} & 0 & 0 \\ 0 & 0 & 0 & 1 \\ 0 & 0 & 0 & -\dfrac{D_2}{M_2} \end{bmatrix} ; C = \begin{bmatrix} q_1 & 0 & -q_1 & 0 \\ 0 & q_2 & 0 & -q_2 \\ 0 & 0 & 0 & 0 \\ 0 & 0 & 0 & 0 \end{bmatrix}$$

;

$$B_1 = \begin{bmatrix} 0 & 0 \\ -\dfrac{1}{M_1} & 0 \\ 0 & 0 \\ 0 & -\dfrac{1}{M_2} \end{bmatrix} ; B_2 = \begin{bmatrix} 0 & 0 \\ \dfrac{K_{f1}}{M_1} & 0 \\ 0 & 0 \\ 0 & \dfrac{K_{f2}}{M_2} \end{bmatrix} D = \begin{bmatrix} 0 & 0 \\ 0 & 0 \\ r_1 & 0 \\ 0 & r_2 \end{bmatrix}$$

Then the above-mentioned problem can be translated to H∞ standard design problem as follows.

Assume the state space of the general controlled plant is:

$$\begin{cases} \dot{x} = Ax + B_1\omega + B_2u \\ Z = C_1x + D_{12}u \end{cases} \tag{5}$$

And the column of the D_{12} is full rank, $(A \quad B_2)$ is stable, then the transfer function of the general controlled plant is

$$G(s) = \begin{bmatrix} A & B & B_1 & B_2 \\ \vdots & \ddots & \vdots & \vdots \\ C_1 & \cdots & 0 & D_{12} \\ I & \cdots & 0 & 0 \end{bmatrix} \tag{6}$$

For the controlled plant (5), the required state feedback controller is

$$u = Kx \tag{7}$$

Such that the close-loop system (6) and (7) are inner stable, and

$$\|T_{z\omega}(s)\|_\infty < 1 \tag{8}$$

For the above-mentioned system, the following theorem exists.

Theorem 1: For the given controlled plant (7), the necessary and sufficient condition which the state feedback matrix K is existed such that the close-loop system (6) and (7) are inner and meet equation (8) is that existing positive defined matrix X meets the Riccati inequation as follows as in [7].

$$A^T X + XA + XB_1 B_1^T X + C_1^T C_1 - \qquad (9)$$
$$(XB_2 + C_1^T D_{12})(D_{12}^T D_{12})^{-1} \times (B_2^T X + D_{12}^T C_1) < 0$$

If the above inequation has positive defined solution X, then to ensure that the close-loop system is stable and Equ.(8) holds, the state feedback matrix is

$$K = -(D_{12}^T D_{12})^{-1}(B_2^T X + D_{12}^T C_1)$$

(10)

We use MATLAB software package to solve (9) and (10), so the required controller K is obtained.

IV. THE SIMULATION RESULTS

Regarding two linear motors with the same parameters:

$$K_f = 25 \text{ N/A}; \ P(s) = \frac{1}{10s + 1.2}.$$

The parameters of IP position controller are:

$$K_s = 11.34; \ K_I = 809.2; \ K_s = 25.62.$$

When $q_1 = q_2 = 5000$ and $r_1 = r_2 = 50$, the obtained H_∞ feedback controller K is

$$K = \begin{bmatrix} -7072.5 & -88.5 & 7072.5 & 88.5 \\ 7072.5 & 88.5 & -7072.5 & -88.5 \end{bmatrix}$$

When $t = 1$s, axis a is added by 100N step disturbance suddenly, When $t = 1.5$s, axis b is added by 50N step disturbance suddenly, the position output responses of two axis to the position command of the step commands are shown in Fig.3. The synchronous error curve of the synchronous servo system is shown in Fig.4. It is obvious that the scheme presented in the paper has good tracking performance and synchronous performance.

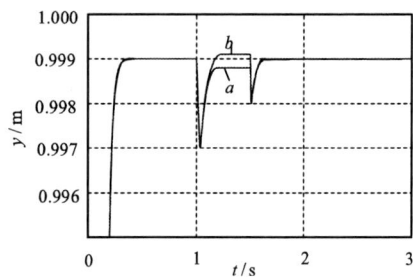

Fig.3. positions output responses of the dual axes

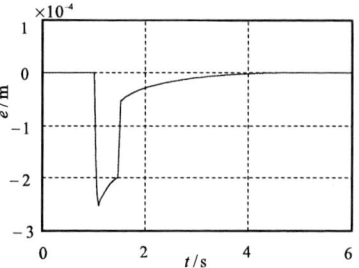

Fig.4. position synchronous error curve

V. CONCLUSION

The problem of H_∞ control for synchronous drive with dual linear motors is researched in this paper. The position synchronous drive controller based on the H_∞ control theory is presented with regarded the dual linear motors as a whole plant. And the H_∞ feedback controller is combined with IP position controller so as to can meet the position tracking performance and ensure the synchronous performance of the system. The simulation results have proved that the presented scheme is effective.

REFERENCES

[1] Q.D. Guo, C.Y. Wang, *Precision control technology for linear AC servo system*, Beijing: Machine Industry Press, 2000.

[2] J.G. Fu, Q.D. Guo, G.P. Tang. "H_∞ robust performance design of position controller for linear permanent magnet synchronous servo motor." Trans. of China Electro-technology Society, Vol.16, No.3, pp.16-19, 2001

[3] Q.D. Guo, L.C. Chi, C.Y. Wang. "H_∞ control for linear permanent magnet synchronous AC servo system with a command-compensator based on artificial neural network". Control and Decision, Vol.15, No.1, pp75-78, 2000

[4] Q.D. Guo, Y. Zhou, W. Guo. "Robust position controller design for high-precision permanent magnet linear synchronous motor." Electric Machines and Control, Vol.2, No.4, pp:208-212, 1998

[5] M.N. Lee, J.H. Moon, K.B. Jin, et al. "Robust H∞ control with multiple constraints for the track-following system of an optical disk drive." IEEE Trans. on Ind. Electric, Vol.45, No.8, pp:638-645, 1998

[6] Li Yi, Masayoshi Tomizuka. "Two-degree-of-freedom control for hard disk servo systems." IEEE Trans on Mechatronics, Vol.4, No.3, pp:17-24, 1999

[7] T.L. Shen. *H_∞ control theory and applications*. Beijing: Tinghua University Press, 1996

Research on Linear Motor Driving System Based on Wavelet Transform

Cui Jiefan, Zhao Lijun, Wang Hemin, Wan Junzhu and Jiang Lili
Shenyang University of Technology
College of Electrical Engineering, P.R. China 110023
E-mail: zhlj1163@163.com

Abstract—End-effect is a main reason of influence on the characteristic of linear motor driving system, its direct impact is nonstable and undulate edges magnetic field. The wavelet transform is applied to analyze the linear motor driving system performance in this paper. The improving thrust response system has been presented based on choice wavelet function and wavelet transform. Simulation results indicate that the proposed control strategy can abate the thrust ripple problem caused by end effect in linear motor control system, and make the system have good performances.

Keywords-Wavelet transform; Linear motor; End effect ; Direct thrust control system

I. INTRODUCTION

France physicist Morlet first applied the wavelet to analyzing the partial characteristic of earthquake wave in 1984. Because wavelet transform is a kind of analysis method for time-scale (frequency) signal, in time-domain and frequency-domain, which has the ability of exploring signal features with partial characteristic. Therefore, in the last years, the special analysis method has made itself theories rapid development and extensive application, especially in the aspects of signal analyzing and image processing areas.

Linear motor has longitudinal edge and transverse edge, for this reason the linear motor exist special end-effect. Longitudinal end-effect is caused by finite length primary iron-core. Transverse end-effect is caused by finite width of primary and secondary, secondary current and secondary plates affecting the air-gap magnetic field. This is the main difference between the linear motor and the rotating machine. Longitudinal end-effect not only causes motor losses, lower electric efficiency and thrust, but also leads to motor work characteristic aggravation. Therefore, key factor is to analyze the longitudinal end effect in this paper.

Traditional analysis method, Fourier transform, has localization contradiction of time-domain and frequency-domain, some messages while analyzing nonstable signal will be usually lost. Therefore, it is necessary to research

a new method that can solve this problem reasonably and effectively so as to improve linear motor driving system performance.

Single dimension continuous wavelet transform has higher sensitivity, stronger ability of denoising and lower demand of the input signal, and doesn't need objects mathematic model. The single dimension continuous wavelet transform is used for analysis the performance of linear motor driving system.

II. SINGLE DIMENSION CONTINUOUS WAVELET TRANSFORM

The continuous wavelet sequence can be described as,

$$\psi_{a,\tau}(t) = \frac{1}{\sqrt{|a|}}\psi\left(\frac{t-\tau}{a}\right) \qquad a, \tau \in \mathrm{R}; a \neq 0$$

(1)

where a is scale parameter，τ is shift parameter. The square-integrable function $\psi(t)$ is called Mother Wavelet. A wavelet sequence can be obtained by dilation and shift transformation of the mother wavelet $\psi(t)$. The continuous wavelet transform of arbitrary function $x(t) \in L^2(R)$ is expressed by (2).

$$WT_x(a,\tau) = \int_{-\infty}^{\infty}\frac{1}{\sqrt{a}}\overline{\psi\left(\frac{t-\tau}{a}\right)}x(t)\mathrm{d}t$$

(2)

The mother wavelet $\psi(t)$ formed wavelet sequence has an observable window function, so $\psi(t)$ should satisfy following constraint condition:

$$\int_{-\infty}^{\infty}|\psi(t)|\mathrm{d}t < \infty$$

(3)

$\hat{\psi}(\omega)$ is a continuous function which must be zero at the initial point for satisfying (2), then,

$$\hat{\psi}(0) = \int_{-\infty}^{\infty}\psi(t)\mathrm{d}t = 0$$

(4)

And its Fourier transform is $\hat{\psi}(\omega)$, when $\hat{\psi}(\omega)$ fulfills the admissible condition:

$$C_{\psi} = \int_{R}\frac{|\hat{\psi}(\omega)|^2}{|\omega|}\mathrm{d}\omega < \infty$$

(5)

---

[a] The project is supported by the Province Education Department(2004D049) and Province Nature Science Foundation (20052040)

It is shown that single dimension continuous wavelet transform uses $\psi(t)$ both scale a dilations and time τ shifts to analyze the signal. The signal is expanded in window area $[\ \tau-\delta,\tau+\delta\ ]*[\ \omega-\varepsilon,\omega+\varepsilon\ ]$, where δ and ε represent the time-span and the frequent-span of window respectively. The time-frequency analysis is multi-resolution if varying the window area. The high frequency signal is suitably analyzed by the gradually exquisite time step, and the low frequent signal is finely analyzed by the exquisite frequency step. The windows of the time and frequency are adjusted through changing the signal frequency. Time-frequency localization analysis of signal can be achieved.

III. CORRECT CHOICE WAVELET FUNCTION

Three wavelet functions in Fig. 1 are respectively selected as mother wavelets, and single dimension continuance wavelet transform is adopted to analyze the magnetic field in Fig.2. Among three mother wavelets, the harmonics of high degree are contained in Haar wavelet, the next is Coiflet 2, and the last one is Daubechies 4.

Haar wavelet function is defined as following:

$$\psi(t)=\begin{cases}+1 & 0\leq x<1/2 \\ -1 & 1/2\leq x<1\end{cases}$$

(7)

Scale function:

$$\phi(x)=\begin{cases}1 & 0\leq x<1 \\ 0 & x<0\ or\ x\geq1\end{cases}$$

(8)

Daubechies, who is a famous wavelet analysis scholar in the word, constructs Daubechies wavelet function, which provided an effective analysis method comparison with Haar wavelet function. Daubechies wavelet function has the basic wavelet named dbN, in which N is order number (N=1, 2, ..., 10). The dbN function is standard orthogonal wavelet, which don't have definite expression besides Haar.

The scholar Daubechies also constructs Coiflet wavelet function, which has a series of basic wavelet coifN (N=1,2,3,4,5).

The traditional analysis of signal is the Fourier transform[3]. Although the method can separately be applied in time-domain or frequency-domain, there is limitation in analysis of the nonstable signal such as the magnetic field in Fig. 2. It is shown in Fig. 3 the result of Fourier transform analyze.

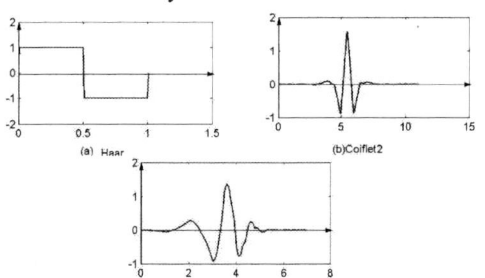

(a) Haar   (b)Coiflet2

(c)Daubechies4

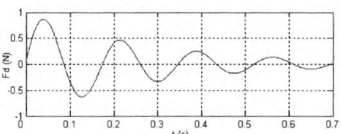

Fig. 1 Selected several mother wavelets

Fig.2 magnetism field fluctuation

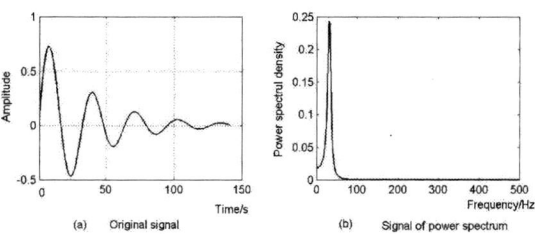

(a)   Original signal   (b)   Signal of power spectrum

Fig. 3 Fourier transform analyze the end efficacy

Any characteristic of frequency-domain doesn't be found according to the original signal. The relation between power spectrum and frequency is constituted after Fourier transform. It is clearly shown in Fig. 3(b) that signal is only made of sine signal under 100Hz. The results don't include messages in time domain.

Simulation results are obtained shown in Fig. 4 while applying the single dimension continuous wavelet transforms to analyze the magnetic field.

The low frequency data are recorded in files (cai.mat, i=1,2,3), high frequency data are recorded in files (cdi.mat, i=1,2,3). So it is helpful to get the high frequency and low frequency decomposition coefficient according to the demand for improving the characteristic of linear motor.

The end effect of the linear motor causes the magnetic field fluctuation, which does not include the harmonic of high degree[4,5]. Different wavelet functions applied in system will obtain different results shown in Fig. 4. The harmonic analysis by Daubechies 4 is different from those by Coiflet 2 and Haar because of lower degree harmonic. In order to analyze the linear motor end effect influence on performances, the frequency spectrum of mother wavelet should be considered. Form Fig. 4, the results of Daubechies 4 harmonic analysis is suitability for analysis end effect influence.

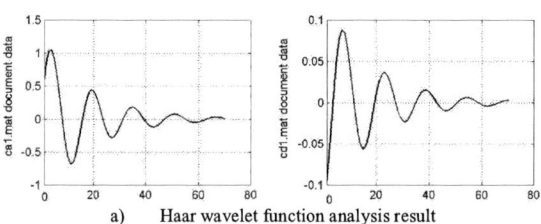

a)   Haar wavelet function analysis result

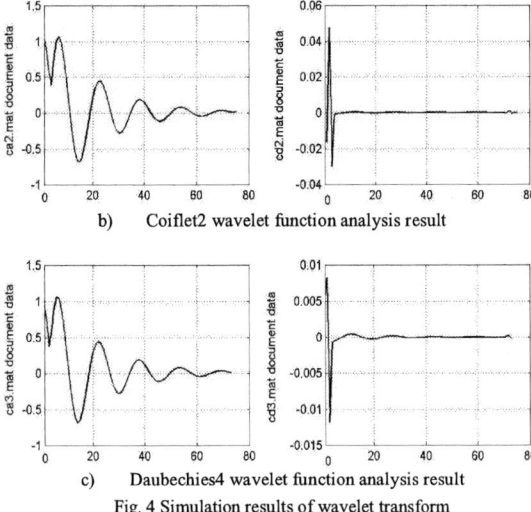

b)   Coiflet2 wavelet function analysis result

c)   Daubechies4 wavelet function analysis result

Fig. 4 Simulation results of wavelet transform

IV.   ANALYSIS END EFFICACY INFLUNCE ON THE DIRECT THRUST CONGTROL SYSTEM

The end-effect of linear motor is a special phenomenon, which is caused by finite primary winding and primary iron core. Its effect is expressed by end efficacy. The end efficacy is not perfect for control system, especial the accurate controlling. Equivalent mathematic model of end efficacy is obtained by analysis and experiment as follow[2]

$$F_d = F_{dm} \cos(\frac{x}{\tau} 2\pi + \theta_0)$$

$$= F_{dm} \cos(\frac{vt}{\tau} 2\pi + \theta_0)$$

(6)

Where F_d is called end efficacy, F_{dm} is called the amplitude of end efficacy, x is called linear motor displacement, θ_0 is called the constant of linear motor, v is velocity.

Fig. 5 shows the simulation of direct thrust control system. In this simulation system adopts the parameter of linear motor as following, the nominal value of primary winding quality M_n =25kg , the nominal value of viscosity friction coefficient B_n =0.2 N·S/m , thrust coefficient k_f =25 N/A , the nominal value of permanent magnet flux linkage ψ_f^* = 0.286Wb , pole span τ_n =36 mm , resistance R =1.2 Ω , d and q axes inductance $L_d = L_q$ =18.74 mH [6]. When t =0.2s add in end efficacy to direct thrust control system.

When t =0.2s adding the end efficacy into direct thrust control system,

$$F_d = 30 \cdot \sin\left(\frac{2\pi}{0.016} x\right) N \tag{9}$$

The simulation of direct thrust control system is shown as Fig. 6, at the same time, observing the thrust influence cased by end efficacy.

When $t = 0.2s$, end efficacy influence is considered in direct thrust control system. Then thrust impulsive is created obviously shown in Fig.7. The system steady state is impacted by the thrust impulsive. Therefore, end effect influence must be compensated in control system, in order to keep control system operating steady.

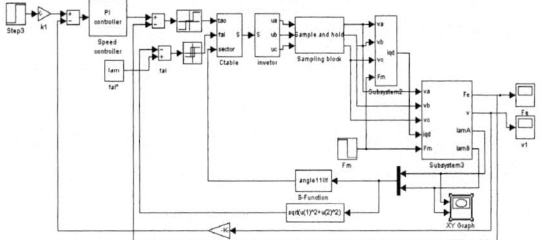

Fig. 5 The simulation scheme of direct thrust control system

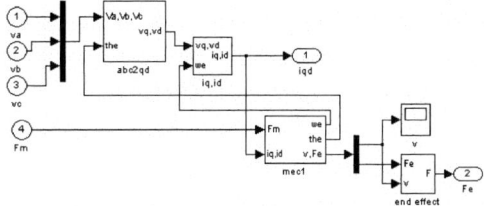

Fig. 6 The simulation considered end efficacy in direct thrust control system

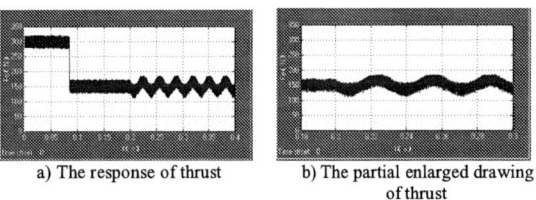

a) The response of thrust   b) The partial enlarged drawing of thrust

Fig. 7 End efficacy influence on the direct thrust control system

Based on direct thrust control system, the control system of end efficacy is created. Fig. 8 shows the simulation scheme of improvement control system, which is based on Daubechies 4.

The decomposition program of wavelet analysis is as following[7],

$$\begin{aligned}
&\{ \text{for } (i = 0 ; \ i <= K ; \ i++) \\
&\quad \{ C_{k+1,j} = 0 ; \ d_{k+1,j} = 0 ; \\
&\quad \text{for } (j = 2i ; \ j <= L + 2i ; \ j++) \\
&\quad \{ C_{k+1,j} = C_{k+1,j} + h_0(j - 2i)C_{k,j} ; \\
&\quad\quad d_{k+1,j} = d_{k+1,j} + h_1(j - 2i)C_{k,j} ; \\
&\quad \} \\
&\quad \} \\
&\}
\end{aligned}$$

1851

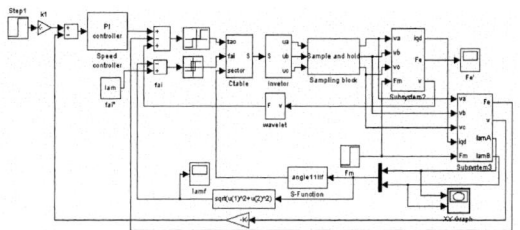

Fig. 8 The improvement control system of linear motor

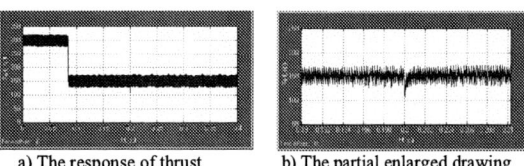

a) The response of thrust   b) The partial enlarged drawing
of thrust

Fig. 9 The improvement thrust response of control system

Comparison Fig.9 with Fig. 7, when $t = 0.2s$, analyzing the end efficacy accurately by Daubechies 4, the thrust fluctuate cased by end efficacy is compensated. Then, the end effect influence on control system thrust response is abated.

V. CONCLUSIONS

Comparing with the Fourier transform analysis method, the wavelet transform technique has the characteristics that it analyzes the signal combining time domain with frequency domain together, for this reason it can effectively solve the problem of time-domain and frequency-domain limitation. It is important to search a new method that can solve the problem in linear motor driving system so as to improve linear motor servo characteristic. Through correct choice wavelet function, it is possible to consider both the frequency spectrum of mother wavelet and the characteristics of original signal. The result of signal analyzed is beneficial for abating the end effect influence on linear motor driving system. Simulation results indicate that the proposed control strategy can reasonably and effectively abate the thrust ripple problem of linear motor control system, and make the control system have good performance.

REFERENCES

[1] Ye. Yunyue, "The principle and application of linear motor", *China Machine Press*, pp.40-48, 2000.

[2] Liu Lili, Xia Jiakuan and Jiang Ping. Study on the end effect and compensation technique of permanent magnetic linear synchronous motor. *Journal of Shenyang University of Technology*. Vol. 27, pp.261-266, 2005.

[3] S. Nornaka, "Simplified Fourier Transform Method of LIM Analyses Based on Space Harmonic Method", *Linear Drives for Industry Application*, pp.187-190, 1998.

[4] Yoshihiko. Mori, "End-effect Analysis of Linear Induction Motor Based on the Wavelet Transform Technique", *IEEE Transactions on magnetics*, 35(5), pp.3739-3741, 1999.

[5] Li. Haodong, "Research on the End-effect and Control of the PM Linear Motor Used in the Electric Discharge Machining", *Shenyang university of technology*, 2003.

[6] Guo. Qingding and Wang. Chengyuan and Zhou. Meiwen and Sun. Yanyu, "Precision control of linear AC serve system", *China Machine Press*, pp.30-37, 2000.

[7] Hu. Changhua, Li. Guohua, and Liu.Tao, "The system analysis and design based on MATLAB6.X- Wavelet transform", *Xi'an university of electron science & technology press*, pp.15-17, 2004.

[8] Liu Lifeng. Research of Direct Thrust Force Control System of Linear Motor Based on DSP. *Shenyang University of Technology*. 2005.

2006 5th International Power Electronics and Motion Control Conference

Study on Rotor Position Detection Error in Sensorless BLDC Motor Drives

Li Qiang, Wang Ruixia

School of Automation, Nanjing University of Science and Technology, Nanjing, 210094, P. R. China
chnliqiang@163.com

Abstract--In this paper, the main reason causing rotor position detection error in sensorless BLDC motor drives that the band pass filters are employed to get the zero-crossings of Back-EMF is presented. The effects on commutation that result from the rotor position detection error are investigated, and the simulation results are given. The correction method is suggested, and the experimental results are provided.

Keywords- BLDCM; Back-EMF ; rotor position detection error; correction; commutation

I. INTRODUCTION

Brushless DC motor is widely used in many fields because of its simple structure, high efficiency and free maintenance and so on. Traditionally, a rotor position sensor is used for detecting the rotor position in BLDC motor drive since it is of simple control and low cost. However these sensors also cause system complexity and low reliability. For example, Hall sensor is not able to operate in the compressor of air-conditioner. It is better to use sensorless controller in the air-conditioner system. In sensorless control of a brushless DC motor drive, a typical method called "Back EMF method" is sampled to detect the EMF zero-crossing of the unexcited winding. However there exists the rotor position detecting error inevitably in the "Back-EMF method" when low-pass or band-pass filters are employed to get zero-crossings. In this paper, the main reason causing rotor position detection error in the zero-crossing detecting Back EMF method in a position sensorless brushless DC motor is analyzed, and the effects on commutation are discussed. The correction method is suggested, and the experimental results are also provided.

II. MAIN REASON OF ROTOR POSITION DETECTION ERROR

The RC low-pass or band-pass filters are used in the back-EMF method to eliminate high frequency harmonic because of PWM in the terminal voltages of the motor. Meanwhile, the low-pass or band-pass filters will cause phase shift of Back EMF wave.

When a three-phase wye-connected BLDC motor operates with six steps mode and only two windings are excited at any time, the winding current is commutated at 30 electrical degrees lagged behind back-EMF zero-crossing. It is known that the interval of two commutation events is 60 electrical degrees according to electrical machine principle. It can be considered that the motor is commutated at the point of $\alpha = 30 + k*60(k = 0,1,2,\cdots)$ electrical degrees lagged behind the back-EMF zero-crossing, and the commutation point is of 90 electrical degrees lagged behind the back-EMF zero-crossing when $k = 1$. The time of the back-EMF zero-crossing is the commutation time theoretically if the phase-shift of the low-pass filters is 90 electrical degrees. The magnitude of the phase-shift through low-pass filters is directly related to the frequency of the Back EMF wave. The frequency is proportional to the speed of the rotor. Therefore, the phase-shift is different in various operation speeds, which results in the rotor position detecting error.

Figure 1 shows the circuit of a passive RC band-pass filter used in position sensorless BLDC motor drives.

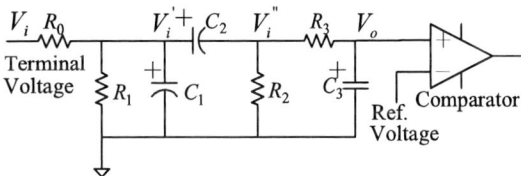

Figure 1. A band-pass filter circuit

V_i is the terminal voltage of the motor and the reference voltage, i.e. equivalent neutral point voltage, is supplied with a wye-connected symmetrical resistance circuit in Figure 1. The voltages and phase-shifts are calculated as follows.

$$V_i' = \frac{R_1}{R_0 + R_1 + j\omega R_0 R_1 C_1} V_i \qquad \alpha_1 = \tan^{-1}(\frac{\omega R_0 R_1 C_1}{R_0 + R_1}) \tag{1}$$

$$V_i'' = \frac{j\omega C_2 R_2}{1 + j\omega R_2 C_2} V_i' \qquad \alpha_2 = -\tan^{-1}(\frac{1}{\omega R_2 C_2}) \tag{2}$$

$$V_o = \frac{1}{1 + j\omega R_3 C_3} V_i'' \qquad \alpha_3 = \tan^{-1}(\omega R_3 C_3) \tag{3}$$

1-4244-0448-7/06/$25.00 ©2006 IEEE   1853

where, $\alpha = 2\pi f$, f is the frequency of the Back-EMF.

The total phase-shift is

$$\alpha = \alpha_1 + \alpha_2 + \alpha_3$$
$$= \tan^{-1}(\frac{2\pi f R_0 R_1 C_1}{R_0 + R_1})$$
$$- \tan^{-1}(\frac{1}{2\pi f R_2 C_2}) + \tan^{-1}(2\pi f R_3 C_3)$$

(4)

When the parameters of resistances and capacitances are fixed in equations (1), (2), (3) and (4), the phase-shift of the signal through low pass filters only relates to the frequency of the Back EMF

$$\alpha = \varphi(f)$$

The phase-shifts are near 90 electrical degrees when the optimized resistance and capacitance values are selected. The phase-shift is 90 electrical degrees on one special frequency under this case. Figure 2 shows the difference between the theoretical and the filtered zero-crossing.

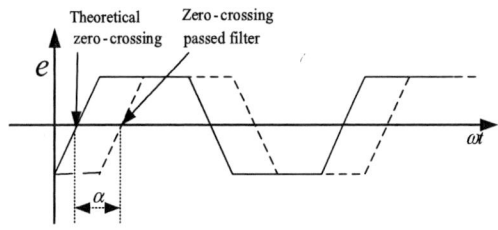

Figure 2.   Differences between actual and filtered zero-crossing

The phase-shift α presents the differences between the theoretical and the filtered zero-crossing. The magnitude of α varies with the frequency, i.e., the speed of the motor. There exists rotor position detecting error if the filtered zero-crossings are used as commutation signals directly. The filtered zero-crossing leads the actual commutation time when α <90 electrical degrees. In the otherwise, the filtered zero-crossing lags behind the actual commutation time.

Figure 3 shows the terminal voltage, the filtered back-EMF and filtered zero-crossing signal when the motor operates at speed of 4000 rpm. The zero-crossing signals are obtained from the filtered terminal voltage and the equivalent neutral point voltage. Comparing with Figure 3(a) and (b), it shows that the filtered zero-crossing lags behind the actual value about 90 electrical degrees.

III.   EFFECT OF ROTOR POSITION DETECTION ERROR ON COMMUTATION

From above, it causes delaying commutation that employs the rotor position information directly, or causes leading commutation that corrects the rotor position

information inappropriately. Delaying commutation or leading commutation causes that the brushless DC motor runs out of phase, and deteriorates the performance.

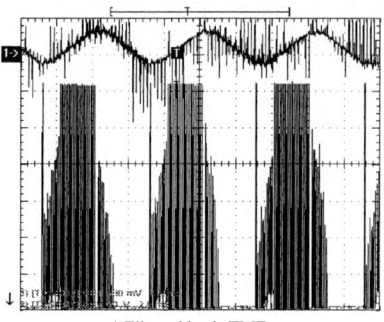

a) Filtered back-EMF

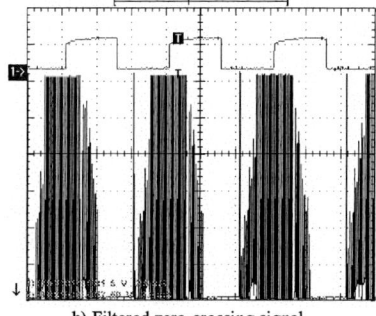

b) Filtered zero-crossing signal

Figure 3.   Filtered back-EMF and zero-crossing signal

A.   Delaying Commutation

The Back EMF of phase-a shown as Figure 4 is given as follows.

$$e_a = \begin{cases} K_e\omega & -\pi/3 < \omega t \le \pi/3 \\ 3K_e\omega - \dfrac{6K_e\omega^2}{\pi}t & \pi/3 < \omega t \le 2\pi/3 \\ -K_e\omega & 2\pi/3 < \omega t \le 4\pi/3 \\ -9K_e\omega + \dfrac{6K_e\omega^2}{\pi}t & 4\pi/3 < \omega t \le 5\pi/3 \end{cases}$$

(5)

where K_e is the EMF constant, and ω is the angular velocity.

The expressions of other phases can be presented similarly and the phase lags $2\pi/3$ electrical radians in turn.

The terminal voltage of phase-a in the period shown in Figure 1 is provided as equation (6). The average voltages are used to substitute for chopped pulses generated by PWM operation in the equation.

1854

$$u_{PWMa} = \begin{cases} \rho V_{DC} & -\pi/3 - \alpha < \omega t \le \pi/3 - \alpha \\ 0 & \pi/3 - \alpha < \omega t \le 2\pi/3 - \alpha \\ -\rho V_{DC} & 2\pi/3 - \alpha < \omega t \le 4\pi/3 - \alpha \\ 0 & 4\pi/3 - \alpha < \omega t \le 5\pi/3 - \alpha \end{cases} \quad (6)$$

where ρ is the duty cycle of PWM wave. V_{DC} is the voltage of DC bus. α is the lagging angle.

Similarly, the terminal voltages of other phases can be given and the phase lags $2\pi/3$ electrical radians in turn.

Supposing that the commutation delays $\pi/3$ electrical radians, i.e., $\alpha = \pi/3$. The phase relation between the Back EMFs and terminal voltages are shown in Figure 1, and the thick lines represent DC supply average voltages of the three phase windings.

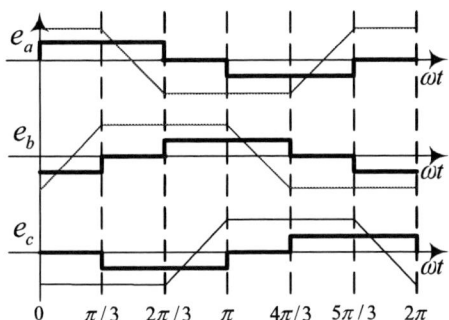

Figure 4. The back EMF and voltage waveforms with commutation delaying $\pi/3$ electrical radians

Referring to Figure 4 and the equations (5), (6), the commutation process around $\omega t = 2\pi/3$ can be analyzed. Before $\omega t = 2\pi/3$, the motor operates in phase period between $\pi/3 \sim 2\pi/3$. Phase-a and phase-c are conducted in this phase period, and the voltage equation can be derived as follows.

$$2ri_{ac} + 2L_\sigma \frac{di_{ac}}{dt}$$
$$= \rho V_{DC} + (4K_e\omega - \frac{6K_e\omega^2}{\pi}t) \quad (\pi/3 < \omega t \le 2\pi/3) \quad (7)$$

When $\omega t = 2\pi/3_-$, the voltage equation can be given as equation (8)

$$2ri_{ac} + 2L_\sigma \frac{di_{ac}}{dt} = \rho V_{DC} \quad (8)$$

During this instantaneous interval, the voltage equation for commutation without delay is given as

$$2ri_{ac} + 2L_\sigma \frac{di_{ac}}{dt} = \rho V_{DC} - 2K_e\omega \quad (9)$$

After $\omega t = 2\pi/3_+$, the motor operates in phase period between $2\pi/3 \sim \pi$. Phase-b and phase-c are conducted in this phase period. The period can be split into two intervals. During the first interval, phase-b and phase-c are conducted and phase-b freewheels through the diode with the final current of the previous phase period as an initial current. During the second interval, the freewheeling

current decays to zero and phase-b and phase-c are conducted only.

It is assumed that the phase current will be decreased to zero when the chopped pulse turns on again. The initial current of freewheeling for the commutation delayed $\pi/3$ electrical radians and without delay can be calculated respectively by equation (10) and (11).

$$i_{ac}(t) = \frac{V_{DC}}{r}(1 - e^{-\frac{r}{L}t}) \quad (10)$$

$$i'_{ac}(t) = \frac{V_{DC} - 2K_e\omega}{r}(1 - e^{-\frac{r}{L}t}) \quad (11)$$

The parameters of the sample BLDC motor are listed as: number of poles $2p$ =4, phase resistance r =0.9Ω, phase inductance L =3.35 mH, K_e =13.2. The carrier frequency of PWM is 5 KHz. The voltage of the DC bus rail is 300 V. The maximum currents occur under ρ =41.7% and n =4200 rpm when chopped pulses turn off.

$$i_{ac}(8.34e-5) = 3.69 \quad (A)$$
$$i'_{ac}(8.34e-5) = 3.01 \quad (A)$$

The phase current is greatly increased because the commutation is delayed $\pi/3$ electrical radians. Meanwhile the initial current of freewheeling is increased correspondingly in the following phase period.

The unexcited phase-a and excited phase-c form a circulation circuit through the lower diode of phase-a and the lower IGBT of phase-c in phase period between $2\pi/3 \sim \pi$. The voltage equation of the circulation circuit is shown as equation (12).

$$2ri_{ac} + 2L_\sigma \frac{di_{ac}}{dt} = e_c - e_a$$
$$= -4K_e\omega + \frac{6K_e\omega^2}{\pi}t \quad (2\pi/3 < \omega t \le \pi) \quad (12)$$

Equation (12) indicates that the circulated current is always exited in the unexcited phase within the phase period. The total current of unexcited phase includes the freewheeling current and the circulating current. In the unexcited phase period, the current is changed from the reverse direction to the forward direction, the circulating current will not be flowed because the circulation circuit can't generate.

Figure 5 shows the simulation voltage and current for commutation delaying $\pi/3$ electrical radians. The result shows that the theoretical analysis is correct.

B. Leading commutation

The Back EMFs, terminal expressions, the voltage and the current equations of three phase windings with the leading commutation are given as alike as that of the delay commutation.

Figure 6 is the Back EMFs and terminal voltages waveforms when commutation leads $\pi/3$ electrical radians, i.e., $\alpha = -\pi/3$ in equation (6). The thick lines represent the DC supply average voltages of three phase windings same as those in Figure 4.

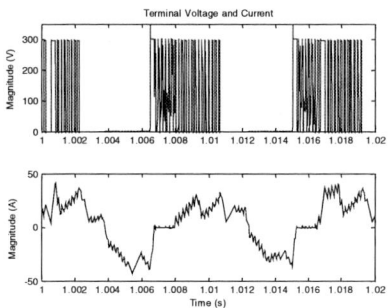

Figure 5.   Simulation result for commutation delaying $\pi / 3$ electrical radians

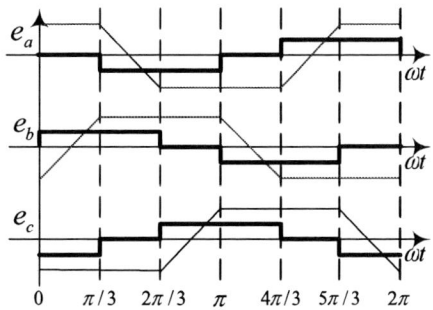

Figure 6.   The back EMFs and terminal voltages with commutation leading $\pi / 3$ electrical radians

Figure 7 shows the simulation result of the terminal voltage and current for commutation leading $\pi / 3$ electrical radians. The current waveform shows that the circulating current occurs in the unexcited phase period when the current is changed from reverse conduction direction to forward conduction direction.

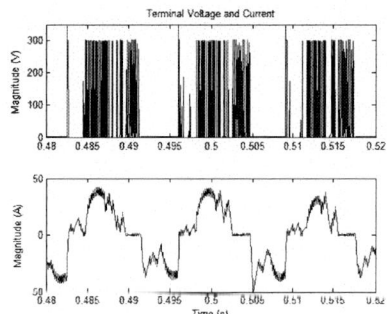

Figure 7.   The simulated terminal voltage and current for commutation leading $\pi / 3$ elec. radians

IV.   CORRECTION METHOD AND EXPERIMENT RESULTS

From above, it shows that the phase-shift of band pass filters causes rotor position detection error in Back EMF zero-crossing of the sensorless BLDC motor drive. The values of phase-shift for the filter circuit in Figure 1 can be calculated according to equation (4). Table I shows the values of phase-shift and corrections under various speeds.

TABLE I.
PHASE-SHIFT AND CORRECTION VALUES

Speed (rpm)	Frequency of B-EMF (Hz)	Shift angle (elec. degrees)	Correction angle (elec. degrees)
1000	33.3	83.50	6.50
2000	66.7	87.00	3.00
3000	100	88.30	1.70
4000	133.3	89.00	1.00
5700	190	89.90	0

In addition, the actual values of phase-shift can be gotten when the motor is driven by another one. The correction values can be given by the theoretical calculation and experimental measurement under various speeds, and these are put in a LUT (look-up table) in the control software. When the BLDC motor operates, the correction values can be searched from the table and used to improve the motor performance.

The experimental terminal voltage and phase current are shown in Figure 8 with no correction, i.e., delaying commutation under about 700 rpm, which is coincide with that showed in Figure 5.

The terminal voltage and phase current are shown in Figure 9 with excessive correction, i.e., leading commutation under about 700 rpm, and the circulating current is similar with that in Figure 6.

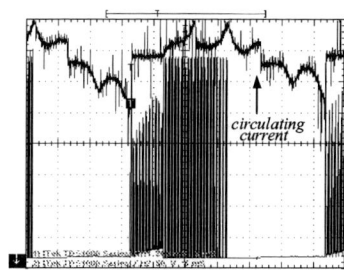

Figure 8.   Experiment result with delaying commutation

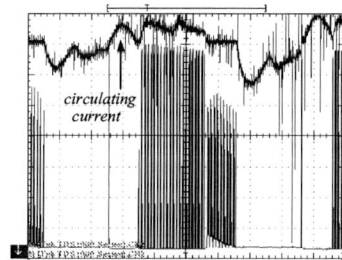

Figure 9.   Experiment result with leading commutation

A terminal voltage and a phase current are shown in Figure 10 with appropriate correction, i.e., normal commutation under about 700 *rpm*. The experiment result indicates that there is no circulating current under this condition, and the phase current is smaller and has little ripple than that under other conditions. The torque ripple will be reduced greatly and the motor operates smoothly with appropriate correction.

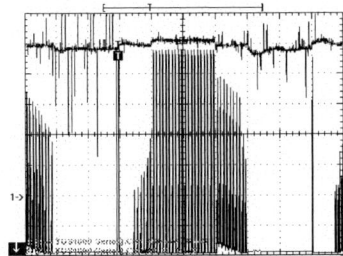

Figure 10. Experiment result with corrected commutation

V. CONCLUSIONS

In this paper, the main reason, phase-shift of the band pass filters, causing rotor position detection error in the zero-crossing detecting Back EMF method in position sensorless brushless DC motor drives is analyzed, and the effects of the detection error to the motor performance are discussed. The error can be corrected with a LUT to improve the operation performance of the motor. The phase current and its ripple are decreased, and the circulating current is eliminated. The experiment, with a Mitsubishi SHB130FRP compressor and a Haixin KFR-4001W/ZBP air conditioner, shows that the result is good at improving the operation performance and reducing the torque ripple of the BLDC motor.

REFERENCES

[1] Zhang Chen, The principle and applications of BLDC motor, China Machine Press, 1996;

[2] Li Zhongming, Rare earth permanent magnet electrical machines, National Defence Press, 1999;

[3] Shen, J.X., Tseng, K.J., "Analyses and compensation of rotor position detection error in sensorless PM brushless DC motor drives", IEEE Transactions on Energy Conversion, Vol. 18, No. 1, March 2003, pp87-93;

[4] Zhang Xiangjun, "Zero-crossing algorithm and phase correction of BEMF in the sensorless control of trapezoidal BLDC motors", Electric drive, 2001(2), pp14-16

[5] Gui-Jia Su, McKeever, J.W., "Low cost sensorless control of brushless DC motors with improved speed range". Applied Power Electronics Conference and Exposition, 2002. APEC 2002. Seventeenth Annual IEEE. 2002:286-292.

[6] Todd D. Batzel, Kwang Y. Lee, "Commutation torque ripple minimization for permanent magnet synchronous machines with Hall effect position feedback". IEEE Transaction on Energy Conversion, Vol. 13, No. 3, Septtmber 1998: 257-262

[7] Byoung-Hee Kang, Choel-Ju Kim, Hyung-Su Mok, Gyu-Ha Choe, "Analysis of torque ripple in BLDC motor with commutation time". ISIE 2001:1044-1048

[8] Li Qiang, Lin Mingyao, Hu Minqiang. Design and Implementation of Filters for Back EMF in a Sensorless BLDC Motor Drive. International Conference on Electrical Machines and Systems 2005, Sept. 27-29, 2005, Nanjing, China: 1899-1902

[9] Lin Mingyao, Li Qiang. Effect of Rotor Position Error on Commutation in Sensorless BLDC Motor Drives. International Conference on Electrical Machines and Systems 2005, Sept. 27-29,2005,Nanjing,China:497-499

2006 5th International Power Electronics and Motion Control Conference

A New Scheme to Direct Torque Control of Interior Permanent Magnet Synchronous Machine Drives for Constant Inverter Switching Frequency and Low Torque Ripple

Jun Zhang, M. Faz Rahman, and Colin Grantham

The University of New South Wales, Australia

School of Electrical Engineering and Telecommunications

Email: f.rahman@unsw.edu.au; jun.zhang@student.unsw.edu.au

Abstract— **This paper presents a new direct torque control (DTC) scheme based on space vector modulation (SVM) for Interior Permanent Magnet Synchronous Machine (IPMSM) drives. The new scheme provides a variable option for the improvement of DTC controlled IPMSM. Closed-loop control of both torque and flux is developed with two proportional-integral (PI) controllers. The stator voltage is generator through SVM unit. The torque and flux ripples are greatly reduced with fixed inverter switching frequency comparing with classical switching-table based DTC scheme. The analysis of the control principle provides a guide to design the controller parameters. Modeling results confirm the effectiveness of the proposed scheme.**

Keywords-Direct torque control; Permanent Magnet Synchronous Machine; Space Vector Modulation

I. INTRODUCTION

Interior Permanent Magnet Synchronous Machine (IPMSM) offers many advantages over the induction machine, such as overall efficiency, effective use of reluctance torque, smaller losses and compact motor size. Moreover, use of flux weakening control based on salient pole behavior supports a wider range of speeds at any given output level. In the late 1990's, direct torque control (DTC) technique for the IPMSM was proposed [1-3]. The advantages of the DTC include fast responses and the elimination of the current controllers, associated coordinate transformation, and the rotor position sensor required for the coordinate transformation. Although DTC has many advantages over vector control, it still has some drawbacks as reported in some literatures. The ripples in torque and flux are relatively higher when compared with those in vector control. Furthermore, the switching frequency of the inverter is not constant which changes with rotor speed, load torque and the bandwidth of the two hysteresis comparators.

In order to improve the performance of the classical DTC [4, 5], there exist different solutions. By using multiple level inverters [6], more control voltage space vectors can be generated to make torque response smooth.

With more power switches needed, the system cost and complexity increase. Predictive algorithms are adopted to calculate the most appropriate voltage space vectors to minimize deviation between the estimated values and actual ones of flux linkage and torque. In such efforts, some modified DTC schemes based on space vector modulation for induction machine drive were reported in [7-11]. With these schemes, fixed switching frequency and lower torque ripple were achieved with the help of proportional-integral (PI) controller and space vector modulation (SVM) technique.

This paper proposed a new DTC-SVM scheme for IPMSM drives. Closed-loop digital control for both torque and flux is implemented with two PI controllers, and the stator voltage is produced by a SVM unit. This scheme features low flux and torque ripples and fixed switching frequency. The analysis of the control principle of proposed scheme is given. Modeling results for a 1kW IPMSM show its good performance for dynamic and steady state with low torque and flux ripples.

This paper is organized as follows. Section II introduced the machine model in the stator flux reference frame. The structure of the proposed DTC-SVM scheme is presented in Section III. Modeling results of the proposed scheme is given and compared with classical switching-table-based DTC (ST-DTC).

II. MACHINE MODEL IN THE STATOR FLUX REFERENCE FRAME

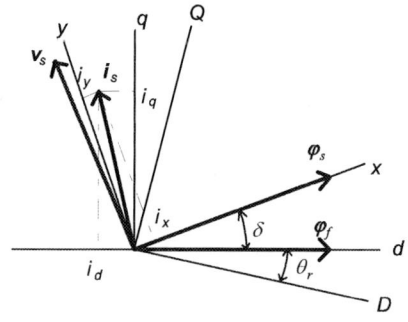

Figure 1 The stator and rotor flux linkages in different reference frames.

1-4244-0448-7/06/$25.00 ©2006 IEEE

As stated in [3], the stator flux linkage vector, φ_s, and rotor (magnet) flux linkage vector, φ_f, can be drawn in the rotor flux (d-q), stator flux (x-y) and stationary (D-Q) reference frames as in Figure 1. The stator flux and torque in the xy reference frame can be express as

$$
\begin{cases}
\begin{bmatrix} \varphi_x \\ \varphi_y \end{bmatrix} = \begin{bmatrix} L_d \cos\delta & L_q \sin\delta \\ -L_d \sin\delta & L_q \cos\delta \end{bmatrix} \begin{bmatrix} \cos\delta & -\sin\delta \\ \sin\delta & \cos\delta \end{bmatrix} \begin{bmatrix} i_x \\ i_y \end{bmatrix} \\
\qquad + \varphi_f \begin{bmatrix} \cos\delta \\ -\sin\delta \end{bmatrix} \\
\varphi_y = 0
\end{cases} \quad (1)
$$

$$
T_e = \frac{3P\varphi_s}{4L_d L_q}[2\varphi_f L_q \sin\delta - \varphi_s (L_q - L_d)\sin 2\delta] \quad (2)
$$

where L_d, L_q are direct and quadrature inductances, ω_r is the electrical rotor speed, T_e is the electromagnetic torque, P is the pole pairs, δ is the load angle, and φ_s, φ_f are stator and magnetic flux linkages.

The angle between the stator and rotor flux linkages δ, is the load angle. In the steady state, δ is constant corresponding to a load torque and both stator and rotor fluxes rotate at the synchronous speed. In transient operation, δ varies and the stator and rotor fluxes rotate at different speeds. Since the electrical time-constant is normally much smaller than the mechanical time-constant, the rotating speed of stator flux with respect to the rotor flux can be easily changed. It will be seen that fast change of torque can be brought about by controlling the change of δ or the rotating speed of the stator flux vector. This is the fundamental idea of the DTC for IPMSM.

The angle δ can be obtained by

$$
\begin{cases}
\delta = \theta_s - \theta_r \\
\theta_s = \omega_s t \\
\theta_r = \omega_r t
\end{cases} \quad (3)
$$

where θ_s and θ_r are the angles of are stator and magnetic flux vector in stationary ($D-Q$) frame, respectively.

As shown in (3), the control of the torque can be eventually realized by varying the angular speed ω_s of the stator flux vector.

Paper [3] didn't provide the voltage equations of the IPMSM in the stator flux reference frame (x-y), which can be derived as

$$
\begin{cases}
V_{sx} = R_s i_x + \dfrac{d\varphi_x}{dt} = R_s i_x + \dfrac{d\varphi_s}{dt} \\
V_{sy} = R_s i_y + \omega_s \varphi_x = R_s i_y + \omega_s \varphi_s
\end{cases} \quad (4)
$$

where R_s is the stator armature resistance.

Equation (4) shows the rotating speed of the stator flux vector can be controlled by appropriate stator voltage vector. It is obvious that the amplitude of stator flux

vector can be regulated by x component of stator voltage directly. And the torque can be indirectly regulated by y component of stator voltage once the load angle δ is regulated by the movement of the stator flux vector.

Note that calculation of the commanded voltage vector by (4) requires the derivation of the stator flux magnitude, which is a dc quantity. Thus, this scheme is less noisy [12].

III. PROPOSED DTC-SVM SCHEME

Based on above analysis, a complete scheme of direct torque and flux control for IPMSM that allows effective torque control has been developed and it is indicated in Figure 2.

The structure of proposed DTC-SVM scheme is similar with the Stator Flux Oriented Control scheme (SFOC). However, they are totally different. Firstly, no current control and decoupling are required with DTC-SVM. Secondly, all the feedback variables are in stationary frame and no rotating frame orientation is required.

As for the basic DTC, relatively higher ripples in flux and torque are produced. These drawbacks of basic DTC are due to the selection of stator voltage vector which is not the appropriate one. In three-phase inverter, a space voltage vector can be composed with the eight basic space voltage vectors with SVM technique. In the proposed scheme, the appropriate space voltage vector can be generated with SVM as long as the reference voltage vector is produced by the PI controllers. Therefore, the stator flux-linkage and torque can be regulated precisely and fixed switching frequency can be obtained.

The voltage vector should be transferred from the stator flux reference frame (x-y) to the stationary frame (D-Q) by (5) before using SVM algorithm.

$$
\begin{bmatrix} V_{sD} \\ V_{sQ} \end{bmatrix} = \begin{bmatrix} \cos\theta_s & -\sin\theta_s \\ \sin\theta_s & \cos\theta_s \end{bmatrix} \begin{bmatrix} V_{sx} \\ V_{sy} \end{bmatrix} \quad (5)
$$

where θ_s is the angle between the stator flux frame ($x-y$) and stationary frame ($D-Q$), i.e. the angle of stator flux linkage vector in the stationary frame.

Then the reference voltage vector is

$$
\vec{V}_{ref} = V_D + jV_Q \quad (6)
$$

With SVM technique, the demand space voltage vector can be composed by two active and one zero voltage vectors, which is illustrated in right part of Figure 3.

For example, when $\vec{V}_{ref}$ locates between $\vec{V}_1$ and $\vec{V}_2$, it can be expressed as

$$
\vec{V}_{ref} = \vec{V}_0 \frac{T_0}{T_s} + \vec{V}_1 \frac{T_1}{T_s} + \vec{V}_2 \frac{T_2}{T_s} \quad (7)
$$

where T_0, T_1, and T_2 are the effective time intervals of $\vec{V}_0$, $\vec{V}_1$ and $\vec{V}_2$ respectively within the sampling period T_s.

1859

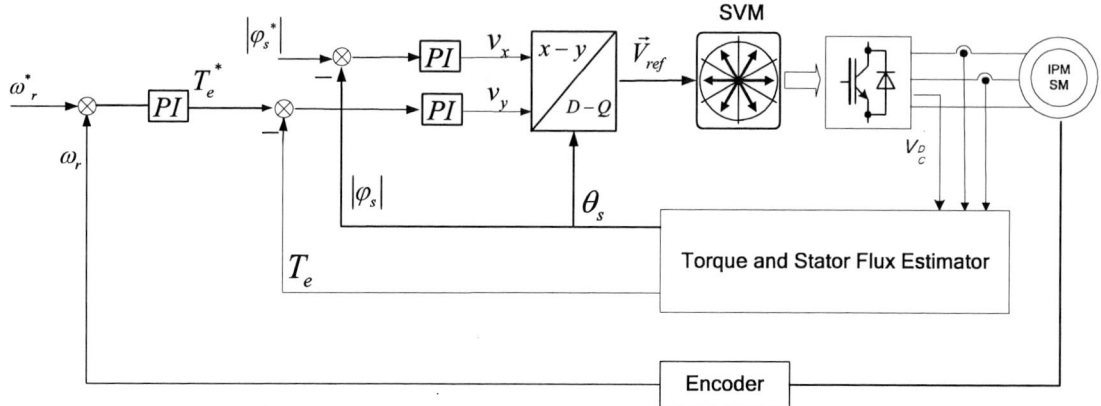

Figure 2 Control structure of DTC-SVM for IPMSM

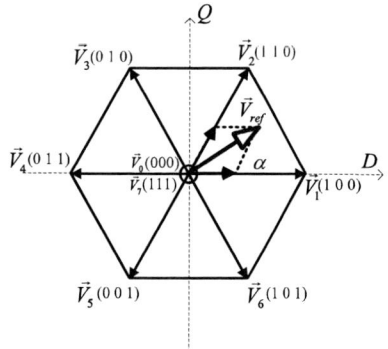

Figure 3 Reference space voltage vector

From Figure 3, it can be obtained

$$\begin{cases} V_{ref}\cos\alpha = V_1\dfrac{T_1}{T_s}+V_2\dfrac{T_2}{T_s}\cos\dfrac{\pi}{3} \\ V_{ref}\sin\alpha = V_2\dfrac{T_2}{T_s}\sin\dfrac{\pi}{3} \end{cases} \quad (8)$$

Thus

$$\begin{cases} T_1 = \dfrac{V_{ref}\sin(\dfrac{\pi}{3}-\alpha)}{V_1\sin\dfrac{\pi}{3}}T_s \\ T_2 = \dfrac{V_{ref}\sin\alpha}{V_2\sin\dfrac{\pi}{3}}T_s \end{cases} \quad (9)$$

then

$$T_0 = T_s - T_1 - T_2 \quad (10)$$

IV. RESULTS

Matlab/Simulink models were constructed to examine the proposed scheme and compare with classical ST-DTC. The parameters of IPMSM are shown in Table 1.

TABLE 1 PARAMETERS OF IPMSM MACHINE

Number of pole pairs	P	2
Stator resistance	R	5.8 Ω
Magnet flux linkage	φ_f	0.533 Wb
d-axis inductance	L_d	0.0448 H
q-axis inductance	L_q	0.1024 H
Phase voltage	V	132 V
Phase current	I	3 A
Base speed	ω_b	1260 rpm
Rated torque	T_b	6 Nm

As indicated in [3], stable torque control can be achieved if

$$\varphi_s < \frac{L_q}{L_q-L_d}\varphi_f \quad (11)$$

$$\delta_m = \cos^{-1}\left(\frac{a/\varphi_s - \sqrt{(a/\varphi_s)^2+8}}{4}\right) \quad (12)$$

where $a = \dfrac{\varphi_f L_q}{L_q - L_d}$

Therefore, the selection of rated stator flux linkage is according to (11).

$$\psi_s = 0.55\,(Wb) \quad (13)$$

Then the maximum load angle is calculated from (12)

$$\delta < 1.982 \quad (rad) \quad (14)$$

In the Simulink models, the sampling time is set as 75 μs for ST-DTC and 150 μs for proposed DTC-SVM. The switching table use in [3] is employed for basic DTC. Both steady state and dynamic behaviors of these two schemes are presented as follows.

Figure 4 (a) and (b) shows steady state of torque and stator flux of ST-DTC and proposed DTC-SVM under full load (6 Nm) when the IPMSM runs at 1200 rpm. Both of the torque and flux has ripples are greatly reduced with proposed DTC-SVM. And the amplitude of the stator flux is kept as constant with proposed control scheme.

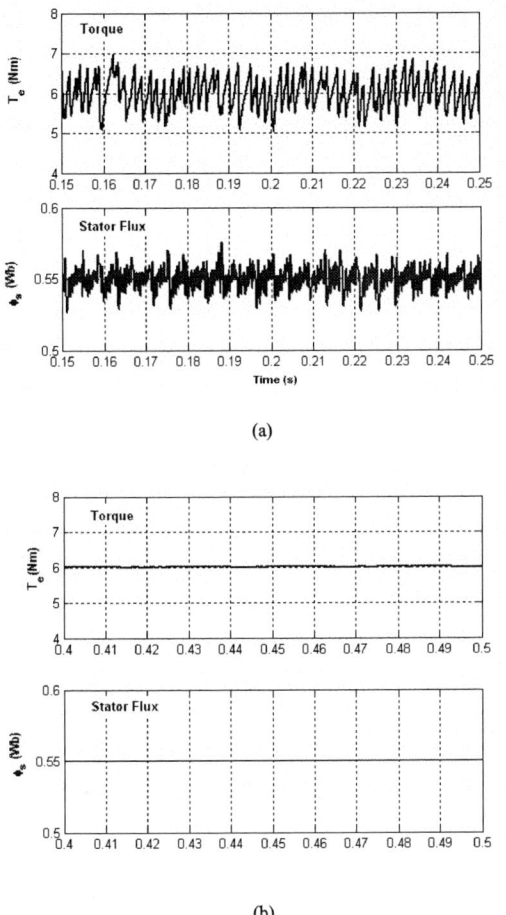

(a)

(b)

Figure 4. The torque and stator flux at 1200 rpm with full load (a) ST-DTC, Ts =75 μs (b) DTC-SVM, Ts =150 μs

Figure 5 shows torque loop dynamic response when a square wave torque reference is applied. As shown in Figure 6 (b), the torque of IPMSM follows the reference smoothly without steady state error with proposed DTC-SVM. Moreover, the fast DTC transient performance is preserved with reduce torque ripple.

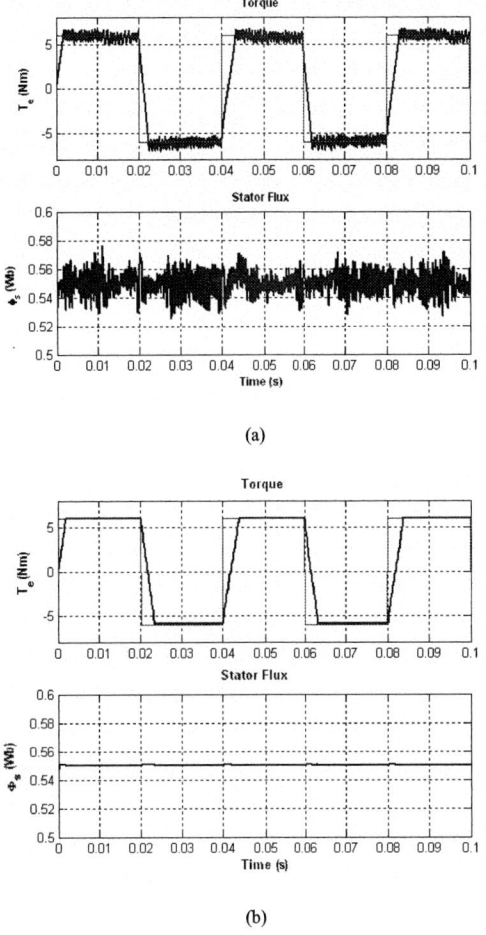

(a)

(b)

Figure 5 Dynamic performance of torque and stator flux with closed torque loop (a) ST-DTC, Ts =75 μs (b) DTC-SVM, Ts =150 μs

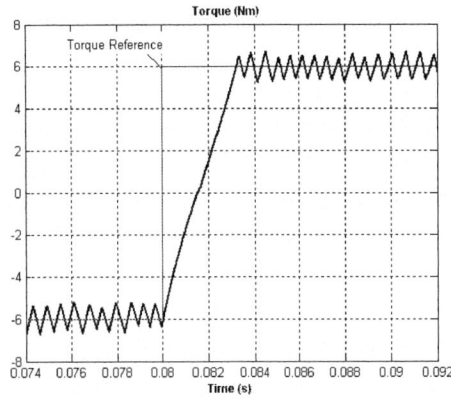

(a)

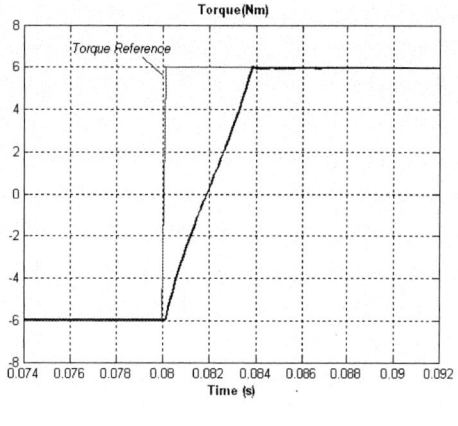

(b)

Figure 6 Dynamic response of torque - zoom in from Figure 5 (a) ST-DTC, Ts =75 μs (b) DTC-SVM, Ts =150 μs

V. CONCLUTION

In this paper, a new direct torque control scheme for IPMSM with space vector modulation is developed. Both torque and flux are controlled to ensure best steady state and dynamic performance. The modeling results confirm that it is an effective solution for the direct torque control of IPMSM drives with constant inverter switching frequency and low torque ripple.

REFERENCES

[1] C. French and P. Acarnley, "Direct torque control of permanent magnet drives," *Industry Applications, IEEE Transactions on*, vol. 32, pp. 1080-1088, 1996.

[2] M. F. Rahman, L. Zhong, and L. Khiang Wee, "A direct torque-controlled interior permanent magnet synchronous motor drive incorporating field weakening," *Industry Applications, IEEE Transactions on*, vol. 34, pp. 1246-1253, 1998.

[3] L. Zhong, M. F. Rahman, W. Y. Hu, and K. W. Lim, "Analysis of direct torque control in permanent magnet synchronous motor drives," *Power Electronics, IEEE Transactions on*, vol. 12, pp. 528-536, 1997.

[4] M. Depenbrock, "Direct self-control (DSC) of inverter-fed induction machine," *Power Electronics, IEEE Transactions on*, vol. 3, pp. 420-429, 1988.

[5] I. Takahashi, and Noguchi, T, "A New Quick-Response and High-Efficiency Control Strategy of an Induction Motor," *Industry Application, IEEE Transactions on*, vol. IA-22, pp. 820-827, 1986.

[6] C. A. Martins, X. Roboam, T. A. Meynard, and A. S. Carvalho, "Switching frequency imposition and ripple reduction in DTC drives by using a multilevel converter," *Power Electronics, IEEE Transactions on*, vol. 17, pp. 286-297, 2002.

[7] D. Swierczynski and M. P. Kazmierkowski, "Direct torque control of permanent magnet synchronous motor (PMSM) using space vector modulation (DTC-SVM)-simulation and experimental results," presented at IECON 02 [Industrial Electronics Society, IEEE 2002 28th Annual Conference of the], 2002.

[8] C. B. Lascu, I.; Blaabjerg, F., "A modified direct torque control for induction motor sensorless drive," *Industry Applications, IEEE Transactions on*, vol. 36, pp. 122-130, 2000.

[9] A. Tripathi, A. M. Khambadkone, and S. K. Panda, "Space-vector based, constant frequency, direct torque control and dead beat stator flux control of AC machines," presented at Industrial Electronics Society, 2001. IECON '01. The 27th Annual Conference of the IEEE, 2001.

[10] T. Lixin, Z. Limin, M. F. Rahman, and Y. Hu, "A novel direct torque controlled interior permanent magnet synchronous machine drive with low ripple in flux and torque and fixed switching frequency," *Power Electronics, IEEE Transactions on*, vol. 19, pp. 346-354, 2004.

[11] T. Lixin, Z. Limin, M. F. Rahman, and H. Yuwen, "A novel direct torque control for interior permanent-magnet synchronous machine drive with low ripple in torque and flux-a speed-sensorless approach," *Industry Applications, IEEE Transactions on*, vol. 39, pp. 1748-1756, 2003.

[12] G. S. Buja and M. P. Kazmierkowski, "Direct torque control of PWM inverter-fed AC motors - a survey," *Industrial Electronics, IEEE Transactions on*, vol. 51, pp. 744-757, 2004.

2006 5th International Power Electronics and Motion Control Conference

A Modified Toque Control for Interior Permanent Magnet Synchronous Motor Drive Without a Speed Sensor

Yanping Xu, Yanru Zhong and Hui Yang

Department of Electrical Engineering, Xi'an University of Technology, Xi'an, China

Abstract—Direct torque control (DTC) can produce quick and robust response, but it has the problems of large torque ripples and inconstant inverter switching frequency. This paper introduces a modified direct torque control based on the space vector pulse width modulation (SVPWM) for interior permanent magnet synchronous motor (IPMSM) drive without a speed sensor. Two PI controllers regulate the flux and torque respectively and the inverter is controlled by the SVPWM technique in the proposed DTC system. Speed estimator is based on the model reference adaptive system (MRAS). Simulation results show that the performance of the proposed DTC system has been improved with respect to the conventional DTC. The DTC system can effectively reduce the flux and torque ripples and speed estimator can accurately track the speed of motor.

Keywords-direct torque control; interior permanent magnet synchronous motor; space vector pulse width modulation; estimator

I. INTRODUCTION

Since the direct torque control (DTC) was proposed in the middle of 1980s, the DTC principle has been widely used for induction motor drives [1][2]. The basic principle of DTC is to directly select stator voltage vector according to the differences between the references of torque and stator flux linkage and their actual values. Compared with vector control, the DTC has many advantages such as less machine parameter dependence, simpler implantation and quicker dynamic torque response. The principle has also been applied to the Interior Permanent Magnet Synchronous Motor (IPMSM) [3]-[6]. Although DTC has many advantages over vector control, it still has some drawbacks. Due to the torque and flux hysteresis controllers, the conventional DTC has large torque and flux ripples, which deteriorate the control system performance, especially at the low speed. Moreover, the inverter has not constant switching frequency.

To solve the problems of large torque ripple and inconstant inverter switching frequency in the conventional direct torque control (DTC), many researchers have given attentions to these problems. A control algorithm based on a discrete space vector modulation (DSVM) technique for induction motors was

reported in [7]. This DSVM uses prefixed time intervals within a cycle period and a higher number of voltage space vectors are used with respect to those used in conventional DTC. However, much more voltage vectors will increase the complexity of control systems. A new DTC scheme for IPMSM with low ripple in flux and torque and fixed switching frequency was reported in [8]. In [9], a fuzzy logic and adaptive control method was used to reduce torque ripples in DTC for induction motors. However, this method is based on expert knowledge.

Another important issue for a sensorless drive is speed estimation. Both open-loop and closed-loop speed and position estimators are widely analyzed in the papers [10][11]. In [10], a novel rotor position sensorless technique for permanent magnet AC motors was present. This technique uses values of flux linkage and the back-EMF functions to estimate rotor position based on the measurement of the phase voltages and currents of the motor. A technique of speed estimation from the stator flux vector and the torque angle for IPMSM was proposed in [11]. This method uses the rotor flux linkage vector to estimate the rotor speed with sufficient accuracy both for the steady-state and dynamic conditions of operation.

In this paper, a modified DTC method for IPMSM is presented based on SVPWM techniques. The block diagram of the proposed IPMSM DTC is given in section II. The principle of SVPWM technique and the speed estimator that based on the MRAS are also introduced in detail. Simulation results in section IV show that the proposed DTC method for IPMSM can effectively reduce the flux and torque ripples and improve the performance of control system.

II. PROPOSED SENSORLESS PMSM DRIVE

The block diagram of the proposed PMSM DTC drive without a speed sensor is shown in Fig.1. The speed controller is a classical proportional-integral (PI) regulator, which produces the reference torque. The calculated flux and torque are compared with the reference flux and torque. Two PI controllers regulate the flux and torque, respectively. Then the inverter is controlled by the space vector pulse width modulation (SVPWM). A DC bus voltage and two phases current are detected to calculate the stator flux linkage and torque.

1-4244-0448-7/06/$25.00 ©2006 IEEE   1863

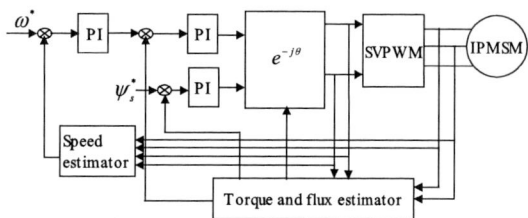

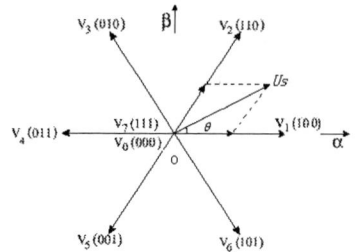

Fig.1. The block diagram of the proposed IPMSM DTC

For the proposed DTC method, the estimation of flux linkage and torque are carried out in the stationary reference frame, as shown in (1)-(4):

$$\psi_d = \int (u_d - R_s i_d) dt + \psi_{d|t=0} \tag{1}$$

$$\psi_q = \int (u_q - R_s i_q) dt + \psi_{q|t=0} \tag{2}$$

$$\psi_s = \sqrt{\psi_d^2 + \psi_q^2} \tag{3}$$

$$T_e = \frac{3}{2} n_p (\psi_s \times i_s) = \frac{3}{2} n_p (\psi_d i_q - \psi_q i_d) \tag{4}$$

Where ψ_d, ψ_q are direct and quadrature axis flux linkage, respectively, u_d, u_q are direct and quadrature axis voltage, i_d, i_q are direct and quadrature axis current, R_s is stator resistance, ψ_s is stator flux linkage, T_e is electromagnetic torque, n_p is pole pairs, i_s is stator current.

A. Space Vector Pulse Width Modulation

The objective of space vector pulse width modulation (SVPWM) technique is to approximate the reference voltage instantaneously by combination of switching states corresponding the basic space vectors. At any time, the stator voltage vector always lies in one of the six sectors, as shown in Fig. 2. For any small period of time, the stator voltage vector can be formed by the combination of adjacent basic space vectors. Suppose that the reference voltage vector lies in the Nth sectors, the respective durations of switching states corresponding to adjacent vectors V_N, V_{N+1} (N=6, $V_{N+1}=V_N$), and the zero vectors V_0, V_7 are shown in (5)~(8).

$$\gamma = \theta - \frac{(N-1)\pi}{3}, 0 \le \gamma < \frac{\pi}{3} \tag{5}$$

$$T_N = T_P \frac{2U_s}{\sqrt{3}V_{dc}} \sin(\frac{\pi}{3} - \gamma) \tag{6}$$

$$T_{N+1} = T_P \frac{2U_s}{\sqrt{3}V_{dc}} \sin \gamma \tag{7}$$

$$T_0 = T_7 = \frac{T_P - T_N - T_{N+1}}{2} \tag{8}$$

At any PWM period T_P, the applied sequence of adjacent vectors V_N, V_{N+1} and the zero vectors V_0, V_7 are V_0-V_N-V_{N+1}-V_7-V_{N+1}-V_N-V_0, and the respective durations are $T_0/2$-$T_N/2$-$T_{N+1}/2$-T_7-$T_{N+1}/2$-$T_N/2$-$T_0/2$, as shown in Fig.3.

Fig.2. Space vector modulation of stator voltage vector

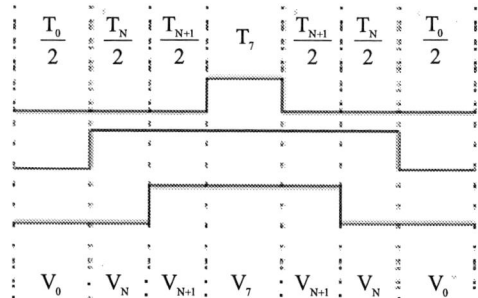

Fig.3. Sequence of Space voltage vectors

B. Speed Estimation

The speed estimator is based on a model reference adaptive system (MRAS) method. MRAS technique uses a speed independent model and a speed dependent model. The error from the two models is forced to zero by an adjusting mechanism, which gives the estimated speed. The stator current equations of the PMSM in the d-q reference frame are shown as (9)(10).

$$\frac{di_d}{dt} = -\frac{R_s}{L} i_d + \omega i_q + \frac{u_d}{L} \tag{9}$$

$$\frac{di_q}{dt} = -\frac{R_s}{L} i_q + \omega i_d - \frac{\psi_r}{L} + \frac{u_q}{L} \tag{10}$$

Where ω is motor speed, L is axis inductance, ψ_r is rotor flux linkage. Equations (9) and (10) show that the motor speed is related to the current models. Therefore, the reference model is the PMSM model and the adaptive model is a current model. Equation (9) and (10) can be written as (11).

$$\frac{d}{dt} \begin{bmatrix} i_d + \frac{\psi_r}{L} \\ i_q \end{bmatrix} = \begin{bmatrix} -\frac{R_s}{L} & \omega \\ \omega & -\frac{R_s}{L} \end{bmatrix} \begin{bmatrix} i_d + \frac{\psi_r}{L} \\ i_q \end{bmatrix}$$

$$+ \frac{1}{L} \begin{bmatrix} u_d + \frac{R\psi_r}{L} \\ u_q \end{bmatrix} \tag{11}$$

If define that

$$i_d^{'} = i_d + \frac{\psi_r}{L} \tag{12}$$

$$i_q^{'} = i_q \tag{13}$$

$$u_d^{'} = u_d + \frac{R_s \psi_r}{L} \tag{14}$$

$$u_q^{'} = u_q \tag{15}$$

then, equation (11) can be written as (16).

$$\frac{d}{dt}\begin{bmatrix} i_d^{'} \\ i_q^{'} \end{bmatrix} = \begin{bmatrix} -\dfrac{R_s}{L} & \omega \\ \omega & -\dfrac{R_s}{L} \end{bmatrix} \begin{bmatrix} i_d^{'} \\ i_q^{'} \end{bmatrix} + \frac{1}{L}\begin{bmatrix} u_d^{'} \\ u_q^{'} \end{bmatrix} \tag{16}$$

Equation (16) can be abbreviated as (17).

$$\frac{d}{dt}i^{'} = Ai^{'} + Bu^{'} \tag{17}$$

The adaptive model is

$$\frac{d}{dt}\begin{bmatrix} \hat{i}_d^{'} \\ \hat{i}_q^{'} \end{bmatrix} = \begin{bmatrix} -\dfrac{R_s}{L} & \hat{\omega} \\ \hat{\omega} & -\dfrac{R_s}{L} \end{bmatrix} \begin{bmatrix} \hat{i}_d^{'} \\ \hat{i}_q^{'} \end{bmatrix} + \frac{1}{L}\begin{bmatrix} u_d^{'} \\ u_q^{'} \end{bmatrix} \tag{18}$$

Equation (18) also can be abbreviated as (19).

$$\frac{d}{dt}\hat{i}^{'} = \hat{A}\hat{i}^{'} + Bu^{'} \tag{19}$$

The state error is shown as (20).

$$e = i^{'} - \hat{i}^{'} \tag{20}$$

The motor speed ω is calculated and corrected by a PI adaptation mechanism, as shown (21).

$$\hat{\omega} = (K_p + K_I \frac{1}{s})e \tag{21}$$

where K_p and K_I are the proportional and integral gains of the PI estimator. The block diagram of the speed estimator is shown in Fig.4. If the errors between the reference model and the adaptive model are tuned to zero, then the motor speed can be identified.

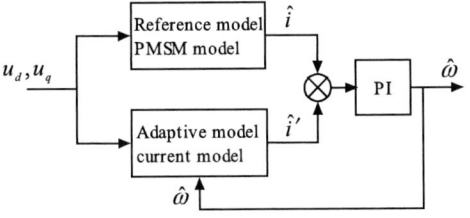

Fig.4. Block diagram of the speed estimator

III. SIMULATION RESIUTS

To study the performance of the proposed modified DTC of PMSM drive without a speed sensor, two MATLAB/Simulink models were developed. One is used for the conventional DTC and the other for the proposed DTC. The IPMSM parameters are $R_s = 2.875\Omega$, $L_d = L_q = 8.5mH$, $\psi_r = 0.3Wb$, $n_p = 2$.

Fig. 5 shows the torque and flux response of the conventional DTC for IPMSM when the load torque has step change between 2 Nm to 5 Nm. Fig. 6 shows the torque and flux response of the proposed DTC under the same conditions. Fig. 5 indicates that the torque and flux ripples are apparent in the conventional PMSM DTC due to the torque and flux hysteresis controllers. Fig. 6 shows that the torque and flux ripples are dramatically reduced in the proposed control systems and the performance of the system has been improved.

To determine the speed estimator performance, the estimator is open-loop operated and the motor speed is controlled using the real speed. A step reference speed from 0 rad/s to 100 rad/s is applied at t=0.

Fig. 7 shows the real and estimated speed in the proposed DTC when the load torque is changed from 1Nm to 3Nm at t=0.07s. Fig. 7 shows that the estimated speed could accurately track the change of the real speed and have fast response. The proposed system is relatively robust with respect to the change of the load torque.

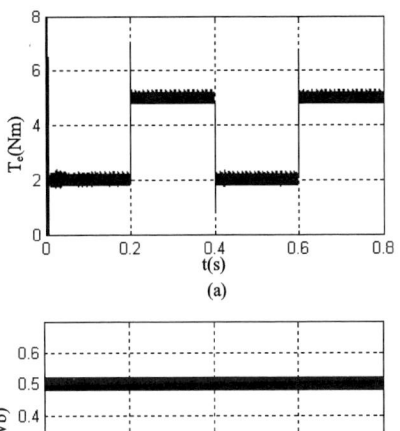

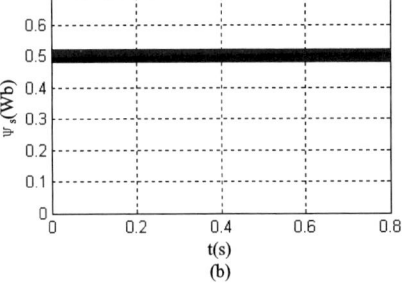

Fig.5 (a) Torque simulation wave of conventional IPMSM DTC and (b) flux linkage simulation wave of conventional IPMSM DTC

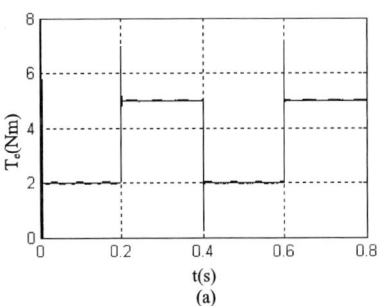

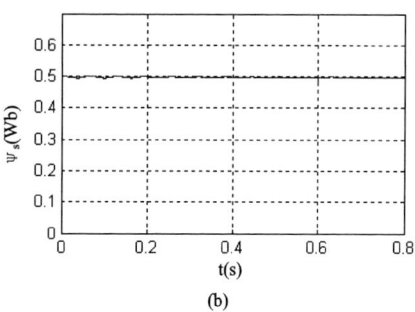

(b)

Fig.6　(a) Torque simulation wave of proposed IPMSM DTC and (b) flux linkage simulation wave of proposed IPMSM DTC

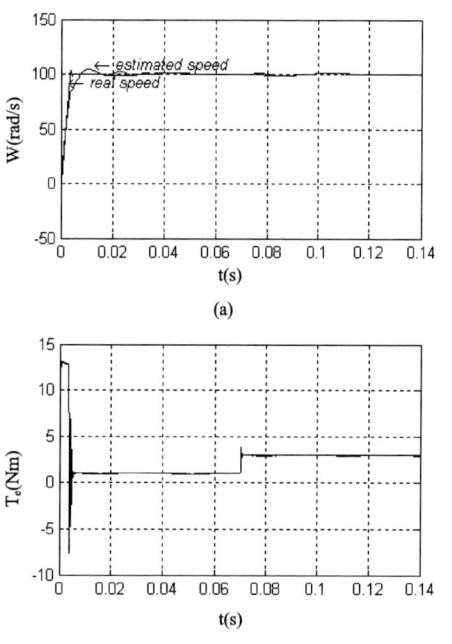

(a)

(b)

Fig.7　(a) Speed estimation wave of proposed IPMSM DTC and (b) torque simulation wave of proposed IPMSM DTC

IV. CONCLUSION

In this paper, a modified direct torque control method for interior permanent magnet synchronous motors without a speed sensor is presented based on the a voltage space vector modulator and a MRAS speed estimator. The proposed direct torque control method apparently reduces the torque and flux ripples while preserves the fast response and robustness merits of the conventional DTC. The proposed method also provide constant inverter

switching frequency and a MRAS speed estimator was applied to estimate speed. Simulation results show that the proposed DTC approach can improve the performance of IPMSM DTC and the MRAS speed estimator is capable of tracking the change of real speed accurately.

REFERENCES

[1] I. Takahashi and T. Naguchi, "A new quick-response and high-efficiency control strategy of an induction motor," *IEEE Trans. Ind. Applicat.*, vol.IA-22, pp.820-827, Sep./Oct.1986.

[2] M. Depenbrock, "Direct self-control of inverter-fed induction machine," *IEEE Trans. Power Electron.*, vol.3, no.4, pp.420-429, Oct.1988.

[3] C. French and P. Acarnley, "Direct torque control of permanent magnet drives," *IEEE Trans. Ind. Applicat.*, vol.IA-32, pp.1080-1088, Sep./Oct.1996.

[4] L.Zhong, M. F. Rahman, W. Y. Hu and K. W. Lim, "Analysis of direct torque control in permanent magnet synchronous motor drives," *IEEE Trans. Power Electron.*, vol.12, no.3, pp.528-535, May.1997.

[5] M. F. Rahman, L.Zhong and K. W. Lim, "A direct torque-controlled interior permanent magnet synchronous motor drive incorporating field weakening," *IEEE Trans. Ind. Applicat.*, vol.34, no.6, pp.1246-1253, Nov./Dec.1998.

[6] L.Zhong, M. F. Rahman, W.Y. Hu, M.A. Rahman, "A direct torque controller for permanent magnet synchronous motor drives, " *IEEE Trans. Energy Conversion*, vol.14, no.3, pp. 637 - 642, May.1999.

[7] Domenico Casadei, Giovanni Serra and Angelo Tani, "Implementation of a DTC algorithm for induction machines based on discrete space vector modulation," *IEEE Trans. Power Electron,*, vol.15, No.4, pp.769-776, July.2000.

[8] Lixin Tang, Limin Zhong, M.F. Rahman and Yuwen Hu, "A novel direct controlled interior permanent magnet synchronous machine drive with low ripple in flux and torque and fixed switching frequency," *IEEE Trans. Power Electron.*, vol.19, No.2, pp.346-354, Mar.2004.

[9] Luis Romeral, Antoni Arias, Emiliana Aldabas and Marcel G. Jayne, "Novel DTC scheme with fuzzy adaptive torque-ripple reduction," *IEEE Trans. Ind. Electron.*, vol.50, No.3,pp.487-492, June.2003.

[10] Li Ying and Nesimi Ertugrul, "A novel robust DSP-based indirect rotor position estimation for permanent magnet AC motors without rotor saliency," *IEEE Trans. Power Electron.*, vol.18, No.2, pp.539 546,March.2003.

[11] Muhammed Fazlur Rahman, L. Zhong, Md. Enamul Haque and M.A. Rahman, "A direct torque controlled interior Permanent-magnet synchronous motor drive without a speed sensor," *IEEE Trans. Energy Conversion.*, vol.18, No.1, pp.17-22, March.2003.

1866

2006 5th International Power Electronics and Motion Control Conference

Direct Torque Control for Interior Permanent Magnet Synchronous Motors Using Matrix Converters

D. Xiao and M. F. Rahman

School of Electrical Engineering and Telecommunications
The University of New South Wales, Sydney, Australia

Abstract—**An improved direct torque control (DTC) scheme for the matrix converter fed Interior Permanent Magnet Synchronous Machine (IPMSM) is presented in this paper. In the proposed scheme, the modified DTC strategy for matrix converter IPMSM drives is achieved using a variable-structure torque and stator flux controller, and indirect space vector modulation (ISVM) technique. It features low torque and flux ripples, sinusoidal input/output currents and unity fundamental displacement factor at the input regardless of the load power factor while maintaining constant switching frequency. Numerical simulations are carried out in both steady-state and transient conditions, verifying the effectiveness of this new proposed scheme.**

Keywords—*direct torque control (DTC), Interior Permanent Magnet Synchronous Machine (IPMSM), space vector modulation (SVM), matrix converter (MC)*

I. INTRODUCTION

The direct torque control (DTC) technique for IPMSM was proposed in the late 1990's [1]. The main advantages of DTC are robust and fast torque response, elimination of coordinate transformation and no requirement of PWM pulses generation and current regulators. However, further research is still being done to adapt DTC to new converters and reduce torque and flux ripples while maintaining fixed switching frequency [2].

Recently, the three-phase to three-phase matrix converter (MC) has received considerable interest as a viable alternative to the conventional voltage-source inverter (VSI) due to a number of advantages, such as inherent four-quadrant operation, sinusoidal input/output waveforms, and controllable input power factor. Furthermore, the absence of bulky dc-link electrolytic capacitors for energy storage allows long lifetime, high integration capability, extreme temperature and critical volume/weight applications [3]-[5]. The basic DTC scheme for matrix converter induction motor drives was initially proposed in [6]. Three hysteresis comparators were applied to control electromagnetic torque, stator flux, and input power factor. At each sampling period, the proper switching configuration is selected to compensate the instantaneous errors of flux magnitude and torque under the constraint of unity input power factor. As the converter doesn't update the switching state till the outputs of hysteresis controllers change states, the ripples in stator flux linkage and torque are relatively high. Furthermore, the switching frequency varies according to rotor speed, load torque and bandwidths of the three hysteresis controllers. In this paper, a new DTC scheme for IPMSM drives fed by matrix converter using Indirect Space Vector Modulation (ISVM) and a variable-structure torque and stator flux linkage controller is presented as shown in Fig. 1, which features low torque and flux ripples, constant switching frequency, unity input power factor, and sinusoidal input/output waveforms while preserving the fast response and robustness of DTC compared with the classical DTC.

II. ISVM FOR MATRIX CONVERTER

An approach widely used in matrix converter is the Indirect Space Vector Modulation (ISVM), which separates the modulation into rectifying and inverting vector modulations [3], [7]-[9]. There are two reference vectors to be synthesized in this method. One is input current vector corresponding to rectification stage, the other is output voltage vector corresponding to inversion stage. The combination of two adjacent active switching vectors and a zero vector determines both direction and magnitude of the reference vector. According to the law of sines, the duty cycles of active switching vectors used to synthesize the input current vector are calculated with (1), (2) while the duty cycles of active vector for the inversion stage are calculated with (3), (4).

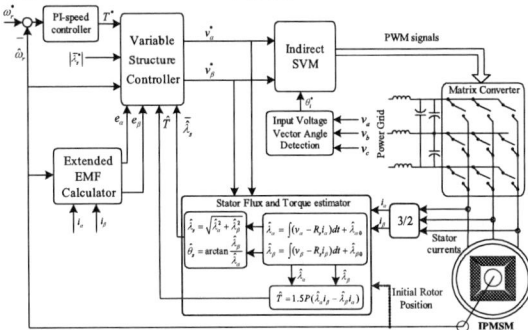

Figure 1.   Block diagram of the VS-DTC of the matrix converter fed IPMSM drive system

$$d_\gamma = T_\gamma / T_s = m_C \cdot \sin(\pi/3 - \theta_i^*) \qquad (1)$$

$$d_\delta = T_\delta / T_s = m_C \cdot \sin(\theta_i^*) \qquad (2)$$

$$d_\alpha = T_\alpha / T_s = m_V \cdot \sin(\pi/3 - \theta_o^*) \qquad (3)$$

$$d_\beta = T_\beta / T_s = m_V \cdot \sin(\theta_o^*) \qquad (4)$$

where m_C and m_V are the current and voltage modulation indexes, θ_i^* and θ_o^* are the reference vectors angles within their respective sectors. To merge the two independent vector modulations into one, the four duty cycles for new active vector pair can now be derived from the product of rectifying duty cycles (1), (2) and inverting duty cycles in (3), (4).

$$d_{\alpha\gamma} = d_\alpha \cdot d_\gamma, d_{\alpha\delta} = d_\alpha \cdot d_\delta, d_{\beta\delta} = d_\beta \cdot d_\delta, d_{\beta\gamma} = d_\beta \cdot d_\gamma$$
$$d_0 = T_0 / T_s = 1 - d_{\alpha\gamma} - d_{\alpha\delta} - d_{\beta\delta} - d_{\beta\gamma} \qquad (5)$$

III. VARIABLE-STRUCTURE DTC FOR MATRIX CONVERTER

The equation of the IPMSM in the rotor reference frame is [10]:

$$\begin{bmatrix} v_d \\ v_q \end{bmatrix} = \begin{bmatrix} R_s + pL_d & -P\omega_r L_q \\ P\omega_r L_d & R_s + pL_q \end{bmatrix} \begin{bmatrix} i_d \\ i_q \end{bmatrix} + \begin{bmatrix} 0 \\ P\omega_r \lambda_f \end{bmatrix} \qquad (6)$$

The model can be described in the stationary reference frame as (7).

$$\begin{bmatrix} v_\alpha \\ v_\beta \end{bmatrix} = \begin{bmatrix} R_s + pL_d & \omega_{re}(L_d - L_q) \\ -\omega_{re}(L_d - L_q) & R_s + pL_d \end{bmatrix} \begin{bmatrix} i_\alpha \\ i_\beta \end{bmatrix} + \begin{bmatrix} e_\alpha \\ e_\beta \end{bmatrix} \qquad (7)$$

The extended EMF (EEMF) is defined as (8).

$$e = \begin{bmatrix} e_\alpha \\ e_\beta \end{bmatrix} = \left[(L_d - L_q)(\omega_{re} i_d - pi_q) + \omega_{re} \lambda_f \right] \begin{bmatrix} -\sin\theta_{re} \\ \cos\theta_{re} \end{bmatrix} \qquad (8)$$

The state equation of the IPMSM can be derived from (7) as (9).

$$\begin{bmatrix} pi_\alpha \\ pi_\beta \end{bmatrix} = \begin{bmatrix} -R_s / L_d & -\omega_{re}(L_d - L_q)/L_d \\ \omega_{re}(L_d - L_q)/L_d & -R_s / L_d \end{bmatrix} \begin{bmatrix} i_\alpha \\ i_\beta \end{bmatrix}$$
$$+ \begin{bmatrix} -1/L_d & 0 \\ 0 & -1/L_d \end{bmatrix} \begin{bmatrix} e_\alpha \\ e_\beta \end{bmatrix} + 1/L_d \begin{bmatrix} v_\alpha \\ v_\beta \end{bmatrix} \qquad (9)$$

The stator flux linkages and torque are given by

$$T = 1.5P(\lambda_\alpha i_\beta - \lambda_\beta i_\alpha) \qquad (10)$$

$$\begin{cases} \dot{\lambda}_\alpha = v_\alpha - R_s i_\alpha \\ \dot{\lambda}_\beta = v_\beta - R_s i_\beta \end{cases} \qquad (11)$$

The switching surface is defined as $S = [S_1 \; S_2]^T$,

$$S_1 = e_T(t) + K_T \int_0^t e_T(\tau)d\tau - e_T(0)$$
$$S_2 = e_\lambda(t) + K_\lambda \int_0^t e_\lambda(\tau)d\tau - e_\lambda(0) \qquad (12)$$

$$\dot{S}_2 = \dot{e}_\lambda + K_\lambda e_\lambda = (\dot{\lambda}^* - \dot{\hat{\lambda}}) + K_\lambda(\lambda^* - \hat{\lambda})$$
$$\dot{S}_1 = \dot{e}_T + K_T e_T = (\dot{T}^* - \dot{\hat{T}}) + K_T(T^* - \hat{T}) \qquad (13)$$

where $e_T = T^* - \hat{T}$, $e_\lambda = \lambda^* - \hat{\lambda} = \lambda_s^{*2}/2 - (\hat{\lambda}_\alpha^2 + \hat{\lambda}_\beta^2)/2$.

Substituting for $\hat{T}$, $\hat{\lambda}$, $\dot{\hat{T}}$ and $\dot{\hat{\lambda}}$ into (13) leads to

$$\dot{\underline{S}} = \underline{F} + \underline{D} \bullet [u_\alpha, u_\beta]^T$$
$$\underline{D} = \begin{bmatrix} -1.5P(i_\beta - \hat{\lambda}_\beta/L_d) & -1.5P(\hat{\lambda}_\alpha/L_d - i_\alpha) \\ -\hat{\lambda}_\alpha & -\hat{\lambda}_\beta \end{bmatrix}$$

$$F_1 = K_T e_T - 1.5P\{\hat{\lambda}_\alpha[\omega_{re}(L_d - L_q)/L_d \cdot i_\alpha - R_s i_\beta/L_d - e_\beta/L_d]$$
$$- \hat{\lambda}_\beta[-R_s i_\alpha/L_d - \omega_{re}(L_d - L_q)/L_d \cdot i_\beta - e_\alpha/L_d]\}$$

$$F_2 = K_\lambda e_\lambda + \hat{\lambda}_\alpha R_s i_\alpha + \hat{\lambda}_\beta R_s i_\beta$$

In order to drive the state trajectory to the intersection of the above switching surfaces, the VS controller generates the command output voltage vector components provided for the inversion stage of ISVM module. The switching control law can be selected as (14) according to Lyapunov approach [10].

$$u_{\alpha,\beta}^* = -D^{-1} \begin{bmatrix} \mu_1 & 0 \\ 0 & \mu_2 \end{bmatrix} \begin{bmatrix} sign(S_1) \\ sign(S_2) \end{bmatrix} \qquad (14)$$

The chattering problem can be remedied by (15). The switching function is replaced by a continuous function around the sliding surface neighborhood by introducing smoothing factors.

$$sign(S_i) = \begin{cases} 1, \; if \; S_i > \lambda_i \\ -1, \; if \; S_i < -\lambda_i \; \; (Smoothing \; factors \; \lambda_i > 0) \\ S_i / \lambda_i, \; if \; |S_i| < \lambda_i \end{cases} \qquad (15)$$

To implement the ISVM for matrix converter, the input reference current vector is given by the input voltage vector for an instantaneous unity power factor. Under balanced condition, the input current vector angle is straightforwardly set to the supply voltage vector angle. An optimized ISVM is applied to minimize total switching losses by selecting the optimized switching sequence and zero vector from a criteria that assures the minimum number of switching transitions [8].

IV. NUMERICAL SIMULATION

A Matlab/Simulink model is constructed to verify the proposed DTC IPMSM drive system. The parameters of the test IPMSM and model are given in Tables I and II. The simulations have been carried out assuming three-phase balanced input voltages and ideal switching devices.

It can be noticed in Figs. 2–8 that the proposed drive system operates very well under the transient conditions. The torque and flux ripples are reduced without compromising the fast dynamic responses to the command changes. The stator and filtered input currents are sinusoidal with nearly unity input power factor. Rated torque reversal at -6 Nm and +6 Nm was tested with stator flux linkage maintained at the rated value 0.55 Wb. Fig. 2 shows the dynamic responses of actual speed, estimated torque, current components in the rotor frame, three-phase stator currents and filtered input currents, respectively. In this case, the motor is accelerated from standstill to 1000 rev/min during positive rated torque command. The motor starts to decelerate after the torque is changed to the negative rated value. The magnitude of stator currents keep constant while the frequency varies with the speed. However, the filtered input currents have a speed-varying magnitude but a constant frequency 50 Hz. The transitions between the two conditions above are shown in Fig. 3. The current waveform is almost sinusoidal immediately after the step command.

Such input side quantities as input phase voltage and filtered input current are also shown to test the performance of the input side of the matrix converter. During motoring operation (from 0 to 0.1 sec.), the filtered input current (i_a) is in phase with the corresponding phase voltage (v_a). Then the input current suddenly reverses the phase 180°, opposite that of the input voltage, during the regenerative breaking. Under these two conditions, the matrix converter drive draws electric power from the mains and then generates electric power back to the mains with sinusoidal waveforms and unity input power factor.

Fig. 4 shows the four-quadrant operation characteristics of the proposed drive system. This is examined by applying ±104.72 rad/sec (±1000 rev/min) speed step reference and 1 Nm load. The motor accelerates from standstill to 104.72 rad/sec and stays in steady-state for 0.12 sec (first quadrant). When the speed step is changed to -104.72 rad/sec, the speed is decelerated until it reaches zero. In this transition from the first to the second quadrant, the motor operates as a generator. Following the reference, the speed reaches -104.72 rad/sec and remains at that speed (third quadrant) until the step command comes again to 104.72 rad/sec. That results in a transition from the third to the fourth quadrant. Eventually, the machine returns to quadrant one and reaches 104.72 rad/sec. The torque remains in saturation limited to the rated value until the speed reaches its final values. The corresponding stator currents have constant magnitude and speed-varying frequency during the transitions but constant magnitude and frequency in the steady-state. The input current frequency remains at 50 Hz all the time, while the magnitude varies with speed during transitions but constant in the steady-state.

Figs. 4 and 6 show stator flux magnitude, flux components on axis α and β in the stationary reference frame, extended back EMF and manifolds of torque and flux for torque and speed step commands. It can be seen that the state variables, torque and flux, are driven towards their own sliding surfaces by discontinuous control signals under these two conditions. In the steady states, the extended EMFs exhibit smooth sinusoidal waveforms. The stator flux follows its reference (0.55 Wb) with low ripples in both steady-state and transient conditions.

Figs. 7 and 8 illustrate the input phase voltage, filtered input current and motor current in the four-quadrant operation in detail. In Fig. 7, the motor operates in steady-state in the first quadrant, i.e. positive torque and positive speed, from 0.18 to 0.24 second. The magnitude and frequency of both stator and input currents are constant. From 0.24 to 0.34 sec, the transition from quadrant one to quadrant two occurs and the machine operates as a

generator. In this condition, the input current reverses its phase from 0° to 180° with respect to the corresponding phase voltage and regenerative energy of the motor is fed back to the supply. The reverse motoring occurs after speed reaches zero at about 0.34 sec. The speed accelerates to -104.72 rad/sec with negative rated torque before it reaches its final value and the steady-state operation is provided in the third quadrant. The input current remains in phase with the input voltage in this quadrant. Similarly, Fig. 8 demonstrates the simulation waveforms for the transition from the third to the fourth quadrant. In this case, the motor operates from reverse motoring in the third quadrant to reverse breaking in the fourth quadrant followed by forward motoring and keeps steady in the first quadrant. Therefore, the system can operate under stable and efficient four-quadrant operation.

The steady-state behavior has been investigated at 1000 rev/min and 3 Nm load. As shown in Figs. 9 and 11, the harmonics in the stator and filtered input currents are low, and the magnitudes of the dominant harmonics, at 5 kHz, are less than 2.5% of those of their fundamentals, respectively. In this case, fixed switching frequency, 5 kHz, is achieved, which is equal to the reciprocal of the sampling period 200 μs. The stator current shows sinusoidal waveform with fundamental magnitude 1.879 A at 100/3 Hz and total harmonic distortion (THD) 1.83% in the steady-state. The filtered input current contains 50 Hz fundamental with its magnitude 1.073 A and THD 2.88%. Fig. 10 shows the output line-to-line voltage and its frequency spectrum of the matrix converter in the range from 0 to 11 kHz. The matrix converter operates at 33.333Hz output frequency with 5 kHz switching frequency.

V. CONCLUSION

In this paper, a new matrix converter fed IPMSM drive scheme using DTC technique has been proposed. The Indirect SVM technique is employed to achieve decoupling control of the matrix converter on the basis of the control requirements of the motor side and performance of the grid side. The stator flux linkage and torque control is realized by variable structure control and SVM in the inversion stage and unity input power factor is achieved in rectification stage.

The effectiveness and feasibility of the proposed drive system has been tested in steady-state and transient conditions under rated torque reversal and four-quadrant operation, carrying out some numerical simulations. The dynamic behaviors have been particularly shown during the transitions between motoring and regenerative breaking operation. The torque and flux ripples are significantly reduced, and also the switching frequency remains constant independent of operating conditions. The filtered input currents, the stator currents and the input line-to-line voltages are improved, of which the dominant harmonics are around a fixed frequency 5 kHz determined by the ISVM sampling period.

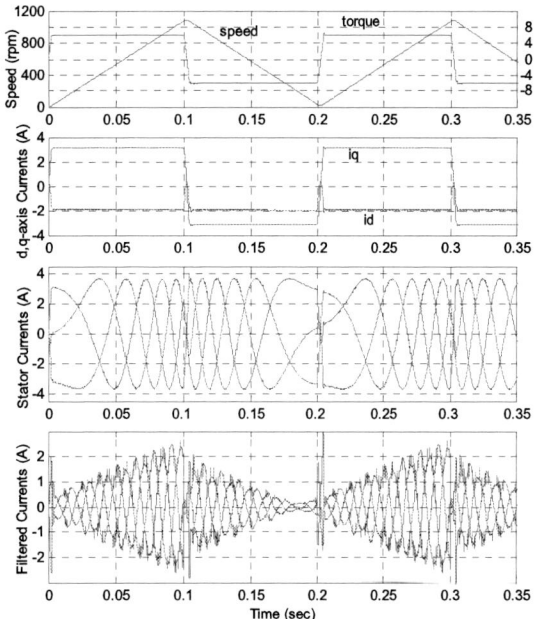

Figure 2. Responses to a step torque command at ±6 Nm

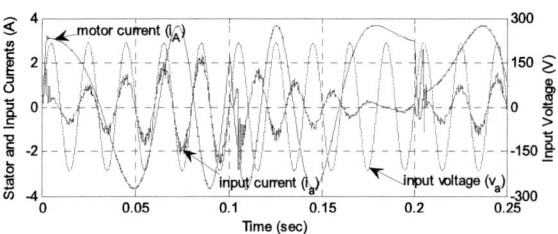

Figure 3. Transition between the first and the second quadrant

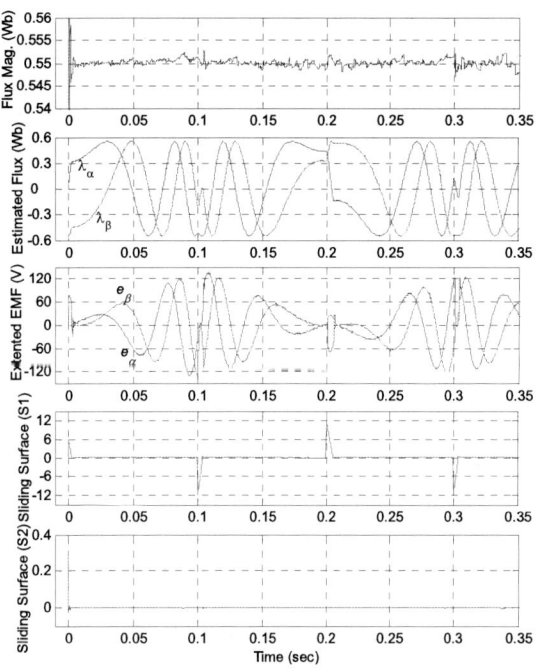

Figure 4. Stator flux, EEMF and sliding surfaces with ±6 Nm step

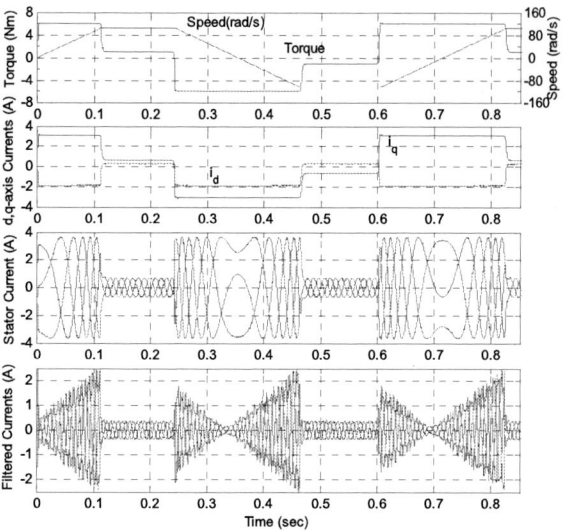

Figure 5. Four-quadrant operation waveforms with light load

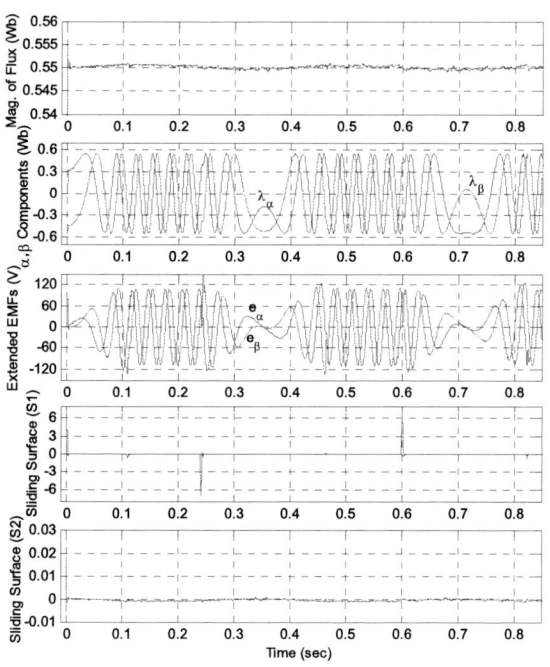

Figure 6. Stator flux, EEMF and sliding surfaces with speed step

TABLE I   PARAMETERS OF THE IPMSM USED IN THIS PAPER

Rated output power (Watt)	P_r	1000
Rated phase voltage/current (V/A)	V / I	132/3
Magnetic flux linkage (Wb.)	λ_f	0.533
Number of pole pairs	P	2
Rated torque (Nm)	T_b	6
Stator resistance (Ω)	R_s	5.8
dq-axis inductances (mH)	L_d , L_q	44.8, 102.7
Rotor inertia (Kg.m^2)	J	0.00529
Friction coefficient (Nm/rad/s)	D	0.0006

TABLE II.   PARAMETERS OF VS-DTC SCHEME

Positive control gains of VS controller	μ_1 , μ_2	3200, 200
Smoothing factors of sliding surfaces	λ_1 , λ_2	0.4, 0.04
Control gains of VS controller	K_T , K_λ	3.0, 5.0
Initial stator flux components (Wb)	$\lambda_{\alpha 0}$, $\lambda_{\beta 0}$	0, -0.533
Sampling period (μs)	T_s	200

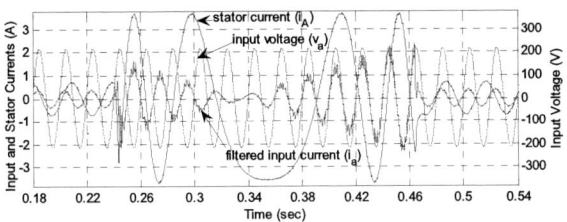

Figure 7.   Transitions from the first to the second quadrant

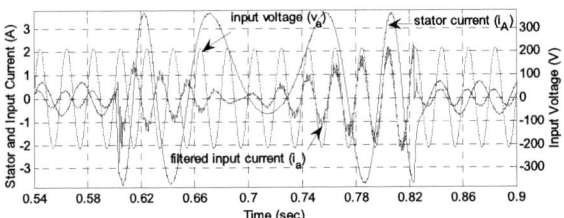

Figure 8.   Transitions from the third to the fourth quadrant

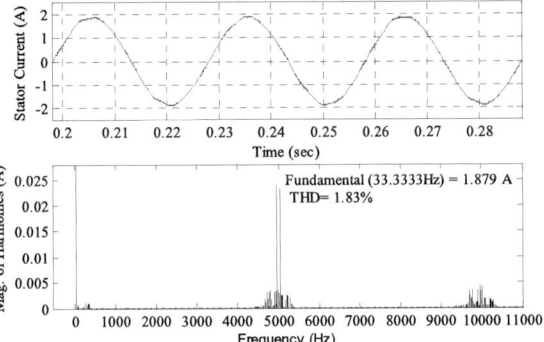

Figure 9.   Stator current and its spectrum at 1000 r/min and 3 Nm

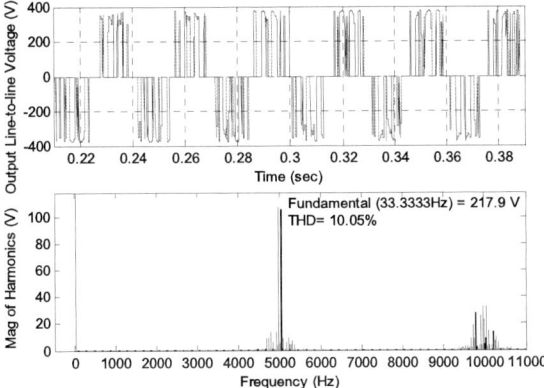

Figure 10.   Output voltage and spectrum at 1000 r/min, 3 Nm

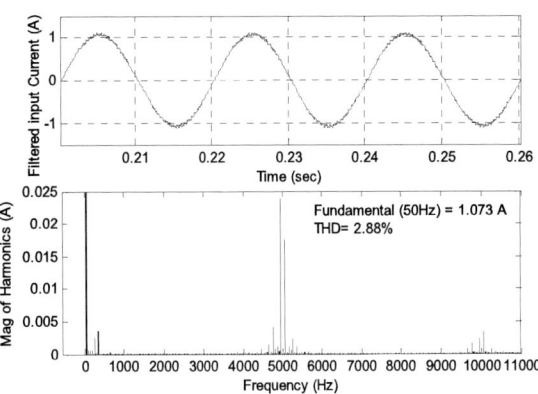

Figure 11.   Filtered input current and spectrum at 1000 r/min, 3 Nm

REFERENCES

[1]   C. French and P. Acarnley, "Direct torque control of permanent magnet drives," *IEEE Trans. Ind. Appl.*, vol. 32, no. 5, pp. 1080–1088, Sept./Oct. 1996.

[2]   L. Tang, L. Zhong, M. F. Rahman, and Y. Hu, "A novel direct torque control for interior permanent-magnet synchronous machine drive with low ripple in torque and flux-a speed sensorless approach," *IEEE Trans. Ind. Appl.*, vol. 39, no. 6, pp. 1748–1756, Nov./Dec. 2003.

[3]   P. W. Wheeler, J. Rodríguez, J. C. Clare, L. Empringham, and A. Weinstein, "Matrix converter: a technology review," *IEEE Trans. Ind. Electron.*, vol. 49, no. 2, pp. 276–288, Apr. 2002.

[4]   M. Venturini, "A new sine wave in sine wave out, conversion technique which eliminates reactive elements," in *Proc. POWERCON 7*, 1980. pp. E3_1–E3_15.

[5]   M. Venturini and A. Alesina, "Analysis and design of optimum-amplitude nine-switch direct AC-AC converters," *IEEE Trans. Power Electron.*, vol. 4, no. 1, pp. 101–112, Jan. 1989.

[6]   D. Casadei, G. Serra, and A. Tani, "The use of matrix converters in direct torque control of induction machines," *IEEE Trans. Ind. Electron.*, vol. 48, no. 6, pp. 1057–1064, Dec. 2001.

[7]   D. Casadei, G. Grandi, G. Serra, and A. Tani, "Space vector control of matrix converters with unity input power factor and sinusoidal input/output waveforms," in *Proc. EPE*, Brighton, U.K., Sept.13–16, 1993, vol. 7, pp. 170–175.

[8]   P. Nielsen, F. Blaabjerg, and J. K. Pedersen, "Space vector modulated matrix converter with minimized number of switchings and a feedforward compensation of input voltage unbalance," in *Proc. IEEE PEDES'98*, vol. 2, 1996, pp. 833–839.

[9]   L. Huber, and D. Borojevic, "Space vector modulated three-phase to three-phase matrix converter with input power factor correction," *IEEE Trans. Ind. Appl.*, vol. 31, no. 6, pp. 1234–1245, Nov./Dec. 1995.

[10]   Z. Xu and M. F. Rahman, "A variable structure torque and flux controller for a DTC IPM synchronous motor drive", in *Proc. IEEE PESC'04*, 2004, vol. 1, pp. 445–450.

2006 5th International Power Electronics and Motion Control Conference

A Neural Network Based Initial Position Detection Method To Permanent Magnet Synchronous Machines

Mengjia Jin[*], P.C.K Luk[**], Jianqi Qiu[*], Cenwei Shi[*] and Ruiguang Lin[*]

[*]Department of Electrical Engineer Zhejiang University, Hangzhou, China

[**]Department of Aerospace,Power & Sensors Cranfield University, Shrivenham, UK

Mengjiajin@hotmail.com

Abstract – **Initial Position is one of the key points in the permanent magnet synchronous machine (PMSM) senseless control. This paper presents a neural network method with currents difference as its inputs and the rotor position angle as its outputs. 256 different 100μs-width voltage vector pulse pairs injects into stator before position calculation. The voltage vector direction distributes symmetrically around an electrical cycle. So 256 current pairs can be obtained. Because of the saturation of the stator, the positive and negative vector pulses have different current responses. 256 current difference data can be got. As there's much noise in the data a neural network is introduced. 256 current difference data are used as the inputs of the neural network, and the electrical angle the rotor position is the outputs. A direct torque controlled (DTC) PMSM experimental system with the proposed initial position detecting method is set up. The scheme is verified by the experimental system with a 600w PMSM. The simulation and experimental results show that the scheme presented in the paper is available to get the initial position with 5.6% error and able to start a PMSM successfully.**

Index terms – PMSM, ANN, initial position

I. INTRODUCTION

Initial position information is essential for a permanent magnet synchronous motor drivers both vector control or direct torque control. The rotor's position information is often got by an encoder in vector controlled PMSM drives. Unfortunately for the cost reason most of the encoder applied in industry is incremental not absolute, so there's some problem to get the initial position. In direct torque control method the rotor's position is not necessary when the motor is running, while it requires the initial position information when starting. Uncertainness of initial position may cause reserve at start or starting failure [1].

PMSMs' starting methods can be divided into three groups [2]. The first is aligning the rotor to a known position by a known stator current vector. It can be realized by a current close loop control or a constant switch pattern [3]. The critical problem of this method is

when the load is unknown the currents required is also unknown. Further more when the rotor is near to the applied current vector the torque will decrease quickly. Such decrease will make it difficult to align accurately. The second is by mathematic procedure. The algorithms such as Kalman and sliding mode can make the oscillation minimized [4][5]. The third is getting initial position by rotor's saliency. It used in interior permanent magnet synchronous machines (IPMs) successfully [6-11]. For surface mounted permanent magnet synchronous machines (SM-PMSMs), stator saturation method is used [12][13].

Artificial neural networks (ANN) have the ability of emulating nonlinear multi-variable function [14]. It is used in this paper to filtrate the noisy signal and pick up the position information. The initial position information can be got from the anisotropies caused by the saturation of stator iron. An electrical cycle is divided into 256 different directions symmetrically and 256 voltage vectors are obtained. Current response pairs can be measured when the voltage vector applied positively and oppositely to the stator coil. The scheme takes the difference of the current pairs as the input data of the ANN and the estimated angle as the outputs. Both the simulation and experimental results are shown.

II. SATURATION OF PMs

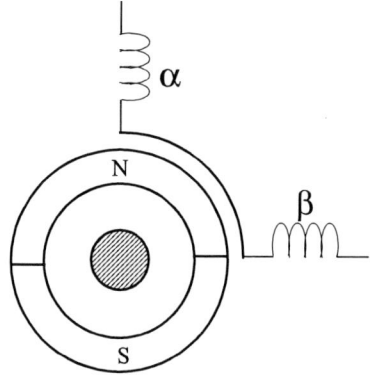

Fig. 1 PMSM sketch in αβ frame

A surface mounted permanent magnet synchronous machine has a structure looks like fig. 1 in αβ frame. When the rotor has the same direction as α axis, the flux in this direction gets maximums. The iron material always works near saturation area for the desire of higher iron material usage factor, shown as fig. 2.

Where, (Iw,Ψw) is the work point. A same current changing ΔI will cause different flux variation as the inference of stator saturation. The increments of flux caused by incremental currents are not so obvious as the decrease of flux for the descendent currents. For a PMSM, Iw is the equivalent current in stator caused by permanent magnets.

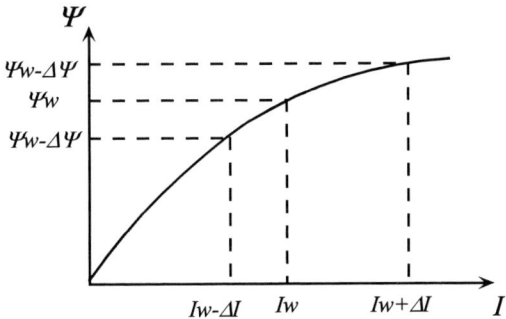

Fig. 2 Stator iron material character considering saturation

ΔI is the actual stator current Is. The inductance Ls is expressed as

$$L_s = \frac{\psi_s}{I_s}$$

So the increasing-flux-currents (currents who increase flux) have less inductance than the decreasing-flux-currents. It doesn't require the structure saliency of the rotor. To a standstill point on the stator, the work point is shifting with the rotation of the rotor. And the saturation of the stator is also changing with the rotor's location. If the voltage (Fig. 3 a) pilots to the stator coil, current difference can be obtained according to the equation

$$U_s = L_s \frac{di_s}{dt}$$

For example when the rotor flux has the same direction with α axis, and the voltage vector pulse has the same direction as rotor flux. Then the inductance of positive current is less than that of the opposite. The voltage pulse shown in fig. 3 (a) will gets a current as fig. 3 (b). $\Delta I\alpha$ is the difference of Iα+ and Iα-, which is caused by stator saturation. Every voltage vector pulse is applied, an Is+ and an Is- pair will be obtained and a ΔI can be got. In this paper, an electrical cycle is divided into 256 voltage vectors. So 256 ΔI can be got.

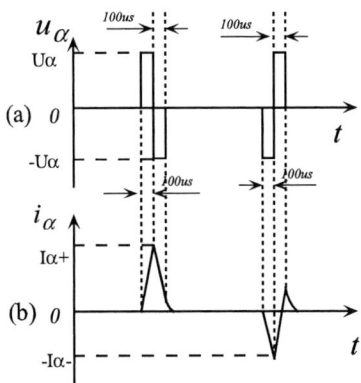

Fig. 3 Voltage pulse and current response

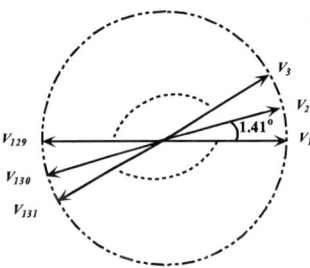

Fig. 4 the voltage vectors applied

Different rotor position produces different ΔI shape. Fig.5 (a) to (h) show ΔI shapes when the rotor stands at 0° and 315° electrical angle respectively.

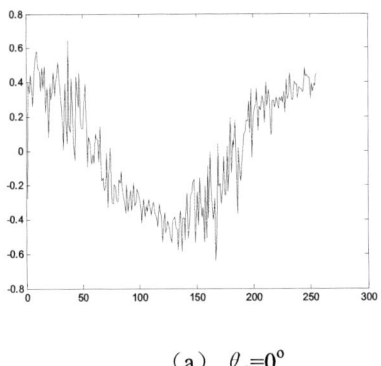

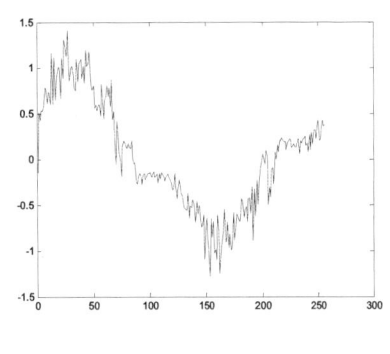

(a) $\theta_r=0°$

(b) $\theta_r=45°$

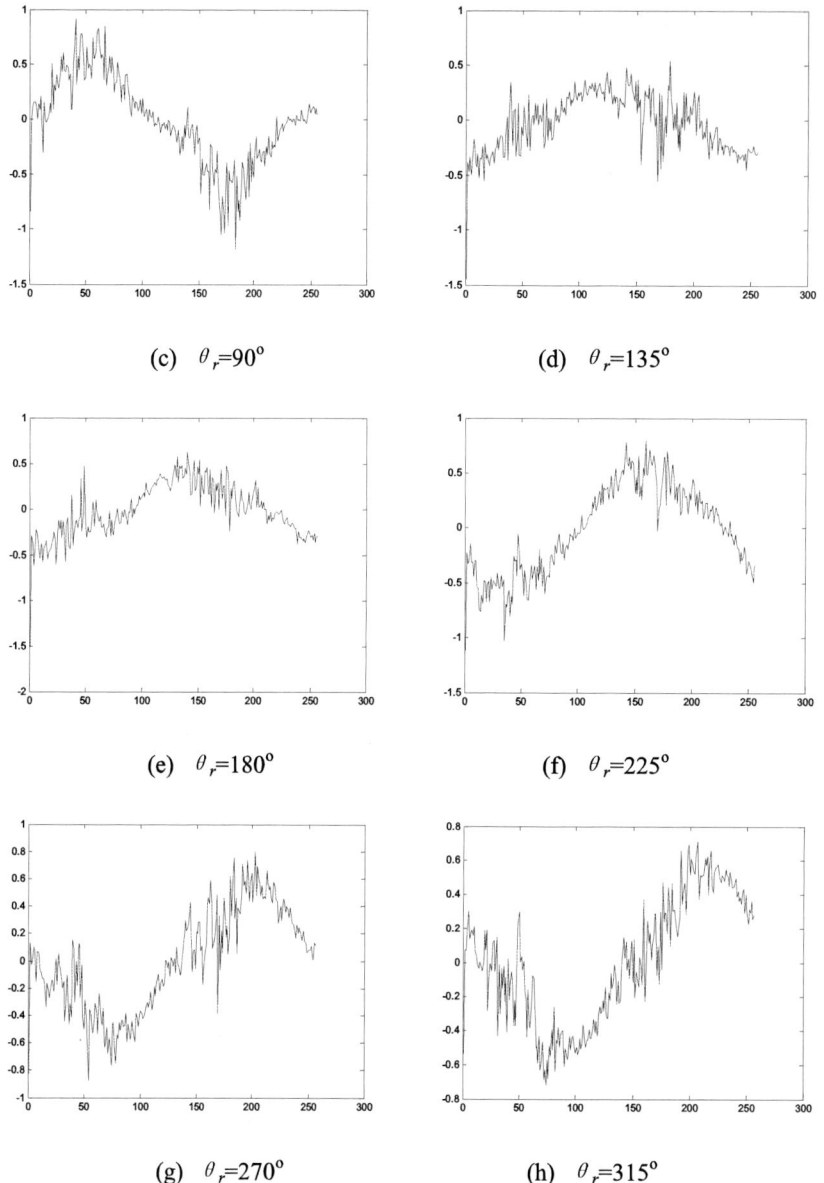

(c) $\theta_r=90^o$ (d) $\theta_r=135^o$

(e) $\theta_r=180^o$ (f) $\theta_r=225^o$

(g) $\theta_r=270^o$ (h) $\theta_r=315^o$

Fig. 5 ΔI sequences when the rotor fixed on different position

Fig. 5 shows that the ΔI shapes contain the rotor initial position information but with much disturbance. These disturbance and uncertainness make the signal not clear. A neural network is applied in the position identification to learn the relationship between the ΔI and the position.

III. BACKPROPAGATION NEURAL NETWORK

A backpropagation neural network [8] is used in the system. The structure of the network is shown in fig.6.

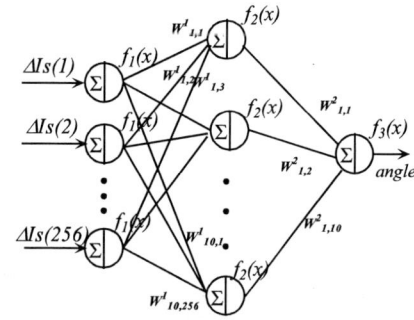

Fig.6 back propagation neural network

The networks have three layers, input layer, one-hidden layer and output layer. The input layer contains 256 input neurons. The hidden layer has 10 neurons and there's one neuron in output layer. The functions of each layer is

$$\begin{cases} f_1(x) = x \\ f_2(x) = \dfrac{1}{1-e^{-x}} - \dfrac{1}{2} \\ f_3(x) = x \end{cases}$$

IV. TRAINING OF THE NETWORK

A. Obtaining The Training Data

The experimental PMSM has 4 pole pairs, and an electrical cycle corresponds to 1/4 mechanical round. The electrical cycle is divided into 32 points. Then 256 currents and a position (an input-output pair) can be got when the rotor fixed at one point. So there are only 32 training data sets. It's not enough for training and needs to be extended. Assuming that the output keeps unchanged when the 256 currents changing in the same scale of 0.98, the process will continue until it decreases to 0.2 of the original data. So every data set generates about 80 data, and 2560 as total. Slight white noise of the input has no effect on the output. A set of white noises with 0.05-0.1 the amplitude that of the signal is added to the data. The increasing step of the noise keeps 0.01 and there are 5 sets of data in the range 0.05-0.1. So there are 128,000 sets training data. The training error is shown as fig.7.

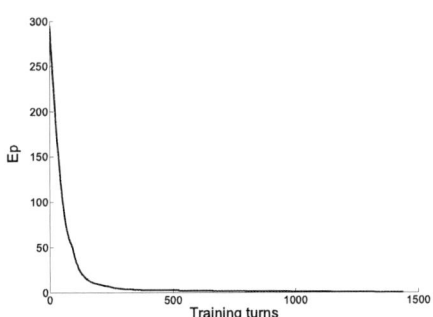

Fig.7 training error

Where, Ep is the sum of square error of the objective output and training output. It achieves 0.0098 after 1300 training.

B. Verification of The Network

The testing data is generated by 0.12 noises and the original data (untrained). The output of the network and actual position are shown in fig.8. The estimated position and the real one is almost agree when the study is finished.

V. EXPERIMENTAL RESULTS

A DTC method is applied in the experimental system that was described in detail in the paper [15]. The parameters of the PMSM used in the experiments are listed in table 1.

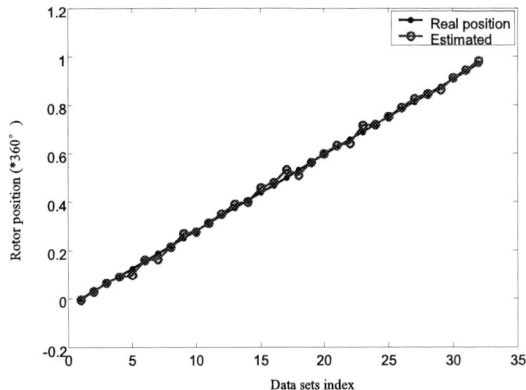

Fig.8 actual and estimated position with 0.12 noises off-line

TABLE 1.PARAMETERS OF PMSM IN SIMULATION AND EXPERIMENTS

Power (Kw)	0.6
Rated line voltage (V)	110
Phase flux of PM (Wb)	0.2144
Pole pairs	4
Rated torque (Nm)	6
Rated speed (rev/min)	1000

When the rotor is fixed at different angle, voltage pairs are piloted into the stator on line. A TI DSP TMS2407A generates the estimated angle. Fig.9 shows the estimated angle and actual one and error between them.

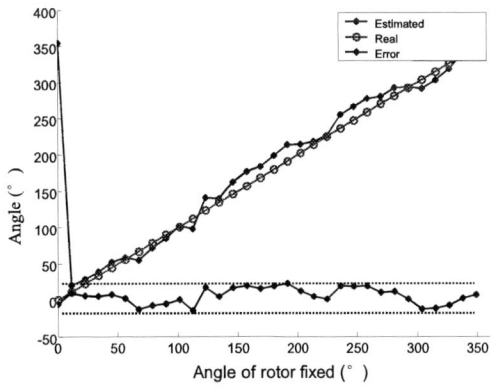

Fig.9 the actual and estimated position on-line

In fig.9, the maximum of the error is 20°(5.6% of 360°). Fig.10 shows the speed and torque curves of the starting procedure with rated load with an assumptive estimated initial angle of 75°, while the rotor stands at 90° (the actual estimated angle may be more accurate than 75°) .So in the worst situation with 20° estimation errors, the machine can start up steadily without any reverse.

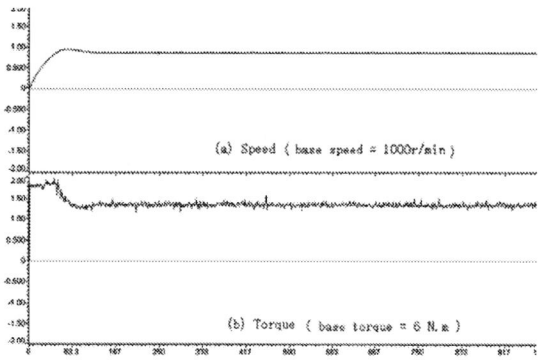

(a) Speed (base speed ≈ 1000r/min)

(b) Torque (base torque = 6 N.m)

Fig.10 Start up with estimated angle of 70° when the rotor fixed at 90°

(sample period 1.5ms)

VI. CONCLUSION

This paper proposed a novel method to detect the initial position of PMSM without the requirement of the mechanical saliency. The saturation of the iron material and neural network are used. The training and test simulation shows the ability of the ANN. The experimental results show that the machine can start up successfully with the permitted error.

REFERENCE

[1] G. Pfaff, A. Weschta, A. Wick, "Design and experimental results of a Brushless AC Servo Drive," IEEE Transactions on Industry Applications, Vol. 20, NO. 4, pp: 814-821, July1984.

[2] N. Matsui, "Sensorless PM Brushless DC motor Drives," IEEE Transactions on Industrial Electronics, Vol. 43, NO. 2, pp: 300-308, April 1996.

[3] N. Matsui, "Sensorless operation of brushless DC motor drives," Proc. of the IEEE-IECON'93, pp: 739-744, 1993.

[4] F. Parasiliti, R. Petrella, M. Tursini, "Sensorless Speed Control of a PM Synchronous Motor by Sliding Mode Observer," Proc. of the IEEE ISIE '97, Guimarles, Portugal, Vol. 3, pp: 1106-1111, Jul. 1997.

[5] F. Parasiliti, R. Petrella, M. Tursini, "Rotor Speed and Position Detection for PM Synchronous Motors Based on Sliding Mode Observer and Kalman Filter," Proc. Of EPE '99, Lausanne, Switzerland, pp: 394-398, September 1999.

[6] M. Tursini, R. Petrella, F. Parasiliti, "Initial rotor position estimation method for PM motors", IEEE Transactions on Industry Applications, Vol. 39, NO. 6, pp: 1630 - 1640, Nov. 2003.

[7] M. Boussak, "Sensorless speed control and initial rotor position estimation of an interior permanent magnet synchronous motor drive" proc. Of IECON 02 2002 Vol. 1, pp: 662 – 667, Nov. 2002.

[8] Matsui N, Takeshita T. "A novel starting method of sensorless salient-pole brushless motor",Conference Record of Industry Applications Society, vol.1, pp:386 – 392,1994.

[9] Yu-seok Jeong, Lorenz R D, Jahns T M, Seung-Ki Sul. "Initial rotor position estimation of an interior permanent-magnet synchronous machine using carrier-frequency injection methods", IEEE Transactions on Industry Applications, Vol. 41, No. 1, pp:38 – 45, Jan. 2005.

[10] Hyunbae Kim, Kum-Kang Huh, Lorenz, R D, Jahns T M. "A novel method for initial rotor position estimation for IPM synchronous machine drives", IEEE Transactions on Industry Applications, Vol. 40, No. 5pp: 1369 – 1378, Sept. 2004.

[11] Jung-Ik Ha, Ide K, Sawa T, Seung-Ki Sul. "Sensorless rotor position estimation of an interior permanent-magnet motor from initial states", IEEE Transactions on Industry Applications, Vol.39, NO. 3, pp: 761 – 767, May 2003.

[12] Tanaka K, Yuzawa T, Moriyama R, Miki I. "Initial rotor position estimation for surface permanent magnet synchronous motor", Record of Thirty-Sixth Industry Applications Conference, Vol. 4, No.30, pp: 2592 - 2597, Oct. 2001.

[13] Yuzawa T, Tanaka K, Moriyama R, Miki I. "An efficient estimation method of sensorless initial rotor position for surface PM synchronous motor", Proceeding of IEEE International Electric Machines and Drives Conference, pp: 44 – 49, 2001.

[14] Martin T.Hagan, Howard B.Demuth and Mark Beale, "Neural network design", THOMSON press pp. 11.1 –11.45, 2002.

[15] L. Zhong, M.F. Rahman, W.Y. Hu, "A direct torque controller for permanent magnet synchronous motor drives", IEEE Transactions on Energy Conversion, Vol.14, No.3, pp:637 - 642, Sept. 1999.

2006 5th International Power Electronics and Motion Control Conference

A New Recurrent Fuzzy Neural Network Sliding Mode Position Controller Based on Vector Control of PMLSM Using SVM

Junyou Yang, Ruijuan Chen, Naiguang Fa

School of Electrical Engineering, Shenyang University of Technology, Shenyang, 110023, China
chenruijuan111@126.com

Abstract—Sliding mode controller using a recurrent fuzzy neural network (RFNN) is presented, in which RFNN is utilized to estimate the real-time lumped uncertainty for the position control of permanent magnet linear synchronous motor (PMLSM) drive system, so that the control effort can be reduced. Considering the convergence rate, global feed-forward RFNN is employed instead of global feedback RFNN. Furthermore, space vector modulation (SVM) that can decrease the vector deviation is adopted. Simulation results show that the proposed new recurrent fuzzy neural network sliding mode position control scheme provides a fast and robust regulation for the mover position.

Keywords—sliding mode control; chattering; recurrent fuzzy neural network (RFNN); PMLSM; vector control; SVM

I. INTRODUCTION

Based on its notable advantages, such as no backlash and less friction, high speed and high precision in long distance locations, high thrust force and reliability, permanent magnet linear synchronous motors (PMLSMs) are widely used in servo drives [1]. However, it is well known that PMLSM is greatly affected by mechanical parameter variations and external load disturbance in the drive system. For high performance, an appropriate control strategy is needed eagerly.

Sliding mode control, which has been employed in the position and speed control of AC servo systems recently, is insensitive to parameter variations and external disturbances once the control system enters the sliding mode. However, its applications are limited by three main problems, i.e. the assumption of known uncertainty bounds, static error and chattering phenomenon. With the help of the designed integral-operation switching surface [2], the second existing problem can be solved, especially on loaded occasion. Furthermore, in order to reduce chattering phenomenon, the sign function in sliding mode controller is replaced by saturation function [3]. However, both of them result in a decreasing robust. To overcome the weakness, second order sliding mode control is emerged [4][5], but it's difficult to come true. Recurrent

fuzzy neural network (RFNN), which is composed of fuzzy control and recurrent neural network (RNN), has dynamic and robust virtues. By means of the nonlinear adaptation and learning ability of RNN, the fuzzy inference mechanism and fuzzy rules in RFNN can be adjusted simply compared with conventional fuzzy controller. To compensate the disadvantages, the combination of sliding mode control and RFNN is expected. Space vector modulation (SVM), which is utilized to reduce thrust force ripple, uses a special scheme to switch the power transistors to generate pseudo sinusoidal currents. The scheme comes from the translation of the voltage reference vector into an amount of commutation (on/off) for each power transistors.

Therefore, a new RFNN sliding mode controller to control the mover position based on vector control of PMLSM is designed, in which the RFNN is utilized to estimate the real-time lumped uncertainty for the position control system, thus the control effort can be reduced. Moreover, since the global feedback RFNN has drawbacks of ripple train, not fast enough convergence rate and large computation cost, a global feed-forward RFNN is applied. In the study, a four-layer recurrent network is investigated, the input signals are the switch surface and its derivation, and the output is the estimated system uncertainty. To compare with the sliding mode controller using the two kinds of RFNN respectively, several simulation systems are built to verify the feasibility of the proposed control scheme.

II. THE MODEL OF PMLSM

Adopting Park vector transformation, and based on the principle of total power invariance, the mathematics model of PMLSM is deduced. In a synchronous rotating reference frame, it is given as follows[6]:

$$u_d = R_s i_d + p\lambda_d - \omega_e \lambda_q, \qquad (1)$$

$$u_q = R_s i_q + p\lambda_q + \omega_e \lambda_d, \qquad (2)$$

where

$$\lambda_d = L_d i_d + \lambda_{PM}, \qquad (3)$$

Supported by National Natural Science Foundation of China (50375102) and Program for Liaoning Excellent Talents in University (RC-04-14)

1-4244-0448-7/06/$25.00 ©2006 IEEE

$$\lambda_q = L_q i_q, \tag{4}$$

$$\omega_e = P_n \pi v / \tau. \tag{5}$$

The electromagnetic force

$$F_e = \pi P_n [\lambda_{PM} i_q + (L_d - L_q) i_d i_q] / \tau, \tag{6}$$

and the mover dynamic equation is

$$F_e = Mpv + Dv + w(t), \tag{7}$$

where u_d, u_q are the d and q axis voltages; i_d, i_q are the d and q axis currents; R_s is the phase-winding resistance; L_d, L_q are the d and q axis inductances; ω_e is the electric velocity of the mover; λ_{PM} is the permanent magnet flux linkage; P_n is the number of primary poles and p denotes the differential operator; v is the linear velocity; τ is the pole pitch; M is the total mass of the moving element system; D is the viscous friction and iron-loss coefficient and $w(t)$ is the external disturbance term.

A. Field-oriented PMLSM Model

With the implementation of vector control, if $i_d = 0$, the mathematics model of PMLSM can be simplified as follows:

$$F_e = K_F i_q^*, \tag{8}$$

$$K_F = \pi P_n \lambda_{PM} / \tau, \tag{9}$$

$$H_p(s) = 1/(Ms + D), \tag{10}$$

where K_F is the thrust coefficient; i_q^* is the command of the thrust current and s is the Laplace operator.

III. New Recurrent Fuzzy Neural Network Sliding Mode Position Controller

Two mathematics models, i.e. the sliding mode controller model and the model with RFNN are built. Then, a comparison is made between global feedback RFNN and global feed-forward RFNN. For a fast control performance, global feed-forward RFNN is chosen finally.

A. The Model of Sliding Mode Position Controller

In nominal condition without parameter variations and external load disturbances, the dynamic equation of PMLSM drive system can be written as [7]:

$$U(t) = \overline{A}_1 \ddot{d}(t) + \overline{B}_1 \dot{d}(t), \tag{11}$$

where

$$\overline{A}_1 = \frac{\overline{M}}{\overline{K}_F}, \overline{B}_1 = \frac{\overline{D}}{\overline{K}_F}, U(t) = i_q^*(t),$$

and d denotes the mover position of PMLSM.

Considering the parameter variations and external load disturbances, we can obtain

$$\tilde{A}_1 \ddot{d}(t) + \tilde{B}_1 \dot{d}(t) + C_1 F_L = U(t), \tag{12}$$

where $C_1 = 1/K_F$, $\tilde{A}_1$ and $\tilde{B}_1$ include the uncertainties, which are introduced by the system parameters $\overline{M}$ and $\overline{D}$, F_L represents the external disturbance to the system.

Moreover, (12) can be rewritten as follows:

$$U(t) = \overline{A}_1 \ddot{d}(t) + \overline{B}_1 \dot{d}(t) + (\Delta \overline{A}_1 \ddot{d}(t) + \Delta \overline{B}_1 \dot{d}(t) + C_1 F_L)$$

$$= \overline{A}_1 \ddot{d}(t) + \overline{B}_1 \dot{d}(t) + \gamma, \tag{13}$$

the lumped real-time uncertainty γ

$$\gamma \equiv \Delta \overline{A}_1 \ddot{d}(t) + \Delta \overline{B}_1 \dot{d}(t) + C_1 F_L. \tag{14}$$

If the positive scalar λ is introduced, according to (13), we modified the reference trajectory equation [8] to:

$$\overline{A}_1 \left(\ddot{d}_d(t) - \lambda \dot{\tilde{d}}(t) \right) + \overline{B}_1 \left(\dot{d}_d(t) - \lambda \tilde{d}(t) \right) + \gamma \overset{\Delta}{=} W(t)\Theta, \tag{15}$$

where

$$W(t) = [\ddot{d}_d(t) - \lambda \dot{\tilde{d}}(t) \quad \dot{d}_d(t) - \lambda \tilde{d}(t) \quad 1] = [W_1(t) \, W_2(t) \, 1],$$

$$\Theta = \left[\overline{A}_1 \, \overline{B}_1 \, \gamma \right]^T,$$

and $d_d(t)$ is the desired mover position signal, the tracking error is defined as $\tilde{d}(t) = d(t) - d_d(t)$, $S(t)$, as the switching function is defined as follow:

$$S(t) = \dot{\tilde{d}}(t) + \lambda \tilde{d}(t). \tag{16}$$

B. Sliding Mode Controller Using RFNN

The RFNN in the presented sliding mode position controller is utilized to estimate the lumped real-time uncertainty γ. Its output $\hat{\gamma}$ is the estimate of γ, thus a new vector $\hat{\Theta}_s = \left[\overline{A}_1 \, \overline{B}_1 \, \hat{\gamma} \right]^T$ can be acquired.

The sliding mode law using RFNN is defined as

$$\Theta_s(t) = -F(t)\hat{\Theta}_s,$$

$$F(t) = \begin{bmatrix} \mathrm{sgn}(f_1(t)) & 0 & 0 \\ 0 & \mathrm{sgn}(f_2(t)) & 0 \\ 0 & 0 & \mathrm{sgn}(S(t)) \end{bmatrix},$$

$$f_i(t) = W_i(t)S(t), \quad i = 1, 2,$$

$$\mathrm{sgn}(f_i(t)) = \begin{cases} 1, & f_i(t) > 0 \\ 0, & f_i(t) = 0 \\ -1, & f_i(t) < 0 \end{cases}, \tag{17}$$

the sliding mode control law is

$$\begin{cases} U(t) = U_S(t) + U_l(t) \\ U_l(t) = -\alpha \dot{\tilde{d}}(t) - \beta \tilde{d}(t) \quad (\alpha, \beta > 0) . \\ U_S(t) = W(t)\Theta_s(t) \end{cases} \quad (18)$$

The inputs of the RFNN are $S(t)$ and its derivative $\dot{S}(t)$.

C. The Structure of RFNN

RFNN has the advantages of both recurrent neural network (RNN) and fuzzy control. Their combination is researched in great demand. Several types of recurrent neural networks have been introduced [9]. Nowadays, two of them are used frequently, i.e. global feedback RNN and global feed-forward RNN, resulting in two kinds of RFNN, i.e. global feedback RFNN and global feed-forward RFNN, respectively. The global feedback RFNN adopted in [7], which adds the overall output to the input layer, makes the computation task huge and all weights should be adjusted when training. For the purpose of simplification, a global feed-forward RFNN is presented.

Fig. 1 shows the structure of a four-layer global feed-forward RFNN, which contains an input layer, a membership layer, a rule layer and an output layer.

Nodes in the input layer are language variables. For every node, the net input and the net output are

$$net_i^1(N) = x_i^1(N) ,$$

$$y_i^1(N) = f_i^1(net_i^1(N)) = net_i^1(N), \quad i = 1,2 , \quad (19)$$

where $x_1^1 = S(t)$, $x_2^1 = \dot{S}(t)$.

In the membership layer, all the six nodes perform a membership function. For the purpose of precision, Gaussian function is chosen as membership function. For

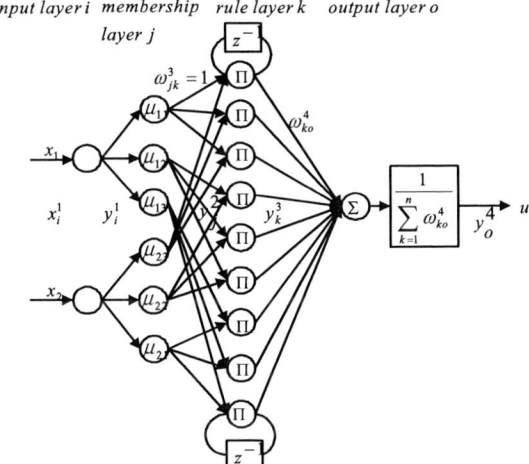

input layer i membership rule layer k output layer o
layer j

Fig.1. Structure of a global feed-forward RFNN

the *jth* node of layer 2, we have

$$net_{ij}^2(N) = -\left(y_{ij}^1 - m_{ij}\right)^2 \Big/ \left(\sigma_{ij}\right)^2 ,$$

$$y_{ij}^2(N) = f_{ij}^2(net_{ij}^2(N)) = \exp(net_{ij}^2(N)), \quad j = 1,2,3 ,$$

$$y_{ij}^2(N) = \mu_{ij}(x_i), \ y_{ij}^1(N) = y_i^1(N), i = 1,2, \ j = 1,2,3 , \quad (20)$$

where m_{ij} and σ_{ij} are the mean and the standard deviations of the *jth* Gaussian function with regard to the *ith* input linguistic variable in the membership layer respectively, and for every input node, there are three linguistic variables.

The third layer, which is used as mechanism inference of fuzzy control, is called rule layer. Each node k in this layer is denoted by $\prod$, that means multiplies the input signals and outputs its product result. Its net input and net output are

$$net_{j_1j_2}^3(N) = \prod_i y_{1j_1}^2(N) y_{2j_2}^2(N) \cdot \omega_k y_k^3(N-1) ,$$

$$y_k^3(N) = f_k^3(net_k^3(N)) = net_k^3(N) ,$$

$$j_1, j_2 = 1,\cdots,3 , \ j_1 j_2 = k = 1,\cdots,l, \ l = 9 , \quad (21)$$

where $y_{1j_1}^2$, $y_{2j_2}^2$ denotes the *jth* input to layer 3; ω_{jk}^3, the weights between the membership layer and the rule layer, are assumed to be 1; ω_k is the recurrent weights; and l is the number of rules when complete rule connection if the same linguistic variables are selected for each input node.

The output layer has only one single node o, and its output is the total sum of all the input signals

$$net_o^4(N) = \sum_k \omega_{ko}^4 y_k^3(N) ,$$

$$\hat{\gamma}(N) = f_o^4(net_o^4(N)) = net_o^4(N), \quad o = 1 , \quad (22)$$

where ω_{ko}^4 is the connection weight between the *kth* rule and the *oth* output, y_k^3 signs the *kth* input to the node of layer 4.

D. The Online Learning Algorithm of The RFNN

The algorithm used is an improved gradient descent search algorithm in the space of network parameters.

IV. THE SPACE VECTOR MODULATION

The SVM, as a method to produce inverter control signals, which is used to approximate the desired stator reference voltage with only eight possible states of switches, is to combine adjacent vectors of the reference voltage and to modulate the application time of each adjacent vector. For an example, modulation period is T_p and the reference voltage vector is V_{ref}, as shown in Fig.2. This vector can be gained by compounding V_1, V_2, and zero space voltage vectors—V_0 and V_7.

1879

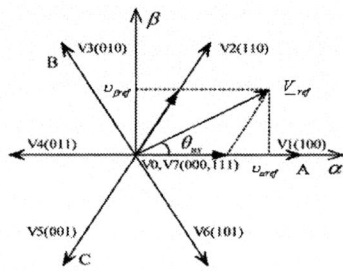

Fig.2. The principle behind reference voltage vector SVM

$$V_{ref}T_p = V_1T_1 + V_2T_2 \, ,$$

$$T_1 = T_p \, 2d \sin\left(\pi/3 - \theta_{us}\right)/\sqrt{3} \, ,$$

$$T_2 = T_p \, 2d \sin\left(\theta_{us}\right)/\sqrt{3} \, ,$$

$$T_0 = T_7 = \left(T_p - T_1 - T_2\right)/2 \, ,$$

$$d = V_{ref} \, V_{ref} \big/ V_{dc} \, , \qquad (23)$$

where V_{dc} is the DC link voltage.

V. SIMULATION RESULTS

The flexibility of the proposed new recurrent fuzzy neural network sliding mode position controller is verified by MATLAB/Simulink simulation results. Meanwhile, its application to vector control for a PMLSM is achieved.

The parameters of the PMLSM in the simulation are $R_s = 18.7 \ \Omega$, $L_d = L_q = L = 26.82 \ \text{mH}$, $\tau = 30 \ \text{mm}$, and $\overline{M} = 11 \ \text{kg}$, $\overline{D} = 1.2 \ \text{Ns/m}$, $\lambda_{PM} = 0.1717 \ \text{wb}$, $P_n = 3$, the control period of SVM $T_s = 100 \mu s$.

The learning of the two kinds of RFNN are carried out respectively, training results are presented in Fig.3 and Fig.4. We can see that global feed-forward RFNN has a more fast convergence rate than global feedback RFNN.

With the designed RFNN sliding mode controller, the simulation results based on vector control of PMLSM are shown in the following figures.

When M changes from $\overline{M}$ to $2\overline{M}$, the contrast of mover position responses of the new RFNN sliding mode control for PMLSM using SVM are shown in Fig.5, curve 1 is the position response when $M = \overline{M}$, and curve 2 is the one when $M = 2\overline{M}$. In Fig.5, the two curves are coincident after 0.05s, which means the parameter variation can't influence the output of the drive system, i.e. the system is insensitive to parameter variations.

Considered an external disturbance, the initial external disturbance is 0N, when t=0.2s it steps to 100N. As shown in Fig.6, the mover position regulation is robust, and both static and dynamic performances are good.

Fig.7 is the block diagram of RFNN sliding mode position controller for PMLSM drive.

If both the parameter variation mentioned above and external disturbance which steps when t=0.3s exist, from Fig.8 to Fig.10 we know that there is no static error, and a fast dynamic response and robustness are obtained. Fig.8 is the current curves, the curve in Fig.9 is the electromagnetic force response, Fig.10 draws its position response.

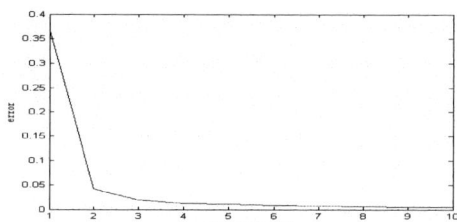

Fig.3. The error curve of global feed-forward RFNN

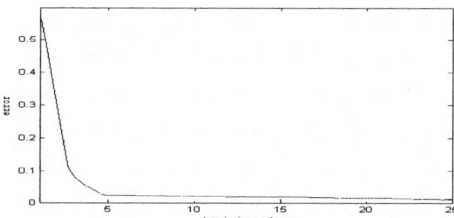

Fig.4. The error curve of global feedback RFNN

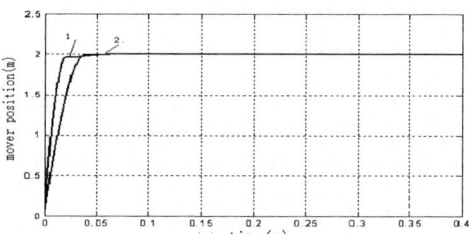

Fig.5. The contrast of position responses curves

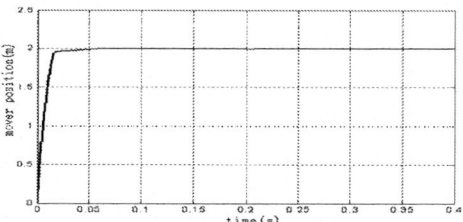

Fig.6. Reference position input is 2m, 100N external disturbance is given when t=0.2s

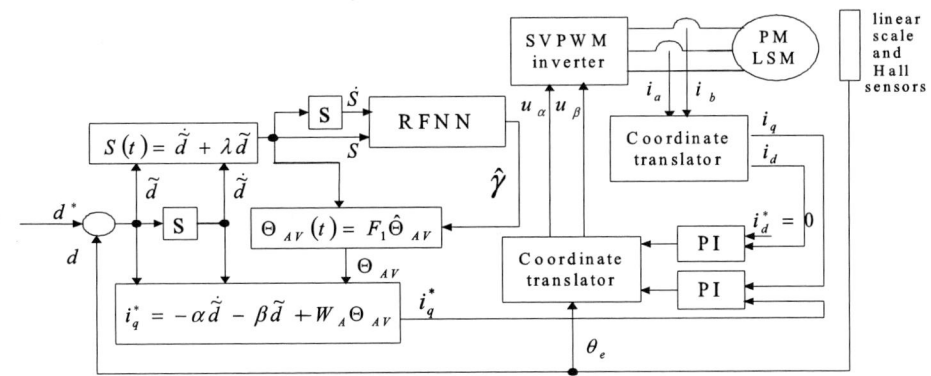

Fig 7. Block diagram of the position controller for PMLSM drive

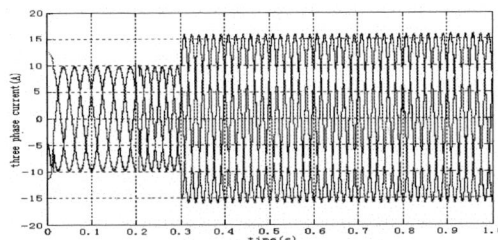

Fig 8. The current curves of the PMLSM

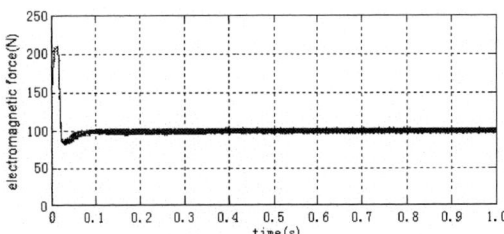

Fig 9. The electromagnetic force response

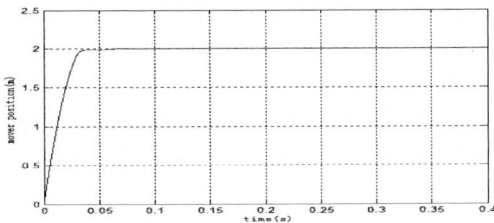

Fig 10. Reference position input is 2m, $M = 2\overline{M}$ and 100N external disturbance is given when t=0.3s

All parameters in the designed controller are obtained by repeated and attempted simulation.

VI. CONCLUSIONS

In the study, a new sliding mode position controller using recurrent fuzzy neural network for PMLSM is designed. The real-time lumped uncertainty including parameter variation and external disturbance are estimated by RFNN. The virtue point of the scheme is that it integrates the main advantages of sliding mode, RNN and fuzzy control. Therefore, it can realize strong robust and reduce chattering phenomenon by the designed position controller with SVM on–line in the condition of system parameter variation and the impact of outside uncertainty factors. Several simulations with SVPWM are done using Simulink and M file，the simulation results verify the feasibility of the proposed control strategy.

REFERENCES

[1] D. L. Trunpher, W. J. Kim, and M. E. Williams, "Design and analysis framework for linear permanent magnet machine," *IEEE Trans. IA.*, vol. 32, no. 2, pp. 371–379, 1996.

[2] Seshagiri S., Khalil H. K., "On introducing integral action in sliding mode control," *Proceedings of the 41st IEEE Conf. Decision and Control*, pp. 1473–1478, February 2002.

[3] Vicente P. V., Gerd H., "Chattering-free sliding mode control for a class of nonlinear mechanical systems," *IEEE Transl. J. Robust and Nonlinear Control*, pp. 1161–1178, November 2001.

[4] T. Floquet, J. P. Barbot and W. Perruquetti, "Second order sliding mode control for induction motor," *Proceedings of the 39st IEEE Conf. Decision and Control*, pp. 1691–1694, December 2000.

[5] G.. Bartolini, A. Ferrara and E. Usai, "On multi-input second order sliding mode control of nonlinear systems with uncertainty," *Proceedings of the 38st IEEE Conf. Decision and Control*, pp. 4245–4250, December 1999.

[6] I. Boldea and S. A. Nasar, *Linear Electric Actuators and Generators*. London, U.K.: Cambridge Univ. Press, 1997.

[7] F. J. Lin, C. H. Lin and P. H. Shen, "Variable-structure control for a linear synchronous motor using a recurrent fuzzy neural network," *IEE Proc. Control Theory Appl.*, vol. 151, no. 4, pp. 395–404, 2004.

[8] YU, H., SENEVIRATNE, L. D., and EARLES, S. W. E., "Exponentially stable robust control law for robot manipulators," *IEE Proc. Control Theory Appl.*, vol. 141, pp. 389–395, 1994.

[9] Cong Shuang, Dai Yi, "Structure of recurrent neural networks," *Computer Application*, vol. 24, no. 8, pp. 18–27, 2004.(in Chinese)

DSP Implementation of Rotor Position Detection Method for Hybrid Stepper Motors

M. Bendjedia*, Y. Ait-Amirat**, B. Walther* and A. Berthon **

*Haute Ecole Arc, engineering, Saint-Imier, Switzerland
** L2ES-UTBM/Universite of Franche-Comte, Belfort, France
Moussa.Bendjedia@he-arc.ch

Abstract— To improve the dynamic performance of the stepper motor it can be driven with a closed loop control. For sensorless control, initial rotor position detection is one of the serious problems. At standstill, if the rotor position is inaccurately detected, the stepper motor can move in the wrong direction or do not start at all. The method that uses inductance saliency to detect the initial rotor position is very suitable for hybrid Stepper motor (HSM) which is highly saturated. This paper will present experimental results for the initial rotor position estimation. At standstill, a sequence of voltage pulses is applied and the measured peak current give information about rotor position.

The DSP System used for the implementation is based on the powerful DS 1103 controller board from dSPACE. It allows programming dynamic real-time systems using Matlab/Simulink environment.

Keywords-Hybrid stepping motor; initial rotor position estimation; dSPACE DS 1103.

I. INTRODUCTION

The Hybrid Stepper Motor (HSM) is used in many applications, such as robotic systems, printers and consumer electronics. The HSM is a doubly salient machine which incorporates a permanent magnet in the rotor as described in [1].

In the open loop control the HSM can use only 50% of its nominal torque. A large torque reserve is required to overcome any load variation. In this classical control scheme there is no feedback of load position to the controller. The motor must responds to each excitation change. If the excitation changes are made too quickly the stepper motor can loose steps and therefore it is unable to move the rotor to the new demanded position. So a permanent error can be introduced between load position and the position expected by the controller. All these limitations can be overcome with a closed loop controller. In this case, the motor requires a rotor position sensor for providing the proper commutation sequence.

Mechanical sensors such as encoder increase cost and size of the motor. These position sensors are temperature sensitive and show lower reliability. To use these sensors, mechanical arrangement must be done. Due to these limitations, sensorless control is more attractive to drive the stepper motor. With today's low cost and high performance DSP's, manufactures of stepping motors are very interested to implement the advanced control algorithms.

Many sensorless methods suffer from start up. If the rotor position is not estimated at standstill, the motor starts from an unknown rotor position. In this case one can obtain a temporary reverse rotation or a starting failure. To overcome these problems, the motor can be started from predetermined rotor position. In the case of hybrid stepper motor, a voltage pulse in the inverter can be applied in order to align the rotor with one phase of the motor. The displacement of the motor before starting is not accepted in many precise motions. The better solution to guarantee a good startup, is the estimation of rotor position with an algorithm at standstill.

In this paper some experimental results will be presented concerning the initial rotor position detection for HSM by the method called "impulse current technique" [2][3]. This method based on the wellknown works of LEI laboratory from EPFL, uses the effect of the inductance saliency. The implementation is based on the powerful DS1103 from dSPACE and minimal external peripheral. The obtained results are good and confirm the effectiveness of the experimental setup.

First this study will start with an overview of the most important sensorless methods .

II. TECHNIQUES OF SENSORLESS ROTOR POSITION ESTIMATION

A. Sensorless methods

There are different methods to extract the rotor position from the Back-EMF. For example, the techniques [1] [2] of zero-crossing phase current extinction and reconstruction of Back-EMF are implemented and compared in [4]. At low speed it is very difficult to get rotor position using these methods.

Kalman filtering is computationally efficient candidate for estimation of speed and rotor position. It can also be used for on-line estimation of system parameters such as load torque and coil resistor. Kalman filter has a low pass filter characteristic which is the greater advantage over other techniques. For HSM, the Extended Kalman Filter (EKF) was implemented in [5] and [6]. The important step in Kalman filter is the choice of filter coefficient values, disturbance covariance Q and measurement covariance R, to obtain the best estimation. In [7] the

EKF was compared with Model Reference Adaptive (MRAS) technique and the results confirm that the disadvantage of EKF is the computational cost due to Jacobian matrix calculations. To overcome this problem we can reduce the order of the system as demonstrated in [8].

The rotor position can be estimated from the flux linkage, which is calculated form stator voltage and phase current [9]. The phase inductance of the motor varies as a function of rotor position. In [10], the phase inductance is calculated from voltage and phase current during different segments of electrical cycle. The calculated value is then used to estimate the rotor position with a lookup table.

B. Standstill methods

The sensorless methods of the previous section are not self-starting. We must use an algorithm to detect the position at standstill to guarantee a good startup. In the case of Back-EMF technique, the motor must first be started accelerated to a certain speed where the Back-EMF can be detected. The variation of inductance with rotor position is the principal effect used in all standstill methods.

The majority of the standstill methods uses an auxiliary signal injection via the PWM inverter. With the measured phase current, we can extract rotor position information. In [11], the author uses sinusoidal signal injection to detect the initial rotor position and continues with Back-EMF technique. In this group of methods, there is an other technique called Indirect Flux Detection by On-line Reactance Measurement (INFORM) [12]. The basic idea of this method is to apply a test voltage space phasor and measure the behaviour of the current space phasor which varies as a function of rotor position.

An other group of standstill methods uses the high frequency content of pulse width modulation (PWM).The rotor position is estimated at every PWM period[13].

The "current impulse technique"[2][3] is very suitable for hybrid stepper motor which is highly saturated. This technique is also able to separate the north pole of the permanent magnet from the south pole as will be shown in the next section.

III.   CURRENT IMPULSE TECHNIQUE

This method is well described by the works of J. Persson in [14].

The phase inductances of hybrid stepper motor vary as a function of the rotor position because of the saturation effect of the core in magnetic circuit. Instead of measuring the inductance, a sequence of voltage pulses is applied at standstill and the measured peak current gives information about initial rotor position.

A positive voltage pulse is applied to the phase A through the driver, with the DC voltage source. The current peak (i_1^+) is hold and memorized. A high value corresponds to the case where the permanent flux has the same direction as the phase current flux. The total flux

increases and saturates the magnetic circuit. In this case the winding inductance is low.

The procedure is repeated with a negative voltage pulse. The current (i_1^-) will give a phase current flux in the opposite direction of the permanent magnet flux. The total flux will decrease and the phase inductance is high.

The two current amplitudes can be then compared.

The sign of these current differences is a function of the total flux. If this difference is positive, the magnet flux has the same direction as that generated by the positive pulse, and consequently the flux is positive. If this difference is negative, the magnet flux has the same direction as that generated by the negative pulse, and then the flux is negative. One phase can inform about the rotor position with an accuracy of 180°. This is also the reason why this method is able to separate the magnetic north pole from the south pole by using both positive and negative currents.

The process is repeated for the second phase B. The values of the current differences Δi_a and Δi_b vary continuously and sinusoidal with the rotor position.

IV.   EXPERIMENTAL SETUP

A. DSP-based drive system

The laboratory setup is presented in Fig.1. The experimental setup consists of a DS1103 board from dSPACE, electronic interface, hybrid stepper motor coupled with a DC motor and an incremental encoder. The dSPACE DS1103 PPC Controller board is a complete real-time control system based on the Motorola PowerPC 604e processor running at 933MHz. The board includes a slave DSP subsystem based on the Texas Instruments TMS320F240 DSP microcontroller.

The most relevant features served by the master PPC are:
-16 multiplexed channels ADC-16 bit ±10 V
- 4 channels ADC-12 bit ± 10V
- 8 channels DAC-14 bit ±10 V
- 32 bit digital I/O
- 7 channels incremental encoder Interface

The most relevant features served by the slave DSP are:
- 16 channels ADC-14 bit [0 5]V
- 18 bit digital I/O
- 10 PWM output signals

The Matlab / Simulink models can be implemented and tested in real-time. The Real-Time Interface (RTI) contains a library of Blocks which connects the Simulink model to the physical world. This library is subdivided in two sublibraries. The Master PPC library contains the blocks served by the Power PC 604e master processor such as Encoder, ADC and DAC. The TMS320F240 DSP uses the slave DSP F240 library which contains the PWM outputs,etc.

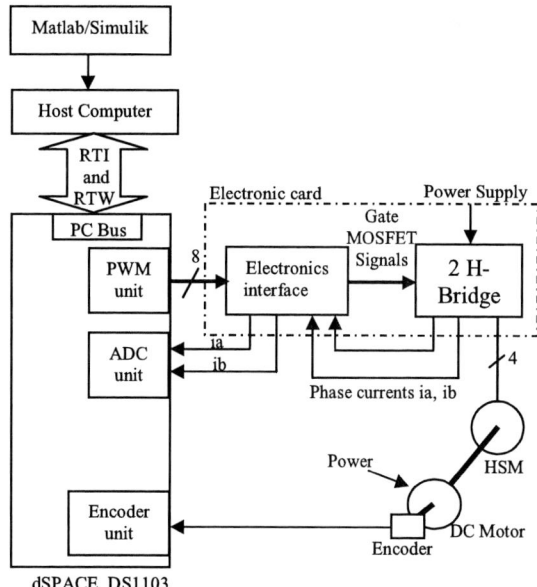

dSPACE DS1103

Figure 1. Experimental setup

TABLE I.
PARAMETERS OF A TEST MOTOR (MODEL 6600-20-2-0.37)

Nominal power	10.5 [W]
Nominal current	3.8 [A]
DC-bus voltage	24 - 60[V]
Stator resistance	0.37 [Ω]
Phase inductance	0.9 [mH]
Rotor inertia	2.48 e-5 [kg m2]
Step angle	1.8[deg]
Holding torque	925[mNm]
Detent torque	37 [mNm]

The Real-Time Workshop (RTW) converts the model to C code. The C code is then automatically compiled to the assembly language of the target processors, assembled, link-edited and downloaded to the PPC Controller Board program memory. Finally, ControlDesk, an experimentation tool, is used to control, tune and monitor the running process. In real time we can capture the signals of the model and change parameters of the system such as PI controller.

The electronic card has been home designed and build(Fig.1). This card contains an electronic interface and two H-bridge converters based on eight MOSFET transistors. These converters form two independently controlled full H-bridges for controlling the current magnitude in the phase winding bi-directionally. The phase currents are measured with shunt resistors attached to each H-bridge. The voltage measured by shunt resistor is amplified and filtered in the "Electronic interface" block. The ADC input is synchronized with the PWM signal using the DS1103SL-DSP-PWMINT block from RTI. This block is configured to generate a trigger signal to ADC input in the OFF-period of PWM signal. Since the PWM blocks are in the slave DSP F240 library the trigger signal is generated by the slave DSP.

The HSM used in the experiment has 200 steps per revolution. Table I shows parameters of the test motor. The HSM is attached to a DC motor equipped with an analog current control loop in order to impose a load torque. An incremental encoder with 2000 counts per revolution is attached to the DC motor.

B. Experimental results

We use four impulses to extract the initial rotor position. When the voltage pulse is applied to the phase, the current peak is hold by the "Maximum" block from Simulink library which memorizes the value. The current differences Δi_a and Δi_b are connected respectively to the Real and imag inputs of "Real-Imag to Complex" block from Simulink library. The output of this Block is connected to "Complex to Magnitude-Angle" block which calculates the initial rotor position.

The DC voltage source used in this laboratory is 30V and the time during which the voltage is applied to a phase is about 0.15ms. The voltage pulses are not width, so the rotor does not move during the standstill algorithm. Fig.2 shows an example of phase current used to detect the initial position.

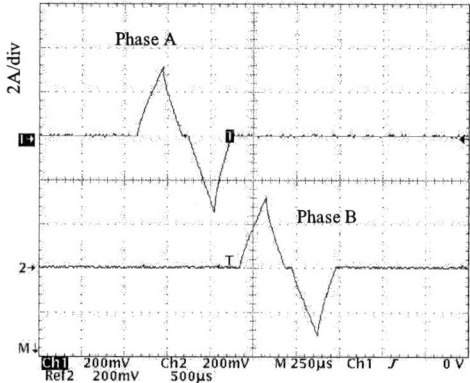

Figure 2. Phase currents to detect the initial position.

Fig.3 shows the measured current differences for 20 different rotor positions. The measurements are distributed in electrical degrees. For comparison, the real rotor position (θ_r) is obtained by an incremental encoder. By calculating Δi_A and Δi_B, two approximately sinusoidal waves which have a periodicity of one electrical period have been carried out..

Fig.3 shows that initial rotor position can be found easily by a trigonometric function as :

$$\theta_0 = a\tan\left(\frac{\Delta i_b}{\Delta i_a}\right) \qquad (1)$$

Fig.4 represents the rotor position estimated by (1) as a function of the real rotor position obtained by a incremental encoder.

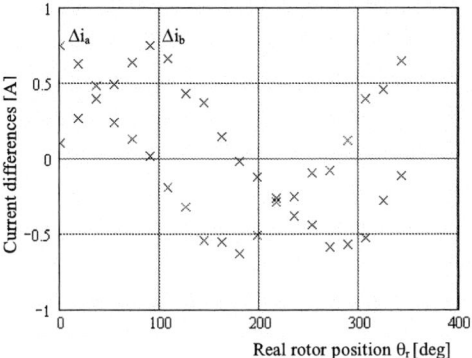

Figure 3. Peak current differences

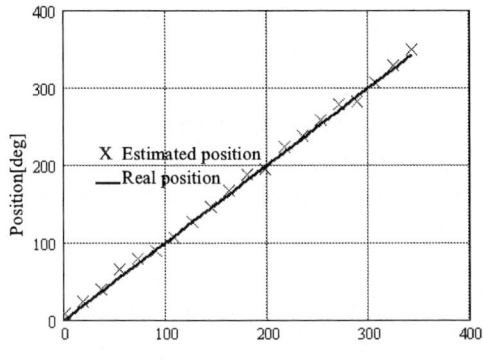

Figure 4. Initial rotor position estimation results

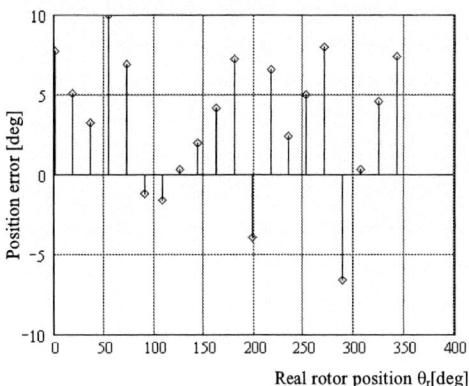

Figure 5.Position estimation error

Fig.5 shows that the maximum position error is about 10°. This error is caused by the current differences shown in Fig.3. which are not sinusoidal.

Precision can be improved by using sinusoidal regression curves of the current differences Δi_a and Δi_b (Fig.6). The new rotor position is obtained using this references .

Using this method the precision is improved as can be seen in Fig 7 and Fig.8.

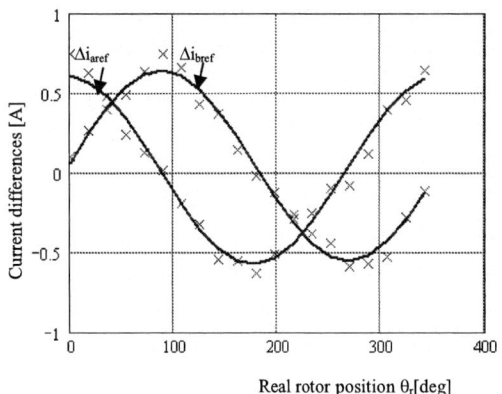

Figure 6. Current differences

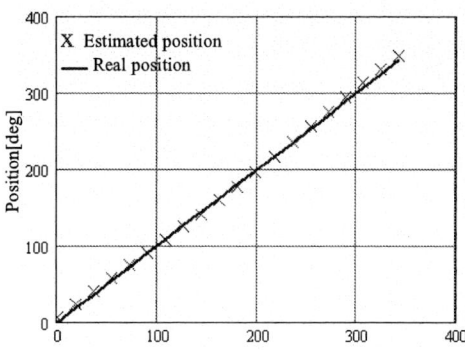

Figure 7. Improving accuracy of rotor position

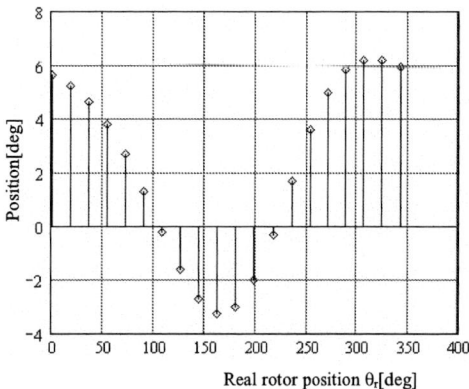

Figure 8. Rotor position error improvement

V. CONCLUSION AND OUTLOUK

In this paper standstill rotor position detection is implemented for the HSM. It guarantees the start-up of the motor in the desired direction. This standstill method utilizes stator inductance variation due to magnetic flux.

The obtained results are good and confirm the effectiveness of the experimental setup.

A possible development of this work will consist in modifying the algorithm to detect the initial rotor position.

With today's low cost and high performance DSP's, we intend to test the advanced control algorithms to detect the speed and the rotor position such as Kalman filter which is itself a low pass filter.

ACKNOWLEDGMENT

The authors gratefully acknowledge the Haute Ecole Arc, engineering School from Saint-Imier (Switzerland) for financial and material supports to this Work.

REFERENCES

[1]   P. P. Acarnley, *"Stepping Motors: A Guide to Modern Theory and Practice"*, 2nd ed. Stevenage, U.K.: Peregrinus, 1984.

[2]   L. Cardoletti, A. Cassat, M. Jufer, "Sensorless position and speed control of a brushless DC motor from start-up to nominal speed", *EPE Journal, Vol. 2, No 1, March 1992, pp 25-34.*

[3]   L. Cardoletti, "Commande et réglage de moteurs synchrones auto-commutés par des capteurs indirects de position ", Thesis, EPFL , Lausanne , 1993.

[4]   F. Bonvin, "Analyse et mise en ouvre de méthodes de commande sans capteur pour moteurs synchrones a aimants permanents", Thesis No 2354, EPFL, Lausanne, 2001.

[5]   C.Obermeier, H.Kellermann and G. Brandenburg, "Sensorless Field Oriented Control of a Hybrid Stepper Motor Using an Extended Kalman Filter", *EPE' 97 European Conference on Power Electronics and Applications, Trondheim 1997, p. 238-43 vol.1*

[6]   J. Persson, Y. Perriard, "An Optimized Extended Kalman Filter Algorithm for Hybrid Stepper Motors", *CD-ROM, IEEE-IECON'03, Roanoke, Virginia, USA, 2-6 November 2003*

[7]   Bilal Akin," A Comparative Study on Non-Linear State Estimators Applied to Sensorless AC Drives: MRAS and Kalman Filter" *The 30th Annual Conference of the IEEE Industrial Electronics Society, November 2 - 6, 2004, Busan, Korea.*

[8]   Y-H. Kim , Y-S. Kook, "High Performance IPMSM Drives without Rotational Position Sensors Using Reduced-Order EKF," IEEE *Trans. on Energy Cons*, Vol 14, No. 4 December 1999.

[9]   N. Ertugrul and P.P. Acarnley, "A New Algorithm for Sensorless Operation of Permanent Magnet Motors", IEEE Transactions on Industry Applications, Vol. 30, No. 1, pp. 126-133, January-February 1994.

[10]   A. B. Kulkarni. M.Ehsani, " A Novel Position Sensor Elimination Technique for the Interior Permanent Magnet Synchronous Motor Drive," *IEEE Transactions on Industry Applications, Vol.28, No. 1 , pp. 144-150, January/February 1992.*

[11]   T. Aihara, A. Toba, T. Yanase, A. Mashimo, and K. Endo, "Sensorless torque control of salient-pole synchronous motor at zero-speed operation," *IEEE Trans. Power Electron.*, vol. 14, pp. 202–208, Jan. 1999.

[12]   M. Schroedl and P. Weinmeier, "Sensorless control of reluctance machines at arbitrary operating conditions including standstill," *IEEE Trans. Power Electron.*, vol. 9, pp. 225–231, Mar. 1994.

[13]   S. Ogasawara and H. Akagi, "Implementation and position control performance of a position-sensorless IPM motor drive system based on magnetic saliency," *IEEE Transactions on Industry Applications, Vol.* 34, NO.4, July/August 1998

[14]   J. Persson," Innovative standstill position detection combined with sensorless control of synchronous motors", Thesis N° 3221, EPFL, Lausanne 2005.

2006 5th International Power Electronics and Motion Control Conference

An In-Wheel Switched Reluctance Motor for Electric Vehicles

P.C.K. Luk, P. Jinupun

Cranfield University/Department of Aerospace Power and Sensors, Swindon, United Kingdom
email: p.c.k.luk@cranfield.ac.uk

Abstract— **An alternative motor technology for electric drive-trains, in particular in use as an in-wheel drive, is presented. The technology is based on a simple double concentric stator structure with a 'yokeless' rotor, which enables the motor to develop constant power with an extended range much closer to zero speed. A distinct feature of the new motor, which is based on switched reluctance, rests on an overlapping magnetic circuitry inherent in the wheel shape structure, resulting in a smooth but exceptionally high torque per current capability with minimum iron losses. Simulation and experimental results will be present to show the potential applications of the motor.**

Keywords - switched reluctance motors; in-wheel motor; electric vehicle

I. INTRODUCTION

Although many drive-train schemes have been proposed for electrically propelled vehicles, the search for a motor technology that enables the drive-train to operate more cost-effectively and efficiently, entirely in constant power to meet operational constraints such as initial acceleration and climb-ability, has never been more compelling. To overcome the stringent torque-speed requirement imposed by these constraints, conventional ICE uses multiple gear transmission. Similarly electric drive-trains also invariably resort to using some mechanical gear system. The potential impact of the proposed alternative motor technology that can achieve a near constant power torque-speed characteristic over a wide speed range extending to standstill condition, is therefore considerable. This will allow a gearless direct drive system to be deployed in a future EV, leading to increased wheel dynamics and drive efficiency. It is felt that the aim to develop and demonstrate a new motor technology is mission critical as it offers the electric vehicle market an attractive and viable alternative, and its apparent increasing dependence on a permanent magnet technology. It also fills a technological gap in motor drives developed by the evolution of two main trends in the industry, where drives either encompass a relatively high-cost/sophisticated motor with a relative simple controller such brushless PM DC drives, or a simple motor with a sophisticated controller such as advanced switched reluctance (SR) motor drives or vector control

drives using the induction motor. The radical design of the proposed motor will result in a drive with a typically simple SR motor structure and a controller very similar to that of a brushless DC. It has a unique direct drive capability that not only removes the gear transmission and hence significant weight advantage, but also offers additional agility, controllability and freewheeling capability that are critical for an independent drive vehicle. Its intrinsic in-wheel structure is also conducive to integration at system and component levels.

This paper reports the design of a novel in-wheel SR motor that has a novel rotor and a stator configuration that uses substantially less back iron than conventional types. The consequence of this unique in-wheel configuration results in an extremely efficient torque generation mechanism and a very high power density motor. Simulation and experimental results are presented to show the potential advantages of the motor over conventional SR motors.

II. NEW MOTOR CONCEPT

The proposed SRM relies on the fundamental mechanism of reluctance torque generation in which the motor's magnetic saliency is exploited to create alignment torque. As the stator is electromagnetically energized, the rotating part moves to restore minimum reluctance or maximum inductance. The novel feature of the new motor, however, rests on a unique stator and rotor configuration that allows flux formation in magnetically isolated yokeless (no yoke) rotor teeth. Alignment forces are produced on the rotor teeth located along the circumference of the yokeless rotor, when flux is established by yoke-less (less yoke) stator pole pairs.

A. The Rotor/Stator Structure

The unique rotor and stator configuration is shown in Fig.1. The rotor base is a slotted disk made of light non-magnetic material. The rotor teeth are housed in the slots to create a seamless rotor structure. The stator consists of multiple pairs of horse-shoe shape poles, with each pair fully aligned with each other but separated by the rim of the rotor. The yoke-less stator pole-pairs are mounted on non-magnetic light material similar to that of the rotor base. Extremely efficient torque generation happens when selected pole-pairs are energized according to their

(a) Yokeless rotor teeth with base rotor

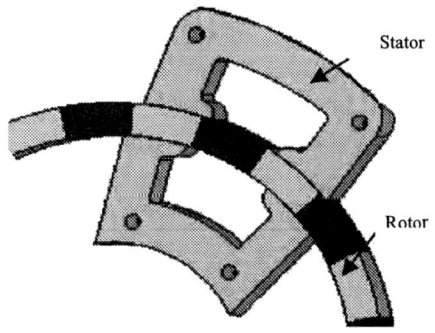

(b) Yoke-less stator core

Fig.1 Conceptual Yokeless SRM

position with the rotor teeth, when the rotor teeth aim to align with the excited stator pole pairs. The torque generation is efficient because the forces are created at the rim of the disc structure, and the magnetic paths are minimized. So torque is maximized for a given set of forces, and iron losses are minimized by short magnetic paths. The main advantage of this structure is very high torque per unit iron mass can be developed, which makes the motor ideal for direct drive.

B. Analytical Equations and Modelling

The new motor has a rotor/stator configuration very different from that of the conventional 6/4 or 8/6 structures. The fundamental torque generation mechanism is however similar. The instantaneous voltage across the terminals of a single phase of an SRM winding is related to the flux linked in the winding by Faraday's law, can be developed into the following power conversion equation.

$$vi = i^2R_m + i\frac{d\phi}{dt} \qquad (1)$$

where, v is the terminal voltage, i is the phase current, R_m is the motor resistance, and ϕ is the flux linked by the winding. The new motor shares a double salient construction in the same way as the conventional SRM. However, the yokeless rotor structure makes it to be saturated more uniformly, thus increasing its torque

density. Eqn (1) disguises the complexity of the motor performance of the motor under saturated and hence highly nonlinear condition. No simplified torque analytical equation may yield satisfactory determination of the generated torque for the motor, leading to the preferred use of energy modeling as in Eqn (2), where the left hand term is the second term of the right hand term of Eqn (1).

$$i\frac{d\phi}{dt} = \frac{dW_m}{dt} + \frac{dW_f}{dt} \qquad (2)$$

where $\frac{dW_m}{dt}$ is the instantaneous mechanical power, and $\frac{dW_f}{dt}$ is the instantaneous power, which is stored in the magnetic field. In this paper, finite element analysis is used to determine the instantaneous mechanical power and hence the generated torque as follows [3].

$$\frac{dW_m}{dt} = T\omega = T\frac{d\theta}{dt} \qquad (3)$$

C. Theory of Operation

Although the basic operation theory is similar to conventional SRMs, the new motor is more like a linear motor in terms of operation, and there are some features specific to this motor. Fig.2 shows the basic electric drive circuit and magnetic circuit for a phase winding. The drive

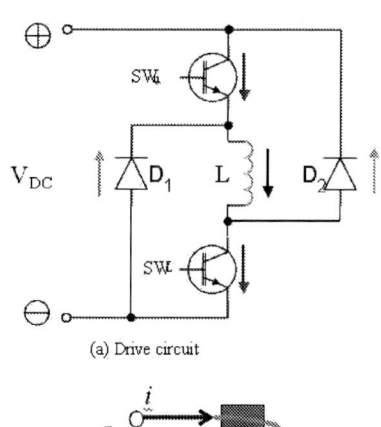

(a) Drive circuit

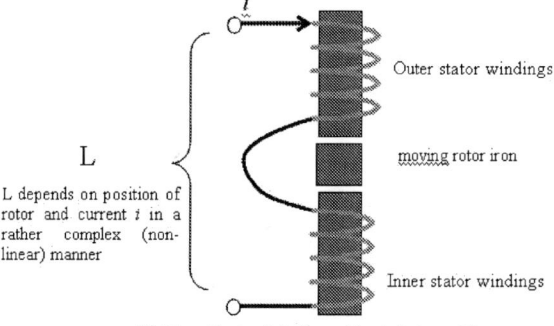

(b) Magnetic circuit (with machine inductance L)

Fig.2 Per Phase Driver Circuit

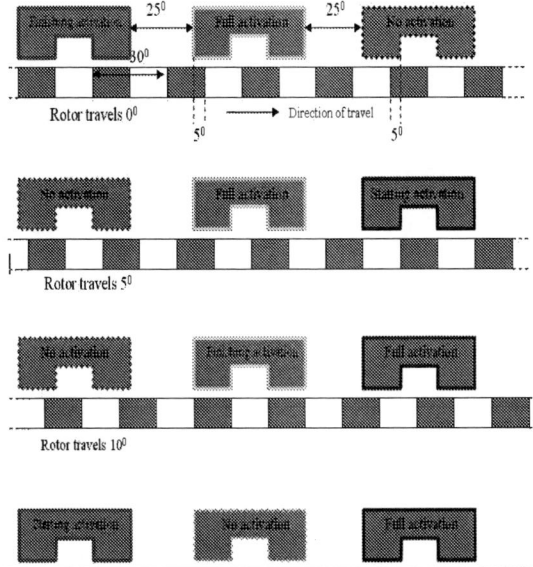

Fig.3 Activation Sequence Diagram

circuit in Fig.2(a) is a typical per phase power inverter circuit for the SRM. When the power switches (SWU and SWL) are on with the same switching signal, current will flow through the stator winding L from the supply VDC via the power switches, producing motoring torque and storing magnetic energy in L. When the power switches are off, current will continue to flow via the diodes (D1 and D2) back to the supply. Since current flows against the supply, it will decrease to zero before end of the cycle. During this period, stored magnetic energy returns to the supply. However, this apparently simple operation is made very complex due to the variation of the value of inductance, due to the excitation current level i and the position of the moving iron rotor, as illustrated in Fig.2 (b). It should be noted that it is true for SRM in general, not just for this design. The variation of L is also highly dependent on the magnetic materials used. Complex control method is often needed to ensure the correct profile of excitation current to produce the required torque level.

Fig.3 shows the switching or activiation sequence, using a developed diagram. It shows only the upper stator windings and the rotor in a 'rolled-out' form for ease of illustration. The stator windings are each separated by 25^0. Thus, in the first diagram of Fig.3, with phase A fully aligned with the rotor, phase B has an 'forward' overlap zone of 5^0 with the rotor, and phase C has a 'backward' overlap zone of 5^0. A 'forward' overlap will generate a forward torque (to the right) when excited, and a 'backward' overlap will generate retarding torque. The choice of these overlaps is important in determining the torque characteristic of the motor, and in particular

minimising torque ripples. The sequence of diagrams in Fig.3 aims to illustrate graphically the switching sequence, or more precisely the activation sequence as will be seen, of the stator windings required to develop a smooth and unidirectional torque to move the rotor to the right. In the top diagram, excitation current flowing in stator winding A must be reduced to zero as soon as possible, as further flow of current will result in pulling the rotor to the left, i.e. a negative torque to the direction of rotation. Winding A is therefore finishing activation. At the same time, winding B is in full activation state as it is the only winding that can produce the main positive torque. However, it does not necessarily mean that during this activation period winding B is fully turned on all the time. The power switches must control the current into winding B at a certain level to develop the appropriate torque.

III. SIMULATION RESULTS

A. FEA Results

Finite element analysis (FEA) has been performed based on the new motor topology for various excitation currents. Fig.4 shows the FEA results and the excitation sequence of the motor flux over a distance of 10^0

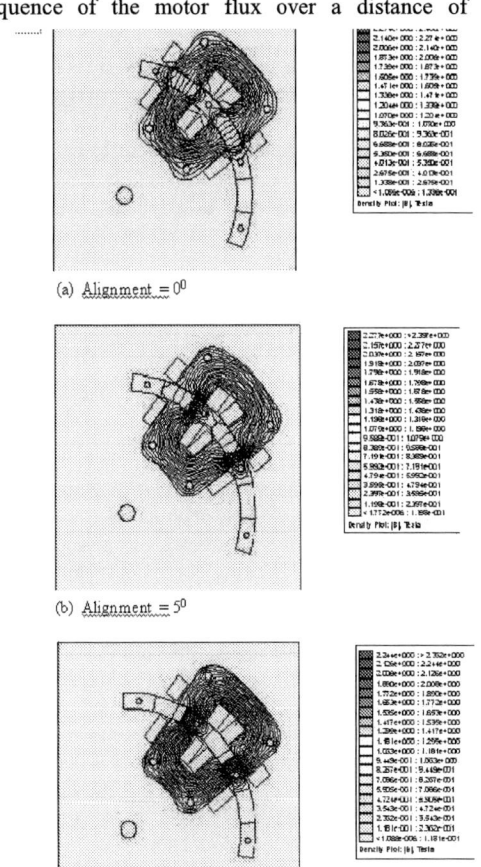

(a) Alignment = 0^0

(b) Alignment = 5^0

(c) Alignment = 10^0

Fig.4 FE Results for typical alignment angles

(mechanical) clockwise displacement. A marker is manually painted on a non-magnetic section the rotor to help show the displacement clearer. It is noted that a relatively high average flux level can be achieved at rated voltage and current level. In Fig.4(a), with the rotor set at an arbitrarily initial position, phase A starts its excitation as the rotor teeth begins aligning with the stator poles while phase B is excited at nearly fully aligned position. Also, phase C is not excited. It is noted that the peak flux level is at 2.1T. Fig.4(b) shows the excitation pattern and the corresponding flux levels after a 5^0 rotor travel. Only phase A is excited at this moment. In Fig.4(c), the rotor has traveled a total of 10^0. Phase A is approaching nearly fully aligned, phase B is now un-excited, and phase C begins its alignment and is now excited. The FEA results are used to calculate the instantaneous torque value.

B. Simulation Results

Simulation results are performed by using a special switched reluctance software program, called *SRSim®*, developed by the authors [4]. It links with the FEA program and allows different switching schemes and control programs to be performed, and can output torque, speed, current and voltage curves, among others. Fig.5 shows the simulation results from *SRSim ®*, when the motor is driving a load of 30Nm at 120rpm at rated voltage. Simple hystersis current control method is used. The upper curve shows the torque and the lower curve is the current from the 3 phases. It can be seen that the torque ripple is relative low for the new SRM compared, even when only a simple control current control is used. It has been demonstrated elsewhere [4] that significant improvement of the torque can be achieved with more advanced control methods.

IV. EXPERIMENTAL RESULTS

A laboratory prototype motor based on the novel SRM concept has been designed and built, which is shown in Fig.6. In order to generate experimental results to be used to validate the simulation model and the FEA results, the motor is setup to drive a controlled load simulated by the braking system of a gokart. Fig.7 shows the whole experimental setup, with the torque sensor and data acquisition system. During the experimental tests, the motor was driving against the braking load applied from the gokart while steady state speed was maintained manually. A relatively low speed of 120rpm was used in the tests, while the load torque was set to about 30Nm. Fig.8 shows the measured torque of the motor. It can be seen that, allowing for some experimental and instrumentation errors, the experimental results are in close agreement with the simulation results. Since the current setup cannot be performed under prolonged testing condition due to the amount of heat generated, further experimental results at higher speeds can only be performed when a more high power dynamometer setup is ready.

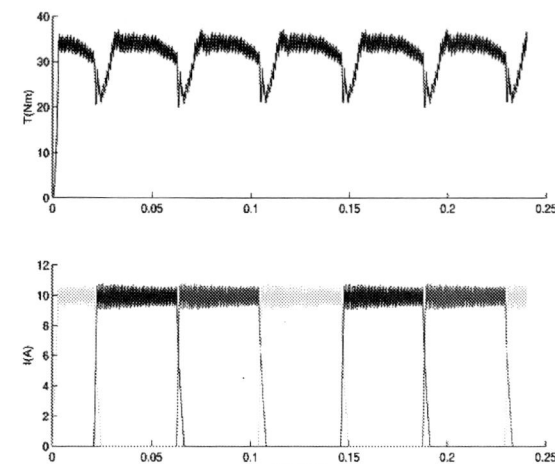

Fig.5 Simulated outputs of the motor with at 120rpm – Total torque (upper curve), phase currents (lower)

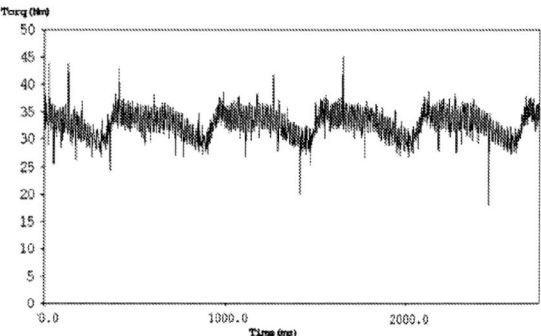

Fig.6 Measured motor torque at 120rpm

V. CONCLUSION

An in-wheel SRM with a yokeless rotor and yoke-less stator is introduced. The apparent advantages of the motor include less iron losses, higher torque density, among others. The basic operations and finite element results of the motor are presented. The FEA results are then used to predict the performance of the new motor using a specialist simulation software. The simulation results are then confronted with the experimental results, which confirm the model of the motor.

REFERENCES

[1] T. J. E. Miller, Switched Reluctance Motors and Their Control. Oxford, U.K.: Oxford Science, 1993.

[2] P. C. Kjaer, J. J. Gribble, and T. J. E. Miller, "High-grade control of switched reluctance machines," IEEE Trans. Ind. Applicat., vol. 33, pp. 1585–1593, Nov./Dec. 1997.

[3] Torrey, D.A. Lang, J.H., 'Modelling a nonlinear variable-reluctance motor drive,' Electric Power Applications, IEE Proceedings B, Sep 1990, Volume: 137, Issue: 5, On page(s): 314-326.

[4] Luk, P.C.K., Jinupun, P, 'Instantaneous thrust control for a linear switched reluctance motor,' Power Electronics Specialists Conference, 2004. PESC 04. 2004 IEEE 35th Annual Publication Date: 20-25 June 2004, Volume: 3, On page(s): 2265- 2269 Vol.3

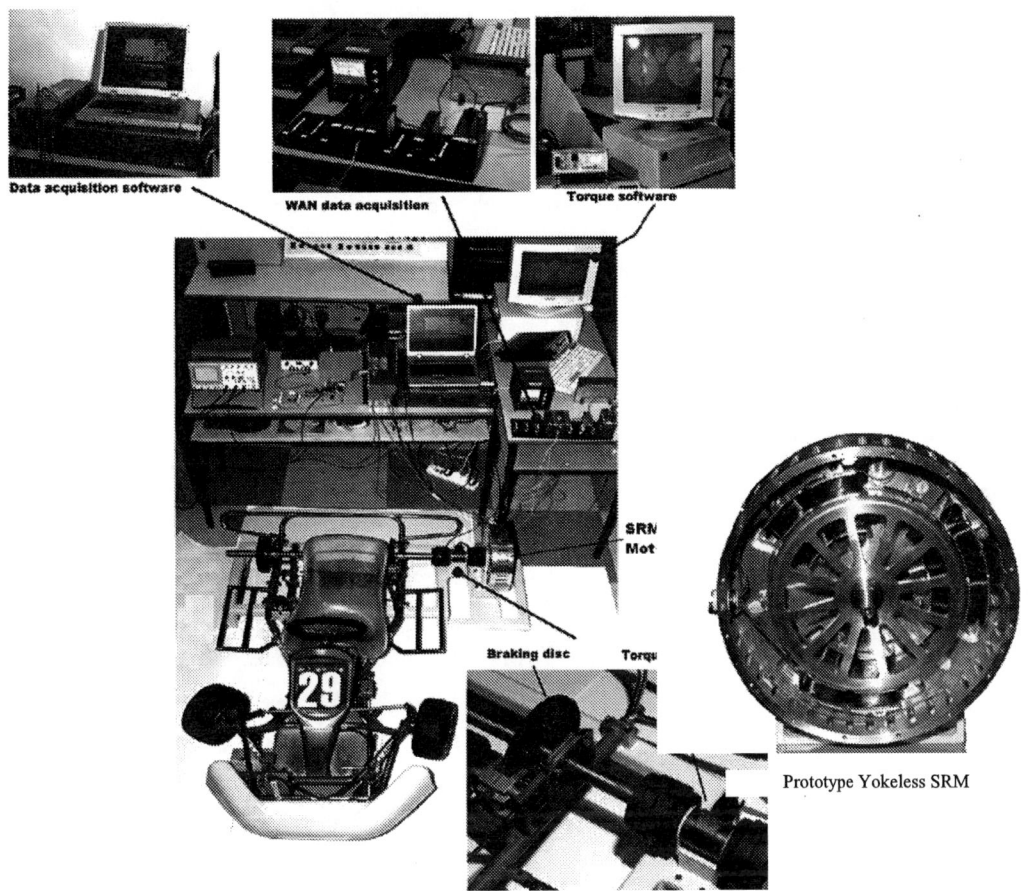

Fig.7 Full experimental Setup

2006 5th International Power Electronics and Motion Control Conference

Speed Sensorless Vector Control of Induction Motor Based on Full-Order Flux Observer

Shanshan Wu, Yongdong Li, Zedong Zheng

Department of Electrical Engineering, Tsinghua University, Beijing, P.R.China

Email: wuss03@mails.tsinghua.edu.cn, liyd@mail.tsinghua.edu.cn, zheng99@mails.tsinghua.edu.cn

Abstract—**This paper presents a DSP-based speed sensorless field oriented vector control system for induction motor. A full-order adaptive observer with stator current and rotor flux as its state variables is used here to estimate the rotor speed. Moreover, the stator resistance is also identified to update the stator resistance value used in the observer so that the rotor speed can be estimated accurately at low speeds. By using the proposed full-order flux observer, the drive system can operate well in a wide range of speed. The validity of the proposed method is confirmed by numerical simulations. Then a prototype is setup and experiments based on a DSP TMS320F2812 prove the excellent performances of the presented control algorithm.**

Keywords-speed sensorless; vector control; full-order adaptive observer; induction motor

I. INTRODUCTION

Recently, speed sensorless vector control of induction motor without a shaft encoder has been developed more and more for its low cost and convenience. In order to obtain good performance of speed control, most researchers focus on the accurate flux and speed estimation and have proposed different methods.

Some methods are open loop schemes based directly on calculation. For example, rotor flux is estimated by pure integration of back electromotive force (EMF) when the classical voltage model is used. But the accuracy of EMF is influenced by mismatch of stator resistance at low speeds. In order to take use of advantage of current models, which doesn't involve stator resistance, voltage model and current model are combined to operate at high frequency and low frequency respectively [1]. Some advanced voltage models are also developed to overcome the problem of pure integration of EMF [2].

Some other methods are closed-loop schemes, which are supposed to be more robust and noise-immune than open loop schemes. Model-reference adaptive system (MRAS) is also developed. Firstly, the voltage model is regarded as the reference model, while the current model which involve rotor speed is regarded as adjustable model [3]. However, the same problems still exist because of using voltage model.

Some papers deal with the problems above by using a full-order speed adaptive flux observer to estimate the

rotor speed and identify stator resistance at the same time [4][5]. In this paper, the stator current and rotor flux are chosen to be state variables of the full-order adaptive observer. The rotor speed is estimated by the observer and is used as a feedback signal for the stator current and rotor flux estimation. Moreover, the stator resistance is also identified to update the incorrect value used in the observer so that the rotor speed can be estimated accurately at low speeds. By using the proposed full-order flux observer, the drive system can operate very well in a wide range of speed. The validity of the proposed method is confirmed by numerical simulations. Then a prototype is setup and experiments based on a digital signal processor (DSP) TMS320F2812 prove the excellent performances of the presented control algorithm.

II. FULL-ORDER SPEED ADAPTIVE FLUX OBSERVER

A. Full-Order Flux Observer of Induction Motor

If the stator current and the rotor flux are selected as the state variables, the state equation of an induction motor can be expressed as (1) in the stationary reference frame.

$$\frac{d}{dt}\begin{bmatrix} i_s \\ \psi_r \end{bmatrix} = \begin{bmatrix} A_{11} & A_{12} \\ A_{21} & A_{22} \end{bmatrix}\begin{bmatrix} i_s \\ \psi_r \end{bmatrix} + \begin{bmatrix} B_1 \\ 0 \end{bmatrix} u_s = Ax + Bu_s \quad (1)$$

where

$$u_s = \begin{bmatrix} u_{s\alpha} & u_{s\beta} \end{bmatrix}^T \text{ stator voltage}$$

$$i_s = \begin{bmatrix} i_{s\alpha} & i_{s\beta} \end{bmatrix}^T \text{ stator current}$$

$$\psi_r = \begin{bmatrix} \psi_{r\alpha} & \psi_{r\beta} \end{bmatrix}^T \text{ rotor flux}$$

$$A_{11} = -\left(R_s/(\sigma L_s) + (1-\sigma)/\sigma\tau_r\right)I = a_{r11}I$$

$$A_{12} = L_m/(\sigma L_s L_r)(1/\tau_r I - \omega_r J) = a_{r12}I + a_{i12}J$$

$$A_{21} = L_m/\tau_r I = a_{r21}I$$

$$A_{22} = -1/\tau_r I + \omega_r J = a_{r22}I + a_{i22}J$$

$$B = 1/(\sigma L_s)\begin{bmatrix} I \\ 0 \end{bmatrix} \quad I = \begin{bmatrix} 1 & 0 \\ 0 & 1 \end{bmatrix} \quad J = \begin{bmatrix} 0 & -1 \\ 1 & 0 \end{bmatrix}$$

$\sigma = 1 - L_m^2/(L_s L_r)$ leakage coefficient

R_s, R_r stator and rotor resistance

L_s, L_r, L_m stator, rotor and mutual inductance

$\tau_r = L_r/R_r$ rotor time constant

ω_r motor angular velocity

1-4244-0448-7/06/$25.00 ©2006 IEEE

The full-order observer, which estimates stator current and rotor flux together, can be written as (2).

$$\frac{d}{dt}\hat{x} = \hat{A}\hat{x} + Bu_s + G(\hat{i}_s - i_s) \qquad (2)$$

where $\wedge$ means the estimated values and G is the feedback gain of the observer. In order to make the observer stable at any speed, the pole position of the observer is assigned as k times of that of the motor. So, G is defined as

$$G = \begin{bmatrix} g_1 & g_2 & g_3 & g_4 \\ -g_2 & g_1 & -g_4 & g_3 \end{bmatrix}^T \qquad (3)$$

where

$$\begin{cases} g_1 = (k-1)(\hat{a}_{r11} + \hat{a}_{r22}) \\ g_2 = (k-1)\hat{a}_{i22} \\ g_3 = (k^2-1)(\hat{a}_{r21} - \rho \cdot \hat{a}_{r11}) + \rho \cdot g_1 \\ g_4 = \rho \cdot g_2 \end{cases}, \; k > 0$$

$$\rho = -(L_s L_r - L_m^2)/L_m$$

B. Adaptive Scheme for Roror Speed and Stator Resistance

If the motor itself is regarded as the reference model and the full-order flux observer is regarded as the adjustable model according to MRAS, the estimation error can be expressed as (4) by subtracting (2) from (1).

$$p\varepsilon = (A + GC)\varepsilon + \Delta A \hat{X} \qquad (4)$$

where

$$\varepsilon = x - \hat{x} \qquad C = [I \quad 0]$$

$$\Delta A = A - \hat{A} = \begin{bmatrix} 0_{2\times2} & J/\rho \\ 0_{2\times2} & J \end{bmatrix}(\omega_r - \hat{\omega}_r)$$

$$+ \begin{bmatrix} -I/\sigma L_s & 0_{2\times2} \\ 0_{2\times2} & 0_{2\times2} \end{bmatrix}(R_s - \hat{R}_s) = \Delta A_1 \Delta \omega_r + \Delta A_2 \Delta R_s$$

In order to derive the adaptive scheme, Lyapunov's theorem is utilized and Lyapunov function is defined as

$$V = \varepsilon^T \varepsilon + (R_s - \hat{R}_s)^2 / \rho_1 + (\omega_r - \hat{\omega}_r)^2 / \rho_2 \qquad (5)$$

where ρ_1 and ρ_2 are positive constants.

After some derivations and mathematical calculations, we can get the following adaptive scheme for the rotor speed and stator resistance.

$$\hat{R}_s = -K_{RP}(\hat{i}_{s\alpha}\varepsilon_1 + \hat{i}_{s\beta}\varepsilon_2) - K_{RI}\int(\hat{i}_{s\alpha}\varepsilon_1 + \hat{i}_{s\beta}\varepsilon_2)dt$$

$$\hat{\omega}_r = K_{\omega P}(\hat{\psi}_{r\beta}\varepsilon_1 - \hat{\psi}_{r\alpha}\varepsilon_2) + K_{\omega I}\int(\hat{\psi}_{r\beta}\varepsilon_1 - \hat{\psi}_{r\alpha}\varepsilon_2)dt$$

where

$$\varepsilon_1 = i_{s\alpha} - \hat{i}_{s\alpha} \qquad \varepsilon_2 = i_{s\beta} - \hat{i}_{s\beta}$$

$$K_{RP}, K_{RI}, K_{\omega P}, K_{\omega I} \quad \text{PI parameters}$$

If the observer feedback gain matrix G is selected as is said in part A, the proposed full-order flux observer is asymptotic stable according to Lyapunov's theorem.

Fig.1 shows the proposed full-order flux observer with both rotor speed and stator resistance estimations.

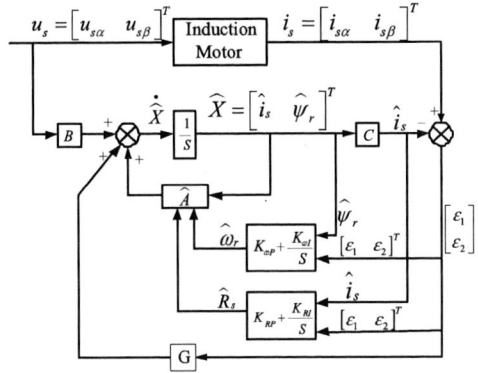

Figure 1. Full-order flux observer with both rotor speed and stator resistance estimations

III. SIMULATIONS

The block diagram of the speed sensorless vector control system is depicted in Fig. 2. Field orientation is achieved by means of the proposed full-order adaptive observer. The proposed system has been simulated. Motor parameters are listed in Table I.

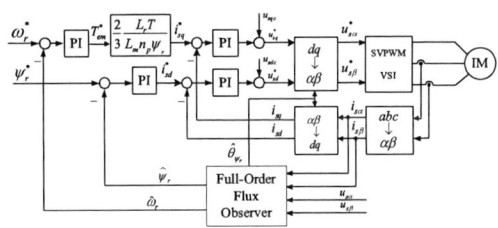

Figure 2. System configuration of speed sensorless vector control using full-order flux observer

Fig. 3 shows the behavior of the estimated rotor speed, with a speed step and a 100% load torque step at 2s and 4s, respectively. The estimated speed follows the real speed very well.

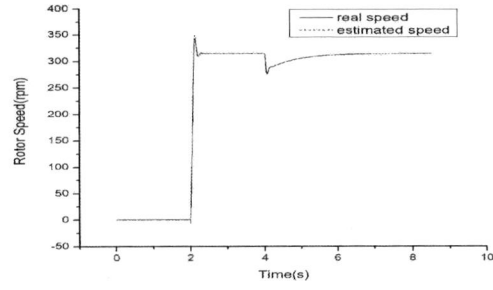

Figure 3. Rotor speed response: speed step (from 0 to 25Hz) at 2s and 100% load torque step at 4s

1893

Fig. 4 shows the quick response of the estimated stator resistance when the real resistance changes.

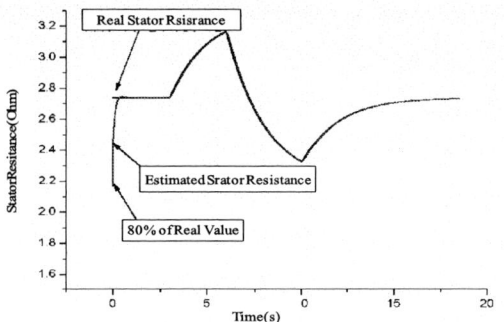

Figure 4. Simulation Result of Stator Resistance Identification

In order to test the influence of the stator resistance variation, the resistance used in the motor model is decreased to 70% of the real value. Fig. 5 shows the simulation results without stator resistance tuning. The estimated values have steady-state errors, so that the stator resistance online identification is proved to be important.

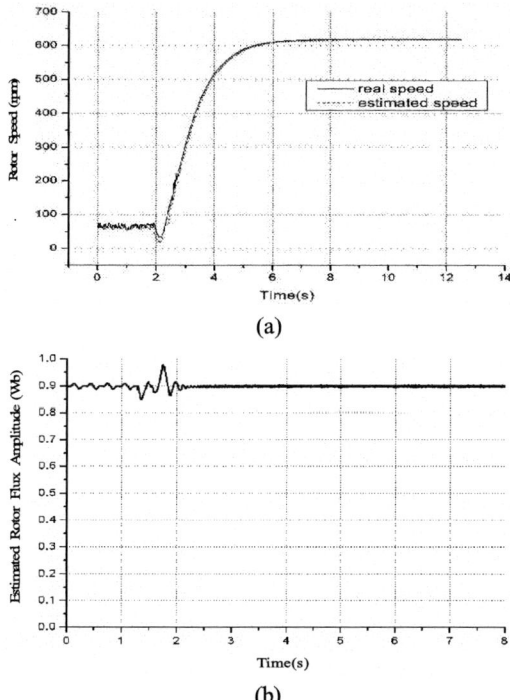

Figure 5. Simulation Results without Stator Resistance Online Tuning(speed steps to 4Hz at 1s, 100% load torque step at 5s): (a) rotor speed   (b) rotor flux amplitude

IV.   EXPERIMENTS

The proposed control system has been implemented on a system based on a DSP TMS320F2812. The sampling frequency is 8 kHz. A dc motor is used to supply load torque.

Fig. 6 shows the responses of the estimated rotor speed and rotor flux during acceleration. The proposed system has good dynamic response, because the estimated parameters could converge to the real values very quickly. Furthermore, both of the estimated values have no steady-state errors.

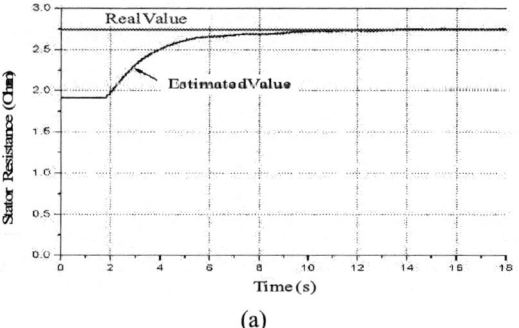

Figure 6. Experimental Waveforms During Acceleration(2Hz →20 Hz): (a) Rotor Speed (b) Estimated Rotor Flux Amplitude

Fig.7 shows that the estimated stator resistance could converge to the real value, whatever the initial value is.

Fig.8 shows the responses of the estimated rotor flux and stator current at very low speed without stator resistance online tuning. The stator resistance used in the observer calculation reduced to 70% of the real value. Visible steady-state errors of the estimated values can be found. It's proved that the stator resistance identification is necessary.

(a)

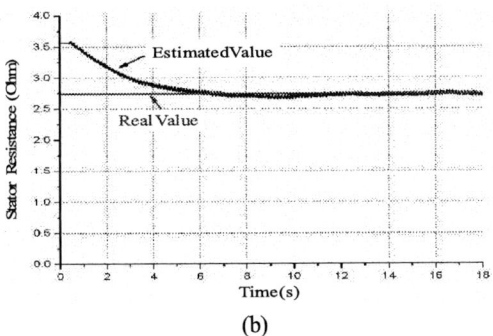

(b)

Figure 7. Experimental Results of Stator Resistance Identification: (a) Initial Value is 70% of The Real Value (b) Initial Value is 130% of The Real Value

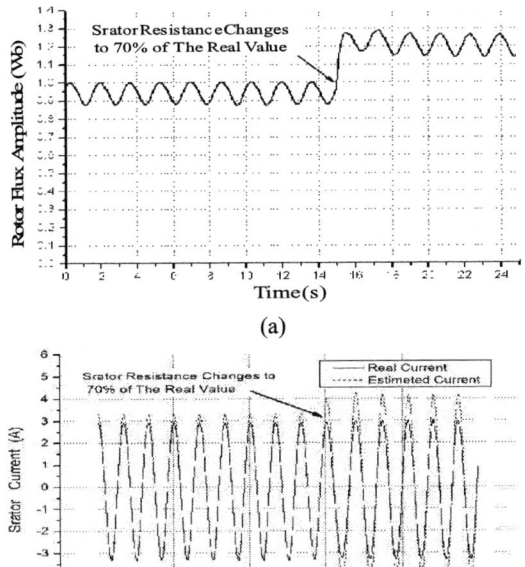

(a)

(b)

Figure 8. Experimental Results at 17rpm without Stator Resistance Online Tuning: (a) Estimated Rotor Flux Amplitude (b) Stator Current

TABLE I.
PARAMETERS OF THE INDUCTION MOTOR

Rated power	2.2kW	R_s	2.74 Ω
Rated voltage	380V	R_r	2.55 Ω
Rated current	4.9A	L_s	0.299H
Number of Poles	4	L_r	0.3128H
Rated Frequency	50Hz	L_m	0.3128H
Rated Speed	1430r/min	J	0.025 kgm²

V. CONCLUSION

This paper presented a speed sensorless vector control system. Field orientation is achieved by means of a full-

order adaptive observer, by which the rotor speed and the stator resistance are both estimated. By using the proposed full-order flux observer, the drive system can operate well in a wide range of speed, especially in low speed range. The validity of the proposed method is confirmed by numerical simulations. Then a prototype is setup and experiments based on a DSP TMS320F2812 prove the excellent performances of the presented control algorithm.

REFERENCES

[1] Shinnaka, "A new hybrid vector control for induction motor without speed and position sensors-frequency hybrid approach using new indirect scheme," IEEE PCC-Nagaoka'97, 1997, 541-548.

[2] Bin Wu and Jun Hu, "New Integration Algorithms for Estimating Motor Flux over a Wide Speed Range," IEEE Trans., VOL.13, NO.5, September 1998: 969-977.

[3] C. Schauder, "Adaptive speed identification for vector control of induction motors without rotation transducers," in Conf. Rec. IEEE/IAS Ann. Mtg., 1989, pp. 493-499.

[4] Hisao Kubota, Kouki Matsuse and Takayoshi Nakano, "DSP-based Speed Adaptive Flux Observer of Induction Motor," IEEE Transactions on Industry Applications, Mar./Apr. 1993, Vol.29, No.2, pp.344-348.

[5] Geng Yang and Tung-Hai Chin, "Adaptive-Speed Identification Scheme for a Vector-Controlled Speed Sensorless Inverter-Induction Motor Drive," IEEE Transactions on Industry Applications, July/August 1993, Vol.29, No.4, p 820-825.

2006 5th International Power Electronics and Motion Control Conference

A Parameter Identification Method for General Inverter-fed Induction Motor Drive

JIANG, Xiaochun   YANG, Geng*   WANG, yunfei

Dept. of Automation, Tsinghua Univ., Beijing 100084, China
* Email：yanggeng@mail.tsinghua.edu.cn

Abstract: It is important to match correctly parameters of controlled induction motor in a controller for some kinds of advanced induction motor control technology, such as sensor-less field-oriented control. Also it is necessary to use a simple estimating algorithm, because a lot of general invertors are equipped with a fixed point CPU. In this paper, a model of an induction motor with single-phase supply is described. Based on the model, an iterative algorithm for parameter identification is presented. The performance of the strategy is demonstrated by both the simulation and the experimental results.

Key words: parameter identification, induction motor, field-oriented control.

I. INTRODUCTION

Due to the high performance, simple structure and low cost, induction motor drives with advanced control strategies, such as vector control, sensorless vector control, etc, have been developed dramatically over the past decades. However, it is important to match correctly parameters of controlled motor in a controller for the advanced technology.

Nowadays, the parameters identification strategies are usually categorized offline and online identification ones. The offline identification strategies include traditional method and modified traditional method. These methods normally need two procedures: stand no-load test and rotor-locked test. However, when the motor have already been coupled to a load, those methods will be limited. So other methods are proposed which can identify the required parameters with the condition of holding the motor at standstill state. There are two types of excitation to be applied in the test. One is DC and another is AC. Procedures described in [1–3] are all based on some tests with single-phase DC excitation to the machine. On the other hand, [4–7] are based on AC excitation. The drawback of these methods is that they require awfully complicated mathematical calculations. As well known, almost all general invertors are equipped with a fixed point 16 bits or 32 bits MCU or DSP. So it is necessary to research a simple but useful estimating algorithm for the inverter-motor system.

This paper proposed a novel identification method which belongs to offline identification strategies and avoids solving the complex nonlinear equations. All the parameters can be estimated only by stator current and voltage while the motor is hold at standstill state. To avoid solving the awfully complicated nonlinear equations, an iterative algorithm is presented. The algorithm is easily implemented and suitable for fixed-DSP (Digital Signal Processor).

The remainder of this paper is organized as follows. Section II discusses the model of the induction motor with single-phase supply. Based on this model, Section III presents a novel parameter identification method with the motor at standstill state. After that, the section proposes an iterative algorithm to avid solving the complex equations. Section IV explains the experimental methods and then gives the simulation and experimental results. Section V presents conclusion.

II. MOTOR MODEL

The diagram of induction motor supplied with a single phase ac voltage is shown in Fig.1. Here, we supposed that the magnetizing curve is in linear region. Then the stator current is only composed of negative and positive sequence components. Then the models about negative and positive sequence currents are shown in Fig.2 (a) and (b), respectively. Where, R_1, L_{l1} are the stator resistance and leakage inductance, respectively; R_2', L_{l2}' are the rotor resistance and leakage inductance, respectively; L_m is the magnetizing inductance, s is the slip ratio, u_+, I_{1+}, I_{2+} are the positive sequence components of the stator voltage, stator current and rotor current, u_-, I_{1-}, I_{2-} are the negative sequence components of the counterpart, respectively.

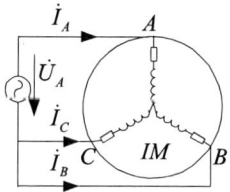

Figure.1. Induction motor with single-phase supply

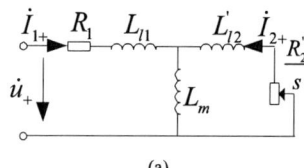

(a)

1-4244-0448-7/06/$25.00 ©2006 IEEE

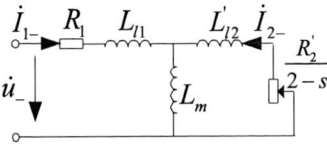

(b)

Figure.2 (a) positive-sequence model, (b) negative-sequence model

Based on the analysis of the models showed in Fig.2 by symmetrical component method, a circuit model of induction motor with single-phase supply is show in Fig.3. Further, Fig.4 is obtained based on Eq. (1) and Fig.3.

$$\begin{cases} L_s = L_r = L_m + L_l \\ L = (L_s L_r / L_m^2 - 1) L_s \\ R = L_s L_r / L_m^2 R_2' \end{cases} \quad (1)$$

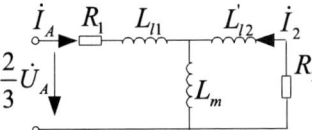

Figure.3. Circuit model of induction motor with single-phase supply

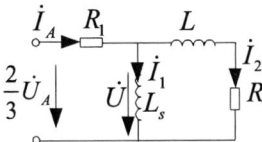

Figure.4. Type □ equivalent circuit

Generally, the equation $L_l = L_{l1} = L_{l2}'$ is supposed [8]. So the relation between Fig. 3 and Fig. 4 is described by Eq.(1) and (2).

$$\begin{cases} L_m = L_s / \sqrt{L / L_s + 1} \\ L_l = L_s - L_s / \sqrt{L / L_s + 1} \\ R_2' = R / (L / L_s + 1) \end{cases} \quad (2)$$

Due to the value of R_1 can be easily obtained by supplying a DC voltage to the stator, we can just put effort to the remainder parameters as L_{l1}, R_2', L_{l2}' and L_m. Then the equivalent circuit model showed in Fig.4 can be simplified to the model in Fig. 5, where the voltage $\dot{U}$ can be calculated by Eq.(3) under steady state.

$$\dot{U} = \frac{2}{3} \dot{U}_A - R_1 \dot{I}_A \quad (3)$$

The model shown in Fig. 5 can be described as

$$\frac{1}{j\omega L_s} + \frac{1}{R + j\omega L} = a + bj = \frac{\dot{I}_A(\omega)}{\dot{U}(\omega)} \quad (4-1)$$

where,

$$a = \frac{R}{R^2 + \omega^2 L^2}$$

$$b = -\frac{1}{\omega L_s} - \frac{\omega L}{R^2 + \omega^2 L^2} \quad (4-2)$$

Figure.5. Simplified equivalent circuit model

The model shown in Fig. 5 can be excited with two sets of single phase voltages with different frequencies, such as $\dot{U}(\omega_1)$ and $\dot{U}(\omega_2)$, and then two deferent currents $\dot{I}_A(\omega_1)$ and $\dot{I}_A(\omega_2)$ can be detected by controller. Consequently, we can get two sets of impedance as follows based on Eq.(4).

$$\begin{cases} \dfrac{1}{j\omega_1 L_s} + \dfrac{1}{R + j\omega_1 L} = a_1 + b_1 j = \dfrac{\dot{I}_A(\omega_1)}{\dot{U}(\omega_1)} \\ \dfrac{1}{j\omega_2 L_s} + \dfrac{1}{R + j\omega_2 L} = a_2 + b_2 j = \dfrac{\dot{I}_A(\omega_2)}{\dot{U}(\omega_2)} \end{cases} \quad (5)$$

where, $\omega_1 > \omega_2$.

Equation (5) can be rewritten as,

$$\begin{cases} \dfrac{R}{R^2 + \omega_1^2 L^2} = a_1 \\ \dfrac{1}{\omega_1 L_s} + \dfrac{\omega_1 L}{R^2 + \omega_1^2 L^2} = -b_1 \\ \dfrac{R}{R^2 + \omega_2^2 L^2} = a_2 \\ \dfrac{1}{\omega_2 L_s} + \dfrac{\omega_2 L}{R^2 + \omega_2^2 L^2} = -b_2 \end{cases} \quad (6)$$

In Eq. (6), there are three unknown variables R, L, L_s and four equations, so Eq. (6) can be solved. So the motor parameters R, L, L_s can be calculated by the measured currents and voltages by Eq.(6).

III. IDENTIFICATION ALGORITHM

Almost all general invertors are equipped with a fixed point 16 bits or 32 bits MCU or DSP. So it is difficult to calculate Eq. (6) with them. To avoid solving Eq. (6), an original iterative algorithm is proposed in this paper.

3.1 IDENTIFICATION ALGORITHM

Traditional iterative algorithm, such as Newton iterative algorithm, is complicated and then is unsuitable for the calculation by a fixed-point DSP. If we assume that the parameters of the induction motor are unchanged during the identification and $\dot{U}(\omega_1)$, $\dot{U}(\omega_2)$, $\dot{I}_A(\omega_1)$, $\dot{I}_A(\omega_2)$ can be acquired from the inverter and relevant sensors, the following steps are designed to estimate the parameters R, L, L_s.

Step1 Set an initial value $L_s = L_s^{(0)}$, based on $\{ L_s^{(0)}, \dot{U}(\omega_1), \dot{I}_A(\omega_1) \}$, $\{ R^{(1)}, L^{(1)} \}$ can be obtained from Eq. (5);

Step2 Based on $\{ R^{(1)}, L^{(1)}, \omega_2, \dot{U}(\omega_2), \dot{I}_A(\omega_2) \}$, $L_s^{(1)}$ is derived from Eq. (5);

Step3 Repeat the step1 and step2. That is, based on $\{ L_s^{(k)}, \omega_1, \dot{U}(\omega_1), \dot{I}_A(\omega_1) \}$, $\{ R^{(k+1)}, L^{(k+1)} \}$ is derived from Eq. (5), and based on $\{ R^{(k+1)}, L^{(k+1)}, \dot{U}(\omega_2), \dot{I}_A(\omega_2) \}$, $L_s^{(k+1)}$ is derived from Eq. (5);

Step4 Repeat step3 until the following absolute values are less than the given value Δ_L or Δ_R, which present the precision of the identification.

$$\left| L_s^{(k+1)} - L_s^{(k)} \right| \leq \Delta_L$$
$$\left| L^{(k+1)} - L^{(k)} \right| \leq \Delta_L \qquad (7)$$
$$\left| R^{(k+1)} - R^{(k)} \right| \leq \Delta_R$$

The process of the iterative algorithm convergence is explained as follows.

$$L_s^{(0)} \xrightarrow{\omega=\omega_1} \{ R^{(1)}, L^{(1)} \} \xrightarrow{\omega=\omega_2} L_s^{(1)}$$
$$\xrightarrow{\omega=\omega_1} \{ R^{(2)}, L^{(2)} \} \xrightarrow{\omega=\omega_2} L_s^{(2)} \cdots \cdots \qquad (8)$$

From Eq. (8), we can get Eq. (9) as follows

$$\{ R^{(1)}, L^{(1)} \} \xrightarrow{\omega=\omega_2} L_s^{(1)}$$
$$\xrightarrow{\omega=\omega_1} \{ R^{(2)}, L^{(2)} \} \qquad (9)$$

In the algorithm, $\{ R^{(k)}, L^{(k)} \}$ are obtained based on $\omega = \omega_1$, while $L_s^{(k)}$ is obtained based on $\omega = \omega_2$. Due to the value of L_s is very small when the frequency is high enough, the value error of L_s has less influence on $\{ R, L \}$ computation when $\omega = \omega_1$. Thus, $\{ R^{(2)}, L^{(2)} \}$ is closer to the true value than $\{ R^{(1)}, L^{(1)} \}$ though they are both derived from the same $L_s^{(1)}$ yet at different $\omega = \omega_1 > \omega_2$ respectively. To the next step, $L_s^{(2)}$ derived from $\{ R^{(2)}, L^{(2)} \}$ is closer to the true value than $L_s^{(1)}$ derived from $\{ R^{(1)}, L^{(1)} \}$. Obviously, $\{ L_s^{(2)}, R^{(2)}, L^{(2)} \}$ is closer to the true value than $\{ L_s^{(1)}, R^{(1)}, L^{(1)} \}$ after one repeat. Repeated the above steps, the absolute errors of $\{ L_s^{(k)}, R^{(k)}, L^{(k)} \}$ between the true values will become less and less until they arrive at the given precision.

3.2 IDENTIFICATION FLOW CHART

Fig.6 is the program flowchart. A DSP is used to generate PWM signals and implement A/D sampling and the proposed algorithm. Firstly, single-phase sine-wave voltages at two different frequencies ω_1, ω_2 are applied to the stators terminals respectively and the stator current and voltage $\dot{I}_A(\omega_1)$, $\dot{I}_A(\omega_2)$ and $\dot{U}_A(\omega_1)$, $\dot{U}_A(\omega_2)$ are recorded. Secondly, $\dot{U}(\omega_1)$ and $\dot{U}(\omega_2)$ are derived based on (3). Finally, according to the iterative algorithm, the induction motor parameters are estimated.

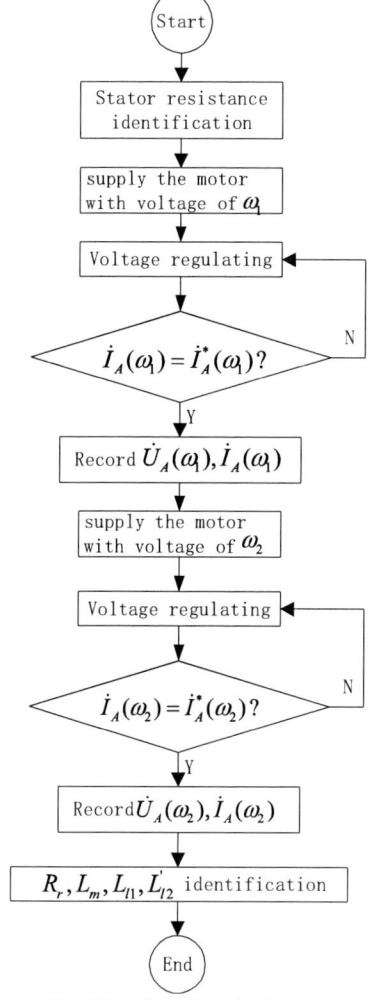

Fig. 6 Flow chart of the algorithm

IV SIMULATION AND EXPERIMENT

Experiments have been applied on two different types of induction motor. AG71-6/B3 is made in China and AB1329R is made in US. Table 1 is the rated parameters of the machines.

TABLE 1
RATED PARAMETERS OF MACHINES

TYPE	Power (KW)	Voltage (V)	Current (A)	Rotor Speed (r/min)
AG71-6/B3	0.25	400	0.98	900
AB1329R	1.50	460	2.30	1736

4.1 SIMULATIONS

Simulations are performed with MATLAB tools. Table 2 is the set parameters of the motor used in simulations. Table 3 is the estimated parameters. Result 1 and result 2 are obtained under the condition that the sampled current with no noise and with 10% white noise respectively. Compared result 1 with result 2, we can conclude that the proposed method is robust to the measure noise.

TABLE 2
MEASURED PARAMETERS OF AG71-6/B3

	R_1 (Ω)	R_2' (Ω)	L_m (mH)	$L_{l1}'L_{l2}'$ (mH)
Parameters	32.3	22.6	928	86

TABLE 3
SIMULATION RESULTS

Results	R_1 (Ω)	R_2' (Ω)	L_m (mH)	$L_{l1}'L_{l2}'$ (mH)
1	32.3	22.6	930	86
2	32.3	22.7	921	87

4.2 EXPERIMENT

The proposed algorithm is implemented in a general voltage source inverter shown as Fig.7, which employs a TMS320LF2407A DSP-based motor controller. The TMS320LF2407A DSP is a 16-bit fixed-point digital signal processor controller. It is equipped with 16 channels 10-bit A/D converter, two of which are used as current sensor by the inverter. The A/D converter is scaled as 250% of the peak-peak value of the rated current of the inverter.

To implement the proposed method, the induction motor is supplied with single-phase sine-wave power at two different frequencies, as shown in Fig. 1. The larger the difference of the two frequencies is, the more rapidly the algorithm converges. Due to the dead-time introduces serious harmonics to the voltage and then the current at low frequency and that will reduce the precision of the identification, the frequency can't be set too low. 4Hz is set in this experiment. On the other hand, the sample rate of A/D converter is designed as 1024 times of the current frequency so the frequency can't be set too high. So 24Hz is selected in this experiment and the sample rate of A/D converter is 24.576 KHz.

If only the magnet circuit is not saturated, the current is expected to big enough to make full use of the A/D converter. When the motor operate with a high frequency ω_1, the current component in magnet circuit is small, so the reference current is set to the rated value of the motor. At the same time, the reference value is set to

60% when the motor operates with a low frequency ω_2.

Fig.7. Experimental system

Because the proposed method is performed under steady state, the motor should be supplied with the current with a long time. However, the motor can not be cooled by the fan at standstill state, the temperature and resistance will rise, especially the stator's temperature. Therefore, the current of the stator terminals could not be supplied in a long time. In the experiment, when the absolute error between RMS of the current and the set value is less than 5% for 10 times, I_A and U_A is recorded as the data for parameter identification.

Since accuracy of the identified parameters are crucial to voltage drop of power electronic devices, inverter dead-time [9] and rotor skin effect, these influences are considered or compensated in the algorithm.

Taking A/D sampling currents as an example, the sampling position of the stator current should be fixed at the points shown by the marks Δ in Fig.8, in order to avoid sampling the harmonic current components. These positions can also reduce the influences of the dynamic waveform which is caused by switching of the power devices and added to the waveform of the fundamental current.

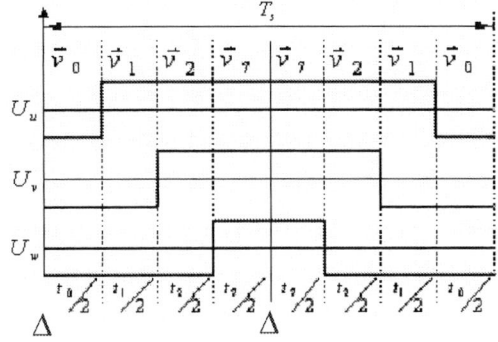

Fig 8 Current sampling point

Table 4 and Table 5 are the experiments results, where data A is the estimated results by the proposed method, data B is the measuring results by a traditional offline identification method where the voltages, currents and the powers are measured with a measurement WT1600 (made by YOKOGAWA Corp). Fig.8 shows the real

1899

current waveforms.

TABLE 4
RESULTS OF AG71-6/B3

Results	R_1 (Ω)	R_2' (Ω)	L_m (mH)	L_{l1}',L_{l2}' (mH)
A	32.3	22.5	928	86
B	32.3	22.3	924	88

TABLE 5
RESULTS OF AB1329R

Results	R_1 (Ω)	R_2' (Ω)	L_m (mH)	L_{l1}',L_{l2}' (mH)
A	5.89	3.70	443	27
B	5.89	3.69	440	27

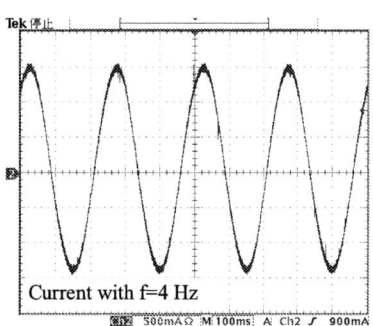

Current with f=4 Hz

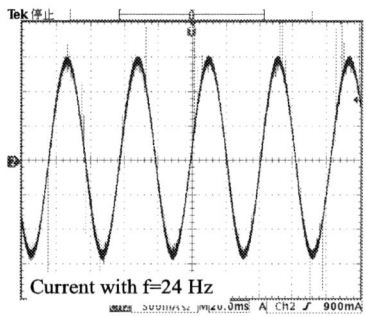

Current with f=24 Hz

Fig.8. current waveforms used for the identification

V. CONCLUSION

This paper proposes a novel parameter identification method for a general inverter-fed induction motor system at standstill state instead of performing locked rotor test and no-load test. Required no extra hardware, this method is effective, simple and practical for the electrical parameter identification of vector controlled induction motor drives.

REFERENCES

[1] P. J. Chrzan and H. Klaassen, "Parameter identification of vector-controlled induction machines," Electrical Engineering, vol. 79, no. 1, pp. 39–46, 1996.

[2] S. I. Moon and A. Keyhani, "Estimation of induction machine parameters from standstill time-domain data," IEEE Trans. Ind. Applicat., vol. 30, pp. 1606–1615, Nov./Dec. 1994.

[3] M. Ruff and H. Grotstollen, "Identification of the saturated mutual inductance of an asynchronous motor at standstill by recursive least squares algorithm," in Proc. Europe. Conf. Power Electron. Applicat., vol. 5, 1993, pp. 103–108.

[4] A. Bünte and H. Grotstollen, "Parameter identification of an inverter-fed induction motor at standstill with a correlation method," in Proc. Europe. Conf. Power Electron. Applicat., vol. 5, 1993, pp. 97–102.

[5] N. R. Klaes, "Parameters identification of an induction machine with regard to dependencies on saturation," in Proc. IEEE Ind. Applicat. Soc. Annu. Meeting, 1991, pp. 21–27.

[6] A. Bünte and H. Grotstollen, "Offline parameter identification of an inverter-fed induction motor at standstill," in Proc. Europe. Conf. Power Electron. Applicat., 1995, pp. 3.492–3.496.

[7] R. J. Kerkman, J. D. Thunes, T. M. Rowan, and D. Schlegel, "A frequency based determination of the transient inductance and rotor resistance for field commissioning purposes," in Proc. IEEE Ind. Applicat. Soc. Annu. Meeting, 1995, pp. 359–366.

[8] A. E. Fitzgerald, Charles Kingsley Jr., Stephen D. Umans, Electric Machinery", 6th edition, The McGraw-Hill Companies, 2003, Chapt. 6.

[9] Yu Gongjun, Zhong Yanru, Yang Geng, 'A compensation method of dead-time for IGBT inverters', Power electronics, China, No. 4, pp.7-9, 1999, In Chinese

Indirect Rotor Field Orientation Vector Control for Induction Motor Drives in the Absence of Current Sensors

Z. S. WANG*, S. L. HO**

* College of Electrical Engineering, Zhejiang University, Hangzhou 310027, China
** Electrical Engineering Department of Hong Kong Polytechnic University, Hong Kong
zjuwzs@hotmail.com, eeslho@inet.polyu.edu.hk

Abstract—**This paper proposes a simplified vector control implementation strategy that can be realized in the absence of current sensors. In order to decouple the torque and flux in the determination of the three-phase voltage reference commands of the SPWM inverter, both stator and rotor currents in the stationary and rotating frames can be derived from the corresponding rotor flux-oriented vector control requirements and motor dynamic equations. A sensitivity analysis to study the effect of parameter deviation or mismatch is also investigated. Simulation results are presented to demonstrate the feasibility and performance of the proposed methodology.**

Keywords- vector control; induction motor control; motor drives

I. INTRODUCTION

The indirect field orientation control (IFOC) strategy is widely used for implementing high performance induction motor drive systems, and has been increasingly adopted as the standard industry solution [1-3]. In this scheme, current sensors are used to measure the motor currents, and two current controllers have to be designed to regulate the motor currents and to infer the M/T-axis voltage reference command. These sensors and controllers not only increase the overall system's cost, but also increase the design complexity in terms of drift compensation and gain correction, particularly if the scheme is to be applicable over the whole speed and torque load ranges.

Conventional slip frequency control scheme does not use voltage and current sensors [4]. However, essentially, it is scalar control. Inevitably, its dynamic performance is poor due to the coupling effect between torque and flux. Yamamura S. etc. integrated the scalar scheme with vector control, and proposed the slip frequency vector control strategy to get good dynamic responds [5]. But in this method, current sensors and their corresponding controllers have to be designed to regulate the motor currents.

This paper presents a novel IFOC implementation method for induction motor drives without the need to use voltage and current sensors. It also eliminates two current feedback loops and their associated controllers, resulting in overall design simplicity and cost reduction for vector controllers. In order to realize the decoupled control between the torque and flux, the stator current is separated into two orthogonal components for torque current and flux magnetizing current, and these components are then regulated independently. Furthermore, it can be shown that if the rotor currents are known values, the voltage applied to the motor can be determined using motor equations. Nevertheless the rotor current cannot be measured directly in induction motors. However, for rotor field orientation drives, one could calculate the rotor current based on the vector control scheme and motor dynamic equations. Thus, the voltage applied to the motor can be controlled precisely. On the other hand, one has to note that as there are no voltage and current feedbacks, the performance of the drive might deteriorate due to parameter deviations/mismatch and disturbances in the input DC link voltage. It is shown from the simulation results that even though there are large value mismatching in parameters, the proposed algorithm could, with the exception of very low speeds, produce good dynamic and steady state performances over a wide range of speed. The effect of DC link voltage disturbance can be neglected in many actual practical application cases.

II. INDUCTION MOTOR MODEL AND THE PROPOSED CONTROL METHODOLOGY

According to the rotor field orientation vector control theory [4] [6], the stator current of squirrel-cage induction motor can be decomposed into two orthogonal components in the synchronous rotating rotor flux-oriented reference frame (M-T frame) which are, namely, the torque current i_{TS}, which is to generate the electromagnetic torque, and the magnetizing current i_{MS}, which is to excite the motor flux. These two current components are separated independently and decoupled each other. Their magnitudes can be calculated according the electromagnetic torque (T_e) and rotor flux level (λ_r), and are governed by the following equations [6]

$$\begin{cases} i_{TS} = \dfrac{4T_e}{3P\lambda_r} \\ i_{MS} = \dfrac{1 + (L_r / R_r)p}{L_M}\lambda_r \end{cases} \quad (1)$$

The subscript s and r indicate the stator and rotor variables, M,T represent the variables in the rotor field orientation rotating reference frame. R_r, L_M, L_S and L_r are the motor rotor resistance, mutual inductance, stator inductance and rotor inductance, respectively. λ_r and T_e are the respective rotor flux and electromagnetic torque. P is the number of pole pairs. p is the d/dt operator.

In most cases, the flux magnitude should be kept at some constant level, particularly when the motor runs below its base speed. So, (1) can be rewritten as

$$\begin{cases} i_{TS} = \dfrac{4T_e}{3P\lambda_r} \\ i_{MS} = \dfrac{\lambda_r}{L_M} \end{cases} \quad (2)$$

The current angle θ_2 in M-T axis is

$$\theta_2 = arctg\,\frac{i_{TS}}{i_{MS}} \quad (3)$$

On the other hand, the slip frequency is [4]

$$\omega_s = \frac{R_r i_{TS}}{L_r i_{MS}} \quad (4)$$

The rotor flux angle in stationary reference frame can be expressed as

$$\theta_1 = \int (\omega_r + \omega_s)dt \quad (5)$$

where ω_r is the rotor mechanical speed. Thus, the stator current can be expressed in the stationary reference frame as:

$$\begin{bmatrix} i_{\alpha s} \\ i_{\beta s} \end{bmatrix} = \begin{bmatrix} cos\,\theta_1 & -sin\,\theta_1 \\ sin\,\theta_1 & cos\,\theta_1 \end{bmatrix} \begin{bmatrix} i_{MS} \\ i_{TS} \end{bmatrix} \quad (6)$$

On the hand, in the rotor flux-orientated reference frame, the rotor side voltage-current equations of squirrel-cage induction motor can be described as [6]

$$\begin{bmatrix} 0 \\ 0 \end{bmatrix} = \begin{bmatrix} L_M P & 0 \\ L_M \omega_s & 0 \end{bmatrix}\begin{bmatrix} i_{MS} \\ i_{TS} \end{bmatrix} + \begin{bmatrix} R_r + L_r p & 0 \\ L_r \omega_S & R_r \end{bmatrix}\begin{bmatrix} i_{Mr} \\ i_{Tr} \end{bmatrix} \quad (7)$$

where i_{Mr}, i_{Tr} are rotor current components in M-T axis.

The first row of (7) can be rewritten as

$$\begin{aligned} 0 &= R_r i_{Mr} + p(L_M i_{Ms} + L_r i_{Mr}) \\ &= R_r i_{Mr} + p\lambda_{Mr} \end{aligned} \quad (8)$$

In the rotor field orientation control, $\lambda_r = \lambda_{Mr}, \lambda_{Tr} = 0$. In most cases, flux magnitude should be kept at some constant levels: $|\lambda_r| = const.$, or $p\lambda_r = 0$. Base on these requirements and the flux definition, the rotor current components in M-T axis, i_{Mr}, i_{Tr}, can be obtained as [6]

$$\begin{cases} i_{Mr} = -p\lambda_{Mr}/R_r = -p\lambda_r/R_r = 0 \\ i_{Tr} = (\lambda_{Tr} - L_m i_{Ts})/L_r = (0 - L_M/L_r)i_{Ts} \\ \quad = -(L_M/L_r)i_{Ts} \end{cases} \quad (9)$$

Using coordinate transformation, the rotor current can be expressed in the stationary reference as

$$\begin{bmatrix} i_{\alpha r} \\ i_{\beta r} \end{bmatrix} = \begin{bmatrix} cos\,\theta_1 & -sin\,\theta_1 \\ sin\,\theta_1 & cos\,\theta_1 \end{bmatrix}\begin{bmatrix} i_{Mr} \\ i_{Tr} \end{bmatrix} \quad (10)$$

The phasor diagram of the stator and rotor current (i_s, i_r) in $\alpha - \beta$ axis and M-T axis is shown in Fig. 1.

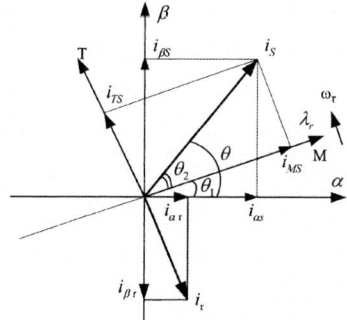

Fig.1 Phasor diagram of stator and rotor current

Based on the motor voltage-current equations in stationary reference frame which are shown below,

$$\begin{aligned} \begin{bmatrix} u_{\alpha s} \\ u_{\beta s} \end{bmatrix} &= \begin{bmatrix} R_S + L_S p & 0 \\ 0 & R_S + L_S p \end{bmatrix}\begin{bmatrix} i_{\alpha s} \\ i_{\beta s} \end{bmatrix} \\ &+ \begin{bmatrix} L_M p & 0 \\ 0 & L_M p \end{bmatrix}\begin{bmatrix} i_{\alpha r} \\ i_{\beta r} \end{bmatrix} \end{aligned} \quad (11)$$

where R_s, L_s are stator resistance and inductance. Using the 2-phase to 3-phase coordinate transformation, the three phase reference command voltage signal can finally be determined.

Thus, the final stator voltage, including frequency, phase and magnitude, can be directly derived from the motor dynamic equations just according to the speed and flux requirements (reference). Any voltage current sensors, signal feedback loops, and corresponding controllers do not need. Moreover, this method does not require complicated algorithm [5][6], and time-consuming parameter online identification technique [6]. So, the proposed control scheme can greatly simplify the system design, and reduce overall cost.

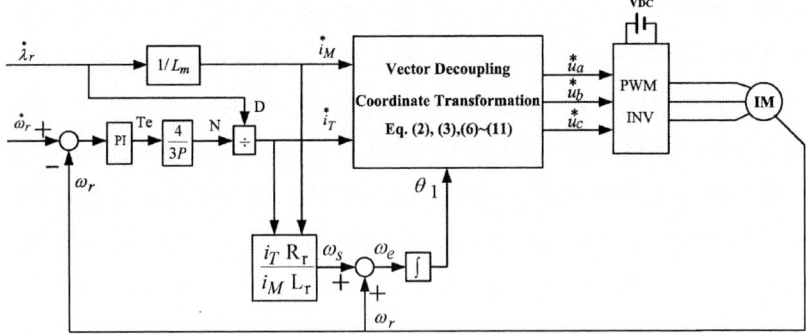

Fig.2 Configuration o the proposed vector control method

III. SIMULATION RESULTS

Digital simulation research is carried out to validate the proposed control method. Fig. 2 shows the system configuration of the vector control: a 2.2 kW induction motor is driven by a SPWM voltage-type inverter, and the three-phase voltage reference command applied to the inverter is derived from the flux and torque references. The simulation investigation is focused on the motor performance with decoupled torque and flux, a sensitivity study to evaluate parameter mismatch or deviation, and the tracking ability of the motor in detuned cases. The motor parameters are $R_S = 3\,\Omega$, $R_r = 3.23\,\Omega$, $L_M = 210mH$, $L_S = L_r = 223mH$, P=2.

A. Performance of the Proposed IFO Conreoller

Detailed simulation tests are carried out to assess the proposed control method.

Figs. 3 and 4 show the torque and flux decoupled performance of the motor. In Fig. 3, the motor is running at a speed of 1000 rpm. It can be seen that as the load torque is suddenly changed from 5.5 Nm to 10.5 Nm (trace a), the rotor flux level hardly changes (trace b, c), although there is a small speed dip but then it is restored quickly (trace d). Fig. 4 shows a different case in that the flux level is suddenly changed at t=5 s by the manipulators (trace a, c), however the torque can be restored to its original level quickly (trace b). Fig. 5 shows the tracking performance of the motor following a trapezoidal speed reference (trace a). Trace b is the speed error, and trace c is the rotor flux in the $\alpha - axis$. It can be seen that the flux level can be kept constant in different speed regions. The M-axis and the T-axis fluxes in trace d illustrate the effect of the rotor field-oriented control: $\lambda_{T\,r} = 0$, $\lambda_{M\,r} = const$. The speed tracking experiment is in a constant load condition (10Nm). This trapezoidal test can be used to evaluate the driver running in bi-direction operation, constant /variable speed and motoring/generative modes.

Fig. 6 shows the speed transient and step torque change responses (trace a). The load torque is proportional to the rotor speed in this case. There are speed reference and load torque step changes at time t=3 s and at t=3.5 s (trace a, c) respectively. Trace d and b show the corresponding flux magnitude and $\alpha - axis$ flux.

Because there are no voltage and current feedbacks in this control method, parameter mismatching or detuned running is a major problem. The following tests are designed to evaluate this performance. In Fig.7, the DC link voltage, which is applied to the inverter, is set to have a $\pm 20\%$ deviation from the rated voltages (the rated voltage is 500V) (trace a). It shows that the steady state torque is not changed (trace c), but the flux level is changed by as much as about $\pm 20\%$. This is still an acceptable result because in practical applications, most motors have their own input voltage ranges and their corresponding flux levels. Moreover, many new generation drivers use power factor correction (PFC) circuits in the first stage, so that their DC link voltage can be maintained at a constant level, and therefore one does not need to take into account this influence.

Fig. 8 shows the flux sensitivity with respect to the inductor parameters (L_M, L_S, L_r) and mismatching in low-speed (100rpm, traces a, b) and high-speed (1200 rpm, trace c, d) running. During t=2s to 3s, $\hat{L}_{MSR} = L^*_{MSR}$. During t=3s to 4s, $\hat{L}_{MSR} = 2L^*_{MSR}$. During t=4s to 5s: $\hat{L}_{MSR} = 0.5L^*_{MSR}$, where L^*_{MSR} means the actual motor L_M, L_S, L_r parameters, and $\hat{L}_{MSR}$ means the corresponding mismatched parameters used in the controller. These mismatching has a slight influence upon the motor performance at high-speed running only, the influence is much more pronounced at low-speeds. The resistance mismatching influence is shown in Fig.9, similarly, it explains the flux level versus different stator and rotor resistances in low-speed (100rpm in trace a, b) and high-speed (1200 rpm in trace in trace d, e) regions. During t=2s to 3s, $\hat{R}_{s\,r} = R^*_{s\,r}$. During t=3s to 4s, $\hat{R}_{s\,r} = 2R^*_{s\,r}$. During t=4s to 5s, $\hat{R}_{s\,r} = 0.5R^*_{s\,r}$, where $R^*_{s\,r}$ means the actual parameters, and $\hat{R}_{s\,r}$ means the mismatched parameters used in the controller. One can

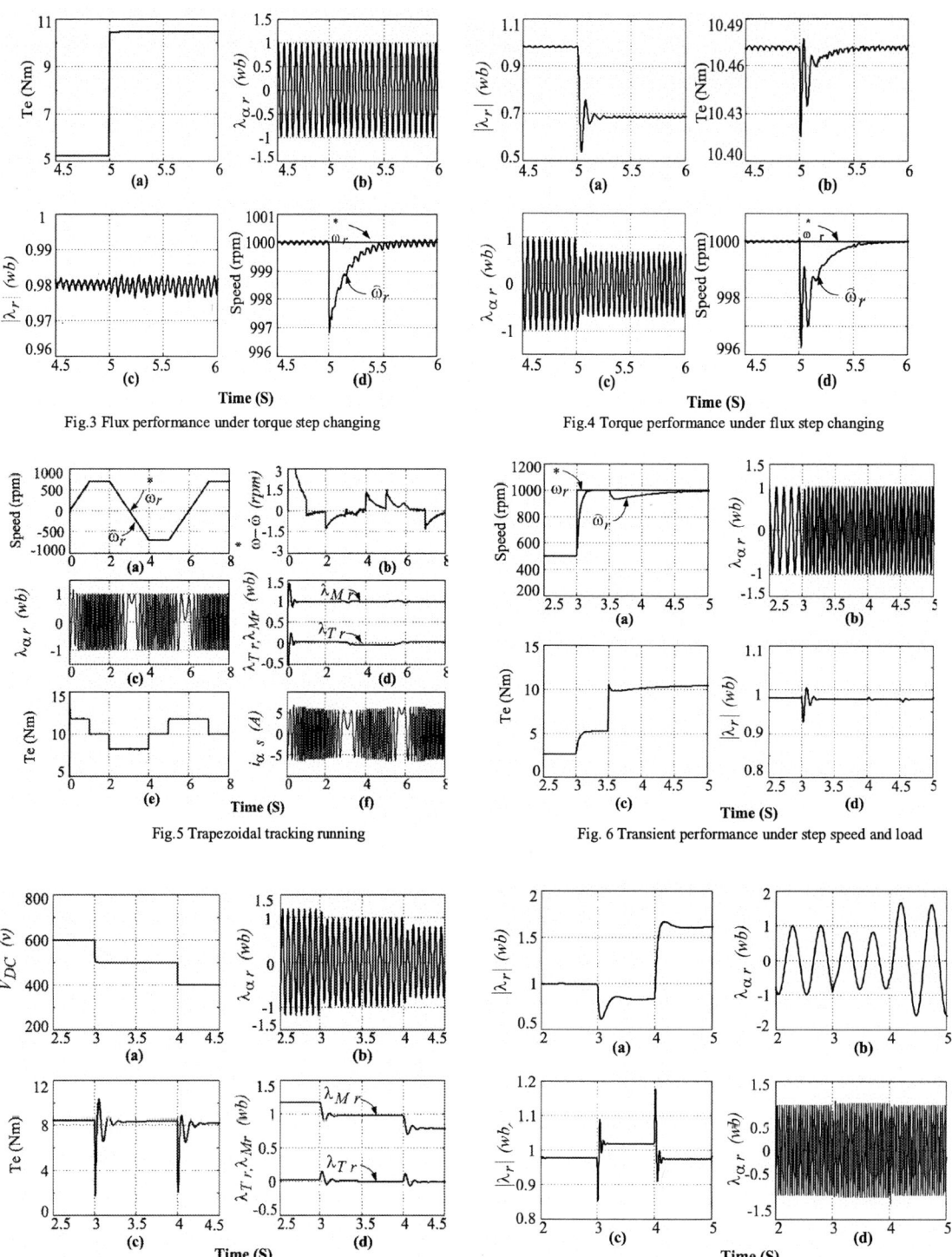

Fig.3 Flux performance under torque step changing

Fig.4 Torque performance under flux step changing

Fig.5 Trapezoidal tracking running

Fig. 6 Transient performance under step speed and load

Fig. 7.Dc link voltage mismatching

Fig.8 Inductor mismatching

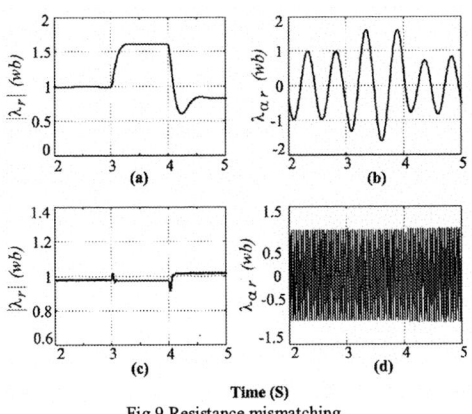

Fig.9 Resistance mismatching

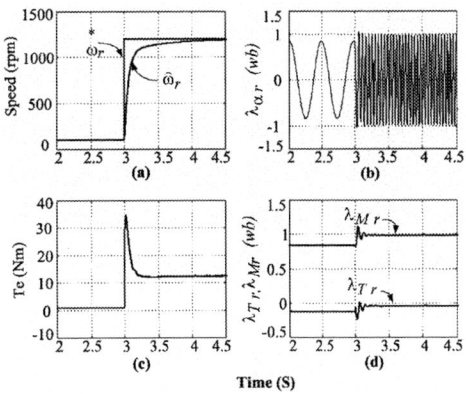

Fig. 10 Synthetic parameter mismatching

see that noticeable influence only appears in case of low speed running.

For the proposed control method, it will have a general tolerance to operate with parameters mismatching to a certain extent. A transient performance with parameter mismatching is shown in Fig. 10. The mismatched parameters are: $R^*_{sr} = 2\hat{R}_{sr}$, $L^*_{MSR} = 1.5\hat{L}_{MSR}$. The load is set to be proportional to the rotor speed. One can see that although the transient time is increased a bit, the steady state flux and torque (speed) are still stable with very small ripples.

B. Dynamic Responds Comparison with Conventional Vector Control

The proposed control algorithm is derived form the basic principle of rotor field orientation control. Consequently, so it is still vector control method. It should be noted that the purpose of this control method is aimed to reduce the money cost in hardware, and simplify the control design. Fig.11 shows the dynamic responds of conventional vector control and proposed control scheme in case of no load ((a)) and 10 Nm load (Trace (b)).

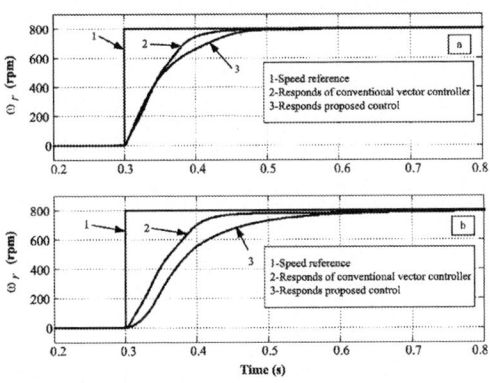

Fig. 11 Step responds of proposed control method and conventional vector control. (a). no load running, (b). running in 10 Nm load

Both trace (a) and (b) show that the proposed control scheme shares comparable good dynamic responds performance with conventional vector controller using current sensors no matter running in no load or in a heavy load.

IV. CONCLUSIONS

A vector control strategy for induction motor drives, which does not use current sensor, is proposed and tested. This scheme can reduce the system cost and simplify the control design. The current feedback loop and controller can be eliminated. It exhibits the decoupled effect between the flux and torque. Sensitivity to the parameter mismatching is evaluated by simulation, and it shows that the proposed algorithm works well over most speed ranges with the exception of low speeds. Moreover, the proposed system can operate stably with significant parameters mismatching as well.

REFERENCES

[1] A. Consoli, G. Scarcella, and A. Testa, "Speed-and Current-Sensorless Field-Oriented Induction Motor Drive Operation at Low Stator Frequencies, " *IEEE Trans. Industry Applications*, Vol.40, No. 1, pp.186-193, Jan./Feb., 2004.

[2] Fang-Zheng Peng, and T. Fukao, Robust Speed Identification for Speed-sensorless Vector Control of Induction Motors, *IEEE Trans. Industry Appl.*, Vol.30, No. 5, pp.1234-1240, Sept./Oct., 1994.

[3] R. Gabriel, W. Leonard, C. Nordby, Field Oriented Control of a Standard AC Motor Using Micro-processors, *IEEE Trans. Industry Applications*, Vol.IA-16, No. 2, pp.186-192, 1980.

[4] Andrzej M. Trzynadlowski, *The Field Orientation Principle in Control of Induction Motors*, Kluwer Academic Publishers, U.S.A.

[5] Yamamura S., Spiral Vector Theory of AC Motor Analysis and Control, *IEEE Industry Applications Society Annual Meeting*, 1991, Conference Record of 1991, pp.79 – 86.

[6] Bose B. K., *Modern Power Electronics and AC Drives*, Prentice-Hall, Englewood Cliffs, New Jersey, 1986.

A Robust Adaptive Sliding-Mode Controller for Slip Power Recovery Induction Machine Drives

J.Soltani, A.Farrokh Payam

Isfahan University of Technology/Department of Electrical and Computer Engineering, Isfahan, Iran
e-mail:j1234sm@cc.iut.ac.ir,amir_farokh@yahoo.com

Abstract-In this paper a robust nonlinear controller is presented for Doubly-Fed Induction Machine (DFIM) drives. The nonlinear controller is designed based on combination of Sliding-Mode (SM) and Adaptive-Backstepping control techniques. Using the fifth order model of DFIM in a stator d,q axis reference frames with stator currents and rotor fluxes as state variables, a SM controller is designed in order to follow a linear reference model. Then, the SM control and adaptive backstepping control approach are combined to design a robust composite nonlinear controller for DFIM, that makes the drive system robust and stable against the parameters uncertainties and external load torque disturbance. In this drive system two back-to-back voltage source two level SVM-PWM inverters are employed in the rotor circuit, to make the drive system capable of operating in motoring and generating modes below and above synchronous speed. Computer simulation results obtained, confirm the effectiveness and validity of the proposed control approach.

Keywords-Doubly fed Induction Machine, Nonlinear Sliding-Mode, Adaptive backstepping, Torque and Flux control.

I. INTRODUCTION

So far, the vector control (VC) and Direct torque control (DTC) methods have been applied to the three phase squirrel cage induction machine drives [1]. Although the field oriented control method in the stator or magnetizing reference frame are applied to the DFIM drives [2,3], little attention has been given to the DTC and flux control of these types of drives.

In field oriented methods applied to DFIM drives, the voltage drop across the stator leakage impedance is neglected. Such an assumption, forces a steady-state error both in motoring and generating modes of operation. In [4] the bang-bang DTC control method is combined with direct rotor flux field oriented control method and is applied to an adjustable doubly fed induction motor drive. In [4], the controller is designed based on neglecting the voltage drop across the rotor resistance. In [5,6], a backstepping tracking controller has been introduced for a DFIM drive. The control method of [5,6] has been applied only for generating mode of operation upon unity of power factor, measured on the stator side of the

machine. This paper describes a nonlinear controller for DFIM drives based on combination of SM control and adaptive backstepping control approaches. The SM control forces the system state nominal trajectories to follow a desired linear reference model DFIM drive while the adaptive backstepping controller makes the drive system performance robust and stable subject to the parameter variations and external load torque. The nonlinear control approach described in this paper has the following advantages:

1) The motor generated torque becomes linear with respect to system control states.

2) The rotor flux can be easily regulated in order to increase the machine efficiency.

3) The system robustness can be achieved against the parameters uncertainty.

The overall system stability is proved by Lyapanouv theory. The nonlinear control method described in [7], uses a third order backstepping approach and a third order linear reference model. In this method, three independent estimation laws are derived for stator and rotor resistance estimation. Such a overparameter estimation laws obtained are not useful to predict the motor resistances. In additions in [7], has not been described how is possible rotor flux to calculate, the rotor fluxes with unknown resistances. The research work presented in this paper is in fact a continuation of research work described in [7], with main objective of achieving a robust DTC and flux control for DFIM drives, using two back-to-back two level SVM-PWM voltage source inverters in the rotor circuit, the proposed control method can be applied for motoring and generating modes of operation below and above synchronous speed. In addition the rotor dc link voltage is maintained constant based on input-output feedback linearization control method, using a rotating reference frame with d axis in coincide with the space voltage vector of the main ac supply.

II. DFIM MODEL

Under assumption of linear magnetic circuits and balanced operating condition, the equivalent two-phase model of the symmetrical DFIM with stator connected to line, represented in fixed stator d-q reference frame is

$$\frac{di_{ds}}{dt} = -(\frac{R_s}{L_\sigma} + \frac{R_r L_m^2}{L_r^2 L_\sigma})i_{ds} + \frac{R_r L_m}{L_r^2 L_\sigma}\psi_{dr} +$$

$$\frac{\omega_r L_m}{L_r L_\sigma}\psi_{qr} + \frac{u_{ds}}{L_\sigma} - \frac{L_m}{L_r L_\sigma}u_{dr}$$

$$\frac{di_{qs}}{dt} = -(\frac{R_s}{L_\sigma} + \frac{R_r L_m^2}{L_r^2 L_\sigma})i_{qs} + \frac{R_r L_m}{L_r^2 L_\sigma}\psi_{qr} -$$

$$\frac{\omega_r L_m}{L_r L_\sigma}\psi_{dr} + \frac{u_{qs}}{L_\sigma} - \frac{L_m}{L_r L_\sigma}u_{qr}$$

$$\frac{d\psi_{dr}}{dt} = \frac{R_r L_m}{L_r}i_{ds} - \frac{R_r}{L_r}\psi_{dr} - \omega_r\psi_{qr} + u_{dr}$$

$$\frac{d\psi_{qr}}{dt} = \frac{R_r L_m}{L_r}i_{qs} - \frac{R_r}{L_r}\psi_{qr} + \omega_r\psi_{dr} + u_{qr} \tag{1}$$

where i_s, ψ_r, u_s, u_r, R and L denote stator current, rotor flux linkage, stator terminal voltage, rotor terminal voltage, resistance and inductance, respectively. The subscripts s and r stand for stator and rotor while subscripts d and q stand for vector component with respect to a fixed stator reference frame. ω_r denotes the rotor electrical angular speed and L_m is mutual inductance. $L_\sigma = L_s(1 - (L_m^2 / L_r L_s))$ is the redefined leakage inductance.

The DFIM generated torque can be expressed in terms of stator currents and rotor flux linkage as follows:

$$T_e = \frac{3P}{2}\frac{L_m}{L_r}(\psi_{dr}i_{qs} - \psi_{qr}i_{ds}) \tag{2}$$

where P is number of poles.

The mechanical equation is given by

$$J\frac{d\omega_m}{dt} + B\omega_m + T_L = T_e \tag{3}$$

where J and B denote the moment of inertia and friction coefficient, respectively, T_L is the external load torque and ω_m is the rotor mechanical angular speed ($\omega_r = (P/2)\omega_m$).

Let

$$x = \begin{bmatrix} i_{ds} & i_{qs} & \psi_{dr} & \psi_{qr} \end{bmatrix}^T \tag{4}$$

be the state vector and for generated torque T_e defined by

$$y = T_e = \frac{3P}{2}\frac{L_m}{L_r}(\psi_{dr}i_{qs} - \psi_{qr}i_{ds}) \tag{5}$$

III. NONLINEAR SLIDING MODE CONTROL

Equation (1) in compact form rewritten as:

$$\dot{x} = f(x) + g_1 u_{dr} + g_2 u_{qr} \tag{6}$$

where x is defined in (4) and

$$f(x) = \begin{bmatrix} -(\frac{R_s}{L_\sigma} + \frac{R_r L_m^2}{L_r^2 L_\sigma})i_{ds} + \frac{R_r L_m}{L_r^2 L_\sigma}\psi_{dr} + \frac{\omega_r L_m}{L_r L_\sigma}\psi_{qr} + \frac{u_{ds}}{L_\sigma} \\ -(\frac{R_s}{L_\sigma} + \frac{R_r L_m^2}{L_r^2 L_\sigma})i_{qs} + \frac{R_r L_m}{L_r^2 L_\sigma}\psi_{qr} - \frac{\omega_r L_m}{L_r L_\sigma}\psi_{dr} + \frac{u_{qs}}{L_\sigma} \\ \frac{R_r L_m}{L_r}i_{ds} - \frac{R_r}{L_r}\psi_{dr} - \omega_r\psi_{qr} \\ \frac{R_r L_m}{L_r}i_{qs} - \frac{R_r}{L_r}\psi_{qr} + \omega_r\psi_{dr} \end{bmatrix} \tag{7}$$

and

$$g_1 = \begin{bmatrix} -\dfrac{L_m}{L_r L_\sigma} & 0 & 1 & 0 \end{bmatrix}^T$$

$$g_2 = \begin{bmatrix} 0 & -\dfrac{L_m}{L_r L_\sigma} & 0 & 1 \end{bmatrix}^T \tag{8}$$

Assuming the machine torque T_e and the norm of the rotor flux modulus $|\psi_r|^2 = \psi_{dr}^2 + \psi_{qr}^2$ are to be the system control outputs. Based on input-output feedback linearization

$$h_1(x) = \frac{3P}{2}\frac{L_m}{L_r}(\psi_{dr}i_{qs} - \psi_{qr}i_{ds})$$

$$h_2(x) = \psi_{dr}^2 + \psi_{qr}^2 \tag{9}$$

Introducing the new variables as:

$$z_1 = h_2(x)$$

$$z_2 = h_1(x) \tag{10}$$

Then, the dynamic model of DFIM in new coordinates is given by:

$$\begin{bmatrix} \dot{z}_1 \\ \dot{z}_2 \end{bmatrix} = \begin{bmatrix} L_f h_2 \\ L_f h_1 \end{bmatrix} + \begin{bmatrix} L_{g_1}h_2 & L_{g_2}h_2 \\ L_{g_1}h_1 & L_{g_2}h_1 \end{bmatrix}\begin{bmatrix} u_{dr} \\ u_{qr} \end{bmatrix} \tag{11}$$

where

$$L_f h_2 = 2\frac{R_r L_m}{L_r}(i_{ds}\psi_{qr} - i_{qs}\psi_{dr}) - 2\frac{R_r}{L_r}(\psi_{dr}^2 + \psi_{qr}^2)$$

$$L_f h_1 = \frac{3P}{2}\frac{L_m}{L_r} \times$$

$$\begin{pmatrix} -(\frac{R_s}{L_\sigma} + \frac{R_r L_m^2}{L_r^2 L_\sigma} + \frac{R_r}{L_r})(i_{qs}\psi_{dr} - i_{ds}\psi_{qr}) - \\ \frac{\omega_r L_m}{L_r L_\sigma}(\psi_{dr}^2 + \psi_{qr}^2) - \omega_r(\psi_{qr}i_{qs} + \psi_{dr}i_{ds}) + \frac{u_{qs}\psi_{dr}}{L_\sigma} - \frac{u_{ds}\psi_{qr}}{L_\sigma} \end{pmatrix}$$

$$L_{g_1}h_2 = 2\psi_{dr}$$

$$L_{g_2}h_2 = 2\psi_{qr}$$

$$L_{g_1}h_1 = \frac{3P}{2}\frac{L_m}{L_r}(i_{qs} + \frac{L_m}{L_r L_\sigma}\psi_{qr})$$

$$L_{g_2}h_1 = -\frac{3P}{2}\frac{L_m}{L_r}(i_{ds} + \frac{L_m}{L_r L_\sigma}\psi_{dr}) \tag{12}$$

Furthermore, a nonlinear state feedback control inputs is employed as:

$$\begin{bmatrix} \hat{u}_{dr} \\ \hat{u}_{qr} \end{bmatrix} = \begin{bmatrix} L_{g1}h_2(x)u_{dr} + L_{g2}h_2(x)u_{qr} \\ L_{g1}h_1(x)u_{dr} + L_{g2}h_2(x)u_{qr} \end{bmatrix} \qquad (13)$$

Then, the dynamic system (11) becomes

$$\begin{bmatrix} \dot{z}_1 \\ \dot{z}_2 \end{bmatrix} = \begin{bmatrix} L_f h_2(x) \\ L_f h_1(x) \end{bmatrix} + \begin{bmatrix} 1 & 0 \\ 0 & 1 \end{bmatrix} \begin{bmatrix} \hat{u}_{dr} \\ \hat{u}_{qr} \end{bmatrix} \qquad (14)$$

From (14) the reference model is introduced as:

$$\dot{z}_m = A_m z_m + B_m u_{ref}$$

$$\begin{bmatrix} \dot{z}_{m1} \\ \dot{z}_{m2} \end{bmatrix} = \begin{bmatrix} -a_{m1} & 0 \\ 0 & -a_{m2} \end{bmatrix} \begin{bmatrix} z_{m1} \\ z_{m2} \end{bmatrix} + \begin{bmatrix} a_{m3} & 0 \\ 0 & a_{m4} \end{bmatrix} \begin{bmatrix} \psi_r^{*2} \\ T_e^* \end{bmatrix} \qquad (15)$$

where a_{m1}, a_{m2}, a_{m3} and a_{m4} are the positive constants.

Using (14) and (15) the system error dynamics is obtained as:

$$\dot{e}_z = A(x) + B(x)\overline{U} \qquad (16)$$

where

$$e_z = \begin{bmatrix} z_1 - z_{m1} & z_2 - z_{m2} \end{bmatrix}^T = \begin{bmatrix} e_{z1} & e_{z2} \end{bmatrix}^T \qquad (17)$$

with

$$A(x) = \begin{bmatrix} L_f h_2(x) \\ L_f h_1(x) \end{bmatrix} \qquad B(x) = \begin{bmatrix} 1 & 0 \\ 0 & 1 \end{bmatrix}$$

$$\overline{U} = \begin{bmatrix} \overline{u}_{dr} \\ \overline{u}_{qr} \end{bmatrix} = \begin{bmatrix} \hat{u}_{dr} + a_{m1}z_{m1} - a_{m3}\psi_r^{*2} \\ \hat{u}_{qr} + a_{m2}z_{m2} - a_{m4}T_e^* \end{bmatrix} \qquad (18)$$

According to the system shown in (16), the sliding switching surfaces are chosen as:

$$\sigma(e_z) = Se_z(x) \qquad (19)$$

where $S \in \Re^{2 \times 2}$ is a constant linear matrix so that the inverse of $SB(x)$ must exist for all x, i.e., $\det (SB(x)) \neq 0$ for all x. Combining (16) and (19), gives

$$\dot{\sigma} = S\dot{e}_z = SA(x) + SB\overline{U} = -Q\operatorname{sgn}(\sigma) - K\sigma \qquad (20)$$

where

$$Q = \begin{bmatrix} q_1 & 0 \\ 0 & q_2 \end{bmatrix} \qquad K = \begin{bmatrix} k_1 & 0 \\ 0 & k_2 \end{bmatrix} \quad q_i, k_i > 0,\ i = 1,2 \qquad (21)$$

where

$$\operatorname{sgn}(\sigma) = \begin{cases} 1, & as \quad \sigma > 0 \\ -1, & as \quad \sigma < 0 \end{cases} \qquad (22)$$

From (20) based on Lyapanouv theory, the SM controller is obtained as:

$$\overline{U} = -(SB)^{-1}[SA(x) + Q\operatorname{sgn}(\sigma) + K\sigma] \qquad (23)$$

Using the control low of (23), the reachability of SM control is guaranteed.

Theorem 1: With the developed nonlinear sliding-mode controller (23) and a stable sliding surface (19), the reaching condition $\sigma^T \dot{\sigma} < 0$ is satisfied, and the controlled system (16) will be stabilized.

Proof: From the designed dynamic sliding surface (22), the following equation can be derived

$$\dot{\sigma}_i = -q_i \operatorname{sgn}(\sigma_i) - k_i \sigma_i \quad i = 1,2 \qquad (24)$$

Multiplying σ_i on the both side of the above equation yields

$$\sigma_i \dot{\sigma}_i = -q_i \sigma_i \operatorname{sgn}(\sigma_i) - k_i \sigma_i^2 = -q_i |\sigma_i| - k_i \sigma_i^2 < 0 \qquad (25)$$

From the above analysis, it is evident that the reaching condition is guaranteed.

Remark 1: Using the model-following nonlinear SM control based on the state-coordinate transformed model for the DFIM, the transient of the generated torque and rotor flux amplitude can be regulated through a linear reference model.

In practical and industrial applications, the rotor and stator resistance R_r and R_s vary in temperature and the magnetic flux saturation. Therefore, a nonlinear SM controller that makes the DFIM stable and robust against the parameters variations is proposed in the next section.

IV. ADAPTIVE BACKSTEPPING SLIDING-MODE CONTROL

Adaptive backstepping control has the features of having more than one estimate per unknown parameter for controlled system with mismatched uncertainties [9].

When the system parameters deviate from the nominal values, especially the resistances R_r and R_s, the tracking error model (17) can be rewritten as

$$\dot{e}_{z1} = L_f h_2(x) + \overline{u}_{dr} + \phi_1 d_1(x)$$

$$\dot{e}_{z2} = L_f h_1(x) + \overline{u}_{qr} + \phi_2 d_2(x)$$

$$\dot{e}_z = [A(x) + \Delta A(x)] + B(x)\overline{U} \qquad (26)$$

where ϕ_i $i = 1,2$ and $\Delta A(x)$ denote in fact uncertainties defined by:

$$\Delta f(x) = \begin{bmatrix} -\Delta\alpha i_{ds} + \Delta\beta\psi_{dr} \\ -\Delta\alpha i_{qs} + \Delta\beta\psi_{qr} \\ \dfrac{\Delta R_r L_m}{L_r} i_{ds} - \dfrac{\Delta R_r}{L_r}\psi_{dr} \\ \dfrac{\Delta R_r L_m}{L_r} i_{qs} - \dfrac{\Delta R_r}{L_r}\psi_{qr} \end{bmatrix}, \Delta A(x) = \begin{bmatrix} L_{\Delta f}h_2(x) \\ L_{\Delta f}h_1(x) \end{bmatrix}$$

$$\alpha = \left(\frac{R_s}{L_\sigma} + \frac{R_r L_m^2}{L_r^2 L_\sigma} \right), \beta = \frac{R_r L_m}{L_r^2 L_\sigma}$$

$$\Delta\alpha = \left(\frac{\Delta R_s}{L_\sigma} + \frac{\Delta R_r L_m^2}{L_r^2 L_\sigma} \right), \Delta\beta = \frac{\Delta R_r L_m}{L_r^2 L_\sigma} \qquad (27)$$

Therefore:

$$L_{\Delta f}h_2(x) = \frac{2\Delta R_r L_m}{L_r}\left[i_{ds}\psi_{dr} + i_{qs}\psi_{qr}\right] - \frac{2\Delta R_r}{L_r}\left[\psi_{dr}^2 + \psi_{qr}^2\right]$$

$$L_{\Delta f}h_1(x) = \frac{3P}{2}\frac{L_m}{L_r}\left(\frac{\Delta R_r}{L_r}(\psi_{dr}i_{qs} \quad \psi_{qr}i_{ds}) \quad \Delta\alpha(i_{qs}\psi_{dr} - i_{ds}\psi_{qr}) \right)$$

$$\phi_1 d_1(x) = \frac{2\Delta R_r}{L_r}\left[L_m(i_{ds}\psi_{dr} + i_{qs}\psi_{qr}) - (\psi_{dr}^2 + \psi_{qr}^2)\right]$$

$$\phi_2 d_2(x) = -\frac{3P}{2}\frac{L_m}{L_r}\left[\frac{\Delta R_r}{L_r} + \Delta\alpha \right]\left[\psi_{dr}i_{qs} - \psi_{qr}i_{ds}\right]$$

$$(28)$$

that is $[\phi_1 d_1(x), \phi_2 d_2(x)]^T = \Delta A(x)$.

Since R_r and R_s are sensitively varied with thermal drift, hence one can assume that $|\phi_i|$ $(i = 1,2)$ is the unknown and bounded constant.

1908

Derivativing the system errors e_z with respect to time t, yields

$$\dot{e}_{z1} = L_f h_2(x) + \bar{u}_{dr} + \hat{\phi}_1 d_1(x) - (\hat{\phi}_1 - \phi_1)d_1(x)$$
$$\dot{e}_{z2} = L_f h_1(x) + \bar{u}_{qr} + \hat{\phi}_2 d_2(x) - (\hat{\phi}_2 - \phi_2)d_2(x) \quad (29)$$

where $\hat{\phi}_i$ $(i=1,2)$ is the estimate of ϕ_i, $k_i, i = 1,2$ is a positive constant feedback gain.

It is obvious that the controllers $\bar{u}_{dr}$ and $\bar{u}_{qr}$ are decoupling with respect to two dynamic models, $[e_{z1}, e_{z2}]$. From (29), the d axes SM control is designed as

$$\bar{u}_{dr} = -L_f h_2(x) - \hat{\phi}_1 d_1(x) - k_1 e_{z1} - \rho_1 \operatorname{sgn}(e_{z1}) \quad (30)$$

where $k_1 > 0$ and ρ_1 is chosen as follows
$\rho_1.d_1(x) \geq |(\hat{\phi}_1 - \phi_1)|.d_1(x)$ when $\hat{\phi}_1$ is in transient and the adaptation law of $\hat{\phi}_1$ is given by

$$\dot{\hat{\phi}}_1 = \gamma_1 e_{z1} d_1(x) \quad (31)$$

where $\gamma_1 > 0$ is the adaptation gain. Similarly, the q axes SM control is designed as:

$$\bar{u}_{qr} = -L_f h_1(x) - \hat{\phi}_2 d_2(x) - k_2 e_{z2} - \rho_2 \operatorname{sgn}(e_{z2}) \quad (32)$$

where k_2 is positive constant feedback gain and

$$\dot{\hat{\phi}}_2 = \gamma_2 e_{z2} d_2(x) \quad (33)$$

where $\gamma_2 > 0$ is the adaptation gain and ρ_2 is chosen as follows
$\rho_2.d_2(x) \geq |(\hat{\phi}_2 - \phi_2)|.d_2(x)$ when $\hat{\phi}_2$ is in transient.
Theorem 2: Using the controller described by (30)-(33), the torque and flux amplitude DFIM is stable and robust subject to the parameters mismatched uncertainties.
Proof: Defining the following Lyapanouv function

$$V_1 = \frac{1}{2}\left[e_{z1}^2 + e_{z2}^2 + \frac{1}{\gamma_1}(\hat{\phi}_1 - \phi_1)^2 + \frac{1}{\gamma_2}(\hat{\phi}_2 - \phi_2)^2 \right] \quad (34)$$

Derivative (34) with respect to time t, using equation (29), gives:

$$\dot{V}_1 = e_{z1}\left[L_{f(x)} h_2(x) + \bar{u}_{dr} + \hat{\phi}_1 d_1(x) - (\hat{\phi}_1 - \phi_1)d_1(x) \right] + $$
$$e_{z2}\left[L_{f(x)} h_1(x) + \bar{u}_{qr} + \hat{\phi}_2 d_2(x) - (\hat{\phi}_2 - \phi_2)d_2(x) \right] + $$
$$\frac{1}{\gamma_1}(\hat{\phi}_1 - \phi_1)\dot{\hat{\phi}}_1 + \frac{1}{\gamma_2}(\hat{\phi}_2 - \phi_2)\dot{\hat{\phi}}_2 \quad (35)$$

Substituting the control laws (30) and (32) and adaptation laws (31) and (33) into (35), the equation it is reduced to

$$\dot{V}_1 = -k_1 e_{z1}^2 - k_2 e_{z2}^2 \leq 0 \quad (36)$$

Defining the following equation

$$M(t) = k_1 e_{z1}^2 + k_2 e_{z2}^2 \geq 0 \quad (37)$$

also,

$$V_1(t) = V_1(e(0), \hat{\phi}(0)) + \int_0^t \dot{V}_1(\tau)d\tau = V_1(e(0), \hat{\phi}(0)) - \int_0^t M(\tau)d\tau \quad (38)$$

where $e = [e_{z1}, e_{z2}]^T$ and $\hat{\phi} = [\hat{\phi}_1, \hat{\phi}_2]^T$. From the definition of the Lyapanouv function $V_1(t) \geq 0$ and the above equation, one can obtain that

$$\lim_{t \to \infty} \int_0^t M(\tau)d\tau \leq V_1(e(0), \hat{\phi}(0)) < \infty \quad (39)$$

Based on the Barbalat's Lemma [8], we can obtain
$$M(t) \to 0 \text{ as } t \to \infty \quad (40)$$
That is, e_{z1} and e_{z2} will converge to zero as $t \to \infty$. Therefore, the proposed controller is stable and robust, even if parameters uncertainties exist.

V. STABILAZATION OF ROTOR DC-LINK VOLTAGE

Using the method described in [9], the d and q axis current equations corresponding to main ac power supply are give by:

$$\frac{di_d}{dt} = \frac{1}{L}(v_d - Ri_d + \omega_e L i_q - v_{d1})$$

$$\frac{di_q}{dt} = \frac{1}{L}(-Ri_q - \omega_e L i_d - v_{q1}) \quad (41)$$

One may note that in this reference frames the d axis is in coincided with the main space voltage vector:
Considering i_d^*, i_q^* as reference currents, therefore:
$$e_1 = i_d - i_d^*$$
$$e_2 = i_q - i_q^* \quad (42)$$

therefore, the system error dynamic is:

$$\dot{e}_1 = \frac{v_d}{L} - \frac{R}{L}i_d + \omega_e i_q - \frac{v_{d1}}{L} - \dot{i}_d^*$$

$$\dot{e}_2 = -\frac{R}{L}i_q - \omega_e i_d - \frac{v_{q1}}{L} - \dot{i}_q^* \quad (43)$$

Defining the ac side inverter reference voltages as:

$$v_{d1} = L(\frac{v_d}{L} - \frac{R}{L}i_d + \omega_e i_q - \dot{i}_d^* - ke_1)$$

$$v_{q1} = L(-\frac{R}{L}i_q - \omega_e i_d - \dot{i}_q^* - ke_2) \quad (44)$$

Linking (43) and (44), gives:

$$\dot{e}_1 = -ke_1$$
$$\dot{e}_2 = -ke_2 \quad (45)$$

Considering a Lyapanouv function as:

$$V = \frac{1}{2}e_1^2 + \frac{1}{2}e_2^2 \quad (46)$$

therefore:

$$\dot{V} = \dot{e}_1 e_1 + \dot{e}_2 e_2 = -k(e_1^2 + e_2^2) < 0 \quad (47)$$

Based on above control strategy the block diagram of rotor dc link voltage controller is depicted in Fig.1.

1909

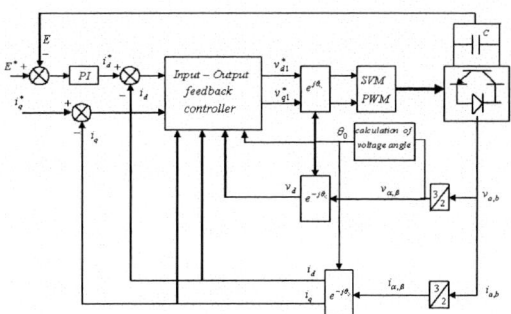

Figure.1. DC link voltage controller

VI. ACTIVE AND REACTIVE STATOR POWER CONTROL IN GENERATING MODE

For electric energy generation, it is usually required to regulate the stator active-reactive power, whose references are assumed to be P_s^* and Q_s^* respectively. These quantities in a special two axis rotating reference frame which is defined in the previous section, referring to [10]:

$$i_q^* = \frac{3}{2}\frac{Q_s^*}{U}$$

$$i_d^* = \frac{3}{2}\frac{P_s^*}{U} \tag{48}$$

using equation (48) flux references are obtained as:

$$\psi_d^* = \frac{1}{\beta\omega_0}\left(-\frac{R_s}{\sigma}i_q^* - \omega_0 i_d^*\right)$$

$$\psi_q^* = \frac{1}{\beta\omega_0}\left(\frac{R_s}{\sigma}i_d^* - \omega_0 i_q^* - \frac{1}{\sigma}U\right) \tag{49}$$

so rotor flux reference and torque reference are calculated as below

$$\psi_r^* = \sqrt{\left(\psi_d^{*2} + \psi_q^{*2}\right)}$$

$$T_e^* = \mu\left(\psi_d^* i_q^* - \psi_q^* i_d^*\right) \tag{50}$$

VII. SYSTEM SIMULATION

The overall block diagram of the proposed control approach is shown in Fig.2. A C^{++} computer program was developed to model this system on P.C.

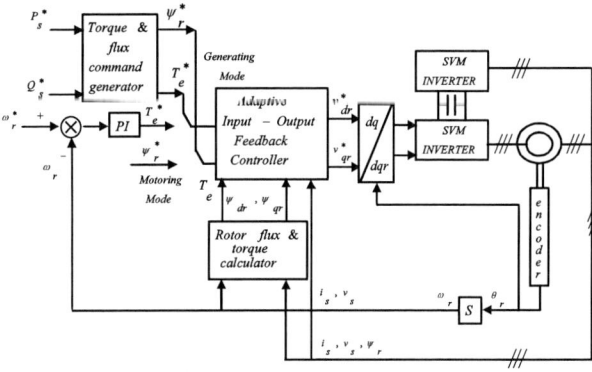

Figure.2. The overall block diagram of the proposed nonlinear controller

In this program, a static runge-kutta fourth order method is used to solve the system equations. The effectiveness and validity of the proposed approach is tested for a three-phase 5 KW, 380 V, six poles,50 Hz DFIM drive [11] by simulation. Simulation results shown in Fig.3 are obtained for motoring mode of operation below and above synchronous speed. These results are obtained in the condition of an exponential speed reference from 0 to 260 (rad/sec) rise up to 350 (rad/sec) at t=2 sec with $\tau_\omega = .07$ sec, an exponential reference flux signal from zero to .5 W-t at t=0 with $\tau_f = .01$ sec, a load torque reference profile which is also shown in Fig.3 and $R_r = 2R_m, R_s = 2R_{sn}$, where R_s and R_r respectively are the stator and rotor resistances. Note the subscript n shows the nominal parameters.

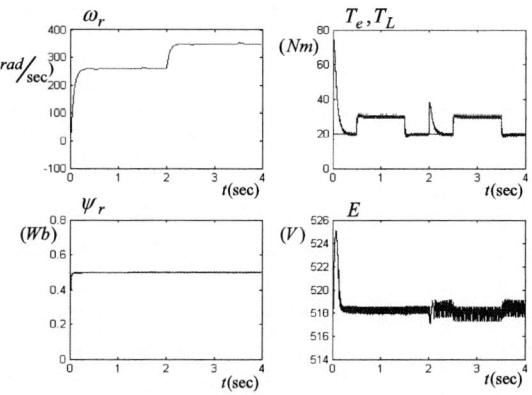

Figure. 3. Drive system motoring mode of operation below and above synchronous speed

Fig.4 shows the drive system performance in the generation mode of operation above synchronous speed. These results are obtained for $R_r = 2R_m, R_s = 2R_{sn}$ and torque reference profile and reference flux signal shown in Fig.4 and a constant $\omega_r = 375$ (rad/sec).

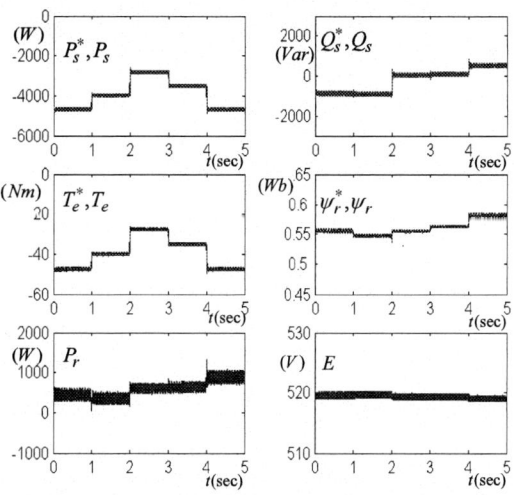

Figure. 4. Drive system performance for the generating mode of operation above synchronous speed

1910

Fig.5 shows the drive system performance in the generation mode of operation below synchronous speed. These results are obtained for the same condition described for Fig.4 and a constant $\omega_r = 280$ ($rad/_{sec}$).

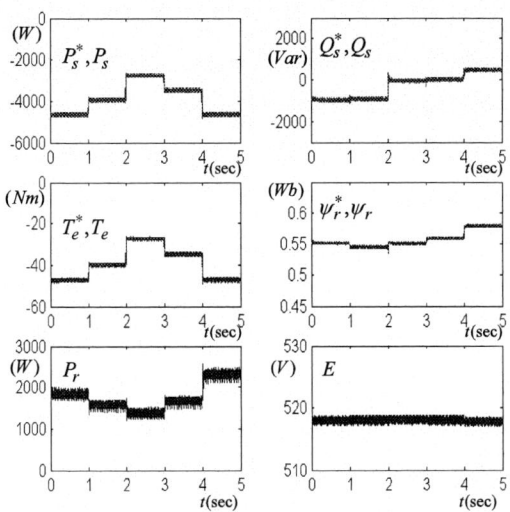

Figure. 5. Drive system performance for the generating mode of operation below synchronous speed

One may note that in Figs.4 and 5, the average active power injected to rotor circuit is positive. That is because of high rotor copper losses obtained by rotor resistance of $R_r = 2R_{rn}$. Also in the above simulation results, T_e is the machine generated torque, ω_r is the rotor speed in electrical ($rad / \sec$), ψ_r is the rotor flux linkage, P_r is the power injected into the rotor circuit, E is the rotor dc link voltage, P_s and Q_s are active and reactive power injected to the stator from ac supply.

VIII. CONCLUSION

In this paper a robust nonlinear controller has been designed for DFIM drives. The proposed controller is derived based on combination of SM control and adaptive input-output control approach. The proposed composite control forces the system states trajectories to follow the nominal states which are obtained by a desired reference model, in spite of stator and rotor resistance uncertainties. The nonlinear control method has been tested for motoring and generating mode of operation bellow and above synchronous speed using two back-to-back two level SVM-PWM voltage sources inverters in rotor circuit. Furthermore the rotor dc link voltage is maintained constant also based on input-output control method, using a rotating synchronous reference frame with d axis coincide with the space voltage vectorin coincide with d axes. Computer simulation results obtained, confirm the validity and capability of the proposed control approach.

REFERENCES

[1] D.Casrdei, F. Profumo, G. Serra and A.Tani, "FOC and DTC:tow viable schemes for induction motors torque control," IEEE Trans. Power Electron., vol.17, no. 5, Sept. 2002, pp. 779-787.

[2] B.Hopfensperger, J.Atkinson, and R.A.Lakin, "Stator-flux-oriented control of a doubly-fed induction machine with and without position encoder," IEE Proc. Electr. Power Appl, vol. 147, no. 4, July. 2000, pp. 241-250.

[3] L.Xu, and W.Cheng, "Torque and reactive power control of a doubly-fed wound rotor induction machine by position sensorless scheme," IEEE Trans, Ind. Appl, 1995, 31(3), pp.636-642.

[4] Z.Wang, F.Wang, M.Zong, and F.Zhang, "A new control strategy by combining direct torque control with vector control for doubly fed machines," in Proc.IEEE-Powercon'2004, Singapore, 21-24 November. 2004, pp. 792-795.

[5] S.Peresada, A.Tilli, and A.Tonielli, "Robust active-reactive power control of a doubly-fed induction generator," in Proc.IEEE-IECON'98, Aachen, Germany, Sept. 1998, pp. 1621-1625.

[6] S.Peresada, A.Tilli, and A.Tonielli, "Indirect stator flux-oriented output feedback control of a doubly fed induction machine," IEEE Trans. Contr Sys, vol. 11, no. 6, November. 2003, pp. 875-888.

[7] H.J.Shieh, K.K.Shyu,"Nonlinear Sliding-Mode Torque Control with Adaptive Backstepping Approach for Induction Motor Drive," IEEE Trans, Ind. Electr, vol. 46, No. 2, April 1999, pp. 380-389.

[8] M.Krstic, I.Kanellakopoulos, and P.Kokotovic, Nonlinear and Adaptive Control Design. New York: Wiley, 1995.

[9] R.Pena, J.C.Clare, and G.M.Asher, "Doubly fed induction generatorusing back to back PWM converter and its application to variable-speed wind energy generator," IEE Proc. Electr. Power. Appl, vol. A3, no. 3, November. 1996, pp. 231-241.

[10] S.Peresada, A.Tilli, and A.Tonielli, "Robust Output Feedback Control of a Doubly Fed Induction Machine," Proc IEEE, 1999, pp. 1348-1354.

2006 5th International Power Electronics and Motion Control Conference

Identification of the Rotor Time Constant in Induction Machines without Speed Sensor

M. Li*, J.N. Chiasson*, M. Bodson**, L.M. Tolbert*

*The University of Tennessee, ECE Department, Knoxville, USA

**The University of Utah, ECE Department, Salt Lake City, USA

Abstract—A differential-algebraic method is used to estimate the rotor time constant T_R of an induction motor without measurements of the rotor speed/position. The method consists of solving for the roots of a polynomial equation in T_R whose coefficients depend only on the stator currents, stator voltages, and their derivatives. Experimental results are presented.

Index Terms—Rotor Time Constant, Sensorless Speed Observer, Induction Motor.

I. INTRODUCTION

Induction motors are very attractive in many applications owing to their simple structure, low cost, and robust construction. Field-oriented control is now used to obtain high performance drive of the induction motor because it gives control characteristics similar to separately excited DC motors. Implementation of a (rotor-flux) field-oriented controller requires knowledge of the rotor speed and the rotor time constant T_R to estimate the rotor flux linkages. There has been considerable work done in the last several years to implement a field-oriented controller without the use of a speed sensor [1][2][3][4][5][6]. However, many of these methods still require the value of T_R, which can change with time due to ohmic heating; that is, to be able to update the value of T_R to the controller as it changes is valuable. The work presented here uses an algebraic approach to identify the rotor time constant T_R without the motor speed information. It is most closely related to the ideas described in [7][8][9][10][11]. Specifically, it is shown that T_R satisfies a polynomial equation whose coefficients are functions of the stator currents, the stator voltages, and their derivatives. A zero of this polynomial is the value of T_R. It is further shown T_R is not identifiable by this technique under steady-state conditions. It is also true (and shown here) that a standard least-squares approach cannot identify T_R under steady-state conditions. In [4], the speed ω and T_R are identified assuming constant but not (sinusoidal) steady state. In [12], the speed is assumed constant, but the flux magnitude is perturbed by a small amplitude sinusoidal signal to identify T_R.

The paper is organized as follows. Section II introduces a space vector model of the induction motor. Section III uses this model to develop a differential-algebraic equation that T_R must satisfy. Section IV shows that in steady state, T_R is not identifiable by either the differential-algebraic method nor a standard linear least-squares method. Section V presents the experimental results, while Section VI gives the conclusions and future work.

II. MATHEMATICAL MODEL OF INDUCTION MOTOR

The starting point of the analysis is a space vector model of the induction motor given by (see e.g., pp. 568 of [13])

$$\frac{d}{dt}\underline{i}_S = \frac{\beta}{T_R}\left(1 - jn_P\omega T_R\right)\underline{\psi}_R - \gamma\underline{i}_S + \frac{1}{\sigma L_S}\underline{u}_S \quad (1)$$

$$\frac{d}{dt}\underline{\psi}_R = -\frac{1}{T_R}\left(1 - jn_P\omega T_R\right)\underline{\psi}_R + \frac{M}{T_R}\underline{i}_S \quad (2)$$

$$\frac{d\omega}{dt} = \frac{n_pM}{JL_R}\operatorname{Im}\left\{\underline{i}_S\underline{\psi}_R^*\right\} - \frac{\tau_L}{J}, \quad (3)$$

where $\underline{i}_S \triangleq i_{Sa} + ji_{Sb}$, $\underline{\psi}_R \triangleq \psi_{Ra} + j\psi_{Rb}$, and $\underline{u}_S \triangleq u_{Sa} + ju_{Sb}$. Here, θ is the position of the rotor, $\omega = d\theta/dt$ is the rotor speed, n_p is the number of pole pairs, i_{Sa}, i_{Sb} are the (two-phase equivalent) stator currents, ψ_{Ra}, ψ_{Rb} are the (two-phase equivalent) rotor flux linkages, R_S, R_R are the stator and rotor resistances, respectively, M is the mutual inductance, L_S and L_R are the stator and rotor inductances, respectively, J is the moment of inertia of the rotor, and τ_L is the load torque. The symbols $T_R = \dfrac{L_R}{R_R}$, $\sigma = 1 - \dfrac{M^2}{L_SL_R}$, $\beta = \dfrac{M}{\sigma L_SL_R}$, $\gamma = \dfrac{R_S}{\sigma L_S} + \dfrac{\beta M}{T_R}$ have been used to simplify the expressions. T_R is referred to as the rotor time constant, while σ is called the total leakage factor.

III. DIFFERENTIAL-ALGEBRAIC APPROACH TO T_R ESTIMATION

The idea of the differential-algebraic approach is to solve (1) and (2) for T_R [14][15]. However, equations (1) and (2) are only four equations while there are six unknowns, namely ψ_{Ra}, ψ_{Rb}, $d\psi_{Ra}/dt$, $d\psi_{Rb}/dt$, ω, and T_R. Equation (3) is not used because it introduces the additional unknown τ_L. To find two more independent equations, equation (1) is differentiated to obtain

$$\begin{aligned}\frac{d^2}{dt^2}\underline{i}_S &= \frac{\beta}{T_R}\left(1 - jn_P\omega T_R\right)\frac{d}{dt}\underline{\psi}_R - jn_P\beta\underline{\psi}_R\frac{d\omega}{dt} \\ &\quad - \gamma\frac{d}{dt}\underline{i}_S + \frac{1}{\sigma L_S}\frac{d}{dt}\underline{u}_S.\end{aligned} \quad (4)$$

Using the (complex-valued) equations (1) and (2), one can solve for $\underline{\psi}_R$ and $\dfrac{d}{dt}\underline{\psi}_R$ in terms of ω, $\underline{i}_S$ and $\underline{u}_S$ and substitute the resulting expressions into (4) to obtain

$$\frac{d^2}{dt^2}\underline{i}_S = -\frac{1}{T_R}\left(1 - jn_P\omega T_R\right)\left(\frac{d}{dt}\underline{i}_S + \gamma\underline{i}_S - \frac{1}{\sigma L_S}\underline{u}_S\right)$$
$$+ \frac{\beta M}{T_R^2}\left(1 - jn_P\omega T_R\right)\underline{i}_S - \gamma\frac{d}{dt}\underline{i}_S + \frac{1}{\sigma L_S}\frac{d}{dt}\underline{u}_S$$
$$- \frac{jn_P T_R}{1 - jn_P\omega T_R}\left(\frac{d}{dt}\underline{i}_S + \gamma\underline{i}_S - \frac{1}{\sigma L_S}\underline{u}_S\right)\frac{d\omega}{dt}.$$
(5)

Solving (5) for $d\omega/dt$ gives

$$\frac{d\omega}{dt} = -\frac{\left(1 - jn_P\omega T_R\right)^2}{jn_P T_R^2} + \frac{1 - jn_P\omega T_R}{jn_P T_R} \times$$
$$\frac{\frac{\beta M}{T_R^2}\left(1 - jn_P\omega T_R\right)\underline{i}_S - \gamma\frac{d}{dt}\underline{i}_S + \frac{1}{\sigma L_S}\frac{d}{dt}\underline{u}_S - \frac{d^2}{dt^2}\underline{i}_S}{\frac{d}{dt}\underline{i}_S + \gamma\underline{i}_S - \frac{1}{\sigma L_S}\underline{u}_S}.$$
(6)

The left-hand side of (6) is real, so the right-hand side must also be real. Note by (1) that $d\underline{i}_S/dt + \gamma\underline{i}_S - \underline{u}_S/(\sigma L_S) = \frac{\beta}{T_R}\left(1 - jn_P\omega T_R\right)\underline{\psi}_R$ so that the right-hand side of (6) is singular if and only if $\left|\underline{\psi}_R\right| = 0$. Other than at startup, $\left|\underline{\psi}_R\right| \neq 0$ in normal operation of the motor. Separating the right-hand side of (6) into its real and imaginary parts, the real part has the form

$$\frac{d\omega}{dt} = a_2\left(u_{Sa}, u_{Sb}, i_{Sa}, i_{Sb}\right)\omega^2 + a_1\left(u_{Sa}, u_{Sb}, i_{Sa}, i_{Sb}\right)\omega$$
$$+ a_0\left(u_{Sa}, u_{Sb}, i_{Sa}, i_{Sb}\right).$$
(7)

The expressions for $a_2\left(u_{Sa}, u_{Sb}, i_{Sa}, i_{Sb}\right)$, $a_1(u_{Sa}, u_{Sb}, i_{Sa}, i_{Sb})$, and $a_0\left(u_{Sa}, u_{Sb}, i_{Sa}, i_{Sb}\right)$ are lengthy in terms of u_{Sa}, u_{Sb}, i_{Sa}, i_{Sb}, and their derivatives as well as of the machine parameters including T_R. As a consequence, they are not explicitly presented here. Their steady-state expressions are given in [6].

On the other hand, the imaginary part of the right-hand side of (6) must be zero. In fact, the imaginary part of (6) is a second degree polynomial equation in ω of the form

$$q(\omega) \triangleq q_2(u_{Sa}, u_{Sb}, i_{Sa}, i_{Sb})\omega^2 + q_1(u_{Sa}, u_{Sb}, i_{Sa}, i_{Sb})\omega$$
$$+ q_0(u_{Sa}, u_{Sb}, i_{Sa}, i_{Sb})$$
(8)

and, if ω is the speed of the motor, then $q(\omega) = 0$. The q_i are functions of u_{Sa}, u_{Sb}, i_{Sa}, i_{Sb}, and their derivatives as well as of the machine parameters including T_R. The expressions for $q_2(u_{Sa}, u_{Sb}, i_{Sa}, i_{Sb})$, $q_1\left(u_{Sa}, u_{Sb}, i_{Sa}, i_{Sb}\right)$, and $q_0(u_{Sa}, u_{Sb}, i_{Sa}, i_{Sb})$ are also lengthy and not explicitly presented here. (Their steady-state expressions are given in [6].) If the speed was measured, then (8) would be equal to zero and could then be solved for T_R. However, in the problem being considered, ω is not known. To eliminate ω, $q(\omega)$ in (8) is differentiated to obtain

$$\frac{d}{dt}q(\omega) = (2q_2\omega + q_1)\frac{d\omega}{dt} + \dot{q}_2\omega^2 + \dot{q}_1\omega + \dot{q}_0$$
(9)

where $dq(\omega)/dt \equiv 0$ if ω is equal to the motor speed. Next, $d\omega/dt$ in (9) is replaced by the right-hand side of (7) so that

(9) may be written as

$$\frac{dq(\omega)}{dt} = g(\omega) \triangleq 2q_2 a_2\omega^3 + (2q_2 a_1 + q_1 a_2 + \dot{q}_2)\omega^2$$
$$+ (2q_2 a_0 + q_1 a_1 + \dot{q}_1)\omega + q_1 a_0 + \dot{q}_0. \quad (10)$$

$g(\omega)$ is a third-order polynomial equation in ω for which the speed of the motor is one of its zeros. Dividing[1] $g(\omega)$ in (10) by $q(\omega)^2$ in (8), $g(\omega)$ may be rewritten as

$$g(\omega) = \frac{1}{q_2}\Big((2q_2 a_2\omega + 2q_2 a_1 - q_1 a_2 + \dot{q}_2)\, q(\omega)$$
$$+ r_1\left(u_{Sa}, u_{Sb}, i_{Sa}, i_{Sb}\right)\omega + r_0\left(u_{Sa}, u_{Sb}, i_{Sa}, i_{Sb}\right)\Big)$$
(11)

$$r_1\left(u_{Sa}, u_{Sb}, i_{Sa}, i_{Sb}\right) \triangleq 2q_2^2 a_0 - q_2 q_1 a_1 + q_2\dot{q}_1 - 2q_2 q_0 a_2$$
$$+ q_1^2 a_2 - q_1\dot{q}_2 \quad (12)$$

$$r_0\left(u_{Sa}, u_{Sb}, i_{Sa}, i_{Sb}\right) \triangleq q_2 q_1 a_0 + q_2\dot{q}_0 - 2q_2 q_0 a_1$$
$$+ q_0 q_1 a_2 - q_0\dot{q}_2. \quad (13)$$

If ω is equal to the speed of the motor, then both $g(\omega) = 0$ and $q(\omega) = 0$, and one obtains

$$r(\omega) \triangleq r_1\left(u_{Sa}, u_{Sb}, i_{Sa}, i_{Sb}\right)\omega + r_0\left(u_{Sa}, u_{Sb}, i_{Sa}, i_{Sb}\right) = 0.$$
(14)

This is now a first-order polynomial equation in ω which uniquely determines the motor speed ω as long as r_1 (the coefficient of ω) is nonzero. (It is shown in Appendix VII-A that $r_1 \neq 0$ in steady state.) Solving for the motor speed ω using (14), one obtains

$$\omega = -r_0/r_1.$$
(15)

Next, replace ω in (8) by the expression in (15) to obtain

$$q_2 r_0^2 - q_1 r_0 r_1 + q_0 r_1^2 \equiv 0.$$
(16)

The expressions for q_i, r_i are in terms of motor parameters (including T_R) as well as the stator currents, voltages, and their derivatives. Expanding the expressions for q_0, q_1, q_2, r_0, and r_1, one obtains a twelfth-order polynomial equation in T_R, which can be written as

$$\sum_{i=0}^{12} C_i\left(u_{Sa}, u_{Sb}, i_{Sa}, i_{Sb}\right) T_R^i = 0.$$
(17)

Solving equation (17) gives T_R. The coefficients $C_i\left(u_{Sa}, u_{Sb}, i_{Sa}, i_{Sb}\right)$ of (17) contain third-order derivatives of the stator currents and second-order derivatives of the stator voltages making noise a concern. For short time intervals in which T_R does not vary, (17) must hold identically with T_R constant. In order to average out the effect of noise on the C_i, (17) is integrated over a time interval $[t_1, t_2]$ to obtain

$$\sum_{i=0}^{12}\left(\frac{1}{t_2 - t_1}\int_{t_1}^{t_2} C_i\left(u_{Sa}, u_{Sb}, i_{Sa}, i_{Sb}\right) dt\right) T_R^i = 0. \quad (18)$$

---

[1]Given the polynomials $g(\omega), q(\omega)$ in ω with $\deg\{g(\omega)\} = n_g$, $\deg\{q(\omega)\} = n_q$, the Euclidean division algorithm ensures that there are polynomials $\gamma(\omega), r(\omega)$ such that $g(\omega) = \gamma(\omega)q(\omega) + r(\omega)$ and $\deg\{r(\omega)\} \leq \deg\{q(\omega)\} - 1 = n_q - 1$. Consequently if, for example, ω_0 is a zero of both $g(\omega)$ and $q(\omega)$, then it must also be a zero of $r(\omega)$.

[2]$q_2 \neq 0$ if ω and the stator electrical frequency ω_S are nonzero, which hold under normal operating conditions. See [6][16].

There are 12 solutions satisfying (18). However, simulation results have always given 10 conjugate solutions. The remaining two solutions include the correct value of T_R while the other one was either negative or close to zero. The method is to compute the coefficients $\frac{1}{t_2-t_1}\int_{t_1}^{t_2} C_i dt$ and then compute the roots of (18). Among the positive real roots is the correct value of T_R. Experimental results using this method are presented in Section V.

IV. IDENTIFIABILITY OF T_R IN STEADY STATE

A. Differential-algebraic approach

The polynomial (18) is now considered with the machine in steady state so that, in particular, the speed is constant. That is, $u_{Sa} + ju_{Sb} = \underline{U}_S e^{j\omega_S t}$ and $i_{Sa} + ji_{Sb} = \underline{I}_S e^{j\omega_S t}$ are substituted into (8) and (14). In steady state, the motor speed in (15) becomes (see Appendix VII-A and [16])

$$\omega = -\frac{r_0}{r_1} = \frac{\omega_S(1-S)}{n_p} \qquad (19)$$

where $S \triangleq (\omega_S - n_p\omega)/\omega_S$ is the normalized slip and ω_S is the electrical frequency. Substituting the steady-state expressions for q_2, q_1, and q_0 as well as the expression (19) for ω into (8), one obtains $q_2\omega^2 + q_1\omega + q_0 =$

$$\frac{n_p^2 T_R^2 |\underline{I}_S|^4 \omega_S^2 L_S(1-\sigma)^2(1-S)}{\sigma(1+S^2\omega_S^2 T_R^2)}\left(\frac{\omega_S(1-S)}{n_p}\right)^2$$

$$+\frac{n_p\omega_S |\underline{I}_S|^4 L_S(1-\sigma)^2\left(1-\omega_S^2 T_R^2(1-S)^2\right)}{\sigma(1+S^2\omega_S^2 T_R^2)}$$

$$\times\left(\frac{\omega_S(1-S)}{n_p}\right) - \frac{|\underline{I}_S|^4\omega_S^2 L_S(1-\sigma)^2(1-S)}{\sigma(1+S^2\omega_S^2 T_R^2)} \equiv 0.$$

That is, in steady state (8) and (14) hold independent of the value of T_R and thus so does (17) making T_R unidentifiable in steady state by this method.

B. Linear least-squares approach

Vélez-Reyes et al [3][4] have used least-squares methods for simultaneous parameter and speed identification in induction machines. In the approach used herein, $d\omega/dt$ is taken to be zero so that a linear (in the parameters) regressor model can be obtained. Specifically, consider the mathematical model of the induction motor in (5). Assuming constant speed, $d\omega/dt = 0$ so that this equation reduces to

$$\frac{d^2}{dt^2}\underline{i}_S = -\frac{1}{T_R}(1-jn_P\omega T_R)\left(\frac{d}{dt}\underline{i}_S + \gamma\underline{i}_S - \frac{1}{\sigma L_S}\underline{u}_S\right)$$

$$+\frac{\beta M}{T_R^2}(1-jn_P\omega T_R)\underline{i}_S - \gamma\frac{d}{dt}\underline{i}_S + \frac{1}{\sigma L_S}\frac{d}{dt}\underline{u}_S \qquad (20)$$

where $\underline{i}_S = i_{Sa} + ji_{Sb}$ and $\underline{u}_S = u_{Sa} + ju_{Sb}$. Decomposing equation (20) into its real and imaginary parts gives

$$\frac{d^2 i_{Sa}}{dt} = \frac{1}{T_R}\left(-\frac{di_{Sa}}{dt} - \frac{R_S}{\sigma L_S}i_{Sa} + \frac{1}{\sigma L_S}u_{Sa}\right)$$

$$+n_p\omega\left(-\frac{di_{Sb}}{dt} - \frac{R_S}{\sigma L_S}i_{Sb} + \frac{1}{\sigma L_S}u_{Sb}\right)$$

$$-\gamma\frac{di_{Sa}}{dt} + \frac{1}{\sigma L_S}\frac{du_{Sa}}{dt} \qquad (21)$$

and

$$\frac{d^2 i_{Sb}}{dt} = \frac{1}{T_R}\left(-\frac{di_{Sb}}{dt} - \frac{R_S}{\sigma L_S}i_{Sb} + \frac{1}{\sigma L_S}u_{Sb}\right)$$

$$-n_p\omega\left(-\frac{di_{Sa}}{dt} - \frac{R_S}{\sigma L_S}i_{Sa} + \frac{1}{\sigma L_S}u_{Sa}\right)$$

$$-\gamma\frac{di_{Sb}}{dt} + \frac{1}{\sigma L_S}\frac{du_{Sb}}{dt}. \qquad (22)$$

The goal here is to estimate T_R without knowledge of ω. So, it is now assumed the motor parameters are all known except for T_R. The set of equations (21) and (22) may then be rewritten in regressor form as

$$y(t) = W(t)K \qquad (23)$$

where $K \in \mathbb{R}^2$, $y \in \mathbb{R}^2$, and $W \in \mathbb{R}^{2\times 2}$ are given by

$$K \triangleq \left[\begin{array}{c} 1/T_R \\ n_p\omega \end{array}\right],$$

$$y(t) \triangleq \left[\begin{array}{c} \dfrac{du_{Sa}}{dt} \quad \sigma L_S\dfrac{d^2 i_{Sa}}{dt} - R_S\dfrac{di_{Sa}}{dt} \\ \dfrac{du_{Sb}}{dt} - \sigma L_S\dfrac{d^2 i_{Sb}}{dt} - R_S\dfrac{di_{Sb}}{dt} \end{array}\right],$$

$$W(t) \triangleq \left[\begin{array}{cc} L_S\dfrac{di_{Sa}}{dt} - u_{Sa} + R_S i_{Sa} & \sigma L_S\dfrac{di_{Sb}}{dt} - u_{Sb} + R_S i_{Sb} \\ L_S\dfrac{di_{Sb}}{dt} - u_{Sb} + R_S i_{Sb} & -\sigma L_S\dfrac{di_{Sa}}{dt} + u_{Sa} - R_S i_{Sa} \end{array}\right].$$

The regressor system (23) is linear in the parameters. The standard linear least-squares approach is to let (i.e., collect data at) $t = 0, T, 2T, \cdots, NT$, multiply (23) on the left by $W^T(nT)$, sum $W^T(nT)y(nT) = W^T(nT)W(nT)K$ from $t = 0$ to $t = NT$, and finally compute the solution to

$$R_W K = R_{YW} \qquad (24)$$

where

$$R_W \triangleq \sum_{n=0}^{N} W^T(nT)W(nT), \quad R_{YW} \triangleq \sum_{n=0}^{N} W^T(nT)y(nT).$$

A unique solution to (24) exists if and only if R_W is invertible. However, R_W is never invertible in steady state as is now shown. To proceed, define

$$D(t) = \left[\begin{array}{cc} i_{Sb}(t) & -i_{Sa}(t) \\ i_{Sa}(t) & i_{Sb}(t) \end{array}\right].$$

In steady state where $u_{Sa} + ju_{Sb} = \underline{U}_S e^{j\omega_S t}$ and $i_{Sa} + ji_{Sb} = \underline{I}_S e^{j\omega_S t}$, $\det(D(t)) = i_{Sa}^2(t) + i_{Sb}^2(t) = |\underline{I}_S|^2$, $D(t)^T D(t) = |\underline{I}_S|^2 I_{2\times 2}$. Multiply both sides of (23) on the left by $D(t)$ to obtain

$$D(t)y(t) = D(t)W(t)K$$

or

$$\left[\begin{array}{c} R_S\omega_S|\underline{I}_S|^2 - \omega_S P \\ \sigma L_S\omega_S^2|\underline{I}_S|^2 - \omega_S Q \end{array}\right] =$$

$$\left[\begin{array}{cc} -\omega_S L_S|\underline{I}_S|^2 + Q & R_S|\underline{I}_S|^2 - P \\ R_S|\underline{I}_S|^2 - P & \sigma L_S\omega_S|\underline{I}_S|^2 - Q \end{array}\right]K \quad (25)$$

where $P \triangleq u_{Sa}i_{Sa} + u_{Sb}i_{Sb}$ and $Q \triangleq u_{Sb}i_{Sa} - u_{Sa}i_{Sb}$ are the real and reactive powers, respectively, whose steady-state expressions are given by (30) and (31) in the Appendix. Using (30) and (31) to replace P and Q in (25), one obtains

$$
\begin{aligned}
\bar{D} &\triangleq D(t) W(t) \\
&= -\frac{|\underline{I}_S|^2 (1-\sigma) \omega_S L_S}{1 + S^2 \omega_S^2 T_R^2} \begin{bmatrix} S^2 \omega_S^2 T_R^2 & S \omega_S T_R \\ S \omega_S T_R & 1 \end{bmatrix}
\end{aligned}
\tag{26}
$$

$$
\begin{aligned}
\bar{Y} &\triangleq D(t) y(t) \\
&= -\omega_S \frac{|\underline{I}_S|^2 (1-\sigma) \omega_S L_S}{1 + S^2 \omega_S^2 T_R^2} \begin{bmatrix} S \omega_S T_R \\ 1 \end{bmatrix}.
\end{aligned}
\tag{27}
$$

That is, in steady state, $\bar{D} \triangleq D(t) W(t) \in \mathbb{R}^{2 \times 2}$ and $\bar{Y} \triangleq D(t) y(t) \in \mathbb{R}^2$ are *constant* matrices. Further, it is easily seen that the determinant of $\bar{D} \triangleq D(t) W(t)$ is zero. Also,

$$
\begin{aligned}
R_{DW} &\triangleq \sum_{n=1}^{N} (D(nT) W(nT))^T (D(nT) W(nT)) \\
&= |\underline{I}_S|^2 \sum_{n=1}^{N} W^T(nT) W(nT) = |\underline{I}_S|^2 R_W.
\end{aligned}
$$

R_{DW} is singular because $D(t) W(t)$ is constant and singular. It then follows that R_W is also singular using steady-state data. Further,

$$
\begin{aligned}
R_{DWY} &\triangleq \sum_{n=1}^{N} (D(nT) W(nT))^T (D(nT) y(nT)) \\
&= |\underline{I}_S|^2 \sum_{n=1}^{N} W^T(nT) y(nT) = |\underline{I}_S|^2 R_{YW}.
\end{aligned}
$$

Thus R_W and R_{YW} are given by

$$
\begin{aligned}
R_W &= R_{DW}/|\underline{I}_S|^2 = N \bar{D}^T \bar{D}/|\underline{I}_S|^2 \\
&= \frac{N |\underline{I}_S|^2 (1-\sigma)^2 \omega_S^2 L_S^2}{1 + S^2 \omega_S^2 T_R^2} \begin{bmatrix} S^2 \omega_S^2 T_R^2 & S \omega_S T_R \\ S \omega_S T_R & 1 \end{bmatrix}
\end{aligned}
\tag{28}
$$

$$
\begin{aligned}
R_{YW} &= R_{DWY}/|\underline{I}_S|^2 = N \bar{D}^T \bar{Y}/|\underline{I}_S|^2 \\
&= \omega_S \frac{N |\underline{I}_S|^2 (1-\sigma)^2 \omega_S^2 L_S^2}{1 + S^2 \omega_S^2 T_R^2} \begin{bmatrix} S \omega_S T_R \\ 1 \end{bmatrix},
\end{aligned}
\tag{29}
$$

where again $\bar{D}$ and $\bar{Y}$ are from (26) and (27), respectively.

By inspection of (28) and (29), $K = \begin{bmatrix} 0 & \omega_S \end{bmatrix}^T$ is one solution to (24). The null space of R_W is generated by $\begin{bmatrix} -1/T_R & S \omega_S \end{bmatrix}^T$ so that all possible solutions are given by $\begin{bmatrix} 0 & \omega_S \end{bmatrix}^T + \alpha \begin{bmatrix} -1/T_R & S \omega_S \end{bmatrix}^T$ for some $\alpha \in \mathbb{R}$. In summary, solving (24) using steady-state data leads to an infinite set of solutions so that T_R is not identifiable using the linear regressor (23) with steady-state data.

V. Experimental Results

To demonstrate the viability of the speed sensorless estimator (18) for T_R, experiments were performed. A three-phase, 0.5 hp, 1735 rpm ($n_p = 2$ pole-pair) induction motor was driven by an Allen-Bradley PWM inverter to obtain the data. Given a speed command to the inverter, it produces PWM

voltages to drive the induction motor to the commanded speed. Here a step speed command was chosen to bring the motor from standstill up to the rated speed of 188 rad/s. The stator currents and voltages were sampled at 10 kHz. The real-time computing system RTLAB from OPAL-RT with a fully integrated hardware and software system was used to collect data [17]. Filtered differentiation (using digital filters) was used for the derivatives of the voltages and currents. Specifically, the signals were filtered with a third-order Butterworth filter whose cutoff frequency was 100 Hz. The voltages and currents were put through a $3 - 2$ transformation to obtain their two-phase equivalent values.

Using the data $\{u_{Sa}, u_{Sb}, i_{Sa}, i_{Sb}\}$ collected between 0.84 sec to 0.91 sec, which includes the time the motor accelerates, the quantities du_{Sa}/dt, du_{Sa}/dt, di_{Sa}/dt, di_{Sb}/dt, d^2i_{Sa}/dt^2, d^2i_{Sb}/dt^2, d^3i_{Sa}/dt^3, d^3i_{Sb}/dt^3 are calculated and used to evaluate the coefficients C_i, $i = 1, 2, \cdots, 12$ in equation (18). Solving (18), one obtains the 12 solutions

$T_{R1} = +0.1064$	$T_{R2} = -0.0186$
$T_{R3} = -0.0576 + j0.0593$	$T_{R4} = -0.0576 - j0.0593$
$T_{R5} = -0.0037 + j0.0166$	$T_{R6} = -0.0037 - j0.0166$
$T_{R7} = -0.0072 + j0.0103$	$T_{R8} = -0.0072 - j0.0103$
$T_{R9} = +0.0125 + j0.0077$	$T_{R10} = +0.0125 - j0.0077$
$T_{R11} = +0.0065 + j0.0018$	$T_{R12} = +0.0065 - j0.0018$.

T_R must be a real positive number, so $T_R = 0.1064$ is the only possible choice. This value compares favorably with the value of $T_R = 0.11$ obtained using the method of Wang et al [18], which requires a speed sensor.

To illustrate the identified T_R, a simulation of the induction motor model was carried out using the measured voltages as input. Then the simulation's output [stator currents computed according to (1) and (2)] are used to compare with the measured (stator currents) outputs. Figure 1 shows the sampled two-phase equivalent current i_{Sb} and its simulated response i_{Sb-sim}. The phase a current i_{Sa} is similar, but shifted by $\pi/(2n_p)$. The resulting phase b current i_{Sb-sim} from the simulation corresponds well with the actual measured current i_{Sb}. Note that in equation (1) $\gamma = \frac{R_S}{\sigma L_S} + \frac{\beta M}{T_R}$ also depends on T_R.

VI. Conclusions and Future Work

This paper presented a differential-algebraic approach to the estimation of the rotor time constant of an induction motor without using a speed sensor. The experimental results demonstrated the practical viability of this method. Though the method is not applicable in steady state, neither is a standard linear least-squares approach. Future work includes studying an on-line implementation of the estimation algorithm and using such an online estimate in a speed sensorless field-oriented controller.

VII. Appendix: Steady-State Expressions

In the following, ω_S denotes the stator frequency and S denotes the normalized slip defined by $S \triangleq (\omega_S - n_p \omega)/\omega_S$. With $u_{Sa} + j u_{Sb} = \underline{U}_S e^{j\omega_S t}$ and $i_{Sa} + j i_{Sb} = \underline{I}_S e^{j\omega_S t}$, it is

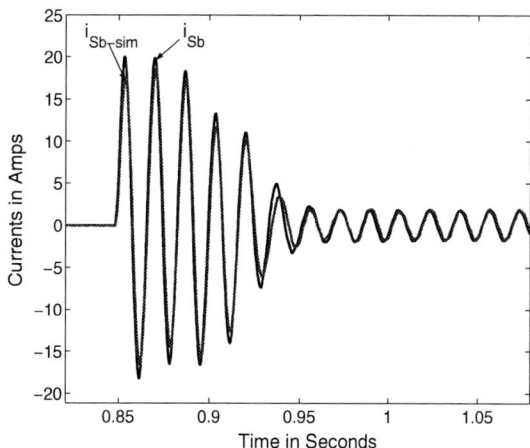

Fig. 1. Phase b current i_{Sb} and its simulated response i_{Sb-sim}.

shown in [19] that under steady-state conditions, the complex phasors $\underline{U}_S$ and $\underline{I}_S$ are related by ($S_p \triangleq \frac{R_R}{\sigma \omega_S L_R} = \frac{1}{\sigma \omega_S T_R}$)

$$
\underline{I}_S = \frac{\underline{U}_S}{R_S + j\omega_S L_S \left(\left(1 + j\frac{S}{S_p}\right) / \left(1 + j\frac{S}{\sigma S_p}\right) \right)}
$$

$$
= \frac{\underline{U}_S}{\left(R_S + \frac{(1-\sigma)S\omega_S^2 L_S T_R}{1 + S^2\omega_S^2 T_R^2}\right) + j\frac{\omega_S L_S \left(1 + \sigma S^2 \omega_S^2 T_R^2\right)}{1 + S^2\omega_S^2 T_R^2}},
$$

and straightforward calculations (see [6]) give

$$
\begin{aligned}
P &\triangleq u_{Sa}i_{Sa} + u_{Sb}i_{Sb} = Re\left(\underline{U}_S \underline{I}_S^*\right) \\
&= |\underline{I}_S|^2 \left(R_S + \frac{(1-\sigma)S\omega_S^2 L_S T_R}{1 + S^2\omega_S^2 T_R^2}\right) \quad (30)
\end{aligned}
$$

$$
\begin{aligned}
Q &\triangleq u_{Sb}i_{Sa} - u_{Sa}i_{Sb} = Im\left(\underline{U}_S \underline{I}_S^*\right) \\
&= |\underline{I}_S|^2 \frac{\omega_S L_S \left(1 + \sigma S^2 \omega_S^2 T_R^2\right)}{1 + S^2\omega_S^2 T_R^2}. \quad (31)
\end{aligned}
$$

A. Steady-State Expression for r_1 and r_0

It is now shown that the steady-state value of r_1 in (12) is nonzero. Substituting the steady-state values of q_2, q_1, q_0, a_2, a_1, and a_0 shown in [6] (noting that $\dot{q}_1 \equiv 0$ and $\dot{q}_2 \equiv 0$ in steady state) into (12) gives

$$
\begin{aligned}
r_1 = & -|\underline{I}_S|^6 \left(\frac{1}{1 + S^2\omega_S^2 T_R^2}\right)^3 \frac{n_p^4 (1-\sigma)^6 L_S^2}{\sigma^4} \times \\
& \omega_S^3 \left(1 + T_R^2 \omega_S^2 (1-S)^2\right)^2 \frac{1}{den} \\
r_0 = & |\underline{I}_S|^6 \left(\frac{1}{1 + S^2\omega_S^2 T_R^2}\right)^3 \frac{n_p^3 (1-\sigma)^6 L_S^2}{\sigma^4} \times \\
& \omega_S^4 (1-S) \left(1 + \omega_S^2 T_R^2 \times (1-S)^2\right)^2 \frac{1}{den}
\end{aligned}
$$

where

$$
\begin{aligned}
den \triangleq & \; n_p T_R |\underline{I}_S|^4 \left(\left(\frac{(1-\sigma)}{\sigma T_R} \frac{1 + S^2\omega_S^2 T_R^2 - S\omega_S^2 T_R^2}{1 + S^2\omega_S^2 T_R^2}\right)^2 \right. \\
& \left. + \left(\frac{(1-\sigma)}{\sigma} \frac{\omega_S}{1 + S^2\omega_S^2 T_R^2}\right)^2 \right). \quad (32)
\end{aligned}
$$

Recall from Section III [following (6)] that $den = 0$ if and only if $\left|\underline{\psi}_R\right| = 0$. It is then seen that $r_1 \neq 0$ in steady state.

REFERENCES

[1] K. Rajashekara, A. Kawamura, and K. Matsuse, Eds., *Sensorless Control of AC Motor Drives - Speed and Position Sensorless Operation*. IEEE Press, 1996.

[2] P. Vas, *Sensorless Vector Control and Direct Torque Control*. Oxford University Press, 1998.

[3] M. Vélez-Reyes, W. L. Fung, and J. E. Ramos-Torres, "Developing robust algorithms for speed and parameter estimation in induction machines," in *Proceedings of the IEEE Conference on Decision and Control*, 2001, pp. 2223–2228, orlando, Florida.

[4] M. Vélez-Reyes, "Decomposed algorithms for parameter estimation," Ph.D. dissertation, Massachusetts Institute of Technology, 1992.

[5] M. Bodson and J. Chiasson, "A comparison of sensorless speed estimation methods for induction motor control," in *Proceedings of the 2002 American Control Conference*, May 2002, pp. 3076–3081, anchorage, AK.

[6] M. Li, J. Chiasson, M. Bodson, and L. M. Tolbert, "Observability of speed in an induction motor from stator currents and voltages," in *IEEE Conference on Decision and Control*, December 2005, pp. 3438–3443, seville Spain.

[7] J. Reger, H. S. Ramirez, and M. Fliess, "On non-asymptotic observation of nonlinear systems," in *Proceedings of the 44th IEEE Conference on Decision and Control*, December 2005, pp. 4219–4224, seville Spain.

[8] M. Diop and M. Fliess, "On nonlinear observability," in *Proceedings of the 1st European Control Conference*. Hermès, Paris, 1991, pp. 152–157.

[9] M. Fliess and H. Síra-Ramirez, "Control via state estimation of some nonlinear systems," in *Symposium on Nonlinear Control Systems (NOLCOS-2004)*, September 2004, stuttgart, Germany.

[10] M. Fliess, C. Join, and H. Síra-Ramirez, "Complex continuous nonlinear systems: Their black box identification and their control," in *14th IFAC Symposium on System Identification (SISYD 2006)*, 2006, newcastle, Austalia.

[11] M. P. Saccomani, "Some results on parameter identification of nonlinear systems," *Cardiovascular Engineering: An Internation Journal*, vol. 4, no. 1, pp. 95–102, March 2004.

[12] I.-J. Ha and S.-H. Lee, "An online identification method for both stator and rotor resistances of induction motors without rotational transducers," *IEEE Transactions on Industrial Electronics*, vol. 47, no. 4, pp. 842–853, August 2000.

[13] J. Chiasson, *Modeling and High-Performance Control of Electric Machines*. John Wiley & Sons, 2005.

[14] M. Diop and M. Fliess, "Nonlinear observability, identifiability and persistent trajectories," in *IEEE Conference on Decision and Control*. Brighton England, 1991, pp. 714–719.

[15] D. Nešić, I. M. Y. Mareels, S. T. Glad, and M. Jirstrand, "Software for control system analysis and design: symbol manipulation," in *Encyclopedia of Electrical Engineering*. John Wiley & Sons, J. Webster, Editor, 2001, available online at http://www.interscience.wiley.com:83/eeee/.

[16] M. Li, "Differential-algebraic approach to speed and parameter estimation of the induction motor," Ph.D. dissertation, The Univesity of Tennessee, 2005.

[17] Opal-RT Technologies, "RT-LAB," see http://www.opal-rt.com.

[18] K. Wang, J. Chiasson, M. Bodson, and L. M. Tolbert, "An on-line rotor time constant estimator for the induction machine," in *Proceedings of the IEEE International Electric Machines and Drives Conference*, May 2005, pp. 608–614, san Antonio TX.

[19] W. Leonhard, *Control of Electrical Drives*. 3rd Edition, Springer-Verlag, Berlin, 2001.

2006 5th International Power Electronics and Motion Control Conference

Adaptive Control of Doubly Fed Field-Oriented Induction Machine Based On Recursive Least Squares Method Taking the Iron Loss Into account

N. R. Abjadi[*], J. Askari[**] and J. Soltani[**]

[*] Islamic Azad University – Majlessi Branch, City of Majlessi, Isfahan, Iran
[**] Faculty of electrical and computer engineering, Isfahan University of Technology, Isfahan, Iran
e-mail: abjadi@ec.iut.ac.ir

Abstract—In this paper, a slip recovery power induction machine drive is introduced which is adaptively vector controlled in the stator flux reference frame. Using this method, the active and reactive powers injected to the stator circuits can be independently controlled so that the drive system is robust and stable against the variations of the electrical parameters and uncertainties. In the proposed control method, the recursive least square (RLS) method is applied to estimate the machine parameters taking into consideration the machine iron loss resistance. The RLS estimator operates in parallel with the system main controllers. The effectiveness and validity of the proposed method is tested by computer simulation.

Keywords- Doubly Fed Induction Machine, Recursive Least squares, Stator Field Oriented Control

I.   INTRODUCTION

The general block diagram of a doubly-fed induction machine (DFIM) drive is depicted in Fig. 1. Having controlled the slip power of this machine, the drive system is capable of operating in motoring and generating modes below and above synchronous speed [1].

The DFIM drive is an alternative for conventional induction motor especially for fans and pumps, which can use smaller converters and reduces the system cost. To get a good performance the vector control or feedback linearization techniques should be used to overcome the nonlinear structure and heavy state coupling of these machines [2]. The mentioned techniques however, require the machine actual electrical parameters.

So far many papers have been published for estimation of squirrel cage induction machine parameters [3-4]. In this research work, our intention is to use the vector control method described in [5]. In addition, to conserve the system robustness and stability subject to the uncertainties and variations of the parameters, a RLS

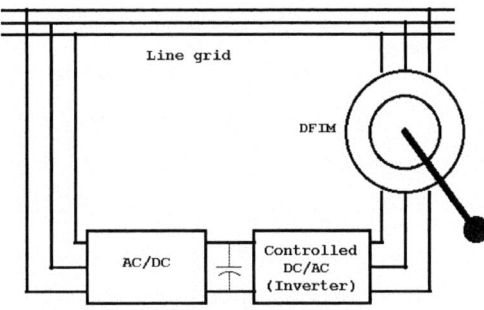

Figure 1. Typical DFIM scheme

estimator is used to on-line estimate the machine electrical parameters including the iron loss resistance.

Based on our little knowledge and search no any paper has been published so far in which the iron loss resistance of this drive has been taken into account.

RLS method is one of the parametric methods for system identification which is attractive for real-time applications. Because the RLS algorithm is recursive, very few data points need to be stored between iterations, and computations occur at specified intervals with a predictable computational complexity [6].

In previous published research works [4], this method has been used to estimate the mechanical parameters of cage induction motors. The purpose of this paper is to estimate the electrical parameters. The RLS method can be used easily for the models which are linear in the parameters. Fortunately in DFIM the voltage equations are linear in the parameters and contain all of the electrical machine parameters even leakage inductances of the machine.

The paper is organized as follow: in section II the DFIM model and control objectives are given. The RLS method and estimation of machine electrical parameters are presented in section III and IV without and with considering the iron loss respectively. Finally in section V simulation results are discussed.

1-4244-0448-7/06/$25.00 ©2006 IEEE

II. DFIM STATOR FIELD ORIENTED CONTROL METHOD

The d and q axis equivalent circuits of a DFIM are shown in Fig. 2. In this figure λ_{ds}, λ_{qs}, λ_{dr} and λ_{qr} are the d- and q- axis flux linkage components of the stator and rotor windings respectively, which are expressed by

$$\lambda_{ds} = L_s\,i_{ds} + L_m\,i_{dr}$$
$$\lambda_{qs} = L_s\,i_{qs} + L_m\,i_{qr}$$
$$\lambda_{dr} = L_r\,i_{dr} + L_m\,i_{ds}$$
$$\lambda_{qr} = L_r\,i_{qr} + L_m\,i_{qs}$$
(1)

From Fig. 2 using (1), the DFIM voltage equations in a general d and q axis rotating reference frame are given by [7]

$$v_{ds} = (R_s + L_s\,p)\,i_{ds} - \omega\,L_s\,i_{qs} + L_m\,p\,i_{dr} - \omega\,L_m\,i_{qr}$$
$$v_{qs} = \omega\,L_s\,i_{ds} + (R_s + L_s\,p)\,i_{qs} + \omega\,L_m\,i_{dr} + L_m\,p\,i_{qr}$$
$$v_{dr} = L_m\,p\,i_{ds} - (\omega_r - \omega)\,L_m\,i_{qs} + (R_r + L_m\,p)\,i_{dr} -$$
$$(\omega_r - \omega)\,L_m\,i_{qr}$$
$$v_{qr} = (\omega_r - \omega)\,L_m\,i_{ds} + L_m\,p\,i_{qs} + (\omega_r - \omega)\,L_m\,i_{dr} +$$
$$(R_r + L_m\,p)\,i_{qr}$$
(2)

where ω is the angular speed of the rotating reference frame, ω_r is the rotor angular speed and $p = d/dt$.

Moreover, the machine electromagnetic torque is given by

$$T_e = \frac{3P\,L_m}{4}(i_{qs}\,i_{qr} - i_{ds}\,i_{dr})$$
(3)

where P is number of machine poles.

Based on the induction machine stator flux field oriented control method, it is evident that by choosing the speed of the stator flux vector as the speed of the reference frame, the d and q axis stator flux linkage components become

$$\lambda_{ds} = \lambda_s$$
$$\lambda_{qs} = 0$$
(4)

where λ_s is the magnitude of the stator flux linkage.

Linking (1) and (4), the stator current component can be obtained as

$$i_{ds} = \frac{\lambda_s - L_m\,i_{dr}}{L_s}$$
$$i_{qs} = \frac{-L_m\,i_{dr}}{L_s}$$
(5)

Substituting (5) in (3), the electromagnetic torque becomes

$$T_e = \frac{-3P\,L_m}{4\,L_s}\lambda_s\,i_{qr}$$
(6)

From (6), for a constant λ_s, the machine torque can be controlled by i_{qr} [2].

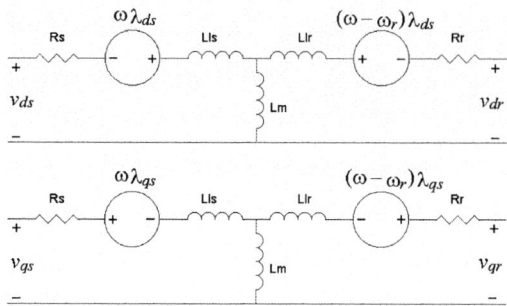

Figure 2. Arbitrary reference-frame equivalent circuits for an induction machine

III. DFIM ELECTRICAL PARAMETERS ESTIMATION

In order to apply the RLS method the model must be written in the form

$$y(t) = \varphi^T(t)\theta$$
(7)

where y is an observed variable, θ is the vector of parameters to be estimated and φ is the vector of known functions which depend on other known variables. The model is indexed by variable t, which often denotes time [8].

Consider the following cost function

$$J = \frac{1}{2}\sum_{i=1}^{t}[y(t) - \varphi^T\theta]^2$$
(8)

In RLS method, the cost function of (8) can be minimized by updating the estimated θ through the following recursive procedure [8]

$$\hat{\theta}(t) = \hat{\theta}(t-1) + K(t)(y(t) - \varphi^T(t)\hat{\theta}(t-1))$$
$$K(t) = P(t-1)\varphi(t)(I + \varphi^T(t)P(t-1)\varphi(t))^{-1}$$
$$P(t) = (I - K(t)\varphi^T(t))P(t-1)$$
(9)

where I is a unit matrix with appropriate dimension and $P(t)$ is a matrix which must be positive definite and at the beginning of the procedure can be selected as a positive multiple of the unit matrix.

In model (7), the vector of parameters is assumed to be constant, but in several cases such as electrical machines, parameters may by vary. To overcome this problem, two methods have been suggested. First is to use a discount factor or a forgetting factor. Second is to reset the matrix $P(t)$ alternatively with a diagonal matrix with large elements [8]. In our simulation, we use the second method.

Now, consider the DFIM equivalent circuits shown in Fig. 2 in attached to the rotor frame, that is $\omega = \omega_r$. One can obtain that

$$v_{ds} - v_{dr} = R_s\,i_{ds} + L_{ls}\,p\,i_{ds} - R_r\,i_{dr} - L_{lr}\,p\,i_{dr}$$
$$- \omega_r(L_{ls}\,i_{qs} + L_m\,i_{qs} + i_{qr})$$
(10)

Comparing (10) with (7), yields

$$y = v_{ds} - v_{dr}$$

$$\varphi^T = [i_{ds}, p\, i_{ds} - \omega_r\, i_{qs}, -i_{dr}, -p\, i_{dr}, -\omega_r\, i_{qs} - \omega_r\, i_{qr}] \quad (11)$$

$$\theta^T = [R_s, L_{ls}, R_r, L_{lr}, L_m]$$

At this stage the RLS method can be applied to estimate the DFIM electrical parameters. To solve the problem of derivatives ($p\, i_{ds}$, $p\, i_{dr}$) in (11), the following first order filter is used

$$H_f(p) = \frac{1}{p + a} \quad (12)$$

For example the term $p\, i_{ds}$ in (11) becomes

$$\frac{p}{p + a} ids = \frac{p + a - a}{p + a} ids = ids - a\, i_{dsf} \quad (13)$$

where i_{dsf} denotes the filtered value of i_{ds}.

IV. DFIM ELECTRICAL PARAMETERS ESTIMATION TAKING THE IRON LOSS RESISTANCE INTO ACCOUNT

Fig. 3 shows the DFIM equivalent circuits described in rotor reference frame considering the iron loss resistance in parallel with magnetizing inductances.

From Fig. 3, it can be seen that

$$L_m p\, i_{qm} = R_i (i_{qs} + i_{qr} - i_{qm}) \quad (14)$$

$$v_{ds} - v_{dr} = R_s\, i_{ds} + L_{ls}\, p\, i_{ds} - L_{ls}\, \omega_r\, i_{qs}$$
$$- L_m \omega_r\, i_{qm} - R_r\, i_{dr} - L_{lr}\, p\, i_{dr} \quad (15)$$

Substituting for i_{qm} from (14), in (15) and multiplying both sides of the resultant equation by $L_m p + R_i$ yields

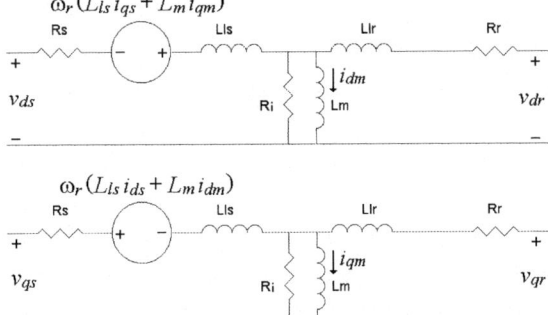

Figure 3. Rotor reference-frame equivalent circuits for an induction machine taking the iron loss into account

$$(v_{ds} - v_{dr}) R_i = -L_m p (v_{ds} - v_{dr}) + R_s L_m p\, i_{ds} +$$
$$R_s R_i\, i_{ds} + L_{ls} L_m p^2 i_{ds} + L_{ls} R_i p\, i_{ds} -$$
$$L_{ls} L_m p\, \omega_r\, i_{qs} - L_{ls} R_i \omega_r\, i_{qs} -$$
$$L_m R_i \omega_r (i_{qs} + i_{qs}) -$$
$$R_r L_m p\, i_{dr} - R_r R_i\, i_{dr} -$$
$$L_{lr} L_m p^2 i_{dr} - L_{lr} R_i p\, i_{dr}$$

$$(16)$$

Comparing (16) with (7), one can obtain that

$$y = v_{ds} - v_{dr}$$

$$\varphi^T = [-p(v_{ds} - v_{dr}), p\, i_{ds}, i_{ds}, p^2 i_{ds} - p\, \omega_r\, i_{qs},$$
$$p\, i_{ds} - \omega_r\, i_{qs}, -\omega_r (i_{qs} + i_{qr}), -p\, i_{dr}, -i_{dr},$$
$$- p^2 i_{dr}] \quad (17)$$

$$\theta^T = [\frac{L_m}{R_i}, \frac{R_s L_m}{R_i}, R_s, \frac{L_{ls} L_m}{R_i}, L_{ls}, L_m, \frac{R_r L_m}{R_i} + L_{lr},$$
$$R_r, \frac{L_{lr} L_m}{R_i}]$$

In order to solve the problem caused by second derivatives in (17), a second order filter is used as

$$H_f(p) = \frac{\omega_n^2}{p^2 + 2\xi\, \omega_n\, p + \omega_n^2} \quad (18)$$

V. SYSTEM SIMULATION

The overall block diagram of the system control is depicted in Fig. 4. The MATLAB/SIMULINK software is used to model this system. Simulation results are obtained for a DFIM with parameters shown in Table I.

Simulation results shown in Fig. 5 correspond to DFIM drive system performance without the iron loss effect taken into account. The same results are shown in Fig. 6 but with iron loss resistance taken into account.

In Figs. 5. At the time t=1.65 s the value of the magnetizing inductance increases 30 %. In this simulation the matrix $P(t)$ has been reset each 0.1 s with a diagonal matrix.

TABLE I. Machine parameters

$P_n = 5.5\, kW$	$L_m = 300\, mH$
$R_s = 1.2\, \Omega$	$L_{ls} = 14\, mH$
$R_r = 0.9\, \Omega$	$L_{lr} = 12\, mH$

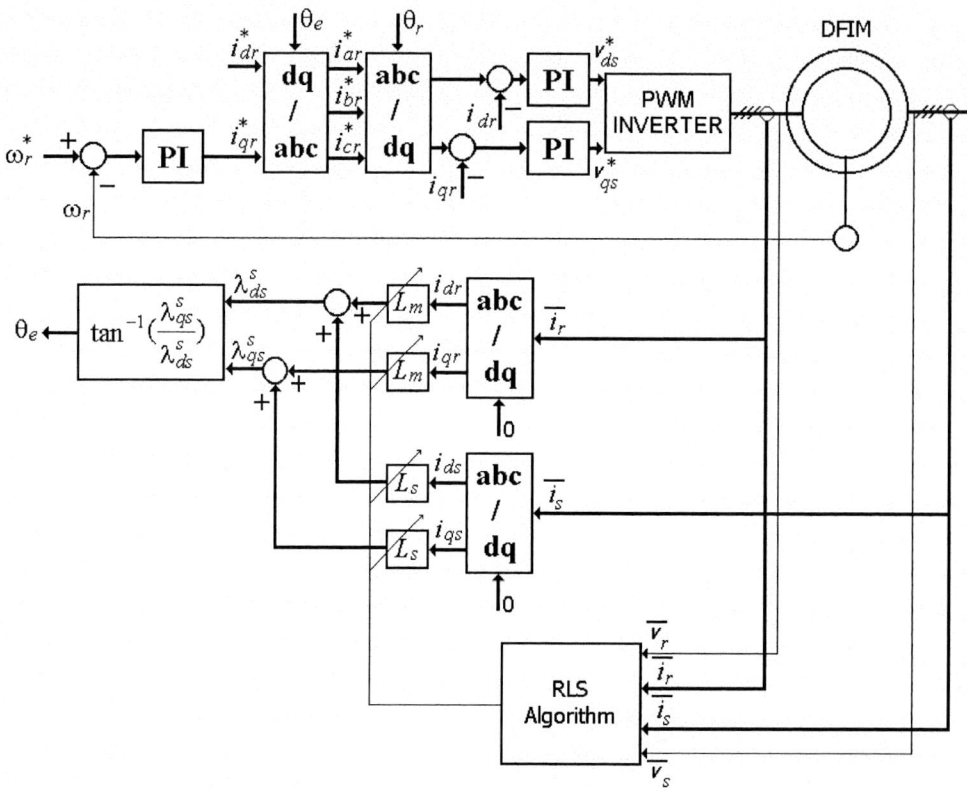

Figure 4. Overall Block diagram of the proposed adaptive DFIM stator field orientation control method

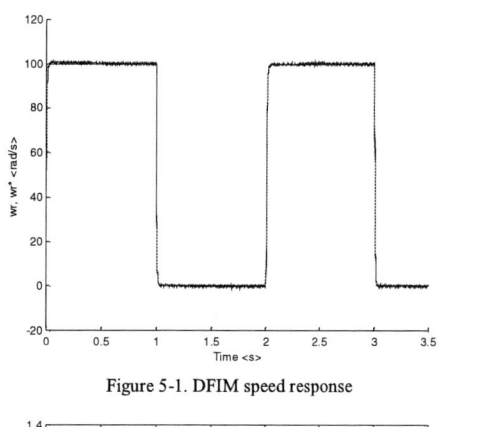

Figure 5-1. DFIM speed response

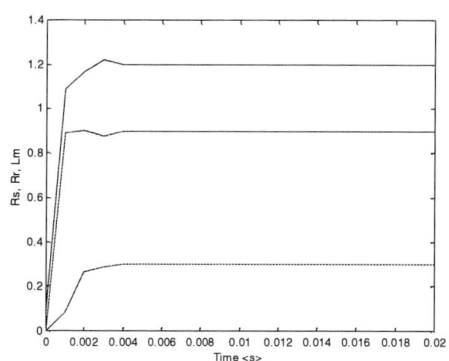

Figure 5-3. The expanded plot of the estimated parameters

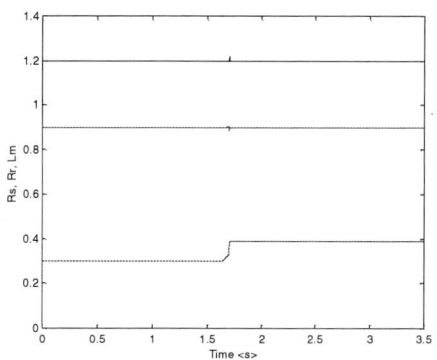

Figure 5-2. Estimated parameters

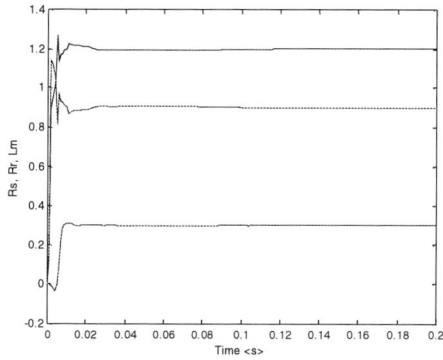

Figure 6. DFIM estimated parameters taking the iron loss into account

VI. CONCLUSION

In this paper it has been shown using the RLS method all of the DFIM electrical parameters can be estimated. In addition the effect of the iron loss in estimation of parameters has been considered. Computer simulations show the effectiveness of the proposed adaptive control method.

REFERENCES

[1] S. Peresada, A. Tilli and A. Tonielli, "Dynamic output feedback linearizing control of a doubly-fed induction motor," *Proceedings of the IEEE International Symposium on Industrial Electronics*, Vol. 3, pp. 1256-1260, 1999.

[2] W. Zheng, W. Fengxiang and Z. Fengge, "Study on stator field orientation control method of doubly fed machine," *IEEE Conf., The 4th International Power Electronics and Motion Control Conference*, Vol. 2, pp. 652-654, IPEMC 2004.

[3] V. Vasic, S. N. Vukosavic and E. Levi, "A Stator Resistance Estimation Scheme for Speed Sensorless Rotor Flux Oriented Induction Motor Drives," *IEEE Trans. On Energy Conversion*, Vol. 18, No. 4, pp. 476-483, Dec. 2003.

[4] F.-J. Lin and H.-M. Su, "A high-performance induction motor drive with on line rotor time-constant estimation," *IEEE Trans. On Energy Conv.*, Vol. 12, No. 4, pp. 297-303, 1997.

[5] B.Hopfensperger, D.J.Atkinson and R.A. Lakin, "Stator-flux-oriented control of a doubly-fed induction machine with and without position encoder," *IEE Proc.-Electr. Power Appl.*, Vol. 147, No. 4, pp. 241-250, July 2000.

[6] D. K. Jackson, S. B. Leeb and S. R. Shaw, "Adaptive control of power electronic drives for servomechanical systems," *IEEE Trans. On Power Elect.*, Vol. 15, No. 6, pp. 1045-1055, Nov. 2000.

[7] P. C. Krause, Analysis of electric machinery, McGraw-Hill Book Company.

[8] K. J. Astrom and B. Wittenmark, Adaptive control, 2nd Ed., Addison-Wesley Longman Publishing Co., Boston, MA, USA, 1994.

2006 5th International Power Electronics and Motion Control Conference

Analysis and Design of PDM Converter with High Frequency Link for HEV Drive System

Ma Xianmin
College of Electrical and Control Engineering,
Xi'an University of Science & Technology, Shaanxi, Xi'an, 710054

Abstract—**A discrete pulse density modulation (PDM) converter with high frequency (HF) link is introduced to improve the performance of hybrid electric vehicle (HEV) drive system requiring high power density. The drive system is made of HF resonant inverter, HF transformer and PDM converter. The inverter changes battery DC voltage into AC signal with 20kHz, which is distributed as power sources for different equipments in HEV by the transformer. By mean of zero-voltage switching technique, the PDM converter transfers the HFAC signal into the single phase AC power with low frequency to drive induction motor. Based on the vehicle dynamic characteristics, the mathematic analysis of the PDM converter is presented, the operation principle and control strategy are described. The theory analysis shows that the proposed system has the advantages of higher energy density, flexibility in the choice of voltage levels in the HEV, and freedom from acoustic noise. The simulation results prove that the system has the merit of easy control, faster system response, and the lower switch losses.**

Keywords-hybrid electric vehicle; high frequency AC; pulse density modulation; zero-voltage switching

I. INTRODUCTION

Electric vehicle (EV) has stood in the spotlight all over the world in recent years, zero or low emission vehicles have now been the newest buzzwords, because the traditional internal combustion engine (ICE) is a major source of urban pollution by giving out black smoke, hydrocarbons, and nitrogen oxides, making urban air contamination much worse. The electric vehicle motor propulsion system has demonstrated its superiority in high power density, efficiency and reliability. Recently, the pulse width modulation (PWM) dc-ac converter has been used in newly developed EVs and seems to be dominant in the propulsion system of EVs. In fact, there are some defects with hard PWM switching technique. Owing to coupled capacitors, the higher *dv/dt* on the output is produced and electrical magnetic interference (EMI) is generated to form a disturbance source for the main circuits. Under the regeneration conditions, the diode reverse recovery and snubber interaction result in very high device stresses, so a larger safe operation area

specification and compromise reliability are needed. Because of high switching losses, the switching frequency is also low. At the inverter switching frequency, acoustic noise can be very obnoxious [1].

In the past years, several soft switching inverters have been proposed to reduce these switching losses and EMI problems, and some of them may be suitable as propulsion drive system for EVs. In fact, these soft switching inverters have emerged as a new family of power converters for electric propulsion drives. Some researchers even claimed that soft switching inverters were the best for power train of EV or HEV.

The paper presents a scheme for the drive system of hybrid electric vehicle. The proposed new topology circuit is made of high frequency resonant inverter, high frequency transformer and soft switching PDM ac-ac converter. Based on vector control principle, the torque and speed can be accurately controlled.

II. PDM CONVERTER SYSTEM FOR HEV

In the development of high power density ac-ac converters, the remarkable progress has been made by using resonant link schemes in the past few years [2]. At zero voltage instant, the switching of the device is made, so these converters can enable the total system to operate at high frequency compared with traditional dc link converters. High frequency switching technique has some advantages such as avoiding acoustic noise, reducing torque and current ripple, making the whole system responding faster [3]. The resulting topology is also suitable to drive system of pure electric vehicles.

The proposed PDM converter with high frequency link for HEV drive system is shown in Figure 1.

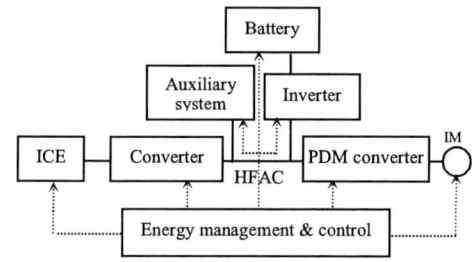

Figure 1. PDM converter drive system for HEV

1-4244-0448-7/06/$25.00 ©2006 IEEE

In this scheme, the proposed system is basically a series hybrid electric vehicle (HEV) in which the battery unit and the power source ICE contribute electrical power together to a common bus, which drives an induction motor through a pulse density modulation (PDM) converter. In fact, the battery unit can also be replaced by other energy storage units such as flywheel, ultra capacitor. Based on same reason, power sources may be fuel cell, gas turbine, or other diesel engine, stirling engine and so on. Therefore, the battery, ultra capacitor and fuel cell can be directly connected to the bus, but the flywheel and engine power sources need generator or converter as interface. It is appeared that the energy management and power flow control in this HEV propulsion power distribution system are very complex, requires powerful microcomputers and digital signal processor.

III. PRINCIPLE OF PROPOSED SYSTEM

The main circuit of the proposed PDM converter with high frequency link for HEV drive system is shown in Figure 2.

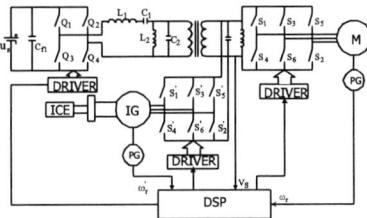

Figure 2. The main circuit of the proposed system for HEV

In Figure 2, a 300V lead-acid battery set is serviced as the energy storage. DC voltage firstly is inverted into quasi-ac square waveform signal that is changed into 20KHz HFAC sinusoidal signal through H-bridge L-C resonant network. The inverter output voltage is boosted to 440V and isolated by HFAC transformer from the PDM converter, then supplied to the converter that feeds the drive motor to push the electrical vehicle. Because the PDM converter is intrinsically a cycloconverter, the switching devices must withstand dual power flow and ac voltage, conduct the current in both directions. The two IGBT with bypass diodes were used as the ac switching, but the two IGBT were connected in inverse-series form.

Although high frequency ac transformer increases the cost and volume, it gives the outstanding benefits as follows:

• It isolates the inverter from the converter, so the optimal choice of battery-inverter and converter-motor is more economical respectively;

• It can raise the high frequency ac link voltage, which makes the terminal voltage of motor higher for the same output power.

The drive motor is controlled according to field orientation control principle, which makes induction motor performance similar to a separately excited dc motor drive. Thus the transient response is very fast and stable. The current I_{qs}^* is proportional to torque as torque command component in the outer loop. In the constant torque region, the rotor flux is controlled by the current I_{ds}^*, which is correspondingly to the flux component. The slip frequency signal ω_{sl} is positive proportion to I_{qs}^* through the slip gain factor K_s. The frequency command ω_s is derived of speed signal ω_r plus the ω_{sl} to produce unit vector signal.

IV. ANALYSIS AND DESIGN

In any ac link system, link voltage should go through zero twice per cycle of the link frequency. If we can limit all switchings in the power converter to these zero voltage points, the switching losses are dramatically reduced because the voltage across the switch is at or near zero during the switching interval. Due to the switching restricted to the zero crossing point of the link voltage, a half cycle of the link voltage becomes the basic unit of synthesis for generating lower frequency signal. Phase voltage V_{ao}, which has a near sinusoidal fundamental component "patched" from the half cycles of the high frequency link voltage V_s, is shown in Figure 3.

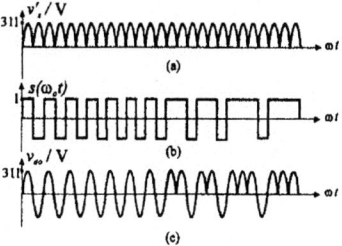

Figure 3. Lower frequency signal synthesized by PDM.

(a) V_s' voltage waveform of half-bridge version of HFAC converter.
(b) switching function S. (c) phase voltage V_{ao}

The principle operation of the PDM converter [4] is shown in Figure 4.

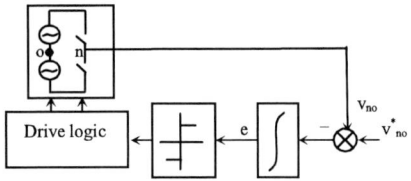

Figure 4. Principle of PDM converter

In Figure 4, the area under a reference voltage V_{no}^* is compared with the area of the synthesized signal V_{no}. If the comparison result shows that the area of the synthesized signal is bigger, thus the controller makes the next half cycle pulse applied in order to decrease the area. This simple mechanism causes the density of the half cycle pulses in the synthesized voltage to be modulated by means of switching function.

Because of the constraint of discrete half cycle zero voltage switching quality, the continuous variation of the output frequency is not possible as the conventional dc-ac inverter. The output frequency f_o is given by where N is integer multiple of three and f_s is the supply

$$f_o = f_s / N \qquad (1)$$

frequency.

PDM scheme has the closed loop nature and simplicity of implementation, and is proportional to the amplitude of the reference signal. Its better stability makes the PDM converter suitable for EV using the high frequency ac link.

At first, it may appear that this system uses a lot of devices making the power circuit complex and unreliable. The device capabilities, however, are the same as those in the convention PWM inverter, because the switching frequency in the PWM inverters is limited by high losses and high device stresses. As a result of zero voltage switching, the safety factor for PDM devices can be smaller than that of PWM devices, the number of devices needed is not crucially a problem, owing to emerging of the integral ac switch modular.

The transfer function of the synthesis sinusoid waveform can be described as

$$H(\omega_o t) = \frac{4}{n\pi}(1 + 2\sum_{k=1}^{k}(-1)^k \cos n(\alpha_k + \theta)) \qquad (2)$$

where $\theta = 0, -\frac{2\pi}{3}, +\frac{2\pi}{3}$.

According to the topology theory, the output voltage can be defined as

$$V_o(\omega_o t) = V_s' H(\omega_o t) \qquad (3)$$

For Y type impedances in three-phase circuit, $V_{an} + V_{bn} + V_{cn} = 0$, and V_{bo} lag V_{ao} to 120 degree, and V_{co} lag V_{bo} 120 degree. So the output neutral voltage can be given as

$$V_{ao} = V_s'H(\omega t + 0) = \frac{4V_s'}{n\pi}(1 + 2\sum_{k=1}^{k}(-1)^k \cos n(\alpha_k + 0)) \qquad (4)$$

$$V_{bo} = V_s'H(\omega t - \frac{2\pi}{3}) = \frac{4V_s'}{n\pi}(1 + 2\sum_{k=1}^{k}(-1)^k \cos n(\alpha_k - \frac{2\pi}{3})). \qquad (5)$$

$$V_{co} = V_s'H(\omega t + \frac{2\pi}{3}) = \frac{4V_s'}{n\pi}(1 + 2\sum_{k=1}^{k}(-1)^k \cos n(\alpha_k + \frac{2\pi}{3})) \qquad (6)$$

According to the relationship of neutral voltage and phase voltage, the phase voltage can described as

$$V_{an} = 2V_{ao}/3 - (V_{bo} + V_{co})/3$$
$$= \frac{16V_s'}{3n\pi}\sum_{k=1}^{k}(-1)^k \cos n\alpha_k(1 - \cos\frac{2}{3}n\pi) \qquad (7)$$

Because n can be divided by 3, so $\cos\frac{2}{3}n\pi = 1$, and $V_{an} = 0$. If n is other odd, then $\cos\frac{2}{3}n\pi = -\frac{1}{2}$, and

$$V_{an} = \frac{8V_s'}{n\pi}\sum_{k=1}^{k}(-1)^k \cos n\alpha_k. \qquad (8)$$

In half-bridge circuit, the Fourier expression of V_s' can be described as

$$V_s' = \frac{2V_m}{\pi} - \frac{4V_m}{\pi}\sum_{m=2,4,6,\cdots}^{\infty} \cos(m\omega_s t)/(m^2 - 1) \qquad (9)$$

The phase voltage can be defined as

$$V_{an} = \frac{16V_m}{n\pi^2}[1 - 2\sum_{m=2,4,6,\cdots}^{\infty} \frac{\cos m\omega_s t}{m^2 - 1}]\sum_{k=1}^{k}(-1)^k \cos na_k \qquad (10)$$

Similarly, the following formula can be deduced as

$$V_{bn} = \frac{16V_m}{n\pi^2}[1 - 2\sum_{m=2,4,6,\cdots}^{\infty} \frac{\cos m\omega_s t}{m^2 - 1}]\sum_{k=1}^{k}(-1)^k \cos n(a_k - \frac{2\pi}{3}) \qquad (11)$$

$$V_{cn} = \frac{16V_m}{n\pi^2}[1 - 2\sum_{m=2,4,6,\cdots}^{\infty} \frac{\cos m\omega_s t}{m^2 - 1}]\sum_{k=1}^{k} \cos n(a_k + \frac{2\pi}{3}) \qquad (12)$$

And the line voltage can be derive of

$$V_{ab} = \frac{16V_s'}{n\pi}\sum_{k=1}^{k}(-1)^{k+1} \sin n(\alpha_k - \frac{\pi}{3})\sin\frac{n\pi}{3} \qquad (13)$$

$$V_{bc} = \frac{16V_s'}{n\pi}\sum_{k=1}^{k}(-1)^k \sin n\alpha_k \sin\frac{2n\pi}{3} \qquad (14)$$

$$V_{ca} = \frac{16V_s'}{n\pi}\sum_{k=1}^{k}(-1)^{k+1} \sin n(\alpha_k + \frac{\pi}{3})\sin\frac{n\pi}{3} \qquad (15)$$

V. SYSTEM OPERATION

To prove the operation of the overall system, the computer simulation models of the PDM converter with the ICE and other system components are now used.

Because the operation of the PDM converter depends upon the generating of the high frequency at the resonant inverter, the delta modulation is used to produce 20KHz sinusoid signal through the L-C hybrid network. The nominal battery voltage 300V was considered, and the HFAC link supply 440V was generated from the resonant inverter. The current i_l of L-C resonant hybrid network is shown in Figure 5.

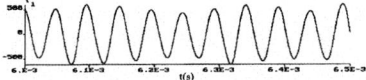

Figure 5. Waveforms of resonant current of L_1 in inverter

Obviously, the resonant current has the sinusoid form. The voltage of HFAC link is shown in Figure 6.

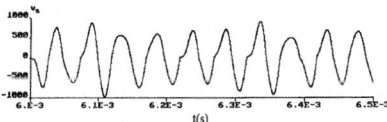

Figure 6. Waveforms of supply of HFAC link

The Figure 6 demonstrates the supply of HFAC link is almost sinusoid, but there is severe distortion. So it is important to restrain harmonics of the link. But further analysis and investigation represent that the Fourier series of V_s is convergent.

The ICE is initially started with the generator that is the same induction motor as the converter motor. During the running time, its speed is maintaining the 5000rpm.

The speed control scheme is also use the field orientation control. Waveforms of given current and actual current of ICE generator are shown in Figure 7.

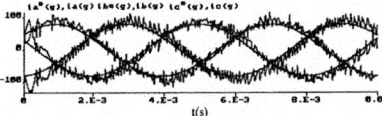

Figure 7. Waveforms of given current and actual current of generator

From Figure 7, the current waveform is almost sinusoid, and the actual current is tracking the command value well. The output low frequency voltage waveform of a phase voltage $V_{a(m)}$ in PDM converter and motor is shown in Figure 8.

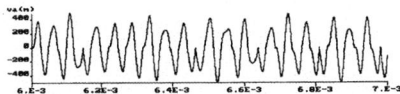

Figure 8. Waveforms of phase voltage in motor

And the phase voltage $V_{a(g)}$ of ICE generator is illustrated in Figure 9.

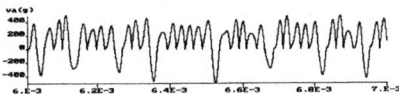

Figure 9. Waveforms of phase voltage in genarator

In both the Figure 8 and Figure 9, there are many half cycle pulses to synthesize the low frequency voltage.

With the technology of pulse density modulation, the PDM ac-ac converter can form the low distortion synthesis of variable amplitude and variable frequency ac signal from the half-cycle pulses of the high frequency voltage, as the input frequency of PDM converter is quite high. Because the switching is restricted to the zero crossing point of the high frequency voltage; the whole switching losses have been minimized. The distortion in the synthesis signals has been demonstrated to be very low.

VI. CONCLUSION

For hybrid electric vehicle drive system, a novel PDM converter with high frequency link has been presented in this paper. With the zero voltage switching, the analysis and design of the proposed PDM converter is described in detail. The results show that system performance is good and suitable for hybrid electric vehicle.

REFERENCES

[1] J.G.Cho, J. Sabate, G. Hua and F.C. Lee, " Zero voltage and zero current switching full bridge PWM converter for high power application" ,*IEEE PESC REC.*, pp.102-108, 1994.

[2] P.M.Espelage, B.K.bose, "High-frequency link power conversion", *IEEE Trans.Ind.App.*,vol. 13,NO.2, pp.384-394,Stept./Oct.,1977.

[3] W.A.Tabisz and F.C.Lee, "Zero-voltage-switching multi-resonant technique-a novel approach to improve performance of high-frequency quasi-resonant converters", Proceeding of PESC'88, pp. 9-17,1988.

[4] P.K.Sood , T.A.Lipo, "Power conversion distribution system using a high-frequency AC link". *IEEE Trans.Ind.App.*,vol. 24,NO.2, pp.288-300,Mrach/April,1988.

2006 5th International Power Electronics and Motion Control Conference

A Multi-Directional Power Converter for a Hybrid Renewable Energy Distributed Generation System with Battery Storage

MEI Qiang[12], WU Wei-yang[2] and XU Zhen-lin[1]
1 Tianjin University, Tianjin, China
2 Yanshan University, Qinhuangdao, China

Abstract—**Power electronics is the key technology for enabling renewable power generation and dispatching Distributed Generation (DG) with improved load-side efficiency. This paper presents a multi-directional power converter with a high-frequency isolation transformer for a stand-alone hybrid renewable energy DG system with battery storage. Full-bridge power converters are employed to realize bi-directional power flow among renewable generator, battery storage and load. Detailed analysis is provided for a four-directional power converter in a wind-photovoltaic distributed generation system. In addition, a two loop PI control scheme is proposed to achieve constant output voltage and exact power flow management. Theoretical analyses are verified by simulations in PSPICE and MATLAB.**

Keywords-renewable distributed generation; multi-directional power converter; high-frequency isolation; power flow control

I. INTRODUCTION

In order to meet the future energy requirement and the pollution reduction demand, applications with clean renewable energy technologies, such as photovoltaic (PV) energy and wind energy, have been vigorously developed over recent decade. In remote rural or energy deficient areas, stand-alone PV and wind power systems are even more attractive choices. However, a substantial energy storage battery bank is required to deliver a reliable power to the load and to draw maximum power from PV arrays or wind turbines since either one has a fitful nature [1], [2]. PV energy and wind energy are complementary since sunny days are usually calm and strong winds are often occurred at cloudy days or at nighttime. Combining a photovoltaic power unit with a wind energy unit will reduce the zero-power intervals. Therefore, a hybrid wind-PV power system has higher availability to deliver continuous power than either individual source and so requires less energy storage devices [3].

Distributed Generation is a small-scale electric power source connected directly to the utility's distribution network or on the customer site of the meter [4]. Distributed Generation (DG) technologies can provide customers with the energy solutions that are more cost-effective, more environmentally friendly, or provide

higher power quality or reliability than conventional solutions, and it provides electric power at a site closer to customers than the central station generation. The premise of DG is to provide electricity to customers at a reduced cost and a higher efficiently while reducing losses than the traditional utility central generating plant with transmission and distribution wires. Other benefits that DG can reduce peak demand charges and overall energy use, and ensure greater power quality and reliability. In addition, the potential for smaller users, such as housing developments and office buildings to switch to on-site power is also high. Various technologies are available for DG, including turbine generators, internal combustion engine, micro-turbines, photovoltaic/solar panels, wind turbines, and fuel cells. The application of renewable energy technologies to advanced distributed generation systems portends the most significant advancement in energy efficiency, conservation and environmental protection for the next decade [5].

Accordingly, it is a good solution for remote rural or energy deficient areas to adopt wind-PV energy in a DG system forming a hybrid renewable energy distributed generation system (HREDG), which possesses merits of both a DG system and a renewable power system. In order to implement such a hybrid renewable power system, several two-input single-output schemes were proposed in [6], [7]. A double-input single-output power converter developed for combined wind-PV power generating system is described in [6]. The PV array and the wind turbine are connected in series and each of them has the same current rating. If one of the dc sources is diminished, it will be very difficult to obtain the regulated voltage output since the input voltage variation is significant In [7], using the high frequency transformer, the circuit has the merits of electrical isolation. In addition, a two-input current-fed full-bridge dc/dc converter with phase-shifted pulse-width-modulation (PWM) control can be achieved to provide a possible approach to combine energy from different sources in a high frequency isolated circuit. However, these circuits have no charging path for the backup battery bank. Thus, they are not suitable for the stand-alone hybrid wind-PV power system

In this paper, the objective is to propose a novel multi-input dc/dc converter, which not only has ability to

1-4244-0448-7/06/$25.00 ©2006 IEEE

combine different renewable sources into one circuit with a high frequency transformer, but also has a battery storage path ensuring the steady-going substitute energy to load/grid in the days of deficit in solar and wind power. According to the aim, the systematic development of a four-directional high frequency isolated converter for the stand-alone hybrid wind-PV power system is presented.

II. MAIN CIRCUIT CONFIGURATION AND OPERATION

Fig.1 shows the schematic diagram of a hybrid renewable energy distributed generation system with battery storage. This system is divided into five main

Power sources of the four-directional converters are various for different operation modes. In operation mode 0, power is transferred from the first input-stage, the second input-stage and the storage-stage to the output-stage; That is, besides the wind turbine and PV panels, the battery bank delivers power to the load/grid via the storage-stage circuit; thus, the battery bank functions as a source. In operation mode 1, power is transferred from the first input-stage and the second input-stage to the storage-stage and the output-stage; here, the four-directional converter supplies power for the load/grid via the output-stage circuit and charges up the battery bank via the

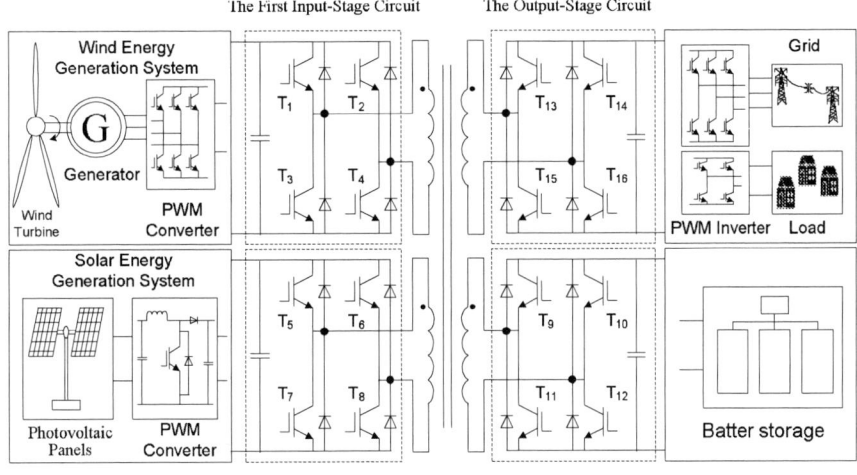

Figure 1. The schematic diagram of hybrid renewable energy distributed generation system

blocks. The First block is wind energy generation system. The second block is solar energy generation system. The third block is battery storage. The fourth block is load side. The fifth block, the central part, is the proposed multi-directional power converter, which consists of two power source input-stage circuits, a four-winding coupled transformer, a battery storage-stage circuit and a common output-stage circuit. In this paper, a four-directional dc/dc converter with high frequency isolated transformer is applied to the central part. Furthermore, a full-bridge converter is introduced into each stage.

A. Operation Modes of Proposed Four-directional Converter

According to whether energy sources (PV panels and Wind turbines) provide power for the load/grid or the storage batter, whether the converter is connected to the load/grid and whether the battery bank supplies or absorbs power, operation states of proposed four-directional converter can be classified into ten possible modes which are listed in Table I. For the PV panels and Wind turbine, supplying power for the load/grid or the battery is denoted by the "1", or by the "0". For the storage battery, the "1" indicates a discharge; whereas, the "0" indicates a charge. Moreover, for the load/grid connecting to the converter is expressed by the "1", or by the "0".

storage-stage circuit. In operation mode 7, the output-stage is disconnected from the load/grid; The PV panels and Wind turbine only export energy through the storage-stage circuit of the four-directional converter i.e. the power sources just charge up the storage battery. By following the above analytic approach, other operation modes also can be educed.

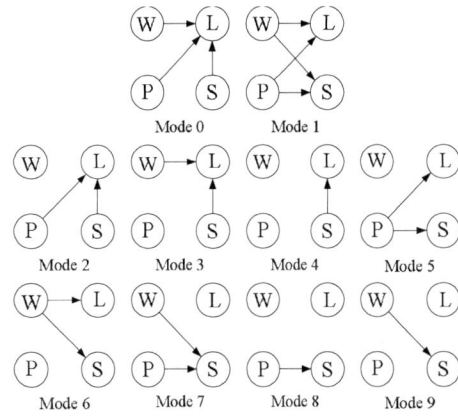

Figure 2. The simplified sketch maps of the hybrid renewable energy distributed generation system under different operation modes

Fig.2 shows the simplified sketch maps of the hybrid renewable energy distributed generation system under different operation modes. Thereinto "W", "P", "S", "L" represent the wind turbine, the PV panels, the storage battery and the load/grid respectively.

B. Equivalent Circuit of the Converter

In terms of the simplified sketch maps of the renewable energy distributed generation system shown in Fig.2; the four-directional converter can be simplified to be various

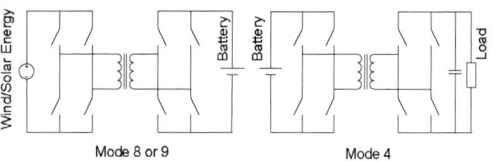

(a) Two-port DC/DC converter

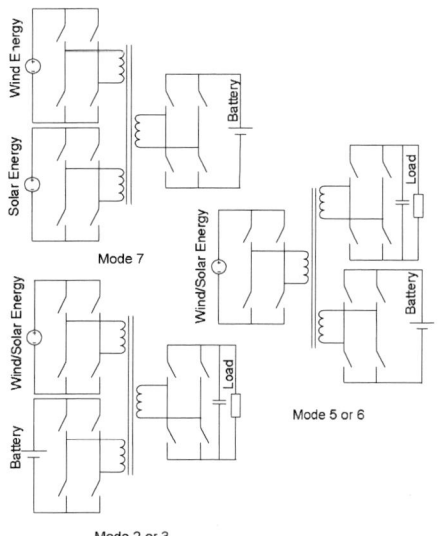

(b) Three-port DC/DC converter

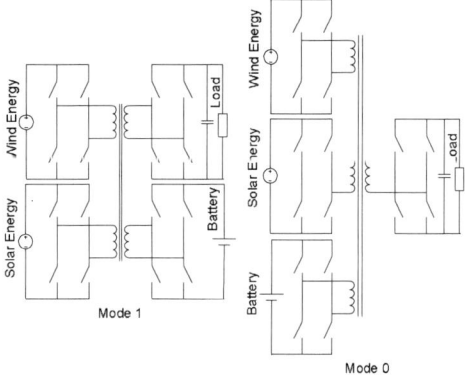

(c) Four-port DC/DC converter

Figure 3. The equivalent circuits of the proposed converter under different operation modes

multi-directional converters. Obviously, the converters under operation modes 4, 8 and 9 can be equivalent to two-port DC-DC converters shown in Fig.3 (a). Under the operation modes 2, 3, 5, 6 and 7, the four-directional converters can be transformed into three-port DC-DC converters shown in Fig.3 (b). However, the structure of converters under operation modes 0 and 1 is not changed, still a four-directional converter shown in Fig.3 (c). Moreover, for a multi-directional converter, the high frequency isolated transformer is one of the most important parts.

Firstly, a turns ratio of $N_1:N_2:N_3:N_4 =1:1:1:1$ is assumed. The equivalent circuit of a four-winding transformer is shown in Fig. 4 (a) where L_{m1} and L_{m2} is the effective magnetizing inductance, and that T_1 and T_2 are two ideal transformers with turn ratios $1: N_3$ and $1: N_4$. The transformer leakage inductances, represented by L_1, L_2, L_3 and L_4, determine the power transfer in connection with the phase displacement of the individual full-bridge converter unit control signals.

In Fig.4, (a) is equivalent circuit of a four-winding transformer; (b) is equivalent circuit of a four-winding transformer used for analyzing the dependency of the power transfer between the ports on the control signal phase displacements; (c) is the simplification of (b); (d) is the tetrahedron equivalent circuit of a four-winding

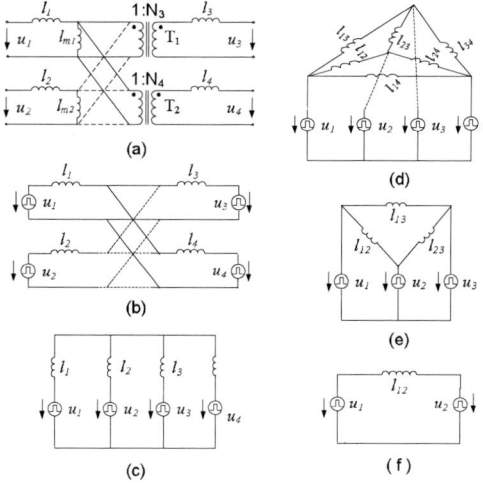

Figure 4. The equivalent circuits of multi-winding transformers

transformer; (e) is the triangle equivalent circuit which is the equivalent circuit of a four-winding transformer when any one of four ports is opened; (d) is equivalent circuit for studying the power flow between two ports when any two of four ports are opened. The parameters of the equivalent circuit can be derived from the measured values of the self inductances Z_1, Z_2, Z_3, Z_4 and the mutual inductances M_{12}, M_{13}, M_{14}, M_{23}, M_{24} and M_{34} of the transformer windings.

$$N_3 = \frac{M_{34}(M_{14}+M_{24})}{M_{14}M_{24}} \qquad L_{m1} = \frac{M_{13}M_{14}}{M_{34}} \qquad L_1 = Z_1 - L_{m1}$$

$$N_4 = \frac{M_{34}(M_{13} + M_{23})}{M_{13}M_{23}} \qquad L_{m2} = \frac{M_{23}M_{24}}{M_{34}} \qquad L_2 = Z_2 - L_{m2}$$

$$L_3 = 2Z_3 - N_3^2 L_{m1} - N_4^3 L_{m2} \qquad L_4 = 2Z_4 - N_4^2 L_{m2} - N_3^3 L_{ml} \qquad (1)\text{-}(8)$$

C. Control Model of Power Flow

In this paper, the control of the power flow in the four-directional converter system is in the simplest case by proper phase shift of the individual full-bridge cells operating in square wave mode. Besides the equivalent circuit of a four-winding transformer, the system control traits are determined based on the power transfer between two ports on the phase displacement of the control signals. The relationship between the bridge phase shift angles and the power flow in the system is found to be

$$P_{13} = S_{13}\varphi_{13}\left(1 - \frac{|\varphi_{13}|}{\pi}\right) \qquad P_{14} = S_{14}\varphi_{14}\left(1 - \frac{|\varphi_{14}|}{\pi}\right)$$

$$P_{23} = S_{23}\varphi_{23}\left(1 - \frac{|\varphi_{23}|}{\pi}\right) \qquad P_{24} = S_{24}\varphi_{24}\left(1 - \frac{|\varphi_{24}|}{\pi}\right)$$

$$P_{34} = S_{34}\varphi_{34}\left(1 - \frac{|\varphi_{34}|}{\pi}\right) \qquad (9)\text{-}(13)$$

$$P_1 = P_{13} + P_{14} \qquad P_3 = P_{14} + P_{24} - P_{34}$$

$$P_2 = P_{23} + P_{24} \qquad P_4 = P_{14} + P_{24} + P_{34} \qquad (14)\text{-}(17)$$

$$S_{13} = \frac{V_1 \tilde{V}_3}{\omega L_{13}} \qquad S_{14} = \frac{V_1 \tilde{V}_4}{\omega L_{14}} \qquad S_{23} = \frac{V_2 \tilde{V}_3}{\omega L_{23}}$$

$$S_{24} = \frac{V_2 \tilde{V}_4}{\omega L_{24}} \qquad S_{34} = \frac{V_3 \tilde{V}_4}{\omega L_{34}} \qquad (18)\text{-}(22)$$

where $\omega = 2\pi f_s$, and f_s is the switching frequency; $\tilde{V}_3$ and $\tilde{V}_4$ are the battery and load voltages referred to the power sources (wind and solar) sides; φ_{13} and φ_{14}, φ_{23} and φ_{24} denote the phase shifts (in radians) of the battery bridge and load bridge with reference to first input-stage bridge and second input-stage bridge, in this paper φ_{13} and φ_{14} are equal to φ_{23} and φ_{24} respectively when two input-stage bridges operate simultaneously; Furthermore, $\varphi_{34} = \varphi_{13} - \varphi_{14}$ or $\varphi_{34} = \varphi_{23} - \varphi_{24}$. P_1 is the power delivered the wind power generator. P_2 is the power delivered by the solar power generator. P_3 is the power stored into the battery (negative sign means that energy is drawn from the battery). P_4 is the power consumed by the load or the power injected into grid.

III. CONTROL SCHEME

According to above equations (9)-(22), different control schemes can be realized. In this paper, a two-loop control strategy is applied in the hybrid renewable energy distributed generation system. One loop is a voltage loop keeping the load voltage V_{load} constant. The other loop is a current loop which controls the directions of power flow among different ports and makes the output power maximum. In order to achieve above aims, two PI controllers are adopted in the control scheme showed in the fig.5. Thereinto, i_1 and v_1 are the feedback current and voltage of the first input-stage; i_2 are the feedback

current and voltage of the second input-stage; i_3 and v_3 are the feedback current and voltage of the storage-stage; i_4 and v_4 are the feedback current and voltage of the output-stage.

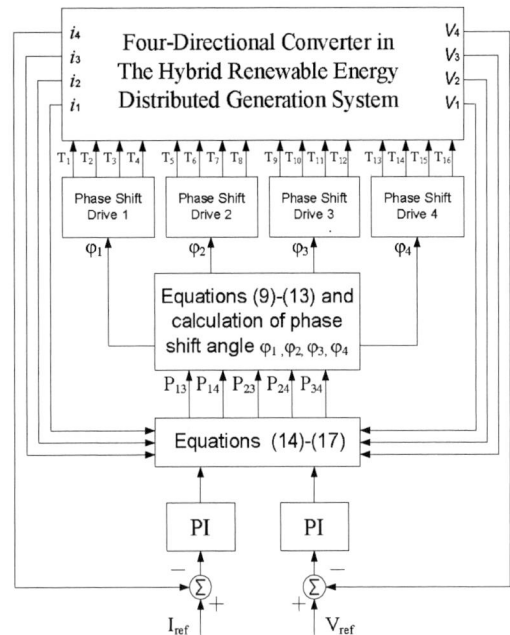

Figure 5. The control scheme of the hybrid renewable energy distributed generation system

IV. SIMULATION ANALYSIS

In order to validate the analytical result, simulations have been done by using the Pspice and MATLAB. The experimentation parameter is as follows:

Switching frequency f_s = 20 KHz

Wind and Photovoltaic power voltage $V_1 = V_2 = 200\text{v}$

Storage voltage $V_3 = 48\text{v}$

Output voltage $V_4 = 200\text{v}$

Inductors $L_1 = 26u\text{H}$, $L_2 = 26u\text{H}$, $L_3 = 1u\text{H}$, $L_4 = 56u\text{H}$

Wind and Photovoltaic power $P_W = P_S = 2\text{kW}$

Storage power P_B between -2kW and 2kW

Load power P_L = 4kW

Transformer turns ratio $N_1 : N_2 : N_3 : N_4 = 19 : 19 : 5 : 19$

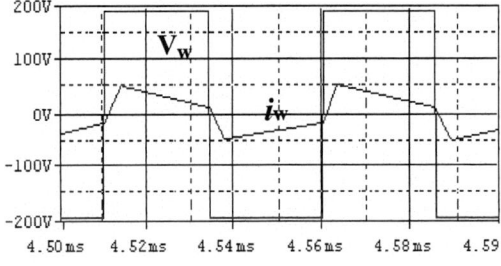

(a) First input-stage voltage and corresponding current waveforms across transformer terminal ($i_w = 5 \ast i_1$)

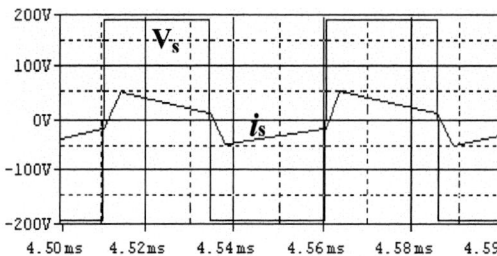

(b) Second input-stage voltage and corresponding current waveforms across transformer terminal ($i_w = 5*i_2$)

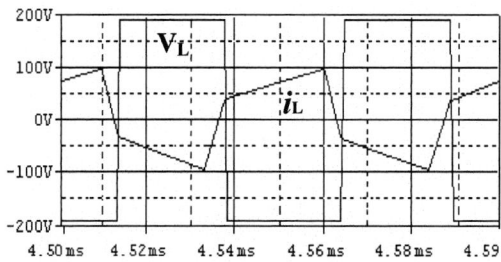

(c) Output-stage voltage and corresponding current waveforms across transformer terminal ($i_w = 5*i_4$)

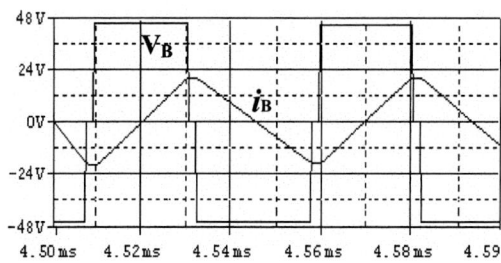

(d) Battery storage-stage voltage and corresponding current waveforms across transformer terminal

Figure 6. Simulation results of the four-directional converter ($\varphi_{13} = 35°$, $\varphi_{14} = 35°$)

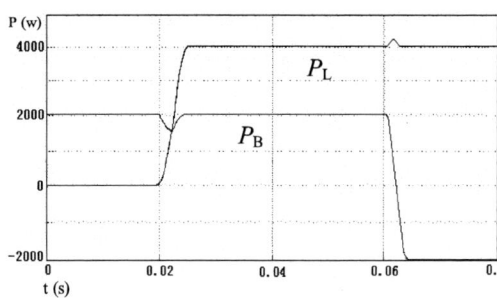

Figure 7. Simulation of the system control behavior

V. CONCLUSION

In this paper, a four-directional DC/DC converter is proposed in the hybrid renewable energy distributed generation system. Due to different combinations of the three possible energy sources and the connection or disconnection of the load/grid, there are ten different operation modes for the proposed four-directional DC/DC converter. Their equivalent circuits are explained and the control mode of the power flow is derived. By adopting phase-shifted PWM and dual PI control scheme, the maximum power and constant output voltage are achieved. The feasibility of the proposed system is verified by computer simulation results in the Pspice and Matlab software. In the next step, a Dsp-controlled 4kW laboratory model of the system shown in Fig.1 will be realized for verifying the proposed control concept. Furthermore, a new type of MPPT (Maximum Power Point Tracing) control method will be considered in the propose system.

REFERENCES

[1] Hao S, Hunter G, Ramsden V, Patterson D, "Control system design for a 20 kW wind turbine generator with a boost converter and battery bank load," Power Electronics Specialists Conference, 2001. PESC'2001 IEEE 32nd Annual Volume 4, pp: 2203 - 2206 vol. 4.

[2] Lessing P.A, "A hybrid photovoltaic-wind power system for the National Data Buoy Center's Coastal Marine Automated Network," OCEANS 2003. Proceedings Volume 4, Sept, 2003, pp: 1997- 2001 Vol.4

[3] Reddy J.B, Reddy D.N, "Probabilistic performance assessment of a roof top wind, solar photo voltaic hybrid energy system," Reliability and Maintainability, 2004 Annual Symposium – RAMS 2004, pp: 654 – 658.

[4] Walid El-Khattam, Y.G.Hegazy and M.M.A Salama, "Stochastic Power Flow Analysis of Electrical Distributed Generation Systems," IEEE Meeting Power Engineering Society 2003, vol. 1, pp. 141–143.

[5] El-Khattam W, Hegazy Y, Salama M.M.A, "Stochastic estimation of the contribution levels of customer operated distributed generation," IEEE Power Engineering Society General Meeting, 2004. pp: 2182 - 2186 Vol.2.

[6] L. Solero, F. Caricchi, F. Crescimbini, o. Honorati and F. Mezzetti, ''Performance of a 10 kW Power Electronic Interface for Combined Wind/PV Isolated Generating Systems," Proceedings of IEEE PESC, 1996, pp. 1027-1032.

[7] Chen Y.M, Liu Y.C, Wu F.Y and Wu T.F, "Multi-input DC/DC converter based on the flux additivity," Industry Applications Conference, 2001. Thirty-Sixth IAS Annual Meeting. Conference Record of the 2001 IEEE, Volume 3, Oct, 2001, pp: 1866 - 1873 vol.3.

2006 5th International Power Electronics and Motion Control Conference

Four-bridge Multilevel Converters Based on Hybrid-clamped Techniques

Xiaofeng Wang Yan Deng Xiangning He

College of Electrical Engineering of Zhejiang University, Hangzhou China

Abstract—In this paper, a four-bridge hybrid-clamped converter topology is proposed based on the analysis of existing multilevel hybrid-clamped techniques. The detailed description is given on the balancing principle of capacitors' voltages and control method of the four-bridge converter. The analysis shows that not only the cost of system is reduced quite a lot due to less active switches and capacitors used in this topology, but also the realization of the balance of capacitors' voltages is easier and more effective compared with that of existing topologies. The four-bridge hybrid-clamped topology is a good topology when both cost and performance are considered. Simulations and experimentations have been carried out and the capacitors' voltages can be balanced well.

Keywords- Multilevel Converter; Hybrid-Clamped; Balance of Capacitor Voltage; Four-Bridge

I. INTRODUCTION

Voltage clamping is to limit voltage at a fixed value across power switches when they are off in voltage-source converter (VSC). For n-level converters, clamping is an important issue in which n-level output voltage generated. Three types of multilevel topologies are popular: diode-clamped [1-3], flying-capacitor-clamped [4-6] and cascaded topology [7-9]. From their names, it is clear that diodes, capacitors and DC sources are the clamping devices for them respectively. When two kinds of devices at least are used for clamping, it is called hybrid clamping. For example, capacitors and diodes or diodes and active switches are used for clamping. In this paper, capacitors, diodes and active switches are combined together forming hybrid clamping. The clamping function requires the balance of capacitors' voltages. For the diode-clamped multilevel topology, the problem of capacitors' voltage balance is complicated when the output level beyond four. That is because in the diode-clamped topology, capacitors across the DC bus are shared by three phases, which results in the charging (discharging) of capacitors influenced by the coupling of three phases. When active power transferred in the system, each DC capacitor has different charging or discharging time which leading to the unbalance of capacitors' voltages and therefore, the distortion of output voltages. The higher the level is, the more serious the unbalance is. Many researchers have done much work [10-11] on this but no effective method especially for topologies higher than five levels until now. This is one of the main reasons that block its industry application.

Hybrid-clamping techniques can be used to realize the balance of DC capacitors' voltages based on the diode-clamped topology. In this paper, a four-bridge multilevel converter based on hybrid-clamping techniques is proposed. This topology uses much less active switches and capacitors, easier control method and better clamping effect achieved than existing topologies proposed before. The general topology in [12] and the improved topology in [13] are mentioned here to compare with the novel topology proposed in this paper. It is proved by theoretical analysis and experimentations that the four-bridge hybrid-clamped topology is a good choice for multilevel converters.

This work is organized as follows: Section II describes the general topology and the improved topology. Section III describes the proposed four-bridge hybrid-clamping topology and relevant analysis. Section IV shows some simulations and experimentations illustrating the new topology. Finally Section V presents the conclusions.

II. TOPOLOGIES BASED ON HYBRID-CLAMPING TECHNIQUES PROPOSED BEFORE

A. General topology

Two topologies are used to compare with the novel topology: the general topology in [12] and the improved topology in [13]. The general topology has self-balancing ability of capacitors' voltages from which the traditional multilevel topologies described above can be derived [12]. Fig.1 shows one bridge of a five-level converter in the general topology. S_{a1}-S_{a4} and $S_{a1'}$-$S_{a4'}$ are main devices and S_{c1}- S_{c12} are clamping devices. The main devices are used to produce the desired levels and the clamping devices are used for clamping and balancing the capacitors' voltages. Each pair of switches

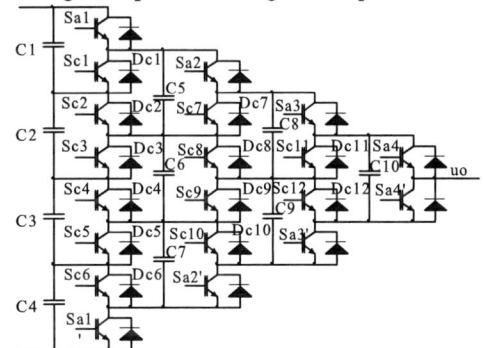

Fig.1 One bridge of five-level converter with general topology in [12]

1-4244-0448-7/06/$25.00 ©2006 IEEE

across a capacitor in a column have complimentary driving pulses. For an n-level topology, it needs (n-1) DC bus capacitors and n(n-1) active switches, (n+1)(n-1)/4 clamping capacitors for each phase. This topology is immune to different load and can be applied in active or passive power conditions. But it needs lots of active switches and capacitors, so it is not cost-effective and seldom used in industry.

B. Improved topology

In order to reduce the number of active switches and capacitors used in the general topology, [13] presents an improved topology based on it. The clamping principle is almost the same with the general topology. As in Fig.2, one bridge of a five-level converter is showed. Sa1-Sa4 and Sa1'-Sa4' are main devices, Sc1-Sc6, Dc7-Dc12 are clamping devices. For an n-level converter, it needs (n-1) DC bus capacitor and (4n-6) active switches, (n-2)(n-3) clamping diodes, (n-2) clamping capacitor for each phase.

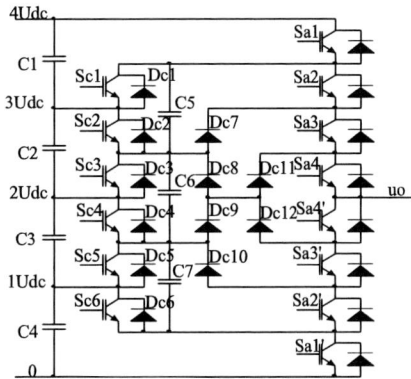

Fig.2 One bridge of five-level converter with improved topology in [13]

So it needs less devices than the general topology. In the improved topology, a pair of switches across each capacitor also has complimentary driving pulses. The clamping devices Sc1, 3, 5 turn on and off at the same time, and Sc2, 4, 6 turn on and off at the same time. Because of the simpler structure, the control circuit is easier than that in the general topology.

Comparing two topologies in Fig.1 and Fig.2, the voltage stress of each power device is one-unit voltage. It is clear the improved one needs 6 active switches, 3 capacitors less and 6 diodes more than those of the general one. So the cost is lower using the improved structure. Although the number of switches used in the improved topology is much less than that in the general topology, it still needs many active switches and capacitors and leads to high cost.

III. FOUR-BRIDGE HYBRID-CLAMPING TOPOLOGY, CLAMPING PRINCIPLE AND RELEVANT ANALYSIS

A. The structure of the four-bridge hybrid-clamped topology

Although both topologies in section II can realize clamping function well, they need many active switches and capacitors for a three-phase system. In this paper, a

novel topology − four-bridge hybrid-clamping topology is proposed that reduces the number of active switches and capacitors quite a lot. The simpler control method is used and better clamping effect is achieved using this topology as showed in Fig.3. Sx1-Sx4 and Sx1'-Sx4'(x=a, b, c) are the main switches, D1-D36, clamping diodes, and Scl1/Dcl1-Scl8/Dcl8, clamping switches. C1-C4 are the DC bus capacitors and C5-C7 are the clamping capacitors. For an n-level converter, it needs (n-1) DC bus capacitors and (n-1)+(n-1)/3 active switches, (n-1)(n-2) clamping diodes, (n-2)/3 clamping capacitors for each phase in average. The number of active switches and capacitors are reduced a lot and therefore the cost reduced. The comparison of the number of devices used in a three-phase five-level converter is as Table. I.

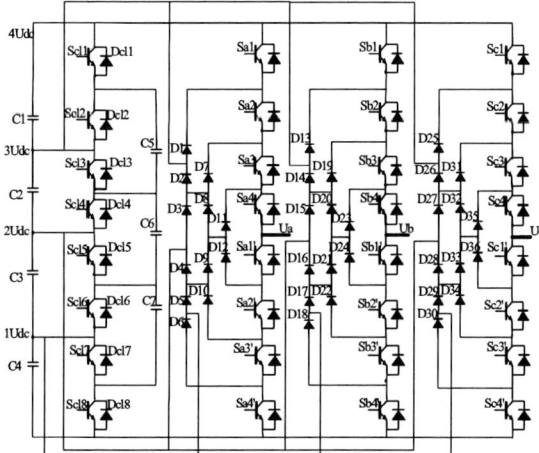

Fig.3 Three-phase topology of the proposed four-bridge hybrid-clamping converter

In this topology, the structure and function of three bridges is the same as the traditional diode-clamped converter, but the forth bridge is different. The forth bridge has the same number of active switches and diodes as in other three bridges, however, the position is different with them. It is located between the DC capacitors and clamping capacitors. The four-bridge topology groups the clamping function into the forth bridge and the voltage clamping of the whole system is realized in this bridge.

TABLE.I Comparison of Devices Number
in Three Kinds of Five-level Hybrid-clamped Topologies

Device \ Topology	Active switch	Clamping diode	Clamping capacitor	DC capacitor
General	36	0	18	4
Improved	18	18	9	4
Four-bridge	8	36	3	4

B. Clamping principle of the novel topology

The control of three diode-clamped bridges can use the methods as the traditional diode-clamped topology. Different with other two topologies described in section

II, the drive signals for the clamping switches have nothing to do with main switches. They can be controlled independently because of its independent structure.

There are two clamping conditions: (1) When Scl1, Scl3, Scl5, Scl7 are turned on, C1, C2, C3 are connected in parallel with C5, C6, C7 respectively (Fig.4a), so Vc1=Vc5, Vc2=Vc6, Vc3=Vc7; (2) When Scl2, Scl4, Scl6, Scl8 are turned on, C2, C3, C4 are connected in parallel with C5, C6, C7 respectively (Fig.4b), so Vc2=Vc5, Vc3=Vc6, Vc4=Vc7. Both conditions are controlled to appear alternatively according to the frequency of clamping switches. So finally, Vc1=Vc2=Vc3=Vc4=Vc5=Vc6=Vc7 achieved, the voltages on all capacitors realize the dynamic balance, the higher

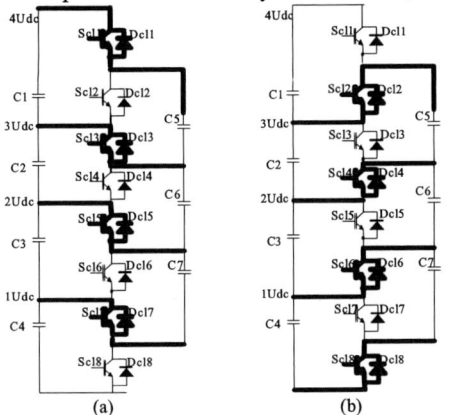

(a)   (b)
Fig.4 two conditions of voltage clamping in four-bridge hybrid-clamping converter

the frequency, the better the clamping performance. All the devices carry one-unit voltage stress. According to the analysis, the frequency of the main switches and clamping switches can be different. The control of this novel topology is much easier and more flexible than that of the former mentioned ones. All control schemes as the traditional diode-clamped multilevel converter can be applied in the novel topology. However, for the general and improved topologies, their control methods limited because the operation of clamping devices is relevant to main switches. For example, in Fig.2, complementary driving signals must be used in Sa1/ Scl1 and Sa1'/Sc6 because Sa1, Scl1 are across C1 and Sa1', Sc6 across C4. So in this topology, in order to realize different levels of output voltage at all modulation indices as well as guarantee the balance of capacitors' voltage, the switching states will be controlled according to the rules which limiting the use of different control methods [13].

C. Relevant analysis for the novel topology

In the four-bridge topology, the control of the clamping switches is more convenient and effective than that in two former mentioned topologies. For both of them, though different objectives of the main switches and clamping switches, they must operate at the same frequency and some of them have complementary driving pulses. In the field of multilevel applications, the

power is usually large and the operating frequency of main power switches used is several kHz. For the four-bridge topology, the clamping switches can be controlled independent of the main switches, and they can operate at higher frequency. Fig.5 is the equivalent circuit when two capacitors connected in parallel and (1)-(4) describe the relationship of power loss P_{loss}, energy loss E_{loss} and voltage difference $\triangle V$, average instantaneous current I_{avg}, the capacitor C, switching frequency f_{sw} [12]. There is the power loss when the clamping switches turn on because there is the voltage error between two parallel capacitors (Fig.5a). In Fig.5b, R is the conducting resistance. Power loss (P_{loss}) is expressed as in (1), which shows that power loss is inverse proportional to C, f_{sw} and irrelevant to R under the presumption of fixed frequency of the main switches. R is the factor influencing the value of current and the during time of charging or discharging (Fig.5c). Equation (2) expresses

$$E_{loss} = C \cdot (\Delta V)^2 / 4 \qquad (1)$$

$$\Delta V = I_{avg} / (C \cdot f_{sw}) \qquad (2)$$

$$E_{loss} = I_{avg}^2 / (4C \cdot f_{sw}^2) \qquad (3)$$

$$P_{loss} = I_{avg}^2 / (4C \cdot f_{sw}) \qquad (4)$$

the voltage error resulting from average charging or discharging current (I_{avg}) in each switching period (f_{sw},

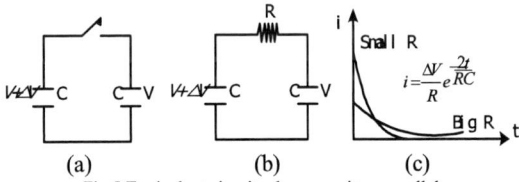

(a)   (b)   (c)
Fig.5 Equivalent circuit when capacitors parallel

switching frequency). Equation (3) and (4) derived from (1) and (2) show the energy loss and power loss of each switch. The power loss can be reduced when bigger capacitor or higher frequency is into use. For the general and improved topology, the frequency of clamping switches is the same as main switches and can't be raised arbitrarily. In the four-bridge topology, the frequency of clamping switches is controlled independently and it can work at higher frequency than the main switches. So the performance of the balance on capacitors' voltages is better and the power loss is smaller.

IV   SIMULATIONS AND EXPERIMENTATIONS

In order to verify the validity of the four-bridge hybrid-clamped topology, simulations are carried out in Pspice and a five-level four-bridge inverter prototype is set up in the lab. Fig.6-7 are simulation results. Fig.8-10 are

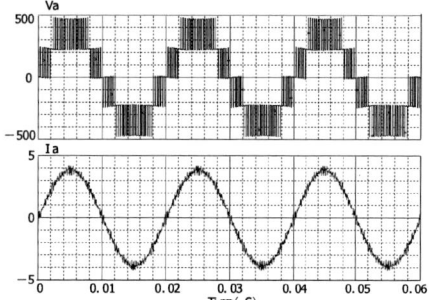

Fig.6 Phase voltage and current at inductive load

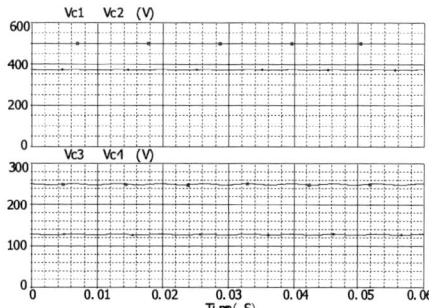

Fig.7 Voltage of four DC bus capacitors

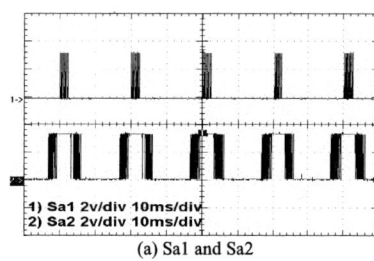

(a) Sa1 and Sa2

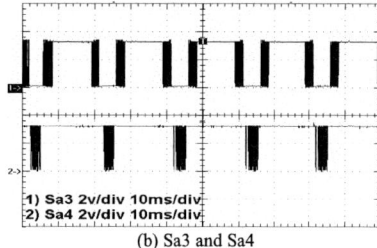

(b) Sa3 and Sa4

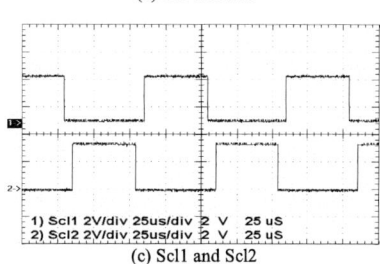

(c) Scl1 and Scl2

Fig.8 Drive pulses of four main switches in phase A and clamping switches

experimental results. Fig.6 shows the voltage and current at inductive load; Fig.7 shows the capacitors' voltages of the DC bus. For experimentation, FPGA (ACEX series EP1K30TC144-3 in Altera company) and some drive chips (M57959L) are used. Fig.8a-8c are driving signals of four switches in one up-bridge and complementary driving signals for the clamping switches; Fig.9 shows the voltage and current at inductive load; Fig.10 shows the capacitors' voltages of the DC bus. It can be seen that the capacitors' voltages can be balanced well and the performance of the output voltage is good. The industry application of the four-bridge hybrid-clamping converter as the main circuit of active power filter is under research and the further results will be present in the near future.

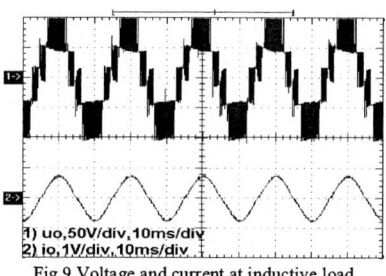

Fig.9 Voltage and current at inductive load

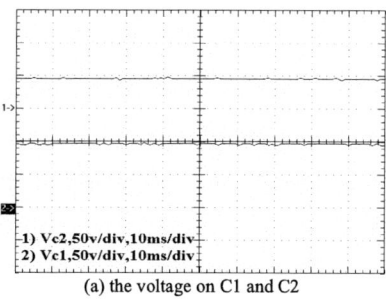

(a) the voltage on C1 and C2

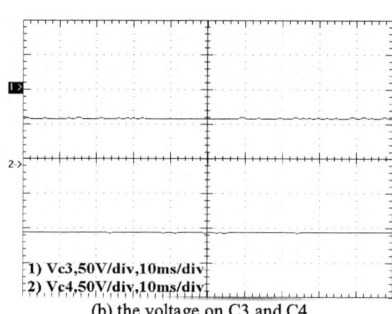

(b) the voltage on C3 and C4

Fig.10 Voltage of four DC bus capacitors

V. CONCLUSION

A novel hybrid-clamped multilevel topology — four-bridge topology is present in this paper. It can reduce large number of active switches and capacitors comparing with the general topology and the improved topology. The forth bridge of the proposed topology functions as clamping bridge. This structure simplifies the control and makes all control schemes applicable.

Moreover, the frequency of clamping switches can be controlled independently, so it has the advantage of reducing the power loss when two capacitors connected in parallel instantaneously by increasing the frequency of clamping switches. Simulations and experimentations verify the validity of this novel topology.

ACKNOWLEDGMENT

The authors would like to thank the financial support of National Nature Science Foundation of China (50307012, 50277035).

REFERENCES

[1] A. Nabae, I. Takahashi, and H. Akagi, "A new neutral-point clamped PWM inverter," *IEEE Trans. Ind. Applicat.*, vol. IA-17, pp. 518–523,Sept./Oct. 1981.

[2] L. Tolbert, F.-Z. Peng, and T. Habetler, "Multilevel converters for large electric drives," *IEEE Trans. Ind. Applicat.*, vol. 35,issue 1, Jan./Feb.1999, pp. 36–44.

[3] Takashi Ishida, Kouki Matsuse, Tetsuya Miyamoto, "Fundamental Characteristics of Five-Level Double Converters With Adjustable DC Voltages for Induction Motor Drives", IEEE *Trans. Ind. Electron*, Vol. 49, No. 4, AUGUST 2002, pp775-782

[4] Keith A. Corzine, and Xiaomin Kou, "Capacitor Voltage Balancing in Full Binary Combination Schema Flying Capacitor Multilevel Inverters", *IEEE Power Electron. Letters,* Vol. 1, No. 1, March 2003, pp2-5.

[5] Miguel F. Escalante, Jean-Claude Vannier, and Amir Arzandé, "Flying Capacitor Multilevel Inverters and DTC Motor Drive Applications", IEEE *Trans on Industrial Electron.,* Vol. 49, No.4, August 2002, pp809-815.

[6] Guillaume Gateau, Maurice Fadel, Pascal Maussion, etc., "Multicell Converters: Active Control and Observation of Flying-Capacitor Voltages", IEEE *Trans on Industrial Electron.,* Vol. 49, No. 5, October 2002, pp998-1008

[7] Keith Corzine, and Yakov Familiant, "A New Cascaded Multilevel H-Bridge Drive", IEEE *Trans on Power Electron.,* Vol. 17, No. 1, January 2002, pp125-131.

[8] Thierry A. Meynard, Henri Foch, Philippe Thomas, etc., "Multicell Converters: Basic Concepts and Industry Applications", IEEE *Trans on Industrial Electron.,* Vol. 49, No.5, October 2002, pp955-963

[9] Donald Grahame Holmes, and Brendan P. McGrath, "Opportunities for Harmonic Cancellation with Carrier-Based PWM for Two-Level and Multilevel Cascaded Inverters", *IEEE Trans. Ind. Applicat.*, Vol. 37, No. 2, March/April 2001, pp574-582

[10] Gautam Sinha, and Thomas A. Lipo, "A Four-Level Inverter Based Drive with a Passive Front End", IEEE *Trans on Power Electron.,* Vol. 15, No. 2, March 2000, pp285-294

[11] Takashi Ishida, Kouki Matsuse, Tetsuya Miyamoto,etc., "Fundamental Characteristics of Five-Level Double Converters With Adjustable DC Voltages for Induction Motor Drives", IEEE *Trans on Industrial Electron.,* Vol. 49, No. 4, August 2002, pp775-782.

[12] F. Z. Peng, "A Generalized Multilevel Inverter Topology with Self Voltage Balancing," IEEE *Trans. on Ind. Applicat.,* Vol.37, No.2, 2001, pp611-618

[13] Alian chen, Xiangning He, "A hybrid clamped multilevel inverter topology with neutral point voltage balancing ability", 35th annual IEEE *Power Electronics Specialist Conference*, Aachen Germany, 2004, pp3952-3956.

2006 5th International Power Electronics and Motion Control Conference

Standardization of Input/Output Impedance Specifications of Buck Converters Based on the System Integration Concept

Tao Wu, Student Member and Xinbo Ruan, Senior Member, IEEE

College of Automation Engineering
Nanjing University of Aeronautics and Astronautics
Nanjing 210016, Jiangsu Province, P.R.China
E-mail: wutaonuaa@yahoo.com.cn

Abstract—Power electronics has evolved to a point where building large-scale power electronics systems has become a reality. However, due to lack of standardized power electronics modules, it is difficult to reduce the costs of building power electronics systems via the system integration approach. In this paper, the input and output impedance of power electronics modules are selected as the objects for standardization at the system level, which can be used to ensure the system stability in the small-signal sense. The characteristics of I/O impedance are analyzed in detail. Further, the influence of three factors including the output power rating, the switching frequency and the control mode on the I/O impedance is discussed.

Keywords-Impedance; system stability; standardization

I. INTRODUCTION

With the evolvement of power processing technologies, building large-scale power electronics systems has become a reality. Especially, distributed power systems recently have drawn a lot of attention, due to the advantages such as high reliability, good modularity and easy maintainability [1]. As illustrated in Fig.1, a typical DC distributed power system (DPS) consists of multiple power electronics modules, which are connected to an intermediated dc power distribution bus. It is expected that power electronics modules, which are connected to the dc bus, have the advantages of high efficiency, high reliability and high power density. More importantly, it is eagerly expected that standardized power electronics module will be used to build such a system via system integration approach. Obviously, it is sensible to take advantage of standardized parts in the form of integrated circuits to build power electronics systems. However, until today power electronics products are still custom-designed, because standardized units or parts are unavailable. This, in turns, results in the long design cycle time and high costs of commercial products. Further, the I/O characteristics of power electronics modules are often unknown to engineers building large-scale power electronics systems. Consequently, the

---
* This paper is supported by Fok Ying Tung Education Foundation under Award Number 91058.

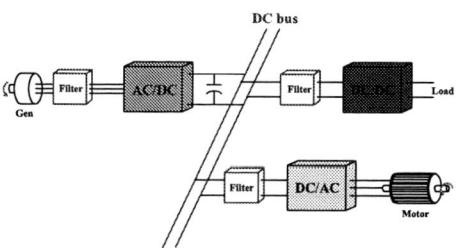

Fig.1. A typical dc distributed power system

sufficient stability margin of DPS is difficult to be ensured [2~4].

In order to reduce the cost of building power electronics systems and improve the reliability of whole systems, US Navy firstly proposed the power electronics building blocks (PEBB) concept [5]. The so-called "PEBB" is a standardized module which integrates power devices, protection and sensing circuit etc. together. It is obvious that the implementation of the PEBB concept gives a new paradigm shift to the development of power electronics and raises the degree of integration to a higher level. Actually, the system integration can be realized at three levels, namely device level, module level and system level [6]. US Navy's main contribution to the system integration is that they proposed a kind of system integration approach at the module level. At the system level, overall power electronics modules in Fig.1 should be regarded as "black boxes". The inner properties such as topologies and control methods are no longer the focus of the research. Instead, the I/O character (input/output impedance) should be given much more attention. Moreover, the input/output impedances are useful to check the stability and the dynamic performance of whole systems. In this paper, therefore, I/O impedances are selected as objects of standardization and the basis for issuing I/O impedance specifications is given.

II. CHARACTERISTICS OF I/O IMPEDANCE OF BUCK-TYPE CONVERTERS

A. Formulas of I/O Impedances

Buck-derived and Boost-derived converters are commonly seen in DPS. In this paper, characteristics of

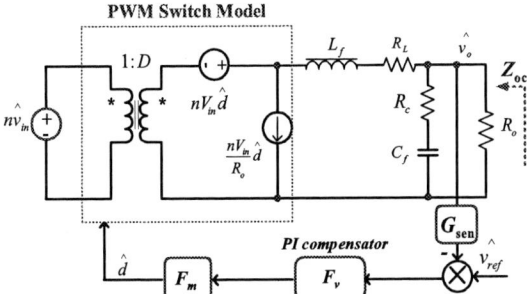

Fig.2. Closed-loop small signal model of the isolated buck converter

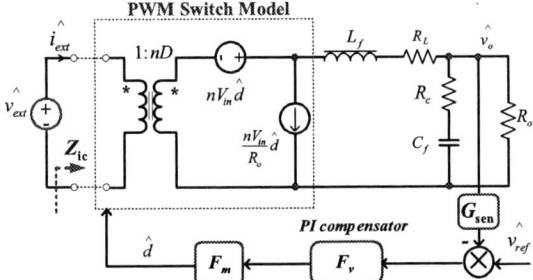

Fig.3. Closed-loop input impedance of the buck converter

I/O impedance of power electronics modules are clearly demonstrated with the Buck-derived converter as an example.

Fig.2 shows the closed-loop small-signal model of an isolated Buck converter working in CCM mode [7]. Based on this model, it is easy to derive the open-loop and closed-loop output impedance

$$Z_o(s) = R_L \frac{(1+s/\omega_c)\cdot(1+s/\omega_L)}{1+s/(Q\cdot\omega_o)+s^2/\omega_o^2} // R_o \qquad (1)$$

where $\omega_c = \dfrac{1}{R_c C_f}$, $\omega_L = \dfrac{R_L}{L_f}$,

$$\omega_o = \frac{1}{\sqrt{L_f C_f}}, Q = \frac{\sqrt{L_f/C_f}}{R_L+R_c}.$$

$$Z_{oc}(s) = \frac{Z_o(s)}{1+T_v(s)} \qquad (2)$$

where T_v is the loop gain.

In order to derive the expression of the input impedance, the external voltage source is put across the input port of the small-signal model, as shown in Fig.3. Based on the definition of closed-loop input impedance, the equation of closed-loop input impedance can be obtained

$$Z_{ic}(s) \overset{\Delta}{=} \frac{\hat{v}_{ext}}{\hat{i}_{ext}} = \frac{Z_i(s)\cdot[1+T_v(s)]}{1-\dfrac{\Delta(s)\cdot T_v(s)}{\alpha(s)}} \qquad (3)$$

where $Z_i(s) = \dfrac{R_o}{n^2 D^2}\cdot\dfrac{\Delta(s)}{\alpha(s)}$,

$$\Delta(s) = nV_{in}/G_{vd}(s),$$

$$\alpha(s) = \frac{1+s(R_o+R_c)C_f}{1+sR_c C_f},$$

n represents the turns ratio of the secondary to primary windings of the transformer.

B. Characteristics of I/O Impedances

In order to get deep insight into characteristics of I/O impedances, an isolated Buck converter is modeled, with the following parameters: V_{in}=400V; V_o=54V; n=0.2; f_s=200kHz; L_f=24μH; R_L=0.01Ω; C_f=620μH; R_c=0.145Ω; R_o=2.916Ω.

Fig.4 shows the open-loop and closed-loop output impedance. It can be shown that the open-loop output impedance appears the resonant peaking and the magnitude of the resonant peak depends on the value of Q. At high frequencies, R_c, the ESR of the output capacitor determines the output impedance. With the voltage loop closed, the output impedance is sharply reduced within the control bandwidth. When frequency goes beyond the crossover frequency f_c, the loop gain T_v becomes little and little. So the closed-loop output impedance will be dominated by the open-loop output impedance. As a result, the ESR also determines the closed-loop output impedance at high frequencies.

Fig.5 shows the open-loop and closed-loop input impedance. It can be seen that the low frequency portion of the input impedance behaves as a negative resistance. For a converter loaded with the constant power, assuming conversion efficiency of the module is 100%, it can be obtained

$$P_o = V_{in}I_{in} \qquad (4)$$

where V_{in} and I_{in} is the average of input voltage and input current, respectively.

After superimposing the small-signal perturbation on the quiescent point, it can be obtained

$$P_o = (V_{in}+\hat{v}_{in})(I_{in}+\hat{i}_{in}) \qquad (5)$$

It is easy to understand from (4), (5) that the close-loop input impedance is

$$Z_{ic} \overset{\Delta}{=} \frac{\hat{v}_{in}}{\hat{i}_{in}} = -\frac{V_{in}^2}{P_o} \qquad (6)$$

The equation (6) means that the magnitude of closed-loop input impedance depends on the input voltage and the output power of the module. When frequency goes beyond the crossover frequency f_c, passive components of the output filter will take place of the voltage loop to dominate the closed-loop input impedance. Thus, at relatively high frequencies the closed-loop output impedance is approximately equal to the open-loop input impedance, as shown in Fig. 5.

III. STANDARDIZATION OF OUTPUT IMPEDANCE SPECIFICATIONS

In the previous section, characteristics of input and output impedance of the Buck converter are analyzed.

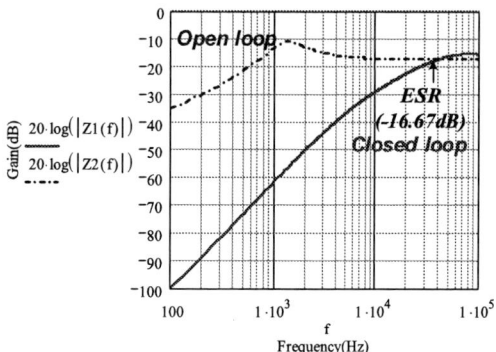

Fig.4. Open-loop and closed-loop output impedance

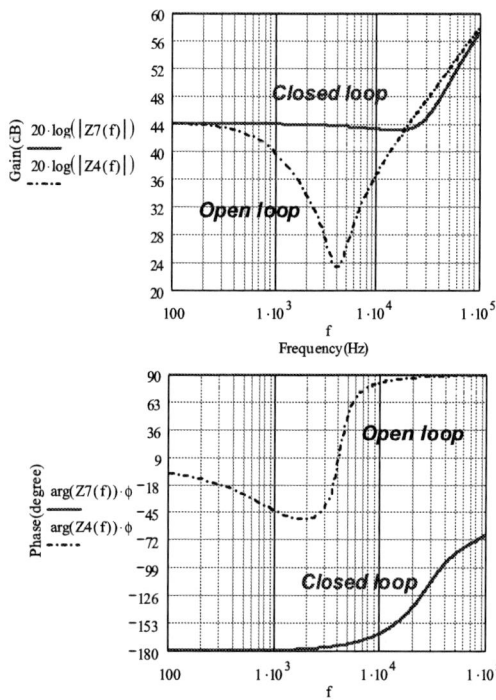

Fig.5. Open-loop and closed-loop input impedance of the Buck converter

Considering the input/output voltage remains constant, the closed-loop output impedance depends on the output power rating, the switching frequency and the control mode (voltage/current mode). So in order to issue specifications of the output impedance, it is necessary to analyze the influence of variations of the output power rating, the switching frequency and the control mode on the closed-loop output impedance. In order to plot the transfer function of the output impedance, the parameters of the isolated buck converter are input voltage V_{in}=400V; output voltage V_o=54V; turns ratio n=0.2.

Obviously, the output filter has close relationships with the open-loop and closed-loop impedance of the Buck converter. So the calculation formula of the output filter is given as follows before analyzing three factors' effect

on the output impedance. Derivation process of (7) and (8) have been given in the appendix.

$$L_f = \frac{V_o^2}{0.2P_{o(nom)}}\frac{1-D}{f_s} \tag{7}$$

where $P_{o(nom)}$ is nominal output power.

$$C_f = \frac{20P_{o(nom)}}{V_o^2 f_s}(\frac{1}{8} + \lambda f_s) \tag{8}$$

where $\lambda = R_c C_f$.

Generally, for certain type of electrolytic capacitors provided by certain vender, the product of the capacitance and the ESR is approximately constant [8]. For the simplicity of analysis, select 380LQ-type electrolytic capacitor provided by CDE company as the output capacitor, whose λ is around 90×10^{-6}. Since the product of the capacitance and the ESR is assumed as a constant, the theoretic value of ESR corresponding to the output capacitor calculated by (8) is given as

$$R_c = \frac{90\times10^{-6}}{C_f}(Ohm) \tag{9}$$

where the unit of C_f is F.

A. Output power rating effect

In order to plot the transfer function of the open-loop and closed-loop output impedance with the power rating as a running parameter, two Buck converters are modeled, with parameters: $f_{s1}=f_{s2}$=200kHz; P_{o1}=1kW, P_{o2}=5kW. And voltage compensator F_v is designed, with following guidelines: crossover frequency f_c is selected as one fifth of the switching frequency (f_c=40 kHz); corner frequency f_z is one tenth of the crossover frequency (f_z=4 kHz). The modulator gain F_m is 0.303 and voltage sensing gain G_{sen} is 0.052.

Fig. 6 shows a set of open-loop and closed-loop output impedance with the power rating as a running parameter. It can be clearly seen that the open-loop output impedance is reduced as the power rating of the module is increased. It is easy to understand from (8), (9) that the theoretic value of ESR is anti-proportional to the output power. Thus, as the power rating is increased, the high frequency gain of the open-loop output impedance, which is determined by the ESR, is reduced. Similar to the open-loop output impedance, the closed-loop output impedance is also reduced with power ratings increased.

B. Switching frequency effect

In order to plot the open-loop and closed-loop output impedance with the switching frequency as a running parameter, two Buck converters are modeled, with $P_{o1}=P_{o2}$=1kW; f_{s1}=50kHz, f_{s2}=200kHz. And design guidelines of the control loop are the same as that of the case of output power ratings' effect.

Fig. 7 shows a set of open-loop and closed-loop output impedance with the switching frequency as a running parameter. As illustrated in Fig.7, it is evident that variations of switching frequencies lead to differences of

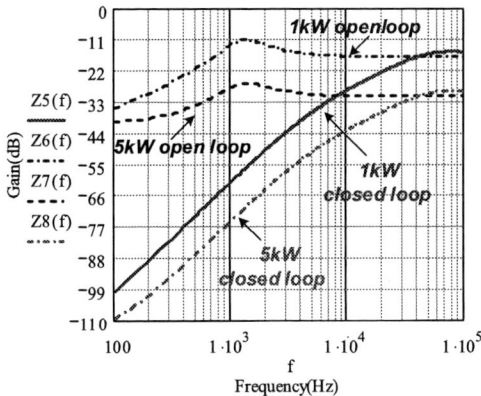

Fig.6. Open-loop and closed-loop output impedance with the power rating as a running parameter

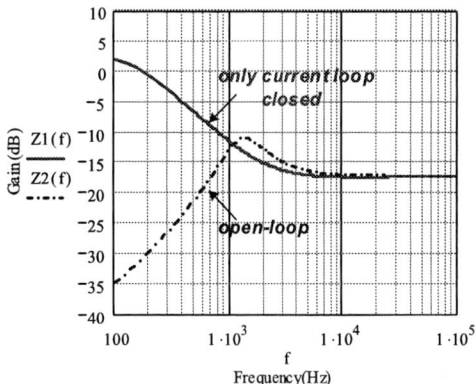

Fig.8. Comparison of open-loop output impedance and output impedance with only current loop closed

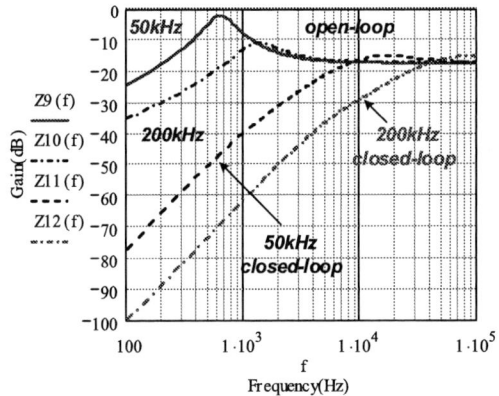

Fig.7. Open-loop and closed-loop output impedance with the switching frequency as a running parameter

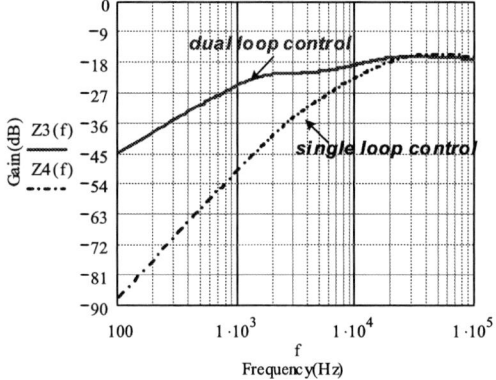

Fig.9. Comparison of Output impedances with single-loop control and dual-loop control

low frequency gains of both open-loop and closed-loop output impedance. For the closed-loop output impedance, this is due to the fact that system bandwidth is increased as the switching frequency is increased, which indicates that the frequency range where closing the voltage loop has the reducing effect on the open-loop output impedance is increased. Thus, the low frequency gain of the closed-loop output impedance, corresponding to the module with a higher switching frequency, is lower.

C. Control mode effect

The analysis results above are obtained with the voltage-mode control. In the following section, the effect of different control modes on the output impedance will be discussed.

Reference [9] gives the equation of the closed-loop output impedance with peak-current mode control

$$
Z_{oc\_dualloop} = \frac{Z_{open} + T_i (Z_{open} - \dfrac{G_{vd} G_{ii}}{G_{id}})}{1 + T_i + T_v} \tag{10}
$$

where T_i is the inner current loop gain, T_v is the outer voltage loop gain.

In order to compare closed-loop output impedances

with different control modes. Another Buck converter, with P_o=1kW and f_s=200kHz, are modeled. For peak-current mode control, the parameters of the inner current loop are inductor current sensing gain R_i=0.02; m_c=4.8. The design guidelines of outer voltage loop are the crossover frequency f_c is selected as one tenth of the switching frequency (20 kHz); corner frequency f_z is one tenth of the crossover frequency (2 kHz).

With the only current loop closed, the output impedance of the module is converted from 2nd order to 1st order. This is due to the current loop, which makes other part of the module function as a current source with almost infinite large output impedance. As shown in Fig.8, current-mode control gives high output impedances at low frequencies, compared with the open-loop output impedance.

Fig.9 compares closed-loop output impedances corresponding to the voltage-mode and peak-current mode control. It can be shown that peak-current mode control gives high output impedances at low frequencies. However, no matter what kind of control-mode will be used, the ESR still determines the output impedance at high frequencies.

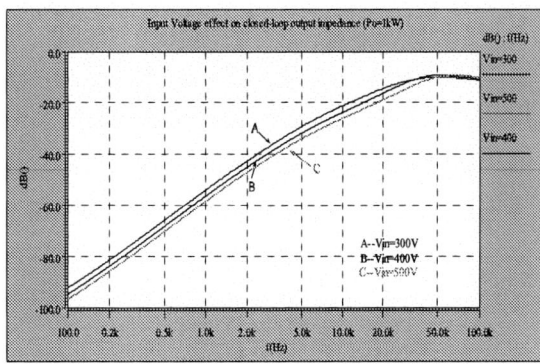

Fig.10. Simulated output impedances with different input voltages

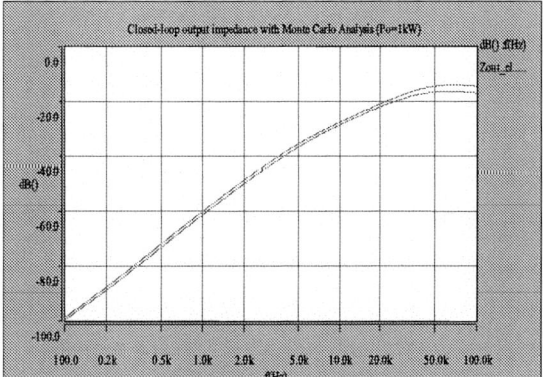

Fig.11. Envelope of output impedance (via Monte Carlo analysis)

D. Practical considerations for deriving the output impedance specifications

Practically, output impedances will be varied as operating points change. When determining the output impedance specification, the output impedance at certain operating point with the worst case magnitude of output impedance should be considered. Further, the characteristic of the output impedance might vary due to the component tolerance. Thus, the output impedance specification should be issued considering the operating region and the component tolerance.

One technique to find out the worst case magnitude of output impedance is to run Monte Carlo analyses by means of SABER. Here, an average model of the Buck converter is created in SABER, with parameters the same as that in section II. Fig. 10 shows the simulated closed-loop output impedance considering the operating region. It can be seen that the worst case output impedance is obtained at the low line condition (V_{in}=300V). For the operating point of V_{in}=300V, perform the Monte Carlo analysis considering filter inductor, filter capacitor and ESR with 30% tolerance. Fig. 11 shows the envelope of output impedance curves obtained by Monte Carlo analysis. From that, the worst case with the highest magnitude of output impedance can be found.

IV. STANDARDIZATION OF INPUT IMPEDANCE SPECIFICATIONS

A dc DPS can be simplified as a cascaded system that

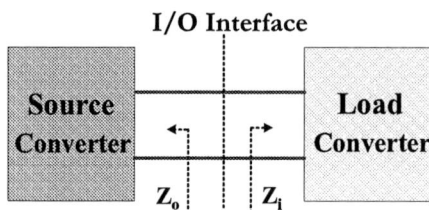

Fig.12. The simplified cascaded power system

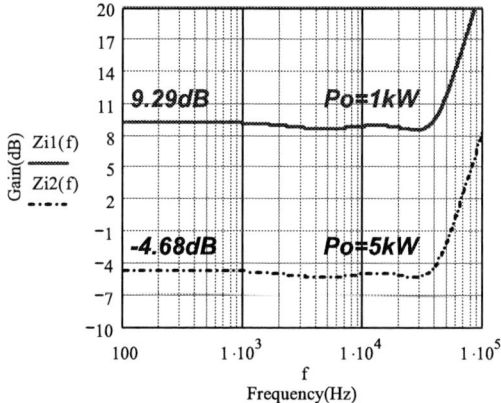

Fig.13. Closed-loop input impedance with different power ratings

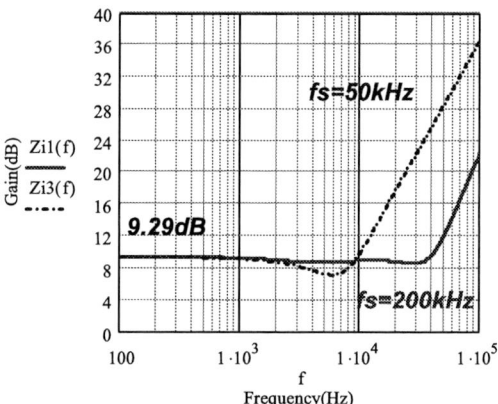

Fig.14. Closed-loop input impedance with different switching frequencies

includes a source converter at the front and a load converter at the end, as illustrated in Fig. 12. Generally, the source converter and the load converter are designed based only on the stability requirement in its stand-alone operation. After system integration, the interaction between two converters may cause system instability. A typical example is the system instability caused by the negative resistive property of a load converter with closed-loop control. The closed-loop controlled converter behaves like a negative resistor. As long as the load converter is powered by an ideal voltage source with very low output impedance, the system is stable. However, in the case that the close loop controlled load converter whose output impedance is larger than the negative resistance in amplitude, the whole system may become unstable.

1940

For checking small-signal system stability of the DPS, Middlebrook established the concept of impedance criterion in 1976. As illustrated in Fig. 12, Z_o and Z_i are the output and input impedances in the I/O interface. The small-signal system stability can be ensured when the curve of Z_o/Z_i does not circle (-1, 0) point on the s-plane. Therefore, to ensure the stability of cascaded systems, not only the output impedance of the source converter should be analyzed, but also the input impedance of the load converter should be checked. Here, the influences of variations of the output power rating and the switching frequency on the closed-loop input impedance are analyzed. Typical buck converters serving as load converters are modeled, with parameters: input voltage V_{in}=54V; output voltage V_o=30V.

A. Output power rating effect

In order to plot the closed-loop input impedance with the power rating as a running parameter, two Buck converters are modeled, with parameters: V_{in}=54V; V_o=30V; $f_{s1}=f_{s2}$=200kHz; P_{o1}=1kW, P_{o2}=5kW. And the design guidelines of the control loop are the same as that of sectionIII. The voltage sensing gain G_{sen} is 0.1.

Fig.13 shows closed-loop input impedances with the power rating as a running parameter. High output power rating gives a low magnitude of input impedance. For load converters which connect to the same DC bus, the module delivering higher output power has lower closed-loop input impedance.

B. Switching frequency effect

Fig.14 shows closed-loop input impedances with the switching frequency as a running parameter. It can be seen that the low frequency gain of two input impedances is the same, because they both have the same input voltage and the output power. The low switching frequency gives a high magnitude of input impedance at relatively high frequencies.

V. CONCLUSIONS

The main contribution by this paper is to select the input/output impedance as objects of standardization from the perspective of the system level. Based on the analysis of characteristics of I/O impedance of Buck converters, output power ratings, switching frequencies and control-modes' effect on the I/O impedance are discussed. The analysis results are beneficial to issuing I/O impedance specifications. It is hoped that this paper may stimulate the researchers' enthusiasm for getting involved in the standardization of power electronics modules.

APPENDIX

This appendix provides the derivation of the calculation of output filter including filter inductor L_f and output capacitor C_f, which are given in the equation of (7) and (8).

When requiring the ripple current of the inductor current is less than 20% of rated load current, the filter inductor can be easily obtained

$$L_f = \frac{V_o^2}{0.2P_{o(nom)}} \frac{1-D}{f_s} \tag{A1}$$

where $P_{o(nom)}$ is nominal output power.

After filter inductor is determined, the ripple current of filter inductor is obtained

$$\Delta I_{L_f} = \frac{V_o(1-D)}{L_f f_s} \tag{A2}$$

Assuming the ripple current of filter inductor is totally assumed by the output filter capacitor, the voltage ripple of the output capacitor can be determined

$$\Delta V_o = \frac{\Delta I_{L_f}}{8C_f f_s} + \Delta I_{L_f} R_c \tag{A3}$$

Substituting (A2) into (A3), the voltage ripple of the output capacitor can be derived

$$\Delta V_o = \frac{(1-D)V_o}{L_f C_f f_s^2} \cdot (\frac{1}{8} + \lambda f_s) \tag{A4}$$

where $\lambda = R_c C_f$.

When requiring that the voltage ripple of output voltage should be limited within 1% of the rated output voltage, the product of filter inductor and filter capacitor can be obtained

$$L_f C_f = \frac{(1-D)}{0.01 f_s^2} \cdot (\frac{1}{8} + \lambda f_s) \tag{A5}$$

Substituting (A1) into (A5), the output filter capacitor can be derived

$$C_f = \frac{20P_{o(nom)}}{V_o^2 f_s} \cdot (\frac{1}{8} + \lambda f_s) \tag{A6}$$

REFERENCES

[1] W.A. Tabisz, M.M.Jovanovic and F.C. Lee, "Present and future of distributed power systems ,'' in Proc. IEEE APEC, 1992, pp.11-18

[2] X.Feng, Z.Ye, K.Xing, F.C.Lee and D.Boroyevich, "Individual load impedance specification for a stable DC distributed power system," in Proc. IEEE APEC, 1999, pp.923-929

[3] X.Feng, Z.Ye, K.Xing, F.C.Lee and D.Boroyevich, "Impedance specification and impedance improvement for DC distributed power system," in Proc. IEEE PESC, 1999, pp.889-894

[4] X.Feng, J. Liu and F.C.Lee, "Impedance specifications for stable DC distributed power systems," IEEE Trans. on Power Electron, Vol.17, NO.2, Mar. 2002

[5] F.C.Lee and D.Peng, "Power electronics building blocks and system integration," in Proc. IEEE IPEMC, 2000, pp.1-8

[6] F.Blaabjerg, A.Consoli, J.A.Ferreira, J.D. van Wyk, "The future of electronic power processing and conversion," IEEE Trans. on Power Electron, Vol.20, NO.3, pp.715-720, May 2005

[7] K.Yao, Y. Meng, P.Xu and F.C.Lee, "Design considerations for VRM transient response based on the output impedance," IEEE Trans. on Power Electron, Vol.18, pp. 1270-1277, Nov. 2003

[8] Abraham I. Pressman, "Switching Power Supply Design", 2nd edition, McGraw-Hill companies published

[9] R.B.Ridley, B.H.Cho and F.C.Lee, "Analysis and interpretation of loop gains of multi-loop controlled switching regulators," IEEE Trans. on Power Electron, Vol.3, pp.489-498, Oct. 1988.

Research on The Magnetic Integration in Three-Level ZCS Quasi-Resonant Buck Converter

Jiang Ying [*], Xiang Hui-jie [**], Yang Yu-gang [*], Liu Nan [*]

[*]Department of Electrical Engineering, Liaoning Technical University, Fuxin 123000, China

[**]Department of Computer and Electrical Engineering, Suzhou Vocational University, Suzhou 215104, China

Abstract—Paper bases on the principles of the coupling inductor to illustrate the planar integration of two resonant inductors and one filter inductor in the ZCS Buck QRC, all three inductors are integrated in one magnetic core, then the decreasing in the volume and height can be achieved, through comparing the close coupling with the loose coupling, how to achieve the soft switching is discussed, and the stable output voltage is guaranteed too. The simulated results of the circuit and the flux line are also included in this paper.

Keywords- magnetic integration；three-level ZCS Buck QRC

I. INTRODUCTION

The current trends for the power electronic converter technology demand the high frequency, small volume and the light weight, and with the development of the micro-electronics' manufactural technology, it is possible to integrate all the active components and all the discrete passive components into one chip[1], and the planar magnetic integration is adopted to decrease the numbers of the discrete inductors and save more volume[2][3]. In order to reduce the voltage stress, the three-level converters which in nature provide more switches to share the voltage stress is widely researched, because of the high voltage the soft switching which need more resonant inductors must be adopted to increase the efficiency[4] [5], this paper take the three-level ZCS Buck convert for example to illustrate the integration for the two resonant inductors and one filter inductor.

II THE OPERATION OF THE THREE-LEVEL ZCS BUCK CONVERTER

The three-level ZCS Buck converter which three inductors is integrated is shown by Fig.2. Parameters are defined as follows: L_{r1} and L_{r2} are the resonant inductors, L_f is the filter inductor, C_{r1} and C_{r2} are the resonant capacitors, C_f is the filter capacitor, M is the mutual inductance, k is the coupling coefficient, V_{in} is the input

voltage, V_0 is the output voltage, u_{Lr} is the voltage of L_r, u_{Lf} is the voltage of L_f, D is the duty cycle .

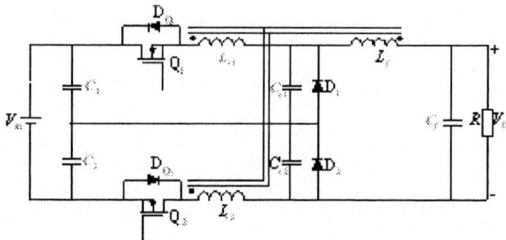

Fig.1 The magnetic integration of the three-level ZCS Buck converter

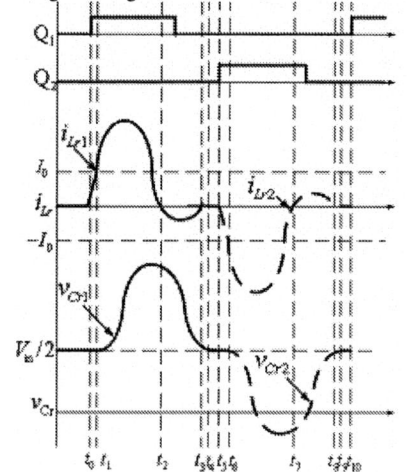

Fig.2 the key waveform of the three-level ZCS Buck converter

The Backward coupling three-level ZCS Buck converter is taken for example to illustrate the magnetic integration and the main circuit is shown as Fig.2 (b).Such assumptions are made: (1) all the switch, diode, inductors and capacitors are ideal; (2) filter inductor L_f more larger than resonant inductor L_{r1} and L_{r2} , and $L_{r1}=L_{r2}=L_r$, $C_{r1}=C_{r2}=C_r$; (3) the capacitor C_f is large enough that it keeps the output voltage V_0 constant.

The operation of the integrated convertor can be divided into ten different intervals for one switching cycle and its key waveform and intervals are shown by the

Fig.2 and Fig.3.

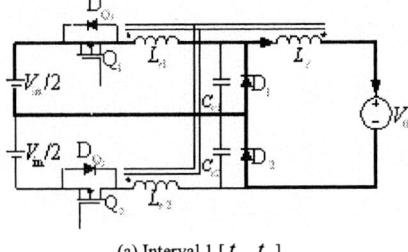

(a) Interval 1 [t_0 , t_1]

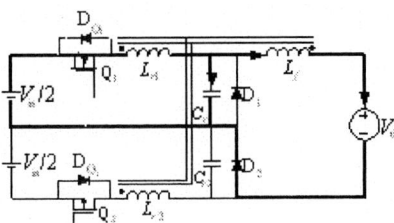

(b)Interval 2 [t_1 , t_2]

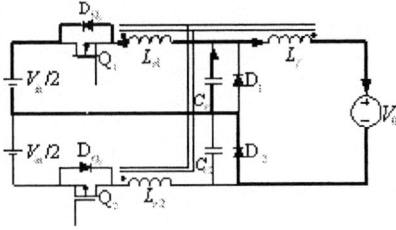

(c)Interval 3[t_2 , t_3]

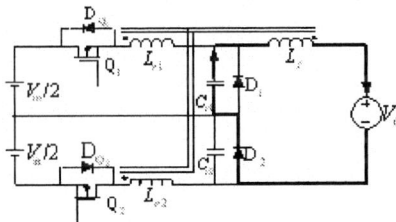

(d)Interval 4[t_3 , t_4]

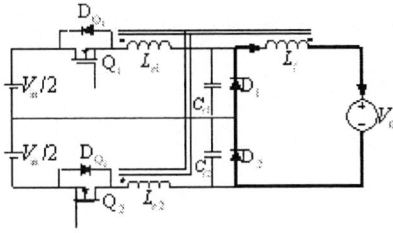

(e) Interval 5 [t_4 , t_5]

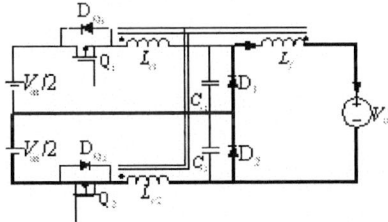

(f) Interval 6 [t_5 , t_6]

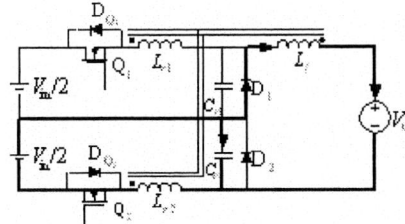

(g) Interval 7 [t_6 , t_7]

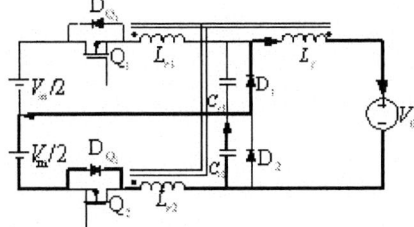

(h) Interval 8 [t_7 , t_8]

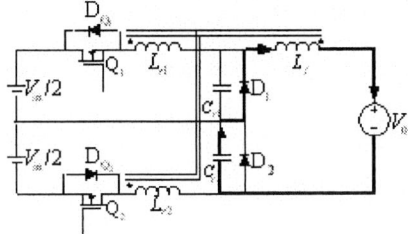

(i) Interval 9 [t_8 , t_9]

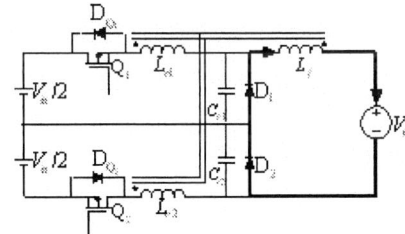

(j)Interval 10 [t_9 , t_{10}]

Fig.3 Equivalent circuit for each stage

Interval 1[t_0 , t_1]:

Because of the clamping by the diode D_1, and the resonant inductor current i_{Lr1} ramps up linearly, the description can be shown below:

$$i_{Lr1}(t) = \frac{v_{in}L_f + v_0 M}{L_{r1}L_f - M^2}(t - t_0) \tag{1}$$

Interval 2[t_1 , t_2]:

L_{r1} and C_{r1} begin to resonate, and because of the coupling between the L_{r1} and the L_f, then the L_f resonates too. The description of the resonant current i_{Lr1} and the resonant voltage V_{Cr1} can be got as follows:

$$\begin{cases} i_{Lr1}(t) = \dfrac{V_{in}/2 - V_0}{Z_1}(t - t_1) + \left(\dfrac{V_{in}/2}{Z_2} + \dfrac{V_0}{Z_3}\right)\sin w(t - t_1) \\ v_{Cr1}(t) = V_{in}/2 - \dfrac{L_r}{Z_1}(V_{in}/2 - V_0) - wL_r\left(\dfrac{V_{in}/2}{Z_2} + \dfrac{V_0}{Z_3}\right)\cos w(t - t_1) \end{cases} \quad (2)$$

where

$$\begin{cases} Z_1 = L_{r1} + L_f - 2M \\ Z_2 = \dfrac{(L_{r1} + L_f - 2M)\sqrt{(L_{r1}L_f C_{r1} - M^2 C_{r1})(L_{r1} + L_f - 2M)}}{(L_{r1} + L_f - 2M)C_{r1} - (L_{r1}L_f C_{r1} - M^2 C_{r1})} \\ Z_3 = \dfrac{(L_{r1} + L_f - 2M)\sqrt{(L_{r1}L_f C_{r1} - M^2 C_{r1})(L_{r1} + L_f - 2M)}}{(L_{r1}L_f C_{r1} - M^2 C_{r1}) - (L_{r1} + L_f - 2M)MC_{r1}} \\ \omega = \sqrt{\dfrac{L_{r1} + L_f - 2M}{L_{r1}L_f C_{r1} - M^2 C_{r1}}} \end{cases} \quad (3)$$

Interval 3 $[t_2, t_3]$:

L_{r1}, C_{r1} and L_f continue to resonate, and its description is the same as above. At t_2, i_{Lr} reaches zero and the anti-parallel diode D_{Q1} turns on, the i_{Lr} goes oppositely, then switch Q_1 can be turned off with zero current. At t_3, i_{Lr1} reaches zero again.

Interval 4 $[t_3, t_4]$:

After t_3, the process of the resonant stops, and the filter inductor current i_{Lf} discharges the C_{r1}, and because $L_f \gg L_r$, i_{Lf} can be regarded as the load I_0, the voltage V_{Cr1} can be derived as follows:

$$v_{Cr1}(t) = v_{Cr1}(t_2) - \frac{I_o}{C_{r1}}(t - t_2) \quad (4)$$

Interval 5 $[t_4, t_5]$:

The filter inductor current i_{Lf} flows through freewheeling diode D_1 until another switch cycle's beginning.

The interval 6-10 is the similar with the interval 1-5, just the Q_1 is substituted with Q_2.

According to

$$k = \frac{M}{\sqrt{L_r L_f}} \quad (5)$$

Then the resonant angular frequency can be described as:

$$\omega = \sqrt{\frac{L_r + L_f - 2kL_r L_f}{L_r L_f C_r - k^2 L_r L_f C_r}} \quad (6)$$

When the coupling is close coupling (the higher value of the coupling coefficient k), the w increases and can't achieve the zero current turning off when the switching cycle is fixed. Therefore, there are two main methods to solve this problem: the switching frequency should be

increased or the coupling coefficient k should be decreased in order to achieve the soft switching.

The operation principle of the forward coupling is similar with that of the backward coupling, when substitute M in the description of the backward coupling for the $-M$, then the description of the forward coupling can be derived

III. THE ANALYSIS FOR THE OUTPUT OF THE THREE-LEVEL ZCS BUCKCONVERTER

In the three-level ZCS Buck converter, there are at most two inductors to operate at the same time, one is the filter inductor and another is the resonant inductor, in order not to influent the operation of the converts, the filter inductor L_f is much larger than the resonant inductor L_r, and its current can be seen as the constant through the switch circle. And the filter inductor connects with the load which is paralleled by filter capacitor C_f which can keep its voltage constant,

$$\begin{cases} u_{Lr} = L_r \dfrac{di_{Lr}}{dt} \pm M \dfrac{di_{Lf}}{dt} \\ u_{Lf} = L_f \dfrac{di_{Lf}}{dt} \pm M \dfrac{di_{Lr}}{dt} \end{cases} \quad (7)$$

where,

$\pm$: forward and backward coupling; i_{Lr}: the current of the resonant inductor; i_{Lf}: the current of the filter inductor; M: mutual inductance.

Because the filter inductor L_f is very large then i_{Lf} can be seen as the constant, then the equation can be abbreviated:

$$\begin{cases} u_{Lr} = L_r \dfrac{di_{Lr}}{dt} \\ u_{Lf} = \pm M \dfrac{di_{Lr}}{dt} \end{cases} \quad (8)$$

Based on the this equation, the voltage of the resonant inductor u_{Lr} is not influenced by the coupling and its equation is the same as the none-coupling's; the voltage of the filter inductor u_{Lf} is influenced by the voltage of mutual inductance which caused by the resonant current i_{Lr}, but filter inductor connects with filter capacitor C_f which can keep the u_{Lf} stable, then coupling don't affect the normal operation of the converts.

IV. THE DESIGN OF MAGNETIC INTEGRATION

In order to attain the magnetic integration of the two resonant inductors and the filter inductor, the coupling inductor should be model precisely, and the two resonant inductors should be decoupled, then the planar **EI** magnetic core is adopted and its model is shown by Fig.4.the middle leg don't has the gap, and both right leg and left leg have gap, then the two resonant inductors can be decoupled, and the deduce of the equivalent circuit is shown by Fig.5.

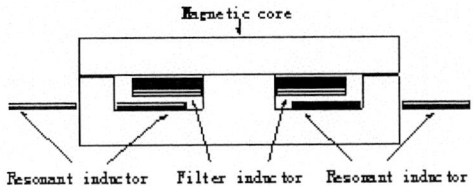

Fig.4 The sketch map of planar magnetic integration

(a) the magnetic circuit

(b) source transfer

(c)decoupling

(d) the equivalent circuit
Fig.5 the deduce of the equivalent circuit

V SIMULATION RESULTS

The resonant current and voltage are simulated by the software Saber, and the parameter values are defined as follows: input voltage V_{in}=20v;output voltage V_0 =10v;output current I_0=2A;filter inductor L_f =100uH; resonant inductor L_r =1.6 uH; filter capacitor L_f =25uF; resonant capacitor C_r=64nF;resonant frequency f_r=500 kHz ;switching frequency f=250 kHz .Fig.5 shows the simulated waveforms that the resonant voltage and current don't change obviously when the coupling is the loose coupling(k=0.3), and the resonant current reaches to

zero before the switch's turning off, then achieve the soft-switching; when the coupling is the close coupling(k =0.9), the resonant frequency increases and the resonant current resonate again after reaching zero, then cant achieve the soft-switching, then the frequency can be adjusted in order to achieve the soft-switching, and when the frequency increases, resonant current reaches zero.

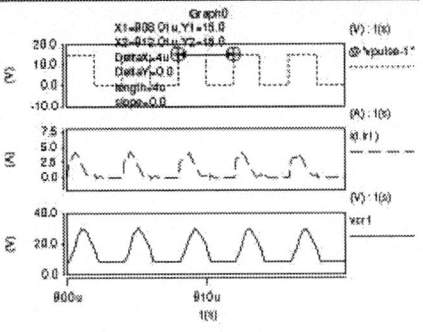

(a) the switch Q_1's waveform of the loose coupling(k =0.3)

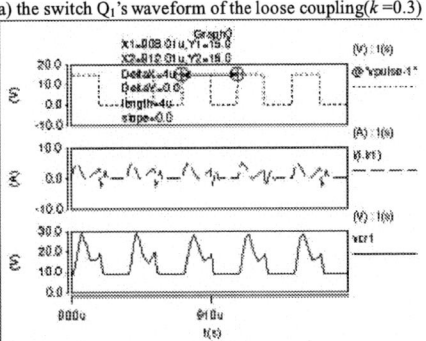

(b) the switch Q_1's waveform of the close coupling(k =0.9)

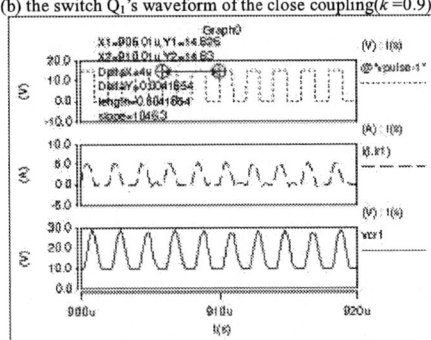

(c) the switch Q_1's waveform of the close coupling when increase frequency

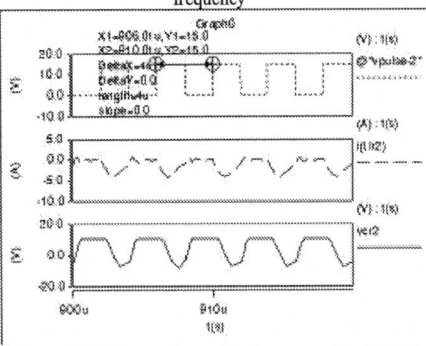

(d) the switch Q_2's waveform of the loose coupling(k =0.3)

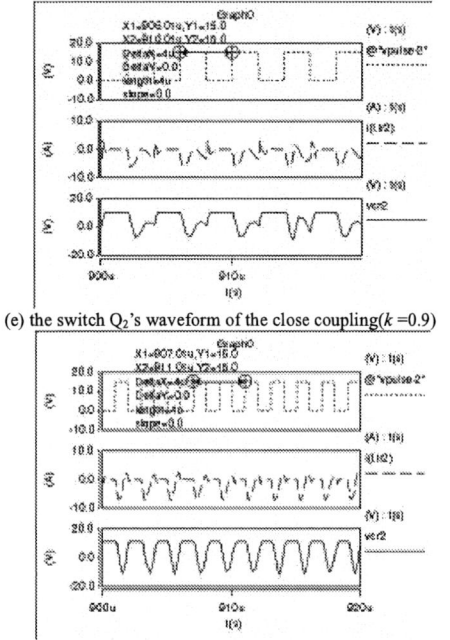

(e) the switch Q_2's waveform of the close coupling(k =0.9)

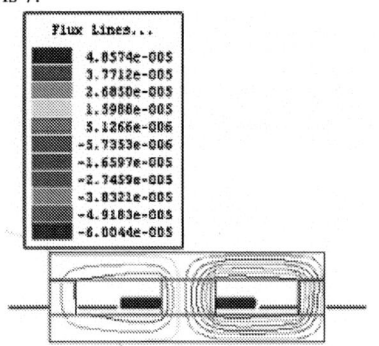

(f) the switch Q_2's waveform of the close coupling when increase frequency

Fig.6　Stimulation waveforms of the coupling

Based on the circuit parameters above, the flux line of the planar magnetic integration is simulated by software Maxwell. And the parameters are defined as follows: both of air gap are 40umm; the width of the conductor is 6mm; the thickness of the conductor is 70um; the dimension of the magnetic core is 32$mm\times$20$mm\times$9mm; the winding turn of resonant inductor is 1; the winding turn of filter inductor is 7.

(a) The flux line when left resonant inductor and filter inductor operates

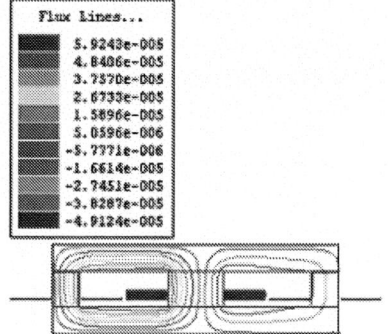

(b) The flux line when right resonant inductor and filter inductor operates

Fig.9 The flux line's distribution

VI CONCLUSION

This paper illustrates the operation of the three-level ZCS Buck converter when three inductors are integrated, and the soft switching can be achieved when the frequency increases or the coupling coefficient decreases. and the model for three inductor's integration is designed, the volume and the height can be decreased by the planar magnetic integration and adapt to the trend of the power elelctronics.

REFERENCES

[1]Daniel A. James; Neil Davey; Leon Gourdeas. A modular integrated platform for micro sensor applications. Conference on Micro electronics: Design, Technology, and Packaging; Dec 10-12,2003; Perth, Australia

[2]Yu-gang Yang: Magnetic Thchology In Modern Power Electrronics, Beijing: Science Techtology Press,2003:207-212

[3]Lingyin Zhao; Johan T.Strydom; J.Daan van Wyk. Design Considerations for an Integrated LC Resonant Module. 2000 CPES Center for Power Electronics Systems Power Electronics Seminar September 17-19, 2000 Blacksburg, VA.

[4]Meynard T A, Foch H. Multi-level conversion: high voltage choppers and voltage-source inverters [A]. IEEE PESC [C].1992,pp.397-403.

[5]Ke Jin, Xinbo Ruan; Zero-voltage-switching Multi-resonant Three-level Converters; Proceedings of the CSEE;2004..pp.156-161

Decoupling Control of Magnetically Levitated Induction Motor with Inverse System Theory

Yang Zhou , Huangqiu Zhu and Tianbo Li
Jiangsu University / School of Electrical and Information Engineering, Zhenjiang 212013, China
zy__88@163.com

Abstract—**A magnetically levitated induction motor is a multivariable, nonlinear and strong coupling system. In order to achieve the rotor suspending and working steadily, it is necessary to realize dynamic decoupling control between torque force and radial suspension forces. In this paper, a method based on inverse system theory is used to study on decoupling control of magnetically levitated induction motors. Firstly, the working principle of radial suspension forces is expounded, and then the state equations of this motor are set up. Secondly, feasibility of decoupling control based on inversion theory for magnetically levitated induction motor is discussed in detail, and the dynamic feedback linearization method of system decoupling and linearizing is used. Finally, linear control system techniques are applied to these linearization subsystems to synthesize and simulate. The simulation results have shown that this kind of control strategy can realize dynamic decoupling control between torque force and radial suspension forces, and the control system has fine dynamic and static performance.**

Keywords-magnetically levitated induction motor; inverse system; dynamic feedback linearization; decoupling control

I. INTRODUCTION

A magnetically levitated induction motor is a multivariable, nonlinear system, and there are couplings between the torque subsystem and flux linkage subsystem, in addition, there are couplings between radial force subsystems themselves. If the motor doesn't be taken some right decoupling control methods, the rotor of motor couldn't be suspended and the motor couldn't work steadily. In this paper, the 2-degree freedom magnetically levitated induction motor is adopted, and the α-th order inverse system method is used to linearize the whole system of the magnetically levitated induction motor, and the dynamic decoupling control algorithm is deduced. Finally, the linearity system theory is used to synthesize and simulate these linearization subsystems.

II. α-TH ORDER INVERSE SYSTEM METHOD

The α-th order inverse system method is an approach of

The project was sponsored by the National Natural Science Foundation of China (50575099), and High Technology Research of Jiangsu Province(BG2005027)

using feedback linearization method to study on the system design theory [1]. The basic idea is: firstly, a α-th order inverse system is constructed, which can be realized by feedback linearization method using the inverse model of system object; secondly the system is transformed to a linear system, namely pseudolinear system; finally, the linearity system theory is used to synthetize the system.

III. DECOUPLING CONTROL OF MAGNETICALLY LEVITATED INDUCTION MOTOR

A. Working Principle of 2-Degree Freedom Magnetically levitated induction motor and Principle of Radial Forces Generation

The stator of magnetically levitated induction motor is wound 2-pole windings and 4-pole windings compoundly. The magnetic field produced by the 2-pole windings and the rotation magnetic field produced by the 4-pole windings affect each other in the air gap. 2-pole windings are called as radial force windings. And the 4-pole windings are called as torque force windings, which produce rotation magnetic field and torque. The electrified torque windings will produce rotation magnetic field when magnetically levitated induction motor work. If the electrified radial windings produce rotation magnetic field and the magnetic field of torque winding satisfy the following three conditions, then the interactional magnetic fields will produce radial suspension forces[2]-[3]: (1) $P_1=P_2\pm1$; (2) The two magnetic fields have the same rotation direction; (3) The currents which produce the magnetic field have the same

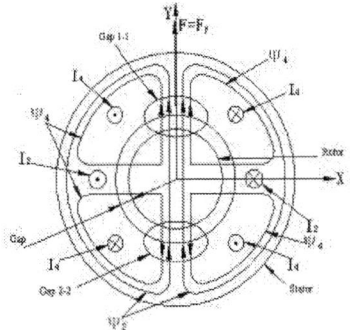

Figure 1. The principle of producing radial suspension forces on magnetically levitated induction motor

frequency.

Fig. 1 shows the working principle of the magnetically levitated induction motor. When 4-pole torque windings and 2-pole radial suspension windings have been electrified by I_4 and I_2 as the figure shown, it will generate the same direction of 4-pole torque flux linkage ψ_4 and 2-pole radial flux linkage ψ_2 in air gap 1-1, and the whole flux linkage will be increased to $\psi_4+\psi_2$, and electromagnetic suction force will also be increased. While in air gap 2-2, ψ_4 and ψ_2 have the opposition direction, and the compound flux linkage will be decreased to $\psi_4-\psi_2$, and electromagnetic suction force will also be decreased, therefore, the rotor will be effected by electromagnetic compound force in the y positive direction. If the direction of current in suspension force windings will be changed, the radial electromagnetic compound force in y opposition direction will be generated. In the same way, electromagnetic compound force will be produced in x direction. So the rotor can be suspended in the balance central position by adjusting the magnitude and direction of the current in radial suspension force windings.

B. Analysis of Suspension Forces on Magnetically levitated induction motor

2-pole radial suspension force windings and 4-pole torque windings are wound compoundly in slots of magnetically levitated induction motor. The electromagnetic couplings of magnetically levitated induction motor are very complex, because there are couplings between the 2-pole windings and 4-pole windings, and there are couplings between the windings themselves [3]. In order to analyse easily, 2-phase windings which has been changed from 3-phase windings in static coordinate through $C_{3/2}$ and $C_{r/s}$ transform is studied. Because the mutual inductance value between 4-pole windings and 2-pole windings is 0, the value of torque windings self-inductance L_{1s} and the value of radial windings self-inductance L_{2s} are constant. The inductance matrix of the motor L can be obtained as [4]

$$ L = \begin{bmatrix} L_{1s} & 0 & -M'\alpha & M'\beta \\ 0 & L_{1s} & M'\beta & M'\alpha \\ -M'\alpha & M'\beta & L_{2s} & 0 \\ M'\beta & M'\alpha & 0 & L_{2s} \end{bmatrix} \tag{1} $$

where α and β are the rotor radial displacement in the direction of x and y. M' is the coefficient of mutual inductance of 4-pole windings and 2-pole windings. The subscript expression s is the component of self-inductance of the stator. According to the relationship of energy conversion, the magnetic energy stored in the windings can be written as

$$ W_m = \frac{1}{2}\boldsymbol{I}^T \boldsymbol{L} \boldsymbol{I} \tag{2} $$

where $\boldsymbol{I} = \begin{bmatrix} i_{d1s} & i_{q1s} & i_{d2s} & i_{q2s} \end{bmatrix}^T$ is current matrix. i_{d1s} and i_{q1s} are the 4-pole windings current component in d-q coordinate,respectively. i_{d2s} and i_{q2s} are the 2-pole windings current component in d-q coordinate, respectively.

Neglecting magnetic saturation, the F_x and F_y are the radial forces in the x- and y-direction, where $F_x=\partial Wm/\partial\alpha$ and $F_y=\partial Wm/\partial\beta$, can be written as

$$ \begin{bmatrix} F_x \\ F_y \end{bmatrix} = \begin{bmatrix} \partial W_m/\partial\alpha \\ \partial W_m/\partial\beta \end{bmatrix} = M'\begin{bmatrix} -i_{d1s} & i_{q1s} \\ i_{q1s} & i_{d1s} \end{bmatrix}\begin{bmatrix} i_{d2s} \\ i_{q2s} \end{bmatrix} \tag{3} $$

C. Electromagnetic Torque of Magnetically levitated induction motor

The electromagnetic forces, which make the rotor suspend and magnetically levitated induction motor operate, are produced by interactional flux linkage of 4-pole windings and 2-pole windings. Because the magnetic field produced by radial force windings is very smaller than the magnetic field produced by torque force windings, neglecting the radial force windings magnetic field, the rotor flux linkage can be satisfied as the following equations

$$ \begin{cases} \dot{\psi}_{dr} = -\dfrac{1}{T_r}\psi_{dr} - \omega_r\psi_{qr} + \dfrac{L_{m1r}}{T_r}i_{d1s} \\ \dot{\psi}_{qr} = -\dfrac{1}{T_r}\psi_{qr} + \omega_r\psi_{d1s} + \dfrac{L_{m1r}}{T_r}i_{q1s} \end{cases} \tag{4} $$

The torque equation for magnetically levitated induction motor is

$$ T_e = p_1\frac{L_{m1r}}{L_r}(\psi_{dr}i_{q1s} - \psi_{qr}i_{d1s}) \tag{5} $$

where ψ_{dr} and ψ_{qr} are the component of rotor flux linkage in d-q coordinate, respectively. ω_r is the speed of the rotor. ψ_{d1s} and ψ_{q1s} are the component of stator torque flux linkage in d-q coordinate,respectively. T_r is the time constant of rotor. p_1 is the pole-pair number of torque windings. L_{m1r} is the mutual inductance between torque windings and rotor.

D. Dynamic Equations of Rotor on Magnetically levitated induction motor

When the magnetically levitated induction motor work, the rotor is effected by the radial forces and the outside forces, the motion equations are as follows [5]

$$ \begin{cases} m\ddot{x} = f_x - F_x \\ m\ddot{y} = f_y - F_y \\ \dfrac{J}{P_1}\dot{\omega}_r = T_e - T_L \end{cases} \tag{6} $$

where m is the mass of the rotor. $f_x = k_s \cdot x$ and $f_y = k_s \cdot y$ are the magnetic tensile forces in the x- and y-direction, respectively. J is the moment of inertia. ω_r is the speed of the rotor. T_e and T_L are the electromagnetic torque and the load torque, respectively. k_s is displacement coefficient.

E. State Equations of Magnetically levitated induction motor

Substituting (3) and (5) into (6), (6) can be written as

$$
\begin{cases}
m\ddot{x} - M'(i_{d1s}i_{d2s} - i_{q1s}i_{q2s}) = f_x \\
m\ddot{y} + M'(i_{q1s}i_{d2s} + i_{d1s}i_{q2s}) = f_y \\
\dfrac{J}{P_1}\dot{\omega}_r = p_1 \dfrac{L_{m1r}}{L_r}(\psi_{dr}i_{q1s} - \psi_{qr}i_{d1s}) - T_L
\end{cases}
\tag{7}
$$

From (4) and (7), we can obtain (8)

$$
\begin{cases}
\dot{\psi}_{dr} = -\dfrac{1}{T_r}\psi_{dr} - \omega_r\psi_{q1s} + \dfrac{L_{m1r}}{T_r}i_{d1s} \\
\dot{\psi}_{qr} = -\dfrac{1}{T_r}\psi_{qr} + \omega_r\psi_{d1s} + \dfrac{L_{m1r}}{T_r}i_{q1s} \\
m\ddot{x} - M'(i_{d1s}i_{d2s} - i_{q1s}i_{q2s}) = f_x \\
m\ddot{y} + M'(i_{q1s}i_{d2s} + i_{d1s}i_{q2s}) = f_y \\
\dfrac{J}{P_1}\dot{\omega}_r = p_1 \dfrac{L_{m1r}}{L_r}(\psi_{dr}i_{q1s} - \psi_{qr}i_{d1s}) - T_L
\end{cases}
\tag{8}
$$

State variables are chosen as

$$
X = \begin{bmatrix} x_1 \\ x_2 \\ x_3 \\ x_4 \\ x_5 \\ x_6 \\ x_7 \end{bmatrix} = \begin{bmatrix} x \\ y \\ \dot{x} \\ \dot{y} \\ \omega_r \\ \psi_{dr} \\ \psi_{qr} \end{bmatrix}
\tag{9}
$$

Input variables are chosen as

$$
U = \begin{bmatrix} u_1 \\ u_2 \\ u_3 \\ u_4 \end{bmatrix} = \begin{bmatrix} i_{d1s} \\ i_{q1s} \\ i_{d2s} \\ i_{q2s} \end{bmatrix}
\tag{10}
$$

Output variables are chosen as

$$
Y = \begin{bmatrix} y_1 \\ y_2 \\ y_3 \\ y_4 \end{bmatrix} = \begin{bmatrix} x \\ y \\ \omega_r \\ \psi_r \end{bmatrix}
\tag{11}
$$

Substituting (9), (10) and (11) into (8), the state equation of the system is written as

$$
\begin{cases}
\dot{x}_1 = x_3 \\
\dot{x}_2 = x_4 \\
\dot{x}_3 = \dfrac{M'}{m}(u_1 u_3 - u_2 u_4) + \dfrac{1}{m}f_x \\
\dot{x}_4 = -\dfrac{M'}{m}(u_1 u_4 + u_2 u_3) + \dfrac{1}{m}f_y \\
\dot{x}_5 = \dfrac{P_1^2 L_{m1r}}{JL_r}(x_6 u_2 - x_7 u_1) - \dfrac{P_1}{J}T_L \\
\dot{x}_6 = -\dfrac{1}{T_r}x_6 - x_5 x_7 + \dfrac{L_{m1r}}{T_r}u_1 \\
\dot{x}_7 = -\dfrac{1}{T_r}x_7 + x_5 x_6 + \dfrac{L_{m1r}}{T_r}u_2
\end{cases}
\tag{12}
$$

Output equation is written as

$$
Y = \begin{bmatrix} y_1 \\ y_2 \\ y_3 \\ y_4 \end{bmatrix} = \begin{bmatrix} x_1 \\ x_2 \\ x_5 \\ \sqrt{x_6{}^2 + x_7{}^2} \end{bmatrix}
\tag{13}
$$

It can be seen from (9)~(13) the magnetically levitated induction motor is a 4-input, 4-output nonlinear, strong coupling system. Fig. 2 shows the state equation structure diagram of magnetically levitated induction motor.

F. Analysis of Decoupling Control

From the state equations of the magnetically levitated induction motor, we can calculate the following parameters: $\dfrac{\partial}{\partial U}\left[y_1^{(2)}\right]$, $\dfrac{\partial}{\partial U}\left[y_2^{(2)}\right]$, $\dfrac{\partial}{\partial U}\left[y_1^{(1)}\right]$, $\dfrac{\partial}{\partial U}\left[y_2^{(1)}\right]$,

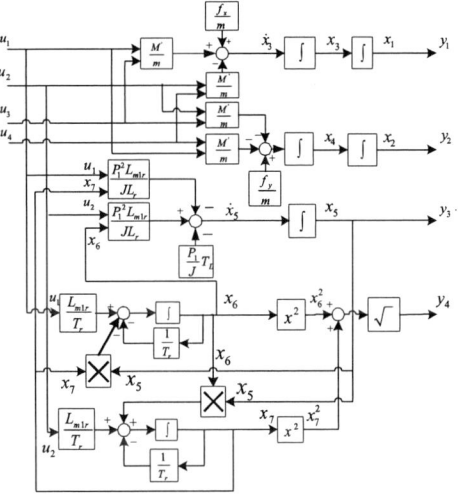

Figure 2. The state equation structure diagram of magnetically levitated induction motor

$\frac{\partial}{\partial U}\left[y_3^{(1)}\right]$, $\frac{\partial}{\partial U}\left[y_4^{(1)}\right]$, $\frac{\partial}{\partial U}[y_1]$, $\frac{\partial}{\partial U}[y_2]$, $\frac{\partial}{\partial U}[y_3]$, $\frac{\partial}{\partial U}[y_4]$, so the relative orders of the system are as follows

$$\alpha = \left(\alpha_1, \alpha_2, \alpha_3, \alpha_4\right) = \left(2, 2, 1, 1\right)$$

It is easy to gain that $\sum_1^4 \alpha_i = 6$. And the order of state equation of the system is 7, so the system is invertible according to theorem [1]. The dynamic feedback linearization method are adopted, now suppose

$$\begin{cases} u_1 u_3 - u_2 u_4 = \phi_1 \\ u_1 u_4 + u_2 u_3 = -\phi_2 \\ x_6 u_2 - x_7 u_1 = \phi_3 \\ x_6 u_1 + x_7 u_2 = \psi_r \phi_4 \end{cases} \quad (14)$$

From (14), the formulas of state feedback arithmetic is as follows

$$\begin{cases} u_1 = -\dfrac{x_7}{\psi_r^2}\phi_3 + \dfrac{x_6}{\psi_r}\phi_4 \\[2mm] u_2 = \dfrac{x_6}{\psi_r^2}\phi_3 + \dfrac{x_7}{\psi_r}\phi_4 \\[2mm] u_3 = \dfrac{u_1}{u_1^2 + u_2^2}\phi_1 - \dfrac{u_2}{u_1^2 + u_2^2}\phi_2 \\[2mm] u_4 = -\dfrac{u_2}{u_1^2 + u_2^2}\phi_1 - \dfrac{u_1}{u_1^2 + u_2^2}\phi_2 \end{cases} \quad (15)$$

Substituting (14) into (12), and from (13), the formulas can be obtained as follows

$$\begin{cases} \ddot{y}_1 = \dot{x}_3 = \dfrac{M'}{m}\phi_1 + \dfrac{1}{m}f_x \\[2mm] \ddot{y}_2 = \dot{x}_4 = \dfrac{M'}{m}\phi_2 + \dfrac{1}{m}f_y \\[2mm] \dot{y}_3 = \dfrac{P_1^2 L_{m1r}}{JL_r}\phi_3 - \dfrac{P_1}{J}T_L \\[2mm] \dot{y}_4 = \dot{\psi}_r = -\dfrac{1}{T_r}\psi_r + \dfrac{L_{m1r}}{T_r}\phi_4 \end{cases} \quad (16)$$

We can see from (16), the system has been turned into linear system and without couplings, as description in (17). Substituting $f_x = k_s \cdot x$ and $f_y = k_s \cdot y$ into (16), from (11), (12) and (13), we can obtain the formula as follows

$$\begin{cases} \ddot{x} = \dfrac{M'}{m}\phi_1 + \dfrac{k_s}{m}x \\[2mm] \ddot{y} = \dfrac{M'}{m}\phi_2 + \dfrac{k_s}{m}y \\[2mm] \dot{\omega}_r = \dfrac{P_1^2 L_{m1r}}{JL_r}\phi_3 - \dfrac{P_1}{J}T_L \\[2mm] \dot{\psi}_r = -\dfrac{1}{T_r}\psi_r + \dfrac{L_{m1r}}{T_r}\phi_4 \end{cases} \quad (17)$$

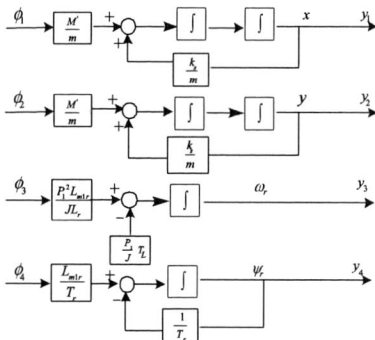

Figure 3. The state equation structure diagram of magnetically levitated induction motor after decoupling control

The state equation structure diagram of magnetically levitated induction motor after decoupling control is shown in Fig. 3.

IV. SYNTHETIZING SYSTEM

A. Synthetizing Position of Rotor System

The normalized linear system described in (17) can be synthesized using the linear system theory. The former two rows of (17) are the displacement subsystems of the rotor of magnetically levitated induction motor, which belongs to the second-order integral system. For example, the transfer function of the displacement system of the rotor is as follows

$$G_k(s) = x(s)/\phi_1(s) = M'/m \cdot s^2 \quad (18)$$

The characteristic equation of the system is as follows

$$s^2 + 2\xi\omega_n s + \omega_n^2 = 0 \quad (19)$$

The parameters ω_n and ξ are chosen as $\omega_n = 800$ rad/s, $\xi = \sqrt{2}/2$, the transfer function of state feedback is as follows

$$a_0 s + a_1 = 2\xi\omega_n m/M' \cdot s + \omega_n^2 m/M' \quad (20)$$

The closed loop transfer function of the system can be obtained as follows

$$G(s) = \frac{6.4 \times 10^5}{s^2 + 1132s + 6.4 \times 10^5} \quad (21)$$

The overshoot of the system is $\sigma = e^{-\frac{\xi}{\sqrt{1-\xi^2}}\pi} = 4.3\%$, the adjusting time is $t_s = 4/\xi\omega_n = 7.06$ ms.

B. Synthetizing Speed System

The third row of (17) is the subsystem of the speed ω_r, which belongs to the first-order integral system. The transfer function of the speed subsystem can be chosen

$$G_k(s) = \frac{\omega_r(s)}{\phi_3(s)} = \frac{P_1^2 L_{m1r}}{JL_r} \cdot \frac{1}{s} \quad (22)$$

1950

The speed adjuster can be choosen as PI adjuster. The transfer function of the system is

$$G_c(s) = \frac{k_1(\tau s + 1)}{\tau s} \tag{23}$$

According to requirement of the design adjuster theory, $G_c(s)$ can be choosen as follows

$$G_c(s) = \frac{2JL_r(\tau s + 1)}{P_1^2 L_{m1r} \tau^2 s} \tag{24}$$

The closed loop transfer function of the rotate speed system is as follows

$$\Phi(s) = \frac{2\tau^{-2}(\tau s + 1)}{s^2 + 2\tau^{-1}s + 2\tau^{-2}} \tag{25}$$

V. SYSTEM SIMULATION

The control strategy can be verified by simulation using the parameters of the designed prototype machine. The parameters of the system are as follows: The stator inductance L_s is 16.31×10^{-2} H; The rotor inductance L_r is 16.778×10^{-2} H; The mutual inductance between stator and rotor L_{m1r} is 15.856×10^{-2} H; The mutual inductance coefficient between stator torque windings and radial force windings M' is 78.2 H/m; The rotor resistance r is 11.48 Ω; The time constant of the rotor T_r is 1.46×10^{-2} s; The quality of rotor m is 2.85 kg; The moment of inertia J is 0.00769 kg·m^2; The pole pairs of torque windings P_1 is 2; The pole pairs of suspension windings P_2 is 3; We can obtain the feedback parameters of radial forces system are: $a_0 = 2\xi\omega_n m/M' = 41.23$, $a_1 = \omega_n^2 m/M' = 23324.81$ and the adjust parameter of torque system, from (23) and (24),

$k_1 = \frac{2JL_r}{P_1^2 L_{m1r}\tau} = 0.041$, where τ is 0.1.

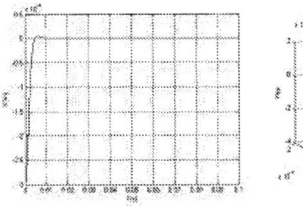

(a) Start up displacement curve in the *x*-direction

(b) The trajectory of the mass center of the rotor

Figure 4. Simulation curves

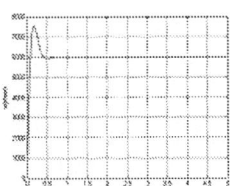

Figure 5. Performance curve of speed subsystem of the magnetically levitated induction motor

From (25) the closed loop transfer function is

$$\Phi(s) = \frac{2\tau^{-2}(\tau s + 1)}{s^2 + 2\tau^{-1}s + 2\tau^{-2}} = \frac{20s + 200}{s^2 + 20s + 200} \tag{26}$$

A. Process of Rotor Rising

When the initialization of x is -0.3 mm, the displacement curve starting up in *x*-direction is shown in Fig. 4 (a). The simulation results have shown that the steady-state error of system approach to 0, the overshoot of system is very small and adjusting time is approach to 0.01 s. When the initialization of x is -0.3 mm and y is -0.4 mm, the trajectory of mass center of rotor is shown in Fig. 4 (b).The rotor position subsystem of decoupling control for magnetically levitated induction motor has fine dynamic and static performance.

B. System of Speed

The response curve of the speed subsystem of magnetically levitated induction motor is shown in Fig. 5. The expectation speed is 6 000 r/min, and the simulation results have shown that the overshoot of the system is less than 5% and the adjusting time is less than 0.5 s, so the speed subsystem has fine performance.

VI. CONCLUSIONS

In the paper, the decoupling control arithmetic adopting α-th order inverse system theory has been educed. From research results, this strategy is succeed in realizing dynamic decoupling control between the radial displacement subsystems and speed subsystem of the magnetically levitated induction motor. The torque and radial suspension subsystems of the magnetically levitated induction motor can be controlled independently and the whole control system has fine dynamic and static performance.

REFERENCES

[1] C. Li and Y. Feng, *The Inverse System Method for Multi-variable Nonlinear Control*, Beijing: Tsinghua University Press, 1991.

[2] X. Liu, Y. Sun, H. Zhu, and et al, "Development, application and prospect of bearingless permanent magnet-type motors," *China Mechanical Engineering*, vol. 15, pp. 1594-1597, September 2004.

[3] Y. He, H. Nian, and B. Ruan, "Optimized air-gap-flux oriented control of an induction-type bearingless motor," *Proceedings of the CSEE*, vol.24, pp. 116-121, March 2004.

[4] A. Chiba, D. T. Power, and M. A. Rahman, "Analysis of no-load characteristics of a bearingless induction motor," *IEEE Trans. on Ind. Appl.*, vol. 31, pp. 77-83, January 1995.

[5] X. Dai, X. Zhang, G. Liu, and et al, "Decoupling control of induction motor based on neural networks inverse," *Proceedings of the CSEE*, vol. 24, pp. 112-117, January 2004.

Fault Detection and Accommodation for Nonlinear Systems Using Fuzzy Neural Networks

H. Xue and J.G. Jiang

Department of Electrical Engineering, Shanghai Jiao Tong University, Shanghai, China
jiang@sjtu.edu.cn

Abstract—**A fault detection and accommodation method based on fuzzy neural networks was presented for nonlinear systems. The fault parameters was designed to detect the fault, adaptive updating method was introduced to estimate and tracking fault, fuzzy neural networks was used to adjust the fault parameters and construct automated fault diagnosis, and the fault compensation control force, which given by fault estimation, was used to realize fault accommodation. This framework leaded to a simple structure and an accurate detection. The simulation results in brushless DC motor showed that it was still able to work well with high dynamic performance and control precision under the condition of motor parameters' variation fault and load torque disturbance.**

Keywords-fuzzy neural network; nonlinear system; fault detection; fault accommodation; adaptive

I. INTRODUCTION

Fault detection and accommodation are becoming more and more important in modern engineering problems. A system failure often causes changes in critical system parameters, or even, changes in the dynamic characters of the system. With the development of the artificial intelligence, fuzzy logic [1] and the method based on neural networks [2] has been introduced into fault detection and accommodation. Both of the two methods have their prominence merits, but they also contain some limits. Fuzzy logic allows for a more natural integration of human knowledge into the fault diagnosis process, but the disadvantage is that if there are too many rules, it becomes a difficult and time-consuming process to get the expert knowledge and obtain the results, and what's more, the standard fuzzy logic method has no capability of self-learning. Neural network has a strong ability of simulating and modeling the nonlinear objects, and can deal with the noises or corrupted data, so neural network is generally regarded as a ideal tool for generating residuals which often used in fault detection [3]. But the main drawback of neural network method is its 'black box' nature, for which it can hardly utilize expert experiences to analyze and depict the fault visually. Combining the advantages of both methods, FNNS (Fuzzy-neural network system) is proposed, which has the adaptation ability of neural

network and the linguistic representation ability of fuzzy logic, so it can represent knowledge in an explicit way, such as rules.

Consequently, Q.S. Zhang [4] introduced the FNNS into fault identification of complex uncertainty system, Jarrah [5-6] proposed to use the identification capability of the fuzzy-neural network for fault detection of non-linear systems, and Polycarpou [7] have designed a general framework for an automatic fault accommodation control using online approximator with fuzzy and neural theory. But the fault characteristics were variable and even the same fault had different representations, so if only depended on the fault diagnosis ability of fuzzy neural network, the incorrect fault detection and missing fault detection may not be avoided, and it would be more difficult to implement fault accommodation.

Based on RBF fuzzy neural network, this note proposed a novel adaptive fault detection method which introduced the adaptive adjustment mechanism for update the fault parameters at real time. The RBF fuzzy neural network was used to get parameters' weights and revise the fault estimation. With the accuracy fault estimation given by RBF fuzzy neural network, compensation controller was designed to implement fault accommodation. Comparing with the prevenient methods, the novel strategy had a simpler structure, and also had a good performance for fault detection and accommodation. The simulation results demonstrated that this method could offer fault detection and precisely tracking for given faults. With the fault accommodation controller, the system still kept stable.

The note was organized as follows. In section II, the structure of the fault system and state observer was given. Section III was devoted to the description of the fuzzy neural network structure, and the section IV interpreted the adaptive mechanism for fault parameter identification. In section V, the proposed fault compensation control was described, and a simulation example was given in Section VI. Finally, some concluding remarks were drawn in Section VII.

II. SYSTEM DESCRIPTION

Consider a nonlinear system described by state-space model of the form

$$\dot{x}(t) = m(x(t), u(t)) + Df(x(t), u(t), \theta)$$
$$y(t) = Cx(t) \qquad (1)$$

---

[a] 973 project *(2005CB221505)*, SRFDP *(20050248058)*.

Where $x(t) \in \Re^n$ denotes the state vector; $u(t) \in \Re^q$ is the vector of measured control input signals; $y \in \Re^p$ is the vector of plant output signals; $m(\cdot)$ is a known smooth nonlinear function, and satisfies $\partial m(x,u)/\partial u \neq 0$; D is system faults directions matrices; $f(\cdot)$ is the failure vector which represents the change in the system due to the presence of unknown failure; $\theta \in \Re^m$ (assume m is known, and satisfy $m \leq n$) denotes the fault parameter, and when $\theta \to 0$, we have $f \to 0$; C is known time-invariant matrices of appropriate sizes.

An observer model can be obtained based on the mathematical model and on the estimate of the fault. The state observer can be formulated as

$$\dot{\hat{x}}(t) = m(x(t),u(t)) + Df(x(t),u(t),\hat{\theta}) + \\ F(\hat{x}(t)-x(t)) + \lambda(t,u) \quad .(2)$$

Where $\hat{x}$ is the estimate state vector; $f(x(t),u(t),\hat{\theta})$ is the estimate of the fault; $\hat{\theta}$ is a parameter vector of the estimate fault; F is positive constant value; $\lambda(t,u)$ is fault compensate control value for fault accommodation.

Assume the target of the control system is to track the ideal output $\tilde{y}(t)$, and the control value without fault is described as

$$u(t) = m^{-1}(C^{-1}\tilde{y}(t), C^{-1}\dot{\tilde{y}}(t)).$$

With (1) and (2), we define the state estimate error as

$$e(t) = \hat{x}(t) - x(t)$$

So it can be obtained that

$$\dot{e}(t) = D(f(x(t),u(t),\hat{\theta}) - f(x(t),u(t),\theta)) \\ - Fe(t) + \lambda(t,u) \quad .(3)$$

III. DESIGN OF FUZZY NEURAL NETWORK FOR FAULT DETECTION

Fault detection is accomplished by fault estimation using adaptive algorithm and parameters emendation given by fuzzy neural network. Generally, the fault is a function of state x and control value u. Fuzzy logic system classifies the system state and input by expert experience, and adaptive algorithm update the fault parameter in real time, we try to use the fuzzy neural network to modify the fault parameter to realize the fault estimation precisely. The structure of RBF fuzzy neural network which consists of five layers was given in Figure 1.

The first layer is input layer. we define the input vector of fuzzy neural network as $X_i = [x_1, x_2, \cdots, x_m, u]^T$, and the output of the node of the first layer is O_i^1.

The second layer is called fuzzification layer. We were choosing Gauss function as membership function

$$\mu(O_i^1) = \exp(-\frac{(O_i^1 - c_i)^2}{\sigma_i^2}), i = 1,2,\cdots,n+1.$$

Where c_i denotes the center of basis function in the input signal space, and we obtain the values by RBF self-adjust law [8]; σ_i^2 denotes the receptive field widths of the neuron given by expert experience, and we define that the

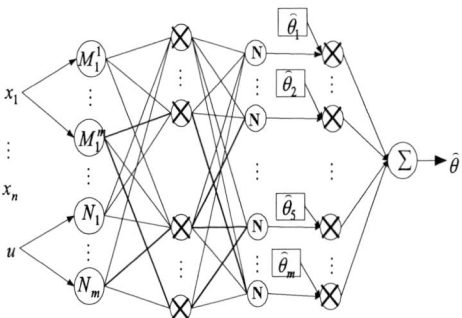

Figure 1. RBF fuzzy neural network model for estimation

output of the second layer as $O_i^2 = \mu(O_i^1)$, and choose the subspace of the output as an equally dissymmetrical space $\{1,2,\cdots,m-1\}$, so the elements of the aggregate is used to partition into m fuzzy subset.

The third layer is rules layer, using the given fuzzy rules-$IF-THEN$, we define the m rules for the output of the second layer as follows,

$RULE\ i$:

$IF\ O_1^2\ is\ M_1^i\ \cdots\ and\ O_n^2\ is\ M_n^i\ and\ u\ is\ N_i$, $THEN\ \ O_i^3 = \overline{w}_i$.

And the output of the third layer is weights of the fault parameters.

The fourth layer is named as parameters adjustment layer. In this layer, the fault parameters of m dimension was adjusted by weights calculation,

$$O_i^4 = \hat{\theta}_i = g(z(\Delta w_i^4, \overline{w}_i), \overline{\theta}_i), i = 1,2,\cdots,m.$$

Where, we define $z(\Delta w_i^4, \overline{w}_i) = 1/(1+e^{-\Delta w_i^5 \overline{w}_i}) \in [0,1)$ as activate function of neural network; and we have

$$g(z(\Delta w_i^4, \overline{w}_i), \overline{\theta}_i) = \eta_\theta \frac{z(\Delta w_i^4, \overline{w}_i)}{\sum\limits_{i=1}^{m} z(\Delta w_i^4, \overline{w}_i)} \cdot \overline{\theta}_i.$$

Where η_θ is fuzzification factor, $\hat{\theta}_i$ ($i=1,2,\cdots,m$) denotes the fault parameters estimation by the node of the forth layer, and $\overline{\theta}_i$ is calculated by adaptive algorithm which given in (6).

The fifth layer is called defuzzification layer. For amendatory fault parameters, we make a information-integration, and given

$$O^5 = \hat{\theta} = \sum_{i=1}^{m} w_i^5 \hat{\theta}_i.$$

In order to cope with model uncertainties and noise input in nonlinear systems, with the definition of the fault parameter and the condition that norm bounded of output of neural network, it can be obtain that there exit a threshold $\beta > 0$, and

$\left\| \hat{\theta} \right\| < \beta$, fault free working mode (non-detectable fault)

$\left\| \hat{\theta} \right\| \geq \beta$, a fault is present

Then, it is known that by comparing the output of the fuzzy neural network and the threshold, the fault can be detected.

We proposed a gradient method for updating the weights of RBF fuzzy neural network. Given l training samples, after making the system discretely, for n steps sampling, we define the performance criteria,

$$J = \frac{1}{2l} \sum_{i=1}^{l} \sum_{j=1}^{n} (\widetilde{\theta}_{ij}(X_i) - \hat{\theta}_{ij}(X_i))^2 .$$

Where $\hat{\theta}_{ij}(X_i)$ is the real output of the system, and $\widetilde{\theta}_{ij}(X_i)$ is the ideal prescribed output.

Given the fourth and fifth weight is adjustable, the weights algorithm is as follows

$$\Delta w_i^4 = -\eta_1 \frac{\partial J}{\partial w_i^4}$$

$$= -\eta_1 \frac{\partial J}{\partial \hat{\theta}_{ij}(X_i)} \cdot \frac{\partial \hat{\theta}_{ij}(X_i)}{\partial g(\overline{w}_i, \hat{\theta}_i)} \cdot \frac{\partial g(\overline{w}_i, \hat{\theta}_i)}{\partial w_i^4}$$

$$= \eta_1 \sum_{k=1}^{l} \sum_{j=1}^{n} \frac{2}{2l} (\widetilde{\theta}_{kj}(X_i) - \hat{\theta}_{kj}(X_i)) \cdot$$

$$\frac{\sum_{j=1}^{i-1} z(\Delta w_j^4, \overline{w}_j) + \sum_{j=i+1}^{m} z(\Delta w_j^4, \overline{w}_j)}{(\sum_{j=1}^{m} z(\Delta w_j^4, \overline{w}_j))^2} \cdot$$

$$w_i^5 \eta_\theta \overline{\theta}_i \cdot z' \cdot \overline{w}_i \qquad ,$$

$$\Delta w_i^5 = -\eta_2 \frac{\partial J}{\partial w_i^5}$$

$$= \eta_2 \sum_{k=1}^{l} \sum_{j=1}^{n} \frac{2}{2l} (\widetilde{\theta}_{kj}(X_i) - \hat{\theta}_{kj}(X_i)) \hat{\theta}_i .$$

Where η_1 , η_2 are leaning rate.

For making the neural network convergence faster, we chose the suitable adjust factor ρ , and we get

$$w_i^4 = w_{i-1}^4 + \Delta w_i^4 + \rho(w_i^4 - w_{i-1}^4) ,$$

$$w_i^5 = w_{i-1}^5 + \Delta w_i^5 + \rho(w_i^5 - w_{i-1}^5) .$$

Then,

$$w_i^4 = \frac{1}{1-\rho} (w_{i-1}^4 - \rho w_{i-1}^4 + \Delta w_i^4), \quad (4)$$

$$w_i^5 = \frac{1}{1-\rho} (w_{i-1}^5 - \rho w_{i-1}^5 + \Delta w_i^5). \quad (5)$$

The note proposed a novel structure of RBF fuzzy neural network, the neurons are adaptive adjustment, and we only need to regulate the weights of the fourth layer and fifth layer, so it can be easy to calculated and ratiocinated.

IV. FAULT PARAMETER ADAPTIVE UPDATING ALGORITHM

Using the state estimation by the observer, the fault parameter adaptive updating algorithm is given as

$$\dot{\hat{\theta}}(t) = \hat{\theta}(t) + Z^T(t, \hat{\theta}(t)) \Gamma_\theta(t)(\hat{x}(t) - x(t)) . \quad (6)$$

with

$$Z(t, \hat{\theta}(t)) = \frac{\partial \eta(t, \hat{\theta}(t))}{\partial \hat{\theta}(t)}$$

$$\dot{\eta}(t, \hat{\theta}(t)) = (I - \Gamma_\eta)\eta(t, \hat{\theta}(t)) + \Gamma_\eta(\dot{e}(t) + Fe(t)) - H(\hat{x}(t) - x(t))$$

Where Γ_η is chosen such that $I - \Gamma_\eta$ has all its eigenvalues inside the unit circle; $H \in \Re^{m \times n}$ is a chosen constant matrix; the gain matrix $\Gamma_\theta(t)$ is chosen as follows,

$$\Gamma_\theta(t) = \hbar(Z(t, \hat{\theta}(t))Z^T(t, \hat{\theta}(t)) + Q)^{-1} . \quad (7)$$

Where $\hbar$ is a positive constant, Q is a positive definite symmetric matrix.

This adaptive mechanism is similar to that used by Caccavale and Villani [9] except that we added a projection operator $He(t)$ to ensure that the fault parameter will converge to real value faster.

V. FAULT COMPENSATION CONTROL DESIGN

For fault accommodation, we design the fault compensation control $\lambda(t, u)$. It has been proved that the RBF fuzzy neural network can approximate arbitrarily closely any non-linear non-singular function [10]. Defining the minimum error of RBF fuzzy Neural network approximating to fault parameter,

$$\overline{\varepsilon} = \theta^* - \hat{\theta} .$$

Where, θ^* is the ideal fault estimation parameter, and defining the fault estimation error as

$$\varepsilon = f(x(t), u(t), \theta^*) - f(x(t), u(t), \theta) .$$

Assuming that when the fault parameter is chosen as θ^* , then $|\varepsilon| < \delta$.

The Taylor series expansion for $f(x(t), u(t), \hat{\theta})$ at ideal weights as

1954

$$f(x(t), u(t), \hat{\theta}) = f(x(t), u(t), \theta^*) + \\ f_\theta'(\theta^* - \hat{\theta}) + O(\bar{\varepsilon}) \qquad (8)$$

For RBF fuzzy neural network approximating characteristic, the higher-order terms in the Taylor series are bounded by

$$\| O(\bar{\varepsilon}) \| \le c_1 + c_2 \| \bar{\varepsilon} \|.$$

Where c_1, c_2 is known constant.

From (3), it can be obtained that

$$\dot{e}(t) = -Fe(t) + D(f(x(t), u(t), \hat{\theta}) - \\ f(x(t), u(t), \theta)) + \lambda(t, u(t)) \qquad (9)$$

Define the Lyapunov function candidate

$$L = \frac{1}{2} e(t) P e^T(t) + \frac{1}{2\gamma} (\theta^* - \hat{\theta})^T (\theta^* - \hat{\theta}).$$

Where P is a symmetric positive definite matrix; γ is a positive constant.

Differentiating yields

$$\dot{L} = e^T(t)\dot{e}(t) + \frac{1}{\gamma}(\theta^* - \hat{\theta})^T \dot{\hat{\theta}} \\
= -Fe^2(t) + (D(f(x(t), u(t), \bar{\theta}) - f(x(t), u(t), \theta)) + \\
\lambda(t, u(t)))e(t) + \frac{1}{\gamma}(\theta^* - \hat{\theta})^T \dot{\hat{\theta}} \\
\le -Fe^2(t) + (D(f_\theta'\bar{\varepsilon} + c_1 + c_2\bar{\varepsilon}) + \lambda(t, u(t))) + \\
\frac{1}{\gamma}(\theta^* - \hat{\theta})^T \dot{\hat{\theta}}$$

So we have

$$\lambda(t, u(t)) = -D(f_\theta' + c_2)\delta - Dc_1 - \\
\frac{1}{\gamma} \delta \dot{\hat{\theta}} \operatorname{sgn}(\bar{\varepsilon}) \qquad (10)$$

Note from (10), if

$$\lambda(t, u(t)) = -D(f_\theta' + c_2)\delta - Dc_1 - \frac{1}{\gamma} \delta \dot{\hat{\theta}} \operatorname{sgn}(\bar{\varepsilon}),$$

we have $\dot{L} \le 0$, so the system error asymptotically converge to zero and the system asymptotically stable.

VI. SIMULATION RESULTS

To test the accuracy of the accommodation controller, with the motor parameters fault, two kinds of methods (MRAC and FNNS) were applied to speed servo system of brushless DC motor(BLDCM) [11]. The parameters are: DC voltage U=36V, resistance of stator phase winding R=4.2Ω, self-inductances of the stator respectively L=0.002H, mutual inductances of stator respectively M=-0.0063H, moment of inertia J=0.0034kg•m, number of pole pairs n_p=1, and reference

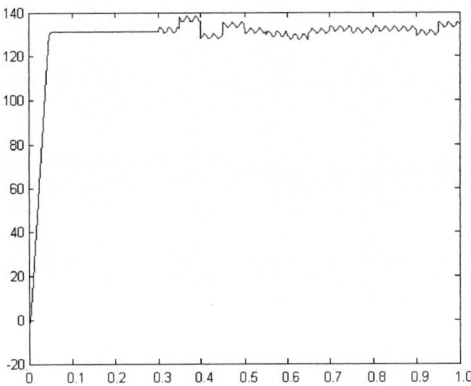

Figure 2. Speed response curve-MRAC
(130 r/m, at 0.3s, Resistance of stator phase winding: 4.2Ω-8.8Ω)

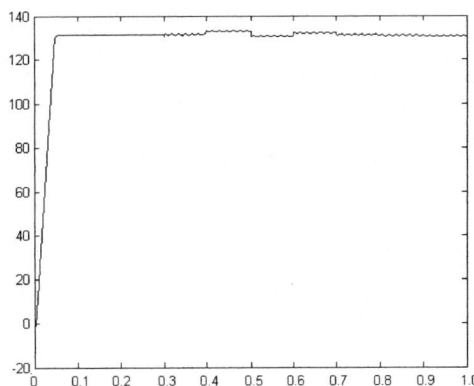

Figure 3. Speed response curve-FNNS
(130 r/m, at 0.3s, Resistance of stator phase winding: 4.2Ω-8.8Ω)

speed n_r=130r/m. Case studies were performed to show the effect of the system fault conditions on the different controller performances. Under the tremendous change of stator resistance (4.2Ω-8.8Ω) and the load torque disturbance (1N•m-5 N•m), two methods were compared with each other.

The speed response using MRAC and FNNS controller respectively are shown in Figure 2-3, where the initial resistance of stator phase winding is 4.2Ω, and at time t=0.3s, the resistance increases to 8.8Ω abruptly. As shown in the waveforms, the MRAC behaves well at high speed effectively, but fails to track after the fault happened, because that the original model is not suit for new operation condition any more. On the other hand, the FNNS has perfect responses to the variable step stator resistance.

Figure 4-5 show the load torque disturbance rejection capability by MRAC controller and FNNS controller. The initial torque is 1N•m, and at time t=0.6s the torque adds to 5N•m. For tracking to the command speed while the load torque is 1N•m, the MRAC has significant overshoot and emerge the steady-state error. Compared to the simulation results of MRAC, FNNS performs much better showing almost no speed change when load disturbed.

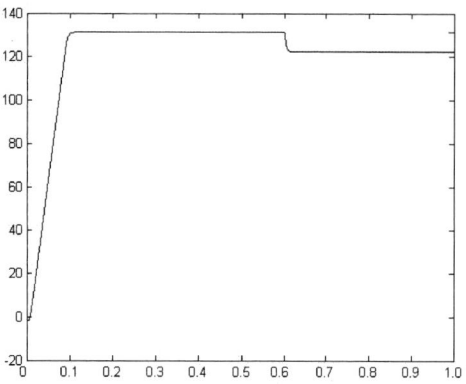

Figure 4. Speed response curve-MRAC
(130 r/m, at 0.6s, Load torque: 1N•m-5N•m)

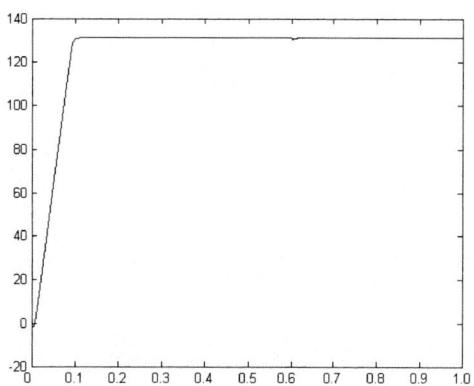

Figure 5. Speed response curve-FNNS
(130 r/m, at 0.6s, Load torque: 1N•m-5N•m)

The simulation results show that the FNNS has much better performance in terms of overshoot, steady-state error and accommodation ability comparing with conventional MRAC, when motor faults happen such as tremendous changes of system parameters or disturbing of outside condition.

VII. CONCLUSION

In this article, a novel fault detection and accommodation method using RBF fuzzy neural network for nonlinear system was proposed. With adaptive fault parameter updating algorithm, RBF fuzzy neural network was used to offer parameters' adjustment, and the fault detection was realized. Furthermore, with Lyapunov stability theory we suggested a fault accommodation control for unpredictable malfunction. The simulation results showed that the RBF fuzzy neural network can track the fault, the system keep stable with fault occurring, and the use of the fuzzy logic system make the system robust to noise measurement. To point the limitation that the approach relies on the existence of the accurate mathematical model of the intact plant, when model changed, the RBF fuzzy neural network should be retrained.

REFERENCES

[1] M.E. Shendy and S.M. Alan, "A fuzzy expert systems for fault detection in statistical process control of industrial process," *IEEE Trans. on Syst., Man, and Cybe.*, vol. 30, pp. 281-289, February 2000.

[2] X.Q. Liu, H.Y. Zhang and J. Liu, "Fault detection and diagnosis of permanent magnet DC motor based on parameter estimation and neural network," *IEEE Trans. on Indu. Elec.*, vol. 47, pp. 1021-1030, May 2000.

[3] A. Bernieri, G. Betta and A. Pietrosanto, "A neural network approach to instrument fault detection and isolation," *IEEE Trans. on Inst. and Meas.*, vol. 44, pp. 747-750, March 1995.

[4] W. Wilson, I. Fathy and G. Farid, "A neuro-fuzzy approach to gear system monitoring," *IEEE Trans. on Fuzzy Syst.*, vol. 12, pp. 710-723, May 2004.

[5] O. Jarrah and M. Alrousan, "Fault detection and accommodation in dynamic systems using adaptive neurofuzzy systems," *Cont. Theo. Appl.*, vol. 148, pp. 283-290, April 2001.

[6] X. Zhao and D.Y. Xiao, "Fuzzy diagnosis of nonlinear systems based on modular fuzzy neural networks," *Cont. Theo. and Appl.*, vol. 18, pp. 395-400, March 2001.

[7] M.M. Polycarpou and J.H. Helmicki, "Automated fault detection and accommodation: a learning systems approach," *IEEE Trans. on Syst., Man, and Cybe.s*, vol. 25, pp. 1447-1458, November 1995.

[8] K. Sukumar, *A new generation of adaptive control: an intelligent supervisory approach.* Toledo, USA: Toledo University, 2004.

[9] F. Caccavale and L. Villani, *Fault diagnosis for industrial robots.* Berlin Heidelberg, Germany: Sping-Verlag, 2003.

[10] M.G. Negorita, V. Palade and D. Neagu, *Computational intelligence engineering of hybrid systems.* Berlin Heidelberg, Germany: Sping-Verlag, 2005.

[11] Z.C. Ji, Y.X. Shen and J.G. Jiang, "A novel method for modeling and simulation of BLDC system based on Matlab," *J. Syst, System Simu.*, vol. 15, pp. 1745-1749, December 2003.

2006 5th International Power Electronics and Motion Control Conference

A Novel Constant Power Control of High Frequency Electronic Ballast Applying the PLL Technique for a Metal Halide Lamp

Chang-Hua Lin, Chung-Lun Ou, Tien-Shuo Liu, and Ken-Chuan Hsu
St. John's University / Department of Computer & Communication Engineering, Taiwan, China
ljh@mail.sju.edu.tw

Abstract—This paper presents a new constant power control of high frequency electronic ballast for a metal halide lamp based on the phase-locked loop (PLL) technique. First, the PLL is employed as a feedback mechanism to improve the resonant frequency variations caused by lamps aging or an impedance difference between used lamps characteristics. The proposed system can adjust the operating frequency to sustain constant power for a metal halide lamp according to the impedance of the adopted lamps. Moreover, the control strategy also can be used to substantially reduce the warm-up process of a metal halide lamp. The design considerations and measured results are detailed to verify the validity of the proposed control strategy.

Keywords-electronic ballast; metal halide lamp; PLL; constant power design

I. INTRODUCTION

Currently, metal halide lamps are being used for many outdoor and indoor applications. They show excellent characteristics by providing color rendering, lighting efficiency, and lamp lifespan [1], [2]. Generally metal halide lamps should be ignited and supplied with ballast to stabilize the lamp power and to limit the lamp current [3]. In principle, a high frequency electronic ballast has many merits: compact size, lighter weight, high efficiency, and can be widely applied to all kind of gas discharge lamps [4]. In addition, lamp aging or variations of lamp impedance may result in over load and then to shorten the lamp lifespan. Therefore, the constant power control is essential for electronic ballast design to retain the system's stability and extend the lamp life span [5]. To achieve the constant power control, both the lamp voltage and the lamp current are usually taken as the feedback parameters to maintain lamp power [5], [6]. However, the design of said controller is too complicated with a high cost. A new control scheme is thus proposed to deal with the aforementioned shortcomings regarding the metal halide lamps. The control scheme utilizes the phase-locked loop (PLL) technique to fast track the optimal operating frequency of the electronic ballast according to the variation of the lamp impedance. In this paper, we capture a phase difference between the driving signal and the lamp current, rather than the traditional-used method regarding lamp current and lamp voltage, as the feedback variable to achieve the constant power control by regulating the operating frequency. Finally, the

proposed control circuit also can substantially reduce the warm-up process of the metal halide lamp.

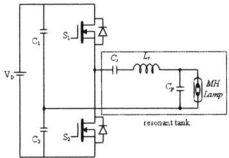

Fig. 1 The schematic of the employed half-bridge resonant inverter.

(a)   (b)

Fig. 2 (a) The equivalent circuit of the employed resonant inverter, (b) the simplified circuit of Fig. 2a.

II. ANALYSIS OF HALF-BRIDGE RESONANT INVERTER

The employed configuration of a half-bridge series-parallel-loaded resonant inverter is shown in Fig. 1, which includes two alternative switching power switches S_1, S_2, and a resonant tank to ignite and drive the metal halide lamp [4], [7]. The equivalent circuit of Fig. 1 in a steady state is shown in Fig. 2a, where the resonant tank consist of a resonant inductor L_r, a series-resonant capacitor C_s, a parallel-resonant capacitor C_p and the equivalent lamp impedance R_ℓ. When the operating frequency of the circuit is close to its resonant frequency, the harmonics and DC component of the current flowing in the circuit will be filtered out due to the high quality factor of the main circuit. Therefore, sinusoidal voltage and current can be produced to start and to drive the metal halide lamp. In addition, the metal halide lamp is replaced with an equivalent resistance R_ℓ for analysis. Since the equivalent lamp resistance R_ℓ in steady state is much lower than the equivalent impedance of the resonant capacitor C_p, that is, $1/\omega C_p \gg R_\ell$ [8], Fig. 2a can be simplified as Fig. 2b. Thus the resonant frequency in steady state is

$$\omega_o = \frac{1}{\sqrt{L_r C_{eq}}} \tag{1}$$

where

$$C_{eq} = \frac{C_s C_p}{C_s + C_p} \tag{2}$$

1-4244-0448-7/06/$25.00 ©2006 IEEE

To simplify the design procedure and easy to analysis the resonant characteristics of Fig. 2, we also define two parameters as follows:

$$\alpha = \frac{C_p}{C_s} \tag{3}$$

$$Q = \frac{R_\ell}{\sqrt{\dfrac{L_r}{C_{eq}}}} = \frac{R_\ell}{\omega_o L_r} = \omega_o C_{eq} R_\ell \tag{4}$$

From Fig. 2b, we can obtain the mathematical models of the frequency characteristics and lamp power versus operating frequency. The transfer function of the resonant tank can be derived by sinusoidal approximation method

$$\frac{v_\ell(j\omega)}{v_d(j\omega)} = \frac{1}{(1+\alpha)\left[1-\left(\dfrac{\omega}{\omega_o}\right)^2\right]+j\dfrac{1}{Q}\left(\dfrac{\omega}{\omega_o}\dfrac{\alpha}{1+\alpha}\right)} \tag{5}$$

The magnitude of equation (5) is

$$\left|\frac{v_\ell(j\omega)}{v_d(j\omega)}\right| = \frac{1}{\sqrt{(1+\alpha)^2\left[1-\left(\dfrac{\omega}{\omega_o}\right)^2\right]^2 + \dfrac{1}{Q^2}\left(\dfrac{\omega}{\omega_o}-\dfrac{\omega_o}{\omega}\dfrac{\alpha}{1+\alpha}\right)^2}} \tag{6}$$

The phase angle of the transfer function is

$$\phi = -\tan^{-1}\left\{\frac{\dfrac{1}{Q}\left(\dfrac{\omega}{\omega_o}-\dfrac{\omega_o}{\omega}\dfrac{\alpha}{1+\alpha}\right)}{(1+\alpha)\left[1-\left(\dfrac{\omega}{\omega_o}\right)^2\right]}\right\} \tag{7}$$

According to (6), the lamp power can be estimated as

$$P_\ell = \frac{v_\ell^2}{R_\ell} = \frac{2V_D^2}{\pi^2 R_\ell\left[(1+\alpha)^2\left[1-\left(\dfrac{\omega}{\omega_o}\right)^2\right]^2 + \dfrac{1}{Q^2}\left(\dfrac{\omega}{\omega_o}-\dfrac{\omega_o}{\omega}\dfrac{\alpha}{1+\alpha}\right)^2\right]} \tag{8}$$

Substituting various values of load into abovementioned equations, the frequency characteristics versus normalized frequency at different loading conditions can be depicted as Fig. 3 and Fig. 4 by Matlab simulation. Fig. 3a and 3b show the magnitudes and the phase angles of the derived transfer function, respectively. Next, the relationships between lamp power and operating frequency under various lamp impedances are shown in Fig. 4. From (6)-(8) and the above simulations shown in Figs. 3-4, we can clearly see that lamp voltage v_ℓ, lamp power P_ℓ, and phase angle ϕ are all much related as closely as the lamp impedance R_ℓ, the operating frequency f, and the parameters of the resonant tank (including resonant inductor L_r, and resonant capacitors C_s, C_p). In other words, the resonant frequency and phase angle will vary with the lamp impedance and hence cannot keep a constant power state. According to (8), the operating frequency must be changed with the lamp impedance to obtain a constant lamp power. Actually, the lamp power variations, caused by lamp aging or lamp impedance difference, cannot be eliminated directly by constant voltage driving or constant current driving. In this paper, we propose a novel controller based on PLL technique to

improve effectively the resonant point variation due to aging or lamp impedance difference. The proposed controller can adjust the operating frequency automatically, when the operating point away from the original one, to maintain constant lamp power and to shorten the warm up phase of metal halide lamp [9].

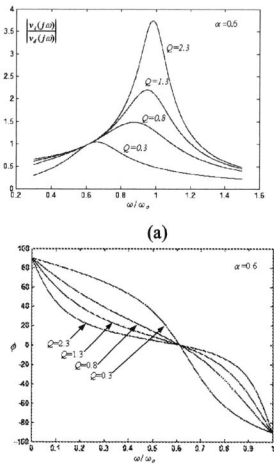

(a)

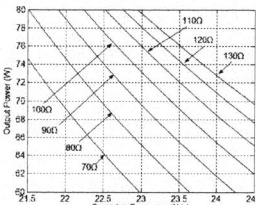

(b)

Fig. 3 Frequency characteristics of the resonant circuit versus normalized frequency at various values of Q (a) lamp voltage (b) phase angle

Fig. 4 Output power versus operating frequency at various value of lamp resistances.

III. IMPLEMENTATION OF CONSTANT POWER CONTROL

In the early days, phase-locked loop technique was primarily applied to signal synchronization in communication engineering and then to the servomotor control. Today it is also used in other electronic products [10]. The basic configuration of PLL is constituted of a phase detector, a low pass filter, and a voltage-controlled oscillator (VCO) [10], [11]. The equivalent impedance variation of the metal halide lamp between the ignition phase and the steady state is much severe. The practical measured lamp impedance during warm up phase is shown in Fig. 5. In fact, the lamp impedance will increase progressively with the lamp aging. Therefore, the resonant frequency and electrical characteristics will also vary with the lamp impedance. Fig. 6 depicts the comparisons of lamp power between constant voltage driving and constant current driving at various lamp impedances. From Fig. 6, the lamp power will increase with the lamp aging when constant current driving is adopted. On the other hand, constant voltage driving will result in that lamp power is inverse proportion to the lamp impedance [6].

1958

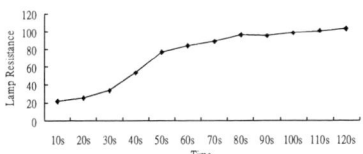

Fig. 5 The measured lamp characteristics during warm up phase.

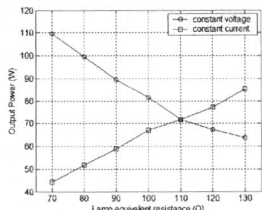

Fig. 6 Comparisons of output power between constant voltage driving and constant current driving at various lamp resistances.

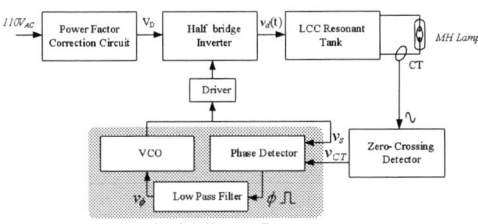

Fig. 7 The proposed metal halide lamp electronic ballast with constant power control.

Hence, this paper presents a controller based on the phase-locked loop technique to get simultaneously the phases of lamp current and driving signal of system for tracking the optimal operating frequency. The block diagram of the complete system is shown in Fig. 7, where the phase detector senses the phase difference, ϕ, between the sampled signal, v_{CT}, of the lamp current, and the driving signal, v_s, of the proposed system. The phase difference ϕ will vary as long as the lamp conditions change. Then, the low pass filter will generate a corresponding dc value to regulate the VCO and thus synchronize the output frequency of the PLL controller. In the adopted PLL module, CD4046BC, the phase detector is applied with the driving signal v_s of system and the sampled signal v_{CT} of lamp current, captured by a current transformer and a zero-crossing detector for detecting their frequency variation and the phase difference between these two signals. When the lamp impedance changes and then results in the deviation of resonant characteristics, the phase detector will generate an phase difference signal ϕ, indicating the variation amount of frequency, or phase, to immediately adjust the output frequency of VCO and, in turn, to regulate the operating frequency of the inverter. Therefore, the phases of v_s and v_{CT} can be shifted back to their original lock-in states. Hence, the output signal v_s of VCO and the sampled signal v_{CT} of lamp current will remain synchronized, with a fixed phase difference, to accomplish the keeping of the constant lamp power. In

addition, with the capture-range setting of PLL, the metal halide lamp can be operated under the preset central frequency during the initial starting period under the condition of an absence of the feedback signal comparison. Moreover, the lock-in range of PLL can be set to the lamp's operating frequency during normal operation. When the resonant characteristics of the system are influenced by the lamp aging, the PLL will quickly track the optimum operating frequency within the lock-in range and maintain the synchronization between v_s and v_{CT} so as to achieve the constant lamp power design.

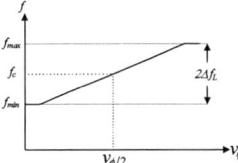

Fig. 8 Typical PLL characteristics of input voltage versus output frequency.

The characteristic curve of the input voltage and the output frequency in PLL is shown in Fig. 8, where f_c is the operating frequency of metal halide lamp in steady state. The frequency deviation Δf of PLL can be determined according to the range of lamp impedance deviation. Hence, the minimum operating frequency f_{min} and the maximum operating frequency f_{max} can be expressed, respectively, as follows:

$$f_{min} = f_c - \Delta f_L \qquad (9)$$

$$f_{max} = f_c + \Delta f_L \qquad (10)$$

From equations (9) and (10), the parameter α is defined as

$$\beta = \frac{f_{max}}{f_{min}} \qquad (11)$$

Next, we can use β and the characteristic curves of CD4046BC to estimate the values of timing resistor in PLL [14].

IV. DESIGN CONSIDERATION AND EXPERIMENTATION

The lamp impedance variation of metal halide lamp is much severe from the ignition state to the steady state [12]. Additionally, the lamp impedance will increase with the used hour, thus it is a key parameter in the constant lamp power design. Two types of metal halide lamp, OSRAM HQI-TS 70W/WDL and 70W/NDL, respectively, are utilized in experimentations to verify the performance of the proposed control strategy. The adopted lamps have rated power of both 70W, operating voltage v_ℓ of 90V$_{rms}$ and 85V$_{rms}$, operating current i_ℓ of 0.8mA$_{rms}$ and 0.85mA$_{rms}$, and starting voltage v_{start} above 2.8kV$_{rms}$, thus the equivalent lamp impedance in steady state can be estimated about 113Ω and 100Ω, respectively. In this paper, we select the starting frequency of 62kHz, the operating frequency of 23.5 kHz, and input voltage V_D=320V$_{DC}$, which is provided from the active PFC. The most important parameters in the proposed circuit structure are the resonant inductor L_r, resonant capacitors C_s and C_p, and the timing resistors of the PLL controller. The design considerations and steps of these parameters are discussed below

1) Deciding the ratio α of resonant capacitors

Because the lamp impedance is high enough before the lamp has been ignited, the resonant frequency of the inverter is approximately as $\omega_p=1/(L_rC_p)^{1/2}$. At this condition, inverter can provide a high voltage to ignite the metal halide lamp. After igniting, the inverter operates the metal halide lamp in the steady state, its resonant frequency is approximately as $\omega_s=1/(L_rC_s)^{1/2}$ due to $1/\omega C_p \gg R_\ell$. In practical design, the values of α is generally within the range of 1/10~1/30 [8], [13]. We choose α to be 1/17 in the actual design.

2) Calculating the resonant components, L_r, C_s, and C_p

In steady state, substituting the operating voltage v_ℓ and the input voltage, v_d, of the resonant tank into (6), we get

$$\left|\frac{v_\ell(j\omega)}{v_d(j\omega)}\right|=\frac{90\sqrt{2}}{\frac{4}{\pi}\times\frac{320}{2}}=0.62 \qquad (12)$$

Substituting the selected value of α, operating frequency of 23.5 kHz and (12) into (4) and (6), we obtain the resonant inductor L_r=1.4mH (use 1.45mH) and the equivalent capacitance C_{eq}=5.1nF. Combining (2) and (3) yields

$$C_{eq}=\frac{C_p}{1+\alpha}=\frac{C_s}{1+\frac{1}{\alpha}} \qquad (13)$$

Solving (13) with the given parameters, we obtain C_s=90.1nF (use 100nF) and C_p=5.4nF (use 5.6nF).

3) Estimating the timing resistors of PLL

The desired values of components R_1, R_2, and C shall be chosen to set the lock-in range of CD4046BC [14]. The central frequency, f_c, of PLL is preset to the operating frequency f= 23.5kHz. The frequency variation, Δf_L, of ±5kHz is substituted into (9) to yield the minimum operating frequency f_{min} of PLL shown below

$$f_{min} = f_c - \Delta f_L \qquad (14)$$
$$= 23.5kHz - 5kHz$$
$$= 18.5kHz$$

Using the characteristic curve of CD4046BC [14] with the frequency of (14) estimates R_2=10kΩ and C=10nF (use 11nF). From (10), the maximum operating frequency of PLL f_{max}=28.5kHz is obtained. Then substituting the minimum and maximum operating frequencies into (11) yields β as follows

$$\beta=\frac{28.5kHz}{18.5kHz}\cong 1.5 \qquad (15)$$

Using the characteristic curve of CD4046BC with β of (16) estimates

$$\frac{R_2}{R_1}=0.8 \qquad (16)$$

Entering (16) into [14] finds $R_1 \cong 12.5$kΩ (use 12.1kΩ).

To verify the validity of the proposed control strategy, we use different metal halide lamps and equivalent high

power resistor as the test loads. First, the lamp of 70W/WDL (R_ℓ=113Ω) is as the test target, Fig. 9 demonstrates the measured waveforms of lamp voltage v_ℓ, lamp current i_ℓ, and input voltage v_d, of the resonant tank, whose phase is the same with the driving signal. At this experimentation, the lamp power is about 72W, the phase difference φ=55°, between v_ℓ and v_d. Next, we replace another lamp of 70W/NDL (R_ℓ=100Ω) as the test target without the proposed PLL controller, Fig. 10 displays the measured waveforms of lamp voltage v_ℓ, lamp current i_ℓ, and input voltage v_d, respectively. In this condition, the lamp power reduces to 64W, and the phase difference φ increases to 64°. When the proposed PLL controller is incorporated in system, the retested results of the same lamp is shown in Fig. 11. We observe that the lamp power is near 71W, and the phase difference φ decreases to 54°.

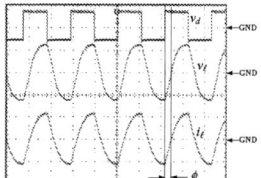

Fig. 9 Measured waveforms of v_d, v_ℓ, and i_ℓ with constant power control at lamp resistance R_ℓ =113Ω (Ver: 250V/div for v_d; Ver: 100V/div for v_ℓ; Ver: 1A/div for i_ℓ; Hor: 400ns/div)

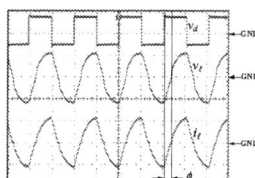

Fig. 10 Measured waveforms of v_d, v_ℓ, and i_ℓ without constant power control at lamp resistance R_ℓ =100Ω (Ver: 250V/div for v_d; Ver: 100V/div for v_ℓ; Ver: 1A/div for i_ℓ; Hor: 400ns/div)

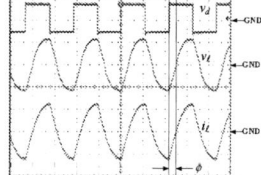

Fig. 11 Measured waveforms of v_d, v_ℓ, and i_ℓ with constant power control at lamp resistance R_ℓ =100Ω (Ver: 250V/div for v_d; Ver: 100V/div for v_ℓ; Ver: 1A/div for i_ℓ; Hor: 400ns/div)

We finally replace the tested lamp with different high power resistor of 70Ω, 90Ω, and 130Ω, respectively. As Fig. 12 and Table I, we observe that all the lamp powers retain at 70W±5%, however, the corresponding operating frequency are 21.8kHz, 22.9kHz, and 24kHz, respectively. Therefore, the proposed PLL controller can immediately adjust the optimal operating frequency of system to maintain the desired constant lamp power.

In this paper, the proposed control strategy not only is able to achieve constant lamp power design, but also offers a better preheating process to shorten the warm up time. Fig. 13 demonstrates the lamp voltage and lamp current during the warm up phase without the PLL controller. The complete warm up process takes about 90 seconds. When the PLL controller is embedded in system, we remeasure these waveforms shown in Fig. 14. From the measured results, the warm up phase effectively decrease to 45 seconds.

TABLE I Tested results with equivalent high power resistors

R_ℓ	70Ω	90Ω	130Ω
$v_\ell(V_{rms})$	70	81	98
$i_\ell(A_{rms})$	0.99	0.87	0.743
$P_\ell(W)$	69	70.5	72.5
$f(kHz)$	21.8	22.85	24
ϕ	62.6°	59°	52.5°

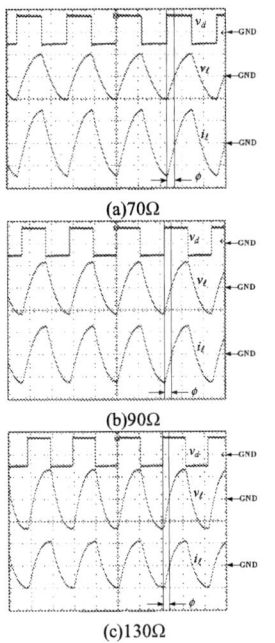

(a)70Ω

(b)90Ω

(c)130Ω

Fig. 12 Measured waveforms of v_d, v_ℓ, and i_ℓ with constant power control at various loads (a) 70Ω (b)90Ω (c)130Ω (Ver: 250V/div for v_d; Ver: 100V/div for v_ℓ; Ver: 1A/div for i_ℓ; Hor: 400ns/div)

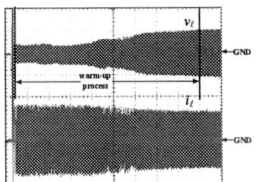

Fig. 13 Lamp voltage and lamp current during warm up process without PLL controller (Ver: 100V/div for v_{comp}; Ver: 10V/div for

v_ℓ; Ver: 1A/div for i_ℓ; Hor: 10s/div)

V. CONCLUSION

A new feedback mechanism of constant lamp power control for a metal halide lamp is proposed in this paper. A half-bridge resonant inverter is incorporated with a PLL controller to adjust the operating frequency to achieve constant lamp power design. In addition, the proposed control strategy also can be used to shorten the warm up process of a metal halide lamp up to 50%.

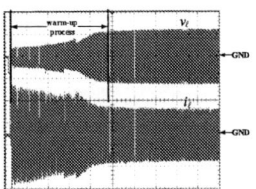

Fig. 14 Lamp voltage and lamp current during warm up process with PLL controller (Ver: 100V/div for v_{comp}; Ver: 10V/div for v_ℓ;

Ver: 1A/div for i_ℓ; Hor: 10s/div)

ACKNOWLEDGMENT

This study is sponsored by the National Science Council (NSC), Taiwan, China, under Research Proposal # NSC94-2213-E-129-015; NSC's financial support to this project is hereby acknowledged.

REFERENCES

[1] M. Sugiura, "Review of Metal-Halide Discharge-Lamp Development 1980-1992," in *Proc. IEE. Science, Measurement and Technology*, vol.140, No.6, Nov. 1993, pp. 443-449.

[2] Sam Ben-Yaakov, M. Gulko, and A. Giter. "The Simplest Electron Ballast for HID Lamps," in *Proc. IEEE APEC'96*, pp. 634-640.

[3] M. Ponce, A. Lopez, J. Correa, J. Arau, and J. M. Alonso, "Electronic ballast for HID lamps with high frequency square waveform to avoid acoustic resonances," in *Proc. IEEE. APEC 2001*, vol.2, pp. 658-663.

[4] R. L. Steigerwald, "A comparison of Half-Bridge Resonant Converter Topologies," *IEEE Trans. Power Electron.*, vol.3, pp. 174-182, Apr. 1988.

[5] Zhang Weiqiang, and Xu Dianguo, "Novel constant power control of electronic ballast for HPS lamps," in *Conf. Rec. IEEE ICIT'02*, vol.1, pp. 129-132, Dec. 2002.

[6] P. Jong-Yeon, and J. Dong-Youl, "Electronic ballast with constant power output controller for 250 W MHD lamp," in *Proc. IEEE ISIE 2001*, vol.1, pp. 46-51.

[7] M. K. Kazimerczuk and W. Szaraniec, "Electronic ballast for fluorescent lamps," *IEEE Trans. Power Electron.*, vol.8, pp. 386-395, Oct. 1993.

[8] Dongyan Zhang, Weiping Zhang, Yuanchao Liu, and Xuesun Zhao, "Design of LCC resonant inverter for metal halide lamp ballast," in *Conf. Rec. IPEMC 2004*, vol.3, pp. 1558-1562

[9] F. Rodriguez-Valdes, and J. H. Avila, "Solution to the warm up problems on HID sodium lamp ballasts controlled by resonant current," in *Conf. Rec. PESC'03*, vol.3, pp. 977-981

[10] G. C. Hsieh and J. C. Hung "Phase-locked loop techniques. A survey," *IEEE Trans. Ind. Electron.*, 1996, vol. 43, pp. 609 -615

[11] Yoko Mizutani, Taiju Suzuki, Jiroaki Ikeda, Hirofumi Yoshida and Shigenobu Shinohara "Frequency control of MOSFET full bridge power inverter for maximizing output power to megasonic transducer at 3 MHz," *IEEE Proc. The 33th Ind. Appli. Conference Record*, 1998, vol.3, pp. 1644-1651

[12] W. W. Byszewski, A. B. Budinger, and Y. M. Li, "HID Starting: Glow Discharge and Transition to the Thermionic Arc," *Journal of the Illuminating Engineering Society*, Summer 1991, pp. 3-9.

[13] C. Branas, F. J. Azcondo, and S. Bracho, "Electronic ballast for 150 W HPS lamps with compensated output power," *IEE Electronics Letters.*, vol.35, 24 June 1999 pp. 1041 – 1043

[14] *Applications Handbook*, National Semiconductor, Nov. 1995.

2006 5th International Power Electronics and Motion Control Conference

The Voltage Stability Research of Ship Electric Power System

Zhaomin, Fanyinhai

Dalian Maritime University, Dalian, China

Abstract—Ship electric power system is an independent electric power system, and is important in ships. At present, with the increasing tonnage of ships and more and more ships propelled by electric power, the capacity of generator in a marine power station is getting larger and larger. Because of electric propulsion technology applied by modern marines, the control system of voltage stability must be with better performance. The electric energy produced by diesel generators is not only transported to each load but also sent to marine propulsion device. The voltage of marine power system must be kept in reasonable range; otherwise it would affect the electrical system of marine greatly. In this paper The definition of voltage stability and its state of study were first summarized, since in voltage stability, any element arouse the tension loses firmly, the load losing accompanies in it, establishing the suitable load model is the key step of voltage stability analysis. The analysis has been underway to the load model building problem in the ship power system. Then according to the problem in voltage stability index aspect, a new index which is quite easy to calculate and can be implemented for online use was proposed. The method does not need complicated matrix computation and other time-consuming calculations, so the solution can be expected to be faster.

Key words-ship electric power system; voltage stability; load model; stability indicator

I. INTRODUCTION

A. The Concept of Ship Electric Power System Voltage Stability

Since the 70's tips, the greater progress was acquired in research of voltage stability problem. The effects of voltage stability key element were gradually understood; and the voltage stability mechanism and nature were comprehended initially Voltage stability is the capability that after the disturbance the system operates at the equilibrium point subjected to the combined action of characteristic of the system and load. The voltage stability analysis ways is divided into two great kinds: one is based on the static analysis of power flow equation, and the other based on dynamic analysis of differential equation.

B. The Present Situation of Voltage Stability Problem Research

At present, the voltage stability research is chiefly divided into two great respects, the first includes of voltage stability problem and how to describe it, and the second consists of research of voltage stability target and prevention methods of voltage collapse and so on. Ordinarily, under the situation that first respect problem has been resolved; the second respect problem can be solved. But until now, first respect problem does not still obtain the very good settlement. Therefore, at the moment two respect problems replace to be underway.

II. THE RESEARCH OF LOAD MODELS BUILDING

A. Voltage Stability together with Load Developments Relationship

In the voltage stability problem, losing stability of voltage can be traced to the load, dynamic characteristic of load-tap changing transformer and the limitation of maximum current of excitation system of ship electric power system. On the basis of dynamic characteristic of three types of elements, the mechanism of losing stability has been interpreted in many literatures. Yet, no matter what reason the losing of stability were caused, the load loses firmly and wholly accompanies in it. So establishing the suitable load model is the most important to analyze the stability of ship electric power system. In the research of voltage stability, the below respects of load dynamic characteristic should be considered [1].

With the dropping of load bus voltage, the load will absorb more reactive power from the system, and then deteriorate the system area equilibrium state of reactive power. It causes the positive feedback mechanism of the voltage dropping.

Dynamic active power recovering property: with the dropping of voltage, different types of active and reactive power load can resume the specified level with either the fast or slow velocity. If under the extreme situation，the voltage will recover to the previous level before the dropping of voltage completely.

To achieve equilibrium of input/output active power, the dynamic load has intrinsic property to adjust admittance. Any dynamic load relate to some kind of active power equilibrium to meet the active power demand, when this kind of equilibrium is wrecked, it should regulate with such inherent means and come into being the various kinds dynamics characteristic.

From the standpoint of nature，the dynamic load characteristic relate to losing of voltage stability is the natural property that in order to maintain equilibrium of active power automatically adjust admittance. This

1-4244-0448-7/06/$25.00 ©2006 IEEE

property has to been reflected on the load model used for ships electric power system voltage stability research.

B. The Analysis of Voltage Stability Load Model Building

In the voltage stability problem, for particular position of load property, it is benefit to solve the problem that via the load model the voltage stability were widely analyzed. In other words, load model building should reflect the nature characteristic that relate to voltage stability.

The dynamic model that adopted in the voltages stability research of ships electric power system may be the below three types:

Reactive power equilibrium model

The static voltage level is chiefly decided by the equilibrium condition of the reactive power, so the voltage collapse may be got in touch with the reactive power disequilibrium. In the voltage stability load model building, the relationship between some variables of the burden node and the reactive power disequilibrium were established. The dynamics of load is described by the following model:

$$\frac{dv}{dt} = \frac{Qd - Qs(V)}{K(V)} \quad (1)$$

That model thought that the voltage alternation was determined by reactive power disequilibrium, and point out that from physics there is not the dynamic procedure which the reactive power equilibrium determined the voltage changes.

Power recovering model

After the disturbance of load, power resuming posses the significant effect to the voltage stabilization. According to the characteristic of dynamic load that under disturbance of step voltage possess approximately single order resumed power, that model pertains to input/output model [2].

$$T_p P_d + P_d = P_s(V) + K_p(V)V \quad (2)$$

$$T_q + Q_d = Q(V) + K_q(V)V \quad (3)$$

In that model active power and reactive power have the analogous recovering property.

Since in the foregoing model after disturbance the bus voltage V, load active power Pd and reactive power Qd would not successively alternate, it is not convenient to analyze voltage stability, so the model may be improved to the following sate variable standard form:

$$T_p P_r + P_r = N_p(V) \quad (4)$$

$$T_q Q_r + Q_r = N_q(V) \quad (5)$$

$$P_d = P_r + P_t(V) \quad (6)$$

$$Q_d = Q_r + Q_t(V) \quad (7)$$

$$N_p(V) = p_s(V) - P_t(V) \quad (8)$$

$$Nq(V) = Q_s(V) - Q_t(V) \quad (9)$$

By means of mutation, Variable P_r, Q_r turns into the successive alternation state variable.

Another kind of universal load model describing load power recovering property is the following form, which also pertains to the sate variable pattern [3].

$$Tp\frac{dx}{dt} = Ps(V) - P \quad (10)$$

$$Tq\frac{dy}{dt} = Qs(V) - Q \quad (11)$$

From the foregoing model, it is known that the input/output model may transformed to the state variable model, when it is transformed, with different state variable chosen, the model may possess the difference form, and the model of recovering power stress on the port property.

Mechanism style model

The mechanism style model is the system model from physics property. In the contemporary ship electric power system, the sort of system dynamic load are more and more, which defer to different type of common property of dynamic load, in other words, equivalent admittance is regulated afterwards equilibrium damaged, the model reflecting the relationship of admittance changing and the power equilibrium may act as below three kinds:

one kind of load model adopt conductance and susceptance as the state variable, such form is following separately:

$$\frac{dG}{dt} = \frac{1}{T_G}(P_O - V_L^2 G) \quad (12)$$

$$\frac{dB}{dt} = \frac{1}{T_B}(Qo - V_L^2 B) \quad (13)$$

$$\frac{dG}{dt} = f_{G2}(G, V_L)(Po - V_L^2 G) \quad (14)$$

$$\frac{dB}{dt} = f_{B2}(B, V_L)(Qo - V_L^2 B) \quad (15)$$

In it, $f_{G2}(G, V_L), f_{B2}(B, V_L)$, as some functions of load parameter, sometimes the reactive power disequilibrium may be replaced by the active power disequilibrium.

The synthetically load dynamic model consisted of static load and OLTC is proposing as follow from the property of OLTC [4].

$$\frac{dG_L}{dt} = -\frac{1}{T}(G_L - f_G(V)) \qquad (16)$$

$$\frac{dB_L}{dt} = -\frac{1}{T}(B_L - f_B(V)) \qquad (17)$$

In it, $f_G(V)$ and $f_B(V)$ is the motionless voltage characteristic of load conductance GL and susceptance BL separately and is the function of ration.

The synthetically load model taking resistance as the sate variable [5].

$$\frac{dR_E}{dt} = -\frac{R_E^{\,2}}{T_E}(P_T - P_E) \qquad (18)$$

The foregoing model can be divided into two types, One type merely has the differential equation that described dynamic changing of conductance G's, Another type has the differential equation that described dynamic changing of both the conductance and the susceptance. The most strongpoint of this type of mechanism style pattern is favor to interpret the voltage stability, and when used to the dynamic procedure it is easy to take shape the evidence. Yet it is alike the other two types of model ahead, the main point of establishing the load model is in view of the voltage stability problem comprehends the voltage stability problem for the sake of the quality.

III. THE SUPERVISION TARGET OF VOLTAGE STABILITY

A. The Proposition of Target

Ship electric power system is an electric network including generator, transmission line, load, voltage control component and so on. First one transmission line of the system that the chart notifies was considered.

Parameter s in the picture expresses sending end node, the r expresses accepting end node, and the total expresses the gross node load.

Next voltage collapse was researched utilizing the concept of the largest transferring power between the network nodes. The fundamental idea is adopting the Dai WeiNan's equivalent electric circuit to the correlation node, Calculation of open circuit voltage and the Dai Weinan's equivalent reactance is for the derivation of voltage stability target, which is the ratio of Dai Weinan's

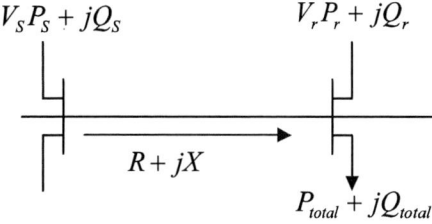

Figure1. Ships transmission line diagram of power system electric network

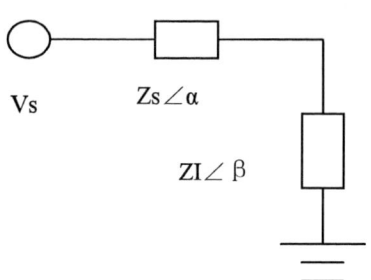

Figure2. Simple power system Dai Weinan of equivalent picture

impedance and load impedance, when ratio is 1, it is known that system is in the largest steady load condition, when ratio is smaller than 1, system voltage is stable, on the contrary it is not the case.

The simple voltage stability derived from the optimal impedance solution of two node system is briefly introduced following:

Using the maximum power method to analyze the voltage stability of simple power system

$$Z_S = R_S + jX_S = Z_S\angle\alpha \qquad (19)$$

$$Z_I = R_I + jX_I = Z_I\angle\beta \qquad (20)$$

The power that the load get is

$$P = I^2 R_f = \frac{V_S^{\,2}}{(R_S + R_L)^2 + (X_S + X_L)^2} R_I \quad (21)$$

In it I is module of load impedance ZI current, in case XI/RI=tgβ, β is constant, the terms that load obtains the maximum power is

$$\frac{dP}{dR_I} = 0 \qquad (22)$$

Straightens it out

$$R_S^{\,2} + X_S^{\,2} = R_I^{\,2} + X_I^{\,2} \qquad (23)$$

The critical condition is

$$Z_S = Z_L \qquad (24)$$

Foundation

$$P = \frac{V_S^{\,2} Z_L \cos\beta}{(Z_I\cos\beta + Z_s\cos\alpha)^2 + (Z_I\sin\beta + Z_s\sin\alpha)^2} \tag{25}$$

Straighten out the style on the critical condition $Z_S = Z_L$ substitute

$$P_{r(\max)} = \frac{V_s^2 \cos\beta}{2Z_s[\cos(\alpha-\beta)+1]} = \frac{V_s^2}{Z_s} \frac{\cos\beta}{4\cos^2\dfrac{(\alpha-\beta)}{2}}$$

(26)

The load voltage is

$$V_{cr} = IZ_l = \frac{V_s Z_l}{\sqrt{(Z_l\cos\beta + Z_s\cos\alpha)^2 + (Z_l\sin\beta + Z_s\sin\alpha)^2}}$$

(27)

Straighten out the style on the critical condition $Z_S = Z_L$ substitute

$$V_{cr} = \frac{V_s}{\sqrt{2[1+\cos(\alpha-\beta)]}} = \frac{V_s}{\cos(\dfrac{\alpha-\beta}{2})}$$ (28)

Here the voltage stabilization is decided by comparison of Zs and ZL.

Considering electric network figure 2, when $Z_S/Z_r = 1$ the power transmitted to the load is the largest, it is also the critical point of the voltage collapse. With simplifying the system network to the simple network with the single transmission line of forgoing parameter, $Z_S/Z_r = 1$ was regarded as forecast target of voltage collapse. When $Z_S/Z_r < 1$, system voltage is stable, when $Z_S/Z_r > 1$, System voltage collapses. As well the value of Pr and P r (max) can be compared; the system stability was measured by means of Pr/ Pr (Max).

B. Derivation of Target

Shown in figure 3, the load of transmission line can be regarded as merely accepting end power by means of the particular line transfer, rather than the gross node load. Zs∠θ It is line impedance, Z r∠δ is relevant load impedance and $\delta = tg^{-1}(Q_r/Pr)$.

In case only the module of load impedance changes, δ maintains constant, the problem may be simplified. In fact, the power factor load of power system ordinarily will be maintained constant, with the increasing of the

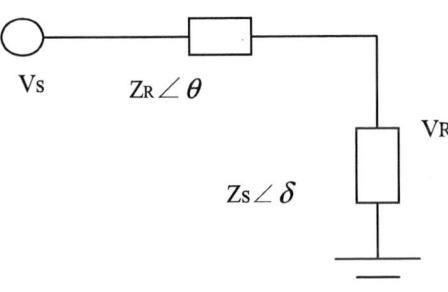

Figure3. Equivalent Dai WeiNan electric network of transmission line

load, Z r decrease, but I add, the accepting end-voltage will drop.

$$I = \frac{V_s}{\sqrt{[(Z_s\cos\theta + Z_r\cos\delta)^2 + (Z_s\sin\theta + Z_r\sin\delta)^2]}}$$

(29)

$$V_r = Z_r I = \frac{Z_r}{Z_S} \frac{V_s}{\sqrt{[1+(Z_r/Z_s)^2 + 2(Z_r/Z_s\cos(\theta-\delta)]}}$$

(30)

The accepting end-power:

$$P_r = V_r\cos\delta = \frac{(V_r)^2}{Z_r}\cos\delta$$ (31)

$$Q_r = V_r I\sin\delta = \frac{(V_r)^2}{Z_r}\sin\delta$$ (32)

$$P_r = \frac{(V_r)^2/Z_s}{1+(Z_r/Z_S)^2 + 2(Z_r/Z_S)\cos(\theta-\delta)} \frac{Z_r}{Z_s}\cos\delta$$

(33)

$$Q_r = \frac{(V_r)^2/Z_s}{1+(Z_r/Z_s)^2 + 2(Z_r/Z_S)\cos(\theta-\delta)} \frac{Z_r}{Z_s}\cos\delta$$

(34)

If $\delta P_r/\delta Z_r = 0$, $Z_r/Z_s = 1$, The accepting end-power is maximum,

Substitute $Z_r/Z = 1$

$$P_r = \frac{(V_r)^2/Z_s}{1+(Z_r/Z_S)^2 + 2(Z_r/Z_S)\cos(\theta-\delta)} \frac{Z_r}{Z_s}\cos\delta$$

(35)

The maximum transfer power $P_{r(\max)}$ is

$$P_{r(\max)} = \frac{V_s^2\cos\delta}{2Z_s[\cos(\theta-\delta)+1]} = \frac{V_s^2}{Z_s}\frac{\cos\delta}{4\cos^2\dfrac{(\theta-\delta)}{2}}$$

(36)

The maximum transfer reactive power is

$$Q_{r(\max)} = \frac{V_s^2}{Z_s}\frac{\sin\delta}{4\cos^2\dfrac{(\theta-\delta)}{4}}$$ (37)

Thus, the fast voltage stabilization target proposed is FVSI(Fast Voltage Stability Index).

$$FVSI(1) = \frac{P_r}{P_{r(\max)}} \qquad (38)$$

$$FVSI(2) = \frac{Q_r}{Q_{r(\max)}} \qquad (39)$$

In it, Pr, Q r may be obtained from the calculation of power flow, Pr depends on the system parameter, network topology structure, system connection and load demand; $P_{r(\max)}$ is decided by sending end-voltage Vs and the equivalent load (the power transfer on transmission line)effect of phase angle δ .

C. The Analysis of Target Feature

Obviously FVSI (1) =FVSI (2), for the sake of convenient of analyses, regard it as FVSI. The target may accurately forecast the voltage collapse.

$$FVSI = \frac{P_r}{P_{r(\max)}} = \frac{P_r}{\dfrac{V_s^2}{Z_s^2} \dfrac{\cos\delta}{4\cos^2\dfrac{(\theta-\delta)}{2}}} \qquad (40)$$

In it Pr determined to the system parameter, network topology structure, system connection and load demand, it is decided by sending end-voltage Vs and the equivalent load (the power transfer on transmission line) effect of phase angle.

The reactive power compensation affects the target.

FVSI is decided by the phase angle δ, when reactive power compensation devices were put in, the generator generates less reactive power. The Qr changes into little thereby cause the phase angle change into little. Finally it causes the transfer power $P_{r(\max)}$ changing into great. Because that target of FVSI changes into little, the reactive power compensation is favor of the voltage stability.

The reactive power limitation of generator affects the target.

When the generator attains the excitation limits, the system will sustain a lower voltage. This causes the power transport to the receiving end decreased and the target will be greater.

The effect of load tap changing transformer to the target.

Just as shunt reactive compensation, load tap changing transformer can improve the static voltage stability of system, when the voltage of secondary side(load side)decrease, OLTC adjust the tap of primary side, change the ratio, then recover the voltage of secondary side, the voltage increase, $P_{r(\max)}$ increase, FVSI value decrease, thus expand the limitation of voltage stability.

IV. CONLUSION

Currently it is in the stage that ships electric power systems regenerate, and the ships electric propulsion are commonly understood. Thus ships electronic power systems are demanded higher; hence it is necessary to research the stability of the ships electric power system. How to establish can choose the dynamic load model suitable for the voltage stability analysis is the key of voltage stability research. The wide-ranging qualitative analysis may be carried on to the voltage stabilization by means of the load model, and the load model structure established reflects the natural dynamic characteristic relate to the voltage losing. Three kinds of dynamic load models are analyzed: reactive power equilibrium model, power recovering model, mechanism style model.

Aiming at the voltage stability security target problem, a new kind of voltage stability supervision target (target of FVSI) was proposed. That target is based on the concept of maximum transfer power, using the equivalent methods, and popularize and apply the previous node equivalent further to each circuit, moreover by means of requesting target of each circuit, judge stability of node and system.

REFERENCES

[1] Prasad G D, Ai-Multhim M A, Ray G D, et al. Comparative Assessment of the Affect of Dynamic load models on voltage stability. Electric Power & Energy System, 1997, 19(5), 305-309

[2] Hill D J. Nonlinear Dynamic Load Models with Recovery for Voltage Stability Studies. IEEE Trans on Power Systems, 1993, 8(1), 166-176

[3] P.Kundur, Power system stability and control, McGraw-Hill, 1994

[4] J.W.Cote and C.C.Liu, Voltage Security Assessment using Generalized Operational Planning Knowledge, IEEE Trans on Power System, Vol. 8, No. 1, Feb 1993, 137-145

[5] IEEE Committee Report, "Voltage stability of power systems: concepts, analytical tools, and industry experience", IEEE/PES 90 TH 0358-2-PWR, 1990.

Parasitic Gate Resistance and Switching Performance

Alan Elbanhawy

Fairchild Semiconductor, San Jose CA, USA

E-mail Address: aelbanhawy@fairchildsemi.com

Abstract—In this paper we discusses in details the effect of the gate equivalent series resistance, ESR, of switching power MOSFETs on different aspects of loss mechanisms when used in DC-DC converters. The loss mechanisms addressed are current rise and fall times, die current distribution and localized shoot through

Keywords-component; Gate ESR; Current Rise and Fall Times; Shoot Through; DC-DC Converter power loss

I. INTRODUCTION

Power MOSFETs' gate effective series resistance (ESR), Rg, plays a major role in determining all aspects of switching losses ranging from the current rise and fall times to shoot through effect as well as the current distribution on the device's dice leading to uneven losses distribution during turn on and off. This paper presents simplified mathematical treatment of all these effects as well as supporting laboratory experimentations performed on synchronous buck converters.

II. DYNAMIC LOSSES

Dynamic losses are the losses encountered during the

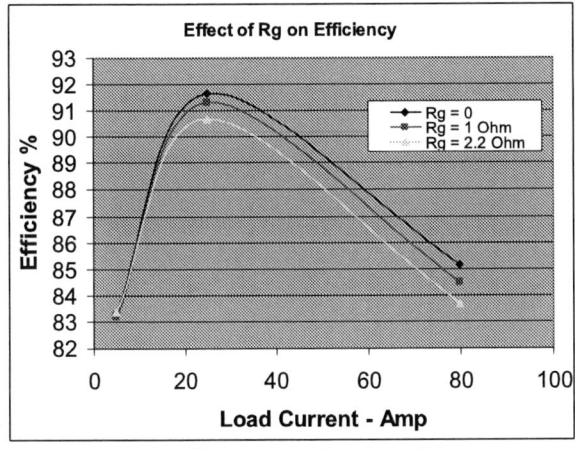

Figure 1. Power efficiency as a function of gate resistance

turn on and turn off of any switching device. "Fig. 1," depicts the reduction in efficiency in a synchronous buck converter as a function of increasing Rg from 0 to 2.2

Ohm. This reduction in efficiency represents an over 9% increase in losses.

The dynamic losses may be calculated from the equation

$$\frac{1}{2} \times Vin \times Id \times fs \times (tri + tfv)$$ Where Vin is the input

voltage, Id is the drain current, fs is the switching frequency and tri and tfv are the current rise time and voltage fall time.

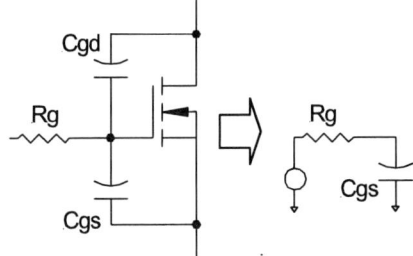

Figure 2. Simplified MOSFET Model

Solving the equations for the equivalent circuit in "Fig. 2," reveals the dependency of the rise and fall times of the current on both Rg and the gate-source capacitance (Cgs). The complete mathematical treatment is found in REF [12].

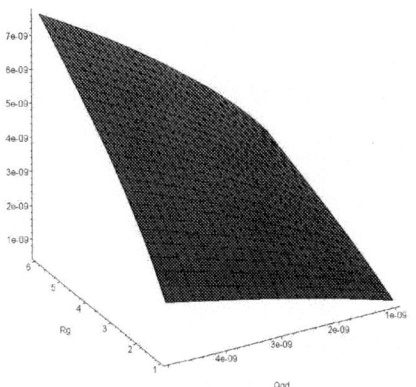

Figure 3. Effect of Rg on drain-source Voltage fall Time

"Fig. 3," depicts the drain voltage fall time as a function of Rg and the gate-drain capacitance Cgd. The voltage fall time may be expressed as follows

$tvf = k \times Rg \times Cgd$ where tvf is the fall time and K is a constant which is different for each case.

III. DISTRIBUTION OF DICE CURRENT

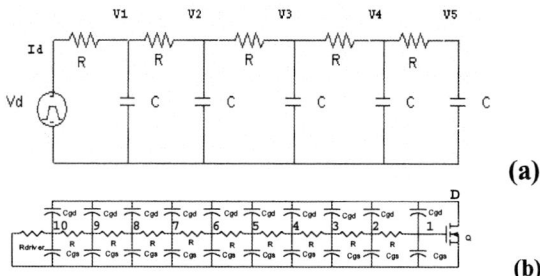

(a)

(b)

Figure 4. Simplified schematic for the mathematical simulation

"Fig. 4a&b," depict the distributed models of the MOSFET used in calculating the current distribution on the dice and the localized shoot through respectively. The dice area is divided in segments with the assumption that within each segment all currents and voltages are identical.

The complete mathematical treatment is found in REF [12] where equations were derived for the drain current in different segments. "Fig. 5," depicts the currents in different segments with the largest current is found in the segment closest to the gate pad.

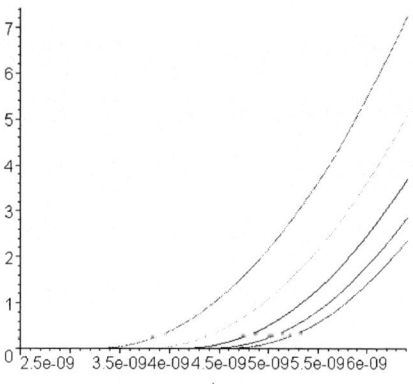

Figure 5. Current in different segments

This uneven distribution results in uneven power dissipation over the dice area and because these events take place in nanoseconds, the heating associated with it is very localized since there is no heat transfer to speak of to the environment during the transition which typically lasts for about 5ns-20ns. This localized heating

will drive the local temperature on the die to a temperature that is higher than the average die temperature hence, care must be exercised when calculating the allowable dynamic losses,

IV. DISTRIBUTED SHOOT THROUGH

MOSFET gate Equivalent Series Resistance, ESR, also plays a major role in determining the shoot through performance of power MOSFETs. Since ESR is a distributed resistive transmission line over the surface of the silicon die, this distribution leads to non uniform and uneven distribution of the shoot through currents and power dissipation over the dice. This means that in areas that are physically close to the gate pad, the power loss will be less than those areas that are the farthest away

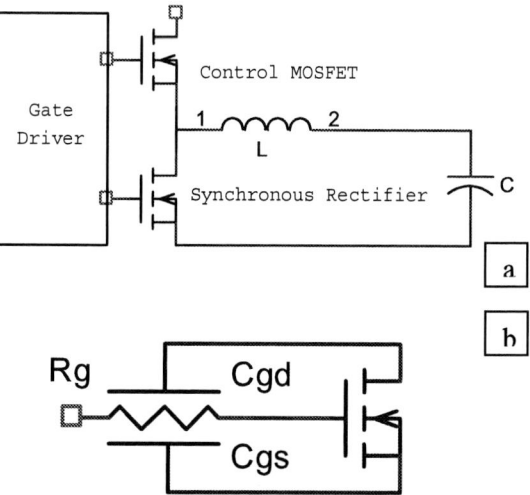

Figure 6. (a) Synchronous Buck Converter and (b) A representative circuit diagram of the distributed ESR (Rg), the gate-source Cgs and the gate-drain Cgd capacitors

The new finding that we are introducing in this paper is the distribution of shoot through current over different parts of the dice as a result of the effect of the distributed gate ESR resistor. "Fig. 6," (b) above shows a representation of Rg, the distributed gate-source capacitance Cgs and the distributed gate-drain capacitor Cgd "Fig 4b,". Based on the results, it is clear that cells or trenches at the end of the gate resistive transmission line has a larger percentage of Rg that connects it to the gate lead which leads to larger amplitudes of gate voltage during shoot through resulting in larger currents and higher power dissipation.

This circuit in "Fig. 6b," can be readily analyzed mathematically but the results will be in terms of the distance along the length of gate trace which may not be

1968

suitable to clearly convey the conclusions of this paper. Instead, we chose to divide the die into ten segments with the assumption that within a segment all conditions of voltages and currents are equal. This way we can see the effects as we move from the gate electrode into the dice. "Fig. 4b," depicts the equivalent circuit for the synchronous rectifier MOSFET (LS) used in the analysis. Notice that the gate driver resistance (Rdriver) is connected to ground indicating that the LS MOSFET is in the off condition.

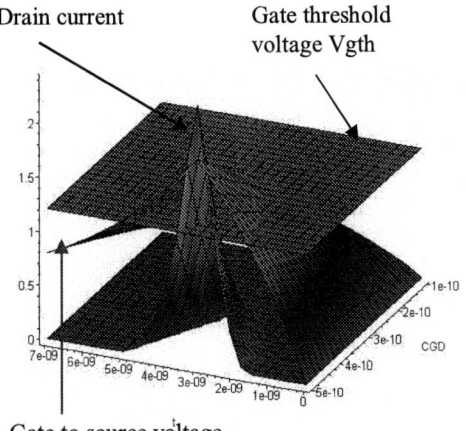

Figure 7. Drain current and gate-source voltage as a function of gate-source Capacitance Cgd

"Fig 7," shows the gate-source voltage of one segment of the synchronous rectifier, the gate threshold voltage plane and the drain current as a function of time and Cgd. The current starts conducting after the gate-source voltage level crosses the gate threshold voltage and spikes rapidly as the voltage increases.

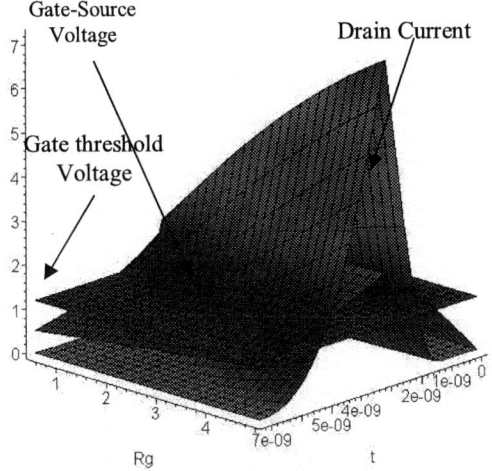

Figure 8. Drain current and voltage of one segment as a function of RG

"Fig. 8," shows the effects of the gate ESR, Rg, on the drain current and the shoot through voltage.

One may observe that while the influence of both Cgd and Cgs is limited in a very narrow window of typical values the effect of Rg is wide reaching along the entire range considered

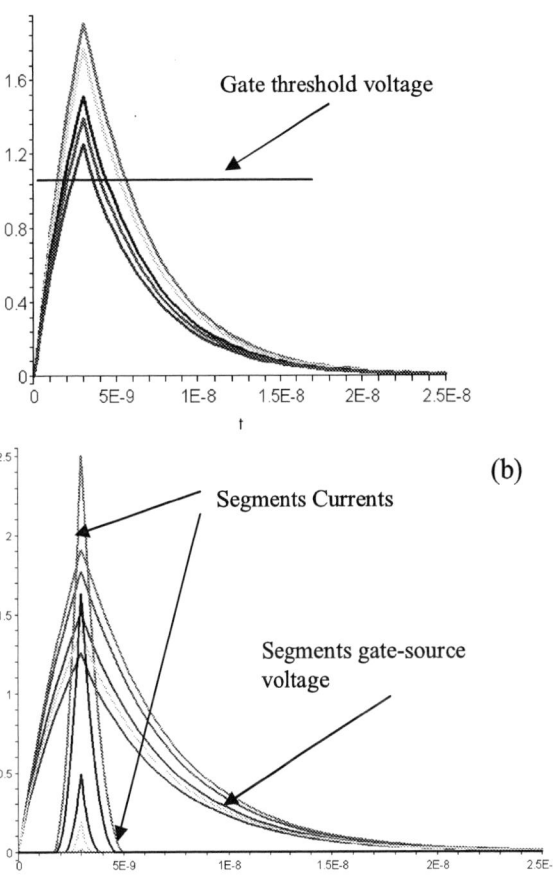

Figure 9. (a) Gate voltage at segments 1, 5, 8, 9 and 10 and (b) the same gate voltages with associated drain currents in the same segments

"Fig. 9," (a) shows the effective gate-source voltage at different segments and exhibits a wide range of voltage levels. The current flowing in each segment may be calculated from the following equation:

$$Isegment := (Vgs\_segment - Vgth)^2 Gm \quad \ldots\ldots\ldots\ldots\ldots (1)$$

Where Vgs_segment = Segment effective gate-source voltage, Vgth = Gate threshold voltage and Gm is the transconductance constant. The nonlinear dependency of the current on the voltage in equation 1 leads to segment currents that widely vary from almost no current to a large percentage of the total shoot through current as can be seen in "Fig. 9," (b).

The complete treatment of the subject may be found in REF[13]

V. CONCLUSION

1. We have shown that the gate ESR plays a major role in determining the losses in a switching device

2. We have shown that gate ESR plays an equal role as either gate-source or gate-drain capacitors

3. With modern technology it is easier to control the gate ESR and reap the benefits of lower losses

4. Considering the fact that the gate ESR, gate-drain and gate-source capacitors are distributed parameters results in a better understanding of the losses due to shoot through and current density distribution across the die

5. Gate ESR has a wide reaching influence on the shoot through behavior and needs to be reduced for improving the shoot through losses and also for improving the dynamic losses

6. Both Cgd and Cgs need to be optimized in conjunction with Rg to eliminate or at least reduce the shoot through

7. The optimization of the parameter in "Fig. 6," above requires an extremely careful consideration of the individual parameter and its effect on all the loss mechanisms in a given power MOSFET and all of these in conjunction with each other and the effect of a given combination of these parameters on the overall performance of the device in the intended application which in our case is a synchronous buck converter

8. With the knowledge that the shoot through losses are very localized, very careful attention must be given to evaluating the propensity of a given MOSFET for cross conduction and every effort must made to completely avoid it

REFERENCES

[1] Evaluating MOSFET Susceptibility for Cross-Conduction, Alan Elbanhawy, Fairchild Semiconductor, PCIM Magazine, Europe
[2] Shoot through analysis with parasitic inductance, Alan Elbanhawy, Fairchild Semiconductor, PCIM Europe Conference 2004
[3] Effect of Gate ESR on MOSFET Switching Losses, Alan Elbanhawy, Fairchild Semiconductor, PCIM Europe Conference 2005
[4] A. Elbanhawy, "Effect of Parasitic Inductance on switching performance" in *Proc. PCIM Europe 2003, pp.251-255*
[5] A. Elbanhawy, "Effect of Parasitic inductance on switching performance of Synchronous Buck Converter" in *Proc. Intel Technology Symposium 2003*

[6] A. Elbanhawy, "Mathematical Treatment for HS MOSFET Turn Off" in *Proc. PEDS 2003*
[7] A. Elbanhawy, "A quantum Leap in Semiconductor packaging" in Proc. *PCIM China, pp. 60-64*
[8] A. Elbanhawy, "The Road to 200 Amps at one Volt VRM" in Proc. PCIM Europe 2004, pp. 54-58
[9] A. Elbanhawy and W. Newberry, "Packaging Parasitic Resistance Frequency Effects" Power Electronics Conference USA 2004 PCIM San Francisco"
[10] A. Elbanhawy and J. Ejury, "Investigations of the Influence of PCB Layout Parasitic Inductances In DC/DC-Converters on the Efficiency" in Proc. PCIM Europe 2004, pp. 31-36
[11] A. Elbanhawy "Are traditional packages suitable for the new generation of DC-DC converter" in Proc. IPEMC China 2004
[12] Effect of gate ESR on MOSFET Switching Losses, Alan Elbanhawy, PCIM Europe Conference 2005
[13] Effect of Gate ESR on Localized Shoot Through on the MOSFET Die, Alan Elbanhawy, IPEC 2005 Conference, Niigata Japan

2006 5th International Power Electronics and Motion Control Conference

PWM Rectifier with DC Reverse-Blocking Diode for High-Reliability Generating Apparatus and Its Application to Gas Heat Pump System.

Akio Toba *, Toshihiro Maeda *, Kouetsu Fujita *, Tomohiko Kato **

* Fuji Electric Group, Japan.
** Aisin Seiki Co. Ltd., Japan.

Abstract—High reliability generating apparatus with a PWM rectifier equipped with a DC reverse-blocking diode is proposed. The PWM rectifier adjusts the power from a permanent magnet synchronous generator by means of position sensorless control. Due to the DC diode, the PWM rectifier is shoot-through free, i.e. the switches in one arm are not broken with their simultaneous turning-on caused by electromagnetic interferences, etc. The apparatus is applied to the Gas Heat Pump system (GHP), which is an efficient air-conditioning system powered by fueling gas, to satisfy its requirement of high reliability to the electrical equipment. There is a control issue in the proposed circuit, that is the DC bus capacitor is electrically separated from the switches when the DC diode is reverse-blocking, which results in the voltage fluctuation at the switches side. This is overcome by adding a stabilization control. The system is tested with a 1.5kW-class experimental setup, which demonstrates desirable performances as expected, and it is confirmed that the rectifier is durable to the simultaneous turning-on of the switches in one arm.

Keywords-Gas heat pump(GHP), Air-conditioning, Generating apparatus, Reliability, PWM rectifier.

I. INTRODUCTION

PWM rectifier is widely used for high power factor AC-to-DC power conversion. One major AC power source to which the PWM rectifier is connected is the electric power grid. Another emerging application is the power controller for an AC rotating generator in the distributed power system as wind turbines, micro-hydroelectric generators, etc. In these systems, the PWM rectifier realizes high power factor, sinusoidal current, and high efficiency, which is ideal to make full use of the latent energy of the power source via the AC generator.

There is one such system in which the high efficiency power conversion is required. That is the Gas Heat Pump system (GHP) with auxiliary generating apparatus. The GHP is an efficient air-conditioning system, in which a compressor is driven by a gas engine, and is broadening its application to the facilities as super-markets, schools, office buildings, etc. The authors have developed the GHP with generating apparatus, of which a sensorless-controlled permanent magnet generator is driven by the gas engine and feeds the electrical equipment in its outdoor unit[1]. Figure 1 shows the overview and system configuration of the outdoor unit of the developed GHP. The purpose of adding the generating apparatus is to operate the system only with the gas, even though the electric power grid is connected for backup as shown in the circuit diagram in Fig. 2.

An important requirement for the generating apparatus is its safety, because the system uses fuelling gas. A critical issue with this aspect is the shoot-through of the

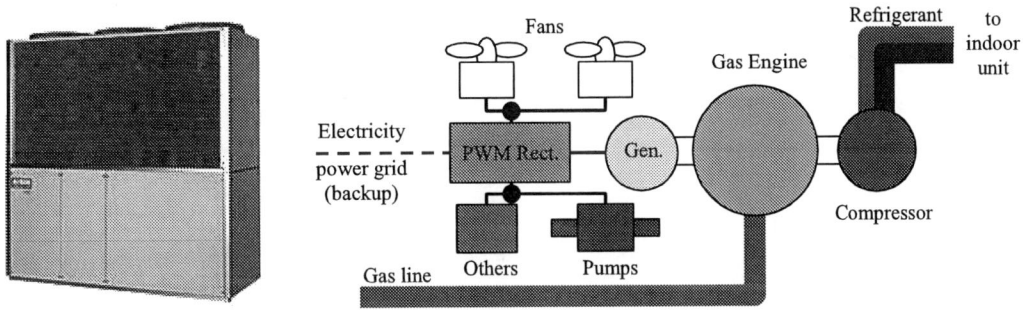

(a) Overview.   (b) System configuration.

Fig. 1. Outdoor unit of the Gas Heat Pump system with generating apparatus.

1-4244-0448-7/06/$25.00 ©2006 IEEE

rectifier arm. That is, the rectifier can be broken if the series connected two switches in one arm are turned-on simultaneously for a small fraction of time as 10 ms. Due to the electromagnetic interference of all kinds, it is still difficult to guarantee a completely shoot-through-free rectifier. One solution for this problem is to put a fuse to the DC bus. However, the dimension of the fuse is usually large and then the packaging problem occurs.

Then the authors propose a PWM rectifier with a reverse-blocking diode at its DC bus. With the DC diode, the shoot-through is no longer occurred by the switching pulse failure. Compared to the fuse solution, the packaging is easier, and the uninterruptible operation can be expected even after a short period of switch-gating failure. Circuit design, a control issue, and experimental results are described in this paper.

II. CIRCUIT

A. Circuit and its Variations

Table 1 summarizes the circuit types that have DC diode(s). Type A2 in Table 1 is chosen for the research, while there can be another variation as Type A1, in which an electrolytic capacitor and a snubber capacitor are both at the switches side of the rectifier. Because of the configuration, control of Type A1 can be conventional. Nevertheless, since Type A1 has no shoot-through durability, because the switches can short-circuit the electrolytic capacitor with large capacitance, it must be

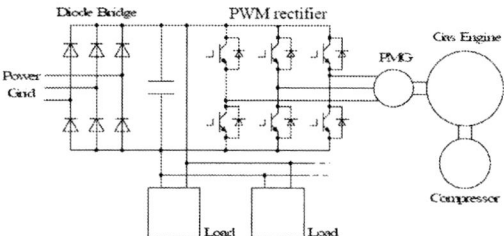

Fig. 2. Circuit diagram of the developed generating system[1].

abandoned. Type A2 is on the other hand shoot-through free due to the small capacitance of the snubber capacitor, while its control must be investigated fully since the electrolytic capacitance can not work to stabilize the operation when the DC diode is reverse-blocking.

Another type of shoot-through free rectifier is shown as Type B in Table 1, which can be found in the literature[2]. This type has only one switch in each arm. Despite the merits as low switching loss and smallest number of switches, unavoidable current distortion is a critical problem in the system that utilizes position sensorless control of the PM generator.

B. Design

Points of the circuit design are followings: 1) The DC diode must be a fast-recovery type, and 2) the capacitance of the snubber capacitor must be the value to avoid damage to switches when the two switches in one arm are turned-on at the same time. Described here is the design consideration for the capacitance:

Assumptions:

1) The energy stored in the snubber capacitor is dissipated and transferred to heat in the IGBTs in one rectifier arm, when a shoot-through is occurred.

2) The temperature rise is so quick that the heat can not spread to the outside of the IGBTs.

Based on these assumptions, the temperature rise ΔT of the IGBT chip can be calculated from the following simple equations.

$$C_h \text{ [J/K]} = m \text{ [g]} \times c \text{ [J/(g K)]} \tag{1}$$

$$\Delta T \text{ [K]} = Q \text{ [J]} / C_h \text{ [J/K]} \tag{2}$$

where C_h; heat capacity, m; mass of an IGBT chip,

c; specific heat, and Q; amount of energy to be dissipated.

Substituting the values in the test system with the snubber capacitance of 4.4 μF, ΔT is calculated to be only 11.6K, which is small enough to avoid the damage of the IGBTs.

Table 1. Comparison of rectifier types with DC reverse-blocking diode(s).

	Type A1	Type A2	Type B
Circuit	(Electrolytic C. Snubber C.)		
Positions of Capacitors	All at switches side	Electrolytic capacitor at load side Snubber capacitor at switches side	- - -
Control	- Good - Conventional control.	- Caution - Unconventional operation mode when DC diode is reverse-blocking.	- Unconventional - Unconventional control. Current is distorted.
Durability to shoot-through	- NG - Switches are broken.	- Good - Shoot-through free.	- Good - Shoot-through free.
Efficiency	- Fair -	- Fair -	- Good - Small number of switches. No DC diode.

III. CONTROL ISSUES

When the DC diode is conducting, the operation of the rectifier type A2 in Table 1 is the same as the conventional one. However, as mentioned above, when the DC diode is reverse-blocking, the electrolytic capacitor is electrically separated from the rectifier and then the voltage at the snubber capacitor fluctuates.

Figure 3 shows the basic control block diagram for the proposed rectifier. The voltage of the electrolytic capacitor E_{dcdet} is detected for feedback control. E_{dcdet} is compared to the reference voltage E_{dcref} and the error is input to the PI regulator to yield the q-axis current reference I_{qref} for the "sensorless current control" block. In the sensorless current controller, there are two current regulators for d- and q- axis currents, i.e. for the magnetizing and the torque components of the generator current, respectively, and for an estimator of the rotor

position[1]. Then the voltage references are produced and modulated to the PWM signals for the switches.

Hence, if the q-axis current becomes positive, which means the power flow is from the rectifier to the generator, e.g. in the case the DC load decreases suddenly, voltage fluctuation occurs as shown in Fig. 4. This phenomenon can be described as follows, with referring the mode numbers in Fig. 4:

[1] By some reasons, the voltage E_{dc1} rises, always accompanied with the rise of E_{dc2}.

[2] When E_{dc2} becomes larger than the reference value E_{dcref}, the output of the PI regulator I_{qref} is limited to the positive limit value I_{LIM-P}. I_{LIM-P} can not be set to zero to absorb the errors that are inherent to the system, otherwise the actual power flow from the generator to the rectifier may not become zero when no power is required in the DC load, resulting in the

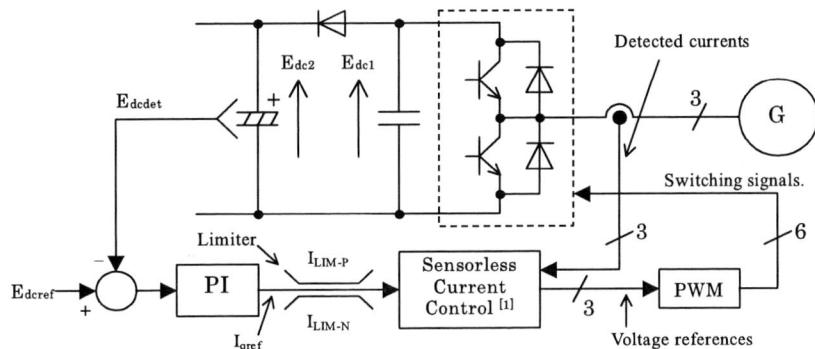

Fig. 3. Basic control block diagram for the proposed rectifier.

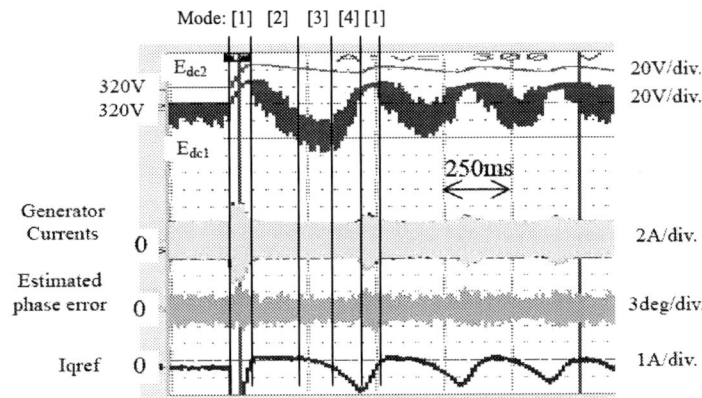

Fig. 4. DC voltage fluctuation with the control configuration in Fig. 3.

Table 2. Specifications of generating system.

GENERATOR	Type	Number of Poles	Revolution range	Rated revolution
	Surface permanent magnet	8	1667 - 4000r/min.	3333r/min.
	Rated output	Rated current		
	1450W	7.2A$_{rms}$		
RECTIFIER	Type	AC input voltage	DC diode	Generator control
	PWM boost rectifier	200Vrms	Fast recovery type	Position sensorless

endless rise of the E_{dc1} and E_{dc2}. As a result of a positive value of I_{qref}, the snubber capacitor is discharged with high rate of change because of its small capacitance, while the electrolytic capacitor is discharged slowly by small load and leak current.

[3] After the voltage E_{dc2} falls below E_{dcref}, I_{qref} becomes negative and the fall of E_{dc1} is stopped.

[4] Despite the rise of E_{dc1}, I_{qref} still decreases until E_{dc1} reaches E_{dcref}. In this interval, the integrator in the PI regulator keep storing its input, i.e. the error between E_{dcref} and E_{dcdet}.

[1] Even after E_{dc1} surpasses E_{dcref}, it still increases until the integrator finishes releasing the stored amount. Then the same sequence is repeated.

This phenomenon can be avoided if the feedback control of the DC voltage is based on the detected voltage of the snubber capacitor, not the electrolytic capacitor as shown in Fig. 3. However, this means that two voltage detectors are required in the system, since the voltage detection of the electrolytic capacitor can not be eliminated to protect the DC bus from an over-voltage, and it should be avoided in terms of cost.

Then a stabilizing element is added to the controller, that is, the integrator is cleared and held to zero when I_{qref} becomes positive. This technique enables the stable operation of the system with one voltage detector as shown in Fig. 3. The effect of this method is described in the next section.

IV. EXPERIMENTS

Table 2 and Figure 5 show the specification and an overview of the experimental system, respectively.

Figure 6 depicts the system response to a speed change with the rated DC load, i.e. the DC diode is always conducting. As is expected, it is confirmed that the system is stable and no difference from the conventional type is observed if the DC diode is conducting. Other tests to investigate the system stability have been carried out, which ensured the stable operation of the proposed circuit. Even with the light or no load, in which the DC

voltage fluctuation occurs, the system never falls into continuous oscillation or divergence.

Figure 7 represents the effect of the stabilizing element mentioned in the previous section. Even after a larger drop of the DC voltage than the case shown in Fig. 4, voltage fluctuation does not occur. This is because the integrator of the PI regulator is held to zero when I_{qref} becomes positive and is not enabled after a certain duration from the time that I_{qref} goes into negative. Due to this operation, there is no overshoot when E_{dc1} reaches E_{dcref}. Through this explanation and discussion, the effectiveness of the stabilization is confirmed.

Figure 8 shows the system operation when the snubber circuit is short-circuited for $10\mu s$, which is long enough to destroy the switches in the conventional rectifier. As to be seen, the snubber capacitor voltage falls to zero and back to the normal value immediately after the short-circuit, while the electrolytic capacitor voltage keeps the reference value. Moreover, the system keeps operating normally after the short-circuit.

From the above tests, the proposed circuit is confirmed to perform as expected and to be durable to the real use.

V. CONCLUSIONS

A new type of PWM rectifier with no shoot-through is proposed, and its application to the Gas Heat Pump air-conditioning system with auxiliary generation apparatus is described. Since the GHP requires high reliability to the equipment, the proposed configuration is desirable. The circuit configurations are investigated and then the positions of the capacitors are determined. A problem in terms of control is pointed out and examined, which is a fluctuation of the DC bus voltage, and its solution is introduced. From the experimental results, it is shown that the proposed configuration operates as expected.

The proposed system can be used not only in the GHP but also in many applications, if the bi-directional power flow is not required to the rectifier.

Several patents are applied with the proposed configuration and its related techniques in a few countries

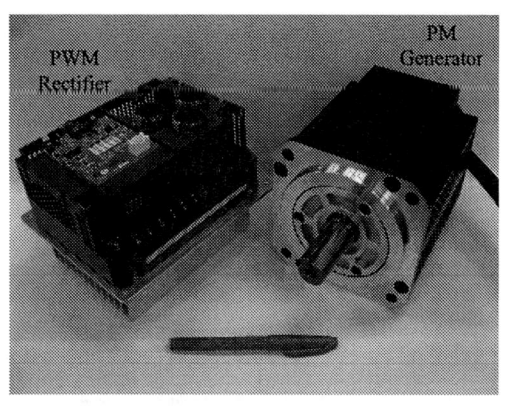

Fig. 5. Overview of the test system.

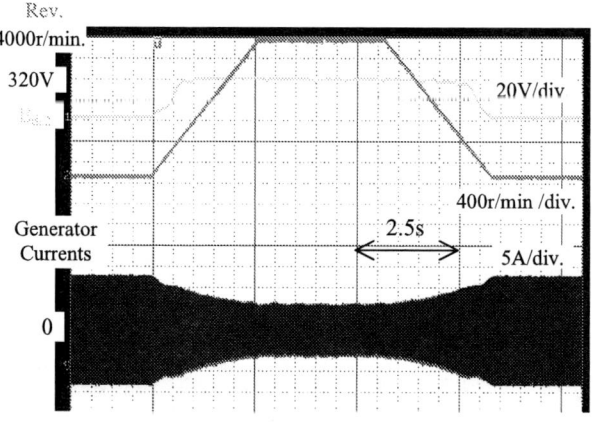

Fig. 6. System response to speed change with rated load.

including Korea and China.

REFERENCES

[1] A. Toba, K. Fujita, T. Maeda, T. Kato, "Generating Apparatus for Gas Heat Pump System using Sensorless-Controlled Permanent Magnet Synchronous Generator," Proceedings of The 2005 IEEJ International Power Electronics Conference, pp. 1549-1554 (2005)

[2] J. Kikuchi, M.D. Manjrekar, T.A. Lipo, "Performance Improvement of Half Controlled Three Phase PWM Boost Rectifier," Proceedings of IEEE Power Electronics Specialists Conference 1999, pp. 319-324. (1999)

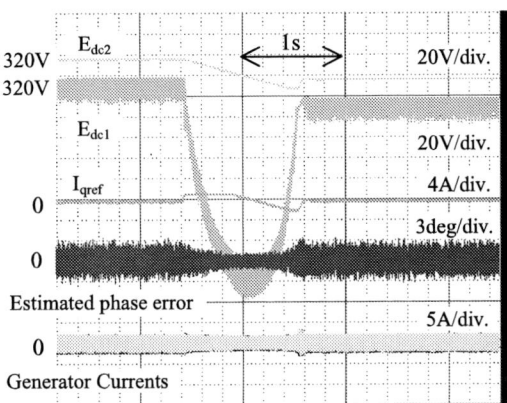

Fig 7. Effect of the DC voltage stabilization.

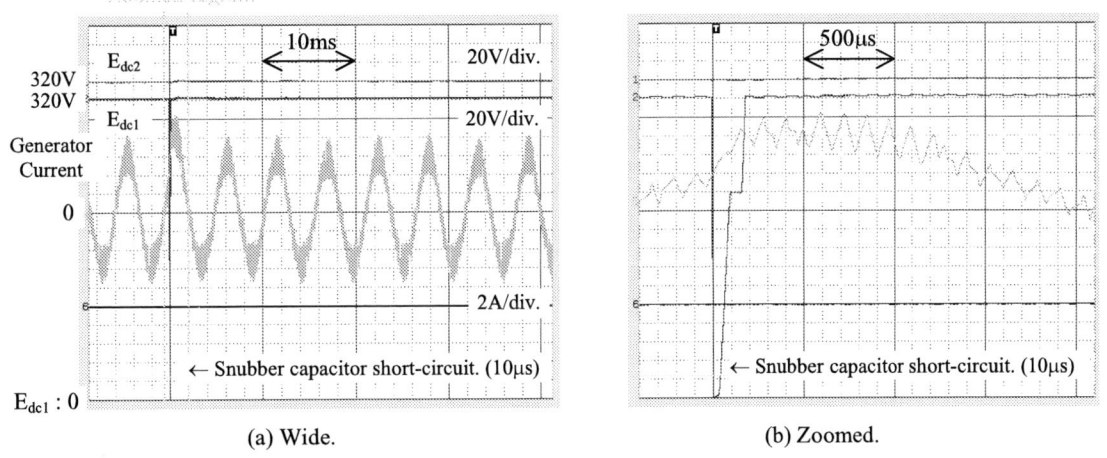

(a) Wide.   (b) Zoomed.

Fig. 8. System response to the snubber capacitor short-circuit.

2006 5th International Power Electronics and Motion Control Conference

A Novel Stator Section Crossing Method of Long Stator Linear Synchronous Motor for Maglev Vehicles

Qian Zhang, Fei Lin, Xiaojie You and Trillion Q. Zheng

School of Electrical Engineering, Beijing Jiaotong University, Beijing, China

Abstract- **This paper proposes a new stator section crossing method that can overcome the drawbacks of the tow-step and three-step methods used in Shanghai maglev railway. This method has no descending of propelling force and idleness of PWM converters. The mathematic model and vector control strategy of the permanent magnetic long stator linear machine (PMLSM) are also researched in this paper. The effectiveness of the proposed section crossing method and control approach are validated by simulation results.**

Keywords- linear synchronous motor; long stator; section crossing method; vector control

I. INTRODUCTION

Linear synchronous motor (LSM) and linear induction motor are both used in Maglev drive system. The typical application of short stator linear induction motor is HSST system of Japan. HSST system is controlled onboard and the train gets power from two power-supplying rails. It is difficult for the train to get power supply from power-supplying rails, so the HSST system is suitable in the case of fairly low speed. The typical applications of LSM are Transrapid of Germany and MLX of Japan. Transrapid and MLX are driven by carriageways, the control and power conversion systems are on ground. The speed of them is able to be 500 km/h, so they are high speed Maglev [1]. The Maglev vehicle has been constructed in Shanghai is Transrapid system. For LSM, to achieve higher efficiency and reduce the reactive power production the stator is arranged by many sections, and we only active the section where the mover is.

LSM, as a core technology of the high speed Maglev system in Shanghai, is a research hotspot nowadays. The excitation of mover of LSM can be electromagnetic, permanent magnetic or the hybrid of the two kinds upwards. It is necessary to keep the vertical

magnetic field steady for an invariable levitating force, so LSM can be simplified to be PMLSM when we research the section crossing technique and the control of horizontal propelling force. This paper research the vector control of PMLSM and presents a novel section crossing method which can overcome the drawbacks of the section crossing methods used in Shanghai maglev railway.

II. CONFIGURATION AND MODEL OF PMLSM

The permanent magnets of PMLSM are underneath the vehicle, and the stator windings are on carriageways. Traveling wave is generated when three-phase currents flow in three-phase stator windings. The interaction of traveling wave and the permanent magnets underneath the vehicle produces a propelling force that leads the vehicle ahead, as fig.1 shows: τ is the distance between two permanent magnet poles; σ is the air gap between stator and mover which is detected and close-loop controlled by the control system of Maglev train. A small air gap is propitious to efficiency and power factor but make the control more difficult. Voltage of power supply and parameters of the PMLSM will determine the propelling force. The propelling force and resistant force such as wind-resistance will determine the acceleration of vehicle. The frequency of power supply will determine the final speed in steady-state.

To save energy and get high efficiency, the stator of PMLSM is arranged by many sections and only the section where the mover is should be drove. Each section is much longer than a vehicle so only some of the stator windings can interact with permanent magnets. The magnetic field inducted by the stator windings which is under the magnetic poles (permanent magnets on board) and the magnetic field inducted by the magnetic poles constitutes the main magnetic field, so the inductance of

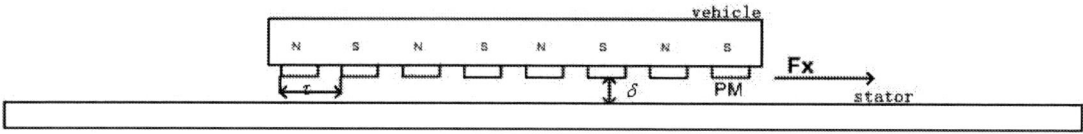

Fig.1 configuration and basic principle PMLSM

this part of stator windings is main-inductance. The magnetic field inducted by stator windings which have no magnetic poles forms loop with the air and constitutes the leak magnetic field, so this part of stator windings is leakage-inductance [2-3].

Ignore the transverse and longitudinal ending effect, under the d-q-0 coordinate, the voltage equation of PMLSM can be given by

$$U_q = Ri_q + \frac{\pi V_s}{\tau}\Psi_d + \frac{d}{dt}\Psi_q \quad (1)$$

$$U_d = Ri_d - \frac{\pi V_s}{\tau}\Psi_q + \frac{d}{dt}\Psi_d \quad (2)$$

The flux linkage expression is given by

$$\Psi_d = i_d(L_\delta + L_{md}) + \Psi_f \quad (3)$$

$$\Psi_q = i_q(L_\delta + L_{mq}) \quad (4)$$

The propelling force on the steady state is

$$F_x = \frac{3}{2}\frac{\pi}{\tau}(\Psi_d i_q - \Psi_q i_d) \quad (5)$$

Where $L_d = L_{md} + L_\sigma$, $L_q = L_{mq} + L_\sigma$, L_d is d-axis inductance which contains L_{md} and L_σ, d-axis main-inductance and d-axis linkage-inductance respectively, and L_q is q-axis inductance which contains L_{mq} and L_σ, q-axis main-inductance and q-axis linkage-inductance respectively. V_s is speed of vehicle or magnetic poles, and Ψ_f is magnetic flux inducted by permanent magnets in stator windings.

A fourth order state equation can be deduced form the series of expressions upwards

$$\dot{i}_d = \frac{1}{L_d}(U_d - Ri_d + \frac{\pi V_s}{\tau}L_q i_q) \quad (6)$$

$$\dot{i}_q = \frac{1}{L_q}(U_q - Ri_q - \frac{\pi V_s}{\tau}L_d i_d - \frac{\pi V_s}{\tau}\Psi_f) \quad (7)$$

$$\dot{V}_s = (F_x - f)/m \quad (8)$$

$$\dot{\theta} = \frac{\pi V_s}{\tau} \quad (9)$$

III. VECTOR CONTROL OF PMLSM

The vector control is a high-dynamic and high performance control strategy for ac motor drive[4-5]. The basic principle of vector control of PMLSM is to decouple control of stator current [6], so the propelling force is only relative to flux and current. This is somewhat like the control of DC motor. High performance can be achieved through vector control.

The vector control of PMLSM is somewhat simpler than linear induction motor. The position of mover magnetic flux is the same as the mechanical position of mover, so the position of mover magnetic flux can be ascertained by detecting the actual position of mover. There phase alternating current, i_a, i_b, i_c, can be converted to two direct currents in synchronous rotating coordinate, i_d, i_q. The active stator winding is much longer than the length of vehicle, so the leakage-inductance is much bigger than the main-inductance. So we can ignore the difference of the d-axis main-inductance L_{md} and q-axis main-inductance L_{mq} for simplicity when we use equivalent circuit of PMLSM to calculate the performance. The following expression of propelling force can be deduced from equation (3), (4) and (5)

$$F_x = \frac{3\pi}{2\tau}[i_d i_q(L_d - L_q) + \Psi_f i_q] \quad (10)$$

It can be deduced that F_x is direct proportion to i_q when L_d is approximately equal to L_q, so the F_x can be controlled by controlling of i_q. When i_d is kept to be 0, we can achieve the maximal propelling force via minimal current magnitude. After i_d, i_q is ascertained, we can convert the i_d, and i_q to i_a, i_b, i_c, which are current commands given to PWM converter through dq_to_abc transformation.

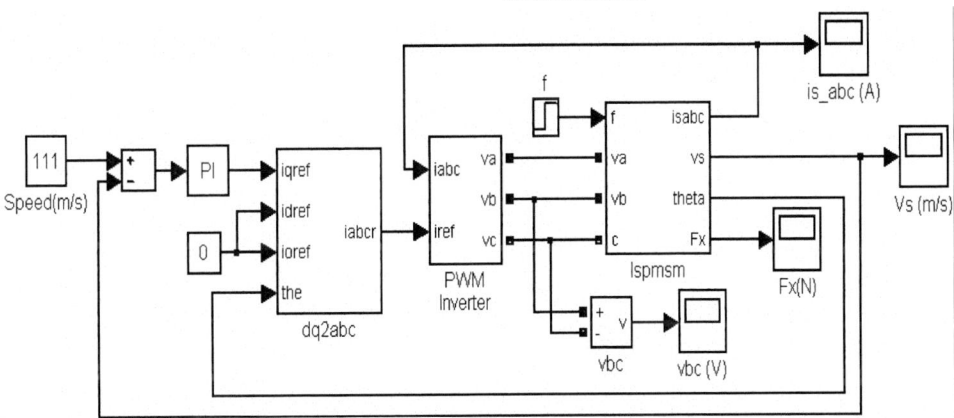

Fig.2 Vector control system of PMLSM

IV. THE SECTION CROSSING METHOD

4.1 Technique background

For the existed high speed maglev railway in Shanghai, the stator is arranged by several electrically independent sections, and we drive only the section in which the mover is to achieve higher efficiency and reduce the reactive power production. Two-step method and three-step method are both used for Shanghai maglev railway. Two-step method is used for pulling the vehicle into rail station or pulling the vehicle outside rail station in low speed. Three-step method is used for drive the vehicle in high speed between two rail stations [1].

Two-step method is to arrange the stator to be interlaced sections, has two PWM converts with two sets of cables to supply power to the two side of carriageway, each side is an independent LSM. When the vehicle in section crossing state, the current of the crossing LSM gradually descends to zero, then turn off the switches which connect the cable and former section and turn on the switches which connect the cable and the latter one. The crossing side cannot propel the vehicle and only another side is propelling the vehicle, this causes the descending of propelling force. For two-step method stator section is arranged as fig.3

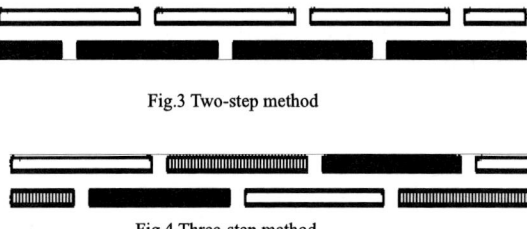

Fig.3 Two-step method

Fig.4 Three-step method

Three-step method is by setting three PWM converters with three sets of cables to resolve the drawback of two-step method, descending of propelling force in section crossing state. When vehicle is in stator crossing state, the current in former section gradually descends to zero and simultaneity the current in latter section gradually ascends. When vehicle is not in section crossing state one of the three PWM converts is in idleness. The idleness of one PWM convert increases the cost of maglev system. With three-step method, when in section crossing state, the vehicle bestrides on two stator sections, to keep the current in these two stator sections invariable is fairly complicated. For three-step method stator sections is arranged as fig.4.

4.2 A novel section crossing method

This paper presents a novel method that causes no descending of propelling force, has no idleness of PWM converts and is simple to be control when in section crossing state. The stator is arranged as many sections, these sections is not electrical independent but consecutive. The sections in the two sides are not interlaced but absolutely symmetrical. Two PWM converters and two sets of cables are needed to supply power to the two symmetrical LSM on both sides of carriageway. It is arranged as fig.5.

We active only the section that has magnetic poles and disable the sections that have not magnetic poles by controlling the switches beside the carriageway. These switches can be mechanical switches or electrical switches. Because three phases electricity is supplied, each phase has its independent switch. The switches connecting cables and stator windings are called as upper-switch, such as S^{n-1}_{upper} and S^n_{upper}, and the switches connecting neutral point and stator windings are called bottom-switches, such as S^{n-1}_{bottom} and S^n_{bottom}. An upper switch and a bottom switch connecting the same section and in the same phase constitute a switch set, for instance S^n_{upper} and S^n_{bottom} constitute switch set n. The distance between an upper switch and bottom switch in the same set should be longer than two times of vehicle's length. The stator windings between an upper switch and bottom switch in the same set is a power-supply section, for example power-supply section n-1 is between the switch set n-1. Adjacent two power-supply sections overlap each other, and the overlapped part is called as a stator-crossing section. For example the overlapped part of power-supply section n-1 and power-supply section n is called as stator crossing section n-1. The stator-crossing section should be longer than the length of vehicle. Because the stator windings on both sides of the carriageway are symmetrical and the configuration of phase B, C is the same as phase A, only one phase in one side is shown in sketch map for concision as fig.5.

When the vehicle is in power-supply section n and not in stator-crossing section n-1 or stator-crossing section n, the switch set belong to the power-supply section n, S^n_{upper} and S^n_{bottom}, are turned on, all other switches are turned off. When the vehicle is in section-crossing section n, the overlapped part of power-supply section n and power-supply section n+1, switch set n turn off and switch set n+1 turn on simultaneous. By this way the vehicle successfully leave the power-supply section n and enter the power-supply section n+1. Via this method only two PWM converters are needed, the same as two-step method. No propelling force loss is achieved, the same as three-step method. We only control the turning on and turning off of the sets of switches, we don't need to control the converters to control the stator current.

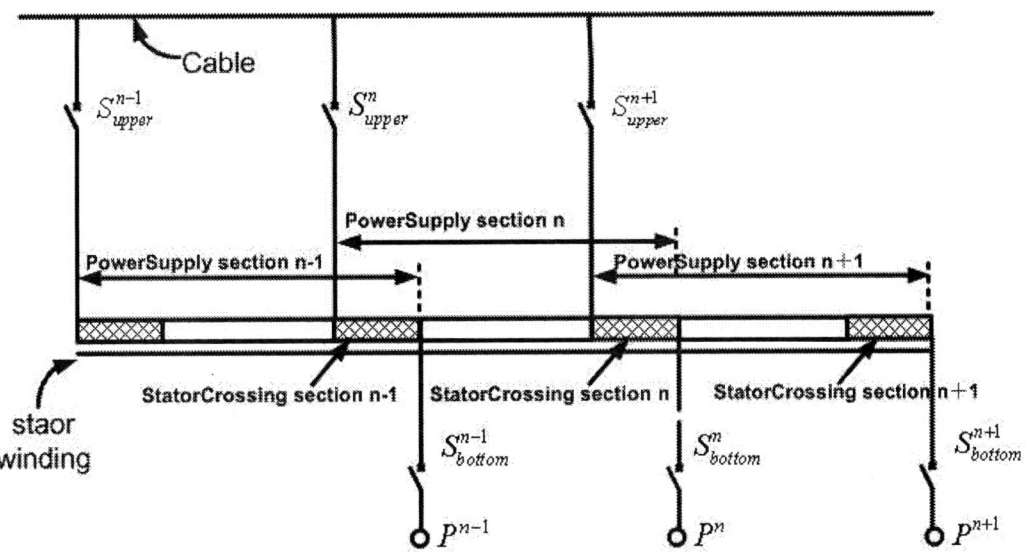

Fig.5 Sketch of the stator crossing method

V. SIMULATION RESULTS

The parameters used in the simulation are as follows, rated velocity is 400km/h, phase rated voltage is 7760 V, rated frequency is 215 Hz, and polar distance is 258mm, stator leakage inductance 1.4974mH, d-axis synchronous inductance 0.8856mH, and q-axis synchronous inductance 0.6986mH, magnetic flux in stator windings inducted by permanent magnets is 0.272Wb.

If we start the motor directly, and give resistant force at 0.07s, the curve of velocity, propelling force and stator current are as follows.

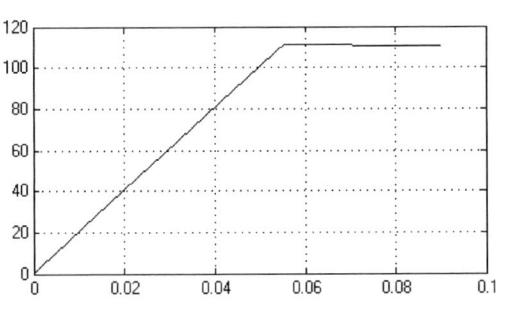

Fig.6 curve of velocity

From the above curve, we can get that velocity in steady state is smooth and can trace the synchronous velocity. Motor is started by unchanged propelling force. In accelerating stage the stator current is fairly big and acceleration is steady. In the steady state and before the resistant force is given, the stator current surge around zero. After the resistant force is given the stator current ascends to a special value and the propelling force trace to equal the resistant force.

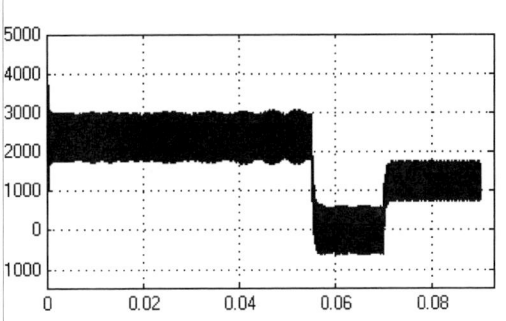

Fig.7 curve of propelling force

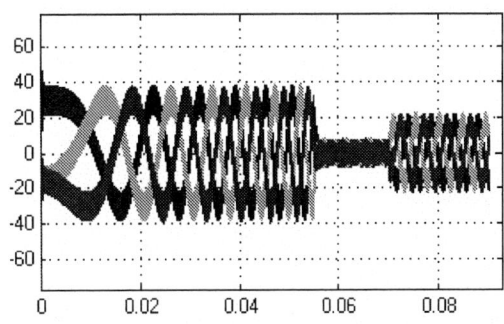

Fig.8 curve of stator current

According to the section crossing method proposed in this paper, from 0.08s to 0.084s, we change the active section from one to another by turning off the former switches when current crosses zero and turn on the latter switches simultaneously. The stator current and vehicle speed curve on the section crossing time are as follows.

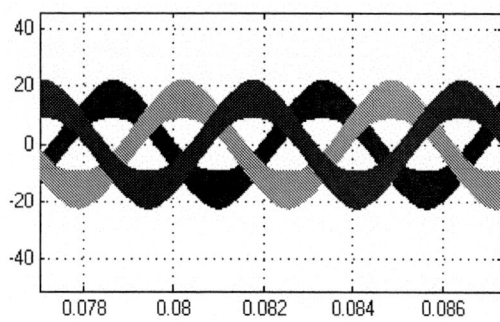

Fig.9 current curve on stator crossing time

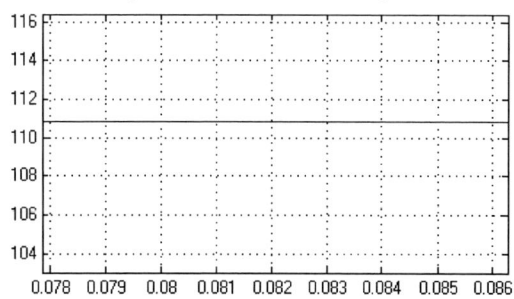

Fig.10 speed curve on stator crossing time

We can get from above two figures that vehicle speed and stator current are both smooth and unchanged before and after the crossing time. So we can have a smooth section crossing via the novel section crossing method.

VI. CONCLUSION

This paper presents a novel section crossing method which can overcome the drawbacks of the section crossing methods used in Shanghai maglev railway. Only two PWM converters are needed, the same as two-step method. No propelling force loss is achieved, the same as three-step method. We don't need to control the converters to regulate the stator current when section crossing. The simulations of the section crossing method under vector control of PMLSM show the effectiveness.

VII. ACKNOWLEDGEMENT

This research has been sponsored by the Important Special Item of High Speed Maglev Transportation Technology in High-tech Research and Development Program of China (contract number: 2005AA505101-629).

REFERENCES

[1]Xiangming Wu. "Maglev Trian". Shanghai scientific and technical publisher.2003

[2] Z. Deng, I. Boldea, and S. A. Nasar "Forces and parameters of permanent magnet linear synchronous machines" *IEEE Trans. Magn.*, vol. 23, pp. 305–309, Jan. 1987.

[3] Bolden.L. Nasar.S.A. "Linear electric actuators and generators," Energy Conversion. IEEE Transactions , Volume,14 Issue: 3 , Sep 1999 page: 712-717

[4] Zhang, B.; Pong, M.H.; "Maximum torque control and vector control of permanent magnet synchronous motor," International Conference on Power Electronics and Drive Systems, 1997. Volume 2, 26-29 May 1997 Page(s):548 - 552

[5] Fei Yu; Xiaofeng Zhang; Huaishu Li; Zhihao Ye; "The space vector PWM control research of a multiphase permanent magnet synchronous motor for electrical propulsion," ICEMS 2003. Volume 2, 9-11 Nov. 2003, Page(s):604 - 607

[6] Roberto Leidhold and Peter Mutschler. "Speed Sensorless Control of a Long-Stator Linear Synchronous-Motor arranged by Multiple Sections". IECON. 2005, 6-10 Nov., 2005 Page(s):1395 - 1400

2006 5th International Power Electronics and Motion Control Conference

Common Mode Current Suppression in Full-Bridge Converter Based on Simulated Annealing Algorithm***

Yonggao Zhang*,**, Kai Zhang*, Yunbin Zhou* and Yong Kang*

* College of Electrical and Electronics Engineering, Huazhong University of Sci. & Tech. Wuhan 430074, P.R China
** School of Electrical and Electronics Engineering, EastChina Jiaotong University, Nanchang 330013, P.R China
E-mail: zhkhust@263.net

Abstract—In PWM converter, the serious electromagnetic interference (EMI) is occurred due to high speed operation of switch devices. In this paper, production mechanism of common mode (CM) current in single phase full-bridge converter is analyzed, and severe affection to CM current owing to different transmission delays of two drive pulses is illustrated. It is proposed a digital CM current suppression method based on optimal drive pulses. A novel evaluation measure for CM noise level based on the energy of CM noise current is proposed. The relationship between CM noise level and the delay time of drive signals is studied. Simulated annealing searching algorithm is used to search for the global optimum of CM current energy. Simulated results show that CM current suppression based on simulated annealing algorithm (SAA) can regulate the delay time of drive signals without getting trapped at local minimums. It suppresses CM current level effectively and improves the EMC performance of PWM converter.

Keywords-PWM converter, common mode current, drive signals, common mode current energy, simulated annealing algorithm

I. INTRODUCTION

With the application of high-speed high-power switch devices, the dynamic performance of power electronic products has been improved and the volumes and weight have become small. However, severe electromagnetic interference (EMI) is produced due to high dv/dt and high di/dt caused by high speed switching of power devices. There are basically two types of EMI in power electronics, namely common mode (CM) current and differential mode (DM) current. DM current flows in and out of a power device through its input power cables. CM current flows in the same direction through two power cables. CM current usually constitutes a major part in total EMI current [1-4].

At present, most study of power devices EMI is concerned on switching power supply and three-phase

PWM inverter system[5-9]. In this paper, CM conducted EMI in single-phase full-bridge converter is analyzed. Compared CM conducted EMI in full-bridge circuit with differential modulate technique. It is conclusion that the conducted CM current with bipolar SPWM modulation technique is smaller than that with unipolar SPWM modulation technique because the two legs provide perfect CM current compensation for each other. However, different transmission delays of two drive signals in bipolar SPWM modulation technique always produce large CM current in real case. It is important for CM current suppression to regulate of time delay of two drive signals.

This paper proposed a novel time delay regulation method with digital closed-loop control. It has the potential of regulating the time delay of two drive signals by using some predictive control algorithm, or it can take advantage of the switching information provided by the main digital controller of the converter, thus eliminating the negative impact of the time delay. Fast CM noise evaluation and regulation of t the time delay of two drive signals are two key techniques in a digital CM current suppression system.

In this paper, CM current level is evaluated by measuring the energy of CM current. The relationship between the energy of CM current and the time delay of two drive signals is studied, and the results reveal more than one local minimums of the CM current energy as a function of the time delay. In order to avoid getting trapped at local minimums, simulated annealing algorithm (SAA) is used to find the optimal time delay. Simulated results show that SAA works well and can effectively reduce CM current energy, suppress the CM current noise level and improve the EMC performance of PWM converter.

II. CM CURRENT PRODUCTION MECHANISM UNDER DIFFERENT MODULATION TECHNOLOGY

A single phase full-bridge converter consisted of IGBTs is shown in Fig.1. There is a thin insulation layer between the device junction and the base plate, which is usually fixed on a heat sink that is grounded for safety reason. This forms a parasitic capacitance (see Cp1 and Cp2 in

---

***This work was sponsored by the National Natural Science Foundation of China (NSFC, under project 50407011) and Delta Electronics Ltd.

1-4244-0448-7/06/$25.00 ©2006 IEEE

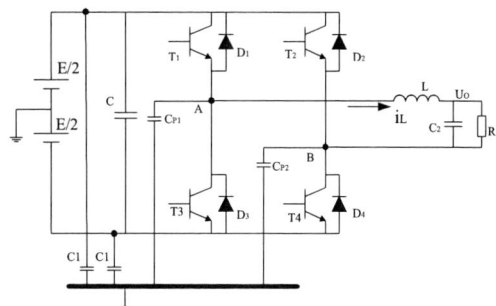

Figure1. Single-phase full-bridge converter

Fig.1) between the midpoints of legs and ground. As the IGBTs switching on and off in high speed, high dv/dt at the midpoints of legs results in charging and discharging currents through the parasitic capacitors. Specifically, these currents flow through the parasitic capacitors and

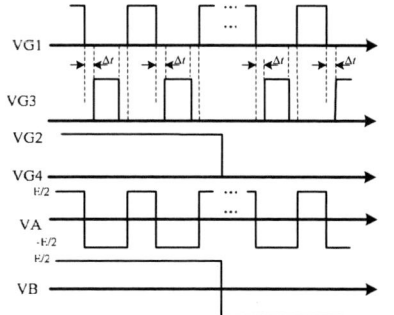

Figure2. Drive signals and midpoint voltages for unipolar SPWM

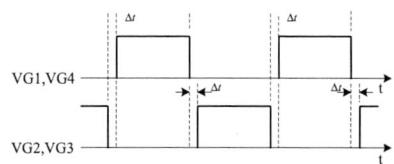

Figure3. Drive signals for bipolar SPWM

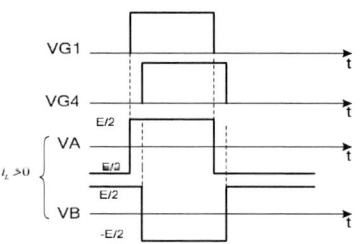

Figure4. Loss of synchronicity between midpoint voltages due to different transmission delays of gating signals

converge into the ground, which is named as CM current. CM current gets its way into the DC bus through some common mode capacitors.

Drive signals ($V_{G1} \sim V_{G4}$) and midpoint voltages (V_A, V_B) of the legs for unipolar SPWM are shown in Fig.2. For convenience, only a small portion around the zero-crossing point of the modulating signal is shown. Δt is

the blank time, E is the voltage of the DC source, and the load current is assumed to be positive. Since V_A changes between ±E/2 in switching frequency and V_B changes between ±E/2 in fundamental frequency, the CM conducted EMI is mainly generated by the left leg. Clearly, with unipolar SPWM, V_A and V_B cannot cancel out each other.

In the case of bipolar SPWM, switch pairs T1-T4 and T2-T3 turn on and off simultaneously, and the two switch pairs operates in a complementary manner. Fig.3 shows drive signals and midpoint voltages of legs for bipolar SPWM. No matter what the actual direction of load current, the two midpoint voltages always change in an opposite way. Usually, the parasitic capacitance between each inverter leg's midpoint and the ground can be considered to be equal because of same parameter of power devices. As a result, two legs of the converter provide perfect compensation for each other and the CM current can be cancelled out automatically.

Unfortunately, in a real converter system the drive signals always have to go through some processing circuits before arriving at the power devices. Such circuits usually include logic gates, voltage comparators, optical couplers, and some application-specific driving circuits. Due to inevitable deviations in signal transmission characteristics of these devices, transmission delay of each drive signal is different. Therefore, the exact synchronicity of switching actions shown in Fig.3 never exists in real case.

An example of this situation is shown in Fig.4. Suppose $i_L>0$, T1 and T4 are all off, and load current flows through D2 and D3. Then T1 turns on ahead of T4 due to faster transmission of its drive signal, load current starts to flow through T1 and D2, and V_A changes from -E/2 to E/2. When T4 turns on, load current flows through T1 and T4, and V_B changes from E/2 to -E/2. When T1 turns off, T4 is still on, load current goes through D3 and T4, and V_A goes back to its initial negative polarity. Finally when T4 turns off, load current flows through D2 and D3, and V_B goes back to positive polarity. Clearly, in this case ($i_L>0$) V_{G1} determines V_A and V_{G4} determines V_B. The two midpoint voltages no longer change in an exactly opposite way due to different transmission delays of gating signal V_{G1} and V_{G4}. Although these differences (usually several hundred ns) are fairly small compared with the pulse widths, it can easily disrupt the automatic compensation mechanism mentioned above, because the oscillation frequency of the CM current can be as high as of MHz level.

III. CM CURRENT LEVEL EVALUATION

The basic idea of a novel time delay regulation method with digital closed-loop control can be illustrated by Fig.5.

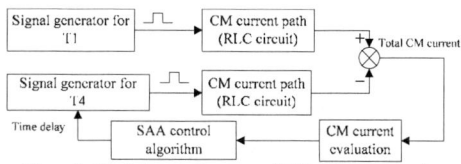

Figure5. Digital control system of CM noise suppression

The path of the CM current can be modeled as an equivalent RLC circuit. For such a system, CM noise evaluation measure and closed-loop control strategy are the two key techniques.

For simplicity, it is desirable to evaluate the CM current with a single-value index. Under such circumstances, the 'energy' of the CM current e_{CM} is adopted here to serve as that index, the calculation of which is given by

$$e_{CM} = \sum_{n=0}^{N-1} \left| i_{CM}[n] \right|^2 \qquad (1)$$

Obviously, it is just a sum of squared values of the sampled CM current i_{CM}. According to Parseval's theorem, the time-domain energy of the CM current as calculated with (1) equals the mean-square value of its magnitude spectrum, i.e.:

$$e_{CM} = \sum_{n=0}^{N-1} \left| i_{CM}[n] \right|^2 = \frac{1}{N} \sum_{k=0}^{N-1} \left| I_{CM}[k] \right|^2 \qquad (2)$$

where I_{CM} is the discrete Fourier transform (DFT) of i_{CM}. The rightmost part of (2) is also called the frequency-domain energy. Apparently this is a good measure of the overall noise spectrum, and its calculation is far more convenient than that of a DFT.

Due to inevitable differences in signal transmission delays, and the high-frequency nature of the CM current, the lead/lag time of the drive signal needs to be fine-tuned, otherwise a CM current will occur, even if power devices have same parameters.

The relationship between the CM current energy and the delay time of the drive signals is studied. The frequency of drive signals is 8 kHz. The duty cycle is 50% and the amplitude is 100V. The parameters of the series RLC circuit is R=150Ω, L=1mH, and C=1nF. The delay time is available from -1.185e-5s to1.2e-5s, with a step size of 1.5e-7s.

The result is shown in Fig.6, which indicates that CM current energy have five minimum value in the range of available time delay (-1.185e-5, -6.15e-6, 0, 6.3e-6, 1.2e-5). Therefore, it is important for the closed-loop control algorithm to be able to search for the global minimum and figure out time delay of a correspondence. Conventional hill climbing search algorithm (HCSA) is easy to trap at a local minimum value. Simulated annealing search algorithm (SASA), which allows up-hill moves, is proposed in this paper in order to search for the global minimum efficiently.

IV. SIMULATED ANNEALING SEARCH ALGORITHM

A. Simulated Annealing Algorithm

Simulated annealing (SA) is an optimization procedure based on the behavior of condensed matter at low temperatures that mirrors the annealing process that takes place in nature. The procedure employs Monte Carlo iterative calculation to search for global optimum solution by simulating the annealing process. Along with

continuous descent of temperature at the condition of a start temperature, simulated annealing algorithm (SAA) searches global optimum solution of an objective function randomly in all solution space. It can probabilistically jump out local optimum solution and drive to global one [10].

SAA implement optimal search following by procedure: 1) preset the variable range of model parameter, select randomly an initial state x0 and compute the value of objective function E(x0). 2) generate a new state x in the domain of current state by randomly and compute the value E(x), figure out △E=E(x)-E(x0). 3) if △E<0 , new state x is accepted. If △E>0, x is accepted in the probability of P=exp(-△E/T), where T is current temperature. At the same time, change current state to new state, that is x0=x, E(x0) =E(x). 4) Repeat the operations of step 2) and 3) for a preset number of inner cycle at current temperature T. 5) decrease temperature slowly in temperature updating functions. 6) Repeat the operation from step 2) to 5) until condition of convergence is satisfied. SAA include two cycles, inner cycle named sampling process produce new state and judge whether the new state is accepted at all temperature, outer cycle is annealing process. So state generating functions, state accepting functions, temperature updating functions, terminating conditions of inner and outer cycle as well as start temperature is the main parts of SAA. In normal, sampling process ending criterion is deterministic sampled number and annealing process is ended at end temperature.

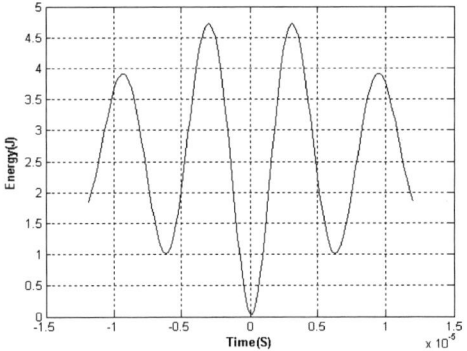

Figure6. The relationship between CM noise energy and delay time of drive signals for T4

B. CM Current Minimum Energy Control

CM current minimum energy control is main goal of closed-loop control system. The total CM current level is reduced by regulating the time delay of drive signals for T4. When time delay of drive signal arrived to T1and T4 is zero, the total current energy is zero. Closed-loop control system regulate the time delay of drive signals in order to make total current energy is minimum.

According to relationship between CM current energy and the time delay of drive signals for T4 mentioned above, there are several local minimum values in the neighborhood of global minimum value. SAA can avoid getting trapped at local minimum value and find the global minimum value in the searching process. The program flow chart of SAA applied in searching CM current minimum energy is showed Fig.7.

To SAA proposed in this paper, it is improved on annealing process and sampling process. In normal SAA, the accepted current state maybe worse than some middle state in searching process because of state accepting in probability. In order to lose the state of 'Best So Far' and improve the searching efficient, it is modified in the ending criterion of annealing process and sampling process. It is illustrated as following: Save the middle optimal state and update at any moment; Set double threshold value, that is to say, sampling process ended when current state is consciously invariable at the times of Step1 in every temperature and annealing process ended when optimal state is continuously invariable at the Step2 times of annealing process.

New state generating function is
$$x1 = x0 + 1.5e - 7 * fix(10 * (rand(1) - 0.5)) \quad (3)$$

New state accepting function is
$$\min\left\{1, e^{\frac{E(S_j) - E(S_i)}{Tk}}\right\} >= rand[0,1] \quad (4)$$

And annealing function or temperature updating functions is
$$T_1 = a * T \quad (5)$$

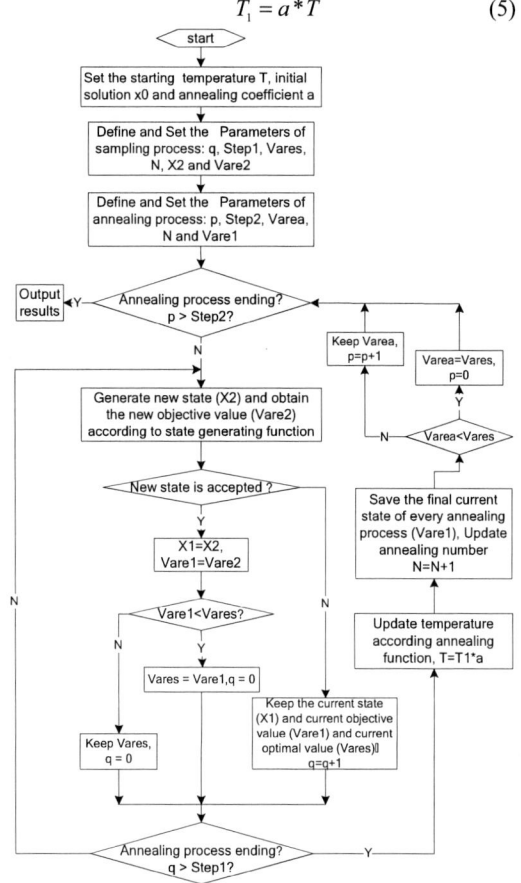

Figure7. SAA program flow chart of minimum energy searching

The sampling process as follows: Firstly, give initial solution of time delay X0 randomly, set current solution X1=X0 and obtain CM current energy at the start temperature Vare1 and set the number of continuously invariable current state q=0. Secondly, set initial optimal solution of sampling process Vares=Vare1, produce new state X2 according to state generating function from current solution, and figure out new current energy Vare2. Thirdly, judge whether the new state is accepted according to new state accepting function, compare the current state and last optimal solution Vares, update the optimal solution and q. Then, judge whether q is larger than Step1or not. If q is larger than Step1, the current optimal solution Vares and current state X1 and Vare1 is saved and turn to annealing process. If q is not larger than Step1, the sampling process is continued.

The annealing process as follows: Firstly, set initial minimum energy of annealing process Varea=Vare1, the annealing number of continuously invariable minimum energy p−0. Secondly, call the sampling process and obtain the new minimum energy Vares and current state X1 and Vare1.Thirdly, compare the Varea and Vares. If Varea is smaller than Vares, the annealing number of continuously invariable minimum energy p is updated to p+1. Otherwise, the CM current minimum energy Varea is updated to Vares and clear p=0. Fourthly, update the temperature according annealing function. Then compare the p and Step2, judge the annealing process is ending or not. If p is larger than Step2, annealing process is ending and output the current optimal solution and it is considered to be searching solution. Otherwise, return to step secondly and continue the annealing process and sampling process.

V. SEARCHING RESULTS AND ANALYSIS

SAA mentioned above and its application of searching the minimum noise energy is implemented by using MATLAB. It is showed that the CM current energy is a global minimum when the delay time of driving signals for T4 is 0 in Fig.6. The simulated results are showed in Fig.8 and in Fig.9.

In SAA, set the initial delay time of driving signal for T4 to be 1e-5s, the threshold number of annealing process ending criterion Step2 to be 25, the threshold number of sampling process ending criterion Step1 to be 5. The initial temperature T=100, annealing coefficient a=0.9.

Fig.8 is the results of CM current minimum energy searching results in every annealing process. The delay time of driving signals is 0s and minimum energy of a correspondence CM current is 0.029475 (J). That is to say, SAA control system regulates the time delay of T4 drive signals to 0s. The current energy is about zero and total noise current is very small, so control system based on SAA suppresses the CM current efficiently.

Fig.9 is the results of delay time of driving signal for T4 in every searching process. It cost about the number of 1400 to searching the global optimal solution. It improves the searching speed and shortens the searching time.

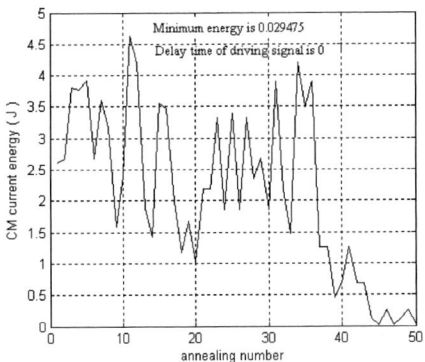

Figure8. Minimum energy searching results

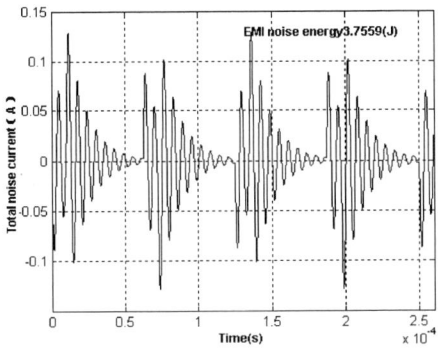

Figure9. Searching results delay time of driving signal for T4

The total CM current before and after SAA control system regulating the phase of compensating current are showed in Fig.10 and Fig.11. The result shows that CM current energy is 3.7559 (J) and current peak value is about 150mA before regulating. And after regulating, CM current energy decreases to 0.029475 and current peak value descent to about 5mA. The results also indicate SAA control system suppresses the CM current level efficiently.

Figure10. Total current before regulating

VI. CONCLUSION

A novel evaluation measure for CM current level based on the energy of CM current is proposed. It is just a sum of squared values of the sampled CM current. The relationship between CM current energy and time delay of drive signals is studied. It is concluded that the energy is an important evaluation measure.

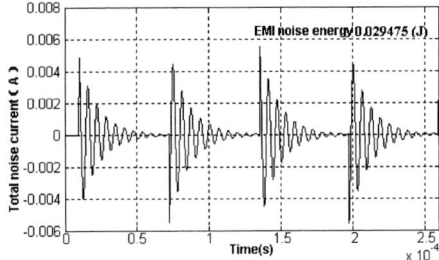

Figure11. Total current after regulating

According to more than one local optimal value in the relationship between CM current energy and time delay of drive signals, simulated annealing random search algorithm is utilized. SAA can search for the global optimal value efficiently and avoid trapping at local optimal value. Simulation results in MATLAB indicate SAA is valid. It can search for global minimum energy efficiently and figure out the phase of compensating current. Closed-loop control system with SAA regulates the phase of compensating circuit and suppresses the CM noise level effectively. Further working is how to apply the SAA into experimental devices.

REFERENCES

[1] Xuejun Pei, Jian Xiong and Yong Kang, etc. Analysis and Suppression of Conducted EMI Emission in PWM inverter [J]. IEMDC'03, pp1787-1792

[2] Xin Wu, N.K. Poon, C.M. Lee, etc, A Study of Common Mode Noise in Switching Power Supply from a Current Balancing Viewpoint [J]. IEEE-PEDS'99, vol.2, pp.621-625

[3] D.H. Liu, J.G. Jiang, Z.M. Zhao. A Systematic Approach to Analyze EMI in Control Circuit of Power Electronic Equipment [J]. IEEE-APEC'01, vol.1, pp.208-212

[4] Chingchi Chen. Novel EMC Debugging Methodologies for High-Power Converters [J]. IEEE International Symposium on Electromagnetic Compatibility, 2000, vol.1, pp.385-390

[5] Alexander L. Julian,"Elimination of Common-Mode Voltage in Three-Phase Sinusoidal Power Converters",IEEE trans. on Power Electronics,Vol.14,NO.5.

[6] Yo-Chan Son and Seung-Ki Sul, " A Novel Active Common-mode EMI Filter for PWM Inverter", IEEE-APEC2002. vol.1, pp.545-549, Mar. 2002

[7] R. Scheich and J. Roudet, S. Bigot, and J. P. Ferrieux. Common mode RFI of a HF power converter: phenomenon, its modeling and its measurement[C]. IEEE EPE, 1993, 164-169.

[8] Teulings, W., Schanen, J. L., Roudet, J. A new technique for spectral analysis of conducted noise of a SMPS including interconnects[C]. IEEE PESC, 1997, 516-521

[9] D. Cochrane, D. Chen, D. Boroyevich. Passive Cancellation of Common-Mode Noise in Power Electronic Circuits [J]. IEEE Trans. on Power Electronics, 2003, 18(3), pp756-763

[10] Ling Wang. Intelligent Optimization Algorithms with Applications [M]. Beijing: Tsinghua University Press, 2001

Summary of Distance Measurement Based on Vision in Localization Technology

Handong Zhang, Gang Wang and Yuwan Cen

School of Electrical Engineering & Information, Anhui University of Technology, Ma'anshan, Anhui, 243002, China
e-mail: zhanghd@ahut.edu.cn

Abstract—In localization technology for a mobile robot, visual locating is widely concerned because of its high precision and large amount of information, et al. Firstly, the paper introduces some popular locating methods and shows the advantages and disadvantages. Secondly, it discusses some technology related to distance measurement in visual locating, such as image segment, feature matching, camera calibration, etc. Development directions of visual locating are also discussed. At last, it analyzes some problems existing in visual locating, some possible schemes to solve them are proposed.

Keywords-localization; binocular stereovision; image segment; feature matching; camera calibration

I. INTRODUCTION

The research on a mobile robot is initiated from 1960s, which is to study a robot's abilities of deducing and planning independently under different environment. Because the mobile robot has the ability of moving and self-rule and can adapt to different surrounding, it was widely used in different fields such as logistics, traffic, medical service and social service. During the development process of the mobile robot, locating and navigating technology is a key one in robot technology. More attention is paid to the localization of a mobile robot in the paper.

II. THE LOCALIZATION TECHNOLOGY FOR A MOBILE ROBOT

Localization is to ascertain the position where the robot works[1]. The localization methods for the mobile robot can be divided into two categories: relative and absolute position measurement. The commonly used localization method includes: an odometer for navigation, inertial navigation, a magnetic compass for navigation, active light tower, global positioning system (GPS), land marks for navigation, map model matching, information fusion, visual localization and so on. Each kind of technique has its respective merit and limitation. Generally several kinds of methods above are synthesized in actual application. It can realize supplementary between the merit and flaw of the methods above to enhance the precision and the reliability of the localization[2].

At present, among localization methods for the mobile robot, the methods researchers pay more attention to include: map model matching, information fusion and navigation based on vision.

The map model matching is a technique that uses the sensors of robot itself to establish one partial condition, and then makes a comparison between the partial map and the overall situation map preserved in the memory, at last, calculates the position in the reality. The current technology is only used in indoor or relatively simple environment [3]. The disadvantage of this method is that the speed of the map model matching technology is slow.

In the information fusion[4], information redundant and supplementary of multi-sensor in the identical condition will be used, and then the information with some fusion methods will be processed to guarantee the precision of the localization information. The multi-sensor information fusion technique is one of the most important development directions in robot localization and navigation technology.

Because the visual localization technology has large amount of information and high measuring precision, it is widely concerned by more and more researchers. But it still has many essential theories and techniques waiting for the solution and the consummation, because the robot vision involves more disciplines and people are insufficient to understand our own visual system. The technology related to visual localization for the mobile robot will be discussed emphatically following.

III. MEASUREMENT TECHNOLOGY AND CORRELATED TECHNOLOGY IN ROBOT VISUAL LOCALIZATION

Robot vision is an important research direction and has longer research history. The robot vision is also known as image understanding. It is a field to study the visual information that a robot needs to complete a task and gain some information from the image. Its basic three goals are as follows:

(1) To calculate the distance between the camera and the goal according to one or more two-dimension projects.

(2) To calculate the movement parameter of the object between the camera and the goal according to one or more two-dimension projects.

(3) To calculate the superficial physical property of the object according to one or more two-dimension projects.

The final goal above is to realize the understanding of three-dimension world, namely, to realize certain functions of human's visual system. That is to say, the visible part of the three-dimension object will be reconstructed by two-dimension project.

As to visual localization for an industry robot, it can be divided into passive vision and active vision according to the different light used in robot's working condition. The merit of passive vision is that it has large information and we can obtain more comprehensive information of the object. But it leads to longer time to process the large amount of information and it is easily influenced by natural light. The processing speed of active vision is quicker, but the information obtained by active vision is not comprehensive. Binocular stereovision is usually involved to survey the information of the workpiece by using the passive vision. For example, EL-Hakin[5] proposed a three-dimension coordinate measurement system of multi-visual sensor based on the principle of binocular vision. The accuracy of this system is 8 um in a surveyed space that is 30cm*30cm*30cm. In addition, there is another measurement technique based on passive vision in which a single camera is used.

A. Monocular vision

The monocular vision is usually used in the aspect of path tracking. Jun Zhou [6] proposed a solution to the relative pose for the wheeled mobile robot applied in the agriculture. Xin Zhou [7] inspected the current divider in high way by using multi-threshold division technology and established conic section model of the divider. Yingming Hao [8] proposed three-dimension visual methods based on two kinds of model—monocular vision based on a model and binocular vision based on a model. The former realized the model restraint through simple restraint and iteration, which started with the physics significance of visual computation. The latter realized the model restraint by fusing in the Least Square Method and the monocular vision based on a model together. Zhihai Zhang [9] proposed a focusing method of monocular vision through analyzing and studying many kinds of position measurement and control system thoroughly. Xiufen Yu [10] considered the delay of visual system in the tracking system based on vision serve for a mobile robot and introduced information measured by monocular vision. She realized the localization of the mobile object in space and gained the movement information with the method of extended and delayed Kalman filter.

B. Binocular vision

The discussion related to three-dimension depth information[11, 12], is mainly launched revolving the measurement of distance. Therefore, three-dimension image is also known as distance chart or depth chart in many papers. Although the binocular parallax method is the most traditional one in the three-dimension visual measuring technique, the research in this aspect is still very few these years. With the development of computer technology and sensor technology, the accuracy of this method is enhanced and the speed of that is enhanced too. Thus it has the prospect to be applied in the military, the aviation, the astronautics, robot guiding, goal recognition and the industry inspecting and so on. Therefore, it is necessary to do further research on the distance measurement based on binocular vision.

Figure 1 shows the principle of distance measurement based on binocular vision. In Figure 1, P is the object point, O_l and O_r is the optical centers of the left and right camera. The two cameras have the same focal length f. P_l and P_r are the points of P in the image, which are formed in the image plane of the left camera and the right one. The length from P to the connected line between the two lens is d. According to triangle similarity we can obtain that:

$$d = \frac{bf}{(m+n)} \qquad (1)$$

In formula (1), m+n is the binocular parallax of the object P in the image plane. Formula (1) indicates that the distance between the camera and the object has the inverse ratio to the binocular parallax, but has the direct radio to the distance between the two cameras and the focal length f. Because b and f can be determined, we can calculate the distance between the cameras and the object as long as we known the binocular parallax. We can acquire the distance information if the point positions of P_l and P_r in the images are known. Therefore, this question is transformed to the question of seeking the matching point in the left image and the right image.

The matching is main step during the process of distance measurement based on binocular parallax. The matching strategies are mainly connected with the restraint in distance measurement. The typical matching methods include the method based on the region and the

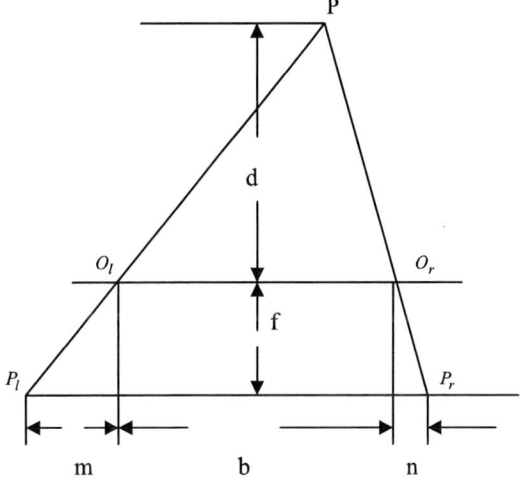

Figure1. the principle of the distance measurement based on binocular vision

one based on the feature. The former method is to consider the gray value's domain of a point in one image as a template, and then search the same (or the similar) gray value distribution in another image, thus realizes the match of the two images. The basic principle of the matching based on the correlation of the gray value is that the image is regarded as two-dimension signal in the view of statistic, then the correlative match between the signals with the method of statistical correlation is sought. The matching based on the feature matching that matches the feature selectively, and then solves the diversity of the matching through emphasizing the structure information of the spatial objects. Baker[13] and Amold[14] proposed matching algorithms of taking edge as the match feature and used dynamic planning at the first time. Ohta[15] proposed dynamic planning algorithm by taking the edge lines as the matching feature. The experiment indicates that the algorithm can get a better match to all kinds of objects. The disadvantage of which is the complexity of the calculation and the realization. Lee[16] proposed matching algorithms of region segmentation while he used the relaxed iterative algorithm.

Image segmentation is another important technology in the distance measurement based on vision. The methods usually used include: image segmentation based on region, image segmentation based on edge, image segmentation based on combination with region and edge, and image segmentation based on some specific theories. Image segmentation based on region is to make use of the similarity of the feature in the region, so that the image can be divided into series of significant regions. The realization of this method is threshold method, region growing and region merging, clustering method and so on. Image segmentation based on edge is to detect the edge between different regions by using different characteristics of the pixels in different regions. The realization of this method is differentiator and series boundary technique. The series boundary technique is to detect the edge and then connect the closed boundary serially, but this method is greatly influenced by the initial point. Edge detection can obtain the partial changing intensity of the gray value and the colored value, while regions segmentation can detect the similarity and the uniformity of the characteristics. To combine with the two methods above, we can avoid the region being over-segmented through the limitation of the edge; at the same time, the outline will be more integrated by regions segmentation. The regions segmentation can reinforce the edges that cannot be detected by edge detection. This method above is namely image segmentation based on combination with region and edge. Image segmentation based on specific theories includes image segmentation based on mathematical morphology, fuzzy set, neural network, and genetic algorithm etc. The principle of image segmentation based on mathematical morphology is to measure and get the corresponding shape in the image through the structure element with

certain shape, so that we can analyze and recognize the image[17]. The main disadvantage of this method is that it cannot solve the problem of time-consuming. Therefore, it is a tendency to combine the method with some one that can save time. Image segmentation based on fuzzy set is widely used in medical image[18]. The principle of image segmentation based on neural network is that, firstly, to get the linear decision functions through training the multilayer perception machines, secondly, to segment the image by classifying the pixels with the decision functions. Yanxin Chen [19] proposed image segmentation with clustering automatically based on competitive Hopfield Neural Network. Genetic algorithm is a method based on evolution theory with natural selection mechanism and the parallel, statistical, stochastic search. Peizheng Wang and Peiming Du [20] proposed muti-threshold image automatic segmentation based on hybrid genetic algorithm. Mingwu Ren [21] proposed a new method for the construction of histogram based on edge pattern.

Another technology in distance measurement based on vision is camera calibration. At present, it can be divided into two categories: traditional camera calibration and camera self-calibration[22]. The principle of the traditional method is that: lay calibration reference in the condition, and then calculate the inside parameter and the outside one of the camera by using some known three dimension coordinates of some points and point coordinates in the image. The difficulty to gain the information of the three-dimension coordinates is to look for and match the feature in the image. The two above are close connected. The question above is referred to as the corresponding question. Qingjie Zhao [23] studied the relationship between image information and the three-dimension information with artificial neural network.

C. Multi-camera vision

In order to solve the deficiencies in binocular vision, Zheng Liu [24] proposed a trinocular vision inspecting system. A line-structured light trinocular vision based on the active vision and combined with coordinate measurement machine is studied. More researches in multi-camera vision are expected in the near future.

IV. PRESENT SITUATION AND DEVELOPING TENDENCY OF THE VISUAL LOCALIZATION IN DOMESTIC AND FOREIGN COUNTRIES

For many years, many researchers have been doing the research on the localization and navigation based on vision. Therefore, the understanding and the solution to the questions in the visual localization have been well developed. Among the ways of localization and navigation based on vision, it is usually used that one or two cameras loaded by a robot based on partial visual navigation. For example, Boley[25] developed a mobile robot with cameras carried by a vehicle, which can be navigated by recognizing the path. Ohya[26] studied the collision problem based on visual navigation system through the cameras and fewer ultrasonic sensors carried by a vehicle. In China，the more successful cases in the mobile robot are: AGV (i.e. Automated Guided Vehicle) and anti-riot robot developed by Shengyang Institute of Automation, navigation system based on vision for a mobile robot developed by Institute of Automation, Chinese Academy of Sciences[27, 28].

According to current developing direction, the development tendencies can be generalized as follows:

(1)Each modules in the stereo visual system will be connected by the comprehensive viewpoint, the full intrinsic information will be excavated;

(2)Each kind of technique and multi-sensor fusion will be comprehensively utilized, including the synthesis use of different matching method and the active vision and passive vision to enhance the reliability of the technique;

(3)The direction from the traditional binocular stereovision to multi-camera vision, from the static vision to dynamic vision will be developed, and the difficulty of the visual computation will be reduced by increasing input of the information;

(4)The algorithm will be more intelligent, the stereovision method based on the knowledge, the model and the rule will be established.

(5)The algorithm will be more parallel, and parallel streamline machine and special-purpose signal processing chip will be applied to enhance the practicability of the system.

(6)The condition and the task restraint will be emphasized, and the optimal choices will be applied to each part, and the visual system with goal and task-oriented will be established.

V. THE PROBLEMS EXISTING IN THE VISUAL LOCALIZATION AND THE POSSIBLE SOLUTION TO THE PROBLEMS

The tracking and localization system based on vision is multi-disciplinary integration of the computer vision, image processing, mathematics, the precise instrument and so on. The measurement based on binocular vision is widely used, and its precision is high, but real-time is worse; the range of the measurement based on the monocular vision is narrow, and the precision of that is lower, so it can be only applied to a mobile robot, but the real-time of the measurement based on the monocular vision is better than that of the binocular vision measurement[29]. From now on, it will be emphasized that to improve the precision of camera calibration, to speed up the feature extraction and the feature matching, at the same time, other factors which influence the system and the solution to the sheltering will be considered.

According to the problems existing in the distance measurement based on vision, our research group proposes some possible solutions to them: (1) the method used will be ameliorated, for example, we can utilize the camera calibration based on neural network to improve the precision of the calibration; (2) the original methods for feature extraction will be synthesized or the new methods for feature extraction will be proposed; (3) the intrinsic restraint information will be excavated according to the installment position of two CCDs, and the matching algorithm already existed will be optimized, at the same time, the speed of matching will be increased and the precision of matching will be improved; (4) with information fusion of multi-sensor, the influence of false image in the process of the image acquisition will be eliminated through the supplementary and redundant among multi-sensor. In one words, although the distance measurement based on vision is becoming more and more consummate, much work needs to do in the aspects of features of the object extraction, speeding up the feature matching, enhancing the precision of the camera calibration etc.

ACKNOWLEDGMENT

This project is supported by the National Natural Science Foundation (50407017) and the Natural Science Foundation of Anhui Education Bureau (2004KJ058，2006KJ019A).

REFERENCES

[1] Y. Guo, Z. Hu and J. Dong, Mobile robot navigation and localization technology[J], *Microcomputer Information,* 2003, 19(8), pp. 10-13. (in Chinese)

[2] J. Borenstein, H. R. Everett, L. Feng and D. Wehe, Mobile robot positioning sensors and techniques[J], *Journal of Robotic Systems,* 1997, 14 (2), pp. 231-249.

[3] Y. Zhang, X. Liu and L. Fu, Extension and application of certainty measure map matching algorithm in car navigation[J], *Application Research of Computers,* 2004, 21(6), pp. 117-119. (in Chinese)

[4] Z. Wang and Y. Guo, Present situation and future development of mobile robot navigation technology [J], *Robot,* 2003, 25(5), pp. 470-474. (in Chinese)

[5] EL-Hakin et al, The VCM automated 3–D measurement system[A], *Theory Application and Performance evaluation*[C], Proc. SPIE, 2000, pp. 460-482.

[6] J. Zhou, C. Zheng and C. Ji, Pose sensing of agricultural wheeled mobile robot[J], *Journal of Image and Graphics,* 2005, 10(3), pp.310-314. (in Chinese)

[7] X. Zhou, Monocular vision navigation in vehicle intelligent auxiliary drive system[J], *Robot,* 2003, 25(4), pp. 289-295. (in Chinese)

[8] Y. Hao, F. Zhu and J. Ou, 3D visual methods for object pose measurement[J], *Journal of Image and Graphics,* 2002, 7A(12), pp. 1248-1251. (in Chinese)

[9] Z. Zhang, Three-Point Method on the 3D position measurement of a robot[J], *Measurement and Control Technology,* 2002, 21(7), pp. 10-13. (in Chinese)

[10] X. Yu, H. Duan and H. Gong, Mobile robot visual locating based on time-delay Kalman filter[J], *Electronics Optics & Control,* 2004, 11(2), pp 29-31. (in Chinese)

[11] R. A. Jarvis, A perspective on range finding techniques for computer vision, *IEEE Trans. On PAMI,* 1996, 5(2), pp. 122-139.

[12] J. Anon, Improved distance sensor[A], *Proc. of the ninth Int. Symposium on Industrial Robots*[C], USA, 1997, pp. 479-481.

[13] H. Baker, et al, Depth from edge and intensity based stereo[A], proc. *the seventh IJCAI*[C], 1998, pp. 631-636.

[14] R D. Aronld, et al, Automated Stereo Perception[M], Standford University Tech. Rep.1996, pp. 83-85.

[15] Y. Ohta, et al, Stereo by intra-and inter-scanline search using dynamic programming, *IEEE Trans .On PAMI*, 1995, 7, pp. 139-154.

[16] H. J. Lee, et al, Region Matching and Depth Finding for 3-D Objects in Stereo Aerial Photographs[J], *Pattern Recognition*, 1998, 23(2), pp. 81-93.

[17] J. Yang, Image Segmentation based on mathematical morphology[J], *Computer Engineering and Science*, 1998, 20(4), pp. 21-27. (in Chinese)

[18] S. Luo, N. Tang, A bias based adaptive fuzzy segmentation algorithm[J], *Journal of Image and Graphics*, 1999, 7(2), pp.111-114. (in Chinese)

[19] Y. Cheng, F. Qi, Image segmentation with clustering automatically based on competitive hopfield neural network[J], *Pattern Recognise and Artificial Intelligent*, 1998, 11(2), pp. 215-221. (in Chinese)

[20] P. Wang and P. Du, A new mixed genetic algorithm for multilevel thresholding[J], *Journal of Image and Graphics*, 2000, 5(1), pp. 44-47. (in Chinese)

[21] M. Ren, J. Yang and H. Sun, A new method for the construction of histogram based on edge pattern[J], *the Research and Development of the Computer*, 2001, 38(8), pp. 972-976. (in Chinese)

[22] M. Qiu, S. Ma and Y. Li, Overview of camera calibration for computer vision[J], *Acta Automatica Sinica*, 2000, 26(l), pp. 43 −55. (in Chinese)

[23] Q. Zhao, Z. Sun and L. Lan, Camera calibration based on artificial neural network[J], *Control and Decision*, 2002, 17(3), pp. 336-339. (in Chinese)

[24] Z. Liu, Multi-Vision Inspecting technique of free-form surface[J], *Journal of Tianjin University*, 2003, 36(2), pp. 148-151. (in Chinese)

[25] D. L. Boley, E. S. Steinmetz and K. T. Sutherland, Robot localization from landmarks using recursive total least squares[A], *Proceedings of the IEEE International Conference on Robotics and Automation*[C]. 1996, pp. 624-630.

[26] A. Ohya, A. Kosaka and A. Kak, Vision-Based Navigation of Mobile Robot with Obstacle Avoidance by Single Camera Vision and Ultrasonic Sensing[A], *Proc. IEEE Int. Conf. Intelligent Robots and Systems*[C].1997, pp.704-711.

[27] Y. Qin, M. Zhao and Y. Qu, Analysis on material handling system of assembly line with AGV and AS/RS[J], *Machinery Design & Manufacture,* 2005, 6, pp.166−169. (in Chinese)

[28] H. Zhang, K. Yuan and Q. Zhou, Visual navigation of a mobile robot based on path recognition[J], *Journal of Image and Graphics*, 2004, 9(7), pp.853-857. (in Chinese)

[29] S. Zhu, G. Sheng, X. Sun and X. Qiang, Analysis of 3-D coordinate vision measuring methods with feature points on workpiece[J], *Optics and Precision Engineering*, 2000, 8(2), pp. 22-25. (in Chinese)

2006 5th International Power Electronics and Motion Control Conference

The studies of Single-phase Inverter Fault Diagnosis Based on D-S Evidential Theory and Fuzzy Logical Theory

Wang Baocheng, Li Danhe, Sun Xiaofeng and Wu Weiyang
Yanshan University, Qinhuangdao city, Hebei province, China
E-mail: ldh820@tom.com

Abstract—A data fusion method for single-phase inverter fault diagnosis based on D-S evidential theory and fuzzy logical theory is presented. By measuring the output voltage、the two bridge-arms voltage and the temperature of MOSFET, the belief function assignment is gotten, and the fusion belief function assignment is gotten by using D-S rule and fuzzy logic, and fault component is found. By comparing the diagnosis results based on separate original data and fused data respectively, it is shown that the latter is more accurate than the former in the fault recognition.

Keywords- data fusion; D-S evidential theory; belief function; fault diagnosis

I. INTRODUCTION

Single-phase inverter has been used in most of industrial applications. When one of fault occurs, the system operation has to be stopped for a non-programmed maintenance schedule. So the fault diagnosis is very important. Though there are many fault diagnosis methods, these methods have some limits. For example, a fault diagnosis method based on expert systems [1] is difficult to maintenance; the method based on neural network [2] is too slow to diagnose; the method based on math model [3] is difficult to make a proper model. Sometimes there may be some misjudge. In this aspect multi-sensor data fusion technique [4] provides a brand new way because of its particular manner to manage multi-dimension information. Data fusion has been used more and more widely. But it has less research that the judgment ability of fault models has been improved by colligated information of multi-sensor data fusion. In this article data fusion technique is used in fault component searching in inverter. This article aims at the diagnosis of four MOSFET damages. By measuring the output voltage、the two bridge-arms voltage and the temperature of MOSFET, the information are gotten, and by using D-S (Dempster-Shafer) evidential theory and fuzzy logic, fault component is found accurately. The diagnosis results on single sensor and ones from fusion technique are compared to find the advantage of the new technique, detailed results and conclusions are provided.

This work is supported by National Natural Science Foundation of China (No. 50237020,50407012).

II. SYSTEM CONSTRUCTION AND FUSION THEORY

The construction of multi-sensor data fusion system [4] is shown in figure1. In figure1, u_0, u_1, $\cdots$, u_i are the recognising fault modes; $m_1(u_0)$, $m_1(u_1)$, $\cdots$, $m_1(u_i)$are the belief function value that sensor 1 endows each fault modes; $m_i(u_0)$, $m_i(u_1)$, $\cdots$, $m_i(u_j)$ are the belief function value that sensor i endows each fault modes; $m(u_0)$, $m(u_1)$, $\cdots$, $m(u_i)$ are the fusion belief function values which endow each fault modes [8]. According to fuzzy logic, for each sensor, the fault possibility of the component tested can be described by a set of membership function values, that may appear two situations: firstly for the mutual affection of each component in inverter, there may be some misjudge when using a signal sensor to distinguish a fault component; secondly if three sensors give different belief function values, it will be much harder to judge a fault component. An approach is to use fuzzy logic and D-S evidential theory, and the input and output data are the meaningful belief function values. In fault diagnosis, using the belief function values of voltage and temperature tested as input data of D-S evidential theory, and output data are the fusion fault belief function values. By using the fusion fault belief function values, the fault component can be found on certain fault determination criterion.

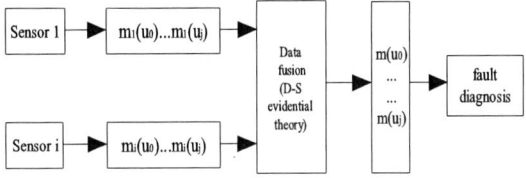

Figure1. Multi-sensor data fusion fault diagnosis system

III. SIMULATION OF SINGLE-PHASE INVERTER FAULT DIAGNOSIS

We simulate using MATLAB, as shown in figure 2. Waveforms of the output voltage and the two bridge-arms voltage can be gotten when the system works well-balanced, as shown in figure 3 and figure 4. We deal with the output voltage using filter and DQ transform, as shown in figure 5 and figure 6. The action of the filter makes the output voltage producing 90^0 phase. By DQ transform, we can get an approximate direct current value. When one MOSFET of fault occurs, waveforms change, as shown in figure 7 and figure 8. The four MOSFET occur fault respectively. The waveforms are shown.

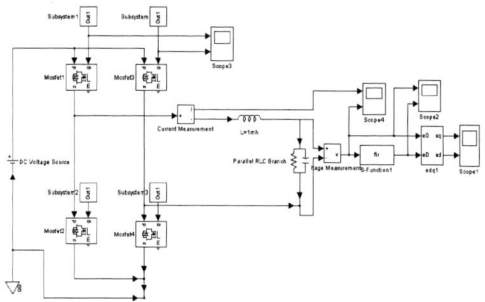

Figure2. The Simulation drawing of Single-phase inverter system

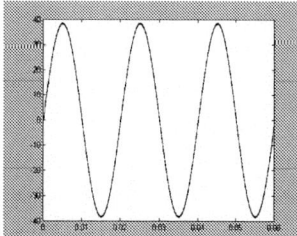

Figure3. Waveforms of the output voltage

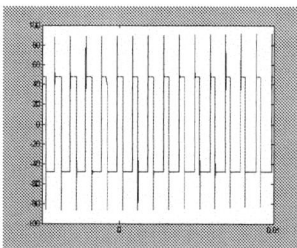

Figure4. Waveforms of the two bridge-arms voltage

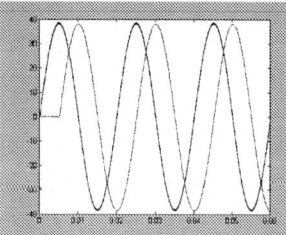

Figure5. Waveforms of the former and after from the filter

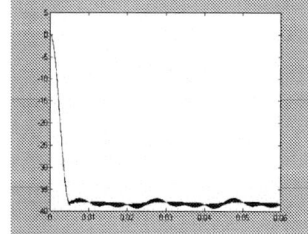

Figure6. Waveforms of the DQ transform output

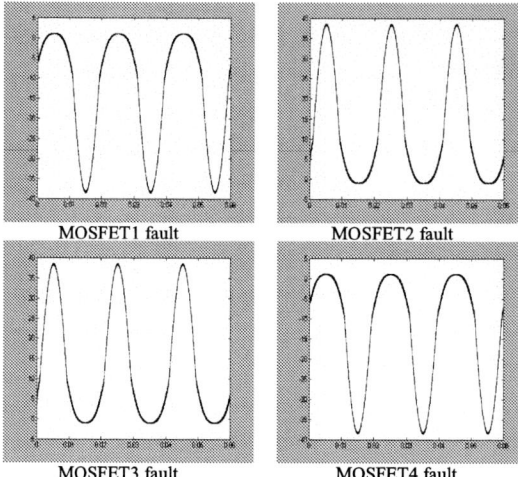

MOSFET1 fault   MOSFET2 fault

MOSFET3 fault   MOSFET4 fault

Figure7. Waveforms of the output voltage

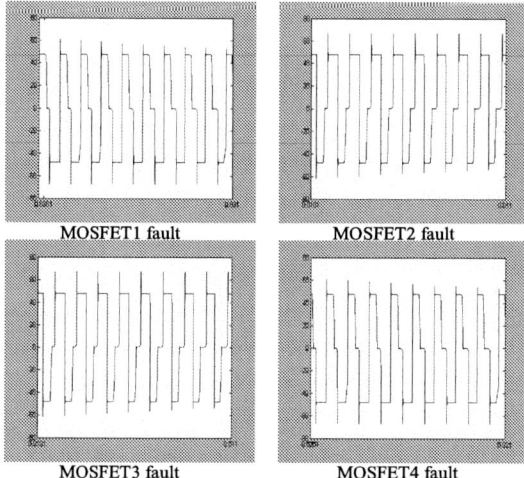

MOSFET1 fault   MOSFET2 fault

MOSFET3 fault   MOSFET4 fault

Figure8. Waveforms of the two bridge-arms voltage

From the waveforms, we can define different membership values. For example, based on the differences between working-well waveform and fault waveform of the output voltage, detecting the differences, we define the membership values as 0 when no fault occurs; 0.6 when MOSFET 1 and 4 fault; 0.5 when MOSFET 2 and 3 fault; because the waveforms of MOSFET 1 and 4 fault is same and the waveforms of MOSFET 2 and 3 fault is same. The concrete values can be shown in the following table. The temperature will be gotten when we do experiments. Firstly, the temperature of each MOSFET is measured when the system is working properly. Secondly, if there is some fault in the circuit, the fault components temperature will change, the new temperature of each component is measured, and the fault membership function value can be calculated using the formula given before. The membership function values are the probability of the fault of MOSFET. After the membership function values gotten, the belief function values can be gotten. Then we can fuse these values. Finally the fusion belief function values are gotten. The fault component can be found on certain fault determination criterion.

IV. MUNTI-SENSOR D-S EVIDENTIAL THEORY DATA FUSION ALGORITHM

A. D-S Evidential Theory

Let Θ be a frame of discernment, including finite basic propositions, noted down $\{u_0, u_1, \cdots, u_i\}$, corresponding basic fault modes in fault diagnosis, which are mutex in Θ. A basic propositions assignment is a function m: $2^{\Theta} \rightarrow [0,1]$ such that [5]

$$\sum_{u \subset \Theta} m(u) = 1 \qquad (1)$$

$$m(\phi) = 0 \qquad (2)$$

in the equation, ϕ is an empty set; m is the belief function assignment in 2^{Θ}; m(u) is the belief function value of u, which expresses the confidence measure of u, namely the degree supporting u.

B. Form of Belief Function

The belief function assignment is a probability reasoning of men endows the supposing object modes and a judgement of men that may be influenced by many factors.

The largest correlative coefficient [6] between sensor i and fault modes is defined by

$$\alpha_i = \max\{C_i(u_j)\}, \quad i=1, 2, \cdots, N_c \qquad (3)$$

The distributed coefficient between sensor i and the respectively correlative coefficient is defined by

$$\beta_i = \frac{\dfrac{N_c W_i}{\sum\limits_{j=1}^{N_c} C_i(u_j)} - 1}{N_c - 1}, \quad N_c \geq 2 \qquad (4)$$

The reliable coefficient of sensor i is defined by

$$R_i = \frac{W_i \alpha_i \beta_i}{\bullet \sum\limits_{k=1}^{N} W_k \alpha_k \beta_k}, \quad k=1, 2, \cdots, N \qquad (5)$$

Then the belief function that sensor i endows the modes u_j is defined by

$$m_i(u_j) = \frac{C_i(u_j)}{\sum\limits_{j=1}^{N_c} C_i(u_j) + N(1 - R_i)(1 - W_i \alpha_i \beta_i)} \qquad (6)$$

The uncertainty belief function of sensor i is defined by

$$m_i(\theta) = \frac{N(1 - R_i)(1 - W_i \alpha_i \beta_i)}{\sum\limits_{j=1}^{N_c} C_i(u_j) + N(1 - R_i)(1 - W_i \alpha_i \beta_i)} \qquad (7)$$

From formula (6) and (7), N and N_c are fixed values to a system; W_i can be confirmed by sensors characteristics and experience; the key vaules are how to confirm the correlative coefficient $C_i(u_j)$. A resolvent is using membership function in fuzzy logic instead of $C_i(u_j)$.

C. Form of Membership Function

Membership function [4] is designed based on the working property of the sensors and the property of the parameters to be tested. For a certain component in inverter system, when the system is working properly, the voltages of component key points should be steady and the temperature should be a fixed value. When there is some fault components in the system, generally the voltage values will diverge from the normal range and the temperature will change. The more the differences are, the more fault possibility will be. For a clear description, we use undetermined coefficient method of membership function in reference and particular testing results to explain the distribution of membership function $u_{ij}(x)$, as shown in figure 9, the formula(1)is the distribution of membership function $u_{ij}(x)$. Thereinto, X_{oij} is the standard value of the component to be tested when the inverter system is working properly; e_{ij} is the normal excursion of the tested component's parameters; t_{ij} is the margin error of the parameter of the component to be tested; $u_{ij}(x)$ is membership function value of the component j to be tested by sensor i; Xi is the practical measuring parameter value from sensor i.

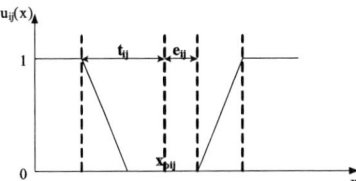

Figure9. The distribution of membership function

$$u_{ij} = \begin{cases} 1 & x_i \leq x_{oij} - t_{ij} \\ -(x_i - x_{oij} + e_{ij})/(t_{ij} - e_{ij}) & x_{oij} - t_{ij} \leq x_i \leq x_{oij} - e_{ij} \\ 0 & x_{oij} - e_{ij} \leq x_i \leq x_{oij} + e_{ij} \\ -(x_i - x_{oij} - e_{ij})/(t_{ij} - e_{ij}) & x_{oij} + e_{ij} \leq x_i \leq x_{oij} + t_{ij} \\ 1 & x_i \leq x_{oij} + t_{ij} \end{cases} \qquad (8)$$

D. The Fusion Rules of D-S Evidential Theory

In the evidential theory [5], suppose m_1 and m_2 are the corresponding belief function assignment value individually on the same identification frame Θ, the focal elements are $A_1, A_2, \cdots, A_k$ and $B_1, B_2, \cdots, B_k$ separately, and suppose:

$$C = \sum_{A_i \cap B_j = \phi} m_1(A_i) m_2(B_j) < 1 \qquad (9)$$

$$m(A) = \begin{cases} \dfrac{\sum\limits_{A_i \cap B_j = A} m_1(A_i) m_2(B_j)}{1 - C} & \forall A \subset \Theta \, and A \neq \phi \\ 0 & A = \phi \end{cases} \qquad (10)$$

E. Fault Component Judge Criterrion

In fault component decision, there are some basic rules. The basic rules [4] are as follows:

1) The fault component should have maximum belief function value.

2) The difference between the belief function values of fault component and other components should be more than a certain value.

3) The uncertainty belief function value should less than a threshold value.

4) The fault component belief function value should greater than the uncertainty belief function value.

V. THE RESULTS OF FUSIN BASED D-S EVIDENTIAL THEORY

We use some data calculating. The data in the table are simulative. From the table, the belief function values

TABLE I.
THE RESULTS OF FAULT DIAGNOSIS USING SINGLE SENSOR AND MULTI-SENSOR

Fault components	Sensor	Fault belief function vaule					Diagnosis results
		m(u1)	m(u2)	m(u3)	m(u4)	m(θ)	
MOSFET1(u1)	output voltage	0.4102	0.0823	0.0157	0.3521	0.3498	uncertainty
	two bridge-arms voltage	0.3998	0.0518	0.0546	0.4561	0.4959	uncertainty
	temperature	0.6154	0.0031	0.0750	0.2748	0.2538	u1 fault
	fusion	0.7951	0.0466	0.0681	0.2540	0.2052	u1 fault
MOSFET2(u2)	output voltage	0.0109	0.3511	0.2013	0.0387	0.4156	uncertainty
	two bridge-arms voltage	0.1000	0.4002	0.3101	0.1109	0.4123	uncertainty
	temperature	0.0101	0.5481	0.2935	0.1410	0.2859	uncertainty
	fusion	0.0040	0.6984	0.1857	0.0801	0.1854	u2 fault
MOSFET3(u3)	output voltage	0.0067	0.2012	0.3901	0.0027	0.3681	uncertainty
	two bridge-arms voltage	0.0133	0.2138	0.6977	0.1081	0.5101	u3 fault
	temperature	0.1011	0.3481	0.6230	0.0729	0.1209	uncertainty
	fusion	0.0021	0.2019	0.7567	0.0460	0.9045	u3 fault
MOSFET4(u4)	output voltage	0.3390	0.0513	0.0433	0.4559	0.4016	uncertainty
	two bridge-arms voltage	0.5677	0.1378	0.0971	0.5569	0.3004	uncertainty
	temperature	0.0210	0.0782	0.1982	0.5101	0.1123	uncertainty
	fusion	0.2560	0.0923	0.1462	0.7193	0.0101	u4 fault

gotten by the sensors respectively are sometimes very similar, the single sensor can not judge the fault component correctly using the judge criterion given before, but fusion belief function values can judge the fault component. In other words, the data fusion algorithm increases the objective's belief function value assignment and decreases the values of the other components. This makes the uncertainty of the diagnosis system decrease greatly. So multi-sensor data fusion based on D-S evidential theory and fuzzy logic improves the analyzable of the system and the judgment ability of fault models, and greatly increases the correction of fault component decision.

VI. CONCLUSION

The technology of data fusion based on D-S evidential theory and fuzzy logic is used in fault diagnosis of single-phase inverter. A kind of belief function formulation and fault component judgement criterion is used. The simulation shows that the problem of uncertainty fault

diagnosis can be solved well and the fusion fault diagnosis is practical and effective.

REFERENCES

[1] K.Debebe, V.Rajagopalan, and T.S.S.SAankar, "Expert systems for fault diagnosis of fault diagnosis of VSI fed AC drives," Conf.Rec.IEEE-IAS Annu. Meeting, 1991, pp. 368-373

[2] Hao Ma, Dehong Xu, "Two fault diagnosis methods comparision of power electronic circuits based on neural network," Mechanism Electric Engineering, 1999(3), pp. 47-49.

[3] Jiang Wang, Longgen Hu, Zhongtang Zhao, "An application of oberserver-based method to fault diagnosis for three phase inverter," Power Technology Applications, 2001(6), pp. 24-27.

[4] Daqi Zhu, Shenglin Yu, "Neural networks data fusion algorithm of electronic equipment fault diagnosis," Processing of the 5th World Congress on Intelligent Control and Automation, 2004, pp. 1815-1818.

[5] Ping Ma, Hailian Du, Feng Lv, "Coal mass estimation of the coal mill based on two-step multi-sensor fusion," Processing of the Fourth International Conference on Machine Learning and Cybernetics, 2005, pp. 1307–1311.

[6] Xinman Zhang, Jiuqiang Han, Xuebin Xu, "Dempster-Shafer reasoning with application to multisensor object recognition system," Processing of the Third International Conference on Machine Learning and Cybernetics, 2004, pp. 975-977.

[7] Bogler P L, "Shafer-Dempster reasoning with applications to multisensor target identification system," [J].IEEE Trans System, Man and Cybernetics, 1987.

[8] Zhu Daqi, Yu Shenglin, "Data Fusion Algorithm Based on D-S Evidential Theory and Its Application for Circuit Fault Diagnosis," ACTA Electrinica Sinica, 2002, pp. 221-223

2006 5th International Power Electronics and Motion Control Conference

A Novel Single-Stage High-Power-Factor Electronic Ballast with Symmetrical Half-Bridge Topology

Chien-Ming Wang[*], *member, IEEE,* Chien-Yeh Ho[**]

[*] National Ilan University /Department of Electrical Engineering, Taiwan, China
[**] Lunghwa University of Science and Technology /Department of Electronic Engineering, Taiwan, China

Abstract—This paper proposes a novel single-stage high-power-factor electronic ballast with symmetrical half-bridge topology for fluorescent lamps. The proposed electronic ballast only uses a symmetrical half-bridge topology to procure the functions of a boost power-factor-correction converter and a half-bridge series parallel-loaded inverter. In spite of its simplicity, an excellent performance concerning load and supply is achieved, ensuring a sinusoidal and in phase supply current. The conventional electronic ballast circuit has larger conduction losses because its power factor correction (PFC) power flow path circuit always includes two diode losses from the front-end bridge rectifier and one power switch loss. The PFC power flow path of the proposed circuit has only two conduction drops in the current flow paths. Therefore, it can provide lower conduction loss than the conventional one. The design equations are derived from the analyzed results based on fundamental approximation, and then an easy-to-use design tool is provided accordingly under considerations of filament heating and ignition. A prototype circuit designed for one 40-W fluorescent lamps operating at 40-kHz switching frequency and 110-V line voltage is built and tested to verify the analytical predictions.

Keywords-electronic ballast; single stage; PFC

I. INTRODUCTION

In resent years, the high frequency electronic ballasts have played a very important role for fluorescent lamps due to the benefits of light weight, small size, high luminous efficiency, and long lamp life. Most electronic ballasts are realized with load resonant inverters since they can provide an appropriate ignition voltage and then a stable arc current with a low crest factor for fluorescent. The peak detection rectifier is traditionally used by the resonant inverter to get the input dc voltage source. Nevertheless, this circuit will cause a large and sharp input current when the input ac source voltage reaches its peak. The harmonics included in the input current is harmful for the other electrical appliances, such as personal computers, and radios. In order to improve this drawback, a power factor correction (PFC) circuit must be attached to the electronic ballast to reduce the input line current harmonics. However, the two-stage solution requires more circuit components, resulting in higher cost and lower efficiency. For simplify the circuit of electronic ballast and reduced its cost, some single-stage electronic ballasts have been proposed by integrating PFC circuit into the inverter stage to perform both functions of the PFC and resonant inverter [1]-[6]. By sharing the active

power switch and the control circuit, the component count can be effectively reduced. However, their circuits have larger conduction losses because their power factor correction (PFC) power flow path circuits always include two diode losses from the front-end bridge rectifier and one power switch loss. For improving this drawback, a novel single-stage high-power-factor electronic ballast with symmetrical half-bridge topology is proposed in this paper. The proposed electronic ballast only uses a symmetrical half-bridge topology to procure the functions of a boost power-factor-correction converter and a half-bridge series parallel-loaded inverter. The proposed circuit is depicted in Fig. 1. The topology shown in Fig. 1(a) is suitable for driving lower voltage lamp and the topology shown in Fig. 1(b) is suitable for driving larger voltage lamp. The PFC power flow path of the proposed circuit has only two conduction drops in the current flow paths. Therefore, it can provide lower conduction loss than the conventional one. A high-power factor can be achieved by operating the converter at discontinuous conduction mode (DCM) at a fixed frequency with a constant duty cycle. In this way, the control circuit is simple, and low-cost commercial ICs are available. The proposed ballast circuit in analyzed and the design rules are listed, and the experimental results verify the analysis.

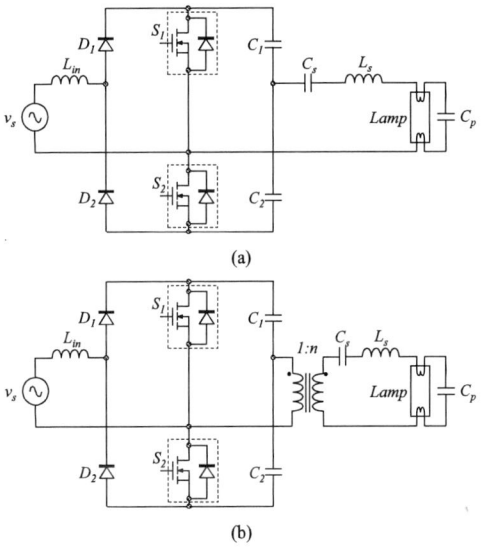

(a)

(b)

Fig. 1 Proposed topology (a) non-isolated (b) isolated.

1-4244-0448-7/06/$25.00 ©2006 IEEE

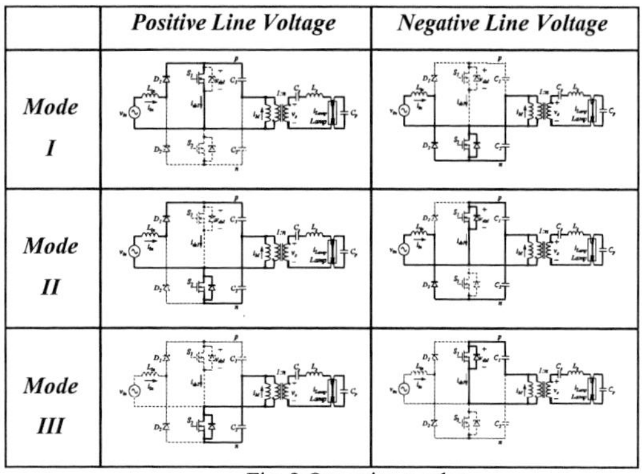

	Positive Line Voltage	*Negative Line Voltage*
Mode I	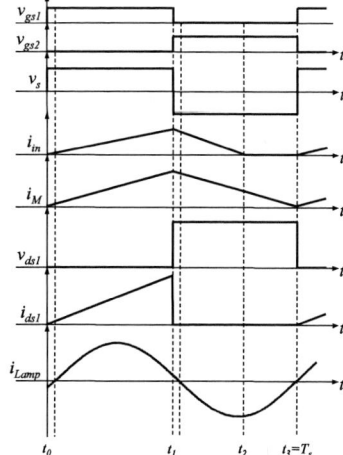	
Mode II		
Mode III		

Fig. 2 Operation modes

Fig. 3 Theoretical waveforms (positive line voltage).

II. CIRCUIT OPERATION

The new proposed single-stage high-power-factor electronic ballast with symmetrical half-bridge topology is shown in Fig. 1. It includes a single-phase voltage supply, a lower conduction losses boost PFC circuit formed by L_{in}, S_1, S_2, C_1, C_2, D_1, and D_2, a half-bridge high-frequency inverter formed by an isolated transformer, S_1, S_2, C_1, C_2, D_1, D_2, C_s, L_s, C_P, and a fluorescent lamp. The power switches S_1 and S_2 are operated at fixed frequency with a constant duty cycle of 50%. The boost PFC circuit is operated in discontinuous conduction mode, the input current naturally follows the sinusoidal waveform of the input voltage, achieving unity power factor to the utility line. The half-bridge isolated high frequency inverter is operated in boundary mode between continuous and discontinuous conduction. A series-resonant tank circuit L_s and C_s in series with the lamp network forms the load resonant circuit of the inverter. In the lamp network, only the capacitor C_P provides the current path for filament heating. Because the power switches S_1 and S_2 are operated at a constant duty cycle of 50% and the turns of primary winding are the same as the turns of secondary winding, v_s is

symmetrical voltage waveform. There should be no dc voltage in the load resonant circuit. In practice, however, a dc voltage may be presented in the resonant circuit due to discrepant duty cycle and winding. This unwanted dc voltage can be easily blocked the capacitor C_s. Such a circuit topology has the advantages of simplicity and high efficiency. Since the circuit operates symmetrically, the circuit operation for the negative half-cycle of the line voltage is identical to that of the positive half-cycle except that the diodes. The circuit operation for a half-cycle can be divided in to three modes in accordance with the conducting power switch during one high-frequency cycle. Fig. 2 shows the operation modes for the positive and negative half-cycle of the line voltage. The input filter is omitted for simplicity. Fig. 3 illustrates the theoretical waveforms for each mode. To achieve a high power factor, the boost PFC circuit is operated in DCM. The resonant inverter is operated at a switching frequency above its resonance. The circuit operation is described as follows:

Mode I $[t_0 < t < t_1]$:

Before $t=t_0$, the switch S_1 is turn-off state and switch S_2 is turn-on state. The energy stored in magnetizing inductor L_m is returned to the boost PFC filter capacitor C_2 through transformer and diode S_2. The voltage V_{C2} across the capacitor C_2 by negative value supplies the output resonant circuit through the second winding of the transformer. When the magnetizing current $i_{LM}(t)$ drop to zero and the switch S_2 is turned off at $t=t_0$. At this instant, the switch S_1 is activated by its gate signal $v_{gs1}(t)$ and this mode begins. The energy stored in the filter capacitor C_1 supplies the output resonant circuit by positive value. Since the load circuit is designed to be inductive, the load resonant current i_{Lamp} is negative at this instant. It increases and then decreases when it arrives its peak value via the resonance of output resonant circuit. Because the boost PFC is operated DCM and the push-pull converter is operated in boundary mode between CCM and DCM, the input current $i_{in}(t)$ and the magnetizing current $i_{LM}(t)$ increase linearly from zero. When the 50% duty ratio is complete, this mode is finished.

Mode II $[t_1 < t < t_2]$:

At $t=t_1$, the switch S_1 is turned off, the switch S_2 is turned on, and this mode begins. The energy stored in the input inductor L_{in} is delivered to the filter capacitors C_1 and C_2. Thus, the input current $i_{in}(t)$ deceases linearly. The energy stored in the capacitor C_2 supplies the magnetizing inductor L_M and the output resonant circuit by negative value. Thus, the magnetizing current $i_{LM}(t)$ also decreases linearly and the lamp current $i_{Lamp}(t)$ decreases continuously toward its negative peak value. When the input current $i_{in}(t)$ drops to zero, the diode D_1 is naturally turned off and this mode ends.

Mode III $[t_2 < t < t_3]$:

During this mode, the voltage V_{C2} across the capacitor C_2 by negative value supplies continuously the second winding of the transformer. The magnetizing

1996

current $i_{LM}(t)$ deceases linearly. The lamp current i_{Lamp} decreases continuously and then increases when it arrives its negative peak value. When the magnetizing current $i_{LM}(t)$ drops to zero and the switch S_1 is turned on again by the gate signal $v_{gs1}(t)$, this mode ends.

After Mode III, the circuit operation is returned to the Mode I of the next cycle. The input current $i_{in}(t)$ and magnetizing current $i_{LM}(t)$ return to zero. The lamp current $i_{Lamp}(t)$ also returns to its initial value. Therefore, the assumption previously made is proven to be valid. Through the analysis presented above, key waveforms of the proposed single-stage high-power-factor electronic ballast with symmetrical half-bridge topology can be plotted as shown in Fig. 3.

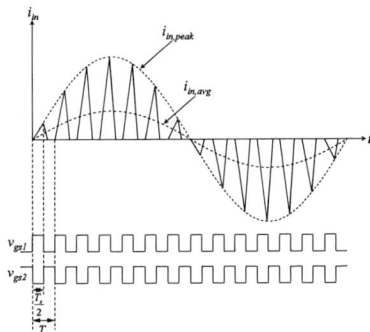

Fig. 4 Conceptual waveform of line current i_{in}.

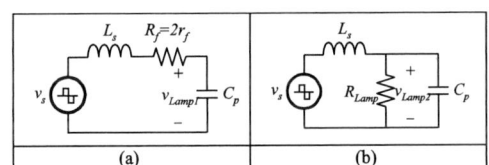

Fig. 5 Equivalent circuit of the load resonant inverter (a) before ignition (b) after started

III. CIRCUIT ANALYSIS

For simplifying the analysis, the following assumptions are made.
1. All components and devices are ideal.
2. The load quality factor of the load resonant circuit is sinusoidal.
3. The filter capacitors C_1 and C_2 are large enough to assume that the voltages V_{C1} and V_{C2} are constant and ripple free. The value of capacitor C_1 equal to value of capacitor C_2, and $V_{C1}=V_{C2}=V_C$
4. The capacitance of C_s in the load resonant circuit is large enough, resulting in zero reactance at switching frequency.
5. The lamp is regarded as an open circuit before ignition and a resistance at the steady state.

Based on the operations described above, the electronic ballast can be treated as two independent stages, the boost PFC and the load resonant inverter.

A. Boost Power Factor Corrector

The electronic ballast is supplied from the ac line voltage source

$$v_{in}(t) = V_m \sin \omega t \qquad (1)$$

where ω and V_m are the angular frequency and the amplitude of the line voltage source, respectively. In practice, the frequency of the line voltage source is much lower than the inverter switching frequency f_s. Under such assumption, the rectified line voltage can be considered as a constant over a high-frequency cycle of the inverter. Since the boost PFC circuit is operated at DCM over an entire line frequency cycle, $i_{in}(t)$ rises from zero at the beginning of Mode I and reaches its peak at the end of Mode I. Then, it drops from its peak value at the beginning of Mode II and reaches zero at the end of Mode II. The waveform of $i_{in}(t)$ is conceptually shown in Fig. 4. Its peaks follow a sinusoidal envelope and can be expressed as

$$i_{in,peak}(t) = \frac{T_s V_m \sin \omega t}{2 L_{in}} \qquad (2)$$

where T_s is the high-frequency switching period. The input current $i_{in,avg}(t)$ is equal to the average of $i_{in}(t)$ by removing its high-frequency contents

$$i_{in,avg}(t) = \frac{1}{T_s} \int_0^{T_s} i_{in}(t)dt = \frac{T_s V_m}{8 L_{in}} (\frac{a \sin \omega t}{a - \sin \omega t}) \qquad (3)$$

where $a = 2V_C / V_m$

The input power can be obtained by taking the average of the instantaneous input power over one line frequency cycle.

$$P_{in} = \frac{1}{\pi} \int_0^\pi v_{in}(t) i_{in,avg}(t) d\omega t = \frac{T_s V_m^2 ay}{8 L_{in} \pi} \qquad (4)$$

where

$$y = \int_0^\pi \frac{\sin^2 \omega t}{a - \sin \omega t} d\omega t$$
$$= -2 - a\pi + \frac{2a^2}{\sqrt{a^2-1}} \left[\frac{\pi}{2} - \tan^{-1}(\frac{-1}{\sqrt{a^2-1}}) \right] \qquad (5)$$

The rms value of input current can be obtained as following.

$$I_{in,rms} = \left[\frac{1}{\pi} \int_0^\pi i_{in,avg}^2(t) d\omega t \right]^{1/2} = \frac{T_s V_m a \sqrt{z}}{8 L_{in} \sqrt{\pi}} \qquad (6)$$

where

$$z = \int_0^\pi \left[\frac{\sin \omega t}{a - \sin \omega t} \right]^2 d\omega t \qquad (7)$$
$$= \pi + \frac{2a}{a^2-1} + \frac{2a(2-a^2)}{\sqrt{(a^2-1)^3}} \left[\frac{\pi}{2} - \tan^{-1}(\frac{-1}{\sqrt{a^2-1}}) \right]$$

Thus, the power factor cab be obtain as

$$PF = \frac{P_{in}}{V_{in,rms} I_{in,rms}} = \frac{y\sqrt{2}}{\sqrt{\pi z}}. \qquad (8)$$

From equation (8), we can find the value of $a = 2V_C / V_m$ is the more large; the power factor is the more good. The input current is very approximate sinusoidal and in phase with the ac line voltage.

B. Series-Resonant Inverter

The voltage $v_s(t)$ applied to the load resonant circuit is a square waveform and can be represented as

$$v_s(t) = \begin{cases} nV_C & 0 \le t \le \pi / \omega_s \\ -nV_C & \pi / \omega_s \le t \le 2\pi / \omega_s \end{cases} \qquad (9)$$

1997

It can also be represented as following by the Fourier series.

$$v_s(t) = \sum_h \left[\frac{4nV_C}{h\pi} \sin(n\omega_s t) \right], h=1, 3, 5, \ldots \ldots \quad (10)$$

With a high quality factor of the load resonant circuit, almost all the harmonic contents of $v_s(t)$ will be filtered out by the resonant circuit.

Because the lamp resistance R_{Lamp} is extremely high before ignition, the equivalent circuit of the load resonant inverter can be regard as a series-resonant series-loaded composed by the filament resistance R_f, the inductor L_s and the capacitor C_p. Thus, the equivalent circuit of the load resonant inverter before ignition can be shown in Fig. 5(a). The natural resonance angular frequency of the series-resonant series-loaded inverter can be denoted as

$$\omega_{r1} = \frac{1}{\sqrt{L_s C_P}} \quad (11)$$

The quality factor at undamped natural frequency is

$$Q_1 = \frac{\omega_{r1} L_s}{R_f} \quad (12)$$

The transfer function of the lamp voltage v_{Lamp1} versus the voltage v_s can be calculated and represented as following.

$$\left| \frac{V_{Lamp1}(j\omega)}{V_s(j\omega)} \right| = \frac{1}{\sqrt{(1 - \omega^2 L_s C_P)^2 + (\omega R_f C_P)^2}} \quad (13)$$

Assuming the lamp is started while the load resonant inverter operates at $\omega = \omega_{r1}$, the peak starting voltage of lamp $V_{Lamp,peak}$ can be evaluated as following.

$$V_{Lamp,peak} = \frac{4nQ_1 V_C}{\pi} \quad (14)$$

After the lamp is started, the fluorescent lamps can be approximated a resistance R_{Lamp}. The equivalent circuit of the load resonant inverter can be regard as a series-resonant parallel-loaded composed by the lamp resistance R_{Lamp}, the inductor L_s and the capacitor C_p. Thus, the equivalent circuit of the load resonant inverter before ignition can be shown in Fig. 5(b). The natural resonance angular frequency of the series-resonant parallel-loaded inverter can be denoted as

$$\omega_{r2} = \frac{1}{\sqrt{L_r \left[\frac{1 + \omega^2 R_{Lamp}^2 C_P^2}{\omega^2 R_{Lamp}^2 C_P} \right]}} \quad (15)$$

The transfer function of the lamp voltage v_{Lamp2} after the lamp is started versus the voltage v_s can be calculated and represented as following.

$$\left| \frac{V_{Lamp2}(j\omega)}{V_s(j\omega)} \right| = \frac{1}{\sqrt{(1 - \omega^2 L_s C_P)^2 + (\frac{\omega L_s}{R_{Lamp}})^2}} \quad (16)$$

We can find the following phenomenon from equation (13). When the circuit operation is more than the resonant frequency, the value of $V_{Lamp1}(j\omega)/\ V_s(j\omega)$ decreases for the circuit operation frequency increases. Thus, we can use this characteristic to warm up gradually the filament. It can avoid the lamp is directly ignited in high voltage and the life of the lamp can be extended.

IV. REALIZATION AND EXPERIMENTATION

An electronic ballast for driving an fluorescent lamp (FL-40D) is illustrated as a design example. The circuit parameters are designed to operate the boost PFC circuit at DCM so that to a high input power factor can be achieved. The controller circuit uses the frequency modulation to ignite the lamp for getting the soft start characteristic so that the life of the lamp will be extended. The design procedure is outlined as follows.

Step 1—Circuit specification.

Input Voltage v_{in}	$110\sqrt{2} \sin[2\pi(60)]t$
Switching Frequency f_s	50kHz
Lamp Power P_{Lamp}	38W
Lamp Voltage $V_{Lamp,max}$	150V
Lamp Equivalent Resistance R_{Lamp}	675Ω
Lamp Filament R_f	50Ω
Input Power Factor PF	>0.95

Step 2—Determine the boost PFC circuit inductor and filter capacitors C_1 and C_2.

For satisfying the circuit requirement $PF>0.95$, the value of $a = V_C/V_m$ must be more than 1.25 from the equation (8). Fortunately, the boost PFC circuit of proposed electronic ballast is operated at fixed frequency with a constant duty cycle of 50% and is operated in discontinuous conduction mode. Thus, the value of $a = V_C/V_m$ approximates 2. Thus, it satisfies the requirement. Assuming a circuit efficiency η is equal to 85% and substituting it and $a=2$ into (4), we can find

$$L_{in} = \frac{T_s V_m^2 a y}{8 P_{in} \pi} = \frac{T_s V_m^2 a y \eta}{8 P_{Lamp} \pi} = 1.527 \times 10^{-3} \quad (17)$$

We use $L_r=1.5mH$. For minimize the ripple voltage of the voltage V_C, we use $C_1=C_2=470\mu F$.

Step 3—Determine the resonant inductor L_s and capacitor C_p.

Assuming the starting peak voltage of the fluorescent lamp is 400V and substituting it and relative circuit specification into (13), (14) and (15), we can obtain

$$\frac{1}{\sqrt{(1 - \omega^2 L_s C_P)^2 + (\omega R_f C_P)^2}} = \frac{400}{\frac{4}{\pi} \times 310} = 1.0134$$
$$(18)$$

and

$$\frac{1}{\sqrt{(1 - \omega^2 L_s C_P)^2 + (\frac{\omega L_s}{R_{Lamp}})^2}} = \frac{150}{\frac{4}{\pi} \times 310} = 0.38 \quad (19)$$

The resonant inductor L_S and capacitor C_P can be calculated as $L_S=5.2mH$ and $C_P=3.896nF$ by using the equations (18) and (19). We use $L_S=5mH$ and $C_P=3.9nF$ in this design example.

In hardware realization, we use MOSFET's IRF740 and S3L60 as the power switches and diodes. This proposed electronic ballast is controlled by the IC TL494 and some logic gate, which drive the active switch at suitable high switching frequencies from starting to steady-state operation. The experimental results are shown in Fig. 6,

7 and 8, respectively. Fig. 6 shows the starting transient voltage waveform of the lamp. The proposed electronic ballast is powered on at a relatively high switching frequency. Then, the switching frequency decreases down to produce the required preheating current. After the preheating voltage has been complete, it decreases further toward the resonant frequency of the load resonant circuit to generate a sufficiently high voltage to ignite the lamp. Once it has been successfully ignited, the lamp current starts to flow. At steady state, the switching frequency is 50kHz to operate the lamp at 38W. Fig. 7 shows the waveforms of the input voltage and current, in which the waveforms of the input voltage and current are almost in phase. The measured power factor is greater than 0.95 and total current harmonic distortion (THD) is less than 8%. Fig. 8 shows the lamp voltage and current waveforms. The measured crest factor of the lamp current is below 1.5.

V. CONCLUSIONS

A novel single-stage high-power-factor electronic ballast with symmetrical half-bridge topology has proposed in this paper with simple and compact configuration. The circuit operation was described and design and the design equations were derived. A prototype circuit of a design example was built and measured to verify the analytical predictions. In spite of the simplicity of circuit configuration, experimental results show that satisfactory performance can be achieved. The power factor is nearly unity and the THD is less than 8%. In addition, the PFC power flow path of the proposed circuit has two conduction drops, the conduction losses can also be reduced. Therefore, it can provide lower conduction loss than the conventional one. Compared with the other electronic ballasts, the performance of the input power factor and lamp crest factor are the same.

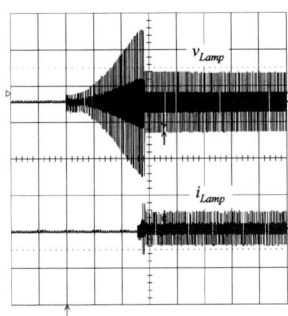

Fig. 6 Starting transient. (v_{Lamp}: 200V/div; i_{Lamp}:2A/div, *time*:0.5s/div).

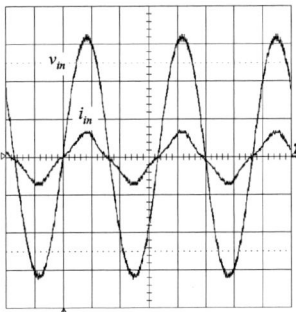

Fig. 7 Input voltage and current after EMI filter. (v_{in}: 50V/div; i_{in}:2A/div, *time*:5ms/div)

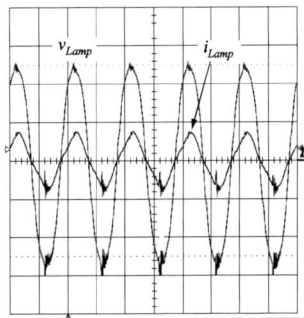

Fig. 8 Lamp voltage and current waveforms (v_{Lamp}: 50V/div; i_{Lamp}:2A/div, *time*:10μs/div)

REFERENCE

[1] J. M. Alnonso, A. J. Calleja, F. J. Ferrero, E. Lopez, J. Ribas, and M. Rico-Secades, "Single-stage constant-wattage high-power-factor electronic ballast with dimming capability," in *IEEE PESC Conf. Rec.* 1998, pp. 2021-2027.

[2] R. de Oliveira Brioschi and J. L. F. Vieira, M "High-power-factor electronic ballast with constant DC-link volatge," in *IEEE Trans. Power Electron.* vol. 13, pp. 1030-1037, 1998.

[3] A. J. Calleja, J. M. Alonso, E. Lopez, J. Ribas, J. A. Martinez, and M. Rico-Secades, " Analysis and experimental results of a single-stage high-power-factor electronic ballast based on flyback converter," in *IEEE Trans. Power Electron.* vol. 14, pp. 998-1006, 1998.

[4] R. N. do Pardo, M. F. da Silva, M. Jungbeck, and A. R. Seidel, "Low cost high-power-factor electronic ballast for compact fluorescent lamps," in *Conf. Rec. IEEE-IAS Annu. Meeting*, 1999, pp. 256-261.

[5] C. S. Lin, and C. L. Chen, "A novel single-stage push-pull electronic ballast with high input power factor," *in IEEE Trans. Industrial Electronics.* vol. 48, no. 4, pp. 770-776, 2001.

[6] H. L. Cheng, C. S. Moo, and W. M. Chen, "A novel single-stage high-stage high-power-factor electronic ballast with sysmmetrical topology," *in IEEE Trans. Industrial Electronics.* vol. 50, no. 4, pp. 759-766, 2001.

2006 5th International Power Electronics and Motion Control Conference

Smoothed-Power Output Supply System for Battery of Stand-alone Renewable Power System Using EDLC

Y. Jia, R. Shibata, N. Yamamura, M. Ishida

Department of Electrical and Electronic Engineering, Mie University, Tsu-city, Mie Pref. Japan
E-Mail: jiayan@cs.elec.mie-u.ac.jp, ryosuke111@hotmail.com, yamamura@elec.mie-u.ac.jp, ishida@elec.mie-u.ac.jp

Abstract—A smoothed-power output topology for battery of Stand-alone Renewable Power System (SARPS) using EDLC (Electric Double Layer Capacitor) is presented. To prolong the service life of battery, decrease maintenance cost and power loss of SARPS by realizing smoothed-power output supply, an EDLC is paralleled with battery as a part of storage system, and connected by a bi-directional Buck/Boost compensation topology. The EDLC stores fluctuant component and compensate battery current through proposed topology and switch control. Whereas the changeful natural condition and initial power condition, an appointed fluctuant power condition as an example to analyze the validity of proposed method in simulation and real-time experiment example firstly. The results show that smoothed-power output for battery can be obtained through proposed topology and satisfy the criterion of smooth precision. And then, the experiment result carried out in actual stand-alone wind power system further shows the feasibility of proposed topology.

Keywords- Smoothed-power;Battery; Service life; Stand-alone Renewable Power System (SARPS); Electric Double Layer Capacitor (EDLC); Bi-directional Buck/Boost compensation topology.

I. INTRODUCTION

Renewable energy resource has become one of effective methods to resolve energy shortage and prevent global warming in now and future. There are two types of combination power and stand-alone power in renewable power system. A stand-alone power system is independent from grid, SARPS has become the main power supply method for the area where grid can't reach, e.g. remote locations, remote monitoring devices etc..

As we know, to ensure the effectively continual power supply throughout the year, storage system such as battery has become one of the necessary parts in SARPS. Now, lead-acid battery is the most common storage equipment in renewable power systems because of lower initial cost and being readily available nearly everywhere in the world. Service life of battery depends on the frequency of recharge and the depth of discharge. In conventional renewable power system, higher harmonic component caused by erratic natural condition still exist and be supplied to battery directly, frequently fluctuation increase the cycle times of recharge and sometimes the peak vibration results in overcharge. This situation hastens the aging of battery and extra maintenance cost is also increased. And simultaneously, increase power [1].

EDLC is often called super-capacitor. It is an energy storage device, which also characterizes battery and capacitor as device and with following characteristic: a virtually unlimited cycle life; low impedance of enhancing load handing when parallelling battery; rapid charging and simple charge method; Substantially higher performance of self-discharge [2].

In this paper, we aim to design a smoothed-power output system for battery to prolong its service life, decrease extra maintenance cost. Therefore, an EDLC, used as compensating element of fluctuant component, is connected with lead-acid battery in parallel as storage system of SARPS, and a bi-directional Buck/Boost compensation topology is proposed. Boost chopper or Buck chopper conducts alternately through PI controller and electronic switch control when natural resource and load change. EDLC absorbs and stores AC component of fluctuant power, ensure constant component, i.e. smooth component is supplied to lead-acid battery.

Because the output dynamics of system varies along with the variety of frequency and amplitude of fluctuant power supply, in this paper, based on an example under an appointed fluctuant condition, the feasibility of proposed topology is verifid by simulation in PSIM and real-time experiment firstly in a short time. The smooth precision is satisfied to the criterion. Simultaneously, in a relative long time, the experiment is carried out in actual stand-alone wind power system, the result further verified the feasibility of proposed topology.

II. STAND-ALONE RENEWABLE POWER SYSTEM (SARPS)

A. Conventional SARPS

Stand-alone power source could be defined as a power source which operates totally independent of grid supplied power, small-scale wind/solar hybrid power system and single wind/solar power system are the common styles of Stand-alone Renewable Power System (SARPS) in the rural and remote locations. The conventional SARPS usually includes following parts: power source (wind turbine generator and photovoltaic), storage system and charge controller, MPPT contoller (Maximum Power-Point Tracking), converter, load etc. It can provide DC or AC current or both to consumer.

A conventional charge controller connects power source to battery. It controls the flow of current to and from the battery subsystem to protect the batteries from overcharge and over discharge. Essential for ensuring obtain maximum state of charge and longest life. But, due to the actual fluctuant power, it is difficult for the conventional charge controller to track and regulate the charge/dischage voltage exactly, overcharge and over-discharge bring on the battery fail prematurely.

B. Configuration of proposed SARPS with EDLC storage system and bi-directional Buck/Boost compensation topology

To ensure supplying smoothed-power output to lead-acid battery, to take advantage of the merits of EDLC, we connect EDLC with battery in parallel as the storage system of proposed SARPS to compensate the fluctuant power. But, we find that EDLC cann't absorb the higher harmonic component of fluctuant power completely and compensate the battery current commendably if it is simply connected in parallel with battery. Look following example, Fig.1. shows a conventional power system with EDLC, assume the fluctuant power supply is $is=10sin(20\pi t)+10[A]$, when voltage of EDLC $V_{ED}=15[V]$, inherent resistance $r_{ED}=23.2[m\Omega]$, battery voltage $V_B=6[V]$, inherent resistance $r_B=0.1[\Omega]$, simulation result as shown in Fig. 2 indicate the ratio of Δib_{max} (the maximum fluctuation range of battery current) to I_{p-p} (the peak-to-peak amplitude of fluctuant power supply) is 19%. It manifests the smooth precision is lower.

Therefore, we adopt a bi-directional Buck/Boost converter to contro the Buck chopper and Boost chopper operates alternately through electronics switching control, to ensure the higher harmonic component to be absorbed by EDLC, but DC component for battery. The proposed SARPS with bi-directional Buck/Boost compensation topology and EDLC storage system is shown in Fig.3. In this paper, the research of smoothed-power output applies the focus to the power pulse that come from wind turbine generator after AC/DC rectifier or solar power system as shown in the dashed line part of Fig.3.

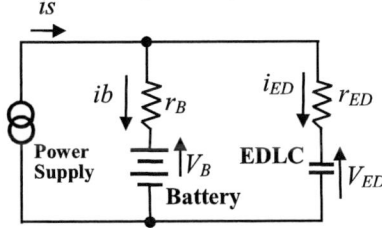

Fig. 1. Main circuit of conventional power system with EDLC

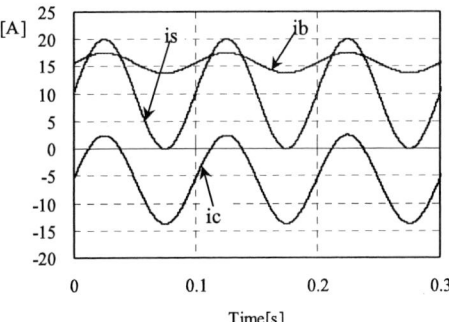

Fig. 2. Simulation result of conventional power system
$is=10sin(20\pi t)+10$ [A], S=19%

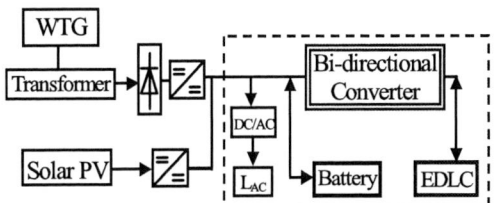

Fig. 3. Stand-alone Renewable Power System (SARPS).

C. Bi-directional Buck/Boost compensation topology

In the main circuit of proposed topology, except the power supply, load and battery, EDLC is connected with battery in parallel through a bi-directional Buck/Boost compensation converter to realize the charge and discharge of EDLC, as shown in Fig.4. In the converter, a low-pass filter (LPF) consisting of L_2 and C_1 is adopted to protect EDLC from transient peak current. Two MOSFET switch is used in experiment. C_2 is a simple snubber of switch. Proposed topology is basically a switched mode circuit, switch S1 and S2 execute alternatively through the active switching control. The defination of main variables are shown in TABLE 1.

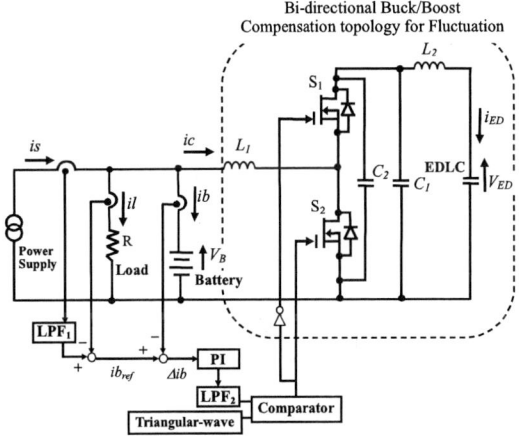

Fig. 4. Proposed Bi-directional Buck/Boost compensation topology

In control circuit block diagram, the DC component of fluctuant power supply is extracted through LPF$_1$ and as

2001

the reference of battery current ib_{ref} is selected. In this paper, we use a proportional & integral (PI) controller to control and adjust the steady state error Δib ($\Delta ib=ib_{ref}-ib$) tend to zero. A LPF$_2$ is placed after PI controller to reduce the influence of oscillation and switching ripples. Assume the duty cycle of switch S2 is d, and then the duty cycle of switch S1 is $(1-d)$, control ON/OFF of switch S1 and S2 by comparing the signal Δd with carrier of triangular wave, realize charge/discharge of EDLC. The cut-off frequency of LPF$_1$ and LPF$_2$ is determined through other system parameters. The frequency of triangular wave is $f_{tri}=31.25$[KHz] in control circuit. In actual condition, switch ON/OFF execute according the actual natural condition (e.g. wind speed or solar radiation) [3], [4], [5].

TABLE I
DEFINATION OF MAIN PARAMETERS

Variables	Meaning
is	current of fluctuant power supply
il	Current of Load
is'	$is - il$
ib	Current of Lead-acid Battery
V_B	Voltage of Lead-acid Battery
ic	Alternating Current Component
i_{ED}	Current of EDLC
V_{ED}	Voltage of EDLC

For example, when rich of natural resource, the Boost chopper conducts, battery bank and EDLC is charged, AC component of fluctuant power with higher harmonic is absorbed and stored by EDLC, but the constant component that is smoothed component is supplied to battery bank. On the other hand, when it weak of natural resource or peak consumption, the Bck chopper conducts, EDLC discharges and supplies power to the battery bank and the load to keep the sustaining power supply to load. In current study, the power supply is regard to be a current source contains higher harmonic fluctuation and the load is a constant resistance.

III. SMOOTH PRECISION AND AN APPOINTED FLUCTUANT POWER CONDITION AS AN EXAMPLE FOR ANALYSIS AN

We evaluate the smooth level of power output using the defined smooth precision, it is evaluated by (1),

$$S = \frac{\Delta ib_{max}}{I_{p-p}} \dots\dots\dots\dots\dots\dots\dots\dots\dots\dots\dots\dots\dots\dots\dots\dots\dots (1)$$

Where, Δib_{max} is the maximum fluctuation range of battery current, and I_{p-p} is the peak-to-peak amplitude of fluctuant power supply. S is the smooth precision, a ratio of Δib_{max} to I_{p-p}.

The smaller S the higher smooth precision can be obtained. The ideal smoothed-power output for battery is DC, i.e. $S=0$. Actually, we purpose to control the battery cuttert fluctuates in a smaller domain and ensure a smooth precision of $S\leq10\%$ through proposed topology.

But, for the complexity of actual condition of renewable resource, frequency and amplitude of fluctuant

input power varies continually, operation characterics of circuit varies too. It become difficult to regulate power output and determine parameters that satisfying design requirement accurately by usual conventional duty modulation method. In this case, to verify the feasibility of proposed topology, assume an appointed fluctuant power condition as an example for analysis firstly.

Parameters in main circuit and PI controller have great influence to output current. Although bigger inductance of L_1 is benefit to the operation of the Buck/Boost choper. The damping efect of input to output has a downward trend with a growth in L_1, smoothing precesion decreases. L_1 has larger effect on higher frequency damping. But then, in actural, unstable phenomena occurs if the L_1 is too small, Boost chopper almost doesn't conduct, smothed-power output cann't obtain. Kp has a deep influence to smooth precision of power output. Damping have a upward trend with a growth in Kp. However, unstable phenomena occures in the larger Kp. L_2 has a deep influence to damping in high-frequency range. Damping has an downword tend with a drop in L_2. C_1 has a few influence on output damping in high-frequency range, but as said foregoing, the cut-off frequency of LPF (L_2, C_1) should far smaller than the switching frequency of the carrier, $f_{tri}=31.25$[KHz], at least is smaller than one-tenth of f_{tri}. And also is far bigger than higher harmonic frequency of fluctuant power in system design.

Considering the smooth output when fluctuant power condition is: Peak to peak amplitude of fluctuant power: $I_{p-p}\leq20$[A]; Higher harmonic frequency: $0.5[Hz]\leq f_h\leq10[Hz]$; and battery voltage, EDLC voltage and resistance of system is $V_B=6$[V] and $R=6[\Omega]$ respectively. Because the limit voltage of existing EDLC in experiment is 24V, considering security of EDLC, we usually think EDLC can act normally when $V_{ED}<20$[V], select $V_{ED}=15$[V], capacitance of EDLC: $C_{ed}=328.9$[F] [6].

IV. SIMULATION EXAMPLES

To verify the feasibility of proposed topology, simulation in PSIM is carried out using the parameters of TABLE II.

TABLE II
PARAMETERS IN SIMULATION

Parameters	Value
L_1	2 [mH]
L_2	0.5 [mH]
C_1	10 [µF]
Kp	3.8
Ki	1

Since PSIM uses the fixed time step during the entire simulation, the time step should be several tens times smaller than the switching period follows the specification of PSIM. Therefore, choose $\tau=10^{-6}$[s] as the time step. Considerring the smooth-power output in short time firstly, Fig. 5, shows the simulation results when power supply is $is=10sin(20\pi t)+10$[A]. Suppose the ratio of Δib_{max} to I_{p-p} is S, S is 4.7% as shown in Fig. 5. The output wave shows AC component in fluctuant power supply can be separated and stored in EDLC ideally and as the compensation power of battery current, but constant component is supplied to battery. Comparing the simulation results of proposed topology in Fig. 5 with that conventional power system without proposed topology, that shown in Fig. 2, it is clarity that the smooth precision of proposed topology is improved greatly.

In additional, if consider the situation of unconstant load, assume resistance is infinite, that is no load. When $is=10sin(20\pi t)+10$[A], simulation is also carried out, the ratio of Δib_{max} to I_{p-p} is 4.8%, smoothed-power for battery can be obtained. Response of system didn't change.In the same way, smoothed-power output also can be obtained when resistance is zero.

For the amplitude of ripple that super-response on top of the output wave that caused by switching control is almost equivalent to 0.056[A], it is much smaller than of I_{p-p} and can be ignored. The system can be regarded as stable under this valid approximation analysis method.

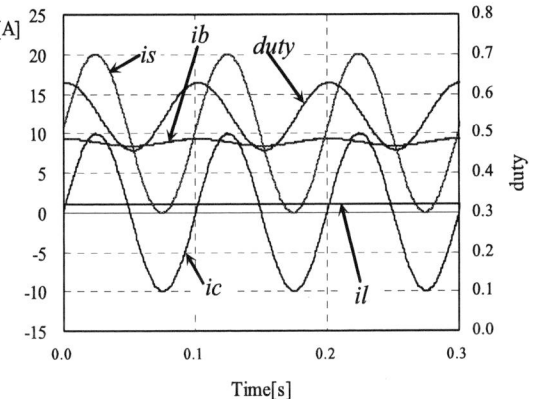

Fig. 5. Simulation result of considering smoothed-power output in a short time
$is=10sin(20\pi t)+10$ [A], S=4.7%

V. EXPERIMENT RESULTS

Simultaneously, real-time experiment is carried out to do further validation. In experiment, the power supply is a sinusoidal current power source, base on the current experiment condition, the peak-to-peak amplitude of fluctuant power source is set in I_{p-p}=12[V]. The gain of PI controller is gained from a DSP (Digital Signal Processing). Other parameters are same with simulation, as shown in TABLE II. The experiment result is shown in Fig. 6. The ratio S of Δib_{max} to I_{p-p} is 6.5% when the power supply is $is=6sin(20\pi t)+6$[A] . The experiment result is better agreed with the foregoing smooth precision target. Validity of proposed topology is verified in experiment too.

In additional, assume the fluctuation component of input current is $is=1.5sin(6\pi t)$ [A], the usefulness of smooth-power output for battery is also validated in

experiment when consider smoothed-power output in a long time through simulating the actural natural power fluctuation. The result is shown in Fig.7., the amplitude of fluctuation component of 3[Hz] can be controlled less than 0.15[A], that is S<10%. It indicate that smothed–power output for battery can be obtained by controlling the compensation current [7]. Assume there isn't proposed topology and EDLC compensation current, and the load current is about 2[A], in this case, current of power source is is supplied into battery directly, datum line of charge/discharge transformation of battery is shown as the line of a-a in Fig. 7., the changeful power source current result in frequently repeating charge/discharge. After proposed smoothed-power output system, observing the charge/discharge times of battery, Fig. 8. is the enlarge figure of from 22[s] to 36[s] of Fig. 7. Fig. 8. shows there is a fewer polarity reverse times of charge/discharge, the recharge cycle time is obviously decreased after propose smothed-power out put supply system.

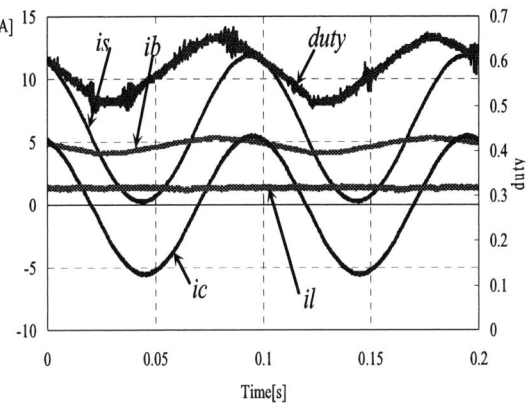

Fig. 6. Experiment result of considering smoothed-power output in a short time
$is=6sin(20\pi t)+6$ [A], S=6.5%

Based on above analysis, we introduce the proposed smoothed-power supply system into actual wind power system and do some experiments. The rated power of wind turbin is 400[W] , rated lead-acid battery voltage V_B=24[V], EDLC voltage V_{ED}=100[V], capacitance of EDLC, C_{ED}=14[F], and other parameters as shown in TABLE II. Fig. 9 shows an example of experiment that load resistance R=0[Ω], and simply consider the feasibility of smooth output through connect wind turbin generator with battery and proposed topology. Fig. 9 shows the smoothed-power output for battery is realized in actual wind power system. During the start-up period wind turbin generator(WTG), as shown from 2[s] to 6[s] in Fig. 9., after the start-up fluctuation current of wind turbin gererator is compensated, the smooth-power flow into battery, it avoid the instantaneous peak current to battery and over charge. During the running period, as shown from 6[s] to 38[s] in Fig. 9., look from the whole output wave, battery current is smooth, peak current and momentary fluctuantion can be smoothed, charge cycle times is decreased greatly than that without smoothed-power output system. And, during the period of stalling, as shown 38[s] to 50[s] in Fig. 9., similar with start-up period, fluctuation current is compensated and smooth current flow in to battery. In additional, due to the compensation of EDLC, battery current decrease slowly with a delay, momentary sharp discharge and over discharge is avoided. Fig. 9. also shows that the step response of proposed topology is better, battery current can come back stable smooth state in several seconds. According to above, it is feasibility that smoothed-power output supply system for battery of SARPS using proposed topology. It will benefite to prolong the service of lead-acid battery of SARPS.

For an actual power system, when load exist and not equal to zero, battery charge/discharge alternatly according to power output of wind turbin generator and natual condition. As shown during the period of from 6[s] to 10[s] in Fig. 9., the output power of wind turbin generator is changeful during this period, assumeing the load current is 4[A]. We can observe that the recharge cycle time is decreased through the smoothed-power output supply system during this period.

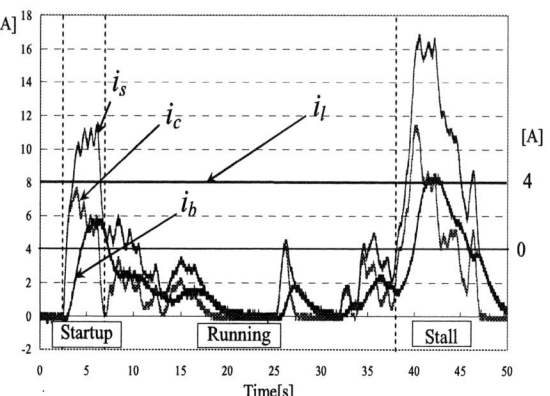

Fig. 9. Experiment result in actural wind power system

VI. CONCLUSION

It is evident that the proposed compensation topology and control method is feasible and effective from the above analysis as well as the results of simulation and experiment. Bi-directional Buck/Boost compensation topology utilize the merite of EDLC effectively through active switch control. As a part of storage system of SARPS, EDLC not only store energy but also absorb and compensate AC component of fluctuant power source to ensure smoothed-power output for lead-acid battery. By that means, ensure to decrease cycle times of recharge and AC power loss. It is benefit for prolonging service life of lead-acid battery and decrease extra maintenance cost. Furthermore, it is verified that the proposed smoothed-power output system is applicable for SARPS.

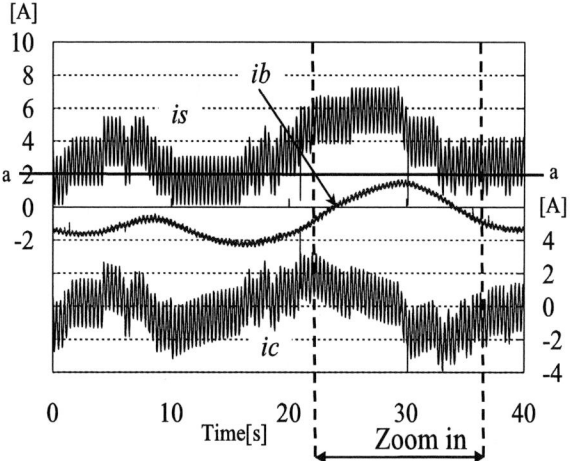

Fig. 7. Experiment result of considering smoothed-power output in
a long time when simulated fluctuation current

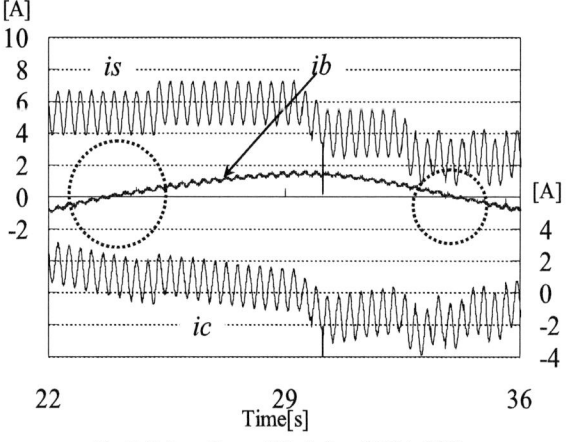

Fig. 8. Enlarge figure of Fig.7. from 22[s] to 36[s]

REFERENCES

[1] The national Renewable Energy Laboratory, Small Wind Energy Systems for the Homeowner, Department of Energy, U.S, January 1997.

[2] Eiyuu TAMURA, MATSUDA etc. The advanced of supercapacitor-Electric Double Lay Capacitor, Function chemical series of electron and ion, Vol.2. ENE.DHI. ESU. Jan. 2002,pp305-327

[3] Georgios A. Papafotiou, and Nikos I. Margaris, calculation and stability Investigation of Periodic Steady State of the Voltage Controlled Buck DC-DC Converter, IEEE Trrans. Power Electron.,vol.19, NO.4, pp959-970, JULY 2004.

[4] M.D.Bernardo, F.Garofalo, L. Glielmo, and F. Vasca, "Switchings bifurcations ans chaos in dc/dc converters," IEEE Trans. Circuits Syst.I, vol. 45, pp. 133-141, Feb. 1998.

[5] Simon S. Ang. Power switching converters. Marcel Dekker Inc,1995.

[6] Y.JIA, R.SHIBATA, N.YAMAMURA, M.ISHIDA, A Control Method of Prolonging the Service Life of Battery in Stand-alone Renewable Energy System using Electric Double Layer Capacitor (EDLC), Proc. Of PEDS2005, IEEE, Distributed Generation 1.

[7] R.SHIBATA, Y.JIA, N.YAMAMURA, M.ISHIDA, A Characteristic of Power Smoothing System using EDLC for Small Stand-alone Power System, Proc. Of The 2006 Annual Meeting IEE Japan, (in Japanese), Vol 7,pp-36-37.

2006 5th International Power Electronics and Motion Control Conference

Supercapacitors characterization for hybrid vehicle applications

F.Rafik[1-2], H.Gualous[2], R. Gallay[3], A.Crausaz[1], A.Berthon[2]

[1] Haute Ecole Arc Ingénierie, CH-2400 LE LOCLE, Switzerland
[2] L2ES-UTBM-Univ.Franche comté, Rue T.MIEG, F90010 BELFORT, France
[3] MAXWELL Technologies, CH-1728 Rossens, Switzerland
Fouad.rafik@he-arc.ch

Abstract— **This paper presents the supercapacitor characterization for hybrid vehicle applications. This characterization takes into account frequency, voltage and temperature variations. The frequencies dependence is mainly caused by the dynamics of the ions in the electrolyte. With high frequencies, the ions do not have time to reach the surfaces not easily accessible, located deeply in the pores which causes a reduction of the capacitance and equivalent series resistance with the frequency. To take in to account these effects in supercapacitor modeling, it is necessary to characterize them in experiments according to the electric, thermal and frequency constraints.**

Keywords: Supercapacitors – Modeling – Characterization – Impedance spectroscopy – Thermal effect

I. INTRODUCTION

Researches about energy means were accentuated these last years to replace fossil energies by renewable energies like wind energy, solar energy, fuel cells....The complexity of integration these alternatives is their applications for the embarked systems as for the case of the hybrid vehicles using the fuel cells for example; the problem with these sources is their limited power. The recent developments in the electrochemical field gave rise to a new element energy storage which proves solution for these power limitation problems. The supercapacitor is the result of these researches during the last decades. It is placed between the batteries and traditional capacitors concerning power and energy. Its capacitance can reach very high values (until 5000F) with a maximum electrolyte voltage up to 2.7V. For its high power, it is shown today like a good alternative allowing an improvement of the power of the system. For example, it can be introduced in parallel with the fuel cells and thus answer the requests of short durations high powers (inferior at one minute). A good energy management available makes it possible to optimize the power consumption of the systems and thus to improve these efficiency. To use efficiently supercapacitors in automotive applications, supercapacitor characterization according to the constraints to which will be subjected is obligatory. These constraints are: temperature, frequency

and charging voltage [1]. So, in this study, we will focus on supercapacitor characterization according to the above mentioned constraints. The characterization method used is the electrochemical impedance spectroscopy applied on the BCAP0010 of 2600F, 2.5V supercapacitor of Maxwell technology.
Different analyses and conclusions are presented.

II. ECTROCHEMICAL IMPEDANCE SPECTROSCOPY

The electrochemical impedance spectroscopy (EIS) is used to characterize the electrochemical behaviour of energy storage devices. It is used to characterize the electrodes materials for supercapacitors, batteries and fuel cells [2]. To characterize a supercapacitor, the sweep in frequency must be done for various voltage levels and also for different temperatures. The EIS permits to study the influence of the frequency on the electrode series resistance set and its specific capacitance contrary to the characterization using constant current with which one can reach significant currents. On the basis of frequencies spectrum, we can determine the physical origin of the obtained impedance.
The measurements have been done on a supercapacitor cell from Maxwell Technologies: BCAP0010 of nominal capacitance 2600F (figure 1), for voltage values in the range from 0 to 2.5 V, and for temperature from -20°C to 60°C. The cell impedances have been measured with the Impedance Spectrometer, IM6 (ZAHNER). A small voltage ripple amplitude (10 mV), with a variable frequency sweeping from 0.01 to 1000 Hz, is applied to the supercapacitor. The current amplitude and phase with respect to the injected voltage permits to determine the module and the phase of the impedance as a function of the frequency.

1-4244-0448-7/06/$25.00 ©2006 IEEE

Figure 1: Test bench overview

The Impedance spectrometer permits to reach the lower frequency and also to charge with a constant current until 40A if we combine it with the Power potentiostat PP240.

III. FREQUENCY SUPERCAPACITOR CHARACTERIZATION

With EIS, the real and the imaginary parts of the impedance can be measured for a given frequency band. Supercapacitor BACP0010 real part impedance and capacitance were determined using the measurements. Different supercapacitor DC bias voltages are applied 0V, 1.25V and 2.5 V. The supercapacitor is based on the activated carbon and organic electrolyte technology. These results show that the series resistance for low frequency is approximately two times higher than that in high frequency (resistance to 1 kHz for example).

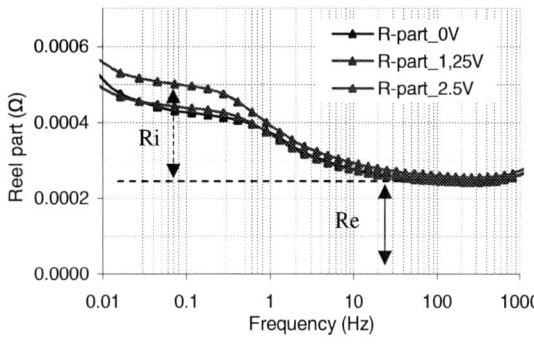

Figure 2: Reel part according to the frequency for a temperature of 20°C

The minimal value of the equivalent series resistance is measured at high frequencies. In the figure 2, this value may be read between 60 and 1000 Hz. It is called, high frequencies resistance (Re) and represents the contact resistance between the activated carbon and the current collector metal as well as the minimal resistance of the electrolyte [3].

These results show also that when the frequency decreases the real part impedance increases. This increase can be justified by the fact that the ions can follow the applied electric field easily, and that their displacements increase the ionic resistance. On the curve, this variation is represented by R_i. It is the electrolyte resistance part [4]. The low frequency equivalent series resistance (ESR), also called DC resistance, is given by Rs = Re + Ri. It's typically visible between 0.01 and 1Hz.

For frequencies lower than 0.02Hz, the real part varies strongly with the frequency. This equivalent resistance variation is mainly driven by the parallel resistance, due to the leakage current through the separator, to the redox reaction of impurities and to the internal charge redistribution.

The capacitance spectrum depends mainly on the properties of the activated carbon. The capacitance is determined by the following expression:

$$C_{diff} = \frac{1}{2\pi \cdot f (2\pi \cdot f \cdot L - \operatorname{Im}(\underline{Z}))}$$

A supercapacitor equivalent inductance can be determined in the high frequency range with the following relation: $L \approx \dfrac{\operatorname{Im}(Z)}{2\pi \cdot f}$

The capacitance is a maximal at very low frequencies when all available surfaces on coal have time to be reached by the ions. So the most part of the surface is located at the bottom of pores, deep and narrow, the capacitance falls quickly as the frequency increases (figure 3).

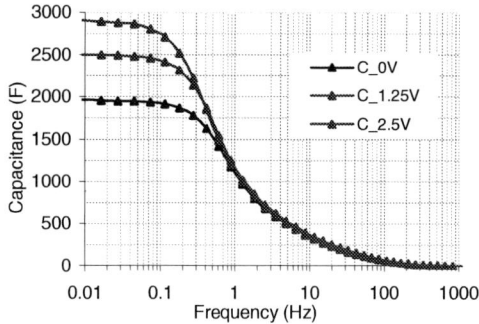

Figure 3: Differential capacitance according to the frequency for a temperature of 20°C

The supercapacitor capacitance variation due to the voltage finds its origin in the physical structure of the supercapacitor. This equivalent capacitance is composed by the two electrode/electrolyte interfaces double layer capacitance. The increase of the capacitance according to the voltage can be interpreted as being due to the increase of the dielectric constant of the electrolyte, or, to the reduction of the distance separating the charges. Inside the electrode, the space charge is created by the charges displacement in the conductor. This charges displacement is at the origin of an electronic capacitance that increases with the density of the electronic states (DOS). These latter increases with the voltage [5].

In the case of a traditional capacitor, the capacitance is constant in the field of voltage and is defined by the relation: Q(U) = C U, where Q(U) is the accumulated charge when the voltage applied at the boundaries of the capacitor is U. For the supercapacitor, the characterization showed that the capacitance varies according to the voltage at the boundaries of the supercapacitor (figure 4). This variation can be considered linear when the supercapacitor is used between its nominal voltage and half of this voltage. Hence, the equivalent capacitance can expressed as following [6]: $C(U) = C_o + K \cdot U$

C_0 is the capacitance value at U=0.

The relation between the current and the electric charge Q always remains given by: $i = \dfrac{dQ}{dt}$

Substituting the Q expression as a function of U and C, taking into account the dependence of C with the voltage, it is easy to show that the current may be given by:

$$i(t) = (C_0 + 2 \cdot K_v \cdot U) \cdot \dfrac{dU}{dt}$$

By analogy with the case of the traditional capacitor, a differential capacitance C_{diff} can be defined to calculate the relation between the current and the voltage. Its expression can be determined as follows: $C_{diff} = C_o + 2 \cdot K \cdot U$

The differential capacitance evolution as a function of the supercapacitor voltage is plotted in figure 4. These experimental results show that the differential capacitance can be considered linear when the supercapacitor is used between its nominal voltage (in order of 2.5V) and the half of this voltage. Hence, in this voltage range, supercapacitor provides 75% of the stored energy.

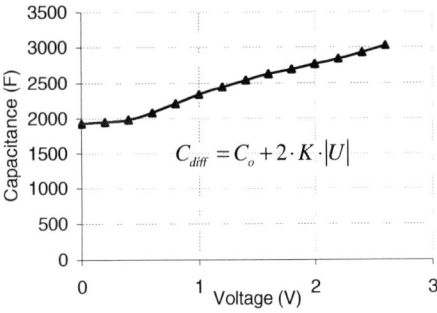

Figure 4: Capacitance as a function of voltage at 0.01 Hz and 20°C.

Table 1 gives the capacitances, differential and of charge for a frequency of 0.01Hz and values of series resistance for a frequency value of 0.1Hz. The capacitances and the series resistance are determined for the operational voltages limit, 1.25V and 2.5V.

Table 1: Parameters determined - series resistance, charge and differential capacitances

Voltage (V)	Resistance (mΩ)	C_{diff}(F)	$C_{diff\text{-}moy}$(F)	k (F/V)	C (F)	C_{moy}(F)
1.25	0.455	2500	2725	220	2225	2337
2.5	0.517	2950		200	2450	

IV. THERMAL SUPERCONDENSATEUR CHARACTERIZATION:

1-Equivalent series resistance

Figure 5 represents the evolution of the impedance real part according to the frequency for several temperatures. The characterization according to the temperature showed that in high frequency range (f > 10Hz), the series resistance variation with temperature can be neglected. In low frequency range, the equivalent series resistance increases when the temperature decreases [7]. This is due to the fact that the electrolyte ionic resistance Ri is influenced strongly by the temperature. Above 0°C Ri varies slowly with the temperature. Below 0°C the temperature dependence is more important. It is due probably to the viscosity of the electrolyte that increases in low temperatures what increases the resistance of the electrolyte.

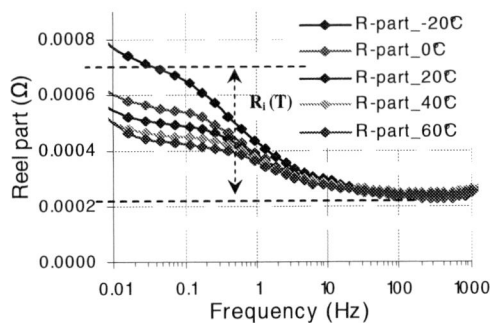

Figure 5: Real part according to the frequency for a DC bias voltage of 2.5V

From the curves represented on the figure 5, we can notice that the variation of series resistance has an exponential form [8].
A relationship between R_i and the temperature has been established from experimental results. It's given by the following expression:

$$R_i = R_{20} \cdot \frac{(1+\exp(-k_T \cdot (T-T_{20}))}{2}$$ where R_{20} is the electrolyte resistance part Ri at 20°C, T the ambient temperature and k_T the temperature coefficient: $k_T = 0.025°C^{-1}$.

The experimental results and the simulations are presented in figure 6.

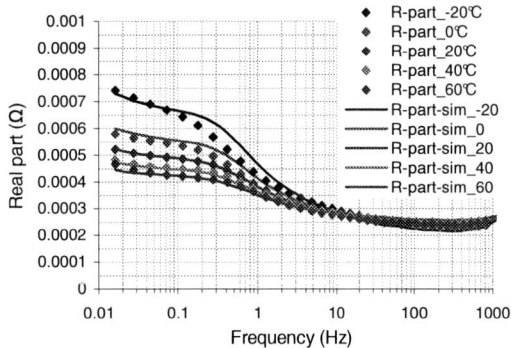

Figure 6: Real part according to the frequency for a DC bias voltage of 2.5V: (Experimental and simulation)

2- Differential capacitance

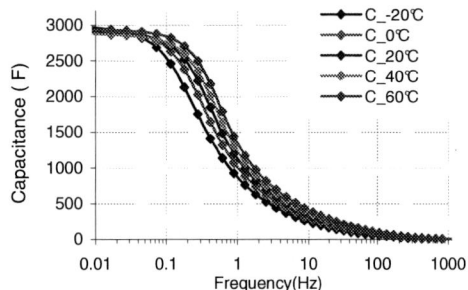

Figure 7: Differential capacitance C_{diff} according to the frequency for a DC bias voltage of 2.5V.

In the case of the capacitance, the experimental result is plotted in figure 7. At very low frequencies (f<0.1Hz), the capacitance is almost constant according to the temperature. It means that the ions penetrate in the depth of the pores of the electrodes regardless of the temperature and there is therefore the same contribution to the capacitance of the double layer for the low temperatures as for higher temperature. The equivalent capacitance is influenced by the temperature only at the frequency range [0.1, 10Hz]. One finds less capacitance at low temperatures and it is due to the series resistance that increases for the low temperatures decreasing the cut off frequencies of the supercapacitor.

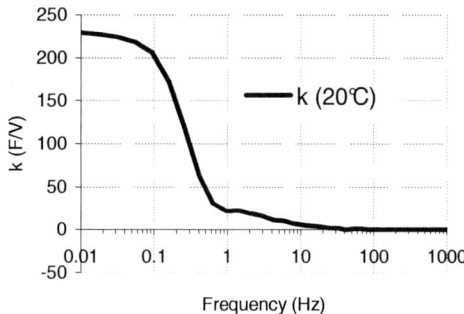

Figure 8: Evolution of K according to the frequency

From the tests carried out, constant K was found for various frequencies. Figure 8 represents the evolution of K according to the frequency for the supercapacitor 2600F. These results show that K is quasi constant for very low frequencies (f<100 mHz).

The linear relation between the capacitance and the voltage is valid only for the low frequencies and when the voltage varies between 1.25 and 2.5 V. Consequently the coefficient K decreases very quickly according to the frequency of use and becomes negligible from about 1Hz; the behavior of the supercapacitor matches a traditional capacitor behavior.

V.   CONCLUSION

This article characterizes the supercapacitor according to the electric, thermal and dynamic constraints in seen to model it. The characterization by using a constant current shows these limits in frequency. With the impedance spectroscopy, one could characterize the supercapacitors for an active field of frequency of 0.1Hz with 1000Hz. This characterization showed that the capacitance of the supercapacitors is strongly influenced by the frequency. For the frequency of 10Hz one loses 90% of the maximum capacitance reached. It has been demonstrated also that the temperature influences the electric behavior of the supercapacitor by increasing his resistance series for the low temperatures. Even with its weak signal characterization, the EIS remains a precise and reliable method comparing with the other characterization methods.

REFERENCES

[1]   B. E. Conway, "Electrochemical supercapacitors: scientific Fundamentals and Technological Applications" Kluwer Academic Press /Plenum Publishers, New York, 1999, pp. 377-556

[2]   S. Buller, E. Karden, D. Kok, R. W. De doncker 'Modeling the Dynamic Behavior of Supercapacitor Using Impedance

Spectroscopy' IEEE Transactions on industry applications, vol. 38, N° 6, 2002, pp. 1622-1626

[3] R. Kötz, M. Carlen, "Principles and applications of electrochemical capacitors" Electrochimica Acta 45, (2000) 2483 – 2498.

[4] Andrew Chu, Paul Braatz, "Comparison of commercial supercapacitors and high-power lithium-ion batteries for power-assist applications in hybrid electric vehicles" I. Initial characterization (Juin 2002), HRL Laboratories, LLC 3011, Malibu Canyon Road, Malibu, CA 90265-4797, USA

[5] V. Hermann, A. Schneuwly and R. Gallay, "High performance double-layer capacitor for power electronic applications", proc. PCIM 2001 in Nürnberg

[6] L. Zubieta, R. Bonert 'Characterization of double-layer capacitors for power electronics applications' IEEE-IAS'98, 1149-1154, 1998

[7] H. Gualous, D. Bouquain, A. Berthon, J.M. Kauffmann "Experimental study of supercapacitor serial resistance and capacitance variations with temperature", Journal of Power Sources 123 (2003) 86–93.

[8] W.Lajnef, O.Briat, S.Azzopardi, E.Woirgard, J.-M.Vinassa "Ultracapacitors electrical modelling using temperature dependent parameters",ESSCAP BELFORT 2004

Power Transfer Maximization and Di/Dt Based Extremum Tracking for a Swing Engine Based Portable Power System

Satish Rajagopalan[1]   Deepak M. Divan[1]   Ronald G. Harley[1]   J. Rhett Mayor[2]

[1]School of Electrical and Computer Engineering
Georgia Institute of Technology
Atlanta, Georgia 30332 USA
rsatish@ece.gatech.edu

[2]Powerix Technologies, LLC
2000 Hogback Road, Suite 13 & 14
Ann Arbor, Michigan 48105 USA

Abstract—Several new distributed generation sources are being proposed today that feature variable frequency multiple unbalanced output voltages. One such example is a portable power system based on a permanent magnet generator (PMG) that is driven by a micro internal combustion swing engine (MICSE). Maximum power transfer in such systems can be challenging and reduces system efficiency by 15 – 30%. Unlike conventional rotating generators, this PMG rotor swings or oscillates through a fraction of a revolution, stops, and swings back to the starting position, where it stops again and so on. This motion is repeated and results in an unbalanced harmonic rich three-phase generator voltage. Discontinuous input currents are common in AC-DC power converters operating from such unbalanced voltages and as a consequence the rms value of the input currents becomes large, thereby limiting the amount of power that can be drawn without tripping the input protection devices. This paper describes power conversion techniques designed to draw the maximum power from such generators. Moreover, key engine controls such as ignition timing are based on the terminal electrical quantities of the PMG which are also simultaneously influenced by the functionality of the power converter. A novel input current di/dt tracker is proposed that indicates the extremum of the engine swings thereby allowing cycle-by-cycle ignition timing adjustments.

Keywords-alternate energy source, variable speed generator

I. DIGEST

A MICSE based portable power system is shown in Fig. 1. The swing engine has four chambers and is designed to be adaptable to a wide range of practical fuel supplies such as butane/propane. The engine oscillates about 70 times a second and drives a PMG.

Three-phase diode bridge rectifiers are often used along with step down (Buck) converters to provide a regulated DC voltage from three-phase inputs. The three-phase voltages in the case of a PMG that is driven by a swing engine are unbalanced and distorted. Figs. 2 and 3 show the unbalanced voltages from the PMG as well as the input current (simulated in SABER) to the power converter consisting of only a three-phase diode rectifier followed by a Buck converter. The rectifier input current in Fig. 3 is discontinuous when this PMG voltage becomes too small. The discontinuous input rectifier currents result in poor input power factor and higher rms values for the input current, thereby limiting the PMG output power.

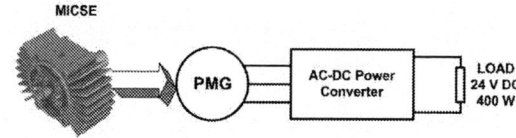

Figure 1.   MICSE based portable power system.

The use of three single-phase ac-dc converters in the three phases has been a commonly used choice to improve power output and power factor in three-phase systems and is well documented [1, 2]. While this topology offers unity input power factor and a superior performance than conventional phase controlled ac-dc topologies, this approach has several drawbacks such as need for split inductors and isolation between phases. The use of isolated single-phase converter stages has also been proposed [2, 3], but these topologies increase the complexity and the cost which may not be an attractive solution for low power converters. Another approach [4] is the use of active power factor correction circuits with a single-switch Boost power factor compensation stage such that the input current is always discontinuous. In this scheme, during the ON period of the Boost switch, all three input phases are shorted forcing the input phase currents to increase in proportion to the phase voltages. However, this scheme has the disadvantage of substantially increasing the current stresses of the switching devices and the high frequency ripple content of the pre-filtered ac input currents [4].

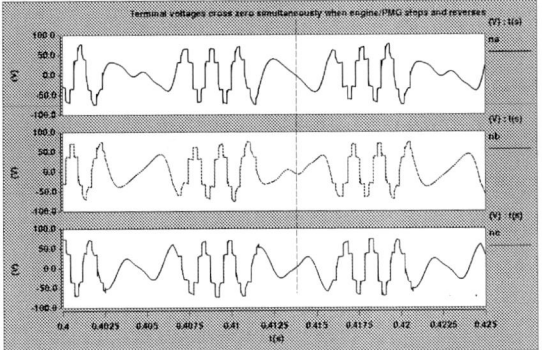

Figure 2. Unbalanced input voltages to power converter.

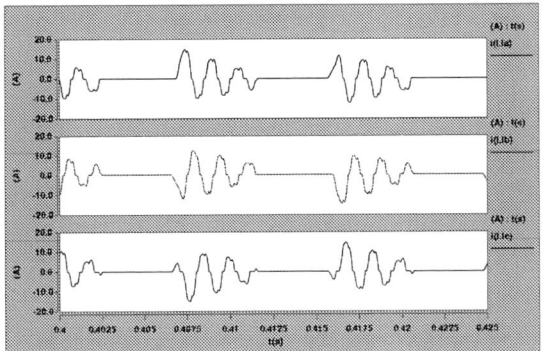

Figure 3. Discontinuous input currents in diode rectifier.

II. AC-DC POWER CONVERTER FOR MAXIMIZING POWER OUTPUT FROM PMG

Since the input voltages to the diode rectifier are unbalanced and harmonic rich, various power maximization schemes have to be considered. One such scheme involves the use of input series capacitors as a passive power maximization scheme. The series capacitors reduce the overall inductance of the PMG windings and improve the output power marginally by decreasing the RMS value of the input currents. However, the input currents are still discontinuous when the input voltages are small. Moreover, the capacitors are bulky and have to be sized to carry a significant rms current.

An alternative scheme has been proposed as shown in Fig. 4 [5]. The variable topology Boost-Buck converter consists of a Boost converter stage in between the diode rectifier and the Buck converter. The Boost converter would operate intermittently during the periods of time when the input current to the diode rectifier is zero, in an attempt to harness more power. The Buck and the Boost converter controls are completely independent and decoupled from each other, thus allowing the Buck converter to perform voltage regulation independent of the output power improvement process of the Boost stage.

Without the new intermittent Boost converter, the rectified DC current I_{dc} is discontinuous when the rectifier

input currents of Fig. 3 are discontinuous. The rectified DC current I_{dc} is therefore a good indicator of rectifier input current discontinuity and is sensed. When I_{dc} reaches zero the Boost switch Q2 is turned ON for a fixed period of time t_{ON}. With Q2 closes the three input phase currents begin to rise in proportion to the voltage from the PMG.

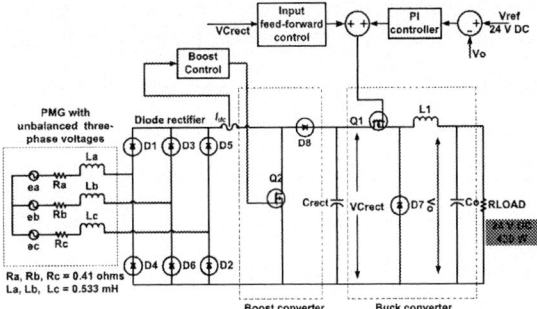

Figure 4. Proposed variable topology Boost-Buck converter with an intermittent Boost converter and a Buck regulator.

A variable topology Boost-Buck converter with feedforward control is simulated in SABER. The voltage feed-forward control (Fig. 4) provides approximately 24 V DC at the Buck output. The maximum output DC power obtained is around 430 W. Simulated rectifier input current waveforms for the converter with and without the Boost stage are shown in Fig. 5 for a fixed Boost switch Q2 ON time of 100 µs. A comparison of the different PMG output power improvement schemes is shown in Table 1. Test case 1 (Table 1) shows that a diode rectifier/buck converter combination without any output power maximization scheme results in an input RMS current of 4.81 A which is almost at the limit of the recommended maximum generator current/phase of 4.9 A. The DC output power of 390 W that can be drawn from the PMG is therefore limited by the high input RMS current. With the addition of series capacitors in the input of the diode rectifier (Test case 2) the input RMS currents drop for the same RLOAD (1.2 ohms) as in Test case 1. This allows slightly more power to be drawn from the machine but at the expense of bulky and heavy series capacitors. Test case 3 show the simulation results for a diode rectifier followed by the variable topology Boost-Buck converter scheme. Among Test cases 1-3, it can be seen that the variable topology converter offers the maximum output power capability.

TABLE 1: COMPARISON OF DIFFERENT POWER MAXIMIZATION SCHEMES

I_a (RMS) A	I_b (RMS) A	I_c (RMS) A	PF	DC Load Power W	Load Current A
Test Case 1: Diode Rectifier/Buck Converter (RLOAD = 1.2 ohms)					
4.81	4.75	4.69	0.74	390	18
Test Case 2: Diode Rectifier/Buck Converter with series capacitor (RLOAD = 1.2 ohms)					
4.47	4.47	4.49	0.79	390	18
Test Case 3: Diode Rectifier/Variable Topology Boost-Buck Converter (RLOAD = 1.08 ohms)					
4.71	4.66	4.65	0.85	430	20

A 24V DC 500 W prototype is built in the lab as shown in Fig. 6. A PMG fabricated during this research and connected to an oscillating drive source is shown in Fig. 7. The output voltage ripple waveform of the variable topology converter is shown in Fig. 8 for an oscillating frequency of 10 Hz. The ripple in the output voltage is less than 2%.

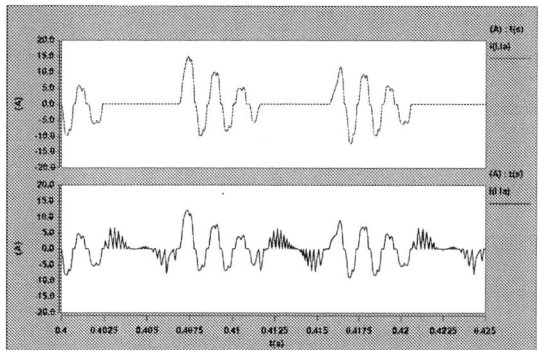

Figure 5. Top: Discontinuous rectifier input current in Phase A (no Boost stage); Bottom: Rectifier input current in Phase A (with Boost converter) shows current drawn during the top waveform's discontinuous period.

Figure 6. 500W Variable Topology Power Converter.

Figure 7. PMG connected to oscillating drive source.

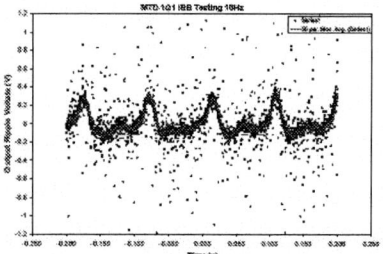

Figure 8. Output voltage ripple during 10 Hz testing < 2%.

The Boost control action is shown in Fig. 9 where the Boost tries to pull current when otherwise the current would be discontinuous. The input L-L voltage and non-discontinuous current are shown in Fig. 10.

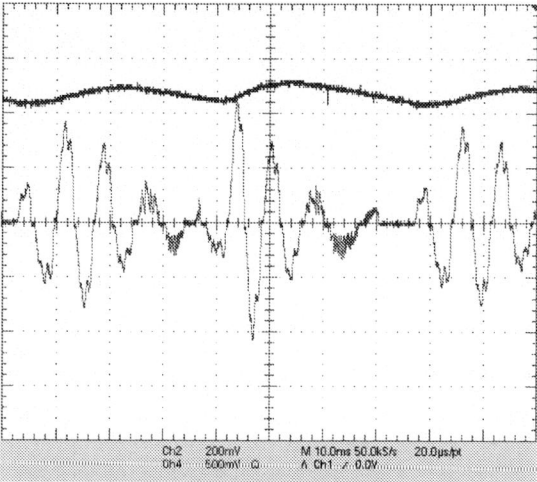

Figure 9. Top (red): Rectified DC voltage; Bottom (Magenta): Input current with diode bridge rectifier and Boost converter demonstrating intermittent Boost switching.

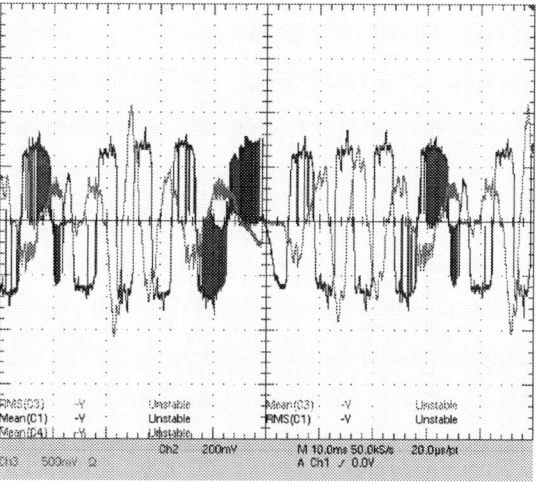

Figure 10. Red: PMG terminal L-L voltage; Green: Non-discontinuous input current with intermittent Boost switching.

III. NOVEL DI/DT TRACKER FOR ENGINE IGNITION TIMING CONTROL

The ignition timing of the MICSE needs to be adjusted periodically to maximize engine efficiency/output. This timing is continuously monitored from the point the engine stops. As the micro-engine oscillates, it comes to a full stop before reversing direction. Without the use of a power maximization converter such as the variable topology Boost-Buck converter, the detection of this stop point is easy. The three back-EMFs (electromotive force) of the PMG (coupled to the engine) go to zero simultaneously when the engine stops. If a simple diode bridge rectifier is used, the terminal generator voltage too is zero (as the input current is zero at low back-EMF voltages) when the back-EMFs are zero (Figs. 2 and 3) and simple comparators can been used to monitor this simultaneous zero crossing of all the three terminal voltages, which in turn is an excellent indicator of the time when the engine stops.

However the use of a Boost converter in the variable topology scheme completely changes this scenario. The Boost converter forces current from the machine even when the back-EMFs are simultaneously decreasing to zero as the engine comes to a stop. This can be seen in Fig. 10 where the input current is no longer discontinuous and the input L-L voltage has no longer a clear zero crossing to indicate engine stop. SABER simulation results in Fig. 11 also show the three phase voltages which no longer have a definite common zero crossing point due to the Boost switching. As the generator terminal voltages no longer go to zero simultaneously, a different technique is needed to detect the time when the engine stops.

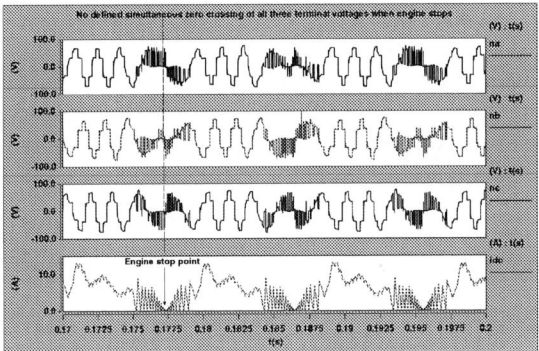

Figure 11. Variable topology Boost-Buck converter: na, nb, nc: three terminal phase voltages of PMG; idc: Rectified DC current.

The simulated PMG back-EMFs and rectified DC waveforms for the variable topology Boost-Buck converter are shown in Fig. 12. It can be seen that the Boost converter forces current when the back-EMFs are simultaneously approaching zero (i.e. when the engine is coming to a stop). When the Boost converter switch is closed for a fixed time, all the three input rectifier currents and consequently the rectified DC current increases (Fig. 13). The increase in the input currents is driven by the amount of back-EMF voltage present during the time. Hence this increase in current gradually decreases as the amount of back-EMF in the three phases decrease. The proposed method shown in Fig. 13 uses a differentiator to calculate the di/dt during the times when the Boost switch is ON. The point where the back-EMFs all go to zero simultaneously is the point where the di/dt is the lowest (ideally close to zero as there is no more back-EMF to drive this current).

A simple op amp differentiator circuit is used to calculate the continuous time di/dt of the rectified DC current. The results of the op amp differentiator are held only for the period when the Boost switch is ON using a Sample/Hold (S/H) circuit with the output filtered using a simple RC filter. The minima of the di/dt curve represent the time when the engine/PMG stops and reverses

direction. The circuit schematic is shown in Fig. 14. The proposed scheme is simulated in SABER and the results obtained are shown in Fig. 15. The middle waveform shows the smoothened di/dt whose minimum point corresponds to the zero crossing of the back-EMFs and can be used to adjust ignition timing on a cycle-to-cycle basis.

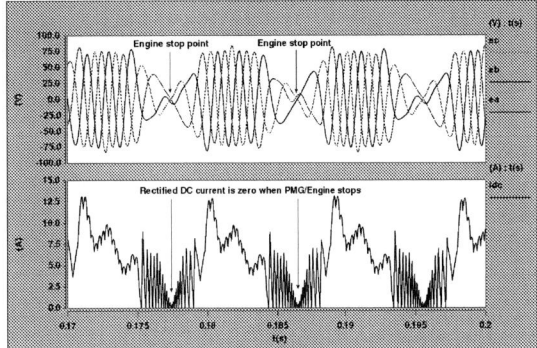

Figure 12. Variable topology Boost-Buck converter: Three back-EMFS ea, eb, ec; rectified DC current (idc).

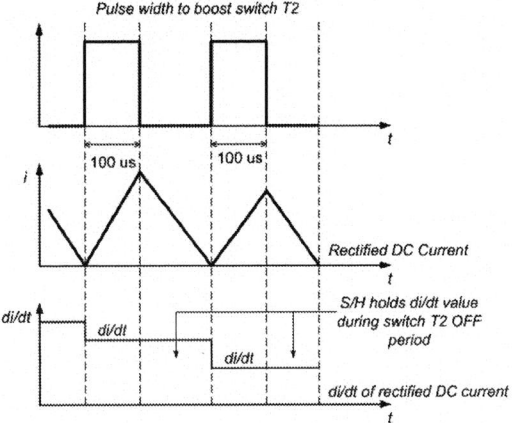

Figure 13. Boost control action and di/dt calculation.

IV. CONCLUSIONS

This paper explains and compares passive and active power maximization schemes for variable speed generators. A variable topology Boost-Buck converter produces a regulated DC output voltage while maximizing the power output from an unbalanced three-phase generator. The Buck converter functions independent of the Boost converter and hence voltage regulation is never compromised. The effects on other auxiliary functions such as engine control of adding such power maximizing converters is explained in detail in the paper. A novel di/dt tracker is proposed that can indicate the zero crossing of the back-EMFs and thereby the position of the engine shaft when the terminal generator voltages do not provide a suitable indication of the zero-crossing. Finally, experimental results are provided that demonstrate the functionality of the variable topology Boost-Buck converter.

REFERENCES

[1] G. Spiazzi and F.C.Lee, "Implementation of Single-phase Boost Power Factor Correction Circuits in Three-phase Applications", *IEEE Trans. Industrial Electronics, 1997,* pp. 365-371.

[2] M. J. Kocher and R. L. Steigerwal, 'An ac-dc converter with high quality input waveform", *IEEE Trans. Industrial Applications,* 1983, pp. 586-589.

[3] R. Ayyanar, N. Mohan, and J. Sun, "Single-Stage Three-phase Power-Factor-Correction Circuit Using Three Isolated Single-

phase SEPIC Converters Operating in CCM," *Proceedings of the 31st Power Electronics Specialists Conference,* 2000, pp. 179-184.

[4] A. R. Prasad, P. D. Ziogas, and S. Manias, "An active power factor correction technique for three-phase diode rectifiers," *IEEE Trans. Power Electronics,* 1991, pp. 83-92.

[5] S. Rajagopalan, D. M. Divan, R. G. Harley, and J. R. Mayor, "Variable Topology Boost-Buck Converter for Maximizing Power Output from Unbalanced Electrical Generators," *to be presented at the IEEE PESC 2006 conference.*

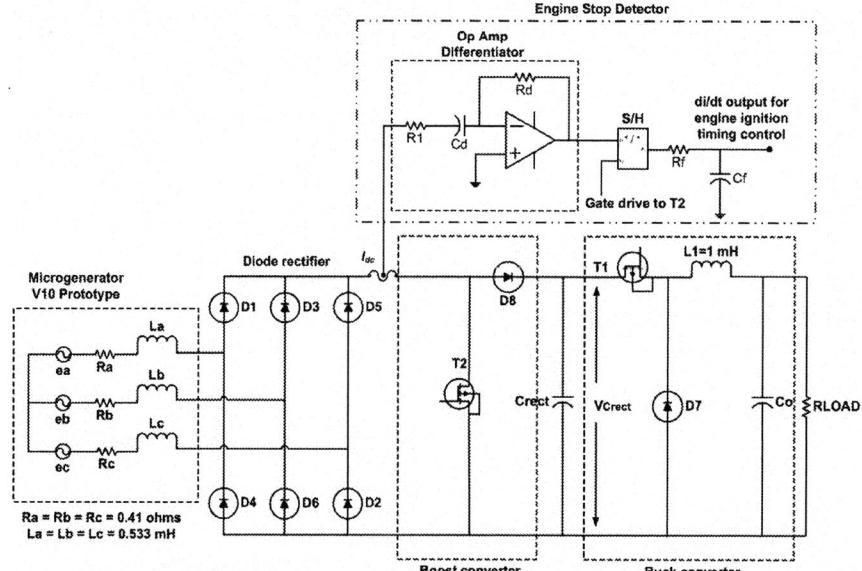

Figure 14. Engine extremum detection using di/dt tracker in a variable topology Boost-Buck converter.

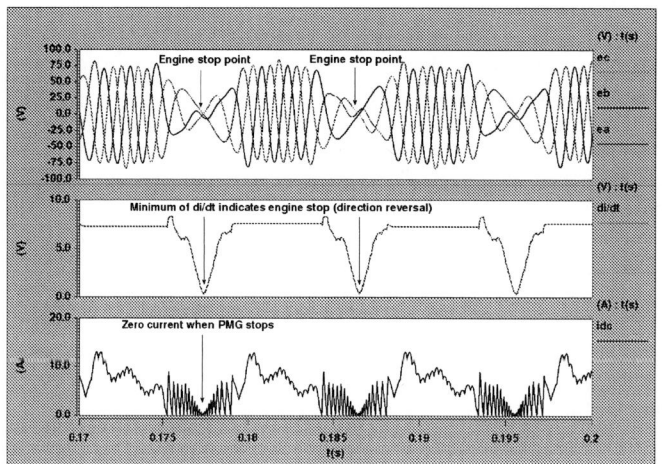

Figure 15. Variable topology Boost-Buck converter: Top: Three back-EMFs; Middle: di/dt of rectified DC current; Bottom: Rectified DC current.

3D FEA of the Stator of the Linear Magnetic Flux Compression Generator

Yanjie Cao, Chengxue Wang

Naval Aeronautical Engineering Institute, Yantai, Shandong Province P.R. China

Email:Cyanjie@263.net

Abstract — The stator of the Linear Magnetic Flux Compression Generator (LMFCG) consists of a helical coil, epoxy (insulator) and a casing, of which the helical coil is the key part. In order to grasp the detailed state about the helical coil under the impact of pulsed current, a 3D solid model of the helical coil was set up and the magnetic-structural coupling analysis of coil was carried out. Some useful conclusions are drawn.

Keywords- linear generator; coupling field; stator; FEA

I. INTRODUCTION

Flux compression generators designs have historically used the energy of highly energetic explosive materials to generator multi-mega-ampere pulses μsec in width. Such kind of pulse current is not suitable for the ETC (Electro-thermal chemical) gun use. A solution to the timing mismatch problem is to substitute a slower burning propellant for the fast burn explosive and to employ an inverse railgun configuration [1]. Although the current pulse and piston acceleration times are better matched to ETC requirements, the inherently low inductance gradient of the inverse railgun geometry creates difficulty in transferring energy from the energy.

Another approach to producing high gradient is to use a coilgun configuration whose output can be tailored by varying the helical pith in a manner similar to that in MCG experiments. Williams was the first to construct a propellant driven, inverse coilgun flux compressor. Kapustjaneko et al. have developed a simplified model for the performance of an inverse coilgun generator driving a rail gun. Mongeau has evaluated the requirement for a repetitive-fire weapon system with an inverse coilgun generator driven by distributed combustion of a fuel-air mixture [1].

E.B. Goldman et al. have constructed a flux compression

power unit for millisecond ETC gun use [2]. In the model, a propellant mixture is used to accelerate a conducting piston into the magnetic field region within a helical coil. An armature integral to the piston makes sliding contact the coil through compliant brushes and provided continuous current flow between the coil and center rod, which acts as a return current path. Interaction between the armature and the magnetic field decelerates the piston and transfers the piston's kinetic energy to the magnetic energy. The initial current and energy in the helical coil are thereby multiplied and delivered to a resistive load.

2D magnetic analysis of the stator in the LMFCG had been accomplished in article [4]. To understand essence of the LMFCG, it is worthy of carrying out the 3D structural analysis of the stator in the LMFCG. In this article, the 3D model of the stator is built up and the magnetic-structural coupling analysis has been accomplished.

II. THE MODEL OF THE LMFCG

The primary component of the LMFCG is illustrated in Fig.1. Prior to firing, the piston integrated with an armature

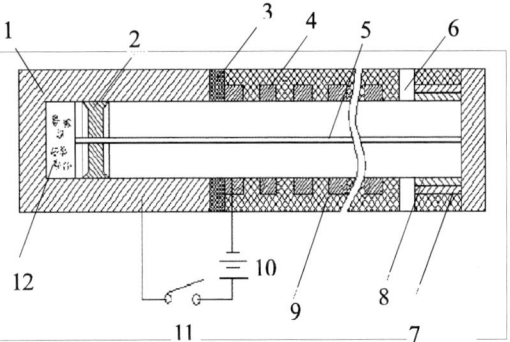

1-body;2-armature;3-insulator;4-helica coil;5-center conducto shaft;6-exhausting hole;7-primary winding of the transformer;8-secondary winding of transformer 9-epoxy;10-charger;11-swith;12-propellant

Figure 1. The model of LMFCG with fundamental parts

ahead is held at the initial position. Solid chemical propellant the same as used in guns is loaded into the chamber. When the propellant is initiated by the ignition of the charge, the pressure within the chamber rises quickly. Once the pressure reaches a certain value, the armature is accelerated forward. Before the armature contacts the helical coil, an initial current in the circuit is established by means of a small pulsed power capacitor, and the initial magnetic flux ($\Phi_0 = L_0 I_0$) is also established within the circuit. As the moving armature (driven by the high-pressure gas) is in continuous contact with the helical coil, the moving armature eliminates the turns of the helical coil from the current path. Ideally, the circuit inductance is forcibly reduced with a consequent increase in the current because the magnetic flux is constant. The decrease in the helical coil inductance must result in the increase of current. This pulse current is transferred to the gun load by the secondary coil of the transformer. In the process, the piston (assembled with an armature) is accelerated, decelerated and stopped, and then returned back to battery position by the magnetic pressure [3]. Then it is ready for the next firing.

III. THE STRUCTURE OF THE STATOR

The stator component of the LMFCG is shown in Fig. 2, it is composed of a helical coil, epoxy (insulator) and a casing. The helical coil has a spindle shape with a rectangular section, it is fabricated from copper. The helical coil ties in with epoxy by moulding and the outside of them is rooted closely by a casing made of steel. For the helical coil, there are 60 turns within a length of 80cm. When the current flows in the circuit, based on the law of eletro-magnetic induction, there must exist a powerful magnetic field within and out of the each turn of the copper coil, and the helical coil should sustain a big electromagnetic force. In order to ensure the LMFCG operating in a good state, the structural intensity of the helical coil is worthy to be concerned about.

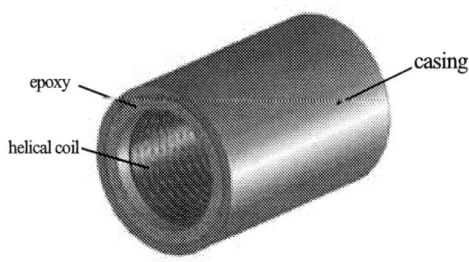

Figure 2 the structure of the stator

IV. THE 3D STRUCTURAL ANALYSIS OF THE STATOR

The process to carry out the structural analysis of stator is

very complex because it involved in magnetic field and structural field. A FEA software which has coupling field analysis function is adopted. Taking the calculation precision and the limit of computer resource into account, the simulation analysis is based on the stator with 10 turns of the helical coil.

A. FEA Model of the stator

Using the certain simulation software, a parameterized simulation program of the stator is created in order to calculate in high efficiency. Method based on element edge and hexahedron element type are adopted. The helical coil takes AZ and VOLT as their degrees of freedom, and insulation and casing takes AZ as its degree of freedom. Get discrete elements by mapped meshing. Fig.3 and Fig.4 show the FEA models of the stator component and the helical coil respectively.

Figure 3   FEA model of the stator

Figure 4   FEA model of the helical coil

B. Electromagnetic Analysis of the stator

On the basis of the FEA models of the stator, a magnetic field analysis of the stator has been carried out. The current waveform applied with 100kA peak value is shown in Fig. 5. Loading procedure has one load step and it is equally divided into 4 sub-steps. Coupling voltage degree of Applying magnetic-flux-parallel condition to the boundary of the model, the boundary condition of magnetic-flux-normal takes effect naturally. ICCG solver is chosen to solve the model. Fig. 6 shows the magnetic density nephogram of the launching coil subassembly at the time of 0.4ms.

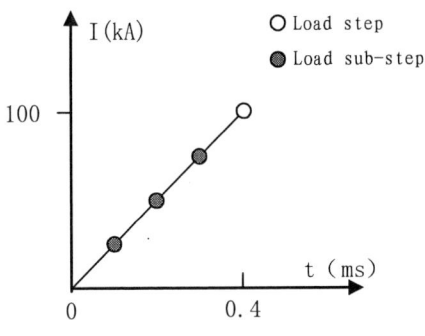

Figure 5   Loa waveform

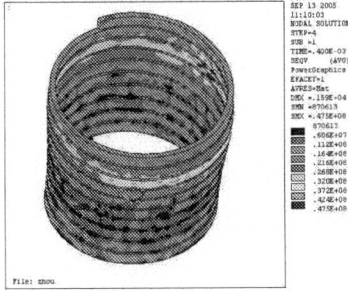

Figure 6   Magnetic density nephogram of the stator at 0.4ms

C.   Stress Field Analysis of the Launching Coil

Magnetic field was coupled with structure. In order to verify the structure intensity of helical coil, structure analysis should be carried out. In this paper, using the magnetic-structural coupling analysis function of the FEA software and taking physical environment method, the structural analysis with stator has been accomplished. Firstly, changing the analysis environment from magnetic field module to structural module and resetting material attribution parameters. Secondly, taking the magnetic force generated in the transient analysis of the magnetic field as the stress load in the structural analysis, and apply (all constraints) to the lower part of the stator subassembly, then ICCG solver is chosen to solve the model. Finally, view and analyze the results, Fig. 7 shows the stress nephogram of the helical coil at 0.4ms.

VI. Summary of the FE A Results and Discussion

A.   Conclusion

The analysis result shows that the magnetic flux density and the current density of the launching coil get thicker and thicker as the current value get higher and higher, at the time of 0.4ms the current reaches its highest value, and so do the magnetic flux density and the current density of the helical coil at the same time. When the current load is 100kA, the maximum stress value is 47.5 MPa. The material of the helical coil is copper, its yielding strength is 280 MPa, so the stress value generated by the launching coil is within the strength limit of the material, and meets the strength requirement.

B. Further Discussion

To understand more about the stator, a 3D FEM analysis of the stator is also carried out with the same method and step when the peak current value is 200kA. The same law of magnetic distribution and stress distribution of the helical coil was obtained. The maximum Von Mises stress value is 201.3MPa when the max. current load is 200kA. If taking the safety fact into account and taking safety coefficient 1.3, the allowable stress [σ] approximate to 215 MPa, the maximum stress is less than the allowable stress[σ], so it can be concluded that the stator part is mechanically sound. But in this case, whether the stator can work properly, the strength of the epoxy is a crucial fact, the result shows that the maximum Von Mises of the epoxy is 100Mpa. So, by choosing the appropriate epoxy, the strength of the stator can be mechanically sound.

References

[1]   Peter Mongeau, Combustion Driven Pulsed Linear Generators for Electric Gun Applications, IEEE Trans. on Mag.,No.1,Janunary 1997.

[2]   Edward B. Goldman et al, Development of a Flux Compression Power Unit for Millisecond ETC Pulsed Power Applications, IEEE Trans. on Mag.Vol.35, No.1,Janunary 1999.

[3]   Ying Wang, J.Li et al   A Mathematical Model of the Reusable Linear Magnetic Flux Compressior,   IEEE Trans. on Mag.Vol.35, No.1,Janunary 2001.

[4]   Yanjie Cao, Shukang Cheng, Structural Analysis of the stator of the Linear Magnetic Flux Compression Generator, The 4th International Power Electronics and Motion Control Conference, Vol2, 2004.08 Xi'an, China.

Figure 7   Stress nephogram of the helical coil at 0.4ms

The Effect of Current Control Strategies on Power Consumption of a Magnetically Levitated Turbomolecular Pump

A.E. Hartavi[*], R.N. Tuncay[*] and M.N. Şahinkaya[**]

[*] Istanbul Technical University / Electrical Machinery Department, Istanbul, Turkey.
E-mail: hartavi@elk.itu.edu.tr
[**] University of Bath / Mechanical Engineering Department, Bath, United Kingdom.

Abstract **This paper presents a novel current control technique called 'Variable Bias Current Control (VBCC)' developed to minimize the power consumption of active magnetic bearings (AMB) used in Turbomolecular pumps. Initially, the mathematical model of the AMBs and the TMP are derived to obtain the simulation model of the system. Then, the effectiveness of bias current on energy consumption is investigated for differential and unidirectional current control strategies. It is shown that it is beneficial to optimize the bias current according to the operating conditions. Finally, the simulation results are utilized to successfully demonstrate the effectiveness of the VBCC technique.**

Keywords- Magnetic bearings, Turbomolecular pump, Power loss.

I. INTRODUCTION

In contrary to conventional bearings, magnetic bearings (MBs) levitate, suspend, and guide rotors without any physical contact. This frictionless operation of MBs offers many advantages such as low vibration, high speed, oil-free and maintenance free operation. As a consequence AMBs are utilized in a wide variety of applications, such as turbomolecular pumps (TMPs), compressors, flywheels, high speed spindles for machine tools, reaction wheels for artificial satellites, blood pumps, gyroscopes, centrifuges, generators etc [1]. Hence, modern TMPs are designed with AMBs [2] to reduce the amount of power losses, to achieve higher rotational speeds, to acquire a contamination free environment and to reduce the level of vibration. However, this unique bearings are inherently unstable and necesssisate a closed loop control [3]. The necessity of the controller appends flexibility to the AMB systems, which can not be achieved with conventional bearings.

Since magnetically levitated TMPs operate in a vacuum environment, the dissipation of the generated heat is extremely hard. Thus, at the design stage all forms of power losses should be regarded. The power losses of AMBs can be classified into three groups: copper loss, iron loss and windage loss. The most significant is the copper loss caused by the flow of currents through the coils of electromagnets. Therefore, stability of the system should be achieved considering the amount of current produced.

In the literature there have been numerous researches that deal with stability problem of AMBs. Only a few of them considers the power losses during the design of the controller. Most of the researchers utilize the Differential Current Control (DCC) approach while few others use Unidirectional Current Control (UCC). In DCC mode of operation, a bias current is fed into the coils, which induce energy consumption, even if no force is required. In UCC strategy a small bias current is used to reduce the amount of power consumed in the system. The magnitude of this bias current is assigned by the static load or the lower limit of the power amplifier [4]. This technique leads to hardships when there are linear controllers or disturbances in the system.

In this research, two current control strategies are considered in terms of the power consumption. The effect of bias current on energy consumption is analysed. A novel current control method called 'Variable Bias Current Control' is successfully developed [5]. In this approach, the bias current is optimized as a function of bearing stiffness and rotor vibration amplitudes. Finally, simulation results of a magnetically levitated TMP are given for different current control techniques to demonstrate the effectiveness of the proposed method.

II. THE SYSTEM MODEL

A. Modeling of AMB

AMBs that are used to levitate the rotor of the TMP have a significant influence on the dynamics of the system. Thus, as an initial attempt the mathematical model of the AMB is obtained. For the sake of simplicity, the model of the MB is obtained only for x-axis.

Figure 1 shows the cross section of an AMB utilized in the TMP. Each of the electromagnets placed opposite to each other to produce counteractive forces. The rotor is made of ferromagnetic material in circular piece to produce an attractive force.

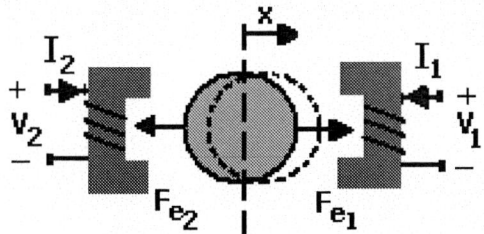

Figure 1   Principle structure of an AMB

Since the two countered electromagnets produce counteractive forces along x-axis, the equation of motion in x-axis can be written by applying Newton's Second Law. The equation of motion for an AMB in x-axis can be given by,

$$m\frac{d^2 x}{dt^2} = F_{e2} - F_{e1} \qquad (1)$$

where F_{e1} and F_{e2} are the forces caused by the right and left electromagnets, m is the mass and x is the displacement of rotor from the equilibrium position. The forces that act upon the rotor are produced by the magnetic field. So, they can be rewritten in a more explicit form by using co-energy principle. The generated electromagnetic force is written as,

$$F_e = -k\frac{i^2}{g_o^2} \qquad (2)$$

where k is the design factor, i is the coil current and g_o is the distance between the rotor and the stator in equilibrium position. By substituting Eq.(2) in Eq.(1), the equation of motion can be written in terms of coil currents and displacement.

$$F = k\left[\left(\frac{i_2}{g_0 - x}\right)^2 - \left(\frac{i_1}{g_0 + x}\right)^2\right] \qquad (3)$$

where i_1 and i_2 are the right and left coil currents, and x is the displacement of rotor from the equilibrium position. As it is perceived from Eq.(3) the forces produced by the electromagnets are proportional to the square of current and inversely proportional to the square of the air gap. This form of the mathematical model is amenable to explain the working principle of the AMB.

B.   Modeling of TMP

The operation principle of the TMP depends on the interaction of a molecule and a moving surface. Hence, the pumping action is caused by a series of blades. Figure 2 shows the schematic view of a magnetically levitated TMP. As it is seen from the figure, TMP-AMB system is a five-axis controlled vertical machine. The rotor is suspended magnetically in the vertical direction by using four radial and one thrust bearing. The rotational motion of the rotor is performed with a Brushless DC motor that is positioned between the upper and lower AMBs. Since the magnetic bearings necessisate a feedback control,

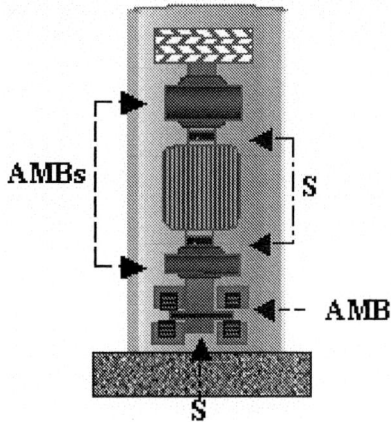

Figure 2   Schematic view of a magnetically levitated TMP

inductive type sensors are utilized to detect the position of the rotor. These sensors are located along and at the bottom of the shaft to measure the radial and axial positions of the rotor.

The equation of magnetically levitated TMP is derived by applying Lagrangian mechanics. First of all, the rotor is modeled as a rigid body. Then, the motion of the rotor is obtained according to the Cartesian coordinates. The origin of the XYZ coordinates is chosen as the geometric center of the rotor. Consequently, the equation of rotational motion of the rotor is written as,

$$[M]\left[\frac{d^2 q}{dt^2}\right] + \Omega[C]\left[\frac{dq}{dt}\right] = Q \qquad (4)$$

where q is the generalized coordinate vector that describes the motion, Q is the generalized input vector consists of forces and moments and M and C are coefficient matrices. The dimension of the matrices and vectors are determined by the number of degree of freedom and they can be written as,

$$q = [x, y, \theta_x, \theta_y, z]^T \qquad (5)$$

$$Q = \left[F_x, F_y, M_x, M_y, F_z\right]^T \qquad (6)$$

$$M = diag(m, m, I_t, I_t, m) \qquad (7)$$

$$C = \begin{bmatrix} 0 & 0 & 0 & 0 & 0 \\ 0 & 0 & 0 & 0 & 0 \\ 0 & 0 & 0 & 0 & I_P \\ 0 & 0 & 0 & -I_p & 0 \\ 0 & 0 & 0 & 0 & 0 \end{bmatrix} \qquad (8)$$

III.   CURRENT CONTROL STRATEGIES

The main motivation of the paper is to reduce the amount of power consumed in the TMP-AMB system. The contactless support of the TMP is provided by the radial and axial MBs. A controller is required to stabilize the system against disturbances and rotor unbalance during the operation of the TMP. The sensors measure the vertical and horizontal displacement of the rotor. The outputs of the sensors are fed into the controller. The current demand that brings the rotor into balance is generated by the controller. Then, they are applied to the coils of the electromagnets through the power amplifiers. Therefore, the amount of power used during the operation of the AMBs is the function of the current that flows through the coils of the electromagnets.

Commonly, a PID type controller is used to produce the control current. Each parameter of the controller functions in a different way. The proportional term causes a control current that results in a restoring force. In other words, it is utilized to balance the negative stiffness of the system. The derivative part applies a force proportional to the speed thus it introduces damping to the system. The integral part is used to set the static position of the system that result in a zero steady state error. Since the proportional term is used to stabilize the system, it is the most important term among those three parameter of the controller. Thus, the control current generated by the controller can be written as,

$$I_c = K_p \, x \tag{9}$$

where I_c is the control current, K_p is the proportional gain and x is the amount of displacement from the equilibrium position.

After defining the control current as a function of rotor position, the power consumption of the system is studied for two different current control strategies.

IV. DIFFERENTIAL CURRENT CONTROL STRATEGY

Differential current control strategy is the most widely used technique utilized in the AMB systems. This mode of operation can be illustrated by Fig.5. To clarify the analysis only the electromagnets that lie along the horizontal axis are considered.

In DCC strategy, initially the maximum allowable current is defined according to the characteristic property of the material used in the electromechanical part of the

bearing. Then, the bias current is selected as a half of the maximum current. The bias current is applied to the electromagnets that are placed opposite of each other. In ideal situations, this will result in a stable levitation. However, when rotor moves from the equilibrium position because of any internal or external disturbance, the rotor starts to accelerate towards one of the electromagnets and stick to it. To prevent the rotor–stator contact and to bring the rotor into the equilibrium position, controller generates a control current. This current is superimposed on the bias current of the right and left electromagnets with opposite signs. Thus, in DCC strategy, the currents flowing through the right and left electromagnets can be written as,

$$I_1 = I_b + I_c \tag{10}$$

$$I_2 = I_b - I_c \tag{11}$$

where I_b is the bias current and I_c is the control current.

V. UNIDIRECTIONAL CURRENT CONTROL STRATEGY

In contrast to DCC strategy, UCC method specifically applies a small bias current. The lower limit of the bias current is set by the static load or the power amplifiers. The controller works on exactly the same principle as DCC but the control current is doubled and applied to the electromagnet which is in the opposite direction of the force acting on the rotor. At that moment, the other electromagnet is just fed by the bias current. Therefore, the currents flowing through the electromagnets in UCC method is written as,

$$I_1 = \begin{cases} I_b, & f > 0 \\ I_b + 2|I_c|, & f < 0 \end{cases} \tag{12}$$

$$I_2 = \begin{cases} I_b + 2|I_c|, & f > 0 \\ I_b, & f < 0 \end{cases} \tag{13}$$

where f>0 corresponds to situation (a) and f<0 corresponds to the situation (b.) This mode of operation is revealed in Fig.4.

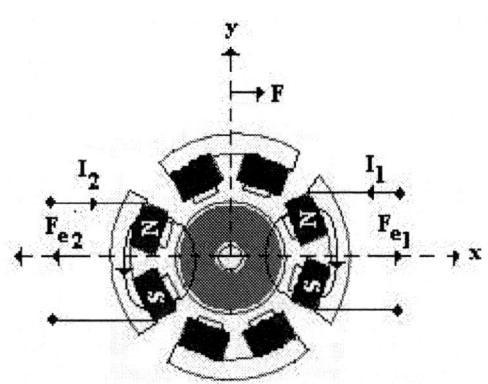

Figure 3   AMB operating with DCC technique

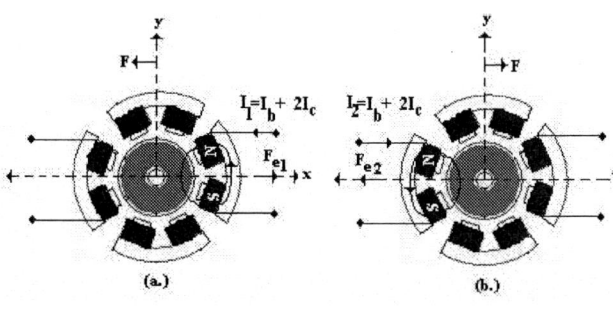

Figure 4   AMB operating with UCC strategy

VI. VARIABLE BIAS CURRENT CONTROL STRATEGY

As a continuation to the previous analysis, in this part of the paper the vibrations caused by the rotation of the rotor are considered. Hence, the copper loss of AMB which is used to suspend a rotor with a rotational speed of Ω can be written as,

$$P = \frac{R}{2\pi} \int_{\theta}^{2\pi-\theta} \left(I_1^2 + I_2^2\right) d(\Omega t) \qquad (11)$$

where R is the resistance of the coil, I_1 and I_2 are the coil currents. If we assume the system stiffness is constant, then the copper loss is the function of coil currents and resistance.

The change of power consumption with respect to bias currents at two vibration levels for both DCC and UCC strategies is shown in Fig.5. During the examination the equivalent stiffness of the system is kept constant; consequently no change in the dynamic response is expected. The result of the analysis reveals that optimizing the bias current according to the operational conditions is beneficial for both unidirectional and bidirectional control strategies. A Variable Bias Current Control (VBCC) strategy is chosen here, which calculates the optimum value of the bias current as a function of the required equivalent stiffness and the measured vibration amplitude.

VII. SIMULATION OF THE SYSTEM

After obtaining the mathematical model of the system that consists of AMBs and rotor dynamics, the simulation model is obtained in Matlab®/Simulink environment. The simulation model is realized by using the parameters given in Table 1. In addition to the system model, a PID controller is also added to achieve the system stability. The proposed bias current control algorithm, VBCC, is also connected in parallel as shown in Fig. 6.

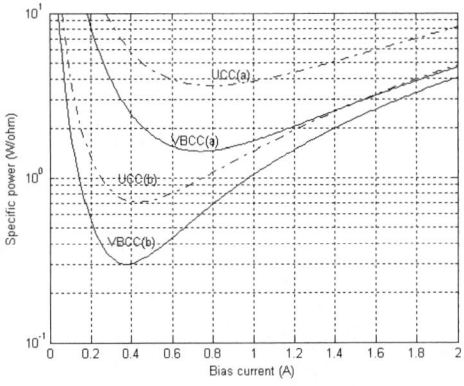

Figure 5   Power consumption versus bias current for a constant equivalent stiffness, at different vibration amplitudes
(a.) 0.4 g_0 , (b.) 0.1 g_0

TABLE I.
DIMENSIONS OF TMP-AMB SYSTEM

Mass of rotor (m)	44	kg
Number of turns per pole (N)	250	-
Air gap clearence (g_o)	0.00028	m
Area of upper mb (A_u)	160×10^{-6}	m^2
Area of lower mb (A_l)	80×10^{-6}	m^2
Magnetic permeability of vacuum (μ_o)	$4\pi \times 10^{-7}$	H/m

The PID controller generates a control current CCC according to rotor position error. The inputs of the VBCC block are the equivalent stiffness and the size of the orbit. The required stiffness (K_{eq}) should be selected according to the working conditions of the system. Low bearing stiffness may result high vibration amplitudes if the system is open to disturbances, which may cause the rotor to contact the auxiliary bearings. The VBCC block produces the optimized value of the bias current and the proportional gain so that the desired equivalent stiffness and the orbit size do not change. As a consequence, AMBs should produce the same performance with lower power consumption.

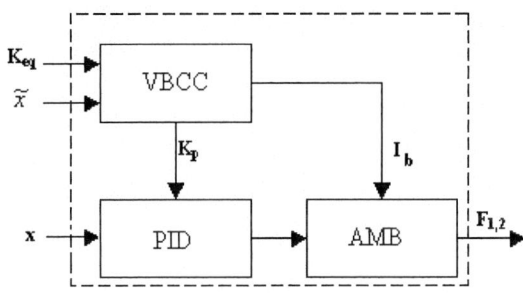

Figure 6   Principle of VBCC strategy

The simulation model is run at a constant speed of 30,000 rpm. An unbalance of 0.75×10^{-5} kgm is added to the system to generate a steady state orbit. Simulations are run for three different current control strategies. The power consumption of the system as about 7.72, 1.53 and 0.9 Watt per axis per one ohm resistance for DCC, UCC and VBCC strategies respectively.

VIII. CONCLUSION

A novel 'Variable Bias Current Control' technique is developed to minimize the energy consumption of a magnetically levitated Turbomolecular pump (TMP). The optimum value of the bias current is determined as a function of equivalent stiffness and rotor vibration amplitudes. Simulated results of the VBCC are compared with two other techniques commonly used in the literature to demonstrate the significant energy savings. This reduction of the power consumption is achieved without any change in the dynamic response of the system.

ACKNOWLEDGMENT

The authors acknowledge the support of BOC Edwards Ltd. for providing the magnetically levitated TMP and thank Dr.Graham Fells and Mr. James Haylock for their technical support.

REFERENCES

[1] F. Matsumura, Y. Okada, T. Namerikawa, "State of art of magnetic bearings", *JSME International Journal, vol.40*, pp.553-559, 1997.

[2] N. S. Harris, "Modern vacuum practice'', *2nd Edition, Bell & Brain Ltd.*, Glasgow, 2001.

[3] S. Earnshaw, "On the nature of molecular forces which regulate the constitution of the luminferous ether''. Trans. Cambridge Phil. Soc., vol 7, *pp 97-112*, 1842.

[4] K. Nonami,, "Recent trend of magnetic bearing and its applications for energy storage flywheel'', *Koyo Engineering Journal,* 164, pp 3-7, 2004.

[5] A. E. Hartavi, "Fuzzy supervisory control and bias current optimization of a Rotor-AMB system'', *Ph.D Thesis*, Istanbul Technical University, Istanbul, Turkey.

2006 5th International Power Electronics and Motion Control Conference

Direct Torque Control of an Interior Permanent Magnet Synchronous Machine fed by a Direct AC-AC Converter

D. Xiao and M. F. Rahman

School of Electrical Engineering and Telecommunications
The University of New South Wales, Sydney, Australia

Abstract—**This paper presents a novel direct torque control (DTC) scheme of an Interior Permanent Magnet Synchronous Motor (IPMSM) fed by a matrix converter. In the proposed scheme, a sliding mode torque and stator flux controller is integrated with a simplified Venturini's modulation algorithm to achieve low torque and flux ripples, sinusoidal input/output currents and unity fundamental displacement factor on the input side regardless of the load power factor while maintaining constant and controllable switching frequency and preserving the fast response and robustness. Numerical simulations are carried out for the classical and modified DTC schemes in both steady-state and transient conditions, verifying the effectiveness of this new proposed scheme which is superior to the classical one.**

Keywords—*direct torque control (DTC), Interior Permanent Magnet Synchronous Machine (IPMSM), matrix converter*

I. INTRODUCTION

Recently, the three-phase to three-phase matrix converter (MC) has emerged to become a viable alternative to the conventional voltage-source inverter (VSI). The matrix converter, as a member of AC-to-AC direct converter family, provides sinusoidal input/output waveforms and allows inherent four-quadrant operation and the adjustment of input power factor on the main side. Furthermore, the absence of bulky dc-link electrolytic capacitors for energy storage allows long lifetime, high integration capability, extreme temperature and critical volume/weight applications [1]-[4].

The basic DTC scheme for matrix converter was initially proposed to apply to induction motor drives [5]. However, some drawbacks, such as large flux and torque ripples and switching frequency variation according to the change of the motor speed and the hysteresis band amplitudes, still exist.

In this paper, a new DTC scheme for IPMSM drives fed by matrix converter is presented as shown in Fig. 1. The variable-structure controller produces the most appropriate stator voltage vectors to track the reference torque and flux. The control stator voltage signals have been limited and converted to the three-phase target voltages before

proceeding to the Venturini's optimum modulation module. The three-phase voltages and instantaneous main voltages are used as the inputs for the modulation algorithm to generate the required duty cycle for each switch in the matrix converter. This method has more advantages over the classical DTC, such as significantly reduced torque and flux ripples and constant switching frequency while preserving the fast response and robustness of DTC. In addition, the quality of input and stator currents can be improved by accurate synthesis of the desired output voltages and input currents.

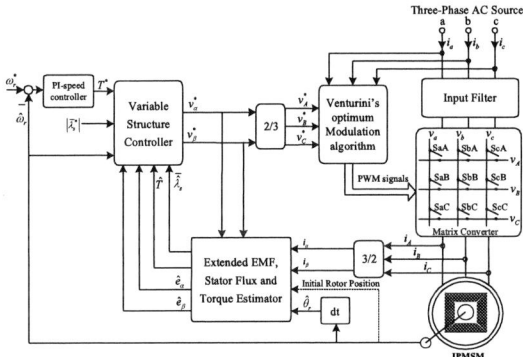

Figure 1. Proposed DTC of the matrix converter fed IPMSM drive

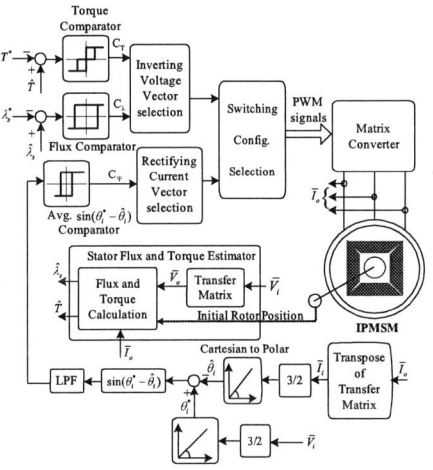

Figure 2. Basic DTC scheme with matrix converter

II. Modulation Algorithm for Matrix Converter

The main task of the modulation algorithm for the matrix converter is to find a modulation matrix $M(t)$ from a set of input voltages and an assumed set of output currents to synthesize the target output voltages and input currents [2], [3]. An injection of a third harmonic of the input and output voltage was proposed in order to obtain a higher voltage transfer ratio [2], [4]. A simplified form of this algorithm with unity input displacement factor is adopted in this work [6]. The nine modulation functions for three-phase outputs are calculated in each sampling interval by measuring any two of three input line-to-line voltages. Then, the magnitudes and positions of the input and target voltage vectors are calculated as

$$V_{im} = \frac{2}{3}(v_{ab}^2 + v_{bc}^2 + v_{ab}v_{bc})^{1/2} \tag{1}$$

$$V_{om} = \sqrt{\frac{2}{3}}(v_A^2 + v_B^2 + v_C^2)^{1/2} \tag{2}$$

$$\omega_i t = \arctan(\frac{\sqrt{3}v_{bc}}{2v_{ab}+v_{bc}}) \tag{3}$$

$$\omega_o t = \arctan(\frac{v_B - v_C}{\sqrt{3}v_A}) \tag{4}$$

Assuming the input power factor is set to one, three triple harmonic terms are defined as

$$K_{31} = \frac{2}{9}\frac{V_{om}/V_{im}}{\sqrt{3}/2}\sin(\omega_i t)\sin(3\omega_i t) \tag{5}$$

$$K_{32} = \frac{2}{9}\frac{V_{om}/V_{im}}{\sqrt{3}/2}\sin(\omega_i t + \frac{2\pi}{3})\sin(3\omega_i t) \tag{6}$$

$$K_{33} = -V_{im}\left[\frac{1}{6}\cos(3\omega_o t) - \frac{1}{2\sqrt{3}}\cos(3\omega_i t)\right] \tag{7}$$

Then, the modulation functions for output phase j ($j = A, B, C$) are given as

$$M_{aj} = \frac{1}{3} + K_{31} + \frac{2}{9V_{im}^2}(v_j + K_{33})(2v_{ab} + v_{bc}) \tag{8}$$

$$M_{bj} = \frac{1}{3} + K_{32} + \frac{2}{9V_{im}^2}(v_j + K_{33})(v_{bc} - v_{ab}) \tag{9}$$

$$M_{cj} = 1 - (M_{aj} + M_{bj}) \tag{10}$$

III. Direct Torque Control for Matrix Converter

A. Basic DTC Using Matrix Converter

The conventional DTC for matrix converter drive can be realized by two stages as represented in Fig. 2. In the inverter stage, the hysteresis flux and torque control can be achieved by selecting one of the six active and two zero voltage vectors with respect to the virtual dc-link voltage, exactly as in VSI. In addition, the direct control of the average value of $\sin(\theta_i^* - \hat{\theta}_i)$ can be completed by choosing a single current vector in the rectifier stage. The stator flux, torque and displacement angle estimators are realized by only measuring the input voltages and stator currents, since the other qualities are calculated on the basis of the switching states in each sampling period.

B. VS-DTC Using Matrix converter

In order to drive the torque and flux to track their desired trajectories, the VS controller is designed in more details in [7]. The procedure can be exhibited by the following equations.

The switching surface is defined as $S = [S_1 \ S_2]^T$,

$$\begin{aligned} S_1 &= e_T(t) + K_T \int_0^t e_T(\tau)d\tau - e_T(0) \\ S_2 &= e_\lambda(t) + K_\lambda \int_0^t e_\lambda(\tau)d\tau - e_\lambda(0) \end{aligned} \tag{11}$$

Differentiating the switching surface vector gives

$$\begin{aligned} \dot{S}_2 &= \dot{e}_\lambda + K_\lambda e_\lambda = (\dot{\lambda}^* - \dot{\hat{\lambda}}) + K_\lambda(\lambda^* - \hat{\lambda}) \\ \dot{S}_1 &= \dot{e}_T + K_T e_T = (\dot{T}^* - \dot{\hat{T}}) + K_T(T^* - \hat{T}) \end{aligned} \tag{12}$$

where $e_T = T^* - \hat{T}$, $e_\lambda = \lambda^* - \hat{\lambda} = \lambda_s^{*2}/2 - (\hat{\lambda}_\alpha^2 + \hat{\lambda}_\beta^2)/2$ K_T and K_λ are positive control gains of flux and torque. Substituting for $\hat{T}$, $\hat{\lambda}$, $\dot{\hat{T}}$ and $\dot{\hat{\lambda}}$ into (12) leads to $\underline{\dot{S}} = \underline{F} + \underline{D} \bullet [u_\alpha, u_\beta]^T$.

$$\underline{D} = \begin{bmatrix} -1.5P\left(i_\beta - \hat{\lambda}_\beta/L_d\right) & -1.5P\left(\hat{\lambda}_\alpha/L_d - i_\alpha\right) \\ -\hat{\lambda}_\alpha & -\hat{\lambda}_\beta \end{bmatrix}$$

$$\begin{aligned} F_1 = K_T e_T - 1.5P\{&\hat{\lambda}_\alpha[\omega_{re}(L_d - L_q)/L_d \cdot i_\alpha - R_s i_\beta/L_d - e_\beta/L_d] \\ &- \hat{\lambda}_\beta[-R_s i_\alpha/L_d - \omega_{re}(L_d - L_q)/L_d \cdot i_\beta - e_\alpha/L_d]\} \end{aligned}$$

$$F_2 = K_\lambda e_\lambda + \hat{\lambda}_\alpha R_s i_\alpha + \hat{\lambda}_\beta R_s i_\beta$$

where R_s is the stator resistance, p is the differential operator, L_d , L_q are direct and quadrature inductances, ω_{re} is the rotor speed in electrical rad/s and e_α and e_β are extended EMFs.

The VS controller generates the command output voltage vector components which are converted to three-phase voltages and provided for the modulation module. The switching control law can be selected as (13) according to Lyapunov approach [7].

$$u_{\alpha,\beta}^* = -D^{-1} \begin{bmatrix} \mu_1 & 0 \\ 0 & \mu_2 \end{bmatrix} \begin{bmatrix} sign(S_1) \\ sign(S_2) \end{bmatrix} \tag{13}$$

The chattering problem can be remedied by (14). The switching function is replaced by a continuous function around the sliding surface neighborhood by introducing smoothing factors.

$$sign(S_i) = \begin{cases} 1, & if\ S_i > \lambda_i \\ -1, & if\ S_i < -\lambda_i \quad (Smoothing\ factors\ \lambda_i > 0) \\ S_i/\lambda_i, & if\ |S_i| < \lambda_i \end{cases} \tag{14}$$

IV. Numerical Simulation

Two Simulink models are built to test the proposed scheme. The steady state and dynamic responses are compared with the basic DTC scheme. The parameters of the tested IPMSM and models are given in Tables I, II and III. The simulations have been carried out assuming three-phase balanced input voltages and ideal switching devices.

Rated torque reversal at -6 Nm and +6 Nm was tested with stator flux linkage maintained at 0.55 Wb. Figs. 3-6 show the dynamic responses of estimated torque, three-phase stator currents and one of three-phase filtered input currents against the corresponding input phase voltage, respectively. It is obvious that the variable-structure DTC drive system operates better than the classical DTC under transient conditions. The torque and flux ripples are significantly reduced by proposed scheme. The stator and filtered input currents are sinusoidal with nearly unity input power factor and smoother than basic DTC as shown in Figs. 3 and 4. During motoring operation (from 0 to 0.1 sec.), the filtered input current is in phase with the corresponding phase voltage. Then the input current suddenly reverses the phase 180°, opposite that of the input voltage, during the regenerative breaking. Figs. 5-7 give the comparison of flux response of the two schemes. It can be noticed that the magnitude of flux stays at rated value immediately after startup. The flux components and extended EMFs exhibit smoother sinusoidal waveforms under VS-DTC. The manifolds of torque and flux are shown in Fig. 6. The states of torque and flux are driven towards their sliding surfaces by discontinuous control signals. The flux is greatly reduced without influencing dynamic response speed under VS-DTC than classical DTC scheme as shown in Fig. 7. Fig. 9 shows that torque steady-state waveform of VS-DTC contains less ripples than that of basic DTC. The proposed scheme still preserves the fast dynamic response as shown in Fig. 8.

Fig. 10 shows the four-quadrant operation characteristics of the proposed drive system. This is examined by applying ±1000 rpm speed step reference and 1 Nm load. Fig. 11 shows stator flux magnitude, flux components in the stationary reference frame, extended back EMFs and manifolds of torque and flux for torque and speed step commands. The torque-speed curve and flux locus for all four-quadrant regions are demonstrated by Fig. 12. The steady-state behavior has been investigated at 1000 rev/min and 3 Nm load under two schemes. From Figs. 13 and 15, high ripples and distortion in currents can be observed and harmonics are distributed over a large frequency span undesirably. The harmonics in the currents are greatly reduced under proposed DTC. The dominant harmonics in currents, around 10 kHz, are less than 0.2% and 3.3% of those of their fundamentals, respectively as shown in Figs. 14 and 16. Figs. 17 and 18 show the output line-to-line voltages and frequency spectrums for these two methods in the range from 0 to 11 kHz. The output line-to-line voltage of VS-DTC contains dominant harmonics around switching frequency while more harmonics are scattered from fundamental to 10 kHz for basic DTC. In this case, a fixed switching frequency 10 kHz is achieved for the VS-DTC.

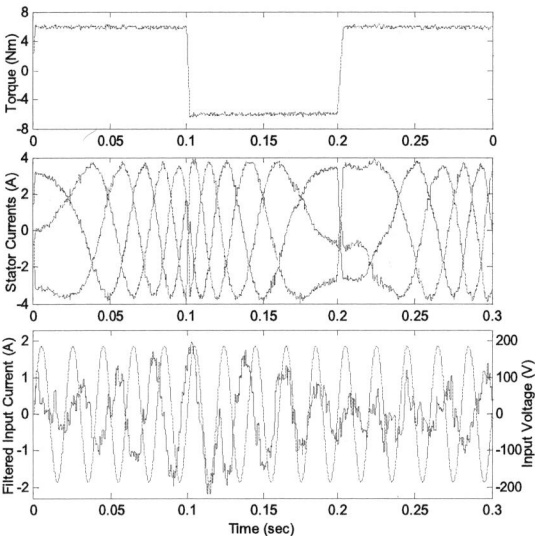

Figure 3. Dynamic response with ±6 Nm command (basic DTC)

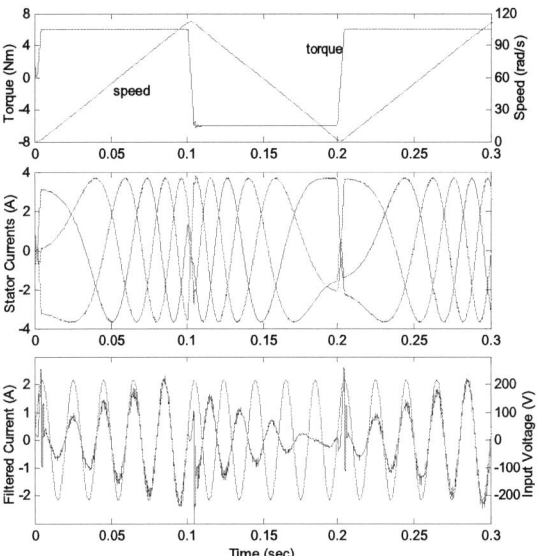

Figure 4. Dynamic response of the VS-DTC with ±6 Nm command

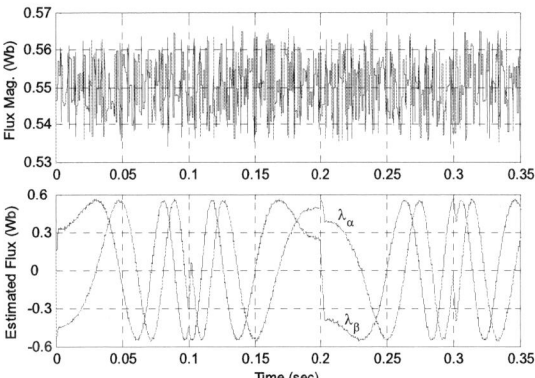

Figure 5. Flux of the basic DTC with ±6 Nm command

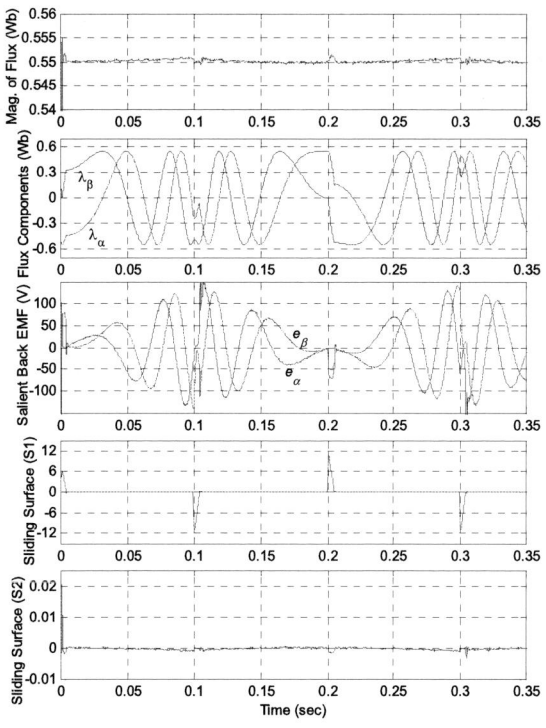

Figure 6. Stator flux, EEMF and sliding surfaces with ±6 Nm step

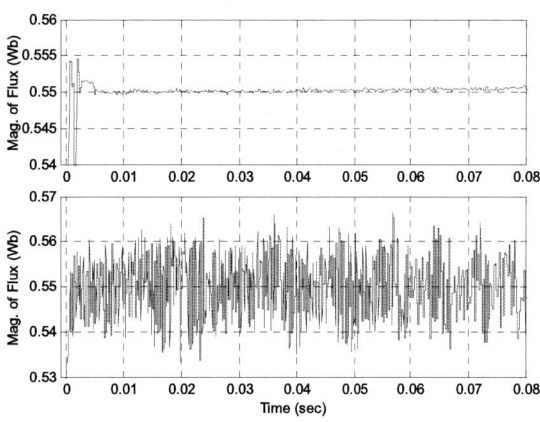

Figure 7. Flux response under VS (a) and basic DTC (b) schemes

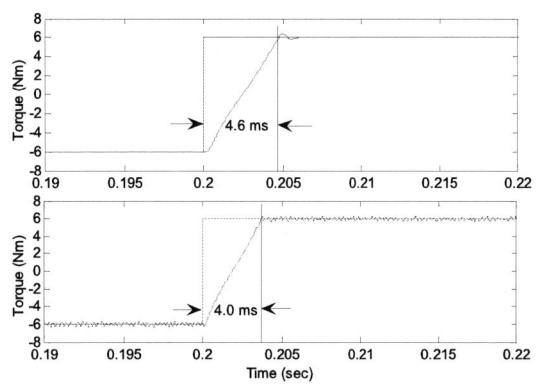

Figure 8. Torque response under VS (a) and basic DTC (b) schemes

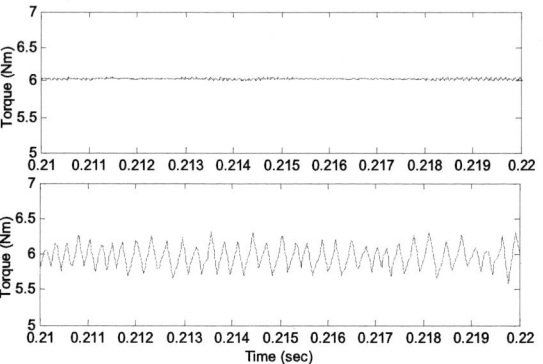

Figure 9. Steady-state torque under VS (a) and basic DTC (b) schemes

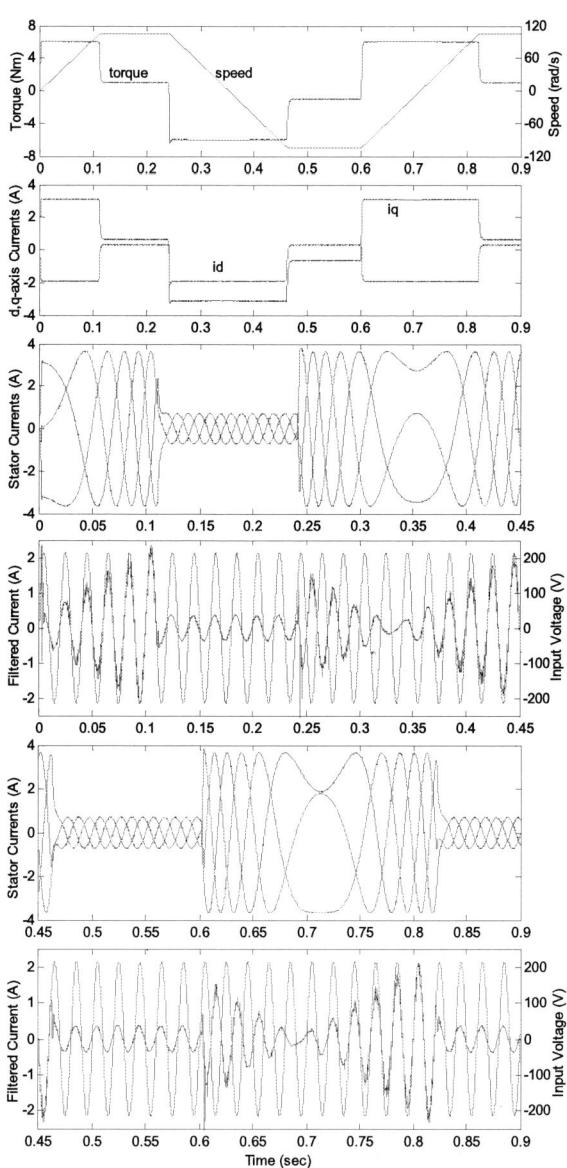

Figure 10. Four-quadrant operation waveforms with light load

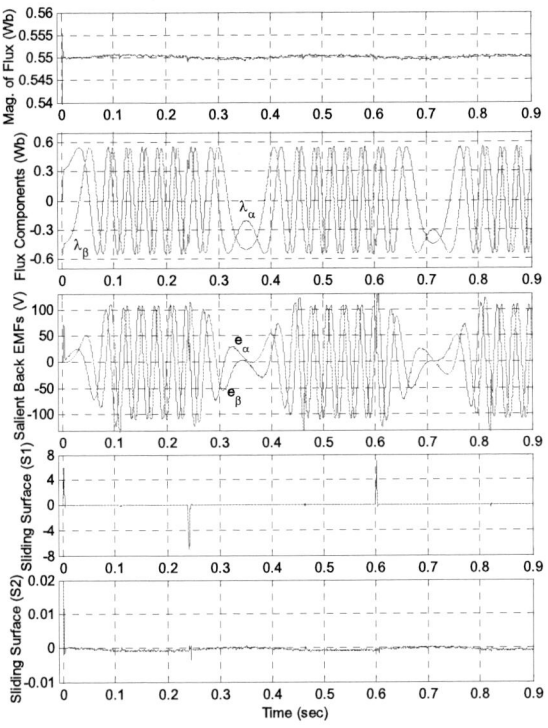

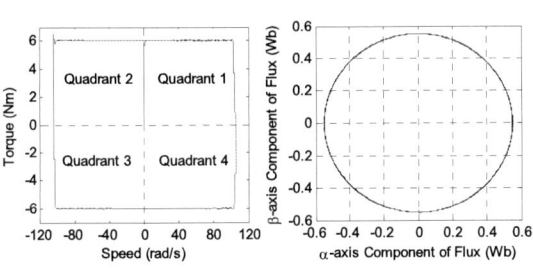

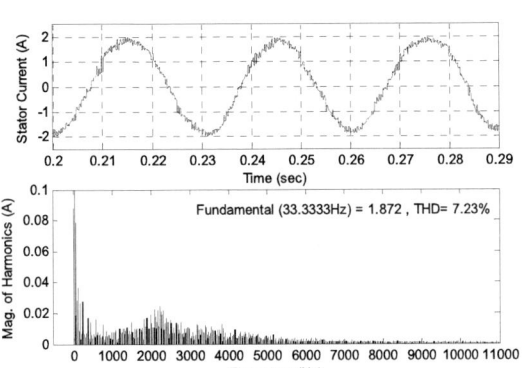

Figure 11. Stator flux, EEMF and sliding surfaces with speed step

Figure 12. Torque-speed characteristic and flux locus

Figure 13. Stator current and spectrum at 1000 r/min and 3 Nm (basic)

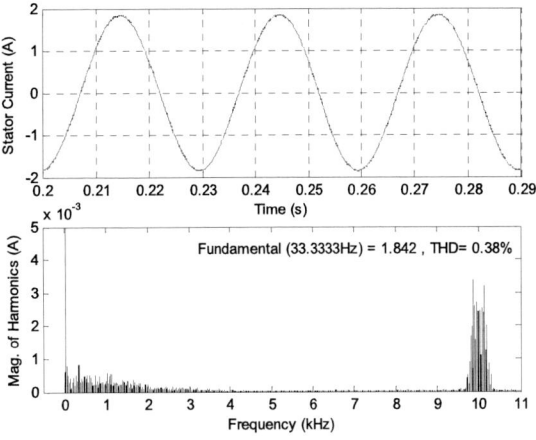

Figure 14. Stator current and its spectrum at 1000 r/min and 3 Nm (VS)

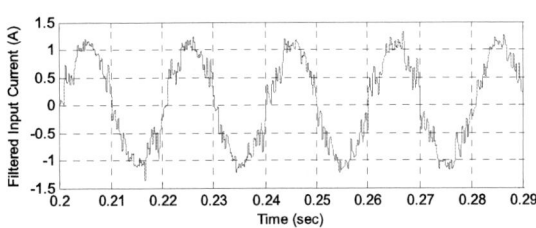

Figure 15. Filtered current and spectrum at 1000 r/min, 3 Nm (basic)

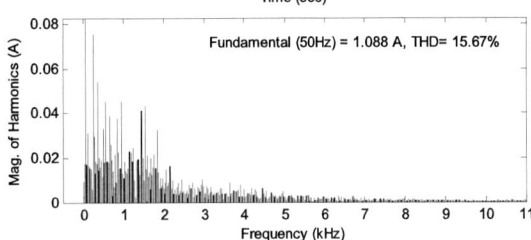

Figure 16. Filtered current and spectrum at 1000 r/min, 3 Nm (VS)

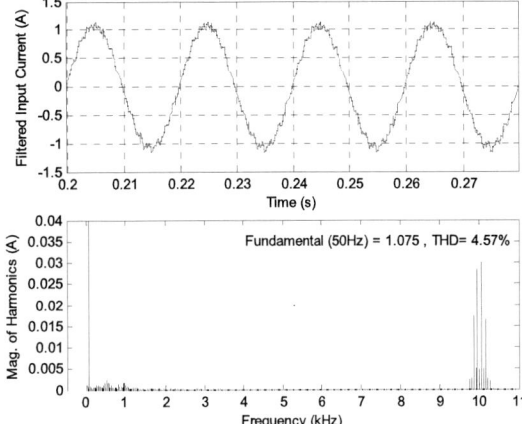

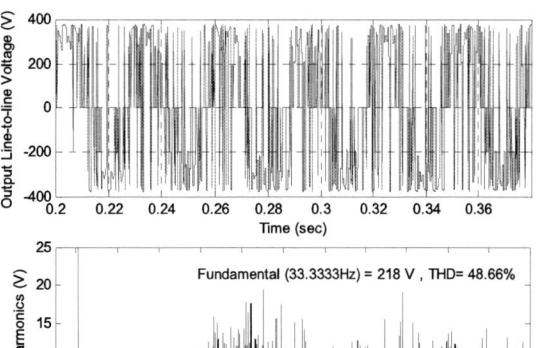

Figure 17. Output voltage and spectrum at 1000 r/min, 3 Nm (basic)

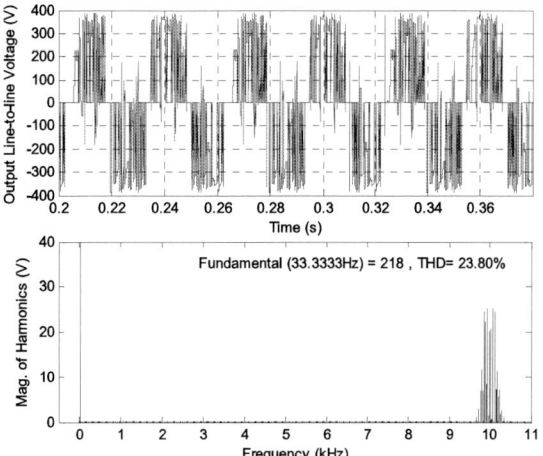

Figure 18. Output voltage and spectrum at 1000 r/min, 3 Nm (VS)

V. CONCLUSION

In this paper, a new matrix converter fed IPMSM drive scheme using DTC technique has been proposed. The Venturini's simplified modulation algorithm is employed to achieve the target output voltages of the matrix converter on the basis of the control requirements of the motor side with unity input power factor on the grid side.

The proposed scheme has been tested in steady-state and transient conditions under rated torque reversal and four-quadrant operation, carrying out some numerical simulations. The numerical simulation results of the proposed drive system are compared with those of the classical DTC. The torque and flux ripples are significantly reduced, and also the switching frequency remains constant independent of operating conditions. In comparison with the basic DTC, the filtered input currents, the stator currents and output line-to-line voltages are improved, of which the dominant harmonics are around a fixed frequency 10 kHz determined by the modulation sampling period.

TABLE I.   PARAMETERS OF THE IPMSM USED IN THIS PAPER

Rated output power (Watt)	P_r	1000
Rated phase voltage/current (V/A)	V / I	132/3
Magnetic flux linkage (Wb.)	λ_f	0.533
Number of pole pairs	P	2
Rated torque (Nm)	T_b	6
Stator resistance (Ω)	R_s	5.8
dq-axis inductances (mH)	L_d , L_q	44.8, 102.7
Rotor inertia (Kg.m^2)	J	0.00529
Friction coefficient (Nm/rad/s)	D	0.0006

TABLE II.   PARAMETERS OF BASIC DTC SCHEME

Stator flux linkage reference (Wb.)	λ_s^*	0.55
Maximum output of speed PI controller (Nm)	T_{max}^*	±6.0
Proportional gain of speed PI controller	K_{SP}	2.0
Integral gain of speed PI controller	K_{SI}	0.05
Electromagnetic torque band (Nm)	B_T	±0.02
Stator flux band (Wb)	B_λ	±0.001
Average $\sin(\theta_i^* - \hat{\theta}_i)$ band	B_ψ	0
Sampling period (μs)	T_s	40

TABLE III.   PARAMETERS OF VS-DTC SCHEME

Positive control gains of VS controller	μ_1 , μ_2	3200, 200
Smoothing factors of sliding surfaces	λ_1 , λ_2	0.4, 0.04
Control gains of VS controller	K_T , K_λ	3.0, 5.0
Initial stator flux components (Wb)	$\lambda_{\alpha 0}$, $\lambda_{\beta 0}$	0, -0.533
Sampling period (μs)	T_s	100

REFERENCES

[1] O.Simon, J. Mahlein, M. N. Muenzer, and M. Bruckmann, "Modern solutions for industrial matrix-converter applications," *IEEE Trans. Ind. Electron.*, vol. 49, no. 2, pp. 401–406, Apr. 2002.

[2] P. W. Wheeler, J. Rodríguez, J. C. Clare, L. Empringham, and A. Weinstein, "Matrix converter: a technology review," *IEEE Trans. Ind. Electron.*, vol. 49, no. 2, pp. 276–288, Apr. 2002.

[3] M. Venturini, "A new sine wave in sine wave out, conversion technique which eliminates reactive elements," in *Proc. POWERCON 7*, 1980. pp. E3_1–E3_15.

[4] M. Venturini, and A. Alesina, "Analysis and design of optimum-amplitude nine-switch direct AC-AC converters," *IEEE Trans. Power Electron.*, vol. 4, no. 1, pp. 101–112, Jan. 1989.

[5] D. Casadei, G. Serra, and A. Tani, "The use of matrix converters in direct torque control of induction machines," *IEEE Trans. Ind. Electron.*, vol. 48, no. 6, pp. 1057–1064, Dec. 2001.

[6] H. Altun and S. Sunter, "A vector controlled matrix converter induction motor drive," PhD Thesis, Department of Electrical and Electronic Engineering, The University of Nottingham, UK, 1993.

[7] Z. Xu and M. F. Rahman, "A variable structure torque and flux controller for a DTC IPM synchronous motor drive", in *Proc. IEEE PESC'04*, 2004, vol. 1, pp. 445–450.

2006 5th International Power Electronics and Motion Control Conference

Control of Distributed Power Systems

Z. Chen [*], Y Hu [**] and F. Blaabjerg [*]

[*]Institute of Energy Technology, Aalborg University, Aalborg, Denmark
[**]School of Science and Technology, North East Wales Institute of Higher Education, Wrexham, UK
zch@iet.aau.dk

Abstract— A large number of distributed generation (DG) units are being integrated into power systems at distribution level. The DG units play an essential role in affecting the efficiency and performance of power systems. This paper describes some possible controlling modes of DG units. The operation and control of the power systems consisting of DG units, in particular separated subsystems, are discussed. Simulation studies have been performed to demonstrate the performance of the control scheme.

Keywords- Distributed generation; control; frequency; voltage

I. INTRODUCTION

The traditional power systems are changing globally. A number of reasons cause these changes, including, load growth and changes in the geographical distribution; new environmental policy with a demand to reduce CO2 emissions, and the usual economic pressures of the marketplace. All this has led to increased interest in the developments of renewable energy generation and Combined Heat and Power (CHP) as distributed generation (DG) units. The DG units, including both renewable and non-renewable sources such as wind turbines, wave generators, Photovoltaic (PV) generators, small hydro, fuel cells and various Combined Heat and Power (CHP) stations, are being actively developed [1, 2, 3]. A wide spread use of DG units in both network connection and autonomous operation is expected.

In conventional generation stations, the generators operate at a fixed speed and thereby with a fixed grid frequency, however, the distributed generation units present quite different characteristics. For example, power electronics are normally used to interface variable speed wind power generators, PV generators and fuel cells. The power electronic technology plays a vital role to meet the requirements of power quality, including frequency, voltage, control of active and reactive power, harmonic minimization etc.

The power systems consisting of the distributed generation units may have frequency and voltage control problems. This paper first describes some operation and control modes of DG units, including the power electronic interfaced units, then discusses the control of the power system with DG units under different conditions, such as the separation of a system into two subsystems.

Simulation examples are presented to illustrate the effects of the control schemes.

II. CONTROL OF DISTRIBUTED GENERATION UNITS

A. Power system load characteristics

Power system load characteristics have a significant effect on system control. In power system analysis, a load model may express the active and reactive powers as functions of the bus voltage and frequency. For example, a load may be represented by the real power (P) and reactive power (Q), which can be related to voltage and frequency. Traditionally, the real power is mainly related to the frequency and the reactive power is closely linked to the system voltage, such load characteristics are shown in Fig. 1 [4].

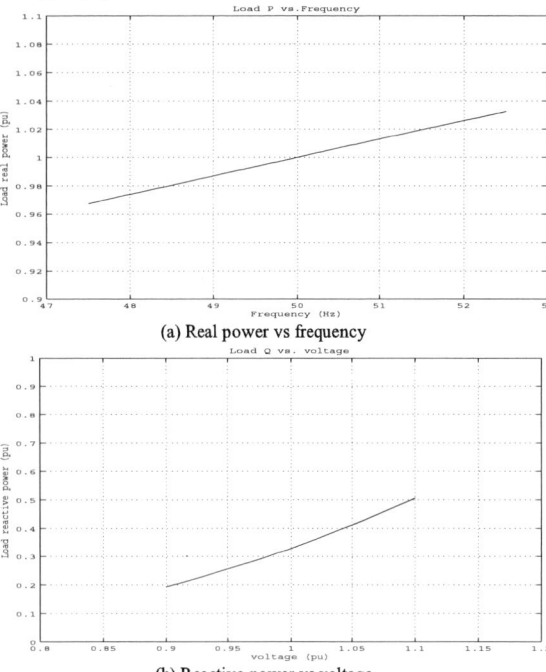

(a) Real power vs frequency

(b) Reactive power vs voltage

Fig. 1. Load real and reactive power characteristics

B. Frequency and voltage drop control

For conventional systems there is a correlation between real power and frequency. If an excess of real power is presented in the system, the speed of generator rotors and the system frequency increase, the generator

1-4244-0448-7/06/$25.00 ©2006 IEEE

controllers will then reduce the real power input to bring the frequency down. On the other hand, if there is a shortage of real power, the frequency will drop. The controller will then increase the input power to raise the frequency. Therefore a power versus frequency droop line can be used to regulate the real power output as shown in Fig. 2(a). Similarly reactive power is typically controlled by a Q vs. V droop line as shown in Fig. 2(b).

The conventional droop method may be described as:

$$f_i = f_o - k_{f,i} P_i \qquad V_i = V_o - k_{v,i} Q_i \qquad (1)$$

Where f_o is the frequency at no load condition, V_o the voltage amplitude at unit power factor. To provide sharing, the slopes $k_{f,i}$ and $k_{v,i}$ are chosen to match the frequency and voltage differences for each node. For a system with DG units in parallel, this approach may be adopted to coordinate all of the DG units to ensure that the power matching and power sharing be achieved.

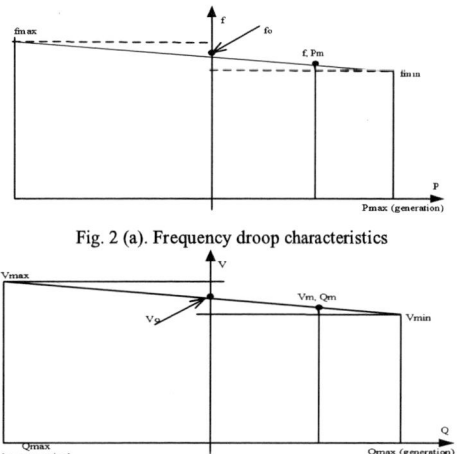

Fig. 2 (a). Frequency droop characteristics

Fig. 2 (b) Voltage droop characteristics

B. Power control of DG units

Not all the DG units have the ability of control their power output as the frequency varies. For example, some renewable energy source powered DG units can only generate the power available at that time. The output power may be reduced but can not be increased easily. For best usage of the energy, the DG units may operate at the optimal power condition. Therefore the fixed power control may be used if the power system constraints are allowed. Power electronics are often used for the integration of such distributed generation units.

Power electronic converters, in particular, voltage source converters (VSC), are widely used for interfacing various types of distributed generation units, including micro turbines, fuel cells, photovoltaic and wind turbines etc. These converters perform the conversions of voltage level and frequency. They can also be designed to optimize the operation of the generation units, such as to capture the maximum available wind power in variable speed wind turbines and to maximize the PV generation by following the maximum power point of a PV system.

A basic VSC structure is shown in Fig. 3. The VSC may be used to control both real and reactive power. One way to implement the power control in a VSC can be based on the dq frame control [4]. With the dq frame control, the control of real power and the control of reactive power are decoupled. The generated real and reactive power references are sent to the VSC controller in dq frame to drive the converter.

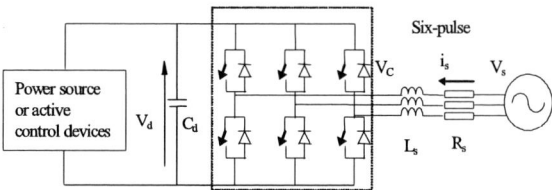

Fig. 3 Voltage source converter (VSC)

C. Voltage control of DG units

The system voltage is closely related to reactive power, therefore the voltage control may be implemented by regulating the reactive power generation. Such a voltage control plus a Q-V droop regulation is shown in Fig. 4, where the voltage-reactive power control is combined with the Q-V droop control. The voltage control follows a Q-V droop characteristic, i.e. the unit will adjust its output voltage based on its reactive generation. Also other generation unit will change their reactive power generation in response to the system voltage variation. Eventually the system voltage will arrive at a new steady state.

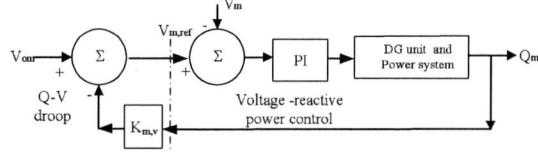

Fig. 4. Q-V droop plus voltage-reactive power controller.

The DG control unit will control the system voltage by

$$V_m = V_{om} - K_{m,v} Q_m$$

$K_{m,v}$ is the reactive power droop coefficient of the DG unit. The reactive power of the unit has to be controlled to achieve the reference voltage.

For a field excited synchronous generator the machine terminal voltage may be directly controlled by regulating the excitation of the field. The block diagram of this type of controller is shown in Fig. 5.

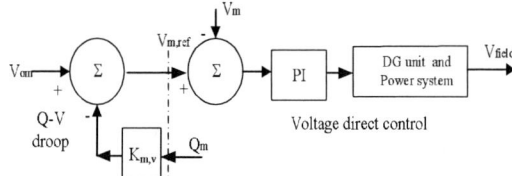

Fig. 5. Q-V droop plus voltage direct controller.

It could be seen that the DG control unit effectively sets the droop line to determine reactive power sharing between the units under steady-state operation. The

2030

voltage will then increase if there is excess reactive power forcing the unit to produce less Q or absorb reactive power, until a new steady-state voltage is reached.

D. Frequency control of DG units

The real power and frequency control may be conducted in the similar way like the reactive and voltage control. The unit responsible for the frequency control will follow a P-f droop characteristic, i.e. the unit will adjust its output fundamental frequency based on its real power output. The frequency variation will enable the other generation units adjust their real power generation correspondingly. After regulation, the system power and frequency will reach a new steady state.

For a frequency control DG unit, the frequency can be calculated as

$$f = f_o - K_{m,f}(s)P_m$$

$K_{m,f}(s)$ is the real power droop coefficient (including a low pass filter function) of the generation unit. The diagram of the frequency control system is shown in Fig. 6.

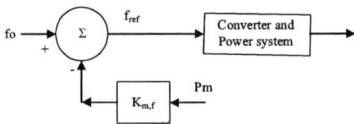

Fig. 6 Block diagram of frequency control of DG unit.

D. Droop characteristics with power controlled DG units

In power control mode, a DG controller will try to follow the real and reactive power references, the real power limit may be set by the available power, the reactive power may be limited by the power electronic capacity.

These references may be determined by

$$P_{DG,i,ref} \cdot = P_{DGo,i} + \Delta P_{DG,i}$$

$$\Delta P_{DG,i} = K_{pi}(s)\Delta f$$

and

$$Q_{DG,i,ref} \cdot = Q_{DGo,i} + \Delta Q_{DG,i}$$

$$\Delta Q_{DG,i} = K_{qi}(s)\Delta V_i$$

Where $P_{DG,i,ref}$, $P_{DGo,i}$ and $\Delta P_{DG,i}$ are real power references, base operating point and share variation respectively; $Q_{DG,i,ref}$, $Q_{DGo,i}$ and $\Delta Q_{DG,i}$ are reactive power reference, base operating point and share variation. $K_{pi}(s)$ and $K_{qi}(s)$ are the real and reactive power share coefficients (including a low pass filter function) of the generation unit i. ΔV_i is terminal voltage variation of generation unit i and Δf is the system frequency variation. Such controllers are shown in Fig. 7 and Fig. 8.

The control mode of a DG unit can be:

- Frequency control mode with P-f droop,
- Real power control mode with P-f droop,
- Fixed real power control mode

and

- Voltage reactive power control mode with P-f droop,
- Fixed terminal voltage control mode,
- Reactive power control mode with P-f droop.

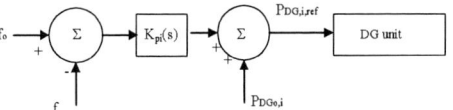

Fig. 7 Real power reference of power controlled VSC unit.

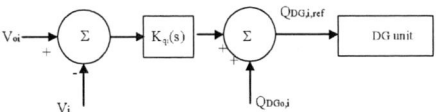

Fig. 8 Reactive power reference of power controlled VSC unit.

The adopted operational modes may be changed during the operation when the system conditions change, consequently the role a DG unit plays may change. For example, the fixed real power control mode may be used for a generation unit operating with certain power constraint conditions, for example, a CHP generator which has to supply a certain heat load or a renewable source powered generation unit whose power is limited by the available renewable energy source at that time.

Reactive power control may also arranged in the similar way such as the real power and frequency control but the power electronic converter normally has the ability of generating reactive power, therefore it may not be so limited as the real power. The system voltage may be controlled by these converters in a distributed meaner as long as these converters are not located too close to compete for the voltage control of the same busbar.

III. SYSTEM VOLTAGE AND FREQUECNY CONTROL

Frequency and voltage are two most important operational indexes for power systems. All the generation units in a synchronised system have to operation at the same frequency. On the other hand, the voltages across the network may be different.

Not all the generation units may be permitted to set frequency and control voltage. Instead, in a synchronised system, only one generation unit in the group should takes care of the direct frequency control; in a group of the closed related DG units, only one generation unit may perform the voltage control. Other generation units work in a real power controlled mode with droop or a fixed real power operation mode to adjust their output real power and in a reactive power controlled mode with droop mode. In this way, the tasks of frequency-power control and voltage–reactive power control are shared. The droop characteristics enable the easy cooperation among the DG units.

The frequency and voltage control is especially important in case of a contingency or incident condition, such as the loss of generation or load, control must be performed to maintain a stable operation. The action of the individual generation units performing the control may have to follow preset characteristics. The power generation must be adjusted to follow the changes of system conditions within the specified time.

The transitions of power system operation states require power re-balance by regulating of DG generations and/or load. The transition may involve significant dynamics and transients, which much be dealt with carefully in order to smoothly transfer the power system from one state to the other. In case that the system is separated into several subsystems, one unit in each group should be take over the frequency control and also the low priority load may have to be cut to maintain reasonable voltage and frequency.

IV. SYSTEM PERFORMANCE STUDY

Simulation studies have been carried out in PSCAD/EMTDC. The studied system is shown in Fig. 9 where a standalone power system consisting of two groups of distributed generation units, G_1 with G_2 and G_3 with G_4 respectively. G_1 and G_3 are the main generation unit for respective DG group. There are four busbars with respective load P_{L1}, Q_{L1}, P_{L2}, Q_{L2}, P_{L3}, Q_{L3}, and P_{L4}, Q_{L4} distributed evenly. Also there is an induction machine load, P_M, Q_M, connected on busbar 4.

G_1 and G_3 are assumed to have relative large capacity therefore can produce large enough real and reactive power for system frequency and voltage control, on the other hand, G_2 and G_4 only have a limited capacity, such as renewable source powered DGs, hence operate with the real power and reactive power control mode as discussed to inject the available power into the system.

In normal operation, the voltage control of the system can be performed in a distributed style, i.e. both G_1 and G_3 may operate in the voltage control mode. However, the only one frequency control unit should exist for a synchronized power system. G_1 is responsible for power system frequency control when the whole system operating as a synchronized system. But when the system is separated, for example, by operating some circuit breakers to clear a power system fault, the separated subsystems have to be able to control their frequency and voltage respectively. In such a case, G_1 and G_3 will perform respectively frequency control for their own systems.

The simulation results of three studied cases are presented in this paper. The common condition is as follows: The system starts with normal operation and load at all busbar are distributed evenly, at 5.0 seconds of the simulation time, some further load at busbars of 3 and 4 are switched in, and at 7.0 seconds, a three phase short circuit fault occurs in the system, the fault lasts for 220

ms and cleared by opening the circuit breakers. The clearance of the fault separates the system into two parts.

Case 1: The short circuit fault occurs in the mid of line 2-3. The clearance of the fault makes the system separated into two parts which have the similar load and generation capacity as normal operation condition. The simulation results are shown in Fig. 10. It can be seen that the voltages of busbar 2 and 3 have larger voltage dips during the fault since they are close to the fault location, consequently the real and reactive load at the two busbars are significantly dropped due the voltage drop caused by the fault. It is also noted that the voltages at busbar 3 and 4 recovered slowly in comparison with the voltages at busbar 1 and 2, because the induction machine load in the subsystem where busbar 3 and 4 are located. The induction machine need large amount of reactive power to reestablish its voltage, the reactive power flow on the circuit may delay the recovery of the system voltage. A further delay of the fault clearance may even result in an over speed of the induction machine, consequently the cut off of the induction machine. Also the voltage of the subsystem supported by DG group 2 is lower than the value at normal operation. The frequency is dropped as well, the DGs in the subsystem-2 respond by increasing the real and reactive power, especially the unit of G_3 since G_4 is capacity limited. The frequencies of the two separated subsystems are different. The subsystem of the group 2 DG units has a reduced frequency. The frequency of the group 2 DG units is performed by the unit G_3, which changes to the frequency control mode from a power control mode after the system is separated.

Case 2: The system experiences similar events as above, but the fault location is at the mid of line 1-2. In this case, the clearance of the fault makes the system separated into two subsystems which have quite different load. The simulation results are shown in Fig. 11. It can be seen that the voltage of subsystem 2 (busbars 2, 3 and 4) has the reduced voltage after the fault is cleared, especially the voltage at busbar 2, the far end of the separated subsystem with DG units G_3 and G_4. Although the induction machine is still recovered, the system suffers a low voltage and low frequency situation. As expected, the unit G_3 in the subsystem 2 has large pressure to supply the required real and reactive power.

Case 3: The case has the similar situation as *Case 2*, but some low priority load in the subsystem (busbars 2, 3 and 4) has been disconnected at 8.0 seconds in response to the low frequency and voltage. The simulation results are shown in Fig. 12. This action helps to increase the frequency and voltage so that the frequency and voltage in the subsystem are better than that in Case 2.

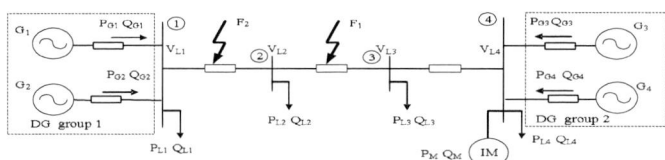

Fig. 9 The studied power system

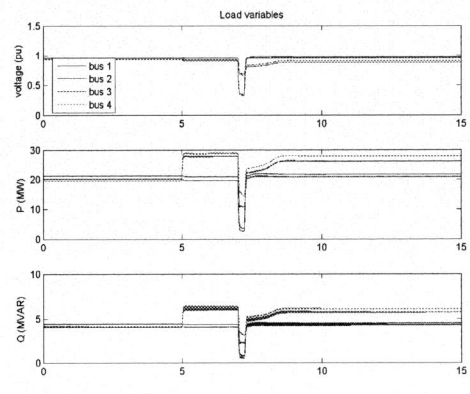

(a) busbar voltage, load P and Q

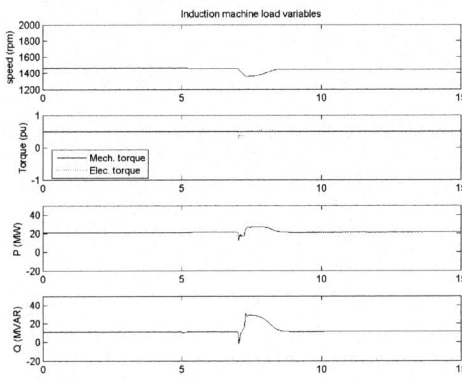

(b) Induction machine speed, load torque and electromagnetically torque, P and Q

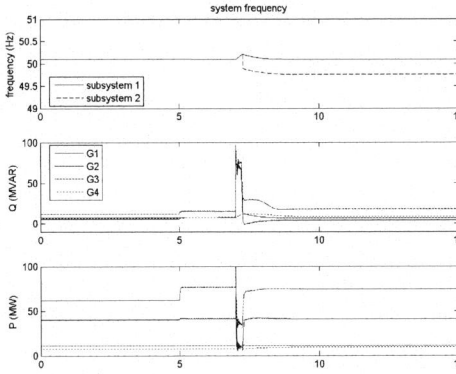

(c) System frequencies and generator P and Q
Fig. 10 Simulation results of case 1

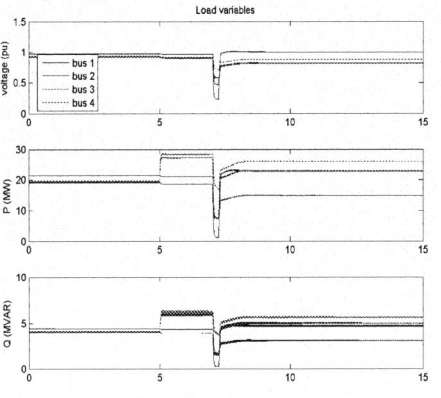

(a) busbar voltage, load P and Q

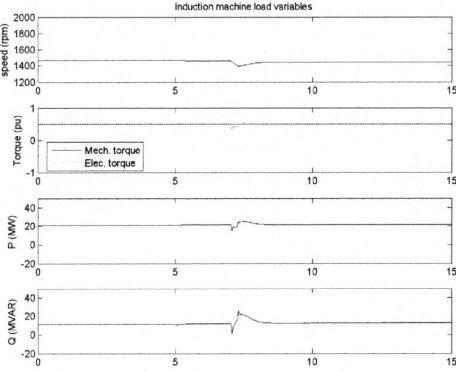

(b) Induction machine speed, load torque and electromagnetically torque, P and Q

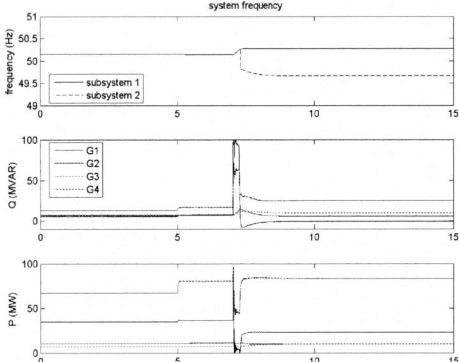

(c) System frequencies and generator P and Q
Fig. 11 Simulation results of case 2

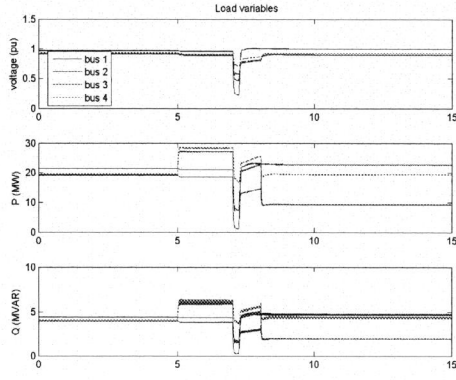

(a) busbar voltage, load P and Q

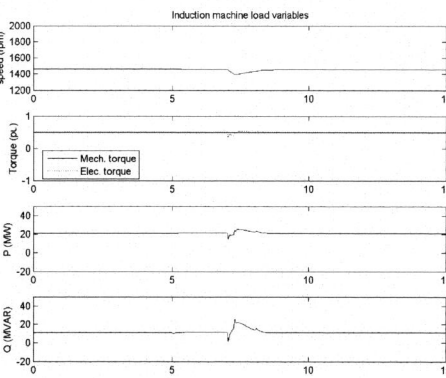

(b) Induction machine speed, load torque and electromagnetically torque, P and Q

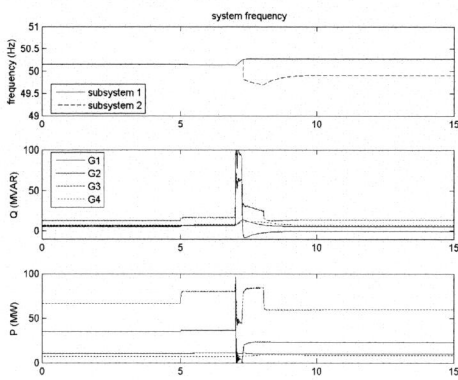

(c) System frequencies and generator P and Q

Fig. 12 Simulation results of case 3

V. CONCLUSIONS

In this paper, the operation and control of distributed generation units, including power electronic interfaced DG units, have been discussed. The method of controlling the frequency and voltage has been developed in order to achieve a satisfactory operation of the distributed generation system.

Different operational modes of the distributed generation units are discussed. The system control will be performed by coordinating the DG units in various operation modes and transferring from one mode to another if necessary.

In a synchronized system, one DG unit is responsible for setting the frequency. The other DG units will operate in a power controlled mode, i.e. the power output will be controlled in response to the variation of system frequency. The system voltage may be controlled in a distributed style, i.e. the voltages at different locations in a system may be controlled by different units.

The system behavior during the transitions is studied by simulations. The key parameters, voltage and frequency, are monitored. The transfer. from an autonomous distribution system into further separated smaller subsystems has been studied. The effectiveness of control (including DG control and load control) has been illustrated by the simulation results.

REFERENCES

[1] G. Strbac, "Impact of dispersed generation on distribution systems: a European perspective" Power Engineering Society Winter Meeting, 2002. IEEE , Volume: 1 , 2002 Page(s): 118 – 120.

[2] M. Dussart, P. Lauwers; S. Magnus, Y. Laperches, "Connection requirements for dispersed generation: evolutions of existing requirements and need for further standardization" Electricity Distribution, 2001. Part 1: Contributions. CIRED. 16th International Conference and Exhibition on (IEE Conf. Publ No. 482), Volume: 4 , 2001 Page(s): 8 pp. vol.4.

[3] J.A.P. Lopes, "Integration of dispersed generation on distribution networks-impact studies", Power Engineering Society Winter Meeting, 2002. IEEE , Vol: 1 , 2002, Page(s): 323 –328.

[4] Z Chen, Y. Hu, "Control of Power Electronic Converters for Distributed Generation Units", Proc. of the 31st Annual Conference of the IEEE Industrial Electronics Society, IECON 2005, pp. 1317-1322, ISBN: 0-7803-9253-1.

[5] Z. Chen, E. Spooner. "Voltage Source Inverters for High-Power, Variable-Voltage DC Power Sources", IEE Proc. –Generation, Transmission and Distributions, Vol. 148, No. 5, September 2001, pp. 439-447.

[6] R. Lasseter and P. Piagi, "Providing premium power through distributed resources," in *Proc. 33rd Hawaii Int. Conf. System Sciences*, Jan. 2000, pp. 1437–1445.

[7] M. Prodanovic, T.C. Green, H. Mansir, "A survey of control methods for three-phase inverters in parallel connection"; Eighth International Conference on Power Electronics and Variable Speed Drives, 2000. (IEE Conf. Publ. No. 475) 18-19 Sept. 2000 Page(s):472 – 477.

[8] J. Lang, T.C. Green, G. Weiss, Q.-C. Zhong, "Hybrid control of multiple inverters in an island-mode distribution system", IEEE 34th Annual Power Electronics Specialist Conference, 2003. PESC '03. 2003, Volume 1, 15-19 June 2003 Page(s):61 - 66 vol.1.

Author Index

A

Abbasian, M.A. 1043
Abdelhamid, T. H. 411
Abedini, A. 224
Abjadi, N. R. 1917
Abo-Khalil, Ahmed G. 1477
Abramovitz, A. 1412
Agarwal, Anant K. 157
Agarwal, Vivek 281
Ahmed, Nabil A. 242
Ahn, C. H. 1198
Aide, Xu 1162
Ai-Juan, Jin 1421, 1426
Ait-Amirat, Y. 1882
Ajjarapu, Venkataramana 505
Akagi, Hirofumi 23
Akimasa, Koji 1613
Andersen, Henrik Rosendal 1032
Ando, Tatsuo 947
Arpilliere, M. 249
Ashida, M. 2000
Askari, J. 1917

B

Badica, M. 1751
Bai, Haijun 219, 826
Bai, Zhifeng 1581
Baihua, ...161
Balda, J.C. 1353
Banaei, M. R. 759, 764
Bao, G.Q. 813
Baocheng, Wang 569, 1991
Baoming, Ge 918
Barsoum, ...1148
Bendjedia, M. 1882
Berthon, A. 515, 1882, 2005
Bhattacharya, Subhashish 1450
Bin, Su 406
bin, Wu 1368
Binder, A. 842
Bing, Chen 1401
Bisogno, F.E. 1117
Blaaberg, F. 46, 1107, 2029
Bo, Chen 122
Böcker, Joachim 1112
Bodson, M. 1912
Bojoi, R. 1651
Boroyevich, D. 249
Boroyevich, Dushan 1836
Bréhaut, Stéphane 92
Brouji, H. El 1663

C

C, Sreekumar 281
Cailin, Wang 1167
Calderon-Lopez, G. 1328
Calverley, S.D. 977
Camara, M.B. 515
Câmpeanu, A. 1751
Cao, Binggang. 1581
Cao, R. X. 510
Cao, Yanjie 2015
Carazo, A. V. 1117
Cartes, David 774
Cen, Yuwan 1986
Chan, C.C. 57
Chang, Chung-Hsing 291
Chang, Duan Qi 1551
Chang, H.-H. 1343
Chang, Jie (Jay) 102
Chang, Lon-Kou 417
Chang, Yuan 1722
Changhong, Wang 1793
Changzheng, Zhang 489, 739
Chau, K. T. 1788
Chen, Bin 1450
Chen, C.-C. 117
Chen, Cheng-Hu 913
Chen, Chern-Lin 332
Chen, Chien-An 967
Chen, Guiyou 1202
Chen, Guocheng 1560
Chen, Guozhu 794
Chen, H. 933
Chen, H. G. 1129
Chen, J. 194
Chen, Jian 1218
Chen, Jiann-Fuh 361, 1178
Chen, Jiaxin 346, 831
Chen, Jie 113
Chen, Jun-Ning 199, 286, 1283, 1392
Chen, Min 442
Chen, Qiaoliang 433, 642
Chen, Rui 1171
Chen, Ruijuan 1253, 1877
Chen, Tso-Min 332
Chen, Wei 171, 1081
Chen, Xi 1507
Chen, Xiangjun 607
Chen, XuWu 438
Chen, Y.-M. 108, 117
Chen, Yao 1454
Chen, Yen-Ming 1763
Chen, Yuan-rui 1397
Chen, Yunpeng 236
Chen, Z. 49, 499, 1773, 2029
Chen, Zongxiang 142
Chenchen 386
Cheng, Chun-An 1178
Cheng, Ming 1746, 1815
Cheng, Ming-Yang 913

Author Index

Cheng-ning, Zhang 1027
Chengsheng, Wang 589
Cherifi, A. 574
Chi, Song 890, 1825
Chiang, Huann-Keng 967
Chiasson, J. N. 1703, 1912
Chiu, Huang-Jen 291
Cho, Yun-hyun 1238, 1784
Choi, E.S. 1382
Chongjian, Li 589
Chun, Dong 1623
Chun, YonDo 1784
Chung, Jung Kee 1736
Chunjiang, Zhang 554, 559, 1473, 1618
Corzine, Keith A. 637
Costa, François 92
Crausaz, A. 2005
Cui, Bo 798
Cui, Jiefan 657
Cui, Junwei 1436
Cvetkovski, G. 254

D

Dai, Ke 789
Dai, Renchang 1122
Dai, Yue-Hua 199
Da-ming, Liu 1674
Danhe, Li 1991
De Doncker, R. W. 31
Deng, Jianming 923
Deng, Yan 1931
Dianguo, Xu 301, 1713
Ding, Xiaoyu 1560
Ding, Ye 1223
Divan, D. 16
Divan, Deepak M. 2010
Doi, Toshimitsu 356, 1302, 1307
Dong, Jiang 880
Donghua, Luo 1634
Dongsheng, Zuo 468
Dong-Shoutian, ...484
DongYu, ...1623
Dou, Sen 537
Du, Guiping 316
Du, Zhong 1450
Duan, Baoxing 70
Duan, Huijuan 798
Duan, Shan Xu 1522
Duan, Shanxu 1218
Duarte, Jorge L. 784

E

Ebrahimi, Yousef 779
Eiuo, Bin 1358
El Din, Ashraf Salah El Din Zein 847
Elbanhawy, Alan 342. 1967

Endo, Tsunehiro 947
Ertugrul, Nesimi 147, 962

F

Fa, Naiguang 1253, 1877
Fang, Liang 817
Fang, Xin 789
Fang, Xu-Peng 166
Fang, Yu 1406
Fang, Zhuo 122, 1542
Fathy, Khairy 356, 1302, 1307, 1358, 1363
Fei, Wanmin 1138
Fei-peng, Xu 647
Feng, D. 499
Feng, Huang 1401
Feng, L. 842
Feng, Zhao 622, 679
Feng, Zheng 585
Fengxiang, Wang 449, 903
Feyzi, M. Reza 204, 1228
Forrest, S. J. 977
Forsyth, A. J. 1323, 1328
Francis, Jerry 1836
Friedrichs, Peter 132
Fröhleke, Norbert 1112
Fuchs, F.-W. 325
Fujita, Kouetsu 1971
Fukuda, S. 1468
Fukushima, Kentaro 1333
Funian, Hua 449
Furuya, Atsushi 1598
Fu-sheng, Wang 463
Futami, M. 1468

G

Gallay, R. 2005
Gang, Ma 378
Gao, F. 1107
Gao, Yan 157
Gao, Yang 113, 1159
Gao, Yong 1198
Gao, Z. Y. 1071
Garinto, Dodi 306
Ge, L.S. 1458
Ge, Lu-sheng 1368, 1576
Ge, Qiongxuan 1171
Geng, Pan 789
Goharrizi, A. Yazdanpanah 1697
Gong, Yu 1223, 1788
Grabner, C. 999
Graczkowski, J. J. 1096
Grantham, Colin 1207, 1858
Gruenberger, Hans Pert 219, 826
Grundmann, Frank 870, 1442
Gu, G. 175
Gu, Herong 473, 1585

Author Index

Gu, Yilei 171, 276
Gualous, H. 515, 2005
Guan, Xiaohan 688
Guang, Zeng 734, 1669
Guangzheng, NI 1091
Guenther, D. 842
Guilan, Chen 554, 1006, 1049
Gui-xin, Shao 1027
Guiyou, Chen 1630, 1634
Guo, Hongche 612, 853
Guo, Qingding 612, 853, 896
Guo, Wei 952
Guo, Xin 337
Guo, Youguang 346, 831
Guobiao, Gu 808
Guocheng, San 1473
Guojun, Lu 802
Guoxin, Zhu 1802
Gustin, F. 515

H

Habetler, Thomas G. 836
Haibing, Hu 937
Haijie, Xu 937
Haiping, Xu 684, 1298
Haitao, Zhang 161
Halász, S. 693
Han, B.D. 1143
Han, Chong 1450
Han, Chong Zhao 652
Han, F. T. 1071
Han, S.K. 1382
Hang-Tian, Li 1421, 1426
Harley, Ronald G. 836, 2010
Hartavi, A.E. 2018
Hashimoto, Takayoshi 1333
He, Guofeng 657
He, Junping 1081
He, Shijie 1463
He, Xiangning 83, 1931
He, Zhongyi 1537
Hemin, Wang 1849
Heming, Li 458
Hendrix, Marcel A. M. 784
Herong, Gu 559, 1618
Hirao, Mitsuhiro 1768
Ho, Chien-Yeh 1527, 1995
Ho, S. L. 1901
Hong, Peng 899
Hong, Shen 559, 1273, 1618
Hong-mei, LI 463
Hongren, Yin 401
Hori, Yoichi 1797
Hosseini, S. H. 759, 764, 1697
Hosseini, Seyyed Hossein 753, 779, 1679
Howe, D. 908. 928, 1841

Hsu, Kai-Sheng 967
Hsu, Ken-Chuan 1957
Hsu, W.P. 718
Hu, D.Q. 1143
Hu, Haibing 127, 1183
Hu, Jiangang 703
Hu, Qing 1806
Hu, Qingbo 526
Hu, Songqin 351
Hu, Weihao 321, 585
Hu, Wenhua 397
Hu, Xuezhi 438
Hu, Y. 2029
Hu, Z. L. 1708
Hu, Zongbo 316
Hua, Li 729, 1571
Hua, Wei 1746, 1758
Huade, Li 401
Huang, Zhenyue 1213
Huang, Alex Q. 113, 157, 1159, 1450
Huang, Chien-Lan 748, 1278
Huang, Congsheng 1288
Huang, Jin 1017
Huang, Xuwen 1288
Huang, Yafeng 542
Huang, Yi 1076
Huang, Yuehui 580, 1512
Hui, Li 729
Hui, Wu 862
Hui, Zhang 1032, 1492
Hui-jie, Xiang 1942

I

Ichinose, M. 1468
Inoue, Kaoru 1233, 1613
Inoue, Shigenori 23
Iov, F. 46
Iwanski, G. 494

J

Jang, Jeong-Ik 1482
Jangwanitlert, A. 1353
Járdán, R.K. 1338
Jeon, K. S. 1198
Jewell, G. W. 977
Ji, Yanchao 627, 1507
Jia, C. 1323
Jia, Y. 2000
Jia, Y.P. 1143
Jiag, Maoh-Chin 1527
Jian, Chen 74, 769
Jian, Cui 1431
Jian, Liu 272
Jian, Wu 1713
Jiang , Chang 1213
Jiang-Hui, Chen 1401

Author Index

Jiang, J. Z. 1788
Jiang, J.G. 1458, 1952
Jiang, J.J. 152
Jiang, J.Z. 813
Jiang, Jianguo 1081
Jiang, Jianzhong 1223
Jiang, Xianglong 1608
Jiang, Xiaochun 1896
Jianguo, Jiang 468
Jianlin, Zhu 1557
Jianru, Wan 1273
Jian-Ru, Wan 1431
Jian-wen, Zhang 1657
Jianze, Wang 1693
Jiarong, Kan 1532
JiaYi, Yuan 808
Jie, Shuo 1806
Jie, Wang 569
Jiefan, Cui 862, 1849
Jin, Jianxun 831
Jin, Mengjia 885, 1872
Jin, Shun 617
Jin, Tianjun 1183
Jin, Wenxi 127
Jin, Xin Min 1454
Jing, Liu 88
Jin-gang, Li 549
Jing-Gang, Zhang 1669
Jinjun, Liu 1061, 1492, 1722
Jinlong, Zhang 401
Jinupun, P. 1887
Jiqiang, Wang 903
Jiuhe, Wang 401
Jo, WonYoung 1784
Johal, H. 16
Johnson, C. M. 977
Joseph, Alan 1076
Jou, H.L 718
Jun, Liu 1012
Jun, Wang 899
Jung, Kun-seok 1238
Junjuan, Sun Xiaofeng Wu 569
Junmin, Zhang 1684
Junzhu, Wan 1849
Jwo, Ko-Wen 1590

K

Kaijie, Feng 1741
Kaipei, Liu 209, 1684
Kang, B.W. 1143
Kang, Ju-Sung 1358
Kang, Y. 194
Kang, Yong 97, 564, 789, 1218, 1522, 1981
Karimi, E. 1697
Kato, Tomohiko 1971
Kato, Toshiji 1233, 1613

Ke, Dao-Ming 199
Ke, Fu-Jing 1154
Ke, Yi-Jing 1392
Kerkman, Russel J. 1054
Kesong, Ye 699, 1832
Khaehintung, Noppadol 137, 368
Khajee, M. Darkalee 759
Khan, Mahamnad Mansoor 386
Kim, E. D. 1198
Kim, Jang-Hwan 662
Kim, Joo Han 1736
Kim, Young-Sin 1482
Kimura, Noriyuki 1768
Kiranon, Wiwat 137
Kita, H. 1468
Koczara, W. 494
Konghirun, M. 972
Koo, DaeHyun 1784
Kou, X. 1096
Krishnaswami, Sumi 157
Ku, Chung-Ping 1590
Kumar, Pavan 537
Kun, Li 1447
Kunakorn, Anantawat 368
Kuo, J. -S. 1343

L

L., M. 1148
Lai, Ching-Ming 1590
Lai, Stephen L. 1502
Lai, Y. M. 296
Lang, Yongqiang 708
Lee, Chi-Yang 1192
Lee, Dong-Choon 1477, 1482
Lee, Fred C. 1
Lee, Hyun Woo 392, 1302, 1307, 1358, 1363, 1372, 1377
Lee, Se-Hyun 1477
Lemberg, Nicholas 989
Li, Chongjian 995
Li, Dong 1202
Li, Dongsheng 947
Li, F. 152
Li, H. 1773
Li, Han 1006, 1049
Li, Hongtao 688, 1248
Li, M. 1703, 1912
Li, Ma 88, 209
Li, Mingzhu 1537
Li, Min-zu 97
Li, Qi 79
Li, Qunzhan 423
Li, Rongyuan 1112
Li, Shijie 1171
Li, Tianbo 1947
Li, Wen 1797

Author Index

Li, Wenguang 1815
Li, Xia 1674
Li, Y.W. 1101
Li, Yaohua 995
Li, Yong 428
Li, Yongbin 942
Li, Yongdong 1892
Li, Zhanlong 1416
Li, Zhaoji 70, 79
Li, Zheng-Ping 286
Li, Zhou 1630, 1634
Liang, L. 1129
Liang, Tsorng-Juu 361, 1178
Liang, Zhonghua 607
Liao, Changming 102
Li-Jiahui, ...484
Lijie, Chen 1595
Li-jun, Hang 406
Lijun, Zhao 862, 1849
Lili, Jiang 1849
Liming, Liu 74
Lin, Bor-Ren 748, 967, 1278
Lin, Chang-Hua 1957
Lin, Fei 184, 1976
Lin, Liangrui 1243
Lin, Li-Wei 291
Lin, Ray-Lee 361
Lin, Ruan 808
Lin, Ruiguang 885, 1872
Lin, W.-C. 108
Lin, Yang-Sheng 1178
Lin, Ying-De 1154
Lin, Yu-Tzung 1192
Ling, Xia 899
Lingjie, Meng 674
Lipo, T.A. 989
Liqiang, Yuan 161
Liu, Cheng-Tsung 1763
Liu, Ching-Hsiung 361
Liu, Guiqiu 896
Liu, Hongwei 798
Liu, Hsing-Fu 417
Liu, Jian 1267
Liu, Jianqiang 184
Liu, Jingbo 703
Liu, Jinjun 713, 1726
Liu, K. 1071
Liu, Kaipei 453
Liu, Shu-Lin 1267
Liu, Tien-Shuo 1957
Liu, Wei-Shih 361, 1178
Liu, Wenhua 542
Liu, Wenji 1248
Liu, Xiang 229
Liu, Xiaodong 351
Liu, Xinhua 1223

Liu, Yuanchao 236, 1248
Liu, Zhengang 688
Liu-Xueli, ...484
Liwei, Zhang 1012
Loh, P. C. 1107
Lorenz, L. 39
Lou, Z. L. 373
Lu, Bin 836
Lu, Bing 1
Lu, Cheng 489, 739
Lu, Haihui 1054
Lu, P.-C. 117
Lu, Shuai 637
Lu, Xiaodong 83
Lu, Zhengyu 127, 171, 276, 526, 1183
Luk, P.C.K 478, 1872, 1887
Luo, Fang 789, 1522
Lyons, James 1122

M

Ma, Hao 1312, 1637
Ma, Hongfei 708
Ma, Wenchuan 627
Ma, Xiangfei 1836
Ma, Xuejun 438, 1288
Maeda, Toshihiro 1971
Manmek, Thip 1207
Mansouri, O. 574
Mao, Hong 1267
Mathew, Anu 442
Matsumoto, Shuji 1598
Matsuse, Kouki 1598
Mayor, J. Rhett 2010
Meghriche, K. 574
Member, Student 1825
Meng, Zheng 236
Mi, Chris 942
Miao, Guan 744
Miao, Zhao 549
Miller, Nicholas 1122
Ming, Cheng 1758
Ming, Zhou 1431
Ming, Zong 449, 903
Ming-fu, Zhao 1623
Mingli, Ding 1793
Min-qian, Ke 734
Miyatake, Masafumi 242
Moghbelli, H. 597
Mohr, M. 325
Molinas, Marta 63
Moon, G.W. 1382
Morimoto, Keiki 356, 1302, 1307
Morizane, Toshimitsu 1768
Mou, Shann-Chyi 291
Mu, Gang 542
Mudannayake, Chathura P. 1207

Author Index

N

N., N. 1148
Na, He 1713
Nagy, I. 1338
Naidu, S. R. 1731
Nakaoka, Mutsuo 356, 392, ... 1302,
1307, 1358, 1363, 1372, 1377
Nakayama, Y. 1468
Nan, C. H. 1708
Nan, Liu 214, 1942
Nan, Zhao 918
Nasiri, A. 224
Neff, K. L. 1096
Niasar, A. Halvaei 597
Ning, Gaidi 1463
Ninomiya, Tamotsu 1333
Nishimae, Kazuya 1233
Nittayarumphong, S. 1117
Niu, Shuangxia 1788
Nolle, Eugen 219, 826
Nondahl, Thomas A. 1054
Notohara, Yasuo 947
Nozawa, Yusuke 1598
Nuttall, D. R. 1328

O

Ogiwara, Hiroyuki 1307, 1358
Ohara, S. 1468
Oka, Kazuo 1598
Okude, Takaaki 1363
Oleschuk, V. 1651
Omata, Ryuji 1598
Omori, Hideki 392, 1358, 1363, 1372, 1377
Ou, Chung-Lun 1957
Ouyang, Wen 989

P

Pan, Junmin 142, 267, 1348
Pan, Ming-Ho 1590
Pan, Sanbo 1348
Pang, Da-Chen 1763
Paponpen, K. 972
Park, J. D. 1198
Paweletz, A. 842
Payam, A. Farrokh 1906
Pedersen, John K. 1773
Pei, Yunqing 321
Peng, Fang Z. 1076
Pengcheng, Zhu 74
Petchjatuporn, Panom 137
Petkovska, L. 254
Piwko, Richard 1122
Poon, N. K. 1502
Poure, P. 1663
Prado, R. N. do 1117

Q

Pratt, Annabelle 537
Profumo, F. 1651

Qi, Feng 1637
Qi, Wang 1793
Qian, Lewei 774
Qian, Zhaoming 127, 171, 276, 1076, 1183
Qiang, Li 1853
Qiang, Mei 1926
Qiao, Ermin 337
Qiao, Wei 836
Qiaofu, Chen 489, 739
Qi-gang, Fu 734
Qing, Sun 862
Qingding, Guo 1802, 1846
Qingdong, Zhou 1793
Qingfan, Zhang 1634
Qinglin, Zhao 532, 1387, 1517
Qingyu, Yang 699, 1832
Qinmu, Wu 1820
Qiu, Dongyuan 316, 1293
Qiu, Jianqi 885, 1872
Qiu, Zhiling 794
Qizhi, Zhan 1542

R

Radecker, M. 1117
Rafik, F. 2005
Ragon, S. 249
Rahman, M. F. 983, 1646, 1867, 2023
Rahman, M. Faz 1858
Rahnavard, Reza 779
Rajagopalan, Satish 2010
Ren, Hai Peng 652
Ren, Shi 699, 1832
Rentschler, A. 842
Rhyu, Se Hyun 1736
Rosado, Sebastian 1836
Rosario, L.C. 478
Ruan, L. 175
Ruan, Xinbo 1936
Ruixia, Wang 1853
Ruliang, Zhang 1167
Ruxi, Wang 1061

S

Saadate, S. 1663
Sabahi, Mehran 753, 1228, 1679
Sabzali, A. 411
Saha, Bishwajit 392, 1372
Sahinkaya, M.N. 2018
Sakamoto, Kiyoshi 947
Sanchez-Gasca, Juan 1122
Scozzie, Charles 157
Segawa, Takeshi 1333

Author Index

Shancheng, Xing 1023
Shanxu, Duan 769
Shao, Changhong 607
Shao-De, Zhang 1447
Shaojun, Xie 1532
Shao-Long, Li 1421, 1426
Sharifian, M. B. B. 204
Shen, Guoqiao 1566
Shen, Hong 1603
Shen, J.X. 908, 928
Shen, Miaosen 1076
Shen, W. 249
Sheng, K. 1188
Sheng, Weihui 995
Shergin, V. V. 1133
Shi, Cenwei 885, 1872
Shi, Y. F. 1841
Shiang, J. -Z. 1086
Shibata, R. 2000
Shi-feng, Zhang 1447
Shiri, A. 821
Shoulaie, A. 821
Shu, Mantang 1560
Shu, Zhibing 1811
Shun, Jin ....
Shutong, Qiao 468
Shyu, Kuo-Kai 1590
Sibo, Ge 699, 1832
Sirisuk, Phaophak 137, 368
Skorokhod, Y. Y. 1133
Sneineh, Anees Abu 1318
Soltani, J. 1038, 1043, 1906, 1917
Song, Wenchao 1450
Song, Wenxiang 1560
Songboonkaew, J. 1353
Songhua, Shen 744
Soong, Wen Liang 962
Souza, E. V. N. 1731
Stankovic, A.M. 1651
Stefanovic, V. 249
Su, Hongsheng 423
Sugimoto, Hidehiko 1258
Sugimura, Hisayuki 392, 1372, 1377
Sul, Seung-Ki 662
Sun, Chin 417
Sun, Jia-E 199
Sun, Jian 442
Sun, Sizhou 351
Sun, Wei 866
Sun, Xiaofeng 674
Sun, Yuxin 179
Sunat, Khamron 137
Sung, Ha Kyeong 1736
Suul, Jon Are 63
Suzuki, Takahiro 947

T

Tai, Wei-Chih 913
Takahashi, Toshio 428
Tan, Guang-Hui 627, 1507
Tan, Ruimin 1032
Tan, Siew-Chong 296
Tanaka, Chikara 947
Tang, Yan 866
Tang, Yupeng 1416
Tang, Yu-peng 669
Taniguchi, Katsunori 1768
Tao, Haimin 784
Tao, Liu 122
Tenconi, A. 1651
Teodorescu, R. 46
Tezcan, Ibrahim 1546
Thammasiriroj, W. 1353
Tian, Kai 1318
Toba, Akio 1971
Tolbert, L. M. 1703, 1912
Tongjing, Sun 1630
Tsai, C.-T. 108
Tsai, Ming-Fa 1154
Tsay, Shuh-Chuan 1278
Tse, Chi K. 296, 580, 1512
Tseng, S. -Y. 1086, 1343
Tseng, S.-H. 1086
Tuncay, R.N. 2018
Tzou, Ying-Yu 1192

U

Undeland, Tore 63

V

Vahedi, A. 597, 821
van der Broeck, Heinz 1546
Vansencc, Flalph 875
Varjasi, I. 693
Venkataramanan, Giri 259
Volskiy, S. I. 1133

W

Walther, B. 1882
Wan, Deyu 1585
Wan, Shuyun 1608
Wang, Bin 674
Wang, Bingsen 259
Wang, Changkun 1258
Wang, Chengxue 2015
Wang, Chien-Ming 1527, 1995
Wang, Deyu 473
Wang, F. 249
Wang, Fred 1836
Wang, Gang 1986
Wang, Hua 1507

Author Index

Wang, J. 977
Wang, J.K. 813
Wang, Jianhui 1436
Wang, Jian-quan 229
Wang, Juan 798
Wang, Jui-Kum 1154
Wang, Li-Li 1283
Wang, Linbing 311
Wang, Liqiao 632, 724
Wang, Ming-Yan 1318
Wang, Pei-zhen 1576
Wang, Qingyi 1608
Wang, Qun-jing 857
Wang, Shuo 1
Wang, Xiaofeng 1931
Wang, Xiaoyu 713, 1726
Wang, Y.F. 1458
Wang, Yaonan 866
Wang, Yue 1463
Wang, Yunfei 1896
Wang, Z. A. 1708
Wang, Z. S. 373, 1901
Wang, Zhaoan 321, 433, 642, 713, 1463, 1726
Wang, Zhixin 520, 1487
Watkins, S. J. 1551
Wei, Dong 189
Wei, Guo 880
Wei, Liu 1473
Wei, Shi 1061
Wei, Wen 684, 1298
Wei, Xueliang 789, 1522
Weibin, Cheng 602
Weiguo, Liu 679
Wei-ping, Zhou 1674
Weiyang, Wu 189, 532, 554, 559, 569, 1273, 1387, 1473, 1517, 1618, 1991
Wei-Yang, Wu 1926
Wen , Z. 175
Wen, H.-T. 1343
Wen, Xuhui 337, 622
Wenjuan, Dong 1542
Wenlang, Deng 1557
Wenlong, Qu 378
Wenqing, Shi 684, 1298
Wenxi, Yao 937
Wetzel, Hermann 1112
Wiseman, J. 1101
Wu, B. 1101
Wu, Bin 397
Wu, C.-Y. 117
Wu, Chih-Yu 417
Wu, Hongxia 438, 1288
Wu, J.C 718
Wu, Jiaju 1258
Wu, Jiande 83
Wu, Li 1487

Wu, Li 520
Wu, Q. P. 1071
Wu, Shanshan 1892
Wu, T.-F. 108, 117
Wu, Tao 1936
Wu, Wei-yang 1603
Wu, Weiyang 473, 632, 674, 724, 1585
Wu, Wilson 537
Wu, Yong 1608

X

Xi, Zhai 122
Xia, Kun 857
Xiangjun, Zhang 301
Xiangrong, Li 301
Xiangyun, Fu 1693
Xianmin, Ma 1922
Xianmin, Mu 1693
Xiao, D. 983, 1867, 2023
Xiao, G. C. 1708
Xiao, Lei 209
Xiao, Wenxun 1293
Xiao, Zheng 1447
Xiaobo, Yang 1273
Xiaofeng, Sun 189, 1991
Xiaofeng, Zhang 958, 1626
Xiaohuan, Wang 1273
Xiaojie, Wu 468
Xiao-ping, Yang 1718
Xiaoqiang, Guo 532
Xiaotan, Zhao 589
Xiaoxia, Wei 1693
Xiaoyi, Jin 189
Xiaoyu, Wang 1722
Xie, Jian 870, 1442
Xie, S.S. 1143
Xie, Yong 1406
Ximei, Zhao 1802, 1846
Xindong, Tian 808
Xing, Yan 1406, 1537
Xinming, Huang 1492
Xinxin, Wang 1162
Xu, Cai 1657
Xu, D. 1101
Xu, Dehong 1566
Xu, Dianguo 708
Xu, Jianping 1066
Xu, Jia-peng 669
Xu, Jinbang 1608
Xu, Longya 703, 890, 1779
Xu, Ming 1
Xu, Wancai 627
Xu, Y. N. 194
Xu, Yanping 1863
Xu, Longya 1825
Xuan-fang, Yang 1674

Author Index

Xue, H. 1952
Xue, Shan 622
Xuhui, Wen 684, 1006, 1012, 1049, 1298
Xun, Li 769

Y

Yabin, LI 458
Yamamura, N. 2000
Yan, Caizhong 1811
Yan, Chen 1623
Yan, Gangui 542
Yan, Wang 1718
Yanchao, Ji 1693
Yanfeng, Wu 729
Yang, Bo 83, 311
Yang, Chun-Sheng 1278
Yang, Geng 1896
Yang, Hui 1863
Yang, J.J. 718
Yang, Jia-qiang 1017
Yang, Junyou 657, 1253, 1877
Yang, R. 152
Yang, S. Y. 510
Yang, Sheng 505
Yang, X.J. 1458
Yang, Xiao-bo 1603
Yang, Xi-jun 229
Yang, Xing-hua 229
Yang, Xu 433, 642
Yang, Zhaoning 1450
Yang, Zhongping 184
Yang, Zilong 473
Yanhui, He 1061
Yanliang, Xu 1741
Yan-min, Su 734, 1669
Yan-ru, Zhong 382, 549, 602, 1718
Yansong, Hou 1571
Yao, Duan 880
Yao, Lei 1463
Yao, Ruoping 989
Yao, Tianjun 127
Yao, Yue-feng 1397
Yaogang, ...386
Yaohua, Li 589
Yao-Xin, ...484
Yazdanpanah, R. 1038
Ye, Min 1581
Ye, Pengsheng 142
Ye, Peng-sheng 229
Yeic, Zhuliang 875
Yesong, Li 1820
Yeung, Heidi H.T. 1502
Yi, Qin 1820
Yi, Wen 1387
Yi, Zhang 862
Yidan, Sun 449, 903

YII, ...1148
Yin, Qiang 1054
Ying, Jiang 1942
Ying, Li 147
Yinhai, Fan 1162
Yong, Gao 88, 1167, 1595
Yong, Kang 74, 769
Yong, Wang 744
Yongchang, Zhang 161
Yonglong, Peng 458
Yoon, H.K. 1382
You, Keping 1646
You, Xiaojie 1976
Youbin, Zhao 739
Yougui, Guo 1557
Youn, M.J. 1382
Yu, Dongmei 1806
Yu, Hongxiang 627
Yu, L.C. 1188
Yu, Y. H. 1129
Yu, Zhang 489
Yuan, Chang 713, 1726
Yuan, Xiaoming 593, 1122, 1566
Yuan, Yang 1595
Yuan, Zhou 647
Yuanbin, Wang 272
Yuanfang, Wen 802
Yuanyuan, Liu 378, 1162
Yuda, Chen 739
Yue, Wang 1061
Yue-feng, Yang 406
Yu-fan, Xi 1669
Yu-gang, Yang 1942
Yugang, Yang 214
Yun-qing, Pei 585
Yun-Xiang, Xie 1401

Z

Zargari, N. 1101
Zeng, Fanpeng 1507
Zeng, Guohong 1689
Zhai, Xiaohua 866
Zhang , B. 152
Zhang, Bo 70, 316, 505, 1293
Zhang, C. L. 1198
Zhang, C. W. 510
Zhang, D. 813
Zhang, Dong 1788
Zhang, Dongyan 236, 1243
Zhang, Fengge 219, 826
Zhang, Hairong 1811
Zhang, Handong 1986
Zhang, Hongyan 794
Zhang, Hui 453
Zhang, Jia 1689
Zhang, Jianzhong 1746

Author Index

Zhang, Jun 1858
Zhang, Kai 564, 1981
Zhang, L. 1551
Zhang, Luan-guo 229
Zhang, Qian 1976
Zhang, Qiang 774
Zhang, Shifu 219, 826
Zhang, Tengchao 179
Zhang, Weiping 236, 688, 1243, 1248
Zhang, X. 510
Zhang, Xi 267
Zhang, Xianmiao 1183
Zhang, Xiaoqiang 1248
Zhang, Xueguang 708
Zhang, Yanli 1138
Zhang, Yingchao 952
Zhang, Yingqi 593
Zhang, Yonggao 564, 1981
Zhang, Yongping 316
Zhang, Yu 1218
Zhang, Yuan 1779
Zhao, Wenxiang 1746, 1815
Zhao, Xiaotan 995
Zhao, Xusen 688, 1243
Zhao, Y. 194
Zhao, Zhengming 952
Zhaoan, Wang 122, 1061, 1542, 1722
Zhaomin, Fanyinhai 1962
Zhao-ming, Qian 406
Zhao-yong, Zhou 647
Zhao-Yulin, ...484
Zhe, Chen 802, 1387
Zhe, Zhang 559, 1618
Zheng- guo, Wu 1674
Zheng, Shi-cheng 1576
Zheng, Trillion Q. 184, 1012, 1976
Zheng, Zedong 1892
Zhengfeng, Ming 382, 1091
Zhengguo, Wu 729, 1023
Zhengming, Zhao 161, 880
Zheng-Na, ...484
Zhengyu, Lu 406, 937, 958, 1626
Zhen-lin, Xu 1926
Zhi, Na 1198
Zhili, Tan 769
Zhi-Qiang, Wei 1431
Zhi-yuan, Zhang 1623
Zhong, Yanru 1863
Zhong, Yan-ru 617
Zhongmin, Wang 1630
Zhongnan, Guo 554
Zhongying, Chen 1517
Zhou, Qian-zhi 1368
Zhou, Qianzhi 397
Zhou, Tao 1066
Zhou, Wenqi 1312, 1637

Zhou, Y. M. 1129
Zhou, Yang 923, 1947
Zhou, Yu-Fei 286, 1283, 1392
Zhou, Yunbin 564, 1981
Zhu, Guo-rong 97
Zhu, Huangqiu 179, 817, 923, 1213, 1947
Zhu, Jianguo 346, 831
Zhu, Jingwei 962
Zhu, Xiaoyong 1746, 1815
Zhu, Yuran 798
Zhu, Yu-wu 1238
Zhu, Z.Q. 908, 928, 1841
Zou, X. D. 194

A-10

CURRAN ASSOCIATES INC.
proceedings
.com

9781424404483